Spin-Label Electron Paramagnetic Resonance Spectroscopy

Spin-Label Electron Paramagnetic Resonance Spectroscopy

Derek Marsh
Max-Planck-Institut für biophysikalische Chemie, Göttingen, Germany

CRC Press
Taylor & Francis Group
Boca Raton London New York

CRC Press is an imprint of the
Taylor & Francis Group, an informa business

CRC Press
Taylor & Francis Group
6000 Broken Sound Parkway NW, Suite 300
Boca Raton, FL 33487-2742

Printed on acid-free paper

International Standard Book Number-13: 978-1-4822-2089-6 (Hardback)
International Standard Book Number-13: 978-1-032-33729-6 (Paperback)

Visit the Taylor & Francis Web site at
http://www.taylorandfrancis.com

and the CRC Press Web site at
http://www.crcpress.com

Dedicated to the memory of
Harden M. McConnell – an inspiration to us all.

Contents

Preface

This book is exactly what the title says it is: a textbook on the EPR spectroscopy of nitroxide spin-label probes. It is not a book on spin labelling nor is it a book on applications. With the exception of the final chapter on site-directed spin labelling, applications are included only sporadically as illustrations. My aim is to help understand the underlying magnetic resonance aspects of spin-label methods: providing the necessary background for current applications and treating EPR-properties that could aid in developing new applications.

In the spirit of a textbook, I make no attempt at comprehensive referencing nor systematically to assign priority. Instead, I refer to papers that I have found particularly helpful and those with clear exposition didactically. Of course, I also refer to my own work, because this is something with which I have reasonable familiarity. Sometimes, a particular section is based essentially on a single paper that naturally I refer to: also in the relatively infrequent cases where I quote a result without any derivation. Some of my work for the book resulted in original publications, and I list these at the end of this preface.

As an EPR textbook, this book is perhaps unusual, indeed I hope so. My original aim was a book for practitioners (or would-be practitioners) in the field. However, a reviewer of the draft proposal said that such a book could be used as text for a graduate course. Although far from my intention, encouraged by the publishers, I decided to take up this challenge. The result is a long series of Appendices A–V at the end of the book, which I divide somewhat arbitrarily into Fundamental Appendices (A–M) and Specialist Appendices (N–V). I hope sincerely that this experiment will be of lasting benefit to the EPR field as a whole. Individual chapters also have appendices, either to amplify points in the main text or to provide useful reference data, e.g., spin-Hamiltonian parameters. In this way, I intend the book also to have some aspects of a handbook.

It is my great pleasure to acknowledge Frau Inge Dreger for her invaluable and untiring help in the preparation of this book, throughout all of its varying stages.

Derek Marsh,
Göttingen, Germany, 2019.

PUBLICATIONS

Intermediate dipolar distances from spin labels. *J. Magn. Reson.* 238:77–81, 2014.

EPR moments for site-directed spin-labelling. *J. Magn. Reson.* 248:66–70, 2014.

Bonding in nitroxide spin labels from ^{14}N electric-quadrupole interactions. *J. Phys. Chem.* A 119:919–921, 2015.

Nuclear spin-lattice relaxation in nitroxide spin-label EPR. *J. Magn. Reson.* 272:166–171, 2016.

Cross relaxation in nitroxide spin labels. *J. Magn. Reson.* 272:172–180, 2016.

Coherence transfer and electron T_1-, T_2-relaxation in nitroxide spin labels. *J. Magn. Reson.* 277:86–94, 2017.

Spin-label order parameter calibrations for slow motion. *Appl. Magn. Reson.* 49:97–106, 2018.

Correction to: “Spin-label order parameter calibrations for slow motion”. *Appl. Magn. Reson.* 50:1241–1243, 2019.

Molecular order and T_1-relaxation, cross-relaxation in nitroxide spin labels. *J. Magn. Reson.* 290:38–45, 2018.

Author

Derek Marsh works at the Max Planck Institute for Biophysical Chemistry, Göttingen. Dr. Marsh obtained his B.A. degree in physics from the University in Oxford in 1967 and his D.Phil. degree from the same institution in 1971. He worked subsequently at the Astbury Department of Biophysics, University of Leeds; at the Biology Division of the National Research Council of Canada, Ottawa; at the Max Planck Institute in Göttingen and at the Biochemistry Department of the University of Oxford, before moving permanently to Göttingen in 1975.

Dr. Marsh's research interests centre around studies of the structure and dynamics of biological membranes and of lipid bilayer model membranes, using different biophysical techniques, the principal being spin-label electron paramagnetic resonance spectroscopy. Amongst books and numerous other publications, he is the author of the *Handbook of Lipid Bilayers* (2013).

Introduction

1

1.1 INTRODUCTION

Electron paramagnetic resonance (EPR) – otherwise called electron spin resonance (ESR) – is a highly selective spectroscopy, because it requires an unpaired electron spin. In molecular systems, this restricts us to free radicals, transition metal compounds and the few paramagnetic molecules such as oxygen that have triplet ground states. Many free radicals are short-lived, and others are highly delocalized. Nitroxides, on the other hand, are stable free radicals in which the unpaired electron spin is localized to the N–O bond. Nearly all spin labels are nitroxides: *N*-oxyl derivatives of secondary amines, >N–O, in various nitrogen-containing heterocyclic rings. The nitroxide is protected against disproportionation reactions by immediately flanking methyl groups. This gives a simple EPR spectrum, where the hyperfine structure is dominated by the nitrogen nuclear spin.

Localization of the unpaired electron results in an angularly anisotropic spectrum that is an excellent reporter of molecular orientation and rotational dynamics. The full range of EPR properties that make nitroxides useful as spin labels is listed in Table 1.1. Linewidths (or T_2-relaxation) are determined by rapid rotational motions, in the nanosecond regime or faster. Power-saturation properties (or T_1-relaxation) are sensitive to slow rotation, with correlation times in the microsecond range. Absolute values of hyperfine couplings and g-values are sensitive to polarity of the immediate environment. The exchange interaction between spin labels can be used to measure collision rates and rates of translational diffusion. The magnetic dipole interaction between spin labels, or between spin labels and transition metal ions, can be used to measure distances in the range up to 8 nm, or possibly further.

Nitroxide free radicals are not only inherently stable but also can participate in a versatile chemistry. This contributes greatly to their value as spin labels. Figure 1.1 shows some typical spin labels: these divide themselves into small molecules, which are used as dilute probes, and nitroxide-bearing reagents that can be used for labelling macromolecules. The probes are either simple soluble nitroxides or small molecules such as lipids, ligands and enzyme substrates that are spin-labelled. Prominent among the spin-label reagents are those for covalently modifying protein side chains. These are analogues of the classical reagents of protein chemistry. Although many applications of spin labelling are in the fields of molecular biophysics and structural biology, these EPR techniques are equally applicable to other areas of physical and macromolecular chemistry, soft-matter physics and materials science.

TABLE 1.1 EPR properties of spin labels

EPR parameter	*physical property*	*biological application*
^{14}N-hyperfine anisotropy $\left(A_{zz}-A_{xx}\approx 2.5\text{ mT}\right)$ CW-EPR; ESE	rotational motion on *ns-timescale*; libration (*sub-ns*)	lipids and protein side chains
hyperfine couplings a_o, A_{zz}; and g-values g_o, g_{xx} (from HF-EPR)	polarity and H-bonding (H_2O)	membrane polarity profiles; water accessibility in proteins
remote nuclear couplings ESEEM; ENDOR	accessibility; proximity; H-bonding; distances	membrane D_2O-penetration profile; ligand conformation
spin–spin exchange interactions $\approx k_{HE}c_{SL}$ CW-EPR	collision frequencies $k_{HE}c_{SL}$~$10^7\,s^{-1}$	lipid translation diffusion; bimolecular reactions; oxygen accessibility
anisotropic spin–lattice relaxation (~1 μs) ST-EPR; 3-PULSE ESE	rotational diffusion in *sub-millisecond* range.	membrane proteins; supramolecular assemblies (e.g., muscle)
paramagnetic relaxation enhancement, $\Delta(1/T_1)$ POWER SATURATION	accessibility to fast relaxers	SDSL; "membrane-bound" O_2, aqueous paramagnetic ion complexes, other spin labels
dipolar spin–spin interactions ~ $1/r^6$, $1/R^n$ CW-EPR; pELDOR; T_1-enhancement	distances, intermolecular and intramolecular separations	SDSL, molecular structure; immersion depths in proteins and membranes

Abbreviations: CW-EPR, conventional continuous-wave EPR (Chapters 3–9); ESE, electron spin echo EPR (Chapter 13); HF-EPR, high field/frequency EPR; ESEEM, electron spin-echo envelope modulation (Chapter 14); ENDOR, electron–nuclear double resonance (Chapter 14); ST-EPR, saturation transfer EPR (Chapter 11); pELDOR (or DEER), pulsed electron–electron double resonance (Chapter 15); SDSL, site-directed spin labelling (Chapter 16).

TEMPO

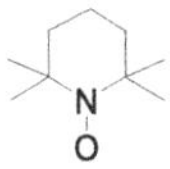

TEMPOL

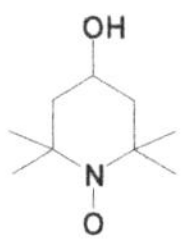

TEMPONE

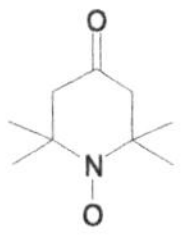

TEMPO-NH_2

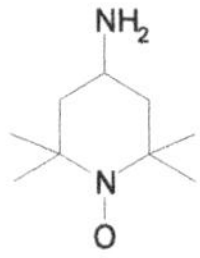

TOAC

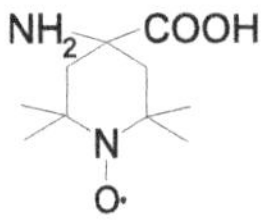

PROXYL-COOH

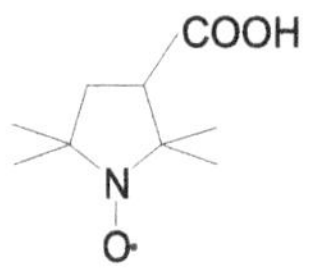

$PYCONH_2$

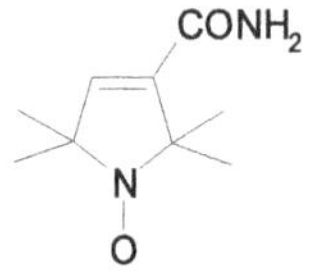

DTBN

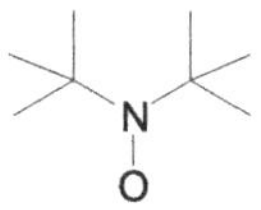

TEMPO-Mal

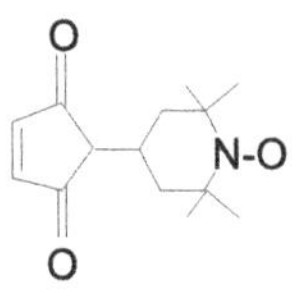

FIGURE 1.1 Nitroxide spin labels. TEMPO derivatives substituted at 4-position of piperidinyl ring; PROXYL derivatives at 3-position of pyrrolidinyl ring; PY derivatives at 3-position of pyrrolinyl ring; DOXYL derivatives at 2-position of oxazolidinyl ring. See Table A1.1 for chemical names.

(Continued)

TEMPO-NCS

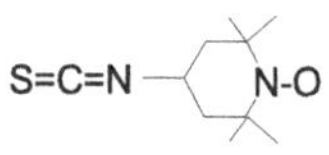

MTSSL

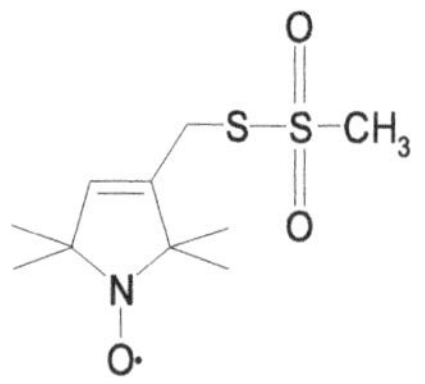

5-DOXYL stearic acid

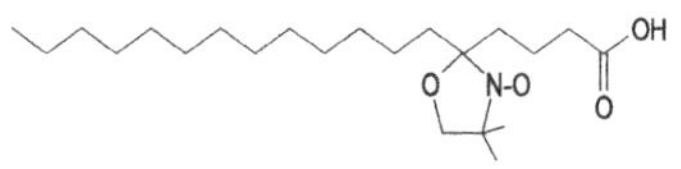

5-DOXYL phosphatidylcholine

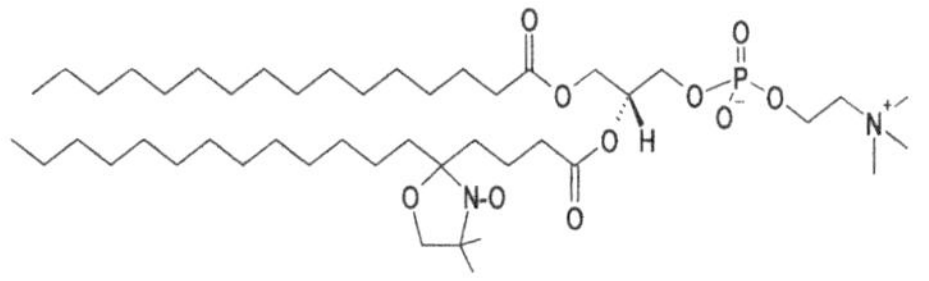

3-DOXYL cholestane

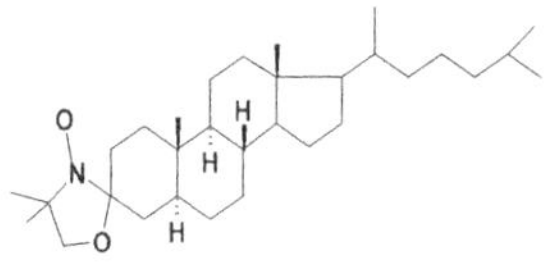

3-DOXYL androstanol

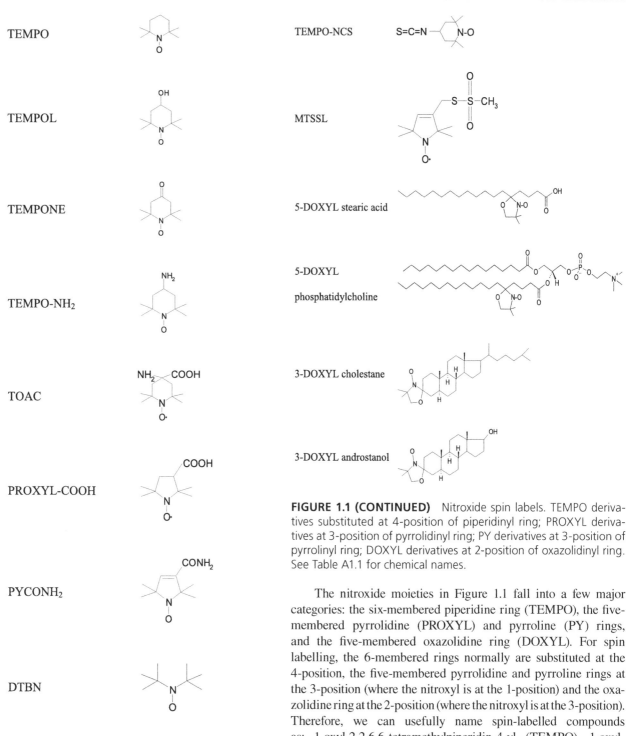

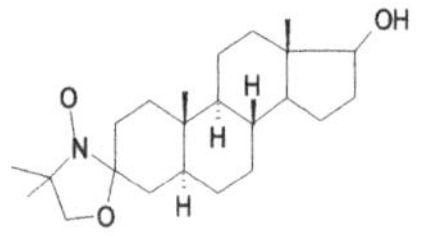

FIGURE 1.1 (CONTINUED) Nitroxide spin labels. TEMPO derivatives substituted at 4-position of piperidinyl ring; PROXYL derivatives at 3-position of pyrrolidinyl ring; PY derivatives at 3-position of pyrrolinyl ring; DOXYL derivatives at 2-position of oxazolidinyl ring. See Table A1.1 for chemical names.

The nitroxide moieties in Figure 1.1 fall into a few major categories: the six-membered piperidine ring (TEMPO), the five-membered pyrrolidine (PROXYL) and pyrroline (PY) rings, and the five-membered oxazolidine ring (DOXYL). For spin labelling, the 6-membered rings normally are substituted at the 4-position, the five-membered pyrrolidine and pyrroline rings at the 3-position (where the nitroxyl is at the 1-position) and the oxazolidine ring at the 2-position (where the nitroxyl is at the 3-position). Therefore, we can usefully name spin-labelled compounds as: 1-oxyl-2,2,6,6-tetramethylpiperidin-4-yl (TEMPO), 1-oxyl-2,2,5,5-tetramethylpyrrolidin-3-yl (PROXYL), 1-oxyl-2,2,5,5-tetramethylpyrrolin-3-yl (PY) and 3-oxyl-4,4-dimethyloxazolidin-2-yl (DOXYL) derivatives. The six-membered ring is conformationally flexible, whereas the five-membered rings are rigid. In particular, the DOXYL ring is incorporated structurally into the labelled molecule: it derives from the corresponding ketone. Additionally, the piperidine-ring nitroxides are more susceptible to chemical reduction than are the five-membered rings, particularly the PROXYLs.

Abbreviations, chemical names and molecular weights of some spin labels are listed in Table A1.1, at the end of the chapter.

1.2 ELECTRON SPIN, MAGNETIC MOMENT AND RESONANCE ABSORPTION

An unpaired electron possesses a magnetic moment, μ_e, that is associated with the electron angular momentum $\hbar\mathbf{S}$, where $\mathbf{S}$ is the quantized spin vector and $\hbar$ is Planck's constant divided by 2π (i.e., $\hbar \equiv h/2\pi$). The constant of proportionality between the magnetic moment and angular momentum is called the gyromagnetic ratio, γ_e:

$$\mu_e = \gamma_e \hbar \mathbf{S} = -g\beta_e \mathbf{S} \tag{1.1}$$

The second equality on the right provides an alternative expression for the magnetic moment in terms of the electron g-value ($g_e = 2.002319$ for a free electron), and the Bohr magneton $\beta_e \left(\equiv e\hbar/2m_e\right)$, where $-e$ and m_e are the electron charge and mass, respectively. The electron magnetic moment, unlike some nuclear magnetic moments (e.g., ^{14}N), is negative. Therefore, as defined in Eq. 1.1, the Bohr magneton and electron g-value are assumed to be positive, whereas the electron gyromagnetic ratio is negative.

The energy of interaction of the electron magnetic moment with a magnetic field (strictly speaking the magnetic induction), $\mathbf{B}$, is determined by its projection on the field direction (see Section B.1 in Appendix B):

$$\mathcal{H} = -\mu \cdot \mathbf{B} \tag{1.2}$$

where $\mathcal{H}$ is the Hamiltonian operator, the eigenvalues of which (see Appendix C) are the energies of the electron moment in the field. If the static magnetic field defines the z-direction, then the Hamiltonian for the electron spin is:

$$\mathcal{H} = g\beta_e B_z S_z \tag{1.3}$$

which we call the Zeeman interaction. An unpaired electron has a spin $S = \frac{1}{2}$, which is quantized such that the z-projection, S_z, can take only the values $M_S = \pm\frac{1}{2}$ (see Appendix E.3). These are conventionally referred to as spin-up and spin-down, and sometimes are designated as α and β for the magnetic quantum numbers $M_S = +\frac{1}{2}$ and $M_S = -\frac{1}{2}$, respectively.

According to Eq. 1.3, the static magnetic field splits the energy levels of the $M_S = \pm\frac{1}{2}$ electron spin states by an amount $\Delta E = g\beta_e B_z$. We can induce transitions between these spin states by applying electromagnetic radiation of frequency ν such that the microwave energy quantum $h\nu$ exactly matches the energy-level splitting ΔE:

$$h\nu = g\beta_e B_o \tag{1.4}$$

where B_o is the static magnetic field strength. This is the resonance condition for EPR absorption. An alternative form is $\omega = \gamma_e B_o$, where $\omega(\equiv 2\pi\nu)$ is the angular frequency of the microwaves. In practice, we use a fixed microwave frequency and scan the static magnetic field to achieve resonance. Table 1.2 lists the resonance fields B_o that match the different microwave frequency bands used in spin-label EPR spectroscopy. By far the most common microwave frequency is 9 GHz which is used with standard EPR spectrometers that have electromagnets. High-field, or equivalently high-frequency, (HF) EPR designates spectrometers that require superconducting magnets: starting with microwave frequencies of 94 GHz upwards.

TABLE 1.2 EPR fields B_o and microwave frequencies ν_e

ν_e (GHz)	B_o (T)	ν_N ^{14}N (MHz)	ν_N ^{15}N (MHz)	ν_N ^{1}H (MHz)	ν_D ^{2}H (MHz)
9.4	0.335	1.031	1.446	14.257	2.189
35	1.247	3.837	5.383	53.085	8.149
94	3.348	10.306	14.456	142.571	21.885
150	5.343	16.445	23.069	227.506	34.924
260	9.262	28.505	39.986	394.344	60.534
360	12.824	39.469	55.365	546.015	83.817

ν_N is the nitrogen nuclear magnetic resonance (NMR) frequency in field B_o and ν_H/ν_D that of hydrogen/deuterium.
$\nu_e = \omega_e/2\pi$, and so on.

When we include the hyperfine interaction $A\mathbf{I}\cdot\mathbf{S}$ of the unpaired electron spin with the nitrogen nuclear spin $\mathbf{I}$, the Hamiltonian for a nitroxide spin becomes:

$$\mathcal{H} = g\beta_e B_z S_z + AI_z S_z \tag{1.5}$$

where A is the strength of the coupling between electron and nuclear spins. Because the hyperfine interaction is much weaker than the electron Zeeman interaction, the magnetic field still defines the z-direction for quantization of the electron spin. Including hyperfine structure, the EPR transitions then become (cf. Eq. 1.4):

$$h\nu = g\beta_e B_o + Am_I \tag{1.6}$$

where m_I is the nuclear spin projection quantum number. Figure 1.2 shows the energy levels and EPR transitions when we scan the magnetic field for a normal ^{14}N-nitroxide (i.e., $m_I = 0, \pm 1$). Transitions are allowed when the electron spin flips but the nuclear spin does not: $\Delta M_S = \pm 1$ and $\Delta m_I = 0$ (see later in Section 1.5). We get a three-line hyperfine structure from ^{14}N, where the low-field line corresponds to $m_I = +1$, the central line to $m_I = 0$ and the high-field line to $m_I = -1$. The hyperfine splitting between the three lines is $A/g\beta_e$ in magnetic field units. For more details, see Appendices B and E. Note that EPR spectra usually are displayed as the first derivative, because we use field modulation with phase-sensitive detection.

1.3 ANGULAR ANISOTROPY OF THE NITROXIDE SPECTRUM

The most important feature of a nitroxide EPR spectrum is its dependence on the angle that the magnetic field makes with the molecular axes. Both the g-value (which specifies the field about

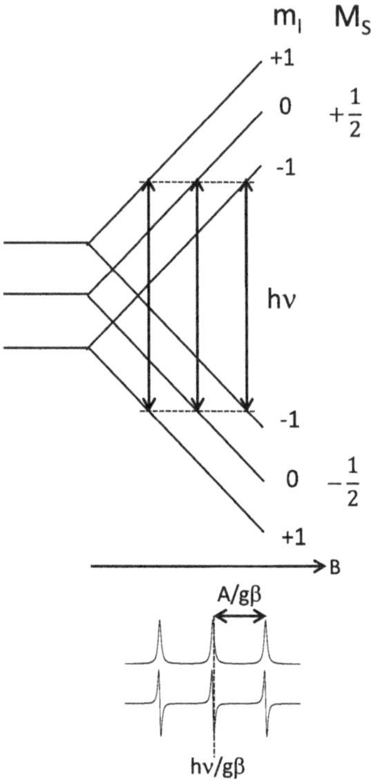

FIGURE 1.2 Energy levels of ^{14}N-nitroxyl free radical, in magnetic field *B*. In absence of a field, energy levels split by hyperfine interaction of electron spin $\left(S=\frac{1}{2}\right)$ with nitrogen nuclear spin $(I=1)$. In an applied field, each hyperfine level splits linearly by Zeeman interaction with two electron-spin orientations: $M_S=\pm\frac{1}{2}$. Magnetic quantum numbers of electron and nuclear spins (M_S and m_I, respectively) for each energy level are given on the right. Allowed EPR transitions $\left(\Delta M_S=\pm 1, \Delta m_I=0\right)$, indicated by vertical arrows, give three-line absorption spectrum with splitting $A/g\beta_e$ (A is hyperfine coupling constant in energy units). Absorption and conventional first-derivative spectra are shown.

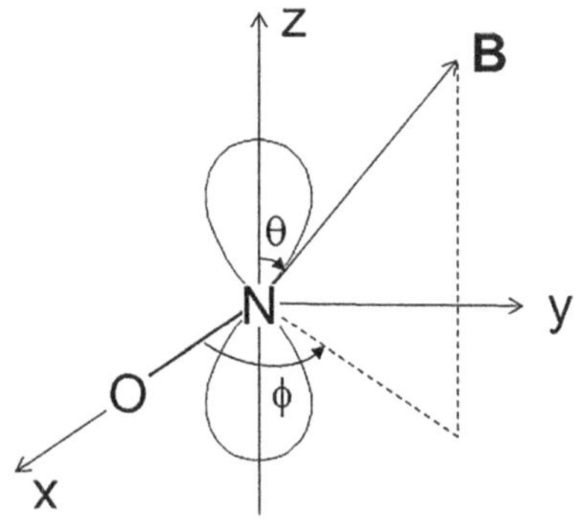

FIGURE 1.3 Axes for nitroxide spin label. *z*-axis defined by nitrogen 2*p* atomic orbital that participates in nitroxide π-bond (which contains unpaired electron). *x*-axis lies along N–O bond. *g*-value is g_{zz} for magnetic field B along nitroxide *z*-axis; g_{xx} when B lies along *x*-axis; g_{yy} when B lies along *y*-axis. Similarly, hyperfine coupling A has values A_{xx}, A_{yy}, A_{zz} when B is along *x,y,z*, respectively. Electron spin always remains quantized along magnetic-field direction.

which the spectrum is centred) and the hyperfine coupling A (which determines the hyperfine splittings) depend on angular orientation. The *g*-factor and hyperfine coupling are, in fact, tensors (see Section B.4 of Appendix B).

Figure 1.3 shows the principal magnetic axes for a nitroxyl free radical. In this axis system, we say that the *g*- and *A*-tensors are diagonal. When we put the magnetic field **B** along the *x*-axis, the *g*-value is g_{xx} and the hyperfine coupling is A_{xx}. Similarly, with **B** along the *y*- or *z*-axis, we get g_{yy}, A_{yy} or g_{zz}, A_{zz}, respectively. As we shall see in Chapter 2, for any intermediate field direction (θ,ϕ), we can express the resonance field and hyperfine coupling in terms of these two sets of three principal values.

We show the effects of hyperfine and *g*-value anisotropy on nitroxide EPR spectra in Figure 1.4. At the normal microwave frequency of 9 GHz (upper panel), spectral anisotropy is dominated by that of the nitrogen hyperfine structure. Consider first a single-crystal sample, where all nitroxides have the same orientation. The A_{zz}-splitting, with B along the nitroxide *z*-axis is much larger than the A_{xx}- or A_{yy}-splitting with B along the *x*- or *y*-axis. Note that the hyperfine anisotropy displays axial symmetry: $A_{xx}\cong A_{yy}$, because it is dipolar in origin (see Section 3.4 in Chapter 3). Compared with the A_{zz}-hyperfine splitting, spectra

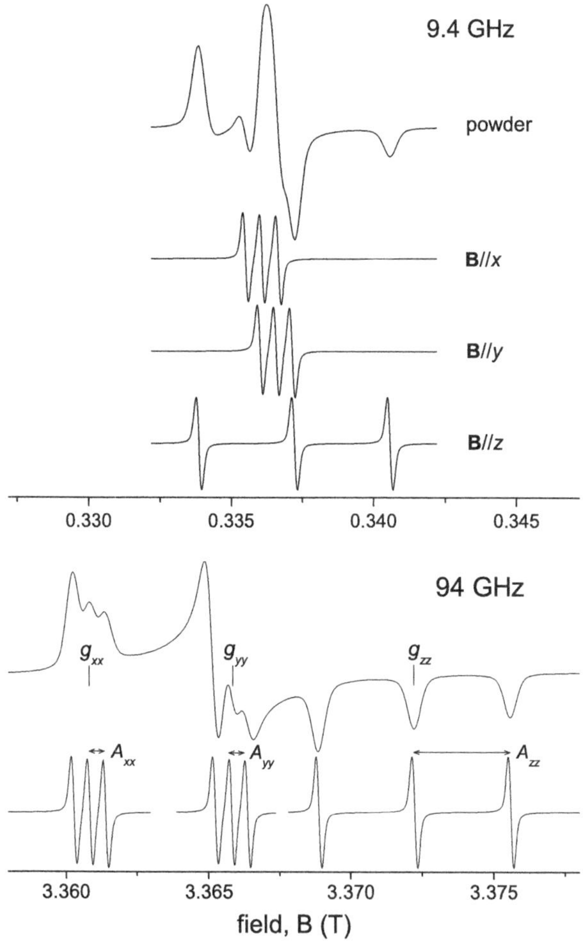

FIGURE 1.4 Hyperfine (A_{ii}) and *g*-value $\left(g_{ii}, i=x,y,z\right)$ spectral anisotropy for ^{14}N-nitroxide spin labels at EPR frequencies of 9.4 GHz (top panel) and 94 GHz (bottom panel). Top spectrum in each panel from randomly-oriented powder sample; lower spectra from uniformly-oriented samples with magnetic field **B** directed along nitroxide *x*-, *y*- or *z*-axis. Total field width displayed: 20 mT for both cases; spectra are conventional first derivative.

with different orientations of the magnetic field show only small differences in their centre field. The effect of g-value anisotropy is relatively unimportant at magnetic fields corresponding to a microwave frequency of 9 GHz.

When our sample is a powder, nitroxides are oriented in all possible directions relative to the static magnetic field. The EPR spectrum is a superposition of overlapping components contributed by all different orientations of the nitroxide axes. We call this a powder pattern, shown as the topmost spectrum in Figure 1.4. The orientational resolution that we get in a single crystal is lost. All hyperfine splittings are present in the powder spectrum, but we resolve only the largest, $2A_{zz}$, because this defines the outer wings of the spectrum. The lineshape no longer consists of typical first-derivative lines, because it is the first derivative of the envelope formed by summing all the constituent absorption spectra. We deal with powder lineshapes in more depth in Section 2.4 of Chapter 2.

At the high frequency of 94 GHz (lower panel in Figure 1.4), the spectral anisotropy is dominated by that of the g-tensor. From Eq. 1.6, we see that effects of g-tensor anisotropy depend linearly on the resonance field, whereas the hyperfine interaction is independent of field. The differences in field centre, for B oriented along the x-, y- and z-axes, are now much larger than any of the hyperfine splittings. Also, we see that the g-tensor clearly is not axial. For B along the x-axis, the spectrum is centred well to low field – and for B along the z-axis well to high field – of the spectrum with B along the y-axis. In principle, HF-EPR gives us a considerable increase in resolution for powder spectra. In practice, however, line broadening may mean that A_{xx}- and A_{yy}-splittings cannot be resolved fully.

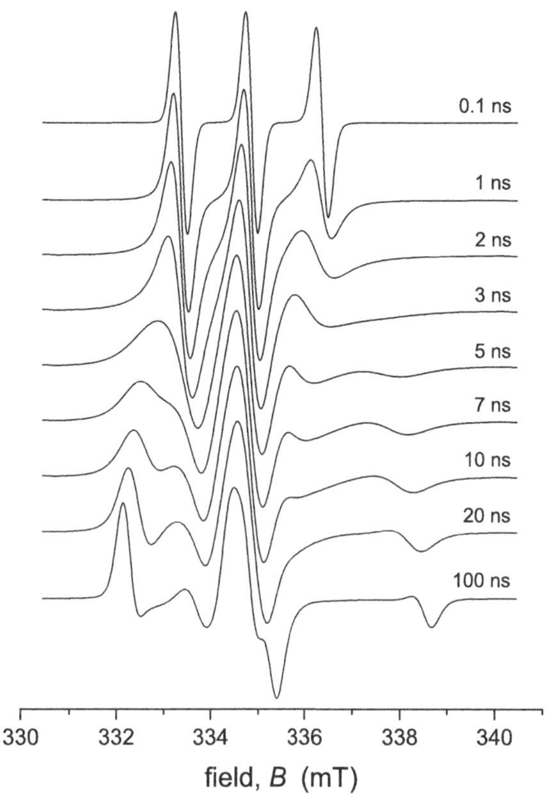

FIGURE 1.5 Spin-label EPR spectra for varying rates of isotropic rotation. From top to bottom, rotational correlation times τ_R (ns) increase, i.e., motion gets slower. EPR frequency = 9.4 GHz. Spectra simulated using stochastic Liouville equation with inhomogeneous line broadening of 0.3 mT, as described in Sections 6.7–6.10 of Chapter 6.

1.4 ROTATIONAL DYNAMICS AND SPIN-LABEL LINESHAPES

A fundamental property of magnetic resonance spectra is that their lineshapes change remarkably when exchange takes place between two species that have different resonance frequencies (or fields). This is typified by the classical case of two-site exchange, which we describe in greater detail in Section 6.3 of Chapter 6. The changes in lineshape are determined by the ratio of the exchange frequency to the difference in resonance frequency of the two species exchanging.

Starting from no exchange, the resonance lines first broaden, then begin to move together and eventually collapse to a single broad resonance, as the exchange frequency increases. This defines the slow-motion regime. Further increase in exchange frequency results in progressive narrowing of the single collapsed line. This defines the fast (or motional narrowing) regime.

Applied to spin-label dynamics, the different resonances correspond to different angular orientations (within the powder pattern), and rotational motion exchanges these resonance positions. We characterize the rate of rotation by a correlation time τ_R. For Brownian diffusion, τ_R specifies the mean-square angular rotation in time t: $\langle \theta^2 \rangle \approx t/3\tau_R$ (see Section 7.2 in Chapter 7). At 9 GHz, the spectra are dominated by the nitrogen hyperfine structure, and its anisotropy $A_{zz} - A_{xx}$ determines the timescale of rotational sensitivity. Roughly speaking, the spectra are most sensitive to correlation times close to the inverse of the anisotropy in angular frequency units: $\tau_R = 1/\gamma_e(A_{zz} - A_{xx}) \approx 2$ ns. This is near the upper border of the fast-motional regime.

Figure 1.5 shows 9-GHz nitroxide EPR spectra simulated for Brownian rotational diffusion with rotational correlation times ranging from the very slow regime ($\tau_R > 100$ ns) to the extreme-narrowing regime ($\tau_R < 0.1$ ns). At the longest correlation time, the spectrum strongly resembles a rigidly immobilized powder pattern. We detect residual motion only from line broadening $\Delta\Delta B$ of the outer peaks, relative to the rigid limit. The increase in linewidth is related to the correlation time by the expression for lifetime broadening:

$$\Delta\Delta B_{1/2} \approx \frac{1}{\gamma_e \tau_R} \tag{1.7}$$

which applies to the half-width of the absorption spectrum at half-height, $\Delta B_{1/2}$. Decreasing the correlation time from $\tau_R = 100$ ns to $\tau_R = 20$ ns causes further broadening of the outer peaks and these move in from the rigid-limit splitting. From the two-site exchange model, the decrease in the hyperfine coupling A_{zz} for the outer lines is:

$$\Delta A_{zz} \approx \frac{1}{A_{zz}}\left(\frac{1}{\gamma_e \tau_R}\right)^2 \tag{1.8}$$

Further decrease to $\tau_R = 10$ ns produces qualitatively the same effect in the outer wings of the spectrum. Beyond this, the overall lineshape changes. Collapse of the spectral anisotropy in the low-field $(m_I = +1)$ hyperfine manifold occurs between $\tau_R = 7$ ns and $\tau_R = 5$ ns, whereas that in the high-field $(m_I = -1)$ hyperfine manifold occurs between $\tau_R = 5$ ns and $\tau_R = 3$ ns. All spectral anisotropy has collapsed at correlation time $\tau_R = 2$ ns. The spectrum then consists of just three first-derivative lines, of different widths, at the average positions for the three $m_I = 0, \pm 1$ hyperfine manifolds. (Differences between high- and low-field manifolds arise from non-vanishing effects of g-value anisotropy.) In this regime, we get progressive motional narrowing, where the linewidth is directly proportional to the rotational correlation time. The line broadening for hyperfine manifold m_I is then:

$$\Delta\Delta B_{m_I} \approx \frac{1}{\gamma_e}\left\langle \Delta\omega_{m_I}^2 \right\rangle \tau_R \tag{1.9}$$

where $\left\langle \Delta\omega_{m_I}^2 \right\rangle$ is the second moment of the spectral anisotropy (see Appendix U). This leads to differential broadening of the three motionally-narrowed hyperfine lines according to:

$$\Delta\Delta B_{m_I} \approx \left(A + Bm_I + Cm_I^2\right) \times \tau_R \tag{1.10}$$

where A, B and C are the different contributions to $\left\langle \Delta\omega_{m_I}^2 \right\rangle$. For example, the C-term, which broadens low- and high-field lines equally, is proportional to $\left(A_{zz} - A_{xx}\right)^2$. See Sections 7.7, 7.8 and 7.14 of Chapter 7 for more details.

The spectra shown in Figure 1.5 correspond to spin labels whose rotational motion is unrestricted in its angular extent. All orientations are accessible, and rapid motion produces an isotropic average. If the motion is restricted only to certain angles, then the spectral anisotropy is not completely averaged out, no matter how fast the rotation. We then expect a powder spectrum from samples that are randomly oriented to the magnetic field direction, but the extent of spectral anisotropy is now reduced compared with a true powder. The reduction in spectral extent depends on the angular amplitude of motion. Figure 1.6 shows a simple example. The nitroxide z-axis is tilted at an angle β_{eff} to a fixed (||) axis about which the spin label rotates rapidly. The correlation time for rotation about this axis is short: $\tau_{R\parallel} = 0.3$ ns (cf. Figure 1.5), but that for rotation of the axis itself is long enough not to reduce the remaining spectral anisotropy $(\tau_{R\perp} > 30$ ns$)$. For $\beta_{eff} = 0$, there is no motional averaging and a full powder pattern results. As β_{eff} increases, the outer hyperfine peaks that are split by $2A_{\parallel}$ move inwards, and a pair of inner hyperfine peaks emerges from the centre of the spectrum (see vertical ticks for $\beta_{eff} = 31°$). The inner peaks are of opposite amplitude to the outer peaks and have a splitting that is related to the hyperfine splitting $2A_{\perp}$ for the magnetic field perpendicular to the rotation axis. As the outer peaks move in, the inner peaks move out with further increase in β_{eff}. Thus, the total spectral anisotropy, given by $A_{\parallel} - A_{\perp}$, progressively decreases. Finally, outer and inner hyperfine peaks merge to give a spectrum that is similar to the isotropic average at $\tau_R = 1$ ns in Figure 1.5. This occurs at the so-called magic angle: $\beta_{eff} = 47.54°$ that corresponds to $\cos^2\beta_{eff} = \frac{1}{3}$. As we shall see later, in Chapter 8, the residual spectral anisotropy is related to the order parameter of the nitroxide z-axis:

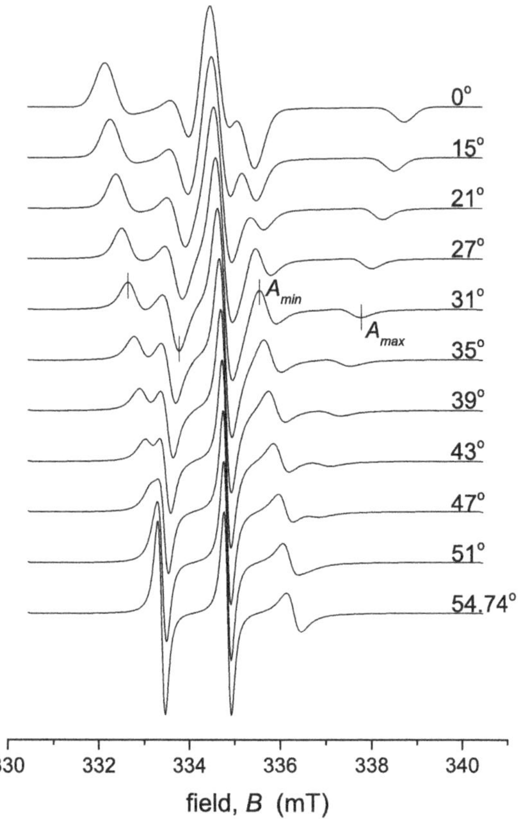

FIGURE 1.6 Spin-label EPR spectra for restricted angular amplitude of motion. Rotation is about a single axis with nitroxide z-axis inclined at fixed angle $\beta_{eff} = 0-54.74°$. From top to bottom, β_{eff} increases corresponding to increments of 0.1 in order parameter S_{zz} from 1 to 0. EPR frequency = 9.4 GHz. Spectra simulated using stochastic Liouville equation with correlation time $\tau_{R\parallel} = 0.3$ ns for uniaxial rotation (inhomogeneous line broadening 0.08–0.5 mT depending on β_{eff}). Outer ticks (for $\beta_{eff} = 31°$) correspond to hyperfine splitting $2A_{max} = 2A_{\parallel}$ and inner ticks to $2A_{min}$ that is related to $2A_{\perp}$.

$$S_{zz} \equiv \tfrac{1}{2}\left(3\cos^2\beta_{eff} - 1\right) = \left(A_{\parallel} - A_{\perp}\right)/\Delta A \tag{1.11}$$

where $\Delta A \equiv A_{zz} - \frac{1}{2}\left(A_{xx} + A_{yy}\right)$ is the value of $A_{\parallel} - A_{\perp}$ for $S_{zz} = 1$ (i.e., $\beta_{eff} = 0$). In fact, the irregularly spaced values of β_{eff} in Figure 1.6 correspond to equal increments of 0.1 in S_{zz} over the complete range 0–1. The partially averaged spectral anisotropy is the same for equal order parameter, no matter how this is achieved. In general, we have $S_{zz} \equiv \frac{1}{2}\left(3\left\langle\cos^2\beta\right\rangle - 1\right)$, where angular brackets indicate an average over the ensemble of accessible orientations β (see Section 8.3 in Chapter 8).

1.5 TRANSITION PROBABILITIES AND SELECTION RULES

We now ask how the transitions in electron spin orientation are brought about by the microwave radiation field. The answer comes from time-dependent perturbation theory in quantum mechanics, which we deal with in Appendix M. The result is known as Fermi's golden rule. Our time-dependent perturbation comes from interaction of the electron magnetic moment, $\boldsymbol{\mu} = -g\beta_e \mathbf{S}$, with the microwave magnetic field. It varies sinusoidally with time t: $V(t) = g\beta_e \mathbf{S}\cdot\mathbf{B}_1 \times 2\cos\omega t$, where $2\mathbf{B}_1$ is the magnetic field vector, and ω the angular frequency, of the microwave radiation. From Eq. M.11 of Section M.2, we see that the rate $W_{m,m'}$ at which the microwave magnetic field induces transitions between the $M_S = m$ and $M_S = m'$ spin orientations depends on the square of the matrix element $\langle m'|\mathbf{S}\cdot\mathbf{B}_1|m\rangle$ of the interaction that connects these states:

$$W_{m,m'} = 2\pi\gamma_e^2 \left|\langle m'|\mathbf{S}\cdot\mathbf{B}_1|m\rangle\right|^2 \delta\left(\omega_{m,m'} - \omega\right) \tag{1.12}$$

where $\omega_{m,m'}$ is the angular frequency of the transition, and the Dirac δ-function is defined by: $\int_{-\infty}^{\infty}\delta(x)\,\mathrm{d}x = 1$ for $x = 0$ but is zero otherwise. This corresponds to an infinitely narrow line centred at the angular frequency: $\omega = \omega_{m,m'}$. Note that interaction with the radiation field causes up and down transitions to occur at the same rate, because $W_{m,m'} = W_{m',m}$ from Eq. 1.12. These transitions correspond to absorption and stimulated emission, respectively.

We know from Appendix E on angular momentum that only the S_x or S_y component of the spin operator $\mathbf{S}$ couples states with different values of M_S (see Table E.1). Therefore, to induce transitions between the spin levels, the microwave magnetic field $\mathbf{B}_1$ must be perpendicular to the static magnetic field $\mathbf{B}_o$, because $\mathbf{B}_o$ splits the energy levels of different spin orientations and defines the z-quantization axis. Microwave resonators in EPR spectrometers are arranged to fulfill the condition $\mathbf{B}_1 \perp \mathbf{B}_o$. Let us put $\mathbf{B}_1$ along the x-direction, relative to $\mathbf{B}_o$, then $V(t) = 2g\beta_e B_{1,x}\cos\omega t \times S_x$. From Eqs. E.11 to E.13 in Appendix E, we see that the only non-zero matrix elements of S_x join states with the same total spin quantum number S and spin-projection quantum numbers that differ by one. Thus, the selection rule for magnetic dipole transitions of the electron spin is: $\Delta S = 0$, $\Delta M_S = \pm 1$. The δ-function in Eq. 1.12 gives us another selection rule, that of energy balance which corresponds to the resonance condition: $\omega = \omega_{m,m'}$. Therefore, in the presence of hyperfine structure, we have the further selection rule: $\Delta m_I = 0$ for the nuclear spin projection. This is because the nuclear resonance frequency $(\omega_n = \gamma_n B_o)$ is very different from the electron resonance frequency $(\omega_e = \gamma_e B_o)$, at the same magnetic field strength B_o.

The selection rule for EPR transitions is met automatically by the $S = \frac{1}{2}$ two-level system of a single unpaired electron spin (see Figure 1.2). From Eq. 1.12, the rate of microwave-induced transitions then becomes simply:

$$W_{+1/2,-1/2} = \tfrac{1}{2}\pi\gamma_e^2 B_{1,x}^2 \delta\left(\omega_{+1/2,-1/2} - \omega\right) \tag{1.13}$$

where we take the matrix element $\left|\langle +\tfrac{1}{2}|S_x|-\tfrac{1}{2}\rangle\right| = \tfrac{1}{2}$ from Table E.1. An important feature of Eq. 1.13 is that the transition probability W is proportional to $B_{1,x}^2$, i.e., to the microwave power. However, EPR spectrometers normally detect the microwave field strength (strictly speaking the microwave electric-field strength), not the microwave power. Therefore, the EPR signal detected is directly proportional to the square root of the microwave power, i.e., to $\sqrt{W}$, in the absence of microwave saturation.

Equation 1.13, as also Eq. 1.12, applies to an infinitely sharp line. For absorption lines of non-vanishing width, we must replace the delta function in Eq. 1.13 with the envelope function $g(\omega_o - \omega)$ that defines the actual lineshape. The rate of induced transitions for a $S = \frac{1}{2}$ system then becomes:

$$W = \tfrac{1}{2}\pi\gamma_e^2 B_{1,x}^2 g(\omega_o - \omega) \tag{1.14}$$

where $\omega_o \equiv \omega_{+1/2,-1/2}$ is the angular resonance frequency, and the lineshape is normalized: $\int_{-\infty}^{\infty} g(\omega_o - \omega)\cdot \mathrm{d}\omega = 1$.

1.6 SPIN–LATTICE RELAXATION AND SATURATION

We learnt in the preceding section that the microwave B_1-field induces transitions in a $S = \frac{1}{2}$ system, at a rate W that is the same in both directions (see left side in Figure 1.7). For simplicity, we let the relative populations in the upper and lower levels be $1 - n$ and $1 + n$, respectively, where n is proportional to the population difference between the two levels. The rate equation for the population in the lower level is then:

$$\frac{\mathrm{d}n}{\mathrm{d}t} = W(1-n) - W(1+n) = -2Wn \tag{1.15}$$

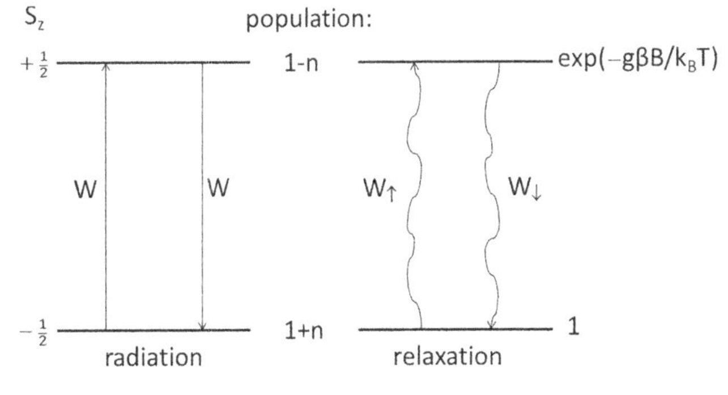

FIGURE 1.7 Radiation-induced transitions at rate W (*left side*) and spontaneous transitions from spin–lattice relaxation at rates $W_\downarrow$ and $W_\uparrow$ (*right side*), for $S = \frac{1}{2}$ electron spin system in magnetic field B. Instantaneous relative populations of upper and lower levels $(M_S = \pm\frac{1}{2})$ are $1 - n$ and $1 + n$, respectively. Relative populations at thermal equilibrium indicated on the right.

We obtain the same result for the upper level. We see immediately that this rate equation applies equally well to the absolute population difference between the two levels. Integration of Eq. 1.15 shows that the population difference decays exponentially to zero with time: $n(t) = n(0)\exp(-2Wt)$, because transitions are induced with equal probability in both directions.

We get a non-vanishing EPR absorption because, in practice, the spin system is not isolated. It is coupled to the so-called lattice, which maintains thermal equilibrium during microwave irradiation by acting as a sink for the microwave energy absorbed. The corresponding transitions in the spin system, which tend to return the population difference to its equilibrium value, are known as spin–lattice relaxation. At thermal equilibrium, the population of the upper state, relative to the lower state, is given by the Boltzmann factor $\exp\left(-g\beta_e B_o/k_B T\right)$ where k_B is Boltzmann's constant and T is the absolute temperature. To maintain this population difference, the rate of spin–lattice transitions from upper to lower level, $W_\downarrow$, must be faster than that from lower to upper level, $W_\uparrow$ (see right side of Figure 1.7). The rate equation for the population in the lower level is given by:

$$\frac{dn}{dt} = W_\downarrow(1-n) - W_\uparrow(1+n) = W_\downarrow - W_\uparrow - \left(W_\downarrow + W_\uparrow\right)n \quad (1.16)$$

in the absence of microwave irradiation. At thermal equilibrium $dn/dt = 0$, and from Eq. 1.16 the equilibrium population difference is then: $n_o = \left(W_\downarrow - W_\uparrow\right)/\left(W_\downarrow + W_\uparrow\right)$. We therefore can rewrite Eq. 1.16 as:

$$\frac{dn}{dt} = \left(W_\downarrow + W_\uparrow\right)\left(n_o - n\right) = \frac{n_o - n}{T_1} \quad (1.17)$$

where the spin–lattice relaxation time T_1 is defined by: $1/T_1 \equiv W_\downarrow + W_\uparrow$. According to Eq. 1.17, the population difference increases exponentially from zero to the equilibrium value n_o: $n(t) = n_o\left(1 - \exp\left(-t/T_1\right)\right)$, with time t after applying a static magnetic field.

We are now in a position to obtain the rate equation for the population difference under microwave radiation, in the presence of spin–lattice relaxation. Combining Eqs. 1.15 and 1.17 we get:

$$\frac{dn}{dt} = -2Wn + \frac{n_o - n}{T_1} \quad (1.18)$$

The steady-state population difference under continuous microwave irradiation therefore becomes:

$$n = \frac{n_o}{1 + 2WT_1} \quad (1.19)$$

where we use $dn/dt = 0$ in Eq. 1.18. This is the fundamental equation that describes saturation of the integrated intensity of absorption in magnetic resonance. When $2W << 1/T_1$, the population difference remains at the equilibrium value n_o. When, however, the microwave power is increased such that $2W >> 1/T_1$, spin–lattice relaxation cannot keep up with transitions induced by the microwave radiation and the population difference tends to zero. Absorption of microwaves and the corresponding EPR absorption signal then become saturated.

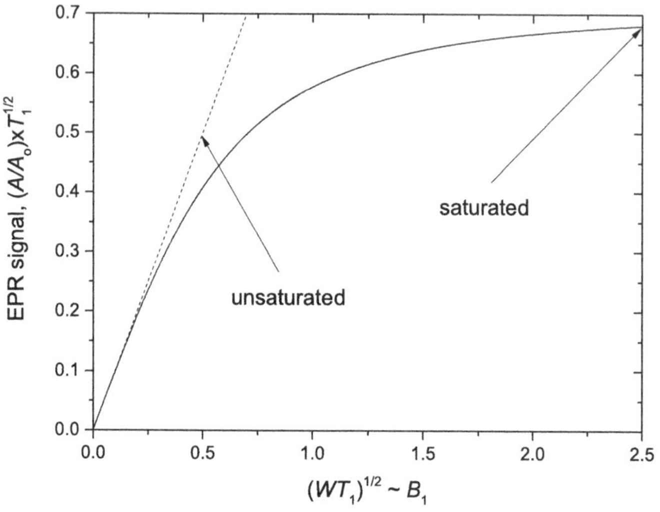

FIGURE 1.8 Saturation of EPR signal: $A(W)$ saturates with increasing microwave power $B_1^2 \propto W$, according to Eq. 1.20 (solid line). W: rate EPR transitions are induced by microwave radiation; T_1: spin–lattice relaxation time. Dashed line: dependence when EPR transition does not saturate.

The rate of absorption of microwave energy is given by $dE/dt = Wn \times \hbar\omega_o$, where $\hbar\omega_o$ is the microwave quantum absorbed per transition. However, as we said in Section 1.5, the signal amplitude, A, from the EPR spectrometer is proportional to the microwave field, not to the power absorbed. The EPR signal is therefore proportional to $\sqrt{Wn}$ and from Eq. 1.19 becomes:

$$A(W) = \frac{A_o\sqrt{W}}{\sqrt{1 + 2WT_1}} \quad (1.20)$$

where $A_o\sqrt{W}$ is the EPR signal at low microwave powers. Figure 1.8 shows the dependence of the EPR signal on W that we get from Eq. 1.20 for a progressive-saturation experiment in continuous-wave (CW) EPR. Initially, the EPR signal increases linearly with $\sqrt{W}$, i.e., with the square root of the microwave power. As saturation sets in, the signal increases less rapidly with increasing $\sqrt{W}$, finally levelling off to a constant value. This is the saturation behaviour that we expect for the integrated intensity of the EPR absorption spectrum (see Eq. 11.11 in Section 11.2 of Chapter 11). In Eq. 1.20, we tacitly assume an infinitely narrow line, i.e., W is given by Eq. 1.13. For saturation of the amplitude $A(\omega)$ within an EPR line of shape function $g(\omega_o - \omega)$, we get from Eqs. 1.14 and 1.19:

$$A(\omega) = \frac{A_o B_1 g(\omega_o - \omega)}{1 + \pi\gamma_e^2 B_1^2 T_1 g(\omega_o - \omega)} \quad (1.21)$$

assuming a linear detector, as in standard EPR spectrometers.

1.7 SPIN–SPIN RELAXATION AND LINEWIDTHS

Spin–lattice relaxation, which we considered in the previous section, limits the lifetime, τ, of the spin system in the upper or excited state. This broadens the spectral lines according to the Heisenberg uncertainty relation: $\Delta E \cdot \tau \approx h$, where ΔE is the uncertainty in energy of the upper state. However, lifetime broadening is not the only contribution to spectral linewidths and mostly is not the dominant one. Other time-dependent interactions modulate the relative energies of the spin levels and cause line broadening that is characterized by the transverse or spin–spin relaxation time, T_2.

The spin–lattice relaxation time, T_1, determines the rate at which the longitudinal spin-magnetization, M_z, that lies along the static magnetic field returns to its equilibrium value M_o. (Recall that magnetization, and relaxation, are properties of an ensemble of spins, because an individual electron spin can take up only one of two – quantized – orientations.) The T_2-relaxation time, on the other hand, determines the rate at which the transverse spin magnetization, M_x and M_y, relaxes to zero. Spin–spin interactions are not the only contributions to transverse relaxation nor the most common. Any time-dependent interaction that modulates the local magnetic field experienced by the electron spin in question can cause T_2-relaxation.

The EPR signal is determined by the transverse spin magnetization, M_x, that results from the oscillating microwave magnetic field, $B_{1,x}$, which is perpendicular to the static magnetic field, $B_{o,z}$. The time dependence of the spin magnetization is determined by the Bloch equations (see Section 6.2 in Chapter 6). In the absence of saturation, solution of the Bloch equations predicts a Lorentzian lineshape for the EPR absorption:

$$A_L(B) = \frac{\gamma_e (\pi T_2)^{-1}}{(1/T_2)^2 + \gamma_e^2 (B - B_o)^2} \tag{1.22}$$

where $B_o (= \omega_o/\gamma_e)$ is the resonant magnetic field and γ_e is the electron gyromagnetic ratio. Note that the intensities in Eq. 1.22 are scaled to give a normalized integral: $\int_{-\infty}^{\infty} A_L(B) \cdot dB = 1$.

The Lorentzian absorption lineshape and its first derivative (which is the normal EPR display) are shown in the upper panel of Figure 1.9. The absorption line is centred on the resonant field, B_o, and has half-width at half-height given by:

$$\Delta B_{\frac{1}{2}}^L = \frac{1}{\gamma_e T_2} \tag{1.23}$$

which is half the full-width at half-maximum, FWHM $= 2\Delta B_{\frac{1}{2}}^L$. Thus, the EPR linewidth is determined by the transverse or spin–spin relaxation time, T_2. The peak-to-peak linewidth of the first-derivative EPR spectrum is:

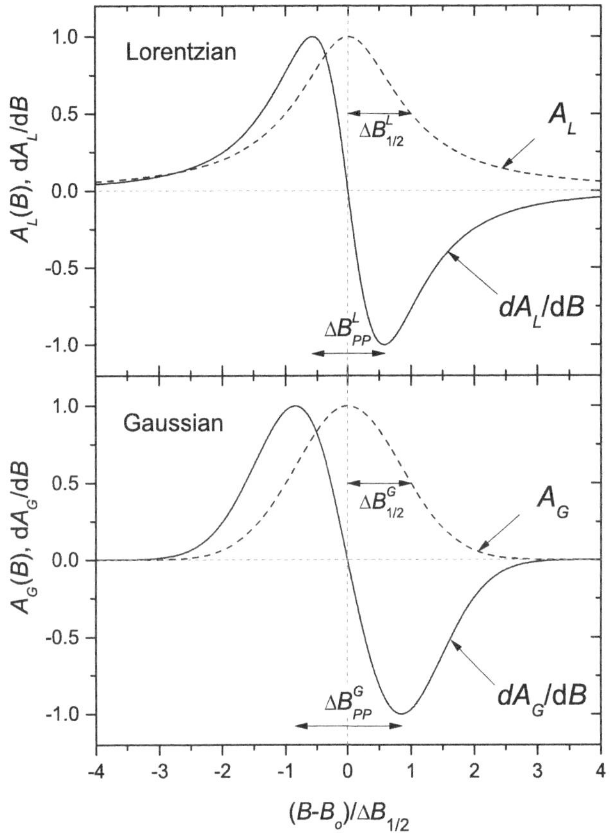

FIGURE 1.9 Lorentzian (*upper panel*) and Gaussian (*lower panel*) lineshapes. EPR absorption (dashed lines) and first derivative (solid lines). Normalized field-scan axis, $(B - B_o)/\Delta B_{\frac{1}{2}}$, gives equal half-widths at half-height, $\Delta B_{\frac{1}{2}}$, for the absorption. Gaussian lineshape represents an envelope of unresolved Lorentzian components with narrower widths.

$$\Delta B_{pp}^L = \frac{2}{\sqrt{3}} \frac{1}{\gamma_e T_2} \tag{1.24}$$

for a purely Lorentzian line.

Lorentzian broadening, the so-called homogeneous broadening, determines the natural width of an EPR line. The linewidth observed experimentally may be larger than the homogeneous width because of so-called inhomogeneous broadening. This arises when the EPR line consists of a distribution of unresolved Lorentzian components. For spin labels, the most common source of inhomogeneous broadening is unresolved hyperfine structure from protons attached to the nitroxide ring (see Section 2.5 in Chapter 2). Often we can represent inhomogeneously broadened lines by a Gaussian shape for the overall envelope. A normalized Gaussian absorption lineshape is given by:

$$A_G(B) = \frac{1}{\sigma_G \sqrt{2\pi}} \exp\left(-\frac{(B - B_o)^2}{2\sigma_G{}^2} \right) \tag{1.25}$$

where B_o is the centre field and σ_G^2 is the variance (or second moment) of the Gaussian distribution. A Gaussian absorption lineshape and its first derivative are shown in the lower panel of

Figure 1.9. The absorption is centred on the field B_o and has half-width at half-height: $\Delta B_{1/2}^G = \sigma_G \sqrt{2\ln 2}$, which is determined by the standard deviation, σ_G. Correspondingly, the peak-to-peak linewidth of the first-derivative spectrum is: $\Delta B_{pp}^G = 2\sigma_G$, for a purely Gaussian line.

We see from Figure 1.9 that the wings of a Lorentzian absorption line extend out much further than do those of a Gaussian line with the same width at half-height. Correspondingly, for identical absorption linewidths $\Delta B_{1/2}$, the peak-to-peak width ΔB_{pp} of a Gaussian first-derivative line is 1.47 times that of a Lorentzian.

APPENDIX A1: SPIN LABELS

TABLE A1.1 Molecular weights of some nitroxyl spin-labels

nitroxide	*formula*	*molecular weight*
TEMPO, 1-oxyl-2,2,6,6-tetramethylpiperidine	$C_9H_{18}NO$	156.25
TEMPONE, 1-oxyl-2,2,6,6-tetramethylpiperidin-4-yl oxide	$C_9H_{16}NO_2$	170.23
$[^2H]_{16}$-TEMPONE	$C_9{}^2H_{16}NO_2$	186.33
TEMPOL, 1-oxyl-2,2,6,6-tetramethylpiperidin-4-ol	$C_9H_{18}NO_2$	172.24
TEMPO-NH_2, 1-oxyl-2,2,6,6-tetramethylpiperidin-4-ylamine	$C_9H_{19}N_2O$	171.26
TEMPO-COOH, 1-oxyl-2,2,6,6-tetramethylpiperidin-4-yl carboxylic acid	$C_{10}H_{18}NO_3$	200.25
TOAC, 1-oxyl-2,2,6,6-tetramethyl-4-aminopiperidin-4-yl carboxylic acid	$C_{10}H_{19}N_2O_3$	215.27
PROXYL-NH_2, 1-oxyl-2,2,5,5-tetramethylpyrrolidin-3-ylamine	$C_8H_{17}N_2O$	157.23
PROXYL-COOH, 1-oxyl-2,2,5,5-tetramethylpyrrolidin-3-yl carboxylic acid	$C_9H_{16}NO_3$	186.23
PYCONH_2, 1-oxyl-2,2,5,5-tetramethylpyrrolin-3-ylamide	$C_9H_{15}N_2O_2$	183.23
PYCOOH, 1-oxyl-2,2,5,5-tetramethylpyrrolin-3-yl carboxylic acid	$C_9H_{14}NO_3$	184.21
PYMeOH, 1-oxyl-2,2,5,5-tetramethylpyrrolin-3-yl methanol	$C_9H_{16}NO_2$	170.23
DTBN, di-*tert*-butyl nitroxide	$C_8H_{18}NO$	144.23
PADS.H_2, peroxylamine disulphonic acid	$H_2NO_7S_2$	192.15
protein reagents		
TEMPO-Mal, *N*-(1-oxyl-2,2,6,6-tetramethylpiperidin-4-yl)maleimide	$C_{13}H_{19}N_2O_3$	251.30
PROXYL-Mal, *N*-(1-oxyl-2,2,5,5-tetramethylpyrrolidin-3-yl)maleimide	$C_{12}H_{17}N_2O_3$	237.27
PYMeMal, *N*-[(1-oxyl-2,2,5,5-tetramethylpyrrolin-3-yl)-methyl]maleimide	$C_{13}H_{17}N_2O_3$	249.29
TEMPO-NHCOCH_2I, *N*-(1-oxyl-2,2,6,6-tetramethylpiperidin-4-yl)iodoacetamide	$C_{11}H_{20}N_2O_2I$	339.19
PROXYL-NHCOCH_2I, *N*-(1-oxyl-2,2,5,5-tetramethylpyrrolidin-3-yl)iodoacetamide	$C_{10}H_{18}N_2O_2I$	325.17
TEMPO-NCS, 1-oxyl-2,2,6,6-tetramethyl-piperidin-4-yl-isothiocyanate	$C_{10}H_{17}N_2OS$	213.32
PROXYL-MeNCS, 1-oxyl-2,2,5,5-tetramethylpyrrolidin-3-ylmethylisothiocyanate	$C_{10}H_{17}N_2OS$	213.32
TEMPO-NHCOPhHgCl, *N*-(1-oxyl-2,2,6,6-tetramethylpiperidin-4-yl) *p*-chloromercuribenzamide	$C_{16}H_{22}ClN_2O_2Hg$	510.40
MTSSL, (1-oxyl-2,2,5,5-tetramethylpyrrolin-3-yl)methylmethanethiosulfonate	$C_{10}H_{18}NO_3S_2$	264.34
PROXYL-COOSucc, *N*-(1-oxyl-2,2,5,5-tetramethylpyrrolidin-3-yl carboxy)-succinimide	$C_{13}H_{19}N_2O_5$	283.30
PYCOOSucc, *N*-(1-oxyl-2,2,5,5-tetramethylpyrrolin-3-yl carboxy)-succinimide	$C_{13}H_{17}N_2O_5$	281.28
lipids		
DOXYL (additional)	$C_4H_6NO_2$	100.10
n-DOXYL SA, stearic acid	$C_{22}H_{42}NO_4$	384.57
n-DOXYL PC.H_2O, phosphatidylcholine	$C_{46}H_{94}N_2O_{11}P$	882.22
3-DOXYL-cholestane	$C_{31}H_{53}NO_2$	471.76
3-DOXYL-androstan-17β-ol	$C_{23}H_{38}NO_3$	376.55

Lipids. PC: palmitoyl-stearoyl phosphatidylcholine. DOXYL ≡ 3-oxyl-4,4-dimethyloxazolidin-2-yl.

RECOMMENDED READING

Atherton, N.M. 1993. *Principles of Electron Spin Resonance*, Chichester, U.K.: Ellis Horwood Ltd.

Carrington, A. and McLachlan, A.D. 1969. *Introduction to Magnetic Resonance: With Applications to Chemistry and Chemical Physics*, New York: Harper and Row.

Knowles, P.F., Marsh, D., and Rattle, H.W.E. 1976. *Magnetic Resonance of Biomolecules*, London: Wiley-Interscience.

Marsh, D. 1981. Electron spin resonance: spin labels. In *Membrane Spectroscopy*, Molecular Biology, Biochemistry and Biophysics, Vol. 31, ed. Grell, E., 51–142. Berlin-Heidelberg-New York: Springer-Verlag.

Marsh, D. and Horváth, L.I. 1989. *In Advanced EPR, Applications in Biology and Biochemistry*, ed. Hoff, A.J., 707–752. Amsterdam-New York: Elsevier Science Publishers B.V.

Pake, G.E. 1962. *Paramagnetic Resonance: An Introductory Monograph*, New York: W.A. Benjamin.

Slichter, C.P. 1996. *Principles of Magnetic Resonance*, Berlin: Springer.

Wertz, J.E. and Bolton, J.R. 1972. *Electron Spin Resonance: Elementary Theory and Practical Applications*, New York: McGraw-Hill Inc.

The Nitroxide EPR Spectrum 2

2.1 INTRODUCTION

This chapter deals with the basic features of the nitroxide EPR spectrum and the associated spectral analysis. Of first importance are the positions of the spectral lines and, most crucially, their angular variation with direction of the spectrometer magnetic field. These are specified by the spin Hamiltonian and the associated energies of the unpaired electron spin. The second aspect concerns the spectra of randomly oriented samples: their so-called powder lineshapes and how the spectral splittings and linewidths can be obtained from them. The third aspect deals with inhomogeneous broadening of EPR lines that arises from unresolved proton hyperfine structure.

Orientational dependence is part of the basic theory of magnetic resonance, which can be understood mostly in terms of angular geometry, without special recourse to quantum mechanics. Alternative approaches to some of these problems that use quantum mechanical and matrix methods are included in Appendices H and I at the end of the book.

2.2 NITROXIDE SPIN HAMILTONIAN

The spin Hamiltonian contains only those energy terms that depend on the electron or nuclear spins. The spins are represented by angular-momentum operators (**S** and **I**, respectively) that act upon the spin wave functions. We label these wave functions with the magnetic quantum numbers, M_S and m_I, which are the observable values of the z-projection of the angular momentum. For the unpaired electron with spin quantum number $S=\frac{1}{2}$, these are $M_S=+\frac{1}{2}$ and $-\frac{1}{2}$, corresponding to spin-up and spin-down orientations. For the nitrogen ^{14}N nucleus with spin $I=1$, the spin projections are $m_I=+1$, 0 and -1. Please refer to Appendix B for an introduction to vectors, matrices and tensors; Appendix C for the basics of quantum mechanics and Appendix K for an introduction to the electronic structure of atoms and molecular bonding.

The spin Hamiltonian for an isolated nitroxide contains two principal energy terms. At conventional spectrometer frequencies (and higher), by far the larger term is the Zeeman interaction of the electron magnetic moment with the magnetic field, **B**, of the spectrometer. We describe this with the electron g-tensor, **g**. The second term is the hyperfine interaction of the electron spin, **S**, with the nuclear spin, **I**, of the nitrogen atom in the nitroxide. We describe this with the hyperfine tensor, **A**. The energy levels of the electron spin are then specified by the following spin Hamiltonian:

$$\mathcal{H}_S=\beta_e\mathbf{B}\cdot\mathbf{g}\cdot\mathbf{S}+\mathbf{I}\cdot\mathbf{A}\cdot\mathbf{S} \tag{2.1}$$

where $\beta_e\left(=e\hbar/2m_e=9.274009\times10^{-24}\ \mathrm{J\cdot T^{-1}}\right)$ is the Bohr magneton. Strictly speaking, **B** is the magnetic induction (in Tesla), but in magnetic resonance we conventionally refer to it as the magnetic field. Note also that **B** and **I** on the left of the **g** and **A** square-matrix tensors are row vectors (strictly speaking transposed columns, $\mathbf{B}^T$ and $\mathbf{I}^T$), whereas **S** on the right is a normal column vector (see Appendix B.4). In Eq. 2.1, the spin **S** is an effective spin, because the true electron spin **s** has the free-electron g-value ($g_e=-2.0023193$), which does not depend on the orientation of the magnetic field. As we explain later, the magnetic moment of the unpaired electron contains small contributions from its orbital angular momentum that are not quenched entirely by chemical bonding between the nitrogen and oxygen. We represent these orbital contributions by the anisotropy of the g-tensor acting together with the effective spin **S** (see Section 3.12). For this reason, we use the term paramagnetic resonance (i.e., EPR), rather than simply spin resonance (ESR). Finally, we should note that the Zeeman interaction with the nitrogen nucleus – which is isotropic – is omitted from Eq. 2.1 but must be included at higher fields ($B\geq1.25$ T, i.e., microwave frequencies of 35 GHz or higher) and when dealing with hyperfine spectroscopy (see Chapter 14). Table 2.1 gives representative values of the g-tensor and hyperfine tensor elements for nitroxide spin labels. Values for further spin labels are listed in the Appendix at the end of this chapter.

The g- and A-tensors are diagonal when referred to the nitroxide axes, which are shown in Figure 2.1. Expressed in terms of these principal axes, the spin Hamiltonian then becomes:

$$\begin{aligned}\mathcal{H}_S&=\beta_e\left(g_{xx}B_xS_x+g_{yy}B_yS_y+g_{zz}B_zS_z\right)\\&\quad+A_{xx}I_xS_x+A_{yy}I_yS_y+A_{zz}I_zS_z\end{aligned} \tag{2.2}$$

where g_{xx}, g_{yy}, g_{zz} and A_{xx}, A_{yy}, A_{zz} so-defined are the principal elements of the g- and A-tensor. We specify the components of the static magnetic field, $\mathbf{B}\equiv\left(B_x,B_y,B_z\right)$, by the cosines

TABLE 2.1 Magnetic tensor elements for DOXYL (androstanol), pyrrolinyl (PYMeOH) and piperidinyl (TOAC, TEMPONE) nitroxides in solid host matrices. ^{14}N hyperfine tensor and g-tensor elements are given by A_{ii} and g_{ii}, with $i \equiv x, y, z$ as defined in Figure 2.1[a]

	DOXYL androstanol[b]	*PYMeOH*[c]	*TOAC*[d]	*TEMPONE*[e]
g_{xx}	$2.00913 \pm 3 \times 10^{-5}$	$2.00905 \pm 2 \times 10^{-5}$	2.00915	$2.0096 \pm 2 \times 10^{-4}$
g_{yy}	$2.00610 \pm 3 \times 10^{-5}$	$2.00617 \pm 2 \times 10^{-5}$	2.00666	$2.0063 \pm 2 \times 10^{-4}$
g_{zz}	$2.00231 \pm 3 \times 10^{-5}$	$2.00227 \pm 2 \times 10^{-5}$	2.00269	$2.0022 \pm 1 \times 10^{-4}$
g_o	$2.00585 \pm 3 \times 10^{-5}$	$2.00583 \pm 2 \times 10^{-5}$	2.00617	$2.0060 \pm 2 \times 10^{-4}$
A_{xx}	14.9 ± 0.6 MHz (0.53 ± 0.02 mT)	13.60 ± 0.15 MHz (0.484 ± 0.005 mT)	19.26 MHz (0.685 mT)	11.5 ± 1.4 MHz (0.41 ± 0.05) mT
A_{yy}	13.8 ± 0.6 MHz (0.49 ± 0.02 mT)	13.50 ± 0.15 MHz (0.481 ± 0.005 mT)	12.44 MHz (0.443 mT)	17.1 ± 1.4 MHz (0.61 ± 0.05) mT
A_{zz}	90.8 ± 0.6 MHz (3.24 ± 0.02 mT)	93.20 ± 0.15 MHz (3.326 ± 0.005 mT)	100.9 MHz (3.60 mT)	93.7 ± 0.6 MHz (3.34 ± 0.02) mT
a_o	39.8 ± 0.6 MHz (1.42 ± 0.02 mT)	40.10 ± 0.15 MHz (1.430 ± 0.005 mT)	44.2 MHz (1.58 mT)	40.8 ± 1.1 MHz (1.45 ± 0.04) mT

[a] $g_o = \frac{1}{3}(g_{xx} + g_{yy} + g_{zz})$
$a_o = \frac{1}{3}(A_{xx} + A_{yy} + A_{zz})$

[b] 3-DOXYL-5α-androstan-17β-ol in *o*-xylene at −135°C (Smirnova et al. 1995).

[c] 1-oxyl-2,2,5,5-tetramethylpyrrolin-3-yl methanol in *ortho*-terphenyl at −93°C (Savitsky et al. 2008).

[d] 1-oxyl-2,2,6,6-tetramethyl-4-aminopiperidin-4-yl carboxylic acid (TOAC) in the nonapeptide Fmoc-Aib$_8$-TOAC-OMe (where Aib is α-aminoisobutyric acid and Fmoc is fluoren-9-ylmethyloxycarbonyl) in acetonitrile at −193°C (Carlotto et al. 2011).

[e] 1-oxyl-2,2,6,6-tetramethyl-piperidin-4-yl oxide in toluene at −124°C (Hwang et al. 1975).

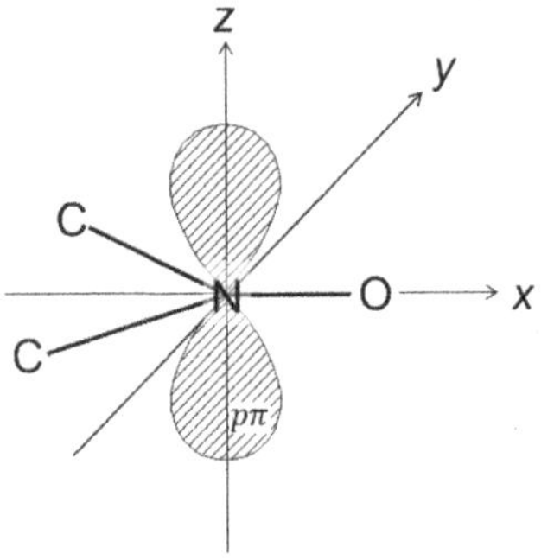

FIGURE 2.1 Principal axis system for a nitroxide. *z*-axis lies along nitrogen 2*p* atomic orbital of nitroxide *p*π-bond that contains the unpaired electron. *x*-axis lies along N–O bond; *y*-axis lies in C–NO–C plane.

$(l_x = \cos\theta_x, l_y = \cos\theta_y, l_z = \cos\theta_z)$ of the angles that the field makes with the *x*-, *y*- and *z*-axes of the nitroxide (see Figure 2.2). The Zeeman part of the spin Hamiltonian, which is by far the dominant term in Eq. 2.2, then becomes:

$$\mathcal{H}_{Zeeman} = \beta_e B\left(g_{xx} l_x S_x + g_{yy} l_y S_y + g_{zz} l_z S_z\right) \quad (2.3)$$

Because the *g*-value anisotropy is relatively small, we can assume that the electron spin is quantized along the magnetic field direction. The maximum discrepancy in effective *g*-value that results from this assumption is approximately 2 ppm. Even at a microwave operating frequency of 360 GHz ($B = 12.9$ T), this corresponds to a maximum error in field position that is only of the order of 10 μT.

If S_z' is the electron spin component in the direction of quantization, then – just as for the magnetic field itself – the

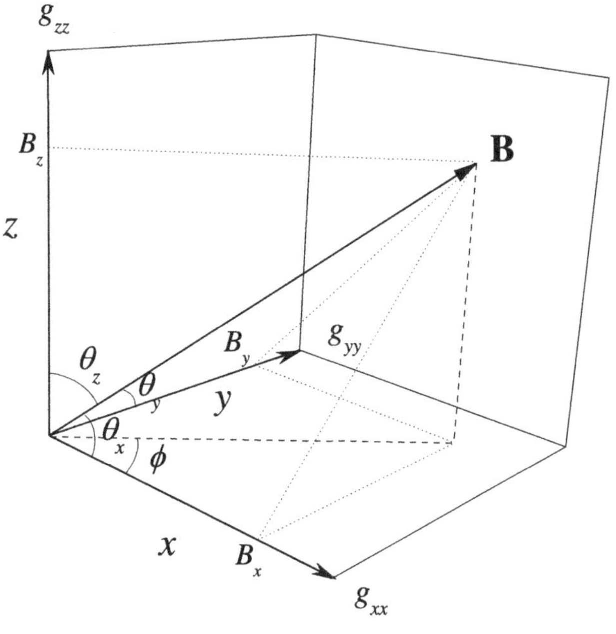

FIGURE 2.2 Orientation of magnetic-field vector **B** to the nitroxide *x,y,z*-axes. Direction cosines of **B**: $(l_x, l_y, l_z) \equiv (\cos\theta_x, \cos\theta_y, \cos\theta_z)$; polar coordinates (B, θ_z, ϕ). Direction cosines satisfy: $l_x^2 + l_y^2 + l_z^2 = \cos^2\theta_x + \cos^2\theta_y + \cos^2\theta_z = 1$ and relate to polar coordinates by: $l_x = \sin\theta_z \cos\phi$, $l_y = \sin\theta_z \sin\phi$ and $l_z = \cos\theta_z$.

components of the electron spin in the original nitroxide axis system are given by: $(S_x, S_y, S_z) = (l_x, l_y, l_z) S_z'$ (cf. Figure 2.2). Rewritten in terms of quantization along the field direction, Eq. 2.3 therefore becomes:

$$\mathcal{H}_{Zeeman} = \beta_e B\left(g_{xx} l_x^2 + g_{yy} l_y^2 + g_{zz} l_z^2\right) S_z' = g\beta_e B S_z' \quad (2.4)$$

where the angular-dependent effective g-value in the new system is:

$$g = g_{xx}l_x^2 + g_{yy}l_y^2 + g_{zz}l_z^2 \tag{2.5}$$

Together with Eq. 2.5, the right-hand equality in Eq. 2.4 gives the electron Zeeman energies. In effect, we have diagonalized the Zeeman Hamiltonian (approximately) by changing to a new axis system where the z-axis is the magnetic field direction.

Now we must transform the electron-spin components that appear in the hyperfine interaction in Eq. 2.2 to the new quantization axes. In first order, only the hyperfine terms that contain S_z' interest us; we call these the secular (I_zS_z') and pseudosecular (I_xS_z', I_yS_z') terms. To this level of approximation, the hyperfine part of the spin Hamiltonian in the S_z' axis system is:

$$\mathcal{H}_{hfs} = (l_xA_{xx}I_x + l_yA_{yy}I_y + l_zA_{zz}I_z)S_z' \tag{2.6}$$

where we ignore off-diagonal terms in S_x' and S_y' (see Appendix J later, for discussion of second-order hyperfine shifts). Unlike the electron spin, **S**, the nitrogen nuclear spin, **I**, is not quantized along the external spectrometer field direction. This is because the internal hyperfine field at the nucleus that is produced by the magnetic moment of the electron is much greater than the external magnetic field. For a ^{14}N nucleus, the maximum hyperfine coupling is $A_{zz} \approx 90-100$ MHz, which corresponds to an effective magnetic field of 29–32 T at the nitrogen nucleus. For comparison, the nitroxide electron Larmor frequency at this field is 820–910 GHz, considerably higher than any practical spin-label EPR frequency.

The coefficients multiplying the nuclear spin components I_x, I_y and I_z in Eq. 2.6 are direction cosines of the internal hyperfine field (Abragam and Bleaney 1970). Therefore, we write the nuclear part of the first-order hyperfine Hamiltonian in Eq. 2.6 in terms of this second new axis system as:

$$\mathcal{H}_{hfs} = A\left(\frac{l_xA_{xx}}{A}I_x + \frac{l_yA_{yy}}{A}I_y + \frac{l_zA_{zz}}{A}I_z\right)S_z' = AI_z'S_z' \tag{2.7}$$

where the nuclear spin is quantized along the new z-axis with direction cosines: $(l_x', l_y', l_z') \equiv (l_xA_{xx}, l_yA_{yy}, l_zA_{zz})/A$. The trigonometrical identity: $l_x'^2 + l_y'^2 + l_z'^2 = 1$ for these direction cosines then requires that the angular-dependent effective hyperfine coupling, A, is given by:

$$A^2 = l_x^2A_{xx}^2 + l_y^2A_{yy}^2 + l_z^2A_{zz}^2 \tag{2.8}$$

Essentially, we have diagonalized the hyperfine Hamiltonian by rotating the nuclear spin to the direction of the vector **A·S**. We illustrate this in Figure 2.3, for the two-dimensional case of an external field confined to the x,z-plane. The alternative approaches of rotating the axes explicitly, or of diagonalizing the hyperfine Hamiltonian directly, are described in Appendix I and H, respectively, at the end of the book.

We have now transformed the spin Hamiltonian to a simple diagonal form, where the g-value (Eq. 2.5) and the first-order hyperfine coupling, A (Eq. 2.8) both depend explicitly on the magnetic field direction:

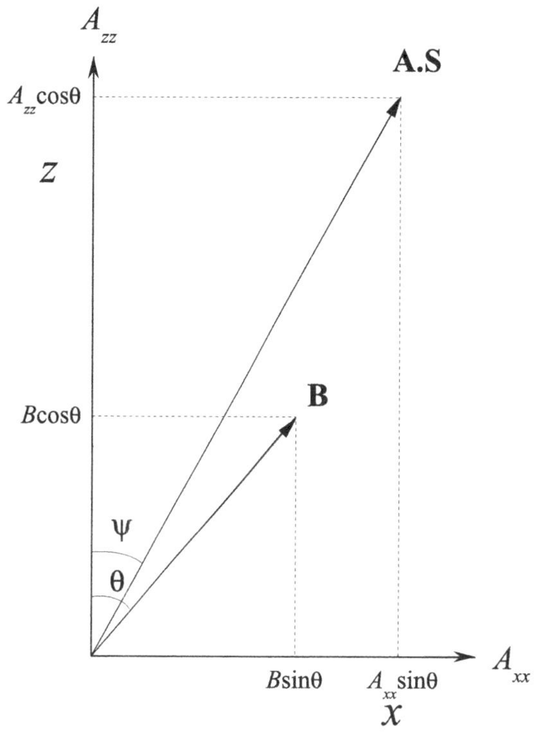

FIGURE 2.3 Relative orientations of internal hyperfine field $\mathbf{A}\cdot\mathbf{S}$ and external spectrometer magnetic field **B**. Electron-spin z-component lies along magnetic field, in x,z-plane inclined at angle θ to nitroxide z-axis. Nitrogen nuclear-spin z-component lies along vector $\mathbf{A}\cdot\mathbf{S}$ that makes angle ψ with nitroxide z-axis: $\tan\psi = (A_{xx}/A_{zz})\tan\theta$.

$$\mathcal{H}_S = g\beta_eBS_z + AI_zS_z \tag{2.9}$$

The primes now have no special significance for I_z and S_z, and therefore we have dropped them. We ignore second-order contributions to the hyperfine interactions because the Zeeman interaction is much larger than the hyperfine interaction (especially at high field). Remembering that the eigenvalues of S_z are $M_S = \pm\frac{1}{2}$ and those of I_z are m_I, the allowed EPR transitions $(\Delta M_S = \pm 1, \Delta m_I = 0)$ are given to first order from Eq. 2.9 by:

$$h\nu = E_{+1/2} - E_{-1/2} = g\beta_eB + Am_I \tag{2.10}$$

where ν is the microwave frequency of the EPR spectrometer and h (= 6.626069 × 10^{-34}J s) is Planck's constant.

2.3 ANGULAR VARIATION OF NITROXIDE EPR SPECTRA

For practical purposes, it is more convenient to express the magnetic field direction in terms of the polar angles, θ and ϕ (see Figure 2.2 with $\theta \equiv \theta_z$), instead of by the three direction cosines, one of which is redundant. From Eqs. 2.5 and 2.8, the effective g-value and hyperfine coupling are then given by:

$$g(\theta,\phi) = \left(g_{xx}\cos^2\phi + g_{yy}\sin^2\phi\right)\sin^2\theta + g_{zz}\cos^2\theta \quad (2.11)$$

$$A(\theta,\phi) = \sqrt{\left(A_{xx}^2\cos^2\phi + A_{yy}^2\sin^2\phi\right)\sin^2\theta + A_{zz}^2\cos^2\theta} \quad (2.12)$$

From Eq. 2.10, the angular dependence of the nitroxide EPR resonance field, $B_{m_I}(\theta,\phi)$, is then given by:

$$B_{m_I}(\theta,\phi) = \frac{h\nu}{g(\theta,\phi)\beta_e} - \frac{A(\theta,\phi)}{g(\theta,\phi)\beta_e} m_I \quad (2.13)$$

where the explicit dependence on the polar angles θ and ϕ of the field relative to the nitroxide principal axes is defined by Eqs. 2.11 and 2.12.

Figure 2.4 gives an example of the angular variation of the resonance positions for a single crystal of horse oxyhaemoglobin that is spin-labelled with a nitroxide derivative of maleimide (TEMPO-Mal). The anisotropy in the g-tensor determines the angular dependence of the central resonance, and the anisotropy of the A-tensor determines that for the splittings of the outer resonances. The maxima and minima in the angular variation coincide with the **a**- and **b**-axes of the monoclinic haemoglobin crystal. This suggests that the magnetic principal axes of the nitroxide are oriented along the two perpendicular crystal axes of the monoclinic crystal. From the values of the hyperfine splittings, we find that the nitroxide z-axis is parallel to the **b**-axis, and the x-axis is parallel to the **a**-axis. The corresponding tensor components are $A_{zz} = 91.3 \pm 3$ MHz and $A_{xx} = 28.0 \pm 3$ MHz, and the associated g-tensor elements are $g_{zz} = 2.00270 \pm 0.0002$ and $g_{xx} = 2.00783 \pm 0.0002$ (Ohnishi et al. 1966) – cf. Table 2.1. The solid lines in Figure 2.4 show that Eqs. 2.11–2.13, together with these magnetic parameters, can describe the angular dependence of the EPR resonance positions to within the experimental accuracy. Libertini and Griffith (1970) find this also for measurements at higher resolution with the small di-*tert*-butyl nitroxide spin label alone in a single crystal of tetramethylcyclobutadione. Spin Hamiltonian tensors for the latter are given in the Appendix at the end of the chapter (Table A2.3).

It is a particularly fortunate circumstance in Figure 2.4 that the nitroxide principal axes coincide with the crystallographic axes. This requires that the nitroxide attached at residue Cys-93 must be very specifically immobilized within a hydrophobic pocket in the β-subunit (Ohnishi et al. 1966). Also, only then are the single labels that are attached to the two β-chains of a single haemoglobin molecule likely to be equivalent: because they are related by the twofold molecular symmetry axis. This is required for both labelled chains to give identical EPR spectra at all field orientations, as we see in Figure 2.4. Generally, we do not expect the magnetic principal axes of a spin-labelled protein to lie along the crystal axes. Then we must perform angular variations about three independent crystal axes. The hyperfine and g-tensors obtained by this procedure will contain off-diagonal elements (g_{xy}, A_{xy}, etc.), and we must diagonalize them to get the principal values (g_{xx}, A_{xx}, etc.) and to find the directions of the principal axes relative to the crystal axes. Details are given in standard EPR text books (Carrington and McLachlan 1969; Wertz and Bolton 1972). In certain situations, e.g., membranes and liquid crystals, molecular motion can result in dynamically averaged hyperfine and g-tensors that have axial symmetry. Such cases considerably simplify the interpretation of spectra from aligned samples (see further in Chapter 8).

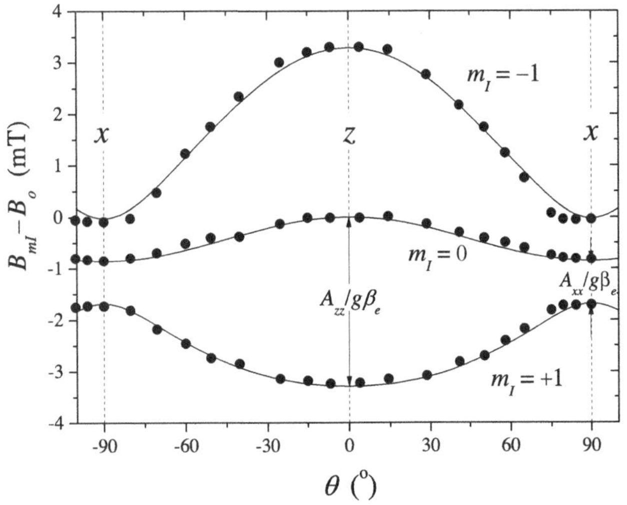

FIGURE 2.4 Angular variation of resonance-field position B_{m_I}, for $m_I = 0$ and ± 1 ^{14}N-hyperfine lines. *N*-(1-oxyl-2,2,6,6-tetramethylpiperidin-4-yl) maleimide at Cys-93 of the β-chains in monoclinic single crystal of horse oxyhaemoglobin (Ohnishi et al. 1966). Magnetic field is rotated in (001)-plane, i.e., about **c***-axis. Nitroxide z-axis is parallel to crystallographic **b**-axis ($\theta = 0°$), which coincides with twofold symmetry axis of $\alpha_2\beta_2$ haemoglobin tetramer; nitroxide x-axis is parallel to crystallographic **a**-axis ($\theta = 90°$). Spectra recorded at 9 GHz; resonance positions referred to central resonance ($B_o = h\nu/g_{zz}\beta_e$) with magnetic field along the z-axis. Solid lines: Eqs. 2.11–2.13, where (001)-plane corresponds to $\phi = 0°$.

2.4 EPR POWDER PATTERNS

In most cases that we encounter, samples are unlikely to be specifically aligned, certainly not single crystals. Instead, they usually are randomly oriented solutions, dispersions or glassy/amorphous solids. For these samples, all possible orientations of the nitroxide axes relative to the magnetic field are present and give rise to so-called powder spectra. The EPR spectral intensity distribution, $I(B_m)\cdot \mathrm{d}B_m$, is related directly to the orientational distribution of the nitroxide axes, $p(\theta,\phi)\cdot \mathrm{d}\theta \cdot \mathrm{d}\phi$ by the formal identity:

$$I(B_m)\cdot \mathrm{d}B_m = p(\theta,\phi)\cdot \mathrm{d}\theta \cdot \mathrm{d}\phi \quad (2.14)$$

Consider an axially symmetric system (specified by an average over ϕ in Eqs. 2.11–2.13), in which the unique axis is randomly oriented. Then the orientational distribution is

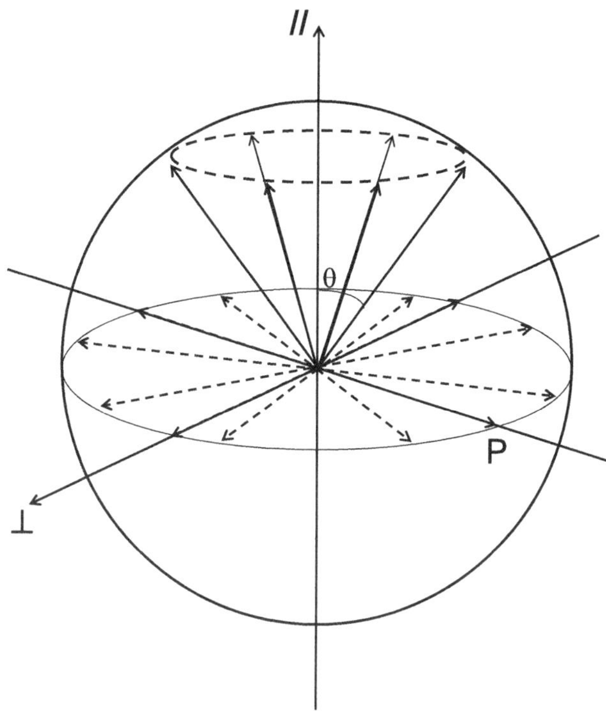

FIGURE 2.5 Angular degeneracy for axial symmetry. Far more angular vectors lie in the perpendicular plane (⊥), compared to the unique parallel direction (||). Probability of orientation θ is proportional to perimeter of dashed circle, i.e., $p(\theta)\cdot \mathrm{d}\theta = \sin\theta \cdot \mathrm{d}\theta$.

$p(\theta,\phi)\cdot \mathrm{d}\theta \cdot \mathrm{d}\phi \equiv \sin\theta \cdot \mathrm{d}\theta$ (see Figure 2.5), and the spectral intensity distribution of the powder pattern is given by:

$$I(B_m) = \frac{\sin\theta}{\mathrm{d}B_m/\mathrm{d}\theta} \tag{2.15}$$

where we get the derivative $\mathrm{d}B_m/\mathrm{d}\theta$ from Eq. 2.13. With axial symmetry, the equivalents of Eqs. 2.11 and 2.12 for the effective g-value and hyperfine coupling are:

$$g(\theta) = g_{\parallel}\cos^2\theta + g_{\perp}\sin^2\theta \tag{2.16}$$

$$A(\theta) = \sqrt{A_{\parallel}^2\cos^2\theta + A_{\perp}^2\sin^2\theta} \tag{2.17}$$

where the unique axis (z) is designated || and the two equivalent axes (x,y) are designated ⊥.

At the standard EPR frequency of 9 GHz, the Zeeman anisotropy is small, and primarily we can consider the hyperfine anisotropy that is given by Eq. 2.17. The normalized intensity distribution in the powder pattern for the absorption spectrum of each outer hyperfine line is then given from Eqs. 2.13, 2.15 and 2.17 by:

$$I(B_m) = \frac{1}{\sqrt{B_{\parallel}^2 - B_{\perp}^2}} \cdot \frac{B_m}{\sqrt{B_m^2 - B_{\perp}^2}} \tag{2.18}$$

where $B_{\parallel} = A_{\parallel}m_I/g\beta_e$ and $B_{\perp} = A_{\perp}m_I/g\beta_e$, with g assumed to be isotropic. Note that $I(B_m)$ has units of inverse field, so that the normalized integral: $\int_{B_{\perp}}^{B_{\parallel}} I(B_m)\cdot \mathrm{d}B_m = 1$ is dimensionless. The envelope pattern described by Eq. 2.18 is dominated by the divergence at $B_m(\theta = \pi/2) = B_{\perp}$ (see Figure 2.6, top panel). This corresponds to the degeneracy of orientations in the perpendicular plane (see Figure 2.5), as opposed to the unique direction of the parallel axis, which gives the step at $B_m(\theta = 0) = B_{\parallel}$. The perpendicular discontinuity becomes less sharp when we take into account the width of the lines at each orientation. The middle

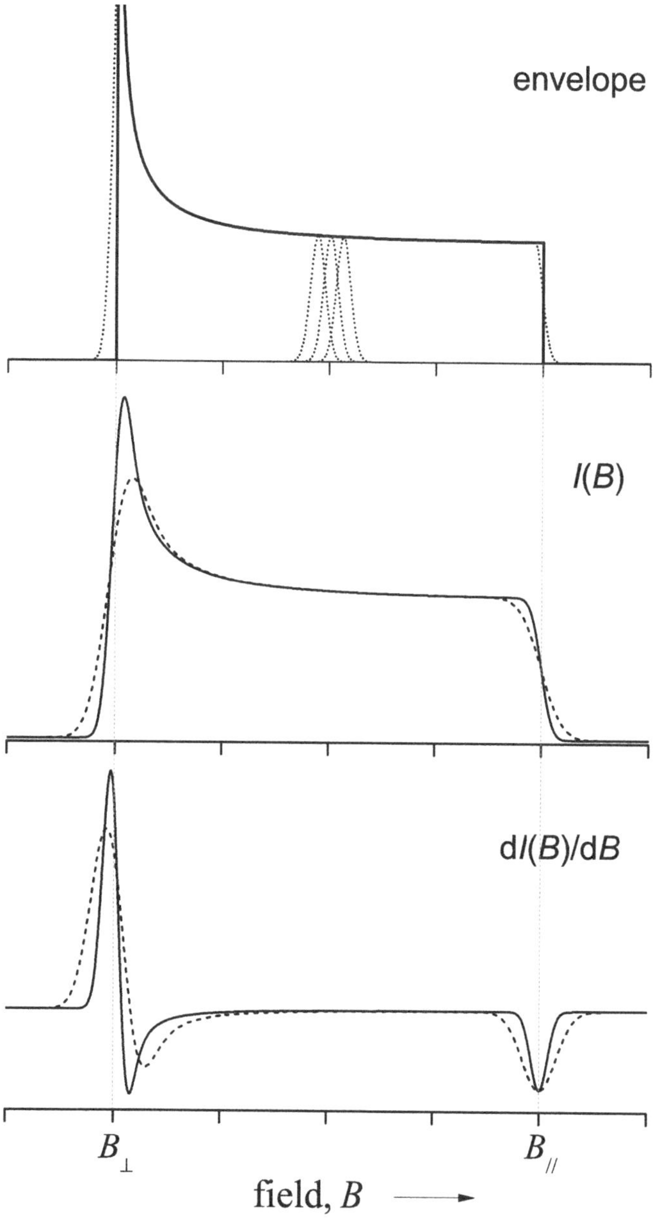

FIGURE 2.6 Powder-spectrum envelope for axial hyperfine anisotropy from Eq. 2.18 (*top panel*). Only high-field manifold ($m_I = -1$) of ^{14}N-nitroxide is shown, extending from $B_{\perp} = A_{\perp}/g\beta_e$ to $B_{\parallel} = A_{\parallel}/g\beta_e$ where g is assumed isotropic. *Middle panel*: spectrum obtained by convoluting with absorption lineshape; *bottom panel*: first derivative. Dashed lines: convolutions with double the linewidth used for solid lines.

panel in Figure 2.6 shows the result of convoluting an absorption lineshape with the envelope $I(B_m)$ (see Section 2.6 below and Appendix T.2, for details of convolution). Lineshapes with intrinsic widths differing by a factor of two are illustrated. The bottom panel of Figure 2.6 gives the usual, first-derivative, EPR lineshape that is recorded by a conventional spectrometer with field modulation. We obtain this spectrum either simply by differentiating the absorption lineshape given in the middle panel or directly by convoluting the envelope function from the top panel with the intrinsic first-derivative lineshape (see Eq. 2.40, later below). Independent of the linewidth, the parallel peak in the first-derivative spectrum is located exactly at the $B_{\parallel}$-turning point. This is not true, however, of the first-derivative perpendicular peak, which corresponds only approximately to the $B_{\perp}$-turning point and deviates more from this for the larger linewidth (see Figure 2.6, bottom panel).

A significant feature of the parallel turning point at $B_{\parallel}$ is that the first-derivative spectrum gives the true *absorption* lineshape for the $\theta = 0$ orientation of the magnetic field (Weil and Hecht 1963; Hubbell and McConnell 1971). When we convolute the distribution in $B_m(\theta)$ with the intrinsic lineshape $a(B-B_m)$ for a particular orientation θ, the absorption lineshape becomes:

$$I(B) = \int_0^{\pi/2} a(B-B_m)\sin\theta \cdot \mathrm{d}\theta \tag{2.19}$$

where $a(B-B_m)$ is centred about the resonance position $B_m(\theta)$. We then get the first-derivative spectrum by differentiating the absorption spectrum with respect to the magnetic field, B:

$$\frac{\mathrm{d}I(B)}{\mathrm{d}B} = \int_0^{\pi/2} \frac{\mathrm{d}a(B-B_m)}{\mathrm{d}B}\sin\theta \cdot \mathrm{d}\theta \tag{2.20}$$

Near $\theta = 0$, we approximate Eq. 2.17 for the hyperfine splitting by: $A(\theta) \approx A_{\parallel}\cos\theta$, and consequently the resonance field is given by: $B_m \approx B_{\parallel}\cos\theta$. In the region of $\theta = 0$, Eq. 2.20 therefore becomes:

$$\frac{\mathrm{d}I(B)}{\mathrm{d}B} = \frac{1}{B_{\parallel}} \int_{\pi/2}^{0} \frac{\mathrm{d}a(B-B_m)}{\mathrm{d}B} \cdot \mathrm{d}B_m \tag{2.21}$$

which we can integrate directly by change of variables from B and B_m to $(B-B_m)$. Thus, around $B = B_{\parallel}$, the first-derivative powder lineshape is well approximated by:

$$\frac{\mathrm{d}I}{\mathrm{d}B} = \frac{1}{B_{\parallel}} a(B-B_{\parallel}) \tag{2.22}$$

which is simply the absorption lineshape for the magnetic field oriented parallel to the nitroxide z-axis. This general result has important implications for the analysis of spin-label spectra. The linewidths are determined by rotational dynamics, and therefore we can use the outer wings of the powder spectra to determine rotational correlation times. Further details are given in Chapters 7 and 8.

At high EPR fields/frequencies, the spin-label spectrum is dominated by the g-value anisotropy, which is non-axial (see Eq. 2.11). Nonetheless, the g-value anisotropy, $\Delta g(\theta,\phi) = g(\theta,\phi) - g_{zz}$, is small compared with the overall g-value, e.g., g_{zz}. Therefore, we can linearize the resonance anisotropy by using the relation: $\delta B/B \approx -\delta g/g$, to give:

$$\Delta B(\theta,\phi) = B_z - B(\theta,\phi) \approx \left(\Delta B_x \cos^2\phi + \Delta B_y \sin^2\phi\right)\sin^2\theta \tag{2.23}$$

where $\Delta B_x = (g_{xx} - g_{zz})B_z/g_{zz}$, $\Delta B_y = (g_{yy} - g_{zz})B_z/g_{zz}$ and $B_z = h\nu/g_{zz}\beta_e$. When we hold angle θ fixed and only vary ϕ, the envelope function is given by:

$$I(\Delta B,\theta) = \frac{1/(2\pi)}{\mathrm{d}\Delta B/\mathrm{d}\phi} = \frac{1}{4\pi\sqrt{\left(\Delta B_x \sin^2\theta - \Delta B\right)\left(\Delta B - \Delta B_y \sin^2\theta\right)}} \tag{2.24}$$

We then obtain the full envelope by integrating over the allowed range of θ:

$$I(\Delta B) = \int_0^{\pi/2} I(\Delta B,\theta)\sin\theta \cdot \mathrm{d}\theta \tag{2.25}$$

We can express Eq. 2.25 as an elliptic integral $K(k)$ (Bloembergen and Rowland 1953):

$$I(\Delta B) = \frac{1}{\pi\sqrt{\Delta B_y}} \frac{1}{\sqrt{\Delta B_x - \Delta B}} K(k) \tag{2.26}$$

for $\Delta B_y > \Delta B > 0$, and

$$I(\Delta B) = \frac{1}{\pi\sqrt{\Delta B_x - \Delta B_y}} \frac{1}{\sqrt{\Delta B}} K(1/k) \tag{2.27}$$

for $\Delta B_x > \Delta B > \Delta B_y$, where

$$k^2 = \frac{\Delta B_x - \Delta B_y}{\Delta B_y} \cdot \frac{\Delta B}{\Delta B_x - \Delta B} \tag{2.28}$$

The complete elliptic integral of the first kind is defined by:

$$K(k) \equiv F(k,\pi/2) = \int_0^{\pi/2} \frac{\mathrm{d}t}{\sqrt{1-k^2\sin^2 t}} = \int_0^1 \frac{\mathrm{d}x}{\sqrt{1-x^2}\sqrt{1-k^2x^2}} \tag{2.29}$$

where $K(0) = \pi/2$ and $K(1) = \infty$.

Figure 2.7 shows the resulting powder lineshapes, omitting N-hyperfine structure, that we obtain with the typical g-value anisotropy for a nitroxyl spin label. The envelope diverges at the intermediate y-turning point that is specified by g_{yy}, whereas the outer turning points specify the step functions at the x- and z-edges (top panel of Figure 2.7). Convolution with an intrinsic

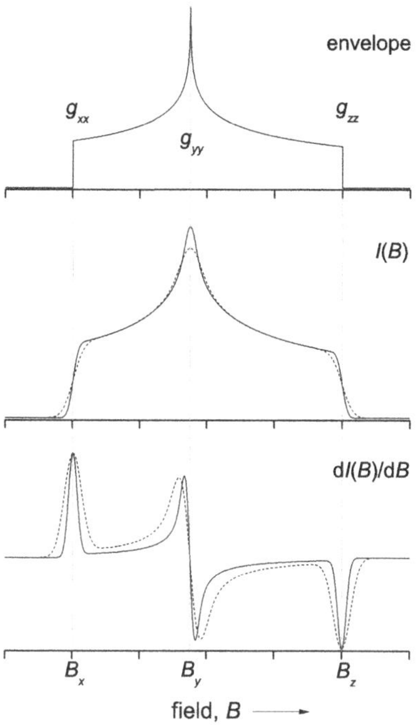

FIGURE 2.7 High field/frequency EPR powder spectra for nitroxide spin labels. N-hyperfine structure not shown, because smaller than Zeeman anisotropy. *Top panel*: powder envelope from Eqs. 2.26–2.28 ($B_x = h\nu/g_{xx}\beta_e$, $B_y = h\nu/g_{yy}\beta_e$, $B_z = h\nu/g_{zz}\beta_e$). *Middle panel*: convolution with absorption lineshape; *bottom panel*: first derivative. Dashed lines: convolutions with double the linewidth used for solid lines.

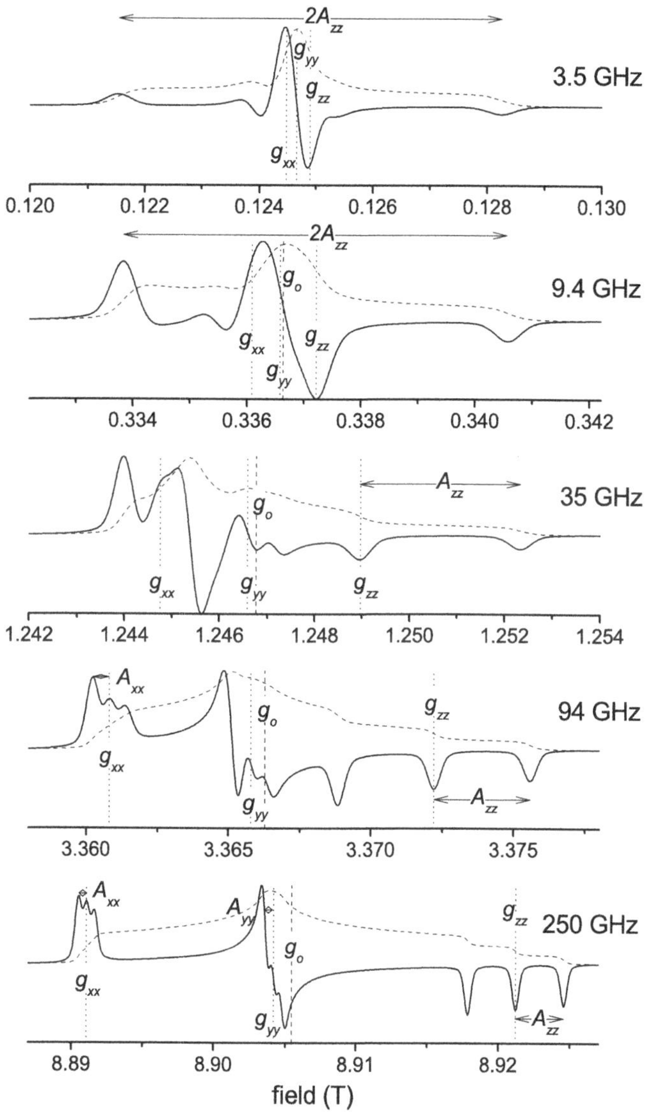

FIGURE 2.8 ^{14}N-nitroxide powder spectra at different EPR frequencies, 3.5–250 GHz. Note different field ranges. First-derivative intrinsic lineshapes are summed numerically (solid lines) for different orientations θ, ϕ of magnetic field to nitroxide principal axes. Absorption spectra (dashed lines) from integrating first-derivative spectrum or summing individual absorption lineshapes. Hyperfine splittings A_{xx}, A_{yy} and A_{zz} in field units; *g*-values g_{xx}, g_{yy} and g_{zz} indicate corresponding resonance-field positions. g_o corresponds to isotropic *g*-value.

absorption lineshape broadens the discontinuity and outer edges of the envelope function (middle panel). The first-derivative spectrum is characterized by a derivative-like feature at B_y and oppositely oriented absorption-like peaks at B_x and B_z in the outer wings (bottom panel). The positions of the outer peaks determine g_{xx} and g_{zz} exactly, and the centre of the derivative feature also corresponds to the g_{yy}-position. As for the axial case, the outer peaks in the derivative spectrum reflect the true absorption lineshapes at B_x and B_z.

In practice, we simulate spin-label EPR powder spectra by using numerical methods. We calculate the resonance fields for a grid of the angles θ and ϕ, and assign an intrinsic lineshape with each of these resonance positions as centre. Summing all these lines then gives us the powder lineshape. Figure 2.8 shows powder lineshapes calculated for a representative ^{14}N-nitroxide at different EPR frequencies. Different features of the hyperfine and *g*-tensors are resolved depending on the microwave operating frequency. Better resolution than that indicated in Figure 2.8 is achieved by using deuterated spin labels. For the lower frequency range, 3.5–35 GHz, the only parameters that we can read immediately from the spectrum, without simulation, are A_{zz} and g_{zz}. At 3.5 GHz, the outer peaks are split by $2A_{zz}$ and are centred about the field corresponding to g_{zz}. The same applies for 9.4 GHz, but the negative peak at the foot of the central line then also corresponds to the g_{zz}-position. At 35 GHz, the g_{zz}-peak is now moved out to high field and the high-field A_{zz}-splitting is clearly resolved. The remainder of the 35-GHz spectrum is characterized by a high degree of spectral overlap. First at high fields (i.e., 94 GHz), can we read the g_{xx}-position directly from the spectrum, and with less certainty the g_{yy}-position. At 250 GHz, the g_{yy}-position is well defined. Whether the minor elements of the *A*-tensor are well resolved at high field/frequency depends very much on the degree of line broadening. We can achieve good resolution of the A_{xx}- and A_{yy}-splittings at 94 GHz by using deuterated spin labels in glass-forming solvents.

The resonance position that corresponds to the isotropic *g*-value, $g_o = \frac{1}{3}(g_{xx} + g_{yy} + g_{zz})$, is also indicated in Figure 2.8. This is significant for measurements at low/intermediate frequency because it corresponds to the mean field position in the absorption powder spectrum, i.e., the point about which the first moment becomes zero (see Appendix U). Sometimes known as the first-moment theorem, this is an approximate result that holds good to better than $\delta g_o = 10^{-4}$ for nitroxides (Hyde and Pilbrow 1980). The value of g_{zz} is found directly from the powder spectrum. (Unfortunately, this is the least interesting g-tensor element because it lies close to the free-electron value – see Section 3.12 in Chapter 3.) Determining g_o from the first-moment theorem then immediately gives the sum: $g_{xx} + g_{yy}$, and if we know one of these two elements, we may calculate the third.

2.5 INHOMOGENEOUS BROADENING

Inhomogeneous (i.e., non-Lorentzian) broadening arises from unresolved proton hyperfine structure in each of the ^{14}N-hyperfine components of the nitroxide spectrum. For equivalent protons, the unresolved hyperfine pattern consists of equally spaced Lorentzian lines whose intensities are given by the binomial distribution (see Figure 2.9). In practice, we find that we can approximate inhomogeneous broadening, from all sources, by a Gaussian distribution of Lorentzian lines.

For a given ^{14}N-hyperfine component, the total peak-to-peak, first-derivative linewidth ΔB_{pp}^{tot} is related to the peak-to-peak Gaussian $\left(\Delta B_{pp}^{G}\right)$ and Lorentzian $\left(\Delta B_{pp}^{L}\right)$ linewidths by the following relation (Dobryakov and Lebedev 1969):

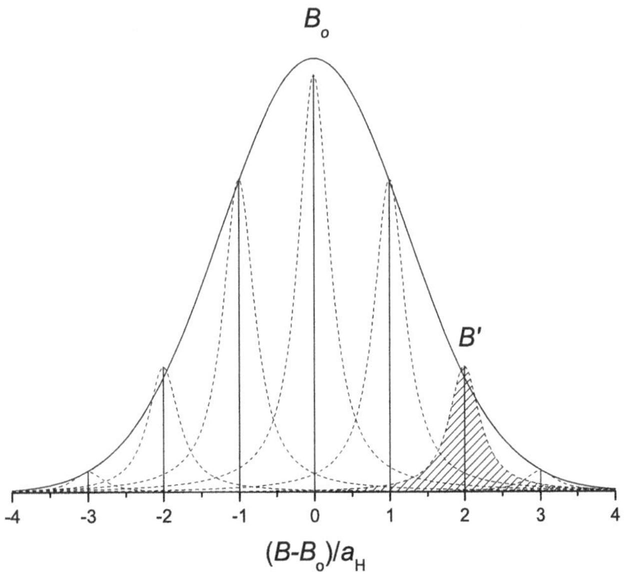

FIGURE 2.9 Hyperfine structure pattern for $N = 6$ equivalent protons ($I = \frac{1}{2}$) with coupling constant a_H (dashed lines). Intensities have binomial distribution in ratios 1:6:15:20:15:6:1. Solid line: Gaussian lineshape (Eq. 1.25) centred at field B_o, with first-derivative peak-to-peak width $\Delta B_{pp}^{G} = 2\sigma_G = \sqrt{N} a_H$, where σ_G^2 is second moment. One of the intrinsic Lorentzian lineshapes (Eq. 1.22) associated with the proton hyperfine multiplet (that centred at field B') is shaded.

$$\left(\frac{\Delta B_{pp}^{G}}{\Delta B_{pp}^{tot}}\right)^2 + \frac{\Delta B_{pp}^{L}}{\Delta B_{pp}^{tot}} = 1 \tag{2.30}$$

This accidental property was discovered by numerical integration of the first-derivative Voigt lineshape. We can use it to extract the Lorentzian linewidth from the measured peak-to-peak linewidth, if the Gaussian broadening is known. For hyperfine multiplets of N_j equivalent nuclei of spin I_j, the Gaussian linewidth is given from Eq. U.23 in Appendix U by:

$$\left(\Delta B_{pp}^{G}\right)^2 = \left(\frac{4\alpha}{3}\right)\sum_j N_j I_j \left(I_j + 1\right) a_j^2 \tag{2.31}$$

where a_j is the hyperfine coupling constant of the jth multiplet, and α is the factor relating the square of the width of the multiplet to the square of the width of a Gaussian line of equal second moment (see, e.g., Figure 2.9). By simulation we find that, within the range of likely proton hyperfine splittings for nitroxide spin labels, the effective Gaussian broadening is not sensitively dependent on the detailed multiplet pattern. Therefore, we can express the Gaussian linewidth in terms of an effective hyperfine coupling, a_e, for $n = \sum N_j$ equivalent protons (Bales 1980):

$$\Delta B_{pp}^{G} = (n\alpha)^{1/2} a_e \tag{2.32}$$

where we find $\alpha \cong 1.06$ from simulation with $I_j = \frac{1}{2}$ for protons.

We give some representative values of n, a_e and ΔB_{pp}^{G} in Table 2.2. Note that, if all protons are equivalent, a_e is given by the true hyperfine coupling constant a_H, and the value of α is accurately determined. We should also remember that the proton coupling will depend on solvent polarity, as does the nitrogen hyperfine coupling (see Section 3.8 in Chapter 3). The values given for guidance in Table 2.2 are means of different determinations of the proton hyperfine couplings in different solvents and at different temperatures. The 6-membered piperidine-ring nitroxides with a tetrahedral carbon at the 4-position and a chair/boat conformation (see Section 3.7 and Table 3.4 in Chapter 3) have similar Gaussian broadenings, in the region of 0.14 mT. On the other hand, TEMPONE with a trigonal carbon at the 4-position and a twisted ring conformation has a much smaller inhomogeneous broadening of 0.03 mT. The 5-membered pyrroline-ring nitroxides ($PYCONH_2$) have a reduced Gaussian linewidth, relative to the TEMPO-like nitroxides, in the region of 0.10 mT. Correspondingly, a comparable Gaussian linewidth of $\Delta B_{pp}^{G} = 0.104 \pm 0.001$ mT is found for the 5-membered pyrrolidine-ring nitroxide PROXYL-COOH (Bales et al. 1998). For oxazolidine-based nitroxides, the two methyl groups and the single methylene of the DOXYL ring have small proton hyperfine couplings, as evidenced by $\Delta B_{pp}^{G} = 0.04$ mT for DOXYL-Me_2. For this class of spin labels, inhomogeneous broadening is contributed mostly by protons from the molecule to which the DOXYL is attached, see, e.g., DOXYLcHx in Table 2.2. Effective Gaussian linewidths in the region of 0.18 mT are estimated for DOXYL-labelled steroids. For comparison,

TABLE 2.2 Effective proton hyperfine couplings a_e, total numbers of protons involved n and resulting Gaussian inhomogeneous linewidths ΔB_{pp}^G, as defined by Eqs. 2.31 and 2.32 with $\alpha \cong 1.06$, for piperidinyl (TEMPO-), pyrrolinyl (PY-) and DOXYL nitroxides

nitroxide[a]	n	a_e (μT)	ΔB_{pp}^G (μT)
TEMPO	18	26.3 ± 1.0	114.9 ± 4.6
TEMPOL	17	33.1 ± 1.5	140.6 ± 6.2
TEMPO-NH_2	17	33.7 ± 1.0	142.9 ± 4.0
TEMPO-Mal	17	32.1	136.4
TEMPONE	16	8.2 ± 2.0	33.8 ± 8.1
PYCONH_2	13	25.9 ± 1.6	96.1 ± 5.8
DTBN	18	9.3 ± 2.0	40.5 ± 8.7
DOXYL-Me_2	14	11.3	43.4
DOXYL cholestane	13	48.1	178.5
DOXYL androstanol	13	48.1	178.5
DOXYL cHx	16	48.9	201.2

Note: standard deviations represent both range of proton hyperfine couplings in different solvents and different determinations. See Appendix A3.2 at the end of Chapter 3.

[a] TEMPO, 1-oxyl-2,2,6,6-tetramethylpiperidine;
TEMPOL, 1-oxyl-2,2,6,6-tetramethylpiperidin-4-ol;
TEMPO-NH_2, 1-oxyl-2,2,6,6-tetramethylpiperidin-4-ylamine;
TEMPO-Mal, *N*-(1-oxyl-2,2,6,6-tetramethylpiperidin-4-yl)maleimide;
TEMPONE, 1-oxyl-2,2,6,6-tetramethylpiperidin-4-yl oxide;
PYCONH_2, 1-oxyl-2,2,5,5-tetramethylpyrrolin-3-ylamide (or other 3-substituents);
DTBN, di-*tert*-butyl nitroxide;
DOXYL-Me_2, 2,2,4,4-tetramethyl-oxazolidine-3-oxyl (2-DOXYL propane);
DOXYL cholestane, 3-(3′-oxyl-4′,4′-dimethyloxazolidin-2′-yl)-5α-cholestane;
DOXYLandrostanol, 3-(3′-oxyl-4′,4′-dimethyloxazolidin-2′-yl)-5α-androstan-17β-ol;
DOXYL cHx, 3-oxyl-4,4-dimethyloxazolidin-2-yl cyclohexane.

Gaussian linewidths of $\Delta B_{pp}^G = 0.080 \pm 0.002\,\text{mT}$ are found for DOXYL-labelled alkyl chains (Bales et al. 1987). The small, noncyclic nitroxide DTBN is characterized by a small Gaussian broadening of ca. 0.04 mT – similar to that of TEMPONE.

In many cases, we may not know the proton hyperfine couplings accurately or they may be subject to exchange narrowing (Sachse et al. 1987). Bales (1982) has devised a simple but effective method for separating Gaussian and Lorentzian contributions to the total linewidth. This uses relative amplitudes in the wings of the spectrum, as shown in Figure 2.10. The diagnostic region is at magnetic-field positions $\pm 1.32\Delta B_{pp}$ distant from the centre of the line. At this point, the amplitude in the wings, h_{wing}, relative to the total peak-to-peak amplitude, h_{pp}, differs considerably between Gaussian and Lorentzian lineshapes (see Figure 1.9). The diagnostic ratio, defined by:

$$\Psi = \frac{h_{wing}}{h_{pp}} \tag{2.33}$$

is $\Psi^G = 0.067$ for a pure Gaussian and is $\Psi^L = 0.213$ for a pure Lorentzian (see Figure 2.10). Empirical calibration of the ratio of Gaussian to Lorentzian linewidths comes from simulating a 19-proton multiplet with differing values of hyperfine coupling and intrinsic Lorentzian width (see Figure 2.11). Fitting with a

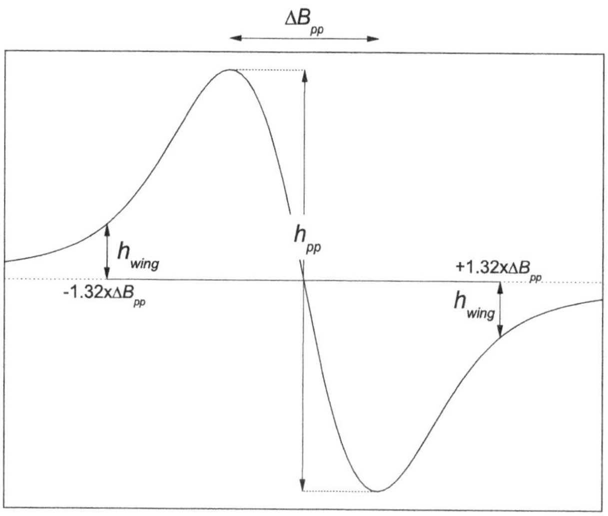

FIGURE 2.10 Single ^{14}N-hyperfine line showing position in spectral wings at $\pm 1.32 \times \Delta B_{pp}$ that defines diagnostic amplitude h_{wing} for Gaussian–Lorentzian decomposition (Bales 1982). Total peak-to-peak amplitude: h_{pp}.

third-order polynomial gives the following dependence on diagnostic ratio Ψ (Bales 1982):

$$\frac{\Delta B_{pp}^G}{\Delta B_{pp}^L} = 16.01 - 236.5\Psi + 1263.5\Psi^2 - 2372.3\Psi^3 \tag{2.34}$$

The fit is shown by the solid line in Figure 2.11. This calibration is valid in the range $\Delta B_{pp}^G / \Delta B_{pp}^L = 0$ to 2.04, the latter limit corresponding to the point at which the multiplet structure no longer remains unresolved. Combination of Eq. 2.34 with Eq. 2.30 then lets us determine both Gaussian and Lorentzian linewidths.

A further calibration, applicable to nitroxides with proton superhyperfine structure that remains unresolved beyond effective Gaussian to Lorentzian ratios of $\Delta B_{pp}^G / \Delta B_{pp}^L = 2.04$, is (Bales 1989):

$$\frac{\Delta B_{pp}^G}{\Delta B_{pp}^L} = \frac{-0.7624\Psi^2 + 0.4091\Psi - 0.0527}{\Psi^2 - 0.2591\Psi + 0.0075} \tag{2.35}$$

This calibration is shown by the dotted line in Figure 2.11. For the lower extents of Gaussian broadening, numerically it does not differ greatly from Eq. 2.34.

2.6 VOIGT LINESHAPES AND SPECTRAL CONVOLUTION

A more complete approach to inhomogeneous broadening, which is not restricted to sharp first-derivative lines, is to use the entire lineshape. The normalized Voigt absorption lineshape is a Gaussian convolution of Lorentzian absorption lines. We must add (i.e., integrate) the contribution at field B from each

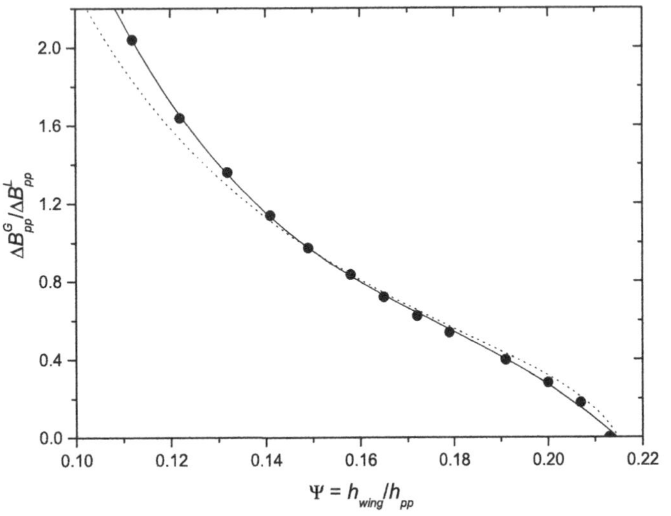

FIGURE 2.11 Dependence of ratio of Gaussian to Lorentzian broadening, $\Delta B^G_{pp}/\Delta B^L_{pp}$, on diagnostic parameter $\Psi = h_{wing}/h_{pp}$ defined in Figure 2.10. Solid circles: from summing 19 binomially-distributed Lorentzian lines (Bales 1982). Solid line: least-squares fit (Eq. 2.34); dotted line: Eq. 2.35 (Bales 1989).

Lorentzian, for every position B' in the Gaussian envelope about which a Lorentzian component is centred (see Figure 2.9):

$$A_V(B) = \frac{1}{\sqrt{2\pi^3}}\left(\frac{\Delta B^L_{1/2}}{\sigma_G}\right) \times \int_{-\infty}^{\infty} \frac{\exp\left(-(B'-B_0)^2/2\sigma_G^2\right)}{\left(\Delta B^L_{1/2}\right)^2 + (B-B')^2} \mathrm{d}B' \qquad (2.36)$$

where $\Delta B^L_{1/2}$ is the half-width at half-height of each Lorentzian absorption component, σ_G^2 is the variance (or second moment) of the Gaussian distribution of Lorentzian components and B_o is the centre of the Gaussian distribution (see Section 1.7 in Chapter 1). The scaling factor in Eq. 2.36 is such that the lineshape is normalized to unit integral: $\int A_V(B)\cdot \mathrm{d}B = 1$. This absorption form of the Voigt lineshape is directly applicable to the outer peaks of the first-derivative powder spectrum because they have pure absorption lineshapes (Guzzi et al. 2009; and see Section 2.4). The first-derivative Voigt lineshape is given from Eq. 2.36 by:

$$\frac{\mathrm{d}A_V}{\mathrm{d}B} = -\sqrt{\frac{2}{\pi^3}}\left(\frac{\Delta B^L_{1/2}}{\sigma_G}\right) \times \int_{-\infty}^{\infty} \frac{(B-B')\exp\left(-(B'-B_o)^2/2\sigma_G^2\right)}{\left(\left(\Delta B^L_{1/2}\right)^2 + (B-B')^2\right)^2} \mathrm{d}B' \qquad (2.37)$$

which is a Gaussian convolution of first-derivative Lorentzian lines. Equation 2.37 may be transformed on integrating by parts to give:

$$\frac{\mathrm{d}A_V}{\mathrm{d}B} = \frac{-1}{\sqrt{2\pi^3}}\left(\frac{\Delta B^L_{1/2}}{\sigma_G^3}\right) \times \int_{-\infty}^{\infty} \frac{(B'-B_o)\exp\left(-(B'-B_o)^2/2\sigma_G^2\right)}{\left(\Delta B^L_{1/2}\right)^2 + (B-B')^2} \mathrm{d}B' \qquad (2.38)$$

which is a Lorentzian convolution of first-derivative Gaussian lines. The latter form is more stable for numerical evaluations when the Gaussian width is greater than the Lorentzian, because the narrower line is then the absorption not the derivative (Bales 1989).

The Voigt lineshape is one particular example of the convolution of an intrinsic lineshape $f(B-B')$ with a broadening function $p(B')$:

$$A(B) = f(B) * p(B) \equiv \int f(B-B')p(B')\cdot \mathrm{d}B' \qquad (2.39)$$

where the asterisk symbol indicates the convolution integral (see Appendix T.2). In Eq. 2.36 for the Voigt lineshape, $f(B-B')$ is the Lorentzian lineshape and $p(B')$ is the Gaussian distribution. Note, however, that convolution is commutative and could equally be viewed as the contribution at field B from a Gaussian lineshape $p(B-B')$ that is broadened by a Lorentzian centred at B_o. The identity of the two forms of the first derivative of the Voigt lineshape that is given by Eqs. 2.37 and 2.38 is therefore a general property of the convolution integral:

$$\frac{\mathrm{d}}{\mathrm{d}B}(f * p) = \left(\frac{\mathrm{d}f}{\mathrm{d}B}\right) * p = f * \left(\frac{\mathrm{d}p}{\mathrm{d}B}\right) \qquad (2.40)$$

This relation is important not only for obtaining the first-derivative spectrum but also for determining partial derivatives that we need in nonlinear least-squares fitting, e.g., by the Levenberg–Marquardt algorithm (Smirnov and Belford 1995). For the latter, we choose the function with an analytical derivative, hence avoiding the need for numerical differentiation.

Spectral convolution is simplified greatly by Fourier transformation (see Appendix T), because the Fourier transform of a convolution integral is just equal to the product of the Fourier transforms of the spectral functions being convoluted:

$$\mathcal{F}\{f * p\} = \mathcal{F}\{f(B)\} \times \mathcal{F}\{p(B)\} \qquad (2.41)$$

where the Fourier transform, $\mathcal{F}(t)$, is defined by:

$$\mathcal{F}\{f(B)\}(t) = \int_{-\infty}^{\infty} f(B)\exp(-iBt)\cdot \mathrm{d}B \qquad (2.42)$$

The inverse Fourier transform is correspondingly obtained from:

$$f(B) = \frac{1}{2\pi}\int_{-\infty}^{\infty} \mathcal{F}(t)\exp(iBt)\cdot \mathrm{d}t \qquad (2.43)$$

This order of the Fourier transform pair is the inverse of that encountered in pulse EPR, or in relaxation theory, where Eq. 2.43 is considered as the forward transform, which produces the spectrum in the frequency domain from the time-domain signal. Note that the Fourier transform of a Lorentzian function is an exponential function: $\mathcal{F}\left\{A_L(B-B_o)\right\} = \exp\left(-\Delta B^L_{1/2} t\right)$, and the Fourier transform of a Gaussian function is itself Gaussian: $\mathcal{F}\left\{A_G(B-B_o)\right\} = \exp\left(-\sigma_G^2 t^2/2\right)$. We obtain the Fourier transform of the first derivative function, df/dB, from that of the original function by using the general rule: $\mathcal{F}\{df/dB\} = it \times \mathcal{F}\{f(B)\}$ (see Appendix T).

We can convolute the intrinsic lineshape $f(B)$ with more than one envelope or broadening function. Possibilities are hyperfine stick patterns representing (partially) resolved proton splittings such as in Figure 2.9 (Smirnova et al. 1995), or instrumental broadening from over modulation (Hyde et al. 1990) or too severe filtering. We see from Eq. 2.41 that we can perform these further convolutions in Fourier space simply by multiplying together the Fourier transforms of the individual contributions, $p_i(B)$:

$$\mathcal{F}\{f * p_1 * p_2 * \ldots\} = \mathcal{F}\{f(B)\} \times \mathcal{F}\{p_1(B)\} \times \mathcal{F}\{p_2(B)\} \times \ldots \tag{2.44}$$

The way in which convolution can be combined with fast Fourier transform procedures for least-squares fitting of experimental spectra is described by Smirnov and Belford (Smirnov and Belford 1995).

A particularly valuable application of convolution with least-squares fitting is for determining additional Lorentzian broadening caused by Heisenberg spin exchange, e.g., with molecular oxygen. In this case, we use the unbroadened spectrum (in the absence of oxygen) as the $p(B)$ lineshape. The $f(B)$ lineshape, which we then obtain by least-squares optimization, gives the additional Lorentzian broadening arising from Heisenberg exchange (Smirnov and Belford 1995). Figure 2.12 shows data

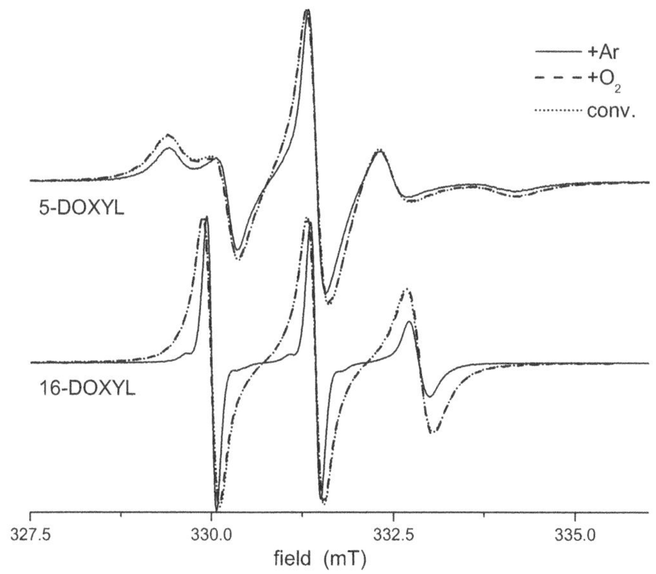

FIGURE 2.12 Convolution (dotted line) of Lorentzian broadening function with spectrum of deoxygenated sample (solid line), fitted to spectrum of same sample in presence of oxygen (dashed line). Convolution and experimental spectrum in oxygen are almost coincident. Lorentzian broadening from Heisenberg exchange between nitroxide spin and paramagnetic oxygen. 9-GHz EPR spectra: 5-DOXYL phosphatidylcholine (PC) (upper row), 16-DOXYL PC (lower row) in fluid phospholipid membranes at 39°C. Lorentzian broadening: $\Delta\Delta B^L_{1/2} = 55 \pm 3$ and $110 \pm 6\,\mu$T for 5-DOXYL PC and 16-DOXYL PC, respectively. (Data from Dzikovski et al. 2003.)

from this type of experiment for spin-labelled lipid chains in the hydrophobic interior of fluid lipid membranes. The spectral lineshapes differ greatly depending on the position of the spin label in the lipid chain (5-DOXYL PC or 16-DOXYL PC). Nevertheless, convolution (dotted line) of the spectrum in the absence of oxygen (solid line) with a Lorentzian absorption lineshape accounts quantitatively for the broadening induced by oxygen (dashed line) throughout the whole of the spectrum, in both cases. We can use a similar approach to allow for inhomogeneous broadening in spectral simulations for spin-label rotational dynamics (cf. Chapter 7). We convolute the simulated spectrum with a Gaussian lineshape, which we then include in the fitting procedure.

2.7 CONCLUDING SUMMARY

1. *Spin-Hamiltonian Tensor Axes:*
   The hyperfine- (A_{xx}, A_{yy}, A_{zz}) and g-tensor (g_{xx}, g_{yy}, g_{zz}) are diagonal in the nitroxide axis system, where the z-axis lies along the symmetry axis of the nitrogen $2p\pi$-orbital and the x-axis along the N–O bond.
2. *Angular Dependence:*
   The dependence of the resonance-field position $B_{m_I}(\theta,\phi)$, of the m_I hyperfine manifold, on orientation of the static magnetic field is given by:

$$B_{m_I}(\theta,\phi) = \frac{h\nu}{g(\theta,\phi)\beta_e} - \frac{A(\theta,\phi)}{g(\theta,\phi)\beta_e} m_I \tag{2.45}$$

   where θ, ϕ are polar angles of the field relative to the nitroxide axes, and ν is the microwave frequency. The angular-dependent g-value and hyperfine coupling are:

$$g(\theta,\phi) = \left(g_{xx}\cos^2\phi + g_{yy}\sin^2\phi\right)\sin^2\theta + g_{zz}\cos^2\theta \tag{2.46}$$

$$A(\theta,\phi) = \sqrt{\left(A_{xx}^2\cos^2\phi + A_{yy}^2\sin^2\phi\right)\sin^2\theta + A_{zz}^2\cos^2\theta} \tag{2.47}$$

   For axial symmetry, these become:

$$g(\theta) = g_{\parallel}\cos^2\theta + g_{\perp}\sin^2\theta \tag{2.48}$$

$$A(\theta) = \sqrt{A_{\parallel}{}^2\cos^2\theta + A_{\perp}{}^2\sin^2\theta} \tag{2.49}$$

3. *Powder Patterns:*
   The outer extrema in the first-derivative spectra of a randomly oriented sample are given by the corresponding EPR absorption lineshapes. At 9 GHz, these correspond to the $m_I = \pm 1$ hyperfine lines with the magnetic field along the nitroxide z-axis, which are split by $2A_{zz}$ and centred about the g_{zz} position. At 95 GHz, these correspond to the g_{xx} resonance at low field and the g_{zz} resonance (which is split by $\pm A_{zz}$) at high field.

4. *Inhomogeneous Broadening:*
   If the Gaussian inhomogeneous broadening ΔB_{pp}^{G} is known, the intrinsic Lorentzian broadening ΔB_{pp}^{L} is obtained from the measured peak-to-peak linewidth ΔB_{pp} by:

$$\left(\frac{\Delta B_{pp}^{G}}{\Delta B_{pp}}\right)^2 + \frac{\Delta B_{pp}^{L}}{\Delta B_{pp}} = 1 \qquad (2.50)$$

   For symmetrical first-derivative lines, the ratio of Gaussian to Lorentzian broadening $\Delta B_{pp}^{G}/\Delta B_{pp}^{L}$ can be obtained from a diagnostic line height in the wings as defined by Bales (see Section 2.5 and Eqs. 2.34, 2.35).
5. *Convolution – Voigt Lineshape:*
   An inhomogeneously broadened lineshape is given by a Gaussian convolution of Lorentzian lines with half-width $\Delta B_{1/2}^{L}$ at half-height in absorption. This normalized Voigt absorption lineshape is:

$$A_V(B) = \frac{1}{\sqrt{2\pi^3}}\left(\frac{\Delta B_{1/2}^{L}}{\sigma_G}\right) \times \int_{-\infty}^{\infty} \frac{\exp\left(-(B'-B_o)^2/2\sigma_G^2\right)}{\left(\Delta B_{1/2}^{L}\right)^2 + (B-B')^2}\,\mathrm{d}B' \qquad (2.51)$$

where σ_G^2 is the variance (or second moment) of the Gaussian distribution of Lorentzian components, and B_o is the centre of the Gaussian distribution. If the Gaussian width is greater than the Lorentzian, the first-derivative Voigt is given by:

$$\frac{\mathrm{d}A_V}{\mathrm{d}B} = \frac{-1}{\sqrt{2\pi^3}}\left(\frac{\Delta B_{1/2}^{L}}{\sigma_G^3}\right) \times \int_{-\infty}^{\infty} \frac{(B'-B_o)\exp\left(-(B'-B_o)^2/2\sigma_G^2\right)}{\left(\Delta B_{1/2}^{L}\right)^2 + (B-B')^2}\,\mathrm{d}B' \qquad (2.52)$$

and by Eq. 2.37 for the reverse situation.

The Fourier transform of the first derivative of a spectrum $f(B)$ convoluted with a broadening function $p(B)$ is given by the product of the Fourier transform of the first-derivative spectrum with the Fourier transform of the broadening function:

$$\mathcal{F}\left\{\mathrm{d}(f * p)/\mathrm{d}B\right\} = \mathcal{F}\{\mathrm{d}f/\mathrm{d}B\} \times \mathcal{F}\{p(B)\} \qquad (2.53)$$

The first-derivative broadened spectrum is then recovered by taking the inverse Fourier transform.

APPENDIX A2: SPIN-HAMILTONIAN TENSORS

TABLE A2.1 Magnetic tensor elements for DOXYL (3-oxyl-4,4-dimethyloxazolidin-2-yl) nitroxides in solid host matrices. ^{14}N-hyperfine tensor and *g*-tensor elements are given by A_{ii} and g_{ii}, with $i \equiv x, y, z$ as defined in Figure 2.1[a]

	DOXYL androstanol[b]	*DOXYL cholestane*[c]	*DOXYL cholestane*[d]	*DOXYL-Me$_2$*[e]	*5-DOXYL SA*[f]	*5-DOXYL PC*[g]	*14-DOXYL PC*[g]
g_{xx}	$2.00913 \pm 3{\times}10^{-5}$	$2.00855 \pm 3{\times}10^{-5}$	$2.0089 \pm 1{\times}10^{-3}$	$2.0088 \pm 5{\times}10^{-4}$	$2.00895 \pm 5{\times}10^{-5}$	2.00857	2.00889
g_{yy}	$2.00610 \pm 3{\times}10^{-5}$	$2.00574 \pm 3{\times}10^{-5}$	$2.0058 \pm 1{\times}10^{-3}$	$2.0058 \pm 5{\times}10^{-4}$	$2.00617 \pm 5{\times}10^{-5}$	2.00586	2.00587
g_{zz}	$2.00231 \pm 3{\times}10^{-5}$	$2.00221 \pm 3{\times}10^{-5}$	$2.0021 \pm 1{\times}10^{-3}$	$2.0022 \pm 5{\times}10^{-4}$	$2.00257 \pm 5{\times}10^{-5}$	2.00197	2.00201
g_o	$2.00585 \pm 3{\times}10^{-5}$	$2.00550 \pm 3{\times}10^{-5}$	$2.0056 \pm 1{\times}10^{-3}$	$2.0056 \pm 5{\times}10^{-4}$	$2.00590 \pm 5{\times}10^{-5}$	2.00547	2.00559
A_{xx}	14.9 ± 0.6 MHz (0.53 ± 0.02 mT)	20.0 ± 0.6 MHz (0.71 ± 0.02 mT)	16.3 ± 1.4 MHz (0.58 ± 0.05 mT)	16.6 ± 1.4 MHz (0.59 ± 0.05 mT)	–	–[h]	–[h]
A_{yy}	13.8 ± 0.6 MHz (0.49 ± 0.02 mT)	12.6 ± 0.6 MHz (0.45 ± 0.02 mT)	16.3 ± 1.4 MHz (0.58 ± 0.05 mT)	15.2 ± 1.4 MHz (0.54 ± 0.05 mT)	18.5 ± 0.6 MHz (0.66 ± 0.02 mT)	–[h]	–[h]
A_{zz}	90.8 ± 0.6 MHz (3.24 ± 0.02 mT)	84.6 ± 1.1 MHz (3.02 ± 0.04 mT)	86.3 ± 1.4 MHz (3.08 ± 0.05 mT)	92.2 ± 1.4 MHz (3.29 ± 0.05 mT)	94.5 ± 0.6 MHz (3.37 ± 0.02 mT)	95.5 MHz (3.41 mT)	91.6 MHz (3.27 mT)
a_o	39.8 ± 0.6 MHz (1.42 ± 0.02 mT)	39.1 ± 0.7 MHz (1.39 ± 0.03 mT)	39.6 ± 1.4 MHz (1.41 ± 0.05 mT)	41.3 ± 1.4 MHz (1.47 ± 0.05 mT)	43.8 ± 1.1 MHz (1.56 ± 0.04 mT)		

[a] $g_o = \frac{1}{3}(g_{xx} + g_{yy} + g_{zz})$

$a_o = \frac{1}{3}(A_{xx} + A_{yy} + A_{zz})$.

[b] 3-(3′-oxyl-4′,4′-dimethyloxazolidin-2′-yl)-5α-androstan-17β-ol in *o*-xylene at −135°C (Smirnova et al. 1995).

[c] 3-(3′-oxyl-4′,4′-dimethyloxazolidin-2′-yl)-5α-cholestane in toluene at −123°C (Earle et al. 1993).

[d] 3-(3′-oxyl-4′,4′-dimethyloxazolidin-2′-yl)-5α-cholestane in single crystal of cholesteryl chloride (Hubbell and McConnell 1971).

[e] 2-(3′-oxyl-4′,4′-dimethyloxazolidin-2′-yl)propane in single crystal of 2,2,4,4-tetramethyl-1,3-cyclobutadione at 23°C (Jost et al. 1971).

[f] 5-(3′-oxyl-4′,4′-dimethyloxazolidin-2′-yl) stearic acid in dry dimyristoyl phosphatidylcholine:myristic acid 1:2 mol/mol mixture at room temperature (Gaffney and Marsh 1998).

[g] 1-acyl-2-[*n*-(3′-oxyl-4′,4′-dimethyloxazolidin-2′-yl)stearoyl]-*sn*-glycero-3-phosphocholine (*n*-DOXYL PC) in dimyristoyl phosphatidylcholine+40 mol% cholesterol bilayer membranes at 30°C (Livshits et al. 2004). Values of A_{zz} determined at −100°C by selecting positional isomers that have the same value of g_o as that for the positional isomer in question at 30°C.

[h] Values in the region of 0.5 mT but not determined with any certainty from the powder patterns (Livshits et al. 2004).

TABLE A2.2 Magnetic tensor elements for piperidinyl nitroxides (6-membered ring) in solid host matrices. ^{14}N-hyperfine tensor and *g*-tensor elements are given by A_{ii} and g_{ii}, with $i \equiv x, y, z$ as defined in Figure 2.1[a]

	TEMPONE[b]				*TEMPO*[c]			*TEMPOL*[e]	
	toluene	*acetone*	*EtOH*	*85% aq. glycerol*	*toluene*	*oleic acid-oil*	*TOAC*[d]	*crystal*	*toluene*
g_{xx}	2.00936 ± 3×10^{-5}	2.0095 ± 3×10^{-4}	2.0092 ± 4×10^{-4}	2.0084 ± 2×10^{-4}	2.00980 ± 7×10^{-5}	2.00955 ± 5×10^{-5}	2.00915	2.0099	2.00988
g_{yy}	2.00633 ± 3×10^{-5}	2.0062 ± 3×10^{-4}	2.0061 ± 4×10^{-4}	2.0060 ± 2×10^{-4}	2.00622 ± 7×10^{-5}	2.00632 ± 5×10^{-5}	2.00666	2.0061	2.00614
g_{zz}	2.00233 ± 3×10^{-5}	2.0022 ± 2×10^{-4}	2.0022 ± 3×10^{-4}	2.0022 ± 1×10^{-4}	2.00220 ± 7×10^{-5}	2.00240 ± 5×10^{-5}	2.00269	2.0024	2.00194
g_o	2.00601 ± 3×10^{-5}	2.0060 ± 3×10^{-4}	2.0058 ± 4×10^{-4}	2.0055 ± 2×10^{-4}	2.00607 ± 7×10^{-5}	2.00609 ± 5×10^{-5}	2.00617	2.0061	2.00599
A_{xx}	11.5 ± 0.6 MHz (0.41 ± 0.02 mT)	13.5 ± 1.4 MHz (0.48 ± 0.05 mT)	13.4 ± 1.7 MHz (0.47 ± 0.06 mT)	15.5 ± 1.4 MHz (0.55 ± 0.05 mT)	16.9 ± 0.3 MHz (0.60 ± 0.01 mT)	19.2 ± 1.4 MHz (0.684 ± 0.050 mT)	19.3 MHz (0.68 mT)	14.9 MHz (0.53 mT)	20.5 ± 1 MHz (0.73 ± 0.04 mT)
A_{yy}	14.3 ± 0.6 MHz (0.51 ± 0.02 mT)	15.2 ± 1.4 MHz (0.54 ± 0.05 mT)	15.9 ± 1.7 MHz (0.56 ± 0.06 mT)	16.0 ± 1.4 MHz (0.57 ± 0.05 mT)	20.5 ± 0.3 MHz (0.73 ± 0.01 mT)	19.2 ± 0.6 MHz (0.684 ± 0.020 mT)	12.4 MHz (0.44 mT)	19.7 MHz (0.70 mT)	21.75 ± 1 MHz (0.77 ± 0.04 mT)
A_{zz}	94.2 ± 0.6 MHz (3.36 ± 0.02 mT)	95.3 ± 0.8 MHz (3.40 ± 0.03 mT)	98.4 ± 1.1 MHz (3.51 ± 0.04 mT)	100.3 ± 0.8 MHz (3.58 ± 0.03 mT)	95.8 ± 0.8 MHz (3.42 ± 0.03 mT)	98.1 ± 0.6 MHz (3.50 ± 0.02 mT)	100.9 MHz (3.60 mT)	98.1 MHz (3.50 mT)	98 ± 1 MHz (3.50 ± 0.04 mT)
a_o	40.0 ± 0.6 MHz (1.43 ± 0.02 mT)	41.3 ± 1.2 MHz (1.47 ± 0.04 mT)	42.5 ± 1.5 MHz (1.52 ± 0.05 mT)	43.9 ± 1.2 MHz (1.57 ± 0.04 mT)	44.4 ± 0.5 MHz (1.58 ± 0.02 mT)	45.5 ± 0.8 MHz (1.62 ± 0.03 mT)	44.2 MHz (1.58 mT)	44.2 MHz (1.58 mT)	46.8 ± 1 MHz (1.67 ± 0.04 mT)

[a] $g_o = \frac{1}{3}(g_{xx} + g_{yy} + g_{zz})$

$a_o = \frac{1}{3}(A_{xx} + A_{yy} + A_{zz})$.

[b] 1-oxyl-2,2,6,6-tetramethylpiperidin-4-yl oxide in: toluene at below −130°C (Budil et al. 1993), frozen acetone, frozen ethanol or 85% glycerol-water at −62°C (Hwang et al. 1975).

[c] 1-oxyl-2,2,6,6-tetramethylpiperidine in toluene glass (Ondar et al. 1985) or in oleic acid-canola oil (2:1 v/v) at −135°C (Smirnov and Belford 1995).

[d] 1-oxyl-2,2,6,6-tetramethyl-4-aminopiperidin-4-yl carboxylic acid (TOAC) in nonapeptide Fmoc-Aib$_8$-TOAC-OMe (where Aib is α-aminoisobutyric acid and Fmoc is fluoren-9-ylmethyloxycarbonyl) in acetonitrile at −193°C (Carlotto et al. 2011).

[e] 1-oxyl-2,2,6,6-tetramethylpiperidin-4-ol in single crystal of 2,2,6,6-tetramethyl-piperidin-4-ol at 23°C (Tabak et al. 1983) or in toluene at −233°C (Florent et al. 2011).

TABLE A2.3 Magnetic tensor elements for pyrrolinyl nitroxide (5-membered ring) and for di-*tert*-butyl nitroxide (DTBN) in solid host matrices. ^{14}N hyperfine tensor and *g*-tensor elements are given by A_{ii} and g_{ii}, with $i \equiv x, y, z$ as defined in Figure 2.1[a]

	PYMeOH[b]			
	o-terphenyl	*glycerol*	*bi-MTSSL*[c]	*DTBN*[d]
g_{xx}	$2.00905 \pm 2\times10^{-5}$	$2.00841 \pm 2\times10^{-5}$	$2.0086 \pm 2\times10^{-4}$	$2.00872 \pm 5\times10^{-5}$
g_{yy}	$2.00617 \pm 2\times10^{-5}$	$2.00604 \pm 2\times10^{-5}$	$2.0067 \pm 1\times10^{-4}$	$2.00616 \pm 5\times10^{-5}$
g_{zz}	$2.00227 \pm 2\times10^{-5}$	$2.00223 \pm 2\times10^{-5}$	$2.0023 \pm 1\times10^{-4}$	$2.00270 \pm 5\times10^{-5}$
g_o	$2.00583 \pm 2\times10^{-5}$	$2.00556 \pm 2\times10^{-5}$	$2.0059 \pm 1\times10^{-4}$	$2.00586 \pm 5\times10^{-5}$
A_{xx}	13.60 ± 0.15 MHz (0.484 ± 0.005 mT)	14.70 ± 0.15 MHz (0.523 ± 0.005 mT)	16.7 ± 0.3 MHz (0.593 ± 0.011 mT)	21.34 ± 0.14 MHz (0.759 ± 0.005 mT)
A_{yy}	13.50 ± 0.15 MHz (0.481 ± 0.005 mT)	14.70 ± 0.15 MHz (0.524 ± 0.005 mT)	11.3 ± 0.3 MHz (0.401 ± 0.011 mT)	16.71 ± 0.14 MHz (0.595 ± 0.005 mT)
A_{zz}	93.20 ± 0.15 MHz (3.326 ± 0.005 mT)	101.40 ± 0.15 MHz (3.618 ± 0.005 mT)	105.0 ± 0.5 MHz (3.747 ± 0.019 mT)	89.08 ± 0.14 MHz (3.178 ± 0.005 mT)
a_o	40.10 ± 0.15 MHz (1.430 ± 0.005 mT)	43.60 ± 0.15 MHz (1.555 ± 0.005 mT)	44.4 ± 0.4 MHz (1.580 ± 0.014 mT)	42.38 ± 0.14 MHz (1.511 ± 0.005 mT)

[a] $g_o = \frac{1}{3}(g_{xx} + g_{yy} + g_{zz})$
$a_o = \frac{1}{3}(A_{xx} + A_{yy} + A_{zz})$.

[b] 1-oxyl-2,2,5,5-tetramethylpyrrolin-3-yl methanol in *o*-terphenyl or glycerol at −93°C (Savitsky et al. 2008).

[c] 2,2,5,5-tetramethylpyrroline-1-oxyl-3,4-bismethyl(methanethiosulphonate) cross-linked to T4 lysozyme (Fleissner et al. 2011).

[d] di-*tert*-butyl nitroxide in single crystal of 2,2,4,4-tetramethyl-1,3-cyclobutadione at 23°C (Libertini and Griffith 1970).

TABLE A2.4 Magnetic tensor elements for ^{15}N-nitroxide in solid host matrix. ^{15}N hyperfine tensor and *g*-tensor elements are given by A_{ii} and g_{ii}, with $i \equiv x, y, z$ as defined in Figure 2.1[a]

	TEMPOL[b]	
	90% glycerol	$A_{ii} \times \gamma_n(^{14}\mathrm{N})/\gamma_n(^{15}\mathrm{N})$[c]
g_{xx}	2.00834	–
g_{yy}	2.00575	–
g_{zz}	2.0020	–
g_o	2.00536	–
A_{xx}	27.43 MHz (0.976 mT)	19.56 MHz (0.696 mT)
A_{yy}	26.53 MHz (0.945 mT)	18.92 MHz (0.674 mT)
A_{zz}	141.00 MHz (5.032 mT)	100.51 MHz (3.587 mT)
a_o	65.06 MHz (2.318 mT)	46.37 MHz (1.652 mT)

[a] $g_o = \frac{1}{3}(g_{xx} + g_{yy} + g_{zz})$
$a_o = \frac{1}{3}(A_{xx} + A_{yy} + A_{zz})$.

[b] 1-oxyl-2,2,6,6-tetramethylpiperidin-4-ol in 90% aqueous glycerol at −34°C (Haas et al. 1993).

[c] $\gamma_n(^{14}\mathrm{N})/\gamma_n(^{15}\mathrm{N}) = 0.7129$.

REFERENCES

Abragam, A. and Bleaney, B. 1970. *Electron Paramagnetic Resonance of Transition Ions*, Oxford: Oxford University Press.

Bales, B.L. 1980. A simple accurate method of correcting for unresolved hyperfine broadening in the EPR of nitroxide spin probes to determine the intrinsic linewidth and Heisenberg spin exchange frequency. *J. Magn. Reson.* 38:193–205.

Bales, B.L. 1982. Correction for inhomogeneous line broadening in spin labels, II. *J. Magn. Reson.* 48:418–430.

Bales, B.L. 1989. Inhomogeneously broadened spin-label spectra. In *Biological Magnetic Resonance*, Vol. 8, Spin Labeling: Theory and Applications, eds. Berliner, L.J. and Reuben, J., 77–130. New York: Plenum Publishing Corporation.

Bales, B.L., Schumacher, K.L., and Harris, F.L. 1987. Correction for inhomogeneous line broadening in spin labels. III. Doxyl-labeled alkyl chains. *J. Magn. Reson.* 72:364–368.

Bales, B.L., Peric, M., and Lamy-Freund, M.T. 1998. Contributions to the Gaussian line broadening of the proxyl spin probe EPR spectrum due to magnetic-field modulation and unresolved proton hyperfine structure. *J. Magn. Reson.* 132:279–286.

Bloembergen, N. and Rowland, T.J. 1953. On the nuclear magnetic resonance in metals and alloys. *Acta Metallurgica* 1:731–746.

Budil, D.E., Earle, K.A., and Freed, J.H. 1993. Full determination of the rotational diffusion tensor by electron paramagnetic resonance at 250 GHz. *J. Phys. Chem.* 97:1294–1303.

Carlotto, S., Zerbetto, M., Shabestari, M.H. et al. 2011. *In silico* interpretation of cw-ESR at 9 and 95 GHz of mono- and bis- TOAC-labeled Aib-homopeptides in fluid and frozen acetonitrile. *J. Phys. Chem. B* 115:13026–13036.

Carrington, A. and McLachlan, A.D. 1969. *Introduction to Magnetic Resonance. With Applications to Chemistry and Chemical Physics*, New York: Harper and Row.

Dobryakov, S.N. and Lebedev, Y. 1969. Analysis of spectral lines whose profile is described by a composition of Gaussian and Lorentz profiles. *Sov. Phys – Dokl.* 13:873–875.

Dzikovski, B.G., Livshits, V.A., and Marsh, D. 2003. Oxygen permeation profile in lipid membranes: non-linear spin-label EPR. *Biophys. J.* 85:1005–1012.

Earle, K.A., Budil, D.E., and Freed, J.H. 1993. 250-GHz EPR of nitroxides in the slow-motional regime - models of rotational diffusion. *J. Phys. Chem.* 97:13289–13297.

Fleissner, M.R., Bridges, M.D., Brooks, E.K. et al. 2011. Structure and dynamics of a conformationally constrained nitroxide side chain and applications in EPR spectroscopy. *Proc. Natl. Acad. Sci. USA* 108:16241–16246.

Florent, M., Kaminker, I., Nagaranjan, V., and Goldfarb, D. 2011. Determination of the ^{14}N quadrupole coupling constant of nitroxide spin probes by W-band ELDOR-detected NMR. *J. Magn. Reson.* 210:192–199.

Gaffney, B.J. and Marsh, D. 1998. High-frequency, spin-label EPR of nonaxial lipid ordering and motion in cholesterol-containing membranes. *Proc. Natl. Acad. Sci. USA* 95:12940–12943.

Guzzi, R., Bartucci, R., Sportelli, L., Esmann, M., and Marsh, D. 2009. Conformational heterogeneity and spin-labelled -SH groups: pulsed EPR of Na,K-ATPase. *Biochemistry* 48:8343–8354.

Haas, D.A., Sugano, T., Mailer, C., and Robinson, B.H. 1993. Motion in nitroxide spin labels - direct measurement of rotational correlation times by pulsed electron double resonance. *J. Phys. Chem.* 97:2914–2921.

Hubbell, W.L. and McConnell, H.M. 1971. Molecular motion in spin-labelled phospholipids and membranes. *J. Am. Chem. Soc.* 93:314–326.

Hwang, J.S., Mason, R.P., Hwang, L.P., and Freed, J.H. 1975. Electron spin resonance studies of anisotropic rotational reorientation and slow tumbling in liquid and frozen media. 3. Perdeuterated 2,2,6,6-tetramethyl-4-piperidone *N*-oxide and analysis of fluctuating torques. *J. Phys. Chem.* 79:489–511.

Hyde, J.S., Pasenkiewicz-Gierula, M., Lesmanowicz, A., and Antholine, W.E. 1990. Pseudo field modulation in EPR spectroscopy. *Appl. Magn. Reson.* 1:483–496.

Hyde, J.S. and Pilbrow, J.R. 1980. A moment method for determining isotropic g-values from powder EPR. *J. Magn. Reson.* 41:447–457.

Jost, P.C., Libertini, L.J., Hebert, V.C., and Griffith, O.H. 1971. Lipid spin labels in lecithin multilayers. A study of motion along fatty acid chains. *J. Mol. Biol.* 59:77–98.

Libertini, L.J. and Griffith, O.H. 1970. Orientation dependence of the electron spin resonance spectrum of di-t butyl nitroxide. *J. Chem. Phys.* 53:1359–1367.

Livshits, V.A., Kurad, D., and Marsh, D. 2004. Simulation studies on high-field EPR of lipid spin labels in cholesterol-containing membranes. *J. Phys. Chem. B* 108:9403–9411.

Ohnishi, S., Boeyens, J.C.A., and McConnell, H.M. 1966. Spin-labeled hemoglobin crystals. *Proc. Natl. Acad. Sci. USA* 56:809–813.

Ondar, M.A., Grinberg, O.Ya., Dubinskii, A.A., and Lebedev, Ya.S. 1985. Study of the effect of the medium on the magnetic-resonance parameters of nitroxyl radicals by high-resolution EPR spectroscopy. *Sov. J. Chem. Phys.* 3: 781–792.

Sachse, J.-H., King, M.D., and Marsh, D. 1987. ESR determination of lipid diffusion coefficients at low spin-label concentrations in biological membranes, using exchange broadening, exchange narrowing, and dipole-dipole interactions. *J. Magn. Reson.* 71:385–404.

Savitsky, A., Dubinskii, A.A., Plato, M. et al. 2008. High-field EPR and ESEEM investigation of the nitrogen quadrupole interaction of nitroxide spin labels in disordered solids: toward differentiation between polarity and proticity matrix effects on protein function. *J. Phys. Chem. B* 112:9079–9090.

Smirnov, A.I. and Belford, R.L. 1995. Rapid quantitation from inhomogeneously broadened EPR spectra by a fast convolution algorithm. *J. Magn. Reson.* 113:65–73.

Smirnova, T.I., Smirnov, A.I., Clarkson, R.B., and Belford, R.L. 1995. W-Band (95 GHz) EPR spectroscopy of nitroxide radicals with complex proton hyperfine structure: fast motion. *J. Phys. Chem.* 99:9008–9016.

Tabak, M., Alonso, A., and Nascimento, O.R. 1983. Single crystal ESR studies of a nitroxide spin label. I. Determination of the G and A tensors. *J. Chem. Phys.* 79:1176–1184.

Weil, J.A. and Hecht, H.G. 1963. On the powder lineshape of EPR spectra. *J. Chem. Phys.* 38:281–282.

Wertz, J.E. and Bolton, J.R. 1972. *Electron Spin Resonance. Elementary Theory and Practical Applications*, New York: McGraw-Hill Inc.

Hyperfine Interactions and g-Values

3

3.1 INTRODUCTION

This chapter describes the origin of the magnetic spin-Hamiltonian parameters, the g-values and hyperfine couplings, which determine the line positions, splittings and also widths. For many applications of spin-label EPR (e.g., to molecular mobility), we need only know the values of the spin-Hamiltonian parameters, without enquiring into their origin. For other applications, notably those that exploit the sensitivity to environmental polarity (see Chapter 4), some understanding of what determines these values is most helpful. We therefore include a relatively detailed account of the EPR parameters of nitroxide free radicals here, in the hope that this will stimulate development of new applications of spin-labelling, which exploit some of these further properties. In common with other atomic and molecular phenomena, we must describe these EPR properties in terms of quantum mechanics. A short primer on vectors and matrices, quantum mechanical methods, spin angular momentum, and atomic and molecular structure is given in Appendices B–E and K at the end of the book.

3.2 HÜCKEL MOLECULAR ORBITALS FOR A NITROXIDE

The nitrogen atom (configuration $1s^2 2s^2 2p_x p_y p_z$) has seven electrons: two are in the $1s$ atomic orbital; three participate in sp^2 hybridized σ molecular orbitals of the R_1–N, R_2–N and N–O bonds; and the two remaining electrons are in the $2p_z\pi$ molecular orbital. The oxygen atom (configuration $1s^2 2s^2 2p_x p_y^2 p_z$) has eight electrons: two in the $1s$ atomic orbital, one in the $2sp_x$ hybridized σ molecular orbital of the N–O bond, a lone pair in the non-bonding $2sp_x$ hybrid atomic orbital, a lone pair in the $2p_y$ atomic orbital and a single electron in the $2p_z\pi$ molecular orbital. An alternative, and more realistic, hybridization scheme for oxygen is to have the lone pairs in two equivalent sp^2 orbitals, with the third sp^2 hybrid forming the N–O σ-bond (see Section 3.12 later). See Figure 2.1 for definition of the nitroxide x,y,z-axes.

The EPR-active unpaired electron resides primarily in the two-centre π-system, for which the Hückel molecular orbitals are linear combinations of $2p_z$ atomic orbitals, φ_N and φ_O, from the nitrogen and oxygen, respectively:

$$\Psi = c_N \varphi_N + c_O \varphi_O \tag{3.1}$$

where c_N and c_O are the admixture coefficients (see Figure 3.10 given later). Normalization requires that $c_N^2 + c_O^2 = 1$, if overlap between the nitrogen and oxygen atomic orbitals is neglected (cf. Appendix K.2). The unpaired electron density on the nitrogen is then given by c_N^2, and that on the oxygen is similarly $c_O^2 \left(=1-c_N^2\right)$.

Application of the variational principle to the expectation value for the energy, E, of the π-system leads to the following linear equations for the admixture coefficients (Griffith et al. 1974; Reddoch and Konishi 1979):

$$\begin{pmatrix} \alpha_N - E & \beta_{NO} \\ \beta_{NO} & \alpha_O - E \end{pmatrix} \begin{pmatrix} c_N \\ c_O \end{pmatrix} = 0 \tag{3.2}$$

where we neglect overlap between the nitrogen and oxygen atomic orbitals, i.e., $\langle \varphi_N | \varphi_O \rangle \cong 0$ (cf. Eqs. K.10 and K.11). In Eq. 3.2, we define the coulomb integral for the oxygen by:

$$\alpha_O = \langle \varphi_O | \mathcal{H} | \varphi_O \rangle \tag{3.3}$$

where $\mathcal{H}$ is the one-electron Hamiltonian. That for the nitrogen is similarly:

$$\alpha_N = \langle \varphi_N | \mathcal{H} | \varphi_N \rangle \cong \alpha_O + h_{NO}\beta_{NO} \tag{3.4}$$

where the second relation on the right refers the nitrogen coulomb integral to that for oxygen. We approximate the h_{NO} parameter by: $h_{NO} = X_N - X_O = -0.5$, where X_N and X_O are the Pauling electronegativities of nitrogen and oxygen, respectively (Streitwieser 1961). Finally, the resonance integral in Eq. 3.2 is defined by:

$$\beta_{NO} = \langle \varphi_N | \mathcal{H} | \varphi_O \rangle \tag{3.5}$$

where again we use the oxygen atom as reference.

Solution of the secular equation corresponding to the linear equations in Eq. 3.2 gives the energy levels of the nitroxyl molecular orbitals:

$$E^{\pm} = \alpha_O + \frac{\beta_{NO}}{2}\left(h_{NO} \pm \sqrt{4 + h_{NO}^2}\right) \tag{3.6}$$

where the upper sign corresponds to the bonding molecular orbital ($E^{(b)}$), and the lower sign corresponds to the antibonding

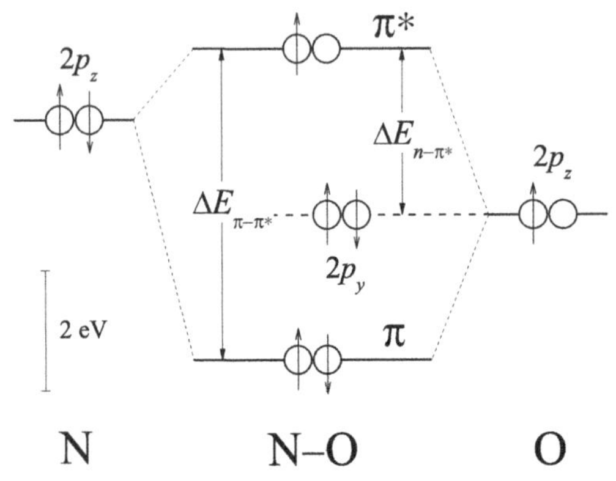

FIGURE 3.1 Energy level diagram for $2p_z\pi$ molecular orbitals of the N–O bond. π and π^* are bonding $(\Psi^{(b)})$ and anti-bonding (Ψ^*) π-molecular orbitals of the N–O bond, where constituent atomic orbitals, φ_N and φ_O, are $2p_z$ orbitals on N and O, as indicated. Energy level of the non-bonding $2p_y$ orbital that contains one of the lone pairs (n) on oxygen is also indicated (dashed line). Excitation energies of allowed π–π^* and overlap-forbidden n–π^* optical transitions are indicated. More generally, the two lone pairs reside in equivalent $2sp^2$ hybrid orbitals on the oxygen. Energy level of the nitrogen $2p_z$ orbital is indicated only qualitatively.

molecular orbital (E^*), because β_{NO} is negative. From Figure 3.1, we see that the energy of the π–π^* optical transition is given by:

$$\Delta E_{\pi-\pi^*} = E^* - E^{(b)} = -\beta_{NO}\sqrt{4+h_{NO}^2} \tag{3.7}$$

The π–π^* optical absorption band lies at 240 nm, corresponding to an excitation energy of $\Delta E_{\pi-\pi^*} = 5.17\,\text{eV}$ (Brière et al. 1965; Kikuchi 1969), and therefore we can approximate the resonance integral by: $\beta_{NO} = -2.5\,\text{eV}$ from Eq. 3.7 (Griffith et al. 1974).

Substituting the solutions given in Eq. 3.6 into Eq. 3.2 now yields the values for the admixture coefficients c_N and c_O, which because of the parametrization in Eq. 3.4 do not depend on β_{NO}. The ratio of admixture coefficients becomes (cf. Eq. K.16 in Appendix K):

$$\frac{c_O}{c_N} = \frac{E^{\pm} - \alpha_N}{\beta_{NO}} = \frac{1}{2}\left(-h_{NO} \pm \sqrt{4+h_{NO}^2}\right) \tag{3.8}$$

which together with the normalization condition gives the individual coefficients.

The antibonding (Ψ^*) and bonding ($\Psi^{(b)}$) molecular orbitals are then of the form (cf. Eqs. G.18 and G.19 in Appendix G, and Griffith et al. 1974):

$$\Psi^* = c_N\varphi_N - c_O\varphi_O \tag{3.9}$$

$$\Psi^{(b)} = c_O\varphi_N + c_N\varphi_O \tag{3.10}$$

where the admixture coefficients interchange and their relative sign reverses on going from the antibonding to the bonding orbital. From the φ_N admixture coefficient in Eq. 3.9, the unpaired spin density on the nitrogen is given by:

$$\rho_\pi^N = |c_N|^2 = \frac{1}{2\left(1+\frac{1}{2}h_{NO}\sqrt{1+\left(\frac{1}{2}h_{NO}\right)^2}+\left(\frac{1}{2}h_{NO}\right)^2\right)} \tag{3.11}$$

Correspondingly, the unpaired spin density on the oxygen is: $\rho_\pi^O = 1-|c_N|^2$, where the unpaired electron resides in the antibonding orbital (see Figure 3.1). With the value of $h_{NO} \cong -0.5$ approximated from electronegativities (see above), this predicts that the spin density on the nitrogen is: $\rho_\pi^N \cong 0.62$ and correspondingly $\rho_\pi^O \cong 0.38$. Clearly, these values depend sensitively on the value chosen for h_{NO} (Griffith et al. 1974). Taking the value $h_{NO} = -0.3\,\text{eV}$ estimated in Appendix K.2, we get instead $\rho_\pi^N \cong 0.57$. From analysis of the polarity dependence of nitroxide hyperfine splittings (see Table 3.1 in Section 3.4 below), we deduce a value of $\rho_\pi^N = 0.61 \pm 0.01$ from an isotropic coupling of $a_o^N = 1.55\,\text{mT}$, which corresponds to moderate-polarity solvents (cf. Chapter 4). This reduces to $\rho_\pi^N = 0.55 \pm 0.01$ for low-polarity environments, where $a_o^N = 1.40\,\text{mT}$.

Polarized-neutron diffraction maps the spin-density distribution in crystals of nitroxides at high magnetic fields and low temperature. For neat nitroxides that are not hydrogen bonded, the unpaired spin density is localized largely to the N–O $p\pi$-bond and is distributed equally between nitrogen and oxygen (Bordeaux et al. 1993; Brown et al. 1983; Pontillon et al. 1999). Normalized spin densities are given for the six-membered ring nitroxides TEMPONE and TEMPOL in Table A3.1 of Appendix A3.1 at the end of the chapter. In TEMPONE, 53% of the N–O spin density is centred on nitrogen and 47% on oxygen. For TEMPOL, this distribution changes to 62%/38% because spin density is shifted onto the nitrogen by hydrogen bonding of the oxygen with the 4-OH hydroxyl from the neighbouring molecule in the crystal. Appendix A3.1 also contains an experimentally parameterized expression (viz. Eq. A3.1) for the molecular orbital of the unpaired electron in a six-membered ring, TEMPO-like nitroxide. This molecular orbital reproduces the spin-density projection obtained from polarized-neutron diffraction (Brown et al. 1983) and predicts that 57% of the unpaired spin density in the N–O bond is on the nitrogen and 43% on the oxygen.

3.3 ISOTROPIC NITROGEN HYPERFINE COUPLINGS

The common ^{14}N-isotope of nitrogen (natural abundance 99.636%) has nuclear spin I = 1 and gyromagnetic ratio relative to the electron of $g_n\beta_n/g_e\beta_e = -1.098202 \times 10^{-4}$. On the other hand, the rare ^{15}N-isotope (natural abundance 0.364%) has $I = \frac{1}{2}$ and 40% larger gyromagnetic ratio of opposite sign $g_n\beta_n/g_e\beta_e = 1.540508 \times 10^{-4}$. The hyperfine coupling constants of ^{15}N are therefore 1.403 × larger than those of ^{14}N.

Magnetic hyperfine couplings arise from interaction of the magnetic moment from the electron spin, $\boldsymbol{\mu}_S = g\beta_e\mathbf{S}$, with that from the nuclear spin, $\boldsymbol{\mu}_I = g_n\beta_n\mathbf{I}$. As we shall see in the following section, magnetic dipole–dipole interactions have no

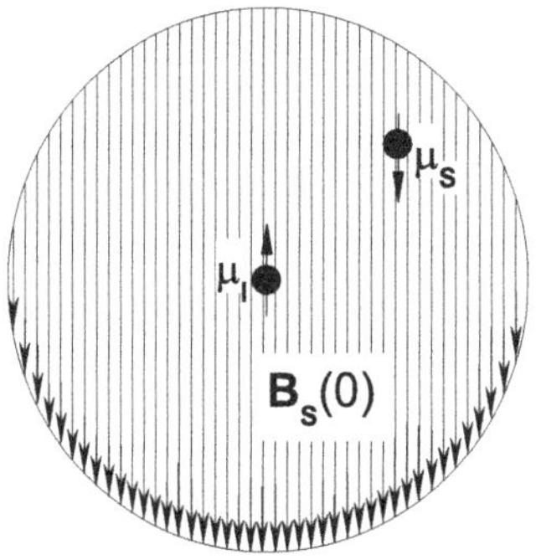

FIGURE 3.2 Fermi contact interaction that gives isotropic hyperfine coupling. Nuclear magnetic moment μ_I interacts with distributed magnetic moment μ_S from the s-electron spin density, which gives uniform magnetic induction $\mathbf{B}_S(0)$ at the nucleus.

isotropic component. Therefore, isotropic hyperfine couplings come only from unpaired electrons that have non-vanishing probability density, $|\psi(0)|^2$, at the N-nucleus and so give rise to the Fermi contact interaction (see Figure 3.2). This means that only unpaired electrons with s-orbital character will contribute to the isotropic hyperfine coupling.

In a semi-classical approach, we assume that the s-electron contributes a uniform magnetization, $\mathbf{M_S}(0) = g\beta_e\mathbf{S}|\psi(0)|^2$, per unit volume at the nucleus. For a spherically symmetric electron distribution, the magnetic induction at the centre is given from classical magnetostatics by: $\mathbf{B_S}(0) = (2/3)\mu_o\mathbf{M_S}(0)$, irrespective of radius. The interaction energy for a nucleus at the centre of the s-electron cloud is then given by the hyperfine Hamiltonian:

$$\mathcal{H}_{hfs}^{iso} = -\mu_I \cdot \mathbf{B_S}(0) = a_o\mathbf{I} \cdot \mathbf{S} \tag{3.12}$$

with isotropic hyperfine coupling constant:

$$a_o = \frac{2\mu_o}{3} g\beta_e g_n\beta_n |\psi(0)|^2 \tag{3.13}$$

where μ_o (= $4\pi \times 10^{-7}$ N A^{-2}) is the permeability of free space. This semi-classical approach is justified by a more rigorous calculation using Dirac's relativistic Hamiltonian (Fermi 1930).

In nitroxide radicals, the unpaired electron occupies a $2p_z\pi$ orbital, which – in common with all π-radicals – has a node at the nucleus (see Figure 2.1). Isotropic hyperfine coupling, therefore, only can come from admixture of nitrogen s-electron orbitals into the $2p_z\pi$ orbital. This occurs by configuration interaction, which arises from quantum-mechanical exchange between the unpaired 2pπ electron and the nitrogen 2s electrons that participate in the N–O σ-bond. In atoms, the exchange integral (see Sections 9.6 and 9.7) is positive, so parallel spins are lower in energy than anti-parallel spins (the first of Hund's rules). Preferential admixture of the σ-orbital with spin parallel to that in the $2p_z\pi$ orbital then creates positive unpaired spin density at the nucleus. Another way of picturing the situation is that exchange interaction with the unpaired p-electron reduces electrostatic repulsion with the s-electron of parallel spin, relative to that with antiparallel spin. This results in different radial distributions of the s-electrons, causing a net positive spin density at the nucleus, which we refer to as core polarization.

One could ask why there is an interaction between atomic configurations. The answer is that the s- and p-configurations, and so on, are not eigenstates of the complete Hamiltonian for the nucleus plus electrons. They are stationary states (i.e., eigenstates) of a Hamiltonian with a spherically symmetrical, or central, potential (e.g., that for a closed electron shell). It is the non-spherically symmetric part of the electrostatic electron–electron repulsion that gives rise to the exchange interaction between electrons in different configurations (see Appendix K.1).

The unpaired spin density at the nitrogen nucleus that results from σ–π interactions between configurations will be directly proportional to the unpaired π-electron spin density, ρ_π^N, on the nitrogen, and possibly also to that on the oxygen, ρ_π^O. In analogy with isotropic proton and ^{13}C-hyperfine splittings of π-radicals (McConnell 1956; Karplus and Fraenkel 1961), the isotropic nitrogen hyperfine coupling is therefore given by (Lemaire and Rassat 1964; Ayscough and Sargent 1966):

$$a_o^N = Q_N^N\rho_\pi^N + Q_{ON}^N\rho_\pi^O \tag{3.14}$$

where Q_N^N and Q_{ON}^N are constants. Note that Q_N^N contains contributions from the nitrogen 1s electrons, as well as from the 2s electrons (Karplus and Fraenkel 1961), although the contribution of the latter is the more significant. Because the unpaired electron is essentially confined to the N–O π-orbital, we have: $\rho_\pi^N + \rho_\pi^O \approx 1$, and thus Eq. 3.14 becomes:

$$a_o^N = \left(Q_N^N - Q_{ON}^N\right)\rho_\pi^N + Q_{ON}^N \tag{3.15}$$

which gives the isotropic coupling solely in terms of the unpaired spin density on the nitrogen. For hyperfine splittings from ^{13}C, the effect of π-spin densities on adjacent atoms is important (Karplus and Fraenkel 1961), but for nitrogen it is found to be small, i.e., $Q_N^N >> Q_{ON}^N$ (Lemaire and Rassat 1964; Ayscough and Sargent 1966; and see Sections 3.4 and 3.6).

3.4 ANISOTROPIC NITROGEN HYPERFINE COUPLINGS

The anisotropic part of the ^{14}N-hyperfine tensor arises from interaction of the nitrogen nuclear magnetic moment, $\mu_I = g_n\beta_n\mathbf{I}$, with the magnetic field, $\mathbf{B}_{dip}$, that is associated with the electron spin dipole, $\mu_S = g\beta_e\mathbf{S}$. This is a classical magnetic dipole–dipole interaction, the geometry for which we show in Figure 3.3. From conventional magnetostatics, the electron dipolar magnetic field at vector position $\mathbf{r}$ is given by (Eq. F.12 in Appendix F):

$$\mathbf{B}_{dip} = \frac{\mu_o}{4\pi}\left(\frac{3(\mu_S \cdot \mathbf{r})\mathbf{r}}{r^5} - \frac{\mu_S}{r^3}\right) \tag{3.16}$$

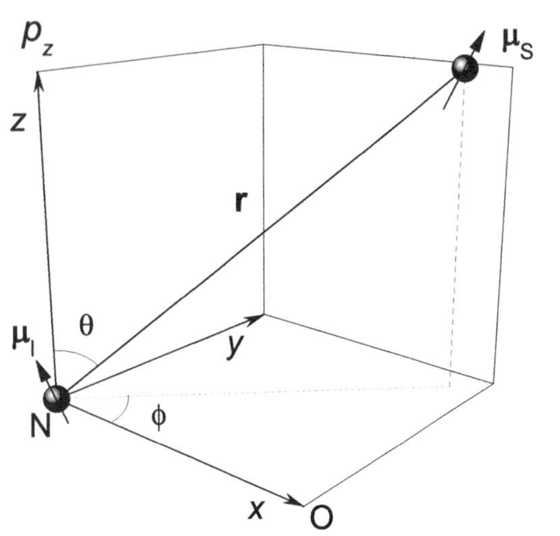

FIGURE 3.3 Magnetic dipole–dipole interaction between unpaired electron spin *S* and nitrogen nuclear spin *I*, which gives anisotropic N-hyperfine splittings. Dipolar hyperfine tensor is diagonal in the nitroxide *x,y,z*-axis system indicated (cf. Figure 2.1) and axially symmetric about the nitrogen $2p_z$ orbital.

where μ_o is the permeability of free space. Therefore, the dipolar contribution to the hyperfine Hamiltonian is:

$$\mathcal{H}_{hfs}^{dip} = -\mathbf{\mu}_I \cdot \mathbf{B}_{dip} = -\left(\frac{\mu_o}{4\pi}\right) g\beta_e g_n \beta_n \left\langle \frac{\mathbf{I}\cdot\mathbf{S}}{r^3} - \frac{3(\mathbf{I}\cdot\mathbf{r})(\mathbf{S}\cdot\mathbf{r})}{r^5} \right\rangle \quad (3.17)$$

where the angular brackets represent averaging over the unpaired electron distribution. Note that the dipolar hyperfine tensor $\mathbf{T}^d$ is traceless, i.e., $\mathrm{Tr}\left(\mathbf{T}_{ij}^d\right) = T_{xx}^d + T_{yy}^d + T_{zz}^d = 0$. This is because the sum of the coefficients of I_xS_x, I_yS_y and I_zS_z of the final term on the right in Eq. 3.17 is: $3\left(x^2+y^2+z^2\right)/r^5 = 3/r^3$, which cancels with the sum of the coefficients of the **I**·**S** term.

The dipolar contribution to the nitrogen hyperfine tensor comes from the unpaired electron in the $2p_z$-orbital (cf. Figure 3.3). This orbital is axially symmetric and, therefore, the cross-terms (*xy*, *yz* and *zx*) within the angular brackets in Eq. 3.17 average to zero, i.e., the dipolar hyperfine tensor is diagonal in the nitroxide axis system of Figure 2.1. Furthermore, the average values of the x^2 and y^2 terms in Eq. 3.17 are equal because of the axial symmetry. Thus, we can write the dipolar hyperfine Hamiltonian as:

$$\mathcal{H}_{hfs}^{dip} = T_{\parallel}^d I_z S_z + T_{\perp}^d \left(I_x S_x + I_y S_y\right) \quad (3.18)$$

where $T_{\parallel}^d \equiv T_{zz}^d$ and $T_{\perp}^d \equiv T_{xx}^d = T_{yy}^d$. Because the dipolar tensor is traceless, we also have:

$$T_{\perp}^d = -\tfrac{1}{2} T_{\parallel}^d \quad (3.19)$$

and need only a single element to specify the dipolar tensor fully. From Eq. 3.17, the parallel element of the dipolar tensor is:

$$T_{\parallel}^d = \left(\frac{\mu_o}{4\pi}\right) g\beta_e g_n \beta_n \left\langle \frac{3z^2 - r^2}{r^5} \right\rangle \quad (3.20)$$

In polar coordinates, the *z*-coordinate in Eq. 3.17 becomes $z = r\cos\theta$, and the angular part of a p_z-wavefunction is: $\left|p_z\right\rangle = \sqrt{3/2}\cos\theta$ (after integrating over ϕ: see Appendix K, Table K.1). Hence, the angular average that we need in Eq. 3.20 is:

$$\left\langle 3\cos^2\theta - 1\right\rangle = \frac{3}{2}\int_0^{\pi}\left(3\cos^2\theta - 1\right)\cos^2\theta\sin\theta \cdot d\theta = \frac{4}{5} \quad (3.21)$$

From Eqs. 3.20 and 3.21, the dipolar coupling for an unpaired electron in a p_z-orbital then becomes:

$$T_{\parallel}^d = \frac{4}{5}\left(\frac{\mu_0}{4\pi}\right) g\beta_e g_n \beta_n \left\langle \frac{1}{r^3} \right\rangle_{2p,\mathrm{N}} \rho_{\pi}^{\mathrm{N}} \quad (3.22)$$

where $\left\langle 1/r^3 \right\rangle$ is the value of $1/r^3$ averaged over the radial distribution of a nitrogen 2*p* orbital, and ρ_{π}^{N} is the unpaired electron spin density on the nitrogen. Using nitrogen self-consistent-field wave functions to determine $\left\langle 1/r^3 \right\rangle_{2p,\mathrm{N}} = 2.093 \times 10^{31}\,\mathrm{m}^{-3}$ (see Whiffen 1964; Clementi et al. 1962), we find that $T_{\parallel,o}^d = \frac{4}{5}\left(\mu_o/4\pi\right) g_e\beta_e g_n\beta_n \left\langle r_{2p}^{-3}\right\rangle$ is 95.6 MHz (≈3.41 mT) for ^{14}N, and −134.2 MHz (≈ −4.79 mT) for ^{15}N, where $T_{\parallel,o}^d$ is the value of $T_{\parallel}^d$ for $\rho_{\pi}^{\mathrm{N}} = 1$.

We obtain the full hyperfine tensor, $\mathbf{A}_{ij}$, by adding the isotropic coupling constant, a_o^{N}, to each of the diagonal elements of the dipolar hyperfine tensor, $\mathbf{T}_{ij}^d$. The principal *z*-element of the hyperfine tensor then becomes:

$$A_{zz} = a_o^{\mathrm{N}} + T_{\parallel}^d \quad (3.23)$$

Using Eq. 3.15 from the previous section to replace the dependence of $T_{\parallel}^d$ on ρ_{π}^{N}, we find that the A_{zz}-element of the hyperfine tensor is related to the isotropic coupling constant, a_o^{N}, by:

$$A_{zz} = \left(1 + \frac{T_{\parallel,o}^d}{Q_{\mathrm{N}}^{\mathrm{N}} - Q_{\mathrm{ON}}^{\mathrm{N}}}\right) a_o^{\mathrm{N}} - \frac{T_{\parallel,o}^d Q_{\mathrm{ON}}^{\mathrm{N}}}{Q_{\mathrm{N}}^{\mathrm{N}} - Q_{\mathrm{ON}}^{\mathrm{N}}} \quad (3.24)$$

which predicts a linear dependence of A_{zz} on a_o^N.

Griffith et al. (1974) demonstrated an approximately linear relation between A_{zz} and a_o^{N} for DOXYL spin-labelled stearic acids or their methyl esters in a number of glass-forming media. We also see a similar dependence for the 6-membered ring nitroxide TEMPONE, as shown in Figure 3.4. Linear regression to the combined data yields A_{zz} = (2.13 ± 0.10) × a_o^{N} + (0.24 ± 0.16) mT, implying from Eq. 3.24 that $Q_{\mathrm{ON}}^{\mathrm{N}} \approx 0$. From the gradient in Figure 3.4, we estimate that $Q_{\mathrm{N}}^{\mathrm{N}} \approx 85 \pm 7$ MHz (3.0 ± 0.3 mT). Table 3.1 lists values of the mean ratio of anisotropic to isotropic hyperfine couplings, A_{zz}/a_o^{N}, for different nitroxide spin labels in a range of solvents that differ in polarity. Also listed are the values of $Q_{\mathrm{N}}^{\mathrm{N}}$ that we deduce from Eq. 3.24 with $Q_{\mathrm{ON}}^{\mathrm{N}} = 0$ and $T_{\parallel,o}^d = 95.6$ MHz. The precision is considerably higher than the estimate from Figure 3.4, with a mean value from Table 3.1 of $Q_{\mathrm{N}}^{\mathrm{N}} = 72.0 \pm 1.4$ MHz (≅ 2.56 ± 0.05 mT). From analysis of perturbations on complex formation with DTBN or TEMPO, Cohen and Hoffman (1973) obtained a linear dependence in agreement with Eq. 3.24 but with a non-zero value of $Q_{\mathrm{ON}}^{\mathrm{N}}$. The electron transfer complexes investigated by these authors produce a much

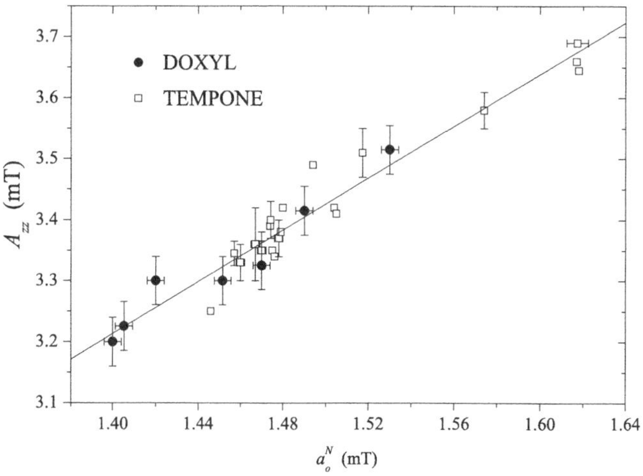

FIGURE 3.4 Correlation of maximum ^{14}N-hyperfine tensor element A_{zz} with isotropic hyperfine coupling constant $a_o^{\rm N}$, for 3-oxyl-4,4-dimethyloxazolidin-2-yl (DOXYL) fatty acid (or methyl ester) nitroxides (solid circles; Griffith et al. 1974), and for perdeuterated 1-oxyl-2,2,6,6-tetramethylpiperdin-4-yl oxide (TEMPONE) (open squares; Zager and Freed 1982), in frozen and fluid media of different polarities. Solid line: linear regression (Eq. 3.24) for combined data set.

TABLE 3.1 Mean ratios of anisotropic to isotropic ^{14}N-hyperfine couplings, $A_{zz}/a_o^{\rm N}$, for nitroxides in solvents of different polarity and proticity, and resulting values of $Q_{\rm N}^{\rm N}$ for σ–π coupling[a]

spin label[b]	$A_{zz}/a_o^{\rm N}$	$Q_{\rm N}^{\rm N}$/MHz (mT)	*Refs.*[c]
DOXYL	2.33 ± 0.05	72 ± 3 (2.57 ± 0.09)	1
	2.29 ± 0.02	74 ± 1 (2.64 ± 0.04)	2
MTSSL	2.34 ± 0.04	71 ± 2 (2.52 ± 0.08)	3
MTSSL/β–SH	2.36 ± 0.02	70 ± 1 (2.50 ± 0.04)	3
TOAC	2.32 ± 0.07	72 ± 4 (2.58 ± 0.13)	1

[a] Deduced from Eq. 3.24 with $Q_{\rm ON}^{\rm N}=0$ and $T_{\parallel,o}^{d}=95.6$ MHz (Whiffen 1964).
[b] DOXYL, 3-oxyl-4,4-dimethyloxazolidin-2-yl (substituted at the 2-position); MTSSL, 1-oxyl-2,2,5,5-tetramethylpyrrolin-3-ylmethyl methanethiosulphonate;
MTSSL/β–SH, β-mercaptoethanol S–S adduct of MTSSL;
TOAC, 1-oxyl-2,2,6,6-tetramethyl-4-aminopiperidin-4-yl carboxylic acid.
[c] References: 1. Marsh and Toniolo (2008); 2. Griffith et al. (1974); 3. Owenius et al. (2001).

larger perturbation of the unpaired electron spin distribution than do the polarity-induced shifts considered here. Therefore, we prefer a value of $Q_{\rm ON}^{\rm N}\approx 0$ in Eq. 3.15 for the unperturbed state. We come to a similar conclusion from analysis of ^{13}C-hyperfine interactions that we give later in Section 3.6.

3.5 PARAMAGNETIC SHIFTS IN NITROXIDE NMR

Before discussing hyperfine couplings other than those of nitrogen, it is useful to consider the shifts in NMR resonance position that are induced by the unpaired electron spin. These paramagnetic shifts determine hyperfine couplings with better resolution than does EPR and also give the sign of the hyperfine interaction. The EPR spectrum, on the other hand, does not depend on the sign of the hyperfine coupling.

In a strong magnetic field, B, the isotropic terms in the spin Hamiltonian that depend on the nuclear spin, I_z, are:

$$\mathcal{H}_n=-g_n\beta_n B\cdot I_z+a_o^{(\rm N)}\langle S_z\rangle I_z \tag{3.25}$$

where g_n and $a_o^{(\rm N)}$ are respectively the nuclear g-factor and isotropic hyperfine coupling of the nucleus (N) concerned, $\beta_n\left(\equiv e\hbar/2M_p=5.050783\times10^{-27}\,{\rm J.T^{-1}}\right)$ is the nuclear magneton, and $\langle S_z\rangle$ is the averaged electron spin component at the nucleus. The upfield shift, ΔB, in nuclear resonance field, B, caused by interaction of the nuclear spin with the unpaired electron spin is therefore:

$$g_n\beta_n\Delta B=a_o^{(\rm N)}\langle S_z\rangle \tag{3.26}$$

It now remains only to calculate the value of $\langle S_z\rangle$.

Electron spin relaxation is very much faster than that of the nucleus. Therefore, the average value, $\langle S_z\rangle$, is different from zero only because the populations, $N_\pm$, of the two electron spin orientations $\left(M_S=\pm\frac{1}{2}\right)$ at thermal equilibrium are slightly different. The mean value of S_z is thus given by:

$$\langle S_z\rangle=\frac{1}{2}\frac{N_+-N_-}{N_++N_-}\cong\frac{1}{4}\left(\frac{N_+}{N_-}-1\right) \tag{3.27}$$

In a magnetic field B, the Zeeman energy levels of the electron spin are: $E_\pm=\pm\frac{1}{2}g_o\beta_eB\ll k_BT$, where k_B is Boltzmann's constant and T is the absolute temperature. From Boltzmann statistics, the ratio of spin populations is given by: $N_+/N_-=\exp\left(-g_o\beta_eB/k_BT\right)\cong 1-g_o\beta_eB/k_BT$, and Eq. 3.27 becomes:

$$\langle S_z\rangle=-\frac{g_o\beta_eB}{4k_BT} \tag{3.28}$$

This corresponds to Curie's Law for the paramagnetic susceptibility of an $S=\frac{1}{2}$ system. Combining Eqs. 3.26 and 3.28 then gives the relative paramagnetic chemical shift:

$$\delta_{pm}=\frac{\Delta B}{B}=-\frac{g_o\beta_e}{g_n\beta_n}\frac{a_o^{(\rm N)}}{4k_BT} \tag{3.29}$$

We sometimes call this the temperature-dependent paramagnetic shift, to distinguish it from general contributions to the chemical shielding that arise from circulation of the paired electrons. Equation 3.29 lets us determine the sign of the hyperfine coupling, $a_o^{(\rm N)}$, because the paramagnetic shift, δ_{pm}, may be either upfield or downfield (Brown et al. 1960). We determine the paramagnetic shift either relative to the chemical shift of the diamagnetic hydroxylamine, which is prepared by mild chemical reduction of the nitroxide, or relative to the parent amine before oxidation to give the paramagnetic nitroxide.

3.6 ^{13}C-HYPERFINE COUPLINGS

The ^{13}C-isotope has nuclear spin $I = \frac{1}{2}$ and gyromagnetic ratio relative to the electron of $g_n\beta_n / g_e\beta_e = -3.8210 \times 10^{-4}$. At natural abundance (1.07%), ^{13}C-hyperfine structure is resolved as satellites in the wings of the ^{14}N-hyperfine lines, when the nitroxide tumbles rapidly in a non-viscous solvent. Because the four protecting methyl groups that flank the nitroxide at the β-position in stable spin labels are often equivalent, these tend to dominate the ^{13}C hyperfine structure. Frequently, the isotropic hyperfine coupling of the β-carbons is comparable to that of the α-carbons (Brière et al. 1971 and see Table 3.2). This is also the case for the more familiar β-proton splittings in π-radicals (McLachlan 1958; Heller and McConnell 1960), and has been suggested to indicate hyperconjugation, i.e., direct electron transfer from carbon to the nitrogen (Colpa and de Boer 1964).

We list Cα and Cβ ^{13}C-hyperfine coupling constants for a variety of nitroxide spin labels in Table 3.2. Where the values have a sign, the coupling constants are determined by ^{13}C-NMR. The Cα coupling constants are negative, as expected for parallel σ–π polarization of the nitrogen σ-orbital in the N–C bond by the unpaired spin in the nitrogen $2p_z\pi$-orbital (see Section 3.3). This is then paired with the opposite spin in the carbon σ-orbital, which results in negative spin density at the carbon nucleus (see Figure 3.5). The situation is completely analogous to the more familiar case of α-proton hyperfine couplings in carbon-based π-radicals (McConnell 1956). The Cβ coupling constants are all positive, as is the case also for β-protons (Heller and McConnell 1960). This is consistent with a small direct admixture of the Cβ σ-orbital into the $2p_z\pi$-orbital of the nitroxide (i.e., hyperconjugation), as well as with direct σ–π polarization.

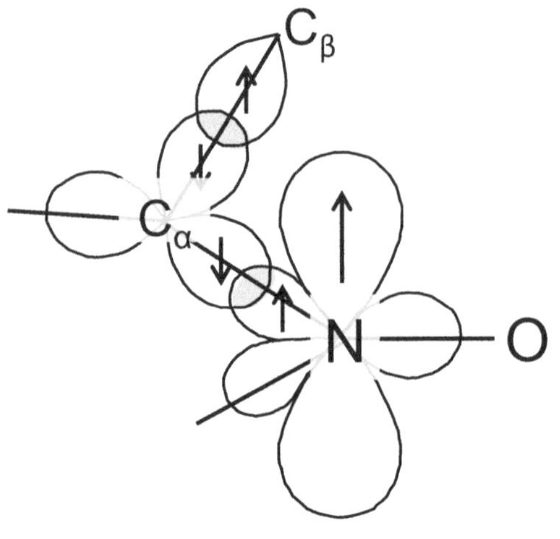

FIGURE 3.5 Electron spin polarizations and orbital overlap in nitrogen and carbon σ-bonds of a nitroxide. Unpaired electron resides in nitrogen $2p_z\pi$-orbital. Exchange polarization (σ–π) favours parallel spin alignment in nitrogen sp^2 σ-orbitals. Spin pairing in N–Cα σ-bonds gives antiparallel spin alignment in Cα sp^2-orbitals; that in Cα–Cβ σ-bonds gives parallel spin alignment in Cβ sp^2-orbitals. Direct nitrogen $2p_z$-orbital admixture (hyperconjugation) also results in parallel spin alignment in Cβ sp^2-orbitals.

Because the unpaired electron is confined essentially to the N–O bond, the isotropic hyperfine coupling from the α-carbon nucleus is given by:

$$a_o^{C\alpha} = Q_{NC\alpha}^{C} \rho_\pi^{N} \tag{3.30}$$

TABLE 3.2 Isotropic ^{13}C-hyperfine couplings of nitroxides

	$a_o^{C\alpha}$		$a_o^{C\beta}(CH_3)$		$a_o^{C\beta}(CH_2)$		$a_o^{C\gamma}$		
nitroxide[a]	(MHz)	(mT)	(MHz)	(mT)	(MHz)	(mT)	(MHz)	(mT)	*Refs.*[d]
DTBN	−12.4	−0.44	+11.8	+0.42					1
	12.3 ± 0.1	0.438 ± 0.005	12.2 ± 0.3	0.435 ± 0.012					2
TEMPO	−10.1	−0.36	+13.8	+0.49	+2.3	+0.082	−0.9	−0.032	1
			16.0 ± 2.8	0.57 ± 0.1					3
TEMPOL	−10.7	−0.38	+10.4	+0.37	+3.4	+0.12	−0.1	−0.002	1
	11.2[b]	0.400[b]	17.8	0.640					4
TEMPONE	−14.3	−0.51	+16.0	+0.57	+6.5	+0.23	−0.11	−0.038	1
			15.7 ± 0.6	0.56 ± 0.02					3
$PYCONH_2$			17.0	0.607					4
			18.0 ± 0.6	0.64 ± 0.02					3
PROXYL[c]			19.1 ± 0.6	0.68 ± 0.02					5

[a] TEMPO, 1-oxyl-2,2,6,6-tetramethylpiperidine;
TEMPOL, 1-oxyl-2,2,6,6-tetramethylpiperidin-4-ol;
TEMPONE, 1-oxyl-2,2,6,6-tetramethylpiperidin-4-yl oxide;
$PYCONH_2$, 1-oxyl-2,2,5,5-tetramethylpyrrolin-3-ylamide;
DTBN, di-*tert*-butyl nitroxide;
PROXYL, 1-oxyl-2,2,5,5-tetramethylpyrrolidin-3-yl derivatives (alcohol, PROXYL-OH; oxime, PROXYL-NOH; ketone, PROXYL-O).
[b] Assignment by comparison with ref. 1.
[c] Identical values for PROXYL-OH, PROXYL-NOH and PROXYL-O (see ref. 5).
[d] References: 1. Hatch and Kreilick (1972); 2. Brière et al. (1968); Brière et al. (1971). 3. Brière et al. (1965); 4. Ottaviani (1987); 5. Dupeyre et al. (1964).

where $Q^{C}_{NC\alpha}$ is the contribution to the hyperfine interaction from polarization by the unpaired spin on the nitrogen, which is practically the sole contribution. Substituting in Eq. 3.30 for ρ^{N}_{π} from Eq. 3.15 leads to the following relation between the carbon and nitrogen coupling constants:

$$a_o^{N} = \frac{\left(Q_N^N - Q_{ON}^N\right)}{Q^{C}_{NC\alpha}} a_o^{C\alpha} + Q^N_{ON} \tag{3.31}$$

Figure 3.6 (solid circles) shows the linear dependence of a_o^N on $a_o^{C\alpha}$ that is found, in accordance with Eq. 3.31, for di-*t*-butyl nitroxide in solvents of increasing polarity (cf. Chapter 4). Linear regression to the data for the α-carbon yields $a_o^N = (2.02 \pm 0.13) \times \left|a_o^{C\alpha}\right| + (0.65 \pm 0.06)$ mT. The relatively small intercept implies that $Q^N_{ON} \approx 0.6$ mT in Eq. 3.31, which is in qualitative agreement with the result found for DOXYL nitroxides and TEMPONE in Figure 3.4 (see Section 3.4). From the gradient in Figure 3.6, we see that the increase in isotropic ^{13}C-hyperfine coupling of the α-carbon is half that of the ^{14}N nitrogen nucleus. More specifically, the Q-value for the α-carbon becomes: $Q^{C}_{NC\alpha} \approx \frac{1}{2}\left(Q^N_N - Q^N_{ON}\right) = 35.6 \pm 3.0$ MHz ($\cong 1.27 \pm 0.11$ mT).

The ^{13}C-hyperfine coupling of the β-carbons (e.g., tetramethyl groups) also increases with increasing polarity. Correlation with the ^{14}N-coupling of the nitrogen is given by the open diamonds in Figure 3.6. Again we find a linear dependence, for which regression gives: $a_o^N = (3.51 \pm 0.18) \times a_o^{C\beta} + (0.03 \pm 0.08)$ mT. In this case, the intercept at $a_o^{C\beta} = 0$ is essentially zero (Lemaire and Rassat 1964). Because $a_o^{C\beta}$ also depends linearly on the unpaired spin density, ρ^N_π (cf. Section 3.7 below), a relation similar to Eq. 3.31 for the α-carbon holds equally for the β-carbon. Therefore, the vanishingly small a_o^N-intercept for $a_o^{C\beta} = 0$ implies that $Q^N_{ON} \approx 0$ (Lemaire and Rassat 1964). From the steeper gradient in Figure 3.6, we see that the isotropic ^{13}C-hyperfine

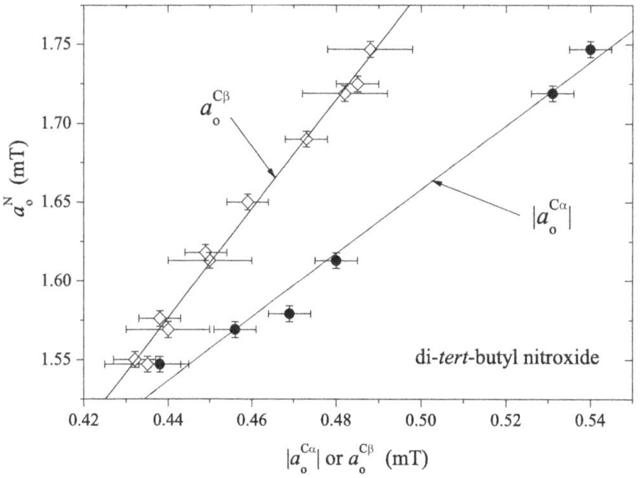

FIGURE 3.6 Correlation of isotropic ^{14}N-hyperfine coupling constant a_o^N with absolute value of (negative) ^{13}C-hyperfine coupling constant $a_o^{C\alpha}$ for α-carbon atom of di-*tert*-butyl nitroxide (solid circles) in solvents of different polarities (Brière et al. 1971). Solid line: linear regression (Eq. 3.31). Open diamonds: corresponding correlation for ^{13}C-hyperfine couplings from methyl groups, i.e., β-carbons (Brière et al. 1971; Lemaire and Rassat 1964).

coupling of the β-carbon is 0.28 × that of the ^{14}N nitrogen nucleus. More specifically, the Q-value for the β-carbon is given by: $Q^{C}_{NC\beta} \approx 0.28 \times \left(Q^N_N - Q^N_{ON}\right) = 20.5 \pm 1.5$ MHz ($\cong 0.73 \pm 0.05$ mT).

3.7 ^{13}C-HYPERFINE COUPLINGS OF β-C ATOMS

The ^{13}C-hyperfine coupling of nitroxide β-carbons (viz., tetramethyl groups and adjacent ring methylenes) not only depends on the unpaired spin density, ρ^N_π, on the nitrogen but also is sensitive to the orientation of the Cα–Cβ bond. The latter property makes β-carbon hyperfine splittings particularly useful. In analogy with hyperfine splittings of β-protons in π-radicals (Heller and McConnell 1960; Stone and Maki 1962), the isotropic ^{13}C-hyperfine coupling for β-carbon atoms is given by:

$$a_o^{C\beta} = \left[Q^C_{NC\beta}(90^\circ) + \left(Q^C_{NC\beta}(0^\circ) - Q^C_{NC\beta}(90^\circ)\right)\cos^2\vartheta\right]\rho^N_\pi \tag{3.32}$$

where ϑ is the dihedral angle between the nitroxide z-axis (i.e., the $2p_z\pi$ orbital) and the Cα–Cβ bond (see Figure 3.7). $Q^C_{NC\beta}(\vartheta)$ gives the contribution to the Cβ-hyperfine interaction from polarization by the unpaired spin on the nitrogen, for the conformation with dihedral angle ϑ. We expect that $Q^C_{NC\beta}(90^\circ)$ is small compared with $Q^C_{NC\beta}(0^\circ)$, because in the perpendicular orientation there is no possibility of overlap between the β-carbon σ-orbitals and the nitrogen $2p_z\pi$-orbital that would favour both hyperconjugation and direct σ–π polarization. For just this reason, we choose a dependence on $\cos^2\vartheta$. The quadratic dependence means that the sign of the dihedral angle is unimportant, because solely the inclination of the Cα–Cβ bond is determining, not its absolute direction.

To apply Eq. 3.32 in conformational analysis, we must correct for possible differences in ρ^N_π. These can arise from differences in environmental polarity (see Figure 3.6), and more importantly from differences in spin label structure, namely, between six-membered ring and five-membered ring nitroxides, and also DTBN. Because we found that $Q^N_{ON} \approx 0$ in Eq. 3.14, we can correct simply by normalizing the carbon hyperfine coupling with respect to that for nitrogen, giving:

$$a_o^{C\beta}/a_o^N = C_0 + C_2\cos^2\vartheta \tag{3.33}$$

where the dimensionless calibration constants are: $C_0 \equiv Q^C_{NC\beta}(90^\circ)/Q^N_N$ and $C_2 \equiv \left(Q^C_{NC\beta}(0^\circ) - Q^C_{NC\beta}(90^\circ)\right)/Q^N_N$. We give values of the normalized β-carbon hyperfine coupling for different nitroxides in Table 3.3. Differences in polarity and in structure of the nitroxide ring are clearly reflected by the values of the nitrogen hyperfine coupling, a_o^N, that we include in this table. In the final column of Table 3.3, we see that normalization effectively removes the dependence of the β-carbon couplings on environmental polarity. Hence, this normalization will correct similarly for intrinsic differences between various chemical structures, the remaining differences then being attributable to the molecular conformations.

TABLE 3.3 Isotropic ^{13}C-hyperfine couplings for methyl β-carbons of different nitroxides, normalized to the nitrogen spin density

nitroxide[a]	*solvent*	$a_o^{C\beta}(CH_3)$ (mT)	a_o^N (mT)	$a_o^{C\beta}(CH_3)/a_o^N$
DTBN	benzene	0.435 ± 0.012	1.547 ± 0.005	0.281 ± 0.009
	water	0.482 ± 0.012	1.719 ± 0.005	0.280 ± 0.008
TEMPOL	pyridine	0.640	1.556	0.411
	EtOH	0.682	1.609	0.423
TEMPONE	ethylene glycol	0.56 ± 0.02	1.51 ± 0.01	0.371 ± 0.016
	water	0.59 ± 0.02	1.60 ± 0.01	0.369 ± 0.015
TEMPO-NOH	ethylene glycol	0.55 ± 0.02	1.54 ± 0.01	0.357 ± 0.015
	water	0.58 ± 0.02	1.63 ± 0.01	0.356 ± 0.014
$PYCONH_2$	pyridine	0.607	1.455	0.417
	EtOH	0.619	1.501	0.412
	ethylene glycol	0.64 ± 0.02	1.49 ± 0.01	0.430 ± 0.016
	water	0.66 ± 0.02	1.60 ± 0.01	0.412 ± 0.015
PROXYL[b]	ethylene glycol	0.68 ± 0.02	1.53 ± 0.01	0.444 ± 0.016

For references, see Table 3.2.
[a] TEMPO-NOH, 1-oxyl-2,2,6,6-tetramethylpiperidin-4-yl oxime;
TEMPOL, 1-oxyl-2,2,6,6-tetramethylpiperidin-4-ol;
TEMPONE, 1-oxyl-2,2,6,6-tetramethylpiperidin-4-yl oxide;
$PYCONH_2$, 1-oxyl-2,2,5,5-tetramethylpyrrolin-3-ylamide;
DTBN, di-*t*-butyl nitroxide;
PROXYL, 1-oxyl-2,2,5,5-tetramethylpyrrolidin-3-yl derivatives (alcohol, PROXYL-OH; oxime, PROXYL-NOH; ketone, PROXYL-O).
[b] Similar values for PROXYL-OH, PROXYL-NOH and PROXYL-O.

We obtain values of the calibration constants C_2 and C_0 by applying Eq. 3.33 to conformationally predictable nitroxides (Brière et al. 1965; Hatch and Kreilick 1972). In Table 3.3, di-*t*-butyl nitroxide (DTBN) has the lowest value of $a_o^{C\beta}/a_o^N$ (= 0.280 ± 0.009). Assuming free rotation of the β-methyl groups about the Cα–N bond, an average value of $\cos^2\vartheta$:

$$\left\langle \cos^2\vartheta \right\rangle = \frac{1}{2\pi}\int_0^{2\pi} \cos^2\vartheta \cdot d\vartheta = \frac{1}{2} \tag{3.34}$$

is appropriate to DTBN in Eq. 3.33. The largest values of $a_o^{C\beta}/a_o^N$ in Table 3.3 are for the five-membered ring nitroxides, the pyrrolidines PROXYL, PROXYL-OH, PROXYL-NOH and PROXYL-O, and the pyrroline $PYCONH_2$ (with mean $a_o^{C\beta}/a_o^N = 0.418 \pm 0.009$). Assuming a planar conformation for the five-membered ring, the β-methyl groups have a dihedral angle of $\vartheta = 30°$ (see Figure 3.7, left), and we obtain $\cos^2\vartheta = 3/4$ in Eq. 3.33 for the pyrrolidines and $PYCONH_2$. Combining these two sets of data then gives $C_2 = 0.552 \pm 0.072$ and $C_0 = 0.004 \pm 0.045$ for the calibration constants in Eq. 3.33. This confirms that

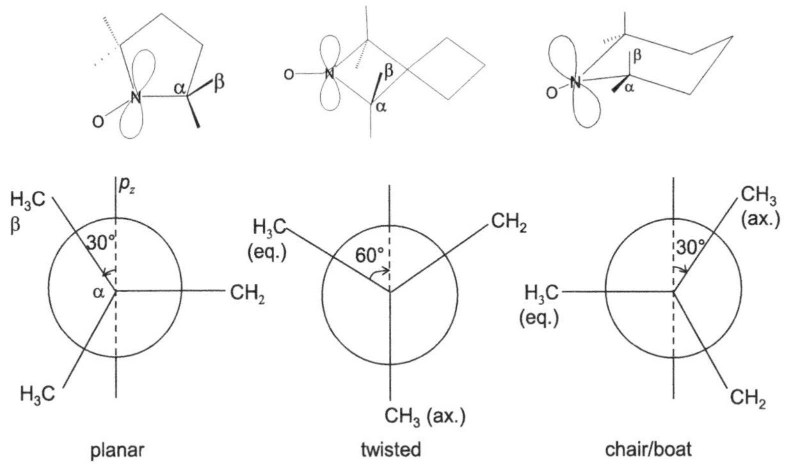

FIGURE 3.7 Nitroxide ring conformations (*upper* row) and dihedral angle, ϑ, of Cα–Cβ bond relative to nitrogen $2p_z$ orbital (*lower* row). *Left*: planar five-membered ring (pyrrolidines and pyrrolines); *centre*: twisted conformation of six-membered (piperidine) ring; *right*: chair (or boat) conformation of six-membered (piperidine) ring. Lower row: projection along Cα–N bond with N $2p_z$-orbital axis vertical. Dihedral angle indicated for one of the β-methyl groups. *Planar*: $\vartheta = 30°$ for both methyls, $\vartheta = 90°$ for methylene. *Twisted*: $\vartheta = 60°$ for equatorial methyl and methylene, $\vartheta = 0°$ for axial methyl. *Chair/boat*: $\vartheta = 30°$ for axial methyl and methylene, $\vartheta = 90°$ for equatorial methyl.

the first term in Eq. 3.33, C_0, is small and justifies not considering the β-methylene group, which has $\vartheta = 90°$ (see Figure 3.7, left), for pyrrolidines and pyrrolines in Table 3.2.

Now we can apply these results to conformational analysis of the six-membered ring nitroxides (Brière et al. 1965). For these piperidine nitroxides, the six-membered ring may assume a chair or boat conformation, or the intermediate twisted conformation. For one form of the chair or boat, the axial methyl and the β-methylene have a dihedral angle of $\vartheta = 30°$, and the equatorial methyl has $\vartheta = 90°$ (see Figure 3.7, right). For the inverse form of the boat or chair, the dihedral angle of the β-methylene remains the same, but the axial and equatorial methyls interchange. If ring inversion is slow, two carbons will have a normalized ^{13}C-coupling of $a_o^{C\beta}(30°)/a_o^N = 0.41 \pm 0.05$ and the third will have almost zero ^{13}C-coupling. If ring inversion is rapid, the two methyl carbons will have the averaged coupling: $\left(a_o^{C\beta}(30°) + a_o^{C\beta}(90°)\right)/2a_o^N = 0.21 \pm 0.03$ and the methylene coupling will remain at $a_o^{C\beta}(30°)/a_o^N = 0.41 \pm 0.05$. For intermediate inversion rates, the methyl carbon resonances may be broadened beyond detection and then only the methylene with $a_o^{C\beta}(30°)/a_o^N = 0.41 \pm 0.05$ will be observed.

In the twisted conformation, the equatorial methyl and the β-methylene have a dihedral angle of $\vartheta = 60°$, and the axial methyl has $\vartheta = 0°$ (see Figure 3.7, centre). Ring inversion again results in exchange of the axial and equatorial methyls, whereas the dihedral angle of the β-methylene remains unchanged. For slow ring inversion, two carbons will have a normalized coupling of $a_o^{C\beta}(60°)/a_o^N = 0.21 \pm 0.02$ and the third will have a coupling of $a_o^{C\beta}(0°)/a_o^N = 0.55 \pm 0.07$. For rapid ring inversion, the two methyl carbons will have an averaged coupling: $\left(a_o^{C\beta}(0°) + a_o^{C\beta}(60°)\right)/2a_o^N = 0.35 \pm 0.05$ and the methylene will remain with a coupling of $a_o^{C\beta}(60°)/a_o^N = 0.21 \pm 0.02$. For broadened methyl resonances at intermediate inversion rates, only the methylene with $a_o^{C\beta}(60°)/a_o^N = 0.21 \pm 0.02$ might be observed.

We must distinguish piperidine nitroxides with a tetrahedral carbon at position 4 of the ring (e.g., TEMPOL) from those with a trigonal carbon at position 4 (TEMPONE and TEMPO-NOH). From Table 3.3, we see that the methyl β-carbons of TEMPOL have resolved isotropic ^{13}C-hyperfine coupling constants that are in the region of $a_o^{C\beta}/a_o^N = 0.41$. TEMPONE and the oxime piperidine-nitroxide (TEMPO-NOH), on the other hand, have resolved β-couplings that are in the region of $a_o^{C\beta}/a_o^N = 0.35 - 0.37$. The TEMPOL coupling is therefore consistent with a chair (or boat) conformation, although possible overlap with the large central hyperfine line (from ^{12}C with $I = 0$) makes it difficult to decide on the rate of ring inversion. The $a_o^{C\beta}$ coupling constants observed for TEMPONE and TEMPO-NOH are consistent with rapid exchange between two equivalent twisted conformations. A twisted conformation is expected for nitroxides with a trigonal carbon at the 4-position, in analogy with cyclohexanedione (Groth and Hassel 1964; Mossel et al. 1963). Slow exchange between twisted conformations is excluded by the hyperfine couplings for TEMPONE and TEMPO-NOH.

Natural abundance ^{13}C satellites do not allow resolution of smaller hyperfine couplings, but this could be remedied by isotopic substitution with ^{13}C, hence allowing a more complete conformational analysis. For reference, we list determinations of nitroxide ring conformations from x-ray diffraction of single crystals in Table 3.4.

TABLE 3.4 Nitroxide conformation in tetramethyl piperdine, pyrrolidine or pyrroline derivatives, and in a dimethyloxazolidine nitroxide, determined from single-crystal x-ray diffraction

nitroxide[a]	*ring conformation*	*N–O bond length* (nm)	*C–N–C bond angle* (°)	*N–O angle to C–N–C plane* (°)	*Refs.*[b]
TEMPO	chair	0.1296(5)	125.5(3)	19.4 ± 1.2	1
	chair	0.1283(9)	123.6(3)	22	2
TEMPOL	chair	0.1291(7)	125.4(5)	15.8(8)	3
TEMPONE	twist	0.1276(3)	123.5(2)	0	4
TOAC	twist-boat	0.128	124.4	0–5	5,6
PROXYL-COOH	twist	0.1272(3)	116.2(2)	3.3	7
PYCOOK	planar	0.1277(8)	114.7(7)	0	8
$PYCONH_2$	planar	0.1267(5)	114.8(5)	0	9
MTSSL	planar	0.12767(14)	114.96(10)	2.9	10
DOXYLcHx	planar	0.1259(4)	112.9(3)	0	11

[a] TEMPO, 1-oxyl-2,2,6,6-tetramethylpiperidine;
TEMPOL, 1-oxyl-2,2,6,6-tetramethylpiperidin-4-ol;
TEMPONE, 1-oxyl-2,2,6,6-tetramethylpiperidin-4-yl oxide;
TOAC, 1-oxyl-2,2,6,6-tetramethyl-4-aminopiperidin-4-yl carboxylic acid;
PROXYL-COOH, 1-oxyl-2,2,5,5-tetramethylpyrrolidin-3-yl carboxylic acid;
PYCOOK, 1-oxyl-2,2,5,5-tetramethylpyrrolin-3-yl carboxylate potassium salt;
$PYCONH_2$, 1-oxyl-2,2,5,5-tetramethylpyrrolin-3-ylamide;
MTSSL, 1-oxyl-2,2,5,5-tetramethylpyrrolin-3-ylmethyl methanethiosulphonate;
DOXYLcHx, 3-oxyl-4,4-dimethyloxazolidin-2-yl cyclohexane.

[b] References: 1. Bordeaux et al. (1973). 2. Capiomont and Lajzérowicz-Bonneteau (1974). 3. Berliner (1970). 4. Bordeaux and Lajzérowicz (1974). 5. Crisma et al. (2005). 6. Flippen-Anderson et al. (1996). 7. Wetherington et al. (1974). 8. Boeyens and Kruger (1970). 9. Turley and Boer (1972). 10. Zielke et al. (2008). 11. Bordeaux and Lajzérowicz-Bonneteau (1974).

3.8 PROTON HYPERFINE COUPLINGS

The proton (i.e., ^{1}H-isotope, natural abundance 99.9885%) has nuclear spin $I = \frac{1}{2}$ and gyromagnetic ratio relative to the electron of $g_n\beta_n/g_e\beta_e = -1.519270\times10^{-3}$. The deuteron (^{2}H-isotope, natural abundance 0.0115%), on the other hand, has $I = 1$ and much smaller gyromagnetic ratio $g_n\beta_n/g_e\beta_e = -2.332173\times10^{-4}$. To stabilize against disproportionation reactions in nitroxides used for spin labelling, there are no hydrogen atoms close to the unpaired electron orbital. The nearest-neighbour hydrogens are three bonds removed from the nitrogen. Thus, in spite of the large gyromagnetic ratio, proton hyperfine couplings in nitroxyl spin labels are relatively small. Nonetheless, isotropic proton hyperfine couplings can be partially resolved in deoxygenated solutions; they also contribute directly to ENDOR and ESEEM frequencies, and quite generally determine inhomogeneous broadening in spin-label EPR.

Figure 3.8 shows the correlation between proton hyperfine couplings, $|a_o^{\mathrm{H}}|$, and nitrogen coupling, a_o^{N}, for 1-oxyl-2,2,6,6-tetramethylpiperidine (TEMPO) in solvents of different polarity. For methylene protons attached to the γ-carbon at position 4 of the ring, the proton coupling increases with increasing solvent polarity, i.e., with increasing a_o^{N}. Anomalously, the absolute value of the proton coupling from the methyl groups, which are attached to the α-carbons at positions 2 and 6 of the ring, decreases with increasing a_o^{N}, rather as if this depends on the spin density on the oxygen, i.e., on ρ_π^{O} not ρ_π^{N}. In both cases, however, the dependence on a_o^{N} is approximately linear. Results of linear regression are: $|a_o^{\mathrm{H}}| = -(24.2\pm5.8)\times10^{-3}\times a_o^{\mathrm{N}} + (60.2\pm9.2)\,\mu\mathrm{T}$ and $a_o^{\mathrm{H}} = (23.5\pm1.5)\times10^{-3}\times a_o^{\mathrm{N}} - (18.8\pm2.3)\,\mu\mathrm{T}$ for methyls and γ-methylene, respectively. On the other hand, the isotropic coupling for the methylene protons of the β-carbon at positions 3 and 5 of the piperidine ring is considerably larger but varies little with the polarity of the solvent, having a mean value of $a_o^{\mathrm{H}} = -39.2\pm0.3\,\mu\mathrm{T}$ for the different values of a_o^{N}.

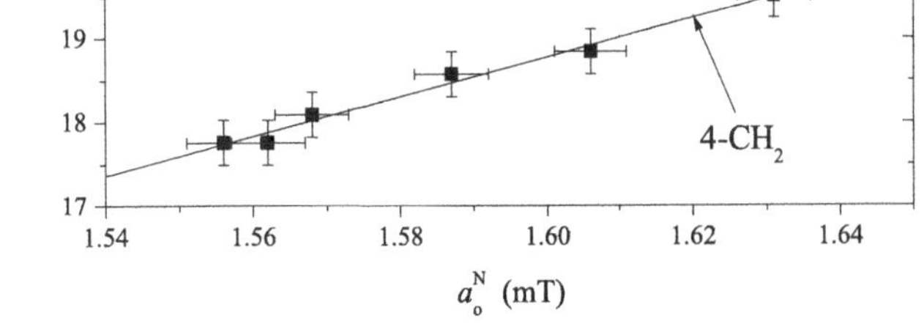

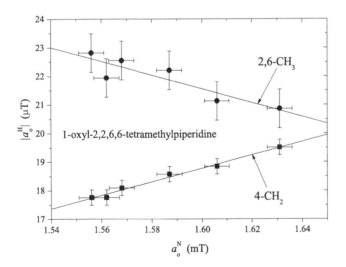

FIGURE 3.8 Correlation of isotropic proton hyperfine coupling constant $|a_o^{\mathrm{H}}|$ with ^{14}N-coupling constant a_o^{N} for protons on tetramethyl groups (circles) and 4-position methylene (squares) of 1-oxyl-2,2,6,6-tetramethylpiperidine in solvents of different polarities (Brière et al. 1967). Coupling constant is negative for methyl protons and positive for γ-methylene protons. Solid lines are linear regressions.

Table 3.5 gives proton hyperfine couplings, and numbers of equivalent protons, for the 5-membered pyrroline-ring nitroxide PYCONH$_2$, the 6-membered piperidine-ring nitroxides TEMPOL, TEMPO-NH$_2$, TEMPO and TEMPONE, and for the simple di-*tert*-butyl nitroxide (DTBN). As expected, all protons of DTBN are equivalent. For PYCONH$_2$, all methyl protons are equivalent because of the reflection symmetry of the planar 5-membered ring. For the 4-substituted piperidine-ring nitroxides TEMPOL and TEMPO-NH$_2$, there are two inequivalent sets of methyl protons and also of β-methylene protons. This clearly indicates two conformations of the 6-membered ring that are in slow exchange. In Section 3.7, we concluded from ^{13}C-hyperfine couplings that this corresponds to ring inversion of the boat/chair conformation. For the non-substituted parent TEMPO nitroxide, all methyl protons are equivalent, as are all β-methylene protons. This arises from averaging of the hyperfine couplings by fast ring inversion. Significantly, the proton hyperfine couplings of both methyls and β-methylenes of TEMPO are close to the mean of the corresponding inequivalent couplings of TEMPOL or TEMPO-NH$_2$. For TEMPONE, which has trigonal symmetry at the substituted 4-position, all methyl protons and all methylene protons are also equivalent. This is consistent with fast ring inversion, as deduced already from the ^{13}C-couplings. The proton hyperfine couplings of TEMPONE differ considerably, however, from those of TEMPO, although these also are in fast exchange. Clearly the conformation of TEMPONE is different from that of TEMPO. In Section 3.7 we concluded that the TEMPONE ring has a twisted conformation, not the chair/boat conformation that is allowed with tetrahedral symmetry at the 4-position of the ring.

A more complete list of proton hyperfine couplings for different nitroxides in solvents of various polarities is given in Appendix A3.2 at the end of this chapter.

3.9 ^{17}O-HYPERFINE COUPLINGS

The normal ^{16}O-isotope of oxygen (natural abundance 99.757%) has no nuclear spin. The ^{17}O-isotope (natural abundance 0.038%) has nuclear spin $I = 5/2$ and gyromagnetic ratio relative to the electron $g_n\beta_n/g_e\beta_e = 2.06039\times10^{-4}$. For ^{17}O-enriched nitroxides, each nitrogen hyperfine line is split into six ^{17}O-hyperfine lines.

Figure 3.9 shows correlations of the isotropic ^{17}O-hyperfine coupling constants a_o^{O} with those of ^{14}N and ^{13}C (a_o^{N} and $a_o^{\mathrm{C\alpha}}$) for ^{17}O-enriched 1-oxyl-2,2,6,6-tetramethylpiperidin-4-yl oxide (TEMPONE) in hydrogen-bonding solvents of different polarity. The ^{17}O-coupling increases as the ^{14}N- and ^{13}C-couplings decrease. This is to be expected because the ^{17}O-coupling depends primarily on the unpaired spin density, ρ_π^{O}, on the

TABLE 3.5 Isotropic proton hyperfine couplings, a_o^H, for 5-membered pyrroline- and 6-membered piperidine-ring nitroxides. Data from (Windle 1981)

		$a_o^H(\beta Me)$ (μT)		$a_o^H(\beta CH_2)$ (μT)		
nitroxide[a]	*solvent*	*equatorial*	*axial*	*equatorial*	*axial*	$a_o^H(\gamma CH_2)$ (μT)
DTBN	CCl_4	11.0 (18H)				
	H_2O	6.0 (18H)				
$PYCONH_2$	CCl_4	24.0 (12H)		44.0 (1H)		
	H_2O	20.0 (12H)		35.0 (1H)		
TEMPOL	CCl_4	46.0 (6H)	3.0 (6H)	43.0 (2H)	29.0 (2H)	7.0 (1H)
	H_2O	44.0 (6H)	6.0 (6H)	45.0 (2H)	36.0 (2H)	15.0 (1H)
$TEMPO-NH_2$	CCl_4	51.0 (6H)	4.0 (6H)	42.0 (2H)	29.0 (2H)	4.0 (1H)
	H_2O	43.6 (6H)	3.8 (6H)	45.0 (2H)	31.5 (2H)	9.0 (1H)
TEMPO	CCl_4	23.0 (12H)		39.0 (4H)		18.0 (2H)
	H_2O	17.5 (12H)		40.0 (4H)		16.0 (2H)
TEMPONE	CCl_4	11.5 (12H)		2 (4H)[b]		
	H_2O	6.0 (12H)		– (4H)		

Note: Values in parentheses are numbers of equivalent protons with the hyperfine coupling indicated. Equatorial and axial orientations apply only to TEMPOL and $TEMPO-NH_2$ 6-membered rings.
[a] DTBN, di-*t*-butyl nitroxide;
$PYCONH_2$, 1-oxyl-2,2,5,5-tetramethylpyrrolin-3-ylamide;
TEMPO, 1-oxyl-2,2,6,6-tetramethylpiperidine;
TEMPOL, 1-oxyl-2,2,6,6-tetramethylpiperidin-4-ol;
$TEMPO-NH_2$, 1-oxyl-2,2,6,6-tetramethylpiperidin-4-ylamine;
TEMPONE, 1-oxyl-2,2,6,6-tetramethylpiperidin-4-yl oxide.
[b] Value determined in benzene by NMR (Brière et al. 1967).

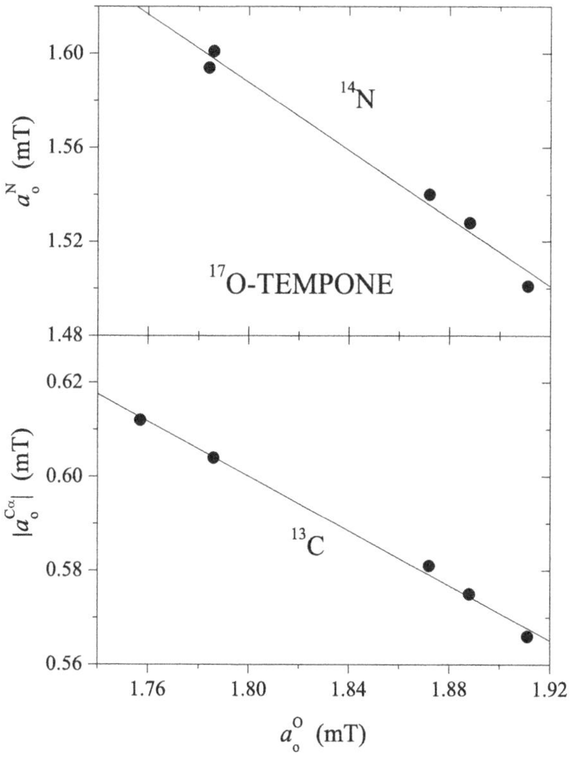

FIGURE 3.9 Correlation of isotropic ^{14}N-coupling constant a_o^N (upper panel) and ^{13}C-coupling constant $|a_o^{C\alpha}|$ (lower panel), with ^{17}O-hyperfine coupling constant a_o^O of 1-oxyl-2,2,6,6-tetramethylpiperidin-4-yl oxide (TEMPONE) in protic solvents of different polarities (Hayat and Silver 1973). Solid lines: linear regressions (Eq. 3.37 and equivalent).

oxygen, whereas the ^{14}N- and ^{13}C-couplings depend on ρ_π^N, the spin density on the nitrogen, where $\rho_\pi^O + \rho_\pi^N \approx 1$. In analogy with Eq. 3.14 of Section 3.3, the isotropic ^{17}O-coupling constant is given by (Karplus and Fraenkel 1961):

$$a_o^O = Q_O^O \rho_\pi^O + Q_{NO}^O \rho_\pi^N \tag{3.35}$$

where Q_O^O and Q_{NO}^O are the contributions to σ–π spin polarization at the oxygen, from unpaired spin density on the oxygen and nitrogen, respectively. Because $\rho_\pi^O + \rho_\pi^N \approx 1$, we can write Eq. 3.35 solely in terms of the unpaired spin density on the nitrogen:

$$a_o^O = Q_O^O - \left(Q_O^O - Q_{NO}^O\right)\rho_\pi^N \tag{3.36}$$

which is equivalent to Eq. 3.15 for the nitrogen coupling constant. Using the latter then gives the following relation between the isotropic coupling constants of oxygen and nitrogen:

$$a_o^O = Q_O^O - \frac{\left(Q_O^O - Q_{NO}^O\right)}{Q_N^N} a_o^N \tag{3.37}$$

where we assume that $Q_{ON}^N \approx 0$, as found in Sections 3.4 and 3.6. From Eq. 3.30, an exactly equivalent version of Eq. 3.37 relates the ^{13}C-coupling constant, $a_o^{C\alpha}$, to a_o^O. The linear relations implied by Eq. 3.37 and its equivalent hold reasonably well for TEMPONE, as seen in Figure 3.9. Note that we find a shallower linear dependence for aprotic solvents from that shown for protic solvents in Figure 3.9. See Chapter 4 for discussion of the dependence of ^{17}O-hyperfine couplings on polarity and hydrogen bonding (Section 4.4.2).

According to Eq. 3.37 and its equivalent for ^{13}C, we can obtain the principal Q-value for oxygen, Q_O^O, from the intercept on the a_o^O-axis in Figure 3.9. We find mutually consistent values of $Q_O^O = 3.96 \pm 0.15$ mT and 3.86 ± 0.08 mT from the ^{14}N- and ^{13}C-coupling constants, respectively. To estimate Q_{NO}^O from Eq. 3.37, we need values of $Q_N^N = 2.56 \pm 0.05$ mT and $Q_{NC\alpha}^C = 1.27 \pm 0.11$ from Sections 3.4 and 3.6. This results in $Q_{NO}^O = 0.48 \pm 0.5$ and $Q_{NO}^O = -0.61 \pm 0.6$ mT from the ^{14}N- and ^{13}C-coupling constants, respectively. These values are reasonably consistent and suggest that Q_{NO}^O is small in comparison with Q_O^O, as we found in the analogous case for the ^{14}N-coupling constant in Sections 3.4 and 3.6.

We also get information from the ^{17}O-hyperfine anisotropy (cf. Section 3.4). The parallel element of the dipolar ^{17}O-hyperfine tensor has been determined for ^{17}O-TEMPONE in frozen diethyl ether: $T_{\parallel}^d = T_{\parallel,o}^d \rho_\pi^O = 5.06$ mT, where the isotropic coupling in non-frozen solution is $a_o^O = 1.784$ mT (Hayat and Silver 1973). The corresponding intrinsic value of $T_{\parallel}^d$ for $\rho_\pi^O = 1$ is $T_{\parallel,o}^d = 288$ MHz ($\cong$ 10.28 mT) – see Whiffen (1964). Ignoring the Q_{NO}^O term in Eq. 3.35, we estimate the leading term as: $Q_O^O \approx a_o^O / \left(T_{\parallel}^d / T_{\parallel,o}^d\right) = 3.62$ mT, which is reasonably close to the values deduced above from Figure 3.9.

3.10 ^{14}N-QUADRUPOLE COUPLINGS

The distribution of electrical charge, $+Ze$, in the nucleus is not spherically symmetric. No nucleus has an electric dipole moment, but nuclei with spins $I > \frac{1}{2}$ (^{14}N but not ^{15}N) have a non-vanishing quadrupole moment, eQ. The nuclear electric quadrupole moment does not interact with the magnetic moment of an electron spin and therefore does not contribute in first order to the EPR spectrum. The quadrupole moment interacts with the electric field gradient produced by the electronic environment of the nucleus. This interaction contributes in first order to the nuclear transition frequencies that we observe in ESEEM or ENDOR (see Chapter 14).

Consider the interaction energy of the nuclear charge distribution of density $\rho(\mathbf{r})$ with the electrostatic potential $V(\mathbf{r})$ from the electron cloud:

$$W_{el} = \int \rho(\mathbf{r}) V(\mathbf{r})\, d^3 r \qquad (3.38)$$

where the integral is over the volume of the nucleus. We can expand the potential $V(\mathbf{r})$ in a Taylor series about $\mathbf{r} = 0$, the centre of the nucleus. The first term is the simple electrostatic energy $V(0)\int \rho(\mathbf{r})\, d^3 r$. The second term is zero because it contains the electric dipole moment of the nucleus $\int \mathbf{r}\rho(\mathbf{r})\, d^3 r = 0$, interacting with the electric field $\mathbf{E}(\mathbf{r}) = \partial V(\mathbf{r})/\partial \mathbf{r}$. The third term is the electric quadrupole interaction, which is the product of the second moment of the charge distribution $\int \mathbf{r}^2 \rho(\mathbf{r})\, d^3 r$ with the electric field gradient $\partial \mathbf{E}(\mathbf{r})/\partial \mathbf{r} = \partial^2 V(\mathbf{r})/\partial \mathbf{r}^2$.

Referred to the principal axes x,y,z of the electric field gradient, the quadrupole interaction energy is:

$$W_Q = \frac{1}{2}\int \rho(x,y,z)\left(x^2\frac{\partial^2 V}{\partial x^2} + y^2\frac{\partial^2 V}{\partial y^2} + z^2\frac{\partial^2 V}{\partial z^2}\right) dx dy dz \qquad (3.39)$$

from the Taylor expansion of Eq. 3.38. We can define an axially symmetric electric field gradient by a single parameter:

$$eq \equiv \frac{\partial^2 V}{\partial z^2} = -2\frac{\partial^2 V}{\partial x^2} = -2\frac{\partial^2 V}{\partial y^2} \qquad (3.40)$$

because Laplace's equation requires that $\partial^2 V/\partial x^2 + \partial^2 V/\partial y^2 + \partial^2 V/\partial z^2 = 0$, assuming that the electron density at the nucleus is zero. The quadrupolar energy then becomes:

$$W_Q = \frac{eq}{2}\sum_i e_i\left(z_i^2 - \tfrac{1}{2}\left(x_i^2 + y_i^2\right)\right) = \frac{eq}{4}\sum_i e_i\left(3z_i^2 - r_i^2\right) \qquad (3.41)$$

where we replace the integration by a summation over all charges e_i in the nucleus. Based on Eq. 3.41, we define the nuclear quadrupole moment by the quantum mechanical matrix element:

$$eQ = \langle I,I|\sum_i e_i\left(3z_i^2 - r_i^2\right)|I,I\rangle \qquad (3.42)$$

Our problem is to evaluate Eq. 3.42 for the nuclear spin states $|I,m_I\rangle$. As for other electrostatic interactions in EPR, e.g., in ligand- or crystal-field theory, we resort to the Wigner–Eckart theorem (see Section Q.4 in Appendix Q). The angular part of the Cartesian expression in Eq. 3.42 is the spherical harmonic $Y_{2,0}$ (see Appendices K and V), and therefore is equivalent to an angular momentum operator (see Appendix K). We obtain this operator equivalent by replacing the coordinates $(x/r, y/r, z/r)$ with the corresponding components of the nuclear spin (I_x, I_y, I_z). Hence, matrix elements of the Cartesian expression in Eq. 3.42 are given by those of the spin tensor operator:

$$Q_2^0 = \langle I\|\alpha\|I\rangle\left(3I_z^2 - I(I+1)\right) \qquad (3.43)$$

where the so-called reduced matrix element $\langle I\|\alpha\|I\rangle$ is a constant of proportionality that does not depend on m_I. We determine $\langle I\|\alpha\|I\rangle$ by evaluating the matrix element $\langle I,I|Q_2^0|I,I\rangle$ and equating this with eQ. Equation 3.43 then becomes:

$$Q_2^0 = \frac{eQ}{I(2I-1)}\left(3I_z^2 - I(I+1)\right) \qquad (3.44)$$

From Eq. 3.41, with the operator replacement given by Eq. 3.44, we can finally write the spin Hamiltonian for the nuclear quadrupole interaction as:

$$\mathcal{H}_Q = \frac{e^2 qQ}{4I(2I-1)}\left(3I_z^2 - I(I+1)\right) \qquad (3.45)$$

for axial symmetry.

TABLE 3.6 Nuclear electric quadrupole coupling tensors for ^{14}N-nitroxide and ^{2}H-water

P_{xx} (MHz)	P_{yy} (MHz)	P_{zz} (MHz)	e^2qQ/h (MHz)	η
PYMeOH/*o*-terphenyl[a]				
+1.26 ± 0.03	+0.53 ± 0.03	−1.79 ± 0.04	−3.58 ± 0.08	−0.41 ± 0.03
$(COOD)_2 \cdot 2D_2O$[b]				
−0.061 ± 0.001	−0.054 ± 0.001	+0.115 ± 0.001	+0.230 ± 0.002	0.062 ± 0.006
−0.063 ± 0.002	−0.046 ± 0.001	+0.109 ± 0.002	+0.219 ± 0.002	0.16 ± 0.03

[a] 1-oxyl-2,2,5,5-tetramethylpyrrolin-3-yl methanol in *o*-terphenyl or glycerol at 180 K; referred to conventional nitroxide *x,y,z*-axes (Savitsky et al. 2008).
[b] D_2O in oxalic acid crystal at −70°C; *z*-axis along OD bond (Chiba 1964).

An alternative way to write Eq. 3.45 is in terms of a traceless quadrupole tensor (cf. Eq. 3.40): $\mathcal{H}_Q = \mathbf{I}^T \cdot \mathbf{P} \cdot \mathbf{I}$, where $\mathbf{P} = \left(-\frac{1}{2}, -\frac{1}{2}, +1\right) P_{\parallel}$ with $P_{\parallel} \equiv e^2qQ/(2I(2I-1))$. For non-axial symmetry, the quadrupole interaction tensor is $\mathbf{P} = \left(P_{xx}, P_{yy}, P_{zz}\right)$, with $P_{xx} + P_{yy} + P_{zz} = 0$ to satisfy Laplace's equation. In addition to eq, we then need a second parameter $\eta \equiv \left(P_{xx} - P_{yy}\right)/P_{zz}$ to specify the asymmetry in electric field gradient. The nuclear quadrupole Hamiltonian hence becomes:

$$\mathcal{H}_Q = \frac{e^2qQ}{4I(2I-1)}\left(3I_z^2 + \eta\left(I_x^2 - I_y^2\right) - I(I+1)\right) \tag{3.46}$$

for the general non-axial case.

Principal elements of the ^{14}N-quadrupole tensor for the pyrroline nitroxide PYMeOH (1-oxyl-2,2,5,5-tetramethylpyrrolin-3-yl methanol) are given in Table 3.6. This nitroxide is the core of the methane thiosulphonate spin label MTSSL that is used routinely in site-directed spin labelling. The principal axes are collinear with those of the *A*- and *g*-tensor (see Table 2.1), where the $2p\pi$-orbital is along the *z*-axis and the N–O bond along the *x*-axis. Table 3.6 also gives the ^{2}H-quadrupole tensor for D_2O that is used to study water accessibility by ESEEM spectroscopy.

3.11 ELECTRIC FIELD GRADIENTS AND *p*-ORBITAL OCCUPANCIES

The electric field gradient at the nucleus is determined by contributions from the *p*-orbitals of the valence electrons, because *s*-orbitals are spherically symmetric and do not contribute to the field gradient. This is also true of all closed electron shells, because they too are spherically symmetric. Each *p*-orbital is symmetrical about its unique axis. Therefore, the components $\left(q_{xx}, q_{yy}, q_{zz}\right)$ of the field gradient from a p_z-electron, for example, are:

$$q_{xx}\left(p_z\right) = q_{yy}\left(p_z\right) = -\tfrac{1}{2}q_{zz}\left(p_z\right) \tag{3.47}$$

where the last equality is required to satisfy Laplace's equation, viz., $q_{xx} + q_{yy} + q_{zz} = 0$. Similar expressions hold for the contributions from a p_x- or p_y-electron:

$$q_{xx}\left(p_x\right) = q_{yy}\left(p_x\right) = -\tfrac{1}{2}q_{zz}\left(p_x\right) \tag{3.48}$$

$$q_{xx}\left(p_y\right) = q_{yy}\left(p_y\right) = -\tfrac{1}{2}q_{zz}\left(p_y\right) \tag{3.49}$$

For an isolated atom, the *x*-, *y*- and *z*-directions are equivalent, therefore we also have:

$$q_{xx}\left(p_x\right) = q_{yy}\left(p_y\right) = q_{zz}\left(p_z\right) \equiv q_{np} \tag{3.50}$$

where q_{np} is the field gradient along the symmetry axis of a *p*-orbital in the *n*th shell.

Let n_x, n_y and n_z be the occupation numbers for electrons in the p_x, p_y and p_z-orbitals, respectively, of the central atom. Then the resultant electric field gradient along the *z*-direction is:

$$q_{zz} = n_x q_{zz}\left(p_x\right) + n_y q_{zz}\left(p_y\right) + n_z q_{zz}\left(p_z\right) \tag{3.51}$$

Using Eqs. 3.47–3.50, this becomes:

$$q_{zz} = \left(n_z - \tfrac{1}{2}\left(n_x + n_y\right)\right) q_{np} \tag{3.52}$$

This is the value of the field-gradient parameter q that appears in Eqs. 3.45 and 3.46. Expressions analogous to Eq. 3.52 hold for the transverse components of the field gradient, q_{xx} and q_{yy}. The asymmetry in the field gradient is therefore given by:

$$q_{xx} - q_{yy} = \tfrac{3}{2}\left(n_x - n_y\right) q_{np} \tag{3.53}$$

In terms of Eqs. 3.52 and 3.53, the asymmetry parameter in Eq. 3.46 is: $\eta \equiv \left(q_{xx} - q_{yy}\right)/q_{zz}$. Note that because we have only two independent parameters q_{zz} and η, we need information additional to the quadrupole couplings, if we are to obtain the three orbital occupancies n_x, n_y and n_z (Marsh 2015). As we shall see below, this comes from the unpaired spin density on the nitrogen, ρ_π^N, that we determine from the dipolar nitrogen hyperfine coupling (see Section 3.4).

We also need the electric field gradient q_{np} for an atomic *p*-electron. The electrostatic potential of an electron at radial distance $r = \sqrt{x^2 + y^2 + z^2}$ from the nucleus is $V = -e/(4\pi\varepsilon_o r)$. The field gradient is then: $eq_{zz} \equiv \partial^2 V/\partial z^2 = -e\left(3z^2/r^2 - 1\right)/\left(4\pi\varepsilon_o r^3\right)$. To obtain the field gradient from a *p*-electron, we must then average over the *np*-orbital:

$$q_{np} = \frac{-1}{4\pi\varepsilon_o}\left\langle 3\cos^2\theta - 1\right\rangle_p \left\langle \frac{1}{r^3}\right\rangle_{np} \tag{3.54}$$

where in polar coordinates: $z = r\cos\theta$. The angular average for a p-electron is given by Eq. 3.21 in Section 3.4. Also using the same value for $\langle 1/r^3 \rangle_{2p,\mathrm{N}}$ as for the dipolar hyperfine tensor (see Section 3.4) and quadrupole moment $Q = 2.044 \times 10^{-30}$ m^2 for ^{14}N, we find that $e^2 q_{2p} Q/h = -11.9$ MHz for a nitrogen $2p$-orbital.

The $2p\pi$-bond that contains the unpaired electron is established by transferring covalent electron density π_c from the doubly occupied nitrogen $2p_z$-orbital to the corresponding singly occupied $2p_z$-orbital on the oxygen (cf. Figure 3.1). Thus for the nitrogen p_z-orbital, we have a net occupation number:

$$n_z = 2 - \pi_c \tag{3.55}$$

This creates unpaired electron spin density ρ_π^{N} on the nitrogen that is equal to the covalent transfer from the nitrogen p_z-orbital, i.e., $\pi_c \equiv \rho_\pi^{\mathrm{N}}$. Thus, we can obtain a consistent value for π_c from the dipolar component of the nitrogen hyperfine interaction: $T_\parallel^d = A_{zz} - a_o^N$, which is related to ρ_π^{N} by Eq. 3.22. From hyperfine data for the pyrroline nitroxide PYMeOH in o-terphenyl (see Table 2.1), we obtain $\rho_\pi^{\mathrm{N}} = 0.555 \pm 0.002$, and hence the electron occupancy of the nitrogen $2p_z$-orbital becomes $n_z = 1.445 \pm 0.002$.

Now we can use the experimental quadrupole couplings for PYMeOH in Table 3.6 to obtain the occupancies, n_x and n_y, of the two other nitrogen $2p$-orbitals. From Eqs. 3.52 and 3.53, the principal values of the quadrupole tensor become:

$$P_{zz}/P_{2p} = n_z - \tfrac{1}{2}(n_x + n_y) \tag{3.56}$$

$$(P_{xx} - P_{yy})/P_{2p} = \tfrac{3}{2}(n_x - n_y) \tag{3.57}$$

where $P_{2p} \equiv e^2 q_{2p} Q/(2I(2I-1)h) = -6.0$ MHz is the value of P_{zz} for a ^{14}N nitrogen $2p$-orbital. Together with the known value of n_z, we then find the nitrogen $2p_x$- and $2p_y$-orbital populations: $n_x = 1.11 \pm 0.01$ and $n_y = 1.19 \pm 0.01$.

Finally, we relate the electron occupancies n_x and n_y on the nitrogen to the imbalance of electron pairing in the N–O and N–C σ-bonds of the nitroxide radical. In terms of the bonding orbitals, we can write the electron configuration of the central nitrogen as: $1s^2 2(sp^2)_x 2(sp^2)_{\mathrm{C1}} 2(sp^2)_{\mathrm{C2}} 2p_z^2$, where the three hybrid σ-bonding sp^2-orbitals are singly occupied. To obtain the electron populations of the nitrogen $2p_x$- and $2p_y$-orbitals, we must take the hybridization into account. From Eqs. K.23, K.25 and K.26 of Appendix K, the three sp^2 nitrogen valence orbitals are:

$$\psi_{\mathrm{N-O}} = \sqrt{1 - 2c_s^2}\,|s\rangle + \sqrt{2c_s^2}\,|p_x\rangle \tag{3.58}$$

$$\psi_{\mathrm{N-C_1}} = c_s|s\rangle - \sqrt{\frac{1 - 2c_s^2}{2}}\,|p_x\rangle - \frac{1}{\sqrt{2}}|p_y\rangle \tag{3.59}$$

$$\psi_{\mathrm{N-C_2}} = c_s|s\rangle - \sqrt{\frac{1 - 2c_s^2}{2}}\,|p_x\rangle + \frac{1}{\sqrt{2}}|p_y\rangle \tag{3.60}$$

where, as we see from Eqs. 3.59 and 3.60, c_s^2 is the fraction of s-electron character in the two hybrid orbitals that are directed along the N–C bonds.

We quantify the population imbalance of the two shared electrons in a σ-bond by the fractional ionic character i_σ of the covalent bond, where positive values correspond to electron excess on the nitrogen. The population of the $\psi_{\mathrm{N-O}}$ orbital given by Eq. 3.58 is therefore $1 + i_\sigma(\mathrm{NO})$, and that of each $\psi_{\mathrm{N-C}}$ orbital given by Eqs. 3.59 and 3.60 is $1 + i_\sigma(\mathrm{NC})$. Summing normalized contributions from Eqs. 3.59 and 3.60 (one half from each), the electron population of the nitrogen p_y-orbital is therefore given by:

$$n_y = 1 + i_\sigma(\mathrm{NC}) \tag{3.61}$$

Correspondingly for the nitrogen p_x-orbital, we sum weighted contributions from Eqs. 3.58–3.60:

$$n_x = 1 + 2c_s^2 i_\sigma(\mathrm{NO}) + (1 - 2c_s^2) i_\sigma(\mathrm{NC}) \tag{3.62}$$

where c_s^2 is related to the C–N–C bond angle θ_{CNC} by Eq. K.27 of Appendix K.

Taking $\theta_{\mathrm{CNC}} = 115.0°$ for the closely related pyrroline nitroxide MTSSL from Table 3.4, we calculate that the s-electron admixture in PYMeOH is $c_s^2 = 0.297$. From Eqs. 3.61 and 3.62, we then deduce that $i_\sigma(\mathrm{NC}) = 0.19 \pm 0.01$ and $i_\sigma(\mathrm{NO}) = 0.05 \pm 0.03$, in accord with the relative electronegativities of carbon and oxygen. The value for $i_\sigma(\mathrm{NC})$ approaches the empirical correlation with Pauling electronegativities (Gordy and Cook 1984): $i_\sigma(\mathrm{NC}) = \frac{1}{2}(X_{\mathrm{N}} - X_{\mathrm{C}}) = 0.25$. The positive charge on the nitrogen is $\delta q^+ = \pi_c - i_\sigma(\mathrm{NO}) - 2i_\sigma(\mathrm{NC}) = 0.13 \pm 0.04$. Empirically correcting for the charge on the nitrogen atom by replacing $P_{2p,\mathrm{N}}$ with $P_{2p,\mathrm{N}}(1 + \varepsilon\delta q^+)$, where the shielding factor is $\varepsilon = 0.30$ for nitrogen (Dailey and Townes 1955; Gordy and Cook 1984), reduces this to $\delta q^+ = 0.10$, with only small changes in the occupancies n_x and n_y. The effective group electronegativity of the nitroxide orbital to which the oxygen is bonded is defined from the empirical correlation: $X_g = X_{\mathrm{O}} + 2i_\sigma(\mathrm{NC})$. This becomes $X_g = 3.6 \pm 0.1$, which is considerably larger than the averaged value of 3.0 for nitrogen, because in nitroxides the latter bears a net positive charge.

From the charge transfer from nitrogen to oxygen, we also can predict the electric dipole moment of the N–O bond:

$$p_{el} = e(\pi_c - i_\sigma(\mathrm{NO}))r_{\mathrm{NO}} \tag{3.63}$$

Taking the N–O bond length r_{NO} for MTSSL from Table 3.4, we get $p_{el} = (1.0 \pm 0.07) \times 10^{-29}$ C·m. This is of a similar magnitude to the experimental estimate of $\Delta p_{el} = (0.8 \pm 0.03) \times 10^{-29}$ C·m for the contribution to the total dipole moment from the N–O bond (Vasserman et al. 1965, and see Table 4.1 in Chapter 4).

3.12 *g*-TENSOR OF NITROXIDES

The g-values of nitroxides are anisotropic and differ from the free-spin value ($g_e = 2.0023193$) because of small contributions from the orbital angular momentum, **l**, of the unpaired electron.

These contributions arise from admixtures of higher- and lower-lying molecular orbitals into the $2p_z\pi$-orbital of the N–O bond by spin-orbit coupling. Note that the orbital angular momentum of the ground state is quenched because chemical bonding removes the degeneracy between the atomic p_x, p_y and p_z orbitals, and for each of these alone the orbital angular momentum is zero, e.g., $\langle p_z|l_i|p_z\rangle = 0$, for $i = x,y,z$ – see Table E.1 in Appendix E.

We can view spin-orbit coupling as an interaction of the spin magnetic moment with the magnetic field that is generated by the orbiting electron. It therefore depends on the product of the two angular momenta: ~**l**·**s**, where **s** is the real spin angular momentum (as opposed to that of the effective spin, **S**, which we use in the spin Hamiltonian). However, we cannot take the classical analogy too far, because electron spin is a purely relativistic effect. In fact, the correct relativistic calculation produces a result for the spin-orbit coupling constant, ζ, that differs by a factor of two from the classical model.

Spin-orbit coupling together with the interaction of the orbital and spin magnetic moments with the external magnetic field, **B**, can be treated quantum mechanically as a perturbation. The corresponding perturbation Hamiltonian takes the form:

$$\mathcal{H}_{pert} = \sum_k \zeta_k \mathbf{l}^{(k)} \cdot \mathbf{s} + \beta_e \mathbf{B} \cdot (\mathbf{l} + g_e \mathbf{s}) \tag{3.64}$$

where $\mathbf{l}^{(k)}$ is the orbital angular momentum of the unpaired electron at atom k, ζ_k is the spin-orbit coupling constant for this electron in atom k, and $\mathbf{l}$ is the angular momentum of the unpaired electron in its molecular orbital. The values of the spin-orbit coupling constant are $\zeta_N = 76\ \text{cm}^{-1}$ and $\zeta_O = 151\ \text{cm}^{-1}$ for $2p$-orbitals of the nitrogen and oxygen atoms, respectively (Morton et al. 1962, quoted in Gordy 1980).

Stone's theory, which allows specifically for the gauge invariance in this multicentre problem, gives the g-tensor for π-radicals (Stone 1963). Gauge invariance means that the results must be independent of the origin to which the angular momenta, $\mathbf{l}^{(k)}$ and $\mathbf{l}$, are referred. This requires addition of further terms to the perturbation Hamiltonian in Eq. 3.64 (Stone 1963). Nonetheless, the main result is simply a second-order perturbation expression (see Appendices G and L) for the diagonal elements, g_{ii}, of the g-tensor that involves cross terms between the magnetic field interaction and the spin-orbit coupling. Neglecting overlap between atomic orbitals, this is given by (cf. Appendix L):

$$g_{ii} = g_e + 2\sum_m^{\sigma,n} \frac{\langle \Psi_{\pi^*}|l_i^{(N)} + l_i^{(O)}|\Psi_m\rangle\langle \Psi_m|\zeta_N l_i^{(N)} + \zeta_O l_i^{(O)}|\Psi_{\pi^*}\rangle}{E_{\pi^*} - E_m}$$

$$= g_e + 2\sum_m^{\sigma,n} \frac{\sum_{k,k'}^{N,O} \langle \psi_{\pi^*}^{(k')}|l_i^{(k')}|\psi_m^{(k')}\rangle \zeta_k \langle \psi_m^{(k)}|l_i^{(k)}|\psi_{\pi^*}^{(k)}\rangle}{E_{\pi^*} - E_m} \tag{3.65}$$

where π^* is the antibonding π-orbital that contains the unpaired electron and m are the other occupied and unoccupied orbitals. Note that spin-orbit coupling mixes neither fully-occupied nor totally unoccupied π-orbitals into the π-orbital that contains the unpaired electron. As in Eq. 3.64, the superscripts k, k' in Eq. 3.65 refer to the atom centres N and O. For example, $\psi_m^{(k)}$ is the contribution of the atomic orbitals from atom k to the molecular orbital m; the coefficients giving the weighting for these atomic orbitals are designated $c_k^{(m)}$. We neglect any small contribution from N–C σ-bonds.

Figure 3.10 shows the atomic orbitals involved in g-tensor calculations for nitroxide spin labels. The unpaired electron is located in an antibonding $2p_z\pi^*$ orbital (see Eq. 3.1):

$$\Psi_{\pi^*} \equiv \psi_{\pi^*}^{(N)} + \psi_{\pi^*}^{(O)} = c_N|p_z\rangle_N + c_O|p_z\rangle_O \tag{3.66}$$

where $\rho_\pi^N \equiv c_N^2$, $\rho_\pi^O \equiv c_O^2$ and $\rho_\pi^N + \rho_\pi^O \approx 1$. Equation 3.65 therefore reduces to:

$$g_{ii} = g_e + 2\sum_m^{\sigma,n} \frac{\sum_{k,k'}^{N,O} c_{k'} c_k \zeta_k \langle p_z^{(k')}|l_i^{(k')}|\psi_m^{(k')}\rangle\langle \psi_m^{(k)}|l_i^{(k)}|p_z^{(k)}\rangle}{E_{\pi^*} - E_m} \tag{3.67}$$

which restricts the p-orbitals in $\psi_m^{(k)}$ and $\psi_m^{(k')}$ that can contribute. Those that we need are the bonding and antibonding σ-orbitals (b and a) of the N–O bond and the lone-pair orbitals (n) on the oxygen (Kawamura et al. 1967; and see Figure 3.10). Oxygen has two lone-pair orbitals in the outer shell. Repulsion between lone pairs favours occupation of two equivalent sp^2 hybrids, with the third hybrid participating in the N–O bond, much as for the nitrogen. An sp^2 hybridization scheme also is suggested by hydrogen bonding of two water molecules to the nitroxide (see Figure 4.13 and Table 4.8 in Chapter 4). The electron configuration of oxygen thus becomes: $1s^2 2(sp^2)_x 2(sp^2)_{n1}^2 2(sp^2)_{n2}^2 2p_z$, and the hybrid lone-pair orbitals $n1$, $n2$ are (cf. Eqs. 3.59 and 3.60):

$$\psi_n^{(O)} = c_{O,s}^{(n)}|s\rangle_O + c_{O,x}^{(n)}|p_x\rangle_O \pm c_{O,y}^{(n)}|p_y\rangle_O \tag{3.68}$$

where $c_{O,x}^{(n)} = \sqrt{\frac{1}{2} - \left(c_{O,s}^{(n)}\right)^2}$ and $c_{O,y}^{(n)} = 1/\sqrt{2}$. The bonding and antibonding N–O σ-orbitals, σ_b and σ_a, are linear combinations of

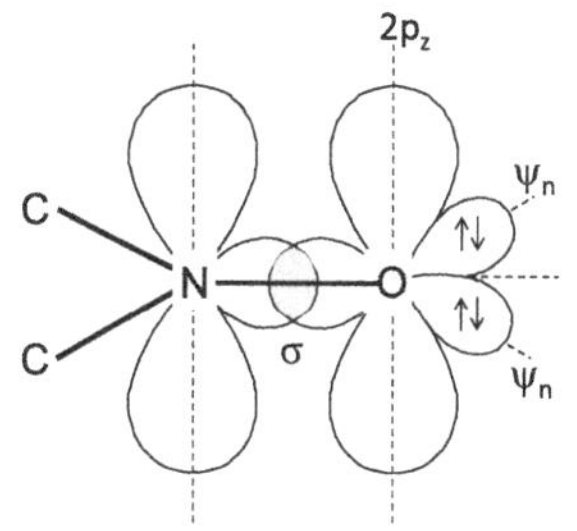

FIGURE 3.10 Nitrogen and oxygen orbitals involved in g-tensor calculations for nitroxides (Eq. 3.67). $2p_z\pi$-orbital (Eq. 3.66) contains unpaired electron, oxygen ψ_n-orbitals (Eq. 3.68) contain lone pairs and σ-orbitals (Eq. 3.69) contain valence electrons. Lone-pair orbitals and σ-orbitals lie in C–NO–C plane; $2p_z$-orbitals are perpendicular to this plane.

the nitrogen $(sp^2)_x$ hybrid orbital (viz., Eq. 3.58) with that of the oxygen:

$$\Psi_{\sigma_a,\sigma_b} \equiv \psi^{(N)}_{\sigma_a,\sigma_b} + \psi^{(O)}_{\sigma_a,\sigma_b}$$
$$= c^{(a,b)}_{N,s}|s\rangle_N + c^{(a,b)}_{N,x}|p_x\rangle_N + c^{(a,b)}_{O,s}|s\rangle_O + c^{(a,b)}_{O,x}|p_x\rangle_O \quad (3.69)$$

where $c^{(b)}_{N,s} = \sqrt{\frac{1}{2} - c_s^2}$, $c^{(b)}_{N,x} = c_s$, $c^{(b)}_{O,s} = \sqrt{\frac{1}{2} - \left(c^{(n)}_{O,s}\right)^2}$ and $c^{(b)}_{O,x} = c^{(n)}_{O,s}$, with c_s as defined in Eqs. 3.59, 3.60 and related to the C–N–C bond angle by Eq. K.27 of Appendix K. Admixture coefficients for the corresponding antibonding molecular orbitals are defined by interchanging nitrogen and oxygen, together with reversing their relative sign as in Eqs. 3.9 and 3.10. Neglecting overlap terms, Eq. 3.69 is normalized with these expressions for the admixture coefficients.

Because *s*-electron orbitals are spherically symmetrical, they do not contribute to *g*-tensor anisotropy. They have only indirect effect on the *p*-orbital admixtures that results from normalization of the hybrid orbitals. The *p*-orbitals are eigenfunctions of the l_z angular momentum operator with eigenvalues given by: $l_z|p_i\rangle = 0$ where $i \equiv x, y, z$. Thus, we see immediately from Eq. 3.67 that the *zz*-principal element of the *g*-tensor is simply:

$$g_{zz} = g_e \quad (3.70)$$

and does not deviate from the *g*-value of a free electron (see also Appendix L).

The l_x angular momentum operator, which specifies the g_{xx}-element, couples only p_y and p_z orbitals (see Table E.1 in Appendix E). Thus, the g_{xx}-tensor element is given from Eqs. 3.67 and 3.68 by:

$$g_{xx} = g_e + \frac{2\zeta_O \rho^O_\pi}{E_{\pi^*} - E_n} \quad (3.71)$$

where $E_{\pi^*} - E_n$ is the n–π^* excitation energy. We get Eq. 3.71 by summing over the two equivalent $2sp^2$-hybrid lone-pair orbitals, because $2\left(c^{(n)}_{O,y}\right)^2 = 1$. Note that we get the same result by assuming that one lone pair is confined to the oxygen $2p_y$-orbital (i.e., $c^{(n)}_{O,y} = 1$) with the other occupying the non-bonding $2sp_x$ hybrid. This is what we expect because hybridization simply corresponds to a change in basis set of orbitals (see Appendix K.3).

The l_y angular moment operator, which specifies the g_{yy}-element, couples only p_x and p_z orbitals (see Table E.1 in Appendix E). The g_{yy}-tensor element therefore comes from $2p_x$-orbitals contained both in the lone pairs and in the N–O σ-bond. From Eqs. 3.67–3.69, it becomes (Kawamura et al. 1967):

$$g_{yy} = g_e + \frac{2\zeta_O \left(C^{(n)}_{O,x}\right)^2 \rho^O_\pi}{E_{\pi^*} - E_n}$$
$$+ 2\sum_m^{a,b} \frac{\zeta_N \left(c^{(m)}_{N,x}\right)^2 \rho^N_\pi + \zeta_O \left(c^{(m)}_{O,x}\right)^2 \rho^O_\pi + (\zeta_N + \zeta_O) c^{(m)}_{N,x} c^{(m)}_{O,x} p_{NO}}{E_{\pi^*} - E_m} \quad (3.72)$$

where $\left(C^{(n)}_{O,x}\right)^2 = \sum_n \left(c^{(n)}_{O,x}\right)^2$ with summation over all lone-pair orbitals *n*, and $p_{NO} (\equiv c_N \times c_O)$ is the partial bond order of the unpaired electron orbital. For the lone-pair contribution from the $2sp^2$ hybrids, we have $\left(C^{(n)}_{O,x}\right)^2 = 2\left(c^{(n)}_{O,x}\right)^2 = 1 - 2\left(c^{(n)}_{O,s}\right)^2$. We see immediately that this is less than the corresponding contribution to g_{xx}. If we take instead the non-bonding $2sp_x$ hybrid: $\psi^{(O)}_n = c^{(n)}_{O,s}|s\rangle_O - \sqrt{1 - \left(c^{(n)}_{O,s}\right)^2}|p_x\rangle_O$ (together with pure $2p_y$), the lone-pair contribution is $\left(C^{(n)}_{O,x}\right)^2 = \left(c^{(n)}_{O,x}\right)^2 = 1 - \left(c^{(n)}_{O,s}\right)^2$. As expected, this is equivalent to the $2sp^2$ case, because the admixture coefficient $c^{(n)}_{O,s}$ (and $c^{(n)}_{O,x}$) is $\sqrt{2}$ times larger for $2sp_x$ hybrids.

We list representative values of the *g*-tensor elements in the Appendix to Chapter 2. Mean *g*-values from there are: $g_{xx} = 2.00898 \pm 4.0 \times 10^{-4}$, $g_{yy} = 2.00602 \pm 2.2 \times 10^{-4}$ and $g_{zz} = 2.00219 \pm 2.1 \times 10^{-4}$ ($N = 29$), where standard deviations reflect the range of nitroxide structures and solvent polarities. Comparing with Eqs. 3.70–3.72 shows, as predicted, that the g_{zz}-element has the smallest value, closest to that of a free electron. The largest deviation from the free-electron *g*-value is for the g_{xx}-element. Thus, the contribution of $|p_y\rangle_O$ to the lone-pair orbitals is greater than that of $|p_x\rangle_O$, and also the contribution of $|p_y\rangle_O$ to the *g*-tensor is greater than that from the N–O bond (which contributes only to g_{yy}). The latter fulfills our expectation that the lone pairs lie closer in energy to the unpaired electron orbital than does the strong N–O σ-bonding orbital (see Figure 3.1 and Kikuchi 1969).

Assuming that $\rho^O_\pi = 0.45$ (cf. Section 3.2), the value of g_{xx} given by Eq. 3.71 implies that the energy separation of the lone-pair orbitals is $E_{\pi^*} - E_n \approx 20{,}400\ \mathrm{cm^{-1}}$. For di-*tert*-butyl nitroxide and TEMPONE, the weak n–π^* absorption band lies at 466 and 450 nm, respectively (Kawamura et al. 1967; Briere et al. 1965), corresponding to excitation energies of 21,460 and 22,220 cm^{-1} that are comparable with our estimate from g_{xx}.

The experimental ratio $(g_{yy} - g_e)/(g_{xx} - g_e) = 0.556$ exceeds $\left(C^{(n)}_{O,x}\right)^2 = 1 - 2\left(c^{(n)}_{O,s}\right)^2 = 0.40$ predicted for the lone-pair contribution alone, from the angle θ_{nOn} between equivalent sp^2 hybrids (see later in Section 4.4.4). We must attribute the remaining g_{yy}-anisotropy $(\delta g_{yy} = 0.0010)$ to the N–O σ-bond, in spite of its higher energy denominator.

3.13 CONCLUDING SUMMARY

1. *Isotropic ^{14}N-hyperfine coupling:*
   The isotropic hyperfine coupling a^N_o is directly proportional to the unpaired spin density ρ^N_π on the nitrogen, as given by Eq. 3.15. An approximate version is:

$$a^N_o \approx Q^N_N \rho^N_\pi \quad (3.73)$$

   where $Q^N_N \approx 2.56$ mT (72 MHz).

2. *Dipolar ^{14}N-hyperfine coupling:*
The dipolar hyperfine coupling tensor $\left(T_{\perp}^{\mathrm{d}},T_{\perp}^{\mathrm{d}},T_{\parallel}^{\mathrm{d}}\right)$ is axial and traceless: $T_{\perp}^{\mathrm{d}}=-\frac{1}{2}T_{\parallel}^{\mathrm{d}}$.
Each element is directly proportional to the unpaired spin density on the nitrogen:

$$T_{\parallel}^{d}=T_{\parallel,o}^{d}\rho_{\pi}^{\mathrm{N}} \tag{3.74}$$

where $T_{\parallel,o}^{d}\approx 3.41\,\mathrm{mT}$ (96 MHz). The complete hyperfine tensor $\left(A_{xx},A_{yy},A_{zz}\right)$ is given by:

$$A_{zz}=a_o^{\mathrm{N}}+T_{\parallel}^{d} \tag{3.75}$$

$$A_{xx}=A_{yy}=a_o^{\mathrm{N}}-\tfrac{1}{2}T_{\parallel}^{d} \tag{3.76}$$

3. *[13]C-hyperfine couplings:*
Isotropic ^{13}C-hyperfine couplings a_o^{C} also are proportional to ρ_{π}^{N}:

$$a_o^{\mathrm{C}}=Q_{\mathrm{NC}}^{\mathrm{C}}\rho_{\pi}^{\mathrm{N}} \tag{3.77}$$

Experimentally, isotropic couplings $a_o^{\mathrm{C}\alpha}$ of Cα carbon atoms are $\approx\frac{1}{2}a_o^{\mathrm{N}}$, i.e., $Q_{\mathrm{NC}\alpha}^{\mathrm{C}}\approx 1.27\,\mathrm{mT}$ (36 MHz). For Cβ carbon atoms: $Q_{\mathrm{NC}\beta}^{\mathrm{C}}\approx 0.73\,\mathrm{mT}$ (21 MHz).
Isotropic couplings of β-carbon atoms are conformationally sensitive:

$$a_o^{\mathrm{C}\beta}/a_o^{\mathrm{N}}\approx C_2\cos^2\vartheta \tag{3.78}$$

where ϑ is the dihedral angle between the nitroxide z-axis and the Cα–Cβ bond, and $C_2\approx 0.55$.

4. *[14]N-quadrupole couplings:*
The nuclear electric quadrupole coupling tensor $\left(P_{xx},P_{yy},P_{zz}\right)$ is traceless: $P_{xx}+P_{yy}+P_{zz}=0$. The principal elements are related to the electron occupation numbers n_x, n_y and n_z in the nitrogen p_x, p_y and p_z-orbitals:

$$P_{zz}/P_{2p}=n_z-\tfrac{1}{2}\left(n_x+n_y\right) \tag{3.79}$$

$$\left(P_{xx}-P_{yy}\right)/P_{2p}=\tfrac{3}{2}\left(n_x-n_y\right) \tag{3.80}$$

where $P_{2p}=-6.0$ MHz is the value of P_{zz} for a ^{14}N nitrogen $2p$-orbital.

5. *g-Values:*
The nitroxide g-tensor is non-axial with $g_{xx}>g_{yy}>g_{zz}\approx g_e$, where g_e is the free-electron g-value. The largest g-tensor element is given by:

$$g_{xx}=g_e+\frac{2\zeta_{\mathrm{O}}\rho_{\pi}^{\mathrm{O}}}{E_{\pi^*}-E_n} \tag{3.81}$$

where $\rho_{\pi}^{\mathrm{O}}\approx 1-\rho_{\pi}^{\mathrm{N}}$ is the unpaired electron spin density on the oxygen, $\zeta_{\mathrm{O}}=151\,\mathrm{cm}^{-1}$ is the spin-orbit coupling constant of the oxygen $2p$-orbital and $E_{\pi^*}-E_n\approx 22{,}000\,\mathrm{cm}^{-1}$ is the n–π* excitation energy.

APPENDIX A3.1: ELECTRON SPIN DENSITIES OF NITROXIDES

TABLE A3.1 Normalized spin populations in TEMPONE and TEMPOL crystals from polarized-neutron diffraction (Bordeaux et al. 1993)

atom	*TEMPONE*[a]	*TEMPOL*[b]
O	0.371 ± 0.018	0.355 ± 0.012
N	0.420 ± 0.019	0.544 ± 0.012
C2	−0.059 ± 0.021	0.018 ± 0.012
Me2 (equatorial)	0.129 ± 0.022	0.041 ± 0.012
Me2′ (axial)	0.026 ± 0.018	0.006 ± 0.012
C6	0.027 ± 0.023	≡ C2
Me6 (equatorial)	−0.038 ± 0.019	≡ Me2
Me6′ (axial)	0.034 ± 0.015	≡ Me2′
C3	0.013 ± 0.018	0.006 ± 0.012
C5	0.053 ± 0.026	≡ C3
C4	0.038 ± 0.021	−0.053 ± 0.018
O4	−0.016 ± 0.013	0.024 ± 0.012
H	–	−0.012 ± 0.012

[a] TEMPONE, 1-oxyl-2,2′,6,6′-tetramethylpiperidin-4-yl oxide. Ring atoms and substituents are numbered from N-atom as "1". *Twisted* ring-conformation in the crystal results in all heavy atoms being inequivalent.
[b] TEMPOL, 1-oxyl-2,2′,6,6′-tetramethylpiperidin-4-ol. Atoms are numbered as for TEMPONE. *Chair* ring-conformation has a mirror plane containing N–O, i.e., C5≡C3, C6≡C2 (and corresponding methyls).

Molecular orbital for unpaired electron in a TEMPO nitroxide.
The projection map of the spin density in di(1-oxyl-2,2,6,6-tetramethylpiperidin-4-yl) suberate allows construction of the $2p\pi^*$ antibonding molecular orbital for the unpaired electron:

$$\psi(\mathbf{r})=c_{\mathrm{N},z}\left|p_z\right\rangle_{\mathrm{N}}+c_{\mathrm{N},x}\left|p_x\right\rangle_{\mathrm{N}}-c_{\mathrm{O},z}\left|p_z\right\rangle_{\mathrm{O}} \tag{A3.1}$$

The nitrogen-centred part of the orbital contains a p_x-admixture, because the N–O bond makes an angle of 18° with the C–N–C plane (Brown et al. 1983, and cf. Table 3.4). The nitrogen p-orbital axis bisects this angle. In Eq. A3.1, normalized Slater-type functions for the atomic p-orbitals are:

$$\left|p_z\right\rangle_{\mathrm{N}}=\sqrt{\frac{\alpha_{\mathrm{N}}^5}{\pi}}z\exp(-\alpha_{\mathrm{N}}r_{\mathrm{N}}) \tag{A3.2}$$

$$\left|p_x\right\rangle_{\mathrm{N}}=\sqrt{\frac{\alpha_{\mathrm{N}}^5}{\pi}}x\exp(-\alpha_{\mathrm{N}}r_{\mathrm{N}}) \tag{A3.3}$$

$$\left|p_z\right\rangle_{\mathrm{O}}=\sqrt{\frac{\alpha_{\mathrm{O}}^5}{\pi}}z\exp(-\alpha_{\mathrm{O}}r_{\mathrm{O}}) \tag{A3.4}$$

expressed in atomic coordinates of the N- and O-atoms, where r_{N} and r_{O} are the radial distances from the respective atom centres. The Slater exponents α_{N} and α_{O} must not be confused with the conventional symbols that are used for the coulomb integrals of Hückel molecular orbital theory in Section 3.2. Fitting the experimental spin-density projection gives (Brown et al. 1983): $c_{N,z} = 0.74$, $c_{N,x} = 0.03$ with $\alpha_N = 33.8\ \mathrm{nm}^{-1}$ and $c_{O,z} = 0.64$ with $\alpha_O = 43.3\ \mathrm{nm}^{-1}$. Note that $|c_{N,z}|^2 + |c_{N,x}|^2 + |c_{O,z}|^2 = 0.96$, which is less than unity because of non-vanishing spin density on other atoms in the nitroxide ring. The fraction of unpaired spin density on the nitrogen is: $|c_{N,z}|^2 + |c_{N,x}|^2 = 0.55$, and that on the oxygen is: $|c_{O,z}|^2 = 0.41$. Thus, 57% of the N–O bond spin density is on the nitrogen and 43% on the oxygen.

APPENDIX A3.2: PROTON HYPERFINE COUPLINGS OF NITROXIDES

TABLE A3.2 Isotropic proton hyperfine couplings, a_o^H, for 6-membered piperidine- (TEMPO-) ring nitroxides

			$a_o^H(\mathrm{Me})$ (μT)		$a_o^H(\beta\mathrm{CH_2})$ (μT)		$a_o^H(\gamma\mathrm{CH_2})$ (μT)	
nitroxide[a]	*solvent*[b]	*T* (K)	*equatorial* (6H)	*axial* (6H)	*equatorial* (2H)	*axial* (2H)	(1H)	*Refs.*[c]
TEMPOL	toluene	185	−46.3	–	−57.0	−30.3	–	1
	DTBN	297	−36	−2	−48	−31	+7	2
	benzene	RT	−45	−2	−48	−31	+7	3
		RT	45	2	48	31	7	5
	CCl_4	295	46.0	3.0	43.0	29.0	7.0	6
	acetone	RT	−45	−2	−48	−21	+7	4
	EtOH	258	47.0	2.0	48.0	33.3	8.0	7
		278	46.0	2.0	48.0	32.5	7.0	7
		298	45.2	2.0	47.8	32.0	7.0	7
	pyridine	258	46.0	2.0	47.3	32.2	8.0	7
		278	45.4	2.0	47.0	31.8	8.0	7
		298	44.3	2.0	46.9	31.2	7.0	7
	H_2O	295	44.0	6.0	45.0	36.0	15.0	6
		293	44.5(1)	9.79(8)	41.7(4)	37.9(2)	17.27(9)	8
TEMPO-NH_2	neat	333	−46.2	−4.7	−49	−30	+9.3	9
	CCl_4	295	51.0	4.0	42.0	29.0	4.0	6
	MeOH	253	46.2	4.7	49	30	9.3	9
	1,3-Pr$(OH)_2$	RT	46.7	4.0	43.6	30.4	6.8	6
	H_2O	295	43.6	3.8	45.0	31.5	9.0	6
		363	−46.3	−3.8	−45.0	−31.5	+8.8	6
TEMPO-COOMe	toluene	185	−46.7	–	−57.0	−30.6	–	1
TEMPO-OOCMe	benzene	RT	−43	0	–	−31	+8	3
		318	−41	0	–	−29	+9	10
TEMPO-NHOCPhOC_8	MeOH	328	−41.0	0	−49.0	−31.0	–	11
	EBBA	363	43.0	0	49.0	31.0	–	12
TEMPOL-Me	benzene	RT	−40	0	0	−27		3
TEMPO-Mal	NaP buffer	293	43.4	–	52.8	18.2	–	13
TEMPOL-Ph	acetone	RT	−44	−8	−50	−31		4
TEMPO-SO_3PhBr	acetone	RT	−41	−4	–	−32	+6	3
TEMPO-NMe_2 C_{12}	H_2O	370	−42.5	–	–	−30	–	14
TEMPO	CCl_4	RT	−23 (12H)		−39 (4H)		+18.0 (2H)	3
		295	23.0 (12H)		39.0 (4H)		18.0 (2H)	6
	benzene	281	23 (12H)		39 (4H)		18 (2H)	5
		318	−22 (12H)		−39 (4H)		+18 (2H)	10
	MeOH	RT	−21 (12H)		−40 (4H)		20.0 (2H)	3
	H_2O	295	17.5 (12H)		40.0 (4H)		16.0 (2H)	6

(Continued)

TABLE A3.2 (Continued) Isotropic proton hyperfine couplings, a_o^H, for 6-membered piperidine- (TEMPO-) ring nitroxides

			a_o^H(Me) (μT)		a_o^H(βCH_2) (μT)		a_o^H(γCH_2) (μT)	
nitroxide[a]	*solvent*[b]	*T* (K)	*equatorial* (6H)	*axial* (6H)	*equatorial* (2H)	*axial* (2H)	(1H)	*Refs.*[c]
TEMPONE	toluene	195	−10.7(12H)		– (4H)			1
	benzene	RT	−12 (12H)		−2 (4H)			3
		318	−11 (12H)		−1 (4H)			10
	CCl_4	295	11.5 (12H)		– (4H)			6
		298	10 (12H)		– (4H)			15
	n-BuOH		10 (12H)		– (4H)			16
	H_2O	273	6 (12H)		– (4H)			15
		295	6.0 (12H)		– (4H)			6
		373	8 (12H)		– (4H)			15

Note: Values in parentheses, e.g., (6H), are numbers of equivalent protons with the hyperfine coupling indicated; these numbers are given at the top of the columns (except for TEMPO and TEMPONE, where axial and equatorial are averaged and numbers of H are specified within the body of the table). Signed couplings are from NMR or ENDOR; unsigned couplings are absolute values obtained from EPR by spectral simulation.

[a] TEMPO, 1-oxyl-2,2,6,6-tetramethylpiperidine;
TEMPOL, 1-oxyl-2,2,6,6-tetramethylpiperidin-4-ol;
TEMPO-NH_2, 1-oxyl-2,2,6,6-tetramethylpiperidin-4-ylamine;
TEMPO-Mal, *N*-(1-oxyl-2,2,6,6-tetramethylpiperidin-4-yl) maleimide;
TEMPONE, 1-oxyl-2,2,6,6-tetramethylpiperidin-4-yl oxide;
TEMPO-COOMe, 1-oxyl-2,2,6,6-tetramethylpiperidin-4-ylcarboxymethyl ester;
TEMPO-OOCMe, 1-oxyl-2,2,6,6-tetramethylpiperidin-4-yl acetate;
TEMPOL-Me, 1-oxyl-2,2,4,6,6-pentamethylpiperidin-4-ol;
TEMPOL-Ph, 1-oxyl-2,2,6,6-tetramethyl-4-phenylpiperidin-4-ol;
TEMPO-SO_3PhBr, 1-oxyl-2,2,6,6-tetramethylpiperidin-4-ylsulpho(*p*-phenylbromide);
TEMPO-NMe_2C_{12}, *N*-(1-oxyl-2,2,6,6-tetramethylpiperidin-4-yl)-*N*,*N*-dimethyldodecylamine;
TEMPO-NHOCPhOC_8, *N*-(1-oxyl-2,2,6,6-tetramethylpiperidin-4-yl)-(*p*-octyloxy)benzoylamine.

[b] DTBN, di-*tert*-butyl nitroxide; EtOH, ethanol; MeOH, methanol; 1,3-Pr$(OH)_2$, 1,3-propanediol; EBBA, *N*-(*p*-ethoxybenzylidene)-*p*-butylaniline; NaP, sodium phosphate buffer; *n*-BuOH, *n*-butanol.

[c] References: 1. Kirste et al. (1982); 2. Kopf and Kreilick (1969); 3. Brière et al. (1967); 4. Brière et al. (1970); 5. Whisnant et al. (1974); 6. Windle (1981); 7. Ottaviani (1987); 8. Nielsen et al. (2004); 9. Labsky et al. (1980); 10. Kreilick (1967); 11. Barbarin et al. (1978b); 12. Barbarin et al. (1978a); 13. Beth et al. (1980); 14. Fox (1976); 15. Jolicoeur and Friedman (1971); 16. Poggi and Johnson Jr. (1970).

TABLE A3.3 Isotropic proton hyperfine couplings, a_o^H, for 5-membered pyrroline- (PY-) and oxazolidine- (DOXYL) ring nitroxides

					a_o^H(βCH_2) (μT)[c]		a_o^H(γCH_2) (μT)[c]		
nitroxide[a]	*solvent*[b]	*T* (K)	a_o^H(Me) (μT)	a_o^H(βCH_2) (μT)	*equatorial*	*axial*	*equatorial*	*axial*	*Refs.*[d]
PYCONH$_2$	CCl_4	295	24.0 (12H)	44.0 (1H)					1
	pyridine	258	23.5 (12H)	50.8 (1H)					2
		278	23.5 (12H)	49.2 (1H)					2
		298	23.4 (12H)	48.5 (1H)					2
	EtOH	258	23.5 (12H)	49.0 (1H)					2
		278	23.4 (12H)	48.0 (1H)					2
		298	23.2 (12H)	47.3 (1H)					2
	H_2O	295	20.0 (12H)	35.0 (1H)					1
PYCOOH	MeOH	210	−23.5 (12H)	−54.1 (1H)					3
PYCOOMe	toluene	185	−23.5 (12H)	−52.0 (1H)					4
DOXYLMe$_2$	$CDCl_3$	RT	−12 (6H)	−5 (2H)					5
DOXYLcHx	$CDCl_3$	RT	−12 (6H)	−6 (2H)	−58 (2H)	−68 (2H)	+101 (2H)	+21 (2H)	5
	$CDCl_3$	RT	12 (6H)	7 (2H)	58 (2H)	68 (2H)	101 (2H)	21 (2H)	5
DOXYLcHxMe$_2$	$CDCl_3$	RT	−13 (6H)	−6 (2H)	−61 (2H)	−70 (2H)		+13 (2H)	5
DOXYLcHx*t*Bu	$CDCl_3$	RT	−13 (6H)	−5 (2H)	−65 (2H)	−70 (2H)	+106 (2H)	+17 (2H)	5

Note: Values in parentheses are numbers of equivalent protons with the hyperfine coupling indicated. Signed couplings are from NMR or ENDOR; unsigned couplings are absolute values obtained from EPR by spectral simulation.

[a] PYCONH$_2$, 1-oxyl-2,2,5,5-tetramethylpyrrolin-3-yl amide; PYCOOH, 1-oxyl-2,2,5,5-tetramethylpyrrolin-3-yl carboxylic acid; PYCOOMe, 1-oxyl-2,2,5,5-tetramethylpyrrolin-3-ylcarboxymethyl ester; DOXYLMe$_2$, 3-oxyl-2,2,4,4-tetramethyloxazolidine; DOXYLcHx, 3-oxyl-4,4-dimethyloxazolidin-2-yl cyclohexane; DOXYLcHxMe$_2$, 1-(3-oxyl-4,4-dimethyloxazolidin-2-yl)-3,5-dimethylcyclohexane; DOXYLcHx*t*Bu, 1-(3-oxyl-4,4-dimethyloxazolidin-2-yl)-4-*tert*-butylcyclohexane.

[b] EtOH, ethanol.

[c] Cyclohexane ring methylenes.

[d] References: 1. Windle (1981); 2. Ottaviani (1987); 3. Mustafi and Joela (1995); 4. Kirste et al. (1982); 5. Michon and Rassat (1974).

TABLE A3.4 Isotropic proton hyperfine couplings, a_o^H, for di-*tert*-butyl nitroxide (DTBN)

nitroxide	*solvent*[a]	*T* (K)	a_o^H(Me) (μT)	a_o^N (mT)[b]	*Refs.*[c]
DTBN	CCl_4	295	11.0 (18H)	1.536	1
	benzene	RT	12 (18H)	1.536	2
	n-C_5H_{12}	189	11 (18H)	1.54	3
	neat		−10.7 (18H)		4
	i-PrOH	253–273	9.5 (18H)	1.587	5
	ethyleneglycol	253–273	8 (18H)	1.628	5
	MeOH		8 (18H)	1.62	6
			−9 (18H)		6
	H_2O	258	9.1 (18H)	1.73	7
		290	6.0 (18H)	1.721	3
		295	6.0 (18H)	1.714	6

Note: Values in parentheses are numbers of equivalent protons with the hyperfine coupling indicated. Signed couplings are from NMR; unsigned couplings are absolute values obtained from EPR by simulation.
[a] *n*-C_5H_{12}, *n*-pentane; *i*-PrOH, isopropanol; MeOH, methanol.
[b] ^{14}N-hyperfine coupling.
[c] References: 1. Windle (1981); 2. Faber et al. (1967); 3. Plachy and Kivelson (1967); 4. Hausser et al. (1966); 5. Poggi and Johnson Jr. (1970); 6. Lim et al. (1976); 7. Ahn (1976).

REFERENCES

Ahn, M. 1976. Electron spin relaxation of di-tertiary-butyl nitroxide in supercooled water. *J. Chem. Phys.* 64:134–138.

Ayscough, P.B. and Sargent, F.P. 1966. Electron spin resonance studies of radicals and radical ions in solution. V. Solvent effects on spectra of mono- and di-phenyl nitric oxide. *J. Chem. Soc.* 907–910.

Barbarin, F., Chevarin, B., Germain, J.P., Fabre, C., and Cabaret, D. 1978a. Electron spin resonance study of macroscopic and microscopic orientations of isotropic and nematic phases of E.B.B.A by means of deuterated nitroxide radical probe. *Mol. Cryst. Liq. Cryst.* 46:181–194.

Barbarin, F., Chevarin, B., Germain, J.P., Fabre, C., and Cabaret, D. 1978b. Synthèse d'une sonde nitroxyde deuterée de forme allongée. *Mol. Cryst. Liq. Cryst.* 46:195–207.

Berliner, L.J. 1970. Refinement and location of the hydrogen atoms in the nitroxide 2,2,6,6-tetramethyl-4-piperidinol-1-oxyl. *Acta Cryst. B* 26:1198–1202.

Beth, A.H., Perkins, R.C., Venkataramu, S.D. et al. 1980. Advantages of deuterium modification of nitroxide spin labels for biological EPR studies. *Chem. Phys. Lett.* 69:24–28.

Boeyens, J.C.A. and Kruger, G.J. 1970. Remeasurements of the structure of potassium-2,2,5,5-tetramethyl-3-carboxypyrroline-1-oxyl. *Acta Cryst. B* 26:668–672.

Bordeaux, D., Lajzérowicz, J., Brière, R., Lemaire, H., and Rassat, A. 1973. Détermination des axes propres et des valeurs principales du tenseur *g* dans deux radicaux libres nitroxydes par étude de monocristaux. *Org. Magn. Reson.* 5:47–52.

Bordeaux, D. and Lajzérowicz-Bonneteau, J. 1974. Structure cristalline du spiro[cyclohexane-1,2'-(4',4'-dimethyloxazolidine-*N*-oxyl)]. *Acta Cryst. B* 30:2130–2132.

Bordeaux, D. and Lajzérowicz, J. 1974. Polymorphisme du radical libre nitroxyde tétraméthyl-2,2,6,6 piperidinone-4 oxyl-1. Affinement de la structure de la phase orthorhombique. *Acta Cryst. B* 30:790–792.

Bordeaux, D., Boucherle, J.X., Delley, B. et al. 1993. Experimental and theoretical spin densities in two alkyl nitroxides. *Z. Naturforsch.* 48a:117–119.

Brière, R., Lemaire, H., and Rassat, A. 1965. Nitroxydes. XV. Synthèse et étude de radicaux libres stables piperidiniques et pyrrolidinique. *Bull. Soc. Chim. France* 3273–3283.

Brière, R., Lemaire, H., Rassat, A., Rey, P., and Rousseau, A. 1967. Nitroxydes. XXIV. Résonance magnétique nucléaire de radicaux libres nitroxydes pipéridiniques. *Bull. Soc. Chim. France* 4479–4484.

Brière, R., Lemaire, H., and Rassat, A. 1968. Nitroxides. XXV. Isotropic splitting constant for ^{13}C in the tertiary carbon position of di-*tert*-butyl nitroxide. *J. Chem. Phys.* 48:1429–1430.

Brière, R., Rassat, A., and Dunand, J. 1970. Étude de l'interconversion d'un radical libre nitroxyde piperidinique, par RMN à 310 MHz. *Bull. Soc. Chim. France* 4220–4226.

Brière, R., Chapelet-Letourneux, G., Lemaire, H., and Rassat, A. 1971. Nitroxydes. XXXV. Étude de l'interaction hyperfine électron-carbone-13; Radicaux nitroxydes sélectivement marqués en α de l'azote. *Mol. Phys.* 20:211–224.

Brown, T.H., Anderson, D.H., and Gutowsky, H.S. 1960. Spin densities in organic free radicals. *J. Chem. Phys.* 33:720–726.

Brown, P.J., Capiomont, A., Gillon, B., and Schweizer, J. 1983. Experimental spin density in nitroxides. A polarized neutron study of the tanol suberate. *Mol. Phys.* 48:753–761.

Capiomont, A. and Lajzérowicz-Bonneteau, J. 1974. Étude du radical nitroxyde tétraméthyl-2,2,6,6-piperidine-1 oxyle-1 ou `tanane'. II. Affinement de la structure cristallographique de la forme quadratique désordonée. *Acta Cryst. B* 30:2160–2166.

Chiba, T. 1964. Deuteron magnetic resonance study of some crystals containing an O-D…O bond. *J. Chem. Phys.* 41:1352–1358.

Clementi, E., Roothaan, C.C.J., and Yoshimine, M. 1962. Accurate analytical self-consistent field functions for atoms. II. Lowest configurations of the neutral first row atoms. *Phys. Rev.* 127:1618–1620.

Cohen, A.H. and Hoffman, B.M. 1973. Hyperfine interactions in perturbed nitroxides. *J. Am. Chem. Soc.* 95:2061–2062.

Colpa, J.P. and de Boer, E. 1964. Hyperfine coupling constants of CH_2 protons in paramagnetic aromatic systems - spin polarization versus hyperconjugation. *Mol. Phys.* 7:333–348.

Crisma, M., Deschamps, J.R., George, C. et al. 2005. A topographically and conformationally constrained, spin-labeled, α-amino acid: crystallographic characterization in peptides. *J. Peptide Res.* 65:564–579.

Dailey, B.P. and Townes, C.H. 1955. The ionic character of diatomic molecules. *J. Chem. Phys.* 23:118–123.

Dupeyre, R.M., Lemaire, H., and Rassat, A. 1964. Nitroxydes (VII): radicaux libres stables pyrrolidiniques. *Tetrahedron Lett.* 27–8:1781–1785.

Faber, R.J., Markley, F.W., and Weil, J.A. 1967. Hyperfine structure in the solution ESR spectrum of di-*tert*-butyl nitroxide. *J. Chem. Phys.* 46:1652–1654.

Fermi, E. 1930. Über die magnetischen Momente der Atomkerne. *Z. Phys.* 60:320–333.

Flippen-Anderson, J.L., George, C., Valle, G. et al. 1996. Crystallographic characterization of geometry and conformation of TOAC, a nitroxide spin-labelled $C^{\alpha,\alpha}$-disubstituted glycine, in simple derivatives and model peptides. *Int. J. Pept. Protein Res.* 47:231–238.

Fox, K.K. 1976. Isotropic proton hyperfine coupling constants of two cationic nitroxides. *J. Chem. Soc. Faraday Trans. II* 72:975–983.

Gordy, W. 1980. *Theory and Applications of Electron Spin Resonance*. Techniques of Chemistry, Vol. XV. New York: Wiley Interscience.

Gordy, W. and Cook, R.L. 1984. *Microwave Molecular Spectra*. Techniques of Chemistry, Vol. XVIII. New York: Wiley Interscience.

Griffith, O.H., Dehlinger, P.J., and Van, S.P. 1974. Shape of the hydrophobic barrier of phospholipid bilayers. Evidence for water penetration in biological membranes. *J. Membrane Biol.* 15:159–192.

Groth, P. and Hassel, O. 1964. Conformation of the cyclohexane-1.4-dione molecule in an addition compound. *Tetrahedron Lett.* 65–68.

Hatch, G.F. and Kreilick, R.W. 1972. NMR of some nitroxide radicals: ^{13}C coupling constants. *J. Chem. Phys.* 57:3696–3699.

Hausser, K.H., Brunner, H., and Jochims, J.C. 1966. Nuclear magnetic resonance of organic free radicals. *Mol. Phys.* 10:253–260.

Hayat, H. and Silver, B.L. 1973. Oxygen-17 and nitrogen-14 σ-π polarization parameters and spin density distribution in nitroxyl group. *J. Phys. Chem.* 77:72–78.

Heller, C. and McConnell, H.M. 1960. Radiation damage in organic crystals. II. Electron spin resonance of $(CO_2H)CH_2CH(CO_2H)$ in β-succinic acid. *J. Chem. Phys.* 32:1535–1539.

Jolicoeur, C. and Friedman, H.L. 1971. Effects of hydrophobic interactions on dynamics in aqueous solutions studied by EPR. *Ber. Bunsenges. Phys. Chem.* 75:248–257.

Karplus, M. and Fraenkel, G.K. 1961. Theoretical interpretation of carbon-13 hyperfine interactions in electron spin resonance spectra. *J. Chem. Phys.* 35:1312–1323.

Kawamura, T., Matsunami, S., and Yonezawa, T. 1967. Solvent effects on the *g*-value of di-*t*-butyl nitric oxide. *Bull. Chem. Soc. Japan* 40:1111–1115.

Kikuchi, O. 1969. Electronic structure and spectrum of the H_2NO radical. *Bull. Chem. Soc. Japan* 42:47–52.

Kirste, B., Kruger, A., and Kurreck, H. 1982. ESR and ENDOR investigations of spin exchange in mixed galvinoxyl/nitroxide biradicals. Syntheses. *J. Am. Chem. Soc.* 104:3850–3858.

Kopf, P.W. and Kreilick, R.W. 1969. Magnetic resonance studies of some phenoxy and nitroxide biradicals. *J. Am. Chem. Soc.* 91:6569–6573.

Kreilick, R.W. 1967. NMR studies of a series of aliphatic nitroxide radicals. *J. Chem. Phys.* 46:4260–4264.

Labsky, J., Pilar, J., and Lovy, J. 1980. Magnetic resonance study of 4-amino-2,2,6,6-tetramethylpiperidine-*N*-oxyl and its deuterated derivatives. *J. Magn. Reson.* 37:515–522.

Lemaire, H. and Rassat, A. 1964. Structure hyperfine due à l'azote dans les radicaux nitroxydes. *J. Chim. Phys.* 61:1580–1586.

Lim, Y.Y., Smith, E.A., and Symons, M.C.R. 1976. Solvation spectra. Part 51. Di-*t*-butyl nitroxide as a probe for studying water and aqueous solutions. *J. Chem. Soc. Faraday Trans. I* 72:2876–2892.

Marsh, D. 2015. Bonding in nitroxide spin labels from ^{14}N electric-quadrupole interactions. *J. Phys. Chem. A* 119:919–921.

Marsh, D. and Toniolo, C. 2008. Polarity dependence of EPR parameters for TOAC and MTSSL spin labels: correlation with DOXYL spin labels for membrane studies. *J. Magn. Reson.* 190:211–221.

McConnell, H.M. 1956. Indirect hyperfine interactions in the paramagnetic resonance spectra of aromatic free radicals. *J. Chem. Phys.* 24:764–766.

McLachlan, A.D. 1958. Hyperconjugation in the electron resonance spectra of free radicals. *Mol. Phys.* 1:233–240.

Michon, P. and Rassat, A. 1974. Nitroxides. LVIII. Structure of steroidal spin labels. *J. Org. Chem.* 39:2121–2124.

Morton, J.R., Rowland, J.R., and Whiffen, D.H. 1962. *Natl. Phys. Lab. Rep.* BPR:13-

Mossel, A., Romers, C., and Havinga, E. 1963. Investigations into the conformation of non-aromatic ring compounds XII (1). Cyclohexanedione-1,4 and related compounds. *Tetrahedron Lett.* 1247–1249.

Mustafi, D. and Joela, H. 1995. Origin of the temperature-dependent isotropic hyperfine coupling of the vinylic proton of oxypyrrolinyl nitroxyl spin labels. *J. Phys. Chem.* 99:11370–11375.

Nielsen, R.D., Canaan, S., Gladden, J.A. et al. 2004. Comparing continuous wave progressive saturation EPR and time domain saturation recovery EPR over the entire motional range of nitroxide spin labels. *J. Magn. Reson.* 169:129–163.

Ottaviani, M.F. 1987. Analysis of resolved ESR spectra of neutral nitroxide radicals in ethanol and pyridine - the dynamic behavior in fast motion conditions. *J. Phys. Chem.* 91:779–784.

Owenius, R., Engström, M., Lindgren, M., and Huber, M. 2001. Influence of solvent polarity and hydrogen bonding on the EPR parameters of a nitroxide spin label studied by 9-GHz and 95-GHz EPR spectroscopy and DFT calculations. *J. Phys. Chem. A* 105:10967–10977.

Plachy, W. and Kivelson, D. 1967. Spin exchange in solutions of di-tertiary-butyl nitroxide. *J. Chem. Phys.* 47:3312–3318.

Poggi, G. and Johnson Jr., C.S. 1970. Factors involved in the determination of rotational correlation times for spin labels. *J. Magn. Reson.* 3:436–445.

Pontillon, Y., Ishida, T., Lelievre-Berna, E. et al. 1999. Spin density of a ferromagnetic tempo derivative. *Mol. Cryst. Liq. Cryst.* 334:359–367.

Reddoch, A.H. and Konishi, S. 1979. The solvent effect on di-*tert*-butyl nitroxide. A dipole-dipole model for polar solutes in polar solvents. *J. Chem. Phys.* 70:2121–2130.

Savitsky, A., Dubinskii, A.A., Plato, M. et al. 2008. High-field EPR and ESEEM investigation of the nitrogen quadrupole interaction of nitroxide spin labels in disordered solids: toward differentiation between polarity and proticity matrix effects on protein function. *J. Phys. Chem. B* 112:9079–9090.

Stone, A.J. 1963. *g*-factors of aromatic free radicals. *Mol. Phys.* 6:509–515.

Stone, E.W. and Maki, A.H. 1962. Hindered internal rotation and ESR spectroscopy. *J. Chem. Phys.* 37:1326–1333.

Streitwieser, A. 1961. *Molecular Orbital Theory for Organic Chemists*, New York: John Wiley.

Turley, J.W. and Boer, F.P. 1972. The crystal structure of the nitroxide free radical 2,2,5,5-tetramethyl-3-carbamidopyrroline-1-oxyl. *Acta Cryst. B* 28:1641–1644.

Vasserman, A.M., Buchachenko, A.L., Rozantsev, E.G., and Neiman, M.B. 1965. Dipole moments of molecules and radicals. Di-*tert*-butyl nitrogen oxide. *J. Struct. Chem.* 6:445–446.

Wetherington, J.B., Ament, S.S., and Moncrief, J.W. 1974. The structure and absolute configuration of the spin-label *R*-(+)-3-carboxy-2,2,5,5-tetramethyl-1-pyrrolidinyloxy. *Acta Cryst. B* 30:568–573.

Whiffen, D.H. 1964. Information derived from anisotropic hyperfine couplings. *J. Chim. Phys.* 61:1589–1591.

Whisnant, C.C., Ferguson, S., and Chesnut, D.B. 1974. Hyperfine models for piperidine nitroxides. *J. Phys. Chem.* 78:1410–1415.

Windle, J.J. 1981. Hyperfine coupling constants for nitroxide spin probes in water and carbon tetrachloride. *J. Magn. Reson.* 45:432–439.

Zager, S.A. and Freed, J.H. 1982. Electron-spin relaxation and molecular dynamics in liquids. 1. Solvent dependence. *J. Chem. Phys.* 77:3344–3359.

Zielke, V., Eickmeier, H., Hideg, K., Reuter, H., and Steinhoff, H.J. 2008. A commonly used spin label: *S*–(2,2,5,5–tetramethyl–1–oxyl–Δ^3-pyrrolin-3-ylmethyl) methanethiosulfonate. *Acta Cryst. B* 64:o586–o589.

Polarity Dependence

4

4.1 INTRODUCTION

Chapter 3 deals with the origins of the spin Hamiltonian parameters: principally the hyperfine tensors and *g*-values, but also the quadrupole tensor. All these turn out to be sensitive to solvent polarity and to hydrogen bonding, making them excellent reporters of the local environment, particularly in biological systems. We cover this aspect in depth in the present chapter.

We can visualize the distribution of unpaired electron spin density in the nitroxide in terms of the two resonance-hybrid structures:

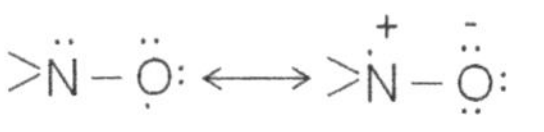

The dipolar resonance structure, on the right, with the unpaired electron spin on the nitrogen atom is favoured by polar environments. Consequently, the nitrogen hyperfine coupling constant is greater in environments of higher polarity. On the other hand, reduction of unpaired spin density on the oxygen in high-polarity media decreases the *g*-value, as also does the solvatochromic blue shift of the n–π^* transition and delocalization of the lone pairs from the oxygen.

Because hydrogen bonding to the –NO moiety produces pronounced *g*-shifts and hyperfine shifts, the polarity sensitivity of nitroxyl spin labels can be divided into two regimes (Steinhoff et al. 2000; Plato et al. 2002): that in aprotic environments and that in protic environments. We consider these two cases separately in this chapter. Then follows an application to biological membranes, including a comparison with the complementary EPR techniques: ESEEM and T_{1e}-relaximetry that we cover later in Chapters 14 and 10–12, respectively.

Much of the theory of solvent polarity effects is based on classical electrostatics. The elements of vector calculus that we need for the electrostatic calculations in this chapter are summarized in Appendix B. The question of units in electromagnetism is dealt with in Appendix A.

4.2 APROTIC ENVIRONMENTS

4.2.1 Reaction Fields

The polarity dependence of EPR spectra from nitroxides in aprotic dielectric media comes from the reaction field, $\mathbf{E}_R$. This electric field results from polarization of the dielectric environment by the permanent electric dipole moment, $\mathbf{p}_{el}$, of the nitroxide (see Figure 4.1).

The strength of the reaction field is directly proportional to the total electric dipole moment, $\mathbf{m}_{el}$, of the nitroxide and inversely proportional to the cube of its effective radius, r_{eff}:

$$\mathbf{E_R} = \frac{-\mathbf{m}_{el}}{4\pi\varepsilon_o r_{eff}^3} f\left(\varepsilon_{r,B}\right) \tag{4.1}$$

where $\varepsilon_{r,B}$ is the relative dielectric permittivity of the environment, i.e., bulk dielectric constant of the medium, and $\varepsilon_o\left(\equiv 1/\left(\mu_o c^2\right) = 8.854187\ldots\times 10^{-12}\ \text{Fm}^{-1}\right)$ is the permittivity of free space. The functional form of $f\left(\varepsilon_{r,B}\right)$ depends on the model that we assume for the radial distribution of the dielectric permittivity (see section immediately following). The simplest is that of Onsager which assumes that the bulk value of ε_r is maintained up to the outer radius of the nitroxide.

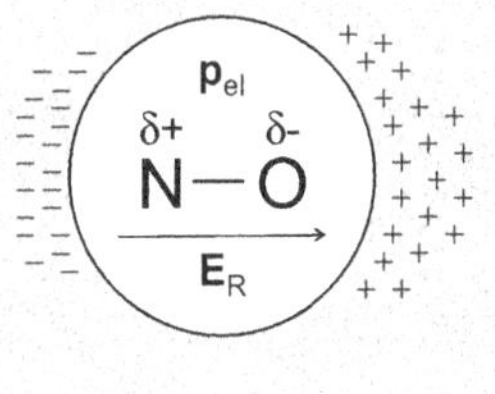

FIGURE 4.1 Permanent electric dipole $\mathbf{p}_{el}$ of the nitroxide polarizes surrounding dielectric giving rise to reaction field, $\mathbf{E}_R$. In turn, $\mathbf{E}_R$ polarizes the nitroxide, augmenting the permanent dipole and increasing unpaired spin density on nitrogen. Extent of induced polarization depends on polarizability α_D of nitroxide, which is related to its refractive index n_D.

The permanent electric dipole, $\mathbf{p}_{el}$, is augmented by polarization of the nitroxide bond by the reaction field (see Figure 4.1):

$$\mathbf{m}_{el} = \mathbf{p}_{el} - \alpha_D \mathbf{E_R} \tag{4.2}$$

where α_D is the electric polarizability of the nitroxide. Combining Eqs. 4.1 and 4.2 gives the following expression for the reaction field:

$$\mathbf{E_R} = \frac{-\mathbf{p}_{el}}{4\pi\varepsilon_o r_{eff}^3} \cdot \frac{f(\varepsilon_{r,B})}{1 - \dfrac{\alpha_D}{4\pi\varepsilon_o r_{eff}^3} f(\varepsilon_{r,B})} \tag{4.3}$$

We can express the electric polarizability, α_D, in terms of the refractive index, n_D, of the pure nitroxide by using the Lorenz–Lorentz relation:

$$\frac{n_D^2 - 1}{n_D^2 + 2} = \frac{\alpha_D}{4\pi\varepsilon_o r_{eff}^3} \tag{4.4}$$

Equation 4.4 comes from standard dielectric theory (Böttcher 1973). Combining Eqs. 4.3 and 4.4 then gives the following expression for the reaction field:

$$\mathbf{E}_R = \frac{-\mathbf{p}_{el}}{4\pi\varepsilon_o r_{eff}^3} \cdot \frac{f(\varepsilon_{r,B})}{1 - f(\varepsilon_{r,B}) \dfrac{n_D^2 - 1}{n_D^2 + 2}} \tag{4.5}$$

which is expressed in terms of the permanent electric dipole moment, refractive index and molecular radius of the nitroxide.

Experimental values for the dipole moment of nitroxide spin labels are given in Table 4.1. The refractive index of di-*tert*-butyl nitroxide is: $n_D = 1.432$ (Reddoch and Konishi 1979), and the average value for DOXYL-fatty acid methyl esters is $n_D = 1.45$ (Seelig et al. 1972). We can estimate effective radii from group contributions to the molecular volume of the spin label: $v_{mol} = (4/3)\pi r_{eff}^3$ (Bondi 1964), which are listed in Table 4.2.

TABLE 4.1 Electric dipole moments (p_{el}) of nitroxides

nitroxide	p_{el} (C.m)	*Refs.*[a]
di-*tert*-butyl nitroxide (DTBN)	1.03×10^{-29} (3.08 D)[b]	1
	0.90×10^{-29} (2.70 D)[b]	2
1-oxyl-2,2,6,6-tetramethylpiperidine (TEMPO)	1.05×10^{-29} (3.14 D)[b]	3
1-oxyl-2,2,6,6-tetramethylpiperidin-4-ol (TEMPOL)	1.04×10^{-29} (3.12 D)[b]	3
diphenylnitroxide	1.00×10^{-29} (3.00 D)[b]	2

Note: Dipole moment of the N–O bond is deduced to be $p_{NO} = 0.8 \times 10^{-29}$ C.m (2.4 ± 0.1 D) (Vasserman et al. 1965).

[a] References: 1. Murata and Mataga (1971); 2. Vasserman et al. (1965); 3. Rosantzev (1970).

[b] Dipole moment in Debyes.

TABLE 4.2 Group contributions $v_{mol,i}$ to molecular volumes for spin labels (Bondi 1964; Marsh and Toniolo 2008)

group, i[a]	*molecular volume,* $v_{mol,i}$ (nm^3)
$-CH_3$	2.270×10^{-2}
$>CH_2$	1.699×10^{-2}
>CH–	1.126×10^{-2}
>C<	0.553×10^{-2}
–CH=	1.406×10^{-2}
>C=CH–	2.240×10^{-2}
>CO	1.943×10^{-2}
>O	0.863×10^{-2}
=O	1.113×10^{-2}
–OH	1.335×10^{-2}
>S	1.793×10^{-2}
>NH	1.342×10^{-2}
>NO	1.832×10^{-2}

[a] Decrement per cyclohexyl or cyclopentyl ring = 0.189×10^{-2} nm^3; decrement per single bond adjacent to carboxyl or amide group = 0.037×10^{-2} nm^3.

4.2.2 Onsager and Block–Walker Models for the Reaction Field

The model that we use for calculating the reaction field is a permanent point dipole, $\mathbf{p}_{el}$, situated at the centre of a spherical cavity of radius r_{eff} immersed in a continuous dielectric medium (Bell 1931; Onsager 1936) – see Figure 4.2. Onsager's original model assumes a homogeneous dielectric of constant relative dielectric permittivity $\varepsilon_{r,B}$, whereas later modifications allow for radial dependence of the relative dielectric permittivity, $\varepsilon_r(r)$, which

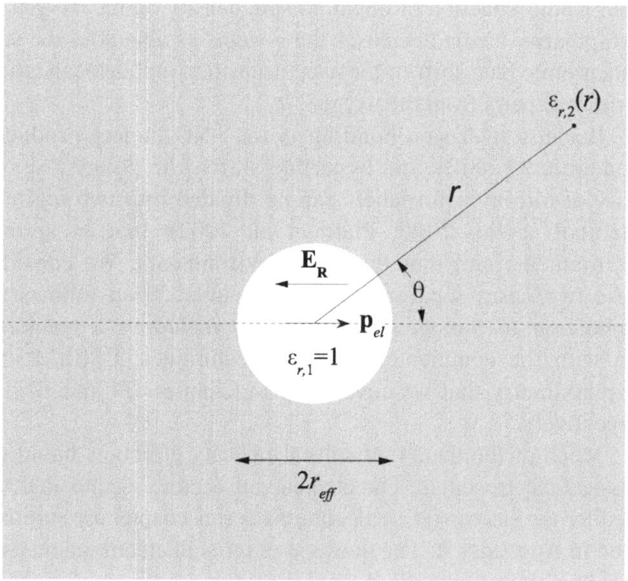

FIGURE 4.2 Model used to calculate reaction field for point dipole $\mathbf{p}_{el}$ in inhomogeneous dielectric medium of radially-dependent relative permittivity $\varepsilon_r(r)$. $\mathbf{E}_R$ is negative when $\mathbf{p}_{el}$ is positive (see Figure 4.1).

reaches the bulk value $\varepsilon_{r,B}$ only at distances far from $r = r_{eff}$ (see Figure 4.3). A survey of different nitroxides in a wide range of aprotic solvents, both polar and apolar (Marsh 2008), reveals that the exponential-inverse dependence proposed by Block and Walker (1973) best describes the polarity dependence of spin-label EPR parameters.

Gauss' theorem for the electric displacement, **D**, in a radially inhomogeneous dielectric is (see Appendix B.2 for vector calculus):

$$\nabla \cdot \mathbf{D} = \varepsilon_o \nabla \varepsilon_r(r) \cdot \mathbf{E} = 0 \tag{4.6}$$

in the absence of excess charge, where **E** is the electric field. In terms of the electrostatic potential, V ($\mathbf{E} = -\nabla V$), Eq. 4.6 becomes:

$$\nabla^2 V + \frac{1}{\varepsilon(r)} \nabla \varepsilon(r) \cdot \nabla V = 0 \tag{4.7}$$

In polar coordinates (see Figure 4.2), we write Eq. 4.7 as:

$$\frac{\partial}{\partial r}\left(r^2 \frac{\partial V}{\partial r}\right) + \frac{1}{\varepsilon(r)} \frac{\partial \varepsilon}{\partial r} \frac{\partial V}{\partial r} + \frac{\partial}{\partial(\cos\theta)}\left[\left(1-\cos^2\theta\right)\frac{\partial V}{\partial(\cos\theta)}\right] = 0 \tag{4.8}$$

for an axially symmetric dipole. For a dipolar system (or a fixed applied electric field), the angular part of the solution is given by the first-order Legendre polynomial: $P_1(\cos\theta) = \cos\theta$. Therefore, writing the separation of variables as $V(r,\theta) = V(r)\cdot\cos\theta$, we obtain the radial solution from:

$$\frac{\partial}{\partial r}\left(r^2 \frac{\partial V(r)}{\partial r}\right) + \frac{1}{\varepsilon(r)} \frac{\partial \varepsilon(r)}{\partial r} \frac{\partial V(r)}{\partial r} - 2V(r) = 0 \tag{4.9}$$

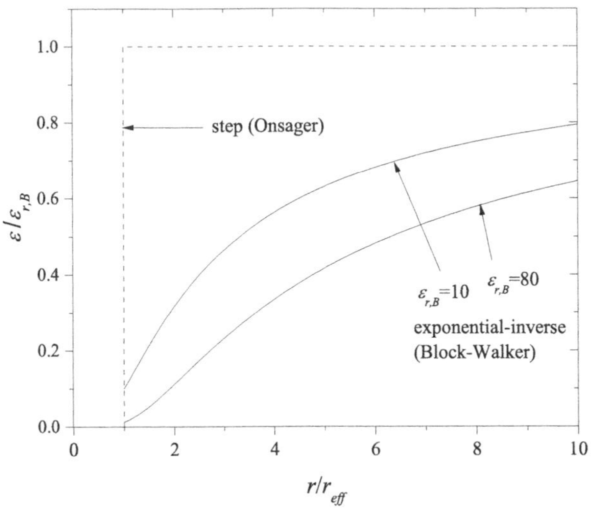

FIGURE 4.3 Radial dependence of relative dielectric permittivity ε for two models of the reaction field. *Dashed line*: step function, Onsager (1936). *Solid line*: exponential-inverse transition (Eqs. 4.18, 4.19), Block and Walker (1973). r_{eff}, effective molecular radius of spin label; starting level: $\varepsilon_r = 1$, for $r/r_{eff} \le 1$. Bloch–Walker model for two bulk dielectric permittivities attained at large distances: $\varepsilon_{r,B} = 10$ and $\varepsilon_{r,B} = 80$.

This differential equation is a general result for a radially inhomogeneous dielectric. To proceed further, we must assume a model for the radial dependence of the dielectric permittivity, $\varepsilon(r)$. First we use the simple Onsager model, and then later introduce the Block–Walker model.

Onsager Model:

In the Onsager model, the dielectric medium is homogeneous, i.e., $\partial\varepsilon(r)/\partial r = 0$ (see Figure 4.3), and Eq. 4.9 simply becomes the radial part of the Laplace equation. Solutions in the regions both inside (1) and outside (2) the cavity are of the general form:

$$V_{1,2}(r) = -\left(\frac{A_{1,2}}{r^2} + B_{1,2} \cdot r\right) \tag{4.10}$$

where $A_{1,2}$ and $B_{1,2}$ are constants that we determine from the boundary conditions:

$$V_1 \to \frac{p_{el}}{4\pi\varepsilon_o r^2} \quad \text{for } r \to 0 \tag{4.11}$$

$$V_2 \to 0 \quad \text{for } r \to \infty \tag{4.12}$$

The two solutions are therefore:

$$V_1(r) = \frac{p_{el}}{4\pi\varepsilon_o r^2} - E_R \cdot r \quad \text{for } r \le r_{eff} \tag{4.13}$$

where the first term on the right is the bare potential due to the fixed dipole, and E_R is the reaction field in the cavity (cf. Figure 4.2), and

$$V_2(r) = \frac{A_2}{r^2} \quad \text{for } r \ge r_{eff} \tag{4.14}$$

for the potential outside the cavity. Both the potential, V, and the radial component of the electric displacement vector, which is given by $\varepsilon_o\varepsilon_r(\partial V/\partial r)$, must be continuous at the surface of the cavity:

$$V_1\left(r_{eff}\right) = V_2\left(r_{eff}\right) \tag{4.15}$$

$$\left(\frac{\partial V_1}{\partial r}\right)_{r=r_{eff}} = \varepsilon_{r,B}\left(\frac{\partial V_2}{\partial r}\right)_{r=r_{eff}} \tag{4.16}$$

Using these boundary conditions (i.e., Eqs. 4.15 and 4.16) to eliminate A_2 between Eqs. 4.13 and 4.14, finally gives the reaction field for the Onsager model:

$$E_R(\text{Onsager}) = \frac{p_{el}}{4\pi\varepsilon_o r_{eff}^3} \frac{2\left(\varepsilon_{r,B} - 1\right)}{2\varepsilon_{r,B} + 1} \tag{4.17}$$

where we assume that the dipole is not polarizable (cf. Eq. 4.1).

Block–Walker Model:

In the Block–Walker model, the dielectric permittivity is inhomogeneous and undergoes an exponential transition to the bulk dielectric constant, $\varepsilon_{r,B}$, that depends inversely on the radial distance, r (Block and Walker 1973):

$$\varepsilon_r(r) = \varepsilon_{r,B} \exp\left(\frac{-\kappa}{r}\right) \tag{4.18}$$

The boundary condition that $\varepsilon_r = 1$ at the molecular surface $(r = r_{eff})$ fixes the exponential decay constant:

$$\kappa = r_{eff} \ln \varepsilon_{r,B} \tag{4.19}$$

and therefore we introduce no extra parameters, relative to the Onsager model. From Eq. 4.18, the explicit form of Eq. 4.9 in the dielectric (i.e., $r \geq r_{eff}$, region 2) becomes:

$$\frac{\partial}{\partial r}\left(r^2 \frac{\partial V(r)}{\partial r}\right) + (2r + \kappa)\frac{\partial V(r)}{\partial r} - 2V(r) = 0 \tag{4.20}$$

In the absence of an applied field, the solution of Eq. 4.20 is of the form (Block and Walker 1973):

$$V_2(r) = C\left[2\kappa^{-1}r + 1 - \left(2\kappa^{-1}r - 1\right)\exp(\kappa/r)\right] \tag{4.21}$$

where we have used the boundary condition given by Eq. 4.12, and C is a constant to be determined.

Within the cavity (i.e., $r \leq r_{eff}$), the situation remains the same as in the Onsager model with the solution, $V_1(r)$, of the form given by Eq. 4.13. Using the boundary conditions, Eqs. 4.15 and 4.16, to eliminate C between Eqs. 4.13 and 4.21, then gives the reaction field for the Block–Walker model:

$$E_R(\text{Block–Walker}) = \frac{p_{el}}{4\pi\varepsilon_o r_{eff}^3}\left(\frac{3\varepsilon_{r,B}\ln\varepsilon_{r,B}}{\varepsilon_{r,B}\ln\varepsilon_{r,B} - \varepsilon_{r,B} + 1} - \frac{6}{\ln\varepsilon_{r,B}} - 2\right) \tag{4.22}$$

where again we assume a non-polarizable dipole (cf. Eq. 4.1).

The function $f(\varepsilon_{r,B})$ in Eq. 4.5 that we need for the reaction field of a *polarizable* nitroxide is then given from Eq. 4.17 by:

$$f_O(\varepsilon_{r,B}) = \frac{2(\varepsilon_{r,B} - 1)}{2\varepsilon_{r,B} + 1} \tag{4.23}$$

for the Onsager model. Similarly, from Eq. 4.22, the corresponding expression is:

$$f_{BW}(\varepsilon_{r,B}) = \frac{3\varepsilon_{r,B}\ln\varepsilon_{r,B}}{\varepsilon_{r,B}\ln\varepsilon_{r,B} - \varepsilon_{r,B} + 1} - \frac{6}{\ln\varepsilon_{r,B}} - 2 \tag{4.24}$$

for the Block–Walker model. Figure 4.4 compares the dependences on bulk dielectric permittivity for the two models. In the Onsager model, $f(\varepsilon_{r,B})$ and the reaction field saturate too quickly with increasing $\varepsilon_{r,B}$, giving strong sensitivity to polarity only for low dielectric constants. The Block–Walker model, on the other hand, predicts a much more gradual dependence of $f(\varepsilon_{r,B})$ – and hence of the reaction field – on $\varepsilon_{r,B}$ (see also Barone 1996).

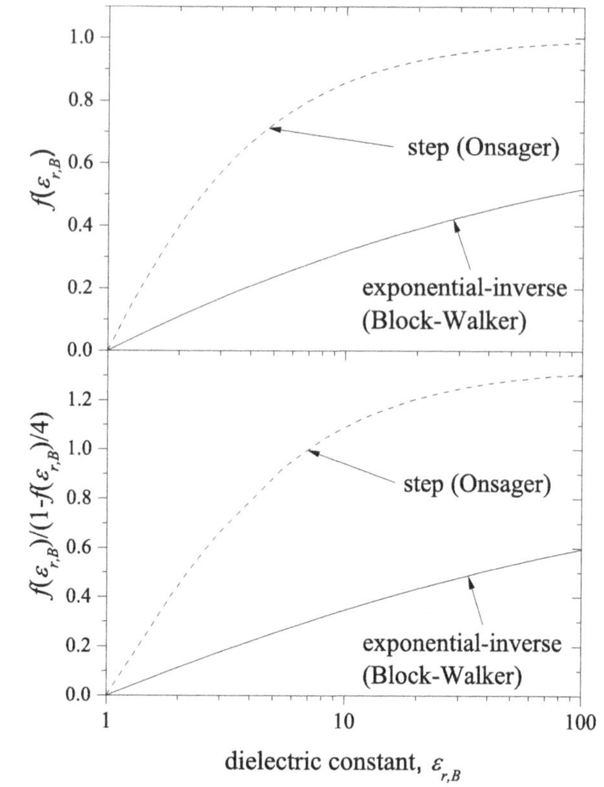

FIGURE 4.4 Dependence of $f(\varepsilon_{r,B})$ that determines reaction field for an unpolarizable nitroxide (upper panel) and of $f(\varepsilon_{r,B})/\left(1-\frac{1}{4}f(\varepsilon_{r,B})\right)$ from Eq. 4.5 that determines reaction field of a polarizable nitroxide (lower panel), on bulk dielectric permittivity $\varepsilon_{r,B}$ of the medium. *Dashed line*: step transition in ε_r (Onsager 1936), Eq. 4.23; *solid line:* exponential-inverse transition in ε_r (Block and Walker 1973), Eq. 4.24.

4.2.3 Influence of Local Electric Fields on Nitroxides

We can analyse the effect of local electric fields (including the reaction field) on the spin-density distribution in nitroxides semi-quantitatively by using molecular orbital theory (Griffith et al. 1974; Reddoch and Konishi 1979). Clearly, the one-electron model that we used in Section 3.2 will not give the correct sign, because an electric field directed from N to O moves electron density off the nitrogen and onto the oxygen. We are therefore interested only in the magnitude of the predicted shift. For the sign, we must resort to the valence-bond picture introduced in Section 4.1.

The one-electron Hamiltonian for the $2p\pi$-system of the nitroxide becomes:

$$\mathcal{H} = \mathcal{H}_o - eV_{loc} \tag{4.25}$$

where $\mathcal{H}_o$ is the unperturbed one-electron Hamiltonian (see Chapter 3, Section 3.2), $-e$ is the charge of the single electron and V_{loc} is the electrostatic potential created by the local electric field: $E_{loc} = -\partial V_{loc}/\partial r$. When we have a local field, Eq. 3.2 of Chapter 3 thus becomes, using the Hückel matrix representation:

$$\begin{pmatrix} \alpha_N - eV_N - E & \beta_{NO} \\ \beta_{NO} & \alpha_O - eV_O - E \end{pmatrix} \begin{pmatrix} c_N \\ c_O \end{pmatrix} = 0 \tag{4.26}$$

where matrix elements of the local electrostatic potential are: $V_N = \langle \varphi_N | V_{loc} | \varphi_N \rangle$ and $V_O = \langle \varphi_O | V_{loc} | \varphi_O \rangle$. Because φ_N and φ_O are the atomic orbitals of nitrogen and oxygen, V_N and V_O are local electrostatic potentials at the nitrogen and at the oxygen, respectively.

To get the unpaired spin density ρ_π^N on the nitrogen atom, we need the coefficient c_N of the antibonding molecular orbital: $\Psi^* = c_N\varphi_N - c_O\varphi_O$ (see Eq. 3.9 in Section 3.2). Diagonalizing the 2×2 matrix in Eq. 4.26 gives the solution that we need (Reddoch and Konishi 1979, and cf. Eq. 3.11):

$$\rho_\pi^N = c_N^2 = \frac{1}{2\left(1 + \frac{1}{2}h_{eff}\sqrt{1 + \left(\frac{1}{2}h_{eff}\right)^2} + \left(\frac{1}{2}h_{eff}\right)^2\right)} \tag{4.27}$$

where $h_{eff} = h_{NO} - e(V_N - V_O)/\beta_{NO}$ and, as in Eq. 3.4, $h_{NO} \equiv (\alpha_N - \alpha_O)/\beta_{NO}$. Assuming that $h_{NO}^2 << 4$ (see Section 3.2) and that the local field is a small perturbation, we can approximate the shift in unpaired spin density by:

$$\Delta\rho_\pi^N \cong \frac{\partial \rho_\pi^N}{\partial h_{eff}}\Delta h_{eff} = -\frac{1}{4}\left(1 - \frac{3}{8}h_{NO}^2\right) \times \frac{e(V_O - V_N)}{\beta_{NO}} \tag{4.28}$$

where the final term on the right is Δh_{eff} arising from the local field. To a first approximation, the spin density shifts linearly with the local electric field:

$$\Delta\rho_\pi^N \cong \left(\frac{er_{NO}}{4\beta_{NO}}\right)E_{loc,x} \tag{4.29}$$

where $E_{loc,x} \approx -(V_O - V_N)/r_{NO}$ is the local field component along the N–O bond, and r_{NO} is the N–O bond length. With a value of $\beta_{NO} = -2.5\,\text{eV}$, as estimated in Section 3.2, and $r_{NO} = 0.13\,\text{nm}$ from Table 3.4, Eq. 4.29 predicts a shift in spin density of absolute size $|\Delta\rho_\pi^N| = 1.3 \times 10^{-11} E_{loc,x}\left(\text{Vm}^{-1}\right)$.

4.2.4 Isotropic Hyperfine Couplings in Aprotic Media

We learnt in Section 3.3 that the isotropic ^{14}N-hyperfine coupling, a_o^N, of nitroxyl spin labels depends linearly on the unpaired electron spin density, ρ_π^N, on the nitrogen atom (see Eq. 3.15). From Eq. 4.29 of the previous section, we therefore expect the reaction field arising from the polar environment to induce linear shifts in hyperfine coupling:

$$\Delta a_o^N = C_E E_R \tag{4.30}$$

Using Eq. 4.5, the isotropic coupling constant then depends on dielectric permittivity of the environment according to (Marsh and Toniolo 2008):

$$a_o^N = a_o^{\varepsilon=1} + K_v \frac{f(\varepsilon_{r,B})}{1 - \frac{n_D^2 - 1}{n_D^2 + 2} f(\varepsilon_{r,B})} \tag{4.31}$$

where $a_o^{\varepsilon=1}$ is the isotropic hyperfine coupling extrapolated to a medium of relative dielectric permittivity $\varepsilon_r = 1$. The coefficient, K_v, of the polarity-dependent term is given by (see Eq. 4.5):

$$K_v = \left(\frac{C_E}{4\pi\varepsilon_o}\right)\frac{p_{el}}{r_{eff}^3} \tag{4.32}$$

which is a constant for a particular nitroxide.

The lower panel of Figure 4.4 shows the dependence of the function $f(\varepsilon_{r,B})/\left(1 - \frac{1}{4}f(\varepsilon_{r,B})\right)$, which determines the polarity sensitivity of the isotropic hyperfine coupling on bulk dielectric constant $\varepsilon_{r,B}$. Here, we assume that $n_D^2 \approx 2$ in Eq. 4.31, as found experimentally for nitroxides (see Section 4.2.1). The Block–Walker model for the reaction field predicts an approximately logarithmic dependence on dielectric permittivity, as we find experimentally with non-polar solvents (see Figure 4.5 and Marsh 2008; Marsh and Toniolo 2008).

Table 4.3 lists calibration constants $a_o^{\varepsilon=1}$ and K_v from Eq. 4.31, for some nitroxides relevant to biological spin labelling. This assumes the Block–Walker model (i.e., Eq. 4.24). We use the DOXYL moiety in site-specific spin labelling of lipid chains, which is important for determining transmembrane polarity profiles. Other nitroxides are involved mostly in spin labelling proteins. MTSSL is the standard reagent, in combination with cysteine-scanning mutagenesis, for site-directed spin labelling of proteins (see Chapter 16). TOAC is a nitroxyl amino acid that we introduce into the protein backbone by peptide synthesis.

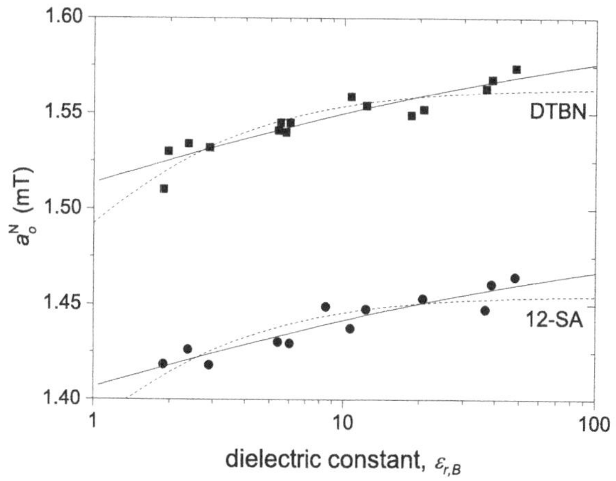

FIGURE 4.5 Dependence of isotropic hyperfine coupling a_o^N on relative dielectric permittivity $\varepsilon_{r,B}$ of the solvent. Squares: di-*tert*-butyl nitroxide (Griffith et al. 1974; Reddoch and Konishi 1979); circles: 12-DOXYL stearic acid (Marsh and Toniolo 2008). Least-squares fits of Onsager (1936) model (step transition in ε_r; dashed line) and Block–Walker (1973) model (exponential-inverse transition in ε_r; solid line). $\varepsilon_{r,B}$-axis is logarithmic.

TABLE 4.3 Dependence of isotropic nitrogen hyperfine couplings, a_o^N, and *g*-values, g_o, of nitroxyl spin labels on solvent polarity, $f_{BW}(\varepsilon_{r,B})/(1-\frac{1}{4}f_{BW}(\varepsilon_{r,B}))$, in aprotic media, according to Eqs. 4.31, 4.38 and 4.24 for the Block–Walker model

spin label[a]	K_v (μT)	$a_o^{\varepsilon=1}$ (mT)	$K_{v,g}\times 10^3$	$(g_o^{\varepsilon=1}-g_e)\times 10^3$	*Refs.*[b]
lipid spin labels:					
DOXYL	87 ± 12	1.410 ± 0.005	−0.19 ± 0.06	3.60 ± 0.02	1
protein spin labels:					
MTSSL	81 ± 55	1.421 ± 0.017	−0.22 ± 0.09	3.65 ± 0.03	1
MTSSL/β-SH	63 ± 31	1.437 ± 0.010	−0.36 ± 0.15	3.69 ± 0.04	1
peptide spin labels:					
TOAC	89 ± 16	1.464 ± 0.006	−0.37 ± 0.06	3.91 ± 0.02	1
	71 ± 20	1.464 ± 0.008			
piperidinyl nitroxides:					
TEMPO-NH_2	98 ± 11	1.524 ± 0.004			2
TEMPOL	99 ± 13	1.519 ± 0.006			2
TEMPO-NMe_3	96 ± 13	1.513 ± 0.006			2
pyrrolinyl nitroxides:					
$PYCONH_2$	109 ± 13	1.404 ± 0.004			2
	99 ± 14	1.410 ± 0.006			
PYCOOH	154 ± 21	1.400 ± 0.009			2
DTBN:	115 ± 11	1.509 ± 0.004	−0.38 ± 0.05	3.84 ± 0.02	1

[a] DOXYL, 2-substituted 3-oxyl-4,4-dimethyl-oxazolidine;
MTSSL, (1-oxyl-2,2,5,5-tetramethylpyrrolin-3-ylmethyl)methanethiosulphonate;
MTSSL/β-SH, β-mercaptoethanol S–S adduct of MTSSL;
TOAC, 1-oxyl-2,2,6,6-tetramethyl-4-aminopiperidin-4-yl carboxylic acid;
TEMPO-NH_2, 1-oxyl-2,2,6,6-tetramethylpiperidin-4-ylamine;
TEMPOL, 1-oxyl-2,2,6,6-tetramethylpiperidin-4-ol;
TEMPO-NMe_3, 1-oxyl-2,2,6,6-tetramethylpiperidin-4-yltrimethylamine;
$PYCONH_2$, 1-oxyl-2,2,5,5-tetramethyl-pyrrolin-3-ylamide;
PYCOOH, 1-oxyl-2,2,5,5-tetramethyl-pyrrolin-3-yl carboxylic acid;
DTBN, di-*tert*-butyl nitroxide.

[b] References: 1. Marsh and Toniolo (2008); 2. Marsh (2008).

The values of K_v in Table 4.3 reflect the relative sizes of the different nitroxides, as predicted by Eq. 4.32. Taking p_{el} = 1.10^{-29}C.m (3 Debye) (see Table 4.1) and C_E = 3.5 × 10^{-11} mT/Vm^{-1} (see following Section 4.3.1 and Schwartz et al. 1997), we obtain effective molecular radii of: r_{eff} = 0.34 ± 0.10 and 0.37 ± 0.07 nm for MTSSL and MTSSL/β–SH, respectively, as compared with r_{eff} = 0.31 ± 0.01 nm for the smaller di-*tert*-butyl nitroxide (DTBN) (Marsh and Toniolo 2008). For both the DOXYL and TOAC nitroxides, we obtain an effective radius of 0.33 ± 0.02 nm, which must correspond only to a segment of the entire spin-labelled lipid or peptide. Effective radii estimated from group contributions to the molecular volume, v_{mol} (see Table 4.2), are: $r_{eff} = (3v_{mol}/4\pi)^{1/3}$ = 0.385, 0.386 and 0.341 nm for MTSSL, MTSSL/β–SH and DTBN, respectively. These are in qualitative accord with the values deduced from the EPR data. For TOAC, the effective radius deduced from the molecular volume of the TOAC ring alone is r_{eff} = 0.335 nm, which is close to the effective experimental value. The DOXYL unit has a considerably smaller molecular volume: we need to include two methylene groups on either side of the point of chain attachment to bring the effective radius, r_{eff} = 0.341 nm, close to that deduced from Eq. 4.32.

From the molecular volumes of MTSSL, MTSSL/β–SH, DTBN and the peptide moiety of TOAC, the mean experimental polarity sensitivity of the isotropic hyperfine coupling is given by: $K_v = (4.21 \pm 0.39\ \mu\text{T.nm}^3) \times (1/r_{eff}^3)$ (Marsh and Toniolo 2008). This relation, together with Table 4.2, can prove useful for predicting the polarity dependence of other nitroxides.

4.3 LOCAL FIELDS FROM FIXED ELECTRIC CHARGES OR DIPOLES

4.3.1 Effect of Fixed Electric Charges on Hyperfine Couplings

An alternative way to determine the effect of local electric fields is provided by fixed electrical charges. Here, we use spin labels with a protonatable group so that the charge can be switched on and off by pH titration (Schwartz et al. 1997).

Figure 4.6 shows the model used for calculating the electrostatic potential arising from a fixed charge. We assume that the spin label occupies a spherical cavity immersed in a homogeneous dielectric (cf. the Onsager model). The volume of the cavity equals the molecular volume of the entire spin-label group $\left(v_{mol} = (4/3)\pi r_{eff}^3\right)$, and we position the oxygen of the N–O bond at the outer circumference, i.e., $r_O = r_{eff}$. The charge q is situated at distance d from the centre of the sphere, where $d < r_O$ in all cases. On the other hand, the radial position, r_N, of the N–O nitrogen atom may be either outside $(d < r_N)$ or inside $(d > r_N)$ that of the fixed charge.

As for the reaction field (see Section 4.2.2), we need solutions of the Poisson equation $\left(\nabla^2 V = -\rho_{el}/\varepsilon_o\varepsilon_r\right)$ but this time in the presence of a net charge, which often lies off-axis with respect to the N–O bond. Quite generally, the angular dependence of the electrostatic potential $V(r,\theta)$ at polar position (r, θ) – relative to q at $(d,0)$ – is determined by the Legendre polynomials, $P_l(\cos\theta)$ (see Appendix V.1). The complete solution of the Poisson equation inside the cavity, at radial distances outside the fixed charge, is (Schwartz et al. 1997):

$$V_1(r,\theta) = \frac{q}{4\pi\varepsilon_o\varepsilon_{r,1}}\sum_{l=0}^{\infty}\left(\frac{d^l}{r^{l+1}} + A_l r^l\right)P_l(\cos\theta) \quad d < r < r_{eff} \tag{4.33}$$

where $\varepsilon_{r,1}$ is the (low) dielectric constant in the cavity and A_l are constants that we must determine. Correspondingly, the potential in the region of higher dielectric constant, $\varepsilon_{r,2}$, outside the cavity is given by (Schwartz et al. 1997):

$$V_2(r,\theta) = \frac{q}{4\pi\varepsilon_o\varepsilon_{r,1}}\sum_{l=0}^{\infty}\left(\frac{d^l}{r^{l+1}} + B_l r^{-(l+1)}\right)P_l(\cos\theta) \quad r_{eff} < r < \infty \tag{4.34}$$

where B_l are also constants to be determined. In Eqs. 4.33 and 4.34, the first term on the right is the potential from the fixed charge q, and the second term is the contribution from polarization of the surrounding dielectric.

We determine the values of A_l and B_l by using the same general boundary conditions as for the reaction field: $V_1\left(r_{eff}\right) = V_2\left(r_{eff}\right)$ and $\varepsilon_{r,1}\left(\partial V_1/\partial r\right)_{r=r_{eff}} = \varepsilon_{r,2}\left(\partial V_2/\partial r\right)_{r=r_{eff}}$ (cf. Eqs. 4.15 and 4.16). We need only the values of A_l to determine the potential at the nitroxide (assuming that $d < r_N$). The explicit solution corresponding to Eq. 4.33 then becomes (Schwartz et al. 1997):

$$V_1(r,\theta) = \frac{q}{4\pi\varepsilon_o\varepsilon_{r,1}}\sum_{l=0}^{\infty}\left(\frac{d^l}{r^{l+1}} + \frac{(\varepsilon_{r,1} - \varepsilon_{r,2})(l+1)}{l\varepsilon_{r,1} + (l+1)\varepsilon_{r,2}}\cdot\frac{d^l r^l}{r_{eff}^{2l+1}}\right) \times P_l(\cos\theta) \quad d < r < r_{eff} \tag{4.35}$$

As in Section 4.2.3, the component of the local field at the nitroxide that is directed along the N–O bond (i.e., nitroxide x-axis) then becomes:

$$E_{loc,x} = -\frac{\partial V_{loc}}{\partial x} = \frac{V_1(r_N,\theta_{NO}) - V_1(r_O,\theta_{NO})}{r_{NO}} \tag{4.36}$$

where r_{NO} is the N–O bond length, and the values of V_1 at $r = r_O$ and $r = r_N$ are given by Eq. 4.35 (assuming that $d < r_N$). If the nitrogen atom of the N–O bond lies within the radius of the fixed charge $(d > r_N)$, then the local potential $V_1(r_N,\theta_{NO})$ at the N-atom is given by Eq. 4.35 but with the first term on the right, d^l/r^{l+1}, replaced by its inverse, r^{l+1}/d^l (Schwartz et al. 1997). Note that Eq. 4.35 without modification is always the form appropriate to the oxygen atom.

Table 4.4 gives values of the local field at the N–O position that we get from Eq. 4.35 (or equivalent) and Eq. 4.36, for a range of protonated cationic spin labels and a single deprotonated anionic spin label (Schwartz et al. 1997). The measured values of the difference, Δa_o^N, in isotropic ^{14}N-hyperfine coupling correlate reasonably well, both in size and sign, with the predicted values of $E_{loc,x}$. Taking the mean value of $\Delta a_o^N/E_{loc,x}$, the calibration factor for the sensitivity of the isotropic coupling to electric fields in Eq. 4.30 becomes $C_E = (3.5 \pm 0.4)\times 10^{-11}$ mT/Vm^{-1}. Combining this with $Q_N^N - Q_{ON}^N = 2.56$ mT from Table 3.1 predicts a value of $(1.4 \pm 0.2)\times 10^{-11}$ V^{-1}m for the proportionality constant in Eq. 4.29 of Section 4.2.3, which is of similar order to the estimate from molecular orbital theory.

These results imply that an electric field of strength $E_x = 10^9$ Vm^{-1} produces a change in isotropic hyperfine coupling of $\Delta a_o^N = 35$ μT (cf. Griffith et al. 1974; Schwartz et al. 1997). *In vacuo*, quantum chemical calculations with density functional theory (DFT) predict a similar shift for the pyrroline nitroxide PYMeOH (Savitsky et al. 2008; and see Table 4.4).

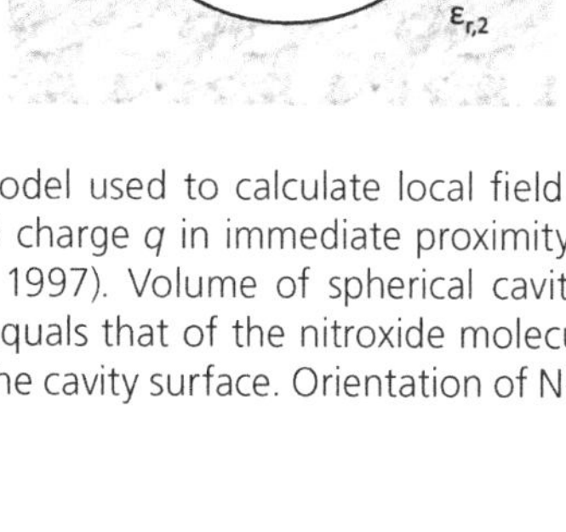

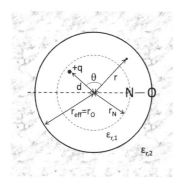

FIGURE 4.6 Model used to calculate local field E_{loc} at the N–O bond, from fixed charge q in immediate proximity to the nitroxide (Schwartz et al. 1997). Volume of spherical cavity with dielectric permittivity $\varepsilon_{r,1}$ equals that of the nitroxide molecule; N–O oxygen atom is fixed at the cavity surface. Orientation of N–O bond is given by $\theta = \theta_{NO}$.

4.3.2 Effect of Fixed Electric Dipoles on Hyperfine Couplings

A local-field effect similar to that of fixed charges is produced also by the electric dipole moment of neutral polar substituents at the 4-position of 1-oxyl-2,2,6,6-tetramethylpiperidine or at the

TABLE 4.4 Local electric fields, $E_{loc,x}$, calculated according to Eqs. 4.35 and 4.36, and charge-induced shifts in isotropic hyperfine coupling, Δa_o^N, for protonatable spin labels in water (Schwartz et al. 1997)[a]

spin label[b]	r_{eff} (nm)	d (nm)	θ_{NO} (°)	$E_{loc,x}$ $(V.m^{-1})$	Δa_o^N (mT)[c]	$\Delta a_o^N / E_{loc,x}$ $(mT/V.m^{-1})$
RH1⁺	0.341	0.161	142	-2.7×10^9	−0.088	$+3.3 \times 10^{-11}$
RH2⁺	0.390	0.159	112	-3.3×10^9	−0.136	$+4.1 \times 10^{-11}$
RH3⁺	0.358	0.155	138	-2.8×10^9	−0.125	$+4.5 \times 10^{-11}$
					−0.138[d]	$+4.9 \times 10^{-11}$
RH4⁺	0.337	0.162	146	-2.7×10^9	−0.072	$+2.7 \times 10^{-11}$
PYCOO⁻	0.347	0.292	130	$+6.3 \times 10^8$	+0.0188	$+3.0 \times 10^{-11}$
					+0.012[d]	$+1.9 \times 10^{-11}$
DFT[e]				$+5.14 \times 10^9$	+0.181	$+3.5 \times 10^{-11}$

[a] Cavity radii, r_{eff}, are determined from group contributions to the molecular volume (Bondi 1964) (see Table 4.2). Dielectric constant of nitroxide-filling cavity assumed to be $\varepsilon_{r,1} = 2$. For the oxygen atom: $r_O = r_{eff}$, and for the N-atom: $r_N = r_O - r_{NO} = r_{eff} - 0.130$ nm. Position (d) of fixed charge, and angle θ_{NO}, determined from optimized geometry produced by semiempirical SCF-LCAO molecular orbital calculations (Schwartz et al. 1997). $E_{loc,x}$, component of local electric field that is directed along N–O bond, from N to O.

[b] RH1, 1-oxyl-2,2,4,5,5-pentamethylimidazoline; RH2, 1-oxyl-2,2,3,5,5-pentamethyl-4-phenylimidazolidine; RH3, 1-oxyl-2,2,3,4,5,5-hexamethyl-imidazolidine; RH4, 1-oxyl-2,2,5,5-tetramethyl-4-aminoimidazoline; PYCOO⁻, 1-oxyl-2,2,5,5-tetramethylpyrrolin-3-yl carboxylate.

[c] Experimental isotropic hyperfine couplings in protonated and unprotonated states of aqueous RH1–RH4 from Khramtsov et al. (1982), and of aqueous PYCOOH from Schwartz et al. (1997), unless noted otherwise.

[d] Data from Gullá and Budil (2001).

[e] Calculations *in vacuo* using density functional theory (DFT) from Savitsky et al. (2008).

3-position of 1-oxyl-2,2,5,5-tetramethylpyrrolidine nitroxides (Brière et al. 1965; Seelig et al. 1972). Figure 4.7 shows the dependence of the isotropic nitrogen hyperfine coupling, a_o^N, on the electric dipole moment of a double bond (e.g., C=O), or the resultant dipole moment of two single bonds (e.g., H–C–O), at the 4-C atom of the piperidine ring. The hyperfine coupling decreases approximately linearly – from TEMPO (H–C–H) to TEMPONE (C=O) – with the component, p_{bond}, of the electric dipole from the substituent bond that is oriented in the direction N to O of the 1-oxyl-piperidine. The decrease in a_o^N is less for the pyrrolidine derivatives than for the corresponding piperidines, because the resultant substituent bond moment at the 3-position is no longer aligned along the N–O direction (Dupeyre et al. 1964). Using the piperidine (6-membered ring) substituents as reference, the dependence of the hyperfine coupling on the resultant bond dipole moment at the 4-position is given from Figure 4.7 approximately by: $a_o^N = -(1.15 \pm 0.08) \times 10^{28}\, p_{bond} + 1.614 \pm 0.004$ mT.

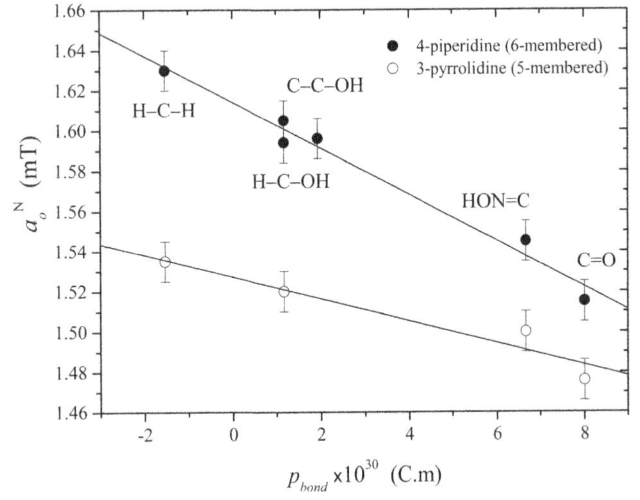

FIGURE 4.7 Dependence of isotropic ^{14}N-hyperfine coupling of 1-oxyl-2,2,6,6-tetramethylpiperidin-4-yl-*X* nitroxides in diethylene glycol (solid circles) on bond electric dipole moment, p_{bond}, of substituent *X* at 4-position of the piperidine ring (Brière et al. 1965). From left to right, substituents are: none, 4-hydroxy (and 4-acetoxy), 4-methyl-4-hydroxy, 4-oxime and 4-oxide. For single-bond substituents, p_{bond} is resultant dipole component directed along the symmetry axis at the 4-position. Open circles: data for 1-oxyl-2,2,5,5-tetramethylpyrrolidin-3-yl-*X* nitroxides (Dupeyre et al. 1964; Lemaire and Rassat 1964). In this case, p_{bond} is inclined at an angle to the N–O direction.

4.3.3 Effect of Polarity and Fixed Charges on Isotropic *g*-Values

The *g*-factors of nitroxides depend on the unpaired spin density, ρ_π^O, on the oxygen atom and additionally on the energies and distribution of the lone-pair orbitals (see Section 3.12 in Chapter 3). Therefore, the isotropic *g*-values respond to environmental polarity, but the sign of the polarity dependence is opposite to that for the hyperfine coupling. The major contribution to the polarity dependence of the nitroxide *g*-tensor comes from the g_{xx} element as expressed by generalization of Eq. 3.71:

$$g_{xx} = g_e + \frac{2\zeta_O \left(C_{O,y}^{(n)}\right)^2 \rho_\pi^O}{\Delta E_{n-\pi^*}} \tag{4.37}$$

where $g_e = 2.002319$ is the free-electron *g*-value, ζ_O is the spin-orbit coupling of oxygen, $\left(C_{O,y}^{(n)}\right)^2 = \sum_n \left(c_{O,y}^{(n)}\right)^2$ for coefficients $c_{O,y}^{(n)}$ of the oxygen $2p_y$ orbital in the lone-pair orbitals n, and $\Delta E_{n-\pi^*}$ is the $n \rightarrow \pi^*$ excitation energy. We introduce $\left(C_{O,y}^{(n)}\right)^2 \le 1$ specifically into Eq. 4.37 to allow for the possibility that lone-pair density is moved off the oxygen by hydrogen bonding (see Section 4.4.4 later). See Section 3.12 for the relation to lone-pair sp^2 hybrids.

Figure 4.8 shows the dependence of $\Delta E_{n-\pi^*}$ on solvent polarity deduced from the optical absorption of di-*tert*-butyl nitroxide at around 466 nm, which is assigned to the $n \to \pi^*$ transition (Kikuchi 1969; Briere et al. 1965). For aprotic solvents, the excitation energy increases approximately linearly with solvent polarity, where the latter corresponds to the energetic contributions of the reaction fields in the ground and Franck–Condon states (Marsh 2009). This increase in $\Delta E_{n-\pi^*}$ results in a decrease in g-value with increasing polarity that is additional to that from the decrease in unpaired spin density, ρ_π^O, on the oxygen (see Eq. 4.37). The maximum change for the protic solvents in Figure 4.8 is 11%, compared with only 3% for aprotic solvents.

The g_{yy} tensor element is considerably less sensitive to polarity than is the g_{xx} element because the contribution from p_x-character of the lone-pair orbital is small in effect, and the energy denominators involving the bonding and antibonding N–O σ-orbitals do not lie as close in energy to the unpaired electron orbital as does the lone pair (see Chapter 3, Sections 3.2 and 3.12). The g_{zz} tensor element is practically insensitive to polarity, because $g_{zz} \approx g_e$.

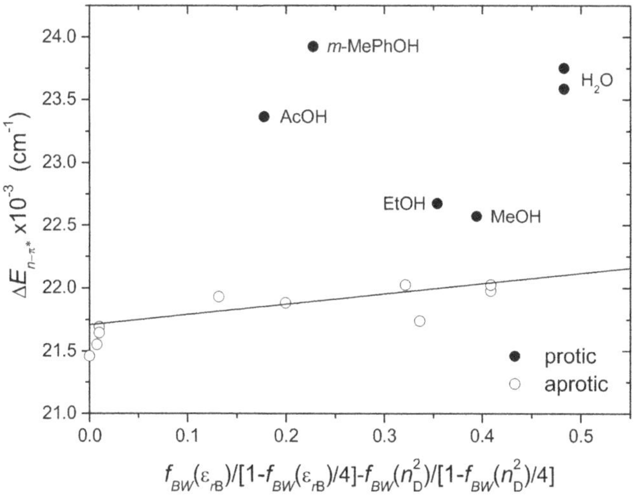

FIGURE 4.8 Dependence of $n-\pi^*$ excitation energy $\Delta E_{n-\pi^*}$ on solvent polarity for di-*tert*-butyl nitroxide in protic (solid circles) and aprotic (open circles) media. Environmental polarity parameterized by: $f_{BW}(\varepsilon_{r,B})/\left[1-\frac{1}{4}f_{BW}(\varepsilon_{r,B})\right]-f_{BW}(n_D^2)/\left[1-\frac{1}{4}f_{BW}(n_D^2)\right]$ for the Block–Walker model (Eq. 4.24); $\varepsilon_{r,B}$ and n_D are dielectric constant and refractive index of the solvent (Marsh 2009). Optical absorption data from Murata and Mataga (1971), Kawamura et al. (1967).

The g-factor, like the hyperfine coupling, also responds approximately linearly to local electric fields at the N–O bond (Plato et al. 2002; Gulla and Budil 2001). Table 4.5 gives results of g-value measurements with spin labels that acquire an electric charge on pH titration, as described for isotropic hyperfine couplings of the same spin labels in Section 4.3.1. The mean experimentally determined dependence on electric field is given by: $\Delta(g_{xx}-g_{zz}) \approx 21\times10^{-14}E_{loc,x}\left(\mathrm{Vm^{-1}}\right)$, $\Delta(g_{yy}-g_{zz}) \approx 4.5\times10^{-14}E_{loc,x}\left(\mathrm{Vm^{-1}}\right)$ and $\Delta g_o \approx 7\times10^{-14}$ $E_{loc,x}\left(\mathrm{Vm^{-1}}\right)$. Estimates made by using density functional theory (DFT) agree at least qualitatively with these values, although the sensitivity of g_{xx} appears underestimated (see Table 4.5).

The g-factor therefore responds to the polarization reaction field in a way analogous to that for the hyperfine coupling. Consequently, the polarity dependence of the isotropic g-value in aprotic environments becomes similarly (cf. Eq. 4.31):

$$g_o = g_o^{\varepsilon=1} + K_{v,g}\frac{f(\varepsilon_{r,B})}{1-\dfrac{n_D^2-1}{n_D^2+2}f(\varepsilon_{r,B})} \tag{4.38}$$

where $g_o^{\varepsilon=1}$ is the isotropic g-factor in a medium of relative dielectric permittivity $\varepsilon_r = 1$ and $K_{v,g}$ is a constant for a particular nitroxide. Table 4.3 given already includes linear regression parameters $K_{v,g}$ and $g_o^{\varepsilon=1}$ for spin labels of biological interest in aprotic solvents.

4.4 PROTIC ENVIRONMENTS

4.4.1 Isotropic Hyperfine Couplings in Protic Media

For protic media, the effects of hydrogen bonding on the hyperfine couplings mostly far outweigh those of solvent polarity. Table A3.1 of Appendix A3.1 to Chapter 3 reports

TABLE 4.5 Dependence of nitroxide g-tensor elements on local electric field, $E_{loc,x}$, from measurements on charged spin labels (Gullá and Budil 2001)[a]

spin label[b]	$E_{loc,x}\left(\mathrm{V\,m^{-1}}\right)$[c]	$\Delta(g_{xx}-g_{zz})$[d]	$\Delta(g_{xx}-g_{zz})/E_{loc,x}$ ($\mathrm{V^{-1}m}$)	$\Delta(g_{yy}-g_{zz})$	$\Delta(g_{yy}-g_{zz})/E_{loc,x}$ ($\mathrm{V^{-1}m}$)	Δg_o[e]	$\Delta g_o/E_{loc,x}$ ($\mathrm{V^{-1}m}$)
RH3$^+$	-2.8×10^9	$+65\times10^{-5}$	$-(23\pm1)\times10^{-14}$	$+9\times10^{-5}$	$-(3.3\pm0.5)\times10^{-14}$	$+26\times10^{-5}$	$-(9.1\pm0.5)\times10^{-14}$
PYCOO$^-$	$+6.3\times10^8$	-12×10^{-5}	$-(19\pm3)\times10^{-14}$	-4×10^{-5}	$-(6\pm2)\times10^{-14}$	-3×10^{-5}	$-(5\pm2)\times10^{-14}$
DFT[f]	$+5.14\times10^9$	-48×10^{-5}	-9×10^{-14}	-23×10^{-5}	-4×10^{-14}	-23×10^{-5}	-4×10^{-14}

[a] $E_{loc,x}$, component of local electric field directed along the N–O bond, from N to O.
[b] RH3, 1-oxyl-2,2,3,4,5,5-hexamethylimidazolidine; PYCOO$^-$, 1-oxyl-2,2,5,5-tetramethylpyrrolin-3-yl carboxylate.
[c] For charged spin-label species in water, from Table 4.4.
[d] Difference in $g_{xx}-g_{zz}$ between charged and neutral species in water. g_{xx} and g_{yy} are referred to g_{zz}, because $g_{zz} \approx g_e$ and is insensitive to local fields.
[e] Difference in isotropic g-value g_o between charged and neutral species in water.
[f] Calculations *in vacuo* using density functional theory (DFT) from Savitsky et al. (2008).

unpaired-electron distributions deduced from polarized-neutron diffraction. These show that hydrogen bonding to the nitroxyl oxygen in crystals of TEMPOL induces large shifts in spin density off oxygen onto nitrogen, which will markedly increase N-hyperfine couplings. Figure 4.9 shows the dependence of the isotropic ^{14}N-hyperfine coupling constants of a DOXYL fatty acid and a TOAC-containing dipeptide on the dielectric permittivity of protic solvents with differing polarities. For the more *apolar* protic solvents, the dependence on polarity obtained with the Block–Walker model is approximately linear, in accordance with Eq. 4.31 for aprotic media. The slopes of the dependence, K_v, are comparable to those obtained with aprotic solvents, but the intercepts, $a_o^{\varepsilon=1}$, are considerable larger (see Marsh and Toniolo 2008). As the proton donor concentration increases at higher polarities, the dependence on polarity shows a steep nonlinearity when hydrogen bonding contributions to a_o^N come to overwhelm those from polarization of the medium. The polarity dependence levels off for the more apolar protic solvents to the left in Figure 4.9 because the bulkier alcohol molecules form hydrogen bonds less efficiently.

In the presence of a proton donor, chemical exchange takes place between free and hydrogen-bonded nitroxides, which have isotropic ^{14}N-hyperfine couplings $a_{o,o}^N$ and $a_{o,h}^N$, respectively. Because exchange is fast compared with the difference in hyperfine couplings, the isotropic coupling constant that we observe experimentally is the weighted average (see, e.g., Marsh 2002a, and Eq. 6.24 in Section 6.3):

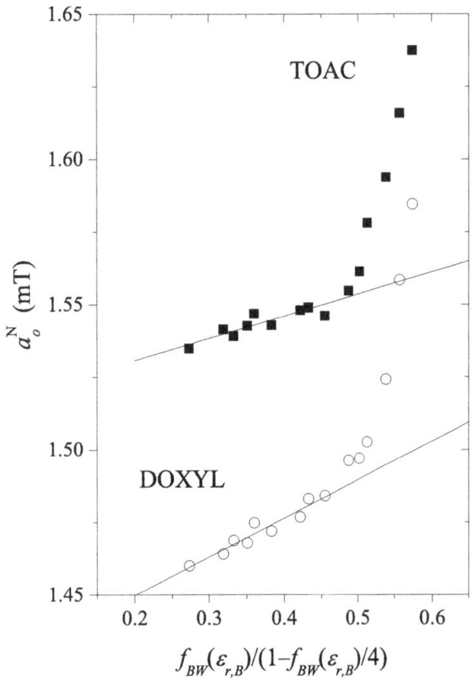

FIGURE 4.9 Dependence of isotropic hyperfine coupling a_o^N on solvent polarity: $f_{BW}(\varepsilon_{r,B})/\left(1-\frac{1}{4}f_{BW}(\varepsilon_{r,B})\right)$ in the Block–Walker model (Eq. 4.24). DOXYL (circles) and TOAC (squares) spin labels in protic solvents. Solid lines: linear regressions for more *apolar* protic media (Marsh and Toniolo 2008).

$$a_o^N([\mathrm{OH}]) = (1-f_h)a_{o,o}^N + f_h a_{o,h}^N \tag{4.39}$$

where $f_h([\mathrm{OH}])$ is the fractional population of hydrogen-bonded nitroxides. The latter depends directly on the concentration of proton-donor –OH groups in the different solvents: $f_h([\mathrm{OH}]) = K_{A,h}[\mathrm{OH}]$, where $K_{A,h}$ is an effective association constant. Figure 4.10 illustrates the linear dependence of a_o^N on molar concentration, [OH], of hydroxyl groups that we get for DOXYL and TOAC spin labels in alkanol solvents and their mixtures with water. Linear-regression parameters for these and other spin labels are given in Table 4.6. From the calibrations in Table 4.3, intercepts $a_{o,o}^N$ for non-hydrogen-bonded nitroxides are close to the hyperfine couplings that we predict for an aprotic solvent of polarity $f(\varepsilon_{r,B})/\left(1-\frac{1}{4}f(\varepsilon_{r,B})\right) \approx 0.5$. Gradients of the dependences on –OH concentration, $\partial a_o^N/\partial[\mathrm{OH}]$, in Table 4.6 are similar for different nitroxides. Density functional theory (DFT) calculations for MTSSL (Owenius et al. 2001) predict values of $a_{o,h} - a_{o,o} \approx 82$ and 145 μT for one and two hydrogen bonds, respectively, which imply that $K_{A,h} \sim 0.02$–0.03 M^{-1} (cf. Eq. 4.39). The standard state to which $K_{A,h}$ refers here is that of a pure hydrogen bonding solvent.

Much higher association constants for hydrogen bonding are expected and found with nitroxides as acceptors in aprotic media (Marsh 2002a; Marsh 2002b). Figure 4.11 shows the dependence of isotropic ^{14}N-hyperfine splitting on concentration of H-bond donor trifluoroethanol (TFE) in apolar solvents toluene or benzene, for a DOXYL- and a TEMPO-based nitroxide. These have the clear appearance of binding curves. Applying the law of mass action, together with Eqs. 4.39 and 4.31, the dependence on proton donor concentration becomes (Marsh 2002a):

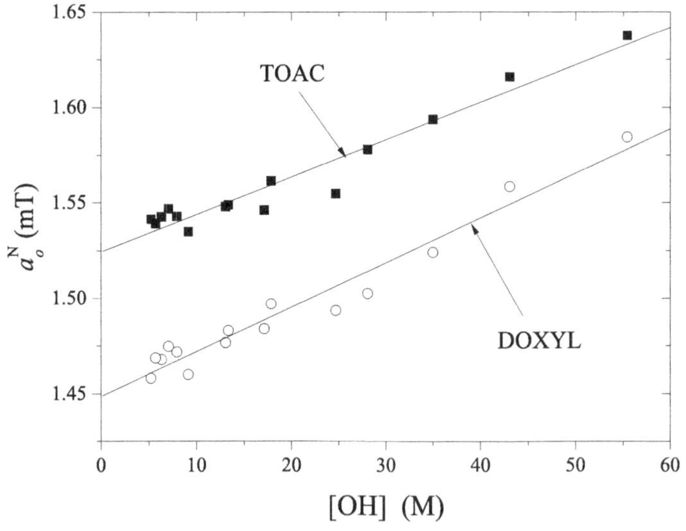

FIGURE 4.10 Dependence of isotropic hyperfine coupling a_o^N on concentration [OH] of hydroxyl proton-donor groups in alkanols and their mixtures with water. Spin labels: DOXYL (circles) and TOAC (squares). Solid lines: linear regressions – see Eq. 4.39 (Marsh and Toniolo 2008).

TABLE 4.6 Slopes and intercepts from linear dependence of isotropic hyperfine splittings, a_o^N, and *g*-values, g_o, on –OH concentration for different spin labels in protic media (see Eqs. 4.39 and 4.41) (Marsh and Toniolo 2008)

spin label[a]	$\partial a_o^N/\partial[\mathrm{OH}](\mu T.M^{-1})$	$a_{o,o}^N$ (mT)	$\partial g_o/\partial[\mathrm{OH}](\mathrm{M}^{-1})$	$g_{o,o}-g_e$	$\partial g_o/\partial a_o^N(\mathrm{T}^{-1})$
DOXYL	2.3 ± 0.1	1.449 ± 0.003	–(5.9 ± 0.3) × 10^{-6}	(3.543 ± 0.007) × 10^{-3}	–2.52 ± 0.11
MTSSL	2.2 ± 0.3	1.472 ± 0.007	–(5.7 ± 0.7) × 10^{-6}	(3.544 ± 0.017) × 10^{-3}	–2.52 ± 0.12
MTSSL/β–SH	2.2 ± 0.2	1.484 ± 0.005	–(5.5 ± 0.5) × 10^{-6}	(3.510 ± 0.011) × 10^{-3}	–2.46 ± 0.10
TOAC	2.0 ± 0.1	1.524 ± 0.003	–(6.6 ± 0.4) × 10^{-6}	(3.709 ± 0.010) × 10^{-3}	–3.33 ± 0.14

[a] DOXYL, 2-substituted 3-oxyl-4,4-dimethyl-oxazolidine; MTSSL, (1-oxyl-2,2,5,5-tetramethylpyrrolin-3-ylmethyl)methanethiosulphonate; MTSSL/β-SH, β-mercaptoethanol S–S adduct of MTSSL; TOAC, 1-oxyl-2,2,6,6-tetramethyl-4-aminopiperidin-4-yl carboxylic acid.

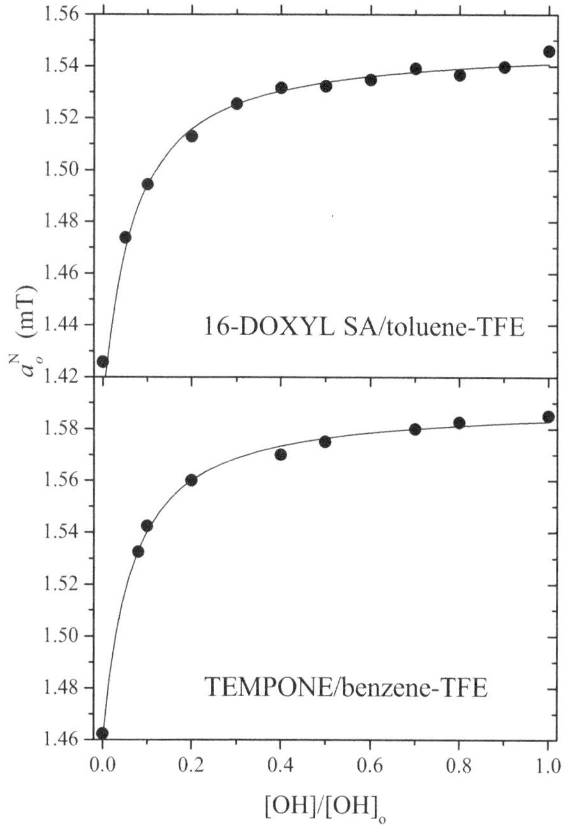

FIGURE 4.11 Dependence of isotropic hyperfine coupling a_o^N of 16-DOXYL stearic acid (top) and TEMPONE (bottom) spin labels on proton-donor concentration [OH], in mixtures of trifluoroethanol (TFE) with toluene or benzene. Solid lines: fits of Eq. 4.40 (incorporating Onsager model), yielding $K_{A,h}$ = 0.96 ± 0.14 and 1.01 ± 0.07 M^{-1} for H-bonding of TFE with DOXYL and TEMPONE, respectively (Marsh 2002a; Marsh 2002b).

$$a_o^N([\mathrm{OH}]) = \frac{a_{o,o}^{\varepsilon=1} + K_v \dfrac{f(\varepsilon_{r,B})}{1-\frac{1}{4}f(\varepsilon_{r,B})} + a_{o,h}^N K_{A,h}[\mathrm{OH}]}{1+K_{A,h}[\mathrm{OH}]} \tag{4.40}$$

where $\varepsilon_{r,B}$ is the dielectric permittivity of the mixture of protic and aprotic solvents, which also depends on [OH]. The solid lines in Figure 4.11 are non-linear least-squares fits of Eq. 4.40, which yield association constants of $K_{A,h}$ ~ 1.0 M^{-1} for H-bonding of TFE with both DOXYL and TEMPONE. (The corresponding values of $a_{o,h} - a_{o,o}$ are ≈ 120 and 130 μT for DOXYL and TEMPONE, respectively.) The association constant for hydrogen bonding with water is likely to be larger than $K_{A,h}$ ~ 1.0 M^{-1}, because assuming this value also for water predicts effective internal water concentrations in lipid membranes that are rather high compared with those expected from the solubility of water in oil (Marsh 2002b).

4.4.2 Isotropic ^{17}O-Hyperfine Couplings in Protic Media

Effects of hydrogen bonding will be felt most directly at the nitroxide oxygen because this is the H-bond acceptor. As we saw in Section 3.9, isotropic ^{17}O-hyperfine couplings are determined predominantly by unpaired π-electron spin density on the oxygen, ρ_π^O. Figure 4.12 shows correlations of isotropic ^{17}O-hyperfine couplings with ^{14}N- and ^{13}C-couplings for ^{17}O-enriched TEMPONE, in both protic and aprotic solvents. We see a clear biphasic dependence, where the ^{17}O-coupling decreases more rapidly (relative to that of ^{14}N or ^{13}C) in hydrogen bonding solvents than in aprotic solvents. This contrasts with the correlation between ^{14}N- and ^{13}C-couplings (see, e.g., Figure 3.6), or between isotropic *g*-values and nitrogen hyperfine couplings (Griffith et al. 1974; Marsh and Toniolo 2008; Owenius et al. 2001), where there is no clear difference in dependences for protic and aprotic solvents. Presumably, hydrogen bonding to the nitroxide oxygen affects σ–π polarization and, thus, changes the value of Q_O^O that relates ^{17}O-hyperfine couplings to unpaired spin density, ρ_π^O, on the oxygen (see Section 3.9, Eq. 3.35). This then gives enhanced sensitivity to ^{17}O-couplings in protic environments.

4.4.3 Isotropic *g*-Values in Protic Media

As with hyperfine couplings, the effects of hydrogen bonding on *g*-values in protic media considerably outweigh those of polarity. Because line shifts arising from *g*-value differences are small compared with the magnitude of the overall resonance field, fast chemical exchange between free and hydrogen-bonded

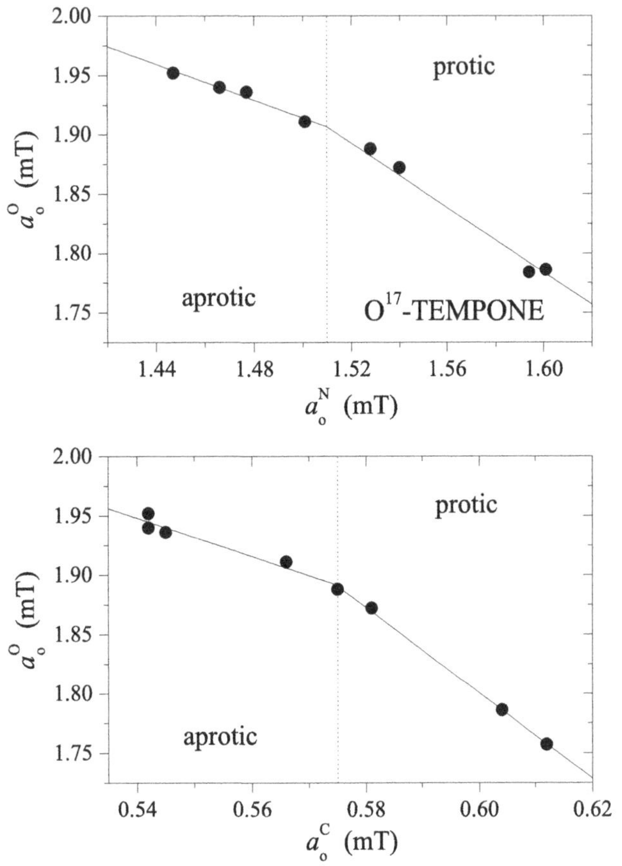

FIGURE 4.12 Correlation of isotropic [17]O-hyperfine coupling constant a_o^O with ^{14}N-hyperfine coupling constant a_o^N (upper panel), and ^{13}C-hyperfine coupling constant a_o^C (lower panel), of 1-[[17]O]oxyl-2,2,6,6-tetramethylpiperidin-4-yl oxide (TEMPONE) in aprotic (left) and protic (right) solvents of various polarities (Hayat and Silver 1973). Solid lines: linear regressions (Eq. 3.37 and equivalent for ^{13}C).

nitroxides effectively averages the g-values. An expression similar to that for hyperfine couplings therefore holds for isotropic g-values in protic solvents (cf. Eq. 4.39):

$$g_o([\mathrm{OH}]) = \left(g_{o,h} - g_{o,o}\right) f_h([\mathrm{OH}]) + g_{o,o} \tag{4.41}$$

where $g_{o,h}$ and $g_{o,o}$ are the isotropic g-values of hydrogen-bonded and free nitroxides, respectively, and $f_h([\mathrm{OH}]) = K_{A,h}[\mathrm{OH}]$ is again the fraction of nitroxides that are hydrogen bonded.

As for hyperfine couplings, isotropic g-values depend linearly on concentration of hydroxyl groups in hydrogen bonding solvents, in accordance with Eq. 4.41. We give data for various spin labels in Table 4.6, introduced already. From DFT calculations, we can estimate that $\left(g_{o,h} - g_{o,o}\right) \approx -1.7\times10^{-4}$ and -3.2×10^{-4} for one and two hydrogen bonds, respectively (Owenius et al. 2001). With the values for $\partial g_o/\partial[\mathrm{OH}]$ in Table 4.6, this gives $K_{A,h} \sim 0.02$–$0.04\ \mathrm{M}^{-1}$, in agreement with the corresponding estimate from the isotropic hyperfine couplings. Table 4.6 also lists values of $\partial g_o/\partial a_o^N$ from correlations of isotropic g-values and hyperfine couplings. These are relevant to discussions of hydrogen bonding in protic media (see following section). Omitting TOAC, but including DTBN (from Marsh and Toniolo 2008), the mean value of this gradient is: $\partial g_o/\partial a_o^N = -2.45\pm0.10\ \mathrm{T}^{-1}$. Assuming that only Δg_{xx} contributes significantly to Δg_o and that A_{xx}, A_{yy} and A_{zz} are all proportional to ρ_π^N, then we have also that $\partial g_{xx}/\partial A_{zz} = \partial g_o/\partial a_o^N \times\left(1+\left(A_{xx}+A_{yy}\right)/A_{zz}\right) \cong -3.2\pm0.1\ \mathrm{T}^{-1}$ for protic solvents.

4.4.4 *g*-Tensor Anisotropy in Protic Media

To exploit the polarity dependence of the g_{xx} tensor element (see Eq. 4.37) fully in spin-label spectroscopy, we must use high-field EPR spectrometers (Marsh et al. 2002). The advantage over hyperfine couplings is the enhanced sensitivity to hydrogen bonding. Figure 4.13 shows the hydrogen-bond geometry for the two water molecules that associate strongly with the nitroxyl oxygen. The inter-bond angle $\theta_{nOn} \approx 115°$ obtained from DFT-optimization (see Table 4.7) indicates that the oxygen lone pairs occupy $2sp^2$ hybrid orbitals (see Eq. 3.68 in Section 3.12). As in Section 3.11 for N-orbital hybrids, we therefore expect that the fractional s-electron character in the equivalent lone-pair orbitals is $\left(c_{O,s}^{(n)}\right)^2 \approx 0.30$ (cf. Eq. K.27). Note that direct experimental

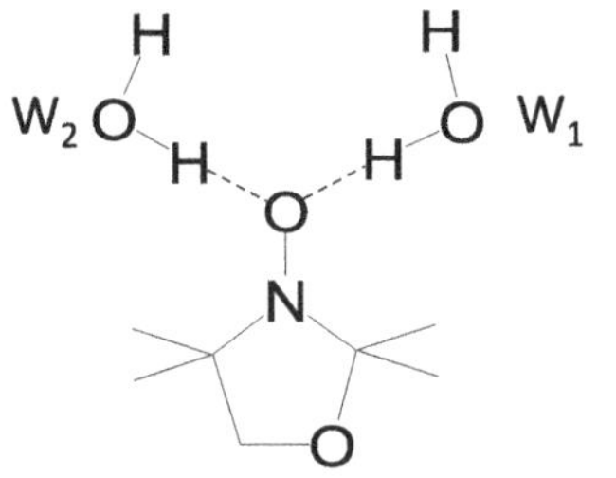

FIGURE 4.13 Pattern of hydrogen bonding between H_2O molecules and a DOXYL (or PROXYL, imidazoline) nitroxide. Geometrical parameters refined by density functional theory are given in Table 4.7. (See Erilov et al. 2005.)

TABLE 4.7 Bond distances (nm) and bond angles (°) for hydrogen bonding of water (W_1, W_2) to DOXYL (Erilov et al. 2005), PROXYL (Pavone et al. 2006) and imidazoline (Improta et al. 2001) nitroxides (NO), from DFT-optimized geometries[a]

	DOXYL[b]	*PROXYL*	*imidazoline*
N–O	0.1282	0.1263	0.1270
HW_1–O	0.1899	0.1875	0.1851
HW_2–O	0.1901	0.1876	0.1854
HW_1–O–N	122.4	124.7	122.6
HW_2–O–N	122.2	124.7	122.5
HW_1–O–HW_2	115.4	110.6	114.8

[a] For geometrical definitions see Figure 4.13.

[b] DOXYL, 2-substituted 3-oxyl-4,4-dimethyl-oxazolidine; PROXYL, 1-oxyl-2,2,5,5-tetramethyl-pyrrolidine; imidazoline, 1-oxyl-2,2,4,5,5-pentamethyl-imidazoline (RH1).

evidence for hybridization of the oxygen lone-pair on hydrogen bonding comes from pulsed ENDOR (see Sections 14.14.1 and 14.14.2 in Chapter 14).

From Eq. 4.37, changes in the g_{xx}-element, $\delta\Delta g_{xx}$, are related to the terms dependent on polarity and hydrogen bonding by (cf. Ondar et al. 1985):

$$\frac{\delta\Delta g_{xx}}{\Delta g_{xx}} = \frac{\delta\rho_\pi^{\mathrm{O}}}{\rho_\pi^{\mathrm{O}}} - \frac{\delta\Delta E_{n-\pi^*}}{\Delta E_{n-\pi^*}} + \frac{\delta\left(C_{\mathrm{O},y}^{(n)}\right)^2}{\left(C_{\mathrm{O},y}^{(n)}\right)^2} \tag{4.42}$$

where $\Delta g_{xx} \equiv g_{xx} - g_e$. Figure 4.8 shows that the positive $\delta\Delta E_{n-\pi^*}$ is much greater in protic solvents than in aprotic solvents. Hydrogen bonding will also decrease the occupation of the lone-pair orbitals, i.e., $\left(C_{\mathrm{O},y}^{(n)}\right)^2$. In contrast, changes in hyperfine coupling depend only on the unpaired spin density (see Eqs. 3.14 and 3.22 in Sections 3.3, 3.4). Therefore, g-values are preferentially sensitive to hydrogen bonding, as compared with hyperfine couplings. From Eq. 4.42, the polarity sensitivities of g-tensor element, g_{xx}, and hyperfine tensor element, A_{zz}, are related by (Möbius et al. 2005):

$$\frac{\delta\Delta g_{xx}}{\Delta g_{xx}} = -\frac{\rho_\pi^{\mathrm{N}}}{\rho_\pi^{\mathrm{O}}}\frac{\delta A_{zz}}{A_{zz}} - \frac{\delta\Delta E_{n-\pi^*}}{\Delta E_{n-\pi^*}} + \frac{\delta\left(C_{\mathrm{O},y}^{(n)}\right)^2}{\left(C_{\mathrm{O},y}^{(n)}\right)^2} \tag{4.43}$$

because $A_{zz} \propto \rho_\pi^{\mathrm{N}}$ and $\rho_\pi^{\mathrm{N}} + \rho_\pi^{\mathrm{O}} \cong 1$ and where we have $\rho_\pi^{\mathrm{N}}/\rho_\pi^{\mathrm{O}} \approx 1$ for nitroxides.

The gradient $\partial g_{xx}/\partial A_{zz}$ proves to be a useful diagnostic indicator for distinguishing protic from aprotic environments (Plato et al. 2002). Figure 4.14 shows the correlation between g_{xx} and A_{zz} for MTSSL spin label at different amino-acid residue positions in site-directed cysteine mutants of the transmembrane protein bacteriorhodopsin (Steinhoff et al. 2000; Wegener et al. 2001). Nearly all positions lie on a straight line, with gradient $\partial g_{xx}/\partial A_{zz} = -2.5\ \mathrm{T}^{-1}$, that joins the apolar, aprotic environment of toluene/polystyrene with the polar, hydrogen bonding environment of water. This shows a close correlation between polarity and proticity in the proton channel of bacteriorhodopsin. Correspondingly, we find a similar gradient of $\partial g_{xx}/\partial A_{zz} = -2.4 \pm 0.1\ \mathrm{T}^{-1}$ between DOXYL labels situated near to the polar interface and those near to the hydrophobic interior of lipid membranes (Kurad et al. 2003), indicating a dominant contribution of water penetration to the polarity profile in biological membranes. Quantum-chemical (DFT) calculations of the incremental changes, ΔA_{zz} and Δg_{xx}, in the hyperfine and g-tensors that are induced by hydrogen bonding of water to the nitroxide are given in Table 4.8, together with direct measurements at 244 and 275 GHz. These data also support the assignment of g_{xx} vs. ΔA_{zz} gradients of this overall size to protic environments.

Open diamonds and open squares in the upper panel of Figure 4.14 show DFT predictions for $g_{xx} - A_{zz}$ correlations when the spin label is not H-bonded and is singly H-bonded, respectively. The diamonds therefore correspond to a completely aprotic environment. Polarity is varied from *in vacuo* to an Onsager environment with the dielectric constant of water, and further by addition of a local polarizing field $E_{loc,x} = 10^6\ \mathrm{Vm}^{-1}$ (Savitsky et al. 2008). The gradient is much reduced to $\partial g_{xx}/\partial A_{zz} = -1.0\ \mathrm{T}^{-1}$ in both cases, when the degree of hydrogen bonding is held fixed. Experimental measurements on resolved H-bonding states, at higher frequencies, confirm the reduction in gradient when H-bonding does not change (Smirnova et al. 2007; Gast et al. 2014; and see below). DFT calculations reproduce trends but not absolute values of g_{xx} and A_{zz}. Following Plato et al. (2002), they are shifted in Figure 4.14 such that the *in vacuo*, non-H-bonded DFT result coincides with measurements for the highly hydrophobic toluene/polystyrene environment. The dashed line for one H-bond then lies above the measurement for water. Extrapolation from the 0 HB and 1 HB lines hence predicts that the nitroxide

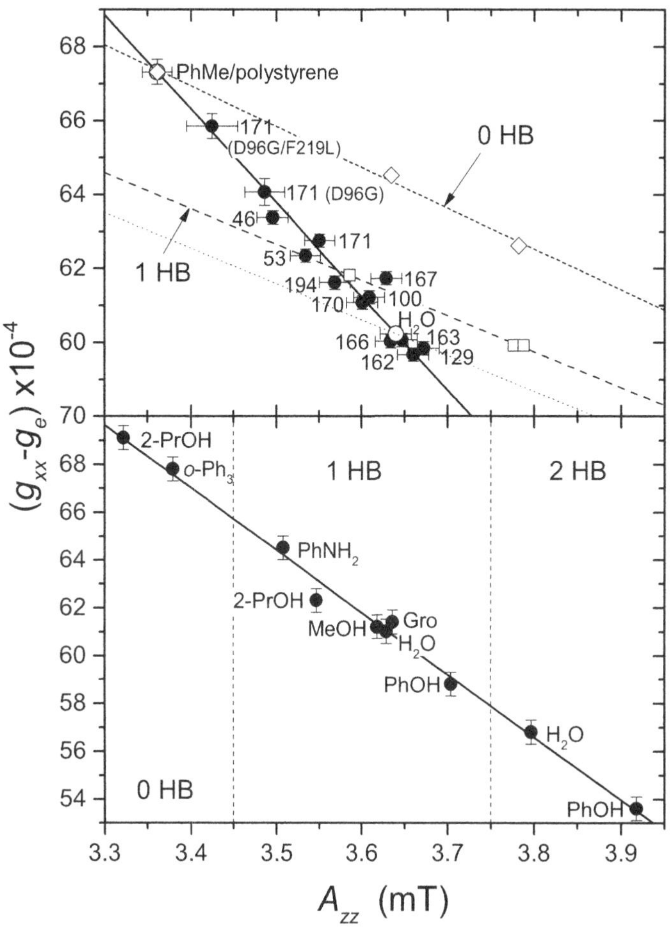

FIGURE 4.14 Correlation of g-tensor element g_{xx} with hyperfine tensor element A_{zz}: MTSSL spin label at different amino-acid residue positions in bacteriorhodopsin (top) and PYMeOH with 0, 1 or 2 hydrogen bonds (HB) in glassy solvents (bottom). Measurements on frozen samples at 95 GHz (top; Steinhoff et al. 2000; Wegener et al. 2001) or 244 GHz (bottom; Nalepa et al. 2019). Open circles, top: MTSSL in water or toluene/polystyrene. DFT predictions (top) for polarity dependence with 0 HB (diamonds/short-dash) and 1 HB (squares/dash) from Savitsky et al. (2008). DFT data shifted so that 0 HB *in vacuo* coincides with experimental toluene/polystyrene.

TABLE 4.8 Contributions of water hydrogen bonding to g_{xx} (and A_{zz}) from DFT calculations for MTSSL model nitroxides

	Δg_{xx}		$\Delta g_{xx}/\Delta A_{zz}\ (\mathrm{T}^{-1})$		
medium	$1H_2O$	$2H_2O$	$1H_2O$	$2H_2O$	*Refs.*[a]
vacuum[b]	-4.4×10^{-4}	-8.2×10^{-4}	−2.0	−2.3	1
vacuum	-5.5×10^{-4}	–	−2.4	–	2
water[c]	-4.6×10^{-4}	–	−3.2	–	2
expt.[d]	$-(6.2\pm0.8)\times10^{-4}$	$-(10.6\pm0.9)\times10^{-4}$	-3.9 ± 0.7	-3.4 ± 0.5	3
	$-(6.1\pm0.4)\times10^{-4}$	$-(10.8\pm0.4)\times10^{-4}$	-3.9 ± 0.5	-3.4 ± 0.3	4

[a] References: 1. Owenius et al. (2001); 2. Savitsky et al. (2008); 3. Gast et al. (2014); 4. Nalepa et al. (2019).
[b] *g*-tensor for nitroxide in vacuum, but *A*-tensor calculated with polarizable medium by using Onsager reaction field.
[c] Simulated by using Onsager reaction field.
[d] Mean Δg_{xx} for MTSSL (cf. Table 4.6) bound to 8 single-Cys mutants of sensory rhodopsin (ref. 3) or PYMeOH in glycerol (ref. 4); $\Delta A_{zz} = 0.157\pm0.010$ mT from difference between 1 and 2 H-bonds of PYMeOH (cf. Table 4.9) in 5vol% aqueous glycerol.

has on average 1.36 H-bonds, when in water. The maximum value is two (cf. Figure 4.13 and Table 4.8).

A valuable feature of high-field EPR is the ability to detect *g*-strain that arises from a heterogeneous population of hydrogen-bonded spin labels, and to resolve discrete levels of hydrogen bonding (Kurad et al. 2003; Gast et al. 2014). Figure 4.15 shows 94- and 360-GHz spectra that illustrate the polarity-associated *g*-strain for a spin label at the $n = 5$ position of the lipid chains (5-DOXYL PC) in hydrated membranes. We get a single, inhomogeneously broadened peak with frequency-dependent width in the g_{xx}-region at the low-field side of the spectrum from 5-DOXYL PC. In contrast, at the $n = 9$ position (i.e., 9-DOXYL PC), we see much reduced broadening of the g_{xx}-feature in the 94-GHz spectrum such that the ^{14}N-hyperfine splitting $(2A_{xx})$ becomes partially resolved. Relative to 9-DOXYL PC (for which $\Delta B = 47\,\mu$T), the inhomogeneous broadening of 5-DOXYL PC increases almost fourfold between 94 and 360 GHz (from $\delta\Delta B = 41$ to $153\,\mu$T), i.e., scales directly with the microwave frequency. The equivalent g_{xx}-strain corresponds to a distribution width of $\delta\Delta g_{xx} \approx 2\times10^{-4}$. This represents a statistical distribution of water molecules in the region of the membrane close to the top of the chain, whereas water is almost absent towards the centre of the membrane. The $n = 8$ segment of the chain (i.e., 8-DOXYL PC) represents a transition region, where the EPR spectrum consists of a superposition of partially resolved components corresponding to the two flanking regions for $n<8$ and $n>8$.

Measurements at the higher frequency of 275 GHz resolve three distinct peaks in the g_{xx}-region for MTSSL (1-oxyl-2,2,5,5-tetramethylpyrrolin-3-ylmethyl-methanethiosulphonate) reacted with different single-Cys mutants of sensory rhodopsin. These correspond to discrete states of hydrogen bonding, yielding $g_{xx} = 2.00917\pm0.00003$, 2.00855 ± 0.00004 and 2.00811 ± 0.00006 for zero, one and two H-bonds, respectively (Gast et al. 2014). ^{14}N-hyperfine couplings of the related PYMeOH (1-oxyl-2,2,5,5-tetramethylpyrrolin-3-yl methanol) in 5vol% aqueous glycerol are correspondingly found to be $A_{zz} = 3.629$, 3.786 ± 0.005 mT for hydrogen bonding to one and two water molecules, respectively.

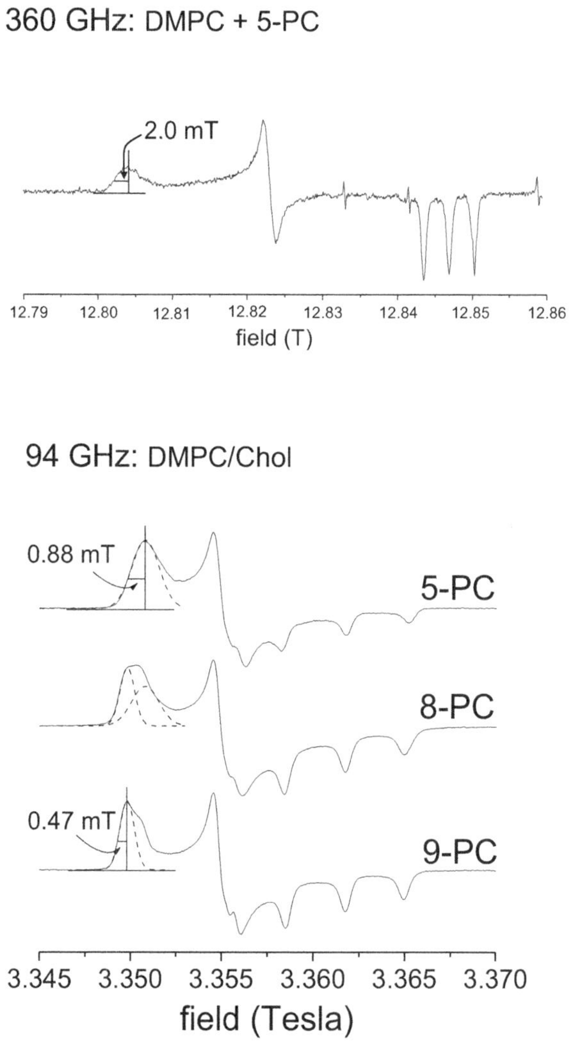

FIGURE 4.15 *Upper*: 360-GHz EPR spectrum of 5-DOXYL phosphatidylcholine (PC) in membranes of dimyristoyl PC at −100°C (see Marsh et al. 2005). *Lower*: 94-GHz EPR spectra of *n*-DOXYL PCs in membranes of dimyristoyl PC+40 mol% cholesterol at −100°C. (Data from Kurad et al. 2003.) Dashed lines in g_{xx}-region: individual hyperfine component in non-H-bonded state (9-PC); inhomogeneous broadening from distribution of H-bonding states (5-PC); coexistence of these two states (8-PC).

Results of DFT calculations, which we present in Table 4.8, are reasonably in agreement with these 275-GHz measurements. They suggest that hydrogen bonding of one water molecule causes a g-shift of $\Delta g_{xx} = -(4.4 - 5.5)\times 10^{-4}$, compared with the experimental value of $-(6.2 \pm 0.8) \times 10^{-4}$. The g-shift measured with spin-labelled lipids between the outer (5-DOXYL PC) and inner (10-DOXYL PC) regions of a membrane is similarly $\Delta g_{xx} \approx -6.2\times 10^{-4}$ (Kurad et al. 2003), which equals the difference between two resolved components at 240GHz (Dzikovski et al. 2012). Thus, the mean number of water molecules that are hydrogen bonded to a spin label positioned in the upper part of the lipid chain is ~1, and the distribution width in this region (see above) is approximately ±0.5 water molecules. Given that a nitroxide can have up to two strong H-bonds with water, these estimates are not unreasonable.

The bottom panel of Figure 4.14 shows the $g_{xx} - \Delta A_{zz}$ correlation for MTSSL model compound PYMeOH under defined H-bonding states, resolved at 244GHz in glassy solvents (Nalepa et al. 2019). In the top panel, distinct states were not resolved in the 95-GHz spectra and the quasi-continuous $g_{xx}/\Delta A_{zz}$ variation comes from spectral overlap and inhomogeneous g-strain. Relative proportions of specific states differ between H-bond donors in the bottom panel. For water, we get 66% 1 HB and 34% 2 HB, giving 1.36 H-bonds on average, similar to that estimated for MTSSL from the top panel. Solvents used are: 1:1 v/v mixture of *ortho*-terphenyl (o-Ph_3) and benzene; 10% aniline ($PhNH_2$) or phenol (PhOH) in this 1:1 mixture; 2-propanol (2-PrOH); methanol (MeOH); glycerol (Gro) and water (H_2O) with 5% or 10% glycerol. Both solvents and spin label are deuterated to improve spectral resolution. The spread of values in g_{xx} and ΔA_{zz} arises mostly from different numbers of H-bonds, different types of H-bond (amine and alcohol) and aromatic vs. aliphatic alcohols. Here, nitroxides are electron donors which form complexes with acceptors such as phenol (Murata and Mataga 1971). We then see a clear linear correlation with gradient $\partial g_{xx}/\partial A_{zz} = -2.6 \pm 0.1\ \mathrm{T}^{-1}$, close to that in the top panel but for reasons different from those suggested above. In the latter case, we assumed that the steep gradient arises from different degrees, on average, of H-bonding to water or other proton donors (cf. Gast et al. 2014).

4.5 QUADRUPOLE INTERACTIONS AND POLARITY: ^{14}N-ESEEM OF NITROXIDES

As explained in Section 3.10 of Chapter 3, nuclear electric quadrupole interactions depend on the electric field gradient at the nucleus and hence on the polarity of the spin-label environment. Quadrupole splittings do not contribute in first order to the CW-EPR spectrum, and consequently we need nuclear/hyperfine spectroscopy (ESEEM or ENDOR – see Chapter 14) to measure them. Using the site selection of high-field EPR, we can determine the P_{xx} and P_{yy} principal values of the ^{14}N-quadrupole tensor by ESEEM (Savitsky et al. 2008) and the P_{zz}-element by using ELDOR-detected NMR at high turning angles (Florent et al. 2011; Nalepa et al. 2014). Experimental values of the quadrupole tensor for various nitroxides in environments of different polarity and proticity are listed in Table 4.9. The P_{yy}-element of the quadrupole coupling tensor is more sensitive to environmental polarity than is the P_{xx} element, because the electric field gradient along x is dominated by the N–O bond (note that $P_{xx} + P_{yy} + P_{zz} = 0$). Using P_{yy} (or P_{zz}) as a polarity index is more robust than the usual hyperfine and g-tensor parameters A_{zz} and g_{xx}, because it is less sensitive to temperature and structural variations (Savitsky et al. 2010). For distinguishing between the effects of polarity and hydrogen bonding, P_{yy} is also a useful addition to the other two parameters (cf. Marsh 2010). From means of intercomparisons within Table 4.9, we can assign $P_{zz} = -1.77 \pm 0.01$ MHz to no hydrogen bonds, $P_{zz} = -1.61 \pm 0.05$ MHz to one hydrogen bond and $P_{zz} = -1.49 \pm 0.05$ MHz to two hydrogen bonds (cf. Gast et al. 2014; Nalepa et al. 2019). The dependence of P_{zz} on A_{zz} for PYMeOH is approximately linear, with gradient $\partial P_{zz}/\partial A_{zz} = 0.020 \pm 0.001$ when both are in MHz (Table 4.9, and cf. Nalepa et al. 2019). Correspondingly, the mean ratios of P_{zz}/A_{zz} are -0.00188 ± 0.0004, -0.00157 ± 0.0006 and -0.00138 ± 0.0004 for 0, 1 and 2 H-bonds, respectively.

Table 4.10 gives the nitrogen $2p$-orbital occupancies that we deduce from the quadrupolar tensors in Table 4.9 by combining with the dipolar nitrogen hyperfine tensor, as described in Section 3.11 of Chapter 3 (Marsh 2015). Corresponding values for the degree of covalent transfer and ionicities (see Eqs. 3.55, 3.61 and 3.62) are also listed in Table 4.10. Particularly relevant is PYMeOH (which constitutes the core of the MTSSL reagent used for site-directed spin labelling) in hydrophobic and protic media. The positive charge on the nitrogen increases from $\delta q^+ = \pi_c - i_\sigma(\mathrm{NO}) - 2i_\sigma(\mathrm{NC}) = 0.13 \pm 0.04$ in o-terphenyl to 0.23 ± 0.04 in glycerol. Empirically correcting for the charge on the nitrogen atom by replacing $P_{2p,\mathrm{N}}$ with $P_{2p,\mathrm{N}}(1 + \varepsilon\delta q^+)$, where the shielding factor is $\varepsilon = 0.30$ for nitrogen (Dailey and Townes 1955; Gordy and Cook 1984), reduces these values to $\delta q^+ = 0.10$ and 0.19, respectively, with only small changes in the other quantities. Comparing the two sets of values for PYMeOH, we see that the greatest sensitivity to polarity and proticity comes from the increase of covalent transfer in the π-bond, followed by a decrease in ionicity of the N–O σ-bond. Both have the effect of increasing the electric dipole of the N–O bond. The N–C σ-bonds, on the other hand, remain almost unaffected.

4.6 MEMBRANE POLARITY PROFILES

The permeability barrier of biological membranes provides a good illustration of how to study environmental polarity by spin-label EPR. We can determine transmembrane polarity profiles at

TABLE 4.9 Nuclear electric quadrupole coupling tensors for ^{14}N-nitroxides

P_{xx} (MHz)	P_{yy} (MHz)	P_{zz} (MHz)	e^2qQ/h (MHz)	η
PYMeOH-d$_{15}$/*o*-terphenyl[a]				
+1.26 ± 0.03	+0.53 ± 0.03	−1.79 ± 0.04	−3.58 ± 0.08	−0.41 ± 0.03
PYMeOH-d$_{15}$/glycerol[a]				
+1.23 ± 0.03	+0.36 ± 0.03	−1.59 ± 0.04	−3.18 ± 0.08	−0.55 ± 0.03
PYMeOH-d$_{15}$/2-propanol[b]				
–	–	−1.77 ± 0.03	−3.54 ± 0.06	
–	–	−1.62 ± 0.03	−3.24 ± 0.06	
PYMeOH-d$_{15}$/*o*-terphenyl-benzene (1:1 v/v)[b]				
		−1.74 ± 0.03	−3.48 ± 0.06	
PYMeOH-d$_{15}$/5vol% aq. glycerol[c]				
–	–	−1.54 ± 0.05	−3.08 ± 0.1	
–	–	−1.49 ± 0.05	−2.98 ± 0.1	
PYMeOH-d$_{15}$/methanol[d]				
		−1.60 ± 0.03	−3.20 ± 0.06	
PYMeOH-d$_{15}$/aniline[d]				
		−1.64 ± 0.03	−3.28 ± 0.06	
PYMeOH-d$_{15}$/phenol[d]				
		−1.57 ± 0.03	−3.14 ± 0.06	
		−1.46 ± 0.03	−2.92 ± 0.06	
TEMPOL/*o*-terphenyl[e]				
+1.22 ± 0.03	+0.54 ± 0.03	−1.76 ± 0.04	−3.52 ± 0.08	−0.39 ± 0.03
TEMPOL/toluene[f]				
+1.29	+0.48	−1.77 ± 0.1	−3.54 ± 0.2	−0.46
TEMPOL-d$_{15}$/2-propanol[g]				
		−1.65	−3.30	
		−1.53	−3.06	
PROXYL-COOH/glycerol[h]				
+1.56 ± 0.6	+0.11 ± 0.8	−1.67 ± 0.3	−3.34 ± 0.6	

[a] 1-oxyl-2,2,5,5-tetramethylpyrrolin-3-yl methanol in *o*-terphenyl or glycerol at 180 K (Savitsky et al. 2008).
[b] 1-oxyl-2,2,5,5-tetramethylpyrrolin-3-yl methanol in 2-propanol-d$_8$ or *o*-terphenyl-d$_{14}$/benzene-d$_6$, at 50 K; for isopropanol, two-component spectrum with 0, 1 H-bond in ratio 2:3 (Nalepa et al. 2014).
[c] 1-oxyl-2,2,5,5-tetramethylpyrrolin-3-yl methanol in D_2O-5vol% glycerol-d$_8$ at 50 K; two-component spectrum for 1, 2 H-bond in ratio 1:2 (Gast et al. 2014).
[d] 1-oxyl-2,2,5,5-tetramethylpyrrolin-3-yl methanol in methanol-d$_4$, aniline-d$_7$ or phenol-d$_6$ at 50 K; for phenol two-component with 1,2 H-bond in ratio 84:16 (Nalepa et al., 2019).
[e] 1-oxyl-2,2,6,6-tetramethylpiperidin-4-ol in *o*-terphenyl at 90 K (Savitsky et al. 2010).
[f] 1-oxyl-2,2,6,6-tetramethylpiperidin-4-ol in toluene at 40 K (Florent et al. 2011).
[g] 1-oxyl-2,2,6,6-tetramethylpiperidin-4-ol in 2-propanol-d$_8$ at 50 K; two-component spectrum in ratio 1:1 (Nalepa et al. 2014).
[h] 1-oxyl-2,2,5,5-tetramethylpyrrolidin-3-yl carboxylic acid in glycerol at 40 K (Florent et al. 2011).

TABLE 4.10 Nitrogen 2*p*-orbital occupancies, n_x, n_y and n_z; covalent transfer, π_c, in nitroxide $2p_z\pi$-bond; and ionicities, i_σ, of N–O and N–C σ-bonds; deduced from ^{14}N quadrupolar and hyperfine tensors

n_x	n_y	n_z	$\pi_c \equiv \rho_\pi^N$	i_σ (NO)	i_σ (NC)
PYMeOH/*o*-terphenyl[a]					
1.11 ± 0.01	1.19 ± 0.01	1.445 ± 0.002	0.555 ± 0.002	0.05 ± 0.03	0.19 ± 0.01
PYMeOH/glycerol[a]					
1.08 ± 0.01	1.18 ± 0.01	1.396 ± 0.002	0.604 ± 0.002	0.02 ± 0.03	0.18 ± 0.01
TEMPOL/*o*-terphenyl[b]					
1.14 ± 0.01	1.22 ± 0.01	1.472 ± 0.002	0.528 ± 0.002	0.11 ± 0.02[c]	0.22 ± 0.01
TEMPOL/toluene[d]					
1.12 ± 0.03	1.21 ± 0.03	1.46 ± 0.01	0.54 ± 0.01	0.09 ± 0.05[c]	0.21 ± 0.03

[a] 1-oxyl-2,2,5,5-tetramethylpyrrolin-3-yl methanol in *o*-terphenyl or glycerol at 180 K (Savitsky et al. 2008).
[b] 1-oxyl-2,2,6,6-tetramethylpiperidin-4-ol in *o*-terphenyl at 90 K (Savitsky et al. 2010).
[c] Approximate because C–NO–C is not planar for TEMPOL (see Table 3.4).
[d] 1-oxyl-2,2,6,6-tetramethylpiperidin-4-ol in toluene at 40 K (Florent et al. 2011).

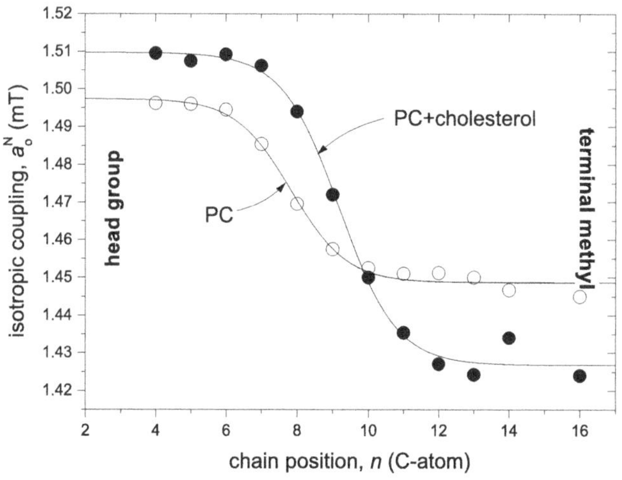

FIGURE 4.16 Polarity profiles for isotropic ^{14}N-hyperfine coupling constant a_o^N of *n*-DOXYL phosphatidylcholine (PC) spin labels in fluid bilayer membranes of dipalmitoyl PC, with (solid circles) and without (open circles) 50 mol% cholesterol. Lines: least-squares fits of Eq. 4.44; fitting parameters in Table 4.11 (Marsh 2001).

high spatial resolution by using lipids systematically spin-labelled with the DOXYL moiety at single sites down their hydrocarbon chains (see Figure 1.1). Besides using polarity-sensitive parameters discussed in this chapter, we can map water penetration profiles directly from ESEEM spectroscopy of D_2O, as described later in Chapter 14. Also, we can obtain penetration profiles of hydrophobic paramagnetic gases, such as molecular oxygen, from enhancements of spin-label T_{1e}-relaxation measured by saturation techniques, as described in Chapters 10–12.

Figure 4.16 shows the transmembrane profile of the isotropic hyperfine coupling constant a_o^N from phospholipid probes (*n*-DOXYL PC), spin labelled at specific positions *n* down their hydrocarbon chains, in lipid-bilayer membranes. Values of the isotropic coupling a_o^N come from the anisotropic hyperfine splittings $A_\parallel$ and $A_\perp$, as described in Chapter 8. We see significant quantitative differences between the polarity profiles of dipalmitoyl phosphatidylcholine membranes with and without cholesterol. Nevertheless, both profiles have the same overall trough-like shape that delineates the permeability barrier of the membrane to polar solutes. We can describe all such profiles with a Boltzmann sigmoidal dependence on *n* (Marsh 2001):

$$I(n) = \frac{I_1 - I_2}{1 + e^{(n-n_o)/\lambda}} + I_2 \tag{4.44}$$

where I_1 and I_2 are the limiting values of the measured EPR-quantity I at the polar head group and terminal methyl ends of the lipid chain, respectively, n_o is the value of n at the point of maximum gradient, corresponding to $I(n_o) = \frac{1}{2}(I_1 + I_2)$ and λ is an exponential decay length. This equation corresponds to a two-compartment distribution between outer $(n < n_o)$ and inner $(n > n_o)$ regions of the membrane, in which the free energy of transfer, $(n - n_o)k_BT/\lambda$, increases linearly with distance, $n - n_o$, from the dividing plane.

Table 4.11 lists the parameters n_o and λ that characterize transmembrane profiles obtained from the different EPR techniques, including D_2O-ESEEM which directly reflects the profile of water penetration. Mostly, the profiles are rather similar, the decay length is λ ~1–1.5 CH_2 groups, with the exception of the data from frozen samples (i.e., g_{xx}, A_{zz} and D_2O-ESEEM) for which a sharper profile is found. The midpoint of the profile is at the n_o ~8–10 position of the lipid chain, with a tendency to the higher end of this range for membranes containing cholesterol.

TABLE 4.11 Parameters fitting the polarity profiles recorded by *n*-DOXYL PC spin-labelled phosphatidylcholine chains in different lipid membranes, according to Eq. 4.44[a]

lipid[b]	*indicator*[c]	n_o	λ	*Refs.*[d]
$(14{:}0)_2$PC	a_o^N	8.00 ± 0.06	0.44 ± 0.06	1
	O_2, saturation	9.8 ± 0.7	1.4 ± 0.6	2
	O_2, linewidth	9.0 ± 0.1	1.4 ± 0.1	2
	O_2, saturation	7.8 ± 0.2	0.5 ± 0.2	3
	NO, saturation	10.2 ± 0.3	1.4 ± 0.3	3
$(14{:}0)_2$PC +chol	a_o^N	9.37 ± 0.09	0.83 ± 0.08	1
	g_{xx}	7.63 ± 0.08	0.25 ± 0.05	4
	A_{zz}	7.60 ± 0.09	0.36 ± 0.06	4
$(16{:}0)_2$PC	a_o^N	7.8 ± 0.1	0.8 ± 0.1	1
$(16{:}0)_2$PC +chol	a_o^N	9.2 ± 0.1	0.8 ± 0.1	1
	D_2O-ESEEM	7.6 ± 0.2	0.4 ± 0.1	5
(16:0/18:1)PC	a_o^N	8.35 ± 0.14	1.03 ± 0.13	1
(16:0/18:1)PC +chol	a_o^N	9.38 ± 0.14	1.25 ± 0.13	1
$(18{:}1)_2$PC	a_o^N	8.24 ± 0.24	0.96 ± 0.22	1
$(18{:}1)_2$PC +chol	a_o^N	10.0 ± 0.1	1.0 ± 0.1	1

[a] Expressed in CH_2 units for C-atom positions in *sn*-2 chain of phosphatidylcholine.
[b] $(14{:}0)_2$PC, dimyristoyl phosphatidylcholine; $(16{:}0)_2$PC, dipalmitoyl phosphatidylcholine; (16:0/18:1)PC, 1-palmitoyl-2-oleoyl phosphatidylcholine; $(18{:}1)_2$PC, dioleoyl phosphatidylcholine; chol, cholesterol (40 or 50 mole%).
[c] a_o^N, isotropic ^{14}N-hyperfine coupling; O_2 and NO, relaxation enhancements by molecular oxygen and nitric oxide, respectively (see Chapters 10–12); g_{xx}, anisotropic *g*-tensor element; A_{zz}, anisotropic ^{14}N-hyperfine tensor element; D_2O-ESEEM, electron spin-echo envelope modulation by D_2O (see Chapter 14).
[d] References: 1. Marsh (2001); 2. Dzikovski et al. (2003); 3. Nedieanu et al. (2004); 4. Kurad et al. (2003); 5. Erilov et al. (2005).

4.7 CONCLUDING SUMMARY

1. *Dielectric constant and aprotic environments:*
In aprotic environments, isotropic hyperfine couplings and *g*-values depend on dielectric permittivity, $\varepsilon_{r,B}$, of the bulk medium via the reaction field, according to expressions of the form:

$$a_o^N = a_o^{\varepsilon=1} + K_v \frac{f_{BW}(\varepsilon_{r,B})}{1 - \frac{1}{4} f_{BW}(\varepsilon_{r,B})} \tag{4.45}$$

where $f_{BW}(\varepsilon_{r,B}) = 3\varepsilon_{r,B}\ln\varepsilon_{r,B}/(\varepsilon_{r,B}\ln\varepsilon_{r,B} - \varepsilon_{r,B} + 1) - 6/\ln\varepsilon_{r,B} - 2$. For a_o^N: K_v ~80–100 μT, and $a_o^{\varepsilon=1}$ depends on the particular spin label (see Table 4.3). For g_o: $K_{v,g}$ ~ $-(0.2\text{–}0.4)\times 10^{-3}$, and $g_o^{\varepsilon=1}$ is given in Table 4.3.
2. *Local electric fields:*
Shifts in isotropic hyperfine coupling are related to local electric-field components along the N–O bond, via redistribution of the unpaired spin density:

$$\Delta a_o^N(\text{mT}) = 3.5\times 10^{-11} E_{loc,x}\left(\text{Vm}^{-1}\right) \tag{4.46}$$

and corresponding expressions for the *g*-values, particularly $\Delta g_{xx} \approx 2\times 10^{-13} E_{loc,x}\left(\text{Vm}^{-1}\right)$.
3. *Hydrogen bonding and protic environments:*
In protic environments, isotropic hyperfine couplings and *g*-values depend linearly on the concentration of hydroxyl groups, [OH], according to expressions of the form:

$$a_o^N = a_{o,o}^N + \left(\frac{\partial a_o^N}{\partial[\text{OH}]}\right)\cdot[\text{OH}] \tag{4.47}$$

For a_o^N: $\partial a_o^N/\partial[\text{OH}]$ ~2.2 μT.M^{-1}, and $a_{o,o}^N$ depends on the particular spin label (see Table 4.6). For g_o: $\partial g_o/\partial[\text{OH}]$ ~ $-(5.5\text{–}6.6)\times 10^{-6}$ M^{-1}, and $g_{o,o}$ is given in Table 4.6.
4. *Hyperfine:g-value correlations:*
The *g*-value vs. hyperfine coupling gradients, $\partial g_o/\partial a_o^N$ and $\partial g_{xx}/\partial A_{zz}$, are diagnostic for hydrogen bonding environments, with values of $-(2\text{–}3.5)$ T^{-1} when spanning protic and aprotic media, and considerably smaller in either protic or aprotic media alone. A hydrogen bond to one water molecule produces a *g*-shift of $\Delta g_{xx} \approx -6\times 10^{-4}$, and -4.5×10^{-4} for the second water (Table 4.8).
5. *Polarity profiles:*
The profiles of polarity, and permeation of O_2 and H_2O, in lipid bilayer membranes follow a Boltzmann sigmoid with respect to the C-atom position, *n*, in the lipid chains:

$$I(n) = \frac{I_1 - I_2}{1 + \exp\left((n - n_o)/\lambda\right)} + I_2 \tag{4.48}$$

where λ ~1–1.5 CH_2 groups and the midpoint of the profile is at n_o ~8–10 (Table 4.11). This profile is established by measurement of: a_o^N, g_{xx} and A_{zz}; relaxation enhancements by O_2; and D_2O-ESEEM amplitudes.

REFERENCES

Barone, V. 1996. Electronic, vibrational and environmental effects on the hyperfine coupling constants of nitroside radicals. H_2NO as case study. *Chem. Phys. Lett.* 262:201–206.

Bell, R.P. 1931. The electrostatic energy of dipole molecules in different media. *Trans. Faraday Soc.* 27:797–805.

Block, H. and Walker, S.M. 1973. Modification of Onsager theory for a dielectric. *Chem. Phys. Lett.* 19:363–364.

Bondi, A. 1964. van der Waals volumes and radii. *J. Phys. Chem.* 68:441–451.

Böttcher, C.J.F. 1973. *Theory of Electric Polarization*, Vol. 1. Amsterdam: Elsevier Scientific Publishing Company.

Brière, R., Lemaire, H., and Rassat, A. 1965. Nitroxydes. XV. Synthèse et étude de radicaux libres stables piperidiniques et pyrrolidinique. *Bull. Soc. Chim. France* 3273–3283.

Dailey, B.P. and Townes, C.H. 1955. The ionic character of diatomic molecules. *J. Chem. Phys.* 23:118–123.

Dupeyre, R.M., Lemaire, H., and Rassat, A. 1964. Nitroxydes (VII): radicaux libres stables pyrrolidiniques. *Tetrahedron Lett.* 27–8:1781–1785.

Dzikovski, B.G., Livshits, V.A., and Marsh, D. 2003. Oxygen permeation profile in lipid membranes: non-linear spin-label EPR. *Biophys. J.* 85:1005–1012.

Dzikovski, B., Tipikin, D., and Freed, J. 2012. Conformational distributions and hydrogen bonding in gel and frozen lipid bilayers: a high frequency spin-label ESR study. *J. Phys. Chem. B* 116:6694–6706.

Erilov, D.A., Bartucci, R., Guzzi, R. et al. 2005. Water concentration profiles in membranes measured by ESEEM of spin-labeled lipids. *J. Phys. Chem. B* 109:12003–12013.

Florent, M., Kaminker, I., Nagaranjan, V., and Goldfarb, D. 2011. Determination of the ^{14}N quadrupole coupling constant of nitroxide spin probes by W-band ELDOR-detected NMR. *J. Magn. Reson.* 210:192–199.

Gast, P., Herbonnet, R.T.L., Klare, J. et al. 2014. Hydrogen bonding of nitroxide spin labels in membrane proteins. *Phys. Chem. Chem. Phys.* 16:15910–15916.

Gordy, W. and Cook, R.L. 1984. *Microwave Molecular Spectra*, Techniques of Chemistry, Vol. XVIII. New York: Wiley Interscience.

Griffith, O.H., Dehlinger, P.J., and Van, S.P. 1974. Shape of the hydrophobic barrier of phospholipid bilayers. Evidence for water penetration in biological membranes. *J. Membrane Biol.* 15:159–192.

Gullá, A.F. and Budil, D.E. 2001. Orientation dependence of electric field effects on the *g* factor of nitroxides measured by 220 GHz EPR. *J. Phys. Chem. B* 105:8056–8063.

Hayat, H. and Silver, B.L. 1973. Oxygen-17 and nitrogen-14 σ-π polarization parameters and spin density distribution in nitroxyl group. *J. Phys. Chem.* 77:72–78.

Improta, R., Scalmani, G., and Barone, V. 2001. Quantum mechanical prediction of the magnetic titration curve of a nitroxide 'spin probe'. *Chem. Phys. Lett.* 336:349–356.

Kawamura, T., Matsunami, S., and Yonezawa, T. 1967. Solvent effects on the *g*-value of di-*t*-butyl nitric oxide. *Bull. Chem. Soc. Japan* 40:1111–1115.

Khramtsov, V.V., Weiner, L.M., Grigoriev, I.A., and Volodarsky, L.B. 1982. Proton exchange in stable nitroxyl radicals. EPR study of the pH of aqueous solutions. *Chem. Phys. Lett.* 91:69–72.

Kikuchi, O. 1969. Electronic structure and spectrum of the H_2NO radical. *Bull. Chem. Soc. Japan* 42:47–52.

Kurad, D., Jeschke, G., and Marsh, D. 2003. Lipid membrane polarity profiles by high-field EPR. *Biophys. J.* 85:1025–1033.

Lemaire, H. and Rassat, A. 1964. Structure hyperfine due à l'azote dans les radicaux nitroxydes. *J. Chim. Phys.* 61:1580–1586.

Marsh, D. 2001. Polarity and permeation profiles in lipid membranes. *Proc. Natl. Acad. Sci. USA* 98:7777–7782.

Marsh, D. 2002a. Polarity contributions to hyperfine splittings of hydrogen-bonded nitroxides - the microenvironment of spin labels. *J. Magn. Reson.* 157:114–118.

Marsh, D. 2002b. Membrane water-penetration profiles from spin labels. *Eur. Biophys. J.* 31:559–562.

Marsh, D. 2008. Reaction fields and solvent dependence of the EPR parameters of nitroxides: the microenvironment of spin labels. *J. Magn. Reson.* 190:60–67.

Marsh, D. 2009. Reaction fields in the environment of fluorescent probes: polarity profiles in membranes. *Biophys. J.* 96:2549–2558.

Marsh, D. 2010. Spin-label EPR for determining polarity and proticity in biomolecular assemblies: transmembrane profiles. *Appl. Magn. Reson.* 37:435–454.

Marsh, D. 2015. Bonding in nitroxide spin labels from ^{14}N electric-quadrupole interactions. *J. Phys. Chem. A* 119:919–921.

Marsh, D., Kurad, D., and Livshits, V.A. 2002. High-field electron spin resonance of spin labels in membranes. *Chem. Phys. Lipids* 116:93–114.

Marsh, D., Kurad, D., and Livshits, V.A. 2005. High-field spin label EPR of lipid membranes. *Magn. Reson. Chem.* 43:S20–S25.

Marsh, D. and Toniolo, C. 2008. Polarity dependence of EPR parameters for TOAC and MTSSL spin labels: correlation with DOXYL spin labels for membrane studies. *J. Magn. Reson.* 190:211–221.

Möbius, K., Savitsky, A., Schnegg, A., Plato, M., and Fuchs, M. 2005. High-field EPR spectroscopy applied to biological systems: characterization of molecular switches for electron and ion transfer. *Phys. Chem. Chem. Phys.* 7:19–42.

Murata, Y. and Mataga, N. 1971. ESR and optical studies on the EDA complexes of di-*t*-butyl-*N*-oxide radical. *Bull. Chem. Soc. Japan* 44:354–360.

Nalepa, A., Möbius, K., Lubitz, W., and Savitsky, A. 2014. High-field ELDOR-detected NMR study of a nitroxide radical in disordered solids: towards characterization of heterogeneity of microenvironments in spin-labeled systems. *J. Magn. Reson.* 242:203–213.

Nalepa, A., Möbius, K., Plato, M., Lubitz, W., and Savitsky, A. 2019. Nitroxide spin labels – Magnetic parameters and hydrogen-bond formation: a high-field EPR and EDNMR study. *Appl. Magn. Reson.* 50:1–16.

Nedieanu, S., Páli, T., and Marsh, D. 2004. Membrane penetration of nitric oxide and its donor *S*-nitroso-*N*-acetylpenicillamine: a spin-label electron paramagnetic resonance spectroscopic study. *Biochim. Biophys. Acta* 1661:135–143.

Ondar, M.A., Grinberg, O.Ya., Dubinskii, A.A., and Lebedev, Ya.S. 1985. Study of the effect of the medium on the magnetic-resonance parameters of nitroxyl radicals by high-resolution EPR spectroscopy. *Sov. J. Chem. Phys.* 3:781–792.

Onsager, L. 1936. Electric moments of molecules in liquids. *J. Am. Chem. Soc.* 58:1486–1493.

Owenius, R., Engström, M., Lindgren, M., and Huber, M. 2001. Influence of solvent polarity and hydrogen bonding on the EPR parameters of a nitroxide spin label studied by 9-GHz and 95-GHz EPR spectroscopy and DFT calculations. *J. Phys. Chem. A.* 105:10967–10977.

Pavone, V., Sillanpaa, A., Cimino, P., Crescenzi, O., and Barone, V. 2006. Evidence of variable H-bond network for nitroxide radicals in protic solvents. *J. Phys. Chem. B* 110:16189–16192.

Plato, M., Steinhoff, H.J., Wegener, C. et al. 2002. Molecular orbital study of polarity and hydrogen bonding effects on the *g* and hyperfine tensors of site directed NO spin labelled bacteriorhodopsin. *Mol. Phys.* 100:3711–3721.

Reddoch, A.H. and Konishi, S. 1979. The solvent effect on di-*tert*-butyl nitroxide. A dipole-dipole model for polar solutes in polar solvents. *J. Chem. Phys.* 70:2121–2130.

Rosantzev, E.G. 1970. *Free Nitroxyl Radicals*, New York-London: Plenum.

Savitsky, A., Dubinskii, A.A., Plato, M. et al. 2008. High-field EPR and ESEEM investigation of the nitrogen quadrupole interaction of nitroxide spin labels in disordered solids: toward differentiation between polarity and proticity matrix effects on protein function. *J. Phys. Chem. B* 112:9079–9090.

Savitsky, A., Plato, M., and Möbius, K. 2010. The temperature dependence of nitroxide spin-label interaction parameters: a high-field EPR study of intramolecular motional contributions. *Appl. Magn. Reson.* 37:415–434.

Schwartz, R.N., Peric, M., Smith, S.A., and Bales, B.L. 1997. Simple test of the effect of an electric field on the ^{14}N-hyperfine coupling constant in nitroxide spin probes. *J. Phys. Chem. B* 101:8735–8739.

Seelig, J., Limacher, H., and Bader, P. 1972. Molecular architecture of liquid crystalline bilayers. *J. Am. Chem. Soc.* 94:6364–6371.

Smirnova, T.I., Chadwick, T.G., Voinov, M.A. et al. 2007. Local polarity and hydrogen bonding inside the Sec14p phospholipid-binding cavity: high-field multi-frequency electron paramagnetic resonance studies. *Biophys. J.* 92:3686–3695.

Steinhoff, H.J., Savitsky, A., Wegener, C. et al. 2000. High-field EPR studies of the structure and conformational changes of site-directed spin labeled bacteriorhodopsin. *Biochim. Biophys. Acta* 1457:253–262.

Vasserman, A.M., Buchachenko, A.L., Rozantsev, E.G., and Neiman, M.B. 1965. Dipole moments of molecules and radicals. Di-*tert*-butyl nitrogen oxide. *J. Struct. Chem.* 6:445–446.

Wegener, C., Savitsky, A., Pfeiffer, M., Möbius, K., and Steinhoff, H.J. 2001. High-field EPR-detected shifts of magnetic tensor components of spin label side chains reveal protein conformational changes: the proton entrance channel of bacteriorhodopsin. *Appl. Magn. Reson.* 21:441–452.

Spin Relaxation Theory

5

5.1 INTRODUCTION

This chapter deals with spin–lattice (T_1) and spin–spin (T_2) relaxation and their mechanisms in the liquid state. Relaxation is induced by fluctuating local magnetic fields at the unpaired electron that arise from rotational diffusion of the spin-labelled molecule. We restrict ourselves here to the motional narrowing region, where molecular rotation is sufficiently rapid to average out the spectral anisotropies $\Delta\omega_{aniso}$ that produce the fluctuating local fields. This requires that $\Delta\omega_{aniso}\tau_c << 1$, where τ_c is the characteristic time scale of molecular motion. We can use time-dependent perturbation theory (see Appendix M) for this fast-motional regime. Each hyperfine state then contributes a single Lorentzian line to the EPR spectrum, where the Lorentzian width is determined by the transverse or spin–spin relaxation time, T_2. For unrestricted rotational motion, these lines are centred at the isotropic resonance positions specified by the hyperfine coupling constant a_o and g-value g_o. We treat spectral lineshapes for slow rotation (outside the motional narrowing regime) in Chapter 6 and restricted rotational motion in Chapter 8.

Fluctuating local fields perpendicular to the static magnetic field of the spectrometer induce transitions between electron spin orientations, just as does the microwave radiation field (see Chapter 1 and Section M.3 of Appendix M). Therefore, these perpendicular fluctuating fields, which arise from the non-secular (i.e., S_x and S_y) terms in the spin Hamiltonian, are responsible for spin–lattice relaxation. To be most effective in T_1-relaxation, they must fluctuate at the EPR resonance frequency, ω_e. Local fields that are parallel to the static field cause fluctuations in separation of the electron-spin energy levels, which broadens the lines and thus contributes to transverse relaxation. These parallel fields arise from the secular (i.e., S_z) terms in the spin Hamiltonian and are most effective in T_2-relaxation when they fluctuate with low frequency. Spin–lattice relaxation causes lifetime broadening of the EPR lines, but this is less effective than the secular broadening mechanism, except for very fast motions when $T_2 = T_1$.

Note that, throughout this chapter, we assume that hyperfine constants are given in angular frequency units (rad s^{-1}).

5.2 TIME-DEPENDENT PERTURBATION THEORY FOR RANDOM FLUCTUATIONS: CORRELATION FUNCTIONS

In fast-motion relaxation theory, the time-dependent spin Hamiltonian is composed of a static part, $\mathcal{H}_o$, and a time-dependent random perturbation, $\mathcal{H}_1F(t)$:

$$\mathcal{H}(t) = \mathcal{H}_o + \mathcal{H}_1F(t) \tag{5.1}$$

We need the wave functions, $\psi(t)$, of the time-dependent Schrödinger equation (see Appendix D.1) that correspond to this Hamiltonian. We expand $\psi(t)$ in terms of the eigenfunctions, $u_m\exp(-i\omega_m t)$, of the unperturbed time-dependent Schrödinger equation (see Appendix M.1):

$$\psi(t) = \sum_m c_m(t)u_m\exp(-i\omega_m t) \tag{5.2}$$

where the spin states $u_m\left(\equiv|M_S,m_I\rangle\right)$ are eigenfunctions of the stationary system with energy eigenvalues $\hbar\omega_m$ (i.e., $\mathcal{H}_o u_m = \hbar\omega_m u_m$).

As described in Appendix M.1, the rate equation for the expansion coefficients in Eq. 5.2 is given by (see Eq. M.6):

$$\frac{\partial c_{m'}(t)}{\partial t} = \frac{-i}{\hbar}\sum_m \langle m'|\mathcal{H}_1|m\rangle F(t)\exp(i\omega_{m'm}t)c_m(t) \tag{5.3}$$

where $\omega_{m'm} \equiv \omega_{m'} - \omega_m$ is the transition frequency and matrix elements of the perturbation are defined by: $\langle m'|\mathcal{H}_1|m\rangle \equiv \int u^*_{m'}\mathcal{H}_1u_m\cdot \mathrm{d}^3r$, with integration over all space. For transitions from the system originally in a single state m, i.e., $c_{m'}(0) = \delta_{m',m}$, the approximate solution of Eq. 5.3 is:

$$c_{m'}(t) = \frac{-i}{\hbar}\langle m'|\mathcal{H}_1|m\rangle\int_0^t F(t')\exp(i\omega_{m'm}t')\cdot\mathrm{d}t' \tag{5.4}$$

This corresponds to first-order time-dependent perturbation theory – see Eq. M.7 in Appendix M.1.

The probability of finding the system in state m' after time t is $|c_{m'}(t)|^2 = c^*_{m'}c_{m'}$ and the corresponding transition probability (per unit time) is:

$$W_{m'm} = \frac{\partial c^*_{m'}}{\partial t} c_{m'} + c^*_{m'} \frac{\partial c_{m'}}{\partial t} \tag{5.5}$$

From Eqs. 5.3 and 5.4, we have:

$$\frac{\partial c^*_{m'}}{\partial t} c_{m'} = \frac{1}{\hbar^2} \left|\langle m'|\mathcal{H}_1|m\rangle\right|^2 F^*(t)\exp(-i\omega_{m'm}t) \times \int_0^t F(t')\exp(i\omega_{m'm}t')\cdot \mathrm{d}t'$$

$$= \frac{1}{\hbar^2} \left|\langle m'|\mathcal{H}_1|m\rangle\right|^2 \int_0^\tau F^*(t)F(t-\tau)\exp(-i\omega_{m'm}\tau)\cdot \mathrm{d}\tau \tag{5.6}$$

where we change variables: $t' = t - \tau$ for the second equality. We get an expression analogous to Eq. 5.6 for $c^*_{m'}\partial c_{m'}/\partial t$ but with integration limits from $-\tau$ to zero.

When $F(t)$ fluctuates randomly about zero in a stationary fashion (see Figure 5.1), the autocorrelation function defined by the ensemble average:

$$G(\tau) = \langle F^*(t)F(t-\tau)\rangle \tag{5.7}$$

has a fixed, non-vanishing value that is independent of t. Note that the following general relations: $G(\tau) = G^*(\tau) = G(-\tau)$, hold for correlation functions. The rate of transition from initial state m to final state m' is then given from Eqs. 5.5 and 5.6 and equivalent, by:

$$W_{m'm} = \frac{1}{\hbar^2} \left|\langle m'|\mathcal{H}_1|m\rangle\right|^2 J(\omega_{m'm}) \tag{5.8}$$

where an ensemble average over $\left|\langle m'|\mathcal{H}_1|m\rangle\right|^2$ is now implied. The spectral density, $J(\omega_{m'm})$, in Eq. 5.8 is the Fourier transform of the correlation function, $G(\tau)$:

$$J(\omega_{m'm}) = \int_{-\infty}^{+\infty} G(\tau)\exp(-i\omega_{m'm}\tau)\cdot \mathrm{d}\tau \tag{5.9}$$

Extension of the limits of integration ($\tau \rightarrow \infty$) is allowed in Eq. 5.9 because correlations of stationary random functions persist only over a limited range. Often we use an alternative definition for the spectral density: $j(\omega) = \frac{1}{2}J(\omega)$, which we get by restricting the integration range in Eq. 5.9 to between zero and infinity.

For stationary random functions, the variable t is redundant in Eq. 5.7. Therefore, we can express the correlation function more succinctly by a further change of variable: $\tau = t$. This then depicts the time dependence of the autocorrelation function as:

$$G(t) = \langle F^*(t)F(0)\rangle \tag{5.10}$$

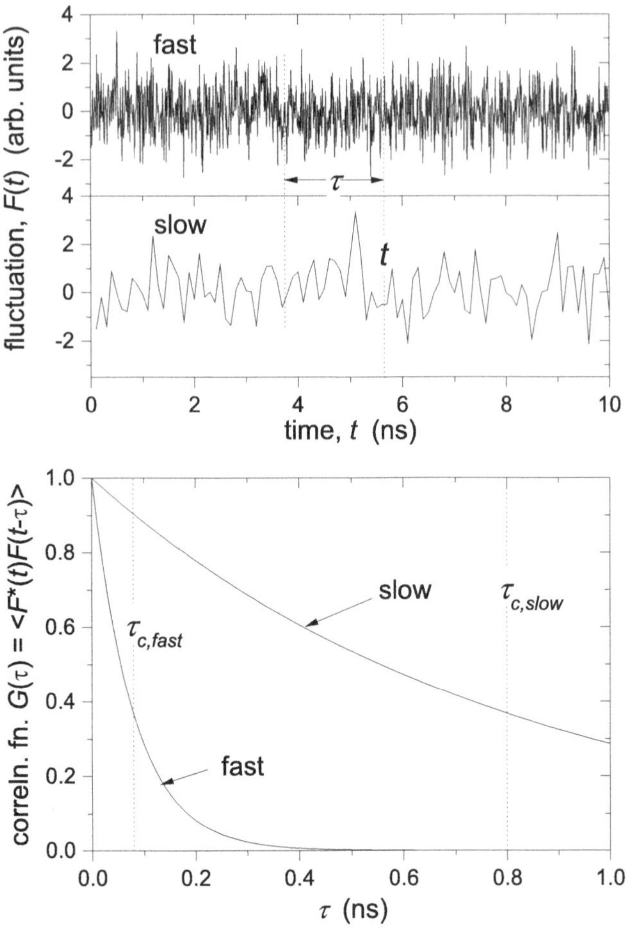

FIGURE 5.1 Fast (*top*) and slow (*middle*) random fluctuations $F(t)$ in angular terms of the spin Hamiltonian (Eq. 5.1). *Bottom*: normalized correlation functions $G(\tau)$ (Eq. 5.7) for fast and slow fluctuations. Correlation functions are exponential (Eq. 5.12) with correlation times $\tau_{c,\mathrm{fast}}$ and $\tau_{c,\mathrm{slow}}$, respectively.

where $t = 0$ is an arbitrarily chosen time.

The relation between the correlation function and conditional probability $P(\Omega_o;\Omega,t)$ is crucial to relaxation calculations. For rotational diffusion, $P(\Omega_o;\Omega,t)$ is the probability that the spin label will have orientation Ω at time t, if it has orientation Ω_o at time $t = 0$. Because $G(t)$ is an ensemble average, it is related to averages over Ω and Ω_o by:

$$\langle F^*(t)F(0)\rangle = \iint F^*(\Omega)F(\Omega_o)P(\Omega_o;\Omega,t)P(\Omega_o)\mathrm{d}\Omega_o\,\mathrm{d}\Omega \tag{5.11}$$

where $P(\Omega_o)$ (=$1/4\pi$) is simply the orientational probability for a random distribution (at $t = 0$). Equation 5.11 is how we usually calculate correlation functions. This involves determining $P(\Omega_o; \Omega,t)$ for a given model describing the dynamic fluctuations – see, e.g., Section 7.5 for isotropic rotational diffusion.

For several stationary random processes of interest, including especially Brownian rotational diffusion (see Eq. 7.24 in Chapter 7), the correlation functions are exponential in time (see bottom panel of Figure 5.1):

$$G(t) = G(0)\exp(-|t|/\tau_c) \tag{5.12}$$

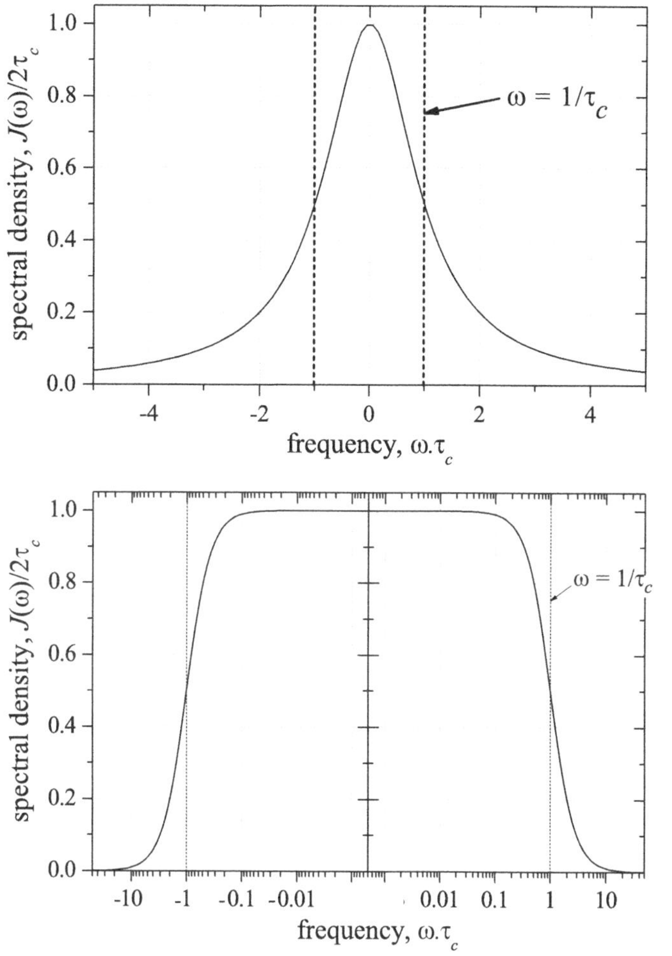

FIGURE 5.2 Angular-frequency dependence of spectral density $J(\omega)$ for exponential correlation functions (Eq. 5.12). Spectral density normalized with respect to correlation time τ_c; angular frequency scaled by τ_c. Fluctuation amplitude in Eq. 5.13 is independent of t and normalized to unity: $\langle |F(0)|^2 \rangle = 1$. Lower panel: frequency axis is logarithmic.

where $G(0) \equiv \langle |F(0)|^2 \rangle$. The corresponding spectral densities are then Lorentzian (see Table T.1 in Appendix T):

$$J(\omega) = \langle |F(0)|^2 \rangle \frac{2\tau_c}{1+\omega^2\tau_c^2} \tag{5.13}$$

where τ_c is the correlation time and angular brackets indicate an ensemble average. For rotational diffusion, $G(0) \equiv \langle |F(0)|^2 \rangle$ is an average over the angular dependence of the spin-label spectrum (cf. Sections 2.2 and 2.3 in Chapter 2).

Figure 5.2 shows the frequency dependence of the spectral density for exponential correlation functions. In this figure, we assume that the mean-square fluctuation amplitude is normalized to $\langle |F(0)|^2 \rangle = 1$. A steep loss of intensity occurs at a characteristic frequency, $\omega \approx 1/\tau_c$, that is determined by the correlation time, τ_c. For short correlation times, the spectral density is concentrated at low frequencies. For long correlation times, on the other hand, the spectral density spreads out over a wide frequency range.

5.3 SPIN HAMILTONIAN AND ANGULAR AVERAGES

For a rapidly tumbling nitroxide, the time-independent spin Hamiltonian is:

$$\mathcal{H}_o = g_o\beta_e \mathbf{B}_o \cdot \mathbf{S} + a_o \mathbf{I}\cdot\mathbf{S} \tag{5.14}$$

where $g_o = \frac{1}{3}(g_{xx}+g_{yy}+g_{zz})$ and $a_o = \frac{1}{3}(A_{xx}+A_{yy}+A_{zz})$ are the isotropic g-value and isotropic nitrogen hyperfine coupling, respectively. The orientation-dependent magnetic interactions, on the other hand, contribute time-dependent terms in $\theta(t)$ and $\phi(t)$ to the spin Hamiltonian (McConnell 1956, and see also Section B.4 in Appendix B):

$$\begin{aligned}
\mathcal{H}_1(t) &= \sum_q \mathcal{H}_1^{(q)} F^{(q)}(t) \\
&= \frac{1}{3}\left(\Delta g\beta_e B_o + \Delta A I_z\right)\left(3\cos^2\theta(t)-1\right)S_z \\
&+ \frac{1}{2}\Delta A \sin\theta(t)\cos\theta(t)\left(I_+e^{-i\phi(t)} + I_-e^{i\phi(t)}\right)S_z \\
&+ \frac{1}{2}\left(\Delta g\beta_e B_o + \Delta A I_z\right)\sin\theta(t)\cos\theta(t)\left(S_+e^{-i\phi(t)} + S_-e^{i\phi(t)}\right) \\
&+ \frac{1}{4}\Delta A \sin^2\theta(t)\left(I_+S_+e^{-i2\phi(t)} + I_-S_-e^{i2\phi(t)}\right) \\
&- \frac{1}{12}\Delta A\left(3\cos^2\theta(t)-1\right)\left(I_+S_- + I_-S_+\right)
\end{aligned} \tag{5.15}$$

where we assume axial symmetry, with $\Delta g \equiv g_\parallel - g_\perp$ and $\Delta A \equiv A_\parallel - A_\perp$. The direction of the static magnetic field, $\mathbf{B}_o$, specifies the z-quantization axis in Eq. 5.15. We write off-diagonal terms using shift operators (see Appendix E): $S_\pm \equiv S_x \pm iS_y$ and $I_\pm \equiv I_x \pm iI_y$ for electron and nuclear spins, respectively, and the ϕ-dependence is written as $\exp(\pm in\phi) = \cos n\phi \pm i\sin n\phi$. The corresponding perturbation Hamiltonian for non-axial tensors is given by Kivelson (1971), Wilson and Kivelson (1966).

The polar angles $\theta(t)$ and $\phi(t)$ in Eq. 5.15 specify the direction of the principal axis (‖) of the A- and g-tensor (see Figure 5.3). These angles vary randomly, as the spin label tumbles by rotational diffusion. Note that the angular terms in Eq. 5.15 appear as second-rank spherical harmonics, $Y_{2,M}(\theta,\phi)$ (see Appendix V.2). From Eq. 5.15, the angular ensemble averages that we need for calculating $\langle |F(0)|^2 \rangle$ in Eq. 5.13 are:

$$\langle |F^{(0)}(0)|^2 \rangle \equiv \langle (3\cos^2\theta - 1)^2 \rangle$$

$$= \frac{\int_0^\pi\int_0^{2\pi} (3\cos^2\theta-1)^2 \sin\theta\cdot d\theta\cdot d\phi}{\int_0^\pi\int_0^{2\pi} \sin\theta\cdot d\theta\cdot d\phi} = \frac{4}{5} \tag{5.16}$$

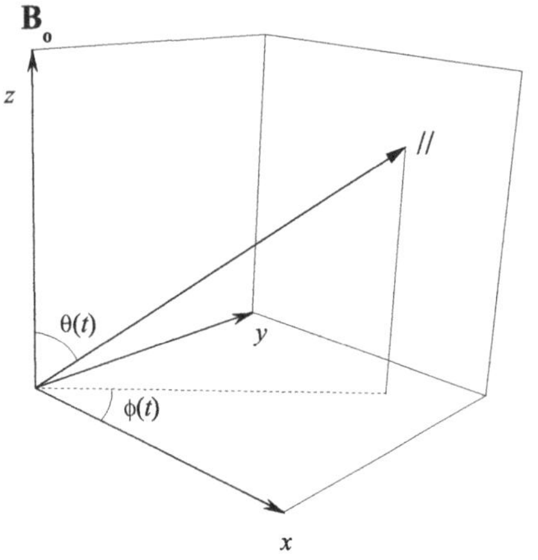

FIGURE 5.3 Orientation of nitroxide magnetic symmetry axis (||), relative to static magnetic field $\mathbf{B}_0$ that specifies z-axis of spin quantization. In Brownian rotational diffusion, locus of nitroxide ||-axis performs random walk on the surface of a sphere.

$$\left\langle \left| F^{(1)}(0) \right|^2 \right\rangle \equiv \left\langle \sin^2\theta \cos^2\theta \right\rangle = \frac{2}{15} \tag{5.17}$$

$$\left\langle \left| F^{(2)}(0) \right|^2 \right\rangle \equiv \left\langle \sin^4\theta \right\rangle = \frac{8}{15} \tag{5.18}$$

for unrestricted Brownian diffusion (see Figure 5.3). These are directly proportional to averages of the squares of the spherical harmonics $Y_{2,0}$, $Y_{2,1}$, $Y_{2,2}$, respectively, i.e., $F^{(M)}(0) \propto Y_{2,M}$. Cross terms in the ensemble averages are zero, because spherical harmonics are orthogonal, i.e., $\iint Y^*_{L,M} Y_{L',M'} \sin\theta \cdot d\theta \cdot d\phi = 0$, unless $L = L'$ and $M = M'$ (see Appendix V.2).

5.4 TRANSITION PROBABILITIES FOR A NITROXIDE

Equation 5.8 in Section 5.2 gives the rate for a particular two-level transition. Figure 5.4 shows the possible transitions for a ^{14}N-nitroxide ($I = 1$), where the three hyperfine levels are labelled by the magnetic quantum number $m_I = 0, \pm 1$. The probability of electron-spin transitions is W_e, that of nuclear-spin transitions is W_n and that of cross relaxation is W_{x1} and W_{x2}. We know from Section 1.6 in Chapter 1 that the spin–lattice relaxation time $T_{1,e}$, for the two-level $S = \frac{1}{2}$ system, depends on the rate of transition in both directions, viz. $T_{1,e} \equiv 1/\left(W_\uparrow + W_\downarrow\right) = 1/(2W_e)$ (see Eq. 1.17). For a spin $I = \frac{1}{2}$ (^{15}N), the nuclear relaxation time is correspondingly related to the rate of backward and forward nuclear transitions by: $T_{1,n} = 1/(2W_n)$. Spin–lattice relaxation is more complicated for three-level systems, however. In Section 12.3, we see that the nuclear spin–lattice relaxation time for ^{14}N-nitroxides with $I = 1$ is best defined by $T_{1,n} = 1/W_n$ (see Figure 12.3). This is the definition that we shall adhere to here for ^{14}N-nitroxides, whilst retaining the usual definition for ^{15}N-nitroxides because the latter is correct for $I = \frac{1}{2}$ nuclei. Note, however, that it is the rate constant W_n, and not the nuclear spin–lattice relaxation time, that is relevant to EPR (see Chapters 10 and 12).

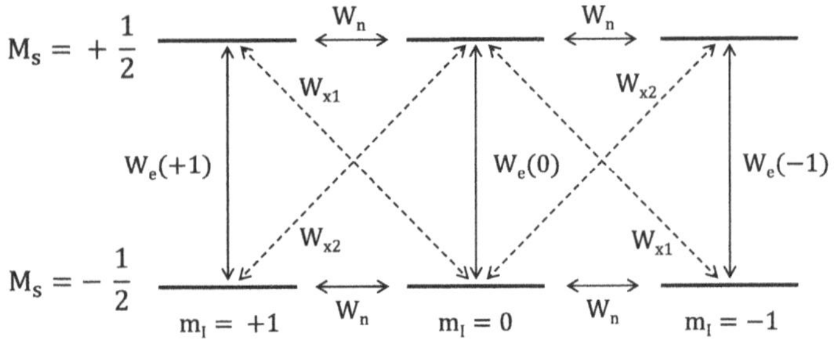

FIGURE 5.4 Energy levels and transitions for ^{14}N-nitroxide $\left(S = \frac{1}{2}, I = 1\right)$ in a magnetic field. Spin states labelled by electron and nuclear magnetic quantum numbers, M_S and m_I, respectively. Transition probabilities for leaving a given state: W_e for electron transitions (Eq. 5.19); W_n for nuclear transitions (Eq. 5.20); W_x for cross relaxation (Eqs. 5.21, 5.22).

To calculate the relaxation rates W_e, W_n, W_{x1} and W_{x2}, we must take into account the various contributions from the different terms in the perturbation Hamiltonian. As is common in spin-label EPR, we now define compound spectral densities $j^{AA}(\omega)$, $j^{gg}(\omega)$ and $j^{gA}(\omega)$ to include the coefficients that pre-multiply the products of spin operators (specifically, $S_\pm I_z$), or products of spin operator and magnetic field (specifically, $S_\pm B_o$), in Eq. 5.15. This results in more compact expressions, but at some cost to transparency, as we shall try to explain. Derivations in terms of conventional spectral densities are given for transverse relaxation in Sections O.3 and O.6 in Appendix O, at the end of the book.

5.4.1 Electron Spin Relaxation $\left(\Delta M_S = \pm 1, \Delta m_I = 0\right)$

Electron spin transitions are induced by the non-secular terms involving the hyperfine and Zeeman operators $S_\pm I_z$ and $S_\pm B_o$ that appear in the third line on the right-hand side of Eq. 5.15. The matrix elements that we need for the transition probability in Eq. 5.8 are $\left|\left\langle \pm\frac{1}{2}, m_I \middle| S_\pm \middle| \mp\frac{1}{2}, m_I \right\rangle\right|^2 = 1$, $\left|\left\langle \pm\frac{1}{2}, m_I \middle| S_\pm I_z \middle| \mp\frac{1}{2}, m_I \right\rangle\right|^2 = m_I^2$, etc. (see Appendix E).

In terms of the modified spectral densities, $j(\omega)$, the electron-spin transition probabilities are then (Freed 1979):

$$W_{e,m_I} \equiv W\left(\mp\tfrac{1}{2}, m_I \to \pm\tfrac{1}{2}, m_I\right)$$
$$= 2j^{gg}(\omega_e)B_o^2 + 4j^{gA}(\omega_e)B_o m_I + 2j^{AA}(\omega_e)m_I^2 \tag{5.19}$$

where j^{gg} and j^{AA} arise respectively from the Zeeman and hyperfine terms alone in Eq. 5.15, and $2j^{gA}$ is from the cross term. The electron Larmor frequency is $\omega_e \cong g_o\beta_e B_o/\hbar$. Note that the common factor of two on the right-hand side of Eq. 5.19 comes from the definition of $j(\omega)$, which equals half $J(\omega)$ (cf. Eq. 5.9 and Section 5.2).

In the following, instead of defining more spectral densities, we express all further transition probabilities in terms of j^{AA} as defined by Eq. 5.19.

5.4.2 Nuclear Spin Relaxation $(\Delta M_S = 0, \Delta m_I = \pm 1)$

Nuclear spin transitions are induced by the pseudo-secular terms involving the hyperfine operator $S_z I_\pm$ that appears in the second line on the right-hand side of Eq. 5.15. The required matrix element is $\left|\langle \pm\frac{1}{2}, m_I \pm 1 | S_z I_\pm | \pm\frac{1}{2}, m_I \rangle\right|^2 = \frac{1}{4}[I(I+1) - m_I(m_I \pm 1)]$. This has the value $\frac{1}{2}$ for a ^{14}N-nitroxide ($I = 1$) and $\frac{1}{4}$ for a ^{15}N-nitroxide $(I = \frac{1}{2})$.

In terms of the modified spectral density for the hyperfine anisotropy $j^{AA}(\omega)$ defined by Eq. 5.19, the nuclear spin transition probabilities are then (Freed 1979):

$$W_n(I) \equiv W\left(\pm\tfrac{1}{2}, m_I \to \pm\tfrac{1}{2}, m_I \pm 1\right)$$
$$= \tfrac{1}{2} j^{AA}(\omega_a)[I(I+1) - m_I(m_I \pm 1)] \quad (5.20)$$

where $\omega_a = \frac{1}{2}a_o$ (in angular frequency units). The nuclear Larmor frequency, ω_n, is small compared with a_o, and we neglect it. The nuclear transition probability designated W_n in Figure 5.4 for a ^{14}N-nitroxide is given from Eq. 5.20 by: $W_n \equiv j^{AA}(\omega_a)$. With a similar definition, the nuclear transition probability for a ^{15}N-nitroxide is correspondingly $W_n \equiv \frac{1}{2} j^{AA}(\omega_a)$.

Note that a different definition of W_n was used for ^{14}N-nitroxides in the CW-ELDOR literature (Hyde et al. 1968) and taken over for analysis of progressive saturation in CW-EPR (Marsh 1992). There, W_n was reserved for the nuclear relaxation rate in $I = \frac{1}{2}$ spin systems, and therefore the rate in $I = 1$ spin systems, specifically ^{14}N-nitroxides, became $2W_n$. In this book, we adhere to the definition given by Eq. 5.20, i.e., W_n is the polarization transfer rate for the system under consideration, irrespective of the nuclear spin. This accords with the definition used in the saturation-recovery ELDOR literature (see Chapter 12), which now is the main source of data on nitrogen nuclear relaxation in spin labels.

5.4.3 Cross Relaxation $(\Delta M_S = \pm 1, \Delta m_I = \pm 1 \text{ or } \mp 1)$

Combined electron and nuclear spin transitions are induced by the non-secular terms involving the hyperfine operators $S_\pm I_\pm$ and $S_\pm I_\mp$ that appear in the fourth and fifth lines, respectively, on the right-hand side of Eq. 5.15. We refer to this as cross relaxation, following the definition used in Freed (1979) and Dalton et al. (1976). The matrix elements that we need are: $\left|\langle \pm\frac{1}{2}, m_I \pm 1 | S_\pm I_\pm | \mp\frac{1}{2}, m_I \rangle\right|^2 = I(I+1) - m_I(m_I \pm 1)$ and $\left|\langle \pm\frac{1}{2}, m_I \mp 1 | S_\pm I_\mp | \mp\frac{1}{2}, m_I \rangle\right|^2 = I(I+1) - m_I(m_I \mp 1)$. Both have the value 2 for a ^{14}N-nitroxide ($I = 1$) and 1 for a ^{15}N-nitroxide $(I = \frac{1}{2})$.

In terms of the spectral density for the hyperfine anisotropy, $j^{AA}(\omega)$, the combined electron- and nuclear-spin transition probabilities then become (Freed 1979):

$$W_{x1} \equiv W\left(\mp\tfrac{1}{2}, m_I \to \pm\tfrac{1}{2}, m_I \pm 1\right)$$
$$= 2 j^{AA}(\omega_e)[I(I+1) - m_I(m_I \pm 1)] \quad (5.21)$$

and

$$W_{x2} \equiv W\left(\mp\tfrac{1}{2}, m_I \to \pm\tfrac{1}{2}, m_I \mp 1\right)$$
$$= \tfrac{1}{3} j^{AA}(\omega_e)[I(I+1) - m_I(m_I \mp 1)] \quad (5.22)$$

Note that $j^{AA}(\omega)$ defined in Eq. 5.19 depends on the average $\frac{1}{4}\langle \sin^2\theta \cos^2\theta \rangle$, whereas the spectral densities appropriate to Eqs. 5.21 and 5.22 depend on $\frac{1}{16}\langle \sin^4\theta \rangle$ and $\frac{1}{144}\langle (3\cos^2\theta - 1)^2 \rangle$, respectively (see Eq. 5.15). To correct for this, $2j^{AA}(\omega)$ must be scaled by factors of 1 and 1/6 in Eqs. 5.21 and 5.22, respectively, as we see from Eqs. 5.16 to 5.18. Here, we neglect the hyperfine and nuclear Larmor frequencies, in comparison with the electron Larmor frequency, ω_e. The transition probabilities for cross relaxation in Figure 5.4 are: $W_{x1} = 4j^{AA}(\omega_e)$ and $W_{x2} = \frac{2}{3} j^{AA}(\omega_e)$ for a ^{14}N-nitroxide. For a ^{15}N-nitroxide, they are instead: $W_{x1} = 2j^{AA}(\omega_e)$ and $W_{x2} = \frac{1}{3} j^{AA}(\omega_e)$. For the electron–nuclear dipolar (END) mechanism assumed here, the two cross-relaxation rates are related by: $W_{x1} = 6W_{x2}$.

5.5 SPECTRAL DENSITIES FOR A NITROXIDE

For Brownian rotational diffusion, the correlation functions are exponential in time (see Section 7.5 in Chapter 7). We assume isotropic rotation for simplicity. The spectral density is then given by Eq. 5.13, where the ensemble averages $\langle |F(0)|^2 \rangle$ are given by the angular averages in Eqs. 5.16–5.18. (This equivalence between time average and ensemble average is demonstrated explicitly by the autocorrelation function given in Eq. 7.24 of Chapter 7.) The rotational correlation time is: $\tau_R = 1/(6D_R)$, where D_R is the rotational diffusion coefficient (see Eq. 7.22 for $L = 2$). Generalizations for axially anisotropic rotational diffusion are given in the Appendix to this chapter, and see also Section 7.6 in Chapter 7.

5.5.1 Hyperfine Anisotropy (END)

Anisotropy of the hyperfine interaction arises from magnetic interaction between the electron and nuclear spin dipoles (see Section 3.4 in Chapter 3), and conventionally is referred to as the electron-nuclear dipolar (END) interaction. From Eqs. 5.13, 5.15

and 5.17, the spectral density for the END mechanism therefore becomes (Freed 1965; Freed 1979):

$$j^{AA}(\omega)=\frac{1}{4}\left[(\Delta A)^2+3(\delta A)^2\right]\cdot\langle\sin^2\theta\cos^2\theta\rangle\frac{\tau_R}{1+\omega^2\tau_R^2}$$
$$=\frac{1}{30}\left[(\Delta A)^2+3(\delta A)^2\right]\cdot\frac{\tau_R}{1+\omega^2\tau_R^2} \quad (5.23)$$

We define the hyperfine anisotropies by:

$$\Delta A=A_{zz}-\tfrac{1}{2}\left(A_{xx}+A_{yy}\right) \quad (5.24)$$

and

$$\delta A=\tfrac{1}{2}\left(A_{xx}-A_{yy}\right) \quad (5.25)$$

where the hyperfine tensor (A_{xx},A_{yy},A_{zz}) is in angular frequency units. The axial approximation that we assumed in Eq. 5.15 corresponds to $\delta A=0$. Note that this spectral density contributes terms to the transition probabilities that are independent of the magnetic field strength (see Eq. 5.19).

For the special case of isotropic rotation, we can express Eq. 5.23 in more compact form by using the identity $\sum(A_{ii}-a_o)^2=\tfrac{2}{3}\left((\Delta A)^2+3(\delta A)^2\right)$:

$$j^{AA}(\omega)=\frac{1}{20}\sum_{i=x,y,z}(A_{ii}-a_o)^2\cdot\frac{\tau_R}{1+\omega^2\tau_R^2} \quad (5.26)$$

which emphasizes that it is the square of the amplitude of fluctuation in hyperfine splitting that determines the transition probability.

5.5.2 *g*-Value Anisotropy (Zeeman Interaction)

Anisotropy of the electron Zeeman interaction arises from *g*-tensor anisotropy. From Eqs. 5.13, 5.15 and 5.17, the spectral density then becomes (Freed 1965; Freed 1979):

$$j^{gg}(\omega)=\frac{1}{30}\left[(\Delta g)^2+3(\delta g)^2\right]\cdot\left(\frac{\beta_e}{\hbar}\right)^2\frac{\tau_R}{1+\omega^2\tau_R^2} \quad (5.27)$$

We define the *g*-value anisotropies by:

$$\Delta g=g_{zz}-\tfrac{1}{2}\left(g_{xx}+g_{yy}\right) \quad (5.28)$$

and

$$\delta g=\tfrac{1}{2}\left(g_{xx}-g_{yy}\right) \quad (5.29)$$

Again, the axial approximation assumed in Eq. 5.15 corresponds to $\delta g=0$. Note that this spectral density contributes a term to the transition probability that is proportional to $B_o{}^2$, the static magnetic field strength squared (see Eq. 5.19).

For the special case of isotropic rotation, we can express Eq. 5.27 in the compact form:

$$j^{gg}(\omega)=\frac{1}{20}\sum_{i=x,y,z}(g_{ii}-g_o)^2\left(\frac{\beta_e}{\hbar}\right)^2\frac{\tau_R}{1+\omega^2\tau_R^2} \quad (5.30)$$

which emphasizes that the square of the amplitude of fluctuation in Zeeman splitting determines the transition probability.

5.5.3 END-Zeeman Cross Term

With the same notation as above, the spectral density that arises from the cross term between the hyperfine and *g*-value anisotropies is (Freed 1965; Freed 1979):

$$j^{gA}(\omega)=\frac{1}{30}\left[\Delta g\Delta A+3\delta g\delta A\right]\cdot\left(\frac{\beta_e}{\hbar}\right)\frac{\tau_R}{1+\omega^2\tau_R^2} \quad (5.31)$$

Again this comes from Eqs. 5.13, 5.17 and 5.15, where the latter corresponds to the axial approximation. This spectral density contributes a term to the transition probability that is directly proportional to the strength of the static magnetic field, B_o (see Eq. 5.19).

For the special case of isotropic rotation, we can express Eq. 5.32 in the form:

$$j^{gA}(\omega)=\frac{1}{20}\sum_{i=x,y,z}(A_{ii}-a_o)(g_{ii}-g_o)\left(\frac{\beta_e}{\hbar}\right)\frac{\tau_R}{1+\omega^2\tau_R^2} \quad (5.32)$$

which emphasizes that the cross product of fluctuation amplitudes in Zeeman and hyperfine splittings determines the corresponding transition probability.

5.5.4 Spin-Rotation Interaction

There is an additional mechanism that contributes a transition probability which is independent of the magnetic field strength, but in this case also is independent of the nuclear spin. This is the spin-rotation interaction, which arises from interaction of the electron spin with a transient magnetic moment generated by inertial rotation of the spin-labelled molecule (Atkins and Kivelson 1966). Spin-rotation interaction is not included in the spin Hamiltonian given by Eqs. 5.14 and 5.15. It makes the additional contribution:

$$\mathcal{H}_{SR}=\mathbf{J}\cdot\mathbf{C}\cdot\mathbf{S} \quad (5.33)$$

where $\mathbf{J}\hbar$ is the angular momentum of the rotating molecule and $\mathbf{C}$ is the spin-rotation coupling tensor. The spin-rotation interaction causes relaxation because the angular velocity of the molecule, $\boldsymbol{\omega}_R$, is modulated by random molecular collisions. (The angular momentum is given by: $\mathbf{J}\hbar=\mathbf{I}^{inert}\cdot\boldsymbol{\omega}_\mathbf{R}$, where $\mathbf{I}^{inert}$

is the moment-of-inertia tensor of the molecule.) For Brownian dynamics, the correlation functions of the angular momentum components are exponential (Hubbard 1963; and see also Eq. 7.69 in Section 7.12):

$$\langle J_i(t)J_i(0)\rangle = \left(k_B T I_{ii}^{inert}/\hbar^2\right)\exp\left(-t/\tau_{J,i}\right) \tag{5.34}$$

where $i \equiv x, y, z$, and I_{ii}^{inert} are the principal elements of the inertial tensor in the molecular axis system. The correlation time for spin-rotation coupling in Eq. 5.34 is related to the correlation time, $\tau_{R,i}$, for Brownian rotational diffusion by (Hubbard 1963):

$$\tau_{J,i} \equiv I_{ii}^{inert}/f_{R,i} = I_{ii}^{inert}/\left(6k_B T\tau_{R,i}\right) \tag{5.35}$$

where $i \equiv x, y, z$; and $f_{R,i}$ are the components of the rotational friction coefficient that enter into the Langevin equation (Eq. 7.65 in Section 7.12) and are given by: $f_{R,i} \equiv k_B T/D_{R,i} = 6k_B T\tau_{R,i}$ (see Sections 7.3 and 7.5: Eqs. 7.9 and 7.22). We assume here that the principal axes of the inertial and friction tensors are collinear. The inverse relation between $\tau_{J,i}$ and $\tau_{R,i}$ in Eq. 5.35 reflects the fact that rotation is slowed down in viscous media, whereas the angular velocity is decelerated more rapidly.

For anisotropic rotation, the spectral density corresponding to the correlation functions of Eq. 5.34 takes the form (McClung and Kivelson 1968; Goldman et al. 1973):

$$j^{SJ}(\omega) = -\frac{2k_B T}{3\hbar^2}\sum_{i=x,y,z} C_{ii}^2 I_{ii}^{inert}\frac{\tau_{J,i}}{1+\omega^2\tau_{J,i}^2} \tag{5.36}$$

where C_{ii} are the principal elements of the spin-rotation coupling tensor in the molecular axis system. The angular momentum $\mathbf{J}$ in Eq. 5.33 is actually the sum of the rigid-rotator momentum that is determined by the moment of inertia, plus the residual orbital angular momentum, $\mathbf{L}$, of the unpaired electron (cf. Section 3.12). A perturbation treatment involving the spin-orbit coupling, similar to that leading to Eq. 3.65 for the effective g-tensor (cf. Appendix L), also can be applied to the spin-rotation coupling (Atkins and Kivelson 1966). Consequently, the spin-rotation coupling tensor is directly proportional to the g-tensor, having principal elements (Nyberg 1967):

$$C_{ii} = -\hbar\left(g_{ii} - g_e\right)/I_{ii}^{inert} \tag{5.37}$$

where g_e (=2.0023193) is the g-value of the free electron. For simplicity, we assume that the axes of the g-tensor and inertial tensor are collinear. Substituting Eq. 5.37 in Eq. 5.36 then gives the following expression for the spectral density:

$$j^{SJ}(\omega) = -\frac{2}{3}k_B T\sum_{i=x,y,z}\frac{\left(g_{ii}-g_e\right)^2}{I_{ii}^{inert}}\frac{\tau_{J,i}}{1+\omega^2\tau_{J,i}^2} \tag{5.38}$$

From Eq. 5.35 we find that $\tau_{J,i} << \tau_{R,i}$ and consequently that $\omega_e^2\tau_{J,i}^2 << 1$, i.e., that spin-rotation coupling corresponds to extreme motional narrowing. Therefore using Eqs. 5.35 and 5.38, we can write the spectral density in terms of the correlation times for uniaxial rotation as:

$$j^{SJ}(\omega) = -\frac{1}{9}\left(\frac{\left(g_{zz}-g_e\right)^2}{\tau_{R\parallel}} + \frac{\left(g_{xx}-g_e\right)^2 + \left(g_{yy}-g_e\right)^2}{\tau_{R\perp}}\right) \tag{5.39}$$

where $\tau_{R\parallel} = 1/(6D_{R\parallel})$ and $\tau_{R\perp} = 1/(6D_{R\perp})$ for rotation around and perpendicular to the nitroxide z-axis, respectively (see Section 7.6). The spectral density for isotropic spin-rotation interaction is then simply:

$$j^{SJ}(\omega) = -\frac{1}{9\tau_R}\sum_{i=x,y,z}\left(g_{ii}-g_e\right)^2 \tag{5.40}$$

where τ_R is the correlation time for isotropic rotation. Because of the extreme motional narrowing condition (i.e., j^{SJ} is independent of ω), longitudinal and transverse relaxation times are equal: $T_1^{SJ} = T_2^{SJ}$, for relaxation by the spin-rotation interaction.

5.6 SPIN–LATTICE RELAXATION RATES

As explained at the beginning of Section 5.4, the electron transition probabilities, W_e, for spin–lattice relaxation are related directly to the relaxation time, T_{1e}, by: $2W_e = 1/T_{1e}$ (see also Chapter 1). With the definition of W_n given in Section 5.4, a similar relation holds for nuclear relaxation of ^{15}N-nitroxides, because this is also a two-level $I = \frac{1}{2}$ spin system. Although often assumed in the literature, this is not true however for the three-level $I = 1$ nuclear spin system of ^{14}N-nitroxides. As explained in Section 12.3 of Chapter 12, the nuclear spin–lattice relaxation time of ^{14}N-nitroxides is better described by $W_n = 1/T_{1n}$, where W_n is defined by Eq. 5.20. This is the definition that we adhere to in this book. Important in EPR, however, are the rate constants W_n and W_x and not the relaxation times themselves. As we see from Eqs. 5.20 to 5.22, the relation between transition probabilities and spectral densities differs by a factor of a half for ^{15}N-nitroxides, compared with ^{14}N-nitroxides, but is identical for the relaxation times.

5.6.1 Electron Spin–Lattice Relaxation

From Eqs. 5.19, 5.23, 5.27, 5.31 and 5.40, the electron spin–lattice relaxation rate is given by:

$$W_e(m_I) \equiv \frac{1}{2T_{1,e}(m_I)} = \frac{1}{18}\sum_{i=x,y,z}\frac{\left(g_{ii}-g_e\right)^2}{\tau_R}$$
$$+\frac{1}{15}\left[\left(\left(\Delta g\right)^2 + 3\left(\delta g\right)^2\right)\cdot\left(\frac{\beta_e}{\hbar}\right)^2 B_o^2 + 2\left(\Delta g\Delta A + 3\delta g\delta A\right)\right.$$
$$\left.\cdot\left(\frac{\beta_e}{\hbar}\right)B_o m_I + \left(\left(\Delta A\right)^2 + 3\left(\delta A\right)^2\right)m_I^2\right]\frac{\tau_R}{1+\omega_e^2\tau_R^2} \tag{5.41}$$

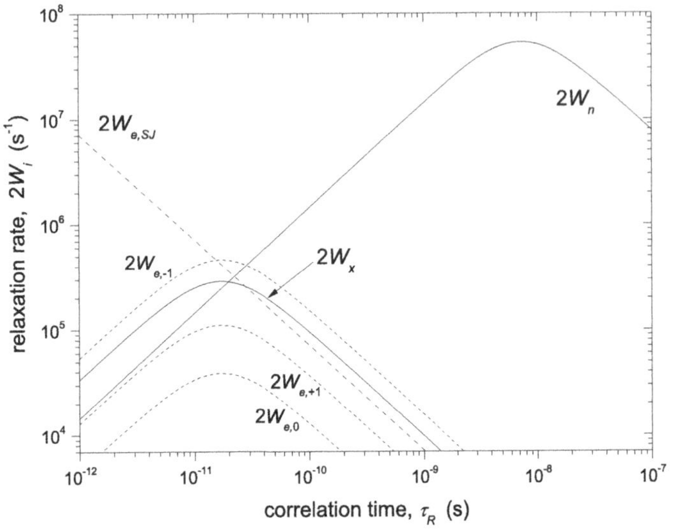

FIGURE 5.5 Contributions to spin–lattice and cross relaxation rates $2W_i$ of ^{14}N-nitroxide, as function of correlation time τ_R for isotropic rotational diffusion. Electron spin–lattice relaxation from g- and hyperfine-tensor anisotropies $(2W_{e,m_I})$, and spin-rotation $(2W_{e,SJ})$, come from Eq. 5.41; mean cross-relaxation rate $(2W_x = W_{x1} + W_{x2})$ from Eqs. 5.43, 5.44; nuclear spin–lattice relaxation rate $(2W_n)$ from Eq. 5.42. Spin-Hamiltonian parameters for TEMPOL from Percival et al. (1975); EPR frequency 9.3 GHz.

for each of the nitrogen hyperfine lines, m_I. The predicted dependence of relaxation rate on rotational correlation time is shown in Figure 5.5. Contributions from modulation of the hyperfine- and g-tensors (light dashed lines) are given separately for the different hyperfine lines $m_I = 0, \pm 1$ of a ^{14}N-nitroxide. All have the same biphasic dependence on correlation time, with a characteristic maximum at $\tau_R = 1/\omega_e$, which corresponds to $\tau_R = 1.7 \times 10^{-11}$ s for an EPR frequency of 9.3 GHz. The contribution from spin-rotation does not depend on m_I, nor on EPR frequency, and is given by the heavy dashed line. It dominates for very fast rotation and decreases monotonically with increasing τ_R. Comparison with experimental data from saturation-recovery EPR is given later in Chapter 12.

5.6.2 Nuclear Spin–Lattice Relaxation

From Eqs. 5.20 and 5.23, the nitrogen nuclear relaxation rate for ^{14}N-nitroxides is given by:

$$W_n \equiv \frac{1}{T_{1,n}} = \frac{1}{30}\left[(\Delta A)^2 + 3(\delta A)^2\right]\frac{\tau_R}{1+\omega_a^2\tau_R^2} \tag{5.42}$$

However, for ^{15}N-nitroxides the right-hand side of Eq. 5.42 gives twice the relaxation rate: $2W_n \equiv 1/T_{1n}$ for $I = \frac{1}{2}$ (see Section 5.4). The ratio of nitrogen nuclear relaxation times for ^{15}N- and ^{14}N-nitroxides is therefore: $T_{1n}(^{15}\mathrm{N})/T_{1n}(^{14}\mathrm{N}) = (\gamma_N(^{14}\mathrm{N})/\gamma_N(^{15}\mathrm{N}))^2 = 0.5082$ for a pure END mechanism, where γ_N is the nitrogen nuclear gyromagnetic ratio and $\omega_a^2\tau_R^2 << 1$. We show the dependence of $2W_n$ on rotational correlation time by a solid line in Figure 5.5, with a maximum at $\tau_R = 1/\omega_a$ that corresponds to $\tau_R = 7.4 \times 10^{-9}$ s for a ^{14}N-nitroxide.

5.6.3 Cross Relaxation

From Eqs. 5.21 and 5.23, the W_{x1} cross-relaxation rate for ^{14}N-nitroxides is given by:

$$W_{x1} = \frac{2}{15}\left[(\Delta A)^2 + 3(\delta A)^2\right]\frac{\tau_R}{1+\omega_e^2\tau_R^2} \tag{5.43}$$

From Eqs. 5.21 and 5.22, the W_{x2} cross-relaxation rate is given by:

$$W_{x2} = \frac{1}{6}W_{x1} \tag{5.44}$$

where W_{x1} is given by Eq. 5.43, for ^{14}N-nitroxides. For ^{15}N-nitroxides, however, the right-hand side of Eq. 5.43 gives twice the relaxation rate: $2W_{x1}$, and correspondingly Eq. 5.44 then gives $2W_{x2}$ (see Section 5.4).

From Eqs. 5.42 and 5.43, we see that the cross-relaxation rate is related to the nuclear relaxation rate by: $W_{x1} = 4W_n(1+\omega_a^2\tau_R^2)/(1+\omega_e^2\tau_R^2)$. Only for very fast motion, when $\omega_e^2\tau_R^2 < 1$, does cross-relaxation become comparable in rate to nuclear relaxation. However, cross-relaxation rates are comparable to the pure END contribution to electron relaxation. From Eqs. 5.41 and 5.43, we find that $W_{x1} = 2W_e(END)$ for the $m_I = \pm 1$ manifolds of ^{14}N-nitroxides, and $W_{x1} = 4W_e(END)$ for ^{15}N-nitroxides. Therefore, we should take cross relaxation into account when analysing electron spin–lattice relaxation of spin labels from saturation-recovery (SR) measurements (Robinson et al. 1994; Robinson et al. 1999). This is clear from Figure 5.5, which shows the dependence of the mean cross-relaxation rate $2W_x = W_{x1} + W_{x2}$ on rotational correlation time as a solid line. As we shall see later in Chapter 12, it is the mean value $2W_x$ that contributes to recovery rates in SR-EPR.

Equations for spin–lattice relaxation rates (and cross-relaxation rates) when rotational diffusion is axially anisotropic are given in the Appendix at the end of the chapter.

5.7 SECULAR, PSEUDOSECULAR AND NON-SECULAR TRANSVERSE RELAXATION

Because of the uncertainty principle, the non-secular and pseudosecular transitions considered in Section 5.4 cause a lifetime broadening, and hence contribute to transverse relaxation, i.e., to T_2. However, there is a more important contribution to line broadening. This arises from modulation of the energies of the spin states between which the EPR transitions take place. Such modulation smears out the transition frequencies and therefore broadens the spectral lines. It comes from the secular terms ($\Delta M_S = 0$, $\Delta m_I = 0$) containing S_z and I_z alone, which occur in the first line on the right-hand side of the time-dependent spin Hamiltonian, $\mathcal{H}_1 F(t)$, in Eq. 5.15 and are associated with the spectral density, $j(0)$, at zero frequency. These low-frequency spectral densities do not contribute to spin–lattice relaxation.

We see the effects of secular T_2-relaxation readily by imposing a small-amplitude random fluctuation $\Delta B_{o,fl}$ on the static magnetic field B_o. This causes random fluctuations $\Delta\omega_{o,fl}$ in the electron Larmor frequency ω_o. Concentrating only on the fluctuation, the equation of motion for the transverse spin magnetization $\widehat{M}_+ = \widehat{M}_x + i\widehat{M}_y$ in the rotating frame is simply (see, e.g., Eq. 6.7 of Chapter 6):

$$\frac{\mathrm{d}\widehat{M}_+}{\mathrm{d}t} = -i\Delta\omega_{o,fl}\widehat{M}_+ \tag{5.45}$$

Formal integration gives the solution:

$$\widehat{M}_+(t) = \widehat{M}_+(0) - i\int_0^t \Delta\omega_{o,fl}(t')\widehat{M}_+(t')\mathrm{d}t' \tag{5.46}$$

We get a first-order solution by substituting $\widehat{M}_+(0)$ for $\widehat{M}_+(t')$ in the integral, which then vanishes because the fluctuations $\Delta\omega_{o,fl}$ are random. Therefore, we must proceed to second order by substituting the first-order solution for $\widehat{M}_+(t')$ into the integral, giving:

$$\widehat{M}_+(t) = \widehat{M}_+(0) - \widehat{M}_+(0)\int_0^t \mathrm{d}t'\Delta\omega_{o,fl}(t')\int_0^{t'} \mathrm{d}t''\Delta\omega_{o,fl}(t'') \tag{5.47}$$

Taking the time derivative and making a change of variable then gives us:

$$\frac{\mathrm{d}\widehat{M}_+}{\mathrm{d}t} = -\widehat{M}_+(0)\int_0^t \Delta\omega_{o,fl}(t)\Delta\omega_{o,fl}(t-\tau)\,\mathrm{d}\tau \cong -\widehat{M}_+(t)\left\langle \Delta\omega_{o,fl}^2\right\rangle j(0) \tag{5.48}$$

where the spectral density $j(0)$ for a stationary fluctuation is defined by: $\left\langle \Delta\omega_{o,fl}^2\right\rangle j(0) = \int_0^\infty \Delta\omega_{o,fl}(t)\Delta\omega_{o,fl}(t-\tau)\,\mathrm{d}\tau$. Equation 5.48 predicts exponential relaxation of $\widehat{M}_+(t)$ at the secular rate:

$$\frac{1}{T_2^{\mathrm{sec}}} = \left\langle \Delta\omega_{o,fl}^2\right\rangle j(0) \tag{5.49}$$

where $\left\langle \Delta\omega_{o,fl}^2\right\rangle$ is the second moment of the fluctuation in Larmor frequency (see Appendix U).

Generalizing Eq. 5.49, secular contributions to the linewidth come from the diagonal matrix elements of the perturbation Hamiltonian, $\mathcal{H}_1 F(t)$, and are of the form (Abragam 1961, Chapter X):

$$\begin{aligned}\frac{1}{T_2^{\mathrm{sec}}} &= \frac{1}{\hbar^2}\left\langle\left(\left\langle +\tfrac{1}{2},m_I\right|\mathcal{H}_1\left|+\tfrac{1}{2},m_I\right\rangle - \left\langle -\tfrac{1}{2},m_I\right|\mathcal{H}_1\left|-\tfrac{1}{2},m_I\right\rangle\right)^2\right\rangle j(0)\\ &= \frac{1}{\hbar^2}\left\langle\left|\left\langle m_I\right|\mathcal{H}_1^o\left|m_I\right\rangle\right|^2\right\rangle j(0)\end{aligned} \tag{5.50}$$

where $\mathcal{H}_1 \equiv \mathcal{H}_1^o S_z$. This secular relaxation rate corresponds to the familiar approximation for the motionally narrowed linewidth: $\Delta\Delta\omega_{m_I} \approx \left\langle \Delta\omega_{m_I}^2\right\rangle\tau_R$ that we give in Eq. 1.9 of Chapter 1 (cf. also Eq. 5.49).

We now proceed further and add transverse components $\Delta B_{+,fl}$ to the randomly fluctuating field in our illustrative model. These induce spin–lattice relaxation, but they also contribute to transverse relaxation by the lifetime broadening mechanism. Simultaneous solution of the equations of motion for both transverse and longitudinal spin magnetization then leads to a net transverse relaxation rate that is given by (see, e.g., Carrington and McLachlan 1969):

$$\frac{1}{T_2} = \left\langle \Delta\omega_{o,fl}^2\right\rangle j(0) + \tfrac{1}{2}\left\langle \Delta\omega_{+,fl}^2\right\rangle j(\omega_o) = \frac{1}{T_2^{\mathrm{sec}}} + \frac{1}{2T_1} \tag{5.51}$$

where $\left\langle \Delta\omega_{+,fl}^2\right\rangle \equiv \gamma_e^2\left\langle \Delta B_{+,fl}^2\right\rangle$ and $1/T_1$ is the rate of spin–lattice relaxation induced by random fluctuations of the transverse magnetic field. As we saw in Sections 5.2 and 5.6, spin–lattice relaxation depends on the probability W_{mk} of the transition from state m to state k. Generalizing Eq. 5.51, the transverse relaxation rate for the $M_S = -\frac{1}{2}$ to $M_S = +\frac{1}{2}$ transition then becomes (see Eq. O.86 derived in Appendix O):

$$\frac{1}{T_2} = \frac{1}{T_2^{\mathrm{sec}}} + \frac{1}{2}\left[\sum_{k\neq m'} W_{m'\to k} + \sum_{k\neq m} W_{m\to k}\right] \tag{5.52}$$

where $m' \equiv \left|+\tfrac{1}{2},m_I\right\rangle$, $m \equiv \left|-\tfrac{1}{2},m_I\right\rangle$ and k corresponds to these electron states but with different values of m_I. The second term in Eq. 5.52 comprises the pseudosecular ($\Delta M_S = 0$, $\Delta m_I = \pm 1$) and non-secular ($\Delta M_S = \pm 1$) contributions. Whenever $T_1 >> T_2$, the secular contribution to T_2-relaxation dominates.

5.8 NITROXIDE TRANSVERSE (SPIN–SPIN) RELAXATION RATES

The secular contribution to the T_2-relaxation rate of a nitroxide is given by Eq. 5.50. In terms of the modified spectral densities defined in Section 5.5, this becomes (Freed 1979):

$$\frac{1}{T_2^{\mathrm{sec}}} = -\frac{1}{2}j_1^{SR}(0) + \frac{8}{3}j^{gg}(0)\cdot B_o^2 + \frac{16}{3}j^{gA}(0)\cdot B_o m_I + \frac{8}{3}j^{AA}(0)\cdot m_I^2 \tag{5.53}$$

Note that $j^{gg}(\omega)$, $j^{gA}(\omega)$ and $j^{AA}(\omega)$ introduced in Eq. 5.19 depend on the angular average $\frac{1}{4}\left\langle \sin^2\theta\cos^2\theta\right\rangle$, whereas the spectral densities appropriate to Eq. 5.50 depend on $\frac{1}{9}\left\langle\left(3\cos^2\theta-1\right)^2\right\rangle$ (see Eq. 5.15). To correct for this, $j^{AA}(\omega)$ etc. must be scaled by a factor of 8/3 (see Eqs. 5.16–5.18).

The pseudosecular and non-secular contributions to the T_2-relaxation rate of a nitroxide are given by the second term on the right of Eq. 5.52. With reference to Figure 5.4 for a ^{14}N-nitroxide, the sum of transition probabilities that we need is $\frac{1}{2}\Sigma(0)=W_e(0)+4W_n+2W_{x1}+2W_{x2}$ for the $m_I=0$ ^{14}N-hyperfine line, and $\frac{1}{2}\Sigma(\pm1)=W_e(\pm1)+2W_n+W_{x1}+W_{x2}$ for the $m_I=\pm1$ ^{14}N-hyperfine lines. However, we would prefer an expression that holds for a general hyperfine state m_I and nuclear spin I (^{14}N and ^{15}N). Equation 5.19 for the electron transitions is already in this form, but the situation for the nuclear transitions is slightly less straightforward. As we see from Eq. 5.15, spin operators $I_\pm$ that induce nuclear transitions occur in pairs (Hermitean conjugates) – as they must to ensure that the overall Hamiltonian is Hermitean. Therefore, we should combine the corresponding pair of transition probabilities. For the pseudosecular pair $S_zI_+ + S_zI_-$, we get from Eq. 5.20:

$$W\left(\pm\tfrac{1}{2},m_I\rightarrow\pm\tfrac{1}{2},m_I+1\right)+W\left(\pm\tfrac{1}{2},m_I\rightarrow\pm\tfrac{1}{2},m_I-1\right)$$
$$=\left(I(I+1)-m_I^2\right)j^{AA}(\omega_a) \tag{5.54}$$

Similarly for the cross-relaxation pairs, we get from Eqs. 5.21 and 5.22:

$$W\left(-\tfrac{1}{2},m_I\rightarrow+\tfrac{1}{2},m_I+1\right)+W\left(+\tfrac{1}{2},m_I\rightarrow-\tfrac{1}{2},m_I-1\right)$$
$$=4\left(I(I+1)-m_I^2\right)j^{AA}(\omega_e) \tag{5.55}$$

$$W\left(-\tfrac{1}{2},m_I\rightarrow+\tfrac{1}{2},m_I-1\right)+W\left(+\tfrac{1}{2},m_I\rightarrow-\tfrac{1}{2},m_I+1\right)$$
$$=\tfrac{2}{3}\left(I(I+1)-m_I^2\right)j^{AA}(\omega_e) \tag{5.56}$$

Summing contributions from Eqs. 5.53–5.56 and 5.19 by using Eq. 5.52, the net transverse relaxation rate including secular, pseudosecular and non-secular contributions finally becomes:

$$\frac{1}{T_2(m_I)}=-\frac{1}{2}\left(j_1^{SR}(0)+j_1^{SR}(\omega_e)\right)+2\left(\frac{4}{3}j^{gg}(0)+j^{gg}(\omega_e)\right)B_o^2$$
$$+I(I+1)\left(j^{AA}(\omega_a)+\frac{7}{3}j^{AA}(\omega_e)\right)+4\left(\frac{4}{3}j^{gA}(0)+j^{gA}(\omega_e)\right)B_om_I$$
$$+\left(\frac{8}{3}j^{AA}(0)-j^{AA}(\omega_a)-\frac{1}{3}j^{AA}(\omega_e)\right)m_I^2 \tag{5.57}$$

This applies to both ^{14}N-nitroxides ($I=1$) and ^{15}N-nitroxides $\left(I=\frac{1}{2}\right)$. We derive the axial version of Eq. 5.57 in Section O.6 of Appendix O, at the end of the book, by using the relaxation matrix. From Eq. 5.57, the dependence of the transverse relaxation rate on the nitrogen hyperfine manifold, m_I, takes the form:

$$\frac{1}{T_2(m_I)}=A+Bm_I+Cm_I^2 \tag{5.58}$$

This relation holds quite generally for fast anisotropic rotation, as well as for isotropic rotation, and for rapid angular motion in an orienting potential. The coefficients A, B and C depend on the correlation times (and order parameters). Note that A and B, but not C, depend on the magnetic field strength (or EPR frequency) – see Eq. 5.57 and Eqs. 5.59–5.62 immediately following. We use Eq. 5.58 extensively for analysing EPR linewidths in Chapter 7 (see Sections 7.7–7.11).

Substituting the spectral densities for isotropic rotational diffusion from Section 5.5 into Eq. 5.57, the A, B and C coefficients for the transverse relaxation rate in Eq. 5.58 are (Wilson and Kivelson 1966; Kivelson 1971; Kooser et al. 1969):

$$A=\frac{4}{45}\left((\Delta g)^2+3(\delta g)^2\right)\left(\frac{\beta_e}{\hbar}\right)^2B_o^2\left(1+\frac{3}{4}\frac{1}{1+\omega_e^2\tau_R^2}\right)\tau_R$$
$$+\frac{1}{30}I(I+1)\left((\Delta A)^2+3(\delta A)^2\right)\left(\frac{1}{1+\omega_a^2\tau_R^2}+\frac{7}{3}\frac{1}{1+\omega_e^2\tau_R^2}\right)\tau_R \tag{5.59}$$

$$A'=\frac{1}{9}\sum_{i=x,y,z}\left(g_{ii}-g_e\right)^2\frac{1}{\tau_R}+A_{resid} \tag{5.60}$$

$$B=\frac{8}{45}\left(\Delta g\Delta A+3\delta g\delta A\right)\left(\frac{\beta_e}{\hbar}\right)B_o\left(1+\frac{3}{4}\frac{1}{1+\omega_e^2\tau_R^2}\right)\tau_R \tag{5.61}$$

$$C=\frac{4}{45}\left((\Delta A)^2+3(\delta A)^2\right)\left(1-\frac{3}{8}\frac{1}{1+\omega_a^2\tau_R^2}-\frac{1}{8}\frac{1}{1+\omega_e^2\tau_R^2}\right)\tau_R \tag{5.62}$$

In Eqs. 5.59 and 5.60, we split the total observed A-parameter into the contribution, A, from rotational modulation of the g-tensor and hyperfine tensor, and the remainder, A'. The remainder comes from spin-rotation interaction plus the residual T_2-relaxation rate, A_{resid}, that is not attributable to any of these mechanisms. Note that in Eq. 5.61, B turns out to be negative, therefore, the high-field line ($m_I=-1$) is broadest. We give corresponding expressions for axially anisotropic rotational diffusion in the Appendix to this chapter.

If we limit the lowest applicable rotational correlation time by neglecting the non-secular spectral densities, Eqs. 5.59, 5.61 and 5.62 then become:

$$A=\frac{4}{45}\left((\Delta g)^2+3(\delta g)^2\right)\left(\frac{\beta_e}{\hbar}\right)^2B_o^2\tau_R$$
$$+\frac{1}{30}I(I+1)\left((\Delta A)^2+3(\delta A)^2\right)\frac{\tau_R}{1+\omega_a^2\tau_R^2} \tag{5.63}$$

$$B=\frac{8}{45}\left(\Delta g\Delta A+3\delta g\delta A\right)\left(\frac{\beta_e}{\hbar}\right)B_o\tau_R \tag{5.64}$$

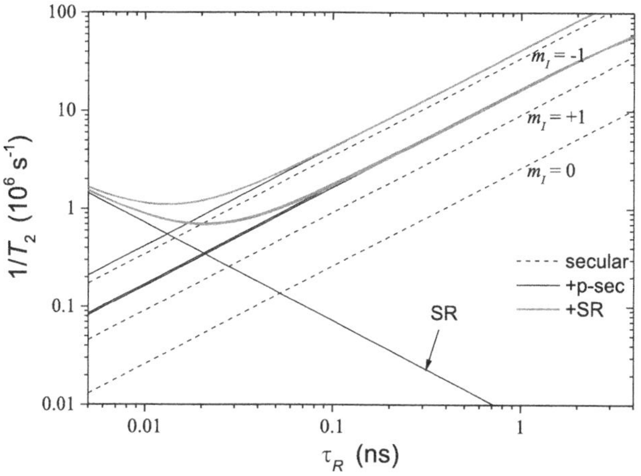

FIGURE 5.6 Secular contributions (dashed lines) and secular-plus-pseudosecular contributions (solid lines) to transverse relaxation rate T_2^{-1} from modulation of ^{14}N-hyperfine and *g*-tensor anisotropies, as function of isotropic rotational correlation time τ_R (see Eqs. 5.57, 5.58 and 5.63–5.65). Grey lines include spin-rotation contributions (SR). Tensor values and EPR frequency as in Figure 5.5.

$$C = \frac{4}{45}\left((\Delta A)^2 + 3(\delta A)^2\right)\left(1 - \frac{3}{8}\frac{1}{1+\omega_a^2\tau_R^2}\right)\tau_R \tag{5.65}$$

where we assume that $\omega_e^2\tau_R^2 >> 1$. This requires correlation times longer than $\approx 5\times10^{-11}$ s for a spectrometer frequency of 9 GHz but proportionately shorter at higher operating frequencies/fields. Including the pseudosecular spectral densities, $j(\omega_a) \propto \tau_R/\left(1+\omega_a^2\tau_R^2\right)$, in Eqs. 5.63 and 5.65 lets us apply motional-narrowing theory up to the limit of its validity, e.g., for correlation times up to 3×10^{-9} s at 9 GHz. Often, the pseudosecular spectral densities are approximated simply by taking $j(\omega_a) \propto \tau_R$, which requires that $\omega_a^2\tau_R^2 << 1$ and restricts correlation times to $\tau_R = 1\times10^{-9}$ s or less. Like the *B*-coefficient (Eq. 5.64), the *A* and *C* relaxation-rate parameters then become linear in τ_R.

Figure 5.6 shows the dependence of the secular and pseudosecular contributions to the transverse relaxation rate on rotational correlation time. Clearly, the pseudosecular terms make an important contribution to the line broadening. For isotropic rotation, the pseudosecular contributions result in very similar values of T_2 for the $m_I = 0$ and $m_I = +1$ hyperfine lines, because then $|C/B| \cong 1$ in Eq. 5.58. Spin-rotation makes little contribution to transverse relaxation, except for very short correlation times, where the non-secular terms must also be included at 9 GHz EPR frequencies. Practically, spin-rotation contributions may be difficult to resolve because of contributions to the *A*-term from residual broadening (see Eq. 5.60).

5.9 CONCLUDING SUMMARY

1. *Transition probabilities and spectral densities:*
   For fast motion, relaxation is induced by the time-dependent random perturbation $\mathcal{H}_1 F(t)$ contributed by the angular anisotropy of the spin Hamiltonian. The rate of transition from state m to state m' is then:

$$W_{m'm} = \frac{1}{\hbar^2}\left|\langle m'|\mathcal{H}_1|m\rangle\right|^2 J(\omega_{m'm}) \tag{5.66}$$

   where $\left|\langle m'|\mathcal{H}_1|m\rangle\right|^2$ are matrix elements of spin operators. For the exponential correlation functions of Brownian diffusion, the spectral density is Lorentzian:

$$J(\omega) = \left\langle |F(0)|^2 \right\rangle \frac{2\tau_R}{1+\omega^2\tau_R^2} \tag{5.67}$$

   where τ_R is the rotational correlation time and $\left\langle |F(0)|^2 \right\rangle$ is an average over the angular anisotropy (see Eqs. 5.16–5.18).
2. *Spin–lattice relaxation rates:*
   The electron spin–lattice relaxation rate $2W_e(m_I) \equiv 1/T_{1,e}(m_I)$ for the m_I-hyperfine line is:

$$\begin{aligned} 2W_e(m_I) = {} & \frac{1}{9}\sum_{i=x,y,z}\frac{(g_{ii}-g_e)^2}{\tau_R} \\ & + \frac{2}{15}\Bigg[\left((\Delta g)^2 + 3(\delta g)^2\right)\cdot\left(\frac{\beta_e}{\hbar}\right)^2 B_o^2 \\ & + 2(\Delta g\Delta A + 3\delta g\delta A)\cdot\left(\frac{\beta_e}{\hbar}\right)B_o m_I \\ & + \left((\Delta A)^2 + 3(\delta A)^2\right)m_I^2\Bigg]\frac{\tau_R}{1+\omega_e^2\tau_R^2} \end{aligned} \tag{5.68}$$

   where $\Delta g = g_{zz} - \frac{1}{2}(g_{xx}+g_{yy})$, $\delta g = \frac{1}{2}(g_{xx}-g_{yy})$ and similarly for the *A*-tensor; $\omega_e \cong g_o\beta_e B_o/\hbar$ is the electron Larmor frequency and B_o is the static magnetic field strength.

   The nitrogen nuclear relaxation rate $W_n \equiv 1/T_{1,n}$ for ^{14}N-nitroxides is:

$$W_n = \frac{1}{30}\left[(\Delta A)^2 + 3(\delta A)^2\right]\frac{\tau_R}{1+\omega_a^2\tau_R^2} \tag{5.69}$$

   where $\omega_a = \frac{1}{2}a_o$ (in angular frequency units). For ^{15}N-nitroxides, the right-hand side gives twice the relaxation rate: $2W_n$.

   The W_{x1} cross-relaxation rate for ^{14}N-nitroxides is:

$$W_{x1} = \frac{2}{15}\left[(\Delta A)^2 + 3(\delta A)^2\right]\frac{\tau_R}{1+\omega_e^2\tau_R^2} \tag{5.70}$$

   and the W_{x2} cross-relaxation rate is $W_{x2} = W_{x1}/6$. For ^{15}N-nitroxides, however, the right-hand side gives twice the relaxation rate: $2W_{x1}$. Cross relaxation contributes significantly to saturation-recovery rates. Results for axially anisotropic rotational diffusion are given in the Appendix.

3. *Tranverse relaxation and linewidths:*
Secular relaxation at rate $1/T_2^{sec}$ is the most important contributor to linewidths or transverse relaxation. It arises from modulation of the energy levels between which transitions take place and depends on the spectral density at zero frequency: $j(0)=\tau_R$. The full transverse relaxation rate includes additionally lifetime broadening from spin–lattice type transitions:

$$\frac{1}{T_2}=\frac{1}{T_2^{sec}}+\frac{1}{2}\left[\sum_{k\neq m'}W_{m'\to k}+\sum_{k\neq m}W_{m\to k}\right] \tag{5.71}$$

where $m'\equiv\left|+\frac{1}{2},m_I\right\rangle$, $m\equiv\left|-\frac{1}{2},m_I\right\rangle$ and k corresponds to these electron states but with different values of m_I. The secular term generally dominates and $T_1>>T_2$, except for very fast motion.
4. *Differential line broadening and rotational correlation times:*
The transverse relaxation rate depends on the nitrogen hyperfine manifold, m_I, according to:

$$\frac{1}{T_2(m_I)}=A+Bm_I+Cm_I^2 \tag{5.72}$$

which holds quite generally for fast rotation, i.e., correlation times up to 3×10^{-9} s at 9 GHz. Coefficients A and B but not C depend on the static field strength. The dependence on rotational dynamics, for isotropic diffusion, is given by Eqs. 5.59–5.62. By limiting to correlation times $\tau_R\geq5\times10^{-11}$ s at 9 GHz, and proportionately shorter at higher operating frequencies, we get:

$$A=\frac{4}{45}\left((\Delta g)^2+3(\delta g)^2\right)\left(\frac{\beta_e}{\hbar}\right)^2B_o^2\tau_R+\frac{1}{30}I(I+1)\left((\Delta A)^2+3(\delta A)^2\right)\frac{\tau_R}{1+\omega_a^2\tau_R^2} \tag{5.73}$$

$$B=\frac{8}{45}(\Delta g\Delta A+3\delta g\delta A)\left(\frac{\beta_e}{\hbar}\right)B_o\tau_R \tag{5.74}$$

$$C=\frac{4}{45}\left((\Delta A)^2+3(\delta A)^2\right)\left(1-\frac{3}{8}\frac{1}{1+\omega_a^2\tau_R^2}\right)\tau_R \tag{5.75}$$

where the B-coefficient is linear in τ_R. If we approximate pseudosecular spectral densities by: $j(\omega_a)\approx\tau_R$, all three coefficients become linear in τ_R, but this restricts correlation times to $\tau_R=1\times10^{-9}$ s or less. Results for axially anisotropic rotational diffusion are given in the Appendix.

APPENDIX A5: TRANSITION PROBABILITIES AND TRANSVERSE RELAXATION RATES FOR FAST ANISOTROPIC ROTATIONAL DIFFUSION

We assume axially anisotropic Brownian diffusion, where the rotational diffusion tensor has principal coefficients $D_{R\parallel}$ and $D_{R\perp}$, corresponding to rotation about and perpendicular to the principal (||) axis (see Section 7.6 in Chapter 7). Rotational correlation times, $\tau_{2,0}$ and $\tau_{2,2}$, corresponding to second-rank spherical tensors are defined by:

$$\tau_{2,0}^{-1}=6D_{R\perp} \tag{A5.1}$$

$$\tau_{2,2}^{-1}=2D_{R\perp}+4D_{R\parallel} \tag{A5.2}$$

– see Eqs. 7.28–7.30 in Chapter 7. We assume that the principal rotation axis (||) is parallel to the nitroxide z-axis. Expressions for the transition probabilities and linewidths can be obtained from the appropriate spectral densities (see Section 7.6) and are given in Dalton et al. (1976). Spin-rotation contributions for the axial case are given by Eq. 5.39 in the main text. The EPR linewidths also are given in Goldman et al. (1972). We obtain corresponding expressions for the principal rotation axis (||) directed along the nitroxide x- or y-axis by cyclic permutation of the x, y, z indices in Eqs. 5.24, 5.25, 5.28 and 5.29 for ΔA, δA, Δg and δg, respectively.

A5.1 Electron Spin–Lattice Relaxation

$$W_e(m_I)=\frac{1}{18}\left[\frac{(g_{zz}-g_e)^2}{\tau_{R\parallel}}+\frac{(g_{xx}-g_e)^2+(g_{yy}-g_e)^2}{\tau_{R\perp}}\right]+\frac{1}{15}\left[(\Delta g)^2\frac{\tau_{2,0}}{1+\omega_e^2\tau_{2,0}^2}+3(\delta g)^2\frac{\tau_{2,2}}{1+\omega_e^2\tau_{2,2}^2}\right]\times\left(\frac{\beta_e}{\hbar}\right)^2B_o^2+\frac{2}{15}\left[\Delta g\Delta A\frac{\tau_{2,0}}{1+\omega_e^2\tau_{2,0}^2}+3\delta g\delta A\frac{\tau_{2,2}}{1+\omega_e^2\tau_{2,2}^2}\right]\times\left(\frac{\beta_e}{\hbar}\right)B_om_I+\frac{1}{15}\left[(\Delta A)^2\frac{\tau_{2,0}}{1+\omega_e^2\tau_{2,0}^2}+3(\delta A)^2\frac{\tau_{2,2}}{1+\omega_e^2\tau_{2,2}^2}\right]\times m_I^2 \tag{A5.3}$$

where $\tau_{R\parallel}^{-1}=6D_{R\parallel}$ and $\tau_{R\perp}^{-1}=6D_{R\perp}$ in the spin-rotation contribution (cf. Section 7.6); the ||-axis is the nitroxide z-axis but can be redefined by cyclic permutation.

A5.2 Nuclear Spin–Lattice Relaxation

$$W_n = \frac{1}{60}\left[(\Delta A)^2 \frac{\tau_{2,0}}{1+\omega_a^2\tau_{2,0}^2} + 3(\delta A)^2 \frac{\tau_{2,2}}{1+\omega_a^2\tau_{2,2}^2}\right] \tag{A5.4}$$

for ^{14}N-nitroxides, and half this for ^{15}N-nitroxides (see Section 5.6).

A5.3 Cross Relaxation

$$W_{x_2} = \frac{1}{15}\left[(\Delta A)^2 \frac{\tau_{2,0}}{1+\omega_e^2\tau_{2,0}^2} + 3(\delta A)^2 \frac{\tau_{2,2}}{1+\omega_e^2\tau_{2,2}^2}\right] \tag{A5.5}$$

$$W_{x_1} = \frac{1}{6} W_{x_2} \tag{A5.6}$$

for ^{14}N-nitroxides, and half that of Eq. A5.5 for ^{15}N-nitroxides (see Section 5.6).

A5.4 Electron Transverse Relaxation

$$\frac{1}{T_2(m_I)} = (A + A') + Bm_I + Cm_I^2 \tag{A5.7}$$

$$A = \frac{4}{45}\left[\left(1+\frac{3}{4}\frac{1}{1+\omega_e^2\tau_{2,0}^2}\right)(\Delta g)^2 \tau_{2,0} + 3\left(1+\frac{3}{4}\frac{1}{1+\omega_e^2\tau_{2,2}^2}\right)(\delta g)^2 \tau_{2,2}\right] \times \left(\frac{\beta_e}{\hbar}\right)^2 B_o^2$$
$$+ \frac{1}{30} I(I+1)\left[\left(\frac{1}{1+\omega_a^2\tau_{2,0}^2} + \frac{7}{3}\frac{1}{1+\omega_e^2\tau_{2,0}^2}\right)(\Delta A)^2 \tau_{2,0} + 3\left(\frac{1}{1+\omega_a^2\tau_{2,2}^2} + \frac{7}{3}\frac{1}{1+\omega_e^2\tau_{2,2}^2}\right)(\delta A)^2 \tau_{2,2}\right] \tag{A5.8}$$

$$A' = \frac{1}{9}\left[\frac{(g_{zz}-g_e)^2}{\tau_{R\parallel}} + \frac{(g_{xx}-g_e)^2 + (g_{yy}-g_e)^2}{\tau_{R\perp}}\right] + A_{resid} \tag{A5.9}$$

$$B = \frac{8}{45}\left[\left(1+\frac{3}{4}\frac{1}{1+\omega_e^2\tau_{2,0}^2}\right)\Delta g \Delta A \tau_{2,0} + 3\left(1+\frac{3}{4}\frac{1}{1+\omega_e^2\tau_{2,2}^2}\right)\delta g \delta A \tau_{2,2}\right] \times \left(\frac{\beta_e}{\hbar}\right) B_o \tag{A5.10}$$

$$C = \frac{4}{45}\left[\left(1-\frac{3}{8}\frac{1}{1+\omega_a^2\tau_{2,0}^2} - \frac{1}{8}\frac{1}{(1+\omega_e^2\tau_{2,0}^2)}\right)(\Delta A)^2 \tau_{2,0} + 3\left(1-\frac{3}{8}\frac{1}{1+\omega_a^2\tau_{2,2}^2} - \frac{1}{8}\frac{1}{(1+\omega_e^2\tau_{2,2}^2)}\right)(\delta A)^2 \tau_{2,2}\right] \tag{A5.11}$$

In Eq. A5.9: $\tau_{R\parallel}^{-1} = 6D_{R\parallel}$ and $\tau_{R\perp}^{-1} = 6D_{R\perp}$ for the spin-rotation contribution (cf. Section 7.6); the ||-axis is the nitroxide z-axis but can be redefined by cyclic permutation of x,y,z. Note that $T_1^{SR} = T_2^{SR}$ for the spin-rotation contribution.

REFERENCES

Abragam, A. 1961. *The Principles of Nuclear Magnetism*, Oxford: Oxford University Press.

Atkins, P.W. and Kivelson, D. 1966. ESR linewidths in solution. II. Analysis of spin-rotational relaxation data. *J. Chem. Phys.* 44:169–174.

Carrington, A. and McLachlan, A.D. 1969. *Introduction to Magnetic Resonance. With Applications to Chemistry and Chemical Physics*, New York: Harper and Row.

Dalton, L.R., Robinson, B.H., Dalton, L.A., and Coffey, P. 1976. Saturation transfer spectroscopy. *Adv. Magn. Reson.* 8:149–259.

Freed, J.H. 1965. Theory of saturation and double-resonance effects in ESR spectra. *J. Chem. Phys.* 43:2312–2332.

Freed, J.H. 1979. Theory of multiple resonance and ESR saturation in liquids and related media. In *Multiple Electron Resonance Spectroscopy*, eds. Dorio, M.M. and Freed, J.H., 73–142. New York and London: Plenum Press.

Goldman, S.A., Bruno, G.V., and Freed, J.H. 1973. ESR studies of anisotropic rotational reorientation and slow tumbling in liquid and frozen media. 2. Saturation and nonsecular effects. *J. Chem. Phys.* 59:3071–3091.

Goldman, S.A., Bruno, G.V., Polnaszek, C.F., and Freed, J.H. 1972. An ESR study of anisotropic rotational reorientation and slow tumbling in liquid and frozen media. *J. Chem. Phys.* 56:716–735.

Hubbard, P.S. 1963. Theory of nuclear magnetic relaxation by spin-rotational interactions in liquids. *Phys. Rev.* 131:1155–1165.

Hyde, J.S., Chien, J.C.W., and Freed, J.H. 1968. Electron-electron double resonance of free radicals in solution. *J. Chem. Phys.* 48:4211–4226.

Kivelson, D. 1971. Electron spin relaxation in liquids. Selected examples. In *Electron Spin Relaxation in Liquids*, eds. Muus, L.T. and Atkins, P.W., 213–277. New York-London: Plenum Press.

Kooser, R.G., Volland, W.V., and Freed, J.H. 1969. ESR relaxation studies on orbitally degenerate free radicals. I. Benzene anion and tropenyl. *J. Chem. Phys.* 50:5243–5247.

Marsh, D. 1992. Influence of nuclear relaxation on the measurement of exchange frequencies in CW saturation EPR studies. *J. Magn. Reson.* 99:332–337.

McClung, R. and Kivelson, D. 1968. ESR linewidths in solution. V. Studies of spin-rotational effects not described by rotational diffusion theory. *J. Chem. Phys.* 49:3380–3391.

McConnell, H.M. 1956. Effect of anisotropic hyperfine interactions on paramagnetic relaxation in liquids. *J. Chem. Phys.* 25:709–711.

Nyberg, G. 1967. Spin-rotational relaxation in solution ESR. *Mol. Phys.* 12:69–81.

Percival, P.W., Hyde, J.S., Dalton, L.A., and Dalton, L.R. 1975. Molecular and applied modulation effects in electron-electron double resonance. 5. Passage effects in high-resolution frequency and field swept ELDOR. *J. Chem. Phys.* 62:4332–4342.

Robinson, B.H., Haas, D.A., and Mailer, C. 1994. Molecular dynamics in liquids: spin-lattice relaxation of nitroxide spin labels. *Science* 263:490–493.

Robinson, B.H., Reese, A.W., Gibbons, E., and Mailer, C. 1999. A unified description of the spin-spin and spin-lattice relaxation rates applied to nitroxide spin labels in viscous liquids. *J. Phys. Chem. B* 103:5881–5894.

Wilson, R. and Kivelson, D. 1966. ESR linewidths in solution. I. Experiments on anisotropic and spin-rotational effects. *J. Chem. Phys.* 44:154–168.

EPR Lineshape Theory

6

6.1 INTRODUCTION

Electron paramagnetic resonance (EPR) spectroscopy of spin-labelled molecules proves extremely useful for studying molecular mobility. As we shall see, conventional 9-GHz spin-label spectra are sensitive to motions on a timescale that is determined by the anisotropy of the ^{14}N-hyperfine splitting of the nitroxide free radical group: $\tau_{motion} \sim \hbar/(A_{zz} - A_{xx}) \approx 2\times10^{-9}$ s. If motions are 100× slower than this, they appear almost static on the conventional EPR time scale. The spectra are then a superposition of all those from the different steps in the molecular motion. Gradually increasing rates of motion give rise to broadening of these individual spectral components, which then finally collapse to the time-average mean value at around the characteristic τ_{motion}-timescale. Beyond this, the averaged lines narrow progressively until the motion becomes 100× faster, when motional broadening is no longer perceptible. This principle of motional averaging underlies the sensitivity of spin-label EPR lineshapes to molecular dynamics. We can describe it most simply by using the classical Bloch equations that we introduce in the next section.

The two-site exchange model displays all essential motionally-sensitive features of the EPR lineshapes, when incorporated in the Bloch equations. Motion that is fast relative to the spectral anisotropy causes motional averaging that does not depend on the mechanism of rotational diffusion. We treat this most readily by using perturbation theory, which we covered in Chapter 5. Fast motion simply causes Lorentzian line broadening, and this depends on rotational rate according to Eqs. 5.58–5.62. We can extend the two-site model used in the exchange-coupled Bloch equations to cover slow rotational diffusion, either by large uncorrelated jumps or in small random steps that characterize Brownian motion. This allows simulation of lineshapes that are valid in the slow-motional regime but not in the fast regime. Alternatively, we can describe rotational diffusion quantum mechanically by using the density matrix. This leads to the stochastic Liouville equation for the spin dynamics, which is valid for all motional regimes. The stochastic Liouville equation has proved very powerful for analysing spin-label EPR lineshapes, particularly in the slow-motional regime (i.e., for rotational correlation times in the range 3–300 ns). We devote the later parts of this chapter to describing these stochastic-Liouville lineshape simulations in some detail.

6.2 BLOCH EQUATIONS AND LORENTZIAN LINESHAPE

The Bloch equations combine the classical equation of motion for the macroscopic electron spin magnetization or bulk magnetic moment $\mathbf{M} = (M_x, M_y, M_z)$ in an external magnetic field $\mathbf{B} = (0,0,B_o)$ with phenomenological expressions for relaxation of the transverse (M_x, M_y) and longitudinal (M_z) components. An individual magnetic moment $\boldsymbol{\mu}$ is related to the associated angular momentum $\mathbf{J}$ by $\boldsymbol{\mu} = \gamma_e \mathbf{J}$, where γ_e is the gyromagnetic ratio of the electron (cf. Eq. 1.1 in Chapter 1). Classically, the rate of change in angular momentum is given by the torque $\boldsymbol{\mu} \times \mathbf{B}$ experienced by the magnetic moment (see Appendix B.1 for vector algebra). Therefore, summing over individual moments, the equation of motion for the magnetization is:

$$\frac{\mathrm{d}\mathbf{M}}{\mathrm{d}t} = \gamma_e(\mathbf{M} \times \mathbf{B}) \tag{6.1}$$

In the absence of a time-varying field, the magnetization precesses about the direction of the static field, $\mathbf{B}_o$, with an angular frequency $\omega = -\gamma_e \mathbf{B}_o$ (see Figure 6.1). We call this Larmor precession. Note that, to incorporate relaxation, which is a property of an ensemble of spins, we need to deal with the

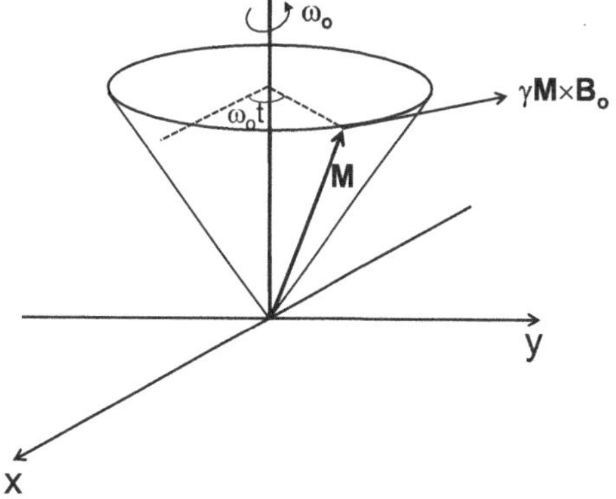

FIGURE 6.1 Free precession of electron spin magnetization **M** about static field $\mathbf{B}_o$, at Larmor angular frequency ω_o. Precessional motion occurs because the field exerts torque $\gamma_e \mathbf{M} \times \mathbf{B}_o$ on the magnetization.

macroscopic magnetization, not just with an individual magnetic moment (see Section 1.6 in Chapter 1).

For unsaturated EPR lineshapes, only transverse components of the magnetization interest us. We assume that spin–lattice (i.e., longitudinal) relaxation is rapid relative to the rate at which spin transitions are induced by the radiation magnetic field $\mathbf{B}_1$. The time dependence of the transverse components of the magnetization, including relaxation, is given by (Bloch 1946):

$$\frac{dM_x}{dt} = \gamma_e (\mathbf{M}\times\mathbf{B})_x - \frac{M_x}{T_2} \tag{6.2}$$

$$\frac{dM_y}{dt} = \gamma_e (\mathbf{M}\times\mathbf{B})_y - \frac{M_y}{T_2} \tag{6.3}$$

where T_2 is the transverse or spin–spin relaxation time. We get EPR absorption when a small oscillatory magnetic field $B_x = 2B_1 \cos\omega t$, applied perpendicular to the static field, is at the resonance frequency $\omega_o = \gamma_e B_o$ of the precessing magnetization. Absorption takes place only from the circularly polarized component of the linearly polarized B_1-field that rotates in the same sense as the Larmor precession.

It helps to view the magnetization in an axis system that rotates at the angular frequency, ω, of the circularly polarized radiation. If the magnetization were static in this rotating frame, it would appear to vary at a rate $\boldsymbol{\omega}\times\mathbf{M}$ in the laboratory-fixed frame (cf. Eq. R.2 in Appendix R). Therefore, in the rotating frame, Eq. 6.1 becomes:

$$\frac{d\hat{\mathbf{M}}}{dt} = \gamma_e\left(\hat{\mathbf{M}}\times\left(\hat{\mathbf{B}}+\boldsymbol{\omega}/\gamma_e\right)\right) \tag{6.4}$$

where the crescent hat (e.g., $\hat{\mathbf{M}}$) indicates vectors referred to the rotating frame. Comparing with Eq. 6.1, we see that the effective field in the rotating frame is $\hat{\mathbf{B}}_{eff} = \mathbf{B}_o - \boldsymbol{\omega}/\gamma_e$, which is parallel to $\mathbf{B}_o$. For strong radiation fields, e.g., in pulse EPR, the effective field is the vector sum of this offset field with the B_1-field, as shown in Figure 6.2. We shall return to this point in later chapters dealing with pulse methods.

From Eq. 6.4, the Bloch equations for the $\hat{M}_x$- and $\hat{M}_y$-components, u and v, of the magnetization in the rotating frame become:

$$\frac{du}{dt} = (\omega_0 - \omega)v - \frac{u}{T_2} \tag{6.5}$$

$$\frac{dv}{dt} = -(\omega_0 - \omega)u + \gamma_e B_1 M_z - \frac{v}{T_2} \tag{6.6}$$

where we assume that the z-component of the magnetization remains at its non-saturated equilibrium value, $M_z = M_o$. Here, u is the component that oscillates in phase with B_1 and v is the component that lags 90° out-of-phase, corresponding to the dispersive and absorptive interactions with the microwave field, respectively. To solve the simultaneous Eqs. 6.5 and 6.6, we define a complex magnetization: $\hat{M}_+ = u + iv$ for which the equation of motion is:

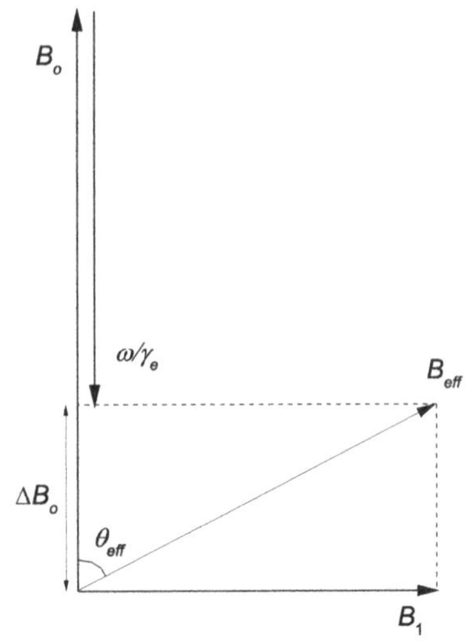

FIGURE 6.2 Effective field $B_{eff}\left(=\sqrt{\Delta B_o^2 + B_1^2}\right)$ in axis frame rotating at angular frequency ω of B_1-radiation field. Effective field is tilted at angle θ_{eff} to static B_o-field, where $\sin\theta_{eff} = B_1/B_{eff}$. ΔB_o: offset from resonance field.

$$\frac{d\hat{M}_+}{dt} = -\left(\frac{1}{T_2} + i(\omega_o - \omega)\right)\hat{M}_+ + i\gamma_e B_1 M_o \tag{6.7}$$

The steady-state solution $\left(d\hat{M}_+/dt = 0\right)$ is then:

$$\hat{M}_+ = \frac{i\gamma_e B_1 M_o}{1/T_2 + i(\omega_o - \omega)} \tag{6.8}$$

Taking the real and imaginary parts of Eq. 6.8, the dispersion (u) and absorption (v) signals, respectively, become:

$$u(\omega) = M_o \frac{\gamma_e B_1 T_2^2(\omega_o - \omega)}{1 + T_2^2(\omega_o - \omega)^2} \tag{6.9}$$

$$v(\omega) = M_o \frac{\gamma_e B_1 T_2^2}{1 + T_2^2(\omega_o - \omega)^2} \tag{6.10}$$

Clearly, the EPR absorption given by Eq. 6.10 has a Lorentzian lineshape (see Eq. 1.22 in Chapter 1). We show both the absorption and dispersion lineshapes in Figure 6.3, together with the usual first derivatives that we get by phase-sensitive detection in spectrometers with field modulation. Note that EPR spectrometers are basically absorption spectrometers. To get the dispersion signal, we must set the reference arm in the microwave bridge 90° out-of-phase with respect to the microwaves reflected from the resonant cavity.

With aqueous samples, we often find small dispersion admixtures to the absorption signal, particularly at higher microwave frequencies. We can describe this as a small phase shift $\Delta\phi$ that gives a composite signal $v\cos\Delta\phi + u\sin\Delta\phi$, where the ratio of dispersion to absorption contribution in the spectrum

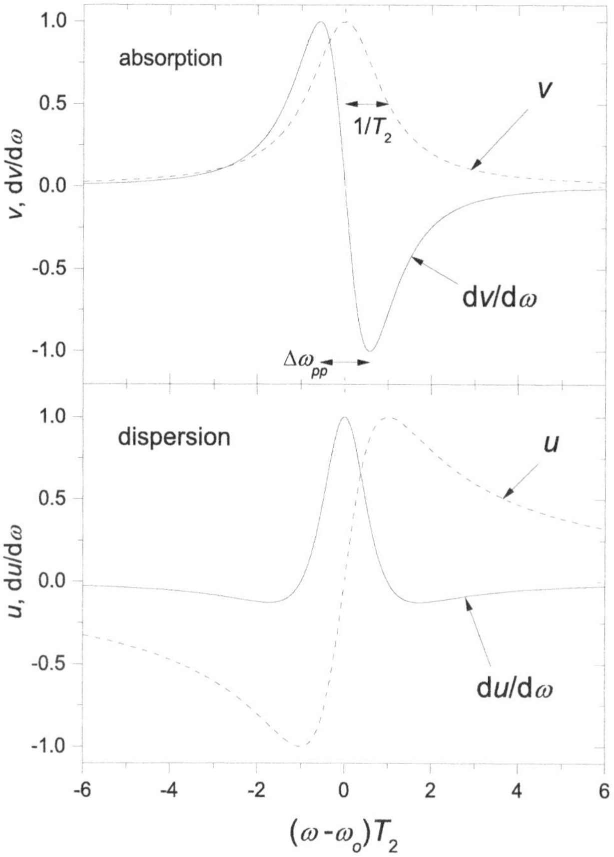

FIGURE 6.3 *Upper panel*: Lorentzian absorption lineshape (----) and its first derivative (—). For normalized integrated intensity, line height has units of (T_2/π) for absorption. Line heights of first derivative relate to those of absorption by factor: $3/(4\Delta\omega_{pp})$. *Lower panel*: Lorentzian dispersion lineshape (----) and first derivative (—). Line heights of first derivative relate to those of dispersion by factor 2×.

is $\tan\Delta\phi$. The resulting first-derivative EPR lineshape then becomes (Smirnova et al. 1995):

$$\frac{dI(B,\Delta\phi)}{dB} = -\frac{\Delta B^L_{1/2}}{\pi} \times \frac{2(B-B_o)^2\cos\Delta\phi + \left(\left(B^L_{1/2}\right)^2 - (B-B_o)^2\right)\sin\Delta\phi}{\left(\left(B^L_{1/2}\right)^2 + (B-B_o)^2\right)^2} \quad (6.11)$$

where $\Delta B^L_{1/2}$ is the half-width at half-height of the Lorentzian absorption spectrum, and the spectrum is written in terms of magnetic field scan B (instead of angular frequency ω) for use in correcting experimental spectra. Note that, under certain circumstances, dispersion admixture is an inevitable feature of spin–spin interactions (see Sections 9.11 and 9.17 in Chapter 9) and even of the nitrogen hyperfine interaction (see Appendix O.4, and also Marsh 2017).

6.3 EXCHANGE-COUPLED BLOCH EQUATIONS

Now we wish to calculate the dynamic effects of exchange between different environments (or orientations) on the EPR lineshapes. We assume a two-site model for physical or chemical exchange:

$$\text{site}(\mathbf{A}) \leftrightarrow \text{site}(\mathbf{B})$$

in conjunction with the Bloch equations (McConnell 1958). The rate equations for the spin magnetization associated with the two environments, $\mathbf{M}_A$ and $\mathbf{M}_B$, respectively, are given by:

$$\frac{d\mathbf{M}_A}{dt} = -\tau_A^{-1}\mathbf{M}_A + \tau_B^{-1}\mathbf{M}_B \quad (6.12)$$

$$\frac{d\mathbf{M}_B}{dt} = \tau_A^{-1}\mathbf{M}_A - \tau_B^{-1}\mathbf{M}_B \quad (6.13)$$

where τ_A^{-1} and τ_B^{-1} are probabilities per unit time of transfer from state **A** to state **B** and vice-versa. At exchange equilibrium: $d\mathbf{M}_A/dt = d\mathbf{M}_B/dt = 0$, and hence from Eqs. 6.12 and 6.13 we get:

$$\tau_A^{-1}/\tau_B^{-1} = M_B^o/M_A^o = (1-f)/f \quad (6.14)$$

where f is the fraction of component **A**. For each of the two sites, we can write a Bloch equation for the complex transverse magnetizations, $\hat{M}_A = u_A + iv_A$ and $\hat{M}_B = u_B + iv_B$, in the rotating frame (cf. Eq. 6.7). Incorporating the rate equations, Eqs. 6.12 and 6.13, then gives us the following expressions for the steady-state complex magnetizations:

$$\left[T_{2,A}^{-1} + \tau_A^{-1} - i(\omega - \omega_A)\right]\hat{M}_A - \tau_B^{-1}\hat{M}_B = i\gamma_e B_1 M_o f \quad (6.15)$$

$$\left[T_{2,B}^{-1} + \tau_B^{-1} - i(\omega - \omega_B)\right]\hat{M}_B - \tau_A^{-1}\hat{M}_A = i\gamma_e B_1 M_o (1-f) \quad (6.16)$$

where ω_A and ω_B are the angular resonance frequencies of components **A** and **B**, respectively, and $T_{2,A}$ and $T_{2,B}$ are their transverse relaxation times. Solving simultaneous Eqs. 6.15 and 6.16 for $\hat{M}_A$ and $\hat{M}_B$, the total transverse magnetization becomes:

$$\hat{M}_+ = \hat{M}_A + \hat{M}_B = -\gamma_e B_1 M_o \times \frac{\left[(\omega-\omega_B) + iT_{2,B}^{-1}\right]f + \left[(\omega-\omega_A) + iT_{2,A}^{-1}\right](1-f) + i\left(\tau_A^{-1} + \tau_B^{-1}\right)}{\left[(\omega-\omega_A) + i\left(T_{2,A}^{-1} + \tau_A^{-1}\right)\right]\cdot\left[(\omega-\omega_B) + i\left(T_{2,B}^{-1} + \tau_B^{-1}\right)\right] + \tau_A^{-1}\tau_B^{-1}} \quad (6.17)$$

where $M_o = M_A^o/f = M_B^o/(1-f)$. As before, the imaginary part of $\widehat{M}_+$: $v(\omega) = v_A + v_B$, gives the amplitude of the resonance absorption.

Figure 6.4 shows spectra simulated for increasing rates of exchange between the two sites. For purposes of illustration, we assume that sites **A** and **B** are equally populated and the only source of line broadening is exchange (i.e., $f = 0.5$ and $T_{2,A}^{-1} = T_{2,B}^{-1} = 0$). From Eq. 6.17, we then get the following simplified expression for the absorption lineshape:

$$v(\omega) = \tfrac{1}{2}\gamma_e B_1 M_o \frac{(\omega_A - \omega_B)^2 \tau_{A,B}^{-1}}{(\omega-\omega_A)^2(\omega-\omega_B)^2 + (2\omega - \omega_A - \omega_B)^2 \tau_{A,B}^{-2}} \tag{6.18}$$

where $\tau_{A,B}^{-1} \equiv \tau_A^{-1} = \tau_B^{-1}$. We give both absorption and first-derivative EPR lineshapes in Figure 6.4, for the complete range of values of the normalized exchange rate, $\tau_{A,B}^{-1}/|\omega_A - \omega_B|$.

For *slow* exchange $\left(\tau_A^{-1}, \tau_B^{-1} << |\omega_A - \omega_B|\right)$, the full lineshape function is given by the following approximation:

$$v(\omega) = \gamma_e B_1 M_0 \times \left[\frac{f\left(T_{2,A}^{-1} + \tau_A^{-1}\right)}{\left(T_{2,A}^{-1} + \tau_A^{-1}\right)^2 + (\omega_A - \omega)^2} + \frac{(1-f)\left(T_{2,B}^{-1} + \tau_B^{-1}\right)}{\left(T_{2,B}^{-1} + \tau_B^{-1}\right)^2 + (\omega_B - \omega)^2}\right] \tag{6.19}$$

In this case, the resonance absorption consists of two Lorentzian lines still centred at the original angular frequencies ω_A and ω_B but with linewidths increased by lifetime broadening. The increase in linewidth is given by the reciprocal of the exchange lifetime, when we express the total linewidth in terms of an effective T_2:

$$T_{2,eff}^{-1}(A) = T_{2,A}^{-1} + \tau_A^{-1} \tag{6.20}$$

$$T_{2,eff}^{-1}(B) = T_{2,B}^{-1} + \tau_B^{-1} \tag{6.21}$$

for components **A** and **B**, respectively.

For slightly faster exchange rates $\left(\tau_A^{-1}, \tau_B^{-1} < |\omega_A - \omega_B|\right)$, the lines begin to move towards each other. In the simplified situation assumed for Eq. 6.18, the line separation in the presence of exchange, $(\omega_A - \omega_B)_{ex}$, then becomes:

$$(\omega_A - \omega_B)_{ex} = (\omega_A - \omega_B)\sqrt{1 - 8\tau_{A,B}^{-2}(\omega_A - \omega_B)^{-2}} \tag{6.22}$$

when $\tau_{A,B}^{-1} < |\omega_A - \omega_B|$.

For *fast exchange* $\left(\tau_A^{-1}, \tau_B^{-1} >> |\omega_A - \omega_B|\right)$, the imaginary part of Eq. 6.17 is given by:

$$v(\omega) = \gamma_e B_1 M_o \frac{fT_{2,A}^{-1} + (1-f)T_{2,B}^{-1} + f^2(1-f)^2(\omega_A - \omega_B)^2(\tau_A + \tau_B)}{\left(fT_{2,A}^{-1} + (1-f)T_{2,B}^{-1} + f^2(1-f)^2(\omega_A - \omega_B)^2(\tau_A + \tau_B)\right)^2 + (f\omega_A + (1-f)\omega_B - \omega)^2} \tag{6.23}$$

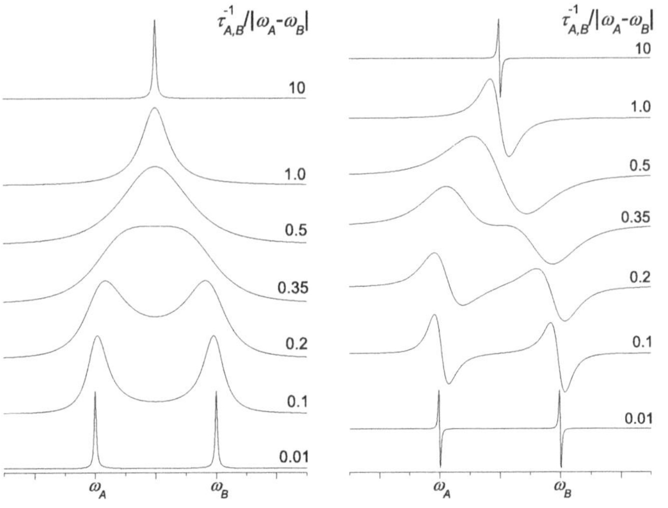

FIGURE 6.4 Absorption (left panel) and first-derivative (right panel) lineshapes for two-site exchange (Eq. 6.18). Exchange rate, $\tau_{A,B}^{-1}$, between equally populated sites A and B increases from bottom to top, as indicated.

The two lines have now collapsed to a single Lorentzian line centred at the mean resonance frequency:

$$\bar{\omega} = f\omega_A + (1-f)\omega_B \tag{6.24}$$

The collapsed line has an effective T_2 determined by the mean value: $\overline{T_2^{-1}} = fT_{2,A}^{-1} + (1-f)T_{2,B}^{-1}$ but is also again subject to a lifetime broadening which is determined by the exchange frequency:

$$T_{2,ex}^{-1} = f^2(1-f)^2(\omega_A - \omega_B)^2(\tau_A + \tau_B) \tag{6.25}$$

This situation of fast exchange is analogous to the motional narrowing condition that we treated in Chapter 5, where the hyperfine lines are at their isotropic positions but subject to a Lorentzian broadening that is given by Eq. 5.57 and depends on the rotational correlation time.

Finally at extremely fast exchange rates, the solution for the absorption lineshape is:

$$v(\omega) = -\gamma_e B_1 M_0 \times \frac{fT_{2,A}^{-1} + (1-f)T_{2,B}^{-1}}{\left(fT_{2,A}^{-1} + (1-f)T_{2,B}^{-1}\right)^2 + (f\omega_A + (1-f)\omega_B - \omega)^2} \tag{6.26}$$

The lineshape remains Lorentzian, centred at the mean angular resonance frequency given by Eq. 6.24, but with a width that is now independent of the exchange frequency and is given by the mean T_2-rate:

$$\overline{T_2^{-1}} = fT_{2,A}^{-1} + (1-f)T_{2,B}^{-1} \tag{6.27}$$

The condition for this is: $(\omega_A-\omega_B)^2(\tau_A+\tau_B) << T_{2A}^{-1}, T_{2B}^{-1}$, i.e., negligible exchange broadening.

Solution of the exchange-coupled Bloch equations embodied in Eqs. 6.19–6.27 summarizes all the effects of molecular dynamics on spin-label EPR spectra (see Figure 6.4). At slow rates of molecular motion (relative to the splittings of the spectral lines that are interchanged by the motion), the lines first broaden whilst remaining at their original positions. As the motional rates increase, the lines move in towards each other and finally collapse to their time-average position. The collapsed lines are broadened to an extent that depends on the rate of motion, and as this rate increases they progressively narrow, finally reaching a width independent of motional rate for extremely fast motions.

A direct application of the two-site exchange model is offered by lipids at the intramembrane interface with membrane-spanning proteins. Depending on lipid/protein ratio, conventional EPR spectra of lipids spin-labelled close to the terminal methyl end of the chain (see Figure 1.1) consist of two components characterized by differing segmental mobility: one from the fluid bilayer regions of the membrane and the other from the lipid–protein boundary (see Marsh and Horvath 1998; Marsh 2018). Two resolved components imply that exchange between the two environments is slow, but spectral simulation using the methods of this section reveals exchange broadening of the narrower component with the expected dependence on lipid/protein ratio and selectivity between different lipid species (Horváth et al. 1988a; Horváth et al. 1988b). Two-site lineshape simulations with exchange rates that are consistent between EPR frequencies of 9 and 35 GHz are also achieved (Horváth et al. 1994).

6.4 BLOCH EQUATIONS WITH SLOW SUDDEN JUMPS

Slow rotational motion corresponds to exchange between different angular orientations, where the spin label will have different resonance frequencies. From the results of Section 6.3 on two-site exchange, we see that the angularly anisotropic EPR lineshape is sensitive to rotation at rates comparable to the hyperfine and g-value anisotropies of the nitroxide. Taking account of this angular dependence, the Bloch equation for the complex transverse magnetization, $\hat{M}_+ = u + iv$, in the rotating frame becomes (cf. Eq. 6.7):

$$\frac{d\hat{M}_+(\omega)}{dt} = -\left(\frac{1}{T_2}+i\left(\omega_{m_I}(\Omega)-\omega\right)\right)\hat{M}_+(\omega)+i\gamma_e B_1 M_o(\Omega) \quad (6.28)$$

where $\omega_{m_I}(\Omega)$ is the resonance frequency for the hyperfine manifold with spin projection quantum number m_I, when the static magnetic field B_o is oriented at polar angles $\Omega \equiv (\theta,\phi)$ to the nitroxide axes. $M_o(\Omega)$ is the equilibrium z-magnetization when the field has this orientation.

The uncorrelated-jump approach used by Livshits (1976) is closest to simple two-site exchange, because the angular jumps are random and thus do not depend on the final orientation. Exchange takes place by sudden jumps from one position to another chosen at random in the (notional) rigid-limit powder lineshape. This is therefore an adiabatic approximation, which is valid in the slow-motion regime, but not in the fast regime. For the latter, we use motional narrowing theory which results in Lorentzian lineshapes, as described in Chapter 5. In this section, we introduce the random-jump model, which is computationally the least intensive of those based on the exchange-coupled Bloch equations.

For sudden random jumps, the probability, $1/\tau_l$, of jumping to a new orientation is the same for all sizes of angular jump (Korst et al. 1972). Here, τ_l is the lifetime between jumps. The equation of motion for the transverse magnetization of nitroxides with orientation Ω is then (cf. Eq. 6.12):

$$\frac{d\hat{M}_+(\Omega)}{dt} = -\tau_l^{-1}\hat{M}_+(\Omega)+\tau_l^{-1}P_o(\Omega)\int \hat{M}_+(\Omega')\cdot d\Omega' \quad (6.29)$$

where the first term on the right is the decrease in magnetization by jumps from orientation Ω to all other orientations, and the second term is the increase in magnetization by jumps to orientation Ω from all other orientations, Ω' (see Figure 6.5). The factor $P_o(\Omega)$ in the second term is the equilibrium orientational distribution ($P_o(\Omega)\cdot d\Omega \equiv (1/4\pi)\sin\theta\cdot d\theta d\phi$, for an isotropic distribution). It satisfies material balance under steady-state conditions: $d\hat{M}_+/dt = 0$ in Eq. 6.29, when angular exchange is at equilibrium.

Incorporating Eq. 6.29 for rotational exchange into Eq. 6.28, the steady-state rate equation for the transverse magnetization becomes:

$$\left(\frac{1}{T_2}+\frac{1}{\tau_l}+i\left(\omega_{m_I}(\Omega)-\omega\right)\right)\hat{M}_+(\Omega,\omega) - \tau_l^{-1}P_o(\Omega)\int \hat{M}_+(\Omega',\omega)\cdot d\Omega' = i\gamma_e B_1 P_o(\Omega) M_o \quad (6.30)$$

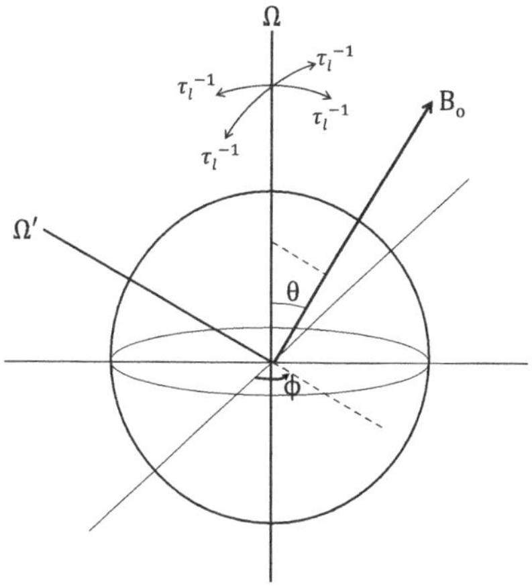

FIGURE 6.5 Uncorrelated sudden-jump model for rotational diffusion. Spin label, with static field B_o orientation $\Omega \equiv (\theta,\phi)$ relative to nitroxide axes, jumps to any other position (where field orientation is Ω') with frequency τ_l^{-1} irrespective of jump size.

which is the equivalent of Eq. 6.15 for two-site exchange. The factor $P_o(\Omega)$ appears on the right because M_o is the total z-magnetization, i.e., $M_o(\Omega) = P_o(\Omega)M_o$. We get the solution for the spectral lineshape, $\hat{M}_+(\omega)$, by integrating Eq. 6.30 over the angular distribution, Ω (Livshits 1976):

$$\hat{M}_+(\omega) \equiv \int \hat{M}_+(\Omega,\omega)\cdot \mathrm{d}\Omega = \frac{i\gamma_e B_1 M_o I_{m_I}(\omega)}{1-\tau_l^{-1} I_{m_I}(\omega)} \tag{6.31}$$

where the integral for each hyperfine manifold m_I is defined by:

$$I_{m_I}(\omega) = \int \frac{P_o(\Omega)\cdot \mathrm{d}\Omega}{T_2^{-1}+\tau_l^{-1}+i(\omega_{m_I}(\Omega)-\omega)} \tag{6.32}$$

and $\int P_o(\Omega)\cdot \mathrm{d}\Omega = 1$ for the normalized distribution. This simple result (i.e., Eq. 6.31), which reduces to a single numerical integration (Eq. 6.32), arises because the angular jumps are uncorrelated. The situation is more complicated for Brownian diffusion, as we see later in Section 6.5, because jumps between adjacent angular sites are then coupled.

Lastly, the angular dependence of the resonance frequency for a nitroxide spin label is given by:

$$\omega_{m_I}(\theta,\phi) = g(\theta,\phi)B_o\beta_e/\hbar + A(\theta,\phi)m_I \tag{6.33}$$

where the angular-dependent g-value $g(\theta,\phi)$ and hyperfine coupling $A(\theta,\phi)$ are given by Eqs. 2.11 and 2.12, respectively (see Chapter 2). We require this expression for the integration in Eq. 6.32. We perform the integration for as many separate values of B_o as needed to define the complete (powder) lineshape. Line shifts as a function of jump lifetime deduced from slow-motional simulations using this model are shown as dotted lines without symbols in Figure 6.7 that is given later, in Section 6.5.

Using the rigid-limit anisotropic resonance frequency, i.e., Eq. 6.33, in the Bloch equations is an adiabatic approximation, because it does not allow for motional averaging of the spin Hamiltonian tensors during rotation. This approximation is acceptable in the slow rotational regime. Non-adiabatic effects appear first in the jump model at times $\tau_l < 50$ ns, with discrepancies in the region of 30% down to 10 ns. The method is not applicable to fast rotation, i.e., $\tau_l < 3$ ns. In the fast regime, motional narrowing produces an isotropic three-line ^{14}N-nitroxide spectrum, with Lorentzian lineshapes differentially broadened by rotational modulation of the hyperfine and Zeeman anisotropies (see Section 5.8 in Chapter 5). We then simply replace $\omega_{m_I}(\Omega)$ in Eq. 6.28 by the isotropic motionally-averaged value $\langle\omega_{m_I}\rangle$, and add the $T_2^{-1}(m_I)$ contribution from time-dependent perturbation theory (i.e., Eq. 5.58) to the intrinsic transverse relaxation rate.

6.5 BLOCH EQUATIONS COUPLED BY SLOW ROTATIONAL DIFFUSION

The opposite situation to the sudden-jump model is Brownian diffusion where rotation takes place in small random angular steps, $\Delta\theta$. McCalley et al. (1972) introduce Brownian rotational diffusion into the Bloch equations by including terms for exchange between adjacent angular sites. We treat this discrete model, which must be solved by numerical methods, in the present section.

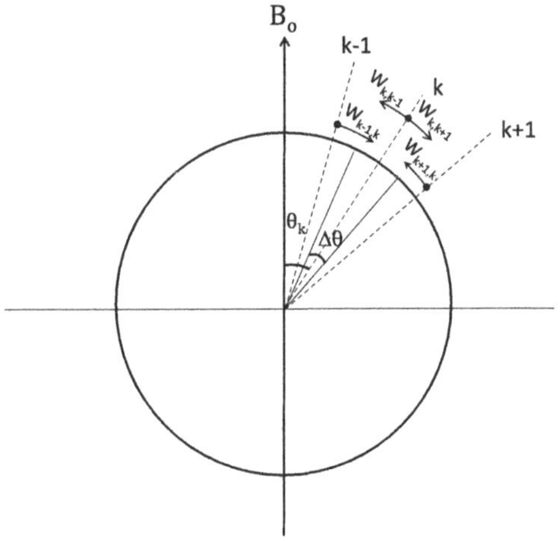

FIGURE 6.6 Small-step Brownian rotational diffusion. Nitroxide in kth angular zone (with orientation θ_k relative to B_o, and width $\Delta\theta$) makes diffusive transitions to adjacent (k ±1)th zones at rates $W_{k,k\pm1}$. Relative population in kth zone is $\sin\theta_k$.

We describe Brownian motion with the rotational diffusion equation (see Sections 7.2 and 7.5 in Chapter 7). The time dependence of the transverse magnetization that comes from molecular motion is then (see Eq. 7.17):

$$\frac{\mathrm{d}\hat{M}_+(\Omega)}{\mathrm{d}t} = -D_R\nabla_\Omega^2\hat{M}_+(\Omega) = -\frac{D_R}{\sin\theta}\frac{\partial}{\partial\theta}\left(\sin\theta\frac{\partial\hat{M}_+(\theta)}{\partial\theta}\right) \tag{6.34}$$

where we assume unrestricted isotropic rotation, and D_R is the rotational diffusion coefficient. We approximate the differentials as small discrete steps $\Delta\theta$:

$$\nabla_\theta^2\hat{M}_+(\theta) = \frac{1}{(\Delta\theta)^2}\left(\frac{\sin\left(\theta+\frac{1}{2}\Delta\theta\right)}{\sin\theta}\hat{M}_+(\theta+\Delta\theta) + \frac{\sin\left(\theta-\frac{1}{2}\Delta\theta\right)}{\sin\theta}\hat{M}_+(\theta-\Delta\theta) - 2\cos\left(\tfrac{1}{2}\Delta\theta\right)\cdot\hat{M}_+(\theta)\right) \tag{6.35}$$

by using the relation: $\partial\hat{M}_+(\theta)/\partial\theta \cong \left[\hat{M}_+\left(\theta+\frac{1}{2}\Delta\theta\right) - \hat{M}_+\left(\theta-\frac{1}{2}\Delta\theta\right)\right]/\Delta\theta$ and equivalents. From this, we define adjacent angular zones k–1, k and k+1 centred about θ_{k-1}, θ_k and θ_{k+1}, respectively, where $\theta_k \equiv \left(k-\frac{1}{2}\right)\Delta\theta$ (see Figure 6.6). In the rotating frame, the complex transverse magnetization of the kth zone is:

$$\hat{M}_k \equiv P_o(\theta_k)\hat{M}_+(\theta_k) = \sin\theta_k\hat{M}_+(\theta_k) \tag{6.36}$$

where $P_o(\theta_k)$ is the angular degeneracy for an isotropic spatial distribution (see Figure 2.5). Using Eq. 6.36 to write Eq. 6.35 in terms of angular zones, the diffusion rate becomes:

$$D_R \nabla_k^2 \widehat{M}_k = W_{k+1,k} \widehat{M}_{k+1} + W_{k-1,k} \widehat{M}_{k-1} - \left(W_{k,k+1} + W_{k,k-1} \right) \widehat{M}_k \tag{6.37}$$

where $W_{k,k+1}$ is the probability of transition from the kth angular zone to the (k+1)th zone, and so on. From Eqs. 6.35 to 6.37, the angular transition probabilities are defined by:

$$W_{k\pm1,k} = \frac{1}{6\tau_R (\Delta\theta)^2} \frac{\sin\left(\theta_k \pm \frac{1}{2}\Delta\theta\right)}{\sin\left(\theta_k \pm \Delta\theta\right)} \tag{6.38}$$

$$W_{k,k+1} + W_{k,k-1} = \frac{1}{3\tau_R (\Delta\theta)^2} \cos\left(\tfrac{1}{2}\Delta\theta\right) \tag{6.39}$$

where $\tau_R \equiv 1/(6D_R)$ is the rotational correlation time for isotropic Brownian diffusion. If $N = \pi/(2\Delta\theta)$ is the total number of angular zones, then $k \geq 1$, $k+1 \leq N$ for Eq. 6.38 and $1 \leq k \leq N-1$ for Eq. 6.39. When $k = N$, Eq. 6.39 is replaced by:

$$W_{N,N-1} = \frac{1}{6\tau_R (\Delta\theta)^2} \frac{\cos(\Delta\theta)}{\cos\left(\frac{1}{2}\Delta\theta\right)} \tag{6.40}$$

to preserve detailed balance, because then $W_{N,N+1} = 0$ by definition.

Incorporating the diffusion equations (Eqs. 6.34 and 6.37) into Eq. 6.28 for the Bloch equations in the rotating frame, we get the rate equation for the magnetization of the kth angular zone:

$$\frac{\mathrm{d}\widehat{M}_k}{\mathrm{d}t} = -\left(\frac{1}{T_2} + W_{k,k+1} + W_{k,k-1} + i(\omega_k - \omega) \right) \widehat{M}_k + W_{k+1,k} \widehat{M}_{k+1} + W_{k-1,k} \widehat{M}_{k-1} + i\gamma_e B_1 M_{o,k} \tag{6.41}$$

where $\omega_k \equiv \omega_{m_I}(\theta_k)$ is given by Eq. 6.33 but with axial g- and A-tensors, corresponding to Eqs. 6.34 and 6.36 which depend only on θ. Separating the real and imaginary parts of the complex transverse magnetization, $\widehat{M}_k = u_k + iv_k$, we get the steady-state relations (McCalley et al. 1972):

$$\left(\frac{1}{T_2} + W_{k,k+1} + W_{k,k-1} \right) u_k - (\omega_k - \omega) v_k - W_{k+1,k} u_{k+1} - W_{k-1,k} u_{k-1} = 0 \tag{6.42}$$

$$\left(\frac{1}{T_2} + W_{k,k+1} + W_{k,k-1} \right) v_k + (\omega_k - \omega) u_k - W_{k+1,k} v_{k+1} - W_{k-1,k} v_{k-1} = \gamma_e B_1 M_{o,k} \tag{6.43}$$

where the z-magnetization $M_{o,k}$ also satisfies a relation like Eq. 6.36. In Eqs. 6.42 and 6.43, the magnetization of the kth angular zone is coupled to that of the adjacent (k − 1)th and (k + 1)th zones, as defined by Eq. 6.37, and so on. We have $2N$ coupled linear simultaneous equations; these are solved numerically for fixed values of the static field, $B\left(\equiv \omega/\gamma_e\right)$, within the powder spectrum. Summing over all v_k, at each B, gives the EPR absorption spectrum as a function of the scan variable B. Summing over the in-phase magnetization, u_k, likewise gives the dispersion spectrum. In the axial approximation, taking $N \geq 30$ angular zones provides sufficient convergence (McCalley et al. 1972).

We use only one angular variable (θ) here, but the simulation should still work well in the outer wings of the spectrum, because these are determined by the hyperfine tensor, which is axial. However, agreement may not be as good in the centre of the spectrum because the g-tensor is completely non-axial. Therefore, we use inward shifts of the outer extrema in the first-derivative spectrum as diagnostic parameters to determine rotational correlation times. Partly, these are analogous to corresponding inward shifts that arise in slow two-site exchange (see Figure 6.4 in Section 6.3). Figure 6.7 shows the inward shifts of low- and high-field outer peaks in the first-derivative spectra of a ^{14}N-nitroxide as a function of the isotropic rotational correlation time. Calculations using the diffusion-coupled Bloch equations are compared with those based on the quantum-mechanical stochastic Liouville equation (Goldman et al. 1972a; Gordon and Messenger 1972), which is described in the remaining sections of this chapter. We see good agreement between the two approaches (see discussion by Gordon and Messenger 1972), and also relatively little dependence on the spin-Hamiltonian parameters, which differ slightly between the various calculations.

For a given rotational correlation time, the shift is greater for the high-field line than for the low-field line, because of the greater spectral anisotropy in the high-field manifold. Shifts of

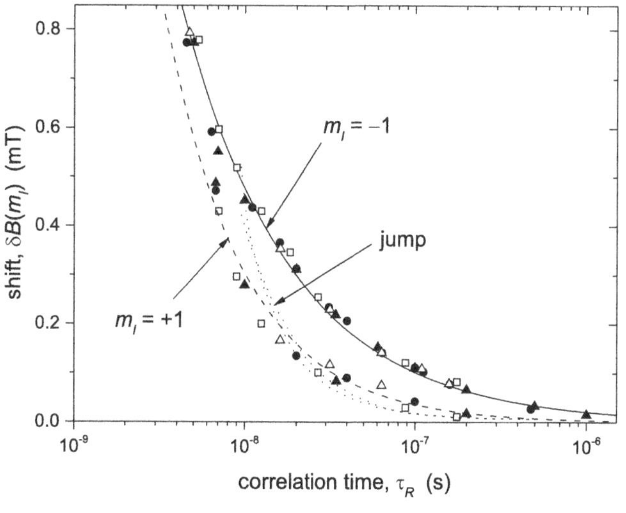

FIGURE 6.7 Inward shifts $\delta B(m_I)$ of low-field ($m_I = +1$; dashed line) and high-field ($m_I = -1$; solid line) outer peaks in first-derivative EPR spectra of ^{14}N-nitroxide, simulated for isotropic Brownian diffusion with rotational correlation times τ_R in slow-motion regime. *Solid symbols*: from Bloch equations (Section 6.5); *open symbols*: from stochastic-Liouville equation or equivalent (Section 6.10). $(A_\perp, A_\parallel) = (0.65, 3.43)$ mT (circles), (0.58,3.08) mT (triangles), (0.6,3.2) mT (squares); $g_\perp - g_\parallel = 0.0051$ (circles), 0.0053 (triangles), 0.0048 (squares); $T_2 = (2.2 - 3.3) \times 10^{-8}$ s. *Dotted lines* (without symbols): strong-jump model (Section 6.4). (Data from McCalley et al. 1972; Gordon and Messenger 1972; Shimshick and McConnell 1972; Livshits 1976.)

the high-field extremum are approximately proportional to $1/\tau_R^{2/3}$. The solid line for $m_I = -1$ in Figure 6.7 is a least-squares fit given by: $\delta B(-1) = 1.974 \times 10^{-6} \tau_R^{-0.673}$, where δB is in mT and τ_R in s. Calibrations such as this allow extrapolation as a function of viscosity to determine the appropriate rigid-limit positions of the extrema (Shimshick and McConnell 1972). The dependence on τ_R in Figure 6.7 differs from that on $\tau_{A,B}$ in Eq. 6.22 of Section 6.3, because the two-site model approximates better to random jumps than to the small-step Brownian diffusion treated here. Simulation results from the sudden-jump model of Section 6.4 are shown in Figure 6.7 by dotted lines without symbols. As expected, the shifts of high- and low-field lines are much more similar in size than found for the Brownian case. Also, shifts are stronger for the jump model: they are approximately proportional to $1/\tau_I^{3/2}$, which is closer to the approximate $1/\tau_{AB}^2$ dependence predicted by Eq. 6.22. More extensive correlation-time calibrations that are based on the outer hyperfine splitting, instead of on the absolute shifts, are given in Section 7.14 of Chapter 7. These latter calibrations are performed with the stochastic Liouville equation.

Application of the present method to determining rotational correlation times of spin-labelled proteins in aqueous sucrose solutions (McCalley et al. 1972; Shimshick and McConnell 1972) confirms that Brownian diffusion is appropriate for these spin-labelled macromolecules. As for the sudden-jump treatment of Section 6.4, this method is an adiabatic approximation because it also uses Eq. 6.33 with rigid-limit spin Hamiltonian tensors. Therefore, it applies only for slow-motion simulations. McCalley et al. (1972) estimate that the adiabatic approximation begins to break down for rotational correlation times $\tau_R < 5$ ns. This is considerably less restrictive than for the jump model of Section 6.4. In practice, we now perform simulations mostly with the stochastic Liouville equation, which is valid over the full range of motional rates. We devote the remainder of the chapter to describing these methods.

6.6 DENSITY-MATRIX METHODS

We see from the classical treatment given above that the EPR experiment detects the macroscopic transverse magnetization, M_x, of the sample. In a quantum mechanical description, the transverse magnetization (or magnetic moment) is directly proportional to the transverse component of the electron spin, S_x (see Chapter 1). The macroscopic magnetization is then determined by the ensemble average of the expectation value, $\overline{\langle S_x \rangle}$, for the entire spin system. As explained in Appendix N, we obtain the ensemble average in a quantum mechanical spin system by using the spin density-matrix, σ:

$$\overline{\langle S_x \rangle} = \mathrm{Tr}(\sigma S_x) \tag{6.44}$$

where the trace Tr is the sum of the diagonal elements of the matrix product σS_x (see Eq. N.4). Because we are dealing with relaxation processes, we are interested primarily in the deviations, $\chi = \sigma - \sigma_o$, of the spin density-matrix from its equilibrium value, σ_o. The macroscopic transverse spin then becomes:

$$\overline{\langle S_x \rangle} = \tfrac{1}{2}\mathrm{Tr}\big(\chi(S_+ + S_-)\big) + \tfrac{1}{2}\mathrm{Tr}\big(\sigma_o(S_+ + S_-)\big) \tag{6.45}$$

where we have introduced the shift operators, $S_\pm$: $S_x = \frac{1}{2}(S_+ + S_-)$ (see Appendix E).

The equilibrium spin density-matrix corresponds to the Boltzmann distribution in statistical mechanics. The energy states that appear in the exponential Boltzmann factors are eigenvalues of the static spin Hamiltonian, $\mathcal{H}_o$, and the equilibrium density-matrix operator is given by (see Eq. N.10 in Appendix N):

$$\sigma_o = \frac{\exp(-\mathcal{H}_o/k_B T)}{\mathrm{Tr}\big[\exp(-\mathcal{H}_o/k_B T)\big]} \tag{6.46}$$

where k_B is Boltzmann's constant and T is the absolute temperature. The denominator in Eq. 6.46 is the statistical-mechanical partition function, which we need to normalize σ_o. Magnetic interactions are small compared with usual thermal energies: $\mathcal{H}_o << k_B T$, and the high-temperature approximation is appropriate. We therefore expand the exponential in Eq. 6.46, keeping only the first terms:

$$\sigma_o = \frac{1}{N_Z}\left(1 - \frac{\mathcal{H}_o}{k_B T} + \ldots\right) \cong \frac{1}{N_Z}\left(1 - \frac{\hbar\omega_o}{k_B T} S_z\right) \tag{6.47}$$

where the partition function in the high-temperature approximation is simply equal to the total number of energy states in the system: $N_Z = (2S+1)(2I+1)$, with $S = \frac{1}{2}$ for the electron spin, and $I = \frac{1}{2}$ or $I = 1$ for ^{15}N or ^{14}N nuclei, respectively. We use the high-field approximation, $\mathcal{H}_o = \hbar\omega_o S_z$, for the right-hand equality of Eq. 6.47. In this case, the eigenfunctions of σ_o are those of the spin operator S_z, and therefore, the second term on the right of Eq. 6.45 does not contribute to the ensemble average $\overline{\langle S_x \rangle}$ in CW-EPR.

Now we can write Eq. 6.45 in terms of matrix elements referred to the usual electron-spin eigenstates, viz., $|M_S = \pm\frac{1}{2}\rangle$:

$$\overline{\langle S_x \rangle} = \tfrac{1}{2}\left(\langle -\tfrac{1}{2}|\chi S_+|-\tfrac{1}{2}\rangle + \langle +\tfrac{1}{2}|\chi S_-|+\tfrac{1}{2}\rangle\right) = \tfrac{1}{2}(\chi_{-+} + \chi_{+-}) \tag{6.48}$$

where we define the matrix elements of the reduced spin density-matrix by: $\chi_{-+} = \langle -\frac{1}{2}|\chi|+\frac{1}{2}\rangle$, and so on. We get this result by operating on the right with $S_\pm$, according to Eqs. E.11 and E.12 of Appendix E. Thus, the EPR signal that we detect from the transverse magnetization is determined by the off-diagonal elements, χ_{+-} and χ_{-+}, of the reduced spin density-matrix. The on-diagonal elements, χ_{++} and χ_{--}, contribute primarily to the saturation properties of the spectrum at high intensities of the radiation field. Because the density matrix is Hermitian (see Appendix N), i.e., $\chi_{+-} = \chi_{-+}^*$ where the asterisk indicates the complex conjugate, we can write Eq. 6.48 as:

$$\overline{\langle S_x \rangle} = \mathrm{Re}(\chi_{-+}) \tag{6.49}$$

where Re denotes the real part of the complex matrix element. To determine the EPR lineshape, we therefore need only evaluate the χ_{-+} off-diagonal matrix element with respect to the electron spin states. We must do this for each hyperfine state, which is labelled with the nuclear magnetic quantum number m_I. The matrix elements of the reduced spin density-matrix that we then need are: $\chi_{-+}^{m_I m_I'}$, which include matrix elements that are off-diagonal in m_I, because we must retain pseudosecular terms (where $m_I' = m_I \pm 1$) when calculating slow-motion spin-label spectra (see Section 6.9 later).

6.7 SLOW-MOTION SIMULATIONS: STOCHASTIC LIOUVILLE EQUATION

The equation of motion for the spin density matrix, σ, is the quantum-mechanical Liouville equation (see Eq. N.18 in Appendix N):

$$\frac{\partial \sigma}{\partial t} = -\frac{i}{\hbar}\left[\mathcal{H}(t), \sigma\right] \tag{6.50}$$

where $\mathcal{H}(t)$ is the randomly varying Hamiltonian that gives rise to spin relaxation. The random variable is the angular orientation, where the probability, P, of finding orientation Ω at time t is given by:

$$\frac{\partial}{\partial t} P(\Omega,t) = -\Gamma_\Omega P(\Omega,t) \tag{6.51}$$

where the Markov operator, Γ_Ω, is independent of time for a stationary process. The equilibrium orientational distribution, $P_o(\Omega)$, is defined by the condition:

$$\Gamma_\Omega P_o(\Omega) = 0 \tag{6.52}$$

The most familiar stochastic process is Brownian rotational diffusion (see Section 7.2). Then Eq. 6.51 is the diffusion equation and Γ_Ω is the diffusion operator, $-D_R \nabla_\Omega^2$, where D_R is the rotational diffusion coefficient and ∇_Ω^2 is the angular part of the Laplacian operator (see, e.g., Eq. 7.17). Combining Eqs. 6.50–6.52 leads to the stochastic Liouville equation (Kubo 1969):

$$\frac{\partial}{\partial t}\sigma(\Omega,t) = -\frac{i}{\hbar}\left[\mathcal{H}(\Omega), \sigma(\Omega,t)\right] - \Gamma_\Omega \sigma(\Omega,t) \tag{6.53}$$

in which the time dependence of the random process is then included solely in the elements of the spin density-matrix, $\sigma(\Omega,t)$. In this case, the reduced spin density-matrix is given by: $\chi(\Omega,t) = \sigma(\Omega,t) - \sigma_o(\Omega)$, where $\sigma_o(\Omega)$ is the equilibrium value of the spin density-matrix. The stochastic Liouville equation then becomes:

$$\frac{\partial \chi(\Omega,t)}{\partial t} + \frac{i}{\hbar}\left[\mathcal{H}(\Omega), \chi(\Omega,t)\right] + \Gamma_\Omega \chi(\Omega,t) = \frac{i}{\hbar}\left[\mathcal{H}(\Omega), \sigma_o(\Omega)\right] \tag{6.54}$$

where we use the equilibrium conditions $\partial \sigma_o / \partial t = 0$ and $\Gamma_\Omega \sigma_o = 0$ in rewriting Eq. 6.53.

The total Hamiltonian, $\mathcal{H}(\Omega)$, consists of an orientation-independent part, $\mathcal{H}_o$, an orientation-dependent part, $\mathcal{H}_1(\Omega)$, and a contribution, $\varepsilon(t)$, from the interaction of the electron spin with the oscillating radiation magnetic field, B_1:

$$\mathcal{H}(\Omega) = \mathcal{H}_o + \mathcal{H}_1(\Omega) + \varepsilon(t) \tag{6.55}$$

By separating orientation-independent and orientation-dependent parts of the Hamiltonian, we avoid the adiabatic approximation (i.e., Eq. 6.33) that is used in Sections 6.4 and 6.5. We see from generalization of Eq. 6.46 that σ_o commutes with $\mathcal{H}_o + \mathcal{H}_1$, i.e., $\left[\sigma_o, \left(\mathcal{H}_o + \mathcal{H}_1\right)\right] = 0$. Therefore, only $\varepsilon(t)$ contributes to the commutator on the right-hand side of Eq. 6.54. In the high-field approximation, the orientation-independent term is given by:

$$\mathcal{H}_o = \hbar \omega_o S_z \tag{6.56}$$

where ω_o is simply the zero-order (i.e., isotropic) angular resonance frequency. For circularly polarized B_1-radiation of angular frequency ω, we have:

$$\varepsilon(t) = \tfrac{1}{2}\gamma_e B_1 \left(S_+ \exp(-i\omega t) + S_- \exp(i\omega t)\right) \tag{6.57}$$

which we can consider as deriving from one of the circular components of the linear oscillating field: $B_{1,x} = 2B_1 \cos \omega t$.

From Eqs. 6.55–6.57 and 6.47, we can write the stochastic Liouville equation for $\chi(\Omega,t)$ (i.e., Eq. 6.54) in terms of the components of $\mathcal{H}(\Omega)$:

$$\frac{\partial \chi(\Omega,t)}{\partial t} + i\omega_o\left[S_z, \chi(\Omega,t)\right] + \frac{i}{\hbar}\left[\mathcal{H}_1(\Omega), \chi(\Omega,t)\right] + \Gamma_\Omega \chi(\Omega,t) = \frac{i\gamma_e B_1 \hbar \omega_o}{2N_Z k_B T} S_- \exp(i\omega t) \tag{6.58}$$

where we use the commutation relations: $\left[S_\pm, S_z\right] = \mp \hbar S_\pm$ (see Appendix E) on the right-hand side. However, we omit the term containing S_+ because we need only the χ_{-+} matrix element to calculate $\overline{\langle S_x \rangle}$ (see Eq. 6.49) and $S_+\left|+\tfrac{1}{2}\right\rangle = 0$. We do not include the commutator $[\varepsilon(t), \chi(\Omega,t)]$ in the left-hand side of Eq. 6.58 because we need this only for the saturation behaviour at high B_1-field strengths (note that $\chi \ll \rho_o$). The solution of Eq. 6.58, in the absence of saturation, is therefore of the form:

$$\chi(\Omega,t) = Z(\Omega,\omega)\exp(i\omega t) \tag{6.59}$$

where $Z(\Omega,\omega)$ is the reduced spin density-matrix in the rotating frame and is independent of t in the steady state. Note that $Z(\omega)$ is complex: $Z \equiv Z' + iZ''$ and is analogous to the complex magnetization, $\hat{M}_+$, that we introduced to solve the Bloch equations in Section 6.2. Here, Z' is the component in phase with the B_1-radiation field, giving the dispersion signal, and Z'' is the component that lags by 90°, giving the absorption signal and the EPR lineshape.

Substituting Eq. 6.59 into Eq. 6.58 and taking the $\langle M_S = -\frac{1}{2} | \ldots | M_S = +\frac{1}{2} \rangle$ matrix element, we then get (Freed et al. 1971):

$$(\omega - \omega_o) Z_{-+} + \frac{1}{\hbar} \left[\mathcal{H}_1(\Omega), Z(\Omega) \right]_{-+} - i \left(\Gamma_\Omega Z(\Omega) \right)_{-+} = \frac{\gamma_e B_1 \hbar \omega_o}{2 N_Z k_B T} \tag{6.60}$$

as the steady-state equation for Z_{-+}. To obtain the total spectral intensity we need to sum $Z_{-+}(\Omega, \omega)$ over the complete distribution of angles $P_o(\Omega)$, as for the macroscopic magnetization (Freed et al. 1971):

$$\bar{Z}_{-+}(\omega) = \int Z_{-+}(\Omega, \omega) P_o(\Omega) \cdot \mathrm{d}\Omega \tag{6.61}$$

where $P_o(\Omega)$ is specified by Eq. 6.52. Performing this operation on Eq. 6.60, therefore, gives:

$$\begin{aligned} &(\omega - \omega_o) \bar{Z}_{-+} + \frac{1}{\hbar} \int \left[\mathcal{H}_1(\Omega), Z(\Omega) \right]_{-+} P_o(\Omega) \mathrm{d}\Omega \\ &- i \int \left(\Gamma_\Omega Z(\Omega) \right)_{-+} P_o(\Omega) \mathrm{d}\Omega = \frac{\gamma_e B_1 \hbar \omega_o}{2 N_Z k_B T} \end{aligned} \tag{6.62}$$

We evaluate the integrals over Ω by expanding $Z_{-+}(\Omega)$ in a complete orthogonal set of eigenfunctions $G_{m'}(\Omega)$ of the Markov operator Γ_Ω, with negative eigenvalues $\tau_{m'}^{-1}$ (Freed et al. 1971):

$$Z_{-+}(\Omega, \omega) = \sum_{m'} \left(C_{m'}(\omega) \right)_{-+} G_{m'}(\Omega) \tag{6.63}$$

where the matrix of expansion coefficients $C_{m'}$ is an operator in spin space that depends on the angular frequency ω but is independent of the orientation Ω. (The prime is for consistency with expansions using explicit eigenfunctions that appear later, in Sections 6.8, 6.9). For rotational diffusion, for instance, the eigenfunctions $G_{m'}(\Omega)$ are the spherical harmonics $Y_{L',M'}(\Omega)$ or Wigner rotation matrices $\mathcal{D}^{L'}_{K',M'}(\Omega)$ (see Sections 7.5 and 7.6 in Chapter 7). From averaging Eq. 6.63 over Ω for an isotropic distribution, we find that the total spectral intensity (i.e., Eq. 6.61) is given simply by the zero-order coefficient:

$$\bar{Z}_{-+}(\omega) = \left(C_0(\omega) \right)_{-+} \tag{6.64}$$

This is because the isotropic equilibrium orientation distribution is $P_o(\Omega) \propto G_0(\Omega)\,(=1)$ that is orthogonal to all $G_{m'}(\Omega)$ with $m' \neq 0$. In particular, the absorption lineshape is given by:

$$v(\omega) = \mathrm{Im}\left(C_o(\omega) \right)_{-+} \tag{6.65}$$

which follows from the complex nature of Z_{-+} that we discussed above, and its relation to the phase and frequency of the B_1-radiation that is defined by Eq. 6.59.

Substituting Eq. 6.63 into Eq. 6.62 yields:

$$\begin{aligned} &(\omega - \omega_o) \bar{Z}_{-+} + \frac{1}{\hbar} \sum_{m'} \int \left[\mathcal{H}_1(\Omega), C_{m'} \right]_{-+} G_{m'}(\Omega) P_o(\Omega) \cdot \mathrm{d}\Omega \\ &= \frac{\gamma_e B_1 \hbar \omega_o}{2 N_Z k_B T} \end{aligned} \tag{6.66}$$

where the third term on the left of Eq. 6.62 vanishes because: $\Gamma_\Omega G_{m'}(\Omega) = \tau_{m'}^{-1} G_{m'}(\Omega)$ and, on integration, orthogonality leaves only $G_0(\Omega)$ $(\propto P_o(\Omega))$ that has the eigenvalue $\tau_0^{-1} = 0$. Finally, pre-multiplying Eq. 6.66 by $G_m^*(\Omega)$ and integrating over Ω to exploit the orthogonality of the eigenfunctions yields:

$$\begin{aligned} &\left(\omega - \omega_o - i \tau_m^{-1} \right) \left(C_m \right)_{-+} + \frac{1}{\hbar} \sum_{m'} \int G_m^*(\Omega) \left[\mathcal{H}_1(\Omega), C_{m'} \right]_{-+} G_{m'}(\Omega) \cdot \mathrm{d}\Omega \\ &= \frac{\gamma_e B_1 \hbar \omega_o}{2 N_Z k_B T} \delta_{m,0} \end{aligned} \tag{6.67}$$

when the distribution is isotropic, i.e., $P_o(\Omega) = 1$, and the eigenfunctions are normalized, viz. $\int G_m^*(\Omega) G_{m'}(\Omega) \cdot \mathrm{d}\Omega = \delta_{m',m}$. The right-hand side of Eq. 6.67 is non-zero only when $m = 0$. For lineshape calculations, the amplitude factor on the right is not important, so in what follows, we simply set: $\gamma_e B_1 \hbar \omega_o / (2 N_Z k_B T) = 1$.

Equation 6.67 yields coupled simultaneous equations for the matrix elements of the coefficients $(C_m)_{-+}$, which we must solve for $(C_0)_{-+}$ that gives the spectral lineshape (see Eq. 6.65). The number of coefficients that we must include to obtain convergence in the calculation increases as the rate of motion decreases. It depends on the product $|\mathcal{H}_1(\Omega)| \tau_{m'}$, because the eigenvalue is inversely related to the correlation time, $\tau_{m'}$. For larger values of $|\mathcal{H}_1(\Omega)| \tau_{m'}$, we need more coefficients $(C_m)_{-+}$. By considering only one order of coefficients beyond the zeroth, we get motional-narrowing theory. Note that, because of the separation of $\mathcal{H}_o$ and $\mathcal{H}_1(\Omega)$ in Eq. 6.55, the treatment is valid for the full range of motional correlation times $\tau_{m'}$ and is not restricted by any adiabatic approximation.

6.8 A SIMPLE EXAMPLE: AXIAL *g*-TENSOR

As an illustration of the stochastic-Liouville method, we consider first a single EPR line that is broadened solely by the secular contribution from modulation of the axial anisotropy $(\Delta g = g_{\parallel} - g_{\perp})$ of the g-tensor. In this axial case, the angular-dependent spin Hamiltonian in Eq. 6.67 is (cf. Eq. 5.15 in Chapter 5):

$$\mathcal{H}_1(\theta) = \tfrac{2}{3} \Delta g \beta_e B_o P_2(\cos\theta) S_z \tag{6.68}$$

where $P_2(\cos\theta) = \frac{1}{2}\left(3\cos^2\theta - 1\right)$ is a second-order Legendre polynomial (see Appendix V.1), and θ is the angle between the static magnetic field B_o and the principal axis of the g-tensor. From

Eq. 6.68, the matrix element of the spin-function commutator on the left side of Eq. 6.67 is:

$$\begin{aligned}[S_z, C_{m'}]_{-+} &\equiv \sum_{M_S=\pm\frac{1}{2}} \langle -\tfrac{1}{2}|S_z|M_S\rangle\langle M_S|C_{m'}|+\tfrac{1}{2}\rangle \\ &- \sum_{M_S=\pm\frac{1}{2}} \langle -\tfrac{1}{2}|C_{m'}|M_S\rangle\langle M_S|S_z|+\tfrac{1}{2}\rangle \\ &= -\tfrac{1}{2}\langle -\tfrac{1}{2}|C_{m'}|+\tfrac{1}{2}\rangle - \tfrac{1}{2}\langle -\tfrac{1}{2}|C_{m'}|+\tfrac{1}{2}\rangle = -(C_{m'})_{-+}\end{aligned} \tag{6.69}$$

because $\langle -\frac{1}{2}|S_z|+\frac{1}{2}\rangle = 0 = \langle +\frac{1}{2}|S_z|-\frac{1}{2}\rangle$.

For isotropic rotational diffusion, the Markov operator is: $\Gamma_\theta \to -D_R\nabla_\theta^2$, eigenfunctions of which are Legendre polynomials:

$$\nabla_\theta^2 \mathrm{P}_L(\cos\theta) = -L(L+1)\mathrm{P}_L(\cos\theta) \tag{6.70}$$

with (negative) eigenvalues $\tau_L^{-1} = L(L+1)D_R$, where D_R is the rotational diffusion coefficient (see also Section 7.5 in Chapter 7). Here ∇_θ^2 is the angular part of the axially symmetric Laplacian operator. The P_L are normalized by using the orthogonality relation: $\int_{-1}^{1} \mathrm{P}_L(x)\mathrm{P}_{L'}(x)\cdot \mathrm{d}x = 2\delta_{L,L'}/(2L+1)$, where $x = \cos\theta$ and $\mathrm{d}x = \sin\theta \cdot \mathrm{d}\theta$ (see Eq. V.4). We can then write Eq. 6.67 by using $\mathrm{P}_{L'}(x)$ for the eigenfunction expansion in $G_{m'}(\Omega)$:

$$\begin{aligned}&\left(\omega - \omega_o - i\tau_L^{-1}\right)(C_L)_{-+} - (2L+1)\frac{\Delta g\beta_e}{6\hbar}B_o \\ &\times \sum_{L'} (C_{L'})_{-+} \int_{-1}^{1} \mathrm{P}_L(x)\left(3x^2-1\right)\mathrm{P}_{L'}(x)\cdot \mathrm{d}x = \delta_{L,0}\end{aligned} \tag{6.71}$$

where we set the amplitude factor equal to unity on the right. We express the argument of the integral on the left-hand side in terms of the recursion relation for Legendre polynomials (Eq. V.5 in Appendix V):

$$(2L+1)x\mathrm{P}_L(x) = (L+1)\mathrm{P}_{L+1}(x) + L\mathrm{P}_{L-1}(x) \tag{6.72}$$

and then evaluate it by using the orthogonality relation. The result is that Eq. 6.71 becomes:

$$\begin{aligned}&\left(\omega - \omega_o - i\tau_L^{-1}\right)(C_L)_{-+} \\ &- \frac{\Delta g\beta_e}{\hbar}B_o\left[a_{L-2}(C_{L-2})_{-+} + a_L(C_L)_{-+} + a_{L+2}(C_{L+2})_{-+}\right] = \delta_{L,0}\end{aligned} \tag{6.73}$$

where:

$$a_{L-2} = \frac{L(L-1)}{2(2L-3)(2L-1)} \tag{6.74}$$

$$a_L = \frac{L(L+1)}{3(2L-1)(2L+3)} \tag{6.75}$$

$$a_{L+2} = \frac{(L+1)(L+2)}{2(2L+3)(2L+5)} \tag{6.76}$$

Note that only values of L differing by 2 are allowed, and consequently L is even. We can readily add a residual Lorentzian line broadening T_2^{-1} that does not depend on motion to Eq. 6.73 by making the substitution: $\omega_o \to \omega_o + iT_2^{-1}$ (see Eq. 6.8).

Equation 6.73 specifies an infinite set of coupled simultaneous equations for the complex coefficients $C_L(\omega)$. We solve these by zero-ing all C_L with $L > n$, which gives 1+(n/2) coupled equations. For slower motion we must retain higher values of L to obtain satisfactory convergence. For $\Delta g\beta_e B_o/\hbar D_R \leq 15$, 150 and 1,500, the values of n at which we can truncate are ≤6, ≈12 and ≈18, respectively (Freed et al. 1971). Note that conventional motional narrowing theory requires that $\Delta g\beta_e B_o/6\hbar D_R \ll 1$.

Figure 6.8 shows simulated first-derivative absorption EPR lineshapes for an axially symmetric g-tensor, with increasing values of the isotropic rotational correlation time, $\tau_R = 1/(6D_R)$. We express the correlation time as the dimensionless quantity $\Delta g\beta_e B_o \times \tau_R$, which here corresponds

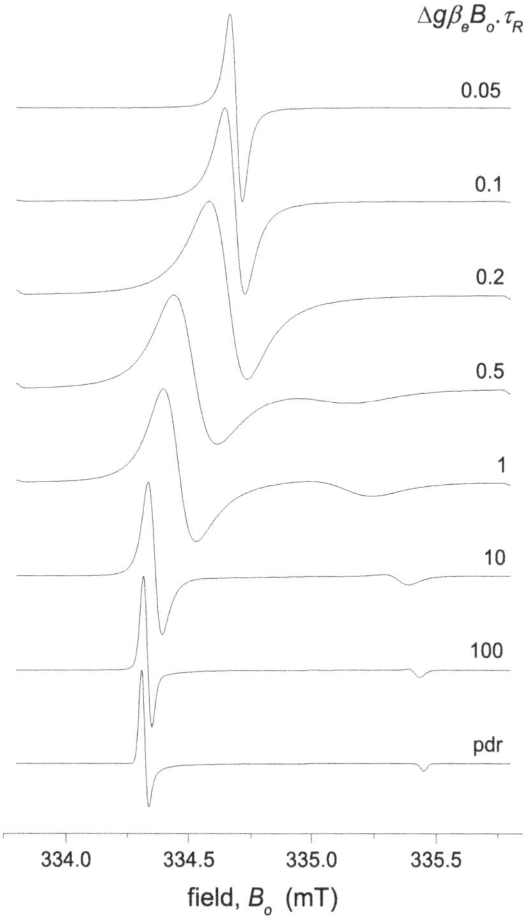

FIGURE 6.8 First-derivative EPR lineshapes for hypothetical spin label with axial g-tensor and no hyperfine structure. Isotropic rotational correlation time τ_R given as $\Delta g\beta_e B_o \times \tau_R$ ($\Delta g = 0.0068$, $B_o = 0.335$ T, appropriate to nitroxide at 9.4 GHz). Spectra simulated with stochastic Liouville equation. "pdr" is rigid-limit spectrum. Intrinsic linewidth = 0.03 mT.

to the range: $\tau_R = 1.5-314$ ns. The spectra in Figure 6.8 display the same systematic changes that we found for increasing rates of two-site exchange in Section 6.3. With increasing rates of rotational diffusion, the outer lines first broaden from their rigid-limit values and then move inwards from their rigid-limit positions. Next, the asymmetry of the lineshape decreases until it collapses to a single line at the isotropic position, which then exhibits motional narrowing on yet further increase in rotational rate. We chose a value of $\Delta g = g_{zz} - g_{xx}$ appropriate to a nitroxide and an EPR frequency of 9.4 GHz for this illustration. The intrinsic linewidth of 0.03 mT is considerably smaller than the inhomogeneous broadening for a typical protonated nitroxide, however, to emphasize the motional broadening.

6.9 EXAMPLE INCLUDING AXIAL ^{15}N-HYPERFINE STRUCTURE

If retaining only secular terms were acceptable for treating nitrogen hyperfine structure, we could perform the lineshape calculations separately for each hyperfine line, as described in the previous section for a single line. However, the pseudosecular hyperfine terms $(I_\pm S_z)$ are important for spin-label lineshapes because the nitrogen nuclear resonance frequency, ω_N, is low (Freed et al. 1971; and see also Figure 5.6 in Chapter 5). Therefore, we must include them in the modulation of the hyperfine anisotropy $(\Delta A = A_\parallel - A_\perp)$. The angular-dependent part of the axial spin Hamiltonian that contains the pseudosecular hyperfine terms is then (see Eq. 5.15 in Chapter 5, and also Eq. S.13 in Appendix S):

$$\mathcal{H}_1(\Omega) = \frac{2}{3}\left(\Delta g\beta_e B_o + \Delta A I_z\right)\mathcal{D}^2_{0,0}S_z - \frac{\Delta A}{\sqrt{6}}\left(\mathcal{D}^2_{+1,0}I_+ - \mathcal{D}^2_{-1,0}I_-\right)S_z \tag{6.77}$$

where the Wigner rotation matrix elements are defined by: $\mathcal{D}^2_{0,0}(0,\theta,0) = \frac{1}{2}\left(3\cos^2\theta - 1\right)$ and $\mathcal{D}^2_{\pm1,0}(\phi,\theta,0) = \mp\sqrt{3/2}\sin\theta \times \cos\theta\exp(\mp i\phi)$. Including the pseudosecular terms means that, instead of the simple Legendre polynomials $P_L(\cos\theta)$ of Section 6.8, we must use spherical harmonics $Y_{L,M}(\theta,\phi)$, which for generality we write here in terms of Wigner rotation matrices with elements: $\mathcal{D}^L_{K,0}(\phi,\theta,0) = \sqrt{4\pi/(2L+1)}Y_{L,-K}(\theta,\phi)$ (see Eq. R.14 in Appendix R). Note that Wigner rotation matrices appear naturally in the angular-dependent Hamiltonian when we express spin interactions by using irreducible spherical tensor operators (see Appendix S).

Wigner rotation matrices, just like spherical harmonics, are eigenfunctions of the isotropic rotational diffusion equation because:

$$\nabla^2_\Omega \mathcal{D}^L_{K,0}(\Omega) = -L(L+1)\mathcal{D}^L_{K,0}(\Omega) \tag{6.78}$$

again with (negative) eigenvalues $\tau_L^{-1} = L(L+1)D_R$ (see also Section 7.5). We normalize $\mathcal{D}^L_{K,0}$ with the orthogonality relation for Wigner rotation matrices (see Eq. R.12):

$$\int \mathcal{D}^{L*}_{K,0}\mathcal{D}^{L'}_{K',0}\cdot d\Omega = \frac{8\pi^2}{2L+1}\delta_{L',L}\delta_{K',K} \tag{6.79}$$

When using $\mathcal{D}^L_{K,0}$ for the $G_m(\Omega)$ eigenfunction expansion in Eq. 6.67, we evaluate the integral on the left-hand side with the standard relation, Eq. R.18 from Appendix R. In the form needed here, this becomes:

$$\int \mathcal{D}^{L*}_{K,0}\mathcal{D}^2_{k,0}\mathcal{D}^{L'}_{K',0}\cdot d\Omega = \frac{8\pi^2}{2L+1}\langle 2L'L;kK'\rangle\langle 2L'L;00\rangle \tag{6.80}$$

where $\langle 2L'L;kK'\rangle \equiv C(2L'L;kK')$ is a Clebsch–Gordan coefficient for which L and L' are even, $L' = L$ or $L \pm 2$ and $K = K' + k$ (see Appendix Q, Sections Q.1, Q.2). (Note that for axial tensors, as in Eq. 6.77: $m = 0 = M' = M$ in the Wigner rotation matrices.) Comparing Eqs. 6.79 and 6.80, we see that the normalization factor cancels out with the multiplying factor in Eq. 6.80, when applying Eq. 6.67. This simplification comes from using Clebsch–Gordan coefficients, although the literature on stochastic-Liouville slow-motion simulations routinely uses Wigner 3-*j* symbols with complicating normalization factors. For the latter reason, we summarize relevant properties of 3-*j* symbols in Section Q.3 of Appendix Q. As noted there, Wigner 3-*j* symbols display inherent symmetries of the vector coupling coefficients more clearly than do the Clebsch–Gordan coefficients.

We need the steady-state Eq. 6.67 of Section 6.7 for matrix elements corresponding to each of the transitions that connect the combined electron and nuclear spin states $\left|-\frac{1}{2},m_I\right\rangle$ and $\left|+\frac{1}{2},m_I'\right\rangle$. Forbidden transitions $(m_I \neq m_I')$ must be included because these connect states that are coupled by the pseudosecular interaction, for which $m_I = m_I' \pm 1$. The allowed and forbidden transitions for a ^{15}N-nitroxide with nuclear spin $I = \frac{1}{2}$ are shown in Figure 6.9. The corresponding angular transition frequencies are:

$$\omega^{++} = \omega_e - \tfrac{1}{2}a_N \tag{6.81}$$

$$\omega^{--} = \omega_e + \tfrac{1}{2}a_N \tag{6.82}$$

$$\omega^{+-} = \omega_e + \omega_N \cong \omega_e \tag{6.83}$$

$$\omega^{-+} = \omega_e - \omega_N \cong \omega_e \tag{6.84}$$

where $\omega_e\left(=\gamma_e B_o\right)$ and $\omega_N\left(=\gamma_N B_o\right)$ are the electron and nuclear angular resonance frequencies, respectively, and a_N is the isotropic hyperfine coupling constant in angular frequency units. The superscripts indicate the nuclear states connected, i.e., ω^{++} and ω^{--} correspond to allowed transitions, and ω^{+-} and ω^{-+} to forbidden nuclear transitions. In each case, the transition from $M_S = -\frac{1}{2}$ to $M_S = +\frac{1}{2}$ electron states is implied, e.g., $\omega^{+-} \equiv \omega^{+-}_{-+}$. Similarly, the matrix elements of the C-coefficients

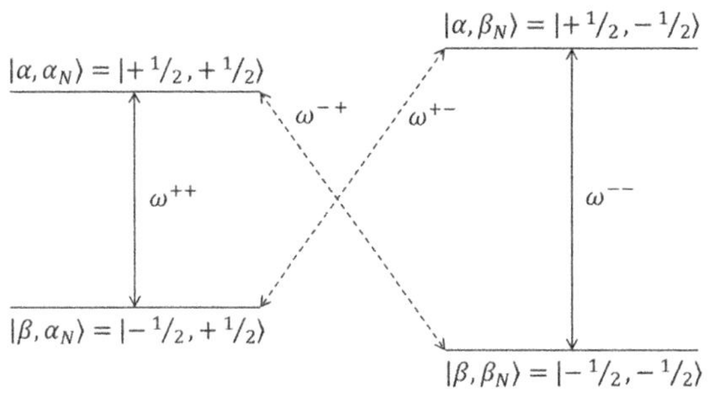

FIGURE 6.9 Energy levels and transitions for ^{15}N-nitroxide ($S=\frac{1}{2}$, $I=\frac{1}{2}$). Electron spin states: α, β for $M_S=\pm\frac{1}{2}$; nuclear spin states: α_N, β_N for $m_I=\pm\frac{1}{2}$. *Solid lines*: allowed transitions; *dashed lines*: forbidden transitions; superscripts indicate nuclear states connected by transition. Angular transition frequencies ω^{++}, ω^{--}, ω^{+-}, ω^{-+} given by Eqs. 6.81–6.84. Matrix elements connecting states corresponding to both sets of transitions are needed to solve the stochastic Liouville equation. Difference between splittings ω^{++} and ω^{--} is exaggerated.

are referred to the nuclear states, and we use the same notation, e.g., $(C_{L,K})^{+-}\equiv(C_{L,K})^{+-}_{-+}$ for $C_{L,K}(\omega^{+-})_{-+}$. Note also that the C-coefficients now depend on both L and K because they are defined in terms of the eigenfunctions $\mathcal{D}^{L}_{K,0}$ (see Eq. 6.63).

We calculate $|M_S,m_I\rangle$-matrix elements of the commutator $\left[\mathcal{H}_1,C_{L',K'}\right]$ in Eq. 6.67 as in the following example for expansion of the product operators, because $\mathcal{H}_1$ is diagonal in $|M_S\rangle$:

$$\left[\mathcal{H}_1,C_{L',K'}\right]^{+-}_{-+}\equiv\langle-,+|\left[\mathcal{H}_1,C_{L',K'}\right]|+,-\rangle$$

$$=\sum_{m_I=\pm\frac{1}{2}}\langle-,+|\mathcal{H}_1|-,m_I\rangle\langle-,m_I|C_{L',K'}|+,-\rangle$$

$$-\sum_{m_I=\pm\frac{1}{2}}\langle-,+|C_{L',K'}|+,m_I\rangle\langle+,m_I|\mathcal{H}_1|+,-\rangle$$

$$=-\tfrac{2}{3}\Delta g\beta_e B_o\mathcal{D}^2_{0,0}\left(C_{L',K'}\right)^{+-}$$

$$+\tfrac{1}{2\sqrt{6}}\Delta A\mathcal{D}^2_{1,0}\left(\left(C_{L',K'}\right)^{++}+\left(C_{L',K'}\right)^{--}\right)\qquad(6.85)$$

where $\langle\pm\frac{1}{2}|I_\pm|\mp\frac{1}{2}\rangle=1$ from the pseudosecular terms contribute matrix elements of $\mathcal{H}_1$ that correspond to nuclear spin flips: $m_I=-\frac{1}{2}$ to $m_I=+\frac{1}{2}$ and vice-versa. Written in matrix form, the commutator for a ^{15}N-nitroxide becomes the column vector that is given by:

$$\left[\mathcal{H}_1,C_{L',K'}\right]=\begin{pmatrix}-\frac{2}{3}(\Delta g\beta_e B_o+\frac{1}{2}\Delta A)\mathcal{D}^2_{0,0} & 0 & \frac{-1}{2\sqrt{6}}\Delta A\mathcal{D}^2_{-1,0} & \frac{1}{2\sqrt{6}}\Delta A\mathcal{D}^2_{1,0}\\ 0 & -\frac{2}{3}(\Delta g\beta_e B_o-\frac{1}{2}\Delta A)\mathcal{D}^2_{0,0} & \frac{-1}{2\sqrt{6}}\Delta A\mathcal{D}^2_{-1,0} & \frac{1}{2\sqrt{6}}\Delta A\mathcal{D}^2_{1,0}\\ \frac{1}{2\sqrt{6}}\Delta A\mathcal{D}^2_{1,0} & \frac{1}{2\sqrt{6}}\Delta A\mathcal{D}^2_{1,0} & -\frac{2}{3}\Delta g\beta_e B_o\mathcal{D}^2_{0,0} & 0\\ \frac{-1}{2\sqrt{6}}\Delta A\mathcal{D}^2_{-1,0} & \frac{-1}{2\sqrt{6}}\Delta A\mathcal{D}^2_{-1,0} & 0 & -\frac{2}{3}\Delta g\beta_e B_o\mathcal{D}^2_{0,0}\end{pmatrix}\begin{pmatrix}C^{++}\\ C^{--}\\ C^{+-}\\ C^{-+}\end{pmatrix}\qquad(6.86)$$

where we use the contraction: $C^{+-}\equiv(C_{L',K'})^{+-}$, et seq. for the matrix elements of the C-coefficients. In superoperator notation, the 4×4 matrix on the right-hand side of Eq. 6.86 represents the operator $\mathcal{H}_1^{\times}$ that is defined by the commutator: $\mathcal{H}_1^{\times}C_{L',K'}\equiv\left[\mathcal{H}_1,C_{L',K'}\right]$. The four versions of Eq. 6.67 for $C_{L,K}$ that correspond to matrix elements for the allowed and forbidden transitions are then (cf. Freed et al. 1971):

$$\left(\omega-\omega^{++}-i\tau_L^{-1}\right)\left(C_{L,0}\right)^{++}-\frac{2}{3}\left(\Delta g\left(\frac{\beta_e B_o}{\hbar}\right)+\tfrac{1}{2}\Delta A\right)$$
$$\times\sum_{L'}\langle 2L'L;00\rangle^2\left(C_{L',0}\right)^{++}-\frac{1}{2\sqrt{6}}\Delta A$$
$$\times\sum_{L'\neq0}\langle 2L'L;00\rangle\langle 2L'L;-1,1\rangle\left[\left(C_{L',1}\right)^{+-}-\left(C_{L',-1}\right)^{-+}\right]=\delta_{L,0}$$
$$(6.87)$$

$$\left(\omega-\omega^{--}-i\tau_L^{-1}\right)\left(C_{L,0}\right)^{--}-\frac{2}{3}\left(\Delta g\left(\frac{\beta_e B_o}{\hbar}\right)-\tfrac{1}{2}\Delta A\right)$$
$$\times\sum_{L'}\langle 2L'L;00\rangle^2\left(C_{L',0}\right)^{--}-\frac{1}{2\sqrt{6}}\Delta A$$
$$\times\sum_{L'\neq0}\langle 2L'L;00\rangle\langle 2L'L;-1,1\rangle\left[\left(C_{L',1}\right)^{+-}-\left(C_{L',-1}\right)^{-+}\right]=\delta_{L,0}$$
$$(6.88)$$

$$\left(\omega-\omega^{+-}-i\tau_L^{-1}\right)\left(C_{L,1}\right)^{+-}-\frac{2}{3}\Delta g\left(\frac{\beta_e B_o}{\hbar}\right)$$
$$\times\sum_{L'\neq0}\langle 2L'L;00\rangle\langle 2L'L;0,1\rangle\left(C_{L',1}\right)^{+-}+\frac{1}{2\sqrt{6}}\Delta A$$
$$\times\sum_{L'}\langle 2L'L;00\rangle\langle 2L'L;1,0\rangle\left[\left(C_{L',0}\right)^{++}+\left(C_{L',0}\right)^{--}\right]=0$$
$$(6.89)$$

$$\left(\omega-\omega^{-+}-i\tau_L^{-1}\right)\left(C_{L,-1}\right)^{-+}-\frac{2}{3}\Delta g\left(\frac{\beta_e B_o}{\hbar}\right)$$
$$\times\sum_{L'\neq0}\langle 2L'L;00\rangle\langle 2L'L;0,-1\rangle\left(C_{L',-1}\right)^{-+}-\frac{1}{2\sqrt{6}}\Delta A$$
$$\times\sum_{L'}\langle 2L'L;00\rangle\langle 2L'L;-1,0\rangle\left[\left(C_{L',0}\right)^{++}+\left(C_{L',0}\right)^{--}\right]=0$$
$$(6.90)$$

where ΔA is now in angular frequency units. Note that $K = 0$ for the allowed transitions, and $K = \pm 1$ for the forbidden transitions. The values of K' come from the selection rule $K = k + K'$ (see above). Because of the triangle rule for Clebsch–Gordan coefficients, the summations are over $L' = L$ and $L \pm 2$, where L and L' are even. We see this explicitly in Eq. 6.73 for the example of the previous section. Consequently, we use the relation $\langle 2L'L;kK'\rangle = \langle 2L'L;-k,-K'\rangle$ in the summations over L'. Expressions needed for the Clebsch–Gordan coefficients are listed in Appendix A6.1 at the end of this chapter. Whereas Eqs. 6.87 and 6.88 include $L = 0$, Eqs. 6.89 and 6.90 require that $L > 0$. Using the definition (Eq. Q.12) and symmetry relations (Eq. Q.14) of Wigner 3-j symbols from Section Q.3 of Appendix Q yields the expressions derived originally by Freed et al. (1971). The latter authors use $C^L_{0,M}$ for our $C_{L,K}$; this comes from adopting a different rotational transformation that interchanges K and M in the $\mathcal{D}^L_{K,M}$ of Eq. 6.77 (see Appendix S). The two treatments become equivalent, however, after exploiting orthogonality of the $\mathcal{D}^L_{K,M}$.

Equations 6.87–6.90 represent four infinite sets of coupled equations, which are coupled amongst themselves by the pseudosecular hyperfine terms. Truncation beyond $L = n$ yields $2(n+1)$ coupled equations to obtain the zero-order coefficients that represent the lineshape:

$$v(\omega) = \mathrm{Im}\left[\left(C_{0,0}\right)^{++} + \left(C_{0,0}\right)^{--}\right] \tag{6.91}$$

– cf. Eq. 6.65. Note that matrix elements corresponding to the forbidden nuclear transitions, $\left(C_{0,0}\right)^{+-}$ and $\left(C_{0,0}\right)^{-+}$ do not enter into the expression for $v(\omega)$.

Figure 6.10 shows first-derivative absorption EPR spectra simulated for an axially symmetric g-tensor and hyperfine tensor, with $I = \frac{1}{2}$. Apart from the restriction to an axial g-tensor (which is less significant at lower EPR frequencies), these lineshapes should reasonably approximate a ^{15}N-nitroxide. At an EPR frequency of 9.4 GHz, the hyperfine anisotropy outweighs the Zeeman anisotropy for nitroxides, and therefore we express the rotational correlation time in terms of the dimensionless quantity $\gamma_e \Delta A \times \tau_R$, in Figure 6.10. The corresponding correlation times are in the range: $\tau_R = 0.14 - 290$ ns. With decreasing values of the isotropic rotational correlation time $\left(\tau_R = 1/(6D_R)\right)$, the spectra display the same systematic changes from a solid-like to a liquid-like spectrum that we found in the previous section for the single EPR line with axial g-tensor.

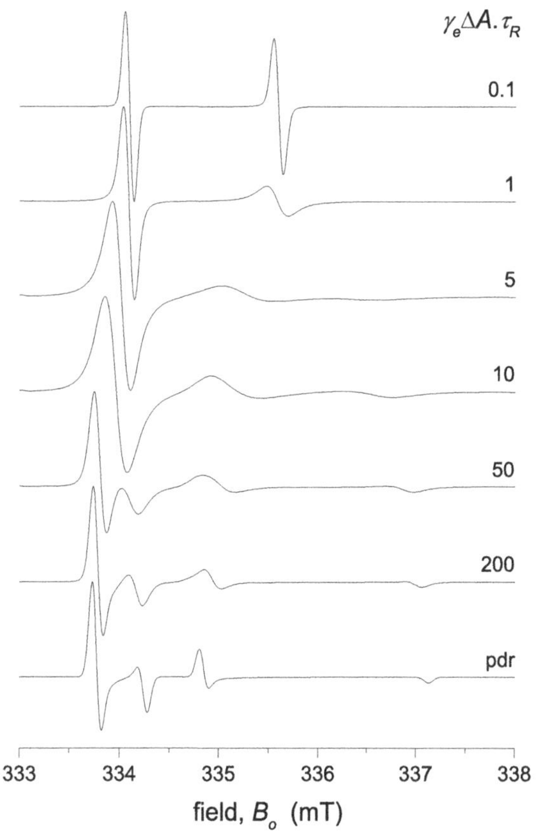

FIGURE 6.10 First-derivative EPR lineshapes for ^{15}N-nitroxide ($S = \frac{1}{2}$, $I = \frac{1}{2}$) with axial g-tensor and hyperfine tensor. Isotropic rotational correlation time τ_R given as $\gamma_e \Delta A \times \tau_R$ ($\Delta A = 3.92$ mT; $\Delta g = 0.0054$, $B_o = 0.335$ T for ^{15}N-nitroxide at 9.4 GHz). Spectra simulated with stochastic Liouville equation. "pdr" is rigid-limit spectrum. Residual Gaussian broadening, FWHM = 0.1 mT.

6.10 SLOW MOTION FOR THE FULL ^{14}N-NITROXIDE

The full angular-dependent part of the spin Hamiltonian, for a ^{14}N-nitroxide (i.e., for $I = 1$) and including non-axial tensors, now becomes (see Eq. S.13 in Appendix S):

$$\begin{aligned}\mathcal{H}_1(\Omega) = {} & \frac{2}{3}\left(\Delta g \beta_e B_o + \Delta A I_z\right)\mathcal{D}^2_{0,0} S_z \\ & + \sqrt{\frac{2}{3}}\left(\delta g \beta_e B_o + \delta A I_z\right)\left(\mathcal{D}^2_{0,2} + \mathcal{D}^2_{0,-2}\right) S_z \\ & - \frac{\Delta A}{\sqrt{6}}\left(\mathcal{D}^2_{+1,0} I_+ - \mathcal{D}^2_{-1,0} I_-\right) S_z \\ & - \frac{\delta A}{2}\left[\left(\mathcal{D}^2_{1,2} + \mathcal{D}^2_{1,-2}\right) I_+ + \left(\mathcal{D}^2_{-1,2} + \mathcal{D}^2_{-1,-2}\right) I_-\right] S_z\end{aligned} \tag{6.92}$$

where $\Delta g = g_{zz} - \frac{1}{2}\left(g_{xx} + g_{yy}\right)$ and $\delta g = \frac{1}{2}\left(g_{xx} - g_{yy}\right)$ and similarly for ΔA and δA. The additional Wigner rotation matrix elements, $\mathcal{D}^L_{K,M}$, that appear in Eq. 6.92 are defined by: $\mathcal{D}^2_{0,\pm 2}(0,\theta,\psi) = \sqrt{3/8}\sin^2\theta \exp(\mp i 2\psi)$, $\mathcal{D}^2_{\mp 1,\pm 2}(\phi,\theta,\psi) = \pm\frac{1}{2}\sin\theta(1-\cos\theta) \times \exp(\pm i(\phi - 2\psi))$ and $\mathcal{D}^2_{\pm 1,\pm 2}(\phi,\theta,\psi) = \pm\frac{1}{2}\sin\theta(1+\cos\theta) \times \exp\left(\mp i(\phi + 2\psi)\right)$ (see Appendix R.3). The Euler angles $\Omega \equiv (\phi,\theta,\psi)$ relate the nitroxide axes to the space-fixed axes that are defined by the magnetic field direction (see Appendix S). Again, we neglect non-secular terms in Eq. 6.92. The reason for expressing the Hamiltonian in terms of Wigner rotation matrices, and the power of the spherical-tensor method, are now apparent. We can apply Eq. R.18, which gives the required angular integrals, just as readily here as for the much simpler Hamiltonian that we treated in

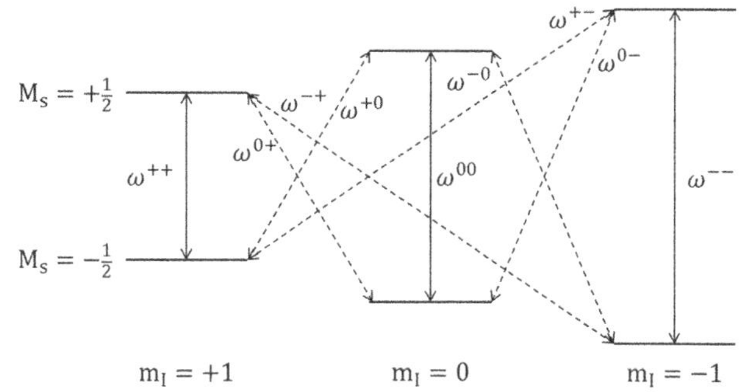

FIGURE 6.11 Energy levels and transitions for ^{14}N-nitroxide ($S=\frac{1}{2}$, $I=1$). Electron spin states: $M_S=\pm\frac{1}{2}$; nuclear spin states: $m_I=0,\pm1$. *Solid lines*: allowed transitions; *dashed lines*: forbidden transitions. Angular frequencies $\omega^{m_I m_I'}$ of transitions connecting m_I and m_I' nuclear states ($\pm$ stands for ±1) are given by Eqs. 6.93–6.97. Differences between ω^{++}, ω^{00}, ω^{--} are exaggerated.

the previous section. Note that the expansion coefficients $C_{L,K}$ of the previous section now become $C^L_{K,M}$, because terms with $M\neq0$ appear for non-axial tensors in the Hamiltonian of Eq. 6.92. As for the previous section, our $C^L_{K,M}$ corresponds to $C^L_{M,K}$ in Freed et al. (1971), because their rotational transformation interchanges K and M in the $\mathcal{D}^L_{K,M}$ of Eq. 6.92 (see Appendix S).

For ^{14}N nuclear spin $I=1$, we now need nine matrix elements: three between states connected by the allowed transitions, $\omega^{m_I m_I}$, and six between those connected by the forbidden transitions, $\omega^{m_I m_I'}$ (see Figure 6.11):

$$\omega^{\pm\pm}=\omega_e\pm a_N \tag{6.93}$$

$$\omega^{00}=\omega_e \tag{6.94}$$

$$\omega^{0\pm}=\omega_e\pm\tfrac{1}{2}a_N\mp\omega_N\cong\omega_e\pm\tfrac{1}{2}a_N \tag{6.95}$$

$$\omega^{\pm0}=\omega_e\pm\tfrac{1}{2}a_N\pm\omega_N\cong\omega_e\pm\tfrac{1}{2}a_N \tag{6.96}$$

$$\omega^{\pm,\mp}=\omega_e\pm2\omega_N\cong\omega_e \tag{6.97}$$

This increases the number of coupled equations for the matrix elements of the $C^L_{K,M}$ coefficients to nine and increases their complexity. But no new principles are involved relative to the situation for $I=\frac{1}{2}$ that we covered in the previous Section 6.9. The matrix elements are larger in number and must now be designated more generally: $C^{m_I m_I'}\equiv\left(C^L_{K,M}\right)^{m_I m_I'}_{-+}$, where the nuclear magnetic quantum numbers are: $m_I,m_I'=-1,\ 0,\ +1$.

We can write the nine equations needed for solely *axial* tensors in terms of the simpler coefficients $\left(C_{L,K}\right)^{m_I m_I'}$ used in the previous section (cf. Freed et al. 1971):

$$\left(\omega-\omega^{\pm\pm}-i\tau_L^{-1}\right)\left(C_{L,0}\right)^{\pm1,\pm1}-\frac{2}{3}\left(\Delta g\left(\frac{\beta_e B_o}{\hbar}\right)\pm\Delta A\right)\times\sum_{L'}\langle 2L'L;00\rangle^2\left(C_{L',0}\right)^{\pm1,\pm1}-\frac{1}{2\sqrt{3}}\Delta A\times\sum_{L'\neq0}\langle 2L'L;00\rangle\langle 2L'L;-1,1\rangle\left[\left(C_{L',+1}\right)^{0,\pm1}-\left(C_{L',-1}\right)^{\pm1,0}\right]=\delta_{L,0} \tag{6.98}$$

$$\left(\omega-\omega^{00}-i\tau_L^{-1}\right)\left(C_{L,0}\right)^{0,0}-\frac{2}{3}\Delta g\left(\frac{\beta_e B_o}{\hbar}\right)\sum_{L'}\langle 2L'L;00\rangle^2\left(C_{L',0}\right)^{0,0}-\frac{1}{2\sqrt{3}}\Delta A\sum_{L'\neq0}\langle 2L'L;00\rangle\langle 2L'L;-1,1\rangle\times\left[\left(C_{L',+1}\right)^{0,+1}-\left(C_{L',-1}\right)^{+1,0}+\left(C_{L',+1}\right)^{0,-1}-\left(C_{L',-1}\right)^{-1,0}\right]=\delta_{L,0} \tag{6.99}$$

$$\left(\omega-\omega^{0\pm}-i\tau_L^{-1}\right)\left(C_{L,+1}\right)^{0,\pm1}-\frac{2}{3}\left(\Delta g\left(\frac{\beta_e B_o}{\hbar}\right)\pm\tfrac{1}{2}\Delta A\right)\times\sum_{L'\neq0}\langle 2L'L;00\rangle\langle 2L'L;01\rangle\left(C_{L',+1}\right)^{0,\pm1}+\frac{1}{2\sqrt{3}}\Delta A\times\left(\sum_{L'}\langle 2L'L;00\rangle\langle 2L'L;1,0\rangle\left[\left(C_{L',0}\right)^{\pm1,\pm1}+\left(C_{L',0}\right)^{0,0}\right]-\sum_{L'\neq0}\langle 2L'L;00\rangle\langle 2L'L;-1,2\rangle\left(C_{L',+2}\right)^{+1,-1}\right)=0 \tag{6.100}$$

$$\left(\omega-\omega^{\pm0}-i\tau_L^{-1}\right)\left(C_{L,-1}\right)^{\pm1,0}-\frac{2}{3}\left(\Delta g\left(\frac{\beta_e B_o}{\hbar}\right)\pm\tfrac{1}{2}\Delta A\right)\times\sum_{L'\neq0}\langle 2L'L;00\rangle\langle 2L'L;0-1\rangle\left(C_{L',-1}\right)^{\pm1,0}-\frac{1}{2\sqrt{3}}\Delta A\times\left(\sum_{L'}\langle 2L'L;00\rangle\langle 2L'L;-1,0\rangle\left[\left(C_{L',0}\right)^{\pm1,\pm1}+\left(C_{L',0}\right)^{0,0}\right]-\sum_{L'\neq0}\langle 2L'L;00\rangle\langle 2L'L;1,-2\rangle\left(C_{L',-2}\right)^{-1,+1}\right)=0 \tag{6.101}$$

$$\left(\omega-\omega^{+,-}-i\tau_L^{-1}\right)\left(C_{L,+2}\right)^{+1,-1}-\frac{2}{3}\Delta g\left(\frac{\beta_e B_o}{\hbar}\right)\times\sum_{L'\neq0}\langle 2L'L;00\rangle\langle 2L'L;0,2\rangle\left(C_{L',+2}\right)^{+1,-1}+\frac{1}{2\sqrt{3}}\Delta A\times\sum_{L'}\langle 2L'L;00\rangle\langle 2L'L;1,1\rangle\left[\left(C_{L',+1}\right)^{0,+1}+\left(C_{L',+1}\right)^{0,-1}\right]=0 \tag{6.102}$$

$$\left(\omega-\omega^{-,+}-i\tau_L^{-1}\right)\left(C_{L,-2}\right)^{-1,+1}-\frac{2}{3}\Delta g\left(\frac{\beta_e B_o}{\hbar}\right)\times\sum_{L'\neq0}\langle 2L'L;00\rangle\langle 2L'L;0,-2\rangle\left(C_{L',-2}\right)^{-1,+1}-\frac{1}{2\sqrt{3}}\Delta A\times\sum_{L'}\langle 2L'L;00\rangle\langle 2L'L;-1,-1\rangle\left[\left(C_{L',-1}\right)^{+1,0}+\left(C_{L',-1}\right)^{-1,0}\right]=0 \tag{6.103}$$

We can exploit the near degeneracy of pairs of forbidden transitions that differ only in the nuclear frequency (see Eqs. 6.95–6.97) to reduce the number of independent equations. To do this, we take linear combinations of the double Eqs. 6.100 and 6.101, and of the single Eqs. 6.102 and 6.103. We define the following linearly independent combinations of coefficients:

$$\left(C_{L',1}\right)^{0,\pm}=\left(C_{L',1}\right)^{0,\pm1}-\left(C_{L',-1}\right)^{\pm1,0} \tag{6.104}$$

$$\left(C_{L',2}\right)^{+,-}=\left(C_{L',2}\right)^{+1,-1}+\left(C_{L',-2}\right)^{-1,+1} \tag{6.105}$$

Then Eqs. 6.98–6.103 reduce to the following six equations, in total:

$$\begin{aligned}&\left(\omega-\omega^{\pm\pm}-i\tau_L^{-1}\right)\left(C_{L,0}\right)^{\pm1,\pm1}\\&-\frac{2}{3}\left(\Delta g\left(\frac{\beta_e B_o}{\hbar}\right)\pm\Delta A\right)\sum_{L'}\langle 2L'L;00\rangle^2\left(C_{L',0}\right)^{\pm1,\pm1}\\&-\frac{1}{2\sqrt{3}}\Delta A\sum_{L'\neq0}\langle 2L'L;-1,1\rangle\langle 2L'L;00\rangle\left(C_{L',1}\right)^{0,\pm}=\delta_{L,0}\end{aligned} \tag{6.106}$$

$$\begin{aligned}&\left(\omega-\omega^{00}-i\tau_L^{-1}\right)\left(C_{L,0}\right)^{0,0}-\frac{2}{3}\Delta g\left(\frac{\beta_e B_o}{\hbar}\right)\\&\times\sum_{L'}\langle 2L'L;00\rangle^2\left(C_{L',0}\right)^{0,0}-\frac{1}{2\sqrt{3}}\Delta A\\&\times\sum_{L'\neq0}\langle 2L'L;-1,1\rangle\langle 2L'L;00\rangle\left[\left(C_{L',1}\right)^{0,+}+\left(C_{L',1}\right)^{0,-}\right]=\delta_{L,0}\end{aligned} \tag{6.107}$$

$$\begin{aligned}&\left(\omega-\omega_e\mp\tfrac{1}{2}a_N-i\tau_L^{-1}\right)\left(C_{L,1}\right)^{0,\pm}-\frac{2}{3}\left(\Delta g\left(\frac{\beta_e B_o}{\hbar}\right)\pm\tfrac{1}{2}\Delta A\right)\\&\times\sum_{L'\neq0}\langle 2L'L;01\rangle\langle 2L'L;00\rangle\left(C_{L',1}\right)^{0,\pm}+\frac{1}{\sqrt{3}}\Delta A\\&\times\left(\sum_{L'}\langle 2L'L;00\rangle\langle 2L'L;1,0\rangle\left[\left(C_{L',0}\right)^{\pm1,\pm1}+\left(C_{L',0}\right)^{0,0}\right]\right.\\&\left.-\tfrac{1}{2}\sum_{L'\neq0}\langle 2L'L;00\rangle\langle 2L'L;-1,2\rangle\left(C_{L',2}\right)^{+,-}\right)=0\end{aligned} \tag{6.108}$$

$$\begin{aligned}&\left(\omega-\omega_e-i\tau_L^{-1}\right)\left(C_{L,2}\right)^{+,-}-\frac{2}{3}\Delta g\left(\frac{\beta_e B_o}{\hbar}\right)\\&\times\sum_{L'\neq0}\langle 2L'L;0,2\rangle\langle 2L'L;00\rangle\left(C_{L',2}\right)^{+,-}+\frac{1}{2\sqrt{3}}\Delta A\\&\times\sum_{L'}\langle 2L'L;1,1\rangle\langle 2L'L;00\rangle\left[\left(C_{L',1}\right)^{0,+}+\left(C_{L',1}\right)^{0,-}\right]=0\end{aligned} \tag{6.109}$$

More general expressions, including non-axial tensors and axially anisotropic rotation, are given in Appendix A6.2 at the end of the chapter. For the allowed transitions, L is even and $K=0$. In general, $L\geq M\geq0$ with M even; and also $K=1$ for the singly forbidden $(\Delta m_I=\pm1)$ transitions, and $K=2$ for the doubly forbidden $(\Delta m_I=\pm2)$ transitions.

Expressed in matrix form with the variables $\left(C_{K,M}^{L}\right)_{-+}^{m_I m_I'}$ arranged in a column vector $\mathbf{C}(\omega)$, the coupled linear equations of the type Eq. 6.98–6.103 become (Freed et al. 1971; Freed 1976):

$$\left(\mathbf{A}+k(\omega)\mathbf{1}\right)\cdot\mathbf{C}(\omega)=\mathbf{U} \tag{6.110}$$

where $\mathbf{U}$ is the column vector that contains the constants on the right-hand side of the equations; $k(\omega)\equiv(\omega-\omega_e)-iT_2^{-1}$ is a constant that contains the spectral sweep ω and multiplies the unit matrix; and $\mathbf{A}$ is a square matrix that contains the coefficients of the $\left(C_{K,M}^{L}\right)_{-+}^{m_I m_I'}$ variables, other than the on-diagonal sweep term, $k(\omega)$. We diagonalize the $\mathbf{A}$ matrix by an orthogonal transformation (see Section B.4 in Appendix B):

$$\mathbf{O}^{-1}\mathbf{A}\mathbf{O}=\mathbf{A}_d \tag{6.111}$$

where $\mathbf{A}_d$ is diagonal with eigenvalues $(A_d)_{jj}$ and $\mathbf{O}^{-1}=\mathbf{O}^{\mathrm{T}}$ (the transpose). Multiplying Eq. 6.110 on the left by $\mathbf{O}^{-1}$ and using Eq. 6.111, we get:

$$\left(\mathbf{A}_d+k(\omega)\mathbf{1}\right)\cdot\mathbf{O}^{-1}\mathbf{C}(\omega)=\mathbf{O}^{-1}\mathbf{U} \tag{6.112}$$

The solution for the $\mathbf{C}$-vector is therefore:

$$\mathbf{C}(\omega)=\mathbf{O}\left(\mathbf{A}_d+k(\omega)\mathbf{1}\right)^{-1}\mathbf{O}^{-1}\mathbf{U} \tag{6.113}$$

Finally, the EPR absorption lineshape again is given by the zero-order coefficients:

$$v(\omega)=\mathrm{Im}\left[\left(C_{0,0}^{0}\right)^{+1,+1}+\left(C_{0,0}^{0}\right)^{0,0}+\left(C_{0,0}^{0}\right)^{-1,-1}\right] \tag{6.114}$$

From Eq. 6.113, we find that this solution corresponds to a superposition of Lorentzians (Freed 1976):

$$\mathrm{Im}\left[\left(C_{0,0}\right)\right]=\mathrm{Im}\left[\sum_j\frac{\left(\mathbf{O}^{\mathrm{T}}\mathbf{U}\right)_j^2}{(A_d)_{jj}+(\omega-\omega_e)-iT_2^{-1}}\right] \tag{6.115}$$

for each ^{14}N-hyperfine manifold.

Guidelines for choosing the number of basis vectors before truncating the infinite set of coupled equations are given by Schneider and Freed (1989). Some examples of the maximum values for L, M and K that we need at different rotational rates are listed in Table 6.1. Excessive truncation leads to spurious oscillations in the simulated spectra. A listing of the FORTRAN IV programme to perform these simulations, including axially anisotropic rotation, is given by Freed (1976). The later version described by Schneider and Freed (1989) incorporates an orientation potential and uses the powerful Lanczos algorithm (see Moro and Freed 1981). The latter reference, which follows Freed (1976), is the last to use their original rotational transformation for

TABLE 6.1 Truncation of basis vectors used in solution of the rotational stochastic-Liouville equation (Vasavada et al. 1987)

$D_R(s^{-1})$	L_{max} (even)	L_{max} (odd)	K_{max}	M_{max}
10^7	6	3	2	2
10^6	14	7	6	2
	14	7	14	2
10^5	30	13	10	2
	30	13	30	2
10^4	54	15	10	2

Isotropic rotational correlation time: $\tau_R = 1/(6D_R)$.
L_{max}, K_{max}, M_{max}: maximum values that specify non-zero $C^L_{K,M}$ expansion coefficients; K = even, $K \le L$ and $M \le 2$. N.B. here L, K, M refer to the software implementation of stochastic-Liouville simulations; they correspond to our L, M, K in Sections 6.9 and 6.10 because Vasavada et al. (1987) label their Wigner rotation matrices $\mathcal{D}^L_{M,K}$ (see Appendix S).
Second of double entries takes y-axis (instead of z-axis) for principal component of hyperfine tensor.
See also Moro and Freed (1981); Schneider and Freed (1989).

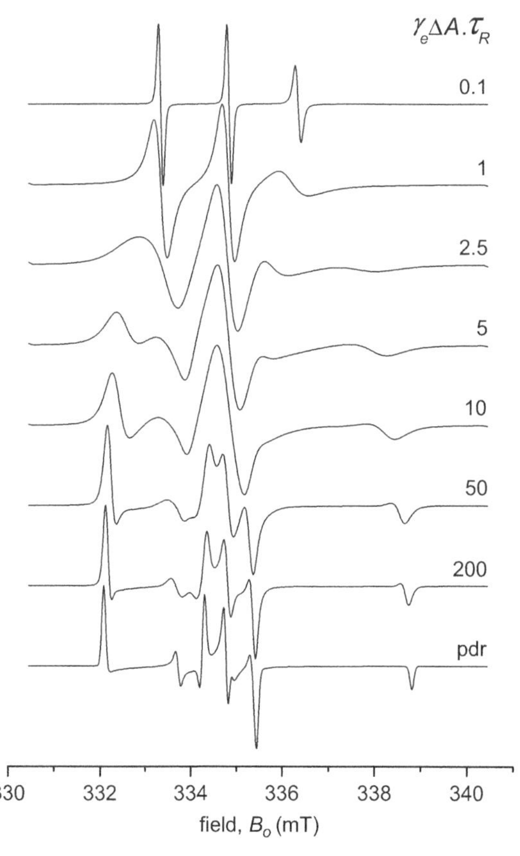

FIGURE 6.12 First-derivative EPR lineshapes for ^{14}N-nitroxide ($S = \frac{1}{2}$, $I = 1$) with non-axial g-tensor and hyperfine tensor. Isotropic rotational correlation time τ_R given as $\gamma_e \Delta A \times \tau_R$ ($\Delta A = 2.79$ mT, $B_o = 0.335$ T for ^{14}N-nitroxide at 9.4 GHz). Spectra simulated with stochastic Liouville equation. "pdr" is rigid-limit spectrum. Residual Gaussian broadening, FWHM = 0.1 mT.

spherical tensors. Subsequent publications (including Schneider and Freed 1989) adopt the transformation that we use (see Appendix S), but interchange K and M, using the notation $\mathcal{D}^L_{M,K}$ instead of $\mathcal{D}^L_{K,M}$ for elements of the Wigner rotation matrices (see Appendix R.3). This double change preserves the original meaning of K and M, which is the reverse of ours.

Figure 6.12 shows 9.4-GHz EPR absorption spectra simulated for isotropic slow rotation of a ^{14}N-nitroxide spin label. Again, we express the rotational correlation time in terms of the dimensionless quantity $\gamma_e \Delta A \times \tau_R$, by scaling with the major hyperfine anisotropy of the ^{14}N-nitroxide. The corresponding correlation times are in the range: $\tau_R = 0.20 - 407$ ns. With decreasing values of the isotropic rotational correlation time $(\tau_R = 1/(6D_R))$, the spectra display systematic changes from a powder-like spectrum to a sharp, three-line, liquid-like spectrum. The inhomogeneous linewidth used here approximates that expected for a perdeuterated spin label in the solid state. Comparable spectra, but simulated with inhomogeneous broadening appropriate to natural protonated spin labels, are shown in Figure 1.5 of Chapter 1. In all cases, any inhomogeneous broadening is added after simulation, by convoluting a Gaussian lineshape with the simulated spectrum, as described in Section 2.6 of Chapter 2.

6.11 COMPOSITE MOTIONS ON DIFFERENT TIMESCALES

We now consider extension to motions that combine different rotational components on different timescales. This approach was applied initially to spin labels in glassy matrices, where the spin label reorients in a local environment that itself relaxes on a longer timescale: the so-called "slowly relaxing local structure" (Polimeno and Freed 1995). More specifically, the treatment applies to segmental motion of spin-labelled molecules whose overall rotation is slower (Lange et al. 1985). Spin-labelled lipid chains in membranes are a prominent example (see Section 8.14 in Chapter 8), as are also spin-labelled macromolecules. In these cases, two separate sets of Euler angles undergo dynamic fluctuations: Ω_{over} specifies the overall orientation of the whole molecule (or environmental cage) and Ω_{loc} gives the local (or internal) orientation of the spin-labelled segment. Both are referred to the static magnetic field (i.e., laboratory axes) or to the director in the case of molecularly ordered systems (cf. Chapter 8).

We illustrate the methods by using the strong-jump model of Section 6.4 as a simple example (Livshits and Marsh 2004). In this context, the Bloch equation for the transverse spin magnetization, $\widehat{M}_+ = u + iv$, in the rotating frame is analogous to that for anisotropic jump diffusion of a symmetric top that was treated by Livshits (1976). We partition the combined motion into jumps in overall orientation Ω_{over} at frequency τ^{-1}_{over} and in local (internal) orientation Ω_{loc} at augmented frequency $\tau^{-1}_{loc} - \tau^{-1}_{over}$. The equation of motion for changes in transverse magnetization by the combined jumps is then (cf. Eq. 6.29):

$$\frac{d\widehat{M}_+(\Omega_{loc},\Omega_{over})}{dt} = -\tau^{-1}_{loc}\widehat{M}_+ + \left(\tau^{-1}_{loc} - \tau^{-1}_{over}\right)\int \widehat{M}_+ \cdot d\Omega'_{loc} + \tau^{-1}_{over} P_o(\Omega_{over})\int \widehat{M}_+ \cdot d\Omega'_{over} \qquad (6.116)$$

where the factor $P_o(\Omega_{over})$ in the final term is the equilibrium distribution of the overall orientation, which in general is governed by a pseudopotential, $U(\Omega_{over})$ (see Section 8.4 in Chapter 8). Combining Eqs. 6.28 and 6.116, in the steady state $\left(d\hat{M}_+/dt = 0\right)$, the modified Bloch equation becomes (cf. Eq. 6.30):

$$\left(\frac{1}{T_{2,m_I}} + \frac{1}{\tau_{loc}} + i\left(\omega_{m_I}\left(\Omega_{loc},\Omega_{over}\right) - \omega\right)\right)\hat{M}_+$$

$$-\left(\tau_{loc}^{-1} - \tau_{over}^{-1}\right)\int \hat{M}_+ \, d\Omega'_{loc} - \tau_{over}^{-1} P_o\left(\Omega_{over}\right)\int \hat{M}_+ \, d\Omega'_{over}$$

$$= i\gamma_e B_1 P_o\left(\Omega_{over}\right) M_o \qquad (6.117)$$

where M_0 is the static z-magnetization, and ω_{m_I} is the angular resonance frequency of the m_I-manifold.

To solve Eq. 6.117 for the jump model, we integrate $\hat{M}_+(\Omega_{loc},\Omega_{over})$ separately over the two sets of orientations Ω_{over} and Ω_{loc} (cf. Livshits 1976). Integration over Ω_{over} covers all orientations, and the resulting integral is the EPR lineshape $v(\omega)$ that is independent of orientation. On the other hand, integration over Ω_{loc} covers only a restricted set of orientations and the resulting integral I_{loc} depends on the remaining set of orientations Ω_{rem} not integrated over. [For the special case of axially anisotropic diffusion, overall rotational jumps in $\Omega_{over} \equiv (\theta,\phi)$ occur at frequency $\tau_\perp^{-1}$, whereas local axial jumps in $\Omega_{loc} \equiv (0,\phi')$ occur at frequency $\tau_\parallel^{-1}$ but only for the ϕ-angle. Therefore, for this case we have: $\Omega_{rem} \equiv (\theta,0)$.]

Returning now to integrating Eq. 6.117, we get (Livshits 1976):

$$v(\omega) \equiv \int \hat{M}_+ \cdot d\Omega_{over}$$

$$= \left(\tau_{over}^{-1} \cdot v(\omega) + i\gamma_e B_1 M_o\right)\int \frac{P_o\left(\Omega_{over}\right)\cdot d\Omega_{over}}{i\left(\omega_{m_I} - \omega\right) + T_{2,m_I}^{-1} + \tau_{loc}^{-1}}$$

$$+\left(\tau_{loc}^{-1} - \tau_{over}^{-1}\right)\int \frac{I_{loc}\left(\Omega_{rem}\right)\cdot d\Omega_{rem}}{i\left(\omega_{m_I} - \omega\right) + T_{2,m_I}^{-1} + \tau_{loc}^{-1}} \qquad (6.118)$$

with

$$I_{loc}\left(\Omega_{rem}\right) \equiv \int \hat{M}_+ \, d\Omega_{loc}$$

$$= \left(\tau_{over}^{-1} \cdot v(\omega) + \left(\tau_{loc}^{-1} - \tau_{over}^{-1}\right) I_{loc}\left(\Omega_{rem}\right) + i\gamma_e B_1 M_o\right) T_{loc}\left(\Omega_{rem}\right) \qquad (6.119)$$

where:

$$T_{loc}\left(\Omega_{rem}\right) \equiv \int \frac{d\Omega_{loc}}{i\left(\omega_{m_I} - \omega\right) + T_{2,m_I}^{-1} + \tau_{loc}^{-1}} \qquad (6.120)$$

Note that $v(\omega)$ appears on the right-hand sides of Eqs. 6.118 and 6.119. Substituting from Eq. 6.119 into Eq. 6.118 and rearranging, the EPR lineshape finally becomes (cf. Eq. 6.31):

$$v(\omega) = i\gamma_e B_1 M_o \int \frac{T'}{1 - \tau_{over}^{-1} \cdot T'} d\Omega_{rem} \qquad (6.121)$$

where:

$$T' = \int \frac{T_{loc}}{1 - \left(\tau_{loc}^{-1} - \tau_{over}^{-1}\right)\cdot T_{loc}} P_o\left(\Omega_{over}\right)\cdot d\Omega_{over} \qquad (6.122)$$

and T_{loc} is given by Eq. 6.120 (see Livshits 1976).

For very slow global tumbling, where overall motion is too slow to affect the EPR spectrum, Eq. 6.121 for the lineshape reduces to:

$$v(\omega) = i\gamma_e B_1 M_o \int T' \cdot d\Omega_{rem} \qquad (6.123)$$

where T' is given by Eq. 6.122. The EPR lineshape is the sum over spectra corresponding to different orientations, Ω_{loc}, of the local director relative to the magnetic field. These spectra depend on the internal (or local) motion, which, in general, is anisotropic in the molecular frame and occurs in an orienting potential. The opposite limiting case is very fast internal motion. Here, we expand the integral in Eq. 6.120 up to second order in τ_{loc}. The EPR lineshape is given by Eq. 6.121 but now with the following reduced form for Eq. 6.122:

$$T' = \int \frac{d\Omega_{over}}{i\left(\left\langle\omega_{m_I}\right\rangle_{loc} - \omega\right) + T_{2,m_I}^{-1} + \left\langle\left(\omega_{m_I} - \left\langle\omega_{m_I}\right\rangle_{loc}\right)^2\right\rangle_{loc} \tau_{loc} + \tau_{over}^{-1}} \qquad (6.124)$$

Rapid local (internal) motion partially averages the resonance field to $\left\langle\omega_{m_I}\right\rangle_{loc}$ and introduces the angular-dependent line broadening $\left\langle\left(\omega_{m_I} - \left\langle\omega_{m_I}\right\rangle_{loc}\right)^2\right\rangle_{loc} \tau_{loc}$ predicted by fast-motional theory (see Eqs. 1.9 and 5.49).

Polimeno and Freed (1995) give the complete treatment for Brownian diffusion, using the stochastic Liouville equation with augmented Smoluchowski operator. Liang and Freed (1999) analyse the validity of different limiting models by simulations using this approach, with particular reference to multifrequency EPR (see also Section 8.15). Stochastic-Liouville treatment of an explicit model for slow long-axis fluctuation accompanied by rapid segmental isomerism of spin-labelled lipid chains in membranes is described in Section 8.14 of Chapter 8.

6.12 CONCLUDING SUMMARY

1. *Sensitivity to rotational motion:*
   Spin-label EPR spectral lineshapes are sensitive to rotation on timescales that are determined by the angular anisotropies of the nitroxide hyperfine- and g-tensors.
2. *Analogy with two-site exchange:*
   At slow rates of molecular rotation, spin-label EPR lines broaden but remain at their original positions. As the rotation rate increases, the anisotropically dispersed

lines move in towards each other, finally collapsing to their isotropic average position. The averaged lines are motionally broadened. With increasing rotational rate, they progressively narrow, finally reaching their intrinsic widths.

3. *Motional-narrowing regime:*
   Fast-motion lineshapes are Lorentzian, independent of rotational mechanism. The Lorentzian widths decrease with faster rotation and are given by perturbation theory.
4. *Slow motion and Bloch equations:*
   Slow-motion lineshapes depend on rotational mechanism. For strong-jump and Brownian motion, they can be simulated by using the rotationally coupled Bloch equations. These simulations use the adiabatic approximation and are not valid for fast motion.
5. *Stochastic-Liouville simulations:*
   The quantum-mechanical stochastic Liouville equation is a general and versatile method for simulating the effects of motion on spin-label EPR lineshapes. It is valid for all rotational rates: see Figure 6.12 for Brownian rotational diffusion. It is an indispensable analytical tool in spin-label EPR spectroscopy.
6. *Composite motions:*
   Combined motions on different timescales occur often. At least one component is in the slow regime. If the second motional component is in the fast regime, e.g., for segmental motions, a single-component slow-motion simulation can use motionally averaged parameters from the fast component. Otherwise, we need a more involved approach to coupled motions, as in Section 6.11.

APPENDIX A6.1: CLEBSCH–GORDAN COEFFICIENTS $\langle 2L'L;m,M\rangle$ NEEDED FOR SLOW-MOTION CALCULATIONS[a]

C-G coefficient	*value*
$\langle 2LL;0,M\rangle$	$\dfrac{3M^2-L(L+1)}{\sqrt{L(L+1)(2L+3)(2L-1)}}$
$\langle 2LL;1,M\rangle$	$-(2M+1)\sqrt{\dfrac{3(L+M+1)(L-M)}{L(2L+3)(2L+2)(2L-1)}}$
$\langle 2LL;2,M\rangle$	$\sqrt{\dfrac{3(L+M+1)(L+M+2)(L-M-1)(L-M)}{L(2L+3)(2L+2)(2L-1)}}$
$\langle 2,L+2,L;0,M\rangle$	$\sqrt{\dfrac{3(L+M+2)(L+M+1)(L-M+2)(L-M+1)}{(L+1)(2L+5)(2L+4)(2L+3)}}$
$\langle 2,L+2,L;1,M\rangle$	$-\sqrt{\dfrac{(L+M+2)(L-M+2)(L-M+1)(L-M)}{(L+1)(L+2)(2L+3)(2L+5)}}$
$\langle 2,L+2,L;2,M\rangle$	$\sqrt{\dfrac{(L-M-1)(L-M)(L-M+1)(L-M+2)}{(2L+5)(2L+4)(2L+3)(2L+2)}}$
$\langle 2,L-2,L;0,M\rangle$	$\sqrt{\dfrac{3(L-M-1)(L-M)(L+M-1)(L+M)}{2L(L-1)(2L-1)(2L-3)}}$
$\langle 2,L-2,L;1,M\rangle$	$\sqrt{\dfrac{(L+M-1)(L+M)(L+M+1)(L-M-1)}{L(L-1)(2L-1)(2L-3)}}$
$\langle 2,L-2,L;2,M\rangle$	$\sqrt{\dfrac{(L+M-1)(L+M)(L+M+1)(L+M+2)}{4L(L-1)(2L-1)(2L-3)}}$

Note: $\langle 2L'L;-m,-M\rangle=(-1)^{L'-L}\langle 2L'L;m,M\rangle$.

[a] See Eqs. 6.87–6.90, 6.106–6.109

APPENDIX A6.2: SOLUTION OF STOCHASTIC LIOUVILLE EQUATION FOR AXIALLY ANISOTROPIC ROTATION

The Wigner rotation matrices $\mathcal{D}^L_{K,M}$ are eigenfunctions of the rotational diffusion equation for a symmetric top (see Section 7.6 in Chapter 7). The corresponding eigenfunctions are given by:

$$1/\tau_{L,M} = D_{R\perp}L(L+1)+(D_{R\parallel}-D_{R\perp})M^2 \tag{A6.1}$$

where $D_{R\parallel}$ and $D_{R\perp}$ are the diffusion coefficients for rotation about the molecular symmetry axis and perpendicular to it, respectively. We designate matrix elements of coefficients in the expansion with the $\mathcal{D}^L_{K,M}$ eigenfunctions by: $\left(C^L_{K,M}\right)^{m_I m_I'} \equiv \left(C^L_{K,M}\right)^{m_I m_I'}_{-+}$ (see Section 6.10).

To simplify things, we assume that the principal rotation axis (||) coincides with one of the nitroxide axes. If the spin Hamiltonian anisotropies are defined by: $\Delta g = g_{zz}-\frac{1}{2}\left(g_{xx}+g_{yy}\right)$ and $\delta g=\frac{1}{2}\left(g_{xx}-g_{yy}\right)$, and similarly for the A-tensor, then the nitroxide z-axis is the principal rotation axis. Cyclic permutation of the indices makes the nitroxide x- or y-axis into the rotation axis.

The equivalents of Eqs. 6.106–6.109, given in Section 6.10 for isotropic rotational diffusion with axial tensors, then become (cf. Goldman et al. 1972b):

$$\begin{aligned}
&\left(\omega-\omega^{\pm\pm}-i\tau_{L,M}^{-1}\right)\left(\bar{C}^{L}_{0,M}\right)^{\pm1,\pm1}-\frac{2}{3}\left(\Delta g\left(\frac{\beta_e B_o}{\hbar}\right)\pm\Delta A\right)\sum_{L'}\langle 2L'L;00\rangle\langle 2L'L;0M\rangle\left(\bar{C}^{L'}_{0,M}\right)^{\pm1,\pm1}\\
&-\sqrt{\frac{2}{3}}\left(\delta g\left(\frac{\beta_e B_o}{\hbar}\right)\pm\delta A\right)\sum_{L'}\langle 2L'L;00\rangle\left(\langle 2L'L;-2,M+2\rangle\left(\bar{C}^{L'}_{0,M+2}\right)^{\pm1,\pm1}+\langle 2L'L;2,M-2\rangle\left(\bar{C}^{L'}_{0,M-2}\right)^{\pm1,\pm1}\right)\\
&-\frac{1}{2\sqrt{2}}\delta A\sum_{L'\neq0}\langle 2L'L;-1,1\rangle\left(\langle 2L'L;-2,M+2\rangle\left(\bar{C}^{L'}_{1,M+2}\right)^{0,\pm}+\langle 2L'L;2,M-2\rangle\left(\bar{C}^{L'}_{1,M-2}\right)^{0,\pm}\right)\\
&-\frac{1}{2\sqrt{3}}\Delta A\sum_{L'\neq0}\langle 2L'L;-1,1\rangle\langle 2L'L;0,M\rangle\left(\bar{C}^{L'}_{1,M}\right)^{0,\pm}=\delta_{L,0}\delta_{M,0}
\end{aligned}\tag{A6.2}$$

$$\begin{aligned}
&\left(\omega-\omega^{00}-i\tau_{L,M}^{-1}\right)\left(\bar{C}^{L}_{0,M}\right)^{0,0}-\frac{2}{3}\Delta g\left(\frac{\beta_e B_o}{\hbar}\right)\sum_{L'}\langle 2L'L;00\rangle\langle 2L'L;0M\rangle\left(\bar{C}^{L'}_{0,M}\right)^{0,0}\\
&-\sqrt{\frac{2}{3}}\delta g\left(\frac{\beta_e B_o}{\hbar}\right)\sum_{L'}\langle 2L'L;00\rangle\left(\langle 2L'L;-2,M+2\rangle\left(\bar{C}^{L'}_{0,M+2}\right)^{0,0}+\langle 2L'L;2,M-2\rangle\left(\bar{C}^{L'}_{0,M-2}\right)^{0,0}\right)\\
&-\frac{1}{2\sqrt{2}}\delta A\sum_{L'\neq0}\langle 2L'L;-1,1\rangle\left(\langle 2L'L;-2,M+2\rangle\left[\left(\bar{C}^{L'}_{1,M+2}\right)^{0,+}+\left(\bar{C}^{L'}_{1,M+2}\right)^{0,-}\right]+\langle 2L'L;2,M-2\rangle\left[\left(\bar{C}^{L'}_{1,M-2}\right)^{0,+}+\left(\bar{C}^{L'}_{1,M-2}\right)^{0,-}\right]\right)\\
&-\frac{1}{2\sqrt{3}}\Delta A\sum_{L'\neq0}\langle 2L'L;-1,1\rangle\langle 2L'L;0,M\rangle\left[\left(\bar{C}^{L'}_{1,M}\right)^{0,+}+\left(\bar{C}^{L'}_{1,M}\right)^{0,-}\right]=\delta_{L,0}\delta_{M,0}
\end{aligned}\tag{A6.3}$$

$$\begin{aligned}
&\left(\omega-\omega_e\mp\tfrac{1}{2}a_N-i\tau_{L,M}^{-1}\right)\left(\bar{C}^{L}_{1,M}\right)^{0,\pm}-\frac{2}{3}\left(\Delta g\left(\frac{\beta_e B_o}{\hbar}\right)\pm\tfrac{1}{2}\Delta A\right)\sum_{L'\neq0}\langle 2L'L;01\rangle\langle 2L'L;0M\rangle\left(\bar{C}^{L'}_{1,M}\right)^{0,\pm}\\
&-\sqrt{\frac{2}{3}}\left(\delta g\left(\frac{\beta_e B_o}{\hbar}\right)\pm\tfrac{1}{2}\delta A\right)\sum_{L'\neq0}\langle 2L'L;01\rangle\left(\langle 2L'L;-2,M+2\rangle\left(\bar{C}^{L'}_{1,M+2}\right)^{0,\pm}+\langle 2L'L;2,M-2\rangle\left(\bar{C}^{L'}_{1,M-2}\right)^{0,\pm}\right)\\
&+\frac{1}{\sqrt{2}}\delta A\sum_{L'}\langle 2L'L;-2,M+2\rangle\left(\langle 2L'L;10\rangle\left[\left(\bar{C}^{L'}_{0,M+2}\right)^{\pm1,\pm1}+\left(\bar{C}^{L'}_{0,M+2}\right)^{0,0}\right]-\tfrac{1}{2}\langle 2L'L;-1,2\rangle\left(\bar{C}^{L'}_{2,M+2}\right)^{+,-}\right)\\
&+\frac{1}{\sqrt{2}}\delta A\sum_{L'}\langle 2L'L;2,M-2\rangle\left(\langle 2L'L;10\rangle\left[\left(\bar{C}^{L'}_{0,M-2}\right)^{\pm1,\pm1}+\left(\bar{C}^{L'}_{0,M-2}\right)^{0,0}\right]-\tfrac{1}{2}\langle 2L'L;-1,2\rangle\left(\bar{C}^{L'}_{2,M-2}\right)^{+,-}\right)\\
&+\frac{1}{\sqrt{3}}\Delta A\sum_{L'}\langle 2L'L;0M\rangle\left(\langle 2L'L;10\rangle\left[\left(\bar{C}^{L'}_{0,M}\right)^{\pm1,\pm1}+\left(\bar{C}^{L'}_{0,M}\right)^{0,0}\right]-\tfrac{1}{2}\langle 2L'L;-1,2\rangle\left(\bar{C}^{L'}_{2,M}\right)^{+,-}\right)=0
\end{aligned}\tag{A6.4}$$

$$
\begin{aligned}
&\left(\omega-\omega_e-i\tau_{L,M}^{-1}\right)\left(\bar{C}_{2,M}^{L}\right)^{+,-}-\frac{2}{3}\Delta g\left(\frac{\beta_e B_o}{\hbar}\right)\sum_{L'\neq 0}\langle 2L'L;0,2\rangle\langle 2L'L;0,M\rangle\left(\bar{C}_{2,M}^{L'}\right)^{+,-}\\
&-\sqrt{\frac{2}{3}}\delta g\left(\frac{\beta_e B_o}{\hbar}\right)\sum_{L'\neq 0}\langle 2L'L;0,2\rangle\left(\langle 2L'L;-2,M+2\rangle\left(\bar{C}_{2,M+2}^{L'}\right)^{+,-}+\langle 2L'L;2,M-2\rangle\left(\bar{C}_{2,M-2}^{L'}\right)^{+,-}\right)\\
&+\frac{1}{2\sqrt{2}}\delta A\sum_{L'}\langle 2L'L;11\rangle\left(\langle 2L'L;-2,M+2\rangle\left[\left(\bar{C}_{1,M+2}^{L'}\right)^{0,+}+\left(\bar{C}_{1,M+2}^{L'}\right)^{0,-}\right]+\langle 2L'L;2,M-2\rangle\left[\left(\bar{C}_{1,M-2}^{L'}\right)^{0,+}+\left(\bar{C}_{1,M-2}^{L'}\right)^{0,-}\right]\right)\\
&+\frac{1}{2\sqrt{3}}\Delta A\sum_{L'}\langle 2L'L;11\rangle\langle 2L'L;0,M\rangle\left[\left(\bar{C}_{1,M}^{L'}\right)^{0,+}+\left(\bar{C}_{1,M}^{L'}\right)^{0,-}\right]=0
\end{aligned}
\tag{A6.5}
$$

where M is even, and in reducing the total number of equations to six, we use the following definitions (cf. Eqs. 6.104 and 6.105):

$$\left(\bar{C}_{1,M}^{L}\right)^{0,\pm}\equiv\left(\bar{C}_{1,M}^{L}\right)^{0,\pm 1}-\left(\bar{C}_{-1,M}^{L}\right)^{\pm 1,0} \tag{A6.6}$$

$$\left(\bar{C}_{2,M}^{L}\right)^{+,-}\equiv\left(\bar{C}_{2,M}^{L}\right)^{+1,-1}+\left(\bar{C}_{-2,M}^{L}\right)^{-1,+1} \tag{A6.7}$$

It is also understood that:

$$\left(\bar{C}_{K,M}^{L}\right)^{m_I,m_I'}\equiv\tfrac{1}{2}\left(\left(C_{K,M}^{L}\right)^{m_I,m_I'}+\left(C_{K,-M}^{L}\right)^{m_I,m_I'}\right) \tag{A6.8}$$

where $\left(\bar{C}_{K,0}^{L}\right)^{m_I,m_I'}\equiv\left(C_{K,0}^{L}\right)^{m_I,m_I'}$. Solution then proceeds exactly as described in Section 6.10, i.e., according to Eqs. 6.110–6.115. As in all stochastic-Liouville calculations of this chapter, Eqs. A6.2–A6.5 are written using Wigner 3-j symbols in the original literature. To convert from the expressions given by Goldman et al. (1972b), we remove the normalizing factor $1/(2L+1)$, reorder the 3-j symbols by cyclic permutation as in Eq. Q.13: $\begin{pmatrix} L & 2 & L' \\ -M & m & M' \end{pmatrix}=\begin{pmatrix} 2 & L' & L \\ m & M' & -M \end{pmatrix}$, and by using the phase factor in Eq. Q.11 then make the substitution: $\begin{pmatrix} 2 & L' & L \\ k & K' & -K \end{pmatrix}\times\begin{pmatrix} 2 & L' & L \\ m & M' & -M \end{pmatrix}=(-1)^{k+K'}\langle 2L'L;kK'\rangle\langle 2L'L;mM'\rangle$, where $M=m+M'$ is even. The resulting expressions are algebraically simpler. Further, because Goldman et al. (1972b) use the same convention as Freed et al. (1971), we replace K by M and correspondingly interchange subscripts in $C_{K,M}^{L}$ to get consistency with our definition of $\mathcal{D}_{K,M}^{L}$ used in Sections 6.9 and 6.10. Then Eqs. A6.2–A6.5 agree with Eqs. 6.106–6.109, when $M=0$ (and thus $C_{K,M}^{L}\equiv C_{L,K}$) and $\delta g=0=\delta A$.

REFERENCES

Bloch, F. 1946. Nuclear induction. *Phys. Rev.* 70:460–474.

Freed, J.H. 1976. Theory of slowly tumbling ESR spectra for nitroxides. In *Spin Labeling, Theory and Applications*, ed. Berliner, L.J., 53–132. New York: Academic Press Inc.

Freed, J.H., Bruno, G.V., and Polnaszek, C.F. 1971. Electron spin resonance line shapes and saturation in the slow motional region. *J. Phys. Chem.* 75:3385–3399.

Goldman, S.A., Bruno, G.V., and Freed, J.H. 1972a. Estimating slow-motional rotational correlation times for nitroxides by electron spin resonance. *J. Phys. Chem.* 76:1858–1860.

Goldman, S.A., Bruno, G.V., Polnaszek, C.F., and Freed, J.H. 1972b. An ESR study of anisotropic rotational reorientation and slow tumbling in liquid and frozen media. *J. Chem. Phys.* 56:716–735.

Gordon, R.G. and Messenger, T. 1972. Magnetic resonance line shapes in slowly tumbling molecules. In *Electron Spin Relaxation in Liquids*, eds. Muus, L.T. and Atkins, P.W., 341–381. New York-London: Plenum Press.

Horváth, L.I., Brophy, P.J., and Marsh, D. 1988a. Exchange rates at the lipid-protein interface of myelin proteolipid protein studied by spin-label electron spin resonance. *Biochemistry* 27:46–52.

Horváth, L.I., Brophy, P.J., and Marsh, D. 1988b. Influence of lipid headgroup on the specificity and exchange dynamics in lipid-protein interactions. A spin label study of myelin proteolipid apoprotein-phospholipid complexes. *Biochemistry* 27:5296–5304.

Horváth, L.I., Brophy, P.J., and Marsh, D. 1994. Microwave frequency dependence of ESR spectra from spin labels undergoing two-site exchange in myelin proteolipid membranes. *J. Magn. Reson.* B 105:120–128.

Korst, N.N., Kuznetsov, A.N., Lazarev, A.V., and Gordeev, E.P. 1972. Shape of the ESR spectrum of radicals in viscous media. *Teoret. Exp. Kh.* 8:51–57.

Kubo, R. 1969. In Stochastic Processes in Chemical Physics. In *Advances in Chemical Physics*, ed. Schuler, E., 101–127. New York: Wiley.

Lange, A., Marsh, D., Wassmer, K.-H., Meier, P., and Kothe, G. 1985. Electron spin resonance study of phospholipid membranes employing a comprehensive line-shape model. *Biochemistry* 24:4383–4392.

Liang, Z.C. and Freed, J.H. 1999. An assessment of the applicability of multifrequency ESR to study the complex dynamics of biomolecules. *J. Phys. Chem. B* 103:6384–6396.

Livshits, V.A. 1976. Slow anisotropic tumbling in ESR spectra of nitroxyl radicals. *J. Magn. Reson.* 24:307–313.

Livshits, V.A. and Marsh, D. 2004. High-field EPR spectra of spin labels in membranes. In *Biological Magnetic Resonance*, Vol. 22, Very High Frequency (VHF) ESR/EPR, eds. Grinberg, O.Y. and Berliner, L.J., 431–464. New York: Kluwer Academic.

Marsh, D. 2017. Coherence transfer and electron T_1-, T_2-relaxation in nitroxide spin labels. *J. Magn. Reson.* 277:86–94.

Marsh, D. 2018. Distinct populations in spin-label EPR spectra from nitroxides. *J. Phys. Chem. B* 122:6129–6133.

Marsh, D. and Horváth, L.I. 1998. Structure, dynamics and composition of the lipid-protein interface. Perspectives from spin-labelling. *Biochim. Biophys. Acta* 1376:267–296.

McCalley, R.C., Shimshick, E.J., and McConnell, H.M. 1972. The effect of slow rotational motion on paramagnetic resonance spectra. *Chem. Phys. Lett.* 13:115–119.

McConnell, H.M. 1958. Reaction rates by nuclear magnetic resonance. *J. Chem. Phys.* 28:430–431.

Moro, G. and Freed, J.H. 1981. Calculation of ESR spectra and related Fokker-Planck forms by the use of the Lanczos algorithm. *J. Chem. Phys.* 74:3757–3772.

Polimeno, A. and Freed, J.H. 1995. Slow motional ESR in complex fluids - the slowly relaxing local structure model of solvent cage effects. *J. Phys. Chem.* 99:10995–11006.

Schneider, D.J. and Freed, J.H. 1989. Calculating slow motional magnetic resonance spectra: a user's guide. In *Biological Magnetic Resonance*, Vol. 8, Spin-Labeling. Theory and Applications, eds. Berliner, L.J. and Reuben, J., 1–76. New York: Plenum Publishing Corp.

Shimshick, E.J. and McConnell, H.M. 1972. Rotational correlation time of spin-labeled α-chymotrypsin. *Biochem. Biophys. Res. Commun.* 46:321–327.

Smirnova, T.I., Smirnov, A.I., Clarkson, R.B., and Belford, R.L. 1995. W-Band (95 GHz) EPR spectroscopy of nitroxide radicals with complex proton hyperfine structure: fast motion. *J. Phys. Chem.* 99:9008–9016.

Vasavada, K.V., Schneider, D.J., and Freed, J.H. 1987. Calculation of ESR spectra and related Fokker-Planck forms by the use of the Lanczos algorithm. II. Criteria for truncation of basis sets and recursive steps utilizing conjugate gradients. *J. Chem. Phys.* 86:647–661.

Dynamics and Rotational Diffusion

7

7.1 INTRODUCTION

A major application of spin-label EPR is to studying rotational diffusion of molecules. This exploits the angular dependence of the EPR spectrum relative to the unique direction defined by the static spectrometer field, and the ability of molecular rotation on a timescale comparable with the inverse of the spectral frequencies to average this angular anisotropy. As we have seen in Chapters 5 and 6, continuous-wave lineshapes at the conventional EPR frequency of 9 GHz display sensitivity to rotational motions with correlation times ranging from approximately 10^{-11} to 10^{-7} s.

In the fast regime, below 3×10^{-9} s, we can measure Lorentzian linewidths to obtain correlation times directly by using expressions deduced from perturbation theory (see Chapter 5). As the slow-motional regime approaches the rigid limit, we can measure linewidths and line splittings to calculate correlation times from empirical calibrations that we establish by spectral simulation. For the remainder of the slow-motional regime, and its overlap with the fast regime, we must perform detailed lineshape simulations by using the stochastic Liouville equation as described in Chapter 6. Whereas spectral lineshapes in the fast-motional regime are independent of rotational mechanism – they are always Lorentzian – slow-motion lineshapes are sensitive to the mode of rotational reorientation. This offers an opportunity to study mechanisms. Reorientation of spin-labelled macromolecules is by small-step Brownian diffusion. However, for small spin labels of size comparable to the solvent molecules, this must not be the case, and we might also envisage various types of jump reorientation.

In this chapter, we deal with different means of deducing correlation times for spin labels undergoing both isotropic and anisotropic rotational diffusion in isotropic media. We defer treating molecular ordering, as in anisotropic media, to the chapter following this. Because it is crucial background, we begin with general considerations of Brownian rotational diffusion, before going on to derive correlation functions for particular diffusion equations. We then deal with fast motion, including high-frequency EPR, before going on to consider other reorientation mechanisms and the slow-motional regime.

7.2 ROTATIONAL DIFFUSION EQUATION

We describe Brownian rotational diffusion by the equivalent of Fick's law for translational diffusion:

$$J(\Omega) = -D_R \frac{\partial P(\Omega,t)}{\partial \Omega} \tag{7.1}$$

where $J(\Omega)$ is the rate of passage of spin labels through angle Ω and $P(\Omega,t)$ is the probability of finding the spin label with angular orientation Ω at time t. Equation 7.1 is the definition of the rotational diffusion coefficient, D_R. We derive the differential equation that governs rotational diffusion from Fick's law by using the continuity equation: $\partial P/\partial t = -\partial J/\partial \Omega$, which balances the flows in and out of the orientation Ω. Substituting this condition into Eq. 7.1 yields:

$$\frac{\partial P(\Omega,t)}{\partial t} = D_R \frac{\partial^2 P(\Omega,t)}{\partial \Omega^2} \tag{7.2}$$

for the rotational diffusion equation that we need to calculate the orientational correlation functions used in relaxation theory (see following sections).

A useful expression due to Einstein relates the rotational diffusion coefficient to the mean-square angular displacement, $\langle \Omega^2 \rangle$, of the spin label. In Brownian rotational diffusion, the orientation of the spin label undergoes small random changes, $\delta\Omega$, resulting from frequent collisions with other neighbouring molecules. The change in orientational probability after a short time τ is given by: $(\partial P/\partial t)\times\tau$. This must be equal to the sum over the distribution of probabilities that the spin label orientation lies within $\delta\Omega$ of the initial orientation. We obtain this by averaging over a Taylor expansion of the probability distribution $P(\Omega)$:

$$P(\Omega+\delta\Omega) - P(\Omega) = \left(\frac{\partial P}{\partial \Omega}\right)\delta\Omega + \frac{1}{2}\left(\frac{\partial^2 P}{\partial \Omega^2}\right)(\delta\Omega)^2 + \dots \tag{7.3}$$

For a random process, the average displacement is: $\langle \delta\Omega \rangle = 0$, and hence:

$$\frac{\partial P}{\partial t} = \frac{\langle (\delta\Omega)^2 \rangle}{2\tau}\left(\frac{\partial^2 P}{\partial \Omega^2}\right) \tag{7.4}$$

to lowest order. Comparison with Eq. 7.2, then yields (Einstein 1906):

$$\langle \Omega^2 \rangle = 2D_R t' \tag{7.5}$$

for the mean-square angular displacement in time t'. (We have now dropped the "δ" because it has no special significance in this context.)

7.3 ROTATIONAL FRICTION COEFFICIENTS

We define the rotational friction coefficient, f_R, as the ratio of the frictional torque, T_R, to the angular velocity of rotation, ω_R:

$$f_R = T_R/\omega_R \tag{7.6}$$

It represents the frictional drag in rotational diffusion. We get the equilibrium orientational distribution in the presence of thermal rotational fluctuations from Boltzmann's law:

$$P(\Omega) \propto \exp\left(\frac{T_R\Omega}{k_B T}\right) \tag{7.7}$$

where the work done by rotation through angle $d\Omega$ is $-T_R \cdot d\Omega$. Then we extract the frictional torque from differentiating Eq. 7.7 with respect to Ω:

$$T_R = \frac{k_B T}{P(\Omega)}\left(\frac{\partial P}{\partial \Omega}\right) \tag{7.8}$$

The flow corresponding to the frictional torque is the angular rotation frequency multiplied by the orientational probability, where from Eq. 7.6 $P(\Omega)\omega_R = P(\Omega)\left(T_R/f_R\right)$. At equilibrium, this is balanced by the opposing flow from rotational diffusion that comes from Fick's law. Hence from Eqs. 7.1 and 7.8, the diffusion coefficient is related to the friction coefficient by:

$$D_R = \frac{k_B T}{f_R} \tag{7.9}$$

which is the rotational analogue of the Einstein relation for translational diffusion.

According to fluid dynamics, the friction coefficient is determined by the viscosity of the medium, η, and by the shape and size of the rotating species. We normally use sticking boundary conditions, where the fluid at the surface of the rotating particle moves with exactly the same velocity as the surface itself. For a sphere of radius a, the resulting Stokes–Einstein expression for the rotational friction coefficient is:

$$f_R^o = 6\eta V_{sph} = 8\pi\eta a^3 \tag{7.10}$$

where V_{sph} is the volume of the sphere. This represents a much stronger dependence on molecular dimensions than that of the friction coefficient for translational diffusion, f_T (see Eq. 9.142 in Chapter 9).

For non-spherical species, the friction coefficients about different axes are not identical and f_R is, in general, a tensor. For particles with axial symmetry, the friction coefficient has two principal values, $f_{R\parallel}$ and $f_{R\perp}$, which correspond to rotation around and perpendicular to the principal axis, respectively. Perrin (1934) used the hydrodynamic model with stick boundary conditions to derive expressions for the friction coefficients of ellipsoids of revolution with semi-axes a, $b = c$, in isotropic media. The friction coefficients for rotation about (‖) and perpendicular (⊥) to the unique a-axis are related to that of a sphere of equal volume, $V_{sph}\left(=4\pi ab^2/3\right)$, by the shape factors, F_R. For rotation about the a and b semi-axes:

$$f_{R\parallel} = f_R^o F_{R\parallel} \tag{7.11}$$

and

$$f_{R\perp} = f_R^o F_{R\perp} \tag{7.12}$$

respectively. For an inverse axial ratio $\rho \equiv b/a$, the shape factors are given by (Perrin 1934):

$$F_{R\parallel} = \frac{2}{3}\frac{1-\rho^2}{1-S\rho^2} \tag{7.13}$$

$$F_{R\perp} = \frac{2}{3\rho^2}\frac{1-\rho^4}{S\left(2-\rho^2\right)-1} \tag{7.14}$$

For a prolate ellipsoid (rodlike, $\rho < 1$), the elliptical integral S in Eqs. 7.13 and 7.14 is given by:

$$S = \frac{1}{\sqrt{1-\rho^2}}\ln\left[\frac{1+\sqrt{1-\rho^2}}{\rho}\right] \tag{7.15}$$

Correspondingly, for an oblate ellipsoid (disclike, $\rho > 1$):

$$S = \frac{1}{\sqrt{\rho^2-1}}\tan^{-1}\left(\sqrt{\rho^2-1}\right) \tag{7.16}$$

in Eqs. 7.13 and 7.14.

Figure 7.1 shows the dependence of the shape factors, $F_{R\parallel}$ and $F_{R\perp}$, on the axial ratio, a/b, for oblate and prolate ellipsoids. Rotation around either axis of an oblate ellipsoid produces a comparable friction, and is retarded relative to that of a sphere of equal volume (for which $a = b$ and $F_R = 1$). At high asymmetries ($a/b < 0.2$), rotation of an oblate ellipsoid becomes strongly

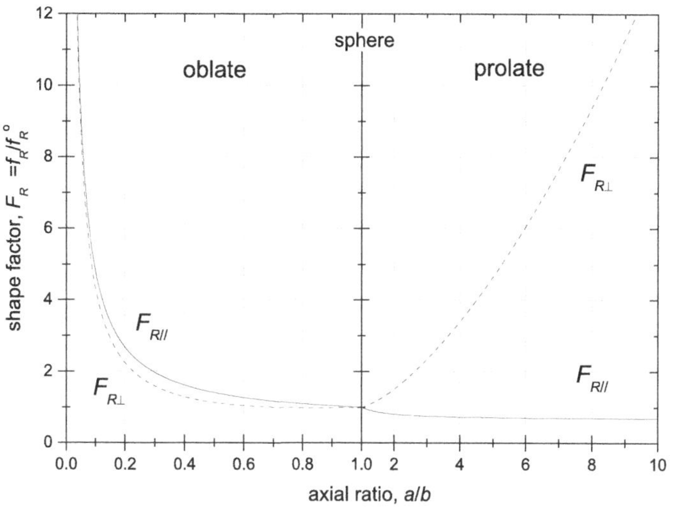

FIGURE 7.1 Rotational friction coefficients, $f_{R\parallel}$ and $f_{R\perp}$, for ellipsoids of revolution. Shape factors, $F_{R\parallel}$ (solid line) and $F_{R\perp}$ (dashed line), as a function of axial ratio, *a/b*, for oblate (left) and prolate (right) ellipsoids of revolution (Eqs. 7.13–7.16). The unique (||) axis lies along the *a* semi-axis. Stick boundary conditions are assumed.

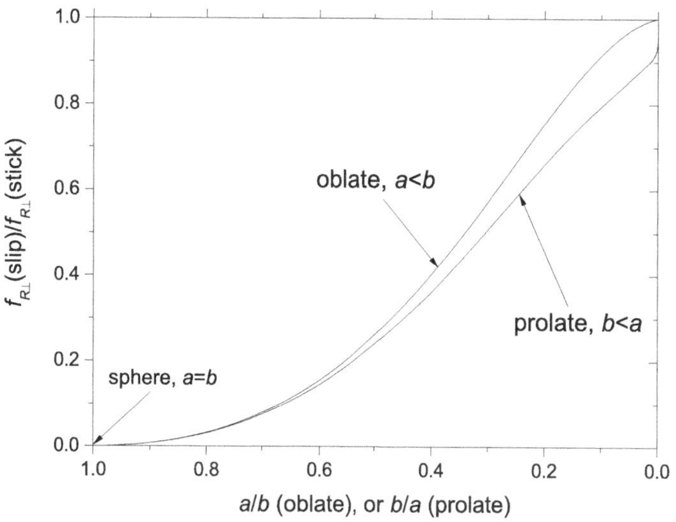

FIGURE 7.2 Rotational frictional ratio, $f_{R\perp}$, for slipping relative to sticking boundary conditions, as function of axial ratio for ellipsoids of revolution, where rotation is about an axis perpendicular to the unique symmetry axis, *a* (Hu and Zwanzig 1974). Axial asymmetry increases from left to right, where the axial ratio for a sphere is one and that for a thin needle or flat disc is zero. For rotation about the unique symmetry axis: $f_{R\parallel} = 0$, under slip boundary conditions.

hindered, and the shape factors tend to a linear dependence: $F_{R\parallel} = F_{R\perp} = 3/(4\pi) \times b/a$, as $a/b \rightarrow 0$ for a disc. Rotation around the long *a*-axis of a prolate ellipsoid is faster than that of the equivalent sphere, reaching a limiting shape factor of $F_{R\parallel} = 2/3$ at high asymmetries for rods. Rotation around the short *b*-axis of a prolate ellipsoid is strongly hindered; the shape factor increases strongly and progressively with increasing asymmetry, tending to the asymptotic relation: $F_{R\perp} = \frac{2}{3}(a/b)^2/\left[2\ln(2a/b) - 1\right]$ at high asymmetries.

7.4 SLIP BOUNDARY CONDITIONS

The friction coefficients and shape factors given above come from conventional fluid dynamics with the so-called sticking boundary condition. In this model, the fluid layer at the surface rotates with the particle. This is a reasonable assumption when the rotating object is much larger than the solvent molecules, e.g., for spin-labelled proteins or other macromolecules. Rough surfaces also tend to promote stick boundary conditions. We find considerable discrepancies, however, when the size of the rotating object is comparable to that of the solvent molecules, e.g., for small spin labels alone. The object then rotates faster than predicted with the stick boundary condition, and a slipping boundary condition may be more appropriate. With slip, there is frictional resistance only if fluid is displaced when the object rotates. There is no tangential component of the normal stress on the surface of the rotating object under slip conditions (see Hu and Zwanzig 1974). Therefore, the friction coefficient is zero for a sphere rotating with slip boundary conditions. Likewise, $f_{R\parallel} = 0$ for rotation about the unique symmetry axis (*a*-axis) of an ellipsoid of revolution with slip. In these cases, the rotational motion is purely inertial without any viscous drag.

Hu and Zwanzig (1974) calculated the friction coefficients, $f_{R\perp}$, for rotation about an axis perpendicular to the unique symmetry axis of ellipsoids of revolution, under slip boundary conditions. Figure 7.2 shows the results of these hydrodynamic calculations for oblate and prolate ellipsoids. The friction coefficient for slip is normalized to that for stick and plotted against the ratio of the length of the shorter axis to that of the longer. A sphere corresponds to an axial ratio of unity and for this: $f_{R\perp}(\text{slip}) = 0$, as already stated. As the axial asymmetry increases, $f_{R\perp}(\text{slip})$ increases at a progressively increasing rate. Initially for modest asymmetries, the ratio $f_{R\perp}(\text{slip})/f_{R\perp}(\text{stick})$ increases to similar extents for both oblate and prolate ellipsoids. But later, the perpendicular friction coefficient with slip approaches that for stick more rapidly for ellipsoids that are oblate. Only at very high asymmetries does $f_{R\perp}(\text{slip})$ for prolate ellipsoids converge to the value for stick boundary conditions. We can approximate this asymptotic form for prolate ellipsoids with $b/a < 0.015$ by the empirical relation: $f_{R\perp}(\text{slip})/f_{R\perp}(\text{stick}) = 1 - 1/\ln\left[4/(b/a)^2\right]$ (Hu and Zwanzig 1974). At the limit of high asymmetry – for thin needles, or flat discs – the perpendicular friction coefficients are the same for slip and stick boundary conditions because they are dominated by the hydrodynamic displacement of fluid by the highly asymmetric rotating object.

The results of hydrodynamic calculations for rotation of a non-axial ellipsoid (semi-axes: $a \neq b \neq c$) with slip boundary conditions (Youngren and Acrivos 1975) are given in Appendix 7.1, at the end of this chapter. Table A7.1 tabulates the shape factors for different combinations of the axial ratios *b*/*a* and *c*/*a*. Which of these axial ratios is *b*/*a* or *c*/*a* is arbitrary.

In general, some combination of stick and slip boundary conditions, or some situation intermediate between them, might be expected. Empirically, we can allow for this by multiplying the friction coefficients, f_R, obtained with stick boundary conditions by a slip coefficient κ, where $0 < \kappa < 1$. From experiment, this works reasonably well, in that the value of κ is independent of temperature (and pressure) for a given solvent but varies between solvents (see, e.g., McClung and Kivelson 1968; Zwanzig and Harrison 1985). Some correlation of κ with solvent size is found, as we might expect from free-volume considerations (Zager and Freed 1982).

7.5 ISOTROPIC ROTATIONAL DIFFUSION

We can describe isotropic Brownian rotational diffusion by the random diffusion of a point, corresponding to the end of a vector fixed in the molecule, over the surface of a unit sphere (see Figure 7.3). The rotational diffusion equation expressed in the spherical polar angles (θ, ϕ) is then (cf. Eq. 7.2):

$$\frac{\partial P}{\partial t} = D_R \nabla_\Omega^2 P = D_R \left[\frac{1}{\sin\theta} \frac{\partial}{\partial\theta} \left(\sin\theta \frac{\partial}{\partial\theta} \right) + \frac{1}{\sin^2\theta} \frac{\partial^2}{\partial\phi^2} \right] P \quad (7.17)$$

where $P(\Omega,t)$ is the instantaneous orientational distribution with respect to the polar angles $\Omega \equiv (\theta, \phi)$ and D_R is the isotropic rotational diffusion coefficient. We can express the general solution to Eq. 7.17 as a sum of spherical harmonics, $\mathrm{Y}_{L,M}(\Omega)$:

$$P(\Omega,t) = \sum_{L,M} c_{LM}(t) \mathrm{Y}_{L,M}(\Omega) \quad (7.18)$$

where $\mathrm{Y}_{L,M}(\theta,\phi) = (-1)^M \sqrt{(2L+1)(L-M)!/4\pi(L+M)!} \times \mathrm{P}_{L,M}(\cos\theta) \cdot e^{iM\phi}$ and $\mathrm{P}_{L,M}(\cos\theta)$ is an associated Legendre function of degree L (see Appendix V.2). Substituting this solution

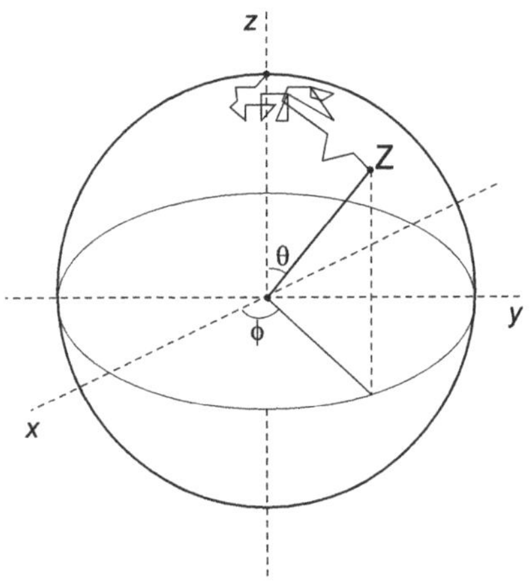

FIGURE 7.3 Rotational diffusion of a macromolecule (or lipid). For isotropic rotation, the locus of the principal molecular axis, *Z*, performs a random walk over the surface of a sphere. For anisotropic rotation, rotations about the *Z*-axis, characterized by the third Euler angle, ψ, must be considered. *x,y,z* are space-fixed axes.

into the diffusion equation gives the following relation for the time dependence of the expansion coefficients:

$$\frac{\mathrm{d}c_{LM}}{\mathrm{d}t} = -D_R L(L+1) c_{LM} \quad (7.19)$$

where we have used the angular part of the solution: $\nabla_\Omega^2 \mathrm{Y}_{L,M}(\Omega) = -L(L+1)\mathrm{Y}_{L,M}(\Omega)$, with L being a positive integer. Solution of Eq. 7.19 for the expansion coefficients is:

$$c_{LM}(t) = c_{LM}(0) \exp(-L(L+1)D_R t) \quad (7.20)$$

where the values of $c_{LM}(0)$ are determined by the initial condition: $P(\Omega,0) = \delta(\Omega - \Omega_o)$. From the closure relation for spherical harmonics (Eq. V.15 in Appendix V): $\delta(\Omega - \Omega_o) = \sum_{L,M} \mathrm{Y}^*_{L,M}(\Omega_o)\mathrm{Y}_{L,M}(\Omega)$, the coefficients at zero time are given by: $c_{LM}(0) = \mathrm{Y}^*_{L,M}(\Omega_o)$.

The conditional probability that a molecule with orientation Ω_o at time $t = 0$ subsequently has orientation Ω at time t therefore becomes:

$$P(\Omega_o;\Omega,t) = \sum_{L,M} \mathrm{Y}^*_{L,M}(\Omega_o)\mathrm{Y}_{L,M}(\Omega) \exp(-t/\tau_L) \quad (7.21)$$

where the time constant τ_L is defined by:

$$1/\tau_L = L(L+1)D_R \quad (7.22)$$

The conditional probability, $P(\Omega_o;\,\Omega,t)$, is exactly what we need to calculate the correlation function, $G(t)$. We do this by integrating over the angular variables Ω and Ω_o (see Eq. 5.11 in Section 5.2):

$$G(t) = \frac{1}{4\pi} \iint F^*(\Omega_0) F(\Omega) P(\Omega_0;\Omega,t) \mathrm{d}\Omega \mathrm{d}\Omega_0 \quad (7.23)$$

In magnetic resonance, the required angular functions, $F(\Omega)$, that appear in the spin Hamiltonian are themselves spherical harmonics, $\mathrm{Y}_{L,M}(\Omega)$ (see, e.g., Eq. 5.15 in Section 5.3). Therefore, from Eqs. 7.21 to 7.23 and the orthogonality of spherical harmonics (Eq. V.14 in Appendix V), the autocorrelation functions for isotropic rotational diffusion are:

$$G_{LM}(t) = \left\langle \left| \mathrm{Y}_{L,M}(\Omega) \right|^2 \right\rangle \exp(-t/\tau_L) \quad (7.24)$$

where the angular brackets indicate an ensemble average over a random angular distribution (cf. Figure 7.3). Note that Eq. 7.24 shows that the correlation functions for rotational diffusion are exponential, and the spectral densities are therefore Lorentzian with correlation times τ_L (cf. Eq. 5.13 in Chapter 5). We used this result extensively in Chapter 5 to obtain expressions for the spin–lattice and spin–spin relaxation rates (see Sections 5.5, 5.6 and 5.8). These relaxation rates are dependent on the $L = 2$ correlation function, with rotational correlation time: $\tau_R \equiv \tau_2 = 1/(6D_R)$ (see Eq. 7.22).

7.6 ANISOTROPIC ROTATIONAL DIFFUSION

For axially anisotropic rotation (i.e., non-spherical molecules or anisotropic media), the diffusion equation is:

$$\frac{\partial P}{\partial t} = -(\mathbf{L}\cdot\mathbf{D}_R\cdot\mathbf{L})P \tag{7.25}$$

where $\mathbf{D}_R$ is the diffusion tensor and $\mathbf{L} = -i\nabla_\Omega$ is the operator that generates an infinitesimal elementary rotation. This is formally equivalent to the dimensionless angular momentum operator (see Appendix R.1). For axially symmetric molecular rotation, we can write the diffusion equation explicitly in terms of the Euler angles, $\Omega \equiv (\phi,\theta,\psi)$ (see Appendix R.2):

$$\frac{\partial P}{\partial t} = D_{R\perp}\left[\frac{1}{\sin\theta}\frac{\partial}{\partial\theta}\left(\sin\theta\frac{\partial}{\partial\theta}\right) + \frac{1}{\sin^2\theta}\frac{\partial^2}{\partial\phi^2} - \frac{2\cos\theta}{\sin^2\theta}\frac{\partial^2}{\partial\psi\,\partial\phi} + \left(\cot^2\theta + \frac{D_{R\parallel}}{D_{R\perp}}\right)\frac{\partial^2}{\partial\psi^2}\right]P \tag{7.26}$$

where $D_{R\parallel}$ and $D_{R\perp}$ are the diffusion coefficients for rotation about the molecular symmetry axis and perpendicular to it, respectively. Here ϕ is a space-fixed azimuthal angle (see Figure 7.3), and ψ is the molecule-fixed azimuthal angle that corresponds to the diffusion coefficient $D_{R\parallel}$. We can write the solution to Eq. 7.26 as a linear combination of normalized Wigner rotation matrices, or generalized spherical harmonics: $\varphi^L_{K,M}(\Omega) = \sqrt{(2L+1)/8\pi^2} \times \mathcal{D}^L_{K,M}(\Omega)$, where $\mathcal{D}^L_{K,M}(\Omega)$ are the elements of Wigner rotation matrices (see Appendix R.3). Formally, these are the eigenfunctions for a symmetric top (see, e.g., Davidov 1965 and Appendix E.4), and for $K = 0$, they are related to the spherical harmonics by Eq. R.13: $\mathcal{D}^L_{0,M}(0,\theta,\psi) = \sqrt{4\pi/(2L+1)} \times Y_{L,-M}(\theta,\psi)$.

We can express the conditional probability in a way similar to the isotropic case (cf. Eq. 7.21):

$$P(\Omega_o;\Omega,t) = \sum_{L,K,M} \varphi^{L*}_{K,M}(\Omega_o)\varphi^L_{K,M}(\Omega)\exp(-t/\tau_{L,M}) \tag{7.27}$$

where the correlation times are now given by the eigenvalues (see Eq. E.17):

$$1/\tau_{L,M} = D_{R\perp}L(L+1) + (D_{R\parallel} - D_{R\perp})M^2 \tag{7.28}$$

of which we need two, e.g., $\tau_{2,0}$ and $\tau_{2,2}$, to define the axial diffusion tensor fully. Here, M is the quantum number for rotation about the molecule-fixed z-axis and K that for the space-fixed z-axis (Brink and Satchler 1968). Other conventions are also used, notably in microwave spectroscopy. Derivation of the autocorrelation functions for axial rotation follows exactly that for isotropic rotation, because the orthogonality properties of the normalized Wigner rotation matrices are similar to those of the spherical harmonics.

Spin–lattice and spin–spin relaxation rates for axially anisotropic rotational diffusion are presented in the Appendix to Chapter 5. These are determined by the $L = 2$ rotational correlation times, $\tau_{2,0}$ and $\tau_{2,2}$, of Eq. 7.28 (cf. Section 5.3). (Note that $\tau_{2,1}$ does not enter, when the diffusion axes are collinear with the spin-label axes.) In analogy with isotropic rotational diffusion, for which the rotational correlation time is related to the rotational diffusion coefficient by: $\tau_R = 1/(6D_R)$, we can define two rotational correlation times for axially anisotropic rotation: one for rotation about the principal axis, $\tau_{R\parallel} = 1/(6D_{R\parallel})$, and the other for rotation about a perpendicular axis, $\tau_{R\perp} = 1/(6D_{R\perp})$. From Eq. 7.28, these are related to the correlation times, $\tau_{2,0}$ and $\tau_{2,2}$, measured from the EPR linewidth coefficients by:

$$\tau_{R\perp} = \tau_{2,0} \tag{7.29}$$

$$\tau_{R\parallel} = \frac{2\tau_{2,0}}{3(\tau_{2,0}/\tau_{2,2}) - 1} \tag{7.30}$$

Correspondingly, the rotational anisotropy is: $\tau_{R\perp}/\tau_{R\parallel} \equiv D_{R\parallel}/D_{R\perp} = \frac{1}{2}(3\tau_{2,0}/\tau_{2,2} - 1)$.

7.7 LINEWIDTHS AND FAST ROTATIONAL MOTION

We can determine the rotational correlation times of relatively small, rapidly tumbling, spin-labelled species in isotropic solvents from the relative line broadening of the three ^{14}N-hyperfine lines in the nitroxide EPR spectrum. For Lorentzian lines, the peak-to-peak linewidth (in magnetic field units) is related to the T_2-relaxation time by Eq. 1.24:

$$\Delta B^L_{pp} = \left(\frac{2}{\sqrt{3}\gamma_e}\right)\frac{1}{T_2} \tag{7.31}$$

where $\gamma_e (\equiv g_o\beta_e/\hbar)$ is the electron gyromagnetic ratio (see Chapter 1). We describe the effects of fast rotational diffusion on the nitroxide T_2-relaxation rate, as determined from time-dependent perturbation theory, in Sections 5.7 and 5.8 of Chapter 5. In the motional narrowing regime, we see from Eq. 5.58 that the Lorentzian EPR linewidths depend on hyperfine manifold, m_I, according to:

$$\Delta B^L_{pp}(m_I) = A_{pp} + B_{pp}m_I + C_{pp}m_I^2 \tag{7.32}$$

where $m_I = +1, 0, -1$ for the low-field, central and high-field lines, respectively, of a ^{14}N-nitroxide. Both linewidth coefficients B_{pp} and C_{pp} depend on the rotational correlation times but in different ways for anisotropic rotational diffusion. Therefore, we can use the relative values of B_{pp} and C_{pp} to diagnose different axial modes of anisotropic rotation diffusion. Note that the high-field line (for which $m_I = -1$) is broadest because hyperfine and g-value anisotropies reinforce; this means that B_{pp} is always negative. Also, it is important not to confuse the B_{pp} linewidth coefficient with the actual linewidth, viz., ΔB_{pp}.

We get the linewidth coefficients B_{pp} and C_{pp} from the Lorentzian contributions to the three linewidths, according to Eq. 7.32:

$$B_{pp} = \frac{1}{2}\left[\Delta B_{pp}^{L}(+1) - \Delta B_{pp}^{L}(-1)\right] \tag{7.33}$$

$$C_{pp} = \frac{1}{2}\left[\Delta B_{pp}^{L}(+1) + \Delta B_{pp}^{L}(-1) - 2\Delta B_{pp}^{L}(0)\right] \tag{7.34}$$

To implement these equations, we must correct for the inhomogeneous broadening of the experimental spectra, as described in Section 2.5 of Chapter 2. If, however, inhomogeneous broadening is negligible compared with the Lorentzian linewidths, we can measure the linewidth coefficients directly from the relative line heights, $h(m_I)$, of the first-derivative absorptions from the three hyperfine lines, according to:

$$B_{pp}^{uncorr} = \frac{1}{2}\Delta B_{pp}^{tot}(0)\left(\sqrt{\frac{h(0)}{h(+1)}} - \sqrt{\frac{h(0)}{h(-1)}}\right) \tag{7.35}$$

$$C_{pp}^{uncorr} = \frac{1}{2}\Delta B_{pp}^{tot}(0)\left(\sqrt{\frac{h(0)}{h(+1)}} + \sqrt{\frac{h(0)}{h(-1)}} - 2\right) \tag{7.36}$$

where $\Delta B_{pp}^{tot}(0)$ is the total peak-to-peak width of the central hyperfine line, and the superscript "*uncorr*" indicates that no correction has been made for inhomogeneous broadening. We can reduce inhomogeneous broadening by using perdeuterated spin labels (e.g., Hwang et al. 1975). Or we can render it less significant by increasing the Lorentzian linewidths via spin exchange, either between spin labels (Sachse et al. 1987) or with molecular oxygen (Kuznetsov et al. 1974), or with saturation broadening (see Chapter 10).

Bales (see Bales 1989; Bales et al. 1987) has established an empirical relationship between the ratios of Lorentzian linewidths, $\Delta B_{pp}^{L}(\pm 1)/\Delta B_{pp}^{L}(0)$, and the corresponding line-height ratios, $h(0)/h(\pm 1)$, of inhomogeneously broadened lines:

$$\frac{\Delta B_{pp}^{L}(\pm 1)}{\Delta B_{pp}^{L}(0)} - 1 = \left(\sqrt{\frac{h(0)}{h(\pm 1)}} - 1\right) \times S(\chi_0) \tag{7.37}$$

where $\chi_0 = \Delta B_{pp}^{G}/\Delta B_{pp}^{L}(0)$ is the ratio of Gaussian to Lorentzian broadening of the central $(m_I = 0)$ hyperfine line, which we determine by the method given in Section 2.5. The calibration function, $S(\chi_0)$, in Eq. 7.37 is given by (Bales 1989):

$$S(\chi_0) = \frac{1 + 1.78\chi_0 + 1.85\chi_0^2}{1 + 2.08\chi_0} \tag{7.38}$$

which applies to two Voigt lineshapes with the same ΔB_{pp}^{G}, and ratio of Lorentzian linewidths in the range: $\Delta B_{pp}^{L}(\pm 1)/\Delta B_{pp}^{L}(0) \leq 2$. From Eqs. 7.33 to 7.37, the linewidth coefficients corrected for inhomogeneous broadening are:

$$B_{pp} = \frac{\Delta B_{pp}^{L}(0)}{\Delta B_{pp}^{tot}(0)} S(\chi_0) \times B_{pp}^{uncorr} \tag{7.39}$$

$$C_{pp} = \frac{\Delta B_{pp}^{L}(0)}{\Delta B_{pp}^{tot}(0)} S(\chi_0) \times C_{pp}^{uncorr} \tag{7.40}$$

where $\Delta B_{pp}^{L}(0)/\Delta B_{pp}^{tot}(0)$ is the ratio of Lorentzian to total peak-to-peak widths of the central hyperfine line. From Eq. 2.30 in Chapter 2, the latter ratio is given by (Dobryakov and Lebedev 1969):

$$\frac{\Delta B_{pp}^{L}(0)}{\Delta B_{pp}^{tot}(0)} = \frac{-1 + \sqrt{1 + 4\chi_0^2}}{2\chi_0^2} \tag{7.41}$$

We get the corrected linewidth coefficients, B_{pp} and C_{pp}, by substituting the uncorrected coefficients from Eqs. 7.35 and 7.36, together with the correcting factors from Eqs. 7.38 and 7.41, in Eqs. 7.39 and 7.40, respectively.

7.8 CORRELATION TIMES FOR FAST ISOTROPIC ROTATIONAL DIFFUSION

The dependence of the T_2-relaxation rate on rotational correlation time for rapid isotropic rotational diffusion is derived in Section 5.8. We must remember that hyperfine constants in expressions for transverse relaxation rates given in Chapter 5 are expressed in angular frequency units. Therefore, we must multiply them by γ_e, the electron gyromagnetic ratio, when given in magnetic field units (e.g., mT). We use hyperfine tensors measured in magnetic field units throughout this chapter. For isotropic rotation, the two m_I-dependent linewidth coefficients are related to the correlation time, τ_R, by Eqs. 5.64 and 5.65. Together with Eqs. 7.31 and 7.32, the linewidth coefficients are then:

$$B_{pp} = \frac{8\gamma_e}{45\sqrt{3}}\left(\Delta g \Delta A + 3\delta g \delta A\right) B_o \tau_R \tag{7.42}$$

$$C_{pp} = \frac{8\gamma_e}{45\sqrt{3}}\left((\Delta A)^2 + 3(\delta A)^2\right)\left(1 - \frac{3}{8}\frac{1}{1 + \omega_a^2 \tau_R^2}\right)\tau_R \tag{7.43}$$

where B_o is the resonance field of the central $(m_I = 0)$ hyperfine line. Here we neglect the non-secular spectral densities at frequency ω_e, which requires that $\omega_e^2 \tau_R^2 >> 1$ (cf. Eqs. 5.61 and 5.62), and correlation times must be longer than $\approx 5 \times 10^{-11}$ s for a spectrometer frequency of 9 GHz (and proportionately lower at higher frequencies/fields – see Section 7.10). If additionally we approximate the pseudosecular spectral densities by $j(\omega_a) \propto \tau_R$, Eq. 7.43 then becomes:

$$C_{pp} = \frac{\gamma_e}{9\sqrt{3}}\left((\Delta A)^2 + 3(\delta A)^2\right) \cdot \tau_R \tag{7.44}$$

where we assume that $\omega_a^2\tau_R^2 << 1$, which restricts correlation times to $\tau_R = 1\times 10^{-9}$ s or less.

Equations 7.42 and 7.44 (or 7.43, for $\tau_R > 1$ ns) give us two independent ways to determine the rotational correlation time. Only if they produce consistent values can we assume that the spin label rotates isotropically (even in an isotropic medium). We collect together values of the tensor anisotropies: ΔA, δA, Δg and δg (see Eqs. 7.51–7.54 given in the next section) for different nitroxide spin labels in Appendix 7.2 at the end of this chapter. Taking mean experimental values for these tensor anisotropies, Eqs. 7.42 and 7.44 yield the following calibrations for determining the isotropic correlation time from the B_{pp} and C_{pp} linewidth parameters:

$$\tau_R = -(3.75 \text{ ns})/B_o(\text{T}) \times B_{pp} \tag{7.45}$$

$$\tau_R = \left(11.2 \text{ ns mT}^{-1}\right) \times C_{pp} \tag{7.46}$$

where B_o is in Tesla, B_{pp} and C_{pp} are in mT, and τ_R is in ns. Consistency with isotropic rotation requires that Eqs. 7.45 and 7.46 yield identical correlation times, i.e., that the absolute ratio of linewidth parameters is: $\left|C_{pp}/B_{pp}\right| \approx 0.335/B_o(\text{T})$. For the standard 9-GHz microwave frequency, the centre-field is: $B_o \approx 0.33$ T and then $\left|C_{pp}/B_{pp}\right| \approx 1$ for isotropic rotation. Equation 7.45 shows that the B_{pp} linewidth parameter becomes sensitive to faster rotation, as the spectrometer frequency/field increases (see Section 7.10, later).

Figure 7.4 shows representative 9-GHz EPR spectra for a nitroxide undergoing isotropic rotational diffusion in the fast motional regime. We see that the spectra consist of three resolved ^{14}N-hyperfine lines, centred at the isotropic resonance positions. The width of each line increases with increasing rotational correlation time, but the high-field line broadens more strongly than the other two hyperfine lines. This is because B_{pp} is negative (see Eq. 7.45), and therefore, B_{pp} and C_{pp} reinforce for the high-field $m_I = -1$ line (cf. Eq. 7.32). Correspondingly, the low-field and central lines have similar widths, because $\left|C_{pp}/B_{pp}\right| \approx 1$ for 9-GHz spectra with isotropic rotation.

As an example of the application of linewidth analysis to isotropic rotation, we use data on the small spin label TEMPONE dissolved in 85% aqueous glycerol from Hwang et al. (1975). This aqueous mixture allows wide variation in solvent viscosity by changing the temperature. Because of its twisted conformation, TEMPONE has relatively small proton hyperfine couplings (see Section 3.8 and Table A3.2 in the Appendix to Chapter 3), and this is reduced yet further by perdeuterating the spin label. Inhomogeneous broadening is therefore very small, allowing precise measurements of homogeneous broadening. The solvent also is deuterated to limit any intermolecular proton hyperfine effects. TEMPONE is a roughly spherical molecule and, therefore, we expect rotation to be almost isotropic.

Figure 7.5 plots the linewidth parameter C_{pp} against B_{pp} for increasing solvent viscosity. We limit the range to correlation times below 1 ns, where we can approximate pseudosecular spectral densities at the hyperfine frequency ω_a by $j(\omega_a) \propto \tau_R$ (see Eqs. 7.43 and 7.44). Over this range, the relation between

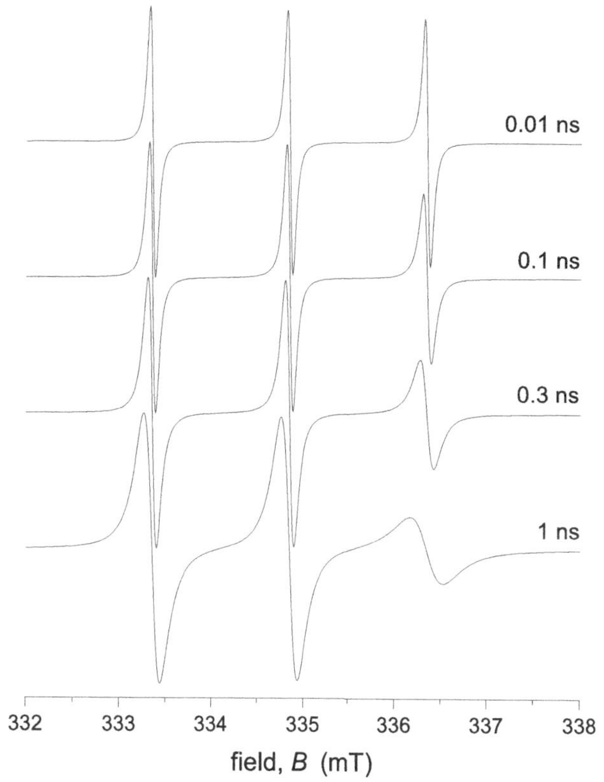

FIGURE 7.4 Simulated 9.4-GHz EPR spectra for fast isotropic rotation with correlation times τ_R given in the figure. Residual Lorentzian linewidth, $2/(\gamma_e T_2) = 0.1$ mT (absorption full-width at half-height). Linewidths are given by Eqs. 5.58 and 5.63–5.65 from Chapter 5. Spin-Hamiltonian tensor anisotropies are mean values from Appendix 7.2.

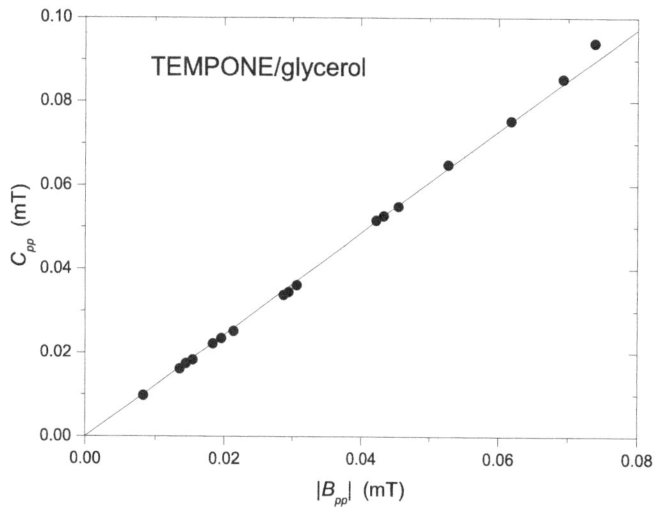

FIGURE 7.5 Correlation of C_{pp} and B_{pp} linewidth parameters, at 9 GHz EPR frequency, for rotational diffusion of TEMPONE (1-oxyl-2,2,6,6-tetramethylpiperidin-4-yl oxide) in 85% glycerol–water at different temperatures (i.e., different viscosities). Both spin label and solvent are perdeuterated. (Data from Hwang et al. 1975.) Solid line: prediction from Eqs. 7.42, 7.44 for isotropic rotation, with magnetic parameters given in Table A7.3 of the Appendix. Note that B_{pp} is negative.

C_{pp} and B_{pp} is linear and agrees quantitatively with Eqs. 7.42 and 7.44, which we can then use to determine the isotropic rotational correlation times, τ_R. For line broadenings greater than those shown in Figure 7.5, we see departures from linearity. These deviations are larger, however, than those predicted for the simple Lorentzian pseudosecular spectral density in Eq. 7.43. They require a non-Debye form: $j(\omega_a) \approx \tau_R / \left(1+\varepsilon'\omega_a^2\tau_R^2\right)$ with $\varepsilon' = 4$ (where $\varepsilon' = 1$ for Lorentzian spectral densities) to describe the dependence of C_{pp} on B_{pp} for correlation times $\tau_R > 1$ ns (Hwang et al. 1975). This is associated with the breakdown of stick boundary conditions for small molecules (cf. Section 7.4) that Hwang et al. (1975) attribute to the influence of fluctuating torques on inertial rotation (see also Section 7.12 below).

Figure 7.6 shows the temperature dependence of the isotropic rotational correlation time for TEMPONE as an Arrhenius plot. The plot is linear over the range where Eqs. 7.42 and 7.44 are valid and yields an activation energy of 53±1 kJ mol^{-1}. Correlation times of >1 ns that we get from motional narrowing theory with the non-Debye spectral density are consistent with the data given in Figure 7.6, and linear extrapolation agrees with measurements made in the slow-motion regime (Hwang et al. 1975).

The activation energy for rotation is comparable to that for the viscosity of the aqueous glycerol solvent $\left(E_a \approx 49 \pm 1 \text{ kJ mol}^{-1}\right)$, but absolute values of the correlation time are much shorter than predicted by the Stokes–Einstein relation for a sphere of this size with stick boundary conditions. The linear dimensions of the TEMPONE molecule are: $\cong 0.84 \times 0.54 \times 0.60\,\text{nm}^3$ (Shibaeva et al. 1972), which gives a geometric mean radius of $\tilde{r} \cong 0.324$ nm. The effective hydrodynamic radius deduced from the viscosity dependence of τ_R is $r_e = 0.164$, however, which corresponds to a slip coefficient of $\kappa \equiv \left(r_e/\tilde{r}\right)^3 = 0.13$ (Hwang et al. 1975). A value of $\kappa = 0.4$ is found with the much less structured solvent toluene, suggesting that TEMPONE in aqueous glycerol rotates in clathrate-type vacancies.

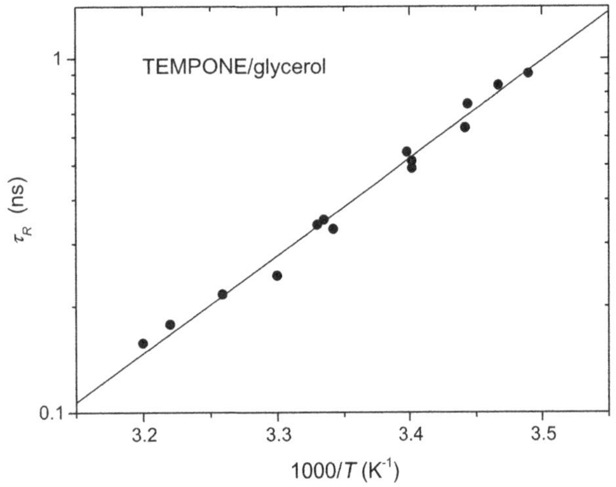

FIGURE 7.6 Temperature dependence of correlation time, τ_R, for isotropic rotation of TEMPONE in 85% aqueous glycerol (Hwang et al. 1975). Data displayed as Arrhenius plot; solid line: linear regression giving activation energy for rotational diffusion: $E_a = 53 \pm 1$ kJ mol^{-1}.

7.9 CORRELATION TIMES FOR FAST ANISOTROPIC ROTATIONAL DIFFUSION

The dependence of the T_2-relaxation rate on rotational correlation time for fast axially anisotropic rotational diffusion is given in the Appendix to Chapter 5. For axial anisotropy, the two m_I-dependent linewidth coefficients are related to the correlation times $\tau_{2,0}$ and $\tau_{2,2}$ of Eq. 7.28 by (Goldman et al. 1972b):

$$B_{pp} = \frac{8\gamma_e}{45\sqrt{3}}\left[(\Delta g\Delta A)\tau_{2,0} + 3(\delta g\delta A)\tau_{2,2}\right]B_o \tag{7.47}$$

$$C_{pp} = \frac{\gamma_e}{9\sqrt{3}}\left[(\Delta A)^2\tau_{2,0} + 3(\delta A)^2\tau_{2,2}\right] \tag{7.48}$$

where both linewidth coefficients, B_{pp} and C_{pp}, and hyperfine anisotropies, ΔA and δA, are in magnetic field units (e.g., mT). We get these expressions from Eqs. A5.10 and A5.11 in the Appendix to Chapter 5 (where the hyperfine anisotropies are in angular frequency units) by using the conditions: $\tau_{2,0}, \tau_{2,2} > 1/\omega_e$ and $< 1/\omega_a$. This restricts correlation times to the range $\approx 5 \times 10^{-11} - 1 \times 10^{-9}$ s at 9 GHz, but extends to proportionately shorter times at higher frequencies/fields (see Section 7.10). From Eqs. 7.47 and 7.48, the rotational correlation times are (Marsh et al. 2007):

$$\tau_{2,0} = \frac{9\sqrt{3}}{8\gamma_e}\frac{1}{\Delta A}\frac{8\delta g \times C_{pp} - \left(5\delta A/B_o\right) \times B_{pp}}{\delta g\Delta A - \Delta g\delta A} \tag{7.49}$$

$$\tau_{2,2} = \frac{3\sqrt{3}}{8\gamma_e}\frac{1}{\delta A}\frac{\left(5\Delta A/B_o\right) \times B_{pp} - 8\Delta g \times C_{pp}}{\delta g\Delta A - \Delta g\delta A} \tag{7.50}$$

where $\gamma_e\left(\approx 2\beta_e/\hbar\right)$ is the electron gyromagnetic ratio and B_o is the resonance field of the central hyperfine line.

For rotation around the nitroxide z-axis as principal (||) axis, the hyperfine anisotropies appearing in Eqs. 7.47 and 7.48 for the linewidth coefficients (and in Eqs. 7.49 and 7.50 for the correlation times) are given by:

$$\Delta A = A_{zz} - \tfrac{1}{2}\left(A_{xx} + A_{yy}\right) \tag{7.51}$$

$$\delta A = \tfrac{1}{2}\left(A_{xx} - A_{yy}\right) \tag{7.52}$$

where A_{xx}, A_{yy} and A_{zz} are the principal elements of the spin-label hyperfine tensor (in magnetic field units). Similar expressions hold for the g-value anisotropies:

$$\Delta g = g_{zz} - \tfrac{1}{2}\left(g_{xx} + g_{yy}\right) \tag{7.53}$$

$$\delta g = \tfrac{1}{2}\left(g_{xx} - g_{yy}\right) \tag{7.54}$$

where g_{xx}, g_{yy} and g_{zz} are the principal elements of the spin-label g-tensor. For rotation around the nitroxide x- or y-axes as principal (||) axes, we permute indices in Eqs. 7.51–7.54 cyclically.

For example, $\Delta A = A_{xx} - \frac{1}{2}(A_{yy} + A_{zz})$, $\delta A = \frac{1}{2}(A_{yy} - A_{zz})$ for rotation around the x-axis, and $\Delta A = A_{yy} - \frac{1}{2}(A_{zz} + A_{xx})$, $\delta A = \frac{1}{2}(A_{zz} - A_{xx})$ for rotation around the y-axis. The different spectral anisotropies that are modulated by preferentially faster rotation about one of the three nitroxide axes give rise to very characteristic differential broadening of the three ^{14}N-hyperfine lines, as we see in Figure 7.7. For z-axis rotation, the high-field line is strongly broadened, whereas the widths of the low-field and central lines are similar. For x-axis rotation, the broadening increases progressively from the low-field, to central, to high-field line. For y-axis rotation, the high-field line is again broadest, but the low-field line is broader than the central line.

We collect values of the spin-Hamiltonian anisotropies: ΔA, δA, Δg and δg for different nitroxide spin labels in Appendix A7.2 at the end of this chapter. These correspond to the spin-Hamiltonian tensors that we give in the Appendix to Chapter 2. We list separate sets of values of ΔA, δA, Δg and δg that are appropriate for rotation preferentially around the z-, x- or y-axes

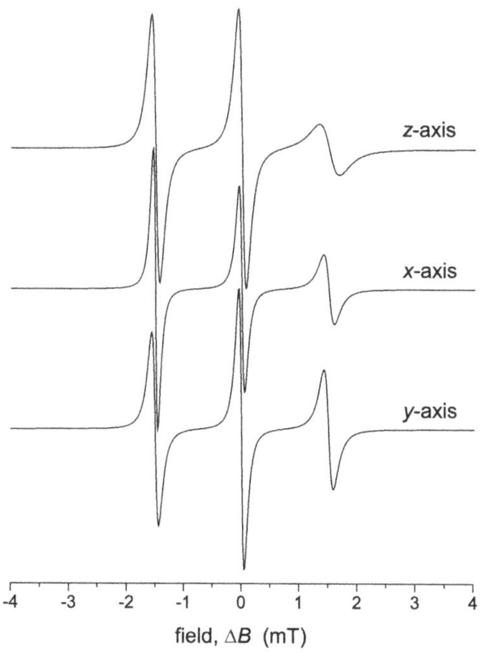

FIGURE 7.7 Simulated 9-GHz EPR spectra for preferential fast rotation about nitroxide z-, x-, or y-axis, with $\tau_{R\parallel} = 0.1$ ns and $\tau_{R\perp}/\tau_{R\parallel} = 10$. Residual Lorentzian linewidth, $2/(\gamma_e T_2) = 0.1$ mT (absorption full-width at half-height). Lorentzian linewidths given by Eqs. A5.7, A5.8, A5.10 and A5.11 from Appendix to Chapter 5. Spin-Hamiltonian tensor anisotropies are mean values from Appendix 7.2.

of the nitroxides. To tabulate coefficients needed for calculating the correlation times, we write Eqs. 7.49 and 7.50 in the following form:

$$\tau_{2,0} = c_1\left(C_{pp} - \frac{c_2}{B_o}B_{pp}\right) \tag{7.55}$$

$$\tau_{2,2} = b_1\left(\frac{b_2}{B_o}B_{pp} - C_{pp}\right) \tag{7.56}$$

where the coefficients c_1, c_2 and b_1, b_2 depend on the mode of anisotropic rotation. Table 7.1 lists numerical values of these calibration coefficients for z-, x- or y-rotation, where B_o is in Tesla, B_{pp} and C_{pp} in mT, and $\tau_{2,0}$ and $\tau_{2,2}$ in ns. These values come from Eqs. 7.49 and 7.50, with cyclic permutations of Eqs. 7.51–7.54, by using the mean experimental tensor anisotropies from Appendix A7.2. Unfortunately, for z-axis rotation, the correlation time $\tau_{2,2}$ depends very strongly on δA which is small (because the hyperfine anisotropy is dipolar and hence almost axial). Therefore, we can determine only the correlation time $\tau_{2,0}(\equiv \tau_{R\perp})$ with reasonable certainty for z-axis rotation (cf. Eq. 7.50). For x- and y-rotation, however, we can get both correlation times, $\tau_{R\parallel}$ and $\tau_{R\perp}$, from the values of $\tau_{2,0}$ and $\tau_{2,2}$ by using Eqs. 7.29 and 7.30.

The fact that correlation times cannot be negative limits the relative values of the linewidth coefficients B_{pp} and C_{pp} in different ways for the three modes of axially anisotropic rotation. Note that experimentally B_{pp} is always negative, whereas C_{pp} is negative only for x-axis rotation. Table 7.1 lists the allowed ranges for the absolute values, $|C_{pp}/B_{pp}|$, of the ratio of linewidth parameters, for z-, x- and y-rotation. As we noted already in the previous section, isotropic rotation requires that $|C_{pp}/B_{pp}| \approx 0.335/B_o(\text{T})$. Thus, we can use the $|C_{pp}/B_{pp}|$ ratio to distinguish different modes of rapid rotation, both isotropic and anisotropic. For instance, at the standard microwave frequency of 9 GHz and $B_o = 0.33$ T, the values of $|C_{pp}/B_{pp}|$ are ≈1, <1, 0.5–1.4 and 0.8–6 for isotropic, z-axis, x-axis and y-axis rotation, respectively. These differences are reflected in the relative line heights of the three hyperfine lines, which we saw already in Figure 7.7.

As an example of linewidth analysis for anisotropic rotation, we use data on the small non-spherical nitroxide PADS (peroxylamine disulphonate), again in 85% aqueous glycerol, from Goldman et al. (1972b). This nitroxide radical contains no protons and therefore inhomogeneous broadening is very small. Figure 7.8 plots C_{pp} against B_{pp} for increasing solvent viscosity in the fast-motion regime. The dependence is linear, but the gradient, $|C_{pp}/B_{pp}| = 1.55 \pm 0.01$, is larger than expected for

TABLE 7.1 Numerical constants relating peak-to-peak linewidth coefficients B_{pp} and C_{pp} (in mT) to rotational correlation times $\tau_{2,0}$ and $\tau_{2,2}$ (in ns) of Eqs. 7.55 and 7.56, and range of validity of the ratio $|C_{pp}/B_{pp}|$ for axially anisotropic rotation about the nitroxide x, y or z-axis

rotation axis	$\|C_{pp}/B_{pp}\|$[a]	c_1 (ns/mT)	c_2 (T)	b_1 (ns/mT)	b_2 (T)
z-axis	$< 0.33/B_o(\text{T})$	11.1	0.004	(–4480)	–0.332
x-axis	$(0.18-0.5)/B_o(\text{T})$	–29.5	–0.460	–24.6	–0.180
y-axis	$(0.26-2)/B_o(\text{T})$	50.9	–0.260	2.30	–1.965

[a] $B_o(\text{T})$ is the centre field in Tesla. Note that for isotropic rotation: $|C_{pp}/B_{pp}| \approx 0.335/B_o(\text{T})$.

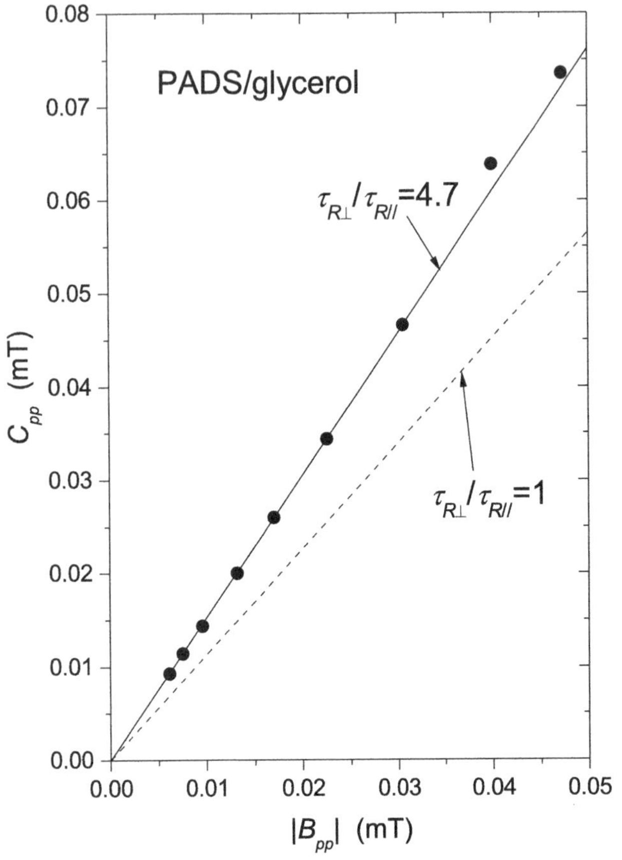

FIGURE 7.8 Correlation of C_{pp} and B_{pp} linewidth parameters, at 9 GHz EPR frequency, for rotational diffusion of peroxylamine disulphonate (PADS) in 85% glycerol–water at different temperatures, (i.e., different viscosities). (Data from Goldman et al. 1972b.) Straight lines: predictions from Eqs. 7.47, 7.48 (with Eqs. 7.29, 7.30) for $\tau_{R\perp}/\tau_{R\parallel}=4.7$ (solid line) and $\tau_{R\perp}/\tau_{R\parallel}=1$ (dashed line). Note that B_{pp} is negative.

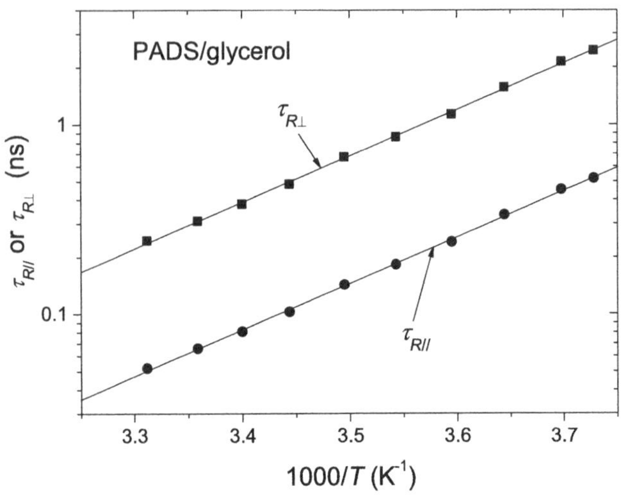

FIGURE 7.9 Temperature dependence of correlation times, $\tau_{R\parallel}$ and $\tau_{R\perp}$, for axial rotation of PADS in 85% aqueous glycerol. (Data from Goldman et al. 1972b.) Rotational anisotropy, $\tau_{R\perp}/\tau_{R\parallel}=4.7$, is independent of temperature (see Figure 7.8); common activation energy: $E_a = 47 \pm 0.5$ kJ mol^{-1}.

isotropic rotation and is consistent exclusively with y-axis rotation, for which we then determine the fixed axial anisotropy to be: $\tau_{R\perp}/\tau_{R\parallel}=4.7\pm0.3$. However, the axial asymmetry of the prolate PADS molecule is relatively modest: $a/b=1.45$ (Howie et al. 1968), and from the frictional ratios in Figure 7.1 this corresponds to only a small rotational anisotropy of $\tau_{R\perp}/\tau_{R\parallel}=1.33$. The much larger experimental asymmetry may arise from strong interaction of the charged SO_3^- groups of $PADS^{2-}$ with solvent. Note also that the rotational anisotropy of PADS in frozen water: $\tau_{R\perp}/\tau_{R\parallel}=2.9\pm0.3$, is considerably smaller than in 85% aqueous glycerol (Goldman et al. 1972b).

Figure 7.9 gives the temperature dependences of the axially anisotropic rotational correlation times, $\tau_{R\parallel}$ and $\tau_{R\perp}$, for PADS as Arrhenius plots. The two plots are linear and parallel over the correlation-time ranges shown and have a common activation energy of 47±0.5 kJ mol^{-1}, because the rotational asymmetry, $\tau_{R\perp}/\tau_{R\parallel}$, does not change with temperature. The activation energy is very similar to that found for isotropic rotation of TEMPONE and for the viscosity of the 85% aqueous glycerol solvent (see Section 7.8). However, as we saw for TEMPONE, the absolute values of the correlation times are much smaller than predicted by the Stokes–Einstein equation with stick boundary conditions. As calculated from the geometric mean of correlation times: $\tilde{\tau}_R=\sqrt{\tau_{R\parallel}\tau_{R\perp}}$, the effective hydrodynamic radius is $r_e=0.11$ nm, whereas the mean radius of PADS from x-ray diffraction is $\bar{r}=0.23$ nm (Howie et al. 1968; Goldman et al. 1972b). This corresponds to an effective slip coefficient of $\kappa=0.11$. Again, as for TEMPONE, there is considerable slip between the small PADS molecule and its solvent cage.

7.10 MEASUREMENTS AT HIGH FIELD/FREQUENCY

From the results of Section 5.8 and the Appendix to Chapter 5, we see that the B_{pp} linewidth coefficient in Eq. 7.32 depends linearly on the magnetic field strength, B_o (see Eqs. 5.61 and A5.10), whereas the C_{pp} coefficient is independent of the field strength (see Eqs. 5.62 and A5.11). Moreover, the A_{pp} coefficient depends quadratically on B_o (see Eqs. 5.59 and A5.8). Going to high EPR operating frequencies, and consequently to high magnetic fields, we therefore expect a dramatic effect on the EPR lineshapes and on their sensitivity to rotational dynamics. There are two major consequences for spin-label dynamics studied at high fields/frequencies (HF). The first is that, in the motional narrowing region, HF-spectra are sensitive to much shorter correlation times. Correspondingly, HF-spectra also enter the slow-motion regime at shorter correlation times. The second consequence is that rotational dynamics come to dominate over all other contributions to the A_{pp} linewidth coefficient in HF-EPR. At conventional EPR frequencies, the residual linewidth contributions, A'_{pp}, make it difficult to use A_{pp} for studying rotational dynamics, but in HF-EPR this problem disappears.

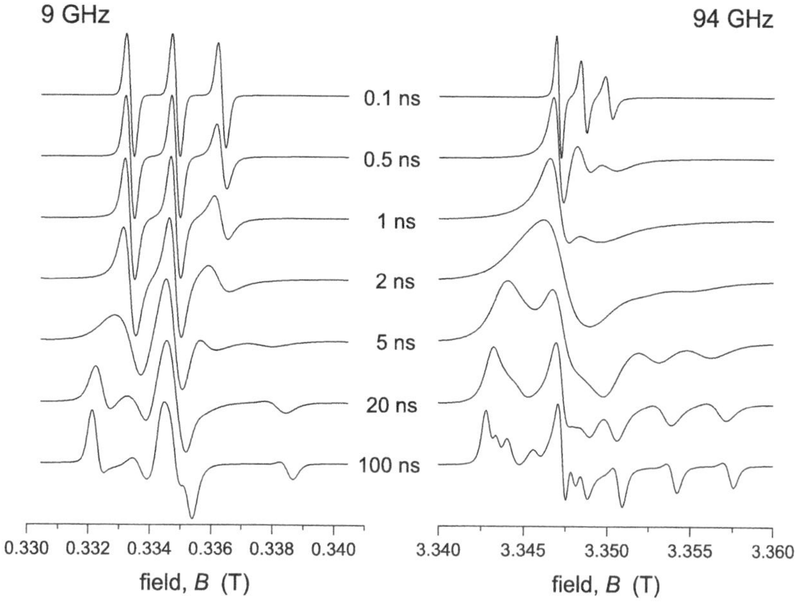

FIGURE 7.10 94-GHz (right column) and 9.4-GHz (left column) EPR spectra of ^{14}N-spin label undergoing isotropic rotational diffusion with correlation times τ_R in ns indicated. Spectra simulated using stochastic Liouville equation (Sections 6.7–6.10); residual Lorentzian linewidth $2/(\gamma_e T_2) = 0.3$ mT (absorption full-width at half-height).

Figure 7.10 illustrates the different motional sensitivities of spin-label EPR at low and high fields. We compare the EPR spectra for a nitroxide spin label at the conventional EPR frequency of 9 GHz with those at a tenfold higher frequency of 94 GHz. The isotropic rotational correlation times τ_R extend from 0.1 ns, characteristic of non-viscous fluids, to 100 ns corresponding to glassy states (see, e.g., Smirnov et al. 1998). For $\tau_R = 0.1$ ns, the spectra consist of three sharp hyperfine lines at both frequencies. At 9 GHz differential line broadening is slight, but at 94 GHz we see strong preferential broadening of the high-field lines, indicating sensitivity to faster rotation at the higher frequency. Also, with increasing correlation time, the spectra broaden more rapidly at 94 GHz than at 9 GHz. Already by $\tau_R = 0.5$ ns, the 94-GHz spectra are in the slow-motion regime, whereas those at 9 GHz are still well within the motional-narrowing region. Only at $\tau_R = 2$ ns do the 9-GHz spectra begin to approach the slow-motion regime. At this stage, the 94-GHz spectrum displays extreme broadening and little structure. By $\tau_R = 5$ ns, the 94-GHz spectrum begins to exhibit the full hyperfine and g-tensor anisotropy, whereas the 9-GHz spectrum is still in an intermediate slow-motion regime where the lineshape remains very sensitive to rotational rate. Finally, the 94-GHz spectra approach the rigid-limit powder lineshape more rapidly with increasing τ_R than do the 9-GHz spectra. At 94 GHz (unlike at 9 GHz), we can resolve all three canonical rigid-limit hyperfine splittings: A_{xx} and A_{yy}, in addition to A_{zz}, if the spin label is perdeuterated. We need these accurate spin-Hamiltonian tensors from rigid-limit spectra to exploit the superior sensitivity of HF-EPR to fast non-axial rotation.

Now we concentrate on the motional narrowing regime. Figure 7.11 shows the A_{pp}, B_{pp} and C_{pp} linewidth coefficients as a function of solvent viscosity to temperature ratio, η/T, for perdeuterated TEMPONE in toluene. Measurements are given for three different EPR frequencies: 250 GHz, 35 GHz and 9.5 GHz (Budil

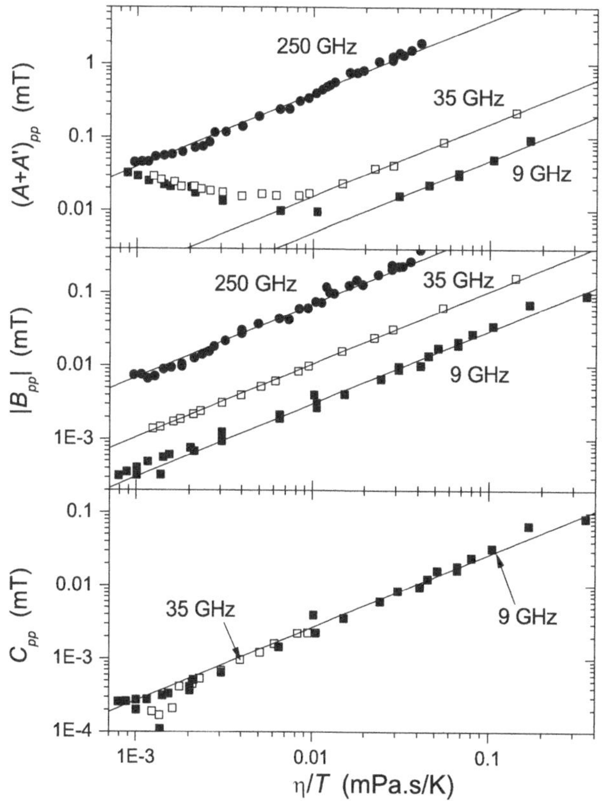

FIGURE 7.11 Dependence of linewidth parameters $(A + A')_{pp}$ (top panel), B_{pp} (middle panel) and C_{pp} (bottom panel) on viscosity to temperature ratio, η/T, for perdeuterated TEMPONE in toluene at EPR operating frequencies: 250 GHz (solid circles), 35 GHz (open squares) and 9.5 GHz (solid squares). (Data from Budil et al. 1993; Hwang et al. 1975.) Both x- and y-axes are logarithmic; straight, solid lines have gradient = +1. Note that B_{pp} is negative.

et al. 1993; Hwang et al. 1975). The solid straight lines have a gradient of +1 in these log-log plots, showing that the linewidth coefficient and η/T are directly proportional. In the motional narrowing regime, the rotational correlation time is therefore directly proportional to η/T.

To within experimental error, values of the C_{pp} coefficient (bottom panel in Figure 7.11) are the same at 9.5 GHz and 35 GHz, as expected from Eq. 7.48, although experimental scatter is greater at 35 GHz. Determining C_{pp} with any degree of precision becomes increasingly difficult at higher EPR frequencies, because the overall line broadening is dominated by the field-dependent A_{pp} and B_{pp} terms. The C_{pp} coefficient is therefore best determined at the standard EPR operating frequency of 9 GHz.

The middle panel in Figure 7.11 shows that the B_{pp} coefficient is directly proportional to the magnetic field strength, as predicted by Eq. 7.47 in the motional narrowing regime. The ratios between intercepts of the different lines are close to the logarithm of the ratio of the corresponding frequencies: log(250/35) and log(35/9.5), respectively. At 9.5 GHz, the data points for low viscosities ($\eta/T < 10^{-2}$ mPa s K^{-1}) lie somewhat above the straight line. This nonlinearity represents the contribution from spectral densities at the Larmor frequency, ω_e, which become appreciable only at very short correlation times. As we mentioned already in Section 7.8 for TEMPONE in aqueous glycerol, these spectral densities are of the non-Debye form: $j(\omega_e) \approx \tau_R / \left(1 + \varepsilon' \omega_e^2 \tau_R^2\right)$ (Hwang et al. 1975). At the higher Larmor frequencies of 35 and 250 GHz, these spectral densities do not contribute appreciably, because then $\omega_e^2 \tau_R^2 >> 1$, even at the lowest viscosities.

The top panel of Figure 7.11 shows the dependence of the A_{pp} coefficient on η/T. At low EPR frequencies, 9 and 35 GHz, the dependence is biphasic (cf. Eqs. 5.59 and 5.60). The high-viscosity region has a log–log slope of +1, as expected from rotational modulation of the hyperfine and g-tensors, whereas the low-viscosity region has the opposite slope, approaching –1. We expect the latter from the inverse dependence on correlation time for relaxation by the spin-rotation mechanism, which is independent of magnetic field strength (see Figure 5.6 and Eqs. 5.39, 5.60). Spin-rotation, in addition to other contributions from the residual linewidth, makes the A_{pp} coefficient unsuitable for detailed study of rotational dynamics at conventional low EPR frequencies. At 250 GHz EPR frequency, however, this is no longer a problem: the dependence of A_{pp} on η/T becomes monophasic with a log–log slope of +1. The strong quadratic dependence of A_{pp} on field strength, B_o, means that the A_{pp} coefficient becomes dominated by rotational modulation of the g-tensor. High-field/frequency EPR spectroscopy therefore provides a further complementary linewidth coefficient, A_{pp}, to study rotational diffusion of spin-labelled molecules.

7.11 FAST NON-AXIAL ROTATION AND HF-EPR

We now exploit measurements of the A_{pp} linewidth coefficient in HF-EPR to determine non-axiality of rapid rotational diffusion. Figure 7.12 shows the dependence of the A_{pp} linewidth coefficient on the B_{pp} coefficient, deduced from the high-field (250 GHz) EPR spectra of perdeuterated TEMPONE in toluene. The relationship is linear, as expected for the fast motional regime (see Eqs. 5.59 and 5.61, Eqs. A5.8 and A5.10). The regression line has a slope of $A_{pp}/B_{pp} = 5.5 \pm 0.1$ and a non-zero intercept that specifies the residual linewidth: $A'_{pp} = 0.075 \pm 0.020$ mT. The A_{pp}/B_{pp} ratio provides a further constraint on the asymmetry of rotational diffusion. In particular, it allows us to determine the complete anisotropy of the rotational diffusion tensor, beyond the axial case that is determined by C_{pp}/B_{pp} at conventional EPR field strengths/frequencies (cf. Section 7.9).

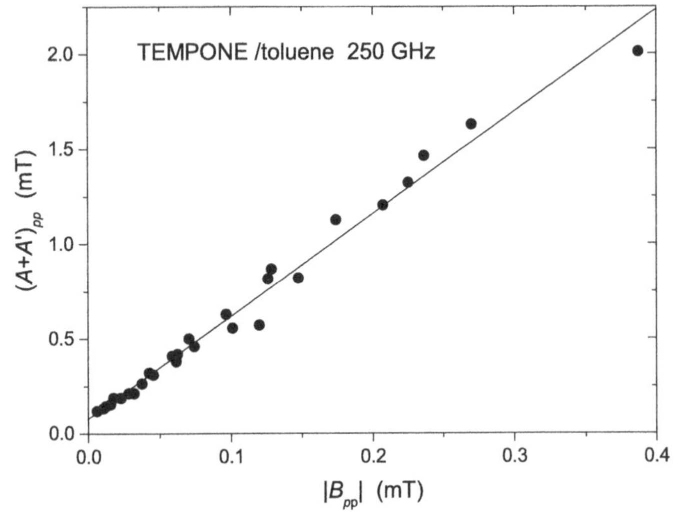

FIGURE 7.12 Correlation of $(A + A')_{pp}$ and B_{pp} linewidth parameters, at 250 GHz EPR frequency, for rotational diffusion of TEMPONE in toluene at different temperatures, (i.e., different viscosities). (Data from Budil et al. 1993.) Gradient of linear regression is $|A_{pp}/B_{pp}| = 5.5 \pm 0.1$ and intercept is $A'_{pp} = 0.075$ mT. Note that B_{pp} is negative.

For completely anisotropic rotational diffusion, all three principal elements of the diffusion tensor $\left(D_{R,x}, D_{R,y}, D_{R,z}\right)$ are different. Expressions for the spectral densities are then rather complex (Freed 1964; Budil et al. 1993). However, if we neglect non-secular spectral densities (i.e., $\omega_e^2 \tau_{eff}^2 >> 1$), and put the pseudosecular frequency ω_a equal to zero, we find rather straightforward relations between the rotational anisotropies ($\rho_x \equiv D_{R,x}/D_{R,z}$ and $\rho_y \equiv D_{R,y}/D_{R,z}$) and the ratios of linewidth coefficients. For the C_{pp}/B_{pp} ratio we get (Kowert 1981):

$$\alpha_{CB}\rho_x = \beta_{CB}\rho_y + \gamma_{CB} \tag{7.57}$$

where

$$\alpha_{CB} = -(\Delta A - 3\delta A)\left[(\Delta g - 3\delta g)B_o\left|\frac{C_{pp}}{B_{pp}}\right| + \frac{5}{8}(\Delta A - 3\delta A)\right] \tag{7.58}$$

$$\beta_{CB} = (\Delta A + 3\delta A)\left[(\Delta g + 3\delta g)B_o\left|\frac{C_{pp}}{B_{pp}}\right| + \frac{5}{8}(\Delta A + 3\delta A)\right] \tag{7.59}$$

$$\gamma_{CB} = 4\Delta A\left[\Delta g B_o\left|\frac{C_{pp}}{B_{pp}}\right| + \frac{5}{8}\Delta A\right] \tag{7.60}$$

with the same definitions of ΔA, δA, etc. as in Eqs. 7.51–7.54 above, expressed in magnetic field units. Similarly, for the A_{pp}/B_{pp} ratio we get (Budil et al. 1993):

$$\alpha_{AB}\rho_x = \beta_{AB}\rho_y + \gamma_{AB} \tag{7.61}$$

where

$$\alpha_{AB} = -(\Delta g - 3\delta g)B_o\left[(\Delta A - 3\delta A)\left|\frac{A_{pp}}{B_{pp}}\right| + \frac{1}{4}(\Delta g - 3\delta g)B_o\right] \tag{7.62}$$

$$\beta_{AB} = (\Delta g + 3\delta g)B_o\left[(\Delta A + 3\delta A)\left|\frac{A_{pp}}{B_{pp}}\right| + \frac{1}{4}(\Delta g + 3\delta g)B_o\right] \tag{7.63}$$

$$\gamma_{AB} = 4\Delta g B_o\left[\Delta A\left|\frac{A_{pp}}{B_{pp}}\right| + \frac{1}{4}\Delta g B_o\right] \tag{7.64}$$

in which A_{pp} is that part of the experimentally measured $(A+A')_{pp}$ parameter which depends upon rotational diffusion (see Section 5.8). Note that B_{pp} is negative, i.e., $A_{pp}/B_{pp} = -|A_{pp}/B_{pp}|$, and similarly for C_{pp}/B_{pp}. The relations given by Eqs. 7.57 and 7.61 are called allowed-value equations, because they specify the combinations of rotational anisotropies, ρ_x and ρ_y, that are possible for a given value of experimental linewidth-parameter ratio. Using both the $|A_{pp}/B_{pp}|$ ratio obtained from high-field/frequency measurements, and the conventional $|C_{pp}/B_{pp}|$ ratio from 9-GHz measurements, then lets us determine ρ_x and ρ_y separately. Figure 7.13 shows the two allowed-value lines that we get from the $|A_{pp}/B_{pp}|$ and $|C_{pp}/B_{pp}|$ ratios for perdeuterated TEMPONE in toluene. The point of intersection of the solid lines gives us the rotational anisotropies: $\rho_x = 1.8 \pm 0.2$ and $\rho_y = 1.5 \pm 0.3$ for TEMPONE (Budil et al. 1993).

We find considerably larger rotational anisotropies than those of TEMPONE for the elongated spin-labelled steroids: 3-DOXYL cholestane ($\rho_x = 0.5 \pm 0.2$ and $\rho_y = 3.0 \pm 0.2$ in toluene; Earle et al. 1993) and 3-DOXYL androstanol ($\rho_x = 1.6 \pm 0.5$ and $\rho_y = 5.8 \pm 1.0$ in *o*-xylene; Smirnova et al. 1995), again from combining high-field and conventional EPR. Interestingly, the allowed-value equation deduced from $|A_{pp}/B_{pp}|$ for the TEMPO spin label in phosphatidylcholine membranes at 94 GHz (dashed line in Figure 7.13) lies very close to that for TEMPONE in toluene (Smirnov et al. 1995). However, we cannot determine the $|C_{pp}/B_{pp}|$ ratio reliably in this case, because the spectrum of membrane-bound TEMPO is not adequately resolved from that of the aqueous spin label at lower EPR frequencies.

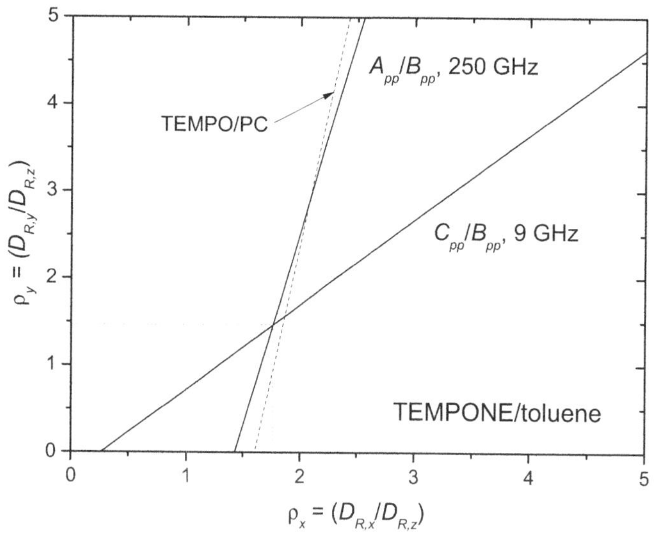

FIGURE 7.13 Allowed combinations of rotational anisotropies $\rho_x \equiv D_{R,x}/D_{R,z}$ and $\rho_y \equiv D_{R,y}/D_{R,z}$ of diffusion tensor for perdeuterated TEMPONE in toluene. Solid straight lines deduced from Eqs. 7.61–7.64 and 7.57–7.60 for linewidth-coefficient ratios A_{pp}/B_{pp} and C_{pp}/B_{pp} determined at 250- and 9.5-GHz EPR frequencies, respectively. (Data from Budil et al. 1993.) The two lines for allowed combinations intersect at the unique solution: $\rho_x = 1.8$ and $\rho_y = 1.5$. Dashed line is for TEMPO in dipalmitoyl phosphatidylcholine lipid membranes, deduced from A_{pp}/B_{pp} at 94 GHz (Smirnov et al. 1995).

7.12 FREE ROTATION AND JUMP DIFFUSION

So far, we have restricted ourselves mostly to the motional narrowing regime and assumed Brownian rotational diffusion. In the fast-motion region, the spectral lineshapes are insensitive to the mechanism of rotation, because the motional contributions to line broadening are always Lorentzian. This no longer holds, however, in the slow-motion regime, and before going on to this, it is useful for us to discuss other types of rotational dynamics. We pointed out already in Section 7.4 that, under slip boundary conditions, the friction coefficient of a sphere goes to zero and likewise for rotation about the unique symmetry axis of a prolate or oblate ellipsoid. Under these circumstances, the molecule rotates freely according to its inertial kinetic energy. We can also envisage alternative rotational mechanisms that occur via sudden large jumps in orientation, with or without intervening inertial rotation.

Brownian rotational diffusion corresponds to rotation in small random steps. It applies to rotating solute molecules that are much larger than the solvent molecules that collide with them. In collision-limited free diffusion, the molecule rotates inertially for a time τ_J and then jumps to a new orientation before resuming free rotation. In jump diffusion, the molecule remains in a fixed orientation for time τ_l (the lifetime) and then jumps instantaneously to a new orientation (see Figure 6.5 in Section 6.4). Free and jump diffusion are more likely when the rotating solute molecule is comparable in size to the solvent molecules or is smaller.

To explore mechanisms, we need an approach that is more fundamental than the diffusion equation; we must go back to the equations of motion for rotation. Applying the analogue of the Langevin equation for linear velocities in the presence of a random force to angular velocities, ω_R, we get the following equation of motion for the angular acceleration (Hubbard 1963; Steele 1963):

$$I^{inert}\frac{d\omega_R}{dt} = -f_R\omega_R + R(t) \tag{7.65}$$

where I^{inert} is the moment of inertia of the rotating molecular species and f_R is the usual rotational friction coefficient. The first term on the right in Eq. 7.65 is the frictional torque experienced by the rotating molecule (see Eq. 7.6), and the second term is a rapidly fluctuating torque, $R(t)$, that arises from random collisions with the surrounding molecules. The random torque has a mean value of zero but a non-vanishing mean-square amplitude, $\langle R^2 \rangle$:

$$\langle R(t)\rangle = 0 \tag{7.66}$$

$$\langle R(t)R(t')\rangle = \langle R^2\rangle \delta(t-t') \tag{7.67}$$

where $\delta(t-t')=1$ for $t=t'$ and is otherwise zero because fluctuations in $R(t)$ are uncorrelated. As usual, the angular brackets in Eqs. 7.66 and 7.67 imply an ensemble average.

Multiplying Eq. 7.65 by $\exp(f_R t/I^{inert})$, we can integrate from 0 to t, giving (Steele 1963):

$$\omega_R(t) = \exp\left(-f_R t/I^{inert}\right) \times\left[\omega_R(0) + \frac{1}{I^{inert}}\int_0^t R(t')\exp\left(f_R t'/I^{inert}\right)\cdot dt'\right] \tag{7.68}$$

Applying this equation twice, and averaging over the distributions in $\omega_R(t)$ and $\omega_R(0)$ by using Eqs. 7.66 and 7.67, we get the autocorrelation function for the angular velocity (Steele 1963; Hubbard 1963):

$$\langle\omega_R(t)\omega_R(0)\rangle = \left(k_B T/I^{inert}\right)\exp\left(-f_R t/I^{inert}\right) \tag{7.69}$$

where k_B is Boltzmann's constant and T is the absolute temperature. The pre-exponential factor in Eq. 7.69 is determined from the equipartition of rotational kinetic energy: $\frac{1}{2}I^{inert}\langle\omega_R^2\rangle = \frac{1}{2}k_B T$, which comes from the Maxwell–Boltzmann distribution of angular velocities and is valid at every t, including the arbitrary $t=0$. This also implies that the mean-square value of the fluctuating torque in Eq. 7.67 is: $\langle R^2\rangle = 2f_R k_B T$ (Steele 1963).

Now we must relate the correlation function for the angular velocities to the angles that appear in the spin Hamiltonian and correspondingly in the rotational diffusion equation. Quite generally, the proportionality constant between the flux, $J(\Omega)$, and the driving force, $\partial P/\partial\Omega$, in relations such as Fick's law (Eq. 7.1) is given by the time integral of the velocity correlation function (Steele 1963; Kubo 1958; Mori et al. 1962):

$$D_R(t) = \int_0^t \langle\omega_R(t)\omega_R(0)\rangle\cdot dt \tag{7.70}$$

Thus, we get the time dependence of the effective diffusion coefficient by substituting Eq. 7.69 in Eq. 7.70 and integrating:

$$D_R(t) = D_R \times\left[1-\exp\left(-t/\tau_J\right)\right] \tag{7.71}$$

where $D_R \equiv k_B T/f_R$ is the usual time-independent rotational diffusion coefficient, and $\tau_J \equiv I^{inert}/f_R$ is the correlation time associated with the angular velocity (or inertial angular momentum). The latter (i.e., τ_J) corresponds to the time of free rotation between molecular collisions. In Section 7.5, we saw that the autocorrelation functions of spherical harmonics are exponential in time when the rotational diffusion coefficient is constant. It therefore comes as no surprise that correlation functions corresponding to a time-dependent diffusion coefficient are not simply exponential functions of time.

We obtain the required diffusion equation for a spherical-top molecule by substituting $D_R(t)$ for D_R in Eq. 7.17. The angular part of the solution is again a sum over spherical harmonics $Y_{L,M}(\Omega)$ (see Eq. 7.18), but the time-dependent expansion coefficients are now given from Eq. 7.19 by:

$$\frac{dc_{LM}}{dt} = -L(L+1)D_R(t)c_{LM}(t) \tag{7.72}$$

Substituting from Eq. 7.71 then gives us the solution:

$$c_{LM}(t) = c_{LM}(0)\exp\left(-L(L+1)D_R\left[t+\tau_J\exp\left(-t/\tau_J\right)-\tau_J\right]\right) \tag{7.73}$$

We see from the development given in Section 7.5 that Eq. 7.73 contains the complete time dependence of the autocorrelation functions that we need. For long times, $t >> \tau_J$, the correlation function becomes a simple exponential: $g_L(t)\approx\exp(-L(L+1)D_R t)$, corresponding to Brownian rotational diffusion, as found already in Section 7.5. At short times, $t << \tau_J$, which correspond to completely free inertial rotation, the correlation function becomes Gaussian: $g_L(t)\approx\exp\left(-L(L+1)D_R t^2/2\tau_J\right)$. Note that the corresponding spectral density: $j_L(\omega)\approx\exp\left(-\omega^2\tau_J/2L(L+1)D_R\right)$, is also Gaussian, because the Fourier transform of a Gaussian function is itself Gaussian (see Table T.1 in Appendix T). From this we infer that the spectral density corresponding to Eq. 7.73 has Lorentzian character at low frequencies, going over to Gaussian character at high frequencies.

A simplified representation of the time dependence in Eq. 7.73: $g(t)\approx\exp\left(-L(L+1)D_R\left(\sqrt{t^2+\tau_J^2}-\tau_J\right)\right)$, is proposed by Egelstaff (1970) for collision-limited free rotational diffusion. This function gives a simple exponential decay at long times but becomes Gaussian at short times. (Note: Klauder and Anderson 1962 give a similar approximate treatment of Lorentzian spectral diffusion; see Appendix A.13 in Chapter 13.) We can approximate the widths at half-height in the corresponding spectral density roughly by an inverse correlation time (Egelstaff 1970):

$$1/\tau_L = \frac{L(L+1)D_R}{\sqrt{1+L(L+1)D_R\tau_J}} \tag{7.74}$$

where τ_L is the exponential correlation time specifying a Lorentzian spectral density of equal width. For comparison, the inverse correlation time for Brownian rotational diffusion is given by Eq. 7.22, namely: $1/\tau_L = L(L+1)D_R$.

Unlike free rotational diffusion, jump diffusion gives rise to exponential correlation functions, where the correlation time depends on the distribution of jump angles (Ivanov 1964). For a particular form of the distribution that is close to random, the correlation time is given by (Egelstaff 1970):

$$1/\tau_L = \frac{L(L+1)D_R}{1+L(L+1)D_R\tau_l} \tag{7.75}$$

where τ_l is the time between jumps (or lifetime), and the effective diffusion coefficient is defined by the mean-square angular jump length, $\langle \vartheta_J{}^2 \rangle$: $D_R \equiv \langle \vartheta_J^2 \rangle / 6\tau_l$. Expressions for the correlation time when the distribution of jump angles is exponential, Gaussian or completely random are given in Goldman et al. (1972b).

Figure 7.14 shows the functional dependence of $1/\tau_L$ on $D_R\tau$ for the three models, Brownian, free and jump diffusion. The lifetime parameter τ has no significance for Brownian diffusion, where $1/\tau_L = L(L+1)D_R$, and therefore we have simply $\tau \equiv 1$ for this case in Figure 7.14. For free or jump diffusion, τ is the duration of free diffusion, τ_J, or the time at a fixed orientation, τ_l, respectively. At low values of the diffusion coefficient, $D_R << 1/\tau$, all three models are equivalent, and both free and jump diffusion approximate to the simple case of Brownian diffusion. As the diffusion coefficient increases, free and jump diffusion begin to diverge from Brownian diffusion, with the difference being larger for jump diffusion. In the limit of $D_R >> 1/\tau$, the effective correlation time for jump diffusion becomes equal to the lifetime: $\tau_L = \tau_l$, for all values of L, as we expect on physical grounds. The different dependence on L for the three models that we see in Figure 7.14 is important in the slow-motion regime. Unlike in the fast-motion regime, the slow-motion lineshapes are dependent on more than a single value of L, and thus we can use them to distinguish between motional models (Goldman et al. 1972b).

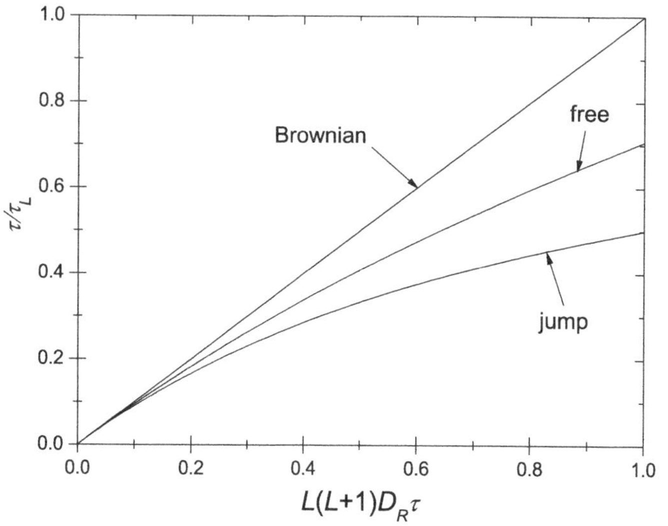

FIGURE 7.14 Dependence of normalized inverse correlation time, $1/\tau_L$, associated with spherical harmonic $Y_{L,M}(\theta,\phi)$, on rotational diffusion coefficient, D_R, for Brownian rotational diffusion (Eq. 7.22), collision-limited free rotational diffusion (Eq. 7.74) and jump rotational diffusion (Eq. 7.75). For Brownian diffusion, $\tau \equiv 1$ (not in the model); for free diffusion, τ is time of inertial rotation τ_J between reorientational collisions; for jump diffusion, τ is lifetime τ_l in fixed orientation before instantaneous transition to a new orientation. For Brownian and free diffusion: $D_R \equiv k_BT/f_R$; for jump diffusion: $D_R \equiv \langle \vartheta_J^2 \rangle / 6\tau$ where ϑ_J is angular jump length.

7.13 SLOW ROTATIONAL DIFFUSION AND ANISOTROPY

Slow-motion simulations for isotropic rotational diffusion, using the stochastic Liouville equation, appear already in Chapters 1 and 6. These illustrate the full range of dynamic sensitivity: from three sharp lines in the extreme motional narrowing region to the rigid-limit powder pattern for extremely slow motion. Figure 7.15 shows comparable 9-GHz spectral simulations for highly anisotropic rotation about the nitroxide x-, y-, or z-axis. Here we set the correlation time about the two non-unique perpendicular axes (e.g., y- and z- for x-axis rotation) to 300 ns, which is too slow to affect conventional EPR lineshapes (see Polnaszek et al. 1981). For slow uniaxial rotation, all three models tend to the rigid limit, just as for isotropic rotation. For fast uniaxial rotation, however, the spectra do not tend to three sharp lines, because only the anisotropy about a single axis is averaged. The result, therefore, is an axial powder pattern with parallel and perpendicular tensor components that are very different for the three models. For say x-axis rotation, we have $A_\parallel = A_{xx}$ and $A_\perp = \frac{1}{2}(A_{yy} + A_{zz})$, and similarly for the g-tensor, in the fast rotational limit. This explains why the lineshapes differ considerably for the three models, before we reach the very slow limit. In particular, the changes in lineshape are rather small for z-axis rotation at 9 GHz, throughout the entire range, because they correspond mostly to averaging of the relatively small Zeeman x–y anisotropy. The fast limit in Figure 7.15 differs from the spectra for axially anisotropic rotation from motional-narrowing theory that were given in Figure 7.7, because both parallel and perpendicular rotations are fast (although different) in the latter case.

7.14 SLOW-MOTION CALIBRATIONS FOR OUTER SPLITTINGS AND LINEWIDTHS

The outer maxima in 9-GHz slow-motion spectra correspond to nitroxides with their z-axes oriented along the magnetic field direction (see Section 2.4). Molecular rotation causes these

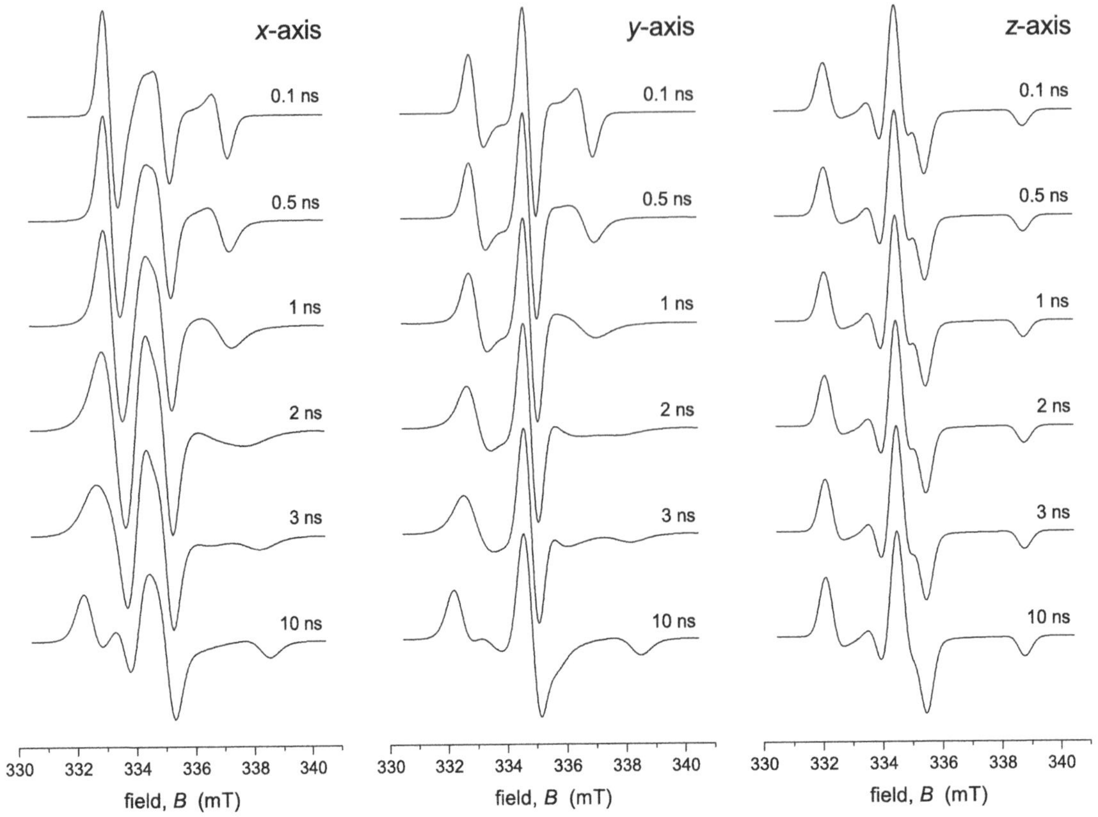

FIGURE 7.15 9.4-GHz EPR lineshapes for anisotropic rotation about *x*-axis (left), *y*-axis (centre) and *z*-axis (right) of the nitroxide. Spectra simulated using stochastic Liouville equation, with additional Gaussian broadening FWHM = 0.5 mT. Axial rotational correlation times $\tau_{R\parallel}$ (in ns) given in the figure; perpendicular value is fixed at $\tau_{R\perp} = 300$ ns (cf. Polnaszek et al. 1981).

nitroxides to exchange with those of different orientation in other parts of the spectrum. In analogy to two-site exchange, as the motional rate of the spin label increases from the rigid limit, the outer hyperfine peaks first broaden and then shift closer together (see Section 6.3). This suggests that we can use the outer hyperfine splitting and linewidths for empirical calibration of rotational correlation times in the slow-motional regime (Goldman et al. 1972a; Mason and Freed 1974; and see also Figure 6.7 in Section 6.5).

For two-site exchange, the increase in linewidth is given by Eqs. 6.20 and 6.21. We express the lifetime broadening as an effective rotational correlation time τ_R, in this simple model. The Lorentzian half-widths at half-height $\Delta B_{m_I} \equiv \left(\Delta B^L_{1/2}\right)_{m_I}$ of the outer extrema then become:

$$\Delta B_{m_I} \approx \Delta B^R_{m_I} + \frac{1}{\gamma_e \tau_R} \tag{7.76}$$

where $\Delta B^R_{m_I}$ is the rigid-limit value of ΔB_{m_I}, with $m_I = +1$ for the low-field peak and $m_I = -1$ for the high-field peak: see Figure 7.16. (Remember that the outer extrema in a first-derivative powder pattern have the straight absorption lineshape – see Eq. 2.22.) Correspondingly, the peak separation $\delta\omega$ given by Eq. 6.22 is reduced by a factor $\sqrt{1-8\tau^{-2}\delta\omega^{-2}} \cong 1-4\tau^{-2}\delta\omega^{-2}$ in the two-site exchange model, where τ is the lifetime and $\delta\omega$ is in angular frequency units. Therefore, we can approximate the splitting $2\langle A_{zz}\rangle$ of the outer hyperfine extrema in this model by:

$$\langle A_{zz}\rangle = A^R_{zz} - \frac{1}{\gamma_e^2 A^R_{zz}}\tau_R^{-2} \tag{7.77}$$

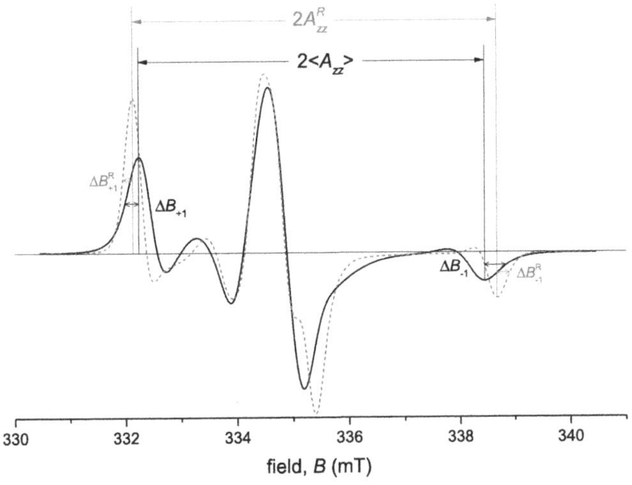

FIGURE 7.16 Slow-motion (solid, black) and rigid-limit (dashed, grey) 9.4-GHz EPR spectra for ^{14}N-nitroxide, showing linewidths ΔB_{m_I} and splittings $2A_{zz}$ used for correlation-time calibrations (Eqs. 7.81, 7.80).

where A_{zz}^R is the rigid-limit value of the partially averaged hyperfine tensor element $\langle A_{zz} \rangle$.

We generalize Eqs. 7.77 and 7.76 to give the following empirical expressions for the outer-splitting constant and linewidths, respectively:

$$\langle A_{zz} \rangle = A_{zz}^R \left[1 - \left(\tau_R / a \right)^{1/b} \right] \tag{7.78}$$

$$\Delta B_{m_I} = \Delta B_{m_I}^R \left[1 + \left(\tau_R / a_{m_I} \right)^{1/b_{m_I}} \right] \tag{7.79}$$

where the fitting parameters a, b and a_{m_I}, b_{m_I} then come from simulations with the stochastic Liouville equation for slow isotropic rotational diffusion (see, e.g., Figures 1.5 and 6.12). We define the a and b parameters in Eqs. 7.78 and 7.79 so as to get simpler expressions for the standard rotational correlation-time calibrations (cf. Goldman et al. 1972a; Mason and Freed 1974):

$$\tau_R = a \left(1 - \frac{\langle A_{zz} \rangle}{A_{zz}^R} \right)^b \tag{7.80}$$

$$\tau_R = a_{m_I} \left(\frac{\Delta B_{m_I}}{\Delta B_{m_I}^R} - 1 \right)^{b_{m_I}} \tag{7.81}$$

Table 7.2 lists calibration parameters for isotropic Brownian rotational diffusion, collision-limited free diffusion and jump diffusion mechanisms of reorientation. Corresponding calibrations of the outer hyperfine splitting (i.e., Eq. 7.78) are also given for slow uniaxially anisotropic rotation about either the y-axis or x-axis, in the Brownian case (see Figure 7.15 and Polnaszek et al. 1981).

In the two-site exchange model, the linewidth calibration constants from Eq. 7.76 are: $b_{m_I} = -1$, and $a_{m_I} = 21.9$ ns for an intrinsic Lorentzian peak-to-peak derivative width of $\Delta B_{pp}^L = 0.3$ mT. These agree reasonably well with the values from slow-motion simulations in Table 7.2, supporting the idea of lifetime broadening. In fact, the increase in linewidth of the two outer extrema is within a factor of 2 (or $\frac{1}{2}$) of $1/\tau_R$, over the entire range of simulations (Mason and Freed 1974). The two-site exchange values for calibration of the outer splitting are: $b_{m_I} = -1/2$, and $a_{m_I} = 1.78$ ns for $A_{zz}^R = 3.2$ mT. These values are closest to the slow-motion calibrations for jump diffusion in Table 7.2, where the exchange analogy is most likely to hold.

Figure 7.17 shows dependences of the outer hyperfine splitting on rotational correlation time that we predict from the calibration constants in Table 7.2. For isotropic rotation, the decrease in splitting is greatest for Brownian diffusion, becoming smaller for collision-limited free diffusion and jump diffusion, in that order. At correlation times, $\tau_R < 7$ ns, the outer lines start to collapse to the isotropic positions, and the calibration is no longer useable. For anisotropic x- and y-rotation, on the other hand, the reduction in outer splitting is less abrupt and the calibrations in Table 7.2 remain valid down to $\tau_R \geq 2$ ns. The upper limit of the calibrations depends on the model, being greatest for isotropic Brownian diffusion, with $\tau_R \leq 300$ ns.

Figure 7.18 shows dependences of the Lorentzian widths of the outer hyperfine peaks on isotropic rotational correlation time. We see immediately that the linewidths are sensitive to slower rotation than is the outer splitting, as expected from the two-site exchange analogy. The high-field $(m_I = -1)$ peak broadens more with slow rotation than does the low-field $(m_I = +1)$ peak,

TABLE 7.2 Parameters fitting outer-splitting and linewidth calibrations (Eqs. 7.80 and 7.81), to give correlation times of slow rotational motion

diffusion model	ΔB_{pp}^L (mT)[a]	A_{zz}^R (mT)	a (ns)	b	m_I	$\Delta B_{m_I}^R$ (mT)[b]	a_{m_I} (ns)	b_{m_I}
isotropic, Brownian	0.03	~3.2	0.257	−1.78				
	0.1		0.590	−1.24	+1	0.080	54.5	−0.999
					−1	0.091	99.5	−1.014
	0.3		0.54	−1.36	+1	0.239	11.5	−0.943
					−1	0.272	21.2	−0.778
	0.5		0.852	−1.16				
	0.8		1.09	−1.05				
isotropic, free	0.03		0.699	−1.20				
	0.1		0.295	−1.68	+1	0.080	53.2	−1.076
					−1	0.091	79.7	−1.125
	0.3		1.10	−1.01	+1	0.239	12.9	−1.033
					−1	0.272	19.6	−1.062
isotropic, jump	0.03		2.46	−0.589				
	0.3		2.55	−0.615				
y-axis, Brownian[c]	0.15–0.3	~3.2	0.2596	−1.396				
x-axis, Brownian[d]	0.15–0.3	3.36	0.56	−0.99				

Calibrations from stochastic-Liouville simulations of 9-GHz spectra for ^{14}N-nitroxides (Goldman et al. 1972a; Mason and Freed 1974)

[a] $\Delta B_{pp}^L = 2/\left(\sqrt{3}\gamma_e T_2\right)$, peak-to-peak first-derivative Lorentzian linewidth (Eq. 1.24) for residual line broadening.

[b] *Note*: in the rigid limit, half-widths at half-height, $\Delta B_{m_I}^R$, of outer hyperfine extrema are: $2\Delta B_{+1}^R = 1.59\Delta B_{pp}^L$ and $2\Delta B_{-1}^R = 1.81\Delta B_{pp}^L$, for low-field and high-field lines, respectively, close to ideal value for a Lorentzian line: $2\Delta B^R = \sqrt{3}\Delta B_{pp}^L$ (see Eqs. 1.23, 1.24).

[c] Polnaszek et al. 1981.

[d] See Figure 7.15.

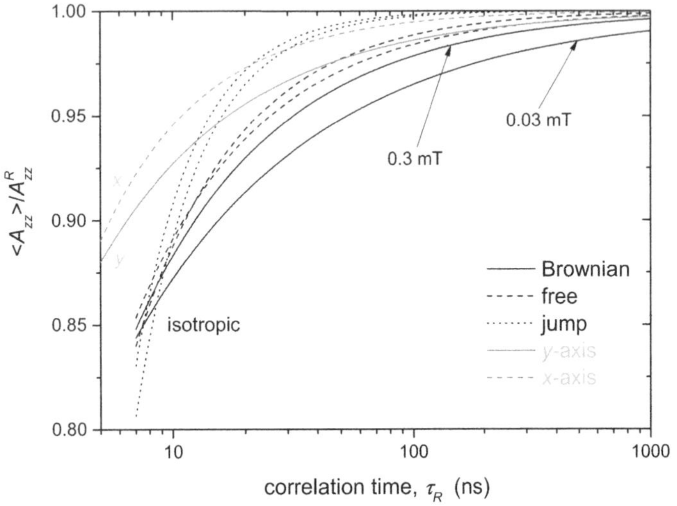

FIGURE 7.17 Dependence of normalized spin-label outer hyperfine splitting, $\langle A_{zz}\rangle / A_{zz}^R$, on correlation time, τ_R, for slow isotropic rotational diffusion (Eq. 7.78, Table 7.2). From 9-GHz spectral simulations for ^{14}N-nitroxide, using stochastic Liouville equation (Goldman et al. 1972a). Solid lines: Brownian diffusion; dashed lines: collision-limited free diffusion; dotted lines: jump diffusion (see Section 7.12). For each pair of curves, upper corresponds to residual linewidth $\Delta B_{pp}^L = 0.3$ mT and lower to $\Delta B_{pp}^L = 0.03$ mT. Grey curves: rotation about nitroxide *y*-axis (solid), or *x*-axis (dashed), alone.

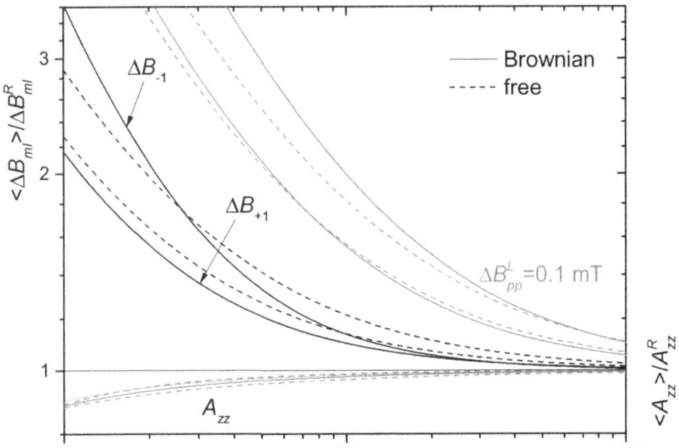

FIGURE 7.18 *Upper curves* (black and grey): dependence of normalized linewidths, $\Delta B_{m_I}/\Delta B_{m_I}^R$, of hyperfine outer extrema ($m_I = \pm 1$), on correlation time, τ_R, for slow isotropic rotational diffusion (Eq. 7.79, Table 7.2). From 9-GHz spectral simulations for ^{14}N-nitroxide, using stochastic Liouville equation (Mason and Freed 1974). For each pair of curves, upper corresponds to high-field peak ($m_I = -1$) and lower to low-field peak ($m_I = +1$). Black lines: residual linewidth $\Delta B_{pp}^L = 0.3$ mT; light-grey lines: $\Delta B_{pp}^L = 0.1$ mT. *Lower curves* (dark grey): corresponding calibrations for normalized outer hyperfine splitting, $\langle A_{zz}\rangle / A_{zz}^R$ (Figure 7.17, Eq. 7.78). For each pair of curves, upper corresponds to residual linewidth $\Delta B_{pp}^L = 0.3$ mT, and lower to $\Delta B_{pp}^L = 0.1$ mT. Solid lines: Brownian diffusion; dashed lines: collision-limited free diffusion. Both axes are logarithmic; minor *y*-axis ticks correspond to increments of 0.2.

because the resonance position changes more rapidly with angle at the high-field turning point than at the low-field turning point (cf. Section 2.3).

At the extreme of the calibrations, choice of rigid-limit parameters A_{zz}^R and $\Delta B_{m_I}^R$ is critical because the difference from these limiting values becomes small for long τ_R. Note that calibration parameters change for different intrinsic line broadenings. We list values for various rigid-limit linewidths in Table 7.2. For many practical situations, choosing the parameters corresponding to $\Delta B_{pp}^L = 0.3$ mT is most appropriate (Freed 1976). In principle, we can combine Eq. 7.79 with the relation between rigid-limit linewidths: $\Delta B_{+1}^R / \Delta B_{-1}^R = 1.59/1.81 = 0.878$ (see footnote b to Table 7.2) to obtain both τ_R and $\Delta B_{\pm 1}^R$ from measurements at low- and high-field.

We can also determine rigid-limit parameters by extrapolating the viscosity dependence to infinity. Figure 7.19 shows the increase in outer hyperfine splitting $2\langle A_{zz}\rangle$ with increasing viscosity for spin-labelled T4 lysozyme. According to Eqs. 7.9 and 7.10, the rotational correlation time depends linearly on viscosity η:

$$\tau_R = \frac{1}{6D_R} = \frac{V_{mol}}{k_B T}\eta \tag{7.82}$$

where V_{mol} is the volume of the protein molecule, which we assume to be roughly spherical. By incorporating Eq. 7.82, a non-linear least-squares fit of Eq. 7.78 to the data of Figure 7.19 yields an extrapolated value of $A_{zz}^R = 3.61 \pm 0.05$ mT. This is less than the true rigid-limit value of $A_{zz}^R = 3.75 \pm 0.02$ mT that is measured in frozen solution at −50°C, indicating limited residual internal motion of the spin-label relative to the protein backbone (Fleissner et al. 2011). The fitted value of the exponent in Eq. 7.78 is $b = -1.82 \pm 0.33$. For Brownian diffusion, this corresponds to a low intrinsic linewidth and implies that $a = 0.257$ ns (see Table 7.2). From Eq. 7.82, the third

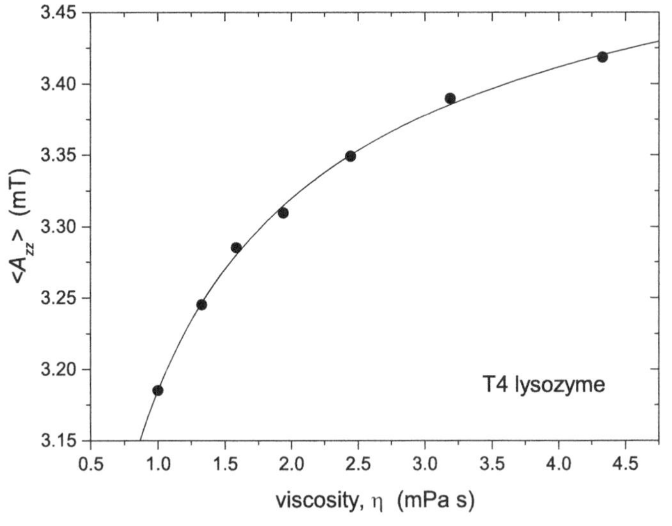

FIGURE 7.19 Dependence of outer hyperfine splitting $2\langle A_{zz}\rangle$ on aqueous viscosity for T4 lysozyme with cross-linking spin label bis-MTSSL (RX) attached at E5C,I9C (Fleissner et al. 2011). Solid line: non-linear least squares fit of Eq. 7.78 for 9-GHz spectra (with Eq. 7.82 for viscosity dependence of correlation time).

non-linear fitting parameter in Figure 7.19 then yields a molecular volume of $V_{mol} = 0.51 \pm 0.26\ \text{nm}^3$, corresponding to a protein radius of $r_{eff} = 2.3 \pm 0.4$ nm. T4 lysozyme has molecular mass $M_r = 18.6$ kDa, which together with a partial specific volume of $\bar{v} = 0.70 \times 10^{-3}\ \text{m}^3\ \text{kg}^{-1}$, gives a molecular volume of $V_{mol} = M_r \times 10^{-3}\bar{v}/N_A = 0.22\ \text{nm}^3$. This is approximately half the value deduced assuming isotropic rotational diffusion. However, T4 lysozyme is not a spherical protein: it consists of two lobes which give an axial ratio of $a/b \approx 2$ (Matthews and Remington 1974; Remington et al. 1978). Thus, the effective correlation time, $\tau_{R\perp}$, could be up to a factor of two greater than that predicted for a sphere (see Eqs. 7.11 and 7.12, and Figure 7.1). Note that, if we know the b calibration parameter, we can get the rigid-limit value of A_{zz}^R simply from extrapolating a linear plot of $\langle A_{zz}\rangle$ against $1/\eta^{1/b}$ (Timofeev and Tsetlin 1983).

We can use a similar approach to the temperature dependence of the rotational correlation time, which is determined by the Arrhenius law:

$$\tau_R = \tau_o \exp\left(+\frac{E_a}{RT}\right) \tag{7.83}$$

where $\tau_o = (V_{mol}/k_B T)\eta_o$ and E_a is the activation energy associated with the effective fluidity $1/\eta$. Figure 7.20 shows the temperature dependence of the outer hyperfine splitting $2\langle A_{zz}\rangle$ for Class II –SH groups of the Na,K-ATPase spin-labelled with either the piperidine (TEMPO-Mal) or pyrrolidine (PROXYL-Mal) nitroxyl maleimide. The x-axis is the reciprocal temperature $1/T$, and therefore the dependence superficially resembles the correlation-time calibration in Figure 7.17. Non-linear least-squares fitting of Eq. 7.78, with Eq. 7.83, to the temperature dependence yields rigid-limit values $A_{zz}^R = 3.50$ and 3.52 mT and

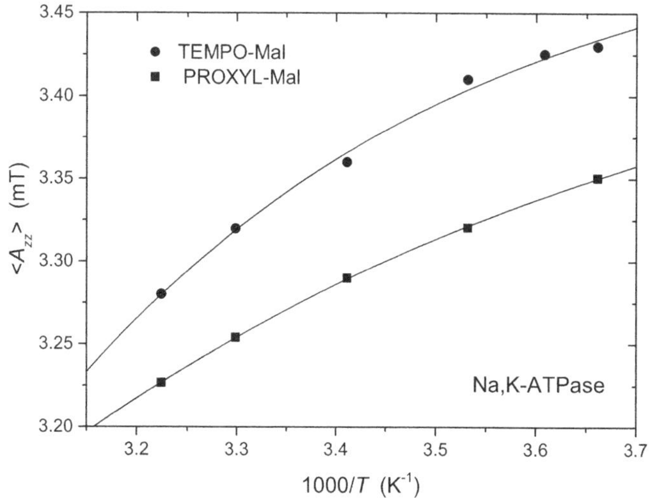

FIGURE 7.20 Temperature dependence of outer hyperfine splitting $2\langle A_{zz}\rangle$ for Na,K-ATPase spin-labelled on Class II –SH groups with PROXYL-Mal (squares) or TEMPO-Mal (circles) nitroxyl maleimides (Esmann et al. 1992). Solid lines: non-linear least squares fits of Eq. 7.78 for 9-GHz spectra (with Eq. 7.83 for temperature dependence of correlation time).

calibration parameters $b = -0.83$ and -0.92, for PROXYL-Mal and TEMPO-Mal, respectively. The corresponding activation energies are $E_a = 7.1$ and 15.7 kJ mol^{-1}. Unfortunately, there is a considerable interdependency of the fitting parameters, which we could reduce by using a wider temperature range for more temperature-stable systems. Note that if, besides the b calibration parameter, we also know the rigid-limit value A_{zz}^R, we can get the activation energy from an Arrhenius plot of $A_{zz}^R - \langle A_{zz}\rangle$ against $1/T$, which immediately gives the value of E_a/b (Esmann et al. 1992).

7.15 CONCLUDING SUMMARY

1. *Brownian diffusion: correlation times.*
   Solution of the rotational diffusion equation yields exponential correlation functions. For isotropic diffusion, the correlation time is:

$$\tau_R \equiv \tau_2 = 1/(6D_R) \tag{7.84}$$

   where $D_R = k_B T/f_R = k_B T/(6\eta V_{sph})$ is the isotropic rotational diffusion coefficient (see Eqs. 7.9 and 7.10). For axially anisotropic diffusion, the correlation times are:

$$\tau_{2,0} = 1/(6D_{R\perp}) \tag{7.85}$$

$$\tau_{2,2} = 1/(2D_{R\perp} + 4D_{R\parallel}) \tag{7.86}$$

   where $D_{R\parallel} = D_R^o/F_{R\parallel}$ and $D_{R\perp} = D_R^o/F_{R\perp}$ are the rotational diffusion coefficients about and perpendicular to the unique (||) axis, respectively, and D_R^o is the diffusion coefficient for a sphere of equal volume. Shape factors, $F_{R\parallel}$ and $F_{R\perp}$, for ellipsoids are given by Eqs. 7.13 and 7.14 and Figure 7.1.
2. *Fast motion.*
   Peak-to-peak, first-derivative, Lorentzian linewidths of the m_I-hyperfine lines in the fast motional regime are given by:

$$\Delta B_{pp}^L(m_I) = A_{pp} + B_{pp}m_I + C_{pp}m_I^2 \tag{7.87}$$

   For isotropic rotation:

$$B_{pp} = \frac{8\gamma_e}{45\sqrt{3}}(\Delta g\Delta A + 3\delta g\delta A)B_o\tau_R \tag{7.88}$$

$$C_{pp} = \frac{\gamma_e}{9\sqrt{3}}\left((\Delta A)^2 + 3(\delta A)^2\right)\cdot\tau_R \tag{7.89}$$

   where the linewidth coefficients and hyperfine anisotropies $\Delta A = A_{zz} - \frac{1}{2}(A_{xx} + A_{yy})$, $\delta A = \frac{1}{2}(A_{xx} - A_{yy})$ are in magnetic field units, e.g., mT (see Tables A7.2–A7.4). (We do not use the A_{pp}-coefficient at 9 GHz because it contains residual broadening that is independent of motion.) We calculate the isotropic rotational correlation time from the experimental linewidth coefficients corrected for inhomogeneous broadening:

$$\tau_R = -(3.75\ \text{ns})/B_o(\text{T}) \times B_{pp} \tag{7.90}$$

$$\tau_R = \left(11.2\ \text{ns mT}^{-1}\right) \times C_{pp} \tag{7.91}$$

where the central field B_o is in Tesla, B_{pp} and C_{pp} are in mT and τ_R is in ns. At 9 GHz, $|C_{pp}/B_{pp}| \approx 1$ for isotropic rotation.

3. *Fast axially anisotropic rotation.*
The axially anisotropic rotational correlation times of Eqs. 7.85 and 7.86 are:

$$\tau_{2,0} = \frac{9\sqrt{3}}{8\gamma_e}\frac{1}{\Delta A}\frac{8\delta g \times C_{pp} - (5\delta A/B_o) \times B_{pp}}{\delta g \Delta A - \Delta g \delta A} \tag{7.92}$$

$$\tau_{2,2} = \frac{3\sqrt{3}}{8\gamma_e}\frac{1}{\delta A}\frac{(5\Delta A/B_o) \times B_{pp} - 8\Delta g \times C_{pp}}{\delta g \Delta A - \Delta g \delta A} \tag{7.93}$$

where the tensor anisotropies ΔA, δA, Δg and δg depend on which of the nitroxide axes is the ||-rotation axis. Table 7.1 lists numerical calibration coefficients for z-, x- and y-rotation. For z-axis rotation, we can only determine $\tau_{2,0} (\equiv \tau_{R\perp})$. At 9 GHz, the diagnostic linewidth-coefficient ratio $|C_{pp}/B_{pp}|$ is ≈1, <1, 0.5–1.4 and 0.8–6 for isotropic, z-axis, x-axis and y-axis rotation, respectively.

4. *Fast non-axial rotation.*
In high-field EPR, the A_{pp} linewidth coefficient of Eq. 7.87 is dominated by rotational modulation of the g-value anisotropy. Then we can combine A_{pp} with B_{pp} and C_{pp} to treat fast anisotropic rotation about three independent axes, with principal elements $(D_{R,x}, D_{R,y}, D_{R,z})$ for the diffusion tensor. The two linewidth-coefficient ratios are related to the rotational anisotropies, $\rho_x \equiv D_{R,x}/D_{R,z}$ and $\rho_y \equiv D_{R,y}/D_{R,z}$, by the allowed-value equations:

$$\alpha_{AB}\rho_x = \beta_{AB}\rho_y + \gamma_{AB} \tag{7.94}$$

$$\alpha_{CB}\rho_x = \beta_{CB}\rho_y + \gamma_{CB} \tag{7.95}$$

for $|A_{pp}/B_{pp}|$ and $|C_{pp}/B_{pp}|$, respectively. The factors α_{AB}, β_{AB} and γ_{AB}, etc. are determined by the respective linewidth-coefficient ratios according to Eqs. 7.62–7.64 and 7.58–7.60. Using the $|A_{pp}/B_{pp}|$ ratio obtained from high-field/frequency measurements, together with the conventional $|C_{pp}/B_{pp}|$ ratio from 9-GHz measurements, lets us determine ρ_x and ρ_y separately.

5. *Slow motion.*
Close to the rigid limit, we can use the linewidths of the outer extrema in 9-GHz spectra, and their splitting, to estimate rotational correlation times τ_R in the 10–300 ns regime. In analogy with two-site exchange, empirical calibrations for the Lorentzian half-widths at half-height ΔB_{m_I} and the outer splitting $2\langle A_{zz}\rangle$ are:

$$\Delta B_{m_I} = \Delta B_{m_I}^R \left[1 + \left(\tau_R/a_{m_I}\right)^{1/b_{m_I}}\right] \tag{7.96}$$

$$\langle A_{zz}\rangle = A_{zz}^R \left[1 - \left(\tau_R/a\right)^{1/b}\right] \tag{7.97}$$

where $\Delta B_{m_I}^R \approx 0.26$ mT is the rigid-limit value of ΔB_{m_I}, with $m_I = +1$ for the low-field peak and $m_I = -1$ for the high-field peak and $A_{zz}^R \approx 3.3$ mT is the rigid-limit value of the partially averaged hyperfine tensor element $\langle A_{zz}\rangle$. Fitting parameters a, b and a_{m_I}, b_{m_I} from simulations for isotropic Brownian diffusion, collision-limited free diffusion and jump diffusion are listed in Table 7.2. Calibrations of the outer hyperfine splitting (i.e., Eq. 7.97) are given also for uniaxial rotation about either the y-axis or x-axis, in the Brownian case. Typical calibrations for isotropic Brownian diffusion are:

$$\tau_R = \begin{pmatrix} 11.5\ \text{ns} \\ 12.1\ \text{ns} \end{pmatrix} \times \left(\frac{\Delta B_{\pm 1}}{\Delta B_{\pm 1}^R} - 1\right)^{\begin{pmatrix} -0.943 \\ -0.778 \end{pmatrix}} \tag{7.98}$$

$$\tau_R = (0.54\ \text{ns}) \times \left(1 - \frac{\langle A_{zz}\rangle}{A_{zz}^R}\right)^{-1.36} \tag{7.99}$$

where upper and lower values in Eq. 7.98 correspond to low- and high-field outer peaks, respectively, and the intrinsic (inhomogeneous) linewidth is appropriate to a protonated nitroxide.

APPENDIX A7.1: SHAPE FACTORS FOR ROTATIONAL DYNAMICS OF A GENERAL ELLIPSOID: $x^2/a^2 + y^2/b^2 + z^2/c^2 = 1$

TABLE A7.1 Shape factors, $F_{R,a}$, $F_{R,b}$ and $F_{R,c}$, for general ellipsoid with semi-axes a, b and c, rotating about each principal axis with *slipping* boundary conditions. From data in Youngren and Acrivos (1975)

b/a / c/a	1.0	0.9	0.8	0.7	0.6	0.5	0.4	0.3	0.2
1.0	$F_{R,a} = 0.00$ $F_{R,b} = 0.00$ $F_{R,c} = 0.00$								
0.9	0.008 0.008 0.000	0.000 0.008 0.008							
0.8	0.033 0.033 0.000	0.008 0.032 0.007	0.000 0.028 0.028						
0.7	0.073 0.073 0.000	0.032 0.068 0.007	0.008 0.063 0.027	0.000 0.058 0.058					
0.6	0.128 0.128 0.000	0.068 0.118 0.007	0.030 0.108 0.025	0.008 0.100 0.053	0.000 0.090 0.090				
0.5	0.197 0.197 0.000	0.122 0.180 0.007	0.067 0.165 0.023	0.028 0.152 0.048	0.007 0.138 0.083	0.000 0.125 0.125			
0.4	0.278 0.278 0.000	0.193 0.255 0.005	0.120 0.232 0.020	0.065 0.210 0.042	0.028 0.188 0.075	0.007 0.168 0.110	0.000 0.148 0.148		
0.3	0.363 0.363 0.000	0.267 0.333 0.005	0.183 0.303 0.017	0.115 0.275 0.035	0.063 0.248 0.063	0.028 0.220 0.093	0.007 0.192 0.128	0.000 0.163 0.163	
0.2	0.450 0.450 0.000	0.342 0.415 0.005	0.250 0.380 0.013	0.172 0.343 0.028	0.110 0.310 0.050	0.060 0.273 0.076	0.027 0.238 0.105	0.007 0.203 0.133	0.000 0.165 0.165

Note: for a sphere $b/a = c/a = 1$; designation of b/a and c/a axial ratios is arbitrary.

APPENDIX A7.2: SPIN-HAMILTONIAN TENSOR ANISOTROPIES

TABLE A7.2 Magnetic tensor anisotropies for DOXYL (4,4-dimethyloxazolidinyl-3-oxyl-2-yl) nitroxides in solid host matrices. ^{14}N-hyperfine tensor and *g*-tensor anisotropies given by ΔA, δA and Δg, δg, respectively[a]

	DOXYL-androstanol[b]	DOXYL-cholestane[c]	DOXYL-cholestane[d]	DOXYL-C_3[e]	5-DOXYL-SA[f]	5-DOXYL-PC[g]	14-DOXYL-PC[g]
	z-axis:						
Δg	$-0.00530 \pm 6\times10^{-5}$	$-0.00494 \pm 6\times10^{-5}$	$-0.0052 \pm 2\times10^{-3}$	$-0.0051 \pm 1\times10^{-3}$	$-0.0050 \pm 1\times10^{-4}$	−0.00525	−0.00537
δg	$0.00151 \pm 3\times10^{-5}$	$0.00141 \pm 3\times10^{-5}$	$0.0016 \pm 1\times10^{-3}$	$0.0015 \pm 5\times10^{-4}$	$0.00139 \pm 5\times10^{-5}$	0.00135	0.00151
ΔA	2.73 ± 0.04 mT	2.44 ± 0.06 mT	2.50 ± 0.10 mT	2.73 ± 0.10 mT			
δA	0.02 ± 0.02 mT	0.13 ± 0.02 mT	0.0 ± 0.05 mT	0.02 ± 0.05 mT			
	x-axis:						
Δg	$0.00493 \pm 6\times10^{-5}$	$0.00458 \pm 6\times10^{-5}$	$0.0050 \pm 2\times10^{-3}$	$0.0048 \pm 1\times10^{-3}$	$0.0046 \pm 1\times10^{-4}$	0.00466	0.00495
δg	$0.00189 \pm 3\times10^{-5}$	$0.00177 \pm 3\times10^{-5}$	$0.0018 \pm 1\times10^{-3}$	$0.0018 \pm 5\times10^{-4}$	$0.00180 \pm 5\times10^{-5}$	0.00195	0.00193
ΔA	−1.34 ± 0.04 mT	−1.03 ± 0.05 mT	−1.25 ± 0.10 mT	−1.33 ± 0.10 mT			
δA	−1.38 ± 0.02 mT	−1.28 ± 0.03 mT	−1.25 ± 0.05 mT	−1.38 ± 0.05 mT			
	y-axis:						
Δg	$0.00038 \pm 6\times10^{-5}$	$0.00036 \pm 6\times10^{-5}$	$0.0003 \pm 2\times10^{-3}$	$0.0003 \pm 1\times10^{-3}$	$0.0004 \pm 1\times10^{-4}$	0.00059	0.00042
δg	$-0.00341 \pm 3\times10^{-5}$	$-0.00317 \pm 3\times10^{-5}$	$-0.0034 \pm 1\times10^{-3}$	$-0.0033 \pm 5\times10^{-4}$	$-0.00319 \pm 5\times10^{-5}$	−0.00330	−0.00344
ΔA	−1.40 ± 0.04 mT	−1.42 ± 0.05 mT	−1.25 ± 0.10 mT	−1.40 ± 0.10 mT			
δA	1.36 ± 0.02 mT	1.16 ± 0.03 mT	1.25 ± 0.05 mT	1.35 ± 0.05 mT			

[a] Tensor anisotropies for axially anisotropic rotation about nitroxide *z*-axis (see Figure 2.1) or for isotropic rotation:

$$\Delta g = g_{zz} - \tfrac{1}{2}\left(g_{xx} + g_{yy}\right)$$

$$\delta g = \tfrac{1}{2}\left(g_{xx} - g_{yy}\right)$$

$$\Delta A = A_{zz} - \tfrac{1}{2}\left(A_{xx} + A_{yy}\right)$$

$$\delta A = \tfrac{1}{2}\left(A_{xx} - A_{yy}\right)$$

For axially anisotropic rotation about nitroxide *x*- and *y*-axes, tensor anisotropies are defined by cyclic permutation of *x*, *y*, *z* indices.

[b] 3-(4′,4′-dimethyloxazolidinyl-3′-oxyl-2′-yl)-17β-hydroxy-5α-androstane in *o*-xylene at −135°C (Smirnova et al. 1995).

[c] 3-(4′,4′-dimethyloxazolidinyl-3′-oxyl-2′-yl)-5α-cholestane in toluene at −123°C (Earle et al. 1993).

[d] 3-(4′,4′-dimethyloxazolidinyl-3′-oxyl-2′-yl)-5α-cholestane in single crystal of cholesteryl chloride (Hubbell and McConnell 1971).

[e] 2-(4′,4′-dimethyloxazolidinyl-3′-oxyl-2′-yl)-propane in single crystal of 2,2,4,4-tetramethyl-1,3-cyclobutadione at 23°C (Jost et al. 1971).

[f] 5-(4′,4′-dimethyloxazolidinyl-3′-oxyl-2′-yl)-stearic acid in dry dimyristoyl phosphatidylcholine:myristic acid 1:2 mol/mol mixture at room temperature (Gaffney and Marsh 1998).

[g] 1-acyl-2-[*n*-(4′,4′-dimethyloxazolidinyl-3′-oxyl-2′-yl)stearoyl]-*sn*-glycero-3-phosphocholine (*n*-DOXYL-PC) in dimyristoyl phosphatidylcholine + 40 mol% cholesterol bilayer membranes at 30°C (Livshits et al. 2004).

TABLE A7.3 Magnetic tensor anisotropies for piperidinyl nitroxides (6-membered ring) in solid host matrices. ^{14}N-hyperfine tensor and g-tensor anisotropies given by ΔA, δA and Δg, δg, respectively[a]

	TEMPONE[b]				TEMPO[c]		*TOAC*[d]	*TEMPOL*[e]
	toluene	acetone	EtOH	85% aq. glycerol	toluene	oleic acid-oil		
	z-axis:							
Δg	$-0.00551 \pm 6\times10^{-5}$	$-0.0057 \pm 5\times10^{-4}$	$-0.0054 \pm 7\times10^{-4}$	$-0.0050 \pm 3\times10^{-4}$	$-0.0058 \pm 1\times10^{-4}$	$-0.0055 \pm 1\times10^{-4}$	−0.0052	−0.0056
δg	$0.00151 \pm 3\times10^{-5}$	$0.0016 \pm 3\times10^{-4}$	$0.0015 \pm 4\times10^{-4}$	$0.0012 \pm 2\times10^{-4}$	$0.00179 \pm 7\times10^{-5}$	$0.00161 \pm 5\times10^{-5}$	0.0012	0.0019
ΔA	2.90 ± 0.04 mT	2.89 ± 0.08 mT	2.99 ± 0.10 mT	3.02 ± 0.08 mT	2.76 ± 0.04 mT	2.82 ± 0.05 mT	3.04 mT	2.88 mT
δA	−0.05 ± 0.02 mT	−0.03 ± 0.05 mT	−0.04 ± 0.06 mT	−0.01 ± 0.05 mT	−0.07 ± 0.01 mT	0.0 ± 0.04 mT	0.12 mT	−0.08 mT
	x-axis:							
Δg	$0.0053 \pm 3\times10^{-4}$	$0.0053 \pm 5\times10^{-4}$	$0.0050 \pm 7\times10^{-4}$	$0.0043 \pm 3\times10^{-4}$	$0.0056 \pm 1\times10^{-4}$	$0.0052 \pm 1\times10^{-4}$	0.0045	0.0057
δg	$0.0020 \pm 1\times10^{-4}$	$0.0020 \pm 2\times10^{-4}$	$0.0019 \pm 3\times10^{-4}$	$0.0019 \pm 1\times10^{-4}$	$0.0020 \pm 7\times10^{-5}$	$0.0020 \pm 5\times10^{-5}$	0.0020	0.0018
ΔA	−1.53 ± 0.04 mT	−1.49 ± 0.09 mT	−1.56 ± 0.11 mT	−1.53 ± 0.09 mT	−1.48 ± 0.03 mT	−1.41 ± 0.07 mT	−1.34 mT	−1.57 mT
δA	−1.42 ± 0.02 mT	−1.43 ± 0.04 mT	−1.47 ± 0.05 mT	−1.51 ± 0.04 mT	−1.35 ± 0.02 mT	−1.41 ± 0.02 mT	−1.58 mT	−1.40 mT
	y-axis:							
Δg	$0.0004 \pm 3\times10^{-4}$	$0.0003 \pm 5\times10^{-4}$	$0.0004 \pm 7\times10^{-4}$	$0.0007 \pm 3\times10^{-4}$	$0.0002 \pm 1\times10^{-4}$	$0.00035 \pm 1\times10^{-4}$	0.0007	−0.0001
δg	$-0.0037 \pm 1\times10^{-4}$	$-0.0036 \pm 2\times10^{-4}$	$-0.0035 \pm 3\times10^{-4}$	$-0.0031 \pm 1\times10^{-4}$	$-0.0038 \pm 7\times10^{-5}$	$-0.0036 \pm 5\times10^{-5}$	−0.0032	−0.0037
ΔA	−1.38 ± 0.04 mT	−1.40 ± 0.09 mT	−1.43 ± 0.11 mT	−1.50 ± 0.09 mT	−1.28 ± 0.03 mT	−1.41 ± 0.05 mT	−1.70 mT	−1.32 mT
δA	1.47 ± 0.02 mT	1.46 ± 0.04 mT	1.52 ± 0.05 mT	1.52 ± 0.04 mT	1.41 ± 0.02 mT	1.41 ± 0.02 mT	1.46 mT	1.48 mT

[a] Tensor anisotropies for axially anisotropic rotation about nitroxide z-axis (see Figure 2.1) or for isotropic rotation:

$$\Delta g = g_{zz} - \tfrac{1}{2}\left(g_{xx} + g_{yy}\right)$$

$$\delta g = \tfrac{1}{2}\left(g_{xx} - g_{yy}\right)$$

$$\Delta A = A_{zz} - \tfrac{1}{2}\left(A_{xx} + A_{yy}\right)$$

$$\delta A = \tfrac{1}{2}\left(A_{xx} - A_{yy}\right)$$

For axially anisotropic rotation about nitroxide x- and y-axes, tensor anisotropies are defined by cyclic permutation of x, y, z indices.

[b] 2,2,6,6-tetramethyl-piperidine-1-oxyl-4-oxo in: toluene at below −130°C (Budil et al. 1993), frozen acetone, frozen ethanol or 85% glycerol-water at −62°C (Hwang et al. 1975).

[c] 2,2,6,6-tetramethylpiperidine-1-oxyl in toluene glass (Ondar et al. 1985) or in oleic acid-canola oil (2:1 v/v) at −135°C (Smirnov and Belford 1995).

[d] 2,2,6,6-tetramethylpiperidine-1-oxyl-4-amino-4-carboxy (TOAC) in nonapeptide Fmoc-Aib$_8$-TOAC-OMe (where Aib is α-aminoisobutyric acid and Fmoc is fluoren-9-ylmethyloxycarbonyl) in acetonitrile at −193°C (Carlotto et al. 2011).

[e] 2,2,6,6-tetramethyl-piperidine-1-oxyl-4-hydroxy in single crystal of 4-hydroxy-2,2,6,6-tetramethyl-piperidine at 23°C (Tabak et al. 1983).

TABLE A7.4 Magnetic tensor anisotropies for pyrroline nitroxides (5-membered ring) and di-*tert*-butyl nitroxide (DTBN) in solid host matrices. ^{14}N-hyperfine tensor and *g*-tensor anisotropies given by ΔA, δA and Δg, δg, respectively[a]

	PYMeOH[b]		bi-MTSSL[c]	DTBN[d]
	o-terphenyl	glycerol		
	z-axis:			
Δg	$-0.00534 \pm 4\times10^{-5}$	$-0.00500 \pm 4\times10^{-5}$	$-0.0053 \pm 2\times10^{-4}$	$-0.0047 \pm 1\times10^{-4}$
δg	$0.00144 \pm 2\times10^{-5}$	$0.00118 \pm 2\times10^{-5}$	$0.0010 \pm 2\times10^{-4}$	$0.00128 \pm 5\times10^{-5}$
ΔA	2.84 ± 0.01 mT	3.10 ± 0.01 mT	3.25 ± 0.03 mT	2.50 ± 0.01 mT
δA	0.001 ± 0.005 mT	0.0 ± 0.005 mT	0.10 ± 0.01 mT	0.082 ± 0.005 mT
	x-axis:			
Δg	$0.00483 \pm 4\times10^{-5}$	$0.00428 \pm 4\times10^{-5}$	$0.0041 \pm 3\times10^{-4}$	$0.0043 \pm 1\times10^{-4}$
δg	$0.00195 \pm 2\times10^{-5}$	$0.00191 \pm 2\times10^{-5}$	$0.0022 \pm 1\times10^{-4}$	$0.00173 \pm 5\times10^{-5}$
ΔA	−1.42 ± 0.01 mT	−1.55 ± 0.01 mT	−1.48 ± 0.02 mT	−1.13 ± 0.01 mT
δA	−1.422 ± 0.005 mT	−1.547 ± 0.005 mT	−1.67 ± 0.02 mT	−1.291 ± 0.005 mT
	y-axis:			
Δg	$0.00051 \pm 4\times10^{-5}$	$0.00072 \pm 4\times10^{-5}$	$0.0012 \pm 2\times10^{-4}$	$0.00045 \pm 1\times10^{-4}$
δg	$-0.00339 \pm 2\times10^{-5}$	$-0.00309 \pm 2\times10^{-5}$	$0.0031 \pm 2\times10^{-4}$	$-0.00301 \pm 5\times10^{-5}$
ΔA	−1.42 ± 0.01 mT	−1.55 ± 0.01 mT	−1.77 ± 0.02 mT	−1.37 ± 0.01 mT
δA	1.421 ± 0.005 mT	1.548 ± 0.005 mT	1.58 ± 0.02 mT	1.210 ± 0.005 mT

[a] Tensor anisotropies for axially anisotropic rotation about nitroxide *z*-axis (see Figure 2.1), or for isotropic rotation:

$$\Delta g = g_{zz} - \tfrac{1}{2}(g_{xx} + g_{yy})$$
$$\delta g = \tfrac{1}{2}(g_{xx} - g_{yy})$$
$$\Delta A = A_{zz} - \tfrac{1}{2}(A_{xx} + A_{yy})$$
$$\delta A = \tfrac{1}{2}(A_{xx} - A_{yy})$$

For axially anisotropic rotation about nitroxide *x*- and *y*-axes, tensor anisotropies are defined by cyclic permutation of *x*, *y*, *z* indices.

[b] 2,2,5,5-tetramethylpyrroline-1-oxyl-3-methylhydroxy in *o*-terphenyl or glycerol at −93°C (Savitsky et al. 2008).

[c] 2,2,5,5-tetramethylpyrroline-1-oxyl-3,4-bismethyl(methanethiosulphonate) cross linked to T4 lysozyme (Fleissner et al. 2011).

[d] di-*tert*-butyl nitroxide in single crystal of 2,2,4,4-tetramethyl-1,3-cyclobutadione at 23° (Libertini and Griffith 1970).

REFERENCES

Bales, B.L. 1989. Inhomogeneously broadened spin-label spectra. In *Biological Magnetic Resonance*, Vol. 8, eds. Berliner, L.J. and Reuben, J., 77–130. New York: Plenum Publishing Corporation.

Bales, B.L., Schumacher, K.L., and Harris, F.L. 1987. Correction for inhomogeneous line broadening in spin labels. III. Doxyl-labeled alkyl chains. *J. Magn. Reson.* 72:364–368.

Brink, D.M. and Satchler, G.R. 1968. *Angular Momentum*, Oxford: Oxford University Press.

Budil, D.E., Earle, K.A., and Freed, J.H. 1993. Full determination of the rotational diffusion tensor by electron paramagnetic resonance at 250 GHz. *J. Phys. Chem.* 97:1294–1303.

Carlotto, S., Zerbetto, M., Shabestari, M.H. et al. 2011. *In silico* interpretation of cw-ESR at 9 and 95 GHz of mono- and bis- TOAC-labeled Aib-homopeptides in fluid and frozen acetonitrile. *J. Phys. Chem. B* 115:13026–13036.

Davidov, A.S. 1965. *Quantum Mechanics*, Oxford: Pergamon Press.

Dobryakov, S.N. and Lebedev, Y. 1969. Analysis of spectral lines whose profile is described by a composition of Gaussian and Lorentz profiles. *Sov. Phys – Dokl.* 13:873–875.

Earle, K.A., Budil, D.E., and Freed, J.H. 1993. 250-GHz EPR of nitroxides in the slow-motional regime - models of rotational diffusion. *J. Phys. Chem.* 97:13289–13297.

Egelstaff, P.A. 1970. Cooperative rotation of spherical molecules. *J. Chem. Phys.* 53:2590–2598.

Einstein, A. 1906. Zur Theorie der Brownschen Bewegung. *Ann. Phys.* 19:371–381.

Esmann, M., Hideg, K., and Marsh, D. 1992. Conventional and saturation transfer EPR spectroscopy of Na,K-ATPase modified with different maleimide-nitroxide derivatives. *Biochim. Biophys. Acta* 1159:51–59.

Fleissner, M.R., Bridges, M.D., Brooks, E.K. et al. 2011. Structure and dynamics of a conformationally constrained nitroxide side chain and applications in EPR spectroscopy. *Proc. Natl. Acad. Sci. USA* 108:16241–16246.

Freed, J.H. 1964. Anisotropic rotational diffusion and electron spin resonance linewidths. *J. Chem. Phys.* 41:2077–2083.

Freed, J.H. 1976. Theory of slowly tumbling ESR spectra for nitroxides. In *Spin Labeling, Theory and Applications*, ed. Berliner, L.J., 53–132. New York: Academic Press Inc.

Gaffney, B.J. and Marsh, D. 1998. High-frequency, spin-label EPR of nonaxial lipid ordering and motion in cholesterol-containing membranes. *Proc. Natl. Acad. Sci. USA* 95:12940–12943.

Goldman, S.A., Bruno, G.V., and Freed, J.H. 1972a. Estimating slow-motional rotational correlation times for nitroxides by electron spin resonance. *J. Phys. Chem.* 76:1858–1860.

Goldman, S.A., Bruno, G.V., Polnaszek, C.F., and Freed, J.H. 1972b. An ESR study of anisotropic rotational reorientation and slow tumbling in liquid and frozen media. *J. Chem. Phys.* 56:716–735.

Howie, R.A., Glasser, L.S.D., and Moser, W. 1968. Nitrosodisulphonates. Part II. Crystal structure of the orange-brown triclinic modification of Fremy's salt (potassium nitrosodisulphonate). *J. Chem. Soc. A: Inorg. Phys. Theor.* 3043–3047.

Hu, C.M. and Zwanzig, R. 1974. Rotational friction coefficients for spheroids with slipping boundary condition. *J. Chem. Phys.* 60:4354–4357.

Hubbard, P.S. 1963. Theory of nuclear magnetic relaxation by spin-rotational interactions in liquids. *Phys. Rev.* 131:1155–1165.

Hubbell, W.L. and McConnell, H.M. 1971. Molecular motion in spin-labelled phospholipids and membranes. *J. Am. Chem. Soc.* 93:314–326.

Hwang, J.S., Mason, R.P., Hwang, L.P., and Freed, J.H. 1975. Electron spin resonance studies of anisotropic rotational reorientation and slow tumbling in liquid and frozen media. 3. Perdeuterated 2,2,6,6-tetramethyl-4-piperidone *N*-oxide and analysis of fluctuating torques. *J. Phys. Chem.* 79:489–511.

Ivanov, E.N. 1964. Theory of rotational Brownian motion. *Sov. Phys. JETP* 18:1041–1045.

Jost, P.C., Libertini, L.J., Hebert, V.C., and Griffith, O.H. 1971. Lipid spin labels in lecithin multilayers. A study of motion along fatty acid chains. *J. Mol. Biol.* 59:77–98.

Klauder, J.R. and Anderson, P.W. 1962. Spectral diffusion decay in spin resonance experiments. *Phys. Rev.* 125:912–932.

Kowert, B.A. 1981. Determination of the anisotropic and non-secular contributions to electron spin resonance linewidths in liquids. *J. Phys. Chem.* 85:229–235.

Kubo, R. 1958. In *Lectures in Theoretical Physics*, 1, ed. University of Colorado, 120–203. New York: Interscience Publishers Inc.

Kuznetsov, A.N., Volkov, A.Y., Livshits, V.A., and Mirzoian, A.T. 1974. Method of studying anisotropic rotation of organic nitroxyl radicals. *Chem. Phys. Lett.* 26:369–372.

Libertini, L.J. and Griffith, O.H. 1970. Orientation dependence of the electron spin resonance spectrum of di-*t*-butyl nitroxide. *J. Chem. Phys.* 53:1359–1367.

Livshits, V.A., Kurad, D., and Marsh, D. 2004. Simulation studies on high-field EPR of lipid spin labels in cholesterol-containing membranes. *J. Phys. Chem. B* 108:9403–9411.

Marsh, D., Jost, M., Peggion, C., and Toniolo, C. 2007. Solvent dependence of the rotational diffusion of TOAC-spin-labelled alamethicin. *Chem. Biodiv.* 4:1269–1274.

Mason, R. and Freed, J.H. 1974. Estimating microsecond rotational correlation times from lifetime broadening of nitroxide ESR spectra near the rigid limit. *J. Phys. Chem.* 78:1321–1323.

Matthews, B.W. and Remington, S.J. 1974. Three-dimensional structure of lysozyme from bacteriophage T4. *Proc. Natl. Acad. Sci. USA* 71:4178–4182.

McClung, R. and Kivelson, D. 1968. ESR linewidths in solution. V. Studies of spin-rotational effects not described by rotational diffusion theory. *J. Chem. Phys.* 49:3380–3391.

Mori, H., Oppenheim, I., and Ross, J. 1962. In *Studies in Statistical Mechanics*, eds. de Boer, J. and Uhlenbeck, G.E. Amsterdam: North-Holland Publishing Company.

Ondar, M.A., Grinberg, O.Ya., Dubinskii, A.A., and Lebedev, Ya.S. 1985. Study of the effect of the medium on the magnetic-resonance parameters of nitroxyl radicals by high-resolution EPR spectroscopy. *Sov. J. Chem. Phys.* 3:781–792.

Perrin, F. 1934. Mouvement brownien d'un ellipsoide - I. Dispersion diélectrique pour des molécules ellipsoidales. *J. Phys. Radium* 5:497–511.

Polnaszek, C.F., Marsh, D., and Smith, I.C.P. 1981. Simulation of the ESR spectra of the cholestane spin probe under conditions of slow axial rotation: application to gel phase dipalmitoyl phosphatidylcholine. *J. Magn. Reson.* 43:54–64.

Remington, S.J., Anderson, W.F., Owen, J., Teneyck, L.F., and Grainger, C.T. 1978. Structure of lysozyme from bacteriophage T4- electron density map at 2.4 Å resolution. *J. Mol. Biol.* 118:81–98.

Sachse, J.-H., King, M.D., and Marsh, D. 1987. ESR determination of lipid diffusion coefficients at low spin-label concentrations in biological membranes, using exchange broadening, exchange narrowing, and dipole-dipole interactions. *J. Magn. Reson.* 71:385–404.

Savitsky, A., Dubinskii, A.A., Plato, M. et al. 2008. High-field EPR and ESEEM investigation of the nitrogen quadrupole interaction of nitroxide spin labels in disordered solids: toward differentiation between polarity and proticity matrix effects on protein function. *J. Phys. Chem. B* 112:9079–9090.

Shibaeva, R.P., Atovmyan, L.O., Neigauz, M.G., Novakovskaya, L.A., and Ginzburg, S.L. 1972. Crystal and molecular structures of 2,2,6,6-tetramethyl-4-piperidinone-1-oxyl. *Zh. Strukt. Khim.* 13:826–830.

Smirnov, A.I. and Belford, R.L. 1995. Rapid quantitation from inhomogeneously broadened EPR spectra by a fast convolution algorithm. *J. Magn. Reson.* 113:65–73.

Smirnov, A.I., Belford, R.L., and Clarkson, R.B. 1998. Comparative spin label spectra at X-band and W-band. In *Biological Magnetic Resonance*, Vol. 14, ed. Berliner, L.J., 83–107. New York: Plenum Press.

Smirnova, T.I., Smirnov, A.I., Clarkson, R.B., and Belford, R.L. 1995. W-Band (95 GHz) EPR spectroscopy of nitroxide radicals with complex proton hyperfine structure: Fast Motion. *J. Phys. Chem.* 99:9008–9016.

Smirnov, A.I., Smirnova, T.I., and Morse, P.D. 1995. Very high-frequency electron paramagnetic resonance of 2,2,6,6-tetramethyl-1-piperidinyloxy in 1,2 dipalmitoyl-*sn*-glycero-3-phosphatidylcholine liposomes – partitioning and molecular dynamics. *Biophys. J.* 68:2350–2360.

Steele, W.A. 1963. Molecular reorientation in liquids. 1. Distribution functions and friction constants. *J. Chem. Phys.* 38:2404–2410.

Tabak, M., Alonso, A., and Nascimento, O.R. 1983. Single crystal ESR studies of a nitroxide spin label. I. Determination of the *G* and *A* tensors. *J. Chem. Phys.* 79:1176–1184.

Timofeev, V.P. and Tsetlin, V.I. 1983. Analysis of mobility of protein side chains by spin label technique. *Biophys. Struct. Mech.* 10:93–108.

Youngren, G.K. and Acrivos, A. 1975. Rotational friction coefficients for ellipsoids and chemical molecules with slip boundary condition. *J. Chem. Phys.* 63:3846–3848.

Zager, S.A. and Freed, J.H. 1982. Electron-spin relaxation and molecular dynamics in liquids. 1. Solvent dependence. *J. Chem. Phys.* 77:3344–3359.

Zwanzig, R. and Harrison, A.K. 1985. Modifications of the Stokes Einstein formula. *J. Chem. Phys.* 83:5861–5862.

Dynamics and Orientational Ordering (Liquid Crystals and Membranes)

8

8.1 INTRODUCTION

Chapter 7 treated isotropic and axially anisotropic rotational diffusion in isotropic media. In both cases, there is full rotation about three independent axes. Now we consider situations where rotation about one or more axes is restricted in amplitude. This results in preferential ordering of the spin-labelled molecule about the complementary (i.e., orthogonal) axes. A common situation is in liquid crystals or lipid membranes. Here, the molecules are partially aligned along a unique axis, the director or membrane normal. They display limited angular excursions about the director but unrestricted rotational motion around this direction. In this case, the system is uniaxial. Another situation of partial ordering is when one end of the spin-labelled molecule is attached to a macromolecule. Limited segmental rotation and steric obstructions at the attachment site then restrict the angular amplitude of the spin label group. Similar situations arise in ligand binding sites or occlusion cavities.

We characterize the degree of orientational ordering by so-called order parameters that are specified by corresponding angular averages. In general, order parameters are determined by an orientational potential. Alternatively, we can relate them to specific geometric or molecular models. For fast rotations, the extent of motional narrowing of the nitroxide hyperfine- and g-tensors depends only on the order parameters. With uniaxial averaging, both tensors become axially symmetric around the director: $\langle \mathbf{A} \rangle = (A_\perp, A_\perp, A_\parallel)$ and $\langle \mathbf{g} \rangle = (g_\perp, g_\perp, g_\parallel)$. The fast-motion lineshapes are again Lorentzian, with widths given by the general relation Eq. 5.58, which now depends on both order parameters and correlation times. In the slow-motion regime, we need spectral simulations: either with a specific angular model or more generally by incorporating an orientational potential in the stochastic Liouville equation (Freed 1976).

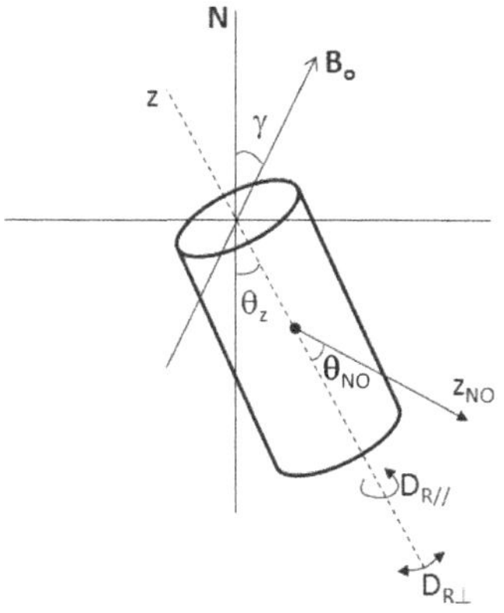

FIGURE 8.1 Orientation of cylindrically symmetric spin-labelled molecule in an ordering environment. **N**, director axis for ordering; z, molecular long axis; z_{NO}, molecule-fixed nitroxide z-axis; $\mathbf{B}_o$, static magnetic field direction.

The ordering director **N** introduces a new axis system. As seen from Figure 8.1, we now need a minimum of three axis systems. The molecular long axis z is partially ordered; it makes an instantaneous angle θ_z with the director. Simultaneously, z is also the principal axis for rotational diffusion of the spin-labelled molecule. Together, random fluctuations in the angle θ_z and rotational diffusion about the molecular axes partially average the spectral anisotropy, and cause line broadening by relaxation. The nitroxide z-axis z_{NO} is fixed in the molecule and makes an angle θ_{NO} with the long axis. A useful simplification is when the nitroxide axis coincides with the molecular axis, as with DOXYL-labelled chains in lipids. Finally, the director makes a fixed angle γ with the static magnetic field $\mathbf{B}_o$. If the sample is not aligned, we get a powder spectrum determined by the partially averaged hyperfine- and g-tensors or more generally by the slow-motion lineshapes for different field orientations.

We begin by assuming fast (partial) motional averaging, and linewidths treated by motional-narrowing theory. It is not until Section 8.12 that we introduce slow motion.

8.2 RESTRICTED TORSIONAL LIBRATION

For simplicity, we consider a libration of the nitroxide z-axis that is restricted to the nitroxide x,z-plane, as shown in Figure 8.2. This corresponds to limited angular oscillations around the nitroxide y-axis, which coincides with the molecular long axis for the 3-DOXYL cholestane and androstanol spin-labelled steroids (cf. Van et al. 1974). The amplitude of the libration is $\phi = \pm\phi_o$ about the space-fixed mean position $\bar{z}$ of the nitroxide

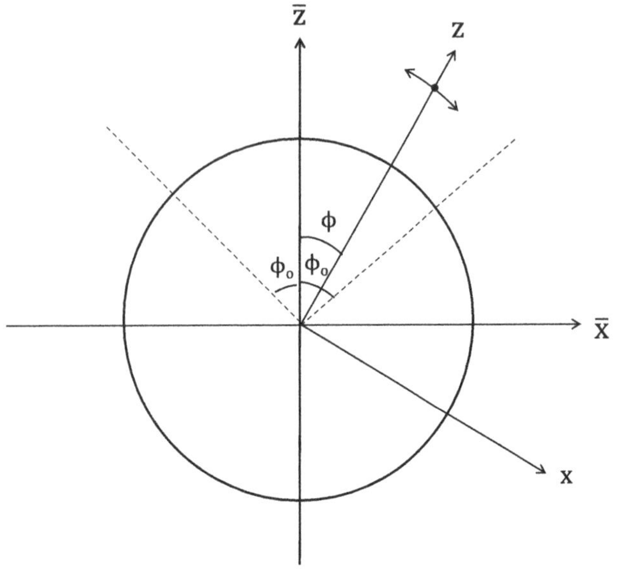

FIGURE 8.2 Limited librational oscillation ϕ of instantaneous nitroxide *z*-axis in nitroxide *x,z*-plane. *z*-axis rotates around fixed nitroxide *y*-axis, with maximum angular amplitude $\pm\phi_o$ about its mean space-fixed direction $\bar{z}$.

z-axis (see Figure 8.2). For purposes of illustration, we consider only the hyperfine tensor, because it is determined more readily than the g-tensor (in 9-GHz EPR) and is sufficient to specify this simple model.

The hyperfine tensor $\mathbf{A} = (A_{xx}, A_{zz})$ is diagonal in the instantaneous nitroxide x,z-axis system. We transform this to the space-fixed (or laboratory) axis-system by rotation $\mathbf{R}(\phi, y)$ about the fixed nitroxide y-axis, as described in Section B.4 of Appendix B (and see also Appendix I). From Eq. B.25, the transformation is given by $\mathbf{A}' = \mathbf{R}\cdot\mathbf{A}\cdot\mathbf{R}^{-1}$, where the rotation matrix $\mathbf{R}$ comes from Eq. B.27. Then we get the motionally narrowed hyperfine tensor by averaging over ϕ:

$$\langle \mathbf{A}' \rangle = \begin{pmatrix} A_{xx}\langle\cos^2\phi\rangle + A_{zz}\langle\sin^2\phi\rangle & (A_{zz}-A_{xx})\langle\sin\phi\cos\phi\rangle \\ (A_{zz}-A_{xx})\langle\sin\phi\cos\phi\rangle & A_{xx}\langle\sin^2\phi\rangle + A_{zz}\langle\cos^2\phi\rangle \end{pmatrix} \tag{8.1}$$

where the average $\langle\sin\phi\cos\phi\rangle = \frac{1}{2}\langle\sin 2\phi\rangle = 0$, because fluctuations are symmetrical about the $\bar{z}$-position (see Figure 8.2).

The motionally averaged hyperfine tensor is therefore diagonal in the space-fixed $\bar{x},\bar{z}$-axis system, with principal elements:

$$\langle A_{\bar{x}\bar{x}} \rangle = A_{zz} - (A_{zz} - A_{xx})\langle\cos^2\phi\rangle \tag{8.2}$$

$$\langle A_{\bar{z}\bar{z}} \rangle = A_{xx} + (A_{zz} - A_{xx})\langle\cos^2\phi\rangle \tag{8.3}$$

Equations 8.2 and 8.3 let us determine $\langle\cos^2\phi\rangle$, the angular average over the librational motion, from experimental measurements of the hyperfine splittings $\langle A_{\bar{z}\bar{z}}\rangle$ and/or $\langle A_{\bar{x}\bar{x}}\rangle$.

The angular average $\langle\cos^2\phi\rangle$ is related to the librational amplitude $\pm\phi_o$ in Figure 8.2 by:

$$\langle\cos^2\phi\rangle = \frac{\int_{-\phi_o}^{\phi_o}\cos^2\phi\cdot d\phi}{\int_{-\phi_o}^{\phi_o} d\phi} = \tfrac{1}{2}\left(1 + \frac{\sin 2\phi_o}{2\phi_o}\right) \cong 1 - \tfrac{1}{3}\phi_o^2 \tag{8.4}$$

where ϕ_o is expressed in radians. The approximation on the extreme right of Eq. 8.4 is for small amplitudes, where $\phi_o << 1$ radian.

8.3 ORIENTATIONAL ORDER PARAMETERS

Instead of just an isolated librational mode, we now consider rotational motion in an anisotropic environment. A typical example is a spin-labelled lipid molecule in a bilayer lipid membrane. Our methods are very similar to those for the simple – essentially one-dimensional – example of the previous section.

The instantaneous orientation of the nitroxide x-, y- and z-axes is defined by angles θ_x, θ_y and θ_z to the director $\mathbf{N}$ (i.e., membrane normal), respectively, as shown in Figure 8.3. We begin with the hyperfine tensor $\mathbf{A} = (A_{xx}, A_{yy}, A_{zz})$, because we can simplify this with the approximation $A_{yy} \cong A_{xx}$. Then a rotation $\mathbf{R}(\theta_z, x')$ by angle θ_z around the x'-axis, which lies in the nitroxide x,y-plane and is orthogonal to $\mathbf{N}$, brings the nitroxide z-axis to lie along the director. This is followed by a rotation $\mathbf{R}(\phi, z')$ by angle ϕ around the new z'-axis, i.e., the director $(z' \equiv \mathbf{N})$, which brings us to a general azimuthal orientation in the uniaxial membrane-fixed axis system (||, ⊥, ⊥). Using the resulting transformation matrix $\mathbf{R} = \mathbf{R}(\phi, z')\mathbf{R}(\theta_z, x')$ (see Appendix I), the motionally averaged A-tensor becomes:

$$\langle \mathbf{A}'(\theta_z,\phi) \rangle = \langle \mathbf{R}(\phi,\theta_z)\cdot\mathbf{A}\cdot\mathbf{R}(\phi,\theta_z)^{-1} \rangle$$

$$= \begin{pmatrix} A_{xx}\langle\cos^2\phi\rangle + (A_{xx}c^2 + A_{xx}s^2)\langle\sin^2\phi\rangle & \frac{1}{2}(A_{xx} - (A_{xx}c^2 + A_{xx}s^2))\langle\sin 2\phi\rangle & (A_{zz} - A_{xx})sc\langle\sin\phi\rangle \\ \frac{1}{2}(A_{xx} - (A_{xx}c^2 + A_{xx}s^2))\langle\sin 2\phi\rangle & A_{xx}\langle\sin^2\phi\rangle + (A_{xx}c^2 + A_{xx}s^2)\langle\cos^2\phi\rangle & (A_{xx} - A_{zz})sc\langle\cos\phi\rangle \\ (A_{zz} - A_{xx})sc\langle\sin\phi\rangle & (A_{xx} - A_{zz})sc\langle\cos\phi\rangle & A_{xx}s^2 + A_{zz}c^2 \end{pmatrix} \tag{8.5}$$

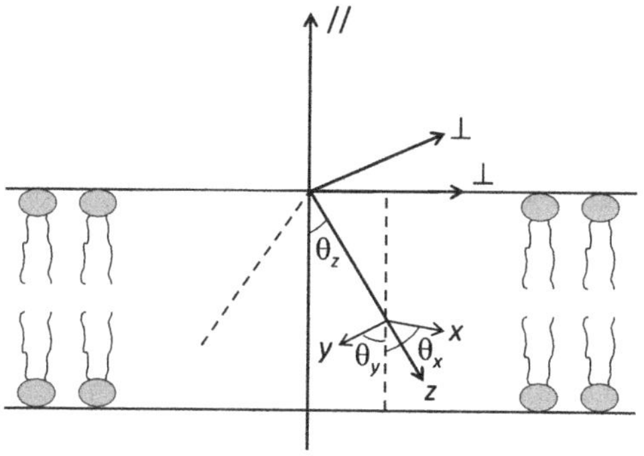

FIGURE 8.3 Instantaneous orientation $\theta_x, \theta_y, \theta_z$ of nitroxide (x,y,z)-axes relative to director of molecular ordering (||), e.g., membrane normal. Molecular rotational diffusion produces time-average axial symmetry about the director.

where $c^2 \equiv \langle \cos^2\theta_z \rangle$, $s^2 \equiv \langle \sin^2\theta_z \rangle = 1 - c^2$ and $sc \equiv \langle \sin\theta_z \cos\theta_z \rangle$. To produce axial symmetry about the director, we average ϕ over the complete range 0 to 2π, which gives $\langle \sin\phi \rangle = \langle \cos\phi \rangle = \langle \sin 2\phi \rangle = 0$. Thus, the motionally averaged hyperfine tensor is diagonal. Further, we have $\langle \cos^2\phi \rangle = 1 - \langle \sin^2\phi \rangle = (1/2\pi) \int_0^{2\pi} \cos^2\phi \cdot d\phi = \frac{1}{2}$, which ensures that the diagonal elements display axial symmetry. The principal elements of the motionally-averaged hyperfine tensor are then:

$$\langle A_\perp \rangle = \tfrac{1}{2}(A_{zz} - A_{xx})\left(1 - \langle \cos^2\theta_z \rangle\right) + A_{xx} \tag{8.6}$$

$$\langle A_\parallel \rangle = (A_{zz} - A_{xx})\langle \cos^2\theta_z \rangle + A_{xx} \tag{8.7}$$

which do not depend on θ_x and θ_y, because we used the approximation $A_{yy} \cong A_{xx}$. Measurement of hyperfine splittings alone (i.e., $\langle A_\parallel \rangle$ and/or $\langle A_\perp \rangle$), therefore gives us the time-average orientation of the nitroxide z-axis, $\langle \cos^2\theta_z \rangle$, relative to the director.

Similarly, Eq. B.25 in Appendix B transforms the g-tensor from the nitroxide axes to the membrane-fixed system: $\mathbf{g}' = \mathbf{R} \cdot \mathbf{g} \cdot \mathbf{R}^{-1}$. Because $\mathbf{g} = (g_{xx}, g_{yy}, g_{zz})$ is completely non-axial, we need a more general form of the rotation matrix $\mathbf{R}$ than for the A-tensor. Performing the transformation, with axial averaging but restricted motional averaging about the membrane normal (||), gives the g-tensor components in the bilayer axis system (Seelig 1970):

$$\langle g_\perp \rangle = \tfrac{1}{2}(g_{xx} - g_{yy})\left(1 - \langle \cos^2\theta_x \rangle\right) + \tfrac{1}{2}(g_{zz} - g_{yy})\left(1 - \langle \cos^2\theta_z \rangle\right) + g_{yy} \tag{8.8}$$

$$\langle g_\parallel \rangle = (g_{xx} - g_{yy})\langle \cos^2\theta_x \rangle + (g_{zz} - g_{yy})\langle \cos^2\theta_z \rangle + g_{yy} \tag{8.9}$$

which depend on the time-average $\langle \cos^2\theta_x \rangle$, in addition to $\langle \cos^2\theta_z \rangle$. Note that only two averages are independent because of the relation between direction cosines: $\langle \cos^2\theta_x \rangle + \langle \cos^2\theta_y \rangle + \langle \cos^2\theta_z \rangle = 1$. Therefore, combining g-value measurements with hyperfine splittings gives us the full set of angular time averages.

A general feature of Eqs. 8.6–8.9 is that the mean values (i.e., traces) of the motionally averaged tensors remain equal to the isotropic values:

$$a_o^N = \mathrm{Tr}(\mathbf{A}) = \tfrac{1}{3}\left(\langle A_\parallel \rangle + 2\langle A_\perp \rangle\right) \tag{8.10}$$

$$g_o = \mathrm{Tr}(\mathbf{g}) = \tfrac{1}{3}\left(\langle g_\parallel \rangle + 2\langle g_\perp \rangle\right) \tag{8.11}$$

This is an important result because the isotropic g-values and hyperfine couplings depend on the polarity of the spin-label environment (see Chapter 4). Therefore, a_o^N and g_o can be used as polarity-sensitive indicators, independent of the motional amplitude.

We often express the angular amplitudes of motion as orientational order parameters, S_{ii}. These relate conveniently to the experimental measurements and summarize all the tensor transformation properties that arise from motional averaging:

$$S_{ii} = \tfrac{1}{2}\left(3\langle \cos^2\theta_i \rangle - 1\right) \tag{8.12}$$

where $i \equiv x, y, z$ corresponds to the order parameters of the nitroxide x-, y- and z-axes, respectively, relative to the director (see Figure 8.3). From Appendix V.1, we recognize that the order parameters are angular averages of a second-order Legendre polynomial, i.e., $S_{ii} \equiv \langle P_2(\cos\theta_i) \rangle$. Note that only two of the three principal order parameters are independent:

$$S_{xx} + S_{yy} + S_{zz} = 0 \tag{8.13}$$

because of the relation between direction cosines: $\cos^2\theta_x + \cos^2\theta_y + \cos^2\theta_z = 1$. For spin labels such as the n-DOXYL fatty acids or phospholipids, the nitroxide z-axis is parallel to the long axis of the lipid molecule. We are then interested most in the S_{zz} order parameter, because this defines the amplitude of motion of the long molecular axis, relative to the director. The second order parameter gives a measure of non-axiality in lateral ordering, $S_{xx} - S_{yy}$. For axially symmetric motion (e.g., of a circular cylinder): $S_{xx} = S_{yy} = -\frac{1}{2}S_{zz}$.

Using Eqs. 8.6 and 8.7, we get the S_{zz} order parameter directly from the extent of averaging of the experimental hyperfine splittings:

$$S_{zz} = \frac{\langle A_\parallel \rangle - \langle A_\perp \rangle}{A_{zz} - A_{xx}} = \frac{\langle A_\parallel \rangle - \langle A_\perp \rangle}{\Delta A} \tag{8.14}$$

where we use the generalization: $\Delta A = A_{zz} - \frac{1}{2}(A_{xx} + A_{yy})$. The S_{xx} order parameter then comes from the experimental g-value $\langle g_\parallel \rangle$ by using Eqs. 8.9 and 8.14:

$$S_{xx} = \frac{3\langle g_\parallel \rangle - (g_{xx} + g_{yy} + g_{zz}) - 2(g_{zz} - g_{yy})S_{zz}}{2(g_{xx} - g_{yy})} \tag{8.15}$$

Practically, this is more convenient than using $\langle g_{\parallel} \rangle - \langle g_{\perp} \rangle$ because we can determine $\langle g_{\parallel} \rangle$ more easily than $\langle g_{\perp} \rangle$. For a 9-GHz axial powder pattern, $\langle g_{\parallel} \rangle$ comes directly from the midpoint of the two outer hyperfine peaks (see Section 2.4 in Chapter 2).

The motionally averaged spin-Hamiltonian parameters in an aligned system behave in exactly the same way as do axial spin-Hamiltonian parameters in a crystal. The spectra depend on the orientation of the magnetic field to the ordering axis, i.e., director, just as described in Sections 2.2–2.4 of Chapter 2. In particular, the effective partially-averaged hyperfine splitting is given from Eq. 2.17 by:

$$\langle A(\gamma) \rangle = \sqrt{\langle A_{\parallel} \rangle^2 \cos^2 \gamma + \langle A_{\perp} \rangle^2 \sin^2 \gamma} \tag{8.16}$$

where γ is the angle between the static magnetic field direction and the director (see Figure 8.1). Figure 8.4 shows the angular dependence of $\langle A(\gamma) \rangle$ for a DOXYL-labelled nitroxide probe in an aligned smectic liquid crystal. This is fitted very well by Eq. 8.16. As seen in Section 2.2, this equation reflects the fact that the nitrogen nuclear spin is not quantized along the magnetic field direction. If it were, Eq. 8.16 becomes analogous to Eq. 2.16 for the angular-dependent g-value:

$$\langle A(\gamma) \rangle_{B_o \to \infty} = \langle A_{\parallel} \rangle \cos^2 \gamma + \langle A_{\perp} \rangle \sin^2 \gamma \tag{8.17}$$

The dashed line in Figure 8.4 shows that the angular dependence is fitted less well by this extreme high-field approximation. Differences between the two fits are not so great for this moderately ordered system; discrepancies between Eqs. 8.16 and 8.17 increase with increasing degree of order, which has a marked effect on angular-dependent linewidths (see Section 8.10, later).

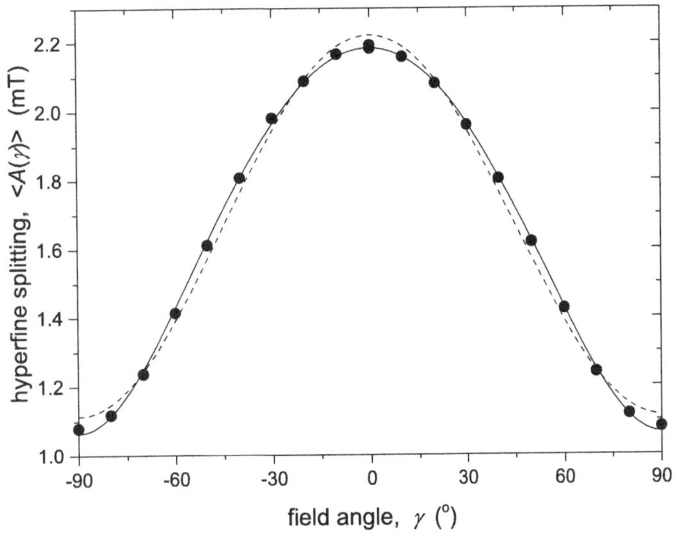

FIGURE 8.4 Dependence of partially-averaged nitrogen hyperfine splitting $\langle \Delta A(\gamma) \rangle$ on angle γ of static magnetic field to director. Spin probe 4-butoxy-4′-ethylDOXYLazobenzene in aligned smectic A phase of 4-butoxy-4′-acetylazobenzene at 104.6°C. Solid line: least-squares fit of Eq. 8.16; dashed line: fit of Eq. 8.17 that assumes nitrogen nuclear spin is quantized along static magnetic field. (Data points from Luckhurst et al. 1987.)

8.4 MOTIONAL MODELS FOR UNIAXIAL ORDER PARAMETERS

If the nitroxide z-axis is oriented at a fixed angle θ_z to the director axis about which rapid ϕ-rotation takes place, the order parameter is given simply by:

$$S_{zz} = \tfrac{1}{2}\left(3\cos^2 \theta_z - 1\right) \tag{8.18}$$

This is a common situation, when the nitroxide is fixed in the spin-labelled molecule and this rotates about an axis that does not coincide with the nitroxide z-axis (see Figure 8.1).

In general, the angles between the nitroxide axes and director fluctuate, contributing to the motional averaging. The angular averages that we need for an axially symmetric system are then:

$$\langle \cos^2 \theta_i \rangle = \frac{\int_0^{\pi} \cos^2 \theta_i \times p(\theta_i) \sin \theta_i \cdot d\theta_i}{\int_0^{\pi} p(\theta_i) \sin \theta_i \cdot d\theta_i} \tag{8.19}$$

where $p(\theta_i)$ is the innate probability of orientation θ_i, and the factor of $\sin \theta_i$ allows for the spatial degeneracy arising from ϕ-rotation about the polar z-axis (see Figure 2.5 in Section 2.4).

In the restricted random walk model (see Figure 8.5), the nitroxide z-axis wobbles randomly within a cone of half-angle θ_C. For this model, Eq. 8.19 becomes:

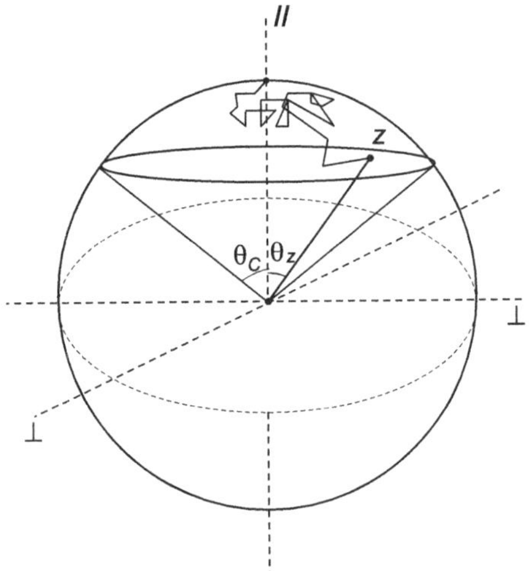

FIGURE 8.5 Restricted-random-walk (or fluctuation-in-a-cone) model of orientational ordering. Nitroxide z-axis performs random walk over sector of spherical surface that is bounded by cone of half-angle θ_C.

$$\langle \cos^2\theta_z \rangle = \frac{\int_0^{\theta_C} \cos^2\theta_z \times \sin\theta_z \cdot d\theta_z}{\int_0^{\theta_C} \sin\theta_z \cdot d\theta_z}$$

$$= \tfrac{1}{3}\left(1 + \cos\theta_C + \cos^2\theta_C\right) \quad (8.20)$$

The order parameter of the nitroxide z-axis is then given by:

$$S_{zz} = \tfrac{1}{2}\cos\theta_C(1 + \cos\theta_C) \quad (8.21)$$

and $S_{xx} = S_{yy}$. Figure 8.6 shows the dependence of the S_{zz} order parameter on angular amplitude θ_C for the cone model. This is compared with the simple case where the nitroxide z-axis is oriented at a fixed angle θ_z to the director axis about which rapid ϕ-rotation takes place (i.e., Eq. 8.18). The angle-dependence is less steep in the case of the cone model, because of angular averaging that results from the fluctuations in θ_z orientation.

A physically realistic alternative to simple geometric models is the mean-field approach. Here, we express orientational probability in terms of an orientation pseudopotential (or potential of mean torque), $U(\theta_i)$:

$$p(\theta_i) = A\exp\left(-U(\theta_i)/k_BT\right) \quad (8.22)$$

where factor A normalizes the distribution, such that: $\int_0^{\pi} p(\theta_i)\sin\theta_i \cdot d\theta_i = 1$. The angular average in Eq. 8.19 then becomes:

$$\langle \cos^2\theta_i \rangle = A\int_0^{\pi} \cos^2\theta_i \sin\theta_i \cdot \exp\left(-U(\theta_i)/k_BT\right)d\theta_i \quad (8.23)$$

Usually, we expand the orientation potential in a series of even-order Legendre polynomials $P_{2n}(\cos\theta_i)$ (see Appendix V.1). Here, we retain only the first term:

$$U(\theta_i) = -k_BT\lambda_2 P_2(\cos\theta_i) \quad (8.24)$$

where the strength of the potential λ_2 is in units of k_BT. This is the classic Maier–Saupe potential for liquid crystals (Maier and Saupe 1959). Using just this term and integrating by parts, we get:

$$\langle \cos^2\theta_i \rangle = \frac{1}{3\lambda_2}\left(\frac{2\exp\left(\frac{3}{2}\lambda_2\right)}{\int_{-1}^{1}\exp\left(\frac{3}{2}\lambda_2 c^2\right)dc} - 1\right) \quad (8.25)$$

from Eq. 8.23, with $c \equiv \cos\theta_i$. Integrating again by parts, we can relate averages of higher even powers of $\cos\theta_i$ to $\langle \cos^2\theta_i \rangle$:

$$\langle \cos^4\theta_i \rangle = \left(1 - \frac{1}{\lambda_2}\right)\langle \cos^2\theta_i \rangle + \frac{1}{3\lambda_2} \quad (8.26)$$

As we shall see later in Sections 8.7–8.9, we need the $\langle \cos^4\theta_i \rangle$ average for relaxation calculations.

Figure 8.7 shows the dependence of the order parameter $S_{zz} \equiv \langle P_2(\cos\theta_z) \rangle$, and also of $\langle P_4(\cos\theta_z) \rangle$, on strength $\lambda_2 \geq 0$ of the Maier–Saupe potential. We determine this from numerical integration of Eq. 8.25, together with Eqs. 8.12, 8.23 and 8.24. In this case, z is the long axis of the molecule, which we assume to be axially symmetric. The order parameters of the molecular x- and y-axes are given simply by $S_{xx} = S_{yy} = -\tfrac{1}{2}S_{zz}$.

If the ordering is no longer uniaxial, i.e., ordering of the molecular x- and y-axes is different, the ordering potential must

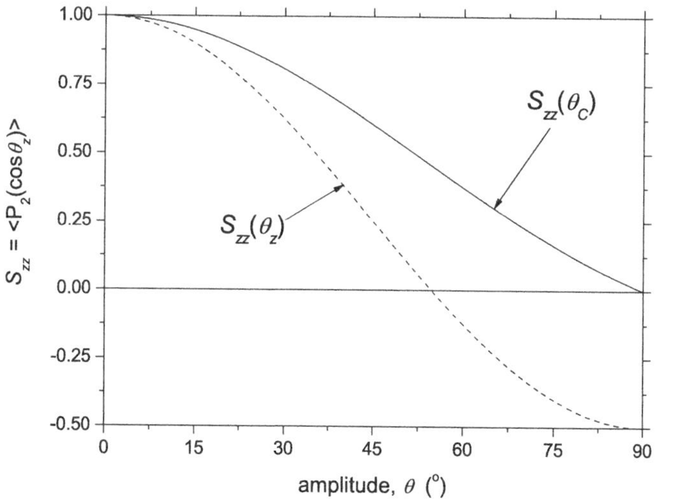

FIGURE 8.6 Dependence of order parameter S_{zz} on angular amplitude of motion, when nitroxide z-axis rotates at fixed angle θ_z to director (dashed line, Eq. 8.18) or wobbles within cone of half-angle θ_C about director as in Figure 8.5 (solid line, Eq. 8.21).

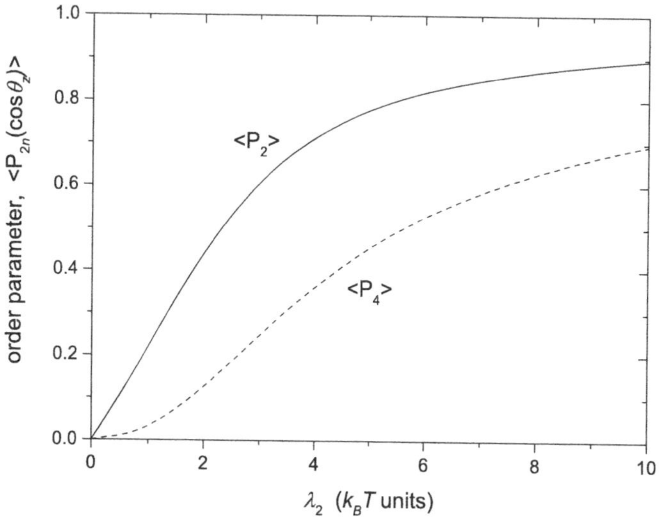

FIGURE 8.7 Dependence of molecular z-axis order parameter $S_{zz} \equiv \langle P_2(\cos\theta_z) \rangle$ (solid line), and of $\langle P_4(\cos\theta_z) \rangle$ (dashed line), on strength λ_2 of Maier–Saupe orientation potential (Eq. 8.24). T, absolute temperature; k_B, Boltzmann's constant.

contain an additional, non-axial term. A form frequently used is (Polnaszek and Freed 1975):

$$U(\theta,\phi) = -k_B T\left(\gamma_2\cos^2\theta + \varepsilon\sin^2\theta\cos 2\phi\right) \tag{8.27}$$

where $\theta \equiv \theta_z$, ϕ is the azimuthal orientation and $\gamma_2 \equiv \frac{3}{2}\lambda_2$ with $|\gamma_2| > |\varepsilon|$. When $\varepsilon > 0$, the molecular y-axis orients preferentially to the x-axis along the director $\left(S_{xx} < S_{yy}\right)$, i.e., the molecular x-axis orients more perpendicular to the director than does the y-axis.

8.5 INDEPENDENT ORDERING COMPONENTS, SEGMENTAL MOTION

We now consider two independent motions, where the nitroxide fluctuates about a local director, which itself fluctuates about the principal director. A typical situation is when the nitroxide z-axis performs limited angular excursions about the long axis of the spin-labelled molecule, and the latter is embedded in a membrane or liquid crystal.

The local motion produces time-averaged axial hyperfine splittings $\langle A'_{zz}\rangle$ and $\langle A'_{xx}\rangle$. From Eq. 8.14, the corresponding local order parameter (relative to the long molecular axis) is:

$$S_{loc} = \frac{\langle A'_{zz}\rangle - \langle A'_{xx}\rangle}{A_{zz} - A_{xx}} \tag{8.28}$$

Motion of the molecular long-axis about the principal director contributes further restricted averaging. This then results in the final axial hyperfine elements $\langle A_{\parallel}\rangle$ and $\langle A_{\perp}\rangle$. The order parameter of the long axis, relative to the principal director, is therefore:

$$S_o = \frac{\langle A_{\parallel}\rangle - \langle A_{\perp}\rangle}{\langle A'_{zz}\rangle - \langle A'_{xx}\rangle} \tag{8.29}$$

From Eqs. 8.14, 8.28 and 8.29, the overall order parameter of the nitroxide z-axis in the principal director system becomes:

$$S_{zz} = \frac{\langle A_{\parallel}\rangle - \langle A_{\perp}\rangle}{A_{zz} - A_{xx}} = S_o \times S_{loc} \tag{8.30}$$

Therefore, the order parameter that results from two independent restricted motions is the product of the two separate order parameters.

Equation 8.30 is a general result that comes from the addition theorem for spherical harmonics (see Eqs. V.17 and V.18 in Appendix V). Averaging over the azimuthal angle $\phi_1 - \phi_2$ in Eq. V.18, we get:

$$\langle P_2(\cos\Theta)\rangle = \langle P_2(\cos\theta_1)\rangle\langle P_2(\cos\theta_2)\rangle \tag{8.31}$$

where Θ is the angle between the principal director (θ_1,ϕ_1) and the nitroxide z-axis (θ_2,ϕ_2). The overall order parameter of the nitroxide z-axis is then: $S_{zz} = \langle P_2(\cos\Theta)\rangle$, relative to the principal director. In terms of Eq. 8.30, θ_1 is the angle between the long molecular axis and the principal director, i.e., $S_o = \langle P_2(\cos\theta_1)\rangle$, and θ_2 is the angle between the nitroxide z-axis and the long molecular axis, i.e., $S_{loc} = \langle P_2(\cos\theta_2)\rangle$.

An important application of Eq. 8.30 is when the nitroxide z-axis is at a fixed angle θ_z to the long molecular axis. Then S_{loc} is given by Eq. 8.18, and S_o gives us information on the ordering of the spin-labelled molecule in its anisotropic environment (e.g., membrane).

A further useful application is to segmental motion in spin-labelled molecules. If motions about adjacent segments are independent, then the corresponding order parameters multiply together as in Eq. 8.30 or 8.31. For m identical segments about which rotation takes place, we can extend Eq. 8.30 to give:

$$S_{zz} = S_o \times S_{seg}^m \tag{8.32}$$

where S_{seg} is the order parameter characterizing the motion of a single segment. Typical examples are lipids spin-labelled at C-atom n of the hydrocarbon chain: the n-DOXYL fatty acids or phospholipids in lipid bilayer membranes.

Figure 8.8 shows order-parameter profiles for chain-labelled lipid spin probes in bilayer membranes. The circles are for spin-labelled fatty acids in a short-chain soap system. These are highly fluid membranes and the partially averaged 9-GHz EPR spectra correspond entirely to the fast-motion regime (Seelig 1970). Complete order parameters of the spin-labelled chains thus come directly from the experimental hyperfine splittings (Eq. 8.14). Although rotations about C–C bonds in hydrocarbon chains are not truly independent, especially in anisotropic

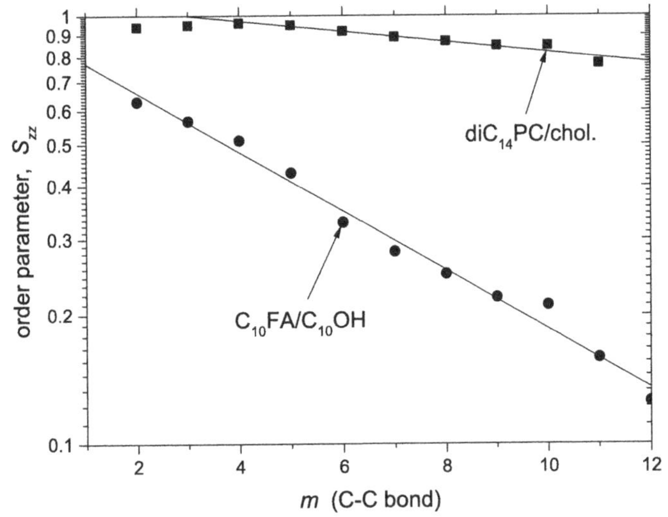

FIGURE 8.8 Order-parameter profile with C–C bond position, m, down the fatty-acid chain for (m + 2)-DOXYL lipids in fluid bilayer membranes. y-axis is logarithmic. *Circles*: fatty-acid spin labels in bilayers of Na decanoate/decanol/water (14:21:15 w/w/w) at 22°C (Seelig et al. 1972). *Squares*: phosphatidylcholine (PC) spin labels in aqueous bilayers of dimyristoyl PC/cholesterol (3:2 mol/mol) at 30°C (Livshits et al. 2004).

environments, the data for the soap system in Figure 8.8 display an order parameter profile consistent with Eq. 8.32 for a freely jointed chain. The corresponding segmental order parameter is $S_{seg} = 0.85$, which relates directly to the probability that the C–C bond is in a *trans* conformation (Hubbell and McConnell 1971). The long-axis order parameter becomes $S_o \approx 0.90$, when this includes the first bond attached to the polar head group, as defined for $m = 0$ in Figure 8.8.

Spin-labelled chains in fluid membranes formed by longer, two-chain phospholipids characteristically have 9-GHz EPR spectra that reflect motion with components in both the fast and the slow motional regimes (Lange et al. 1985; Moser et al. 1989; Schorn and Marsh 1996b). As we shall see later in the chapter, the fast component comes from segmental motion, i.e., *trans-gauche* isomerism. At a higher EPR frequency of 94 GHz, the slow component is driven into the rigid limit, and the $\langle A_{\parallel} \rangle$ hyperfine splitting is determined solely by the fast component. The squares in Figure 8.8 are the fast-motion order parameters from HF-EPR hyperfine splittings, for spin-labelled phosphatidylcholine (PC) in membranes of dimyristoyl (C_{14}) PC with 40 mol% cholesterol. Overall, the order parameters are much higher than for the simple soap system. This is not least because of the high cholesterol content, where the rigid sterol nucleus strongly increases the order of fluid lipid chains (Schreier-Muccillo et al. 1973). Also, the order profile differs from that of the short-chain soap system. The order parameters remain approximately constant at the top of the chain, before the characteristic flexibility gradient sets in towards the terminal methyl end. In this latter region, the segmental order parameter $S_{seg} = 0.97$ is much higher than in the soap system.

8.6 RESTRICTED OFF-AXIS AMPLITUDES AND LATERAL ORDERING (HF-EPR)

Now we treat the situation where the nitroxide z-axis performs limited angular excursions θ_z from the principal director **N**, as in Section 8.3, but there is also lateral ordering in the nitroxide x,y-plane (see Figure 8.9). Incomplete averaging around the nitroxide z-axis accompanies lateral ordering. We describe this by limited excursions of the azimuthal ϕ-angle that are independent of the orientation of the z-axis itself (cf. Section 8.2). The situation is typical of spin-labelled lipid chains in membranes containing cholesterol (Kurad et al. 2001; Kurad et al. 2004).

For n-DOXYL lipid spin labels, the nitroxide z-axis is parallel to the long axis of the lipid chain. The nitrogen hyperfine tensor is almost axially symmetric about the z-principal axis. Therefore, we detect lateral ordering only from the $(g_{xx} - g_{yy})$ Zeeman anisotropy, which means using HF-EPR (94 GHz or higher). For this reason, we concentrate on averaging of the g-tensor by rapid rotation (cf. Livshits and Marsh 2000).

From Eq. 2.11 of Chapter 2, the partially averaged g-value for a general polar orientation (θ,ϕ) of the fixed magnetic field in the instantaneous nitroxide axis system is:

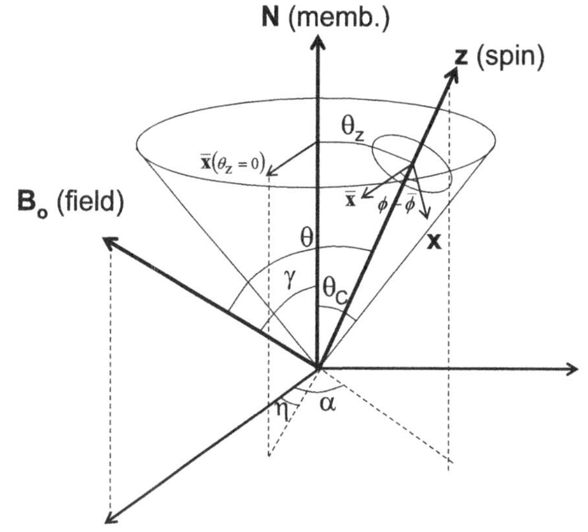

FIGURE 8.9 Relation between instantaneous spin-label axis **z**, magnetic field direction $\mathbf{B}_o$ and director **N** (membrane normal). For spin-labelled lipid chains (n-DOXYL PC), nitroxide z-axis coincides with chain axis, which is uniaxially ordered about **N**. **z** and $\mathbf{B}_o$ are inclined at θ_z and γ, respectively, to **N**. **z** is inclined at θ, with azimuth α in the director system, to $\mathbf{B}_o$. Maximum amplitude of θ_z is θ_C. $\bar{\mathbf{x}}$: space-fixed direction of mean nitroxide x-axis, which performs ϕ-rotations about **z** with mean value $\bar{\phi}$ and fluctuation amplitude $\pm\phi_o$. η: azimuthal orientation of $\bar{\mathbf{x}}$-axis for $\theta_z = 0$, in director system.

$$\langle g(\theta,\phi) \rangle = g_o + \tfrac{1}{3}\left(3\langle \cos^2\theta \rangle - 1\right)\Delta g + \langle \sin^2\theta \cos 2\phi \rangle \delta g \quad (8.33)$$

where g_o is the isotropic g-value and the g-tensor anisotropies are: $\Delta g = g_{zz} - \frac{1}{2}(g_{xx} + g_{yy})$ and $\delta g = \frac{1}{2}(g_{xx} - g_{yy})$. Angular brackets in Eq. 8.33 correspond to averaging over θ and ϕ. These angles fluctuate with changes in orientation of the nitroxide axes that result from molecular rotation.

If the mean ϕ-orientation between nitroxide x-axis and static magnetic field direction is $\phi = \bar{\phi}$ (see Figure 8.9), the non-axial angular average in Eq. 8.33 becomes:

$$\langle \sin^2\theta \cos 2\phi \rangle = \langle \cos 2(\phi - \bar{\phi}) \rangle \langle \sin^2\theta \cos 2\bar{\phi} \rangle \quad (8.34)$$

where the two angular averages on the right-hand side are independent. We use the condition: $\langle \sin 2(\phi - \bar{\phi}) \rangle = 0$, corresponding to symmetric fluctuations about $\bar{\phi}$, to obtain Eq. 8.34.

Now we can relate the field orientation in the nitroxide frame to the angular excursions of the nitroxide axes in the director frame (see Figure 8.9). The nitroxide z-axis is inclined at instantaneous angle θ_z to the director, which is oriented at angle γ to the static-field direction. The magnetic field makes an azimuthal angle η to the mean nitroxide x-direction, in the director axis system. The instantaneous azimuthal orientation of the nitroxide z-direction in the director axis system is given by the angle α, relative to the azimuthal orientation of the magnetic field in the same system. The angular variables appearing in Eqs. 8.33 and 8.34 then are given by (Israelachvili et al. 1975):

$$\cos\theta = \cos\gamma \cos\theta_z - \sin\gamma \sin\theta_z \cos\alpha \quad (8.35)$$

$$\sin^2\theta\cos 2\bar{\phi}$$
$$= \left[(\cos\gamma\sin\theta_z + \sin\gamma\cos\theta_z\cos\alpha)^2 - \sin^2\gamma\sin^2\alpha\right]\cos 2(\alpha-\eta)$$
$$+ 2(\cos\gamma\sin\theta_z + \sin\gamma\cos\theta_z\cos\alpha)\sin\gamma\sin\alpha\sin 2(\alpha-\eta) \quad (8.36)$$

where the angles γ and η are fixed. They specify the orientation of the magnetic field relative to the two director axes of motional averaging, **N** and $\bar{\mathbf{x}}$.

Axially anisotropic rotation of the nitroxide z-axis about the principal director **N** results in averaging over the azimuthal angle α from 0 to 2π. The angular averages in Eqs. 8.33 and 8.34 that we get from Eqs. 8.35 and 8.36 are then:

$$\langle\cos^2\theta\rangle = \tfrac{1}{3}\left[1 + (3\cos^2\gamma - 1)\langle P_2(\cos\theta_z)\rangle\right] \quad (8.37)$$

$$\langle\sin^2\theta\cos 2\bar{\phi}\rangle = \tfrac{1}{6}\sin^2\gamma\cos 2\eta\left[2 + 3\langle\cos\theta_z\rangle + \langle P_2(\cos\theta_z)\rangle\right] \quad (8.38)$$

where $P_2(x) = \frac{1}{2}(3x^2 - 1)$ is the second-order Legendre polynomial (see Appendix V.1). Thus, $\langle P_2(\cos\theta_z)\rangle$ is the conventional order parameter of the nitroxide z-axis. Averaging over θ_z in Eqs. 8.37 and 8.38 is left implicit. We need not only the order parameter $\langle P_2(\cos\theta_z)\rangle$, but also the average $\langle\cos\theta_z\rangle$, to get the composite non-axial average $\langle\sin^2\theta\cos 2\bar{\phi}\rangle$.

The motionally averaged principal g-tensor elements come from Eq. 8.33 by using Eqs. 8.34, 8.37 and 8.38. They are defined by magnetic field orientations (γ, η) = (90°, 0°), (90°, 90°) and γ = 0°, which give the x, y and z diagonal elements, respectively (see Figure 8.9). The g-tensor elements that result are:

$$\langle g_{xx}\rangle = g_o - \tfrac{1}{3}\langle P_2(\cos\theta_z)\rangle\Delta g + \tfrac{1}{6}\langle\cos 2(\phi-\bar{\phi})\rangle\left[2 + 3\langle\cos\theta_z\rangle + \langle P_2(\cos\theta_z)\rangle\right]\delta g \quad (8.39)$$

$$\langle g_{yy}\rangle = g_o - \tfrac{1}{3}\langle P_2(\cos\theta_z)\rangle\Delta g - \tfrac{1}{6}\langle\cos 2(\phi-\bar{\phi})\rangle\left[2 + 3\langle\cos\theta_z\rangle + \langle P_2(\cos\theta_z)\rangle\right]\delta g \quad (8.40)$$

$$\langle g_{zz}\rangle = g_o + \tfrac{2}{3}\langle P_2(\cos\theta_z)\rangle\Delta g \quad (8.41)$$

These three independent g-values depend on three linearly independent terms. We get the isotropic g-value g_o and the conventional order parameter $\langle P_2(\cos\theta_z)\rangle$ directly. The third parameter that we get from the experimental g-values, however, is a composite angular average $\langle\cos 2(\phi-\bar{\phi})\rangle\left[2 + 3\langle\cos\theta_z\rangle + \langle P_2(\cos\theta_z)\rangle\right]$ that depends on both axial and non-axial ordering.

We can extract the angular average $\langle\cos 2(\phi-\bar{\phi})\rangle$ that characterizes non-axial ordering from the composite average, if we assume a model to describe the axial distribution in θ_z. For restricted random walk within a cone of half-angle θ_C (see Figure 8.5), the order parameter S_{zz} comes from Eq. 8.21:

$$\langle P_2(\cos\theta_z)\rangle = \tfrac{1}{2}\cos\theta_C(1 + \cos\theta_C) \quad (8.42)$$

and the other angular average that we need is:

$$\langle\cos\theta_z\rangle = \tfrac{1}{2}(1 + \cos\theta_C) \quad (8.43)$$

in this model. By using Eqs. 8.42 and 8.43, together with Eqs. 8.39–8.41, we determine g_o, θ_C and $\langle\cos 2(\phi-\bar{\phi})\rangle$. If we further model azimuthal x-axis ordering as completely random ϕ-angle fluctuations of maximum amplitude $\pm\phi_o$ about the mean value $\bar{\phi}$ (cf. Figures 8.2 and 8.9), the corresponding order parameter is:

$$\langle\cos 2(\phi-\bar{\phi})\rangle = \sin\phi_o\cos\phi_o/\phi_o \quad (8.44)$$

Thus, we can parameterize the rapid restricted rotational motion with the two independent maximum amplitudes, θ_C of the z-axis tilt, and ϕ_o of the rotation or twist about the z-axis. More generally, but less explicitly, axial ordering is characterized by $\langle P_2(\cos\theta_z)\rangle$ and non-axial or lateral ordering by $\langle\cos 2(\phi-\bar{\phi})\rangle$.

As an example, Figure 8.10 shows the dependence on chain-labelling position, n, of the order parameters $\langle\cos 2(\phi-\bar{\phi})\rangle$ and $\langle P_2(\cos\theta_z)\rangle$, for n-DOXYL PC spin probes in membranes of

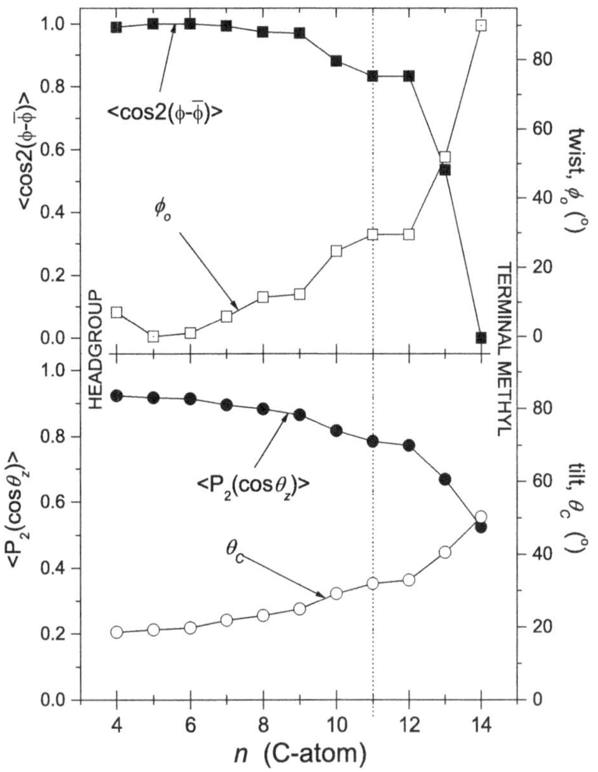

FIGURE 8.10 Dependence of order parameters (solid symbols) and rotational amplitudes (open symbols) on spin-label position n in sn-2 chain of n-DOXYL PC spin probes in membranes of dimyristoyl PC/cholesterol (3:2 mol/mol) at 30°C. *Upper panel*: order parameter $\langle\cos 2(\phi-\bar{\phi})\rangle$ and azimuthal angular amplitude ϕ_o. *Lower panel*: order parameter $\langle P_2(\cos\theta_z)\rangle$ and off-axis angular amplitude θ_C. (HF-EPR data from Kurad et al. 2004; experimental spectra shown later in Figure 8.26).

dimyristoyl phosphatidylcholine (PC) that contain 40 mol% cholesterol. These come from g-value measurements at 94 GHz. Both axial and lateral order remains high and essentially constant up to the C-9 position of the lipid chain. Beyond the C-11 position, the degree of order declines rapidly, more so for the axial rotation determined by $\langle \cos 2(\phi - \bar{\phi}) \rangle$ than for the off-axis fluctuation determined by $\langle P_2(\cos\theta_z) \rangle$. The region of high order corresponds to the location of the rigid steroid nucleus of cholesterol in the membrane, where the lipid chains are ordered not only transversely but also laterally (Kurad et al. 2004). Figure 8.10 also gives the angular amplitudes, θ_C and ϕ_o, derived from Eqs. 8.42 and 8.44. They clearly reflect, in inverse fashion, the dependences on chain-labelling position, n, of the corresponding order parameters.

8.7 MOTIONAL-NARROWING THEORY: LINEWIDTHS AND ORDER

In this section, we use time-dependent perturbation theory to treat relaxation in the motional-narrowing regime, as introduced in Chapter 5. With uniaxial molecular ordering, the time-independent Hamiltonian becomes:

$$\mathcal{H}_o = g_o \beta_e \mathbf{B}_o \cdot \mathbf{S} + a_o \mathbf{I} \cdot \mathbf{S} + \langle \mathcal{H}_1(\theta(t), \phi(t)) \rangle \tag{8.45}$$

where $\langle \mathcal{H}_1(\theta(t), \phi(t)) \rangle$ comes from Eq. 5.15, which assumes axial spin-Hamiltonian tensors. Averaging over the instantaneous angles $\theta(t)$ and $\phi(t)$ in Eq. 5.15, we get:

$$\langle \mathcal{H}_1(\theta(t), \phi(t)) \rangle = \tfrac{1}{3}(3\cos^2\gamma - 1) \langle P_2(\cos\theta_z) \rangle (\Delta g \beta_e B_o + \Delta A I_z) S_z \tag{8.46}$$

where $\langle P_2(\cos\theta_z) \rangle$ is the order parameter and γ is the fixed angle that the static magnetic field $\mathbf{B}_o$ makes with the director $\mathbf{N}$ (cf. Figure 8.9). In Eq. 8.46, we make use of the cosine relation:

$$\cos\theta(t) = \cos\theta_z(t)\cos\gamma + \sin\theta_z(t)\sin\gamma\cos\phi_z(t) \tag{8.47}$$

where $\theta(t)$ is the angle between the magnetic field and the nitroxide z-axis (see Figure 8.9) and the origin for the azimuth $\phi_z(t)$ of the nitroxide z-axis is that of the static field. Also, we ignore the $I_\pm S_\mp$ non-secular terms, which contribute only in second order. Then Eq. 8.46 leads immediately to Eqs. 8.6 and 8.7 for the motionally averaged hyperfine tensor components, when $\gamma = 90°$ and $0°$, respectively. Note that because Eq. 5.15 uses an axial spin Hamiltonian, the motionally averaged g-tensor does not differ here from the hyperfine tensor.

The time-dependent perturbation that we need for the relaxation calculation is:

$$\mathcal{H}_1(t) = \sum_q \mathcal{H}_1^{(q)} F^{(q)}(t) = \mathcal{H}_1(\theta(t), \phi(t)) - \langle \mathcal{H}_1(\theta(t), \phi(t)) \rangle \tag{8.48}$$

We get this from Eqs. 5.15 and 8.46, when we assume axial symmetry for the A- and g-tensors. It is clear from Section 5.3 that the angular averages governing the spectral densities, which come from $\langle |\mathcal{H}_1(t)|^2 \rangle$, will be modified by orientational ordering. Instead of the isotropic averages given in Eqs. 5.16–5.18, we now need the specific angular averages:

$$\begin{aligned} \langle |F^{(0)}(0)|^2 \rangle &\equiv \left\langle \left(3\cos^2\theta - 1 - \langle 3\cos^2\theta - 1 \rangle\right)^2 \right\rangle \\ &= 9\left(\langle \cos^4\theta \rangle - \langle \cos^2\theta \rangle^2 \right) \end{aligned} \tag{8.49}$$

$$\langle |F^{(1)}(0)|^2 \rangle \equiv \langle \sin^2\theta \cos^2\theta \rangle = \langle \cos^2\theta \rangle - \langle \cos^4\theta \rangle \tag{8.50}$$

$$\langle |F^{(2)}(0)|^2 \rangle \equiv \langle \sin^4\theta \rangle = 1 - 2\langle \cos^2\theta \rangle + \langle \cos^4\theta \rangle \tag{8.51}$$

Thus, relaxation in the fast-motion regime depends not only on the $\langle \cos^2\theta \rangle$ average but also on $\langle \cos^4\theta \rangle$. Relaxation measurements therefore provide us with extra information on molecular ordering, beyond that which we get from the motionally averaged line positions or splittings.

We must remember that $\theta(t)$ is the angle between the nitroxide z-axis and the magnetic field, not the director. However, when the magnetic field lies along the director $\mathbf{N}$, Eqs. 8.49–8.51 apply directly to the order of the nitroxide z-axis, i.e., $\theta = \theta_z$. The transverse relaxation rate then depends on hyperfine manifold m_I:

$$1/T_2(m_I) = A + Bm_I + Cm_I^2 \tag{8.52}$$

exactly as for the isotropic case in Chapter 5 (Eq. 5.58). With magnetic field parallel to the director, the linewidth coefficients become (cf. Glarum and Marshall 1967):

$$\begin{aligned} A = \tfrac{1}{9}\left(\frac{\Delta g \beta_e B_o}{\hbar}\right)^2 \left(j_o(0) + \tfrac{9}{2} j_1(\omega_e) \right) \\ + \tfrac{1}{4} I(I+1)(\Delta A)^2 \left(j_1(\omega_a) + \tfrac{1}{18} j_0(\omega_e) + \tfrac{1}{2} j_2(\omega_e) \right) \end{aligned} \tag{8.53}$$

$$B = \left(\frac{\Delta g \beta_e B_o}{\hbar}\right) \Delta A \left(\tfrac{2}{9} j_0(0) + j_1(\omega_e) \right) \tag{8.54}$$

$$C = \tfrac{1}{4}(\Delta A)^2 \left(\tfrac{4}{9} j_0(0) - j_1(\omega_a) - \tfrac{1}{18} j_0(\omega_e) + 2j_1(\omega_e) - \tfrac{1}{2} j_2(\omega_e) \right) \tag{8.55}$$

where the nuclear frequency is given as usual by $\omega_a = \frac{1}{2} a_o$, and all hyperfine constants are in angular frequency units. The spectral densities in Eqs. 8.53–8.55 are related to the angular averages in Eqs. 8.49–8.51 with $\theta = \theta_z$ by (cf. Eq. 5.13):

$$j_q(\omega) = \langle |F^{(q)}(0)|^2 \rangle \frac{\tau_c}{1 + \omega^2 \tau_c^2} \tag{8.56}$$

where τ_c is the correlation time characterizing angular reorientation. Using the isotropic averages: $\langle \cos^2\theta_z \rangle = \frac{1}{3}$ and $\langle \cos^4\theta_z \rangle = \frac{1}{5}$

(see Eqs. 5.16–5.18), Eqs. 8.53–8.56 reduce to Eqs. 5.59, 5.61 and 5.62 for isotropic rotational diffusion, with axial tensors. (Note that the $j_2(\omega_e)$ contribution from cross relaxation in Eqs. 8.53 and 8.55 is twice that given by Glarum and Marshall 1967.)

As we saw in Section 5.8, we can neglect non-secular contributions to the linewidth if $\omega_e^2\tau_R^2 >> 1$, i.e., correlation times are longer than ≈5 × 10^{-11} s at 9 GHz EPR frequency. Conversely if $\omega_a^2\tau_R^2 << 1$, we can put $j_1(\omega_a) = j_1(0)$ in the pseudosecular contributions to Eqs. 8.53 and 8.55, which limits us to correlation times shorter than 1×10^{-9} s (at least for large angular amplitudes). Using these two simplifications, we can investigate the dependence of the linewidth parameters on molecular ordering. We use the relationship between $\langle\cos^2\theta_z\rangle$ and $\langle\cos^4\theta_z\rangle$ averages that is given by Eq. 8.26, which comes from the Maier–Saupe orientation potential. Figure 8.11 shows the dependence of the B and C linewidth coefficients on Maier–Saupe order parameter S_{zz}. Both coefficients are normalized to their values, B_{iso} and C_{iso}, that we get with isotropic rotational diffusion. Within our present approximation, these normalized values are independent of spin-Hamiltonian parameters and rotational correlation time. Both ratios are less than one for moderately high order $(\langle P_2(\cos\theta_z)\rangle > 0.4)$, because restricted rotation is less effective at inducing relaxation than is unrestricted rotation. However, the ratios exceed one (i.e., stronger broadening) at lower order, because the $\langle\cos^2\theta_z\rangle$ and $\langle\cos^4\theta_z\rangle$ averages depend differently on order (see Figure 8.7). A further interesting feature is that the C/C_{iso} ratio becomes negative at high order when $\langle P_2(\cos\theta_z)\rangle > 0.75$, which causes narrowing of the outer lines. This arises because secular and pseudosecular contributions have opposite signs in Eq. 8.55.

In principle, we get the dependence of the linewidth coefficients on magnetic field orientation γ, relative to the director, by substituting Eq. 8.47 for orientation θ, relative to the magnetic field, into Eqs. 8.49–8.51. We do this in the following Section 8.8

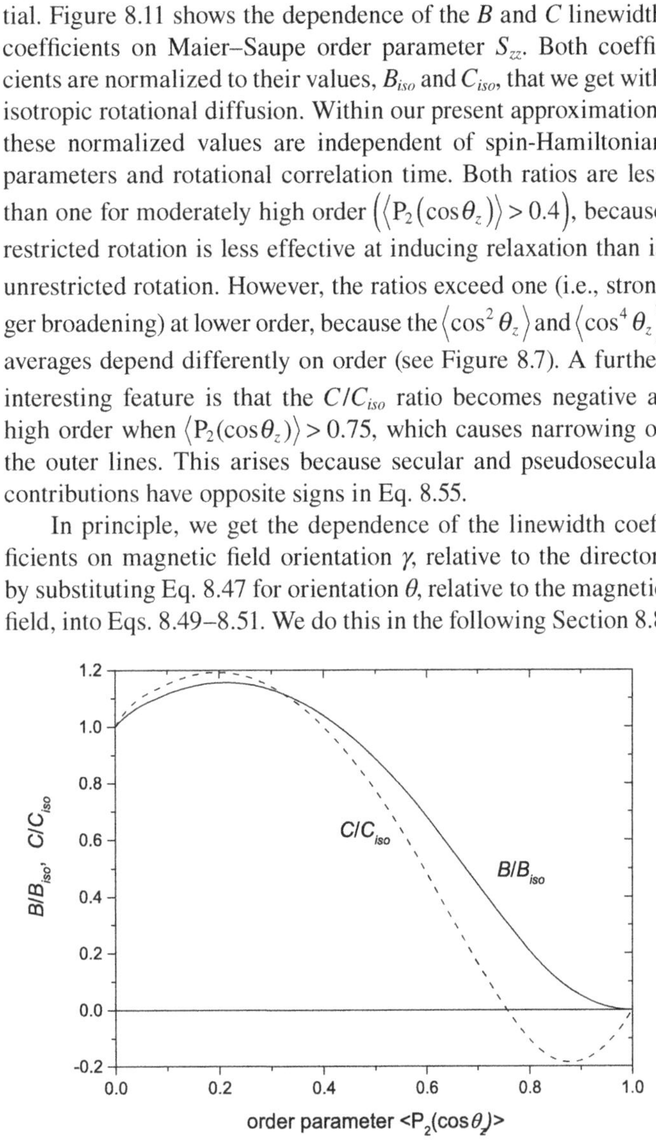

FIGURE 8.11 Linewidth coefficients B and C from Eq. 8.52, as function of order parameter $S_{zz} \equiv \langle P_2(\cos\theta_z)\rangle$. Angular averages determined by Maier–Saupe orientation potential (Eqs. 8.25, 8.26). Non-secular terms neglected, and $\omega_a^2\tau_R^2 << 1$ in pseudosecular spectral densities. Linewidth coefficients normalized to values, B_{iso} and C_{iso}, for isotropic rotational diffusion.

on spin–lattice relaxation, and in Section 8.9 on lateral ordering. In practice, however, transformations between axes are better done with spherical tensor operators and Wigner rotation matrices (i.e., Eq. S.1) than with the Cartesian systems (see Appendices R and S). We describe this general method later, in Section 8.10. For the moment, we only illustrate the dependence on field orientation with a simple model that applies to high (or perfect) order (Hemminga and Berendsen 1972).

Assume that the nitroxide z-axis lies along the director (i.e., $S_{zz} = 1$). The azimuthal angle of the magnetic field in the director system is α; this is the only angle over which motional averaging takes place (cf. Figure 8.9). From Eq. 2.11 in Chapter 2, the effective g-value for polar orientation (γ,α) of the magnetic field is:

$$g(\gamma,\alpha) = (g_{xx}\cos^2\alpha + g_{yy}\sin^2\alpha)\sin^2\gamma + g_{zz}\cos^2\gamma \tag{8.57}$$

The deviation from the motionally averaged value $\langle g(\gamma,\alpha)\rangle$ is (cf. Eq. 8.48):

$$g(\gamma,\alpha) - \langle g(\gamma,\alpha)\rangle = \delta g\cos 2\alpha\sin^2\gamma \tag{8.58}$$

because $\langle\cos^2\alpha\rangle = \frac{1}{2} = \langle\sin^2\alpha\rangle$, and where $\delta g = \frac{1}{2}(g_{xx} - g_{yy})$. If we assume that the nitrogen nucleus is quantized along the magnetic field, a similar expression applies to the hyperfine tensor (but see Luckhurst and Zannoni 1977, and later in Section 8.10). The perturbation Hamiltonian is then:

$$\mathcal{H}_1(t) = (\delta g\beta_e B_o/\hbar + \delta A m_I)\cos 2\alpha(t)\sin^2\gamma \times S_z \tag{8.59}$$

where $\delta A = \frac{1}{2}(A_{xx} - A_{yy})$, in angular frequency units. After averaging over $\alpha(t)$, the secular contribution to the transverse relaxation rate becomes (see Eqs. 5.49, 5.50):

$$\frac{1}{T_2^{sec}(m_I)} = \left\langle\left|\mathcal{H}_1^0(t)\right|^2\right\rangle\tau_{\parallel} = \frac{1}{2}(\delta g\beta_e B_o/\hbar + \delta A m_I)^2\tau_{\parallel}\sin^4\gamma \tag{8.60}$$

where $\tau_{\parallel}$ is the correlation time for rotation around the nitroxide z-axis, which coincides with the molecular long axis and $\langle\cos^2 2\alpha\rangle = \frac{1}{2}$. This expression shows a strong dependence of line broadening on magnetic field orientation γ in ordered systems. With cyclic permutation of the nitroxide axes, Eq. 8.60 applies to the DOXYL steroid labels (where the nitroxide y-axis lies along the long axis) in lipid–cholesterol membranes with high order (Hemminga and Berendsen 1972). The pseudosecular contribution to the tranverse rate is given for this model by Israelachvili et al. (1975).

8.8 SPIN–LATTICE RELAXATION AND CROSS RELAXATION WITH ORIENTATIONAL ORDERING

We can determine the influence of molecular ordering on spin–lattice relaxation rates by using motional narrowing theory as introduced in Chapter 5. The electron spin transition probabilities

$W_e(m_I)$ are given by Eq. 5.19 in Section 5.4, where the spin–lattice relaxation time is $T_{1,e}(m_I) \equiv 1/(2W_e(m_I))$. The corresponding spectral densities $j^{AA}(\omega)$, $j^{gA}(\omega)$ and $j^{gg}(\omega)$ are defined by expressions such as Eq. 5.23 of Section 5.5. These depend on the angular average $\langle \sin^2\theta\cos^2\theta \rangle$ that is defined here by Eq. 8.50. The spin–lattice relaxation rate then becomes (cf. Eq. 5.41):

$$W_e(m_I) = \frac{1}{2}\left[\left((\Delta g)^2 + 3(\delta g)^2\right)\cdot\left(\frac{\beta_e}{\hbar}\right)^2 B_o^2 + 2(\Delta g\Delta A + 3\delta g\delta A)\right.$$
$$\left.\times\left(\frac{\beta_e}{\hbar}\right)B_o m_I + \left((\Delta A)^2 + 3(\delta A)^2\right)m_I^2\right]$$
$$\times\langle \sin^2\theta\cos^2\theta \rangle\frac{\tau_c}{1+\omega_e^2\tau_c^2} \tag{8.61}$$

where τ_c is the rotational correlation time for restricted rotation of the nitroxide z-axis, and θ is the angle that this axis makes with the static magnetic field. For the special case where the magnetic field is parallel to the director, we have simply $\theta = \theta_z$, where θ_z is the angle between the nitroxide z-axis and the director (see Figure 8.3). Initially, we assume this particular orientation of the magnetic field, as in Section 8.7. In addition to aligned samples (including specifically nematic liquid crystals), this orientation also corresponds to the outer wings in an anisotropic powder pattern (see Section 2.4).

We now use Eq. 8.26, which comes from the Maier–Saupe orientational potential, to express the angular average $\langle \sin^2\theta_z\cos^2\theta_z \rangle$ in terms of the order parameter $\langle P_2(\cos\theta_z)\rangle$ for the nitroxide z-axis according to Eq. 8.50:

$$\langle \sin^2\theta_z\cos^2\theta_z \rangle = \frac{3}{2\lambda_2}\langle P_2(\cos\theta_z)\rangle \tag{8.62}$$

where λ_2 represents the strength of the Maier–Saupe potential (see Eq. 8.24). The solid line in the upper panel of Figure 8.12 shows the dependence of the spin–lattice relaxation rate on Maier–Saupe order parameter, when the static field is parallel to the director. We normalize each value to that for isotropic rotational diffusion $W_e(\text{iso})$, given by $\langle \sin^2\theta\cos^2\theta \rangle = \frac{2}{15}$. The normalized values are independent of spin-Hamiltonian parameters, hyperfine manifold and rotational correlation time. As for the linewidth parameters, the ratio is less than one for moderately high order (i.e., $\langle P_2(\cos\theta_z)\rangle > 0.6$), because restricted rotation is less effective at inducing relaxation than is unrestricted rotation. The ratios exceed one (i.e., faster relaxation) at lower order, because the $\langle \cos^2\theta_z \rangle$ and $\langle \cos^4\theta_z \rangle$ averages depend differently on order (see Figure 8.7).

When instead the value of θ_z is fixed, the dependence of spin–lattice relaxation rate on order parameter $S_{zz} = \frac{1}{2}(3\cos^2\theta_z - 1)$ is given by the dashed lines in the upper panel of Figure 8.12. The normalized spin–lattice relaxation rates, for static magnetic field parallel to the director, are then:

$$W_e(S_{zz})/W_e(0) = 1 + S_{zz} - 2S_{zz}^2 \tag{8.63}$$

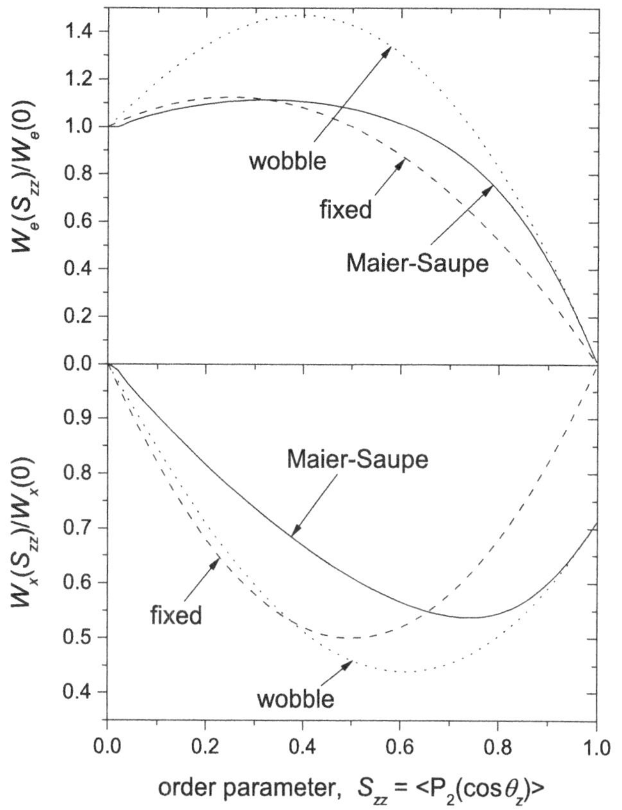

FIGURE 8.12 *Top*: Spin–lattice relaxation rate $W_e = 1/2T_{1e}$ (Eq. 8.61), as function of order parameter $S_{zz} \equiv \langle P_2(\cos\theta_z)\rangle$. Solid line: Maier–Saupe orientation potential (Eq. 8.62); dashed line: fixed θ_z (Eq. 8.63); dotted line: restricted random walk within a cone (Eq. 8.64). *Bottom*: mean cross-relaxation rate $W_x = \frac{1}{2}(W_{x1} + W_{x2})$ (Eqs. 8.65, 8.66) for same three motional models. Relaxation rates $W_e(S_{zz})$, $W_x(S_{zz})$ for static magnetic field parallel to director are normalized to values $W_e(0)$, $W_x(0)$ for $S_{zz} = 0$. For Maier–Saupe potential, normalization is to isotropic rotational diffusion.

where we now normalize to the rate, $W_e(0)$, for $S_{zz} = 0$. Equation 8.63 contrasts with the model-free approach used in NMR where the spin–lattice relaxation rate depends only on S_{zz}^2. However, the effective correlation times in the latter approach depend on angular amplitudes, in addition to rotational rates.

For the wobbling-in-a-cone model illustrated in Figure 8.5, $\langle \cos^2\theta_z \rangle$ is given by Eq. 8.20. Corresponding explicit evaluation of $\langle \cos^4\theta_z \rangle$ then leads to the normalized spin–lattice relaxation rate:

$$W_e(S_{zz})/W_e(0) = 1 + \cos\theta_C - \tfrac{1}{2}(1 + 3\cos\theta_C)\cos^2\theta_C \tag{8.64}$$

where θ_C is the semi-angle of the cone. The dotted line in the upper panel of Figure 8.12 shows the dependence on order parameter for this model, when the magnetic field lies along the director.

The transition probabilities for cross relaxation, which also makes significant contributions in saturation-recovery experiments, depend on angular averages different from those for spin–lattice relaxation. Explicitly, the two cross-relaxation rates are (cf. Eqs. 5.43 and 5.44 and discussion of Eqs. 5.21 and 5.22 in Section 5.4):

$$W_{x1}(m_I)=\frac{1}{8}\left((\Delta A)^2+3(\delta A)^2\right)$$
$$\times\left[I(I+1)-m_I(m_I\pm1)\right]\left\langle\sin^4\theta\right\rangle\frac{\tau_c}{1+\omega_e^2\tau_c^2} \quad (8.65)$$

$$W_{x2}(m_I)=\frac{1}{72}\left((\Delta A)^2+3(\delta A)^2\right)\left[I(I+1)-m_I(m_I\mp1)\right]$$
$$\times\left\langle\left(3\cos^2\theta-1\right)^2\right\rangle\frac{\tau_c}{1+\omega_e^2\tau_c^2} \quad (8.66)$$

which we generalize from the axial spin Hamiltonian given by Eq. 5.15. We can express the angular averages in terms of $\left\langle\cos^2\theta_z\right\rangle$ and $\left\langle\cos^4\theta_z\right\rangle$ (see, e.g., Eq. 8.51), and then determine cross-relaxation rates with the same three models as used for spin–lattice relaxation. The lower panel in Figure 8.12 shows the dependence of the mean cross-relaxation rate, $W_x=\frac{1}{2}\left(W_{x1}+W_{x2}\right)$, on spin-label order parameter, for magnetic field parallel to the director. The combination of angular averages appropriate to the mean rate is $\left\langle\sin^4\theta\right\rangle+\frac{1}{9}\left\langle\left(3\cos^2\theta-1\right)^2\right\rangle$. The mean cross-relaxation rate is the quantity that enters into the rate of saturation recovery (see Section 12.4). Here it is normalized to $W_x(0)$, the mean rate for $S_{zz}=0$, where for the Maier–Saupe potential we use isotropic averages $\left\langle\sin^4\theta\right\rangle=\frac{8}{15}$ and $\left\langle(3\cos^2\theta-1)^2\right\rangle=\frac{4}{5}$ in the normalization.

So far in this section, we have dealt with the special case ($\gamma=0$) where the director axis for ordering lies along the static field. In general, however, the angular averages depend on the angle γ that the magnetic field makes with the director. Therefore, we now use Eq. 8.47 to get the averages $\left\langle\cos^2\theta\right\rangle$ and $\left\langle\cos^4\theta\right\rangle$ as a function of γ:

$$\left\langle\cos^2\theta\right\rangle=\tfrac{1}{2}\left\langle\cos^2\theta_z\right\rangle\left(3\cos^2\gamma-1\right)+\tfrac{1}{2}\left(1-\cos^2\gamma\right) \quad (8.67)$$

$$\left\langle\cos^4\theta\right\rangle=\tfrac{1}{8}\left\langle\cos^4\theta_z\right\rangle\left(3-30\cos^2\gamma+35\cos^4\gamma\right)$$
$$-\tfrac{3}{4}\left\langle\cos^2\theta_z\right\rangle\left(1-6\cos^2\gamma+5\cos^4\gamma\right)+\tfrac{3}{8}\left(1-\cos^2\gamma\right)^2 \quad (8.68)$$

where $\left\langle\cos\phi\right\rangle=0$, $\left\langle\cos^2\phi\right\rangle=\frac{1}{2}$, $\left\langle\cos^3\phi\right\rangle=0$ and $\left\langle\cos^4\phi\right\rangle=\frac{3}{8}$, when averaging over 0 to 2π. By using Eq. 8.50, we then eventually get the dependence of the spin–lattice relaxation rate $W_e(m_I)$ on magnetic field orientation. This is expressed in terms of the angular averages $\left\langle\cos^2\theta_z\right\rangle$ and $\left\langle\cos^4\theta_z\right\rangle$ that we get from one of the three models used above to obtain the relaxation rate when the field lies along the director. It is clear from Eqs. 8.67 and 8.68 that the angular variation of the relaxation rate depends not only on $\cos^2\gamma$ but also on $\cos^4\gamma$. See Marsh (2018a) for further development of this aspect. We treat angular dependence of the linewidths, based on this approach, explicitly in the next section.

8.9 LINEWIDTHS AND LATERAL ORDERING

It is useful to derive linewidth coefficients for the model that includes lateral ordering, which we presented in Section 8.6. Angular averages there come from models (restricted random walk) that are simpler than the Maier–Saupe pseudopotential. We stay within the motional narrowing regime. Calculations are restricted to the leading secular contribution but include non-axial tensors. The linewidth coefficients of Eq. 8.52 then become (Israelachvili et al. 1975):

$$A=\left((\Delta g)^2j_0+2\Delta g\delta g j_1+(\delta g)^2j_2\right)\left(\frac{\beta_eB_o}{\hbar}\right)^2 \quad (8.69)$$

$$B=2\left(\Delta g\Delta A j_0+(\Delta g\delta A+\delta g\Delta A)j_1+\delta g\delta A j_2\right)\left(\frac{\beta_eB_o}{\hbar}\right) \quad (8.70)$$

$$C=(\Delta A)^2j_0+2\Delta A\delta A j_1+(\delta A)^2j_2 \quad (8.71)$$

The corresponding secular spectral densities are:

$$j_0=\left(\left\langle\cos^4\theta\right\rangle-\left\langle\cos^2\theta\right\rangle^2\right)\tau_c \quad (8.72)$$

$$j_1=\left(\left\langle\sin^2\theta\cos^2\theta\cos2\phi\right\rangle-\left\langle\cos^2\theta\right\rangle\left\langle\sin^2\theta\cos2\phi\right\rangle\right)\tau_c \quad (8.73)$$

$$j_2=\left(\left\langle\sin^4\theta\cos^22\phi\right\rangle-\left\langle\sin^2\theta\cos2\phi\right\rangle^2\right)\tau_c \quad (8.74)$$

where τ_c is the correlation time for angular reorientation. For isotropic averaging, the non-zero angular averages are: $\left\langle\cos^2\theta\right\rangle=\frac{1}{3}$, $\left\langle\cos^4\theta\right\rangle=\frac{1}{5}$ and $\left\langle\sin^4\theta\cos^22\phi\right\rangle=\frac{4}{15}$. Equations 8.69–8.74 then produce the secular contributions to the linewidth coefficients for isotropic rotational diffusion that are given by Eqs. 5.59, 5.61 and 5.62 in Chapter 5.

Again, the spectral densities in Eqs. 8.72–8.74 depend on the magnetic field orientation γ, because they are written in terms of the angle θ between field and nitroxide z-axis (see Figure 8.9). For simplicity, we give explicit results on the angular dependence only for two special cases.

Firstly for axial rotations alone, with the nitroxide z-axis oriented along the director, we have: $\theta=\gamma\ (\theta_z=0)$ and $\eta=\bar{\phi}$, which are fixed. Then only the j_2-spectral density is non-zero:

$$j_2=\sin^4\gamma\left(\left\langle\cos^22\phi\right\rangle-\left\langle\cos2\phi\right\rangle^2\right)\tau_\parallel \quad (8.75)$$

where $\tau_\parallel$ is the correlation time for axial rotation. From Eqs. 8.69 to 8.71, the transverse relaxation rate becomes: $1/T_2^{sec}=\left(\delta g\beta_eB_o/\hbar+\delta A m_I\right)^2j_2$. Performing the averages over

$\phi = \bar{\phi} \pm \phi_o$ as in Section 8.6, Eq. 8.75 becomes (Israelachvili et al. 1975):

$$j_2 = \sin^4\gamma\left[\frac{1}{2}\left(1 - \frac{\sin 4\phi_o}{4\phi_o}\right) - \cos^2\eta\frac{\sin 2\phi_o}{2\phi_o}\left(\frac{\sin 2\phi_o}{2\phi_o} - \cos 2\phi_o\right)\right]\tau_{\parallel} \tag{8.76}$$

For full axial rotation $(\phi_o = \pi)$ this gives $j_2 = \frac{1}{2}\sin^4\gamma\tau_{\parallel}$, in agreement with Eq. 8.60 of Section 8.7. Both restricted and unrestricted axial rotation, with complete order, give linewidths that depend on field orientation according to a $\sin^4\gamma$ term alone.

Secondly, for complete axial rotation but restricted random walk of the nitroxide z-axis, azimuthal averages $\langle\ldots\sin 2\phi\rangle = 0$ in Eqs. 8.73 and 8.74 vanish. The j_0 spectral density remains as in Eq. 8.72 and that for j_2 becomes:

$$j_2 = \frac{1}{2}\langle\sin^4\theta\rangle\tau_c \tag{8.77}$$

with correlation time $\tau_c \equiv \tau_\perp$ for the off-axial motion characterized by j_0, but with τ_c representing a hybrid mode in Eq. 8.77. The angular averages over θ are related to the field orientation γ and cone semi-angle θ_C by (Israelachvili et al. 1975):

$$\langle\cos^4\theta\rangle - \langle\cos^2\theta\rangle^2 = \frac{1}{45}(1-\cos\theta_C)^2\left(4 + 7\cos\theta_C + 4\cos^2\theta_C\right) + \cos\theta_C\left(1-\cos^2\theta_C\right)\left[\frac{1}{3}(1+2\cos\theta_C)\sin^2\gamma - \frac{1}{8}(3+5\cos\theta_C)\sin^4\gamma\right] \tag{8.78}$$

$$\langle\sin^4\theta\rangle = \frac{1}{15}(1-\cos\theta_C)^2\left(8 + 9\cos\theta_C + 3\cos^2\theta_C\right) + \cos\theta_C(1+\cos\theta_C)\left[\left(1-\cos^2\theta_C\right)\sin^2\gamma + \frac{1}{8}\left(7\cos^2\theta_C - 3\right)\sin^4\gamma\right] \tag{8.79}$$

We now have a dependence on field orientation that contains terms in both $\sin^2\gamma$ and $\sin^4\gamma$. In this off-axial case, the transverse relaxation rate becomes (see Eqs. 8.69–8.71):

$$\frac{1}{T_2^{sec}(m_I)} = \left(\Delta g\beta_e B_o/\hbar + \Delta A m_I\right)^2 j_0 + \left(\delta g\beta_e B_o/\hbar + \delta A m_I\right)^2 j_2 \tag{8.80}$$

and depends on both γ and θ_C.

Figure 8.13 shows high-field EPR lineshapes simulated according to the fast-motion model for lateral ordering. Line positions are given by Eqs. 8.39–8.41 of Section 8.6 for the partially averaged g-tensor, and equivalents for the hyperfine tensor; and linewidths are determined as described in this section. The left-hand panel gives spectra for increasing amplitudes ϕ_o of rotation about a fixed nitroxide z-axis. Anisotropy between the g_{xx}- and g_{yy}-regions of the spectrum decreases progressively, whereas the g_{zz}-region remains unchanged. For full axial rotation $(\phi_o = 90°)$, the g_{xx}- and g_{yy}-features become completely averaged to a single triplet with ^{14}N-hyperfine splitting $\langle A_\perp\rangle = \frac{1}{2}(A_{xx} + A_{yy})$ centred at the position specified by the mean g-value: $\langle g_\perp\rangle = \frac{1}{2}(g_{xx} + g_{yy})$. Linewidths depend on both amplitude and rate of rotation. The right-hand panel of Figure 8.13 shows what happens when the off-axis rotational amplitude, θ_C, increases simultaneously with the azimuthal amplitude, ϕ_o. In this case, the $(g_{xx} - g_{yy})$ and $g_{zz} - \frac{1}{2}(g_{xx} + g_{yy})$ anisotropies decrease in parallel, with increasing rotational amplitude. The high-field features move to lower field as the g_{xx}-region moves towards g_{yy}.

Lineshapes of the type shown in the right-hand panel of Figure 8.13 correspond to those found frequently for coupled segmental rotation in chain-labelled lipids. We show simulations of such high-field EPR powder spectra using the present model

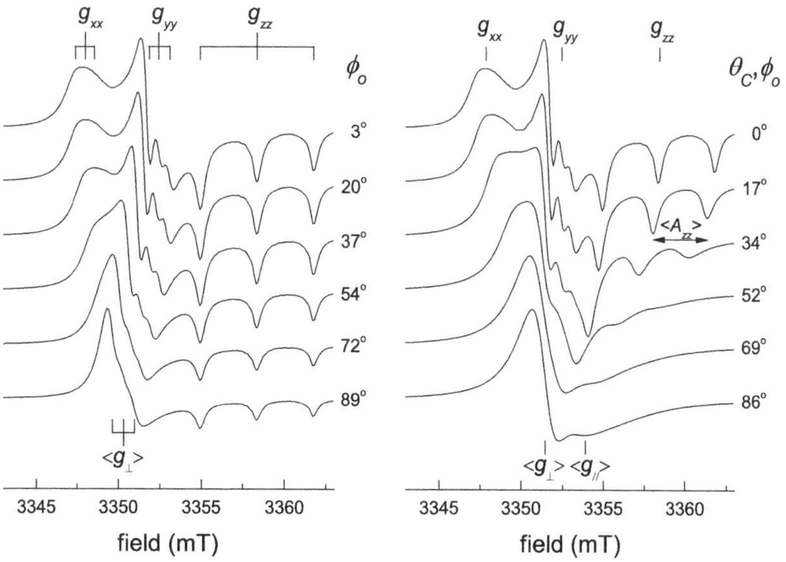

FIGURE 8.13 94-GHz spin-label EPR spectra simulated using motional narrowing theory (Sections 8.6, 8.9), for lateral ordering with different azimuthal rotational amplitudes ϕ_o about the nitroxide z-axis, and fixed correlation time $\tau_c = 1.7$ ns. *Left-hand panel*: without off-axis rotation (i.e., $\theta_C = 0$; $\tau_c = \tau_{\parallel}$); *right-hand panel*: with simultaneous increase in off-axis amplitude (i.e., $\theta_C = \phi_o$; $\tau_\perp = \tau_{\parallel}$). (Data from Livshits and Marsh 2000.)

later in Figure 8.26, which illustrates timescale selection in multifrequency EPR. The experimental 94-GHz spectra shown there are of *n*-DOXYL lipid spin labels in phosphatidylcholine–cholesterol membranes; these yield the lateral and transverse order parameters given already in Figure 8.10.

8.10 ANGULAR-DEPENDENT LINEWIDTHS AND ORDERING: STRONG JUMP MODEL

Now we treat relaxation, still in the fast-motion regime, by using the irreducible spherical tensor formalism (see Appendix S). This method is used routinely for systems with molecular ordering, because transformations with Wigner rotation matrices (Eq. S.1) then become straightforward. Throughout, we assume that the system has complete axial symmetry about the director (i.e., $S_{xx} = S_{yy}$), and that the spin-labelled molecule undergoes rotational diffusion about its long axis (i.e., is also axially symmetric). Also, we assume that the system has reflection symmetry about a plane perpendicular to the director, as in membranes and liquid crystals. As we shall see, these important symmetries limit the number of terms that we must consider. We continue to assume that one of the nitroxide axes (for convenience we take the *z*-axis) coincides with the principal rotation axis (e.g., the long axis) of the spin-labelled molecule. If this is not the case, we need an additional transformation between the nitroxide axes and those of the molecular diffusion tensor. We treat the latter situation later, in Section 8.14.

A word of caution is necessary at the outset. As noted in Appendix S, we use a standard rotational transformation (Eq. S.1) which is not always that adopted in the original EPR literature. In most cases where there are differences, these amount to interchange of indices K and M in $\mathcal{D}^2_{K,M}(\Omega)$.

For simplicity, we begin with the magnetic field oriented along the director axis (cf. Section 8.7). The anisotropic part of the nitroxide spin Hamiltonian is then (see Eq. S.13):

$$\mathcal{H}_1(\Omega) = \frac{2}{3}\left(\Delta g\beta_e B_o + \Delta A I_z\right)\mathcal{D}^2_{0,0}S_z + \sqrt{\frac{2}{3}}\left(\delta g\beta_e B_o + \delta A I_z\right) \times\left(\mathcal{D}^2_{0,2} + \mathcal{D}^2_{0,-2}\right)S_z - \frac{\Delta A}{\sqrt{6}}\left(\mathcal{D}^2_{1,0}I_+ - \mathcal{D}^2_{-1,0}I_-\right)S_z - \frac{\delta A}{2}\left(\mathcal{D}^2_{1,2} + \mathcal{D}^2_{1,-2}\right)I_+S_z + \frac{\delta A}{2}\left(\mathcal{D}^2_{-1,2} + \mathcal{D}^2_{-1,-2}\right)I_-S_z \tag{8.81}$$

where we represent irreducible spherical tensors explicitly in terms of spin-Hamiltonian anisotropies and spin operators, and omit non-secular terms. The whole of the angular dependence is contained in the elements $\mathcal{D}^2_{K,M}(\Omega)$ of the second-rank Wigner rotation matrix (see Appendix R.3). The rotation matrix relates tensors in the nitroxide axis system to those in the director system, along which the static field is aligned. Relaxation is induced by fluctuations in orientation $\Omega(t) \equiv (\phi\theta_z\psi)$ of the nitroxide axes relative to the director (see Figure 8.14). The time-dependent perturbation $\mathcal{H}_1(t)$ is therefore determined by terms such as $\mathcal{D}^2_{K,M}(t) - \langle\mathcal{D}^2_{0,M}\rangle$ (see Eq. 8.48), where $\langle\mathcal{D}^2_{0,M}\rangle$ is the time (or angular) average of the $\mathcal{D}^2_{K,M}$ matrix element over the restricted angular motion. Putting $K = 0$ in the average explicitly indicates complete axial averaging over azimuthal angle ϕ about the director (cf. Eq. R.9). Only the $K = 0$ terms remain after this averaging. Similarly, if diffusion around the molecular long axis (colinear with the nitroxide *z*-axis) is axially symmetric, then averaging over azimuthal angle ψ leaves only the $M = 0$ matrix element non-zero (cf. Figure 8.14). Table 8.1 gives the relation between averages over Wigner rotation matrices and the Saupe order matrix of direction cosines, S_{ij} (Saupe 1964).

We evaluate matrix elements of the spin operators in Eq. 8.81 to determine the secular terms and pseudosecular transition probabilities for transverse relaxation, as in Section 5.8 of Chapter 5. To preserve specific angular averaging, we work with spectral densities $j_{KM}(\omega) = \left\langle\left|\mathcal{D}^2_{K,M}\right|^2\right\rangle \tau_{2,M}/\left(1+\omega^2\tau_{2,M}{}^2\right)$ for the second-rank Wigner rotation matrices $\mathcal{D}^2_{K,M}$, instead of the composite spectral densities used in Chapter 5. From Eqs. 5.50, 5.52 and 5.54, the linewidth coefficients of Eq. 8.52 then become (cf. Nordio et al. 1972; Hemminga 1974):

$$A = \frac{4}{9}\left(\frac{\beta_e B_o}{\hbar}\right)^2\left((\Delta g)^2 j_{00} + 3(\delta g)^2 j_{02}\right) + \frac{1}{6}I(I+1)\left((\Delta A)^2 j_{10} + 3(\delta A)^2 j_{12}\right) \tag{8.82}$$

$$B = \frac{8}{9}\left(\frac{\beta_e B_o}{\hbar}\right)\left(\Delta g\Delta A\cdot j_{00} + 3\delta g\delta A\cdot j_{02}\right) \tag{8.83}$$

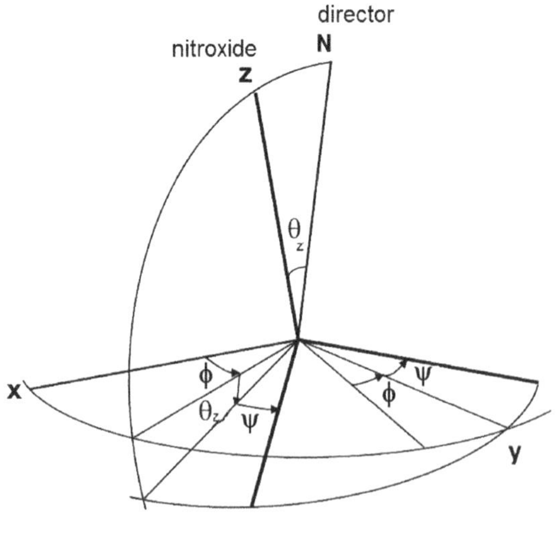

FIGURE 8.14 Euler angles $\Omega \equiv (\phi\theta_z\psi)$ relating director **N** and nitroxide-*x,y,z* axis systems (see Appendix R.2). Axial symmetry about the director refers to azimuthal angle ϕ. Rotational symmetry about nitroxide *z*-axis from molecular diffusion refers to azimuthal angle ψ.

TABLE 8.1 Averages over Wigner rotation matrices $\langle \mathcal{D}^2_{0,M}(\Omega) \rangle$, and corresponding elements of order-parameter matrix S_{ij}

$\langle \mathcal{D}^2_{0,M}(\Omega) \rangle$[a]	S_{ij} $(i,j = x, y, z)$[b]
$\langle \mathcal{D}^2_{0,0} \rangle$	S_{zz}
$\langle \mathcal{D}^2_{0,2} \rangle + \langle \mathcal{D}^2_{0,-2} \rangle$	$\sqrt{2/3}\left(S_{xx} - S_{yy}\right)$
$\langle \mathcal{D}^2_{0,2} \rangle - \langle \mathcal{D}^2_{0,-2} \rangle$	$-i\sqrt{8/3}S_{xy}$
$\langle \mathcal{D}^2_{0,1} \rangle + \langle \mathcal{D}^2_{0,-1} \rangle$	$-i\sqrt{8/3}S_{yz}$
$\langle \mathcal{D}^2_{0,1} \rangle - \langle \mathcal{D}^2_{0,-1} \rangle$	$-\sqrt{8/3}S_{xz}$

[a] Wigner rotation matrix averaged over Euler angles $\Omega = (\phi, \theta, \psi)$ with complete averaging over ϕ but partial averaging over θ, ψ. E.g., $\langle \mathcal{D}^2_{0,2}(0,\theta,\psi) \rangle + \langle \mathcal{D}^2_{0,-2}(0,\theta,\psi) \rangle = \sqrt{3/2}\langle \sin^2\theta \cos 2\psi \rangle$.

[b] Saupe (1964) order matrix.

$$C = \tfrac{4}{9}\left((\Delta A)^2 j_{00} + 3(\delta A)^2 j_{02}\right) - \tfrac{1}{6}\left((\Delta A)^2 j_{10} + 3(\delta A)^2 j_{12}\right) \quad (8.84)$$

where we use the contracted form $j_{KM} \equiv \frac{1}{2}\left(j_{K,+M} + j_{K,-M}\right)$ or $\frac{1}{2}\left(j_{+K,M} + j_{-K,M}\right)$. Note that $j_{K,M} = j_{K,-M} = j_{-K,-M}$, which comes from the symmetry properties of Clebsch–Gordan coefficients (see Appendix Q.2). Secular terms are those with $K = 0$, and pseudosecular terms are those with $K = 1$. For isotropic averaging: $\left\langle \left| \mathcal{D}^2_{K,M} \right|^2 \right\rangle = \frac{1}{5}$ in each case, and Eqs. 8.82–8.84 reduce to Eqs. 5.59, 5.61 and 5.62 for isotropic rotational diffusion (assuming axial tensors and omitting nonsecular terms). All spectral densities in Eqs. 8.82–8.84 become those for zero frequency, $j_{KM} \equiv j_{KM}(0)$, if we simplify the pseudosecular terms ($K = 1$) by putting $\omega_a = 0$. This restricts us to correlation times ≤ 1 ns, when angular amplitudes are large (see Section 5.8).

Now we calculate the $j_{KM}(0)$ spectral densities, neglecting non-secular terms. The autocorrelation functions of Wigner rotation matrix elements are defined by:

$$g_{KM}(t) = \left\langle \left(\mathcal{D}^2_{K,M}(0) - \delta_{K,0}\langle \mathcal{D}^2_{0,M} \rangle \right)\left(\mathcal{D}^2_{K,M}(t) - \delta_{K,0}\langle \mathcal{D}^2_{0,M} \rangle \right)^* \right\rangle$$

$$= \left\langle \mathcal{D}^2_{K,M}(0)\mathcal{D}^{2*}_{K,M}(t) \right\rangle - \delta_{K,0}\left\langle \mathcal{D}^2_{0,M} \right\rangle^2 \quad (8.85)$$

Because it is simplest, we use the strong-jump model, where orientations Ω_o and Ω before and after collision, respectively, are uncorrelated (cf. Section 6.4). For a Poisson process, the probability of collision within time t since the last collision is $1 - e^{-t/\tau}$ (and that of no collision is $e^{-t/\tau}$), where τ is the mean time between collisions. The conditional probability therefore becomes (Luckhurst 1972; cf. also Eq. 13.94 in Section 13.14):

$$P(\Omega_o; \Omega, t) = \delta(\Omega - \Omega_o)e^{-t/\tau} + P_o(\Omega)\left(1 - e^{-t/\tau}\right) \quad (8.86)$$

where $P_o(\Omega)$ is the equilibrium orientational distribution. This leads to the exponential correlation function (see Eq. 5.11 in Section 5.2):

$$g_{KM}(t) = \left(\left\langle \mathcal{D}^2_{K,M}(\Omega)\mathcal{D}^{2*}_{K,M}(\Omega) \right\rangle - \delta_{K,0}\delta_{M,0}\left\langle \mathcal{D}^2_{0,0} \right\rangle^2 \right) e^{-t/\tau} \quad (8.87)$$

where $\mathcal{D}^2_{0,0}(\phi\theta_z\psi) \equiv d^2_{0,0}(\theta_z) = \mathrm{P}_2(\cos\theta_z)$ is a second-order Legendre polynomial (see Eq. R.15). Also, $\mathcal{D}^2_{K,M}(\Omega)\mathcal{D}^{2*}_{K,M}(\Omega) = \left| \mathcal{D}^2_{K,M}(\Omega) \right|^2 \equiv \left| d^2_{K,M}(\theta_z) \right|^2$, which lets us write everything in terms of reduced Wigner rotation matrices $d^2_{K,M}$. We need only the autocorrelation function g_{KM}, because axial symmetry about the director and molecular axis causes cross correlations to vanish. Note that random jumps produce axial symmetry about the molecular long axis (nitroxide z-axis). This results in $M = 0$ when $K = 0$. Integrating Eq. 8.87 over t gives the following zero-frequency spectral densities:

$$j_{KM}(0) = \left(\left\langle \left| d^2_{K,M} \right|^2 \right\rangle - \delta_{K,0}\delta_{M,0}\left\langle d^2_{0,0} \right\rangle^2 \right) \tau_M \quad (8.88)$$

where we give the correlation time τ_M an M-dependence to account empirically for rotational anisotropy about the molecular long axis (Luckhurst and Sanson 1972). Equating pseudosecular spectral densities to those at zero frequency when using Eq. 8.88 for both terms in Eqs. 8.82 and 8.84 assumes that $\omega_a^2 \tau_R^2 \ll 1$ (see Section 8.7), as already mentioned.

We can expand the first average in Eq. 8.88 as a sum of single rotation matrices by using the Clebsch–Gordan series (see Eq. R.17 in Appendix R.3):

$$\left| d^2_{K,M} \right|^2 = (-1)^{K-M} d^2_{K,M} d^2_{-K,-M}$$

$$= (-1)^{K-M} \sum_{L=0,2,4} \langle 22L; K, -K \rangle \langle 22L; M, -M \rangle d^L_{0,0} \quad (8.89)$$

where the first equality exploits symmetry relations for $d^2_{K,M}$ (Eq. R.10), and $\langle L_1 L_2 L; M_1 M_2 \rangle \equiv C(L_1 L_2 L; M_1 M_2)$ is a Clebsch–Gordan coefficient (see Section Q.2 in Appendix Q). Values of L are even because of reflection symmetry through a plane perpendicular to the uniaxial director (Luckhurst and Sanson 1972). Again from Eq. R.15: $d^L_{0,0}(\theta_z) \equiv \mathrm{P}_L(\cos\theta_z)$. The zero-frequency spectral densities then become:

$$j_{KM}(0) = (-1)^{K-M} \sum_{L=0,2,4} \left(\langle 22L; K, -K \rangle \langle 22L; M, -M \rangle \langle \mathrm{P}_L(\cos\theta_z) \rangle \right.$$

$$\left. -\delta_{K,0}\delta_{M,0}\langle \mathrm{P}_2(\cos\theta_z) \rangle^2 \right) \tau_M \quad (8.90)$$

We list numerical values of $\langle 22L; M_2, -M_2 \rangle$ under $M = 0$ in Table A8.2 of Appendix A8.1 at the end of the chapter.

When the magnetic field does not lie along the director, we must transform the spectral density in Eq. 8.88 by using the Wigner rotation matrix with elements $\mathcal{D}^2_{N,K}(\Psi)$, where $\Psi \equiv (0\gamma\alpha)$ are the static polar angles relating the magnetic field direction to that of the director (see Figure 8.9). Axial symmetry about the director renders the azimuthal α-angle redundant. Thus we use simply the reduced Wigner matrix elements $d^2_{N,K}(\gamma) \equiv \mathcal{D}^2_{N,K}(0\gamma 0)$, as listed in Table R.1 of Appendix R, for the transformation. Using the closure rule for matrix multiplication (see Eq. R.24 in Appendix R.3), we replace each rotation matrix element $d^2_{K,M}(\theta_z)$ in Eq. 8.88 by the expansion:

$$d^2_{N,M}(\gamma\theta_z) = \sum_K d^2_{N,K}(\gamma) d^2_{K,M}(\theta_z) \qquad (8.91)$$

The zero-frequency, angular-dependent spectral densities for the Wigner rotation matrices then become:

$$j_{NM}(0,\gamma) = \sum_{K=-2}^{+2} \left| d^2_{N,K}(\gamma) \right|^2 j_{KM}(0) \qquad (8.92)$$

where γ is the angle between the static magnetic field and the director. Note that the $d^2_{N,K}$ matrix elements are real. We evaluate the product of rotation matrices again by using the Clebsch–Gordan series (see Eq. 8.89):

$$\left| d^2_{N,K}(\gamma) \right|^2 = (-1)^{N-K} \sum_{L'=0,2,4} \langle 22L';N,-N \rangle \langle 22L';K,-K \rangle d^{L'}_{0,0}(\gamma) \qquad (8.93)$$

where the Clebsch–Gordan coefficients $\langle 22L';M_2,-M_2 \rangle$ are listed under $M = 0$ in Table A8.2. Again using the identity: $d^{L'}_{0,0}(\gamma) \equiv P_{L'}(\cos\gamma)$, the angular dependence consists of a series of Legendre polynomials of order $L' = 0,2,4$. From Eq. 8.93, we see that the angular dependence of the linewidth coefficients given in Eqs. 8.82–8.84 has the general form:

$$A = A_0 + A_2 P_2(\cos\gamma) + A_4 P_4(\cos\gamma) \qquad (8.94)$$

and similarly for B and C. The Legendre polynomials are defined by: $P_2(\cos\gamma) = \frac{1}{2}(3\cos^2\gamma - 1)$ and $P_4(\cos\gamma) = \frac{1}{8}(35\cos^4\gamma - 30\cos^2\gamma + 3)$ (see Appendix V.1). For static magnetic field parallel to the director ($\gamma = 0$), the B-coefficient for example is: $B = B_0 + B_2 + B_4$, and for field perpendicular to the director ($\gamma = 90°$): $B = B_0 - \frac{1}{2}B_2 + \frac{3}{8}B_4$. Note that the explicit cone model treated in Section 8.9 agrees fully with Eq. 8.94, because the angular dependence of the linewidths is determined there by terms in both $\sin^2\gamma$ and $\sin^4\gamma$.

Figure 8.15 shows the angular dependence of experimental linewidth coefficients A, B and C, for a DOXYL-nitroxide spin probe in an aligned smectic A phase of the parent mesogen, 4-*n*-butyl-4′-acetylazobenzene (Luckhurst et al. 1987). For all three coefficients, the dependence is described well by Eq. 8.94. As we shall see later, this is not always so when the order is higher. Coefficient A is largest, in part because it contains a sizeable contribution from m_I-independent inhomogeneous broadening. This makes A less suitable for analysis than the other two coefficients, unless inhomogenous broadening is corrected for (see Section 2.5 in Chapter 2). Both A and C coefficients are positive, and B is negative as usual. Differential broadening of the outer EPR lines is least when the magnetic field lies either parallel or perpendicular to the director of this smectic mesophase.

We see from Eqs. 8.90, 8.92 and 8.93 that the angular-dependent spectral densities depend on sums over products of four Clebsch–Gordan coefficients, involving both L and L'. From the orthogonality of Clebsch–Gordan coefficients (Eq. Q.6 in Appendix Q), the sum over K collapses the products of the corresponding pair of coefficients to $\delta_{LL'}$. The angular-dependent, zero-frequency spectral densities then become (Luckhurst and Sanson 1972):

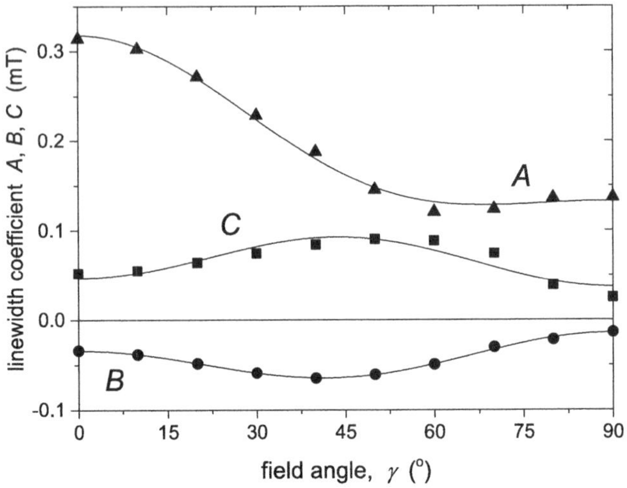

FIGURE 8.15 Dependence of 9-GHz linewidth coefficients A (triangles), B (circles) and C (squares) on angle γ of static magnetic field to the director. Spin probe 4-butoxy-4′-ethylDOXYLazobenzene in aligned smectic A phase of 4-butoxy-4′-acetylazobenzene at 104.6°C. Solid lines: least-squares fits of Eq. 8.94 and equivalents. (Data points from Luckhurst et al. 1987.)

$$j_{NM}(0,\gamma) \equiv \sum_{L=0,2,4} P_L(\cos\gamma) j^L_{NM} \qquad (8.95)$$

where

$$j^L_{NM} \equiv (-1)^{N-M} \langle 22L;N,-N \rangle \langle 22L;M,-M \rangle \times \left(\langle P_L(\cos\theta_z) \rangle - \langle P_2(\cos\theta_z) \rangle^2 \right) \tau_M \qquad (8.96)$$

From this, we see that the angular averages $\langle P_L(\cos\theta_z) \rangle$ (or order parameters) have the same values of L as those that determine the dependence on magnetic field orientation γ. The same holds for the three A_L contributions to the linewidth coefficient A in Eq. 8.94. Substituting Clebsch–Gordan coefficients from Table A8.2 in Eq. 8.96 we get:

$$j^0_{00} = j^0_{10} = \tfrac{1}{5}\left(1 - \langle P_2 \rangle^2\right)\tau_0 \qquad (8.97)$$

$$j^2_{00} = 2 j^2_{10} = \tfrac{2}{7}\left(\langle P_2 \rangle - \langle P_2 \rangle^2\right)\tau_0 \qquad (8.98)$$

$$j^4_{00} = -\tfrac{3}{2} j^4_{10} = \tfrac{18}{35}\left(\langle P_4 \rangle - \langle P_2 \rangle^2\right)\tau_0 \qquad (8.99)$$

$$j^0_{02} = j^0_{12} = \tfrac{1}{5}\tau_2 \qquad (8.100)$$

$$j^2_{02} = 2 j^2_{12} = -\tfrac{2}{7}\langle P_2 \rangle \tau_2 \qquad (8.101)$$

$$j_{02}^4 = -\tfrac{3}{2} j_{12}^4 = \tfrac{3}{35}\langle P_4 \rangle \tau_2 \quad (8.102)$$

where $\langle P_{2n} \rangle \equiv \langle \mathrm{P}_{2n}(\cos\theta_z) \rangle$ and τ_0, τ_2 are the correlation times for $M = 0$ and $M = 2$ in Eq. 8.88. Using Eqs. 8.82–8.84, the angular-dependent linewidth coefficients in Eq. 8.94 and equivalents then become (Luckhurst et al. 1974):

$$A_0 = \tfrac{4}{45}\left[(\Delta g)^2\left(1-\langle P_2 \rangle^2\right)\tau_0 + 3(\delta g)^2 \tau_2\right]\left(\frac{\beta_e B_o}{\hbar}\right)^2 + \tfrac{1}{30} I(I+1)\left[(\Delta A)^2\left(1-\langle P_2 \rangle^2\right)\tau_0 + 3(\delta A)^2 \tau_2\right] \quad (8.103)$$

$$A_2 = \tfrac{8}{63}\langle P_2 \rangle\left[(\Delta g)^2\left(1-\langle P_2 \rangle\right)\tau_0 - 3(\delta g)^2 \tau_2\right]\left(\frac{\beta_e B_o}{\hbar}\right)^2 + \tfrac{1}{42} I(I+1)\langle P_2 \rangle\left[(\Delta A)^2\left(1-\langle P_2 \rangle\right)\tau_0 - 3(\delta A)^2 \tau_2\right] \quad (8.104)$$

$$A_4 = \tfrac{4}{35}\left[2(\Delta g)^2\left(\langle P_4 \rangle-\langle P_2 \rangle^2\right)\tau_0 + (\delta g)^2 \langle P_4 \rangle \tau_2\right]\left(\frac{\beta_e B_o}{\hbar}\right)^2 - \tfrac{1}{35} I(I+1)\left[2(\Delta A)^2\left(\langle P_4 \rangle-\langle P_2 \rangle^2\right)\tau_0 + (\delta A)^2 \langle P_4 \rangle \tau_2\right] \quad (8.105)$$

$$B_0 = \tfrac{8}{45}\left[\Delta g \Delta A\left(1-\langle P_2 \rangle^2\right)\tau_0 + 3\delta g \delta A \tau_2\right]\left(\frac{\beta_e B_o}{\hbar}\right) \quad (8.106)$$

$$B_2 = \tfrac{16}{63}\langle P_2 \rangle\left[\Delta g \Delta A\left(1-\langle P_2 \rangle\right)\tau_0 - 3\delta g \delta A \tau_2\right]\left(\frac{\beta_e B_o}{\hbar}\right) \quad (8.107)$$

$$B_4 = \tfrac{8}{35}\left[2\Delta g \Delta A\left(\langle P_4 \rangle-\langle P_2 \rangle^2\right)\tau_0 + \delta g \delta A \langle P_4 \rangle \tau_2\right]\left(\frac{\beta_e B_o}{\hbar}\right) \quad (8.108)$$

$$C_0 = \tfrac{1}{18}\left[(\Delta A)^2\left(1-\langle P_2 \rangle^2\right)\tau_0 + 3(\delta A)^2 \tau_2\right] \quad (8.109)$$

$$C_2 = \tfrac{13}{126}\langle P_2 \rangle\left[(\Delta A)^2\left(1-\langle P_2 \rangle\right)\tau_0 - 3(\delta A)^2 \tau_2\right] \quad (8.110)$$

$$C_4 = \tfrac{1}{7}\left[2(\Delta A)^2\left(\langle P_4 \rangle-\langle P_2 \rangle^2\right)\tau_0 + (\delta A)^2 \langle P_4 \rangle \tau_2\right] \quad (8.111)$$

These results were derived first by Luckhurst and Sanson (1972), but their expressions for the A-coefficients are in error.

Correlation time τ_0 corresponds to fluctuations of the long molecular axis (nitroxide z-axis), and τ_2 is associated with rotation about this axis. If rotation around the nitroxide z-axis is much faster than rotation of the z-axis, then $\tau_2 \approx 0$. On the other hand, for segmental motion or restricted random walk within a cone, we expect that $\tau_2 \approx \tau_0$. The latter result was found experimentally for spin-labelled fatty acid chains in a lipid bilayer (Schindler and Seelig 1973). For axial tensors: $\delta g = 0 = \delta A$, and the terms in τ_2 disappear from Eqs. 8.103 to 8.111. For isotropic rotation, $\langle \mathrm{P}_2 \rangle = \langle \mathrm{P}_4 \rangle = 0$ and $\tau_0 = \tau_2$. Consequently, only the angle-independent terms A_0, B_0 and C_0 then are non-zero. Correspondingly, Eqs. 8.103, 8.106 and 8.109 reduce to Eqs. 5.59, 5.61 and 5.62 for isotropic rotation in Chapter 5, when non-secular spectral densities are neglected and $\omega_a^2 \tau_R^2 << 1$.

Figure 8.16 shows the dependence of linewidth coefficients B (solid lines) and C (dashed lines) on angle γ between static magnetic field and director that we predict for axial spin-Hamiltonian tensors. The angular-dependent part of the linewidth coefficients, $B - B_o$ and $C - C_o$ (see Eq. 8.94), is plotted from Eqs. 8.107, 8.108 and 8.110, 8.111, respectively. Values are normalized to the linewidth coefficients B_{iso} and C_{iso} for isotropic rotational diffusion that we get from Eqs. 8.106 and 8.109 (with $\langle \mathrm{P}_2 \rangle = 0$). We show results for three different extents of molecular order, characterized by the order parameter $S_{zz} = \langle \mathrm{P}_2(\cos\theta_z) \rangle$ that appears in Eqs. 8.106–8.111. The values of $\langle \mathrm{P}_4(\cos\theta_z) \rangle$ for Eqs. 8.108 and 8.111 are derived from $\langle \mathrm{P}_2(\cos\theta_z) \rangle$ by using Eq. 8.26 for the Maier–Saupe orientation potential (see Figure 8.7). Conversely, we must determine the dependence of the linewidths on magnetic field orientation γ as in Figure 8.15, if we want to measure the $\langle \mathrm{P}_4(\cos\theta_z) \rangle$ order parameter experimentally.

The Maier–Saupe potential, together with the strong-jump model, was used successfully by Schindler and Seelig (1973) to simulate EPR lineshapes of spin-labelled fatty acid chains in a soap bilayer membrane. They relaxed the restriction on axial symmetry of the ordering by allowing $S_{xx} \neq S_{yy}$. As mentioned already in Section 8.5, short-chain soap systems form highly fluid bilayers. We can treat these adequately with the motional narrowing theory used in this and previous sections. A correlation time of $\tau_0 = \tau_2 = 0.23$ ns is found for a C-4 labelled fatty acid in sodium decanoate-decanol bilayers at 21°C (Schindler and Seelig 1973). Longer, two-chain lipids form membranes where important contributions to rotational motion fall in a regime that is slow on the nitroxide EPR timescale. We must treat such

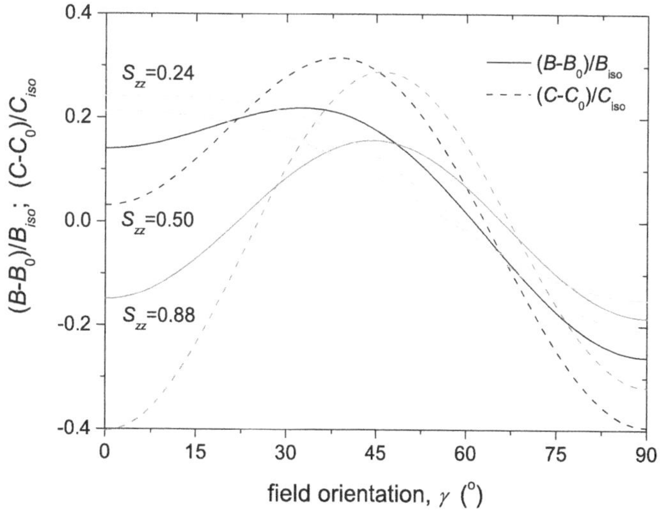

FIGURE 8.16 Angular dependence of linewidth coefficients B (solid lines) and C (dashed lines) – cf. Eq. 8.94, where γ is orientation of static magnetic field to director, for strong-jump model (Eqs. 8.106–8.111). Averages over motional angle θ_z use the Maier–Saupe potential (see Fig. 8.7). Linewidth coefficients are shown for order parameters: $S_{zz} \equiv \langle \mathrm{P}_2(\cos\theta_z) \rangle$ = 0.243 (light grey), 0.496 (black), and 0.884 (grey).

spectra by using the stochastic Liouville equation introduced in Section 6.7 of Chapter 6. This is done for ordered systems later in the chapter – see Section 8.12.

Figure 8.17 shows the angular dependence of the linewidth coefficients for three different nitroxide lipid probes in the same bilayer soap system. Two probes have flexible chains, with different positions of labelling, and the third is a rigid steroid. The theory embodied by Eq. 8.94 does not describe the angular dependence accurately for the probes with high order, such as 5-DOXYL stearic acid and 3-DOXYL cholestane. This is because, as we saw in Section 2.2 of Chapter 2, the nitrogen nuclear spin is not quantized along the magnetic field (see Figure 2.3 and Eqs. 2.6, 2.7), contrary to what we assume in this and previous sections of the present chapter. With proper allowance for the nuclear quantization axis (Luckhurst and Zannoni 1977), the fit to the data given in Figure 8.17 becomes rather good (Luckhurst et al. 1979). This is highly significant because introducing the correct nuclear quantization axis does not involve any additional adjustable parameters. Table 8.2 gives values of the dynamic parameters for the three nitroxide probes. We get both the fourth-rank order parameter $\langle P_4(\cos\theta_z)\rangle$ and the correlation times for strong jumps τ_0 and τ_2 by this procedure. In spite of the better fit, values of the order parameter and correlation times do not change greatly after allowing for the different axis of nuclear quantization. Also, agreement between parameters deduced from independent fits to the B- and C-coefficients is quite good.

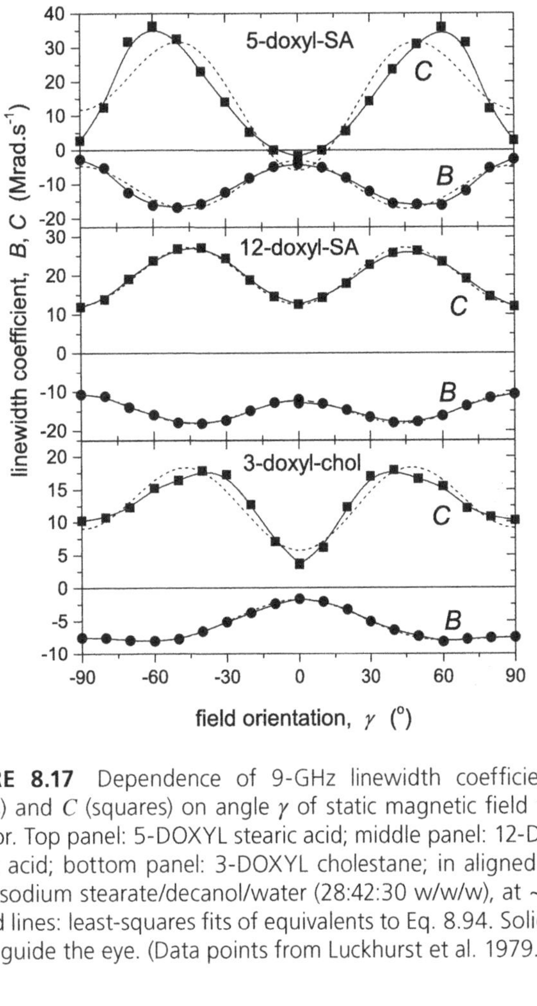

FIGURE 8.17 Dependence of 9-GHz linewidth coefficients B (circles) and C (squares) on angle γ of static magnetic field to the director. Top panel: 5-DOXYL stearic acid; middle panel: 12-DOXYL stearic acid; bottom panel: 3-DOXYL cholestane; in aligned bilayers of sodium stearate/decanol/water (28:42:30 w/w/w), at ~22°C. Dotted lines: least-squares fits of equivalents to Eq. 8.94. Solid lines are to guide the eye. (Data points from Luckhurst et al. 1979.)

Luckhurst and Zannoni (1977) show how to incorporate the correct nitrogen nuclear quantization axis into the strong-jump model. We describe this procedure in Appendix A8.2 at the end of the chapter. Other cases of high order, where this correction cannot be ignored, are afforded by 3-DOXYL cholestane in thiourea inclusions and in the smectic A phase of 4,4′-di-n-heptylazoxybenzene (Luckhurst et al. 1981). Note that, when order is high, motional narrowing obtains up to longer correlation times, because the spectral anisotropy to be averaged is then reduced (see Section 6.3).

8.11 LINEWIDTHS AND ORDERING: BROWNIAN DIFFUSION

Because correlation times may change with the degree of order, relaxation and linewidths depend on the model that we take to describe the restricted motion, even when using motional narrowing theory. This is not true for isotropic rotational diffusion

TABLE 8.2 Order parameter $\langle P_4\rangle$, and strong-jump correlation times τ_0 and τ_2, deduced from angular dependence of linewidth parameters B and C for nitroxide spin-probes in aligned bilayers of 28 wt% sodium decanoate, 42 wt% decanol and 30 wt% water at ~22°C (Luckhurst et al. 1979)

	$\langle P_4\rangle \equiv \langle P_4(\cos\theta_z)\rangle$		τ_0(ns)		τ_2(ns)	
nitroxide[a]	*from B*	*from C*	*from B*	*from C*	*from B*	*from C*
5-DOXYL SA	0.07	0.13	1.45	2.45		
	(0.07)[b]	(0.14)	(1.55)	(2.7)		
12-DOXYL SA	−0.10	−0.08	1.30	1.95		
	(−0.09)[b]	(−0.07)	(1.35)	(2.00)		
3-DOXYL cholestane	0.39	0.48	7.3	9.4	0.41	0.51
	(0.39)[b]	(0.46)	(6.8)	(8.8)	(0.43)	(0.67)

[a] n-DOXYL SA: n-(3′-oxyl-4′,4′-dimethyloxazolidin-2′-yl)-stearic acid; 3-DOXYL cholestane, 3-(3′-oxyl-4′,4′-dimethyloxazolidin-2′-yl)-5α-cholestane.
[b] Values without parentheses come from fitting B,C-versions of Eq. 8.94, together with Eqs. 8.106–8.111. Values in parentheses are calculated from fits (cf. Figure 8.17) that allow for the nitrogen nuclear spin not being quantized along the magnetic field (Luckhurst and Zannoni 1977).

(see Chapter 5); the complication applies only to ordered systems. In the previous section, we used the simplest motional model: that of strong uncorrelated jumps. Brownian rotational diffusion is less straightforward to handle, and in general does not yield simple exponential correlation functions. Solution to the problem was given by Nordio and coworkers (Nordio and Busolin 1971; Nordio et al. 1972) for nematic liquid crystals, and has been applied to a spin-labelled steroid in membranes by Hemminga (1974).

To treat Brownian diffusion in the presence of molecular ordering, we must introduce an orientation potential $U(\theta)$ into the rotational diffusion equation for the orientational probability $P(\Omega,t)$. We do this by twice differentiating the Boltzmann equation for the orientational distribution, Eq. 8.22, which gives:

$$D_{R\perp}\frac{\partial^2 p(\theta)}{\partial\theta^2}=-\frac{D_{R\perp}}{k_BT}\frac{\partial}{\partial\theta}\left(p(\theta)\frac{\partial U}{\partial\theta}\right)\tag{8.112}$$

where $D_{R\perp}$ is the diffusion coefficient for rotation of the molecular long-axis. Because the distribution is axially symmetric, we must allow for the $\sin\theta$-weighting of $p(\theta)$ in the general 3-dimensional case (see Section 2.4). Then adding the contribution $D_{R\perp}\partial^2P/\partial\theta^2$ from Eq. 8.112 to the corresponding term on the right-hand side of the diffusion equation, Eq. 7.26 from Section 7.6, we get (cf. Nordio et al. 1972):

$$\frac{\partial P}{\partial t}=D_{R\perp}\nabla^2_{diff}P-\frac{D_{R\perp}}{k_BT}\frac{1}{\sin\theta}\frac{\partial}{\partial\theta}\left(P\sin\theta\frac{\partial U}{\partial\theta}\right)\tag{8.113}$$

In terms of the Euler angles (ϕ,θ,ψ) of Figure 8.14, the operator $\nabla_{diff}{}^2$ for axial diffusion is given by (see Eq. 7.26):

$$\nabla^2_{diff}=\frac{1}{\sin\theta}\frac{\partial}{\partial\theta}\left(\sin\theta\frac{\partial}{\partial\theta}\right)+\frac{1}{\sin^2\theta}\frac{\partial^2}{\partial\phi^2}-\frac{2\cos\theta}{\sin^2\theta}\frac{\partial^2}{\partial\psi\,\partial\phi}+\left(\cot^2\theta+\frac{D_{R\parallel}}{D_{R\perp}}\right)\frac{\partial^2}{\partial\psi^2}\tag{8.114}$$

where $D_{R\parallel}$ is the diffusion coefficient for rotation about the long molecular (symmetry) axis. Comparing Eqs. 8.113 and 8.114, we see how the orienting torque $-(1/k_BT)\partial U/\partial\theta$ reduces the extent of rotational diffusion in ordered systems.

We use the Maier–Saupe potential from Eq. 8.24; and the torque then becomes: $\partial U/\partial\theta=3k_BT\lambda_2\sin\theta\cos\theta$. As for unrestricted anisotropic diffusion treated in Section 7.6 of Chapter 7, we look for a solution that is an expansion in the $\mathcal{D}^L_{K,M}(\Omega)$ elements of Wigner rotation matrices:

$$P(\Omega,t)=\sum_{L,K,M}c^L_{K,M}(t)\mathcal{D}^L_{K,M}(\Omega)\tag{8.115}$$

because these are eigenfunctions of the Laplacian operator for a symmetric top. The corresponding eigenvalues, $-(D_{R\perp}\tau_{L,M})^{-1}$, are given by Eq. 7.28:

$$1/(D_{R\perp}\tau_{L,M})=L(L+1)+(D_{R\parallel}/D_{R\perp}-1)M^2\tag{8.116}$$

where M takes values $-L$ to $+L$ in integer steps.

As mentioned already, the solution is due to Nordio and Busolin (1971). We substitute Eq. 8.115 in Eq. 8.113 and multiply by $\mathcal{D}^{L*}_{K,M}(\Omega)$ with integration over the full solid angle to exploit the orthogonality of Wigner rotation matrices. Eventually, we get an infinite set of simultaneous differential equations for the coefficients $c^L_{K,M}$:

$$\frac{\partial c^L_{K,M}}{\partial t}=D_{R\perp}\sum_{L'=|L-2|}^{|L+2|}(R_{KM})_{LL'}c^{L'}_{K,M}\tag{8.117}$$

which are coupled by the coefficients $c^{L\pm1}_{K,M}$ and $c^{L\pm2}_{K,M}$. The elements of the matrix $\mathbf{R}_{KM}$ are given by (Nordio et al. 1972):

$$(R_{KM})_{LL'}=-(D_{R\perp}\tau_{L,M})^{-1}\delta_{LL'}+\tfrac{1}{2}k_BT\lambda_2\big(L(L+1)-L'(L'+1)+6\big)\langle 2L'L;0K\rangle\langle 2L'L;0M\rangle\tag{8.118}$$

where $\langle L_1 2L;M_1M_2\rangle=(-1)^{L_1+2-L}\langle 2L_1L;M_2M_1\rangle$ are Clebsch–Gordan coefficients given (for $M_2=0$) in Table A8.1 of the appendix at the end of this chapter. To solve the coupled Eqs. 8.117, we truncate the series in Eq. 8.115 at a value of L that gives sufficient convergence.

Finally, the zero-frequency spectral densities become simply (Nordio and Busolin 1971; Nordio et al. 1972):

$$j_{KM}(0)=\sum_{L\neq0}\left(\langle\mathcal{D}^2_{K,M}\mathcal{D}^{L*}_{K,M}\rangle-\delta_{K,0}\delta_{M,0}\langle\mathcal{D}^2_{0,0}\rangle\langle\mathcal{D}^L_{0,0}\rangle\right)\tau_{LKM}\tag{8.119}$$

These $\omega=0$ spectral densities characterize the secular terms, and the pseudosecular terms when $\omega_a^2\tau_R^2<<1$. The necessary angular averages are (Nordio et al. 1972):

$$\langle\mathcal{D}^L_{0,0}\rangle=-(2L+1)^{-1}(R_{00})_{20}\left(R^{-1}_{00}\right)_{L2}\tag{8.120}$$

$$\langle\mathcal{D}^2_{K,M}\mathcal{D}^{L*}_{K,M}\rangle=(-1)^{K+M}\sum_{L'}\langle L2L';-K,K\rangle\langle L2L';-M,M\rangle\langle\mathcal{D}^{L'}_{0,0}\rangle\tag{8.121}$$

where the latter comes again from the Clebsch–Gordan series (Eq. R.17, together with Eq. R.11), and see Table A8.1. The correlation times in Eq. 8.119 are defined by:

$$\tau_{LKM}=-\tfrac{1}{5}(2L+1)\left(R^{-1}_{KM}\right)_{2L}\tag{8.122}$$

In Eqs. 8.120 and 8.122, the matrix $\mathbf{R}^{-1}_{KM}$ is the inverse of $\mathbf{R}_{KM}$, e.g., $\left(R^{-1}_{KM}\right)_{2L}=(R_{KM})^{L,2}/|\mathbf{R}_{KM}|$, where $(R_{KM})^{L,2}$ is the cofactor of determinant $|\mathbf{R}_{KM}|$, relative to the element $(L,2)$. From the symmetry properties of Clebsch–Gordan coefficients (Eq. Q.8), we get:

$$j_{KM}(0)=j_{-K-M}(0)=j_{K-M}(0)\tag{8.123}$$

for the symmetries of the zero-frequency spectral densities.

Equations 8.119–8.122 let us calculate the zero-frequency spectral densities in terms of the ratio of rotational diffusion coefficients $D_{R\parallel}/D_{R\perp}$ and the strength λ_2 of the orientation potential (or the order parameter S_{zz}). For the model of high ordering with the nitroxide z-axis parallel to the director (i.e., $\langle \mathcal{D}_{0,0}^2 \rangle \approx 1$) that we introduced at the end of Section 8.7, the spectral density reduces to (Nordio et al. 1972; Hemminga 1974):

$$j_{KM}(0) = \left(1-\delta_{K,0}\delta_{M,0}\right)\left\langle \left|\mathcal{D}_{K,M}^2(0)\right|^2 \right\rangle \tau_M = \tfrac{1}{2}\delta_{KM}\left(1-\delta_{K0}\right)\tau_M \tag{8.124}$$

where

$$\tau_M^{-1} = M^2 D_{R\parallel} \tag{8.125}$$

for this simple case, with magnetic field parallel to the director (cf. Eq. 8.60).

The dependence of the spectral densities on orientation γ of the magnetic field to the director follows exactly as in the previous section (see Eqs. 8.92 and 8.93). The angular-dependent, zero-frequency spectral densities become:

$$j_{NM}(0,\gamma) = \sum_{K=-2}^{+2} (-1)^{N-K} j_{KM}(0) \times \sum_{L''=0,2,4} \langle 22L'';N,-N\rangle\langle 22L'';K,-K\rangle \mathrm{P}_{L''}(\cos\gamma) \tag{8.126}$$

where values for the Clebsch–Gordan coefficients are given in Table A8.2 of Appendix A8.1. We thus get:

$$j_{0M}(0,\gamma) = \tfrac{1}{5}\left(j_{0M}+2j_{1M}+2j_{2M}\right) + \tfrac{2}{7}\left(j_{0M}+j_{1M}-2j_{2M}\right)P_2 + \tfrac{6}{35}\left(3j_{0M}-4j_{1M}+j_{2M}\right)P_4 \tag{8.127}$$

$$j_{1M}(0,\gamma) = \tfrac{1}{5}\left(j_{0M}+2j_{1M}+2j_{2M}\right) + \tfrac{1}{7}\left(j_{0M}+j_{1M}-2j_{2M}\right)P_2 - \tfrac{4}{35}\left(3j_{0M}-4j_{1M}+j_{2M}\right)P_4 \tag{8.128}$$

$$j_{2M}(0,\gamma) = \tfrac{1}{5}\left(j_{0M}+2j_{1M}+2j_{2M}\right) - \tfrac{2}{7}\left(j_{0M}+j_{1M}-2j_{2M}\right)P_2 + \tfrac{1}{35}\left(3j_{0M}-4j_{1M}+j_{2M}\right)P_4 \tag{8.129}$$

where $j_{KM} \equiv \tfrac{1}{2}\left(j_{KM}(0) + j_{-KM}(0)\right)$ and $P_{2n} \equiv \mathrm{P}_{2n}(\cos\gamma)$. The first, second and third multiplying factors on the right of Eqs. 8.127–8.129 correspond to the spectral densities j_{KM}^{0}, j_{KM}^{2} and j_{KM}^{4}, respectively, of Eqs. 8.97–8.102 for the strong jump model of the previous section.

For general field orientation γ, the expressions for the linewidth coefficients of Eq. 8.52, derived from Eqs. 8.82 to 8.84, then become (see Hemminga 1974):

$$A(\gamma) = \frac{4}{9}\left(\frac{\beta_e B_o}{\hbar}\right)^2 \left((\Delta g)^2 j_{00}(\gamma) + 3(\delta g)^2 j_{02}(\gamma)\right) + \frac{1}{6}I(I+1)\left((\Delta A)^2 j_{10}(\gamma) + 3(\delta A)^2 j_{12}(\gamma)\right) \tag{8.130}$$

$$B(\gamma) = \frac{8}{9}\left(\frac{\beta_e B_o}{\hbar}\right)\left(\Delta g \Delta A \cdot j_{00}(\gamma) + 3\delta g \delta A \cdot j_{02}(\gamma)\right) \tag{8.131}$$

$$C(\gamma) = \frac{4}{9}\left((\Delta A)^2 j_{00}(\gamma) + 3(\delta A)^2 j_{02}(\gamma)\right) - \frac{1}{6}\left((\Delta A)^2 j_{10}(\gamma) + 3(\delta A)^2 j_{12}(\gamma)\right) \tag{8.132}$$

where hyperfine couplings are in angular frequency units and $j_{KM}(\gamma) \equiv j_{KM}(0,\gamma)$ from Eqs. 8.127 to 8.129. Note that the indices of our j_{KM} are interchanged by Hemminga (1974), presumably because of a different convention for the rotational transformation (see Appendix S). Factors A_0, A_2, A_4 and equivalents that govern angular dependence of the linewidth coefficients (see Eq. 8.94) then come from the first, second and third terms, respectively, for $j_{KM}(0,\gamma)$ in Eqs. 8.127–8.129. We get the constituent values of j_{KM} from Eqs. 8.119 to 8.122.

Figure 8.18 shows EPR spectra simulated in the motional-narrowing regime, for an axially ordered ^{14}N-nitroxide undergoing anisotropic Brownian rotational diffusion in a Maier–Saupe

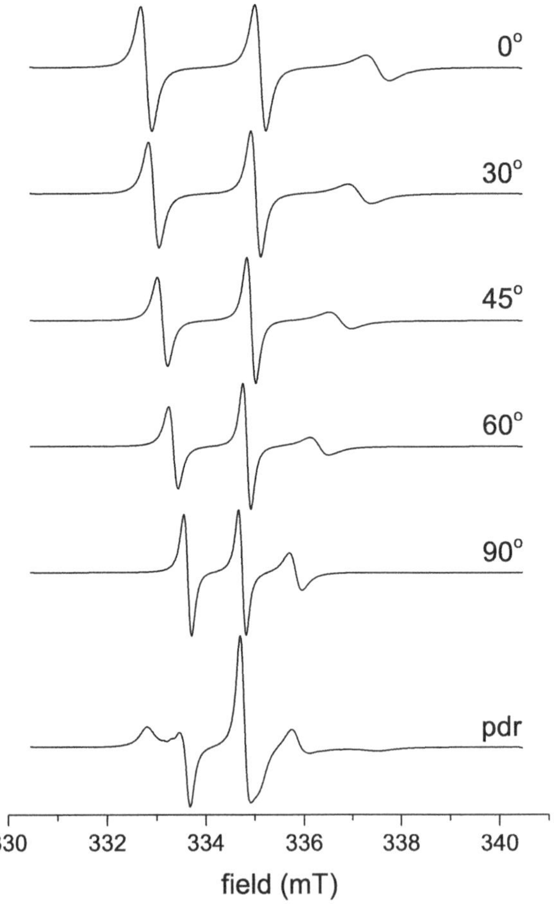

FIGURE 8.18 Fast-motion 9-GHz spectra simulated for Brownian rotational diffusion $\left(D_{R\parallel} = 10D_{R\perp} = 1\times10^9\ \mathrm{s}^{-1}\right)$ with axial order given by Maier–Saupe potential $\left(\lambda_2 = 1.9k_BT;\ S_{zz} = 0.42\right)$. Aligned sample with angles γ of magnetic field to director indicated. Bottom spectrum (pdr): non-aligned powder sample. Residual Gaussian broadening, FWHM = 0.1 mT. Spectra simulated using EasySpin (Stoll and Schweiger 2007).

orientation potential. The bottom spectrum (pdr) is for a non-aligned random sample. The remaining spectra are for an aligned sample with static magnetic field inclined to the direction of axial ordering (i.e., the director) at angles γ given in the figure. Spectra from the aligned sample show a very clear angular dependence of the differential line broadening that is reflected by the relative heights of the three hyperfine lines. As seen also with the linewidth coefficients for the strong-jump model in Figure 8.16, the differential broadening is greater at intermediate angles than for $\gamma = 0°$ and $\gamma = 90°$.

After allowing explicitly for unresolved proton hyperfine structure, Hemminga (1974) achieves quantitative agreement with the lineshapes from 3-DOXYL cholestane in aligned membranes of dipalmitoyl phosphatidylcholine with 50 mol% cholesterol at 49°C, for different orientations γ to the static magnetic field. Using motional narrowing theory for Brownian rotational diffusion, the long-axis order parameter is $S_{ZZ} = 0.86 \pm 0.02$, and the rotational diffusion coefficients are $D_{R\perp} = (3.4 \pm 0.5) \times 10^6\ \mathrm{s}^{-1}$ with $D_{R\parallel}/D_{R\perp} = 40 \pm 4$. Rotation around the molecular long axis, with correlation time $\tau_{R\parallel} = 1.2$ ns, lies within the fast-motion regime. Rotation of the long axis itself is much slower, with $\tau_{R\perp} = 49 \pm 7$ ns. However, as stated already in Section 8.10, molecular ordering extends the range of the motional-narrowing regime. The nitroxide y-axis lies along the steroid long axis of 3-DOXYL cholestane. Rapid rotation about this axis produces a motionally averaged hyperfine element $\langle A_\perp \rangle = \frac{1}{2}(A_{zz} + A_{xx})$ which differs from the long-axis A_{yy} by 1.25 mT. Because of the high ordering, the fraction of this to be motionally averaged is $1 - S_{ZZ} = 0.14$. Converted to an angular frequency, and using the criterion $\Delta\omega_{aniso}\tau_c < 1$, the critical rotational correlation time becomes: $\tau_c < 32$ ns. Thus, we expect incipient slow-motional effects from rotation of the long axis, but a fast-motion approach may still serve as an acceptable approximation. Obviously, the limited motional averaging available at high order precludes accurate determinations of $\tau_{R\perp}$ for 3-DOXYL cholestane.

8.12 ORDERING AND SLOW MOTION

Up until now, we dealt only with motional-narrowing theory, i.e., fast motion. To treat molecular ordering when motion is slow on the nitroxide EPR timescale, we use the stochastic Liouville equation as done already for isotropic rotational diffusion in Sections 6.7–6.10 of Chapter 6. We introduce an orientation potential $U(\theta)$ into the diffusion equation as in Eq. 8.113 of Section 8.11. With the Maier–Saupe potential (Eq. 8.24), we get:

$$\frac{\partial P}{\partial t} = D_{R\perp}\nabla^2_{diff}P + 3\lambda_2 D_{R\perp}\left(\cos\theta\sin\theta\frac{\partial P}{\partial\theta} + \left(3\cos^2\theta - 1\right)P\right)$$

$$\equiv -\Gamma_\Omega P \tag{8.133}$$

where the equilibrium angular probability distribution is:

$$P_o(\theta) = \frac{\exp(\lambda_2 \mathrm{P}_2(\cos\theta))}{\int \exp(\lambda_2 \mathrm{P}_2(\cos\theta))\sin\theta\cdot d\theta} \tag{8.134}$$

Unlike the situation of Sections 6.7–6.10 with no orientational ordering, $P_o(\theta)$ is now no longer a constant. The Markoff operator Γ_Ω of Eq. 8.133 is not Hermitean, but it becomes Hermitean if we use the symmetrizing transformation (Polnaszek et al. 1973):

$$\tilde{\Gamma}_\Omega = P_o^{-1/2}\Gamma_\Omega P_o^{1/2} \tag{8.135}$$

From Eqs. 8.133 to 8.135, we then get (Polnaszek et al. 1973):

$$\tilde{\Gamma}_\Omega = -D_{R\perp}\nabla^2_{diff}$$

$$+ 3D_{R\perp}\lambda_2\left(\tfrac{1}{10}\lambda_2 - \left(1 - \tfrac{1}{14}\lambda_2\right)\mathrm{P}_2(\cos\theta) - \tfrac{6}{35}\lambda_2\mathrm{P}_4(\cos\theta)\right) \tag{8.136}$$

As for the diffusion term alone, the Wigner rotation matrices $\mathcal{D}^L_{K,M}(\Omega)$ are eigenfunctions of this symmetrized Markoff operator $\tilde{\Gamma}_\Omega$, the eigenvalues for which become simply:

$$1/\tau_{L,M} = D_{R\perp}L(L+1) + \left(D_{R\parallel} - D_{R\perp}\right)M^2$$

$$+ 3D_{R\perp}\lambda_2\left(\tfrac{1}{10}\lambda_2 - \left(1 - \tfrac{1}{14}\lambda_2\right)\mathrm{P}_2(\cos\theta) - \tfrac{6}{35}\lambda_2\mathrm{P}_4(\cos\theta)\right) \tag{8.137}$$

The situation is similar to that for axial rotational diffusion in isotropic liquids treated in Appendix A6.2 of Chapter 6. On the left-hand side of the *symmetrized* versions of Eqs. A6.2–A6.5 from Chapter 6, we must now add eigenvalues and corresponding expansions in the $\left\langle \mathcal{D}^L_{K,M} \middle| \ldots \middle| \mathcal{D}^{L'}_{K',M'} \right\rangle$ matrix elements of the term containing λ_2 on the right-hand side of Eq. 8.136. For magnetic field oriented along the director (i.e., $\gamma = 0$), the necessary additions to the first term in each equation are (Polnaszek et al. 1973):

$$-i\tfrac{3}{10}D_{R\perp}\lambda_2^2\left(C^L_{K,M}\right)^{m_I,m_I'}$$

$$+ i3D_{R\perp}\lambda_2\left(1 - \tfrac{1}{14}\lambda_2\right)\sum_{L'}\langle 2L'L;0M\rangle\langle 2LL';0,-K\rangle\left(C^{L'}_{K,M}\right)^{m_I,m_I'}$$

$$+ i\tfrac{18}{35}D_{R\perp}\lambda_2^2\sum_{L'\neq 0}\langle 4L'L;0M\rangle\langle 4LL';0,-K\rangle\left(C^{L'}_{K,M}\right)^{m_I,m_I'} \tag{8.138}$$

where $K = 0$ for Eqs. A6.2 and A6.3, $K = 1$ for Eq. A6.4 and $K = 2$ for Eq. A6.5. When interested only in lineshapes, we need not change terms on the right-hand side of Eqs. A6.2–A6.5. Numerical solution of the coupled equations and calculation of the EPR lineshapes then proceed as in previous cases of slow-motional simulations without molecular ordering (see Section 6.10 in Chapter 6).

Figure 8.19 shows EPR spectra simulated in the slow-motion regime, for an aligned sample with a ^{14}N-nitroxide undergoing anisotropic Brownian rotational diffusion in a Maier–Saupe orientation potential. Simulation parameters are the same as those for the fast-motion simulations in Figure 8.18, except that the rotational diffusion coefficients are five times slower. Now not only do the linewidths depend on magnetic field orientation; the lineshapes

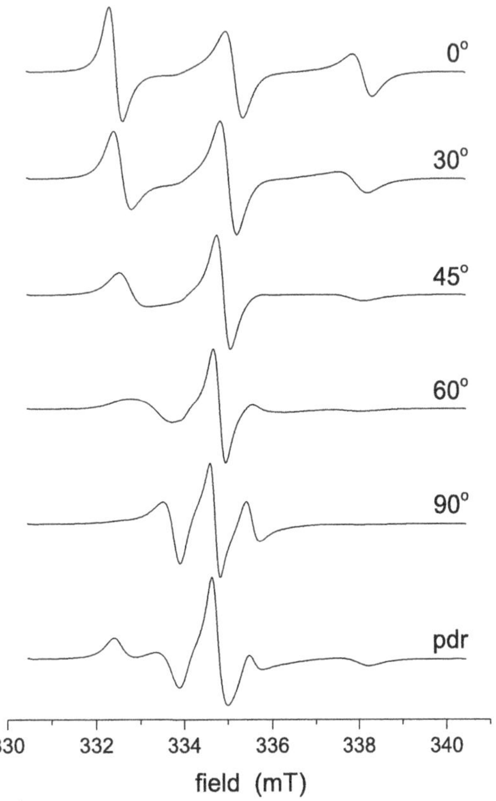

FIGURE 8.19 Slow-motion 9-GHz spectra simulated for Brownian rotational diffusion $\left(D_{R\parallel}=10D_{R\perp}=2\times10^{8}\ \mathrm{s}^{-1}\right)$ with Maier–Saupe ordering potential: $\lambda_2=1.9k_BT\left(S_{zz}=0.42\right)$. Aligned sample with angles γ of magnetic field to director indicated. Bottom spectrum (pdr): non-aligned powder sample. Gaussian broadening, FWHM = 0.1 mT.

differ greatly between the different field orientations, particularly at intermediate angles. These large differences from the fast-motion case become less apparent in the lineshapes of powder spectra from nonaligned samples. Also, slow motion can have a large quantitative effect on line splittings. The outer peaks in the powder spectrum of Figure 8.19 are 1.1 mT further apart than those for the fast-motion case given in Figure 8.18, although the order parameters are identical (cf. Section 7.14 in Chapter 7).

If the orientation potential does not have axial symmetry, e.g., Eq. 8.27 in Section 8.4, we need a more general approach (Polnaszek and Freed 1975). In the presence of an orientation potential, the diffusion equation of Eq. 7.25 written in terms of the operator **L** for an infinitesimal rotation becomes:

$$\frac{\partial P(\Omega,t)}{\partial t}=-\left[(\mathbf{L}\cdot\mathbf{D}_R\cdot\mathbf{L})+\left(\mathbf{L}\cdot\mathbf{D}_R\cdot\mathbf{L}\frac{U(\Omega)}{k_BT}\right)\right]P(\Omega,t)$$
$$\equiv-\Gamma_\Omega P(\Omega,t)\qquad(8.139)$$

Here, the second term is a generalization of Eq. 8.112, and $\mathbf{D}_R$ is the molecular diffusion tensor. From Appendix R.1 we see that $\mathbf{L}=-i\nabla_\Omega$, where ∇_Ω is the gradient vector operator on the surface of a unit sphere, i.e., $|\nabla_\Omega|\equiv\partial/\partial\theta$ in the axial situation above. On making the symmetrization transformation of Eq. 8.135, we then get (Polnaszek and Freed 1975):

$$\tilde{\Gamma}_\Omega=\mathbf{L}\cdot\mathbf{D}_R\cdot\mathbf{L}+\frac{\mathbf{L}\cdot\mathbf{D}_R\cdot\mathbf{L}U}{2k_BT}+\frac{\mathbf{T}_R\cdot\mathbf{D}_R\cdot\mathbf{T}_R}{(2k_BT)^2}\qquad(8.140)$$

where $\mathbf{T}_R=i\mathbf{L}U(\phi\theta\psi)$ is the torque generated by the orientation potential. For the non-axial potential given in Eq. 8.27, the components of the torque become:

$$T_{R,z}/(2k_BT)=\varepsilon\sin^2\theta\sin2\phi\qquad(8.141)$$

$$\left(T_{R,x}\pm iT_{R,y}\right)/(2k_BT)=\pm i\sin\theta\cos\theta\left(\gamma_2\exp(\pm i\phi)-\varepsilon\exp(\mp i\phi)\right)\qquad(8.142)$$

in the molecular axis system. Polnaszek and Freed (1975) give the explicit expression that results for the last two terms in Eq. 8.140, when using Eqs. 8.141 and 8.142. This leads to the extra terms that we must add to the coupled equations for rotational diffusion without an orientation potential, just as in the case of axial ordering considered above (cf. Eq. 8.138).

A general formulation for the stochastic-Liouville calculations, including expansion of the orientation potential and dependence on static-field orientation, is given in the Appendix to Meirovitch et al. (1982). This is coupled with implementation of the highly efficient Lanczos algorithm for tridiagonalization of complex symmetric matrices (Moro and Freed 1981). As referred to already in Section 6.10, the full software implementation is described in detail by Schneider and Freed (1989). Note that a change in convention for rotational transformations is introduced at this point in development of the simulations, with a compensating redefinition of K and M (see Appendix S). Extension to a completely anisotropic rotational diffusion tensor is given in the Appendix to Earle et al. (1993).

8.13 ORDER-PARAMETER CALCULATIONS FROM POWDER SAMPLES

Frequently with lipid-bilayer membranes and other smectic mesophases, we want to extract order parameters from random dispersions of the sample. This is possible when the nitroxide z-axis is parallel to the long axis of the spin-labelled molecule (see Figure 8.3), because the spectra then have their maximum angular anisotropy. We see this from Figures 8.18 and 8.19, which correspond to this particular configuration. The 9-GHz powder spectra there have outer peaks with splitting $2A_{max}$, which correspond to the magnetic field oriented at $\gamma=0°$ to the director of ordering (e.g., membrane normal), and inner peaks with splitting $2A_{min}$ that corresponds to the magnetic field perpendicular to the director $(\gamma=90°)$.

For fast rotation, the motionally averaged hyperfine splitting constants are $\langle A_\parallel\rangle$ and $\langle A_\perp\rangle$ for field orientations $\gamma=0°$ and $\gamma=90°$, respectively. As we saw in Section 2.4 of Chapter 2, the splitting of the outer peaks in a powder pattern then is given precisely by $A_{max}=\langle A_\parallel\rangle$. Even in the fast-motion case, this is not true of the inner splitting, however, because of spectral overlap in the inner regions of the powder pattern. Several corrections have been proposed to relate $\langle A_\perp \rangle$ to A_{min} from the inner splitting (Gaffney 1976; Griffith and Jost 1976; Schorn and Marsh 1997), based on simulations of powder spectra. The first and simplest of these is: $\langle A_\perp \rangle = A_{min} + 0.08$ mT (Hubbell and McConnell 1971).

From Figure 8.19 of the previous section, we see already that things become more complicated when components of the motion are in the slow regime. Although the powder spectrum retains well-defined outer and inner peaks, there are shifts in these relative to the spectrum for fast motion but with the same order parameter (viz., Figure 8.18). As mentioned already, the 9-GHz EPR spectra of nitroxide-labelled lipid chains in biological membranes contain important contributions from slow motion (Lange et al. 1985; Moser et al. 1989; Schorn and Marsh 1996b). We must take this into account when using spin-label order parameters to study chain flexibility.

Figure 8.20 shows slow-motion powder spectra simulated for different values of the S_{zz} order parameter, assuming that the ^{14}N-nitroxide z-axis lies along the long molecular axis of the spin label. This stochastic-Liouville calculation uses the Maier–Saupe ordering potential (Eq. 8.24), as in Figure 8.19. The elements of the axial diffusion tensor are assumed equal $D_{R\parallel} = D_{R\perp}$, which is approximately the case for segmental motion, and have values $D_{R\parallel} = D_{R\perp} = 4 - 7\times 10^7$ s^{-1}, depending on the order parameter. This corresponds to rotational correlation times $\tau_R = 2.4 - 4.2\times 10^{-9}$ s. We list the simulation parameters completely in Table 8.3. The resulting lineshapes resemble those of spin-labelled fatty acids and phospholipids in membranes (see, e.g., Fretten et al. 1980; Pates and Marsh 1987; Rama Krishna and Marsh 1990; Schorn and Marsh 1996a). They are characterized by an outer hyperfine splitting $2A_{max}$ and inner splitting $2A_{min}$, which we mark by ticks on the spectrum for $S_{zz} = 0.5$. At low order parameters, the downward high-field peak becomes visible first for $S_{zz} = 0.2$, as a broad hump superimposed on the outer flank of the spectrum (at higher vertical expansion). At high order parameters, an upward high-field inner peak still remains visible (below the base line) for $S_{zz} = 0.8$, but already the spectra begin to resemble a rigid-limit powder pattern with a central downward peak at the $g_{zz}\,(\equiv g_\parallel)$ position (cf. Figure 2.8).

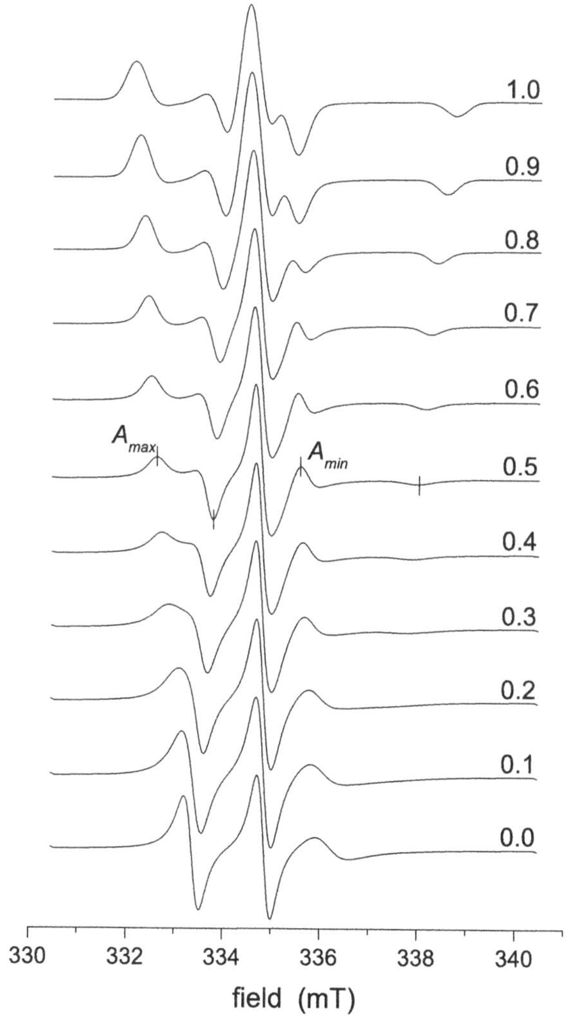

FIGURE 8.20 Slow-motional simulations of ^{14}N-nitroxide 9.4-GHz EPR powder spectra, for values of axial order parameter, $S_{zz} = 0 - 1.0$, indicated. Simulation parameters from Table 8.3. Outer ticks (for $S_{zz} = 0.5$) correspond to outer hyperfine splitting $2A_{max}$ and inner ticks to $2A_{min}$.

Solid symbols in Figure 8.21 show the dependence of A_{max} and A_{min}, measured from the simulated slow-motion spectra in Figure 8.20, on the order parameter S_{zz}. Solid lines show the linear dependence of the motionally-averaged hyperfine constants $\langle A_\parallel \rangle$ and $\langle A_\perp \rangle$ on S_{zz}:

$$\langle A_\parallel \rangle = a_o^{\mathrm{N}} + \tfrac{2}{3}\Delta A\cdot S_{zz} \tag{8.143}$$

$$\langle A_\perp \rangle = a_o^{\mathrm{N}} - \tfrac{1}{3}\Delta A\cdot S_{zz} \tag{8.144}$$

that we get for fast motion from Eqs. 8.6, 8.7 and 8.14. Here, $a_o^{\mathrm{N}} = 1.51$ mT and $\Delta A = 2.74$ mT (see Table 8.3). From the discussion in Section 7.14, we expect that slow motion increases A_{max} relative to $\langle A_\parallel \rangle$ and correspondingly decreases A_{min} relative to

TABLE 8.3 Parameters used in stochastic-Liouville simulations with Maier–Saupe ordering potential (Eq. 8.24), for dependence of 9.4-GHz spin-label spectra on axial order parameter, S_{zz}, of non-aligned random samples (See Figure 8.20)

S_{zz}	$\lambda_2\,(k_BT$ units$)$	$D_{R\perp}\,(\mathrm{s}^{-1})$	ΔB_o^G (mT)
0.0	0.0	6×10^7	0.08
0.1	0.4375	6×10^7	0.08
0.2	0.9125	5×10^7	0.08
0.3	1.348	5×10^7	0.10
0.4	1.807	5×10^7	0.14
0.5	2.325	5×10^7	0.18
0.6	2.963	4×10^7	0.24
0.7	3.865	4×10^7	0.30
0.8	5.495	4×10^7	0.36
0.9	10.395	4×10^7	0.44
1.0	50	4×10^7	0.50

Hyperfine and g-tensors: $(A_{xx}, A_{yy}, A_{zz}) = (0.64, 0.57, 3.33)$ mT and $(g_{xx}, g_{yy}, g_{zz}) = (2.0086, 2.0058, 2.0021)$; and $D_{R\parallel} = D_{R\perp}$. ΔB_o^G: residual Gaussian linewidth (FWHM).

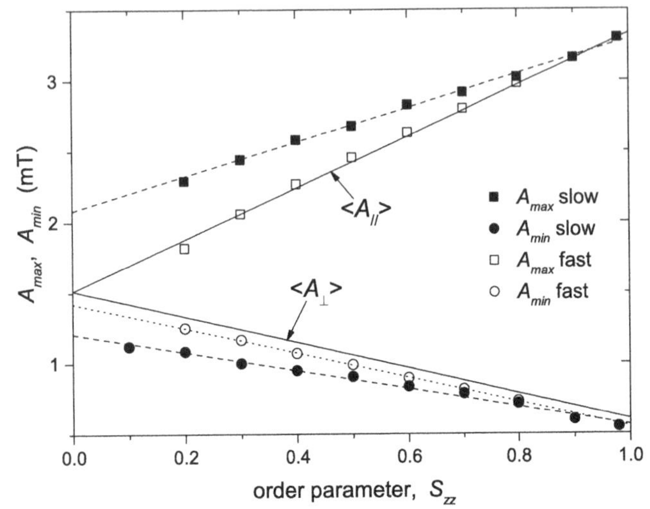

FIGURE 8.21 Outer and inner hyperfine splitting constants, $A_{\max}$ (squares) and $A_{\min}$ (circles), of ^{14}N-nitroxides, as function of axial order parameter S_{zz}. Spectra simulated for slow (solid symbols) and fast (open symbols) rotational diffusion ($D_{R\parallel} = D_{R\perp} = 4-7\times10^{7}\ \mathrm{s}^{-1}$ and $15\times10^{7}\ \mathrm{s}^{-1}$, respectively) in random, non-aligned samples. Other parameters for slow diffusion from Table 8.3. Solid lines: motionally averaged hyperfine constants $\langle A_{\parallel}\rangle$ and $\langle A_{\perp}\rangle$ from Eqs. 8.143 and 8.144. Dashed lines and dotted line: linear regressions.

$\langle A_{\perp}\rangle$. This is just what we find in Figure 8.21, although spectral overlap complicates $A_{\min}$. Open symbols in Figure 8.21 show the dependence of $A_{\max}$ and $A_{\min}$ on S_{zz} simulated for fast motion, where all parameters are the same as in Table 8.3, except that the diffusion constant is now $D_{R\parallel} = D_{R\perp} = 1.5\times10^{8}\ \mathrm{s}^{-1}$. This corresponds to a rotational correlation time $\tau_R = 1.1\times10^{-9}$ s. As predicted above, $A_{\max} = \langle A_{\parallel}\rangle$ for fast motion, but $A_{\min}$ is modified by spectral overlap. The latter effect is smaller, however, than in the slow-motion simulations. In fact, the correction for the fast-motion simulations is almost constant: $\langle A_{\perp}\rangle \cong A_{\min} + 0.07$ mT, as proposed originally from static powder patterns (Hubbell and McConnell 1971).

Values of $A_{\max}$ and $A_{\min}$ measured empirically from the slow-motion simulations (solid symbols in Figure 8.21) show a near-linear dependence on order parameter with the following (dashed) regression lines:

$$A_{\max}(\mathrm{mT}) = \langle A_{\parallel}\rangle + 0.563 - 0.614\times S_{zz} \qquad (8.145)$$

$$A_{\min}(\mathrm{mT}) = \langle A_{\perp}\rangle - 0.328 + 0.328\times S_{zz} \qquad (8.146)$$

Both shifts in splitting reinforce in overestimating the order parameter from $A_{\max} - A_{\min}$ without correction. Substituting from Eqs. 8.145 and 8.146 into Eq. 8.14, we get the following slow-motion calibration for the corrected order parameter:

$$S_{zz} = \left(0.556\ \mathrm{mT}^{-1}\right)\times\left(A_{\max} - A_{\min}\right) - 0.496 \qquad (8.147)$$

in terms of the measurables $A_{\max}$ and $A_{\min}$. This is a very substantial correction and indicates that direct simulation is preferable in this part of the slow-motion regime. By contrast, the calibration deduced from fast-motion simulations (open symbols in Figure 8.21) is:

$$S_{zz} = \left(0.371\ \mathrm{mT}^{-1}\right)\times\left(A_{\max} - A_{\min}\right) - 0.032 \qquad (8.148)$$

which is a much smaller correction.

We also get the (corrected) isotropic hyperfine coupling $a_o^N = \frac{1}{3}\left(\langle A_{\parallel}\rangle + 2\langle A_{\perp}\rangle\right)$ from Eqs. 8.145 and 8.146:

$$a_o^N(\mathrm{mT}) = \tfrac{1}{3}\left(A_{\max} + 2A_{\min}\right) + 0.031 - 0.014\times S_{zz} \qquad (8.149)$$

This depends on S_{zz}, which we must substitute from Eq. 8.147 giving:

$$a_o^N(\mathrm{mT}) = \tfrac{1}{3}\left(0.992A_{\max} + 2.008A_{\min}\right) + 0.038 \qquad (8.150)$$

The a_o-correction for slow motion is small compared with that to the order parameter, because shifts in $A_{\max}$ and $A_{\min}$ almost cancel. Nevertheless, they are important when determining polarity-dependent shifts because these are relatively small (see Sections 4.2 and 4.4 in Chapter 4). Correction of the isotropic hyperfine coupling for fast-motion spectra is correspondingly:

$$a_o^N(\mathrm{mT}) = \tfrac{1}{3}\left(0.968A_{\max} + 2.032A_{\min}\right) + 0.059 \qquad (8.151)$$

because there is no correction for $A_{\max}$ and therefore no compensating cancellation.

When calculating the order parameter, we can use the isotropic hyperfine coupling to correct for differences between the hyperfine tensor of the experimental sample and that used to get ΔA for Eq. 8.14. To do this, we simply scale $\left(A_{\max} - A_{\min}\right)$ in Eq. 8.147 or 8.148 by the factor: $\frac{1}{3}\left(A_{xx} + A_{yy} + A_{zz}\right)/a_o^N = 1.513\ \mathrm{mT}/a_o^N$, where a_o^N comes from Eq. 8.150 or 8.151, respectively. This corrects for differences in environmental polarity and to a first approximation for differences between nitroxide structures (e.g., five- and six-membered rings).

Figure 8.22 gives 9.4-GHz spectra simulated with the stochastic Liouville equation for fixed order parameter $(S_{zz} = 0.5)$ but different values of rotational diffusion coefficient $D_R (\equiv D_{R\perp} = D_{R\parallel})$ extending from the slow- $\left(\le 1.10^{8}\ \mathrm{s}^{-1}\right)$ to fast-motion regimes. These spectra, which incorporate the Maier–Saupe orientation potential, give us some guide as to whether a given experimental powder pattern contains slow-motion components. Figure 8.23 shows the dependence of outer and inner hyperfine splitting constants ($A_{\max}$ and $A_{\min}$) on rotational diffusion coefficient, deduced from the spectra in Figure 8.22. The figure also includes the average value $\frac{1}{3}\left(A_{\max} + 2A_{\min}\right)$ that is related to the isotropic hyperfine coupling. The x-axis in Figure 8.23 is logarithmic and therefore is equivalent to a temperature dependence according to an Arrhenius law for the diffusion coefficient. Both $A_{\max}$ and $A_{\min}$ show an approximately logarithmic (but opposite) dependence on D_R in the slow-motion regime, before becoming constant on reaching the fast regime. Beyond this, $A_{\max} = \langle A_{\parallel}\rangle$ as expected for a fast-motion powder pattern, but $A_{\min}$ is offset from $\langle A_{\perp}\rangle$ because of spectral overlap. The average $\frac{1}{3}\left(A_{\max} + 2A_{\min}\right)$ varies much less (because changes in $A_{\max}$ and $2A_{\min}$ nearly cancel) and reaches a constant value in the fast regime that is offset from a_o^N, as in Eq. 8.151. Experimentally, we use the temperature dependence of the effective isotropic coupling to locate the fast-motion regime,

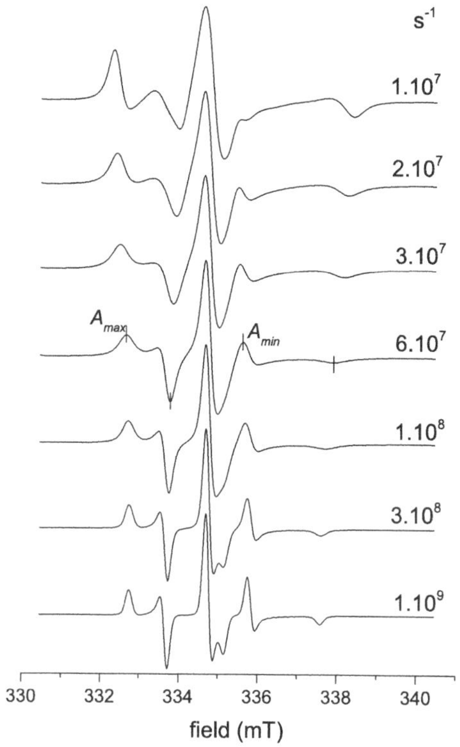

FIGURE 8.22 Stochastic-Liouville simulations of ^{14}N-nitroxide 9.4-GHz EPR powder spectra, for values of rotational diffusion coefficient $\left(D_{R\parallel} = D_{R\perp} = 1-100\times10^7\ \mathrm{s}^{-1}\right)$ indicated, and fixed order parameter $S_{zz} = 0.5$. Other simulation parameters from Table 8.3. Outer ticks (for $D_R = 6\times10^7\ \mathrm{s}^{-1}$) correspond to outer hyperfine splitting $2A_{\max}$ and inner ticks to $2A_{\min}$.

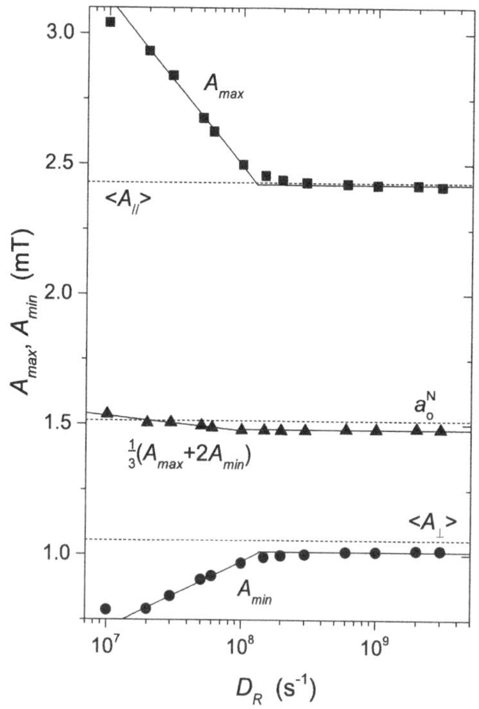

FIGURE 8.23 Dependence of outer ($A_{\max}$, squares) and inner ($A_{\min}$, circles) hyperfine splitting constants, and their average $\left(\frac{1}{3}\left(A_{\max}+2A_{\min}\right)\right.$, triangles) on rotational diffusion coefficient $D_{R\parallel} = D_{R\perp}$, for fixed order parameter: $S_{zz} = 0.5$ (spectra in Figure 8.22). Dashed lines: motionally-averaged hyperfine constants $\langle A_\parallel\rangle$ and $\langle A_\perp\rangle$, and isotropic coupling constant a_o^{N}.

where it becomes constant (Marsh 2001; Marsh et al. 2007). Changes in effective a_o^{N} observed experimentally in the slow regime are larger than those of $\frac{1}{3}\left(A_{\max}+2A_{\min}\right)$ shown in Figure 8.23, because the order parameter also decreases with increasing temperature.

We see from Figure 8.23 that, when the motion is slow, the shifts in $A_{\max}$ and $A_{\min}$ from the motionally-averaged values, $\langle A_\parallel\rangle$ and $\langle A_\perp\rangle$, depend on the value of the diffusion coefficient. Simulated lineshapes, such as those in Figure 8.22, let us make reasonable estimates for D_R. Appendix A8.3 at the end of the chapter lists the calibration constants that we then need to calculate S_{zz} and a_o^{N}, for different diffusion coefficients, by using expressions equivalent to Eqs. 8.147 and 8.149. For more details, see also Marsh (2018b; 2019).

8.14 DYNAMICS OF COMBINED SLOW AND SEGMENTAL MOTIONS: LIPID CHAINS

In Section 8.5, we saw that lipid chain motions have two components: rotation of the chain long axis, plus rotational isomerism of the individual chain segments. In the gel phase of phospholipid bilayers, both components of chain motion are in the slow regime. However, in the physiologically-relevant fluid phase of phospholipid membranes, chain isomerism is fast but long-axis motion of the whole chain can be in the slow-motion regime. We describe first the finite-grid method used by Lange et al. (1985) for treating this combined motion.

Chain isomerism consists of jumps between different configurations of *trans* and *gauche* conformers of the backbone C–C bonds (see Figure 8.24). This process is best treated as discrete jumps of frequency τ_{t-g}^{-1}. The slower whole-body motion of the chain axis, on the other hand, corresponds to restricted Brownian rotational diffusion with anisotropic correlation times $\tau_{R\parallel} = 1/\left(6D_{R\parallel}\right)$, $\tau_{R\perp} = 1/\left(6D_{R\perp}\right)$ and a Maier–Saupe orientation potential. To render the two motional modes compatible, we model Brownian diffusion by transitions on a discrete angular grid, as in Section 6.5 of Chapter 6. But now we incorporate this into the stochastic Liouville equation, instead of into the Bloch equations with adiabatic approximation. Also, we divide not only the θ-rotation with diffusion correlation-time $\tau_{R\perp}$ into an angular mesh, but also the azimuthal ϕ-rotation with correlation time $\tau_{R\parallel}$ for rotation about the chain axis. These two Euler angles relate the principal diffusion axis to the magnetic-field direction in the laboratory axis system (see Figure 8.25). The discrete values are: $\theta_k \equiv \left(k-\frac{1}{2}\right)\Delta\theta$ and $\phi_l \equiv \left(l-\frac{1}{2}\right)\Delta\phi$, where $\Delta\theta = \pi/N_k$, $\Delta\phi = 2\pi/N_l$ are the angular increments and N_k, N_l are the respective total numbers of angular zones.

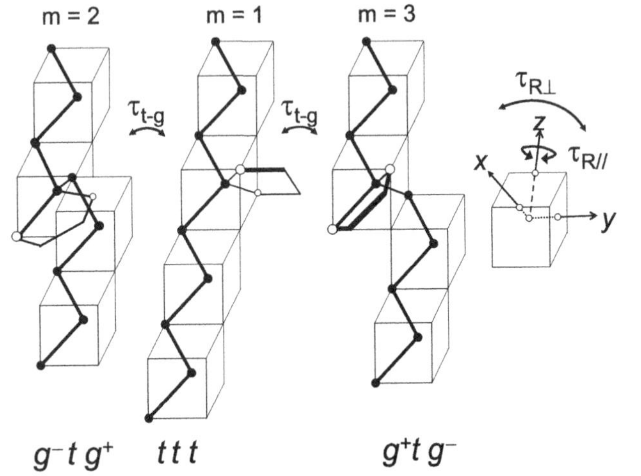

FIGURE 8.24 Conformational states of DOXYL-labelled alkyl chain for three $(m=1-3)$ of the six distinct nitroxide *x,y,z*-axis orientations, relative to the diffusion axis (chain long axis). Sets of Euler angles $(\varphi_m,\vartheta_m,\psi_m)$ corresponding to $m=1-6$ are listed in Table 8.4. Conformations of the bond immediately before the labelled C-atom, and the two immediately following, are indicated. $g^{\pm}g^{\mp}$ combinations are disallowed on intramolecular steric grounds. Conformations change at rate τ_{t-g}^{-1}; long axis *Z* diffuses with correlation times $\tau_{R\parallel}$, $\tau_{R\perp}$.

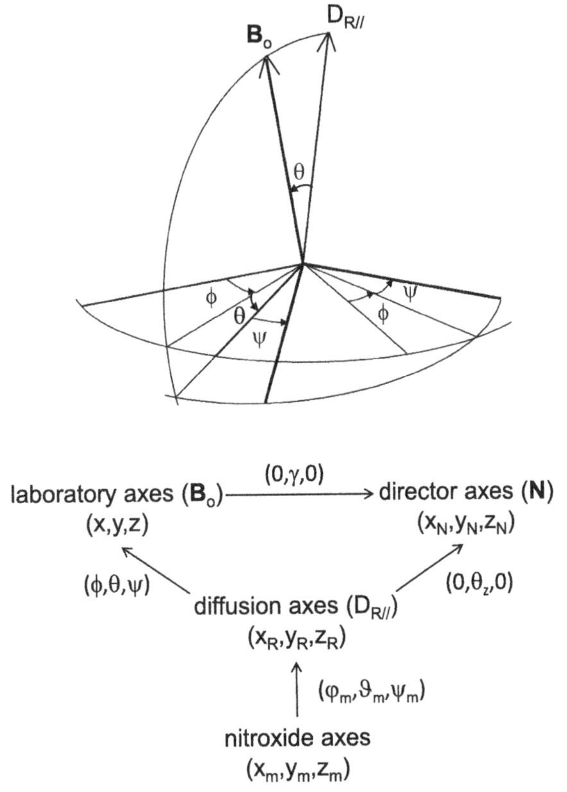

FIGURE 8.25 *Bottom*: Euler transformations between coordinate systems for spin-labelled lipid chains: nitroxide axes (x_m,y_m,z_m); rotational diffusion axes (x_R,y_R,z_R); director axes (x_N,y_N,z_N); and laboratory, or magnetic-field, axes (x,y,z). *Top*: transformation (ϕ,θ,ψ) between diffusion axes $(D_{R\parallel})$ and magnetic-field axes $(\mathbf{B}_o)$.

The nitroxide *x,y,z*-axes are related to the diffusion axis-system by the three Euler angles $(\varphi_m,\vartheta_m,\psi_m)$ that characterize configuration *m* at the chain segment labelled. Euler angles are defined as in Appendix R.2 (see also Figure 8.25). Table 8.4 lists the six sets of angles that give rise to distinct orientations of a DOXYL spin label relative to the diffusion axis (see Figure 8.24). These orientations correspond to the notional H–H vector of the CH_2 group, where we attach the DOXYL ring, being directed along the six edges of a regular tetrahedron. We normalize the populations n_m of the six inequivalent configurations to unity; also *gauche*$^{\pm}$ conformations are equally populated, i.e., $n_2=n_3$ and $n_4=n_5$. This gives $n_6=1-n_1-2(n_2+n_4)$, which means that only three populations are unknown: n_1,n_2 and n_4. We can use these populations to set up a segmental order matrix with elements (see Table 8.1):

$$S_{XX}=\sqrt{3/8}\left(\left\langle\mathcal{D}_{0,2}^{2}\right\rangle+\left\langle\mathcal{D}_{0,-2}^{2}\right\rangle\right)-\tfrac{1}{2}\left\langle\mathcal{D}_{0,0}^{2}\right\rangle \tag{8.152}$$

$$S_{YY}=-\sqrt{3/8}\left(\left\langle\mathcal{D}_{0,2}^{2}\right\rangle+\left\langle\mathcal{D}_{0,-2}^{2}\right\rangle\right)-\tfrac{1}{2}\left\langle\mathcal{D}_{0,0}^{2}\right\rangle \tag{8.153}$$

$$S_{ZZ}=\left\langle\mathcal{D}_{0,0}^{2}\right\rangle \tag{8.154}$$

$$S_{XY}=i\sqrt{3/8}\left(\left\langle\mathcal{D}_{0,2}^{2}\right\rangle-\left\langle\mathcal{D}_{0,-2}^{2}\right\rangle\right) \tag{8.155}$$

$$S_{XZ}=\sqrt{3/8}\left(\left\langle\mathcal{D}_{0,-1}^{2}\right\rangle-\left\langle\mathcal{D}_{0,1}^{2}\right\rangle\right) \tag{8.156}$$

$$S_{YZ}=i\sqrt{3/8}\left(\left\langle\mathcal{D}_{0,1}^{2}\right\rangle+\left\langle\mathcal{D}_{0,-1}^{2}\right\rangle\right) \tag{8.157}$$

where the averages of second-rank Wigner rotation matrix elements (see Appendix R.3) are defined by:

$$\left\langle\mathcal{D}_{K,M}^{2}\right\rangle=\sum_{m=1}^{6}n_m\mathcal{D}_{K,M}^{2}\left(\varphi_m\vartheta_m\psi_m\right) \tag{8.158}$$

Diagonalization of the segmental matrix defined by Eqs. 8.152–8.158 yields the segmental order parameters $S_{Z'Z'}$ and $S_{X'X'}-S_{Y'Y'}$.

TABLE 8.4 Euler angles $(\varphi_m,\vartheta_m,\psi_m)$ relating nitroxide *x,y,z*-axes to the principal diffusion axes for DOXYL spin label in an alkyl chain, and populations n_m for label positions C-*n* in fluid dimyristoyl phosphatidylcholine membranes at 40°C (Moser et al. 1989)

m	1	2	3	4	5	6
config.[a]	*tttt*	*tg⁻tg⁺*	*tg⁺tg⁻*	*g⁺tg⁻t*	*g⁻tg⁺t*	*g⁺g⁺g⁺t*
orientations:						
φ_m	0°	−36.7°	72.7°	−36.7°	72.7°	−162°
ϑ_m	0°	60°	−60°	−60°	60°	−90°
ψ_m	0°	−72.7°	36.7°	−72.7°	36.7°	−18°
populations, n_m:						
C-6	0.50	0.25	0.25	0.00	0.00	0.00
C-10	0.18	0.30	0.30	0.11	0.11	0.00
C-13	0.18	0.24	0.24	0.14	0.14	0.06

[a] Conformation of two bonds immediately before and immediately after labelled C-atom: $t\equiv$ *trans*; $g\equiv$ *gauche*.

These represent the ordering of the most ordered segmental axis Z' and the non-axiality of this order, respectively. For a completely disordered segment, all n_m are 1/6 and $S_{Z'Z'} = 0$; whereas for a fully extended all-*trans* chain, $n_1 = 1$ and $S_{Z'Z'} = 1$.

When discretized, the rate equation for overall and segmental motion that we must incorporate in the stochastic Liouville equation becomes (cf. Eq. 8.139):

$$\frac{\partial P_{klm}(t)}{\partial t} = \left.\frac{\partial P_{klm}(t)}{\partial t}\right|_{over} + \left.\frac{\partial P_{klm}(t)}{\partial t}\right|_{seg} \quad (8.159)$$

where indices k, l refer to orientation of the chain axes, as defined above and m represents the internal chain configuration. We therefore replace the Markov operator by the W-matrix of transition probabilities. The rate equation for overall Brownian diffusion is then:

$$\left.\frac{\partial P_{klm}(t)}{\partial t}\right|_{over} = \sum_{k'}\left(W_{k'kllmm}P_{k'lm}(t) - W_{kk'llmm}P_{klm}(t)\right)$$
$$+\sum_{l'}\left(W_{kkl'lmm}P_{kl'm}(t) - W_{kkll'mm}P_{klm}(t)\right) \quad (8.160)$$

and that for rotational isomerism is:

$$\left.\frac{\partial P_{klm}(t)}{\partial t}\right|_{seg} = \sum_{m'}\left(W_{kkllm'm}P_{klm'}(t) - W_{kkllmm'}P_{klm}(t)\right) \quad (8.161)$$

where P_{klm} is the probability of chain configuration m, with the chain oriented at angles θ_k and ϕ_l.

The transition rate-constants for segmental rotation $m' \to m$ in Eq 8.161 are:

$$W_{kkllm'm} = n_m/\tau_{t-g} \quad (8.162)$$

because, for detailed balance, they depend directly on the equilibrium occupancy of the configuration m to which the transition takes place. The rate constants for Brownian diffusion, $k \to k \pm 1$ and $l \to l \pm 1$, in Eq. 8.160 are given by (cf. Eq. 6.39):

$$W_{k(k+1)llmm} + W_{k(k-1)llmm} = 1/\left(3(\Delta\theta)^2\tau_{R\perp}\right) \quad (8.163)$$

$$W_{kkl(l+1)mm} + W_{kkl(l-1)mm} = 1/\left(3(\Delta\phi)^2\tau_{R\parallel}\right) \quad (8.164)$$

Additionally, for detailed balance, forward and backward diffusion rates ($k \leftrightarrow k'$, $l \leftrightarrow l'$) must be equal:

$$W_{kk'llmm}n_{klm} = W_{k'kllmm}n_{k'lm} \quad (8.165)$$

$$W_{kkll'mm}n_{klm} = W_{kkl'lmm}n_{kl'm} \quad (8.166)$$

Using Eqs. 8.163–8.166, we can express the non-zero elements of W in terms of the correlation times $\tau_{R\parallel}$ and $\tau_{R\perp}$ and the populations n_{klm}.

We express the equilibrium population at a particular site as:

$$n_{klm} = n_{kl}n_m \quad (8.167)$$

where n_{kl} is the population that has orientations θ_k and ϕ_l. We get n_{kl} by integrating the Maier–Saupe distribution (Eq. 8.134):

$$n_{kl} = A_{norm}\int_0^{2\pi} d\psi \int_{\phi_l-\frac{1}{2}\Delta\phi}^{\phi_l+\frac{1}{2}\Delta\phi} d\phi \int_{\theta_k-\frac{1}{2}\Delta\theta}^{\theta_k+\frac{1}{2}\Delta\theta} d\theta \exp\left(\lambda_2 P_2(\cos\theta_z)\right)\sin\theta \quad (8.168)$$

where ψ (a rotation about $\mathbf{B}_o$) is the third Euler angle relating the diffusion axes to the laboratory system (see Figure 8.25), and A_{norm} is the normalization constant. The inclination θ_z of the diffusion axis (i.e., molecular long axis) to the director of ordering comes from the addition theorem for spherical harmonics (Eq. V.18 in Appendix V):

$$P_2(\cos\theta_z) = P_2(\cos\gamma)P_2(\cos\theta) - \tfrac{3}{4}\sin 2\gamma \sin 2\theta \cos\psi$$
$$+\tfrac{3}{4}\sin^2\gamma \sin^2\theta \cos 2\psi \quad (8.169)$$

where γ is the fixed angle between the magnetic field and director.

The spin Hamiltonian used in the quantum-mechanical Liouville equation consists of the secular and pseudo-secular terms:

$$\mathcal{H}_{klm} = \beta_e\left(g_{zz}\right)_{klm} B_o S_z + \left(A_{zz}\right)_{klm} I_z S_z$$
$$+\left(A_{xz}\right)_{klm} I_x S_z + \left(A_{yz}\right)_{klm} I_y S_z \quad (8.170)$$

The tensor elements $\left(g_{zz}\right)_{klm}$, etc. of the **g**- and **A**-tensors are transformed from the nitroxide-axis system to the laboratory system, via the director system, by using the sequential rotation matrix, $\mathbf{R}_{klm} = \mathbf{R}_{lk}(\phi_l\theta_k 0)\mathbf{R}_m(\varphi_m\vartheta_m\psi_m)$, as in Eq. B.25 of Appendix B:

$$\mathbf{g}_{klm} = \mathbf{R}_{klm}\cdot\mathbf{g}\cdot\mathbf{R}_{klm}^{-1}$$
$$= \mathbf{R}_{lk}(\phi_l\theta_k 0)\mathbf{R}_m(\varphi_m\vartheta_m\psi_m)\cdot\mathbf{g}\cdot\mathbf{R}_m^{-1}(\varphi_m\vartheta_m\psi_m)\mathbf{R}_{lk}^{-1}(\phi_l\theta_k 0) \quad (8.171)$$

where $\mathbf{R}_m(\varphi_m\vartheta_m\psi_m)$ is a Cartesian Euler rotation matrix (see Eq. I.21 in Appendix I).

Evaluating spin matrix elements from Eq. 8.170 using the $|M_S, m_I\rangle$-basis functions gives 6×6 combinations for a ^{14}N-nitroxide $\left(M_S = \pm\frac{1}{2};\ m_I = 0, \pm 1\right)$. The total number of points on the orientational grid is $N_k N_l$, for each of which there are $N_m = 6$ conformations. Thus, the stochastic Liouville equation is represented by $216N_kN_l$ coupled differential equations. In the steady state, this becomes four sets of $54N_kN_l$ coupled linear equations, which are solved numerically to give the unsaturated lineshape, essentially as described at the end of Section 6.10 in Chapter 6 (Lange et al. 1985).

Moser et al. (1989) simulated spectral lineshapes of spin-labelled phosphatidylcholine, n-DOXYL PC, in dimyristoyl phosphatidylcholine for different orientations γ of the magnetic field to the membrane normal. Using aligned samples greatly improves the sensitivity to slow-motion components (cf. Figure 8.19). Taking $N_kN_l = 225$ grid points is sufficient for convergence in the fluid phase, where $\tau_{R\parallel} \approx 2.5$ ns. It was assumed that $\tau_{R\perp}/\tau_{R\parallel} = 10$, corresponding to a prolate ellipsoid with axial ratio $a/b \cong 6$, in agreement with previous simulations (Lange et al. 1985). This then lets us determine all other dynamic parameters governing the lipid-chain rotation in the fluid membrane phase by using lineshape simulations (Moser et al. 1989).

Segmental motion, i.e., *trans-gauche* isomerism, lies in the fast regime of conventional nitroxide EPR. This is because all jump lifetimes are too short $(\tau_{t-g} < 0.2 \text{ ns})$ to affect the lineshapes at 9 GHz. Table 8.4 gives populations of the six distinguishable chain configurations in fluid membranes at 40°C, when the DOXYL nitroxide is attached at the C-6, C-10 or C-13 position of the chain. The largest populations are for the straight all-*trans* configurations ($m = 1$ with $\vartheta_1 = 0°$) and for the simple *gauche* configurations ($m = 2$, 3 with $\vartheta_2, \vartheta_3 = \pm 60°$) that introduce a 60°-bend at the label position. Configurations that involve two *gauche* conformations with an intervening *trans* at the label position ($m = 4$, 5 with $\vartheta_4, \vartheta_5 = \mp 60°$) are less populated, and those with adjacent *gauche* conformations ($m = 6$ with $\vartheta_6 = 90°$) are least populated. The corresponding segmental order parameters are: $S_{z'z'} = 0.54$, 0.34 and 0.17 at the C-6, C-10 and C-13 positions of the chain, respectively. On the other hand, the long-axis order parameter remains fixed at $S_{ZZ} = 0.50$ for all chain positions. Table 8.5 gives rotational correlation times and *trans-gauche* lifetimes for the three chain-labelling positions. The correlation time for reorientation of the diffusion long axis is $\tau_{R\perp} = 23$ ns at 40°C and that for rotation about the long axis is ten times shorter: $\tau_{R\parallel} = 2.3$ ns, with a common activation energy of $E_{R\perp} = 49 \pm 2 \text{ kJ mol}^{-1}$ (Lange et al. 1985; Moser et al. 1989).

When segmental motion is well into the motional-narrowing regime, as for lipid chains, we can simply use the segmental-averaged Hamiltonian $\langle \mathcal{H}(\Omega_{over}) \rangle$ in the stochastic Liouville equation, together with the segmental line broadenings $T_2^{-1}(\Omega_{over})$ from fast-motion theory. Here, Ω_{over} represents the angular variables $(\theta_{over}, \phi_{over})$ for the slower overall motion of the lipid chain.

TABLE 8.5 Rotational correlation times ($\tau_{R\parallel}$, $\tau_{R\perp}$) and jump lifetime (τ_{t-g}) from simulating 9-GHz EPR spectra of spin-labelled lipid chains, *n*-DOXYL PC, in aligned membranes of dimyristoyl phosphatidylcholine (Moser et al. 1989)

position, n	*T* (°C)	$\tau_{R\parallel}(\text{s})^a$	$\tau_{R\perp}(\text{s})^a$	$\tau_{t-g}(\text{s})^b$
C-6	25	6.0×10^{-9}	6.0×10^{-8}	1.6×10^{-11}
	35	2.7×10^{-9}	2.7×10^{-8}	1.4×10^{-11}
	45	1.6×10^{-9}	1.6×10^{-8}	1.3×10^{-11}
C-10	25	6.0×10^{-9}	6.0×10^{-8}	7.2×10^{-12}
	35	2.7×10^{-9}	2.7×10^{-8}	6.0×10^{-12}
	45	1.6×10^{-9}	1.6×10^{-8}	6.0×10^{-12}
C-13	25	6.0×10^{-9}	6.0×10^{-8}	2.0×10^{-12}
	35	2.7×10^{-9}	2.7×10^{-8}	1.7×10^{-12}
	45	1.6×10^{-9}	1.6×10^{-8}	1.5×10^{-12}

[a] Overall long-axis motion, independent of spin-label position. Rotational diffusion coefficients: $D_{R\parallel}^{over} = 1/(6\tau_{R\parallel})$ and $D_{R\perp}^{over} = 1/(6\tau_{R\perp})$.
[b] Local *trans-gauche* jump rate τ_{t-g}^{-1} from ^{2}H NMR spectroscopy; spin-label EPR at 9 GHz gives only upper limit: $\tau_{t-g} < 2 \times 10^{-10}$ s.

From Section 6.7 of Chapter 6, the corresponding version of Eq. 6.54 for the overall chain-axis motion becomes:

$$\frac{\partial \chi(\Omega_{over},t)}{\partial t} + \frac{i}{\hbar}\left[\langle \mathcal{H}(\Omega_{over}) \rangle, \chi(\Omega_{over},t)\right] + \left(\Gamma(\Omega_{over}) + T_2^{-1}(\Omega_{over})\right)\chi(\Omega_{over},t) = \frac{i}{\hbar}\left[\langle \mathcal{H}(\Omega_{over}) \rangle, \sigma_o(\Omega_{over})\right] \tag{8.172}$$

where $\Gamma(\Omega_{over})$ is the Smoluchowski diffusion operator for the slow overall motion. In terms of the (segmentally averaged) perturbation Hamiltonian $\langle \mathcal{H}_1(\Omega_{over}) \rangle$, the equivalent of Eq. 6.58 becomes:

$$\frac{\partial \chi(\Omega_{over},t)}{\partial t} + i\omega_o\left[S_z, \chi(\Omega_{over},t)\right] + \frac{i}{\hbar}\left[\langle \mathcal{H}_1(\Omega_{over}) \rangle, \chi(\Omega_{over},t)\right] + \left(\Gamma(\Omega_{over}) + T_2^{-1}(\Omega_{over})\right)\chi(\Omega_{over},t) = \frac{i\gamma_e B_1 \hbar\omega_o}{2N_Z k_B T} S_- \exp(i\omega t) \tag{8.173}$$

with solution in the form given by Eq. 6.59. Treatment of the ordering potential for overall rotation proceeds as described here in Section 8.12. Cassol et al. (1997) used this approach for spin-labelled lipid chains, where they explicitly include the energetics of the *trans-gauche* rotameric states and their transition rates. For 16-DOXYL PC (i.e., labelled at C-16) in dimyristoyl phosphatidylcholine membranes, the rotational correlation time for fluctuations of the long axis is $\tau_{R\perp} = 18 \pm 11$ to 9 ± 2.5 ns over the range 35°C–45°C, and the anisotropy in diffusion rate is $\tau_{R\perp}/\tau_{R\parallel} = 4$. These values are reasonably close to those in Table 8.5, given that the spectra are relatively insensitive to $\tau_{R\parallel}$ when axial diffusion is sufficiently rapid. The corresponding order parameter of the long axis is $S_{zz} = 0.40$ at 40°C, which is lower than that found by Moser et al. (1989). Note, however, that the label in 16-DOXYL PC reaches beyond the end of the C_{14}-chains of the host lipid.

8.15 TIMESCALE SEPARATION AND MULTIFREQUENCY EPR

As mentioned already in Section 8.5, going to higher EPR frequencies can drive slow motions into the rigid limit. Multifrequency EPR therefore offers us a way to distinguish various motional components, based on their differential response to changing microwave frequency.

Table 8.6 lists the longest rotational correlation times that support motional averaging of the Zeeman anisotropy at different EPR operating frequencies. We distinguish between axial rotations around the nitroxide *z*-axis (τ_{axial}) and rotations of the *z*-axis itself $(\tau_{off-axis})$. These are determined by the *g*-value anisotropies $\delta g = \frac{1}{2}(g_{xx} - g_{yy})$ and $\Delta g = g_{zz} - \frac{1}{2}(g_{xx} + g_{yy})$, respectively: $\omega_e \tau_{axial} \le g_e/|\delta g|$ and $\omega_e \tau_{off-axis} \le g_e/|\Delta g|$, where ω_e and g_e are the free-electron Larmor frequency and *g*-value, respectively. For comparison, the maximum

TABLE 8.6 Critical rotational correlation times for motional averaging of *g*-value anisotropy by axial (τ_{axial}) and off-axial ($\tau_{off\text{-}axis}$) rotation, with respect to the nitroxide *z*-axis

EPR frequency (GHz)	τ_{axial} (s)	$\tau_{off\text{-}axis}$ (s)
9.4	$<2.2\times10^{-8}$	$<6.4\times10^{-9}$
34	$<6.2\times10^{-9}$	$<1.8\times10^{-9}$
94	$<2.2\times10^{-9}$	$<6.4\times10^{-10}$
140	$<1.5\times10^{-9}$	$<4.3\times10^{-10}$
260	$<8.2\times10^{-10}$	$<2.3\times10^{-10}$
360	$<5.9\times10^{-10}$	$<1.7\times10^{-10}$

Note: *g*-tensor is for DOXYL nitroxides (see Table A2.1 in Chapter 2 Appendix).

correlation time for off-axis rotational averaging of anisotropy in ^{14}N-hyperfine splitting is $\tau_{off-axis}\le 1/\Delta A\approx 2.1\times10^{-9}$ s. The cut-off for averaging *g*-anisotropy reaches this level first at an EPR frequency of 34 GHz. As the microwave operating frequency increases, the minimum rotational frequency needed for rotational averaging also increases progressively. At EPR frequencies of 94 GHz and above, only subnanosecond off-axis correlation times can support motional averaging. Thus, for composite spin-label motions with different timescales, high-field spectra are sensitive only to the rapid rotational components.

Figure 8.26 illustrates separation of rotational timescales by using 94 GHz and 9 GHz spectra. Spin labels at different positions, *n*, down the lipid chain probe segmental dynamics and overall molecular reorientation in fluid-ordered membranes of a phospholipid–cholesterol mixture. We can simulate the high-field spectra (left panel) satisfactorily by using the fast-motion model with lateral ordering, which we described in Sections 8.6 and 8.9 to specify the line positions and linewidths. Using the same model and parameters does not reproduce the experimental 9-GHz spectra (right panel), because slow overall reorientation, which is frozen out at 94 GHz, contributes significantly to the lineshapes at lower frequency. Note that 9-GHz spin-label spectra are sensitive principally to off-axis excursions of the nitroxide *z*-axis. We see this specifically in the partially averaged values of $\langle A_{zz}\rangle$ deduced from 94-GHz spectra; these are higher than those from 9-GHz spectra, which include additional off-axis contributions from slow motion. We simulate the 9-GHz spectra by using the stochastic Liouville equation with spin-Hamiltonian parameters that we partially average according to Eqs. 8.39–8.41 (and equivalents for the hyperfine coupling), taking angular amplitudes deduced from the 94-GHz spectra (see Figure 8.10). These slow-motion simulations yield the dynamic parameters that govern angular diffusion of the lipid-chain axis in the membrane. Figure 8.27 gives the chain profile of the diffusion coefficient $D_{R\perp}$ (and $D_{R\parallel}$), and order parameter $S_{zz,over}$, for overall rotation. The diffusion coefficients correspond to correlation times in the range $\tau_{R\perp}=32-11$ ns for $n=4-13$, confirming that overall motion is in the slow regime at 9 GHz and would have relatively small effect on the 94-GHz spectra. The order parameter and diffusion coefficient $D_{R\perp}$ for overall rotation remain roughly constant in the upper part of the chain, which corresponds approximately with the location of the rigid steroid nucleus of cholesterol (cf. Kurad et al. 2001). Beyond this however, in the region occupied by the cholesterol hydrocarbon tail, the long axis of the phospholipid chain

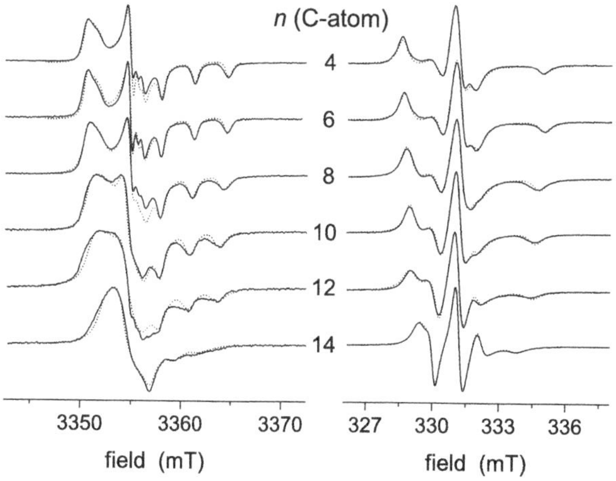

FIGURE 8.26 *Solid lines:* EPR spectra of *n*-DOXYL PC spin-labels in fully hydrated membranes of dimyristoyl PC/cholesterol (3:2 mol/mol) at 30°C. EPR frequency: 94-GHz (*left panel*); 9 GHz (*right panel*). *Dashed lines* (*left*): simulated 94-GHz spectra using motional narrowing theory (Sections 8.6, 8.9) with rigid-limit spin Hamiltonian tensors corrected to 30°C. *Dashed lines* (*right*): simulated 9-GHz spectra using stochastic-Liouville formalism for Brownian diffusion (Eq. 8.173) with motionally averaged spin-Hamiltonian tensors from fast-motion simulations of 94-GHz spectra. (Data from: Livshits et al., 2004; Kurad et al., 2001; Kurad et al. 2004.)

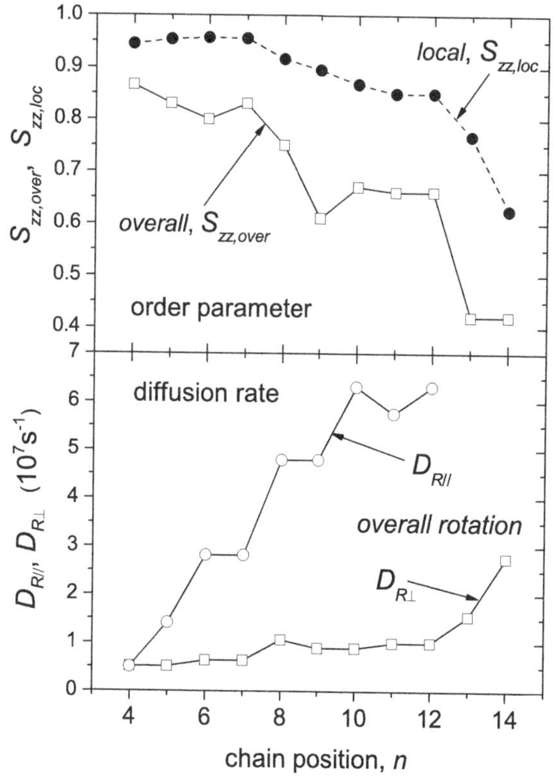

FIGURE 8.27 Dependence of long-axis order parameter $S_{zz,over}$ (upper panel) and rotational diffusion coefficients $D_{R\parallel}$, $D_{R\perp}$ (lower panel) for slow overall motion, on spin-label chain position of *n*-DOXYL PC in dimyristoyl PC/cholesterol (3:2 mol/mol) at 30°C. $S_{zz,loc}$ (dashed line): order parameter for rapid segmental chain motion. From combined simulations of 9- and 94-GHz spectra. (Data from: Livshits et al., 2004; Kurad et al., 2001; Kurad et al. 2004)

no longer approximates a rigid rod and reflects a varying dynamic environment on proceeding deeper into the membrane.

Livshits et al. (2004) discuss means of detecting residual slow-motion contributions in 94-GHz high-field spectra. This consists mainly of comparing effective order parameters deduced from the motionally-averaged hyperfine coupling $\langle A_{zz}\rangle$ with those deduced from the partially averaged g-values $\langle g_{zz}\rangle$. In the absence of slow-motion contributions, these should give the same order parameter $\langle P_2(\cos\theta_z)\rangle$ for rapid off-axial motion. Crucially, simulations of 94-GHz spectra for Brownian diffusion show that $\langle A_{zz}\rangle$ is insensitive to slow off-axis rotation, relative to $\langle g_{zz}\rangle$. Therefore, a lower effective value of $\langle P_2(\cos\theta_z)\rangle$ deduced from $\langle g_{zz}\rangle$ is diagnostic of residual contributions to the 94-GHz spectra from slow overall off-axial tumbling. Such contributions decrease with increasing EPR frequency, approaching progressively closer to the rigid-limit condition.

8.16 CONCLUDING SUMMARY

1. *Molecular ordering*:
Rapid rotation of limited amplitude averages part of the angular anisotropy in nitroxide EPR spectra. In this fast-motion regime, line positions are determined by the partially averaged **A**- and **g**-tensors.
2. *Partial motional averaging*:
Hyperfine splittings in fast-motion spectra (see Eqs. 8.6 and 8.7):

$$\langle A_{\parallel}\rangle = a_o + \tfrac{2}{3}\left(A_{zz} - \tfrac{1}{2}\left(A_{xx} + A_{yy}\right)\right)S_{zz} \quad (8.174)$$

$$\langle A_{\perp}\rangle = a_o - \tfrac{1}{3}\left(A_{zz} - \tfrac{1}{2}\left(A_{xx} + A_{yy}\right)\right)S_{zz} \quad (8.175)$$

are determined by the order parameter of the nitroxide z-axis:

$$S_{zz} \equiv \tfrac{1}{2}\left(3\langle\cos^2\theta_z\rangle - 1\right) = \frac{\langle A_{\parallel}\rangle - \langle A_{\perp}\rangle}{A_{zz} - \tfrac{1}{2}\left(A_{xx} + A_{yy}\right)} \quad (8.176)$$

3. *Order parameters and motional amplitudes*:
Order parameters are determined by the ensemble (or time) average of the angular orientation θ_z of the nitroxide z-axis to the director. For fixed orientation, $\theta_z = \beta$:

$$S_{zz} = \tfrac{1}{2}\left(3\cos^2\beta - 1\right) \quad (8.177)$$

For random wobble in cone of semi-angle θ_C (see Eq. 8.20):

$$S_{zz} = \tfrac{1}{2}\cos\theta_C\left(1 + \cos\theta_C\right) \quad (8.178)$$

Generally, for an orientation potential $U(\theta_z)$ (see Eq. 8.23):

$$\langle\cos^2\theta_z\rangle = \frac{\int \cos^2\theta_z \exp\left(-U(\theta_z)/k_BT\right)\sin\theta_z \cdot d\theta_z}{\int \exp\left(-U(\theta_z)/k_BT\right)\sin\theta_z \cdot d\theta_z} \quad (8.179)$$

where the Maier–Saupe potential is (Eq. 8.24):

$$U(\theta_z)/k_BT = -\tfrac{3}{2}\lambda_2\left(\cos^2\theta_z - \tfrac{1}{3}\right) \quad (8.180)$$

4. *Independent angular motions*:
Order parameters for combined independent motions are multiplicative: $S_{tot} = S_1 \times S_2$ (spherical-harmonic addition theorem). For independent segmental motions with individual segmental order parameters S_{seg}, the order parameter at the mth segment becomes:

$$S_m = S_o \times S_{seg}^m \quad (8.181)$$

5. *Angle-dependent linewidths*:
Linewidths of a nitroxide in the fast-motion regime are given by:

$$1/T_2(m_I) = A + Bm_I + Cm_I^2 \quad (8.182)$$

where coefficients A, B and C depend on orientation γ of the magnetic field to the director of uniaxial ordering according to (Eq. 8.94):

$$A = A_0 + A_2P_2(\cos\gamma) + A_4P_4(\cos\gamma) \quad (8.183)$$

where $P_2(\cos\gamma) = \tfrac{1}{2}\left(3\cos^2\gamma - 1\right)$ and $P_4(\cos\gamma) = \tfrac{1}{8}\left(35\cos^4\gamma - 30\cos^2\gamma + 3\right)$. The linewidth coefficients are determined by secular spectral densities $\left(j(\omega) \propto \tau_c\right)$ that depend on the two order parameters: $\langle P_2(\cos\theta_z)\rangle$ and $\langle P_4(\cos\theta_z)\rangle$.
6. *Ordering and slow motion*:
Slow-motion simulations with molecular ordering require introducing the orientation potential $U(\theta_z)$ into the stochastic Liouville equation, via the resulting torque $\mathbf{L}U(\theta_z)$. Instead of the simple diffusion equation $\partial P/\partial t = (\mathbf{L}\cdot\mathbf{D_R}\cdot\mathbf{L})P(\Omega,t)$, the term $(\mathbf{L}\cdot\mathbf{D_R}\cdot\mathbf{L}U(\Omega)/k_BT)P(\Omega,t)$ must be added, where $\mathbf{D_R}$ is the rotational diffusion tensor (see Eq. 8.139).
7. *Corrections to line splittings in calculating order parameters*:
For anisotropic powder patterns, slow motion causes A_{max} to increase and A_{min} to decrease, relative to the averages for fast motion $\langle A_{\parallel}\rangle$ and $\langle A_{\perp}\rangle$, respectively. Empirical corrections to both A_{max} and A_{min} depend on the motional rate. Calibrations for order parameter S_{zz} and isotropic hyperfine constant a_o^N are given in Appendix A8.3 (see, e.g., Eqs. 8.147 and 8.150, which are representative for lipid chains).
8. *Timescale separation in multifrequency EPR*:
Combined motions on different timescales, e.g., overall and segmental motion, are distinguished with multifrequency EPR. Increasing EPR frequency progressively drives motions into the rigid limit, as the Zeeman anisotropy comes to exceed the minimum motional rate for spectral averaging.

APPENDIX A8.1: CLEBSCH–GORDON COEFFICIENTS $C(L_1M_1L_2M_2;LM) \equiv C(L_1L_2L;M_1M_2) = \langle L_1L_2L;M_1M_2\rangle$

TABLE A8.1 $C(L_1, M-M_2, 2, M_2; LM) \equiv C(L_1 2L; M-M_2, M_2) = \langle L_1 2L; M-M_2, M_2\rangle$

L	$M_2 = \pm 2$	$M_2 = \pm 1$	$M_2 = 0$
4	$\sqrt{\frac{(L_1-1\pm M)(L_1\pm M)(L_1+1\pm M)(L_1+2\pm M)}{(2L_1+1)(2L_1+2)(2L_1+3)(2L_1+4)}}$	$\sqrt{\frac{(L_1+2-M)(L_1+2+M)(L_1\pm M)(L_1+1\pm M)}{(L_1+1)(L_1+2)(2L_1+1)(2L_1+3)}}$	$\sqrt{\frac{3(L_1+1-M)(L_1+2-M)(L_1+1+M)(L_1+2+M)}{(L_1+2)(2L_1+1)(2L_1+2)(2L_1+3)}}$
2	$\sqrt{\frac{3(L_1-1\pm M)(L_1\pm M)(L_1+1\mp M)(L_1+2\mp M)}{L_1(2L_1-1)(2L_1+2)(2L_1+3)}}$	$(1\mp 2M)\sqrt{\frac{3(L_1\pm M)(L_1+1\mp M)}{L_1(2L_1-1)(2L_1+2)(2L_1+3)}}$	$\frac{3M^2-L_1(L_1+1)}{\sqrt{L_1(L_1+1)(2L_1-1)(2L_1+3)}}$
0	$\sqrt{\frac{(L_1-1\mp M)(L_1\mp M)(L_1+1\mp M)(L_1+2\mp M)}{(2L_1-2)(2L_1-1)2L_1(2L_1+1)}}$	$-\sqrt{\frac{(L_1-1-M)(L_1-M)(L_1+1-M)(L_1-1+M)}{(L_1-1)L_1(2L_1-1)(2L_1+1)}}$	$\sqrt{\frac{3(L_1-1-M)(L_1-M)(L_1-1+M)(L_1+M)}{L_1(2L_1-2)(2L_1-1)(2L_1+1)}}$

– see Condon and Shortley (1935).
Note that: $\langle L_1 2L; M-M_2, M_2\rangle = (-1)^{L_1-L}\langle 2L_1L; M_2, M-M_2\rangle$.

TABLE A8.2 $C(2, M-M_2, 2, M_2; LM) \equiv C(22L; M-M_2, M_2) = \langle 22L; M-M_2, M_2\rangle$

L	M	$M_2 = \pm 2$	$M_2 = \pm 1$	$M_2 = 0$
4	± 4	1		
	± 3	$1/\sqrt{2}$	$1/\sqrt{2}$	
	± 2	$\sqrt{3/14}$	$\sqrt{4/7}$	$\sqrt{3/14}$
	± 1	$1/\sqrt{14}$	$\sqrt{3/7}$	$\sqrt{3/7}$
	0	$1/\sqrt{70}$	$\sqrt{8/35}$	$\sqrt{18/35}$
	∓ 1		$1/\sqrt{14}$	$\sqrt{3/7}$
2	± 2	$\sqrt{2/7}$	$-\sqrt{3/7}$	$\sqrt{2/7}$
	± 1	$\sqrt{3/7}$	$-1/\sqrt{14}$	$-1/\sqrt{14}$
	0	$\sqrt{2/7}$	$1/\sqrt{14}$	$-\sqrt{2/7}$
	∓ 1		$\sqrt{3/7}$	$-1/\sqrt{14}$
0	0	$1/\sqrt{5}$	$-1/\sqrt{5}$	$1/\sqrt{5}$

TABLE A8.3 $C(1,0,1,M_2;2,M) \equiv C(112;0,M_2) = \langle 112;0,M_2\rangle$

L	M	$M_2 = \pm 1$	$M_2 = 0$
2	± 1	$1/\sqrt{2}$	$1/\sqrt{2}$
	0	$1/\sqrt{6}$	$\sqrt{2/3}$

– see Condon and Shortley (1935).

APPENDIX A8.2: ALLOWANCE FOR NITROGEN NUCLEAR QUANTIZATION AXIS

As we saw in Section 2.2 of Chapter 2, the nitrogen nuclear spin is not quantized along the static field direction but along that of the hyperfine field **A.S**. The latter is inclined at angle ψ to the director, whereas the static $\mathbf{B}_0$ field makes an angle γ to the director (see Figure A8.1). In Section 8.10, we express the hyperfine Hamiltonian in the magnetic field axis system by using the reduced Wigner rotation matrix with elements $d^2_{N,K}(\gamma)$ (see Eq. 8.91). To express the nuclear spin operators **I** in the axis system where they are quantized, we must rotate by the additional angle $\chi = \gamma - \psi$.

From Figure A8.1, we see that the orientation of the nuclear spin quantization axis to the director is given by: $\sin\psi = \langle A_\perp\rangle \sin\gamma / \langle A(\gamma)\rangle$ and $\cos\psi = \langle A_\parallel\rangle\cos\gamma/\langle A(\gamma)\rangle$, where $\langle A(\gamma)\rangle = \sqrt{\langle A_\parallel\rangle^2\cos^2\gamma + \langle A_\perp\rangle^2\sin^2\gamma}$. We recognize the latter as the effective hyperfine coupling given by Eq. 8.16. The orientation χ of the nuclear quantization axis to the magnetic field is therefore given by:

$$\sin\chi = \frac{(\langle A_\parallel\rangle - \langle A_\perp\rangle)\sin\gamma\cos\gamma}{\sqrt{\langle A_\parallel\rangle^2\cos^2\gamma + \langle A_\perp\rangle^2\sin^2\gamma}} \tag{A8.1}$$

$$\cos\chi = \frac{\langle A_\parallel\rangle\cos^2\gamma + \langle A_\perp\rangle\sin^2\gamma}{\sqrt{\langle A_\parallel\rangle^2\cos^2\gamma + \langle A_\perp\rangle^2\sin^2\gamma}} \tag{A8.2}$$

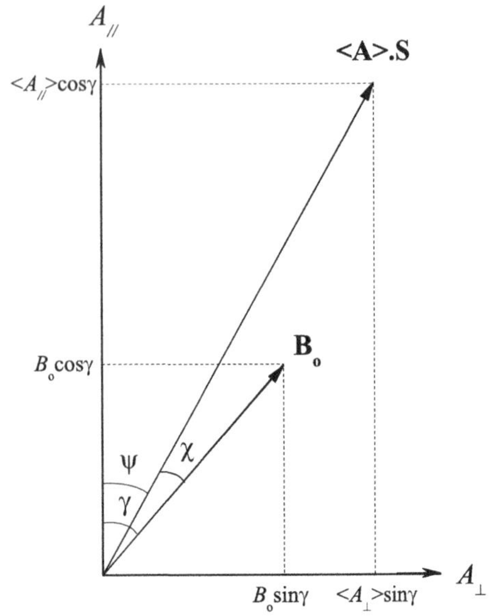

FIGURE A8.1 Relative orientations of hyperfine field, $\mathbf{A}\cdot\mathbf{S}$, and static spectrometer magnetic field, $\mathbf{B_o}$. Electron spin component S_z is directed along the magnetic field, which is inclined at angle γ to the director axis (||). Nitrogen nuclear spin component I_z is directed along $\mathbf{A}\cdot\mathbf{S}$, which makes angle ψ with the director.

The nuclear spin quantization axis becomes parallel to the magnetic field when $\gamma = 0$ and $\gamma = 90°$ but not for intermediate orientations of the magnetic field.

We neglect non-secular terms of the perturbing Hamiltonian (cf. Eq. 8.81), and in the remaining hyperfine terms we must isolate the nuclear spin operators. In the magnetic-field axis frame, the irreducible spherical tensor hyperfine spin operators that interest us are (see Appendix S, Eqs. S.8, S.12 and Table S.2):

$$T_{2,N}(SI) = \langle 112;0N\rangle S_{1,0} I_{1,N} \tag{A8.3}$$

where $I_{1,N}$ are the nuclear spin vectors, and $\langle L_1L_2L;M_1M_2\rangle$ are Clebsch–Gordan coefficients. Necessary values of $\langle 112;0N\rangle$ are given in Table A8.3 of Appendix A8.1. We use Eq. S.1 of Appendix S with reduced Wigner rotation matrix elements $d^1_{m,N}$ to transform the $I_{1,N}$ spin vectors to the nuclear quantization frame:

$$T_{2,N}(SI) = \langle 112;0N\rangle S_{1,0} \sum_{m=0,\pm1} d^1_{m,N}(\chi)\hat{I}_{1,m}$$

$$= \langle 112;0N\rangle \sum_{m=0,\pm1} d^1_{m,N}(\chi)\frac{\hat{T}_{2,m}(SI)}{\langle 112;0m\rangle} \tag{A8.4}$$

where we indicate the nuclear quantization frame with a circumflex hat. In the second equality of Eq. A8.4, we use Eq. A8.3 to express the spin-operator product as an irreducible spherical tensor $\hat{T}_{2,m}(SI)$ in the nuclear quantization frame.

It helps here to write the linewidth coefficients of Eq. 8.52 in terms of composite spectral densities similar to those used in Chapter 5 (see Eq. 5.57):

$$A = \tfrac{8}{3} j^{gg}(0)B_o^2 + I(I+1) j_1^{AA}(0) \tag{A8.5}$$

$$B = \tfrac{16}{3} j^{gA}(0)B_o \tag{A8.6}$$

$$C = \tfrac{8}{3} j_0^{AA}(0) - j_1^{AA}(0) \tag{A8.7}$$

where secular spectral densities have no subscript or $m = 0$, pseudosecular spectral densities have subscript $m = 1$ and we use the approximation $\omega_a = 0$ for the latter. Comparing with Eqs. 8.82–8.84 of Section 8.10, we see that:

$$j^{gg}(0) = \frac{1}{6}\left(\frac{\beta_e}{\hbar}\right)^2\left((\Delta g)^2 j_{00}(0) + 3(\delta g)^2 j_{02}(0)\right) \tag{A8.8}$$

$$j^{gA}(0) = \frac{1}{6}\left(\frac{\beta_e}{\hbar}\right)\left(\Delta g\Delta A\cdot j_{00}(0) + 3\delta g\delta A\cdot j_{02}(0)\right) \tag{A8.9}$$

$$j_m^{AA}(0) = \tfrac{1}{6}\left((\Delta A)^2 j_{m0}(0) + 3(\delta A)^2 j_{m2}(0)\right) \tag{A8.10}$$

where $j_{KM} \equiv \frac{1}{2}\left(j_{K,+M} + j_{K,-M}\right)$ are the spectral densities used in Sections 8.10 and 8.11.

For a general orientation of the magnetic field, the angular-dependent zero-frequency spectral densities that appear in Eqs. A8.5–A8.7 then become (see Luckhurst and Zannoni 1977):

$$j^{gg}(0,\gamma) = \tfrac{1}{6}\left(\beta_e/\hbar\right)^2 \sum_{K=-2}^{+2}\left((\Delta g)^2 j_{K0} + 3(\delta g)^2 j_{K2}\right)\left|d^2_{0K}(\gamma)\right|^2 \tag{A8.11}$$

$$j^{gA}(0,\gamma) = \tfrac{1}{2\sqrt{6}}\left(\beta_e/\hbar\right)\sum_{K,N}\left(\Delta g\Delta A j_{K0} + 3\delta g\delta A j_{K2}\right)$$
$$\times d^2_{0K}(\gamma) d^2_{NK}(\gamma) d^1_{0N}(\chi)\langle 112;0N\rangle \tag{A8.12}$$

$$j_m^{AA}(0,\gamma) = \frac{1}{6\langle 112;m0\rangle^2}\sum_{K,N,M}\left((\Delta A)^2 j_{K0} + 3(\delta A)^2 j_{K2}\right)$$
$$\times d^2_{NK}(\gamma) d^2_{MK}(\gamma) d^1_{mN}(\chi) d^1_{mM}(\chi)\langle 112;0N\rangle\langle 112;0M\rangle \tag{A8.13}$$

where d^L_{KM} are elements of reduced Wigner rotation matrices (see Table R.1 in Appendix R). Summations over N and M are from −1 to +1, in integer steps. As in Section 8.10, the angular-independent spectral densities j_{KM} for the strong-jump model are defined by Eq. 8.90. The $j^{gg}(0,\gamma)$ spectral density in Eq. A8.11 remains unchanged from that specified by Eq. 8.92 of Section 8.10, because it does not involve the nuclear spin.

APPENDIX A8.3: SLOW-MOTION ORDER PARAMETER CALIBRATIONS FOR DIFFERENT D_R

In Section 8.13, we gave expressions to calculate order parameters and isotropic hyperfine couplings by using values of A_{max} and A_{min} that we get from the outer and inner peak splittings in powder spectra of non-aligned samples. For the slow-motion regime, these are given respectively by Eqs. 8.147 and 8.150, which are based on the simulation parameters given in Table 8.3 and the corresponding spectra in Figure 8.20. The dynamic parameters chosen there represent spin-labelled lipid chains in membranes, by matching rotational rates to order parameter. However, in the slow-motion regime, we know from Figures 8.22 and 8.23 that the calibration parameters must depend on the particular value of the rotational diffusion coefficient D_R.

Here, we develop calibrations for different values of D_R in the slow-motion regime. For a fixed value of $D_R \equiv D_{R\perp}(=D_{R\parallel})$, we simulate spectra varying only the order parameter S_{zz}. Gaussian inhomogeneous broadening is kept constant at FWHM = 0.18 mT for all values of D_R. Figure A8.2 shows the dependence of the line shifts $d_{max} = A_{max} - \langle A_\parallel \rangle$ and $d_{min} = \langle A_\perp \rangle - A_{min}$ on S_{zz}, for different values of the diffusion coefficient D_R in the slow regime. This is analogous to Figure 8.21 in Section 8.13. In each case, the dependence is approximately linear over the range $S_{zz} = 0.2 - 0.8$, for both d_{max} and d_{min}. Regression lines are shown on the figure.

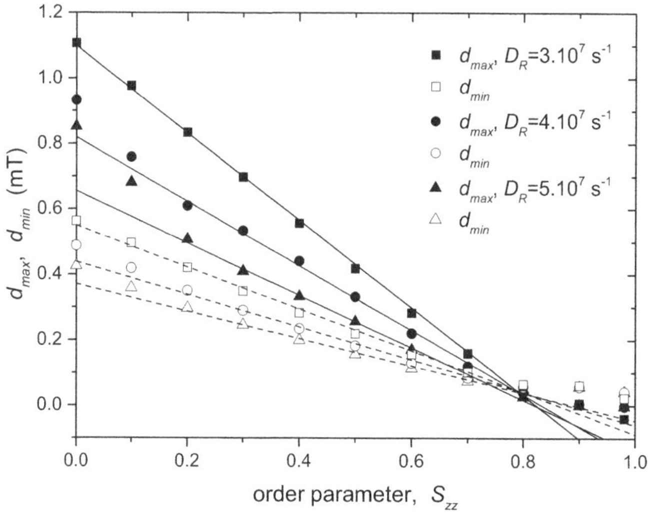

FIGURE A8.2 Dependence on order parameter S_{zz} of shifts, d_{max} (solid symbols) and d_{min} (open symbols), in A_{max} and A_{min} from their motionally averaged values $\langle A_\parallel \rangle$ and $\langle A_\perp \rangle$, because of slow motion and spectral overlap. Spectra simulated as in Figure 8.20 of Section 8.13, but with constant inhomogeneous broadening (FWHM = 0.18 mT), and constant diffusion coefficients D_R with values indicated. Straight lines: linear regressions over range $S_{zz} = 0.2 - 0.8$ (Eqs. A8.14, A8.15).

Directly analogous to Eqs. 8.145 and 8.146 of Section 8.13, empirically measured values of A_{max} and A_{min} thus depend on S_{zz} according to:

$$A_{max} = \langle A_\parallel \rangle + d^o_{max} - d'_{max} \times S_{zz} \tag{A8.14}$$

$$A_{min} = \langle A_\perp \rangle - d^o_{min} + d'_{min} \times S_{zz} \tag{A8.15}$$

Values of the linear regression parameters, over the range $S_{zz} = 0.2 - 0.8$, are given for different values of D_R in Table A8.4. Substituting from Eqs. A8.14 and A8.15 into Eq. 8.14 (in main text), we get the following calibration for the corrected order parameter:

$$S_{zz} = \frac{(A_{max} - A_{min}) - (d^o_{max} + d^o_{min})}{\Delta A - (d'_{max} + d'_{min})} \tag{A8.16}$$

in terms of the measurables A_{max} and A_{min}. We therefore write Eq. A8.16 as:

$$S_{zz} = s_1 \times (A_{max} - A_{min}) - s_o \tag{A8.17}$$

where the calibration parameters s_1 and s_o are given for different values of D_R in Table A8.5.

TABLE A8.4 Regression parameters in Eqs. A8.14 and A8.15 for A_{max} and A_{min}, as function of diffusion coefficient $D_R \equiv D_{R\perp}(=D_{R\parallel})$

$D_R(s^{-1})$	d^o_{max} (mT)	d'_{max} (mT)	d^o_{min} (mT)	d'_{min} (mT)
1.5×10^7	1.349	1.557	0.633	0.672
2.0×10^7	1.319	1.577	0.628	0.702
2.4×10^7	1.250	1.513	0.600	0.684
2.65×10^7	1.195	1.454	0.580	0.667
3.0×10^7	1.100	1.336	0.550	0.638
4.0×10^7	0.820	0.985	0.439	0.499
5.0×10^7	0.656	0.800	0.371	0.418
6.0×10^7	0.495	0.601	0.317	0.352
7.0×10^7	0.330	0.387	0.273	0.295
8.0×10^7	0.225	0.253	0.237	0.249
1.0×10^8	0.198[a]	0.240[a]	0.188	0.188
1.2×10^8	0.131[a]	0.158[a]	0.154	0.145
1.5×10^8	0.084[a]	0.103[a]	0.124	0.105
2×10^8	0.042[a]	0.052[a]	0.098	0.075

Empirical logistic fits (see dotted lines in Figure A8.3):

$$d^o_{max}(\text{mT}) = \frac{1.427}{1 + \left(D_R(s^{-1})/4.613 \times 10^7\right)^{2.782}} + 0.006$$

$$d'_{max}(\text{mT}) = \frac{1.581}{1 + \left(D_R(s^{-1})/4.684 \times 10^7\right)^{3.209}} + 0.065$$

$$d^o_{min}(\text{mT}) = \frac{0.625}{1 + \left(D_R(s^{-1})/5.062 \times 10^7\right)^{2.232}} + 0.065$$

$$d'_{min}(\text{mT}) = \frac{0.664}{1 + \left(D_R(s^{-1})/5.408 \times 10^7\right)^{2.677}} + 0.068$$

[a] Only over range $S_{zz} = 0.5 - 0.8$; otherwise $d^o_{max} \approx 0$, $d'_{max} \approx 0$ ($S_{zz} = 0.1 - 0.4$).

We also get the (corrected) isotropic hyperfine coupling $a_o^{\mathrm{N}} = \frac{1}{3}\left(\langle A_{\parallel}\rangle + 2\langle A_{\perp}\rangle\right)$ from Eqs. A8.14 and A8.15 (cf. Eq. 8.149):

$$a_o^{\mathrm{N}} = \tfrac{1}{3}\left(A_{\max} + 2A_{\min}\right) - \tfrac{1}{3}\left(d_{\max}^{o} - 2d_{\min}^{o}\right) + \tfrac{1}{3}\left(d_{\max}' - 2d_{\min}'\right) \times S_{zz} \tag{A8.18}$$

This depends on S_{zz}, which we must substitute from Eq. A8.16 (or A8.17) giving the following general form (cf. Eq. 8.150):

$$a_o^{\mathrm{N}} = \tfrac{1}{3}\left(f_{\max} A_{\max} + f_{\min} A_{\min}\right) + \delta a_o \tag{A8.19}$$

where the calibration parameters $f_{\max}$, $f_{\min}$ and δa_o are given for different values of D_R in Table A8.5.

Figure A8.3 shows the dependence of the line-shift parameters in Eqs. A8.14 and A8.15 on diffusion coefficient D_R. This is analogous to Figure 8.23 in Section 8.13, but now we use parameters for the S_{zz}-dependence of $d_{\max}$ and $d_{\min}$. From Figure A8.3, we see that the dependence for all four parameters is approximately logarithmic over the range $D_R = (2.65 - 8.0) \times 10^7\ \mathrm{s}^{-1}$, in the slow-motion regime. This lets us express the line-shift parameters directly in terms of D_R:

$$d_{\max}^{o}(\mathrm{mT}) = 16.1 - 2.01 \times \log_{10} D_R\left(\mathrm{s}^{-1}\right) \tag{A8.20}$$

$$d_{\max}'(\mathrm{mT}) = 19.8 - 2.47 \times \log_{10} D_R\left(\mathrm{s}^{-1}\right) \tag{A8.21}$$

$$d_{\min}^{o}(\mathrm{mT}) = 6.00 - 0.730 \times \log_{10} D_R\left(\mathrm{s}^{-1}\right) \tag{A8.22}$$

$$d_{\min}'(\mathrm{mT}) = 7.30 - 0.893 \times \log_{10} D_R\left(\mathrm{s}^{-1}\right) \tag{A8.23}$$

when restricted to this region of the slow-motion regime. We can represent the dependence over the entire range of diffusion coefficients by the empirical logistic equation (dotted lines in Figure A8.3); fits are given in the footnote to Table A8.4. Logistic fits to the calibration parameters for order parameter and a_o are given in the footnote to Table A8.5.

We can extend the analysis by relaxing the restriction that $D_{R\parallel} = D_{R\perp}$, which is an approximation for segmental motion but is not applicable to a non-spherical rigid rotor. Performing simulations with the previous values of $D_{R\perp}$ but now for $D_{R\parallel} = 10D_{R\perp}$, we again find linear dependences according to Eqs. A8.14 and A8.15 over the same range of S_{zz} (Marsh 2018b). Corresponding calibration parameters for S_{zz} and a_o^{N} (see Eqs. A8.17, A8.19) are listed in Table A8.6 for different values of $D_{R\perp}$. Fits of the logistic equation to these calibration parameters are given in the footnote to Table A8.6, for the full range of diffusion coefficients. Comparing with the calibration parameters for $D_{R\parallel} = D_{R\perp}$, we see that increasing the diffusion rate about the long molecular axis by a factor of ten has relatively minor effects on the calibration. These latter calibrations are appropriate to prolate rotors with axial ratios in the region of $a/b \approx 6.5$ (see Section 7.3

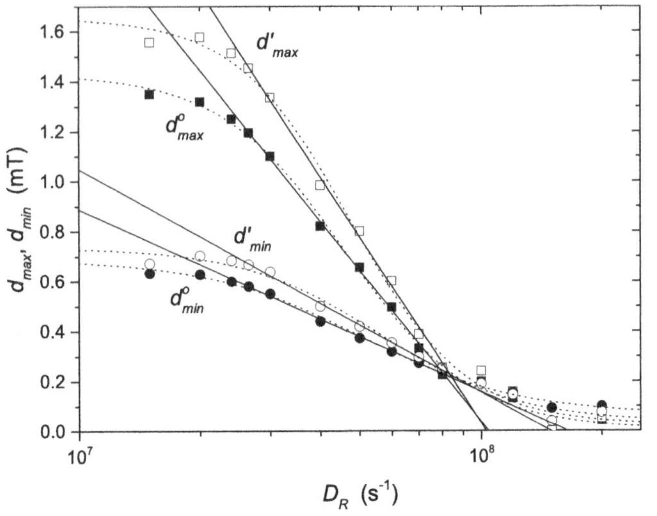

FIGURE A8.3 Dependence on rotational diffusion coefficient $D_R \equiv D_{R\perp} (= D_{R\parallel})$ of linear regression parameters for line shifts $d_{\max}$ and $d_{\min}$ as function of S_{zz} (Figure A8.2 and Table A8.4). Solid lines: linear regressions over range $D_R = (2.65 - 8.0) \times 10^7\ \mathrm{s}^{-1}$ for logarithmic x-axis. Dotted lines: empirical fits with logistic equation (Table A8.4 footnote).

TABLE A8.5 Regression parameters in Eqs. A8.17 and A8.19 for order parameter and isotropic hyperfine coupling, as a function of diffusion coefficient $D_R \equiv D_{R\perp} (= D_{R\parallel})$

$D_R\left(\mathrm{s}^{-1}\right)$	$s_1\left(\mathrm{mT}^{-1}\right)$	s_o	$f_{\max}$	$f_{\min}$	δa_o (mT)
1.5×10^7	2.01438	3.99366	1.14364	1.85636	−0.31248
2.0×10^7	2.24173	4.36568	1.12865	1.87135	−0.2714
2.4×10^7	1.89154	3.49836	1.09147	1.90853	−0.18617
2.65×10^7	1.6547	2.93689	1.06587	1.93413	−0.12903
3.0×10^7	1.33214	2.19694	1.02646	1.97354	−0.04372
4.0×10^7	0.80587	1.01428	0.99621	2.00379	0.02409
5.0×10^7	0.66375	0.68161	0.99214	2.00786	0.03678
6.0×10^7	0.56446	0.45829	0.98075	2.01925	0.06204
7.0×10^7	0.48962	0.29509	0.96673	2.03327	0.0917
8.0×10^7	0.45	0.20786	0.96319	2.03681	0.09983
1.0×10^{8}[a]	0.43535	0.16771	0.98012	2.01988	0.06687
1.2×10^{8}[a]	0.41288	0.11771	0.98171	2.01829	0.06421
1.5×10^{8}[a]	0.39729	0.08246	0.98571	2.01429	0.05752
	(0.38172	0.04724	0.9732	2.0268	0.08581)[b]
2.0×10^{8}[a]	0.38495	0.05415	0.98742	2.01258	0.05318
	(0.37738	0.0371	0.9811	2.0189	0.06739)[b]

Empirical logistic fits:

$$s_1\left(\mathrm{mT}^{-1}\right) = \frac{1.757}{1 + \left(D_R\left(\mathrm{s}^{-1}\right)/3.119 \times 10^7\right)^{5.367}} + 0.443$$

$$s_0 = \frac{4.213}{1 + \left(D_R\left(\mathrm{s}^{-1}\right)/3.043 \times 10^7\right)^{5.160}} + 0.183$$

$$f_{\max} = 0.980 + \frac{0.170}{1 + \left(D_R\left(\mathrm{s}^{-1}\right)/2.65 \times 10^7\right)^{6.53}}$$

$$f_{\min} = 2.020 - \frac{0.170}{1 + \left(D_R\left(\mathrm{s}^{-1}\right)/2.65 \times 10^7\right)^{6.53}}$$

$$\delta a_o(\mathrm{mT}) = 0.068 - \frac{0.399}{1 + \left(D_R\left(\mathrm{s}^{-1}\right)/2.63 \times 10^7\right)^{5.92}}$$

[a] $d_{\max}$ fitted only over range $S_{zz} = 0.5 - 0.8$ (see Table A8.4).

[b] Fits with $d_{\max}^{o} = 0$, $d_{\max}' = 0$.

in Chapter 7). For additional details, see also Marsh (2018b) but note that the calibration parameters given in the latter reference are incorrect (Marsh 2019); the correct values are those given here.

TABLE A8.6 Regression parameters in Eqs. A8.17 and A8.19 for order parameter and isotropic hyperfine coupling, as a function of diffusion coefficient $D_{R\perp}(=0.1\times D_{R\parallel})$

$D_{R\perp}(s^{-1})$	$s_1(mT^{-1})$	s_o	f_{max}	f_{min}	$\delta a_o(mT)$
1.5×10^7	2.25805	4.55239	1.08401	1.91599	−0.17575
2.0×10^7	2.35755	4.62642	1.10419	1.89581	−0.21582
2.4×10^7	1.9727	3.67669	1.07009	1.92991	−0.13875
2.65×10^7	1.69563	3.02791	1.051	1.949	−0.09576
3.0×10^7	1.33364	2.20281	1.01063	1.98937	−0.00925
4.0×10^7	0.82337	1.04946	0.97798	2.02202	0.0629
5.0×10^7	0.6322	0.6186	0.97215	2.02785	0.07795
6.0×10^7	0.55514	0.43971	0.97219	2.02781	0.0793
7.0×10^7	0.48987	0.29547	0.96391	2.03609	0.09822
8.0×10^7	0.47378	0.25449	0.97708	2.02292	0.07194
1.0×10^{8} [a]	0.43282	0.16264	0.97892	2.02108	0.06986
1.2×10^{8} [a]	0.41242	0.11679	0.98246	2.01754	0.06369
1.5×10^{8} [a]	0.39803	0.08341	0.98445	2.01555	0.05915
	(0.3824	0.04808	0.97197	2.02803	0.08735) [b]
2.0×10^{8} [a]	0.38495	0.05415	0.98742	2.01258	0.05318
	(0.37738	0.0371	0.9811	2.0189	0.06739) [b]

Empirical logistic fits:

$$s_1(mT^{-1})=\frac{1.980}{1+\left(D_R/2.979\times10^7\right)^{5.23}}+0.444$$

$$s_0=\frac{4.750}{1+\left(D_R/2.908\times10^7\right)^{5.028}}+0.185$$

$$f_{max}=0.977+\frac{0.116}{1+\left(D_R/2.77\times10^7\right)^{11.67}}$$

$$f_{min}=2.023-\frac{0.116}{1+\left(D_R/2.77\times10^7\right)^{11.67}}$$

$$\delta a_o(mT)=0.071-\frac{0.267}{1+\left(D_R/2.77\times10^7\right)^{10.84}}$$

[a] d_{max} fitted only over range $S_{zz}=0.5-0.8$ (cf. Table A8.5).

[b] Fits with $d^o_{max}=0$, $d'_{max}=0$.

REFERENCES

Cassol, R., Ge, M.T., Ferrarini, A., and Freed, J.H. 1997. Chain dynamics and the simulation of electron spin resonance spectra from oriented phospholipid membranes. *J. Phys. Chem. B* 101:8782–8789.

Condon, E.U. and Shortley, G.H. 1935. *Theory of Atomic Spectra*, Cambridge: Cambridge University Press.

Earle, K.A., Budil, D.E., and Freed, J.H. 1993. 250-GHz EPR of nitroxides in the slow-motional regime: models of rotational diffusion. *J. Phys. Chem.* 97:13289–13297.

Freed, J.H. 1976. Theory of slowly tumbling ESR spectra for nitroxides. In *Spin Labeling, Theory and Applications*, ed. Berliner, L.J., 53–132. New York: Academic Press.

Fretten, P., Morris, S.J., Watts, A., and Marsh, D. 1980. Lipid-lipid and lipid-protein interactions in chromaffin granule membranes. *Biochim. Biophys. Acta* 598:247–259.

Gaffney, B.J. 1976. Practical considerations for the calculation of order parameters for fatty acid or phospholipid spin labels in membranes. In *Spin Labeling. Theory and Applications*, ed. Berliner, L.J., 567–571. New York: Academic Press.

Glarum, S.H. and Marshall, J.H. 1967. Paramagnetic relaxation in liquid-crystal solvents. *J. Chem. Phys.* 46:55–62.

Griffith, O.H. and Jost, P.C. 1976. Lipid spin labels in biological membranes. In *Spin Labeling. Theory and Applications*, ed. Berliner, L.J., 453–523. New York: Academic Press.

Hemminga, M.A. 1974. The angular dependent linewidths of ESR spin probes in oriented smectic systems - application to oriented lecithin-cholesterol multibilayers. *Chem. Phys.* 6:87–99.

Hemminga, M.A. and Berendsen, H.J.C. 1972. Magnetic resonance in ordered lecithin-cholesterol multilayers. *J. Magn. Reson.* 8:133–143.

Hubbell, W.L. and McConnell, H.M. 1971. Molecular motion in spin-labelled phospholipids and membranes. *J. Am. Chem. Soc.* 93:314–326.

Israelachvili, J., Sjösten, J., Eriksson, L.E.G. et al. 1975. ESR spectral analysis of the molecular motion of spin labels in lipid bilayers and membranes based on a model in terms of two angular motional parameters and rotational correlation times. *Biochim. Biophys. Acta* 382:125–141.

Kurad, D., Jeschke, G., and Marsh, D. 2001. Spin-label HF-EPR of lipid ordering in cholesterol-containing membranes. *Appl. Magn. Reson.* 21:469–481.

Kurad, D., Jeschke, G., and Marsh, D. 2004. Lateral ordering of lipid chains in cholesterol-containing membranes: high-field spin-label EPR. *Biophys. J.* 86:264–271.

Lange, A., Marsh, D., Wassmer, K.-H., Meier, P., and Kothe, G. 1985. Electron spin resonance study of phospholipid membranes employing a comprehensive line-shape model. *Biochemistry* 24:4383–4392.

Livshits, V.A., Kurad, D., and Marsh, D. 2004. Simulation studies on high-field EPR of lipid spin labels in cholesterol-containing membranes. *J. Phys. Chem. B* 108:9403–9411.

Livshits, V.A. and Marsh, D. 2000. Simulation studies of high-field EPR spectra of spin-labeled lipids in membranes. *J. Magn. Reson.* 147:59–67.

Luckhurst, G.R. 1972. Electron spin relaxation in liquid crystals. In *Electron Spin Relaxation in Liquids*, eds. Muus, L.T. and Atkins, P.W., 411–442. New York: Plenum.

Luckhurst, G.R. and Sanson, A. 1972. Angular dependent linewidths for a spin probe dissolved in a liquid crystal. *Mol. Phys.* 24:1297–1311.

Luckhurst, G.R., Setaka, M., and Zannoni, C. 1974. An electron resonance investigation of molecular motion in the smectic A mesophase of a liquid crystal. *Mol. Phys.* 28:49–68.

Luckhurst, G.R., Setaka, M., Yeates, R.N., and Zannoni, C. 1979. Orientational order in the lamellar *G* phase of the sodium decanoate-*n*-decanol-water system. An electron resonance investigation. *Mol. Phys.* 38:1507–1520.

Luckhurst, G.R., Setaka, M., and Yeates, R.N. 1981. The effect of nuclear spin quantization on line broadening in the electron resonance spectra of spin probes dissolved in anisotropic media - an experimental investigation. *J. Magn. Reson.* 42:351–363.

Luckhurst, G.R., Smith, S.W., and Sundholm, F. 1987. A spin-labelled liquid crystal. A new spin probe for the study of the orientational order and molecular dynamics of liquid crystals. *Acta Chem. Scand. Ser. A* 41:218–229.

Luckhurst, G.R. and Zannoni, C. 1977. Line broadening in the electron resonance spectra of spin probes dissolved in anisotropic media: the effect of nuclear spin quantization. *Proc. Roy. Soc. A* 353:87–102.

Maier, W. and Saupe, A. 1959. Eine einfache molekular-statistische Theorie der nematischen kristallinflüssigen Phase. Teil 1. *Z. Naturforsch.* 14a:882–889.

Marsh, D. 2001. Polarity and permeation profiles in lipid membranes. *Proc. Natl. Acad. Sci. USA* 98:7777–7782.

Marsh, D. 2018a. Molecular order and T_1-relaxation, cross-relaxation in nitroxide spin labels. *J. Magn. Reson.* 290:38–45.

Marsh, D. 2018b. Spin-label order parameter calibrations for slow motion. *Appl. Magn. Reson.* 49:97–106.

Marsh, D. 2019. Correction to: "Spin-label order parameter calibrations for slow motion". *Appl. Magn. Reson.* 50:1241–1243.

Marsh, D., Jost, M., Peggion, C., and Toniolo, C. 2007. TOAC spin labels in the backbone of alamethicin: EPR studies in lipid membranes. *Biophys. J.* 92:473–481.

Meirovitch, E., Igner, D., Igner, E., Moro, G., and Freed, J.H. 1982. Electron-spin relaxation and ordering in smectic and supercooled nematic liquid crystals. *J. Chem. Phys.* 77:3915–3938.

Moro, G. and Freed, J.H. 1981. Calculation of ESR spectra and related Fokker-Planck forms by the use of the Lanczos algorithm. *J. Chem. Phys.* 74:3757–3772.

Moser, M., Marsh, D., Meier, P., Wassmer, K.-H., and Kothe, G. 1989. Chain configuration and flexibility gradient in phospholipid membranes. Comparison between spin-label electron spin resonance and deuteron nuclear magnetic resonance, and identification of new conformations. *Biophys. J.* 55:111–123.

Nordio, P.L. and Busolin, P. 1971. Electron spin resonance line shapes in partially oriented systems. *J. Chem. Phys.* 55:5485–5490.

Nordio, P.L., Rigatti, G., and Segre, U. 1972. Spin relaxation in nematic solvents. *J. Chem. Phys.* 56:2117–2123.

Pates, R.D. and Marsh, D. 1987. Lipid mobility and order in bovine rod outer segment disk membranes. A spin-label study of lipid-protein interactions. *Biochemistry* 26:29–39.

Polnaszek, C.F., Bruno, G.V., and Freed, J.H. 1973. ESR line shapes in the slow-motional region: Anisotropic liquids. *J. Chem. Phys.* 58:3185–3199.

Polnaszek, C.F. and Freed, J.H. 1975. Electron spin resonance studies of anisotropic ordering, spin relaxation, and slow tumbling in liquid-crystalline solvents. *J. Phys. Chem.* 79:2283–2306.

Rama Krishna, Y.V.S. and Marsh, D. 1990. Spin label ESR and ^{31}P-NMR studies of the cubic and inverted hexagonal phases of dimyristoylphosphatidylcholine/myristic acid (1:2, mol/mol) mixtures. *Biochim. Biophys. Acta* 1024:89–94.

Saupe, A. 1964. Kernresonanzen in kristallinen Flüssigkeiten und kristallin flüssigen Lösungen. *Z. Naturforsch.* 19a:161–171.

Schindler, H. and Seelig, J. 1973. EPR spectra of spin labels in lipid bilayers. *J. Chem. Phys.* 59:1841–1850.

Schneider, D.J. and Freed, J.H. 1989. Calculating slow motional magnetic resonance spectra: a user's guide. In *Biological Magnetic Resonance*, Vol. 8. Spin-Labeling. Theory and Applications, eds. Berliner, L.J. and Reuben, J., 1–76. New York: Plenum Publishing Corp.

Schorn, K. and Marsh, D. 1996a. Lipid chain dynamics and molecular location of diacylglycerol in hydrated binary mixtures with phosphatidylcholine: spin label ESR studies. *Biochemistry* 35:3831–3836.

Schorn, K. and Marsh, D. 1996b. Lipid chain dynamics in diacylglycerol-phosphatidylcholine mixtures studied by slow-motional simulations of spin label ESR spectra. *Chem. Phys. Lipids* 82:7–14.

Schorn, K. and Marsh, D. 1997. Extracting order parameters from powder EPR lineshapes for spin-labelled lipids in membranes. *Spectrochim. Acta A* 53:2235–2240.

Schreier-Muccillo, S., Marsh, D., Dugas, H., Schneider, H., and Smith, I.C.P. 1973. A spin probe study of the influence of cholesterol on motion and orientation of phospholipids in oriented multibilayers and vesicles. *Chem. Phys. Lipids* 10:11–27.

Seelig, J. 1970. Spin label studies of oriented smectic liquid crystals (a model system for bilayer membranes). *J. Am. Chem. Soc.* 92:3881–3887.

Seelig, J., Limacher, H., and Bader, P. 1972. Molecular architecture of liquid crystalline bilayers. *J. Am. Chem. Soc.* 94:6364–6371.

Stoll, S. and Schweiger, A. 2007. EasySpin: simulating CW ESR spectra. In *Biological Magnetic Resonance*, Vol. 27. ESR Spectroscopy in Membrane Biophysics, eds. Hemminga, M.A. and Berliner, L.J., 299–321. New York: Kluwer Publishing.

Van, S.P., Birrell, G.B., and Griffith, O.H. 1974. Rapid anisotropic motion of spin labels. Models for motion averaging of the ESR parameters. *J. Magn. Reson.* 15:444–459.

Spin–Spin Interactions

9

9.1 INTRODUCTION

There are two types of interactions that take place between the electron spins of neighbouring nitroxides. One is the classical dipole–dipole interaction between the magnetic moments that are associated with each electron spin. The other is the Heisenberg exchange interaction, which is an electrostatic interaction that is purely quantum mechanical in origin. Heisenberg exchange manifests itself as an effective spin–spin interaction because of the Pauli principle, which requires that the total wave function (space plus spin) of the electron pair is antisymmetric with respect to interchange of the two electrons. The dipole–dipole interaction depends on the spin separation and is used for distance measurements. Heisenberg exchange essentially is a contact interaction, depending on electron orbital overlap, and is used to measure collision frequencies, second-order rate constants and translational diffusion rates.

Including both dipole–dipole interaction and Heisenberg exchange, we can write the Hamiltonian for spin–spin interaction between electrons 1 and 2 as:

$$\mathcal{H}'_{12} = \mathcal{H}_{dd} + \mathcal{H}_{HE}$$
$$= \frac{\mu_o}{4\pi}\frac{g^2\beta_e^2}{r_{12}^3}\left[\mathbf{s}_1\cdot\mathbf{s}_2 - 3\left(l_x^2 s_{1x}s_{2x} + l_y^2 s_{1y}s_{2y} + l_z^2 s_{1z}s_{2z}\right)\right] + J_{12}\mathbf{s}_1\cdot\mathbf{s}_2 \quad (9.1)$$

where (l_x, l_y, l_z) are the direction cosines of interspin vector $\mathbf{r}_{12}$ in the axis system of the static magnetic field $\mathbf{B}_o$, and J_{12} is the exchange constant. Whereas the exchange interaction is isotropic, the dipole–dipole interaction is totally anisotropic. Isotropic rotational diffusion of sufficiently fast rate averages the dipolar interaction to zero, because then $\langle l_x^2\rangle = \langle l_y^2\rangle = \langle l_z^2\rangle \equiv \int \cos^2\theta \sin\theta\, d\theta = \frac{1}{3}$.

The $\mathbf{s}_1\cdot\mathbf{s}_2$ term in both the dipole–dipole and exchange terms of Eq. 9.1 couples the two individual electron spins to a total spin $\mathbf{S} = \mathbf{s}_1 + \mathbf{s}_2$, where $(\mathbf{s}_1+\mathbf{s}_2)^2 = s_1(s_1+1) + s_2(s_2+1) + 2\mathbf{s}_1\cdot\mathbf{s}_2$, with eigenvalues specified by the quantum number $S = s_1 + s_2 = 1$ or 0 (see Appendices E.3, K.1 and Q.1). In the presence of exchange alone, EPR transitions $(\Delta S = 0, \Delta M_S = \pm 1)$ take place only within the triplet state ($S = 1$) and do not depend on J_{12}. In the presence dipole–dipole coupling, however, the remaining terms in the dipolar Hamiltonian can induce singlet-triplet mixing, and the EPR transitions then involve both the dipole–dipole interaction and spin exchange.

9.2 MAGNETIC DIPOLE–DIPOLE INTERACTIONS – LIKE SPINS AND STRONG COUPLING

The classical expression for the interaction energy of two magnetic dipoles, $\boldsymbol{\mu}_1$ and $\boldsymbol{\mu}_2$, is (see Appendix F.4):

$$\mathcal{H}_{dd} = \frac{\mu_o}{4\pi}\left(\frac{\boldsymbol{\mu}_1\cdot\boldsymbol{\mu}_2}{r_{12}^3} - \frac{3(\boldsymbol{\mu}_1\cdot\mathbf{r}_{12})(\boldsymbol{\mu}_2\cdot\mathbf{r}_{12})}{r_{12}^5}\right) \quad (9.2)$$

where $\mathbf{r}_{12}$ is the vector joining the two dipoles (see Figure 9.1) and μ_o is the permeability of free space ($\mu_o/4\pi = 10^{-7}$ H m^{-1}, see Appendix A). For two electron spins, $\mathbf{s}_1$ and $\mathbf{s}_2$, with dipole moments $\boldsymbol{\mu}_1 = g\beta_e\mathbf{s}_1$ and $\boldsymbol{\mu}_2 = g\beta_e\mathbf{s}_2$, the Hamiltonian for the dipole–dipole interaction is therefore:

$$\mathcal{H}_{dd} = \frac{\mu_o}{4\pi}g^2\beta_e^2\left(\frac{\mathbf{s}_1\cdot\mathbf{s}_2}{r_{12}^3} - \frac{3(\mathbf{s}_1\cdot\mathbf{r}_{12})(\mathbf{s}_2\cdot\mathbf{r}_{12})}{r_{12}^5}\right) \quad (9.3)$$

where β_e is the Bohr magneton. We neglect the small anisotropy of the nitroxide g-values. Expanding the scalar products in Eq. 9.3 (see Appendix B.1) gives:

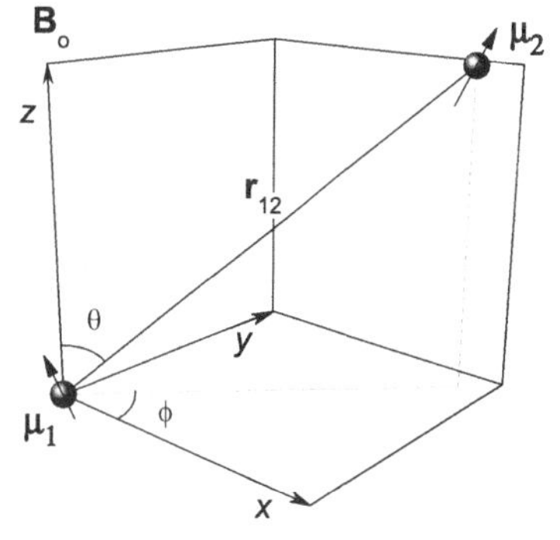

FIGURE 9.1 Coordinate system for two interacting magnetic dipoles, $\boldsymbol{\mu}_1$ and $\boldsymbol{\mu}_2$. Interaction energy of dipole $\boldsymbol{\mu}_2$ with magnetic field $\mathbf{B}_{dip}$ from dipole $\boldsymbol{\mu}_1$ is $\mathcal{H}_{dd} = -\boldsymbol{\mu}_2\cdot\mathbf{B}_{dip}$, and vice-versa. Static laboratory magnetic field, $\mathbf{B}_o$, defines z-direction of spin quantization.

$$\mathcal{H}_{dd} = \frac{\mu_o}{4\pi}\left(\frac{g^2\beta_e^2}{r_{12}^5}\right)\Big[\left(r_{12}^2 - 3x^2\right)s_{1x}s_{2x} + \left(r_{12}^2 - 3y^2\right)s_{1y}s_{2y} + \left(r_{12}^2 - 3z^2\right)s_{1z}s_{2z} - 3xy\left(s_{1x}s_{2y} + s_{1y}s_{2x}\right) - 3yz\left(s_{1y}s_{2z} + s_{1z}s_{2y}\right) - 3zx\left(s_{1z}s_{2x} + s_{1x}s_{2z}\right)\Big] \quad (9.4)$$

Note that the trace (i.e., sum of the diagonal elements) of the dipolar interaction tensor is zero because: $r_{12}^2 = x^2 + y^2 + z^2$. Therefore, static magnetic dipole–dipole interactions are averaged out by rapid rotational diffusion and do not contribute to the line positions or intensities of isotropic EPR spectra. However, dynamic modulation of the dipole–dipole interaction by rotational diffusion gives rise to relaxation and thereby contributes to the EPR linewidths (see Section 9.5.1, later in the chapter).

We now rewrite Eq. 9.4 in terms of polar coordinates instead of Cartesian coordinates (see Figure 9.1) and replace the S_x and S_y components by the shift operators, $S_+ = S_x + iS_y$ and $S_- = S_x - iS_y$:

$$\mathcal{H}_{dd} = \frac{\mu_o}{4\pi}\left(\frac{g^2\beta_e^2}{r_{12}^3}\right)[A + B + C + D + E + F] \quad (9.5)$$

with

$$A = \left(1 - 3\cos^2\theta\right)s_{1z}s_{2z} \quad (9.6)$$

$$B = -\tfrac{1}{4}\left(1 - 3\cos^2\theta\right)\left[s_{1+}s_{2-} + s_{1-}s_{2+}\right] \quad (9.7)$$

$$C = D^{\dagger} = -\tfrac{3}{2}\sin\theta\cos\theta\exp(-i\phi)\left[s_{1z}s_{2+} + s_{1+}s_{2z}\right] \quad (9.8)$$

$$E = F^{\dagger} = -\tfrac{3}{4}\sin^2\theta\exp(-2i\phi)s_{1+}s_{2+} \quad (9.9)$$

where we group the various terms according to their different matrix elements, and the dagger indicates the Hermitean conjugate (see Appendix C.1, C.2). Note that $S_+^{\dagger} = S_-$ and vice-versa. The z-component of the spin is determined by the direction of the static magnetic field, $\mathbf{B}_o$, because the Zeeman energy is much larger than the magnetic dipole–dipole interaction. (The latter corresponds maximally to magnetic field strengths of a few tens of mT, for closely approaching spin labels.) The C, D, E and F terms, which are given by Eqs. 9.8 and 9.9, have off-diagonal matrix elements that couple states separated by the full Zeeman interaction of one or both spins. Therefore, they contribute to the dipolar energy only in second order (cf. Appendix G), and we can neglect them. However, we must retain the B term in the case of strong dipolar coupling, because the states that it couples are then almost degenerate.

The spin Hamiltonian for two spin labels, including dipole–dipole interaction, is thus:

$$\mathcal{H}_{12} = \hbar\left(\omega_1 s_{1z} + \omega_2 s_{2z}\right) + \frac{\mu_o}{4\pi}\left(\frac{g^2\beta_e^2}{r_{12}^3}\right) \times \left(1 - 3\cos^2\theta\right)\left[s_{1z}s_{2z} - \tfrac{1}{4}\left(s_{1+}s_{2-} + s_{1-}s_{2+}\right)\right] \quad (9.10)$$

where ω_1 and ω_2 are the angular resonance frequencies of the two spins, which include both the nitrogen hyperfine structure and the hyperfine and Zeeman anisotropies. Written in matrix form, the spin Hamiltonian is given by (cf. Appendix H):

$\mathcal{H}_{12}$	$\left\vert+\frac{1}{2},+\frac{1}{2}\right\rangle$	$\left\vert+\frac{1}{2},-\frac{1}{2}\right\rangle$	$\left\vert-\frac{1}{2},+\frac{1}{2}\right\rangle$	$\left\vert-\frac{1}{2},-\frac{1}{2}\right\rangle$
$\left\langle+\frac{1}{2},+\frac{1}{2}\right\vert$	$+\frac{1}{2}\hbar(\omega_1+\omega_2)+\frac{1}{4}D_{dd}$			
$\left\langle+\frac{1}{2},-\frac{1}{2}\right\vert$		$+\frac{1}{2}\hbar(\omega_1-\omega_2)-\frac{1}{4}D_{dd}$	$-\frac{1}{4}D_{dd}$	
$\left\langle-\frac{1}{2},+\frac{1}{2}\right\vert$		$-\frac{1}{4}D_{dd}$	$-\frac{1}{2}\hbar(\omega_1-\omega_2)-\frac{1}{4}D_{dd}$	
$\left\langle-\frac{1}{2},-\frac{1}{2}\right\vert$				$-\frac{1}{2}\hbar(\omega_1+\omega_2)+\frac{1}{4}D_{dd}$

where we use $\left\vert m_{s1}, m_{s2}\right\rangle$ as basis functions for the s_1,s_2-spin system, and $D_{dd} \equiv \left(\mu_o/4\pi\right)\left(g^2\beta_e^2/r_{12}^3\right)\left(1 - 3\cos^2\theta\right)$. If the dipolar interaction is larger than the total spectral extent of a single spin label, i.e., $D_{dd} >> \hbar\left\vert\omega_1 - \omega_2\right\vert$, the off-diagonal elements couple the $\left\vert\pm\frac{1}{2},\mp\frac{1}{2}\right\rangle$ states yielding triplet and singlet states with total spin $S = s_1 + s_2 = 1$ or 0, respectively, just as in the case of strong exchange (cf. Section 9.7, Eqs. 9.70–9.73 appearing later). We can see this directly for like spins, where $\omega_1 - \omega_2 = 0$, by diagonalizing the spin-Hamiltonian submatrix directly (see Appendix G).

Alternatively, we can express Eq. 9.10, as far as possible, in terms of the total spin, S. To do this, we write the A and B dipolar terms from Eqs. 9.6 and 9.7 as:

$$A + B = \tfrac{1}{2}\left(3s_{1z}s_{2z} - \mathbf{s}_1\cdot\mathbf{s}_2\right)D_{dd} \quad (9.11)$$

where $s_{1z}s_{2z} = \frac{1}{2}\left(S_z^2 - s_{1z}^2 - s_{2z}^2\right) = \frac{1}{2}\left(S_z^2 - \frac{1}{2}\right)$, and $\mathbf{s}_1\cdot\mathbf{s}_2 = \frac{1}{2}\left(\mathbf{S}^2 - \mathbf{s}_1^2 - \mathbf{s}_2^2\right) = \frac{1}{2}\left(S(S+1) - \frac{3}{2}\right)$ is a constant. The term involving the Zeeman energy is given by: $\omega_1 s_{1z} + \omega_2 s_{2z} = \frac{1}{2}\left(\omega_1 + \omega_2\right)S_z + \frac{1}{2}\left(\omega_1 - \omega_2\right)\left(s_{1z} - s_{2z}\right)$. We then get the Hamiltonian for a strongly dipolar-coupled pair of (*like*) spins by putting $\omega_1 - \omega_2 = 0$:

$$\mathcal{H}_{12} = \tfrac{1}{2}\hbar\left(\omega_1 + \omega_2\right)S_z + \frac{3}{4}\cdot\frac{\mu_o}{4\pi}\left(\frac{g^2\beta_e^2}{r_{12}^3}\right)\left(1 - 3\cos^2\theta\right)S_z^2 \quad (9.12)$$

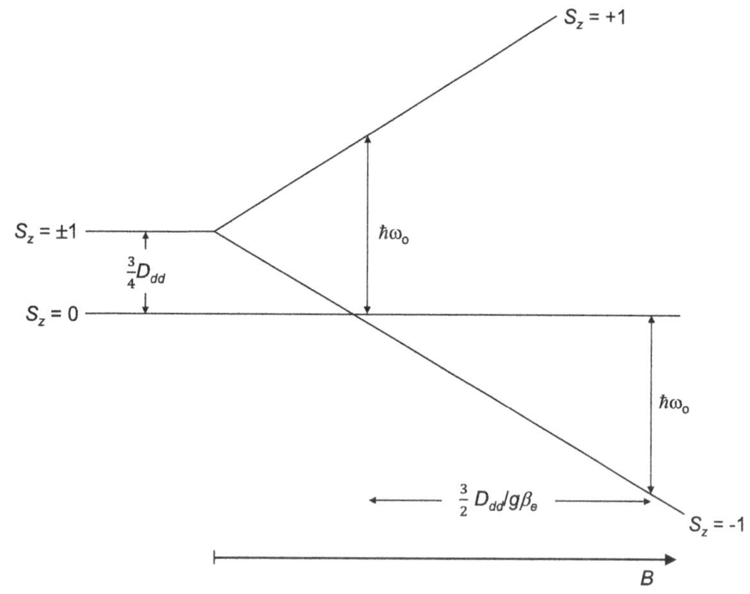

FIGURE 9.2 Triplet energy levels ($S = 1$) for a pair of radicals with electron spins $s_1 = s_2 = \frac{1}{2}$ that are strongly coupled by magnetic dipole–dipole interaction, where $D_{dd} = (\mu_o/4\pi)(g^2\beta_e^2/r_{12}^3)(1-3\cos^2\theta)$. Allowed EPR transitions are governed by selection rules: $\Delta S = 0$, $\Delta M_S = \pm 1$, and are split by $2|\Delta B_{dd}| = \frac{3}{2}D_{dd}/g\beta_e$ along the field axis.

where we are interested only in energy differences within the triplet state ($S = 1$), and therefore neglect all constant terms. Note that, for the same reason, isotropic Heisenberg exchange does not contribute to the spectral splittings for like spins (see Figure 9.11 given later).

The allowed EPR transitions $(\Delta S = 0, \Delta M_S = \pm 1)$ between the triplet energy levels are given from Eq. 9.12 by:

$$\hbar\omega_o = \tfrac{1}{2}\hbar(\omega_1 + \omega_2) \pm \frac{3}{4}\cdot\frac{\mu_o}{4\pi}\left(\frac{g^2\beta_e^2}{r_{12}^3}\right)\left(1 - 3\cos^2\theta\right) \tag{9.13}$$

where ω_o is the angular frequency of the microwave radiation, as indicated in Figure 9.2. Correspondingly, the angular dependence of the dipolar splitting of the resonance field in the EPR spectrum from a pair of like spin labels is given by:

$$\pm\Delta B_{dd}(\theta) = \frac{3}{4}\cdot\frac{\mu_o}{4\pi}\left(\frac{g\beta_e}{r_{12}^3}\right)\left(1 - 3\cos^2\theta\right) \tag{9.14}$$

which varies from $\Delta B_{dd}^{\parallel} = -\frac{3}{2}(\mu_o/4\pi)g\beta_e/r_{12}^3$ to $\Delta B_{dd}^{\perp} = +\frac{3}{4}(\mu_o/4\pi)g\beta_e/r_{12}^3$, for $\theta = 0°$ and $\theta = 90°$, respectively. Here θ is the inclination of the interspin vector $\mathbf{r}_{12}$ to the static field $\mathbf{B}_o$ (see Figure 9.1).

9.3 DIPOLAR POWDER SPECTRA: THE PAKE DOUBLET

For a randomly oriented sample, the dipolar powder spectrum is given from Eq. 2.15 in Chapter 2 by:

$$I(\Delta B_{dd}) = \frac{\sin\theta}{\mathrm{d}\Delta B_{dd}/\mathrm{d}\theta} \tag{9.15}$$

where $I(\Delta B_{dd})$ is the intensity at field position ΔB_{dd}. Using Eq. 9.14 and expressing θ in terms of ΔB_{dd}, one half of the dipolar absorption powder pattern is given by:

$$I(\Delta B_{dd}) = \frac{1}{2\sqrt{3}D_{dd}^o}\frac{1}{\sqrt{\left(1 - \Delta B_{dd}/D_{dd}^o\right)}} \tag{9.16}$$

where $D_{dd}^o = \frac{3}{4}(\mu_o/4\pi)g\beta_e/r_{12}^3$ for like spins, and ΔB_{dd} extends from $-2D_{dd}^o$ to $+D_{dd}^o$, over which range $I(\Delta B_{dd})$ is normalized. We get the other half of the powder pattern by reversing the sign of ΔB_{dd}: it extends from $-D_{dd}^o$ to $+2D_{dd}^o$. The top panel of Figure 9.3 shows the full powder distribution of resonances. We call it the Pake powder pattern, or Pake doublet, after the original analysis of magnetic dipole–dipole interaction in broad-line NMR by Pake (1948). The envelope is dominated by discontinuities that correspond to the perpendicular orientation ($\theta = 90°$); these are split by $2D_{dd}^o = \frac{3}{2}(\mu_o/4\pi)g\beta_e/r_{12}^3$, for like spins. To get the EPR powder lineshape, we must convolute the envelope function with the intrinsic lineshape function at a given orientation, θ. For the absorption EPR lineshape (middle panel of Figure 9.3), we use the intrinsic absorption lineshape for convolution. For the first-derivative EPR lineshape (bottom panel of Figure 9.3), the envelope function remains unchanged but is now convoluted with the first-derivative intrinsic lineshape (see Eq. T.14 in Appendix T). Note that the peaks in the absorption spectrum do not coincide with the discontinuities in the envelope pattern at $\pm D_{dd}^o$. A better approximation to the latter are the points of maximum slope in the absorption, i.e., the major peaks in the derivative spectrum $\mathrm{d}I(B)/\mathrm{d}B$. The parallel edges, which produce absorption

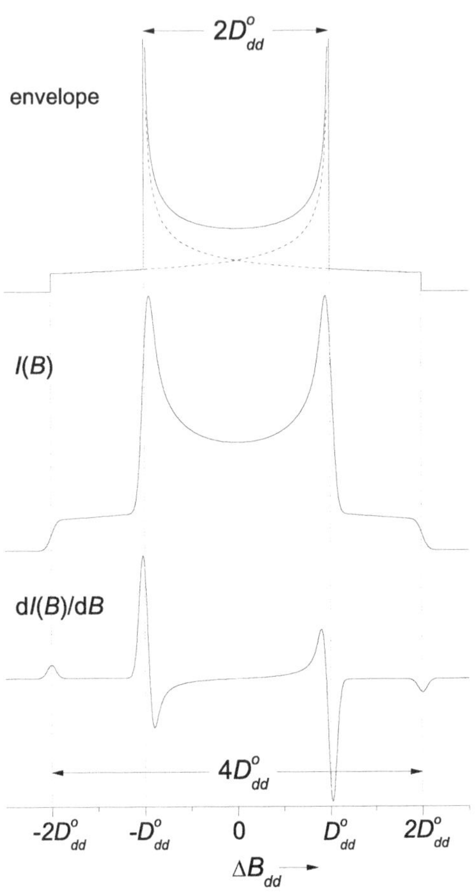

FIGURE 9.3 Dipolar absorption powder pattern envelope (top). Dashed lines: Eq. 9.16 and its mirror image about $\Delta B_{dd} = 0$. Splitting of major peaks for *like* spins is $2D^o_{dd} = \frac{3}{2}\mu_o\left(g\beta_e/4\pi r_{12}^3\right)$ in magnetic-field units, independent of exchange interaction. For *unlike* spins, splitting in field units is $\frac{4}{3}D^o_{dd} - J_{12}/g\beta_e = \mu_o\left(g\beta_e/4\pi r_{12}^3\right) - J_{12}/g\beta_e$, where J_{12} is exchange constant. Middle: envelope convoluted with absorption lineshape (see Eq. 2.39); bottom: first derivative.

lineshapes in the outer wings of the first-derivative spectrum, are split by exactly $\pm 2D^o_{dd}$ (cf. Section 2.4 in Chapter 2). In powder samples, we can use the parallel edges to determine dipolar splittings, if these considerably exceed the total hyperfine anisotropy (Luckhurst 1976; Keana and Dinerstein 1971). However, for smaller dipolar splittings and for distributions of interspin distances, these outer edges tend to be smeared out.

We now introduce hyperfine and g-value anisotropy for a strongly dipolar-coupled pair of spin labels. The conventional $S = s_1 + s_2 = 1$ triplet-state Hamiltonian is:

$$\mathcal{H}_S = \beta_e \mathbf{B}\cdot\mathbf{g}\cdot\mathbf{S} + DS_Z^2 + E\left(S_X^2 - S_Y^2\right) + \tfrac{1}{2}(\mathbf{I}_1 + \mathbf{I}_2)\cdot\mathbf{A}\cdot\mathbf{S} \quad (9.17)$$

where the X,Y,Z-axis system is that of the spin pair, i.e., the Z-axis lies along the interspin vector $\mathbf{r}_{12}$, and the orientation of the magnetic field relative to this Z-axis is given by: $\mathbf{B} = B_o(\sin\theta\cos\phi, \sin\theta\sin\phi, \cos\theta)$ – see Figure 9.1. We write the N-hyperfine interaction for the two spins as: $\mathbf{I}_1\cdot\mathbf{A}\cdot\mathbf{s}_1 + \mathbf{I}_2\cdot\mathbf{A}\cdot\mathbf{s}_2 = \frac{1}{2}(\mathbf{I}_1+\mathbf{I}_2)\cdot\mathbf{A}\cdot\mathbf{S} + \frac{1}{2}(\mathbf{I}_1-\mathbf{I}_2)\cdot\mathbf{A}\cdot(\mathbf{s}_1-\mathbf{s}_2)$ in Eq. 9.17, but omit the term in $\mathbf{s}_1 - \mathbf{s}_2$ because it connects triplet and singlet states, and thus contributes only in second order for strongly coupled systems. Similarly, Heisenberg exchange does not appear in Eq. 9.17 because it contributes only to the singlet-triplet splitting. Both $\mathbf{A}$ and $\mathbf{g}$ are the single-nitroxide values – of the hyperfine and g-tensor, respectively. The dipole–dipole interaction enters as the D- and E-terms in the triplet Hamiltonian: $D = \frac{3}{2}\left(\mu_o/4\pi\right)g_{ZZ}^2\beta_e^2/r_{12}^3$ (see Eq. 9.12 for $\theta = 0$) and $E = \frac{1}{4}\left(\mu_o/4\pi\right)\left(g_{XX}^2 - g_{YY}^2\right)\beta_e^2/r_{12}^3$. The non-axiality of the dipolar interaction, which hitherto we have ignored, is therefore given by: $E/D \approx \frac{1}{3}\left(g_{XX} - g_{YY}\right)/g_{ZZ}$. Taking the maximum g-value anisotropy for a nitroxide spin label (i.e., $g_{xx} - g_{zz} \approx 6.8\times10^{-3}$), this only amounts to $E/D \approx 10^{-3}$.

Figure 9.4 shows representative simulated dipolar powder spectra that include nitroxide hyperfine and g-value anisotropies. The spectra are for spin pairs separated by $r_{12} = 0.6$ nm, which corresponds to strong dipolar coupling ($D = 12.92$ mT) in 9-GHz EPR. In the upper panel ($y_{NO} \parallel \mathbf{r}_{12}$), the principal z-axis of the

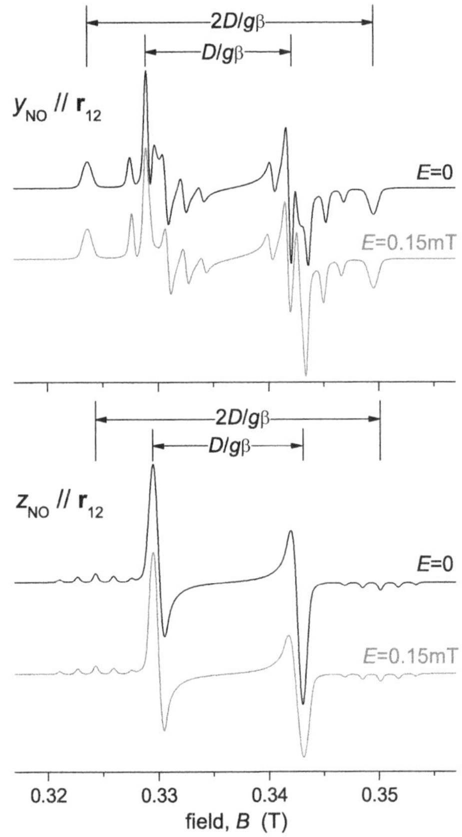

FIGURE 9.4 Dipolar powder spectra for strong coupling in rigid nitroxide biradical; separation between spins $r_{12} = 0.6$ nm. Spectra simulated using triplet spin Hamiltonian (Eq. 9.17). Nitroxide z-axes perpendicular to interspin vector $\mathbf{r}_{12}$ (top panel) and parallel to $\mathbf{r}_{12}$ (bottom panel). Spin-Hamiltonian tensors (top panel): $(g_{XX}, g_{YY}, g_{ZZ}) = (2.00230, 2.00913, 2.00610)$, $(A_{XX}, A_{YY}, A_{ZZ})/g\beta_e = (1.620, 2.65, 2.45)$ mT; (bottom panel): cyclic permutation that gives $A_{ZZ}/g\beta_e = 1.620$ mT.

nitroxide is perpendicular to the Z-axis of the spin pair for both spin labels. This is a common configuration for some spin-label biradicals, e.g., 1,4-bisDOXYL cyclohexane (Michon and Rassat 1975). Hyperfine coupling is resolved only for the field in one of the canonical directions: along the Y-axis of the pair, which corresponds to the nitroxide z-axes. Hyperfine splittings are halved in the coupled system, as seen from the factor of one half appearing in the hyperfine term of Eq. 9.17. Note that the hyperfine structure makes it almost impossible to identify features in the spectrum of the upper panel that we might attribute to the non-axial E-term. Introducing a value of $E = 0.15$ mT, which is ten times that predicted from the g-value anisotropy, does not change the number of major peaks in the spectrum relative to that for $E = 0$. The splitting of the outer, absorption-like peaks of the first-derivative powder pattern is given rigorously by $2D$ in field units; that for one pair of the derivative-like features is close to D, again in field units.

The lower panel of Figure 9.4 ($z_{NO} \parallel \mathbf{r}_{12}$) shows corresponding powder spectra for the situation where both nitroxide z-axes lie in the direction of the interspin vector, i.e., are parallel to the Z-axis. In this case, the outer, absorption-like features are split into five peaks by the hyperfine interaction. In practice, this means that they may be broadened almost beyond detectability. Then we must extract the dipolar coupling solely from the inner derivative-like features, which are broadened only slightly by the minor components of the hyperfine interaction. Note that introducing a non-vanishing value of the dipolar non-axiality, E, has only slight effect on the spectrum: broadening and slightly shifting the derivative-like feature at high field. A good example for this parallel orientation of nitroxide and interspin principal axes is the TOAC spin-label amino acid substituted at residue positions four apart in an α-helix (Hanson et al. 1998). The nitroxide z-axis is then preferentially parallel to the helix axis, and the i, i+4 substitutions position one spin label almost vertically above the other along the helix axis (see Section 16.10).

9.4 DIPOLAR COUPLING OF UNLIKE SPINS

When spins $\mathbf{s}_1$ and $\mathbf{s}'_2$ coupled by magnetic dipole–dipole interaction are distinguishable, the B-term of the dipolar Hamiltonian (given by Eq. 9.7) remains off-diagonal and does not contribute to the interaction energy in first order. We then need consider only the diagonal A-term that is given by Eq. 9.6, and instead of Eq. 9.12, the Hamiltonian for *unlike* spins with magnetic dipole–dipole interaction is:

$$\mathcal{H}_{12} = \hbar\omega_1 s_{1z} + \hbar\omega_2 s'_{2z} + \frac{\mu_o}{4\pi}\left(\frac{g^2\beta_e^2}{r_{12}^3}\right)\left(1-3\cos^2\theta\right)s_{1z}s'_{2z} \quad (9.18)$$

where $\hbar|\omega_1 - \omega_2| >> D_{dd}$, which we call the weak-coupling regime. The situation is much simpler than for the singlet-triplet coupling of like spins. The $\mathbf{s}_1$-spin simply experiences the dipolar field from the $\mathbf{s}'_2$-spin, which is situated at resonance frequency ω_2 well separated from that of the $\mathbf{s}_1$-spin at frequency ω_1. If there is also weak exchange between the unlike spins (weak enough not to cause singlet-triplet coupling), we simply add a term $J_{12}s_{1z}s'_{2z}$ to the interaction Hamiltonian in Eq. 9.18. Because the spins are distinguishable, the exchange coupling J_{12} just adds to the dipolar coupling D_{dd}.

The EPR transition ($\Delta s_{1z} = \pm 1$) of the $\mathbf{s}_1$-spin with quantum number $s_1 = \frac{1}{2}$, is then given by:

$$\hbar\omega_o = \hbar\omega_1 \pm \frac{1}{2}\frac{\mu_o}{4\pi}\left(\frac{g^2\beta_e^2}{r_{12}^3}\right)\left(1-3\cos^2\theta\right) \pm \frac{1}{2}J_{12} \quad (9.19)$$

where ω_o is the angular frequency of the microwave radiation, and $s'_2 = \frac{1}{2}$ for the other spin label. Correspondingly, the angular dependence of the spin–spin splitting of the resonance magnetic field in the EPR spectrum of an unlike spin-label pair is given by:

$$\pm\Delta B_{dd}(\theta) = \frac{1}{2}\frac{\mu_o}{4\pi}\left(\frac{g\beta_e}{r_{12}^3}\right)\left(1-3\cos^2\theta\right) + \frac{1}{2}\frac{J_{12}}{g\beta_e} \quad (9.20)$$

which varies from $\Delta B_{dd}^{\parallel} = -\left(\mu_o/4\pi\right)g\beta_e/r_{12}^3 + \frac{1}{2}J_{12}/g\beta_e$ to $\Delta B_{dd}^{\perp} = +\frac{1}{2}\left(\mu_o/4\pi\right)g\beta_e/r_{12}^3 + \frac{1}{2}J_{12}/g\beta_e$, for $\theta = 0°$ and $\theta = 90°$, respectively. If exchange is negligible (i.e., $J_{12} << D_{dd}$), the spin–spin splittings for unlike spins are 2/3 of those for like spins (cf. Eq. 9.14). In particular, the splitting between the major peaks in the dipolar powder pattern is $2\Delta B_{dd}^{\perp} = \left(\mu_o/4\pi\right)g\beta_e/r_{12}^3$ for unlike spins (cf. Figure 9.3). For the same dipolar splitting, this means in practice that the interspin separation, r_{12}, calculated by assuming like spins is $\sqrt[3]{3/2} = 1.145$ times greater than that calculated assuming unlike spins.

Electron spins are unlike if they are in different hyperfine states, or have different orientations to the magnetic field, and the dipolar coupling is less than the resulting difference in resonance frequency. Differences in orientation arise for asymmetrical rigid biradicals or more commonly from the innate flexibility of attachment of the two spin labels.

Figure 9.5 illustrates the distinction between unlike and like spins for various spin-label spectra. The grey and light grey shaded areas correspond to dipolar splittings of 0.5 and 2 mT, respectively, which correspond to separations between unlike spins of r_{12} = 1.55 and 0.98 nm. The isotropic spectrum in the top panel represents an unusual, but not totally impossible, situation. We would need large-angle segmental motion of the two spin labels, whilst overall macromolecular rotation is too slow to average the dipolar interaction. Here, the dipole–dipole coupling exceeds the width of an individual ^{14}N-hyperfine line but is less than the isotropic hyperfine splitting. For this case, one third of the spins are in the same m_I-hyperfine state and form pairs of like spins, whereas the remaining two-thirds with different values of m_I constitute unlike spin pairs. For a unique value of r_{12}, we would then expect two distinct dipolar powder patterns, with intensities in the ratio 2:1. Note that the situation in the top panel of Figure 9.5 also could apply to a symmetrical rigid biradical, except that the hyperfine splitting would not be isotropic but depend on the orientation of the magnetic field to the interspin vector.

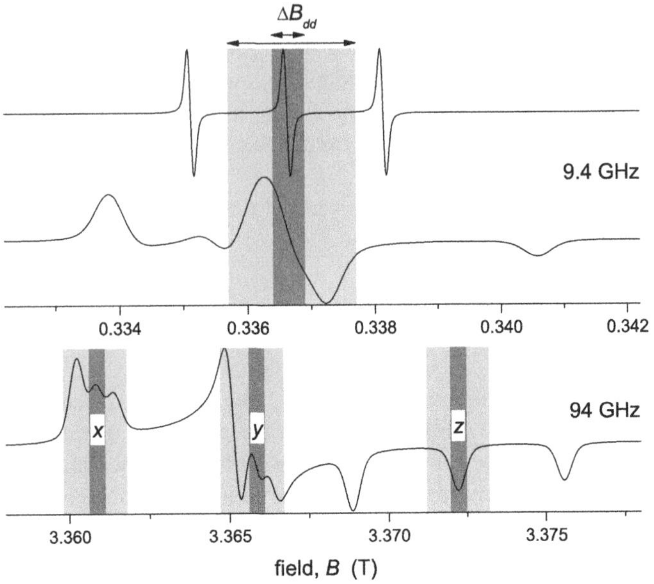

FIGURE 9.5 *Like* and *unlike* nitroxide spins for isotropic spectrum (top), and powder spectra at 9.4 GHz (middle) and 94 GHz (bottom). Grey-shaded area corresponds to dipolar powder pattern of width $\Delta B_{dd} = 0.5$ mT and light-grey area to $\Delta B_{dd} = 2$ mT.

For the powder spectrum at 9.4 GHz that is shown in the middle panel of Figure 9.5, the dipolar splitting of 0.5 mT does not span even the narrowest, central $m_I = 0$ hyperfine manifold. For this we would need a splitting of $\Delta B_{dd} \approx 2$ mT. In the latter case, the $m_I = 0$ spins are like spins, but the $m_I = \pm 1$ spins are not because these two manifolds spread over much wider ranges of spectral anisotropy. Then only one ninth of the spin pairs are between like spins, and the remaining pairs are between unlike spins. The situation is even more extreme if we go to higher frequencies. The powder spectrum at 94 GHz appears in the bottom panel of Figure 9.5 and has a total anisotropic range slightly more than twice that of the 9.4-GHz powder spectrum. At the high frequency, none of the m_I-hyperfine manifolds is grouped together. They are all spread by the g-anisotropy, and the dipolar coupling indicated is sufficient to span only some of the peaks at the x,y,z turning points. In this case, we can consider nearly all spins as unlike. To summarize: whenever the dipolar coupling is considerably smaller than the overall extent of the single-nitroxide powder pattern, the better approximation is to consider the spins as being unlike.

9.5 RELAXATION BY MAGNETIC DIPOLE–DIPOLE INTERACTION

Modulation of the magnetic dipole–dipole interaction between electron spins by either rotational or translational diffusion induces both longitudinal and transverse relaxation (cf. Chapter 5). We treat this here by using motional narrowing theory. Figure 9.6 shows the transitions involved in spin–lattice relaxation for dipolar-coupled *like* and *unlike* electron spins (left and right, respectively). Those with probabilities W_0, W_1 and W_2 are coupled by the B, the C and D, and the E and F terms, respectively, in the dipolar Hamiltonian of Eqs. 9.5–9.9.

The situation is more straightforward for *like* spins, because we need consider only transitions between the $M_S = -1, 0, +1$ triplet states (see left panel in Figure 9.6). Assuming a common spin temperature T_S, we assign populations $N+n$, N and $N-n$ to the three levels (see, e.g., Carrington and McLachlan 1969). These fulfill the requirement that the ratio of populations between levels with the same energy separations are equal, given by a quasi-Boltzmann factor, e.g., $\exp\left(-g\beta_e B/k_B T_S\right)$. In our case: $(N-n)/N = 1 - n/N \approx N/(N+n)$, because n is small. Summing the different relaxation pathways, we find that $dN/dt = 0$ for the $M_S = 0$ level. From either of the two other levels, the rate equation for the population difference is then:

$$\frac{dn}{dt} = -(W_1 + 2W_2)n + \text{constant} \tag{9.21}$$

We need the constant term to maintain detailed balance for the up and down transitions: $W_\uparrow N_- = W_\downarrow N_+$, for each of the W_i s (see Section 1.6 in Chapter 1). In particular, the Boltzmann population difference at equilibrium is $n = n_o$, when $dn/dt = 0$ (cf. Solomon 1955). The rate equation for the population difference n that corresponds to an EPR transition $(\Delta M_S = \pm 1)$ thus becomes:

$$\frac{dn}{dt} = -(W_1 + 2W_2)(n - n_o) \tag{9.22}$$

Comparing with Eq. 1.17 of Chapter 1, we see that the dipolar spin–lattice relaxation rate for *like* spins is:

$$\left(\frac{1}{T_{1dd}}\right)_{like} = W_1 + 2W_2 \tag{9.23}$$

It remains to evaluate the intrinsic transition probabilities W_1 and W_2. These are related to the spectral densities by Eq. 5.8, e.g., $W_1 = \hbar^{-2}\left|\langle \pm 1 | \mathcal{H}_{dd}^o | 0 \rangle\right|^2 J(\omega_{\pm 1,0})$, where $\mathcal{H}_{dd}^o$ is the time-independent part of the fluctuating dipolar Hamiltonian: $\mathcal{H}_{dd} = \mathcal{H}_{dd}^o F(t)$.

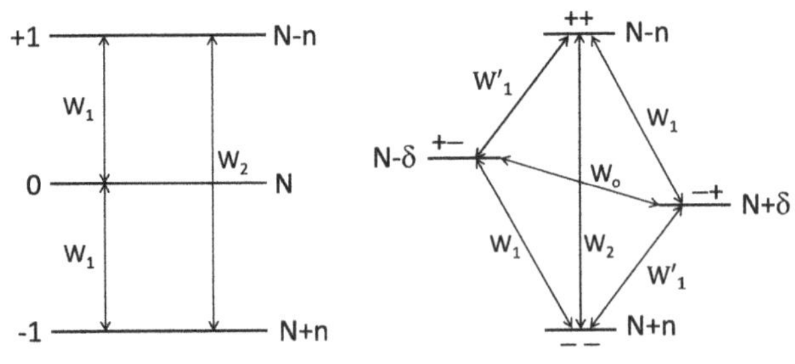

FIGURE 9.6 Energy levels of two dipolar-coupled electron spins $\left(s_1 = s_2 = \frac{1}{2}\right)$ and transitions for spin–lattice relaxation. *Left*: $M_S = 0, \pm 1$ triplet-state levels $(S = s_1 + s_2 = 1)$ for *like* spins. *Right*: $m_{s_1} = \pm\frac{1}{2}$, $m_{s_2} = \pm\frac{1}{2}$ states for *unlike* spins. Relative populations of different levels indicated assume a common spin temperature.

Note that, for *like* spins, we must evaluate the matrix elements of the dipolar Hamiltonian within the triplet state.

To evaluate the spectral densities $J(\omega)$ according to Eq. 5.13 or equivalent, we need the averages $\langle |F(0)|^2 \rangle$. For the different terms, A, B, C, D, E and F (Eqs. 9.6–9.9), in the dipolar Hamiltonian of Eq. 9.5, these are:

$$\left\langle |F^{(0)}(0)|^2 \right\rangle = \left\langle \frac{\left(1-3\cos^2\theta\right)^2}{r_{12}^6} \right\rangle \tag{9.24}$$

$$\left\langle |F^{(1)}(0)|^2 \right\rangle = \left\langle \frac{\sin^2\theta\cos^2\theta}{r_{12}^6} \right\rangle \tag{9.25}$$

$$\left\langle |F^{(2)}(0)|^2 \right\rangle = \left\langle \frac{\sin^4\theta}{r_{12}^6} \right\rangle \tag{9.26}$$

If r_{12} remains constant, we have: $\left\langle |F^{(0)}|^2 \right\rangle = (4/5)\times r_{12}^{-6}$, $\left\langle |F^{(1)}|^2 \right\rangle = (2/15)\times r_{12}^{-6}$ and $\left\langle |F^{(2)}|^2 \right\rangle = (8/15)\times r_{12}^{-6}$ (see Eqs. 5.16–5.18 in Chapter 5). For extreme motional narrowing (i.e., $\omega\tau << 1$), the zero-frequency spectral densities are therefore in the ratio: $J^{(0)}(0) : J^{(1)}(0) : J^{(2)}(0) = \left\langle |F^{(0)}|^2 \right\rangle : \left\langle |F^{(1)}|^2 \right\rangle : \left\langle |F^{(2)}|^2 \right\rangle = 6:1:4$. Note that extreme motional narrowing is independent of the mechanism that produces it.

To get the dependence of the relaxation rate on the spectral densities, we evaluate the matrix elements of $\mathcal{H}_{dd}^{o}$ from the spin-operator coefficients of the angle- and distance-dependent terms in Eqs. 9.6–9.9 (cf. Eqs. 9.24–9.26). For *like* interacting spin labels, with the same nuclear magnetic quantum number $m_I = m$, the dipolar T_1-relaxation rate from Eq. 9.23 becomes (Abragam 1961):

$$\frac{1}{T_{1dd}(m)} = \frac{3}{2}\left(\frac{\mu_o}{4\pi}\right)^2 \gamma_e^4\hbar^2 s(s+1)\left[J^{(1)}(\omega_m) + J^{(2)}(2\omega_m)\right] \tag{9.27}$$

where $s = \frac{1}{2}$ is the spin of each electron, ω_m is the spin label Larmor frequency, γ_e is the electron gyromagnetic ratio and $\hbar$ is Planck's constant divided by 2π. The first term on the right in Eq. 9.27 is the transition probability W_1 between the $M_S = 0$ and ± 1 levels of the triplet with separation $\hbar\omega_m$. The second term is twice that (i.e., W_2) between the $M_S = -1$ and $+1$ levels, which are separated by $2\hbar\omega_m$ (see left panel in Figure 9.6).

Evaluating the dipolar contribution to T_2 is more complicated. We can no longer simply add a secular contribution, as in Eqs. 5.51, 5.52 for a two-level $S = \frac{1}{2}$ system. Solomon (1955) introduces a scheme analogous to that in Figure 9.6 but for transverse components of the spin operators. For *like* interacting spin labels, with the same nuclear magnetic quantum number $m_I = m$, the dipolar T_2-relaxation rate is given by (Solomon 1955; Abragam 1961):

$$\frac{1}{T_{2dd}(m)} = \frac{3}{8}\left(\frac{\mu_o}{4\pi}\right)^2 \gamma_e^4\hbar^2 s(s+1) \times\left[J^{(0)}(0) + 10J^{(1)}(\omega_m) + J^{(2)}(2\omega_m)\right] \tag{9.28}$$

Comparing with Eq. 9.27, we see that the simple relation between T_1 and T_2 for a single spin system no longer holds. For extreme motional narrowing, we find from Eqs. 9.27 and 9.28 that $T_{1dd} = T_{2dd}$ because then $J^{(0)} : J^{(1)} : J^{(2)} = 6:1:4$.

The right panel in Figure 9.6 shows transitions between the $|m_{s1}, m_{s2}\rangle$ product states for *unlike* spins. Now we have more states and more transitions than for like spins. The populations indicated for each state again correspond to a common spin temperature. The population difference that corresponds to z-magnetization of spins 1 is: $[(N+n)+(N+\delta)]-[(N-n)+(N-\delta)] = 2(n+\delta)$ and correspondingly that for spins 2 is $2(n-\delta)$. The rate equations for the populations of the $|++\rangle$ and $|-+\rangle$ levels are, respectively:

$$\frac{d(N-n)}{dt} = W_1(n+\delta) + W_1'(n-\delta) + 2W_2 n + \text{constant} \tag{9.29}$$

$$\frac{d(N+\delta)}{dt} = -W_1(n+\delta) + W_1'(n-\delta) - 2W_0\delta + \text{constant} \tag{9.30}$$

Subtracting Eq. 9.29 from Eq. 9.30, the rate equation for the population difference that corresponds to z-magnetization of spins 1 is:

$$\frac{d(n+\delta)}{dt} = -\left(W_0 + 2W_1 + W_2\right)(n+\delta) - \left(W_2 - W_0\right)(n-\delta) + \text{constant} \tag{9.31}$$

Writing this explicitly in terms of the average S_z-magnetizations of spins 1 and 2, we then get:

$$\frac{dS_{1z}}{dt} = -\left(W_0 + 2W_1 + W_2\right)(S_{1z} - S_{1z}^o) - \left(W_2 - W_0\right)(S_{2z} - S_{2z}^o) \tag{9.32}$$

where S_{1z}^o and S_{2z}^o are the average magnetizations at Boltzmann equilibrium when $dS_{1z}/dt = 0$. This shows that, in general, the longitudinal magnetizations of the two sets of spins are coupled, and we do not expect simple exponential relaxation (see Solomon 1955; Abragam 1961). If the second set of spins stays at equilibrium (e.g., because of fast relaxation), then the first spins relax exponentially with a rate that is given by:

$$\left(\frac{1}{T_{1dd}}\right)_{unlike} = W_0 + 2W_1 + W_2 \tag{9.33}$$

Now we must evaluate the transition probabilities between the product states, e.g., $W_1 = \hbar^{-2}\left|\langle ++|\mathcal{H}_{dd}^{o}|-+\rangle\right|^2 J\left(\omega_{++,-+}\right)$. Otherwise, the calculation proceeds as for like spins.

For *unlike* interacting spin labels, with different values of $m_I = m, m'$, the T_1-relaxation rate of spin labels with $m_I = m$ becomes from Eq. 9.33 (Solomon 1955; Abragam 1961):

$$\frac{1}{T_{1dd}(m,m')} = \left(\frac{\mu_o}{4\pi}\right)^2 \gamma_e^4\hbar^2 s'(s'+1)\left[\tfrac{1}{12}J^{(0)}(\omega_m - \omega_{m'}) + \tfrac{3}{2}J^{(1)}(\omega_m) + \tfrac{3}{4}J^{(2)}(\omega_m + \omega_{m'})\right] \tag{9.34}$$

if the spin labels with $m_I = m'$ are at equilibrium during relaxation of the m-spins. If not, the longitudinal magnetizations of the m- and m'-spins are coupled, as explained above (see Eq. 9.32). For T_2-relaxation this is not the case, the transverse magnetizations precess separately and the relaxation rate is given by (Solomon 1955; Abragam 1961):

$$\frac{1}{T_{2dd}(m,m')} = \left(\frac{\mu_o}{4\pi}\right)^2 \gamma_e^4 \hbar^2 s'(s'+1) \times \left[\tfrac{1}{6} J^{(0)}(0) + \tfrac{1}{24} J^{(0)}(\omega_m - \omega_{m'}) + \tfrac{3}{4} J^{(1)}(\omega_m) + \tfrac{3}{2} J^{(1)}(\omega_{m'}) + \tfrac{3}{8} J^{(2)}(\omega_m + \omega_{m'})\right] \tag{9.35}$$

for a spin label with $m_I = m$. This equation is derived in Section O.2 of Appendix O (see Eq. O.33). The two Larmor frequencies, ω_m and $\omega_{m'}$, are related by $\omega_{m'} = \omega_m + a_N(m' - m)$, where a_N is the nitrogen hyperfine coupling constant expressed in angular frequency units. Note that s' in Eqs. 9.34 and 9.35 is the spin of the electron whose relaxation is not being monitored, which we assign here to a spin label with $m_I = m'$. This is important when relaxation is induced not by an unlike spin label but by a different paramagnetic species, e.g., a transition metal ion. In this case, we must also take into account the different g-value or gyromagnetic ratio (see Section 10.5 in Chapter 10). For extreme motional narrowing, T_1 and T_2 are equal, as for like spins, if the longitudinal magnetizations are uncoupled. Again using $J^{(0)}:J^{(1)}:J^{(2)} = 6:1:4$, Eqs. 9.28 and 9.35 show that $1/T_{2,like} = \frac{3}{2}(1/T_{2,unlike})$ for extreme motional narrowing, just as with static dipolar broadening (see Section 9.4) but not with the second moment (see Section 15.7 in Chapter 15).

Galeev and Salikhov (1996) point out that the traditional assumption that transverse magnetizations of unlike spins are not coupled by dipolar interactions is incorrect. We deal with this point in detail later, in Section 9.17 at the end of the chapter. Experimental evidence for the resulting transfer of transverse magnetization by dipolar interactions comes from dispersion-like contributions to the absorption spectra of interacting spin labels that are of opposite sign to those induced by Heisenberg spin exchange (Peric et al. 2012).

9.5.1 Dipolar Relaxation by Rotational Diffusion

Rotational diffusion induces relaxation by modulating the electron–electron dipolar interaction, in the same way as it does for the electron–nuclear dipolar interaction (i.e., the hyperfine anisotropy) that we treated in Chapter 5. The timescale depends on the size of the dipolar interaction $\Delta\omega_{dd}$, which is determined by the separation r_{12} of the electron spins that we assume are fixed in the rotating molecule. The motional narrowing regime corresponds to the condition $\Delta\omega_{dd}\tau_R << 1$, where τ_R is the rotational correlation time.

We derived the autocorrelation function for isotropic rotational diffusion in Section 7.5 of Chapter 7. Brownian rotational diffusion has exponential correlation functions. Using Eqs. 9.24–9.26, together with the angular averages in Eq. 5.16–5.18 of Chapter 5, the dipolar correlation functions for isotropic rotation are (Abragam 1961):

$$G^{(M)}(t) = \frac{\alpha^{(M)}}{r_{12}^6} \exp(-|t|/\tau_R) \tag{9.36}$$

where $\alpha^{(M)} \equiv \left\langle \left|F^{(M)}(0)\right|^2 \right\rangle = \frac{4}{5}, \frac{2}{15}$ and $\frac{8}{15}$ for $M = 0$, 1 and 2, respectively. Note that $\left\langle \left|F^{(M)}(0)\right|^2 \right\rangle$ is directly proportional to the average over the corresponding spherical harmonic $\left\langle \left|Y_{2,M}\right|^2 \right\rangle$ (see Eqs. 9.24–9.26). The resulting spectral densities are (see Eq. 5.13):

$$J^{(M)}(\omega) = \left(\frac{2\alpha^{(M)}}{r_{12}^6}\right) \frac{\tau_R}{1+\omega^2\tau_R^2} \tag{9.37}$$

where $\tau_R = 1/(6D_R)$ is the rotational correlation time for diffusion coefficient D_R.

For *like* spins with $s = \frac{1}{2}$, the relaxation times from Eqs. 9.27 and 9.28 together with Eq. 9.37 then become:

$$\left(\frac{1}{T_{1dd}}\right)_{rot} = \frac{3}{10}\left(\frac{\mu_o}{4\pi}\right)^2 \frac{\gamma_e^4\hbar^2}{r_{12}^6}\left(\frac{1}{1+\omega_m^2\tau_R^2} + \frac{4}{1+4\omega_m^2\tau_R^2}\right)\tau_R \tag{9.38}$$

$$\left(\frac{1}{T_{2dd}}\right)_{rot} = \frac{3}{20}\left(\frac{\mu_o}{4\pi}\right)^2 \frac{\gamma_e^4\hbar^2}{r_{12}^6}\left(3 + \frac{5}{1+\omega_m^2\tau_R^2} + \frac{2}{1+4\omega_m^2\tau_R^2}\right)\tau_R \tag{9.39}$$

Correspondingly, from Eqs. 9.34 and 9.35 together with Eq. 9.37, the relaxation times for *unlike* spins become:

$$\left(\frac{1}{T_{1dd}}\right)_{rot} = \frac{2}{15}\left(\frac{\mu_o}{4\pi}\right)^2 \frac{\gamma_e^4\hbar^2}{r_{12}^6} s'(s'+1)\left(\frac{1}{1+(\omega_m-\omega_{m'})^2\tau_R^2} + \frac{3}{1+\omega_m^2\tau_R^2} + \frac{6}{1+(\omega_m+\omega_{m'})^2\tau_R^2}\right)\tau_R \tag{9.40}$$

$$\left(\frac{1}{T_{2dd}}\right)_{rot} = \frac{1}{15}\left(\frac{\mu_o}{4\pi}\right)^2 \frac{\gamma_e^4\hbar^2}{r_{12}^6} s'(s'+1)\left(4 + \frac{1}{1+(\omega_m-\omega_{m'})^2\tau_R^2} + \frac{3}{1+\omega_m^2\tau_R^2} + \frac{6}{1+\omega_{m'}^2\tau_R^2} + \frac{6}{1+(\omega_m+\omega_{m'})^2\tau_R^2}\right)\tau_R \tag{9.41}$$

again with the rider concerning coupling of the longitudinal magnetizations for T_1.

When treating T_2-relaxation, i.e., Lorentzian linewidths, we can neglect spectral densities at the electron Larmor frequency, ω_m, in Eqs. 9.39 and 9.41 for EPR frequencies of 9 GHz and above. Also, the spectral density at the hyperfine frequency: $\omega_m - \omega_{m'} \approx a_N$, makes a relatively small contribution to the transverse relaxation rate in Eq. 9.41 for unlike spins (when $\tau_R \geq 8$ ns). Under these circumstances, the relation between the rotation-induced dipolar transverse relaxation rates for like and unlike spins is: $\left(1/T_{2dd}\right)_{unlike} \approx 4/9 \times \left(1/T_{2dd}\right)_{like}$. The corresponding relationship between rotational correlation times deduced from the dipolar broadening is: $\left(\tau_R\right)_{unlike} \approx 9/4 \times \left(\tau_R\right)_{like}$.

Figure 9.7 shows the increase in Lorentzian half-width at half-height, $\Delta\Delta B^{dd}_{1/2,L} = 1/\left(\gamma_e T_{2dd}\right)$, in spin-labelled cysteine double mutants of T4 lysozyme, as a function of the distance, r_{12}, between the two nitroxides (Mchaourab et al. 1997). The values of r_{12} come from energy minimization of the spin labels modelled into the crystal structure of T4 lysozyme; the uncertainty in r_{12} corresponds to spin-label conformers with similar energy. Values for the dipolar broadening come from convoluting a Lorentzian lineshape with the sum of the spectra of the corresponding spin-labelled single mutants. This procedure requires the dipolar broadening to be the same at all positions throughout the spectrum (see Section 2.6 in Chapter 2). From Figure 9.5, we see that this is a reasonable approximation if the dipolar broadening is larger than the width of the sharp features in the single-mutant spectra, where convolution shows its greatest effects. The slope in the log–log plot of $\Delta\Delta B^{dd}_{1/2,L}$ vs. r_{12}, is -6, in accordance with the $1/r_{12}^6$ dependence predicted by Eqs. 9.39 and 9.41. For *like* spins, Eq. 9.39 becomes:

$$\left(\Delta\Delta B^{dd}_{1/2,L}\right)_{rot} \approx \frac{9}{20}\left(\frac{\mu_o}{4\pi}\right)^2 \frac{\gamma_e^3 \hbar^2}{r_{12}^6}\left(\tau_R\right)_{like} \tag{9.42}$$

From Eq. 9.42, together with the intercept of the linear regression to the log–log plot in Figure 9.7, we find that the rotational correlation time of the protein is $\left(\tau_R\right)_{like} = 6$ ns, or $\left(\tau_R\right)_{unlike} = 14$ ns if the spins are *unlike*. For comparison, the rotational correlation time predicted from Debye theory for a spherical protein is: $\tau_{R,o} \equiv 1/6D_{R,o} = M_r \times 10^{-3}\,\bar{v}\eta/(RT) = 5.4$ ns with $M_r = 18.6$ kDa, $\bar{v} = 0.70 \times 10^{-3}$ m^3 kg^{-1}, $\eta = 1$ mPa s (1 cP) and $R = 8.3145$ J mol^{-1} K^{-1} (cf. Eqs. 7.9 and 7.10 in Chapter 7). This corresponds with the value deduced from dipolar relaxation, assuming like spins. However, T4 lysozyme is not a spherical protein: it consists of two lobes which result in an axial ratio of $a/b \approx 2$ (Matthews and Remington 1974; Remington et al. 1978). Thus, the effective correlation time, $\tau_{R\perp}$, could be up to a factor of two greater than that predicted for a sphere (see Eqs. 7.11 and 7.12 and Figure 7.1 in Chapter 7). Consideration of Figure 9.5 suggests that Lorentzian broadening deduced from convolution will be biased towards interactions between unlike spins. Together with the prolate shape of the protein, assuming unlike spins (instead of like spins) is also consistent with the experimental line broadening. In either case, the fit shown in Figure 9.7 provides a useful calibration for determining distances from other double mutants of this protein. Empirically, assuming like spins and taking the Debye correlation time for a sphere produces a good correlation between values of r_{12} from Eq. 9.42 and those found from molecular modelling (Mchaourab et al. 1997).

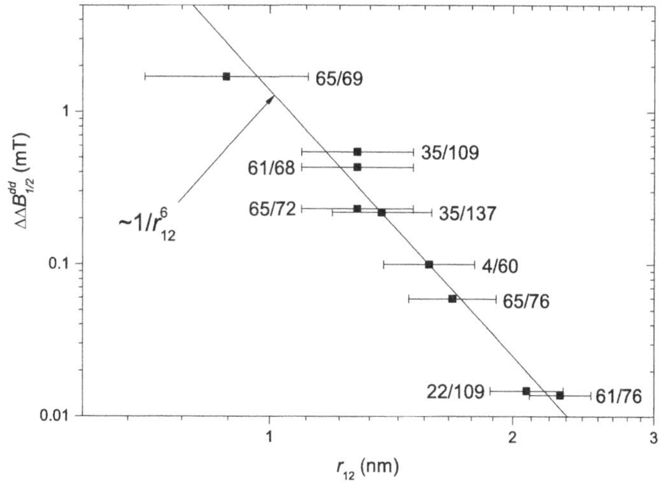

FIGURE 9.7 Lorentzian line broadening, $\Delta\Delta B^{dd}_{1/2,L}$ (half-width at half-height), in spin-labelled cysteine double mutants of T4 lysozyme, relative to single mutants, from spectral convolution. *x*-axis: separation, r_{12}, between nitroxide sites estimated from protein crystal structure. Residues 61–80 constitute an α-helix. (Data from Mchaourab et al. 1997.) Solid line: linear regression to log–log plot, with fixed slope −6 expected for dipolar relaxation.

9.5.2 Dipolar Relaxation by Translational Diffusion

We can evaluate the contribution to dipolar relaxation from translational diffusion by using motional narrowing theory, as described originally for nuclear spin relaxation (Abragam 1961; Hubbard 1966; Berner and Kivelson 1979b). Motional narrowing theory applies in the dynamic limit: $\tau_D/T_{2m} \ll 1$, where the characteristic timescale for translational diffusion is given by: $\tau_D = d_S^2/(2D_T)$, with d_S the distance of closest approach of the two spin dipoles and D_T the translational diffusion coefficient (Sachse et al. 1987; Berner and Kivelson 1979a).

We assume that initially the interspin vectors, $\mathbf{r}_o$, are randomly oriented. From the translational equivalent of Eq. 7.23 for rotational diffusion and Eqs. 9.5–9.9, the time correlation function for dipolar relaxation is given by (Torrey 1953; Abragam 1961):

$$G^{(M)}(t) = \alpha^{(M)} \iint \frac{Y^*_{2,M}\left(\hat{\mathbf{r}}_o\right)}{r_o^3}\frac{Y_{2,M}\left(\hat{\mathbf{r}}\right)}{r^3} P\left(\mathbf{r}_o;\mathbf{r},t\right) P_o\left(\mathbf{r}_o\right)\cdot \mathrm{d}^3\mathbf{r}_o \cdot \mathrm{d}^3\mathbf{r} \tag{9.43}$$

where $\alpha^{(M)} = 16\pi/5, 8\pi/15$ and $32\pi/15$ for $M = 0$, 1 and 2, respectively, is the factor relating spherical harmonics $Y_{2,M}$ to the functions in Eqs. 9.24–9.26: $\left|F^{(M)}(0)\right|^2 = \alpha^{(M)}\left|Y_{2,M}\right|^2$ (see Table V.2 in Appendix V). $P_o\left(\mathbf{r}_o\right)\cdot \mathrm{d}^3\mathbf{r}_o$ is the probability that spin label 2 is located in volume element $\mathrm{d}^3\mathbf{r}_o$ at position $\mathbf{r}_o = \mathbf{r}_{o,2} - \mathbf{r}_{o,1}$ relative to spin label 1, at time zero. For three-dimensional diffusion, $P_o\left(\mathbf{r}_o\right)$ is simply the spin-label density per unit volume, n_{SL}. The conditional probability, $P\left(\mathbf{r}_o;\mathbf{r},t\right)\mathrm{d}^3\mathbf{r}$, is the probability that, at time t, spin label 2 is located in volume element $\mathrm{d}^3\mathbf{r}$ at position $\mathbf{r} = \mathbf{r}_2 - \mathbf{r}_1$ relative to spin label 1, given that it is located at position $\mathbf{r}_o$ at time zero. We get the conditional probability from the translational diffusion equation (see Section 9.13 below):

$$\frac{\partial P(\mathbf{r}_o;\mathbf{r},t)}{\partial t} = 2D_T\nabla^2 P(\mathbf{r}_o;\mathbf{r},t) \tag{9.44}$$

where $2D_T = D_1 + D_2$ is the sum of the translational diffusion coefficients, D_1 and D_2, of the two spin labels. For our purposes, the diffusion equation is solved most conveniently by using integral transform methods (see, e.g., Morse and Feshbach 1953). With the normalized boundary condition $\int P(\mathbf{r}_o;\mathbf{r},t)\cdot \mathrm{d}^3\mathbf{r} = 1$, the solution becomes:

$$P(\mathbf{r}_o;\mathbf{r},t) = \frac{1}{(2\pi)^3}\int \exp\left(-2D_T t\rho^2\right)\exp\left[i\boldsymbol{\rho}\cdot(\mathbf{r}-\mathbf{r}_o)\right]\mathrm{d}^3\boldsymbol{\rho} \tag{9.45}$$

where $\boldsymbol{\rho}$ is the vector conjugate to $\mathbf{r}$, in Fourier space. We can expand the plane-wave functions in Eq. 9.45 in terms of spherical harmonics $\mathrm{Y}_{L,M}(\hat{\mathbf{r}})$ by using the generalization of Bauer's formula (Watson 1945), together with the spherical harmonic addition theorem (see Appendix V.2):

$$\exp(i\boldsymbol{\rho}\cdot\mathbf{r}) = 4\pi\sum_{L,M} i^L \mathrm{Y}^*_{L,M}(\hat{\mathbf{r}})\mathrm{Y}_{L,M}(\hat{\boldsymbol{\rho}})j_L(\rho r) \tag{9.46}$$

where $j_L(\rho r)$ is a spherical Bessel function of the first kind $(j_L(x) \equiv \sqrt{\pi/(2x)}J_{L+1/2}(x)$, where J_L is the standard Bessel function). We obtain an equivalent expression for $\exp(-i\boldsymbol{\rho}\cdot\mathbf{r}_o)$ by taking the complex conjugate of Eq. 9.46. Substituting this and Eq. 9.46 into Eq. 9.45, and then performing the angular part of the integration in Eq. 9.45 by using the orthogonality of the spherical harmonics $\left(\text{i.e.,} \int \mathrm{Y}^*_{L,M}(\hat{\boldsymbol{\rho}})\,\mathrm{Y}_{L',M'}(\hat{\boldsymbol{\rho}})\cdot \mathrm{d}^2\hat{\boldsymbol{\rho}} = \delta_{LL'}\delta_{MM'}\right)$, gives:

$$P(\mathbf{r}_o;\mathbf{r},t) = \frac{2}{\pi}\sum_{L,M}\mathrm{Y}^*_{L,M}(\hat{\mathbf{r}})\mathrm{Y}_{L,M}(\hat{\mathbf{r}}_o)$$
$$\times\int \exp\left(-2D_T t\rho^2\right)j_L(\rho r)j_L(\rho r_o)\cdot\rho^2\cdot \mathrm{d}\rho \tag{9.47}$$

Substituting Eq. 9.47 in Eq. 9.43, and again using the orthogonality properties of the spherical harmonics to perform the angular integrals, yields:

$$G^{(M)}(t) = \frac{2\alpha^{(M)}n_{SL}}{\pi}\int \exp\left(-2D_T t\rho^2\right)\left[\int_{d_S}^{\infty} r^{-1}j_2(\rho r)\cdot \mathrm{d}r\right]^2\rho^2\cdot \mathrm{d}\rho \tag{9.48}$$

where we note that the integrals over r and r_o are identical, with a lower limit $d_S = R_1 + R_2$ that is determined by the collision radii, R_1 and R_2, of the two spin labels. The integral over r in Eq. 9.48 is given by $j_1(\rho d_S)/(\rho d_S)$, and we then get the dipolar spectral density from the Fourier transform of $G^{(M)}(t)$:

$$J^{(M)}(\omega) = \frac{2\alpha^{(M)}}{\pi}\frac{n_{SL}}{d_S D_T}\int_o^{\infty}\left[j_1(x)\right]^2\frac{x^2\mathrm{d}x}{x^4+\omega^2\tau_D^2} \tag{9.49}$$

where we make the substitution $x = \rho d_S$, and define the characteristic time for translational diffusion $\tau_D \equiv d_S^2/(2D_T)$ in accord with Abragam (1961). The integral in Eq. 9.49 is evaluated by Hubbard (1966). We write the spectral density as:

$$J^{(M)}(\omega) = \alpha^{(M)}\frac{n_{SL}}{d_S D_T}I\left(\sqrt{2\omega\tau_D}\right) \tag{9.50}$$

where (Hubbard 1966):

$$I(u) = u^{-5}\left\{u^2 - 2 + e^{-u}\left[\left(u^2-2\right)\sin u + \left(u^2+4u+2\right)\cos u\right]\right\} \tag{9.51}$$

The limit of $I(u)$ as $u \to 0$ is $\frac{2}{15}$, as seen by letting $\omega\tau_D \to 0$ in Eq. 9.49 (Abragam 1961). The spectral density at zero frequency then becomes:

$$J^{(M)}(0) = \frac{2}{15}\alpha^{(M)}\frac{n_{SL}}{d_S D_T} \tag{9.52}$$

which is directly proportional to $\tau_D(\propto 1/D_T)$.

The solid lines in Figure 9.8 show the dependence of the spectral density on the translation correlation time. The upper panel gives the conventional reduced spectral density, $J^{(M)}(\omega,\tau_D)/J^{(M)}(0,\tau_D)$. In the lower panel, the y-axis is multiplied by $\omega\tau_D$ to give the complete dependence on τ_D. The factor of ω in x- and y-axes gives plots that are applicable for all angular resonance frequencies. For comparison, the dotted lines in Figure 9.8 give the corresponding Lorentzian spectral densities for dipolar relaxation by rotational diffusion. Relative to the contribution for pure rotation, the translational spectral density has wider dispersion on a logarithmic scale, with a broad maximum that is located at $\omega\tau_D = 5$. The narrower spectral density for rotation exhibits the familiar maximum at $\omega\tau_R = 1$. The conditions under which the Brownian diffusion model is valid for translational dipolar relaxation are discussed by Torrey (1953), in terms of the theory of random flights.

We obtain the translational relaxation rates by substituting Eq. 9.50 into Eqs. 9.27, 9.28, 9.34 and 9.35. For *like* spins with $s = \frac{1}{2}$:

$$\left(\frac{1}{T_{1dd}}\right)_{trans} = \frac{3\pi}{5}\left(\frac{\mu_o}{4\pi}\right)^2\gamma_e^4\hbar^2\left[I\left(\sqrt{2\omega_m\tau_D}\right)+4I\left(2\sqrt{\omega_m\tau_D}\right)\right]\frac{n_{SL}}{d_S D_T} \tag{9.53}$$

$$\left(\frac{1}{T_{2dd}}\right)_{trans} = \frac{3\pi}{50}\left(\frac{\mu_o}{4\pi}\right)^2\gamma_e^4\hbar^2\left[2+25I\left(\sqrt{2\omega_m\tau_D}\right)\right.$$
$$\left.+10I\left(2\sqrt{\omega_m\tau_D}\right)\right]\frac{n_{SL}}{d_S D_T} \tag{9.54}$$

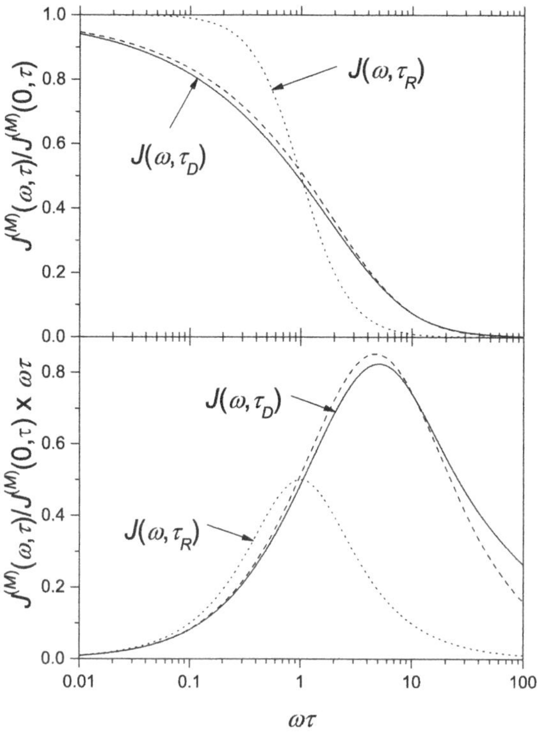

FIGURE 9.8 Spectral density for dipolar relaxation by translational diffusion (solid lines and dashed lines, Eq. 9.50) and by rotational diffusion (dotted lines, Eq. 5.13). Top panel: reduced spectral density, $J^{(M)}(\omega,\tau)/J^{(M)}(0,\tau)$, where $J(0,\tau) \propto \tau$. Bottom panel: reduced spectral density multiplied by $\omega\tau$ to give the full τ-dependence. For translational diffusion: $\tau = \tau_D \equiv d_S^2/(2D_T)$; for rotational diffusion: $\tau = \tau_R \equiv 1/(6D_R)$; where D_T and D_R are the diffusion coefficients. Each τ is scaled by angular resonance frequency ω. Solid lines: classical translational diffusion model (Eq. 9.51); dashed lines: excluded-volume model with reflecting boundary (Eq. 9.57).

where the function $I(u)$ is given by Eq. 9.51. For *unlike* spins with $s = \frac{1}{2}$:

$$\left(\frac{1}{T_{1dd}}\right)_{trans} = \frac{4\pi}{15}\left(\frac{\mu_o}{4\pi}\right)^2 \gamma_e^4\hbar^2 s'(s'+1)\left[I\left(\sqrt{2(\omega_m-\omega_{m'})\tau_D}\right) + 3I\left(\sqrt{2\omega_m\tau_D}\right) + 6I\left(\sqrt{2(\omega_m+\omega_{m'})\tau_D}\right)\right]\frac{n_{SL}}{d_S D_T} \quad (9.55)$$

$$\left(\frac{1}{T_{2dd}}\right)_{trans} = \frac{2\pi}{15}\left(\frac{\mu_o}{4\pi}\right)^2 \gamma_e^4\hbar^2 s'(s'+1) \times\left[\frac{8}{15} + I\left(\sqrt{2(\omega_m-\omega_{m'})\tau_D}\right) + 3I\left(\sqrt{2\omega_m\tau_D}\right) + 6I\left(\sqrt{2\omega_{m'}\tau_D}\right) + 6I\left(\sqrt{2(\omega_m+\omega_{m'})\tau_D}\right)\right]\frac{n_{SL}}{d_S D_T} \quad (9.56)$$

again with the rider on coupling of the longitudinal magnetizations for T_1. As expected, the dipolar relaxation rates are directly proportional to the spin label concentration, n_{SL}, just as are Heisenberg exchange contributions to relaxation by translational diffusion (see Sections 9.14 and 9.15, below).

Ayant et al. (1975) and Hwang and Freed (1975) point out that the classical solution of the diffusion equation given by Eq. 9.47 ignores the finite size of the diffusants, i.e., the excluded volume effect. We do not introduce the cut-off at $r = d_S$ until evaluating the correlation function in Eq. 9.48. Allowance for the excluded volume effect by incorporating a reflecting boundary at $r = d_S$ leads to T_1-relaxation enhancement by translational diffusion that is given by Eqs. 9.53 and 9.55, but with spectral densities now defined by the function (Ayant et al. 1975; Polnaszek and Bryant 1984):

$$I(u) = \frac{4}{27}\frac{1+5u/8+u^2/8}{1+u+u^2/2+u^3/6+4u^4/81+u^5/81+u^6/648} \quad (9.57)$$

The normalized version of this different spectral density function is shown by the dashed lines in Figure 9.8. The absolute spectral density at zero frequency is larger by a factor 10/9 than that for the classical diffusion model, because relaxant accumulates at the reflecting boundary. For T_2-relaxation enhancement with *unlike* spins, we additionally replace $4I(0) = 8/15$ in Eq. 9.56 by 16/27; and for *like* spins, we replace $15I(0) = 2$ in Eq. 9.54 by 20/9.

9.5.3 Translational Diffusion Coefficients and Dipolar Relaxation

Dipolar relaxation contributes to the concentration-dependent linewidths that are used for measuring translation diffusion coefficients by spin-label EPR (cf. Section 9.14, below). Unlike Heisenberg exchange, dipolar relaxation contributes most strongly at low diffusion rates or high viscosity, because it depends inversely on diffusion coefficient (see Eqs. 9.54 and 9.56). Under these conditions, we must correct measurements made using the exchange interaction or even replace them (Devaux et al. 1973; Sachse et al. 1987; Berner and Kivelson 1979a).

Considerable simplifications to the expressions for transverse relaxation ensue if we assume that $\omega_m\tau_D >> 1$, i.e., that $\tau_D >> 2\times10^{-11}$ s at 9 GHz EPR frequency (and proportionally smaller at higher frequencies). Then we can neglect all spectral densities at frequencies ω_m and $\omega_m + \omega_{m'}$ (see Figure 9.8, top). Taking $d_S = 0.64$ nm (the classical diffusion diameter of di-*tert*-butyl nitroxide from Plachy and Kivelson 1967), this requires that diffusion coefficients are in the range $D_T << 1\times10^4$ μm^2 s^{-1}. In practice, this assumption is always acceptable, because dipolar relaxation is ineffective at such high diffusion rates and becomes negligible compared with Heisenberg spin exchange (see Section 9.16).

If we further assume that $a_N\tau_D << 1$, i.e., that $\tau_D << 4\times10^{-9}$s, where a_N is the nitrogen hyperfine constant, then we can put spectral densities at frequencies $\omega_m - \omega_{m'} \approx a_N$ equal to those at zero frequency (see Figure 9.8, top). This requires that the value of the diffusion coefficient is $D_T >> 50\ \mu m^2\ s^{-1}$, which is met by small molecules in non-viscous solution (King and Marsh 1986) but not achieved by spin-labelled lipids in membranes (Sachse et al. 1987; King et al. 1987) or other cases of high viscosity. For lipids, the diffusion coefficient is: $D_T \approx 10\ \mu m^2\ s^{-1}$ in fluid membranes (Marsh 2013), which gives a reduced spectral density $J(\omega)/J(0) \approx 0.16$ at an EPR frequency $\omega/2\pi$ of 9 GHz (see upper panel in Figure 9.8). This becomes even smaller at higher EPR frequencies. The contribution to transverse relaxation is then only 4% of that from the zero-frequency spectral density (cf. Eq. 9.56), and we can neglect it.

For high viscosities ($a_N\tau_D >> 1$), we therefore retain only the spectral density at zero frequency. The transverse relaxation rates from Eqs. 9.54 and 9.56 thus become:

$$\left(\frac{1}{T_{2dd}}\right)_{trans} = \frac{3\pi}{25}\left(\frac{\mu_o}{4\pi}\right)^2 \gamma_e^4\hbar^2 \frac{n_{SL}}{d_S D_T} \tag{9.58}$$

$$\left(\frac{1}{T_{2dd}}\right)_{trans} = \frac{16\pi}{15^2}\left(\frac{\mu_o}{4\pi}\right)^2 \gamma_e^4\hbar^2 s'(s'+1)\frac{n_{SL}}{d_S D_T} \tag{9.59}$$

for like and unlike spins, respectively, provided that slow-motional effects are not important. For motionally narrowed spectra where each nitrogen hyperfine manifold consists of a single line (see Figure 9.5), the probability of collisions between spins of like nuclear magnetic quantum number is $1/(2I+1)$ and that between unlike spins is $2I/(2I+1)$. Here, I is the nitrogen nuclear spin. Thus, the net transverse relaxation rate is given by:

$$\left(\frac{1}{T_{2dd}}\right)_{trans} = \frac{1}{2I+1}\left(\frac{1}{T_{2,like}} + \frac{2I}{T_{2,unlike}}\right) \tag{9.60}$$

From Eqs. 9.58 to 9.60, the transverse relaxation rate induced by dipole–dipole interaction between spins with $s = s' = \frac{1}{2}$ is then:

$$\left(\frac{1}{T_{2dd}}\right)_{trans} = \frac{\pi}{75}\left(\frac{\mu_o}{4\pi}\right)^2 \gamma_e^4\hbar^2 \left(\frac{9+8I}{2I+1}\right)\frac{n_{SL}}{d_S D_T} \tag{9.61}$$

Here n_{SL} is the spin-label concentration in molecules per unit volume (m^3), i.e., $n_{SL} = 10^3 N_A c_M$, where c_M is the concentration in moles/litre and N_A is Avogadro's number.

Equation 9.61 lets us estimate translational diffusion coefficients from the magnetic dipole–dipole contribution to Lorentzian line broadening. For spin-labelled phosphatidylcholine in dimyristoyl phosphatidylcholine membranes, the concentration dependence of line broadening at low temperature in the fluid phase is preponderantly dipolar (see Figure 9.25 given later). From Arrhenius-law temperature dependences fitted in Figure 9.25, the dipolar contribution to the Lorentzian peak-to-peak linewidth at 30°C is $d\Delta B_{pp}^L/dX_{SL} = 3.93$ mT per mole fraction of 16-DOXYL spin-labelled lipid (Sachse et al. 1987). In terms of mole fraction, X_{SL}, the effective spin-label concentration is $n_{SL} = \left(N_A \times 10^3/M_l\bar{v}_l\right)X_{SL}$, where M_l is the relative molecular mass, and $\bar{v}_l$ $\left(m^3/kg\right)$ the partial specific volume, of the lipid (Marsh 2013). By using Eq. 9.61, where $\Delta B_{pp}^L = \left(2/\gamma_e\sqrt{3}\right)T_2^{-1}$ and $I = 1$ for ^{14}N-spin labels, with $d_S = 0.8$ nm for the lateral collision diameter in the membrane, we find a lipid translational diffusion coefficient $D_T = 15\ \mu m^2\ s^{-1}$ at 30°C.

9.6 EXCHANGE INTERACTION AND EXCHANGE INTEGRAL

The Heisenberg exchange interaction is an electrostatic interaction between electrons that is solely of quantum mechanical origin and has no classical analogue. It arises from the indistinguishability of two electrons on close approach, which is a consequence of the Heisenberg uncertainty principle. Exchange manifests itself as an effective spin–spin interaction, because of the Pauli principle. This requires that the overall wave function of particles such as the electron, which have odd-integer spin, must be antisymmetric with respect to exchange of the particles with each other.

Consider the two unpaired electrons, 1 and 2, of nitroxide spin labels a and b (see Figure 9.9). When the spin labels are widely separated, one electron occupies the φ_a-orbital that is centred on a, and the other electron occupies the φ_b-orbital that is centred on b. In the absence of interaction between the two electrons, the product wave function $\varphi_a(1)\varphi_b(2)$ with electron 1 in φ_a and electron 2 in φ_b is a suitable wave function for the combined system, as is equivalently $\varphi_a(2)\varphi_b(1)$ in which the electron labels are exchanged. When the spin labels approach more closely, there is an electrostatic repulsion between the two electrons that is given by the Hamiltonian:

$$\mathcal{H}_{12}' = \frac{e^2}{4\pi\varepsilon_o r_{12}} \tag{9.62}$$

where r_{12} is the distance between the two electrons. We then take the linear combination:

$$\psi = c_1\varphi_a(1)\varphi_b(2) + c_2\varphi_a(2)\varphi_b(1) \tag{9.63}$$

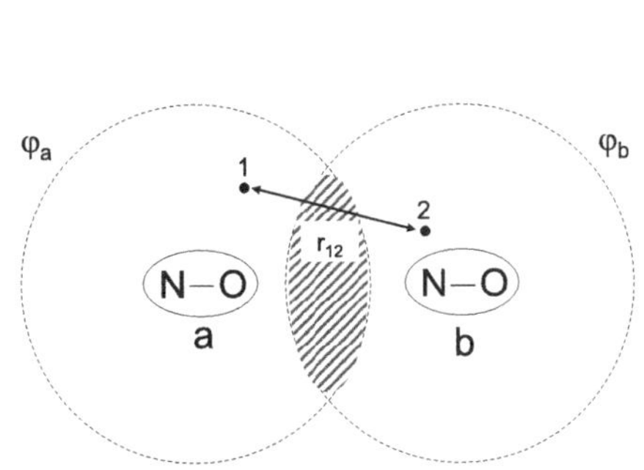

FIGURE 9.9 Schematic representation of electrostatic interaction between unpaired electrons, 1 and 2, of nitroxides a and b. Initially, electron 1 is localized in orbital φ_a of nitroxide a, and electron 2 in orbital φ_b of nitroxide b. On closer approach, orbitals φ_a and φ_b overlap (shaded area), and the electrons are no longer distinguishable.

as a trial wave function. Because the two electrons are indistinguishable, the probability density $|\psi|^2$ must be unchanged when the electron labels are exchanged. Therefore, ψ must remain unchanged or change only in sign when the electrons are exchanged, which requires that $c_2 = \pm c_1$. The two possible (normalized) wave functions are therefore:

$$^1\psi = \frac{1}{\sqrt{2\left(1+|S_{a,b}|^2\right)}}\left(\varphi_a(1)\varphi_b(2)+\varphi_a(2)\varphi_b(1)\right) \tag{9.64}$$

$$^3\psi = \frac{1}{\sqrt{2\left(1-|S_{a,b}|^2\right)}}\left(\varphi_a(1)\varphi_b(2)-\varphi_a(2)\varphi_b(1)\right) \tag{9.65}$$

The superscripts indicate that $^1\psi$ corresponds to a singlet spin state, and $^3\psi$ to a triplet spin state, as will become clear later. In Eqs. 9.64 and 9.65, the overlap integral is defined by:

$$S_{a,b} = \int \varphi_a^*\varphi_b \,\mathrm{d}v \equiv \langle a|b\rangle \tag{9.66}$$

which is an integral over all volume, v. The overlap integral must be included in the normalization of ψ because the wave functions φ_a and φ_b are referred to different centres, and thus are not orthogonal.

The Hamiltonian for electron–electron repulsion that is given by Eq. 9.62 is automatically diagonal when expressed in terms of the $^1\psi$- and $^3\psi$-wave functions that are given by Eqs. 9.64 and 9.65. Alternatively, these wave functions also correspond to the variational minimum for ψ as given in Eq. 9.63. The eigenvalues, or expectation values, of Eq. 9.62 are therefore:

$$E'_\pm \equiv \langle\psi|\hat{\mathcal{H}}'_{12}|\psi\rangle = \frac{K \pm J}{1 \pm |S_{a,b}|^2} \tag{9.67}$$

where the upper sign (+) corresponds to the $^1\psi$-state (i.e., $^1E \equiv E'_+$), and the lower sign (–) corresponds to the $^3\psi$-state (i.e., $^3E \equiv E'_-$). In Eq. 9.67, the direct integral is: $K = \int \varphi_a^*(1)\varphi_b^*(2)\left(e^2/4\pi\varepsilon_o r_{12}\right)\varphi_a(1)\varphi_b(2)\cdot \mathrm{d}v$, which corresponds to a classical coulombic interaction. The exchange integral, on the other hand, is:

$$J = \int \varphi_a^*(2)\varphi_b^*(1)\left(e^2/4\pi\varepsilon_o r_{12}\right)\varphi_a(1)\varphi_b(2)\cdot \mathrm{d}v \tag{9.68}$$

which has no classical analogue. Because it corresponds to an electrostatic repulsion, the exchange integral is positive. (Alternatively, it can be considered as the self-energy of an electron with density distribution $\varphi_a(\mathbf{r})\varphi_b^*(\mathbf{r})$.)

According to Eq. 9.67, the energy levels of the $^3\psi$- and $^1\psi$-states are split by the exchange interaction (see Figure 9.10). The splitting of the ^3E and ^1E energy levels, J_{12}, is given by:

$$J_{12} \equiv {}^3E - {}^1E = 2\frac{K|S_{a,b}|^2 - J}{1-|S_{a,b}|^4} \tag{9.69}$$

Note that for $S_{a,b} = 0$, the $^3\psi$-state lies lower in energy because J is positive. For an atom, this corresponds to the first of Hund's rules because (as we shall see in the next section) $^3\psi$ is the state with higher spin, i.e., the triplet state. In molecular systems, the $^1\psi$-state (corresponding to the singlet) is often lower in energy, indicating that the term involving the overlap integral, $S_{a,b}$, dominates.

9.7 HEISENBERG SPIN EXCHANGE

The spatial wave function $^1\psi$ in Eq. 9.64 is symmetric with respect to exchange of the electrons, because the sign remains unchanged. Therefore, according to the Pauli principle, this must

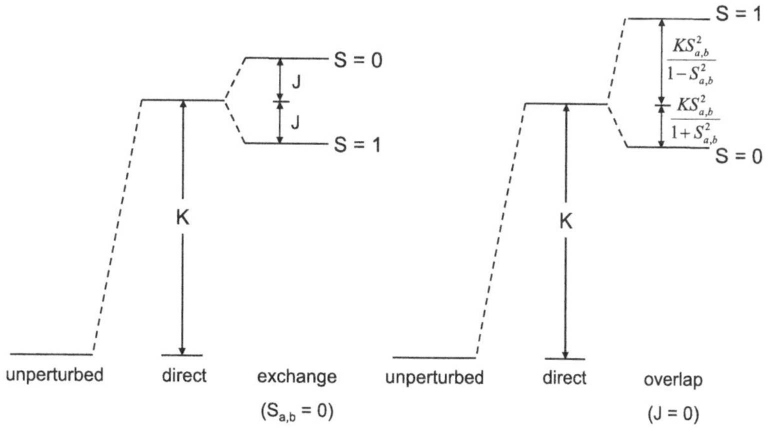

FIGURE 9.10 Energy-level diagrams for electrostatic repulsion between two interacting electrons, e.g., unpaired electrons of nitroxide spin labels. Direct integral K is conventional coulomb repulsion between two electrons; J is non-classical exchange integral. $S_{a,b}$ is overlap integral of the two orbitals in which the electrons reside; $S_{a,b} = 0$ for two electrons on the same atom. *Left hand*: exchange dominates over overlap (i.e., $S_{a,b} \approx 0$); triplet spin state ($S = 1$) lies lowest. *Right hand*: overlap dominates over exchange (i.e., $J \approx 0$); singlet spin state ($S = 0$) lies lowest.

be combined with an antisymmetric spin function in order that the total wave function, which is the product of the space and spin functions, be antisymmetric. There is only one such spin function for two electrons:

$$^{1}\chi_0 = \frac{1}{\sqrt{2}}\left(\alpha(1)\beta(2) - \beta(1)\alpha(2)\right) \tag{9.70}$$

where α and β are the wave functions for electron spin up and spin down, with $m_s = +\frac{1}{2}$ and $m_s = -\frac{1}{2}$, respectively. This corresponds to a singlet spin state with: $S = s_1 + s_2 = 0$ for which $M_S = 0$ (see Appendices E and Q.1). The $^3\psi$ spatial wave function of Eq. 9.65 is antisymmetric and must be combined with a symmetric spin function, of which there are three for two electrons:

$$^{3}\chi_{+1} = \alpha(1)\alpha(2) \tag{9.71}$$

$$^{3}\chi_{-1} = \beta(1)\beta(2) \tag{9.72}$$

$$^{3}\chi_0 = \frac{1}{\sqrt{2}}\left(\alpha(1)\beta(2) + \beta(1)\alpha(2)\right) \tag{9.73}$$

These correspond to a triplet spin state for which: $S = s_1 + s_2 = 1$ with $M_S = \pm 1, 0$ (for Eqs. 9.71, 9.72 and 9.73, respectively). For $^3\chi_{+1}$ both spins point up, for $^3\chi_{-1}$ both point down, and for $^3\chi_0$ the two spins are aligned antiparallel.

The exchange interaction therefore results in an effective coupling of the individual electron spins, $\mathbf{s}_1$ and $\mathbf{s}_2$, in which the triplet state is split from the singlet state by an energy J_{12} that is given by Eq. 9.69 (see Figure 9.10). We can represent this singlet–triplet splitting with a spin Hamiltonian for exchange that has the form:

$$\mathcal{H}_{HE} = J_{12}\mathbf{s}_1 \cdot \mathbf{s}_2 = \tfrac{1}{2}J_{12}\left((\mathbf{s}_1 + \mathbf{s}_2)^2 - \mathbf{s}_1^2 - \mathbf{s}_2^2\right) = J_{12}\left(\tfrac{1}{2}S(S+1) - \tfrac{3}{4}\right) \tag{9.74}$$

where $S(S+1)$ is the eigenvalue of $\mathbf{S}^2$, and $s_1(s_1+1) = s_2(s_2+1) = \frac{3}{4}$ are those of $\mathbf{s}_1{}^2$ and $\mathbf{s}_2{}^2$, respectively (see Appendix E). Equation 9.74 satisfies our requirement that the splitting between triplet ($S = 1$) and singlet ($S = 0$) levels is J_{12}. This is sometimes referred to as the vector model, or the Heisenberg–Dirac–Van Vleck Hamiltonian. Note that, in general, J_{12} contains contributions both from true exchange (J) and from overlap $(S_{a,b})$, according to Eq. 9.69.

9.8 NITROXIDE BIRADICALS

Heisenberg spin exchange in nitroxide biradicals has the effect of coupling the two unpaired electron spins $\left(s_1 = s_2 = \frac{1}{2}\right)$ into triplet ($S = 1$) and singlet ($S = 0$) states. The selection rules for magnetic-dipole transitions are: $\Delta S = 0$, $\Delta M_S = \pm 1$ (see Section 1.5). Thus, for strong exchange, we get EPR transitions only within the triplet (see Figure 9.11). The ^{14}N-nuclear spins ($I_1 = I_2 = 1$) couple to a maximum value of $I = I_1 + I_2 = 2$ and we get five hyperfine lines, with splittings half those of the single spin label and intensities in the ratio 1:2:3:2:1. The latter corresponds to the number of ways that we can combine the individual spin projections, m_{I_1} and m_{I_2}, to give a particular value of $M_I \left(= m_{I_1} + m_{I_2}\right)$.

This relatively simple picture breaks down when the nitrogen hyperfine coupling, a_N, is comparable with the exchange interaction, J_{12}. Electron spins coupled to nuclear spins with different values of m_I then become distinguishable. This induces singlet–triplet mixing, to an extent that depends on the ratio a_N/J_{12}. Assume that the two nitroxides have the same isotropic g-values and hyperfine couplings. Ignoring non-secular hyperfine couplings, the spin Hamiltonian for the biradical is then:

$$\mathcal{H}_{12} = g\beta_e B\left(s_{1z} + s_{2z}\right) + a_N\left(I_{1z}s_{1z} + I_{2z}s_{2z}\right) + J_{12}\mathbf{s}_1 \cdot \mathbf{s}_2 \tag{9.75}$$

It is helpful, as far as is possible, to express Eq. 9.75 in terms of the total spin, $S = s_1 + s_2$, by using Eq. 9.74 for the $\mathbf{s}_1 \cdot \mathbf{s}_2$ term:

$$\mathcal{H}_{12} = g\beta_e BS_z + \tfrac{1}{2}a_N M_I S_z + \tfrac{1}{2}J_{12}\left(S(S+1) - \tfrac{3}{2}\right) + \tfrac{1}{2}a_N \Delta m\left(s_{1z} - s_{2z}\right) \tag{9.76}$$

where $M_I = m_{I_1} + m_{I_2}$ is the magnetic quantum number for the total nuclear spin $(I = I_1 + I_2)$ and $\Delta m = m_{I_1} - m_{I_2}$. The first three terms on the right of Eq. 9.76 describe the coupled spins, s_1 and s_2, in terms of a triplet ($S = 1$) or singlet ($S = 0$) state. Only the final term, which couples states that differ in S by unity, admixes the singlet and triplet states. This occurs because radicals in different hyperfine states ($\Delta m \neq 0$) are no longer equivalent.

Expressed in terms of singlet–triplet quantization, $|S, M_S\rangle$, the matrix form of the spin Hamiltonian given by Eq. 9.76 is:

$\mathcal{H}_{12}$	$\vert 1,+1\rangle$	$\vert 1,-1\rangle$	$\vert 1,0\rangle$	$\vert 0,0\rangle$
$\langle 1,+1\vert$	$g\beta_e B + a_N M_I/2 + J_{12}/4$	0	0	0
$\langle 1,-1\vert$	0	$-g\beta_e B - a_N M_I/2 + J_{12}/4$	0	0
$\langle 1,0\vert$	0	0	$J_{12}/4$	$a_N\Delta m/2$
$\langle 0,0\vert$	0	0	$a_N\Delta m/2$	$-3J_{12}/4$

When $\Delta m \neq 0$, we must diagonalize the 2×2 matrix that connects singlet and triplet states exactly, because the hyperfine coupling and exchange interaction may be comparable in magnitude. The eigenvalues are (see Appendix G, Section G.2):

$$E_{\pm} = -\frac{J_{12}}{4} \pm \frac{a_N}{2}\sqrt{\left(J_{12}/a_N\right)^2 + (\Delta m)^2} \tag{9.77}$$

The corresponding normalized eigenvectors are:

$$|+,0\rangle = \cos\vartheta(\Delta m)\cdot|1,0\rangle + \sin\vartheta(\Delta m)\cdot|0,0\rangle \tag{9.78}$$

$$|-,0\rangle = -\sin\vartheta(\Delta m)\cdot|1,0\rangle + \cos\vartheta(\Delta m)\cdot|0,0\rangle \tag{9.79}$$

where the extent of singlet–triplet admixture is given by:

$$\tan\vartheta(\Delta m) = \frac{-J_{12}/a_N + \sqrt{\left(J_{12}/a_N\right)^2 + (\Delta m)^2}}{\Delta m} \tag{9.80}$$

which determines the relative intensity of the "allowed" and "forbidden" transitions for $\Delta m \neq 0$.

The remainder of the Hamiltonian is diagonal, and the energies of the $|1,\pm1\rangle$ states are given simply by the diagonal elements. The "allowed" EPR transitions, $|1,\pm1\rangle \leftrightarrow |+,0\rangle$ with $\Delta I_z = 0$, which correspond to strong exchange conditions (see Figure 9.11), are therefore given by:

$$h\nu_a^{\pm}(M_I,\Delta m) = g\beta_e B + \tfrac{1}{2}a_N M_I \pm \tfrac{1}{2}J_{12} \mp \tfrac{1}{2}J_{12}\sqrt{1+\left(a_N\Delta m/J_{12}\right)^2} \tag{9.81}$$

with relative intensity, $I_a \propto \cos^2\vartheta(\Delta m)$. Correspondingly, the "forbidden" EPR transitions (see Figure 9.11) are $|1,\pm1\rangle \leftrightarrow |-,0\rangle$ with $\Delta I_z = 0$, which are allowed only for weak exchange, are given by:

$$h\nu_f^{\pm}(M_I,\Delta m) = g\beta_e B + \tfrac{1}{2}a_N M_I \pm \tfrac{1}{2}J_{12} \pm \tfrac{1}{2}a_N\sqrt{(\Delta m)^2 + \left(J_{12}/a_N\right)^2} \tag{9.82}$$

with relative intensity, $I_f \propto \sin^2\vartheta(\Delta m)$. Note that the term involving the square root is identical, apart from the sign, in both Eqs. 9.81 and 9.82. We write it differently in the two cases, to emphasize that the former corresponds to strong exchange and the latter to weak exchange. Equation 9.81 holds also for $\Delta m = 0$ where the transitions are unconditionally allowed but not Eq. 9.82 where these transitions are then strictly forbidden.

Table 9.1 lists the "allowed" hyperfine transition frequencies for the different values of M_I and Δm, and the various combinations of m_{I_1} and m_{I_2} that give rise to them. For $\Delta m = 0$ (i.e., $M_I = \pm2$ and one of the $M_I = 0$ combinations), the positions of the hyperfine lines are independent of the exchange interaction.

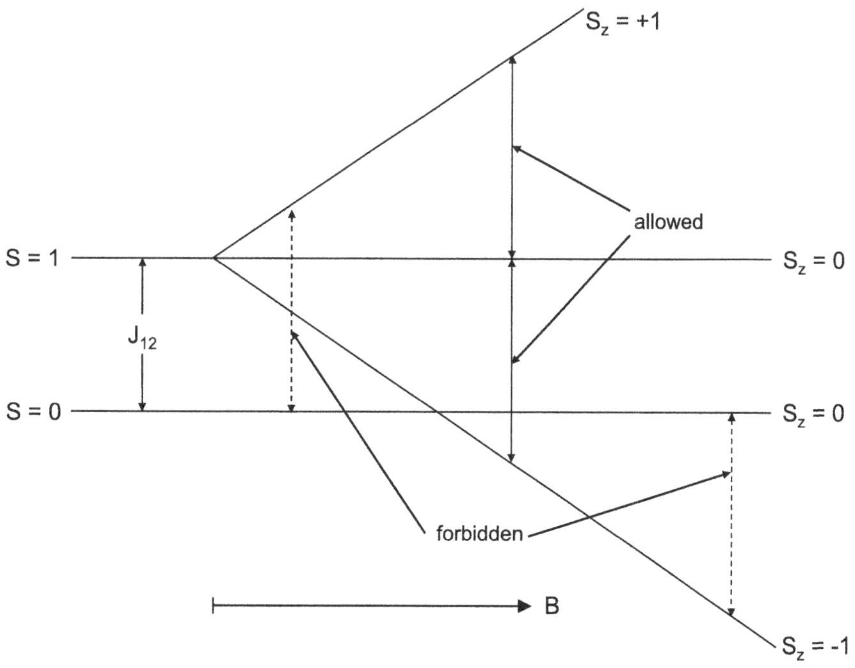

FIGURE 9.11 Triplet and singlet energy levels. For identical radicals $s_1 = s_2 = \frac{1}{2}$, exchange interaction, J_{12}, couples the spins into singlet and triplet manifolds with total spin, $S = 0,1$. Selection rules for magnetic-dipole transitions: $\Delta S = 0$, $\Delta M_S = \pm1$; EPR transitions allowed only within triplet state (solid arrows). "Forbidden" transitions (dashed arrows) allowed only for radicals in different hyperfine states, where they are no longer equivalent.

TABLE 9.1 Hyperfine splittings for allowed EPR transitions of two exchange-coupled nitroxides with $I_1 = I_2 = 1$, where the total nuclear spin projection is $M_I = m_{I_1} + m_{I_2}$, and $\Delta m = m_{I_1} - m_{I_2}$

M_I	m_{I_1}	m_{I_2}	Δm	*allowed transition*[a]	$(J_{12} >> a_N)$[b]
+2	+1	+1	0	$+a_N$	$+a_N$
+1	+1	0	+1	$\frac{1}{2}a_N \mp \frac{1}{2}J_{12}\sqrt{1+(a_N/J_{12})^2} \pm \frac{1}{2}J_{12}$	$\frac{1}{2}a_N \mp \frac{1}{4}a_N^2/J_{12}$
	0	+1	−1	$\frac{1}{2}a_N \mp \frac{1}{2}J_{12}\sqrt{1+(a_N/J_{12})^2} \pm \frac{1}{2}J_{12}$	$\frac{1}{2}a_N \mp \frac{1}{4}a_N^2/J_{12}$
0	0	0	0	0	0
	+1	−1	+2	$\mp \frac{1}{2}J_{12}\sqrt{1+4(a_N/J_{12})^2} \pm \frac{1}{2}J_{12}$	$\mp a_N^2/J_{12}$
	−1	+1	−2	$\mp \frac{1}{2}J_{12}\sqrt{1+4(a_N/J_{12})^2} \pm \frac{1}{2}J_{12}$	$\mp a_N^2/J_{12}$
−1	−1	0	−1	$-\frac{1}{2}a_N \mp \frac{1}{2}J_{12}\sqrt{1+(a_N/J_{12})^2} \pm \frac{1}{2}J_{12}$	$-\frac{1}{2}a_N \mp \frac{1}{4}a_N^2/J_{12}$
	0	−1	+1	$-\frac{1}{2}a_N \mp \frac{1}{2}J_{12}\sqrt{1+(a_N/J_{12})^2} \pm \frac{1}{2}J_{12}$	$-\frac{1}{2}a_N \mp \frac{1}{4}a_N^2/J_{12}$
−2	−1	−1	0	$-a_N$	$-a_N$

[a] Frequency of the corresponding hyperfine line, where the hyperfine constant a_N and exchange constant J_{12} are given in frequency units (Hz). *Note*: for a given M_I, hyperfine lines with non-zero Δm are degenerate.
[b] Approximate hyperfine frequency for strong exchange, from second-order perturbation theory.

These lines have their full intensity, because there are no forbidden transitions. For $\Delta m \neq 0$ (i.e., $M_I = \pm 1$, and the $M_I = 0$ combinations with $\Delta m = \pm 2$), the hyperfine manifolds are split and the extent of the splitting depends on the value of the exchange constant, J_{12}, relative to the hyperfine coupling, a_N. We see this most clearly in the rightmost column of Table 9.1, which gives the second-order shifts in the case of strong exchange. These $\Delta m \neq 0$ lines are reduced in intensity according to $I_a \propto \cos^2 \vartheta(\Delta m)$, because of the "forbidden" transitions.

For a rigid nitroxide biradical, the form of the two-spin spectrum is determined solely by the ratio of the exchange constant to the hyperfine coupling, J_{12}/a_N. Figure 9.12 shows the "allowed" (solid lines) and "forbidden" (dashed lines) EPR lines as a function of J_{12}/a_N. For very strong exchange, $(J_{12}/a_N \to \infty)$, we get just five hyperfine lines $(M_I = 0, \pm 1, \pm 2)$, with splittings $a_N/2$ and relative intensities 1:2:3:2:1, as already noted. For less strong exchange $(J_{12}/a_N \gg 1)$, the $M_I = 0$ and ± 1 manifolds are partly split, giving rise first to inhomogeneous broadening. With decreasing exchange interaction $(J_{12}/a_N > 1)$, the splittings increase and we might resolve the second-order shifts given in Table 9.1. These are largest and most prominent for the central $M_I = 0$ manifold (see Nakajima et al. 1972). As we see from Figure 9.12, the best chance of determining the exchange constant is when it is comparable with the hyperfine splitting: $J_{12}/a_N \approx 1$. Then the "forbidden" transitions in the outer wings are of sufficient intensity that we can obtain J_{12} directly from their position by using Eq. 9.82 (see Ohnishi et al. 1970; Luckhurst and Pedulli 1970). As exchange becomes weaker $(J_{12}/a_N < 1)$, the "forbidden" transitions move inwards and finally all lines group into the three-line hyperfine structure with splittings a_N that is characteristic of a non-interacting nitroxide.

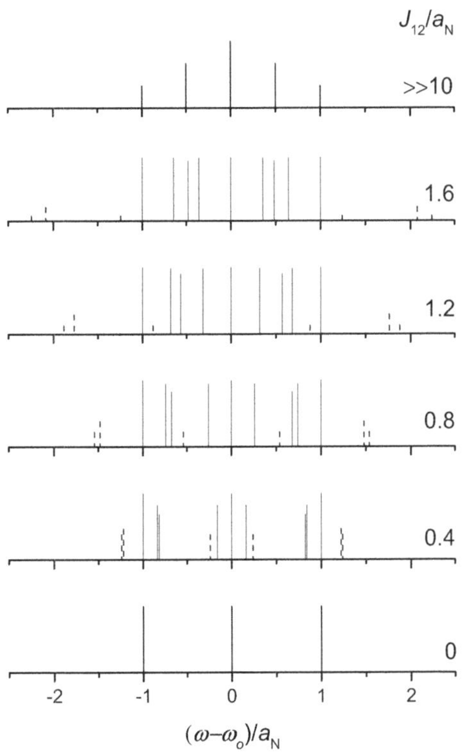

FIGURE 9.12 Stick spectra for symmetric ^{14}N-nitroxide biradical predicted from Eqs. 9.80–9.82, for different exchange coefficients, J_{12}. Solid lines: "allowed" transitions (Eq. 9.81); dashed lines: "forbidden transitions" (Eq. 9.82). Both J_{12} and angular frequency $\Delta\omega$ are normalized to nitrogen hyperfine coupling a_N.

Frequently, biradicals are not rigid but contain a semi-flexible linkage. The biradical can then exist in two or more conformations, where the strength of the exchange interaction differs. If conformational exchange is slow, this will cause further inhomogeneous broadening of the hyperfine manifolds with $\Delta m \neq 0$, because they are sensitive to spin exchange. However, if conformational change is rapid, relative to the differences in J_{12}, modulation of the exchange interaction induces spin relaxation (i.e., homogeneous broadening) of the hyperfine lines that are sensitive to exchange (Luckhurst 1966).

The increase in T_2-relaxation rate for resonances sensitive to exchange is given by $T_{2,ex}^{-1} = (a_N \Delta m)^2 \tau_{J_{12}} / \left(4\langle J_{12}^2 \rangle\right)$, where $\langle J_{12}^2 \rangle$ is the mean-square exchange constant and $\tau_{J_{12}}$ is the correlation time for fluctuations in J_{12} (see Luckhurst 1966; Glarum and Marshall 1967). This assumes fast motion with spectral density $j(J_{12}) = \tau_{J_{12}}$. We then expect differential broadening of the biradical transitions that depends on the value of Δm. Table 9.2 lists the relative broadenings of the various transitions. The hyperfine components ($M_I = 0$ and ± 2) with $\Delta m = 0$ are not broadened. The $M_I = \pm 1$ components with $\Delta m = \pm 1$ are broadened by modulation of the exchange interaction. The $M_I = 0$ components with $\Delta m = \pm 2$ have four times the increase in $M_I = \pm 1$ linewidth. Figure 9.13 shows simulated biradical spectra for motional narrowing with varying values of $\tau_{J_{12}} / \langle J_{12}^2 \rangle$. The $M_I = \pm 1$ components are strongly and variably broadened, relative to the $M_I = 0$ and ± 2 hyperfine components with $\Delta m = 0$. These linewidth features appear in the temperature dependence of experimental spectra from flexible biradicals (Luckhurst and Pedulli 1970). There, the $M_I = 0$ and ± 2 hyperfine components (with $\Delta m = 0$) remain sharp, whereas the $M_I = \pm 1$ components (with $\Delta m = \pm 1$) display a temperature-dependent line broadening.

We can use a simple two-state model to illustrate the effects of conformational transitions on T_2-relaxation of exchange-coupled biradicals. Assume that exchange is extremely strong

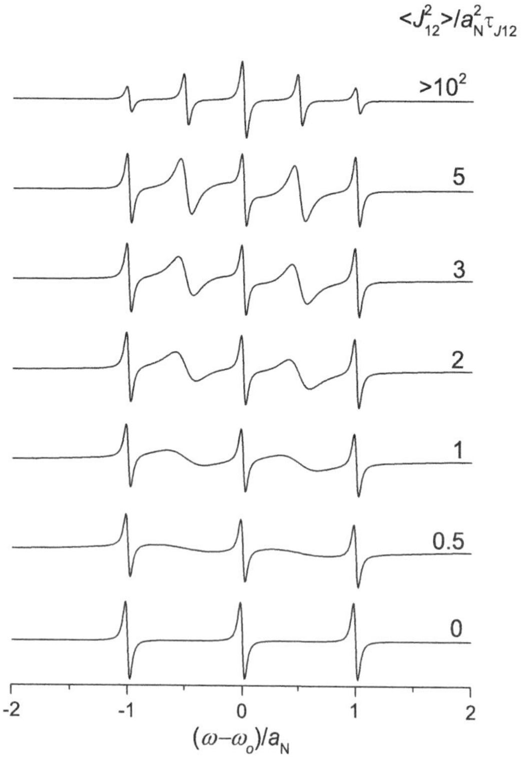

FIGURE 9.13 EPR spectra for flexible ^{14}N-nitroxide biradicals under strong exchange conditions: exchange coefficient J_{12} is modulated by molecular motion with correlation time $\tau_{J_{12}}$. Spectra for values of $\langle J_{12}^2 \rangle / (a_N^2 \tau_{J_{12}})$ indicated; $\langle J_{12}^2 \rangle$: mean-square exchange constant; angular-frequency range $\Delta\omega$ normalized to hyperfine coupling a_N. For two-site exchange (Eq. 9.83): $\langle J_{12}^2 \rangle \equiv 1$; correlation time, $\tau_{J_{12}} \equiv \tau_{conf}$.

in one conformation and zero in the other conformation. The $M_I = \pm 1$ hyperfine lines, at $\pm a_N/2$ under strong exchange, shift to the positions of the $M_I = \pm 2$ lines at $\pm a_N$ or of the $M_I = 0$ lines at 0, when under zero exchange (see Table 9.1 and Figure 9.12). Similarly, the $M_I = 0$ hyperfine lines with $\Delta m = \pm 2$ at position 0, shift to the positions of the $M_I = \pm 2$ lines at $\pm a_N$. The shift of those hyperfine lines that are sensitive to spin exchange is therefore $a_N \Delta m / 2$. For rapid conformational change, the contribution to T_2-relaxation from exchange is then (cf. Eq. 6.25 in Chapter 6):

$$\frac{1}{T_{2,ex}(\Delta m)} = \frac{(a_N \Delta m)^2}{4} \tau_{conf} \tag{9.83}$$

where τ_{conf} is an effective conformational lifetime, and a_N is expressed in angular frequency units. (If τ_A and τ_B are the lifetimes of the conformations with strong exchange and zero exchange, respectively, then for two-site exchange: $\tau_{conf} = \tau_A^2 \tau_B^2 / (\tau_A + \tau_B)^3$). The dependence of the broadening on $(\Delta m)^2$ explains why often only the $M_I = 0$ hyperfine line with $\Delta m = 0$ is observed, because the components with $\Delta m = \pm 2$ are broadened beyond detection. On the other hand, the $M_I = \pm 1$ hyperfine lines (with $\Delta m = \pm 1$) broaden progressively with

TABLE 9.2 Relative line broadening, $T_{2,ex}^{-1}$, of allowed EPR transitions for fluctuating exchange-coupled nitroxides with $I_1 = I_2 = 1$, where the total nuclear spin projection is $M_I = m_{I_1} + m_{I_2}$ and $\Delta m = m_{I_1} - m_{I_2}$

M_I	m_{I_1}	m_{I_2}	Δm	$T_{2,ex}^{-1} / \left(4\langle J_{12}^2 \rangle / a_N^2 j(J_{12})\right)^a$
+2	+1	+1	0	0
+1	+1	0	+1	1
	0	+1	−1	1
0	0	0	0	0
	+1	−1	+2	4
	−1	+1	−2	4
−1	−1	0	−1	1
	0	−1	+1	1
−2	−1	−1	0	0

[a] $\langle J_{12}^2 \rangle$ is the mean-square exchange constant, and $j(J_{12})$ the spectral density of fluctuations in J_{12}. (For the two-site exchange model of Eq. 9.83, $\langle J_{12}^2 \rangle \equiv 1$ and $j(J_{12}) = \tau_{J_{12}}$.)

increasing τ_{conf} (see Figure 9.13). For slow conformational change (cf. Parmon et al. 1973), on the other hand, T_2 is limited just by the conformational lifetime:

$$\frac{1}{T_{2,ex}(\Delta m)} = \frac{1}{\tau_{conf}} \tag{9.84}$$

where, in this case, τ_{conf} is simply the lifetime of the conformation with strong exchange, (i.e., $\tau_{conf} = \tau_A$).

9.9 EXCHANGE PROBABILITY AND EXCHANGE DYNAMICS

The previous sections dealt with a static (or at least equilibrium) picture of exchange. Now we consider the time dependence of spin exchange between nitroxide radicals when they come into proximity.

Consider two nitroxides that originally have oppositely oriented spins, which are characterized by the wave functions $\alpha(1)$ and $\beta(2)$. When the radicals are well separated, we can describe the two-electron system by the simple product wave function $\alpha(1)\beta(2)$, because the notional singlet and triplet states still have the same energy. As the two radicals approach, the exchange interaction removes this degeneracy and the radical pair must be in either a singlet or a triplet state. This requires the radical spins to exchange, because we need mutual spin flips in the $\alpha(1)\beta(2)$ state to produce the $\beta(1)\alpha(2)$ component of the singlet $\left({}^1\chi_0\right)$ and triplet $\left({}^3\chi_0\right)$ states with $M_S = 0$ (see Eqs. 9.70 and 9.73).

We get the rate of spin exchange from the time-dependent Schrödinger equation (see Appendix D.1):

$$i\hbar\frac{\partial\chi}{\partial t} = \mathcal{H}_{HE}\chi \tag{9.85}$$

where $\mathcal{H}_{HE}$ is given by Eq. 9.74. The singlet and triplet states are eigenfunctions (${}^1\chi$ and ${}^3\chi$) of this exchange Hamiltonian with energies ${}^1E \equiv \hbar\omega_S$ and ${}^3E \equiv \hbar\omega_T$, respectively (see Section 9.7). Therefore, a general solution of Eq. 9.85 is a linear combination of singlet and triplet wave functions:

$$\chi(t) = c_S\left({}^1\chi_0\right)e^{-i\omega_S t} + c_T\left({}^3\chi_0\right)e^{-i\omega_T t} \tag{9.86}$$

where the admixture coefficients c_S and c_T are constants to be determined. Rewriting this equation by using Eqs. 9.70 and 9.73, we obtain:

$$\chi(t) = \frac{1}{\sqrt{2}}\Big[\left(c_S e^{-i\omega_S t} + c_T e^{-i\omega_T t}\right)\alpha(1)\beta(2) - \left(c_S e^{-i\omega_S t} - c_T e^{-i\omega_T t}\right)\beta(1)\alpha(2)\Big] \tag{9.87}$$

Because the initial state at time $t = 0$ was $\alpha(1)\beta(2)$, we find that $c_S = c_T = 1/\sqrt{2}$. Hence, the time-dependent wave function during spin exchange is:

$$\chi(t) = \cos\left(J_{12}t/2\hbar\right)\cdot\alpha(1)\beta(2) + i\sin\left(J_{12}t/2\hbar\right)\cdot\beta(1)\alpha(2) \tag{9.88}$$

where the singlet–triplet energy level splitting is: $J_{12} = {}^3E - {}^1E = \hbar\left(\omega_T - \omega_S\right)$, and we replace the exponentials by cosine and sine. Thus, the electrons exchange their spin states in an oscillatory manner, where the frequency of oscillation is determined by the strength of the exchange interaction. The probability density of the spin-exchanged state $\beta(1)\alpha(2)$ is:

$$\rho\left(\beta(1)\alpha(2)\right) = \sin^2\left(J_{12}t/2\hbar\right) \tag{9.89}$$

This equation lets us estimate the probability that spin exchange takes place when two nitroxide monoradicals collide.

After collision, the radical pair dissociates with a unimolecular rate constant, τ_P^{-1}, where τ_P is a measure of the duration of the collision. This is equivalent to assuming an exponential distribution of lifetimes, τ, for the pair:

$$p(\tau) = \exp\left(-\tau/\tau_P\right) \tag{9.90}$$

Hence, the probability, p_{ex}, of spin exchange on collision is given by integrating $\rho(\beta(1)\alpha(2))$ over the distribution, $p(\tau)$, of pair lifetimes:

$$p_{ex} = \frac{\int_0^\infty \sin^2\left(J_{12}t/2\hbar\right)e^{-\tau/\tau_P}\,\mathrm{d}\tau}{\int_0^\infty e^{-\tau/\tau_P}\,\mathrm{d}\tau} \tag{9.91}$$

This leads to the well-known result (Currin 1962; Freed 1966; Johnson 1967):

$$p_{ex} = \frac{1}{2}\left(\frac{\left(J_{12}/\hbar\right)^2\tau_P^2}{1+\left(J_{12}/\hbar\right)^2\tau_P^2}\right) \tag{9.92}$$

For weak exchange (i.e., $\left(J_{12}/\hbar\right)^2\tau_P^2 << 1$), the probability of exchange is: $p_{ex} = \frac{1}{2}\left(J_{12}/\hbar\right)^2\tau_P^2$. For strong exchange (i.e., $\left(J_{12}/\hbar\right)^2\tau_P^2 >> 1$), on the other hand, we get: $p_{ex} = \frac{1}{2}$, as expected from the sinusoidal oscillation of the spin-flip state in Eq. 9.89. The exchange constant for nitroxide radicals is $\left(J_{12}/\hbar\right) \approx 1\times10^{11}$ rad s^{-1} (see later, in Section 9.15). Thus, strong exchange conditions are met for spin-label systems with radical-pair encounter lifetimes of $\tau_P > 1\times10^{-11}$ s. This condition holds for all cases of biological relevance, including diffusion of small molecules in water.

Note that exchange of spin orientations takes place only if the two partners have oppositely oriented spins. If the two spins have the same spin orientation, then no change in spin orientation takes place because the spin system already possesses a triplet wave function, viz., Eq. 9.71 or 9.72. Thus, only half of the possible combinations of radical pairs undergo spin exchange. However, this does not decrease the effects on the EPR spectrum, because the selection rule for each EPR transition is $\Delta M_S = \pm1$ and therefore must involve one state with opposite spins, e.g., $(+,+)\leftrightarrow(-,+)$.

9.10 SPIN-EXCHANGE FREQUENCY AND EPR LINESHAPES

As we learnt in Section 9.9, the Heisenberg spin-exchange interaction between spins S_1 and S_2 reverses the orientation of both spins, when these are oriented mutually antiparallel, i.e., for the combinations $S_{1z}, S_{2z} = +\frac{1}{2}, -\frac{1}{2}$ or $-\frac{1}{2}, +\frac{1}{2}$. If this flip-flop of spin orientation is to affect the EPR spectrum, the resonance frequencies of S_1 and S_2 must be different (cf. Section 6.3 on two-site exchange). For exchange interaction between nitroxides of the same species, this means that S_1 and S_2 must be in different hyperfine states. Let the fractional occupation of spin labels S_1 in hyperfine state $m_I = m$ be f_m, then these undergo effective spin exchange with fraction $1 - f_m$ of partners S_2 that are in hyperfine states $m_I = m'(\neq m)$.

If the pseudo-first order rate constant for exchange of the spin orientation is τ_{HE}^{-1}, the net rate of change in magnetization $\mathbf{M}_{m_I}$ of spins in hyperfine state $m_I = m$ is:

$$\left.\frac{d\mathbf{M}_m}{dt}\right|_{HE} = -(1 - f_m)\tau_{HE}^{-1}\mathbf{M}_m + f_m\tau_{HE}^{-1}\sum_{m'(\neq m)}\mathbf{M}_{m'}$$

$$= -\tau_{HE}^{-1}\mathbf{M}_m + f_m\tau_{HE}^{-1}\sum_{m'}\mathbf{M}_{m'} \tag{9.93}$$

In the first equality, the first term on the right is the decrease in magnetization of the chosen spin orientation (say $S_{1z} = +\frac{1}{2}$) by Heisenberg exchange with oppositely directed spins $\left(S_{2z} = -\frac{1}{2}\right)$ in hyperfine states $m_I = m'$ that are different from m. The second term is the increase in magnetization from exchange interaction of oppositely directed spins $\left(S_{2z} = -\frac{1}{2}\right)$ in the $m_I = m$ hyperfine state with spins of the chosen orientation $\left(S_{1z} = +\frac{1}{2}\right)$ in the other hyperfine states $m'(\neq m)$. In the first equality, the summation is only over hyperfine states with $m' \neq m$, but in the second equality the summation is over all hyperfine states, including $m' = m$. Comparison of the first equality in Eq. 9.93 with Eq. 6.12 of Section 6.3 shows the clear analogy with two-site exchange.

Including Eq. 9.93 in the Bloch equation for complex transverse magnetization $\left(\hat{M}_+ = \hat{M}_x + i\hat{M}_y \equiv u + iv\right)$ in the rotating frame that is given by Eq. 6.7 in Section 6.2, we get the following rate equation for spins in hyperfine state $m_I = m$ (cf. Träuble and Sackmann 1972; Devaux et al. 1973):

$$\frac{d\hat{M}_{+,m}}{dt} = -\left(i(\omega_m - \omega) + \frac{1}{T_{2,m}}\right)\hat{M}_{+,m}$$

$$-\tau_{HE}^{-1}\left(\hat{M}_{+,m} - f_m\sum_{m'}\hat{M}_{+,m'}\right) + i\gamma_e B_1 f_m M_o \tag{9.94}$$

where ω_m and $T_{2,m}$ are the angular resonance frequency and transverse relaxation time, respectively, of spins in hyperfine state $m_I = m$. We assume that the total z-magnetization of the spins M_o, summed over all hyperfine states, remains constant (i.e., no saturation), where B_1 is the microwave magnetic field strength. Equation 9.94 represents a coupled set of linear simultaneous equations for the transverse magnetization of the spins in hyperfine states $m_I = m'$, which include $m' = m$.

We can uncouple the magnetization components in Eq. 9.94 by neglecting all off-diagonal terms, i.e., by neglecting those terms for which $m' \neq m$. This gives us the situation for very slow exchange. The steady-state solution comes from the condition: $d\hat{M}_{+,m}/dt = 0$, giving:

$$\hat{M}_{+,m} \cong \frac{i\gamma_e B_1 f_m M_o}{i(\omega_m - \omega) + 1/T_{2,m} + (1 - f_m)\tau_{HE}^{-1}} \tag{9.95}$$

The EPR absorption lineshape $v(\omega)$ corresponds to the imaginary part of the complex transverse magnetization $\hat{M}_+(\omega)$ (see Section 6.2). Taking the imaginary part of Eq. 9.95, we find that the spectrum for very slow exchange consists of a superposition of Lorentzian absorption lines from each hyperfine state m. Each line has weight f_m and width given by:

$$\frac{1}{T_{2,eff}} = \frac{1}{T_{2,m}} + (1 - f_m)\tau_{HE}^{-1} = \frac{1}{T_{2,m}} + \left(\frac{2I}{2I+1}\right)\tau_{HE}^{-1} \tag{9.96}$$

where I is the nuclear spin that causes the hyperfine splittings (cf. Eqs. 6.20 and 6.21 for two-site exchange). We recognize immediately that the statistical factor $2I/(2I+1)$ in Eq. 9.96 arises because, of the total $2I + 1$ m_I-states of one spin label, $2I$ are different from the m_I-state of the other spin label with which it interacts. The line broadening that results from slow exchange is given by the increase in *half*-width at half-height of the Lorentzian absorption $\left(\Delta\Delta\omega_{HE} \equiv T_{2,HE}^{-1}\right)$:

$$\Delta\Delta\omega_{HE} = \left(\frac{2I}{2I+1}\right)\tau_{HE}^{-1} = \left(\frac{2I}{2I+1}\right)p_{ex}\tau_{coll}^{-1} \tag{9.97}$$

where p_{ex} is the probability of exchange on collision (Eq. 9.92) and τ_{coll}^{-1} is the frequency of bimolecular collisions. The latter is related to the second-order rate constant for collision by: $\tau_{coll}^{-1} = k_{coll}c_M$, where c_M is the molar concentration of spin labels. We can also write Eq. 9.97 in terms of the peak-to-peak linewidth of the first-derivative spectrum, $\Delta\Delta\omega_{pp} = \left(2/\sqrt{3}\right)T_{2,HE}^{-1}$:

$$\Delta\Delta\omega_{pp} = \frac{2}{\sqrt{3}}\left(\frac{2I}{2I+1}\right)\tau_{HE}^{-1} = \frac{2}{\sqrt{3}}\left(\frac{2I}{2I+1}\right)p_{ex}\tau_{coll}^{-1} \tag{9.98}$$

For strong exchange: $p_{ex} = \frac{1}{2}$, and Eq. 9.98 then gives $\tau_{coll}^{-1} = \left(3\sqrt{3}/2\right)\gamma_e\Delta\Delta B_{pp}$ for the field-swept spectrum of a ^{14}N-nitroxide ($I = 1$), where γ_e is the electron gyromagnetic ratio.

At the opposite extreme of very fast exchange, we know from analogy with two-site exchange that the hyperfine structure collapses to a single Lorentzian line at the average frequency (see Section 6.3). Broadening of the lines by exchange is then directly proportional to the square of the hyperfine splitting and inversely proportional to the exchange rate. Generalizing Eq. 6.25 for two-site exchange, the increase in *half*-width at half-height is given by (Molin et al. 1980):

$$\Delta\Delta\omega_{HE} = \left\langle\Delta\omega_{hfs}^{\ 2}\right\rangle\tau_{HE} \tag{9.99}$$

where $\langle \Delta\omega_{hfs}^2 \rangle = \frac{1}{3}I(I+1)a_N^2$ is the second moment of the distribution of hyperfine lines (see Eq. U.23 in Appendix U) and a_N is the nitrogen hyperfine coupling constant in angular frequency units. In fact, Eq. 9.99 is a general result for the secular contribution to T_2-relaxation from motional narrowing theory (see Eq. 5.49 in Chapter 5), because in the fast regime, relaxation rates do not depend on the mechanism of relaxation.

We turn now to the general steady-state solution of Eq. 9.94, which gives us the complex transverse magnetization of spins in hyperfine state $m_I = m$, for all rates of exchange:

$$\widehat{M}_{+,m} = \frac{i\gamma_e B_1 M_o + \tau_{HE}^{-1}\sum_{m'} M_{+,m'}}{i(\omega_m - \omega) + 1/T_{2,m} + \tau_{HE}^{-1}} \times f_m \tag{9.100}$$

For the complete EPR lineshape, we need the total transverse magnetization, which comes from summing over all hyperfine states: $\widehat{M}_+ = \sum_m \widehat{M}_{+,m} \equiv \sum_{m'} \widehat{M}_{+,m'}$. Applying this summation to Eq. 9.100, we then get (cf. Molin et al. 1980):

$$\widehat{M}_+(\omega) = \frac{i\gamma_e B_1 M_o S(\omega)}{1 - \tau_{HE}^{-1} S(\omega)} \tag{9.101}$$

with

$$S(\omega) = \sum_m \frac{f_m}{i(\omega_m - \omega) + 1/T_{2,m} + \tau_{HE}^{-1}} \tag{9.102}$$

where $\sum_m f_m = 1$. This result was obtained originally by Currin (1962), who used a rather less phenomenological approach. Again, we must take the imaginary part of Eq. 9.101 to get the EPR absorption lineshape $v(\omega)$ (see Section 6.2). For Heisenberg exchange of unrestricted frequency τ_{HE}^{-1}, the absorption EPR lineshape then becomes:

$$v(\omega) = \gamma_e B_1 M_o \times \frac{\sigma_o\left(1 - \sigma_o \tau_{HE}^{-1}\right) - \sigma_1^2 \tau_{HE}^{-1}}{\left(1 - \sigma_o \tau_{HE}^{-1}\right)^2 + \sigma_1^2 \tau_{HE}^{-2}} \tag{9.103}$$

where the real summations over the different hyperfine lines, $m_I = m$, are given by:

$$\sigma_o(\omega) = \sum_m \frac{f_m\left(T_{2,m}^{-1} + \tau_{HE}^{-1}\right)}{(\omega - \omega_m)^2 + \left(T_{2,m}^{-1} + \tau_{HE}^{-1}\right)^2} \tag{9.104}$$

$$\sigma_1(\omega) = \sum_m \frac{f_m(\omega - \omega_m)}{(\omega - \omega_m)^2 + \left(T_{2,m}^{-1} + \tau_{HE}^{-1}\right)^2} \tag{9.105}$$

i.e., $S(\omega) \equiv \sigma_o(\omega) + i\sigma_1(\omega)$. Note that σ_o is a sum over Lorentzian absorption lineshapes, and σ_1 is a sum over Lorentzian dispersion lineshapes (cf. Eqs. 6.9, 6.10).

In the specific case of hyperfine structure from a single nitrogen, f_m ($\equiv f$) is the same for each hyperfine component: $f = 1/(2I+1)$ where I is the nitrogen nuclear spin. Simplifications then ensue, if we assume that each line has the same intrinsic width, T_2^{-1}. The absorption EPR lineshape becomes:

$$v(\omega) = \gamma_e B_1 M_o f \times \frac{\sigma'_o\left(T_2^{-1} + \tau_{HE}^{-1}\right)\left[1 - f\sigma'_o\left(T_2^{-1} + \tau_{HE}^{-1}\right)\tau_{HE}^{-1}\right] - f\sigma_1'^2 \tau_{HE}^{-1}}{\left[1 - f\sigma'_o\left(T_2^{-1} + \tau_{HE}^{-1}\right)\tau_{HE}^{-1}\right]^2 + f^2\sigma_1'^2 \tau_{HE}^{-2}} \tag{9.106}$$

with simpler summations over the different hyperfine lines:

$$\sigma'_o(\omega) = \sum_m \frac{1}{(\omega - \omega_m)^2 + \left(T_2^{-1} + \tau_{HE}^{-1}\right)^2} \tag{9.107}$$

$$\sigma'_1(\omega) = \sum_m \frac{\omega - \omega_m}{(\omega - \omega_m)^2 + \left(T_2^{-1} + \tau_{HE}^{-1}\right)^2} \tag{9.108}$$

Corresponding expressions for the first-derivative absorption lineshape of field-swept EPR spectra are derived by Bales and Willett (1984), and by Bales and Peric (1997). We give these analytical results in Appendix A9.1 at the end of the chapter.

In the familiar case of ^{14}N-hyperfine structure: $f = \frac{1}{3}$, for each of the three hyperfine lines. Figure 9.14 shows representative first-derivative absorption lineshapes with increasing

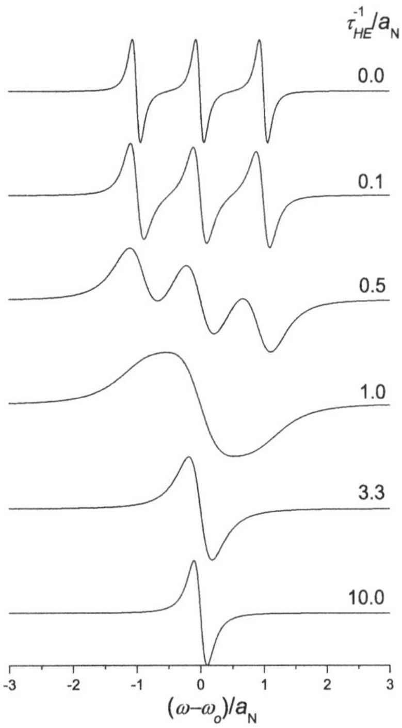

FIGURE 9.14 EPR spectra of freely rotating ^{14}N-nitroxide radicals with increasing spin-exchange frequency, τ_{HE}^{-1} (Eqs. 9.106–9.108, but displayed as first derivative). Exchange frequency scaled to ^{14}N-hyperfine coupling, a_N, in angular-frequency units (rad s^{-1}). Spectra proceed from exchange broadening to hyperfine collapse, and then exchange narrowing, as τ_{HE}^{-1} (i.e., spin-label concentration) increases. Spectra normalized to same line height (cf. Sachse and Marsh 1986).

rates of spin exchange, τ_{HE}^{-1}, from the slow to the fast regime. These derivative lineshapes correspond to Eqs. 9.106–9.108 for the absorption spectrum. The exchange rate is normalized to the isotropic ^{14}N-hyperfine coupling constant, a_N, expressed as an angular frequency. For slow spin exchange (i.e., $\tau_{HE}^{-1} \approx 0.1 \times a_N$), the positions of the hyperfine lines do not change but the lines are broadened. Individual Lorentzian linewidths then give us the exchange rate (see Eqs. 9.97 and 9.98). On increasing the exchange frequency to $\tau_{HE}^{-1} \approx 0.5 \times a_N$, the hyperfine lines become strongly broadened and begin to overlap. At an exchange frequency comparable to the hyperfine coupling (i.e., $\tau_{HE}^{-1} \approx a_N$), the hyperfine structure collapses to an extremely broad single line. Further increase in exchange frequency to $\tau_{HE}^{-1} \approx 3 - 10 \times a_N$ then induces progressive exchange narrowing of the resulting Lorentzian lineshape. In this regime of fast exchange, we again get the exchange rate from the Lorentzian linewidth (see Eq. 9.99). Clearly, we need spectral simulation to determine the exchange rate in the intermediate region of exchange, when $\tau_{HE}^{-1} \approx 0.5 - 2 \times a_N$ (see, e.g., Sachse and Marsh 1986 and Appendix A9.1).

9.11 EXCHANGE-INDUCED DISPERSION

With increasing rates of spin exchange, the lineshapes in Figure 9.14 start as Lorentzian absorptions but then become distorted as we approach incipient exchange narrowing. We know this already for two-site exchange in Section 6.3 (see, e.g., Eq. 6.18). It turns out that we can describe the lineshape distortion rather precisely in this regime as an admixture of dispersion into the Lorentzian absorption (Molin et al. 1980; Bales and Peric 1997). Spectral fitting then lets us identify spin exchange and quantitate it from the extent of dispersion admixture (Bales and Peric 1997; Bales and Peric 2002).

To see this, we first rewrite the steady-state version of Eq. 9.94 by grouping together all magnetization contributions from a particular hyperfine state m:

$$\left(i(\omega_m - \omega) + \frac{1}{T_{2,m}} + (1 - f_m)\tau_{HE}^{-1} \right)\widehat{M}_{+,m} - f_m \tau_{HE}^{-1} \sum_{m'(\neq m)} \widehat{M}_{+,m'} = ic_m \tag{9.109}$$

where the right-hand side $(c_m \equiv \gamma_e B_1 f_m M_o)$ depends only on the equilibrium magnetization M_o. This is an inhomogeneous system of linear equations in which the transverse magnetizations from each hyperfine state are coupled. The general solution to Eq. 9.109 is given by a particular solution, plus the general solution to the homogeneous system of equations that we get by putting the right-hand side equal to zero. The homogeneous system is a set of eigenvalue equations with eigenvectors referred to the transverse magnetizations $\widehat{M}_{+,m}$ as basis functions. This is analogous to the familiar situation for solution of the Schrödinger equation (cf. Appendix G).

Following Molin et al. (1980), we write the complex eigenvalues as: $\omega_{e,m} = \omega'_m + i\Gamma_m$. These are the characteristic roots of a secular determinant made up of the coefficients of $\widehat{M}_{+,m}$ on the left-hand side of Eq. 9.109. Using these characteristic roots, solution of Eq. 9.109 for slow exchange takes the form (Molin et al. 1980):

$$\widehat{M}_{+,m}(\omega) \approx a_m \times \frac{1 + ib_m}{\omega - \omega_{e,m}} \tag{9.110}$$

where a_m and b_m do not depend on ω (cf. Eqs. G.14 and G.15 in Appendix G). To first order, a_m is equal to c_m on the right-hand side of Eq. 9.109, as we shall see later in Section 9.12. We get the absorption lineshape $v_m(\omega)$ for the m_I-manifold, as usual, from the imaginary part of Eq. 9.110:

$$v_m(\omega) \approx \mathrm{Im}\left(\frac{a_m(1 + ib_m)}{\omega - \omega_{e,m}} \right) = a_m \frac{\Gamma_m + b_m(\omega - \omega'_m)}{(\omega - \omega'_m)^2 + \Gamma_m^2} \tag{9.111}$$

The first term in the final equality is a Lorentzian absorption lineshape, and the second term is the corresponding dispersion lineshape (cf. Eqs. 6.10 and 6.9, respectively, in Section 6.2). These components contribute with different weights to the total lineshape. The reasoning leading to Eq. 9.111 is rather general; therefore we expect it to hold for the range of exchange frequencies over which the hyperfine components remain distinguishable.

Figure 9.15 shows a spin-exchanged first-derivative ^{14}N-nitroxide spectrum separated into its absorption and dispersion components. The spectrum in the second row is simulated as described in Section 9.10 (see Eqs. 9.106–9.108), for slow spin exchange at a frequency that maintains the three-line hyperfine structure. Absorption and dispersion components are separated by spectral fitting based on their very different lineshapes (Bales et al. 2009; and see Figure 6.3). The resulting component-spectra appear in the third and fourth rows of Figure 9.15. By fitting concentration-dependent first-derivative experimental spectra, Bales and Peric (1997; 2002) demonstrate that these consist solely of Lorentzian absorption and dispersion components, for both homogeneously and inhomogeneously broadened nitroxides. Inhomogeneous broadening is allowed for by representing the absorption component with a Voigt lineshape or its equivalent, as described in Section 2.6.

Several significant features pertain to Figure 9.15. The first is that the positive dispersion component arises uniquely from Heisenberg spin exchange (Bales et al. 2009). Magnetic dipolar interactions, if present, induce a dispersion component of opposite sign (Peric et al. 2012, and see Section 9.17 later). Secondly, the dispersion component is not influenced by inhomogeneous broadening, which contributes only to the absorption component. Thirdly, there is no exchange-induced dispersion for the $m_I = 0$ hyperfine state. Any dispersion component emerging in the centre of the spectrum comes from instrumental imbalance and the electron-nuclear dipolar (END) part of the hyperfine interaction (Marsh, 2017). To the extent that END can be neglected, we can remove instrumental imbalance by subtracting the $m_I = 0$ contribution from the $m_I = \pm 1$ components. As we shall see later, the exchange-induced $m_I = \pm 1$

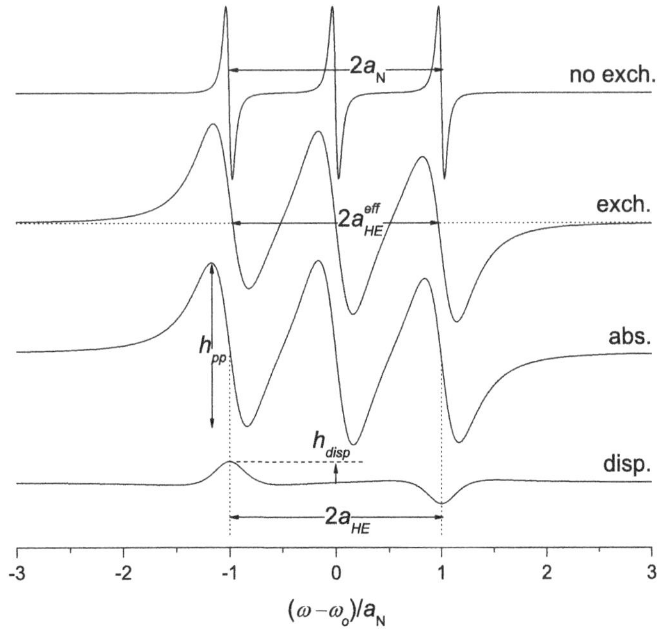

FIGURE 9.15 Lineshape analysis for slow Heisenberg spin-exchange. First-derivative isotropic ^{14}N-nitroxide EPR spectra. *Top to bottom*: 1. no exchange; 2. exchange with frequency $\tau_{HE}^{-1}/a_{\rm N} = 0.29533$ simulated using Eqs. 9.106–108; 3. fitted absorption component of simulated spectrum; 4. fitted dispersion component. (Fitting data from Bales et al. 2009.) Intrinsic ^{14}N-hyperfine coupling constant $a_{\rm N}$ is reduced by exchange to a_{HE} for absorption/dispersion component, whereas effective value a_{HE}^{eff} defined by base-line crossing in original spectrum includes lineshape distortion by exchange.

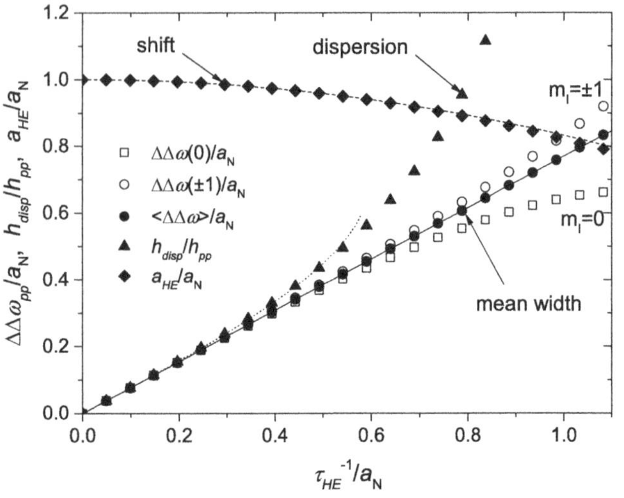

FIGURE 9.16 Exchange broadening $\Delta\Delta\omega_{pp}$ for central ($m_I = 0$, open squares) and outer ($m_I = \pm 1$, open circles) ^{14}N-hyperfine lines, and mean value $\langle \Delta\Delta\omega_{pp} \rangle = \frac{1}{3}[\Delta\Delta\omega(0) + 2\Delta\Delta\omega(\pm 1)]$ (solid circles); relative dispersion (h_{disp}) to absorption (h_{pp}) amplitude (triangles); reduction in isotropic hyperfine splitting constant ($a_{HE}/a_{\rm N}$; diamonds); as function of exchange frequency $\tau_{HE}^{-1}/a_{\rm N}$ for isotropic first-derivative nitroxide spectra simulated using Eqs. 9.106–9.108. (Data from Bales et al. 2009.) Solid straight line: Eq. 9.114 (see text); dotted line: semi-empirical Eq. 9.115; dashed line: Eq. 9.125 for $I = 1$.

dispersion components are equal in magnitude but opposite in sign. Therefore, averaging their absolute values also compensates instrumental imbalance.

Figure 9.16 shows the dependence on Heisenberg exchange frequency of the lineshape characteristics that we measure from Figure 9.15. These are line broadenings, amplitude of the dispersion component, and also the line shifts. Initially, the widths of all three hyperfine lines and the relative amplitude of the dispersion component increase linearly with exchange frequency, at the same rate. Beyond $\tau_{HE}^{-1}/a_{\rm N} \approx 0.3$, the amplitude of the dispersion component begins to increase much more rapidly, and above $\tau_{HE}^{-1}/a_{\rm N} \approx 0.5$, the outer hyperfine lines broaden more rapidly than does the central line. Note, however, that the broadening averaged over all three hyperfine lines still continues to increase with the same linear gradient as that of the individual hyperfine components at lower exchange frequencies. Initially, the inward shift of the absorption resonances $a_{HE}/a_{\rm N}$ is small, as we expect from analogy with two-site exchange. Beyond $\tau_{HE}^{-1}/a_{\rm N} \approx 0.2$, the rate of shift gradually increases but remains less than that of the line broadening. The observed shift $a_{HE}^{eff}/a_{\rm N}$, which we measure directly from the spectrum (see Figure 9.15), is not shown in Figure 9.16. Fortunately, as we shall see later (in Section 9.12), this experimental shift is considerably larger than the intrinsic shift $a_{HE}/a_{\rm N}$, which we can get only from spectral fitting. Tables A9.2.1 and A9.2.2, in Appendix A9.2 at the end of the chapter, list numerical values for the various spectral parameters as a function of exchange frequency simulated for [14]N- and ^{15}N-nitroxides, respectively.

In Figure 9.16, we give both line broadenings and dispersion intensities in terms of peak-to-peak quantities for the usual first-derivative EPR spectra. Differentiating Eq. 9.111 for the absorption, the derivative spectrum becomes:

$$\frac{{\rm d}v_m}{{\rm d}\omega} = a_m \frac{-2\Gamma_m(\omega - \omega'_m) + b_m\left(\Gamma_m^2 - (\omega - \omega'_m)^2\right)}{\left((\omega - \omega'_m)^2 + \Gamma_m^2\right)^2} \tag{9.112}$$

where the first term on the right is the true Lorentzian absorption and the second term is the dispersion admixture. The maximum amplitude of the dispersion component (at $\omega = \omega'_m$) is: $h_{disp} = a_m b_m / \Gamma_m^2$, and the peak-to-peak amplitude of the absorption component (at $\omega = \omega'_m \pm \Gamma_m/\sqrt{3}$) is: $h_{pp} = 3\sqrt{3}a_m/\left(4\Gamma_m^2\right)$. The ratio of dispersion to absorption components is then simply: $h_{disp}/h_{pp} = 4/\left(3\sqrt{3}\right) \times b_m$, where we define h_{disp} and h_{pp} in Figure 9.15. In the following Section 9.12, we derive a value for b_m from perturbation theory. Using Eq. 9.123 from there, the ratio of dispersion to absorption amplitudes of the first derivative is:

$$h_{disp}/h_{pp} = \frac{8}{3\sqrt{3}} \tau_{HE}^{-1} \sum_{m'(\neq m)} \frac{f_{m'}}{\omega_m - \omega_{m'}} \tag{9.113}$$

where $f_m = 1/(2I+1)$ and $\omega_m = a_N m$ for each $m_I = m$. In terms of the hyperfine coupling a_N, the relative amplitude of the dispersion admixture thus becomes:

$$h_{disp}/h_{pp} = \frac{8}{3\sqrt{3}(2I+1)}\frac{\tau_{HE}^{-1}}{a_N}\sum_{m'(\neq m)}\frac{1}{m-m'} = \frac{4}{3\sqrt{3}}\frac{\tau_{HE}^{-1}}{a_N} \quad (9.114)$$

where the second equality applies equally to ^{14}N-nitroxides ($I = 1$) and ^{15}N-nitroxides $\left(I = \frac{1}{2}\right)$. Initially, the dispersion amplitude increases linearly with exchange frequency, as predicted by Eq. 9.114 and shown by the solid line in Figure 9.16. For comparison, we find from Eq. 9.98 that the exchange-induced increase in peak-to-peak linewidth of the first-derivative spectrum is correspondingly: $\Delta\Delta\omega_{pp} = \left(2/\sqrt{3}\right)T_{2,HE}^{-1} = \left(2/\sqrt{3}\right)(1-f)\tau_{HE}^{-1}$. For ^{14}N we have $f = \frac{1}{3}$, which explains why the normalized line broadening $\Delta\Delta\omega_{pp}/a_N$ and relative dispersion amplitude h_{disp}/h_{pp} are superimposable at low exchange frequencies in Figure 9.16 and both have the gradient given by Eq. 9.114. (This is not true for ^{15}N-nitroxides.) At higher exchange frequencies, the dispersion amplitude increases more rapidly. Up to $\tau_{HE}^{-1}/a_N \leq 0.43$, we can approximate the dependence for ^{14}N-nitroxides by (Bales et al. 2009):

$$\tau_{HE}^{-1}/a_N = \frac{3\sqrt{3}}{4}\frac{h_{disp}}{h_{pp}}\left(1 - 0.707\left(\frac{h_{disp}}{h_{pp}}\right)^2\right) \quad (9.115)$$

where the leading term is the linear dependence from perturbation theory in Eq. 9.114, and the correction term is empirical. The calibration curve that we get from Eq. 9.115 is shown as a dotted line in Figure 9.16.

Figure 9.17 shows the relative amplitude of the dispersion component h_{disp}/h_{pp}, and broadening of individual hyperfine lines $\Delta\Delta B_{pp}(m_I)$ in the absorption component, as a function of mean exchange broadening $\left(\langle\Delta\Delta B_{pp}\rangle = 4/(3\sqrt{3}\gamma_e)\times\tau_{HE}^{-1}\right)$, for 16-^{14}N-DOXYL stearic acid methyl ester in ethanol. The exchange frequency τ_{HE}^{-1} is varied either by changing spin-label concentration or by varying temperature, both giving a consistent dependence on exchange broadening (Bales and Peric 2002). The analysis of spectra simulated with Eqs. 9.106–9.108 that we gave in Figure 9.16 describes the experimental data on dispersion admixture and differential line broadening rather well, for this low-viscosity system. Note that, at high exchange frequencies, the central ^{14}N-line is broadened less than the outer hyperfine lines, but the mean broadening retains the linear dependence on exchange frequency. We deal with dispersion contributions from dipolar interactions in higher-viscosity solvents at the end of the chapter, in Section 9.17.

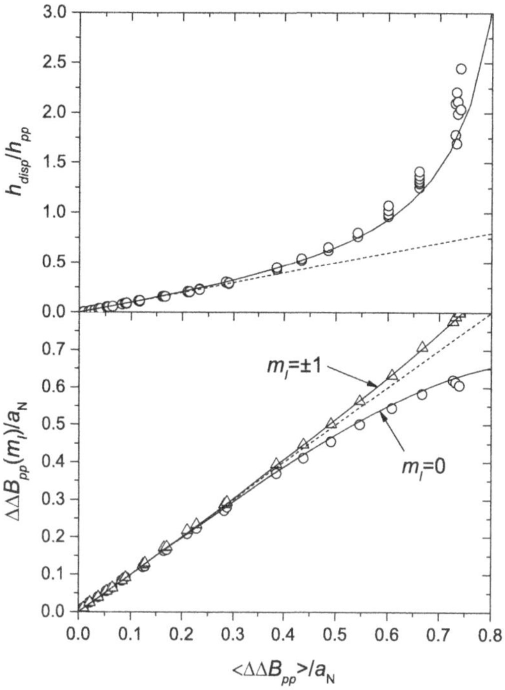

FIGURE 9.17 Amplitude of dispersion component h_{disp}/h_{pp} (top), and broadening of individual hyperfine lines $\Delta\Delta B_{pp}(m_I)$ in absorption component (bottom), for 16-DOXYL stearic acid methyl ester in ethanol (symbols), as function of mean exchange broadening $\langle\Delta\Delta B_{pp}\rangle$ of first-derivative spectrum. (Data from Bales and Peric 2002.) Solid lines correspond to analysis of simulated spectra given in Figure 9.16. Dashed lines have slope +1.

9.12 LINE SHIFTS AND RE-ENCOUNTERS IN SPIN-EXCHANGE SPECTRA

We know from analogy with the situation for two-site exchange in Section 6.3 that slow spin exchange shifts the positions of the resonances inwards towards the centre of the spectrum (see Eq. 6.22). In addition to this resonance shift, the peaks of the absorption spectra (or base-line crossing of the first derivative) shift further inwards because the exchanged-induced admixtures of dispersion distort the Lorentzian lineshapes (Molin et al. 1980; Bales and Peric 1997).

To analyse the line shifts, we follow Molin et al. (1980) and approximate the general expression for the transverse magnetization given in Eqs. 9.94 and 9.100 by taking perturbation theory to second order. The secular equation from the left-hand side of the system of linear equations given by Eq. 9.109 for the transverse magnetization takes the form:

$$\begin{vmatrix} i(\omega_m - \omega) + 1/T_2 + (1-f_m)\tau_{HE}^{-1} & -f_m\tau_{HE}^{-1} \\ -f_{m'}\tau_{HE}^{-1} & i(\omega_{m'} - \omega) + 1/T_2 + (1-f_{m'})\tau_{HE}^{-1} \end{vmatrix} = 0 \quad (9.116)$$

where we give only one of the terms from the sum over m', for illustration. Including first- and second-order terms, the characteristic roots $\omega = \omega_{e,m} \left(\equiv \omega'_m + i\Gamma_m\right)$ are given by (cf. Eq. G.26 in Appendix G):

$$\omega_m - \omega = i\left(\frac{1}{T_2} + \left(1 - f_m\right)\tau_{HE}^{-1}\right) - f_m \tau_{HE}^{-2} \sum_{m'(\neq m)} \frac{f_{m'}}{\omega_{m'} - \omega_m} \quad (9.117)$$

Taking the real part, the direct shift in resonance frequency therefore becomes:

$$\omega'_m - \omega_m = f_m \tau_{HE}^{-2} \sum_{m'(\neq m)} \frac{f_{m'}}{\omega_{m'} - \omega_m} \quad (9.118)$$

which depends on the square of the exchange frequency, τ_{HE}^{-1}.

We deduce the total shift, including lineshape distortion, from the position of the absorption maxima in Eq. 9.111. For small values of dispersion admixture b_m, we find these at spectral frequencies:

$$\omega_{\max,m} \cong \omega'_m + \tfrac{1}{2}\Gamma_m b_m \quad (9.119)$$

for $dv_m/d\omega = 0$ from Eq. 9.112. Comparing this with Eq. 9.118, we see that the shift due to lineshape distortion caused by dispersion admixture is: $\frac{1}{2}\Gamma_m b_m$. The imaginary part of Eq. 9.117 gives us Γ_m. To determine b_m, we need perturbation solutions for the transverse magnetization components $\widehat{M}_{+,m}$. Up to first order, which is sufficient to obtain shifts correct to second order, Eq. 9.109 is given by:

$$\left(i(\omega_m - \omega) + \frac{1}{T_{2,m}} + \left(1 - f_m\right)\tau_{HE}^{-1}\right)\widehat{M}'_{+,m} - f_m \tau_{HE}^{-1} \sum_{m'(\neq m)} \widehat{M}^{(0)}_{+,m'} = ic_m \quad (9.120)$$

where the transverse magnetization $\widehat{M}'_{+,m}$ includes both zeroth-order ($\widehat{M}^{(0)}_{+,m}$) and first-order contributions. The zeroth-order magnetization components in the summation on the left-hand side are given by $\widehat{M}^{(0)}_{+,m'} \cong c_{m'}/\left(\omega_{m'} - \omega_m\right)$. Therefore, Eq. 9.120 reduces to a system of uncoupled simultaneous equations (Molin et al. 1980):

$$\left(\omega_{e,m} - \omega\right)\widehat{M}'_{+,m} = c_m\left(1 - i\tau_{HE}^{-1} \sum_{m'(\neq m)} \frac{f_{m'}}{\omega_{m'} - \omega_m}\right) \quad (9.121)$$

To this particular solution, we must add the general solution to the homogeneous system (see discussion following Eq. 9.109). This corresponds to the secular equation that is given by Eq. 9.116, from which the first-order eigenvectors then become:

$$\widehat{M}_{+,m} = \widehat{M}'_{+,m} - if_m \tau_{HE}^{-1} \sum_{m'(\neq m)} \frac{\widehat{M}'_{+,m'}}{\omega_{m'} - \omega_m} \quad (9.122)$$

(cf. Eqs. G.22 and G.27 in Appendix G). From Eqs. 9.121 and 9.122, we then get the solution for the magnetization components:

$$\widehat{M}_{+,m} \cong c_m \frac{1 - 2i\tau_{HE}^{-1} \sum_{m'(\neq m)} \frac{f_{m'}}{\omega_{m'} - \omega_m}}{\omega_{e,m} - \omega} \equiv a_m \frac{1 + ib_m}{\omega - \omega_{e,m}} \quad (9.123)$$

where $c_m \equiv \gamma_e B_1 f_m M_o$. From Eqs. 9.117–9.119 and 9.123, the total shift in the absorption maximum therefore becomes (Molin et al. 1980):

$$\omega_{\max,m} - \omega_m = \left(1/T_2 + \tau_{HE}^{-1}\right)\tau_{HE}^{-1} \sum_{m'(\neq m)} \frac{f_{m'}}{\omega_{m'} - \omega_m} \quad (9.124)$$

which contains terms both linear and quadratic in the exchange frequency τ_{HE}^{-1}. Comparing with Eq. 9.118, we see that the quadratic term is $1/f_m$ times larger than the direct exchange-induced shift in resonance frequency. For a ^{14}N-nitroxide, this amounts to a factor of three.

Although labelled with the hyperfine state m_I, the foregoing results are quite general. Concretely, for isotropic hyperfine structure from a single nucleus, we have $f_m \equiv f = 1/(2I+1)$ where I is the nuclear spin, and $\omega_m = a_N m$ where the hyperfine coupling constant a_N is expressed in angular frequency units. To derive hyperfine splittings a_{HE} and a_{HE}^{eff} in the presence of exchange (see Figure 9.15), we must distinguish between odd and even nuclear spins I. For even spins, we take half the splitting between the $m_I = +1$ and $m_I = -1$ hyperfine lines, both of which are shifted inwards by the same amount. For odd spins, on the other hand, we take the whole splitting between the $m_I = +\frac{1}{2}$ and $m_I = -\frac{1}{2}$ hyperfine lines, both of which again have equal inward shifts. The summation that we need to calculate the shifts from Eqs. 9.118 and 9.124 is then: $\sum_{m'(\neq m)} 1/(m - m') = 3/(I+1)$ for both odd and even I, when we allow for the double shift with odd I. The hyperfine constants that characterize the exchange shifts defined in Figure 9.15 then become:

$$a_{HE}/a_N = 1 - \frac{3}{(I+1)(2I+1)^2}\left(\frac{\tau_{HE}^{-1}}{a_N}\right)^2 \quad (9.125)$$

$$a_{HE}^{eff}/a_N = 1 - \frac{3}{(I+1)(2I+1)}\left(\frac{1/T_2}{a_N} + \frac{\tau_{HE}^{-1}}{a_N}\right)\frac{\tau_{HE}^{-1}}{a_N} \quad (9.126)$$

for intrinsic and total shifts, respectively. The dashed line in Figure 9.16 comes from Eq. 9.125 for $I = 1$, viz. $a_{HE}/a_N = 1 - \frac{1}{6}\left(\tau_{HE}^{-1}/a_N\right)^2$ without adjustable parameters. Agreement of this perturbation result with exact calculations is good up to quite high exchange frequencies. However, only for peroxylamine disulphonate (PADS^{2-}) do shifts measured experimentally agree with the predictions of Eqs. 9.125 and 9.126 (Bales and Peric 1997). For most other nitroxides, the measured shifts are larger than predicted, whereas we have seen that linewidths, intensities and lineshapes are in good accord with the treatment of exchange given in the two previous sections (Bales and Peric 2002; Bales et al. 2003; Kurban et al. 2008).

The extra shift found experimentally is linear in exchange frequency (Bales and Peric 2002), which is not predicted by Eq. 9.118 for the direct shift in resonance frequency. This suggests a mechanism other than spin exchange during a single, sudden collision. Instead of simply colliding once and then diffusing apart, spin labels may remain within a "cage" and re-encounter several times before finally diffusing apart (Noyes 1954). Using a theory by Salikhov (1985), Bales et al. (2003; 2008) show that such a re-encounter mechanism accounts for the increased linear line broadening induced by exchange.

Whilst in the cage, the interacting pair of spin labels in hyperfine states m and m' undergoes exchange narrowing. If f_{RE} is the mean fraction of time spent in the spin-exchanged state during intervals τ_{RE} between re-encounters, the mean resonance frequency of spin label in hyperfine state m is: $\bar{\omega}_m = (1 - f_{RE})\omega_m + f_{RE}\omega_{m'}$ (cf. Eq. 6.24). Thus the additional shift in resonance frequency, $\delta(\omega_{e,m} - \omega_m)$, that is induced by exchange during re-encounters is:

$$\delta(\omega_{e,m} - \omega_m) = (\omega_{m'} - \omega_m) f_{RE} \tag{9.127}$$

where we assume that the lifetime τ_P of the collision pair is negligible compared with τ_{RE}. Equation 9.127 applies if the distribution of τ_{RE} follows Poisson statistics (Salikhov 1985), but this does not apply to continuous diffusion, which interests us most. For diffusive re-encounters, the additional shift in resonance position is given by (Salikhov 1985; Bales et al. 2003):

$$\delta(\omega_{e,m} - \omega_m) = -\tfrac{1}{2}\tau_{HE}^{-1} \sum_{m'(\neq m)} f_{m'} \sqrt{|\omega_m - \omega_{m'}|\tau_{RE}/2} \tag{9.128}$$

where we assume strong exchange. Adding Eq. 9.128 to Eq. 9.118 introduces a linear term into the shift in resonance frequency, which otherwise is solely quadratic in τ_{HE}^{-1}. Note that Eq. 9.128 adds equally to the total shift of the absorption maximum which is given by Eq. 9.124. In this latter case, however, the effects of re-encounter add to a pre-existing linear term. In terms of the nitrogen hyperfine structure, Eq. 9.128 becomes:

$$\delta(\omega_{e,m} - \omega_m) = -\frac{\tau_{HE}^{-1}\sqrt{a_N \tau_{RE}}}{2\sqrt{2}(2I+1)} \sum_{m'(\neq m)} \sqrt{|m - m'|} \tag{9.129}$$

For a ^{15}N-nitroxide $(I = \frac{1}{2})$, the reduced hyperfine splitting then contains the additional contribution:

$$\delta a_{HE} = -\frac{1}{2\sqrt{2}}\sqrt{a_N \tau_{RE}}\,\tau_{HE}^{-1} \tag{9.130}$$

and for a ^{14}N-nitroxide ($I = 1$):

$$\delta a_{HE} = -\tfrac{1}{6}\left(1 + 1/\sqrt{2}\right)\sqrt{a_N \tau_{RE}}\,\tau_{HE}^{-1} \tag{9.131}$$

either of which adds to both Eq. 9.125 and Eq. 9.126.

Figure 9.18 shows the intrinsic decrease in hyperfine coupling a_{HE} of the absorption component of the spectrum from 16-^{14}N-DOXYL stearic acid methyl ester in ethanol, as a function of the exchange broadening $\left(\langle \Delta\Delta B_{pp}\rangle = 4/\left(3\sqrt{3}\gamma_e\right) \times \tau_{HE}^{-1}\right)$. Combining Eqs. 9.125 and 9.131, we predict the following dependence for a ^{14}N-nitroxide:

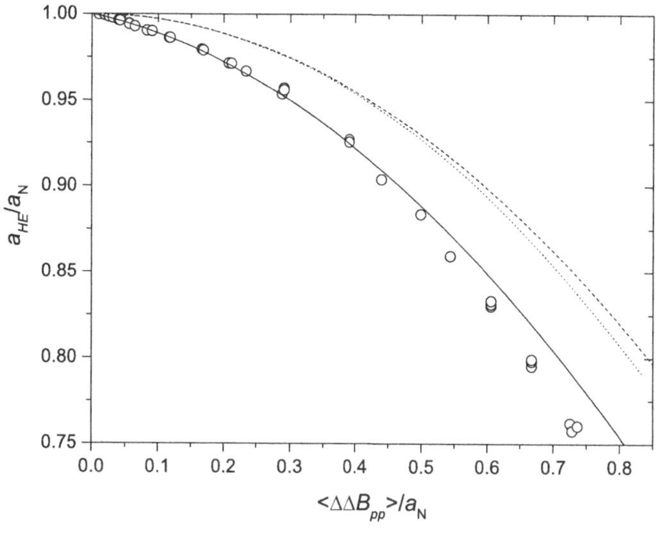

FIGURE 9.18 Intrinsic decrease in hyperfine coupling a_{HE}/a_N of absorption component from spectrum of 16-DOXYL stearic acid methyl ester in ethanol, as function of mean exchange broadening $\langle \Delta\Delta B_{pp}\rangle$ of first-derivative spectrum. Dashed line: prediction from Eq. 9.132 with $\kappa = 0$; dotted line: exact dependence from Figure 9.16. Solid line: least-squares fit of Eq. 9.132 up to $\langle \Delta\Delta B_{pp}\rangle = 0.5a_N$, with linear term as fitting parameter yielding $\kappa = 0.083 \pm 0.002$. Experimental data (circles) from Bales and Peric (2002).

$$a_{HE}/a_N = 1 - \kappa\frac{\langle \Delta\Delta\omega_{pp}\rangle}{a_N} - \frac{9}{32}\left(\frac{\langle \Delta\Delta\omega_{pp}\rangle}{a_N}\right)^2 \tag{9.132}$$

where $\kappa = \left(\sqrt{3}/8\right)\left(1 + 1/\sqrt{2}\right)\sqrt{a_N \tau_{RE}}$. From fitting Eq. 9.132 to the data in Figure 9.18, up to the limits for perturbation theory, we get $\tau_{RE} = 1.9 \times 10^{-10}$ s at 25°C (Bales et al. 2003). According to the treatment by Salikhov (1985), the time between diffusive reencounters is equal to the characteristic diffusion time $\tau_D \equiv d_S^2/(2D_T)$ that we introduced already in connection with dipolar relaxation by translational diffusion (see Section 9.5.2). Using $d_S = 0.64$ nm (see Section 9.5.3) and the Stokes-Einstein prediction for a sphere of this diameter (see following Section 9.13), we estimate $\tau_D = 3.3 \times 10^{-10}$ s in ethanol at 25°C, which is the order of magnitude expected for a low-viscosity solvent. We deal with line shifts from dipolar interactions in higher-viscosity solvents at the end of the chapter, in Section 9.17.

9.13 TRANSLATIONAL DIFFUSION AND FRICTION COEFFICIENTS

In spin-label EPR, the principal application of Heisenberg exchange is for studying translational dynamics. Before continuing with the EPR aspects, we therefore summarize some useful results on translational diffusion in this section.

We describe Brownian translational diffusion by Fick's first law, which states that the flux of spin-labelled molecules is proportional to their concentration gradient:

$$J(x) = -D_T \frac{\partial P(x,t)}{\partial x} \tag{9.133}$$

where $J(x)$ is the rate of flow, per unit area, of spin labels in direction x, and $P(x,t)$ is the probability per unit volume of finding a spin label at position x at time t. Equation 9.133 defines the translational diffusion coefficient, D_T, of the spin-labelled species. We derive the translational diffusion equation from Fick's first law by using the continuity equation: $\partial P/\partial t = -\partial J/\partial x$, which balances the flows in and out of the volume element at position x. Substituting this condition into Eq. 9.133 yields the differential equation:

$$\frac{\partial P(x,t)}{\partial t} = D_T \frac{\partial^2 P(x,t)}{\partial x^2} \tag{9.134}$$

for translational diffusion in one dimension. We can generalize Eq. 9.134 readily to three dimensions by replacing the second-derivative on the right by the operator $\nabla^2 \equiv \partial^2/\partial x^2 + \partial^2/\partial y^2 + \partial^2/\partial z^2$. Normally, we then use spherical polar coordinates.

The translational diffusion coefficient is related to the mean-square displacement, $\langle x^2 \rangle$, of the spin label. In Brownian diffusion, the position of the spin label undergoes small random changes, δx, resulting from frequent collisions with other neighbouring molecules. The change in positional probability after a short time τ is given by: $(\partial P/\partial t) \times \tau$. This must be equal to the sum of probabilities that the spin label lies within δx of the initial position, which we obtain by averaging over a Taylor expansion of the probability distribution:

$$P(x+\delta x) - P(x) = \left(\frac{\partial P}{\partial x}\right)\delta x + \frac{1}{2}\left(\frac{\partial^2 P}{\partial x^2}\right)(\delta x)^2 + \ldots \tag{9.135}$$

For a random process, the average displacement is: $\langle \delta x \rangle = 0$, and hence:

$$\frac{\partial P}{\partial t} = \frac{\langle (\delta x)^2 \rangle}{2\tau}\left(\frac{\partial^2 P}{\partial x^2}\right) \tag{9.136}$$

to second order. Comparison with Eq. 9.134 then yields (cf. Einstein 1905):

$$\langle x^2 \rangle = 2D_T t' \tag{9.137}$$

for the mean-square x-displacement in time t'. Equation 9.137 applies to the one-dimensional case. For translational diffusion in two dimensions, the mean-square displacement is: $\langle r^2 \rangle \equiv \langle x^2 + y^2 \rangle = 4D_T t'$, and for three-dimensions: $\langle r^2 \rangle = 6D_T t'$.

As for rotational diffusion (see Section 7.3), the diffusion coefficient is related to the shape and size of the diffusing molecule, and to the viscosity of the medium, by the frictional coefficient. We define the translational friction coefficient, f_T, as the ratio of the frictional drag, F_i, to the particle velocity, v_i:

$$f_T = F_i/v_i \tag{9.138}$$

In the steady state (i.e., no acceleration), the frictional drag on the diffusing particle is: $F_i = f_T v_i$. The equilibrium distribution of particle positions in the presence of thermal fluctuations is given by the Boltzmann law:

$$P(x_i) \propto \exp\left(\frac{F_i x_i}{k_B T}\right) \tag{9.139}$$

where the work done in translation by dx_i is $-F_i \cdot dx_i$. Differentiating Eq. 9.139 and rearranging, the frictional force is:

$$F_i = \frac{k_B T}{P(x_i)}\left(\frac{\partial P}{\partial x_i}\right) \tag{9.140}$$

The flow corresponding to this frictional force is given by the positional probability multiplied by the velocity, i.e., $P(x_i)v_i = P(x_i)(F_i/f_T)$, which at equilibrium is balanced by the opposing flow from diffusion that is given by Fick's first law. Hence from Eqs. 9.133 and 9.140, the diffusion coefficient is related to the friction coefficient by:

$$D_T = \frac{k_B T}{f_T} \tag{9.141}$$

which is the Einstein relation for translational diffusion (Einstein 1905).

According to fluid dynamics, the frictional coefficient is determined by the viscosity of the medium, η, and by the shape and size of the diffusing species. We normally use sticking boundary conditions, where the fluid at the surface of the diffusing particle moves with exactly the same velocity as the particle itself. For a sphere of radius a, the resulting Stokes-Einstein expression for the frictional coefficient is (Einstein 1905):

$$f_T^o = 6\pi\eta a \tag{9.142}$$

which represents a much weaker dependence on molecular dimensions than that for rotational diffusion. From Eq. 7.10 of Chapter 7, the ratio of rotational to translational frictional coefficients is: $f_R^o/f_T^o = \frac{4}{3}a^2$. For slip boundary conditions, where there is no interaction between the diffusing particle and the fluid molecules, the translational frictional coefficient for a sphere is reduced by one third to: $f_T^o = 4\pi\eta a$ (cf. Section 7.4 in Chapter 7 for rotational diffusion).

The frictional coefficients of non-spherical species are larger than those for a sphere of the same volume. Perrin (1934) used a hydrodynamic model with stick boundary conditions to derive expressions for the friction coefficients of ellipsoids of revolution with semi-axes a, $b = c$, in isotropic media. The translational friction coefficients, f_T, are related to that of a sphere of equal volume, $V_{sph} = 4\pi ab^2/3$, by the shape factors, F_T:

$$f_T = f_T^o F_T \tag{9.143}$$

For inverse axial ratio $\rho \equiv b/a$, the shape factor is given by (Perrin 1934):

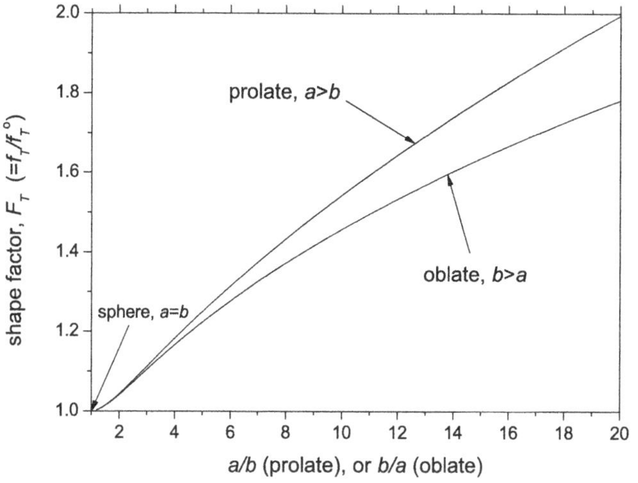

FIGURE 9.19 Translational friction coefficients, f_T, for ellipsoids of revolution. Shape factors, F_T, as function of axial ratio, a/b, for prolate ($a > b$) and oblate ($b > a$) ellipsoids (Eqs. 9.144, 9.145). Stick boundary conditions assumed.

$$F_T = \frac{\sqrt{1-\rho^2}}{\rho^{2/3}\ln\left(\left[1+\sqrt{1-\rho^2}\right]/\rho\right)} \tag{9.144}$$

for a prolate ellipsoid ($\rho < 1$), and by

$$F_T = \frac{\sqrt{\rho^2-1}}{\rho^{2/3}\tan^{-1}\left(\sqrt{\rho^2-1}\right)} \tag{9.145}$$

for an oblate ellipsoid ($\rho > 1$). Figure 9.19 shows the dependence of the shape factors F_T on axial ratio, a/b or b/a, for oblate and prolate ellipsoids. Initially, the shape factors increase rather slowly with increasing asymmetry for both oblate and prolate ellipsoids. For the same axial ratio, the translational shape factors for prolate ellipsoids are greater than for oblate ellipsoids and are much smaller than those for rotational diffusion of the same species (compare Figures 7.1 and 9.19). For a long rod of length $2a$ and radius b, the shape factor for translational diffusion is $F_T = \left(2/3\rho^2\right)^{1/3}/(2\ln(2/\rho)-0.11)$, where $\rho = b/a$ and the volume of the equivalent sphere is $V_{sph} = 2\pi b^2 a$.

9.14 BIMOLECULAR COLLISION RATES AND TRANSLATIONAL DIFFUSION

We can determine collision rate constants for small spin-labelled molecules in non-viscous solvents from the concentration dependence of Lorentzian broadening in the slow-exchange regime. Figure 9.20 shows data from the isotropic three-line spectra of a small ^{14}N-nitroxide-labelled fatty acid in water. We get the Lorentzian, ΔB_{pp}^L, and Gaussian, ΔB_{pp}^G, contributions to the total width, ΔB_{pp}^{tot}, of the inhomogeneously broadened lines by using the methods described in Section 2.5 of Chapter 2. The Lorentzian linewidth increases linearly with increasing spin-label concentration, corresponding to exchange broadening of each nitrogen hyperfine line, whereas the Gaussian linewidth exhibits exchange narrowing of the unresolved proton hyperfine structure. Dipolar interactions do not contribute, because they are averaged out by fast diffusion in non-viscous media (cf. Section 9.5.3, and later below). For slow exchange, the concentration dependence of the spin exchange frequency is related to that of the peak-to-peak Lorentzian linewidth by Eq. 9.98 in Section 9.10. The second-order rate constant for bimolecular collisions then is given by:

$$k_{coll} \equiv \frac{\tau_{coll}^{-1}}{c_M} = \frac{\sqrt{3}}{2p_{ex}}\frac{2I+1}{2I}\gamma_e\left(\frac{d\Delta B_{pp}^L}{dc_M}\right) \tag{9.146}$$

where γ_e is the electron gyromagnetic ratio and c_M is the molar concentration of spin label. The data in Figure 9.20, with $p_{ex} = \frac{1}{2}$ and $I = 1$, yield a collision rate constant $k_{coll} = 4.1\times10^9\ \mathrm{M^{-1}\ s^{-1}}$ (King and Marsh 1986).

We can compare this result with the rate constant for a diffusion-controlled bimolecular reaction between species with collision radii r_1 and r_2, and translational diffusion coefficients $D_{T,1}$ and $D_{T,2}$. From Fick's law (Eq. 9.133 in Section 9.13), the total inward diffusive flux of species 1 (in molecules/s) across a sphere of radius r is:

$$\Phi_1 = J_1 \times 4\pi r^2 = 4\pi D_{T,1} N_A 10^3 r^2 \frac{\partial c_{M,1}}{\partial r} \tag{9.147}$$

where $c_{M,1}$ is the molar concentration of species 1 and N_A is Avogadro's number. In the steady state, Φ_1 is constant. For a diffusion-controlled encounter with species 2, we must integrate

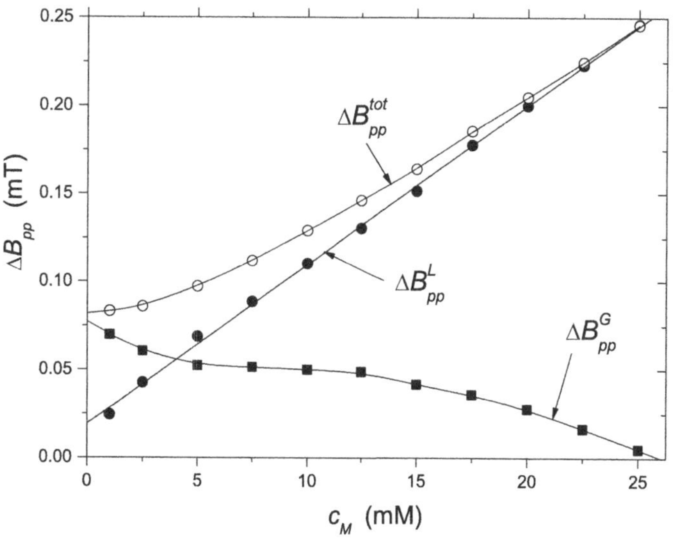

FIGURE 9.20 Concentration dependence of total (open circles), Lorentzian (solid circles) and Gaussian (squares) contributions to peak-to-peak width, ΔB_{pp}, of low-field ($m_I = +1$) hyperfine line in EPR spectrum of spin-labelled valeric acid (4-DOXYL VA) dissolved in water at 20°C. (Data from King and Marsh 1986.)

$\partial c_{M,1}/\partial r\left(\propto r^{-2}\right)$ from the distance of closest approach, $r = r_1 + r_2$, where $c_{M,1} = 0$, to infinity where $c_{M,1}$ reaches its bulk value, $c_{o,1}$. From Eq. 9.147, the (constant) flux is thus:

$$\Phi_1 = 4\pi N_A 10^3 (r_1 + r_2) D_{T,1} c_{o,1} \tag{9.148}$$

The diffusion-controlled collision rate between species 1 and 2 (in M.s^{-1}) is given by: $k_{coll} c_{o,1} c_{o,2} = \Phi_1 c_{o,2}$, and the second-order rate constant (in M^{-1}s^{-1}) becomes (von Smoluchowski 1917):

$$k_{2,diff} = 4\pi N_A 10^3 (r_1 + r_2)(D_{T,1} + D_{T,2}) \tag{9.149}$$

where we replace $D_{T,1}$ by $D_{T,1} + D_{T,2}$ because the origin of r is species 2, which is also diffusing. Taking a value of $r_1 = r_2 = 0.4$ nm for the collision radius of the spin-label group yields a translational diffusion coefficient $D_T = 3.4 \times 10^2\ \mu\text{m}^2\ \text{s}^{-1}$ for spin-labelled valeric acid in water at 20°C. Comparing with the Stokes-Einstein expression for a sphere (Eqs. 9.141 and 9.142): $D_T = k_B T/(6\pi\eta a)$ yields a shape factor in the range $F_T \approx 1.3 - 1.6$, which corresponds to an axial ratio of the equivalent prolate ellipsoid of $a/b \approx 5 - 10$ (King and Marsh 1986).

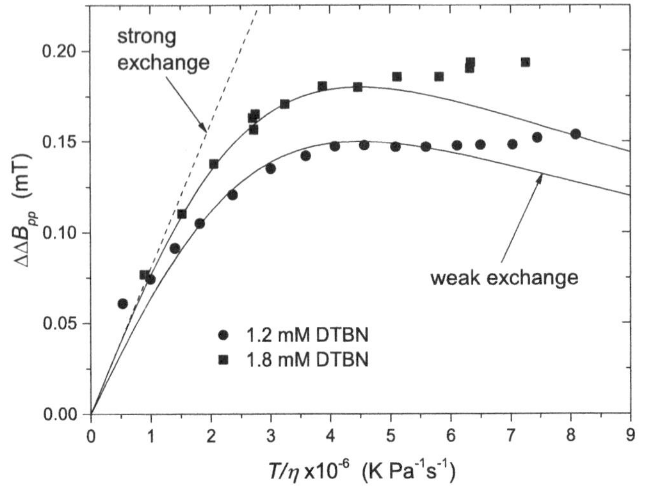

FIGURE 9.21 Dependence of spin-exchange broadening $\Delta\Delta B_{pp}$ on ratio T/η of absolute temperature to solvent viscosity for di-*tert*-butyl nitroxide (DTBN) in liquid propane. *Circles*: [DTBN] = 1.2 mM; *squares*: [DTBN] = 1.8 mM. (Data from Plachy and Kivelson 1967.) *Solid lines*: functional dependence predicted by Eqs. 9.151, 9.153 with temperature-independent jump length λ.

9.15 SPIN EXCHANGE CONSTANT AND BIMOLECULAR COLLISIONS

As stated already, spin exchange during nitroxide collisions is normally strong: $J_{12}^2 \tau_P^2 >> 1$. (From now on, we express the exchange constant J_{12} in angular-frequency units, i.e., units of $\hbar$.) To get weak exchange, the duration of collision τ_P must be very short, which means ultra-low viscosities. Plachy and Kivelson (1967) measured exchange-induced line broadening of the small spin probe di-*t*-butyl nitroxide (DTBN) in short-chain hydrocarbon solvents, propane and *n*-pentane, with increasing temperature. Figure 9.21 shows the dependence of the exchange-induced line broadening $\Delta\Delta B_{pp} \equiv \left(2/\sqrt{3}\gamma_e\right) T_{2,HE}^{-1}$ on the ratio T/η of absolute temperature to solvent viscosity, for DTBN in liquid propane. Linewidths at very low concentration are subtracted from those at higher concentration, in Figure 9.21. (At such low viscosities, the T/η-dependence of the low-concentration linewidths comes from spin-rotation interaction.) With decreasing viscosity, i.e., decreasing τ_P, the exchange linewidth first increases approximately linearly in Figure 9.21 and then reaches a plateau value but does not subsequently decrease as we would expect for the onset of weak exchange according to Eq. 9.92. This is because the mean free step-length λ for diffusive jumps increases with increasing temperature due to thermal expansion.

If X_{SL} is the mole fraction of spin label, the line broadening from exchange is given by (cf. Eq. 9.97 and Eq. 9.160 given later):

$$T_{2,HE}^{-1}/X_{SL} = \frac{2I}{2I+1} p_{ex} z \tau_j^{-1} \tag{9.150}$$

where τ_j^{-1} is the spin label hopping frequency, z is the number of new neighbours encountered at each diffusive jump and p_{ex} is the probability of exchange on collision that is given by Eq. 9.92. The mean duration of collisions is given by: $\tau_P \approx (z_T/z)\tau_j$ where z_T is the total number of nearest neighbours (Plachy and Kivelson 1967). From Eqs. 9.92 and 9.150, we therefore have:

$$T_{2,HE}^{-1}/X_{SL} = \frac{I}{2I+1} z_T J_{12} \left(\frac{1}{1/(J_{12}\tau_P) + J_{12}\tau_P} \right) \tag{9.151}$$

From the Einstein relation for diffusion in three dimensions (cf. Eq. 9.137), the mean free path is related to the time between jumps by:

$$\lambda^2 = 12 D_T \tau_j \tag{9.152}$$

where $2D_T$ is the effective translational diffusion coefficient for bimolecular encounters. The self-diffusion coefficient is given by the Stokes-Einstein relation (Eqs. 9.141 and 9.142): $D_T = k_B T/(6\pi\eta a)$, where a is the effective radius of the diffusant and k_B is Boltzmann's constant. The duration of collision therefore becomes:

$$\tau_P = \frac{z_T}{z} \frac{\pi a}{2k_B} \lambda(T)^2 \left(\frac{\eta}{T} \right) \tag{9.153}$$

where $\lambda(T)$ is the mean jump length, which in general depends on temperature.

The solid lines in Figure 9.21 show the dependence of the line broadening on T/η that we predict from Eqs. 9.151 and 9.153 with a constant value of $\lambda(T)$. Fitting Eq. 9.151 to comparable data for DTBN in *n*-pentane, but including a temperature dependence $\lambda(T) = \lambda_o \left(\rho(T_o)/\rho(T)\right)^{1/3}$ where $\rho(T)$ is the solvent density, yields $z_T J_{12} = 9.9 \times 10^{11}$ s^{-1} (Plachy and Kivelson 1967). For $z_T \approx 8$ nearest neighbours, we then get $J_{12} \cong 1.2 \times 10^{11}$ s^{-1} for the spin-exchange constant of DTBN. The fitting also yields a value of $(z_T/z)a\lambda_o^2 J_{12} = 3.2 \times 10^{10}$ nm^3 s^{-1} from Eq. 9.153 (Plachy and Kivelson 1967). With the classical diffusion radius of DTBN $a \cong 0.32$ nm, and a jump length $\lambda_o \approx 0.5$ nm of the order of a solvent diameter, we then get $z \approx 2.5$. This implies that DTBN encounters two to three new neighbours per jump. Note that our Eqs. 9.92 and 9.152 differ by a factor of a half and two, respectively, from those assumed originally by Plachy and Kivelson (1967).

Alternatively, we can express the dissociation of the collision pair (which takes place at a rate τ_P^{-1}) in terms of continuum theory for translational diffusion. We again integrate Eq. 9.147 for the constant flux of spin label 1 but with the boundary conditions: $c_{M,1} = 0$ at infinity and $c_{M,1} = 1/(N_A 10^3 v_{12})$ at $r = r_1 + r_2$, where $v_{12} = \frac{4}{3}\pi(r_1 + r_2)^3$ is the steric volume occupied by one spin-label molecule. The dissociation rate constant then becomes:

$$\tau_P^{-1} = 4\pi(D_{T,1} + D_{T,2})(r_1 + r_2)\frac{1}{v_{12}} = \frac{2k_B T}{\pi(r_1 + r_2)^3 \eta} \tag{9.154}$$

where we use the Stokes-Einstein expression (Eqs. 9.141 and 9.142) for the diffusion coefficient, with $r_1 = r_2$, for the second equality. Again taking $r_1 \equiv a \cong 0.32$ nm, we get $\tau_P \cong 1.0 \times 10^{-10}$ s for the duration of collisions between DTBN molecules in water at 20°C (where $\eta = 1.0$ mPa s). This easily fulfils the condition for strong exchange $(J_{12}^2 \tau_P^2 \approx 140)$, as already asserted.

Solid symbols in Figure 9.22 show the dependence of exchange broadening $\Delta\Delta B_{pp}/c_M$ on T/η for perdeuterated TEMPONE in water. The approximately linear dependence with positive slope and zero intercept corresponds to strong exchange in this more viscous solvent, as compared with the short-chain alkanes used for Figure 9.21. The rate constant for exchange broadening of TEMPONE in water is $T_{2,HE}^{-1}/c_M = 1.77 \times 10^9$ M^{-1} s^{-1} at 20°C. From Eq. 9.97 this corresponds to a bimolecular collision rate constant: $k_{coll} = 5.3 \times 10^9$ M^{-1} s^{-1}, at 20°C where $p_{ex} = \frac{1}{2}$ for strong exchange. For comparison, the diffusion-controlled collision rate is: $k_{2,diff} = \frac{8}{3} \cdot 10^3 N_A k_B (T/\eta) = 6.5 \times 10^9$ M^{-1} s^{-1} at 20°C, which we get from Eq. 9.149 together with the Stokes-Einstein relation and $r_1 = r_2$. From the gradient in Figure 9.22, the dependence of the collision rate constant of TEMPONE on T/η in water is $\mathrm{d}k_{coll}/\mathrm{d}(T/\eta) = 1.52 \times 10^4$ Pa M^{-1} K^{-1}, which also is comparable with the value that we predict for diffusion-controlled collision, viz. $\frac{8}{3} \cdot 10^3 N_A k_B = 2.2 \times 10^4$ Pa M^{-1} K^{-1}. The latter is given by the dashed line in Figure 9.22.

In contrast to the uncharged TEMPONE, the dependence of exchange broadening of peroxylamine disulphonate (PADS^{2-}) in 50 mM K_2CO_3 on T/η in Figure 9.22 (open circles and squares) displays a modest negative gradient, which corresponds to weak exchange (cf. again Figure 9.21). Electrostatic repulsion between the doubly negatively charged PADS molecules reduces the encounter frequency so much that conditions of strong exchange during collision no longer hold. As seen from the open triangles in Figure 9.22, increasing the ionic strength, which reduces the electrostatic repulsion, results in a positive slope for the T/η-dependence of PADS. Figure 9.23 shows the dependence of the exchange broadening of PADS on ionic strength. Increasing

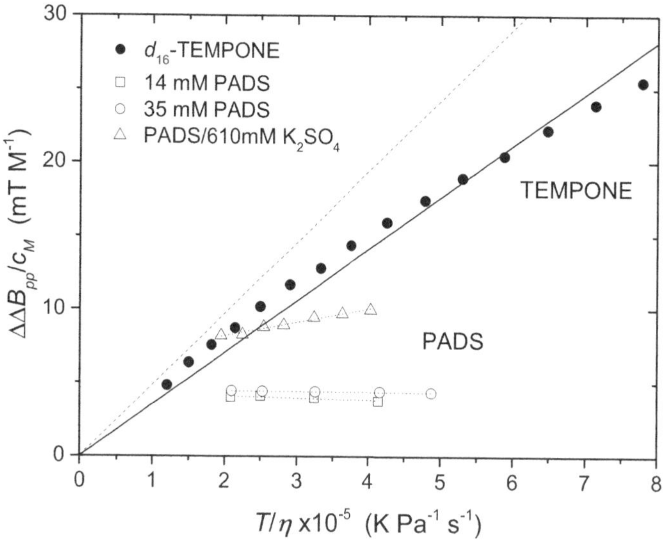

FIGURE 9.22 Dependence of effective rate constant $\Delta\Delta B_{pp}/c_M$ from exchange broadening on T/η for perdeuterated 1-oxy-2,2,6,6-tetramethylpiperidyl-4-oxide (d_{16}-TEMPONE; solid circles; Bales et al. 2009) and peroxylamine disulphonate (PADS; open symbols; Eastman et al. 1970) in water. *Open squares*: 13.9 mM PADS, *open circles*: 35.1 mM PADS; both in 50 mM K_2CO_3. *Open triangles*: 24.4 mM PADS in 50 mM K_2CO_3 plus 560 mM K_2SO_4. *Dashed line*: Stokes-Einstein prediction for diffusion-controlled collisions. *Solid line*: linear regression passing through the origin for solid circles. Dotted lines are linear regressions.

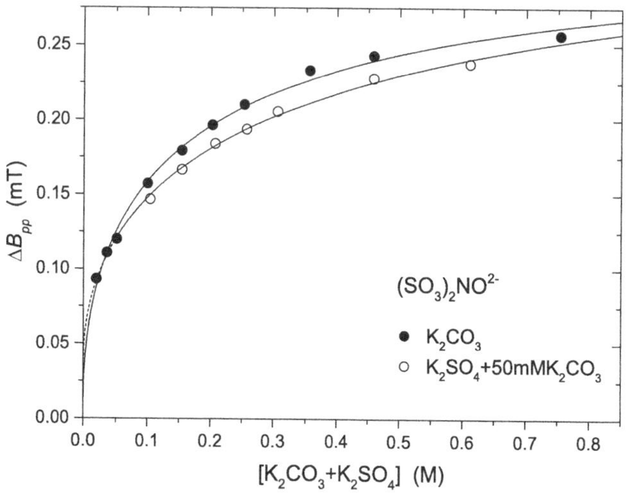

FIGURE 9.23 Linewidth of 24.6 mM aqueous peroxylamine disulphonate (PADS) with increasing ionic strength. *Solid circles*: in K_2CO_3; *open circles*: in K_2SO_4 plus 50 mM K_2CO_3. (Data from Eastman et al. 1970.) Solid lines: fits of dependence ~ $1 - \exp(-m\sqrt{c_I})$, where c_I is the electrolyte concentration.

the electrolyte concentration screens the electrostatic repulsion between PADS molecules, in a way that accords qualitatively with Debye–Hückel theory.

Bales et al. (2011) analyse the ionic strength dependence for exchange broadening of PROXYL-COO$^-$, relative to the uncharged acid form, by using modified Debye theory (Debye 1942). They find that the exchange rate constant is linear in $\sqrt{I}$ up to ionic strengths $I \approx 0.7$. Interestingly, the effective encounter distance $(r_1 + r_2 \approx 0.65 \text{ nm})$ appears temperature dependent, as indeed is the dielectric constant and effective range of the electrostatic repulsion.

Table 9.3 collects together second-order rate constants, both for exchange broadening and bimolecular collision, of various nitroxides in water. We assume strong exchange. Collision rate constants for uncharged spin labels are of a similar magnitude to our estimate for diffusion-controlled encounters. Rate constants for collision between negatively charged spin labels are reduced considerably by electrostatic repulsion but those between the positively charged spin labels less so.

9.16 TRANSLATIONAL DIFFUSION IN MEMBRANES

We can describe the translational diffusion of lipids or other molecules within the plane of a membrane by the two-dimensional diffusion equation (cf. Eq. 9.134):

$$\frac{\partial P}{\partial t} = D_T \nabla^2 P = D_T \left(\frac{\partial^2}{\partial r^2} + \frac{1}{r}\frac{\partial}{\partial r} \right) P \tag{9.155}$$

where $P(r,t)$ is the probability that a given molecule is situated at distance r from the origin at time t, and D_T is the translational diffusion coefficient. Solution of the two-dimensional diffusion equation for the initial boundary condition $P(r,0) = \delta(r)$ gives:

$$P(r,t) = \frac{1}{4\pi D_T t} \exp\left(\frac{-r^2}{4D_T t} \right) \tag{9.156}$$

TABLE 9.3 Rate constants for Heisenberg-exchange broadening, $T_{2,HE}^{-1}/c_M$, of nitroxides in water at ambient temperatures[a]

nitroxide[b]	$T_{2,HE}^{-1}/c_M \times 10^{-9}\ (\text{M}^{-1}\ \text{s}^{-1})$	$k_{coll} \times 10^{-9}\ (\text{M}^{-1}\ \text{s}^{-1})$	*Refs.*[c]
TEMPONE	2.71 ± 0.01	8.13 ± 0.03	1
	1.57	4.71	2
TEMPO	1.79	5.37	3
TEMPO-OOCMe	1.30	3.90	2
TEMPO-OOCCH$_2$I	1.22	3.66	2
TEMPO-OCH$_2$CH$_2$CN	1.36	4.08	2
TEMPENE-COOMe	1.30	3.90	2
4-DOXYL-VA	1.3	4.1	4
DTBN	1.65 ± 0.07	4.95 ± 0.2	5
TEMPO-NH$_3^+$	1.05	3.15	2
PROXYL-NH$_3^+$	1.26	3.78	2
PROXYL-COO$^-$	1.08 ± 0.01	3.24 ± 0.03	6
PYCOO$^-$	0.94	2.82	2
TEMPENE-COO$^-$	0.60	1.80	2
TEMPENE-CH$_2$COO$^-$	0.60	1.80	2
TEMPO-O(CH$_2$)$_2$COO$^-$	0.53	1.59	2
PADS^{2-}, I~0	0.15	0.45	7
PADS^{2-}, 50 mM K$_2$CO$_3$	1.15 ± 0.07	3.45 ± 0.2	8

[a] $T_{2,HE}^{-1} = (\sqrt{3}/2)\gamma_e \Delta\Delta B_{pp}$; $k_{coll} = ((2I+1)/2I)\, p_{ex}^{-1} (T_{2,HE}^{-1}/c_M)$ is the rate constant for bimolecular collision with $p_{ex} = \frac{1}{2}$ (strong exchange) and $I = 1$ (^{14}N-nitroxide), where c_M is the molar spin-label concentration (see Eqs. 9.97 and 9.98).

[b] TEMPONE, 1-oxyl-2,2,6,6-tetramethylpiperdin-4-yl oxide; TEMPO, 1-oxyl-2,2,6,6-tetramethylpiperdine; TEMPO-OOCMe, 1-oxyl-2,2,6,6-tetramethylpiperidin-4-yl acetate; TEMPO-OOCCH$_2$I, 1-oxyl-2,2,6,6-tetramethylpiperidin-4-yl iodoacetate; TEMPO-OCH$_2$CH$_2$CN, 1-oxyl-2,2,6,6-tetramethylpiperidin-4-yl ethoxycyanide; TEMPENE-COOMe, 1-oxyl-2,2,6,6-tetramethylpiperid-3-en-4-yl carboxymethyl ester; 4-DOXYL VA, 4-(3-oxyl-4,4-dimethyloxazolidin-2-yl)-valeric acid; DTBN, di-*tert*-butyl nitroxide (Me$_3$C)$_2$NO; TEMPO-NH$_3^+$, 1-oxyl-2,2,6,6-tetramethylpiperidin-4-ylamine; PROXYL-NH$_3^+$, 1-oxyl-2,2,5,5-tetramethylpyrrolidin-3-ylamine; PROXYL-COO$^-$, 1-oxy-2,2,5,5-tetramethylpyrrolidin-3-yl carboxylate; PYCOO$^-$, 1-oxy-2,2,5,5-tetramethylpyrrolin-3-yl carboxylate; TEMPENE-COO$^-$, 1-oxyl-2,2,6,6-tetramethylpiperid-3-en-4-yl carboxylate; TEMPENE-CH$_2$COO$^-$, 1-oxyl-2,2,6,6-tetramethylpiperid-3-en-4-ylmethyl carboxylate; TEMPO-O(CH$_2$)$_2$COO$^-$, 1-oxyl-2,2,6,6-tetramethylpiperidin-4-ylethoxy carboxylate; PADS, peroxylamine disulphonate (SO$_3^-$)$_2$NO.

[c] References: 1. Keith et al. (1977); 2. Grebenshchikov et al. (1972); 3. Ablett et al. (1975); 4. King and Marsh (1986); 5. Eastman et al. (1970); 6. Bales et al. (2011); 7. Pearson and Buch (1962); 8. Jones (1963).

The corresponding solution for the three-dimensional case is:

$$P(r,t) = \frac{1}{(4\pi D_T t)^{3/2}} \exp\left(\frac{-r^2}{4D_T t}\right) \quad (9.157)$$

When the origin of coordinates is not a fixed position, but is defined by the position of another identical molecule, we must replace D_T in Eqs. 9.156 and 9.157 by $2D_T$. This is the solution needed when we measure the diffusion coefficient by a bimolecular process, such as spin–spin interaction.

We can use either Eq. 9.156 or a two-dimensional random walk to determine the mean-square displacement, $\langle r^2 \rangle$, of the diffusing particle in time t'. This is given by the Einstein relation:

$$\langle r^2 \rangle = 4D_T t' \quad (9.158)$$

where we replace the factor 4 on the right-hand side by 2 or 6 for one-dimensional and three-dimensional diffusion, respectively (see Eq. 9.137). Alternatively, if we use a lattice model for the diffusion process, the diffusion coefficient is given from Eq. 9.158 by:

$$D_T = \tfrac{1}{4}\tau_j^{-1}\lambda^2 \quad (9.159)$$

where λ is the lattice constant (or mean free path) and τ_j^{-1} is the hopping frequency between adjacent lattice sites. Again for diffusive encounters, we must replace D_T by $2D_T$ in Eqs. 9.158 and 9.159.

The frequency of collision is related directly to the translational diffusion coefficient by the lattice model. If a given lipid exchanges with one of its $z_T = 6$ nearest neighbours (see Figure 9.24), it encounters $z = 3$ new nearest neighbours and the collision frequency is given by (Devaux et al. 1973):

$$\tau_{coll}^{-1} = 3\tau_j^{-1} X_{SL} \quad (9.160)$$

where X_{SL} is the mole fraction of spin-labelled lipid molecules (cf. Eq. 9.150). Combining Eqs. 9.159 and 9.160, including the substitution of $2D_T$ for bimolecular collisions, the translational diffusion coefficient then becomes:

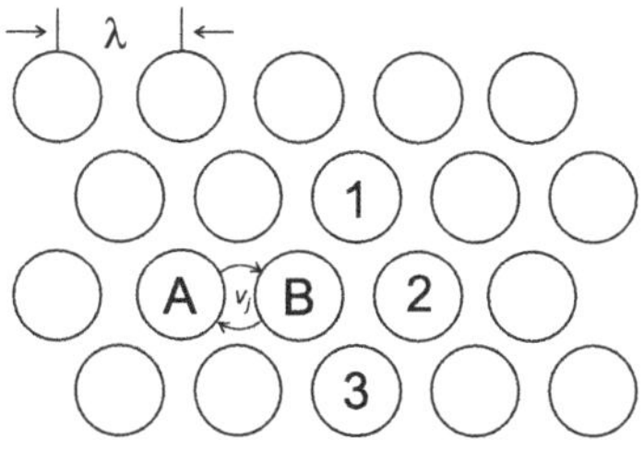

FIGURE 9.24 Lattice model for translational diffusion of lipid molecules in the membrane plane. Translation by interchange of pairs of molecules, such as *A* and *B*, with frequency ν_j. After interchange, one molecule of the pair (*A*) has three new neighbours, 1–3. *B* represents either another lipid molecule or thermally-induced void.

$$D_T = \frac{\lambda^2}{24}\left(\frac{\tau_{coll}^{-1}}{X_{SL}}\right) \quad (9.161)$$

When combined with Eq. 9.97 from Section 9.10, we can use Eq. 9.161 to determine D_T from the dependence of the Heisenberg exchange broadening, $T_{2,HE}^{-1}$, on mole fraction of spin-labelled lipid:

$$D_T = \frac{\lambda^2}{24}\left(\frac{2I+1}{I}\right)\left(\frac{T_{2,HE}^{-1}}{X_{SL}}\right) \quad (9.162)$$

where we assume $p_{ex} = \frac{1}{2}$ for the probability of exchange on collision.

For slow translational diffusion in viscous media, in addition to exchange, the magnetic dipole–dipole interaction (see Section 9.5.3) contributes to the spin–spin interactions. In this case, we distinguish the contributions from the two interactions by their opposing temperature dependences (Scandella et al. 1972). Because the diffusion coefficient is given by the Einstein relation: $D_T = k_B T / f_T$ (see Eq. 9.141), and the friction coefficient, f_T, is directly proportional to the viscosity (see Eq. 9.142), its temperature dependence is of the form:

$$D_T(T) = D_{T,o} T \exp\left(-E_a / k_B T\right) \quad (9.163)$$

where E_a is the activation energy associated with the viscosity.

Figure 9.25 shows the temperature dependence of the concentration-gradient, $\mathrm{d}\Delta B_{pp}^L / \mathrm{d}X_{SL}$, of the peak-to-peak

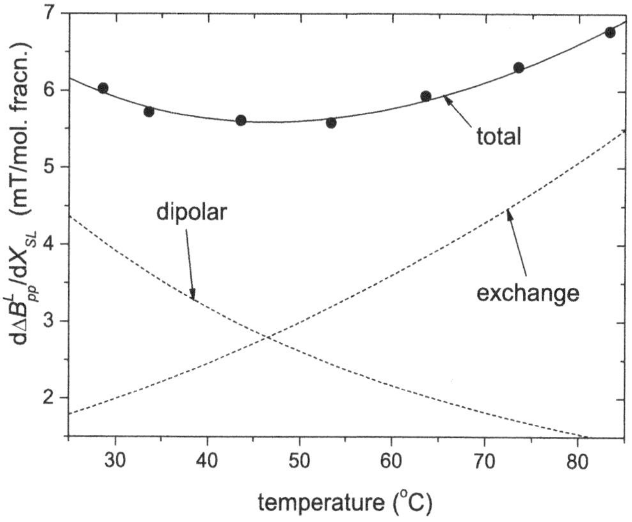

FIGURE 9.25 Temperature dependence for gradient of Lorentzian linewidth, ΔB_{pp}^L, with mole fraction, X_{SL}, of 16-DOXYL PC in dimyristoyl phosphatidylcholine (PC) membranes. *Circles*: experimental peak-to-peak Lorentzian widths of central ($m_I = 0$) hyperfine line (see Section 2.5). *Solid line*: fit for Arrhenius temperature-dependence of diffusion rate (Eq. 9.164); *dashed lines*: contributions from Heisenberg spin exchange and magnetic dipole–dipole interactions (first and second terms in Eq. 9.164). (Data from Sachse et al. 1987.)

Lorentzian linewidth for a spin-labelled lipid in fluid-phase lipid bilayer membranes. The temperature dependence is fitted by the expression:

$$\frac{d\Delta B_{pp}^{L}}{dX_{SL}} = A_{HE}T\exp\left(\frac{-E_a}{k_B T}\right) + \frac{A_{dd}}{T}\exp\left(\frac{+E_a}{k_B T}\right) \quad (9.164)$$

where A_{HE} and A_{dd} are temperature-independent constants characterizing the spin exchange and magnetic dipole–dipole interactions, respectively. From Eqs. 9.162 to 9.164, again with $T_{2,HE}^{-1}/X_{SL} = (\sqrt{3}/2)\gamma_e(d\Delta B_{pp}^{L}/dX_{SL})$, the coefficient for the exchange contribution is:

$$A_{HE} = \frac{48}{\sqrt{3}\gamma_e}\left(\frac{I}{2I+1}\right)\frac{D_{T,o}}{\lambda^2} \quad (9.165)$$

where we need an additional factor of two if there are collisions with labels in the other half of the bilayer. From the analysis given already in Section 9.5.3, the coefficient of the dipolar contribution is correspondingly (see Eq. 9.61):

$$A_{dd} = \frac{2\pi}{75\sqrt{3}}\left(\frac{\mu_o}{4\pi}\right)^2 \gamma_e^3\hbar^2\left(\frac{9+8I}{2I+1}\right)\frac{N_A \times 10^{-3}}{M_l \bar{v}_l d_S}\cdot\frac{1}{D_{T,o}} \quad (9.166)$$

where M_l is the relative molecular mass, and $\bar{v}_l$ (m^3kg^{-1}) the partial specific volume, of the lipid. We assume three-dimensional diffusion within the membrane volume in Eq. 9.166, as described in Section 9.5.3. To treat dipolar relaxation by translational diffusion in two dimensions is complicated because it involves potential divergences (see Korb et al. 1984).

From the fit in Figure 9.25, we get $A_{HE} = 1.60$ mT K^{-1}/mole fraction for the contribution from Heisenberg exchange, with a common activation energy for dipolar and exchange broadening of $E_a = 13.8$ kJ mol^{-1} (Sachse et al. 1987). We now use Eq. 9.165 for the spin exchange coefficient (including the additional 2× factor needed for 16-DOXYL labels), together with $\lambda = 0.8$ nm for the mean jump length of adjacent phospholipids (Träuble and Sackmann 1972) and $I = 1$ for ^{14}N-nitroxides. The temperature dependence of the diffusion coefficient then becomes: $D_T = 9.73 \times T\exp(-1600/T)$ $\mu m^2\,s^{-1}$, which gives $D_T = 12$ $\mu m^2\,s^{-1}$ for dimyristoyl phosphatidylcholine membranes at 30°C. This is comparable to the value estimated from the dipolar contribution to the linewidth in Section 9.5.3, although we approximated the latter by a three-dimensional model.

In principle, we can use the methods of Section 9.11 to help separate magnetic-dipole from exchange contributions to spin–spin interactions, by using the dispersion component of the lineshape. However, complications arise because dipolar interactions also introduce dispersion contributions, albeit of opposite sign, to the overall lineshape (Peric et al. 2012). We deal with this in more detail, in the following Section 9.17.

The analysis from Figure 9.25 is based on exchange broadening of the Lorentzian linewidths of the N-hyperfine lines and is the standard method for measuring translational diffusion coefficients by spin-label EPR. Figure 9.26 illustrates the exchange narrowing of the Gaussian linewidth from the unresolved proton hyperfine structure in each nitrogen manifold, which occurs already at relatively low mole fractions of spin label (cf. Figure 9.20). The square of the Gaussian linewidth is plotted against the square of the mole fraction of spin-labelled lipid, according to Eq. 6.22 for exchange narrowing in Chapter 6:

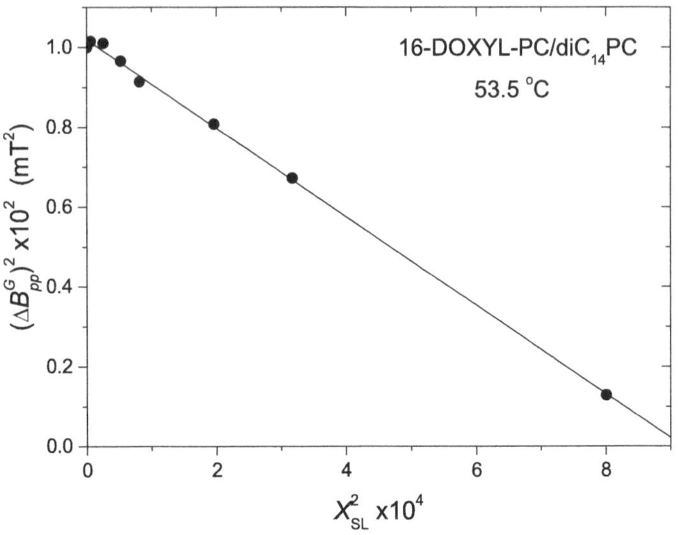

FIGURE 9.26 Dependence of Gaussian peak-to-peak inhomogeneous linewidth, ΔB_{pp}^{G}, on mole fraction, X_{SL}, of 16-DOXYL PC in dimyristoyl phosphatidylcholine membranes at 53.5°C. Highest mole fraction: $X_{SL} = 0.028$. Solid line: linear regression (Eqs. 9.167, 9.168). (Data from Sachse et al. 1987.)

$$\left(\Delta B_{pp}^{G}\right)^2 = \left(\Delta B_{pp}^{G}\right)_o^2 - \frac{8}{\gamma_e^2}\tau_{HE}^{-2} \quad (9.167)$$

For exchange within a single nitrogen hyperfine manifold, we replace $2I/(2I+1)$ in Eqs. 9.96 and 9.97 of Section 9.10 by $1/(2I+1)$. The exchange frequency for motional narrowing of unresolved proton hyperfine structure is then given by the modified version of Eq. 9.162:

$$\tau_{HE}^{-1} = \frac{12}{2I+1}\left(\frac{D_T}{\lambda^2}\right)X_{SL} \quad (9.168)$$

where again we need an additional factor of two if there are collisions with labels in the other half of the bilayer. Therefore, we expect a linear dependence of $\left(\Delta B_{pp}^{G}\right)^2$ on X_{SL}^2, as found in Figure 9.26. From the gradient, the diffusion coefficient estimated by using Eqs. 9.167 and 9.168 is $D_T = 16$ $\mu m^2\,s^{-1}$ at 53.5°C, as compared with $D_T = 19$ $\mu m^2\,s^{-1}$ deduced from exchange broadening at this temperature. The agreement is satisfactory, in view of the approximate nature of the analysis of exchange narrowing.

9.17 DIPOLAR-INDUCED MAGNETIZATION TRANSFER AND DISPERSION LINESHAPES

It is often assumed, as in Sections 9.5.1 and 9.5.2, that magnetic dipole–dipole interactions contribute only to spin dephasing, i.e., to line broadening. However, Galeev and Salikhov (1996) point out that the $S_{1,\pm}S_{2,\mp}$ flip-flop terms and $S_{1,z}S_{2,\pm}$ non-secular terms in the dipolar Hamiltonian (see Eqs. 9.7, 9.8) transfer transverse magnetization between unlike spin labels, just as Heisenberg exchange does. In the presence of dipolar interactions, Eq. 9.93 must be rewritten as (Galeev and Salikhov 1996):

$$\left.\frac{d\mathbf{M}_m}{dt}\right|_{HE+dd} = -\left[(1-f_m)\tau_{HE}^{-1} + \frac{f_m}{T_{2dd,like}} + \frac{1-f_m}{T_{2dd,unlike}}\right]\mathbf{M}_m + f_m\left(\tau_{HE}^{-1} - \frac{1}{T'_{2dd}}\right)\sum_{m'(\neq m)}\mathbf{M}_{m'} \tag{9.169}$$

where the dephasing rates $1/T_{2dd,like}$ and $1/T_{2dd,unlike}$ are given by the general expressions Eq. 9.28 and 9.35, respectively. Specifically for translational diffusion, these become Eqs. 9.54 and 9.56 in Section 9.5.2. The new contribution $1/T'_{2dd}$ in Eq. 9.169 represents the transfer of transverse magnetization (or coherence transfer) by non-secular dipolar interactions. From Eqs. 9.5 to 9.9, we get (Galeev and Salikhov 1996; Salikhov 2010; Marsh, 2017):

$$\frac{1}{T'_{2dd}} = \frac{1}{12}\left(\frac{\mu_o}{4\pi}\right)^2 \gamma_e^4\hbar^2 s(s+1)\left(J^{(0)}(0) + J^{(0)}(\omega_m - \omega_{m'}) + 9J^{(1)}(\omega_m) + 9J^{(1)}(\omega_{m'})\right) \tag{9.170}$$

where the spectral densities for translational diffusion $J^{(M)}(\omega)$ are given by Eqs. 9.50–9.52 and 9.57 in Section 9.5.2. As we see from the second term on the right-hand side of Eq. 9.169, the most important feature of Eq. 9.170 is that coherence transfer by dipole–dipole interaction is of opposite sign to that from Heisenberg exchange. Correspondingly, the dispersion contribution to the lineshape that is induced by dipolar interaction subtracts from that induced by Heisenberg spin exchange (Peric et al. 2012).

Dipolar coherence transfer refers, of course, to unlike spins. The non-secular terms involved are just those that give the enhanced rate of spin dephasing for like spins, relative to unlike spins. We see this immediately from Eq. 9.28 together with Eq. 9.35 for $m' = m$. The difference is: $1/T'_{2dd} = 1/T_{2dd,like} - 1/T_{2dd,unlike} = \frac{1}{6}(\mu_o/4\pi)^2\gamma_e^4\hbar^2 s(s+1)\left[J^{(0)}(0) + 9J^{(1)}(\omega_m)\right]$, which agrees with Eq. 9.170 for $m' = m$.

Substituting for the spectral densities in Eq. 9.170 by using the expressions for translational diffusion, we get:

$$\frac{1}{T'_{2dd}} = \frac{\pi}{15}\left(\frac{\mu_o}{4\pi}\right)^2 \gamma_e{}^4\hbar^2 s(s+1)\left[\frac{8}{15} + 4I\left(\sqrt{2(\omega_m - \omega_{m'})\tau_D}\right) + 6I\left(\sqrt{2\omega_m\tau_D}\right) + 6I\left(\sqrt{2\omega_{m'}\tau_D}\right)\right]\frac{n_{SL}}{d_S D_T} \tag{9.171}$$

where the integral $I(u)$ is given by Eq. 9.51 or 9.57 in Section 9.5.2. Again in Eq. 9.171, n_{SL} is the spin-label number density per unit volume (m^3), d_S is the collision diameter and $2D_T$ the sum of translational diffusion coefficients of the two spin labels, and the characteristic diffusion time is $\tau_D \equiv d_S^2/(2D_T)$. Of course, just like the dephasing rates and spin-exchange frequency, T'^{-1}_{2dd} is directly proportional to the spin-label concentration. If we take the approximation for high viscosities $(a_N\tau_D \gg 1)$ from Section 9.5.3, by retaining only spectral densities at zero frequency (cf. Eq. 9.52), the coherence transfer rate becomes:

$$\frac{1}{T'_{2dd}} = \frac{8\pi}{15^2}\left(\frac{\mu_o}{4\pi}\right)^2 \gamma_e^4\hbar^2 s(s+1)\frac{n_{SL}}{d_S D_T} \tag{9.172}$$

and correspondingly, the dephasing rates $1/T_{2dd,like}$ and $1/T_{2dd,unlike}$ are given by Eqs. 9.58 and 9.59, respectively. The ratio in rate of coherence transfer to that of dephasing by dipolar interaction then becomes: $T'^{-1}_{2dd}/T^{-1}_{2dd,m} = 6/17$ for a ^{14}N-nitroxide (cf. Eq. 9.169), where the total broadening is $T^{-1}_{2dd,m} = f_m T^{-1}_{2dd,like} + (1-f_m)T^{-1}_{2dd,unlike}$ (see Eq. 9.60). For low viscosities $(a_N\tau_D \ll 1)$ on the other hand, the spectral density at frequency $\omega_m - \omega_{m'} \approx a_N$ approaches that at zero frequency, and the ratio increases to $T'^{-1}_{2dd}/T^{-1}_{2dd,m} = 12/19$ for a ^{14}N-nitroxide. The coherence-transfer rate then becomes twice that given in Eq. 9.172.

To calculate spectra that include dipole–dipole interactions, we rewrite Eq. 9.169 in terms of the summation over all magnetization components:

$$\left.\frac{d\mathbf{M}_m}{dt}\right|_{HE+dd} = -\left[\tau_{HE}^{-1} + \frac{1}{T_{2dd,tot}} + \frac{1}{T'_{2dd}}\right]\mathbf{M}_m + f_m\left(\tau_{HE}^{-1} - \frac{1}{T'_{2dd}}\right)\sum_{m'}\mathbf{M}_{m'} \tag{9.173}$$

This is equivalent to the rightmost equality in Eq. 9.93. We then proceed as with the general solution for the lineshape in Section 9.10. Equations 9.101 and 9.102 for the total transverse magnetization $\hat{M}_+ = \hat{M}_x + i\hat{M}_y \equiv u + iv$ become:

$$\hat{M}_+(\omega) = \frac{i\gamma_e B_1 M_o S(\omega)}{1 - \left(\tau_{HE}^{-1} - 1/T'_{2dd}\right)S(\omega)} \tag{9.174}$$

with

$$S(\omega) = \sum_m \frac{f_m}{i(\omega_m - \omega) + 1/T_{2,m} + \tau_{HE}^{-1} + 1/T_{2dd,m} - 1/T'_{2dd}} \tag{9.175}$$

For dipolar interaction alone, the absorption EPR lineshape $v = \mathrm{Im}\left(\hat{M}_+\right)$ then becomes:

$$v(\omega) = \gamma_e B_1 M_o$$

$$\times \frac{\sigma_o\left(T_2^{-1} + T_{2dd}^{-1} - T_{2dd}'^{-1}\right)\left[1 + \sigma_o\left(T_2^{-1} + T_{2dd}^{-1} - T_{2dd}'^{-1}\right)T_{2dd}'^{-1}\right] + \sigma_1{}^2 T_{2dd}'^{-1}}{\left[1 + \sigma_o\left(T_2^{-1} + T_{2dd}^{-1} - T_{2dd}'^{-1}\right)T_{2dd}'^{-1}\right]^2 + \sigma_1{}^2 T_{2dd}'^{-2}} \tag{9.176}$$

where summations over the different hyperfine lines, $m_I = m$, are given by:

$$\sigma_o(\omega) = \sum_m \frac{f_m}{\left(\omega - \omega_m\right)^2 + \left(T_2^{-1} + T_{2dd}^{-1} - T_{2dd}'^{-1}\right)^2} \tag{9.177}$$

$$\sigma_1(\omega) = \sum_m \frac{f_m\left(\omega - \omega_m\right)}{\left(\omega - \omega_m\right)^2 + \left(T_2^{-1} + T_{2dd}^{-1} - T_{2dd}'^{-1}\right)^2} \tag{9.178}$$

Here, we assume that each line has the same intrinsic width, T_2^{-1}. The dipolar broadening T_{2dd}^{-1} depends on f_m, but this is the same for each nitrogen hyperfine line. Inclusion of both dipolar interaction and spin exchange is straightforward; we simply replace $-T_{2dd}'^{-1}$ by $\tau_{HE}^{-1} - T_{2dd}'^{-1}$, at each occurrence in Eqs. 9.176–9.178 (cf. Eq. 9.173).

Figure 9.27 shows first-derivative absorption EPR lineshapes for increasing dipolar interaction strength, T_{2dd}^{-1}, at fixed ratio of the magnetization-transfer rate $T_{2dd}'^{-1}$. These derivative lineshapes correspond to Eqs. 9.176–9.178 for the absorption spectrum of a ^{14}N-nitroxide. We can compare them with the corresponding spectra in Figure 9.14 for increasing frequency of Heisenberg spin exchange. At intermediate extents of line broadening, in the region of $T_{2dd}^{-1}/a_N \approx 0.1 - 0.5$, the lineshapes have an asymmetry that is opposite to that for Heisenberg exchange. This is because dispersion admixture from dipolar magnetization transfer is opposite in sign to that shown in Figure 9.15 for spin exchange. For stronger dipolar interactions, when $T_{2dd}^{-1}/a_N \geq 1$, the spectra begin to collapse towards a single line (cf. Galeev and Salikhov 1996). We see this from the progressive reduction in relative intensity of the central hyperfine component in the bottom two spectra of Figure 9.27. However, the spectra continue to broaden progressively; there is no subsequent spectral narrowing that we get for high spin-exchange frequencies in Figure 9.14. The spectra in Figure 9.27 are calculated with the higher of the two values for the coherence-transfer rate: $T_{2dd}'^{-1}/T_{2dd,m}^{-1} = 12/19$. This corresponds to the situation of rather fast diffusion rates $\left(a_N\tau_D \ll 1\right)$, when Heisenberg exchange is also expected to contribute strongly. For the lower coherence-transfer rate, $T_{2dd}'^{-1}/T_{2dd,m}^{-1} = 6/17$, which corresponds to slow translational diffusion $\left(a_N\tau_D \gg 1\right)$, the spectral lineshapes corresponding to Figure 9.27 are dominated by broadening throughout, with little contribution from dispersion.

For low rates of magnetization transfer, $T_{2dd}'^{-1}$, we can determine dispersion contributions to the lineshape and also line shifts by using perturbation theory, just as in Sections 9.11 and 9.12 for

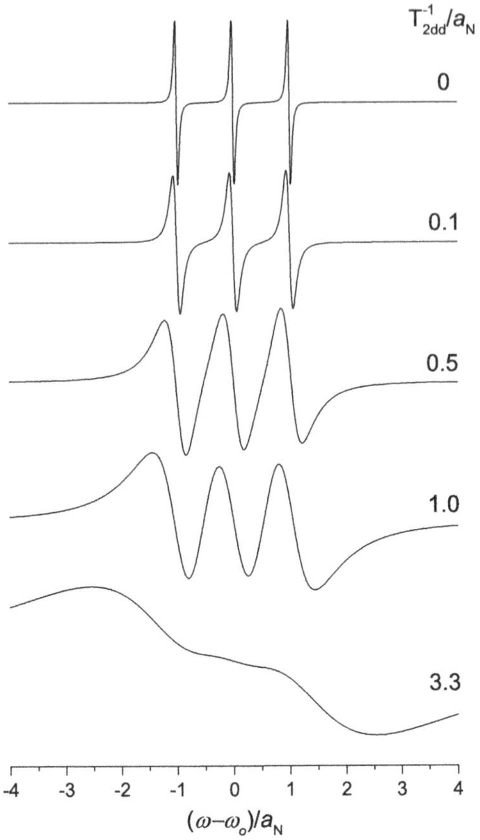

FIGURE 9.27 EPR spectra of freely rotating ^{14}N-nitroxide radicals with increasing dipole–dipole interaction, T_{2dd}^{-1} (Eqs. 9.176–9.178, but displayed as first derivative). Dipolar T_{2dd}^{-1} scaled to ^{14}N-hyperfine coupling a_N in angular-frequency units (rad s^{-1}); rate of coherence transfer fixed at $T_{2dd}'^{-1} = (12/19)T_{2dd}^{-1}$. Spectra normalized to same line height.

(slow) Heisenberg exchange alone. Corresponding to Eqs. 9.113 and 9.114, the ratio of dispersion to absorption amplitudes of the first-derivative spectrum is:

$$h_{disp}/h_{pp} = \frac{8}{3\sqrt{3}}\left(\tau_{HE}^{-1} - T_{2,dd}'^{-1}\right)\sum_{m'(\neq m)} \frac{f_{m'}}{\omega_m - \omega_{m'}} = \frac{4}{3\sqrt{3}}\frac{\tau_{HE}^{-1} - T_{2dd}'^{-1}}{a_N} \tag{9.179}$$

for combined dipolar and exchange interactions, where the second equality applies equally to ^{14}N-nitroxides ($I = 1$) and ^{15}N-nitroxides $\left(I = \frac{1}{2}\right)$. Because $T_{2dd}'^{-1}$ is positive (see Eqs. 9.171 and 9.172), Eq. 9.179 shows clearly that the dispersion contribution from dipolar interaction has the opposite sign to that from exchange shown in Figure 9.15.

Figure 9.28 shows the dependence of the normalized dispersion amplitude on the mean broadening $\left\langle \Delta\Delta B_{pp}\right\rangle = 2/\left(\sqrt{3}\gamma_e\right)\times \left(\left(1 - f_m\right)\tau_{HE}^{-1} + T_{2dd,m}^{-1}\right)$, for increasing concentrations of perdeuterated TEMPONE in the moderately viscous hydrocarbon squalane (Peric et al. 2012). For low rates of magnetization transfer, we expect a linear dependence from Eq. 9.179 at fixed temperature, because τ_{HE}^{-1}, $T_{2dd,m}^{-1}$ and $T_{2dd}'^{-1}$ all increase linearly with spin-label concentration. At higher rates of magnetization transfer, the dispersion contribution increases nonlinearly as indicated by Eq. 9.115 for magnetization transfer by Heisenberg exchange alone. At 303 K, the dispersion contribution increases with increasing broadening because exchange dominates over dipolar interactions at these relatively low viscosities. The increase becomes progressively less steep with decreasing temperature, until it changes sign and becomes a decrease at 253 K. Dipolar interaction now dominates for the higher viscosity at this low temperature. In principle, we can estimate the ratio of dipolar to exchange contributions in the dispersion admixture from the initial linear gradients in Figure 9.28. To do this, we assume a constant ratio, $T_{2dd}^{\prime -1}/T_{2dd,m}^{-1}$, of coherence transfer to dephasing rates for the dipolar interaction. From Eq. 9.179, the ratio of total dispersion amplitude to total broadening is given by:

$$\frac{h_{disp}/h_{pp}}{\langle \Delta\Delta B_{pp}\rangle / a_{\mathrm{N}}} = \frac{2}{3}\frac{\tau_{HE}^{-1} - T_{2dd}^{\prime -1}}{(1-f_m)\tau_{HE}^{-1} + T_{2dd,m}^{-1}} \tag{9.180}$$

when $\Delta\Delta B_{pp}$ and a_{N} are both expressed in field units. We then immediately obtain the ratio of exchange to dipolar rates $\tau_{HE}^{-1}/T_{2dd,m}^{-1}$, for a given value of $T_{2dd}^{\prime -1}/T_{2dd,m}^{-1}$ (cf. above). Note that at low temperatures, the dispersion component in Figure 9.28 extrapolates to a non-vanishing value in the absence of spin–spin interaction. This likely arises from coherence transfer induced by electron–nuclear dipolar interaction within single nitroxides (Marsh, 2017, and see Appendix O.4).

Corresponding to Eq. 9.118, the direct shift in resonance frequency by combined dipolar and exchange interactions is:

$$\omega'_m - \omega_m = f_m\left(\tau_{HE}^{-1} - T_{2dd}^{\prime -1}\right)^2 \sum_{m'(\neq m)} \frac{f_{m'}}{\omega_{m'} - \omega_m} \tag{9.181}$$

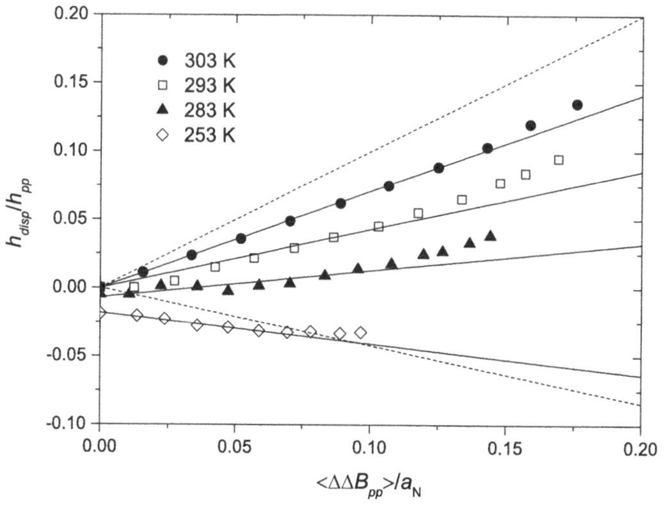

FIGURE 9.28 Dependence of dispersion amplitude h_{disp}/h_{pp} on mean broadening $\langle \Delta\Delta B_{pp}\rangle$ of the three hyperfine lines in the absorption component, for increasing concentrations of perdeuterated ^{14}N-TEMPONE in squalane. Temperature: 303 K (circles), 293 K (squares), 283 K (triangles) and 253 K (diamonds). (Data from Peric et al. 2012.) *Upper dashed line*: prediction of Eq. 9.180 for Heisenberg exchange alone; *lower dashed line*: dipolar interaction alone with $T_{2dd}^{\prime -1}/T_{2dd,m}^{-1} = 12/19$.

which depends on the square of the spin-label concentration, just as for exchange alone. Therefore, the dipolar interaction opposes the direct exchange-induced shift but does not change its sign. Corresponding to Eq. 9.124, the total shift of the maximum in absorption lineshape (or baseline-crossing of the first derivative) becomes:

$$\omega_{\max,m} - \omega_m = \left(1/T_2 + \tau_{HE}^{-1} + T_{2dd,m}^{-1} - f_m T_{2dd}^{\prime -1}\right) \times \left(\tau_{HE}^{-1} - T_{2dd}^{\prime -1}\right) \sum_{m'(\neq m)} \frac{f_{m'}}{\omega_{m'} - \omega_m} \tag{9.182}$$

in the presence of both dipolar and exchange interactions. This contains terms both linear and quadratic in the spin-label concentration, just as for exchange alone. Finally, the effective hyperfine constant that corresponds to Eq. 9.126 for exchange alone becomes:

$$\frac{a_{HE,dd}^{eff}}{a_{\mathrm{N}}} = 1 - \frac{3}{(I+1)(2I+1)} \times \left(\frac{1/T_2 + \tau_{HE}^{-1} + T_{2dd,m}^{-1} - f_m T_{2dd}^{\prime -1}}{a_{\mathrm{N}}}\right)\left(\frac{\tau_{HE}^{-1} - T_{2dd}^{\prime -1}}{a_{\mathrm{N}}}\right) \tag{9.183}$$

when dipole–dipole interaction is included. Because the $T_{2dd}^{\prime -1}$ terms have opposite sign to those of τ_{HE}^{-1}, we see that dipolar interactions increase the effective hyperfine splitting, whereas exchange reduces it.

Figure 9.29 shows the dependence of the normalized effective hyperfine constant on mean total broadening $\langle \Delta\Delta B_{pp}\rangle$, for increasing concentrations of perdeuterated TEMPONE in 70% aqueous glycerol (Bales et al. 2009). At

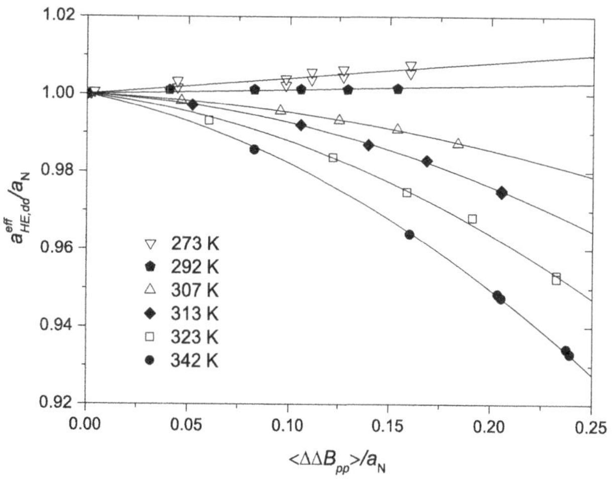

FIGURE 9.29 Dependence of effective hyperfine coupling $a_{HE,dd}^{eff}$ on mean broadening of the three hyperfine lines $\langle \Delta\Delta B_{pp}\rangle$, for increasing concentrations of perdeuterated ^{14}N-TEMPONE in 70% aqueous glycerol. Temperature: 342 K (circles), 323 K (squares), 313 K (diamonds), 307 K (triangles), 292 K (pentagons) and 273 K (inverted triangles). (Data from Bales et al. 2009.)

higher temperatures (≥307 K), the effective hyperfine splitting decreases progressively, at steeper rates with increasing temperature. This resembles the situation for Heisenberg exchange alone (cf. Figure 9.18, and Eqs. 9.125, 9.126 and 9.132), where an additional linear term must be included to account for reencounters (see Section 9.12). At low temperatures, $a_{HE,dd}^{eff}$ increases with increasing total broadening, as expected from Eq. 9.183 when dipolar interactions dominate.

9.18 CONCLUDING SUMMARY

1. *Dipolar coupling for like and unlike spins*:
The first-order dipolar splitting, $\pm\Delta B_{dd}(\theta)$, in the CW-EPR spectrum of a spin-label pair depends on whether the two spin labels are indistinguishable (strong coupling) or distinguishable (weak coupling):

$$\pm\Delta B_{dd}(\theta) = D_{dd}^{\perp}\left(1-3\cos^2\theta\right) \tag{9.184}$$

where $D_{dd}^{\perp} = \frac{3}{4}\left(\mu_o/4\pi\right) g\beta_e/r_{12}^3$ for *like* spins, and $D_{dd}^{\perp} = \frac{1}{2}\left(\mu_o/4\pi\right) g\beta_e/r_{12}^3$ for *unlike* spins. Spins are distinguishable if the dipolar splitting is less than the difference in resonance positions.
2. *Pake Powder Patterns*:
The dipolar spectrum for interspin vectors $\mathbf{r}_{12}$ randomly oriented in θ is an axially symmetric powder pattern with separation $2D_{dd}^{\perp}$ between the principal peaks. This Pake absorption powder pattern is convoluted with the powder spectrum of the single spin label.
3. *Dipolar relaxation:*
Both rotational and translational diffusion induce electron-spin relaxation by modulating the dipole–dipole coupling. For rotational diffusion with correlation time τ_R, the increase in Lorentzian peak-to-peak width is:

$$\Delta\Delta B_{pp,L}^{dd} = \frac{8}{5\sqrt{3}}\gamma_e\left(D_{dd}^{\perp}\right)^2\tau_R \tag{9.185}$$

With the appropriate $D_{dd}^{\perp}$, this applies to both *like* and *unlike* spins, where unlike is the better approximation for powder patterns. For translational diffusion with diffusion coefficient D_T, the increase in Lorentzian peak-to-peak width for *like* spins is:

$$\left(\Delta\Delta B_{pp,L}^{dd}\right)_{like} = \frac{\sqrt{3}\pi}{25}\left(\frac{\mu_o}{4\pi}\right)^2\gamma_e^3\hbar^2\frac{n_{SL}}{d_S D_T} \tag{9.186}$$

where n_{SL} is the spin-label number density per unit volume (m^3), and $d_S \cong 0.64-0.8$ nm is the collision diameter. The corresponding expression for *unlike* spins is 4/9× that for like spins. For a motionally-narrowed (quasi isotropic) spectrum, fraction $1/(2I+1)$ of the collisions are between like and $2I/(2I+1)$ between unlike spins. The net increase in Lorentzian peak-to-peak width is then:

$$\Delta\Delta B_{pp,L}^{dd} = \frac{\pi}{75\sqrt{3}}\left(\frac{\mu_o}{4\pi}\right)^2\gamma_e^3\hbar^2\left(\frac{9+8I}{2I+1}\right)\frac{n_{SL}}{d_S D_T} \tag{9.187}$$

where I is the nitrogen nuclear spin.
4. *Heisenberg spin exchange*:
Heisenberg exchange is a non-classical electrostatic interaction that results in effective spin–spin interaction:

$$\mathcal{H}_{HE} = J_{12}\mathbf{s}_1\cdot\mathbf{s}_2 \tag{9.188}$$

Strong exchange $(J_{12} >> a_N)$ couples the spins to triplet $(S = s_1 + s_2 = 1)$ and singlet $(S = 0)$ states and does not contribute to allowed EPR transitions within the triplet. Moderate exchange $(J_{12} \approx a_N)$ contributes to the spectra of biradicals, from which we can determine J_{12}. Moderate/weak exchange $(J_{12} \le D_{dd})$ adds to the dipolar splitting between *unlike* spins:

$$\pm\Delta B_{dd}(\theta) = D_{dd}^{\perp}\left(1-3\cos^2\theta\right)+\frac{1}{2}\frac{J_{12}}{g\beta_e} \tag{9.189}$$

but not to the dipolar splitting of *like* spins.
5. *Bimolecular collisions and translational diffusion*:
Heisenberg exchange affects the EPR spectrum just as does two-site exchange (cf. Section 6.3). It gives rise to exchange broadening, inward shifts, collapse and exchange narrowing, depending on the exchange frequency, τ_{HE}^{-1} (i.e., on spin concentration). For two (quasi-isotropic) colliding spin labels, the exchange broadening $\left(\Delta\Delta\omega_{HE} \equiv T_{2,HE}^{-1}\right)$ is:

$$\Delta\Delta\omega_{HE} = \left(\frac{2I}{2I+1}\right)p_{ex}\tau_{coll}^{-1} \tag{9.190}$$

where τ_{coll}^{-1} is the collision frequency, and $p_{ex} = \frac{1}{2}$ for strong exchange during collisions. We determine the second-order rate constant for collision $\left(k_{coll} \equiv \tau_{coll}^{-1}/c_M\right)$ from the rate of increase in Lorentzian peak-to-peak linewidth ΔB_{pp}^L with molar concentration, c_M, of spin label:

$$k_{coll} = \sqrt{3}\frac{2I+1}{2I}\gamma_e\left(\frac{d\Delta B_{pp}^L}{dc_M}\right) \tag{9.191}$$

when dipolar contributions are negligible (low viscosity). For translational diffusion of spin-labelled lipids in membranes, the exchange contribution to the Lorentzian linewidth increment is:

$$\frac{d\Delta B_{pp}^L}{dX_{SL}} = \frac{48}{\sqrt{3}\gamma_e}\left(\frac{I}{2I+1}\right)\frac{D_T}{\lambda^2} \tag{9.192}$$

where X_{SL} is the mole fraction of spin-labelled lipid and $\lambda \cong 0.8$ nm is the length of a diffusive jump. Diffusion must be fast enough that dipolar contributions (see Eq. 9.61) are negligible, or the dipolar and exchange contributions separated by their opposite temperature dependences (see Eq. 9.164).

6. *Dispersion contribution to lineshapes, and line shifts:* Transfer of magnetization between unlike spins by Heisenberg exchange admixes dispersion components into the EPR absorption lineshapes and shifts the line positions. For slow exchange, the ratio of dispersion amplitude to peak-to-peak amplitude of the usual first-derivative absorption (see Figure 9.15) is:

$$h_{disp}/h_{pp} = \frac{4}{3\sqrt{3}} \frac{\tau_{HE}^{-1}}{a_N} \tag{9.193}$$

which applies to both ^{14}N- and ^{15}N-nitroxides. The apparent hyperfine coupling, obtained from the positions of base-line crossings in the first-derivative EPR spectrum, is reduced from a_N to a_{HE}^{eff} by slow exchange:

$$a_{HE}^{eff}/a_N = 1 - \frac{1}{2}\left(\frac{1/T_2}{a_N} + \frac{1+1/\sqrt{2}}{3}\sqrt{a_N \tau_{RE}} + \frac{\tau_{HE}^{-1}}{a_N} \right) \frac{\tau_{HE}^{-1}}{a_N} \tag{9.194}$$

for a ^{14}N-nitroxide, where τ_{RE} is the time between immediate diffusive reencounters of the colliding spin labels. At higher viscosities (lower temperatures), magnetic dipole–dipole interactions contribute dispersion admixtures of opposite sign to those from exchange and produce line shifts that increase the apparent hyperfine coupling (see Section 9.17).

APPENDIX A9.1: FIRST-DERIVATIVE ABSORPTION LINESHAPES WITH HEISENBERG SPIN EXCHANGE

From the field-swept versions of Eqs. 9.100 and 9.101, the first-derivative lineshape is given by:

$$\frac{d\tilde{M}_+(B)}{dB} = \frac{1}{\left(1-\left(\tau_{HE}^{-1}/\gamma_e\right)S(B)\right)^2} \times \sum_{m_I} \frac{-if_m}{\left(i(B-B_m)+\gamma_e^{-1}\left(T_{2,m}^{-1}+\tau_{HE}^{-1}\right)\right)^2} \tag{A9.1}$$

where $S(B)$ is defined by Eq. 9.102 in field units, and γ_e is the electron gyromagnetic ratio. Taking the imaginary part of $\tilde{M}_+ = u + iv$ for the EPR absorption, the first-derivative lineshape becomes (Bales and Peric 1997):

$$\frac{dv(B)}{dB} = \frac{DN' - ND'}{D^2} \tag{A9.2}$$

with the real variables:

$$N = \left(1-\left(\tau_{HE}^{-1}/\gamma_e\right)\sigma_2\right)\sigma_2 - \left(\tau_{HE}^{-1}/\gamma_e\right)\sigma_1{}^2 \tag{A9.3}$$

$$D = \left(1-\left(\tau_{HE}^{-1}/\gamma_e\right)\sigma_2\right)^2 + \left(\tau_{HE}^{-1}/\gamma_e\right)\sigma_1{}^2 \tag{A9.4}$$

$$N' = -2\left(1-\left(\tau_{HE}^{-1}/\gamma_e\right)\sigma_2\right)\sigma_5 - 2\left(\tau_{HE}^{-1}/\gamma_e\right)(\sigma_3 - 2\sigma_4)\sigma_1 \tag{A9.5}$$

$$D' = 4\left(\tau_{HE}^{-1}/\gamma_e\right)\left(1-\left(\tau_{HE}^{-1}/\gamma_e\right)\sigma_2\right)\sigma_5 + 2\left(\tau_{HE}^{-1}/\gamma_e\right)^2(\sigma_3 - 2\sigma_4)\sigma_1 \tag{A9.6}$$

The field-dependent sums $\sigma_j(B)$ in Eqs. A9.3–A9.6 are given by:

$$\sigma_1(B) = \sum_m f_m(B-B_m)L_m(B) \tag{A9.7}$$

$$\sigma_2(B) = \sum_m f_m \Delta B_{1/2,m}^L L_m(B) \tag{A9.8}$$

$$\sigma_3(B) = \sum_m f_m L_m(B) \tag{A9.9}$$

$$\sigma_4(B) = \sum_m f_m (B-B_m)^2 L_m(B)^2 \tag{A9.10}$$

$$\sigma_5(B) = \sum_m f_m \Delta B_{1/2,m}^L (B-B_m) L_m(B)^2 \tag{A9.11}$$

with the Lorentzian absorption lineshape defined by:

$$L_m = \left((B-B_m)^2 + \left(\Delta B_{1/2,m}^L\right)^2\right)^{-1} \tag{A9.12}$$

where the Lorentzian half-width at half-height is:

$$\Delta B_{1/2,m}^L = \left(T_{2,m}^{-1} + \tau_{HE}^{-1}\right)/\gamma_e \tag{A9.13}$$

B_m is the resonance field of hyperfine state $m_I = m$, and the sums extend over all hyperfine lines.

APPENDIX A9.2: BROADENING, LINE SHIFTS AND DISPERSION ADMIXTURE BY HEISENBERG SPIN EXCHANGE

TABLE A9.2.1 Relative broadening $\Delta\Delta B_{pp}(m_I)/a_N$, relative dispersion amplitude h_{disp}/h_{pp}, direct hyperfine shifts a_{HE}/a_N and relative intensities $I(\pm 1)/I(0)$, for a ^{14}N-nitroxide ($I=1$), as function of normalized Heisenberg spin-exchange frequency τ_{HE}^{-1}/a_N (Bales et al. 2009)

τ_{HE}^{-1}/a_N	$\Delta\Delta B(0)/a_N$	$\Delta\Delta B(\pm 1)/a_N$	$\langle \Delta\Delta B_{pp}\rangle/a_N$	h_{disp}/h_{pp}	a_{HE}/a_N	$I(\pm 1)/I(0)$
0	0	0	0	0	1	1.00000
0.04922	0.03788	0.0379	0.03789	0.03799	0.9996	0.99757
0.09844	0.0757	0.07582	0.07578	0.07625	0.99838	0.99031
0.14766	0.11339	0.11381	0.11367	0.11508	0.99636	0.97823
0.19688	0.1509	0.15189	0.15156	0.15477	0.99352	0.96138
0.2461	0.18815	0.1901	0.18945	0.19575	0.98985	0.93981
0.29532	0.22507	0.22848	0.22734	0.23829	0.98536	0.91359
0.34454	0.26159	0.26706	0.26524	0.28292	0.98002	0.88283
0.39376	0.29762	0.30588	0.30313	0.32996	0.97382	0.84765
0.44298	0.33306	0.345	0.34102	0.38077	0.96676	0.80820
0.49221	0.36781	0.38446	0.37891	0.4353	0.95881	0.76464
0.54143	0.40175	0.42433	0.4168	0.49497	0.94995	0.71720
0.59065	0.43475	0.46468	0.4547	0.56178	0.94018	0.66614
0.63987	0.46664	0.50558	0.4926	0.63783	0.92946	0.61177
0.68909	0.49725	0.54712	0.5305	0.72426	0.91779	0.55445
0.73831	0.52636	0.58941	0.56839	0.82784	0.90515	0.49465
0.78753	0.55372	0.63258	0.60629	0.95396	0.89153	0.43291
0.83675	0.57904	0.67678	0.6442	1.1147	0.87693	0.36986
0.88597	0.60199	0.72217	0.68211	1.3227	0.86137	0.30630
0.9352	0.62216	0.76895	0.72002	1.6183	0.8449	0.24316
0.98442	0.63913	0.81735	0.75794	2.0847	0.82759	0.18154
1.03364	0.65242	0.86759	0.79587	2.9274	0.80958	0.12269
1.08287	0.66157	0.91992	0.8338	4.9254	0.79107	0.06800

a_N: isotropic hyperfine coupling constant, expressed in same units as the quantity that it normalizes; $\Delta\Delta B \equiv \Delta\Delta B_{pp}$: increase in peak-to-peak linewidth of first-derivative spectrum; $\langle \Delta\Delta B_{pp}\rangle = \frac{1}{3}[\Delta\Delta B(0) + 2\Delta\Delta B(\pm 1)]$: mean broadening of the three hyperfine lines; h_{disp}: maximum amplitude of first-derivative dispersion; h_{pp}: peak-to-peak amplitude of first-derivative absorption; a_{HE}: splitting of absorption maxima from adjacent hyperfine peaks; $I(m_I)$: integrated intensity of m_I-hyperfine line. See Figure 9.15.

TABLE A9.2.2 Relative dispersion amplitude h_{disp}/h_{pp}, and direct hyperfine shifts a_{HE}/a_N, for ^{15}N-nitroxide ($I = 1/2$), as function of normalized Heisenberg spin-exchange frequency τ_{HE}^{-1}/a_N (Bales et al. 2009)

τ_{HE}^{-1}/a_N	h_{disp}/h_{pp}	a_{HE}/a_N
0	0	1
0.01968	0.01516	0.99982
0.03936	0.03033	0.99923
0.05905	0.04553	0.99827
0.07873	0.0608	0.99691
0.09841	0.07612	0.99514
0.11809	0.09155	0.993
0.13778	0.10706	0.99045
0.15746	0.12276	0.98755
0.17714	0.13857	0.98418
0.19683	0.15456	0.98045
0.21651	0.17074	0.97627
0.23618	0.1871	0.97173
0.25588	0.20368	0.96673
0.27555	0.22069	0.96127
0.29523	0.23795	0.95541
0.31492	0.25547	0.94914
0.3346	0.27332	0.94236
0.35429	0.29156	0.93514
0.37397	0.31033	0.92745
0.39364	0.32973	0.91927
0.41334	0.34928	0.91059
0.43301	0.36999	0.90141
0.45269	0.39069	0.89168
0.47238	0.41272	0.88141
0.49206	0.43512	0.87055
0.53143	0.48331	0.84709
0.5511	0.5085	0.83445
0.5708	0.53506	0.82109
0.59047	0.56311	0.80705
0.61015	0.59276	0.79227
0.62984	0.62421	0.77673
0.64952	0.65721	0.76036
0.6692	0.69319	0.74309
0.70856	0.77295	0.70564
0.74793	0.86739	0.66377
0.7873	0.98294	0.61659
0.82666	1.1309	0.56268
0.86603	1.3333	0.5
0.90539	1.6416	0.42458
0.94475	2.2188	0.32778
0.98412	4.2669	0.17753

a_N: isotropic hyperfine coupling constant, expressed in same units as the quantity that it normalizes; h_{disp}: maximum amplitude of first-derivative dispersion; h_{pp}: peak-to-peak amplitude of first-derivative absorption; a_{HE}: splitting of absorption maxima from adjacent hyperfine peaks. See Figure 9.15.

REFERENCES

Ablett, S., Barratt, M.D., and Franks, F. 1975. Self-diffusion and electron spin exchange of hydrophobic probes in dilute solution. *J. Solution Chem.* 4:797–807.

Abragam, A. 1961. *The Principles of Nuclear Magnetism*, Oxford: Oxford University Press.

Ayant, Y., Belorizky, E., Alizon, J., and Gallice, J. 1975. Calcul des densités spectrales résultant d'un mouvement aléatoire de translation en relaxation par interaction dipolaire magnétique dans les liquides. *J. de Phys.* 36:991–1004.

Bales, B.L. and Peric, M. 1997. EPR line shifts and line shape changes due to spin exchange of nitroxide free radicals in liquids. *J. Phys. Chem. B* 101:8707–8716.

Bales, B.L. and Peric, M. 2002. EPR line shifts and line shape changes due to spin exchange of nitroxide free radicals in liquids. 2. Extension to high spin exchange frequencies and inhomogeneously broadened spectra. *J. Phys. Chem. A* 106:4846–4854.

Bales, B.L. and Willett, D. 1984. Electron paramagnetic resonance investigation of the intermediate spin exchange regime. *J. Chem. Phys.* 80:2997–3004.

Bales, B.L., Peric, M., and Dragutan, I. 2003. Electron paramagnetic resonance line shifts and line shape changes due to spin exchange between nitroxide free radicals in liquids. 3. Extension to five hyperfine lines. Additional line shifts due to re-encounters. *J. Phys. Chem. A* 107:9086–9098.

Bales, B.L., Meyer, M., Smith, S., and Peric, M. 2008. EPR line shifts and line shape changes due to spin exchange of nitroxide free radicals in liquids: 4. Test of a method to measure re-encounter rates in liquids employing ^{15}N and ^{14}N nitroxide spin probes. *J. Phys. Chem. A* 112:2177–2181.

Bales, B.L., Meyer, M., Smith, S., and Peric, M. 2009. EPR line shifts and line shape changes due to spin exchange of nitroxide free radicals in liquids: 6. Separating line broadening due to spin exchange and dipolar interactions. *J. Phys. Chem. A* 113:4930–4940.

Bales, B.L., Cadman, K.M., Peric, M., Schwartz, R.N., and Peric, M. 2011. Experimental method to measure the effect of charge on bimolecular collision rates in electrolyte solutions. *J. Phys. Chem. A* 115:10903–10910.

Berner, B. and Kivelson, D. 1979a. Electron spin resonance linewidth method for measuring diffusion - critique. *J. Phys. Chem.* 83:1406–1412.

Berner, B. and Kivelson, D. 1979b. Paramagnetically enhanced relaxation of nuclear spins - measurement of diffusion. *J. Phys. Chem.* 83:1401–1405.

Carrington, A. and McLachlan, A.D. 1969. *Introduction to Magnetic Resonance with Applications to Chemistry and Chemical Physics*, New York: Harper and Row.

Currin, J.D. 1962. Theory of exchange relaxation of hyperfine structure in electron spin resonance. *Phys. Rev.* 126:1995–2001.

Debye, P. 1942. Reaction rates in ionic solutions. *J. Electrochem. Soc.* 82:265–272.

Devaux, P., Scandella, C.J., and McConnell, H.M. 1973. Spin-spin interactions between spin-labeled phospholipids incorporated into membranes. *J. Magn. Reson.* 9:474–485.

Eastman, M.P., Bruno, G.V., and Freed, J.H. 1970. ESR studies of Heisenberg spin exchange. II. Effects of radical charge and size. *J. Chem. Phys.* 52:2511–2522.

Einstein, A. 1905. Über die von der molekularkinetischen Theorie der Wärme geforderte Bewegung von in ruhenden Flüssigkeiten suspendierten Teilchen. *Ann. Phys.* 17:549–560.

Freed, J.H. 1966. On Heisenberg spin exchange in liquids. *J. Chem. Phys.* 45:3452–3453.

Galeev, R.T. and Salikhov, K.M. 1996. Theory of dipole widening of magnetic resonance lines in nonviscous liquids. *Chem. Phys. Reports* 15:359–375.

Glarum, S.H. and Marshall, J.H. 1967. Spin exchange in nitroxide biradicals. *J. Chem. Phys.* 47:1374–1378.

Grebenshchikov, Yu.B., Likhtenshtein, G.I., Ivanov, V.P., and Rozantsev, E.G. 1972. Investigation of electrostatic charges in proteins by the paramagnetic probe method (translated from Russian). *Mol. Biol.* 6: 400–406.

Hanson, P., Anderson, D.J., Martinez, G. et al. 1998. Electron spin resonance and structural analysis of water soluble, alanine-rich peptides incorporating TOAC. *Mol. Phys.* 95:957–966.

Hubbard, P.S. 1966. Theory of electron-nucleus Overhauser effects in liquids containing free radicals. *Proc. Roy. Soc. A* 291:537–555.

Hwang, L.P. and Freed, J.H. 1975. Dynamic effects of pair correlation functions on spin relaxation by translational diffusion in liquids. *J. Chem. Phys.* 63:4017–4025.

Johnson, C.S. 1967. Theory of linewidths and shifts in electron spin resonance arising from spin exchange interactions. *Mol. Phys.* 12:25–31.

Jones, M.T. 1963. Electron spin exchange in aqueous solutions of $K_2(SO_3)_2NO$. *J. Chem. Phys.* 38:2892–2895.

Keana, J.F.W. and Dinerstein, R.J. 1971. New highly anisotropic dinitroxide ketone spin label. Sensitive probe for membrane structure. *J. Am. Chem. Soc.* 93:2808–2810.

Keith, A.D., Snipes, W., Mehlhorn, R.J., and Gunter, T. 1977. Factors restricting diffusion of water-soluble spin labels. *Biophys. J.* 19:205–218.

King, M.D. and Marsh, D. 1986. Translational diffusion of lipid monomers determined by spin label electron spin resonance. *Biochim. Biophys. Acta* 863:341–344.

King, M.D., Sachse, J.-H., and Marsh, D. 1987. Unconstrained optimization method for interpreting the concentration and temperature dependence of the linewidths of interacting nitroxide spin labels. Application to the measurement of translational diffusion coefficients of spin-labeled phospholipids in membranes. *J. Magn. Reson.* 72:257–267.

Korb, J.P., Winterhalter, M., and McConnell, H.M. 1984. Theory of spin relaxation by translational diffusion in two-dimensional systems. *J. Chem. Phys.* 80:1059–1068.

Kurban, M.R., Peric, M., and Bales, B.L. 2008. Nitroxide spin exchange due to re-encounter collisions in a series of *n*-alkanes. *J. Chem. Phys.* 129:064501–1–064501–10.

Luckhurst, G.R. 1966. Alternating linewidths. A novel relaxation process in electron resonance of biradicals. *Mol. Phys.* 10:543–550.

Luckhurst, G.R. 1976. Biradicals as spin probes. In *Spin Labeling. Theory and Applications*, ed. Berliner, L.J., 133–181. New York: Academic Press.

Luckhurst, G.R. and Pedulli, G.F. 1970. Interpretation of biradical electron resonance spectra. *J. Am. Chem. Soc.* 92:4738–4739.

Marsh, D. 2013. *Handbook of Lipid Bilayers*, Boca Raton, FL: CRC Press.

Marsh, D. 2017. Coherence transfer and electron T_1-, T_2-relaxation in nitroxide spin labels. *J. Magn. Reson.* 277:86–94.

Matthews, B.W. and Remington, S.J. 1974. Three-dimensional structure of lysozyme from bacteriophage T4. *Proc. Natl. Acad. Sci. USA* 71:4178–4182.

Mchaourab, H.S., Oh, K.J., Fang, C.J., and Hubbell, W.L. 1997. Conformation of T4 lysozyme in solution. Hinge-bending motion and the substrate-induced conformational transition studied by site-directed spin labeling. *Biochemistry* 36:307–316.

Michon, P. and Rassat, A. 1975. Nitroxides. LXIX. 1,4-Bis(4',4'-dimethyloxazolidine-3'-oxyl)cyclohexane structure determination by electron spin resonance and nuclear magnetic resonance. *J. Am. Chem. Soc.* 97:696–700.

Molin, Y.N., Salikov, K.M., and Zamaraev, K.I. 1980. Spin Exchange. Principles and Applications in Chemistry and Biology, Berlin: Springer Verlag.

Morse, P.M. and Feshbach, H. 1953. *Methods of Theoretical Physics*, New York: McGraw-Hill Book Co. Inc.

Nakajima, A., Ohyanishi, H., and Deguchi, Y. 1972. Magnetic properties of some iminoxyl polyradicals. 3. Exchange interaction in iminoxyl biradicals. *Bull. Chem. Soc. Japan* 45:713–716.

Noyes, R.M. 1954. A treatment of chemical kinetics with special applicability to diffusion controlled reactions. *J. Chem. Phys.* 22:1349–1359.

Ohnishi, S.I., Cyr, T.J.R., and Fukushima, H. 1970. Biradical spin-labeled micelles. *Bull. Chem. Soc. Japan* 43:673–676.

Pake, G.E. 1948. Nuclear resonance absorption in hydrated crystals: fine structure of the proton line. *J. Chem. Phys.* 16:327–336.

Parmon, V.N., Kokorin, A.I., Zhidomirov, G.M., and Zamaraev, K.I. 1973. Evidence for slow exchange in ESR spectra of nitroxide biradicals. *Mol. Phys.* 26:1565–1569.

Pearson, R.G. and Buch, T. 1962. Electron paramagnetic relaxation in ionic solutions. *J. Chem. Phys.* 36:1277–1282.

Peric, M., Bales, B.L., and Peric, M. 2012. Electron paramagnetic resonance line shifts and line shape changes due to Heisenberg spin exchange and dipole-dipole interactions of nitroxide free radicals in liquids. 8. Further experimental and theoretical efforts to separate the effects of the two interactions. *J. Phys. Chem. A* 116:2855–2866.

Perrin, F. 1934. Mouvement brownien d'un ellipsoide - I. Dispersion diélectrique pour des molécules ellipsoidales. *J. Phys. Radium* 5:497–511.

Plachy, W. and Kivelson, D. 1967. Spin exchange in solutions of di-tertiary-butyl nitroxide. *J. Chem. Phys.* 47:3312–3318.

Polnaszek, C.F. and Bryant, R.G. 1984. Nitroxide radical induced solvent proton relaxation - measurement of localized translational diffusion. *J. Chem. Phys.* 81:4038–4045.

Remington, S.J., Anderson, W.F., Owen, J., Teneyck, L.F., and Grainger, C.T. 1978. Structure of lysozyme from bacteriophage T4- electron density map at 2.4 Å resolution. *J. Mol. Biol.* 118:81–98.

Sachse, J.-H., King, M.D., and Marsh, D. 1987. ESR determination of lipid diffusion coefficients at low spin-label concentrations in biological membranes, using exchange broadening, exchange narrowing, and dipole-dipole interactions. *J. Magn. Reson.* 71:385–404.

Sachse, J.-H. and Marsh, D. 1986. Line intensities in spin-exchanged nitroxide ESR spectra. *J. Magn. Reson.* 68:540–543.

Salikhov, K.M. 1985. The contribution from exchange interaction to line shifts in electron spin resonance spectra of paramagnetic particles in solutions. *J. Magn. Reson.* 63:271–279.

Salikhov, K.M. 2010. Contributions of exchange and dipole-dipole interactions to the shape of EPR spectra of free radicals in diluted solutions. *Appl. Magn. Reson.* 38:237–256.

Scandella, C.J., Devaux, P.F., and McConnell, H.M. 1972. Rapid lateral diffusion of phospholipids in rabbit sarcoplasmic reticulum. *Proc. Natl. Acad. Sci. USA* 69:2056–2060.

Solomon, I. 1955. Relaxation processes in a system of 2 spins. *Phys. Rev.* 99:559–565.

Torrey, H.C. 1953. Nuclear spin relaxation by translational diffusion. *Phys. Rev.* 92:962–969.

Träuble, H. and Sackmann, E. 1972. Studies of the crystalline-liquid crystalline phase transition of lipid model membranes. III. Structure of a steroid-lecithin system below and above the lipid-phase transition. *J. Am. Chem. Soc.* 94:4499–4510.

von Smoluchowski, M. 1917. Versuch einer mathematischen Theorie der Koagulationskinetik kolloider Lösungen. *Z. Phys. Chem.* 92:129–168.

Watson, G.N. 1945. *A Treatise on the Theory of Bessel Functions*, Cambridge, UK: Cambridge University Press.

Spin–Lattice Relaxation

10

10.1 INTRODUCTION

This chapter covers the measurable parameters that govern electron spin–lattice relaxation of ^{14}N- and ^{15}N-nitroxides, as studied by continuous-wave (CW) EPR methods. What concerns us here is progressive-saturation EPR and CW electron–electron double resonance (ELDOR). These experiments are characterized by effective spin–lattice relaxation times T_1^{eff} and by ELDOR reduction factors $R_\infty^{p,obs}$, respectively. Progressive-saturation EPR methodology is treated in Chapter 11, and this is by far the more common and straightforward technique. CW-ELDOR has the advantage that it provides information directly about nitrogen nuclear relaxation of the nitroxide, which is almost inextricably mixed in the effective relaxation times determined from progressive-saturation EPR.

Our aim is to express the effective relaxation times and reduction factors in terms of the fundamental processes that determine nitroxide spin–lattice relaxation, which we covered in Chapter 5. This gives us immediate access to dynamic information on the spin-labelled system, in the form of rotational correlation times. The advantage over studies using transverse relaxation, i.e., Lorentzian linewidths, is that we can cover a wider range of rotational correlation times.

Crucial also is the much longer timescale of spin–lattice relaxation (microseconds, as opposed to nanoseconds for transverse relaxation). This allows us to determine relaxation enhancements by much slower dynamic processes and by much weaker spin–spin interactions. Thus, we treat paramagnetic relaxation enhancement in considerable detail, because of the wealth of dynamic and structural information that this can provide, particularly in site-directed spin labelling (see Chapter 16). Also slow two-site exchange and slow/weak Heisenberg spin exchange are important dynamic processes that become accessible to study by CW-saturation EPR.

In Chapter 12, we shall see that saturation-recovery EPR experiments provide a direct time-dependent measurement of the processes covered here. Interpretation of the recovery rate constants is therefore more straightforward than for the effective relaxation times from CW-EPR. However, availability of saturation-recovery equipment is considerably more restricted, and therefore the two techniques provide useful complements in spin-label EPR.

10.2 EFFECTIVE SPIN–LATTICE RELAXATION TIMES

In a CW progressive-saturation EPR experiment, processes other than intrinsic electron spin–lattice relaxation (which occurs at rate $2W_e = 1/T_{1e}$, see Chapter 5) contribute to alleviating saturation. These additional processes contribute to the effective spin–lattice relaxation time T_1^{eff}. They include nuclear spin–lattice relaxation at rate W_n (see Chapter 5) and Heisenberg spin exchange at rate $W_{HE} = \tau_{HE}^{-1}/(2I+1)$ (see Chapter 9).

This is the effective electron T_1 which we measure in experiments that depend upon CW-saturation. If $N_i^{\pm}$ are the populations of the $M_{S,i} = \pm\frac{1}{2}$ levels of the ith electron spin system, the population difference that we need for absorption of microwaves is $n_i = N_i^- - N_i^+$. The steady-state population difference of the pth isolated two-level system under CW-irradiation at resonance frequency ω_p is (see Eq. 1.19 in Chapter 1 and Slichter 1978):

$$n_p = \frac{n_p^o}{1 + 2WT_{1,p}^{eff}} \tag{10.1}$$

where n_p^o is the population difference at Boltzmann equilibrium, and W is the rate at which the microwave B_1-field induces transitions.

Figure 10.1 shows allowed transitions between the $S = \frac{1}{2}$ electron spin states for a ^{14}N-nitroxide with nuclear spin $I = 1$. For simplicity, we neglect cross relaxation from electron–nuclear dipolar interaction (see Section 5.4) and nitrogen quadrupolar relaxation, and assume that all three hyperfine levels have the same intrinsic spin–lattice relaxation rate (cf. Figure 5.4). These are significant restrictions (Marsh 2016a), and we shall return to them later. Ignoring Heisenberg exchange for the moment, the steady-state rate equations for spin population differences $n_{m_I} = N_{m_I}^- - N_{m_I}^+$ in the $m_I = 0, \pm 1$ hyperfine levels are then (Marsh 1992):

$$\frac{dn_0}{dt} = -2W_e\left(n_0 - n_0^o\right) - W_n\left(2n_0 - n_{+1} - n_{-1}\right) - 2Wn_0\delta_{0,p} = 0 \tag{10.2}$$

$$\frac{dn_{\pm1}}{dt} = -2W_e\left(n_{\pm1} - n_{\pm1}^o\right) - W_n\left(n_{\pm1} - n_0\right) - 2Wn_{\pm1}\delta_{\pm1,p} = 0 \tag{10.3}$$

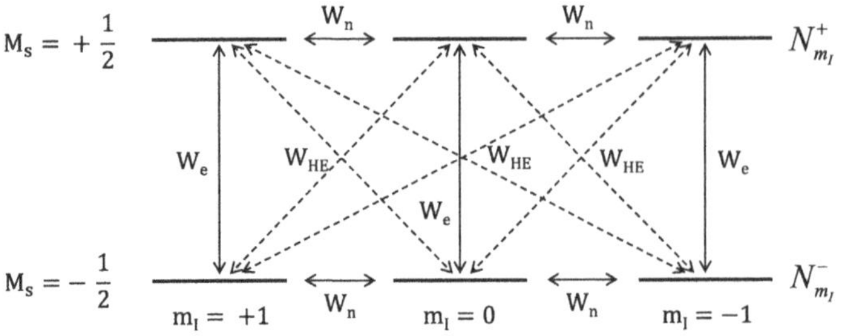

FIGURE 10.1 Energy-level scheme for $S=\frac{1}{2}$, $I=1$ ^{14}N-nitroxide spin label. Populations of states are indicated; $n_{m_I} = N^-_{m_I} - N^+_{m_I}$ is population difference between $M_S = \pm\frac{1}{2}$ levels. Spin–lattice relaxation pathways (solid arrows): W_e for electron transitions, W_n for nuclear transitions. Heisenberg spin exchange $W_{HE}\left(=\frac{1}{6}\tau_{HE}^{-1}\right)$ (dashed lines) does not contribute at low concentrations.

where $n^o_{m_I}$ is the population difference at Boltzmann equilibrium, which is independent of m_I, i.e., $n^o_0 = n^o_{+1} = n^o_{-1}$. The irradiating B_1-field induces transitions at rate W within the $m_I = p$ hyperfine manifold (i.e., $\delta_{m_I,p} = 1$ for $m_I = p$, and is zero otherwise). The transition probability for nuclear spin–lattice relaxation W_n is given by Eq. 5.20 of chapter 5, with $I = 1$ for a ^{14}N-nitroxide. This definition differs by a factor of two from that given for a ^{14}N-nitroxide in Marsh (1992), which stems from the original use in CW-ELDOR (Hyde et al. 1968). The present definition corresponds to that adopted for saturation-recovery studies on nitroxides (see, e.g., Yin and Hyde 1987, and also Chapters 5 and 12), which is the true transition probability for a $I = 1$ nucleus. Further discussion of this point is given in Marsh (2016b).

Solution of Eqs. 10.2 and 10.3 leads to a dependence on the radiation power (i.e., on W) that conforms to Eq. 10.1. The effective spin–lattice relaxation times $T_1^{eff}(m_I)$ that govern the saturation behaviour of the $m_I = 0,\ \pm1$ hyperfine manifolds are then (Marsh 1992):

$$T_1^{eff}(0) = \frac{1+b}{1+3b}\cdot\frac{1}{2W_e} \tag{10.4}$$

$$T_1^{eff}(\pm1) = \frac{1+3b+b^2}{(1+b)(1+3b)}\cdot\frac{1}{2W_e} \tag{10.5}$$

where $b \equiv W_n/(2W_e)\left(=T_{1e}/T_{1n}\right)$ is a ratio of the nuclear to electron spin–lattice relaxation rate. This corresponds to the original definition of b (Hyde et al. 1968; Marsh 1992) but with the new definition of W_n (and T_{1n}).

When $b << 1$, nuclear relaxation is unimportant and each hyperfine state relaxes independently with its intrinsic rate $T_1^{-1}(0) = T_1^{-1}(\pm1) = 2W_e$ (see Figure 10.2). When $b >> 1$, the electron transitions are short-circuited by nuclear relaxation, all then relax at rate $T_1^{-1}(0) = T_1^{-1}(\pm1) = 6W_e$. Otherwise, relaxation of the $m_I = 0$ line is faster than for the $m_I = \pm1$ transitions; the maximum difference in rate is 25%, when $b = 1$ (see dashed line in Figure 10.2). The solid line in Figure 10.2 shows the dependence of effective spin–lattice relaxation time $T_1^{eff}(0)$, for the $m_I = 0$ manifold, on nuclear relaxation time. Except for very short rotational correlation times, the nuclear relaxation time is of the order of or less than the electron relaxation time: $T_{1n} \le T_{1e}$, i.e., $b \ge 1$ (see, e.g., Marsh et al. 1998). Therefore, the true T_{1e} is likely to lie within the range $2T_1^{eff}(0)$ to $3T_1^{eff}(0)$, with the precise value depending on the nuclear relaxation rate and hence on the rotational correlation time (Livshits et al. 1998).

We cover the general case for Heisenberg exchange later, in Section 10.12. For the scheme shown in Figure 10.1, the effective relaxation times that include spin exchange between the different hyperfine manifolds of a ^{14}N-nitroxide are (Hyde et al. 1968):

$$T_1^{eff}(0,b'') = \frac{1+b+b''}{1+3b+3b''}\cdot\frac{1}{2W_e} \tag{10.6}$$

$$T_1^{eff}(\pm1,b'') = T_1^{eff}(0,b'') + \frac{b}{(1+b+3b'')(1+3b+3b'')}\cdot\frac{1}{2W_e} \tag{10.7}$$

Here $b'' \equiv \tau_{HE}^{-1}/(6W_e)\left(=\frac{1}{3}T_1^o\tau_{HE}^{-1}\right)$ is the normalized Heisenberg exchange rate, defined in terms of the pseudo-first order rate constant τ_{HE}^{-1} that we introduced in Eq. 9.93 of Section 9.10. Unlike those of nuclear relaxation, the effects of spin exchange are similar for all three hyperfine manifolds. For $m_I = 0$, spin exchange

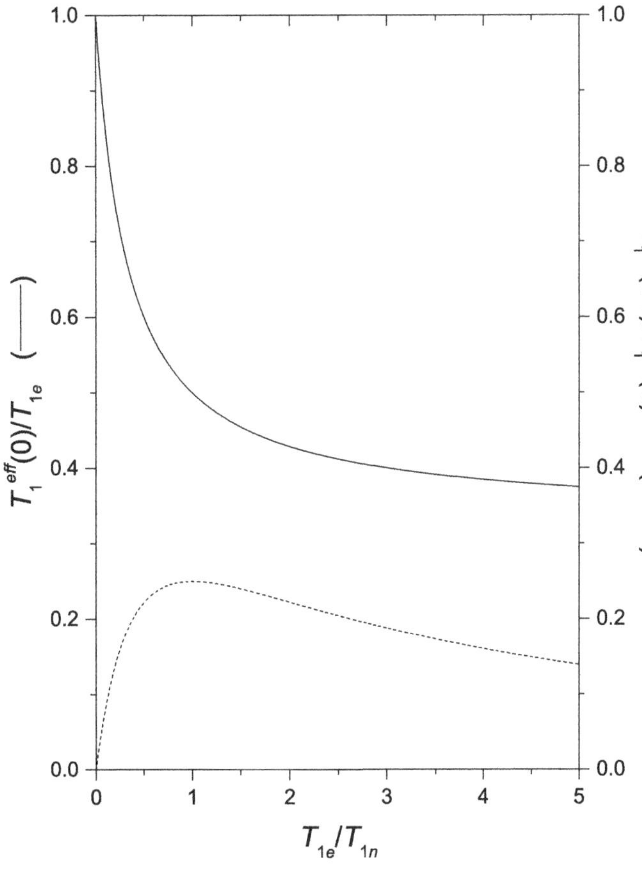

FIGURE 10.2 Dependence on ^{14}N-nuclear spin–lattice relaxation rate $\left(b \equiv W_n/2W_e = T_{1e}/T_{1n}\right)$ of ratio $T_1^{eff}(0)/T_{1e}$ between effective T_1 and intrinsic electron T_{1e} for $m_I = 0$ manifold (solid line); and of normalized difference in effective electron spin–lattice relaxation times $T_1^{eff}(\pm1)/T_1^{eff}(0) - 1$ for $m_I = \pm1$ and $m_I = 0$ hyperfine manifolds (dashed line). From Eqs. 10.4 and 10.5.

contributes to T_1^{eff} just as does nuclear relaxation. For $m_I = \pm 1$, exchange is more effective than nuclear relaxation, because of the less restrictive selection rules (see Figure 10.1). Rate equations for saturation-recovery EPR and ELDOR are given for this case in Section 12.6 of Chapter 12. A treatment analogous to that of Eqs. 10.2 and 10.3 then leads to Eqs. 10.6 and 10.7.

Corresponding results for a ^{15}N-nitroxide (i.e., $I = \frac{1}{2}$) are given in Appendix A10.1, at the end of the chapter. Unlike the treatment here, Appendix A10.1 specifically includes both cross relaxation and different intrinsic spin–lattice relaxation rates for the different hyperfine manifolds (see Section 5.4). Cross relaxation leads to CW saturation where n_p^o in Eq. 10.1 no longer corresponds simply to the equilibrium population difference between the $M_{S,p} = \pm\frac{1}{2}$ levels (Marsh 2016a). This is because longitudinal magnetizations of the electron and nuclear spins are then coupled (cf. Eq. 9.32 for unlike spins in Section 9.5, and also Solomon 1955). Cross relaxation becomes comparable in rate to nuclear relaxation only for very fast motion with rotational correlation times $\tau_R < 2\times 10^{-11}$ s at 9 GHz, and proportionally shorter at higher EPR frequencies (see Section 5.6). However, cross-relaxation rates are comparable to those of intrinsic electron spin–lattice relaxation, and therefore deserve consideration. We shall see in Chapter 12 that this is even more important for saturation-recovery EPR. Results for ^{14}N-nitroxides that include cross relaxation and m_I-dependent intrinsic electron spin–lattice relaxation rates are given in Appendix A10.3 at the end of the chapter (and see also Marsh 2016a).

The expressions for effective T_1-relaxation times (and ELDOR reduction factors) derived in Appendices A10.1 and A10.3 are somewhat more cumbersome than those that neglect cross relaxation and differences in intrinsic T_1. However, they involve no new physical principles. The additional terms covering cross relaxation and m_I-dependent electron spin–lattice relaxation rate still depend only on correlation time and known spin-Hamiltonian parameters. Therefore, there is no reason not to use these more complete expressions for interpreting spin-label results in terms of rotational dynamics.

10.3 REDUCTION FACTORS FOR CW ELECTRON–ELECTRON DOUBLE RESONANCE (ELDOR)

We saw in the last section that nitrogen nuclear spin–lattice relaxation can affect spin-label saturation studies quite significantly. Our principal source of nitrogen nuclear relaxation rates in nitroxide spin labels is from electron-electron double resonance (ELDOR) measurements. Here, we pump with microwaves at one frequency and observe the effects in another part of the spectrum at a different microwave frequency. In continuous-wave ELDOR, we saturate one hyperfine line (or part of the spectrum) by high-power irradiation that is applied continually, whilst we observe the spectrum at a second microwave frequency that corresponds to a different hyperfine line or spectral region. Saturation is transferred from the first hyperfine line to the other hyperfine lines by nuclear relaxation.

If transition p is pumped continuously, and transition obs is observed under non-saturating conditions, we need the solution of rate equations such as Eqs. 10.2 and 10.3 with the steady-state boundary condition $dn_{obs}/dt = 0$. Then the spin population difference at the pump frequency, n_p, is given by Eq. 10.1. The CW-ELDOR reduction factor is given by the ratio of signals observed in the presence and absence of the pumping radiation:

$$R = \frac{n_{obs}(\text{off}) - n_{obs}(\text{on})}{n_{obs}(\text{off})} = 1 - \frac{n_{obs}(\text{on})}{n_{obs}(\text{off})} \tag{10.8}$$

where $n_{obs}(\text{on})$ is the population difference of the observed transition when the pumping microwaves are switched on, and $n_{obs}(\text{off})$ is that when the pump is switched off.

Let us, for example, pump the $m_I = 0$ transition and observe the $m_I = \pm 1$ transition. From Eq. 10.3, we get:

$$\frac{dn_{\pm 1}}{dt} = -(2W_e + W_n)n_{\pm 1} + 2W_e n_{\pm 1}^o + W_n n_0 = 0 \tag{10.9}$$

Substituting n_0 from Eq. 10.1 for the pumped transition, the population difference at the observing frequency becomes:

$$n_{\pm 1}(\text{on}) = \frac{n^o}{2W_e + W_n}\left(2W_e + \frac{W_n}{1 + 2WT_{1,0}^{eff}}\right) \tag{10.10}$$

where $T_{1,0}^{eff}$ is the effective spin–lattice relaxation time for the transition that is pumped, and n^o $\left(\equiv n_0^o = n_{\pm 1}^o\right)$ is the population difference in any one of the m_I hyperfine states at Boltzmann equilibrium. With the saturating microwaves switched off, the population difference is given by Eq. 10.10 but with $W = 0$. The dependence of the CW-ELDOR reduction R on rate W of pumping by the microwaves therefore becomes:

$$R^{-1} = \frac{2W_e + W_n}{W_n}\left(1 + \frac{1}{2WT_{1,0}^{eff}}\right) \tag{10.11}$$

In practice, we extrapolate the CW-ELDOR reduction to infinite microwave power ($W \to \infty$), by plotting R^{-1} against W^{-1} according to Eq. 10.11 (Hyde et al. 1968). For pumping the $m_I = 0$ transition and observing the $m_I = \pm 1$ transition we get, from Eq. 10.11:

$$R_\infty^{0,\pm 1} = b/(1+b) \tag{10.12}$$

where $b \equiv W_n/(2W_e)$. Similarly, for pumping the $m_I = \pm 1$ transitions and observing the $m_I = 0$ or $m_I = \mp 1$ transition, the ELDOR reduction extrapolated to infinite power becomes, respectively (Hyde et al. 1968):

$$R_\infty^{\pm 1,0} = \frac{b(1+b)}{1 + 3b + b^2} \tag{10.13}$$

$$R_\infty^{\pm 1,\mp 1} = \frac{b^2}{1 + 3b + b^2} \tag{10.14}$$

where again $b \equiv W_n/(2W_e)$.

Heisenberg spin exchange connects each of the ^{14}N-nitroxide hyperfine states with the other two (see Figure 10.1). Therefore, it operates exactly as does nuclear relaxation for the $m_I = 0$ hyperfine state. For Heisenberg exchange alone, the CW-ELDOR reduction factor is thus $R_\infty = b''/(1+b'')$ for each combination of pump and observe positions, where once again $b'' \equiv \tau_{HE}^{-1}/(6W_e)$. CW-ELDOR reduction factors that include both nuclear relaxation and Heisenberg spin exchange are given by Eqs. A10.16–A10.18 in Appendix A10.2 at the end of the chapter. Those that include cross relaxation and m_I-dependent intrinsic electron spin–lattice relaxation rates are given in Appendix A10.3. Reduction factors for ^{15}N-nitroxides are given in Appendix A10.1.

10.4 DEPENDENCE OF SPIN–LATTICE RELAXATION ON ROTATIONAL DYNAMICS

We expect the spin–lattice relaxation rates that we determine from saturation studies to depend on molecular dynamics as described in Chapter 5. Figure 10.3 shows the dependence of the spin–lattice relaxation rate for the centre $(m_I = 0)$ hyperfine line of peroxylamine disulphonate (PADS, $(SO_3^-)_2NO$) in aqueous glycerol on the geometric mean correlation time for axial rotational diffusion. Inhomogeneous broadening is unusually small in PADS because there are no protons in this nitroxide, which allows rather precise measurements.

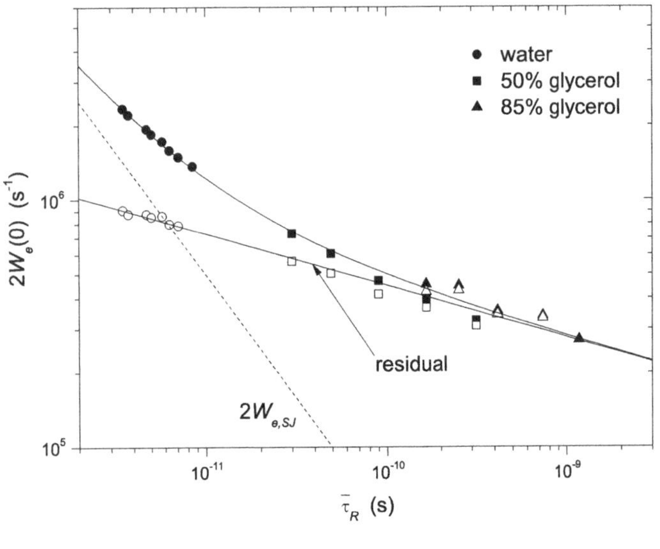

FIGURE 10.3 Spin–lattice relaxation rate $2W_e(0)$ for $m_I = 0$ hyperfine line of peroxylamine disulphonate (PADS) in water (circles), 50% aqueous glycerol (squares) and 85% aqueous glycerol (triangles), as function of correlation time $\bar{\tau}_R = \sqrt{\tau_{R\parallel}\tau_{R\perp}}$ for axial rotational diffusion. Solid symbols: experimental measurements; open symbols: after subtracting spin-rotation contribution $2W_{e,SJ}$ (dashed line). Linear regression to residual rate has gradient −0.2 on log–log scale. (Experimental data from Goldman et al. 1973.)

Rotation of PADS in water alone is very fast $\left(\bar{\tau}_R < 10^{-11}\ \text{s}\right)$, and it is found that $T_2^{-1} = T_1^{-1} \equiv 2W_e(0)$ (Kooser et al. 1969). The data for water in Figure 10.3 are therefore obtained from linewidth measurements. All other data come from CW-EPR progressive saturation determinations of $T_1^{eff}(0)$. Rotational correlation times $\bar{\tau}_R = \sqrt{\tau_{R\parallel}\tau_{R\perp}}$ are derived from EPR linewidths as described in Chapter 7. Nuclear relaxation (and cross relaxation) rates are estimated for the END mechanism, using the correlation times, to obtain values of $2W_e(0)$ which are consistent with measurements of $T_1^{eff}(0)$ from saturation (Goldman et al. 1973).

The dashed line in Figure 10.3 shows the values of $2W_e(0)$ that are predicted for the spin-rotation mechanism of spin–lattice relaxation: $2W_{e,SJ} = \sum (g_{ii} - g_e)^2 / 9\tau_R$ (see Section 5.5.4 and Eq. 5.40). Modulation of the g-value anisotropy by rotational diffusion (see Eq. 5.19) is the only other relaxation mechanism in Chapter 5 that contributes for the $m_I = 0$ hyperfine line. We can neglect this compared with spin rotation, because the maximum contribution is predicted to be $2.3 \times 10^4\,\text{s}^{-1}$, when $\tau_R = 1/\omega_e \approx 1.7 \times 10^{-11}$ s. The open symbols in Figure 10.3 show the relaxation rate that remains after subtracting the contribution of spin rotation. This residual rate depends only weakly on rotational correlation time; linear regression in the log–log plot yields: $W_e(0) \propto \bar{\tau}_R^{-1/5}$. Saturation studies in the slow-motion region reveal the same weak dependence on τ_R, with a low effective activation energy of ≈ 13 kJ mol^{-1} (Goldman et al. 1973).

Equivalent results are found with perdeuterated TEMPONE (1-oxyl-2,2,6,6-tetramethylpiperidin-4-yl oxide) in both apolar and aqueous solvents, except that inhomogeneous broadening by the deuterons leads to greater experimental scatter (Hwang et al. 1975). After subtracting the contribution from spin-rotation (cf. Figure 10.3), the residual relaxation rate of TEMPONE in toluene that we obtain from these studies depends on rotational correlation time according to: $W_e(0) \propto \tau_R^{-0.17\pm0.04}$, which is comparable to the PADS result. More evidence for a mechanism of nitroxide spin–lattice relaxation that depends on a low fractional power of τ_R comes from saturation-recovery measurements, as we shall see later in Chapter 12. This is tentatively ascribed to some form of spectral diffusion, or spin diffusion, which depends on a weak power of the correlation time.

10.5 PARAMAGNETIC SPIN–LATTICE RELAXATION ENHANCEMENT

Coupling a spin label with a fast-relaxing paramagnetic species (e.g., molecular oxygen or a transition-metal ion), either by Heisenberg spin exchange or by a magnetic dipole–dipole interaction, increases the true spin–lattice relaxation rate of the nitroxide, instead of simply effecting cross relaxation. These paramagnetic enhancements in T_1-relaxation are important for determining spin-label accessibilities in the case of Heisenberg exchange, and for distance measurements in the case of dipolar interactions (see Chapter 16).

Because the paramagnetic species produce a true spin–lattice relaxation sink (see Figure 10.4), the steady-state rate equations for population differences between the spin levels are those of Eqs. 10.2 and 10.3 with W_e replaced by $W_e + W_{RL}$, where W_{RL} is the rate of the additional paramagnetic spin–lattice relaxation process. Neglecting cross relaxation, effective spin–lattice relaxation times for the $m_I = 0,\ \pm 1$ hyperfine manifolds, determined by CW-saturation, are then (Marsh 1995a; Marsh 1995b; Subczynski and Hyde 1981):

$$T_1^{eff}(0,b''') = \frac{1+b'''+b}{1+b'''+3b}\cdot\frac{1}{2W_e(1+b''')} \tag{10.15}$$

$$T_1^{eff}(\pm 1,b''') = \frac{(1+b''')^2+3(1+b''')b+b^2}{(1+b'''+b)(1+b'''+3b)}\cdot\frac{1}{2W_e(1+b''')} \tag{10.16}$$

where as before $b \equiv W_n/(2W_e)$, and $b''' \equiv W_{RL}/W_e$ is the normalized paramagnetic relaxation enhancement rate. Thus, although paramagnetic relaxation is a true T_1-process, complications may arise if we cannot ignore nuclear relaxation. If paramagnetic relaxation is faster than nuclear relaxation $(b''' > b)$, and/or nuclear relaxation is very slow compared with electron relaxation ($b << 1$), we get a straightforward relaxation enhancement: $T_1^{-1}(0,b''') = T_1^{-1}(\pm 1,b''') = 2W_e(1+b''')$. Alternatively, if nuclear relaxation is very fast ($b >> 1$), the paramagnetic enhancement is: $T_1^{-1}(0,b''') = T_1^{-1}(\pm 1,b''') = 6W_e(1+b''')$. For intermediate values of nuclear or paramagnetic relaxation, we need to know b from other experiments, such as ELDOR (see Section 10.3 and Chapter 12). However, this applies only if we cannot vary relaxant concentrations sufficiently.

Quantitatively, paramagnetic relaxation enhancement by a magnetic dipole–dipole mechanism depends on the spin–lattice relaxation time $T_{1,R}$, g-value g_R, and total spin S_R, of the relaxant. As we shall see in the next section, relaxation enhancement by spin exchange also may depend on the values of $T_{1,R}$ and S_R (but not on g_R). We list these quantities for various paramagnetic-ion complexes (mostly aqueous) in Table 10.1. For paramagnetic species with relatively long T_1 s at ambient temperatures, the spin–lattice relaxation times come mostly from measurements of EPR linewidths, with the assumption that $T_{1,R} = T_{2,R}$. For species with short T_1 s, where the EPR spectrum is not seen at room temperature ($T_{1,R} < 3\times 10^{-11}$ s at 9 GHz), values of $T_{1,R}$ derive from paramagnetic enhancements of the ligand nuclear relaxation. Paramagnetic broadening of O^{17} NMR linewidths arises primarily from the contact hyperfine interaction, with little contribution from dipolar interaction (see, e.g., Connick and Fiat 1966). Proton spin–lattice relaxation rates and their frequency dispersion, on the other hand, depend more strongly on magnetic dipolar interactions (see, e.g., Hausser and Noack 1964).

For many of the iron-group (3d-series) ions, the orbital angular momentum is quenched by the ligand field (cf. Appendix L). The ground state is then a non-degenerate orbital singlet, and S_R in Table 10.1 is the real spin S of the paramagnetic ion. Co^{2+} and Fe^{2+} are special cases. High-spin Co^{2+} has a spin $S = \frac{3}{2}$, but the ligand field of the hexa-aquo ion leaves a degenerate orbital triplet as the lowest state. Spin-orbit coupling λ splits this by amounts $(J'+1)|\lambda| > k_B T_{room}$, leaving a doublet (with angular momentum $J' = \frac{1}{2}$, but g-value very different from 2) as ground state. In Table 10.1, we describe this doublet by an effective spin $S_R = \frac{1}{2}$ with $\langle g_R \rangle = 13/3$ (see Abragam and Bleaney 1970). Analogously, high-spin Fe^{2+} has spin $S = 2$ and an orbital triplet ground state. Spin-orbit coupling then leaves a $J' = 1$ triplet ground state that we represent by $S_R = 1$ with $\langle g_R \rangle > 3$. The effective spins S_R quoted for lanthanides (4f-series) in Table 10.1 are the quantum numbers J of the real total angular momentum: $\mathbf{J} = \mathbf{L} + \mathbf{S}$, where $\mathbf{L}$ is the unquenched orbital angular momentum. Here, ligand-field splittings are small (relative to $k_B T_{room}$), because the 4f-orbitals lie within the 5s and 5p shells of the central ion, and spin-orbit coupling is strong. Correspondingly, values given for $\langle g_R \rangle$ are the Landé g-factor: $g_J = \frac{3}{2} - [L(L+1) - S(S+1)]/2J(J+1)$ (see, e.g., Abragam and Bleaney 1970). The lanthanide electron relaxation times (except Gd^{3+}) are obtained from paramagnetic relaxation enhancement of water ^{1}H-NMR by assuming a most probable hydration number of nine (Alsaadi et al. 1980).

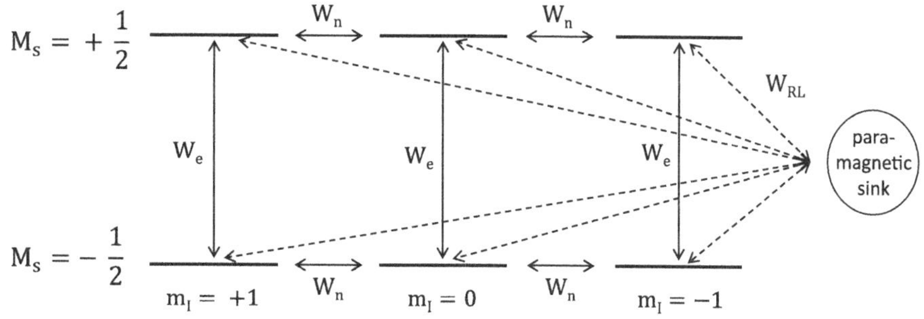

FIGURE 10.4 Energy levels for $S = \frac{1}{2}$, $I = 1$ ^{14}N-nitroxide, in the presence of paramagnetic relaxation sink. Intrinsic electron spin–lattice relaxation rate, W_e, and nuclear relaxation rate, W_n, as in Figure 10.1. Dashed lines: additional relaxation pathways, W_{RL}, contributed by interaction with fast-relaxing paramagnetic species.

TABLE 10.1 Spin–lattice relaxation times, $T_{1,R}$, effective spins, S_R, and g-values, $\langle g_R \rangle$, for paramagnetic-ion complexes at ambient temperatures[a]

ion, R[b]	$\langle g_R \rangle$	S_R	$T_{1,R}$ (s)	Refs.[c]
$Ni(H_2O)_6^{2+}$	2.25	1	3×10^{-12}	1
			4.3×10^{-12}	2
$Co(H_2O)_6^{2+}$	4.33	1/2	$\approx1.5\times10^{-12}$	3, 4
$Co(CH_3OH)_6^{2+}$	2.3	3/2	7×10^{-13}	3
$Cu(H_2O)_6^{2+}$	2.20	1/2	0.9×10^{-9}	5
$Cu(en)_2(H_2O)_2^{2+}$	2.10	1/2	2.7×10^{-9}	5
$Mn(H_2O)_6^{2+}$	1.993	5/2	$2\text{–}3\times10^{-9}$	6, 7
$Cr(H_2O)_6^{3+}$	1.98	3/2	$3\text{–}5\times10^{-10}$	8, 7
$Fe(H_2O)_6^{3+}$	2.0	5/2	6×10^{-11}	7
$Fe(CN)_6^{3-}$	1.79	1/2	$10^{-12}\text{–}10^{-11}$	9
$Fe(H_2O)_6^{2+}$	3.4	1	$<1.4\times10^{-11}$	10
$Ce(H_2O)_9^{3+}$	6/7	5/2	1.1×10^{-13}	11
$Pr(H_2O)_9^{3+}$	4/5	4	0.7×10^{-13}	11
$Nd(H_2O)_9^{3+}$	8/11	9/2	1.5×10^{-13}	11
$Dy(H_2O)_9^{3+}$	4/3	15/2	3.5×10^{-13}	11
$Yb(H_2O)_9^{3+}$	8/7	7/2	1.6×10^{-13}	11
$Gd(H_2O)_n^{3+}$	2.0	7/2	$\geq1.2\times10^{-10}$	12

[a] For lanthanides (Ce^{3+}–Yb^{3+}): $S_R \equiv J$ is the quantum number of the ground-state total angular momentum, with $J = L - S$ in the first half of the series and $J = L + S$ in the second half of the series; and $\langle g_R \rangle \equiv g_J = \frac{3}{2} - [L(L+1) - S(S+1)]/2J(J+1)$ is the Landé g-factor, where L and S are quantum numbers for the orbital angular momentum and true spin, respectively.

[b] en=ethylenediamine, $NH_2CH_2CH_2NH_2$.

[c] References: 1. Friedman et al. (1979); 2. Connick and Fiat (1966); 3. Fiat et al. (1968); 4. Chmelnik and Fiat (1967); 5. Lewis et al. (1966); 6. Bertini et al. (1993); 7. McGarvey (1957); 8. Hausser and Noack (1964); 9. Bertini and Luchinat (1986) (at 77 K $T_{1,R} = 2.3\times10^{-11}$ s; Kulikov and Likhtenstein 1977); 10. Swift and Connick (1962); 11. Alsaadi et al. (1980); 12. Stephens and Grisham (1979).

10.6 PARAMAGNETIC T_1-RELAXATION ENHANCEMENTS BY HEISENBERG EXCHANGE

Spin–lattice relaxation is induced by Heisenberg spin exchange only if the paramagnetic species comes into direct contact with the spin label (see Section 9.6). Relaxation of the paramagnetic species is fast when $T_{1,R} << \tau_P$, where τ_P is the duration of collision between relaxant and spin label that we introduced in Section 9.9, and $T_{1,R}$ is the spin–lattice relaxation time of the paramagnetic species. The criterion for strong exchange is $J_{RL}^2 \tau_P^2 >> 1$ (cf. Eq. 9.92), where J_{RL} is the exchange constant between relaxant and spin label expressed in units of $\hbar$. From Eq. 9.149, we expect a collision lifetime of $\tau_P \approx 1\times10^{-10}$ s for diffusion-controlled collisions in water, where the viscosity is $\eta \approx 1$ mPa s ($\equiv 1$ cP) at 20°C. From Table 10.1, we see that the paramagnetic ions listed, except Mn^{2+}, Cu^{2+}, Cr^{3+} and Gd^{3+}, fulfill the condition for fast spin–lattice relaxation. Also, for molecular oxygen the spin–lattice relaxation time is $T_{1,R} \approx 5-10\times10^{-12}$ s, largely independent of solvent and viscosity (Teng et al. 2001).

For strong exchange with a fast-relaxing paramagnetic species, the enhancement in spin–lattice relaxation rate of the spin label $\left(T_{1,HE}^{-1} \equiv \tau_{HE}^{-1}\right)$ is given simply by the product of the collision rate constant, k_{RL}, and relaxant concentration, c_R. Note that the statistical factor that appears in Eq. 9.97 for exchange between like spins (viz., identical spin labels) is unity for unlike spins such as a spin label and a paramagnetic species. Specifically, the exchange frequency depends on the probability of exchange on collision p_{ex}, and a steric accessibility factor σ_{RL}:

$$\tau_{HE}^{-1} = p_{ex}\sigma_{RL}\tau_{coll}^{-1} = k_{RL}c_R \tag{10.17}$$

where τ_{coll}^{-1} is the frequency of collisions between spin label and relaxant. For strong exchange and fast paramagnetic relaxation, the probability of exchange on collision is $p_{ex} = 1$, because the paramagnetic species acts as part of the lattice. Under these circumstances, Heisenberg exchange contributes equally to T_1- and T_2-relaxation of the spin label, i.e., $\tau_{HE}^{-1} = T_{2,HE}^{-1} = T_{1,HE}^{-1}$, and is independent of the T_1-relaxation time of the paramagnetic species. Thus, saturation experiments are a much more sensitive way to determine paramagnetic relaxation enhancements than are linewidth measurements, because $T_1 >> T_2$ for spin labels. Most notably, we use progressive-saturation EPR for determining accessibilities of site-directed spin labels to different paramagnetic relaxants (see Sections 16.1 and 16.2 in

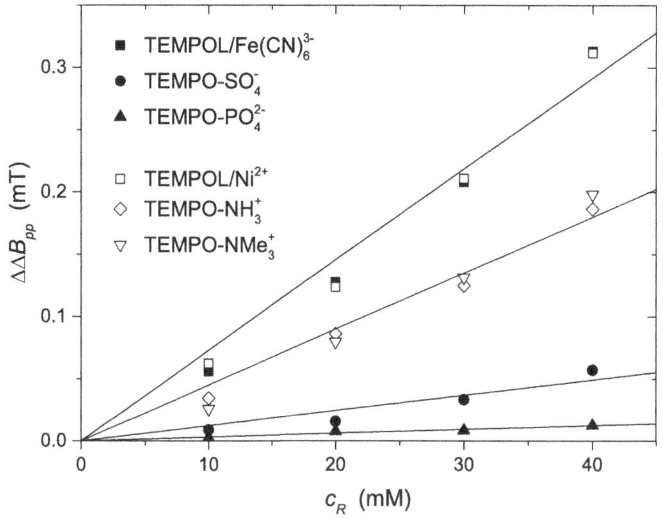

FIGURE 10.5 Line broadening, $\Delta\Delta B_{pp}$, by aqueous paramagnetic relaxants potassium ferricyanide ($K_3(FeCN)_6$, solid symbols) and nickel chloride ($NiCl_2$, open symbols), as function of concentration, c_R. Nitroxides: TEMPOL (squares), TEMPO-SO_4^- (circles), TEMPO-PO_4^{2-} (up triangles), TEMPO-NH_3^+ (diamonds) and TEMPO-NMe_3^+ (down triangles). (Data from Keith et al. 1977.)

Chapter 16). On the other hand, much of what we know about the EPR features of collision-induced exchange, and suitability of paramagnetic relaxants for accessibility studies in SDSL, comes from paramagnetic broadening of small spin labels in solution, particularly those such as TEMPONE where inhomogeneous broadening is small (Molin et al. 1980).

As an example, Figure 10.5 shows the concentration-dependent line broadening of different spin labels by ferricyanide (solid symbols) or Ni^{2+} ions (open symbols), which have short $T_{1,R}$-relaxation times. Exchange with ferricyanide is strong because it displays a strong linear increase with T/η, as the viscosity η of the solution decreases (Grebenshchikov et al. 1972b, cf. Section 9.15 and see later). Relaxation enhancement is much reduced for spin labels that have the same electrical charge as the relaxant ion. We get rate constants for spin exchange with different relaxants by measuring the line broadening and using the relation: $\tau_{HE}^{-1} = \frac{\sqrt{3}}{2}((2I+1)/2I)\gamma_e \Delta\Delta B_{pp}$, where $\Delta\Delta B_{pp}$ is the increase in first-derivative peak-to-peak linewidth, and I is the nitrogen nuclear spin. Rate constants for small molecules in solution that we get in this way are given in the tables of Appendix A10.4 at the end of this chapter.

Strong Heisenberg spin exchange is the dominant mechanism of spin-label relaxation by paramagnetic molecular oxygen (Hyde and Subczynski 1989). For oxygen, the distance of closest approach is $d_{RL} \approx 0.45$ nm (Windrem and Plachy 1980), and the diffusion coefficient in water at 37°C is $D_{O_2} = 2.8 \times 10^{-5}$ cm^2 s^{-1} with $\sigma_{RL} = 1$ (Hyde and Subczynski 1984). In fluid lipid-bilayer membranes, the diffusion coefficient is not much smaller: $D_{O_2} = 1-1.8 \times 10^{-5}$ cm^2 s^{-1} in dimyristoyl phosphatidylcholine at 40°C (Subczynski and Hyde 1981). Thus for oxygen, where exchange is diffusion-controlled (see Section 9.15), the bimolecular rate constant in Eq. 10.17 is in the region of $k_{RL} \approx 10^{10}$ M^{-1} s^{-1} (Subczynski and Hyde 1984).

10.7 PARAMAGNETIC ENHANCEMENT BY WEAK/INTERMEDIATE HEISENBERG EXCHANGE

Collisional exchange between spin labels themselves is nearly always strong (see Section 9.9 in Chapter 9). However, it can be weak when one of the partners is a paramagnetic ion complex, because the ligands may shield the interaction. For rare-earth ions, the $4f$-electrons are shielded additionally by the outer shells of $5s$- and $5p$-electrons. For weak exchange, we can use perturbation theory (Salikov et al. 1971a). The perturbing Hamiltonian is $\mathcal{H}'_{HE}(t) = J_{RL}\mathbf{S}_L \cdot \mathbf{S}_R(t)$, where J_{RL} is the exchange constant, $\mathbf{S}_L$ is the nitroxide spin, and $\mathbf{S}_R(t)$ is that of the relaxant, which is time-dependent because of its fast relaxation. This is scalar relaxation of the second type as treated by Abragam (1961); we give a simplified account here. From Eq. 5.49, the secular contribution to T_2-relaxation of the spin label is: $1/T_{2,L}^{sec} = \langle\Delta\omega^2\rangle_{HE}\tau_c = \frac{1}{3}J_{RL}^2 S_R(S_R+1)T_{1,R}$, where $T_{1,R}$ $(\equiv \tau_c)$ is the spin–lattice relaxation time of the relaxant, which is that appropriate to secular relaxation enhancement (i.e., coming from diagonal terms of the perturbation). The second moment $\langle\Delta\omega^2\rangle_{HE}$ of the perturbation comes from the corresponding expression for the hyperfine interaction in Section U.3.1 of Appendix U (see Eq. U.23). For fast paramagnetic relaxation enhancement, we have: $T_{1,L}^{-1} = T_{2,L}^{-1} = 2(T_{2,L}^{sec})^{-1}$ (cf. Eq. 5.51), which requires that $T_{1,R}^{-1} = T_{2,R}^{-1}$ for the relaxant. The relaxation enhancement of the spin label on collision with the relaxant is therefore:

$$\frac{1}{T_{1,L}} = \frac{1}{T_{2,L}} = \frac{2}{3}J_{RL}^2 S_R(S_R+1)T_{1,R} \tag{10.18}$$

From the longitudinal Bloch equation for the magnetization of an ensemble of spin labels, the probability of spin exchange is: $n/n_o = 1-\exp(-t_P/T_{1,L})$, for a collision of duration t_P (see Eq. 1.17). As for collisional exchange between spin labels (see Section 9.9), we must average over the exponential distribution $\exp(-t_P/\tau_P)$ of collision times. The probability of exchange on collision with the relaxant therefore becomes:

$$p_{ex} = \frac{1}{\tau_P}\int_0^\infty \left(1-\exp(-t_P/T_{1,L})\right)\exp(-t_P/\tau_P)\cdot dt_P = \frac{\tau_P/T_{1,L}}{1+\tau_P/T_{1,L}} \tag{10.19}$$

Substituting from Eq. 10.18 then gives (Salikov et al. 1971a):

$$p_{ex} = \frac{\frac{2}{3}J_{RL}^2 S_R(S_R+1)T_{1,R}\tau_P}{1+\frac{2}{3}J_{RL}^2 S_R(S_R+1)T_{1,R}\tau_P} \tag{10.20}$$

For $J_{RL}^2 S_R(S_R+1)T_{1,R}\tau_P >> 1$, we get $p_{ex} = 1$, which is the general condition for strong exchange between spin labels and a fast-relaxing paramagnetic species. On the other hand, for weak exchange, the exchange probability becomes $p_{ex} = \frac{2}{3}J_{RL}^2 S_R(S_R+1)T_{1,R}\tau_P << 1$.

TABLE 10.2 Exchange constants, J_{RL}, from collisional spin exchange between TEMPONE (or TEMPOL) nitroxide and paramagnetic relaxants R with spin–lattice relaxation times shorter than the duration of collision, i.e., $T_{1,R} < \tau_P$ (cf. Molin et al. 1980)

relaxant, R[a]	S_R	$\tau_{HE}^{-1}/c_R \times 10^{-9}\ (M^{-1}\ s^{-1})$	$T_{1,R}$ (s)	$J_{RL}\ (s^{-1})$	*Refs.*[b]
weak exchange[c]					
$Co(CH_3OH)_6^{2+}$	3/2	0.42	7×10^{-13}	2×10^{10}	1
$Co(acac)_2Py_2$[d]	3/2	0.38 ± 0.01	7×10^{-13}	2×10^{10}	2
$Ni(acac)_2$[d]	1	0.22 ± 0.005	$\sim 10^{-11}$	5×10^{9}	2
$Mn(acac)_3$[d]	2	0.52 ± 0.01	$10^{-11} - 10^{-12}$	$0.4–1 \times 10^{10}$	2
strong exchange[c]					
$Co(H_2O)_6^{2+}$	1/2	0.81 ± 0.07	1.5×10^{-12}	$>4 \times 10^{10}$	3
$CoPy_4Cl_4$	1/2	1.4	$\sim 10^{-12}$	$>7 \times 10^{10}$	1
$Ni(acac)_2(C_6H_5NH_2)_2$	1	1.0	$\sim 10^{-11}$	$>2 \times 10^{10}$	1
$Ni(en)_2(H_2O)_2^{2+}$	1	2.11 ± 0.07	4.3×10^{-12}	$>2 \times 10^{10}$	3
$Ni(H_2O)_6^{2+}$	1	1.86 ± 0.03	4.3×10^{-12}	$>2 \times 10^{10}$	4

[a] acac = acetylacetonate, $CH_3COCHCOCH_3$; en=ethylenediamine, $NH_2CH_2CH_2NH_2$; Py=pyridine, C_5H_5N.

[b] References: 1. Molin et al. (1980); 2. Skubnevskaya et al. (1970); 3. Anisimov et al. (1971); 4. Keith et al. (1977). See Table 10.1 for references to spin–lattice relaxation times $T_{1,R}$ of paramagnetic species.

[c] For weak exchange: $p_{ex} = \frac{2}{3} J_{RL}{}^2 S_R(S_R+1) T_{1,R}\tau_P$, and values of J_{RL} come from Eq. 10.21. For strong exchange: $p_{ex} = 1$ (see Eq. 10.20), and only lower estimates of J_{RL} are possible (see text).

[d] TEMPOL nitroxide (1-oxyl-2,2,6,6-tetramethylpiperidin-4-ol) in chloroform. All others are TEMPONE nitroxide (1-oxyl-2,2,6,6-tetramethylpiperidin-4-yl oxide) in water.

Table 10.2 gives rate constants, τ_{HE}^{-1}/c_R, for spin exchange with paramagnetic relaxants that have short spin–lattice relaxation times, $T_{1,R} < \tau_P$. We can group the relaxants into those for which exchange is strong or weak according to the criteria just given. To do this, we use Eq. 10.17 to estimate the probability of exchange $p_{ex}\sigma_{RL}$ by assuming that the collision rate constant is diffusion controlled, i.e., $\tau_{coll}^{-1}/c_R = k_{2,diff}$. We get $k_{2,diff}$ from Eq. 9.149 in Section 9.14, together with the Stokes-Einstein relation. Exchange is strong if $p_{ex}\sigma_{RL} \approx 1$; otherwise exchange is weak.

When exchange is weak, we can estimate the exchange constant J_{RL} from Eq. 10.20 by using the diffusion-controlled value for the duration of collisions τ_P from Eq. 9.154 in Section 9.15. This requires a value for $r_R + r_L$, the sum of the diffusion radii. We assume $r_R \approx r_L = 0.32$ nm, i.e., the value for DTBN (Plachy and Kivelson 1967). If we take the approximation for weak exchange, the rate constant becomes, from Eqs. 10.17 and 10.20:

$$\frac{\tau_{HE}^{-1}}{c_R} = \frac{2}{3} J_{RL}^2 S_R(S_R+1) T_{1,R}\sigma_{RL}\tau_P \frac{\tau_{coll}^{-1}}{c_R}$$

$$= \frac{8\pi}{9} 10^3 N_A J_{RL}^2 S_R(S_R+1) T_{1,R}\sigma_{RL}(r_R+r_L)^3 \qquad (10.21)$$

where we use diffusion-controlled values from Eqs. 9.149 and 9.154 to get the second equality on the right. This eliminates the viscosity, without using the Stokes-Einstein relation, but we still need $r_R + r_L$. Estimates of the exchange constant J_{RL} that we get from Eq. 10.21 (with $\sigma_{RL} = 1$) are listed for relaxants with weak exchange in Table 10.2, which includes values of $T_{1,R}$ assumed for each relaxant. For strong exchange, we can give only a lower estimate for the exchange constant, which we get from the condition: $\frac{2}{3} J_{RL}^2 S_R(S_R+1) T_{1,R}\tau_P > 1$ (see Table 10.2).

Molin et al. (1980) estimate values of J_{RL} comparable to those in Table 10.2 by using the model of diffusive passage (Waite 1958), instead of Eq. 9.154, to obtain τ_P. This method, which often is used to analyse radical recombination kinetics, requires an estimate for the spatial range of the Heisenberg exchange interaction.

As for collisional exchange between spin labels (see Section 9.15), we can get information on the strength of spin exchange with the relaxant from the dependence of the exchange rate constant τ_{HE}^{-1}/c_R on viscosity of the medium. For strong exchange and diffusion-controlled collisions: $\tau_{HE}^{-1}/c_R = k_{2,diff} = \frac{8}{3} \cdot 10^3 N_A k_B (T/\eta)$, from which we expect a linear dependence on T/η, with a gradient $\frac{8}{3} \cdot 10^3 N_A k_B = 2.22 \times 10^4$ Pa M^{-1} K^{-1}. Figure 10.6 shows the linear dependence of the rate constant for spin exchange of TEMPONE with the nickel ethylenediamine complex $Ni(en)_2^{2+}$ in aqueous glycerol on the ratio T/η (solid symbols). The gradient for $Ni(en)_2(H_2O)_2^{2+}$ is 1.6×10^4 Pa M^{-1} K^{-1}, which is comparable

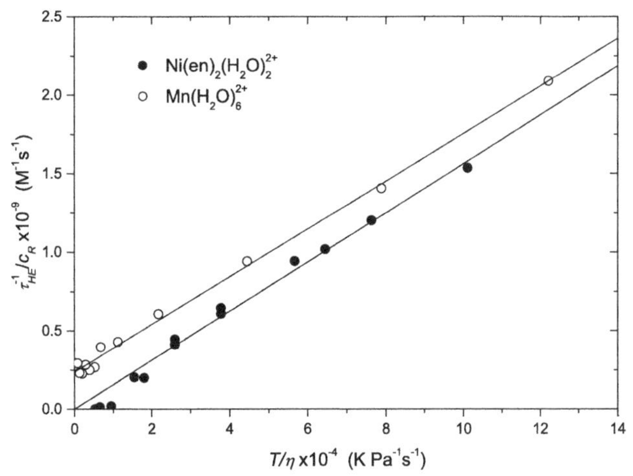

FIGURE 10.6 Dependence of rate constant τ_{HE}^{-1}/c_R for exchange broadening of TEMPONE by $Ni(NH_2CH_2CH_2NH_2)_2$ (solid circles) and Mn^{2+} (open circles) on ratio T/η of absolute temperature to solvent viscosity in water-glycerol mixtures. (Data from Anisimov et al. 1971.) Solid lines are linear regressions.

to the diffusion-controlled estimate. The data in Figure 10.6 therefore demonstrate strong exchange for this paramagnetic relaxant.

In water at 20°C, the diffusion-controlled rate constant for bimolecular collisions calculated as above is $k_{2,diff} = \frac{8}{3} \cdot 10^3 N_A k_B (T/\eta) = 6.50 \times 10^9\ M^{-1}\ s^{-1}$. This is considerably higher than the rate constants τ_{HE}^{-1}/c_R given in Table 10.2 for relaxants inducing strong spin exchange, where we expect $p_{ex} = 1$. We attribute the lower values to steric hindrance $(\sigma_{RL} < 1)$ of access to the localized unpaired electron in the nitroxide. Favouring this interpretation are the considerably higher rate constants for relaxation of the more delocalized radicals bis-diphenyl chromium and diphenylpicryl hydrazine (DPPH) (Skubnevskaya et al. 1966; Molin et al. 1980). Data in Figure 10.6 are for higher viscosities than those in Table 10.2. Note that τ_P increases with increasing viscosity (see Eq. 9.154), which promotes strong exchange. The gradient of the dependence of τ_{HE}^{-1}/c_R on T/η for relaxation enhancement of TEMPONE by $Fe(CN)_6^{3-}$ in glycerol–water mixtures of lower viscosities is 0.4×10^4 Pa M^{-1} K^{-1} (Grebenshchikov et al. 1972b). This is much below that for $Ni(en)_2^{2+}$ in Figure 10.6.

When the spin–lattice relaxation time of the paramagnetic species is long, i.e., $T_{1,R} >> \tau_P$, this does not affect the efficiency of exchange anymore. If the relaxant has a spin $S_R = \frac{1}{2}$, the probability of exchange on collision with a spin label is then given by Eq. 9.92, just as for collision between two spin labels. For relaxants with higher spins, $S_R > \frac{1}{2}$, the total spin of the spin label/relaxant pair becomes: $S = S_R \pm \frac{1}{2}$, on collision. The two total spin states are now split by an amount $J_{RL}\left(S_R + \frac{1}{2}\right)$, instead of simply by J_{RL} when $S_R = \frac{1}{2}$, as is required by the Landé interval rule. For total spin states with a common spin projection quantum number, M, as in the singlet–triplet case treated in Section 9.7, the exchange efficiency thus becomes: $\frac{1}{2} J_{RL}^2 \left(S_{RL} + \frac{1}{2}\right)^2 \tau_P^2 \Big/ \left(1 + J_{RL}^2 \left(S_{RL} + \frac{1}{2}\right)^2 \tau_P^2\right)$. For $S_R = \frac{1}{2}$, this is the sole contribution, but for $S_R > \frac{1}{2}$, there are other combinations of the individual spin projections that have common values of M. Allowing for these other exchange pathways, the probability of exchange on collision of spin label and relaxant becomes (Salikov et al. 1971a; Zitserman 1971):

$$p_{ex} = \frac{2}{3} \frac{S_R(S_R+1) J_{RL}^2 \tau_P^2}{1 + \left(S_R + \frac{1}{2}\right)^2 J_{RL}^2 \tau_P^2} \tag{10.22}$$

which reduces to Eq. 9.92 for $S_R = \frac{1}{2}$. For strong exchange $\left(S_R + \frac{1}{2}\right)^2 J_{RL}^2 \tau_P^2 >> 1$, the exchange efficiency does not reach unity for relaxants with long $T_{1,R}$. Instead, the probability of exchange is: $p_{ex} = \frac{2}{3} S_R(S_R+1) / \left(S_R + \frac{1}{2}\right)^2$ because the individual spin projections are conserved on collision. Table 10.3 lists the exchange probabilities for collisions with strong exchange between a nitroxide $\left(S_L = \frac{1}{2}\right)$ and a slowly relaxing relaxant of spin S_R. The exchange probability increases slowly from $p_{ex} = \frac{1}{2}$ as S_R increases. For weak exchange, on the other hand, the probability of exchange becomes: $p_{ex} = \frac{2}{3} J_{RL}^2 S_R(S_R+1) \tau_P^2$ and depends on both the exchange constant J_{RL} and duration of collision τ_P.

Table 10.4 lists rate constants, τ_{HE}^{-1}/c_R, for spin exchange with paramagnetic relaxants that have long spin–lattice relaxation

TABLE 10.3 Probability p_{ex} of spin exchange induced by strong exchange $(J_{RL}\tau_P >> 1)$ on collision with relaxants of spin S_R and long $(T_{1,R} >> \tau_P)$ or short $(T_{1,R} << \tau_P)$ spin–lattice relaxation times[a]

	p_{ex}	
S_R	$(T_{1,R} >> \tau_P)$[b]	$(T_{1,R} << \tau_P)$[c]
1/2	1/2	1
1	16/27	1
3/2	5/8	1
2	16/25	1
5/2	35/54	1
7/2	21/32	1

[a] τ_P is the duration of a collision.
[b] From Eq. 10.22.
[c] From Eq. 10.20.

TABLE 10.4 Rate constants τ_{HE}^{-1}/c_R for collisional spin exchange between TEMPONE nitroxide and aqueous paramagnetic relaxants R with spin–lattice relaxation times longer than the duration of collision, i.e., $T_{1,R} > \tau_P$

relaxant, R	S_R	$\tau_{HE}^{-1}/c_R \times 10^{-9}\ (M^{-1}\ s^{-1})$	$k_{coll} \times 10^{-9}\ (M^{-1} s^{-1})$[a]	*Refs.*[b]
$Cu(H_2O)_6^{2+}$	1/2	2.66 ± 0.02	5.32 ± 0.04	1
$Cu(NH_2CH_2CH_2NH_2)_2(H_2O)_2^{2+}$	1/2	2.61 ± 0.07	5.2 ± 0.1	2
$CuN(CH_2CH_2NH_2)_3OH^+$	1/2	2.5	5.0	3
$Cu(NH_2CH_2COO)_2$	1/2	2.0	4.0	3
$Cu(NH_2CH_2CH_2N(C_2H_5)_2)_2^{2+}$	1/2	1.0	2.0	3
Cu^{2+} etioporphyrin/pyridine	1/2	2.41 ± 0.12	4.8 ± 0.2	4
$Mn(H_2O)_6^{2+}$	5/2	2.07 ± 0.07	3.2 ± 0.1	2, 5
$Cr(H_2O)_6^{3+}$	3/2	1.24 ± 0.07	2.0 ± 0.1	2
Fe^{3+} haemin/pyridine	1/2	2.40 ± 0.12	4.8 ± 0.2	4
Fe^{3+} etioporphyrin/pyridine	1/2	2.57 ± 0.13	5.1 ± 0.2	4
VO^{2+} etioporphyrin/pyridine	1/2	1.53 ± 0.08	3.1 ± 0.2	4
$VOSO_4$	1/2	1.2 ± 0.2	2.4 ± 0.4	3

[a] Rate constant for bimolecular collisions: $k_{coll} = \tau_{coll}^{-1}/c_R$. Calculated from Eq. 10.17 with $\sigma_{RL} = 1$ and assuming strong exchange, for which: $p_{ex} = \frac{2}{3} S_R(S_R+1)/\left(S_R+\frac{1}{2}\right)^2$; for weak exchange: $p_{ex} = \frac{2}{3} J_{RL}^2 S_R(S_R+1)\tau_P^2$ (see Eq. 10.22).
[b] References: 1. Keith et al. (1977); 2. Anisimov et al. (1971); 3. Molin et al. (1980); 4. Grebenshchikov et al. (1972a); 5. Salikhov et al. (1971b).

times, $T_{1,R} > \tau_P$. This table includes estimates of the rate constants for bimolecular collisions, $k_{2,coll} \equiv \tau_{coll}^{-1}/c_R$, that we obtain by assuming exchange is strong (cf. Eq. 10.22) and the steric factor is $\sigma_{RL} \approx 1$ in Eq. 10.17. Several of these values approach the estimate for diffusion-controlled collisions in water at 20°C: $k_{2,diff} = 6.50 \times 10^9\ M^{-1}\ s^{-1}$, which we get when $r_R \approx r_L$. Values for relaxants with larger ligands progressively fall well below this value and may correspond to weak or sterically hindered exchange.

Figure 10.6 includes data on T/η-dependence of the rate constant for spin exchange of TEMPONE with Mn^{2+} ions, which relax slowly: $T_{1,R} > \tau_P$ (open symbols). The dependence is linear, with a gradient similar to that for the fast-relaxing $Ni(en)_2(H_2O)_2^{2+}$ complex, again suggesting strong exchange. In fact, correcting for $p_{ex} < 1$ gives a gradient for the collision rate constant with Mn^{2+} of 2.3×10^4 Pa M^{-1} K^{-1}, equal to the prediction for diffusion control. However, we get a non-zero intercept in Figure 10.6 for slow-relaxing $Mn(H_2O)_6^{2+}$. This results from dipolar contributions to the line broadening at high viscosities, which depend on τ_P (Anisimov et al. 1971; cf. also Figure 9.25 in Section 9.16). We find the same feature in the viscosity dependence of broadening by the $Cu(en)_2(H_2O)_2^{2+}$ complex that also relaxes slowly (Salikov et al. 1971a). For $Ni(en)_2(H_2O)_2^{2+}$, on the other hand, dipolar broadening is limited by the short $T_{1,R}$, not by the longer τ_P. This gives a zero intercept in Figure 10.6 that is characteristic of exchange interaction alone, for this fast-relaxing complex.

10.8 PARAMAGNETIC ENHANCEMENT BY HEISENBERG EXCHANGE BETWEEN CHARGED SPECIES

For charged systems, the collision rate constant $k_{coll} \equiv \sigma_{RL}\left(\tau_{coll}^{-1}/c_R\right)$ in Eq. 10.17 is given by the Smoluchowski solution of the diffusion equation for bimolecular collisions (see Eq. 9.149), modified by an electrostatic term:

$$k_{coll} = 4\pi\sigma_{RL} d_{RL} D_T \cdot \exp\left(-z_R e \psi_L / k_B T\right) \tag{10.23}$$

where D_T is the translational diffusion coefficient of the relaxant (that of the spin-labelled system is assumed to be negligible by comparison), d_{RL} is the interaction distance between relaxant and spin label, $z_R e$ is the electrical charge on the relaxant and ψ_L is the electrostatic surface potential of the spin-labelled system. Enhancements in relaxation rate by Heisenberg spin exchange are therefore determined by the diffusion-concentration product, plus any electrostatic interactions. In the case of membranes, the local value of the product $D_T(z)c_R(z)$ at position z along the membrane normal is the determining factor.

Figure 10.5, given previously, includes data for paramagnetic broadening of charged spin labels by relaxants bearing an electrical charge with the same sign as that of the spin label. In each case, the relaxation enhancement is less than that of an uncharged spin label, because of electrostatic repulsion between charged relaxant and charged spin label (cf. Eq. 10.23). The effect is greater with the ferricyanide ion, which has a formal charge of −3, than with the nickel ion that has a charge of +2. Figure 10.7 shows the dependence on ionic strength of the rate constant for exchange broadening of the positively-charged TEMPO-NH_3^+ spin label by the negatively-charged ferricyanide ion. The rate constant decreases strongly with increasing ionic strength, exhibiting classical screening of the electrostatic attraction between the two oppositely charged species. At high ionic strength, the exchange rate constant is reduced to the value for the uncharged spin label TEMPONE. We can estimate the ionic-strength dependence of the electrostatic potential of an ion from Debye–Hückel theory for dilute electrolytes. This gives a collision rate constant (see, e.g., Amdur and Hammes 1966):

$$\ln k_{coll}/k_{coll}^o = -\frac{z_R z_L e^2}{4\pi\varepsilon_r\varepsilon_o d_{RL} k_B T} + \frac{z_R z_L e^2 \kappa}{4\pi\varepsilon_r\varepsilon_o k_B T(1+\kappa d_{RL})} \tag{10.24}$$

where $z_L e$ is the electrical charge on the spin label, ε_r is the relative permittivity (effective dielectric constant) of the electrolyte solution, and $\kappa = \sqrt{2N_A 10^3 e^2 I/(\varepsilon_r\varepsilon_o k_B T)}$ is the inverse Debye screening length for an electrolyte of ionic strength $I = \frac{1}{2}\sum z_i^2 c_{M,i}$ that is composed of ions i with charge $z_i e$ at molar concentration $c_{M,i}$. In Eq. 10.24, k_{coll}^o is the value of rate constant k_{coll} without electrostatics. Empirically, we can use Debye–Hückel theory for fitting ionic strength dependences at higher electrolyte concentrations, if we take the distance of closest approach d_{RL} as an adjustable parameter (Robinson and Stokes 1955). We illustrate this by the solid line in Figure 10.7.

Table A10.1, in Appendix A10.4 at the end of this chapter, lists spin-exchange rate constants for different uncharged and

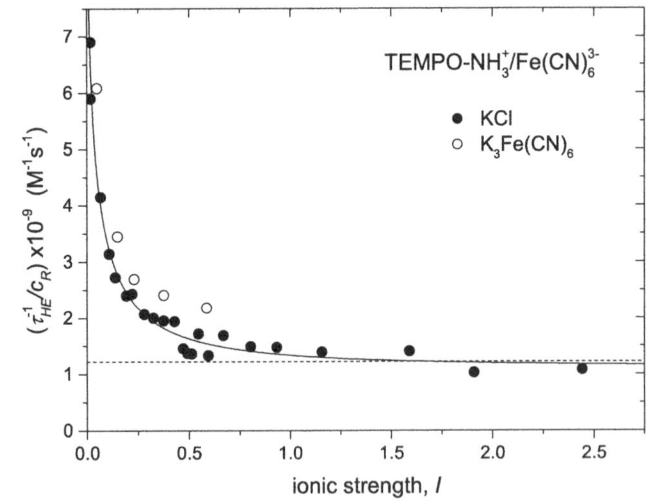

FIGURE 10.7 Ionic strength dependence of rate constant τ_{HE}^{-1}/c_R for exchange broadening of TEMPO-NH_3^+ by ferricyanide $Fe(CN)_6^{3-}$. *Solid circles*: in KCl solutions; *open circles*: with increasing concentrations of $K_3Fe(CN)_6$. (Data from Grebenshchikov et al. 1972b.) Solid line: fit with $\exp\left[-m\sqrt{I}/\left(1+a\sqrt{I}\right)\right]$ dependence, where I is KCl ionic strength and m, a are fitting parameters (Eq. 10.24). Dashed line: rate constant for TEMPONE.

oppositely-charged nitroxides with the charged paramagnetic relaxants ferricyanide $Fe(CN)_6^{3-}$, bis-diphenyl chromium $Cr(C_6H_5 - C_6H_5)_2^+$ and nickel ions Ni^{2+}. Correspondingly, Table A10.2 in Appendix A10.4 lists spin-exchange constants for peroxylamine disulphonate $(SO_3^-)_2NO$, uncharged TEMPONE and TEMPO, and TEMPO-NH_3^+ nitroxides with different paramagnetic metal ion complexes. We see the effects of electrostatic attraction and repulsion clearly in the values of these rate constants.

10.9 PARAMAGNETIC ENHANCEMENT BY STATIC MAGNETIC DIPOLE–DIPOLE INTERACTION

For paramagnetic ions, unlike molecular oxygen, magnetic dipole–dipole interactions – both static and dynamic – can make appreciable contributions to spin-label relaxation enhancements. How effective this mechanism is depends on the spin, magnetic moment and T_1-relaxation time of the paramagnetic ion.

Relaxation enhancement by magnetic dipole interactions between spin label and relaxant depends on their distance apart, and unlike Heisenberg exchange, does not require mutual contact. The static dipolar mechanism dominates over the dynamic mechanism when spin–lattice relaxation of the paramagnetic relaxant is much faster than modulation of the dipolar interaction by translational diffusion of spin label and relaxant. For this, we need $T_{1,R} << \tau_D$, where τ_D is the characteristic dipolar correlation time for translational diffusion that we introduced in Section 9.5.2, and $T_{1,R}$ is the T_1-relaxation time of the paramagnetic relaxant. Fast, random spin flips of the relaxant switch the local dipolar field experienced by the spin label. As we see in the next paragraph, this gives an exponential correlation function, as for rotational diffusion, but with the correlation time equal to $T_{1,R}$. For frozen solutions, this static mechanism is the only source of dipolar relaxation enhancement.

The local field at the spin label is defined by: $\mathcal{H}_{dd} = -\boldsymbol{\mu}_L \cdot \mathbf{B}_{loc}$ where $\mathcal{H}_{dd}$ is the dipolar Hamiltonian given by Eqs. 9.5–9.9 and $\boldsymbol{\mu}_L$ is the magnetic moment of the spin label L. Assume, for simplicity, a two-level system for the paramagnetic relaxant (see Figure 10.8). The relaxant produces a local dipolar field $+B_{z,loc}$ at the spin label when in the upper level, and a local field of $-B_{z,loc}$ when in the lower level. An ensemble of relaxants has relative population $\frac{1}{2}(1+n)$ in the upper level and $\frac{1}{2}(1-n)$ in the lower level. Let all relaxants be initially in the upper level, i.e., $n=1$ at time $t=0$. The local field at the spin label is then $B_{loc}(0) = +B_{z,loc}$. After time t, part of the upper population has relaxed to the lower level, and the net local field is then: $B_{loc}(t) = B_{z,loc} \times n(t)$. We know from Chapter 1 that the population difference in the ensemble decreases exponentially by spin–lattice relaxation: $n(0) - n(t) = n(0)\exp(-t/T_{1,R})$ (see Eq. 1.17). The autocorrelation function for the local field is therefore also exponential:

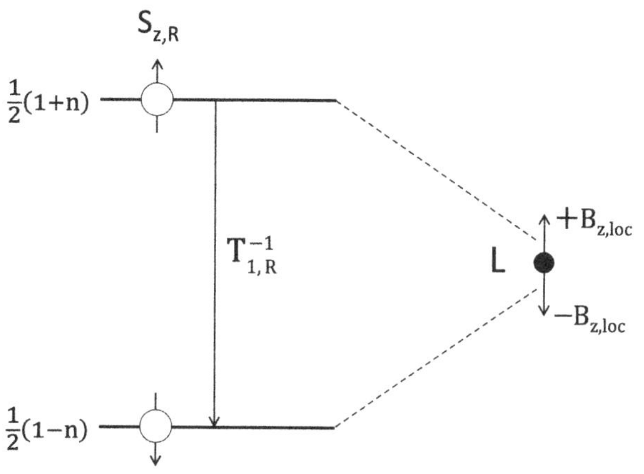

FIGURE 10.8 Relaxation enhancement of spin label L by fast relaxation of paramagnetic relaxant R, at rate $T_{1,R}^{-1}$. Relaxant produces local dipolar field $\pm B_{z,loc}$ at the spin label. Relative population difference between upper and lower levels of the relaxant is n.

$$G(t) \equiv \langle B_{loc}(0)B_{loc}(t)\rangle = (B_{z,loc})^2 \exp(-t/T_{1,R}) \tag{10.25}$$

and the spectral density is correspondingly Lorentzian (see also Slichter 1978).

First, we treat the situation in which spin label L and a single relaxant R are at a fixed distance r_{RL} apart, e.g., attached to the same macromolecule (see left-hand panel in Figure 10.9). The dipolar relaxation rate for *unlike* spins is given by Eq. 9.34 in Section 9.5; the T_1-relaxation rate of the spin label is thus:

$$\frac{1}{T_{1dd}(L)} = \left(\frac{\mu_o}{4\pi}\right)^2 |\mu_R|^2 \gamma_e^2 \left[\tfrac{1}{12} J^{(0)}(\omega_L - \omega_R) + \tfrac{3}{2} J^{(1)}(\omega_L) + \tfrac{3}{4} J^{(2)}(\omega_L + \omega_R)\right] \tag{10.26}$$

where $|\mu_R|^2 = g_R^2 \beta_e^2 S_R(S_R+1)$ is the square of the magnetic moment of the paramagnetic relaxant; ω_L, ω_R are the Larmor frequencies of spin label and relaxant, respectively, and γ_e is the electron gyromagnetic ratio. For a random distribution of spin-label/relaxant pairs, the spectral densities $J^{(M)}(\omega)$ are formally

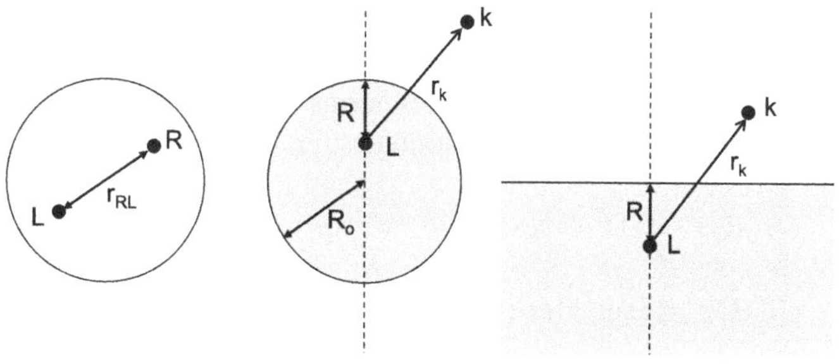

FIGURE 10.9 Spatial relations for relaxation enhancement of spin label L by paramagnetic relaxants. *Left*: fixed relative position r_{RL} of single spin label and relaxant R bound to a macromolecule. *Middle*: spin label at immersion depth R in macromolecule surrounded by solution of relaxants, k. *Right*: spin label at immersion depth R within membrane dispersed in solution of relaxants, k.

equivalent to those given in Section 9.5.1 for isotropic rotational diffusion. This is because we need the same ensemble average over the angular functions $F^{(M)}(0)$ given in Eqs. 9.24–9.26. From Eq. 9.40, we then arrive at the classic Solomon–Bloembergen equation for relaxation enhancement by a single paramagnetic ion (Bloembergen 1949; Solomon 1955):

$$\frac{1}{T_{1,dd}(\text{static})} = \frac{2}{15}\left(\frac{\mu_o}{4\pi}\right)^2 \frac{|\mu_R|^2 \gamma_e^2}{r_{RL}^6} T_{1,R} \times f_1(\omega_L,\omega_R) \quad (10.27)$$

where:

$$f_1(\omega_L,\omega_R) \equiv \frac{1}{1+(\omega_L-\omega_R)^2 T_{1,R}^2} + \frac{3}{1+\omega_L^2 T_{1,R}^2} + \frac{6}{1+(\omega_L+\omega_R)^2 T_{1,R}^2} \quad (10.28)$$

This applies to a random ensemble of inter-dipole vectors. The formal equivalence with isotropic rotational diffusion arises because both have exponential correlation functions. For Ni^{2+} and Co^{2+} ions, $\omega^2 T_{1,R}^2 << 1$ and all three terms in Eq. 10.28 contribute, whereas for Cu^{2+} and Mn^{2+}, $\omega^2 T_{1,R}^2 >> 1$ and the first term in $f_1(\omega_L,\omega_R)$ dominates at 9 GHz.

We now turn to relaxation induced by a solution of paramagnetic species. In this case, the distribution of relaxant relative to the spin label location is important. Typically, we want to determine the immersion depth, R, of the spin label in a membrane or buried within a macromolecule (see middle and right-hand panels in Figure 10.9). We therefore must take specific account of angular relationships. Summing Eq. 10.26 over the distribution of paramagnetic relaxants, k, the net enhancement is (Livshits et al. 2001):

$$\frac{1}{T_{1,dd}(\text{static})} = \left(\frac{\mu_o}{4\pi}\right)^2 |\mu_R|^2 \gamma_e^2 \sum_k \frac{|F_o(\Theta_k)|^2 j(\omega_L-\omega_R) + |F_1(\Theta_k)|^2 j(\omega_L) + |F_2(\Theta_k)|^2 j(\omega_L+\omega_R)}{r_k^6} \quad (10.29)$$

where r_k is the separation of spin label from relaxant, and Θ_k is the angle between the inter-dipole vector $\mathbf{r}_k$ and magnetic field direction. The angular-dependent terms $F_q(\Theta_k)$ are related to absolute values of spherical harmonics:

$$|F_0(\Theta)|^2 = \tfrac{1}{6}\left(1-3\cos^2\Theta\right)^2 \quad (10.30)$$

$$|F_1(\Theta)|^2 = 3\sin^2\Theta\cos^2\Theta \quad (10.31)$$

$$|F_2(\Theta)|^2 = \tfrac{3}{2}\sin^4\Theta \quad (10.32)$$

where the reduced spectral densities are defined by:

$$j(\omega) = \frac{T_{1,R}}{1+\omega^2 T_{1,R}} \quad (10.33)$$

The spin–lattice relaxation time of the paramagnetic relaxant $T_{1,R}$ is assumed to be sufficiently short that it also determines T_2, i.e., $T_{2,R} = T_{1,R}$.

For the moment, we neglect the angular dependence in Eq. 10.29, relative to the steep distance dependence of $1/r^6$ (cf. Hyde and Dalton 1979). Integrating over the paramagnetic relaxant distribution then gives the concentration dependence of the relaxation enhancement (cf. Eq. 10.17), in terms of the distance of closest approach, R. For a membrane-embedded spin label (see right-hand panel in Figure 10.9), area integration over a surface distribution of relaxant gives:

$$\int_A \frac{dA}{r^6} = \int_0^\infty \frac{2\pi y dy}{\left(y^2+R^2\right)^3} = \frac{\pi}{2}\frac{1}{R^4} \quad (10.34)$$

and volume integration for a bulk relaxant gives:

$$\int_V \frac{dV}{r^6} = \int_R^\infty dz \int_0^\infty \frac{2\pi y dy}{(y^2+z^2)^3} = \frac{\pi}{6}\frac{1}{R^3} \quad (10.35)$$

where z lies along the bilayer normal, and y is parallel to the plane of the membrane. Quite generally, the exponent of the R-dependence depends on the dimensionality of the relaxant distribution (Páli et al. 1992).

We now return to the angular dependence in Eq. 10.29. Using polar coordinates defined in Figure 10.10, the angle Θ_k comes from the cosine law:

$$\cos\Theta_k = \cos\theta_k\cos\theta_o + \sin\theta_k\sin\theta_o\cos\phi_k \quad (10.36)$$

where θ_o and θ_k are, respectively, the angles that the magnetic field and interspin vector $\mathbf{r}_k$ make with the membrane normal, and ϕ_k is the azimuthal orientation of $\mathbf{r}_k$ relative to the magnetic field direction. For a bulk solution of relaxants, the volume element dV for integration is that defined by the first equality

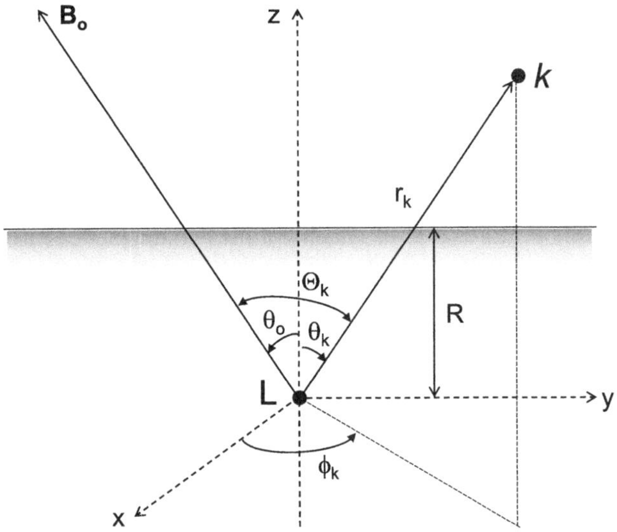

FIGURE 10.10 Angular relations for relaxation enhancement of membrane-embedded spin labels L by bathing solution of paramagnetic relaxants, k, at polar locations (r_k,θ_k,ϕ_k). Static magnetic field $\mathbf{B}_o$ is at angle θ_o to the membrane normal.

in Eq. 10.35, except that now we must replace the factor of 2π by integration over ϕ_k from 0 to 2π. We also need the relation: $\cos^2\theta_k = y^2/(y^2+z^2)$ to implement Eq. 10.36. Integrating over the distribution of relaxants then gives values of T_{1dd}^{-1}(static) that depend on the magnetic field orientation θ_o. The $F_q(\Theta_k)$ terms that we need in Eq. 10.29 then become (Livshits et al. 2001):

$$\int_V \frac{(1-3\cos^2\Theta)^2}{r^6}dV = \frac{\pi}{4R^3}\left(\cos^2\theta_o + \frac{3}{8}\sin^4\theta_o\right) \quad (10.37)$$

$$\int_V \frac{\sin^2\Theta\cos^2\Theta}{r^6}dV = \frac{\pi}{36R^3}\left(1-\frac{3}{8}\sin^4\theta_o\right) \quad (10.38)$$

$$\int_V \frac{\sin^4\Theta}{r^6}dV = \frac{\pi}{36R^3}\left(4-3\cos^2\theta_o + \frac{3}{8}\sin^4\theta_o\right) \quad (10.39)$$

We get analogous expressions for surface distributions of paramagnetic ions.

For macroscopically unoriented membrane dispersions, different parts of the powder pattern will saturate differently, depending on the orientation θ_o of the membrane to the static magnetic field $(0 \le \theta_o \le \pi/2)$. However, we see from Eqs. 10.37 to 10.39 that the residual angular dependence of T_{1dd}^{-1}(static) on θ_o is much weaker than its original dependence on Θ in Eq. 10.29. Therefore, for reasonable estimates of the saturation behaviour of the integrated spin-label EPR intensity, we can average T_{1dd}^{-1}(static) over θ_o. The resulting angular-independent effective value of T_{1dd}^{-1}(static) for a bulk distribution of relaxants is (Livshits et al. 2001):

$$\frac{1}{T_{1,dd}(\text{static})} = \frac{\pi}{45}\left(\frac{\mu_o}{4\pi}\right)^2 \frac{|\mu_R|^2\gamma_e^2}{R^3} T_{1,R}n_R \times f_1(\omega_L,\omega_R) \quad (10.40)$$

where n_R is the paramagnetic-ion concentration, expressed in ions/m^3, and $f_1(\omega_L,\omega_R)$ is again given by Eq. 10.28. When we have a surface distribution of relaxants, $1/R^3$ in Eq. 10.40 is replaced by $3K_s/R^4$, where K_s is a surface binding constant and the surface concentration is $n_{R,s} = K_s n_R$ in ions/m^2. This is exactly the relation between Eqs. 10.34 and 10.35.

Table 10.5 gives numerical estimates of the static dipolar T_1-relaxation enhancements for a spin label situated at $R = 1$ nm in lipid membranes that are immersed in a 30 mM paramagnetic ion solution. For ions with $S_R > \frac{1}{2}$, both estimates of T_{1dd}^{-1} and experimental values of $T_{1,R}$ depend on how we treat zero-field splittings (Livshits et al. 2001). Spectral densities (Eq. 10.28) are calculated as previously for NMR relaxation enhancements (Lindner 1965; Banci et al. 1991). For Ni^{2+}, we use a zero-field splitting of $D = 2.5\,\text{cm}^{-1}$ and average the spin-label relaxation enhancements over the angle between the principal axis of the Ni^{2+}-aquo complex and the interspin vector (Lindner 1965). The increased effective values of ω_R reduce the relaxation enhancement by Ni^{2+}. For high-spin Co^{2+} in octahedral or tetragonally distorted hexa-aquo complexes, the combined effect of large ligand-field splitting and spin-orbit coupling leaves an isolated Kramers doublet as ground state. As already noted, the total spin of Co^{2+} reduces to an effective spin $S_R = \frac{1}{2}$ with g-value $g_R = 4.33$, and no zero-field splitting (see e.g. Abragam and Bleaney 1970). For Mn^{2+}, the zero-field splitting of aquo-complexes is small, and its effect on NMR relaxation dispersion is insignificant at magnetic fields corresponding to 9 GHz EPR (Banci et al. 1986). Therefore we neglect zero-field splitting, which means that enhancements for Mn^{2+} in Table 10.5 are upper estimates. For paramagnetic ions other than Mn^{2+}, we predict rather small values of T_{1dd}^{-1}(static), in the order $Ni^{2+} > Co^{2+} \approx Cu^{2+}$. On the other hand, Mn^{2+} has both long $T_{1,R}$ and a high spin S_R, which give rise to efficient paramagnetic relaxation enhancement. However, Mn^{2+} has an EPR spectrum easily visible at room temperature, which complicates analysis of the saturation behaviour of the spin labels. For strongly absorbed paramagnetic cations, the values of $T_{1,dd}^{-1}$(static) can be increased by up to a factor of ten, because of the smaller average distance from the spin label.

TABLE 10.5 Predicted enhancements in T_1- and T_2-relaxation rates of spin labels in membranes by static dipolar interactions with 30 mM aqueous paramagnetic ions, and R = 1 nm in Eqs. 10.40, 10.28 and Eqs. 10.42, 10.43 for T_{1dd}^{-1} and T_{2dd}^{-1}, respectively (Livshits et al. 2001)

		$T_{1dd}^{-1}(s^{-1})$		$T_{2dd}^{-1}(s^{-1})$	
ion, R[a]	$T_{1,R}$ (s)	*volume*	*surface*[b]	*volume*	*surface*[b]
$Ni(H_2O)_6^{2+}$	3×10^{-12}	0.8×10^4	3.9×10^4	0.9×10^4	4.3×10^4
	5×10^{-12}	1.1×10^4	5.4×10^4	1.3×10^4	6.6×10^4
$Co(H_2O)_6^{2+}$	$\approx10^{-12}$	4.6×10^3	2.2×10^4	4.6×10^3	2.2×10^4
$Cu(H_2O)_6^{2+}$	1×10^{-9}	4×10^3	2.5×10^4	2.4×10^5	1.5×10^6
	3×10^{-9}	1.4×10^3	0.9×10^4	7.3×10^5	4.6×10^6
$Mn(H_2O)_6^{2+}$	1×10^{-9}	1×10^6	1.1×10^7	2.8×10^6	3.1×10^7
	3.5×10^{-9}	1.5×10^6	1.6×10^7	8.9×10^6	9.6×10^7
$Dy(H_2O)_n^{3+}$	3.5×10^{-13}	1.3×10^4	1.9×10^5	1.3×10^4	1.9×10^5
	8×10^{-13}	3.0×10^4	4.5×10^5	3.0×10^4	4.5×10^5

[a] See Table 10.1 for spins S_R, g-values $\langle g_R\rangle$ and references for spin–lattice relaxation times $T_{1,R}$, of paramagnetic ions. Differences in $T_{1,R}$ between aqueous and surface-adsorbed ions are neglected.

[b] Surface concentrations calculated from intrinsic binding constants (i.e., for an uncharged phospholipid membrane): $n_{R,s} = 2.9\times10^{16}$, 2.9×10^{16}, 6.5×10^{16} m^{-2} for Ni^{2+}, Co^{2+}, Mn^{2+} (McLaughlin et al. 1978); $n_{R,s} = 3.8\times10^{16}$, 9×10^{16} m^{-2} for Cu^{2+}, Dy^{3+} (Livshits et al. 2001). See also Marsh (2013).

Static dipolar interactions modulated by fast relaxation of the paramagnetic ion contribute also to T_2-relaxation. From Eq. 9.35 for *unlike* spins, the transverse relaxation rate is:

$$\frac{1}{T_{2dd}(L)}=\left(\frac{\mu_o}{4\pi}\right)^2|\mu_R|^2\gamma_e^2\times\left[\tfrac{1}{6}J^{(0)}(0)+\tfrac{1}{24}J^{(0)}(\omega_m-\omega_{m'})+\tfrac{3}{4}J^{(1)}(\omega_m)+\tfrac{3}{2}J^{(1)}(\omega_{m'})+\tfrac{3}{8}J^{(2)}(\omega_m+\omega_{m'})\right] \quad (10.41)$$

These contributions influence both progressive saturation and also linewidths. In standard Leigh theory (Leigh Jr. 1970), the strong angular dependence of the dipolar relaxation for two isolated spins results in practically complete quenching of all resonances other than those for spin pairs oriented at the magic angle, where $1-3\cos^2\Theta_k=0$. This lets us calibrate the separation of spin label and paramagnetic ion by using the central EPR line height (Leigh Jr. 1970). However, integration over a *distribution* of paramagnetic ions (cf. above), yields a much weaker angular dependence and results in line broadening instead of amplitude quenching. The corresponding effective transverse relaxation rate $T_{2,dd}^{-1}(\text{static})$ for a bulk distribution of ions is (Livshits et al. 2001):

$$\frac{1}{T_{2,dd}(\text{static})}=\frac{\pi}{90}\left(\frac{\mu_o}{4\pi}\right)^2\frac{|\mu_R|^2\gamma_e^2}{R^3}T_{1,R}c_R\times f_2(\omega_L,\omega_R) \quad (10.42)$$

$$f_2(\omega_L,\omega_R)=4+\frac{1}{1+(\omega_L-\omega_R)^2T_{1,R}^2}+\frac{3}{1+\omega_L^2T_{1,R}^2}+\frac{6}{1+\omega_R^2T_{1,R}^2}+\frac{6}{1+(\omega_L+\omega_R)^2T_{1,R}^2} \quad (10.43)$$

In the case of a surface distribution of relaxants, Eq. 10.42 is modified by the same substitution as that in Eq. 10.40 for the T_1-relaxation rate. For Cu^{2+} and Mn^{2+} ions $(T_{1,R}\approx 10^{-9}\text{ s})$, the first term dominates in Eq. 10.43, giving $f_2(\omega_L,\omega_R)\approx 4$. For Ni^{2+}, Co^{2+} and Dy^{3+} ions $(T_{1,R}<<<10^{-9}\text{ s})$, on the other hand, all terms in the spectral density contribute and $f_2(\omega_L,\omega_R)\approx 20$.

Table 10.5 includes numerical estimates of $T_{2,dd}^{-1}(\text{static})$, again for $R=1$ nm and 30 mM bulk concentration of paramagnetic ions. Values for Ni^{2+}, Co^{2+} and Dy^{3+} are comparable to those for $T_{1,dd}^{-1}(\text{static})$, whereas for Cu^{2+} (and to a lesser extent for Mn^{2+}), $T_{2,dd}^{-1}(\text{static})>>T_{1,dd}^{-1}(\text{static})$. For Cu^{2+} ions, $T_{2,dd}^{-1}$ (static) is 50 times higher than for Ni^{2+} ions, and in turn is 10–20 times lower than for Mn^{2+} ions.

10.10 PARAMAGNETIC ENHANCEMENT BY DYNAMIC MAGNETIC DIPOLAR INTERACTION

If translational diffusion is fast, spin-label relaxation is induced by modulation of the dipolar interaction that comes from relative motion of spin label and paramagnetic ion. This is the translational dipolar relaxation mechanism that we treated in Section 9.5.2. The criterion for the dynamic mechanism to dominate over the static mechanism is $\tau_D^{-1}\equiv 2D_T/d_{RL}^2>>T_{1,R}^{-1}$, where D_T is the translational diffusion coefficient and d_{RL} is the distance of closest approach between paramagnetic ion and spin label.

The dipolar enhancement in spin–lattice relaxation rate is given by Eq. 10.26. If we neglect spectral densities at the Larmor frequency and above, the dynamic enhancement in T_1-relaxation becomes:

$$\frac{1}{T_{1dd}(\text{dynamic})}=\frac{1}{12}\left(\frac{\mu_o}{4\pi}\right)^2|\mu_R|^2\gamma_e^2J^{(0)}(\omega_R-\omega_L) \quad (10.44)$$

where the spectral densities for translational diffusion $J^{(M)}(\omega)$ are given in Section 9.5.2. For ions with g-values which differ considerably from that of the spin label: $(\omega_R-\omega_L)^2\tau_D^2>>1$ and the spectral density is $J^{(0)}(\omega_R-\omega_L)\approx 0$. We then expect the dynamic dipolar T_1-relaxation enhancement to be small. For ions with g-values close to that of the spin label, viz., Mn^{2+}: $(\omega_R-\omega_L)^2\tau_D^2<<1$ and the spectral density is $J^{(0)}(\omega_R-\omega_L)\approx J^{(0)}(0)$, where the zero-frequency spectral density is given by Eq. 9.52, i.e., $J^{(0)}(0)=(48\pi/15^2)n_R/(D_Td_{RL})$. The dynamic T_1-relaxation enhancement then becomes:

$$\frac{1}{T_{1dd}(\text{dynamic})}=\frac{4\pi}{15^2}\left(\frac{\mu_o}{4\pi}\right)^2|\mu_R|^2\gamma_e^2\frac{n_R}{D_Td_{RL}} \quad (10.45)$$

where n_R is the paramagnetic-ion concentration, in ions/m^3.

Unlike relaxation by diffusion-controlled Heisenberg spin exchange, the dynamic dipolar interaction is much more significant for T_2-relaxation than for T_1-relaxation, because of the contribution from spectral densities at low frequency. The enhancement in T_2-relaxation rate is given by Eq. 10.41. Assuming again that spectral densities at the Larmor frequency and above contribute negligibly, the dynamic dipolar T_2-relaxation enhancement is:

$$\frac{1}{T_{2dd}(\text{dynamic})}=\left(\frac{\mu_o}{4\pi}\right)^2|\mu_R|^2\gamma_e^2\left[\tfrac{1}{6}J^{(0)}(0)+\tfrac{1}{24}J^{(0)}(\omega_R-\omega_L)\right] \quad (10.46)$$

Substituting the spectral densities, the T_2-relaxation enhancement then becomes:

$$\frac{1}{T_{2dd}(\text{dynamic})}=C\left(\frac{\mu_o}{4\pi}\right)^2|\mu_R|^2\gamma_e^2\frac{n_R}{D_Td_{RL}} \quad (10.47)$$

where $C=2\pi/45$ for $(\omega_R-\omega_L)^2\tau_D^2<<1$ (i.e., specifically for Mn^{2+}), and $C=8\pi/15^2$ for $(\omega_R-\omega_L)^2\tau_D^2>>1$ (i.e., for most other ions). For the former case (i.e., $|g_R-g_L|<<0.01$), the T_1-relaxation rate is $T_{1dd}^{-1}=\frac{2}{5}T_{2dd}^{-1}$ (cf. Eq. 10.45), whereas for the latter case (i.e., $|g_R-g_L|>>0.01$), we have $T_{1dd}^{-1}<<T_{2dd}^{-1}$.

10.11 SPIN–LATTICE RELAXATION ENHANCEMENT BY EXCHANGE PROCESSES

Exchange at rates comparable to that of spin–lattice relaxation alleviates saturation in nitroxide EPR spectra. This results either from transfer of spin polarization, in the case of physical or chemical exchange, or by mutual antiparallel spin flips (i.e., cross relaxation), in the case of Heisenberg exchange (see Figure 10.11). For exchange between two sites, both cases are equivalent as we shall see immediately below. We use rate equations for the spin population differences (cf. Yin and Hyde 1987).

Consider two sites with total spin populations N_i and N_k between which physical or chemical exchange takes place:

$$N_i \underset{\tau_k^{-1}}{\overset{\tau_i^{-1}}{\rightleftarrows}} N_k$$

where τ_i^{-1} is the rate of transfer from site i, and τ_k^{-1} that from site k (see Figure 10.11). At exchange equilibrium, these rate constants are related by detailed balance:

$$\tau_i^{-1}N_i = \tau_k^{-1}N_k \tag{10.48}$$

The rate equation for the spin population difference, $n_i = N_i^- - N_i^+$, at site i is then:

$$\frac{dn_i}{dt} = -\left(\tau_i^{-1}n_i - \tau_k^{-1}n_k\right) \tag{10.49}$$

where N_i^+, N_i^- are the spin populations of the $M_{S,i} = \pm\frac{1}{2}$ states $\left(N_i = N_i^+ + N_i^-\right)$ at site i and similarly for n_k.

On the other hand, we can express Heisenberg spin exchange between the $\pm\frac{1}{2},\mp\frac{1}{2}$ and $\mp\frac{1}{2},\pm\frac{1}{2}$ combinations of spin orientations, $S_{iz}S_{kz}$, for the two sites as:

$$N_i^{\pm} \xleftrightarrow{k_{HE}} N_k^{\mp}$$

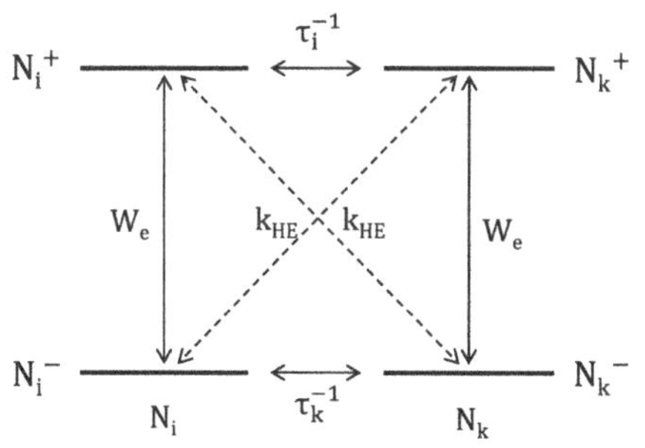

FIGURE 10.11 Energy levels, spin populations $N_i^{\pm}$ and transitions for two spin environments i and k in an EPR lineshape. Spin population difference, $n_i = N_i^- - N_i^+$; transition rate for spin–lattice relaxation, $W_e \equiv 1/\left(2T_1^o\right)$. Rate of chemical (or physical) exchange between spins i and k, $N_i\tau_i^{-1} = N_k\tau_k^{-1}$ at equilibrium; Heisenberg spin-exchange rate (cross relaxation), $2k_{HE}N_i^{\pm}N_k^{\mp}$.

where k_{HE} is the bimolecular rate constant for spin exchange (see Figure 10.11). The corresponding rate equation for the spin population difference at site i is:

$$\frac{dn_i}{dt} = -2k_{HE}\left(N_i^- N_k^+ - N_i^+ N_k^-\right) \tag{10.50}$$

We can rewrite the right-hand side of Eq. 10.50 by using the identity: $\left(N_k^+ + N_k^-\right)\left(N_i^- - N_i^+\right) - \left(N_i^+ + N_i^-\right)\left(N_k^- - N_k^+\right) = 2\left(N_i^- N_k^+ - N_i^+ N_k^-\right)$. The rate equation expressed in terms of population differences n_i, n_k then becomes (Marsh 1992):

$$\frac{dn_i}{dt} = -k_{HE}\left(N_k n_i - N_i n_k\right) \tag{10.51}$$

where N_i, N_k are the total populations, and these remain constant. The pseudo-first order rate constant for spin exchange (or exchange frequency) that we define in Section 9.10 of Chapter 9 is: $\tau_{HE}^{-1} = k_{HE}\left(N_i + N_k\right)$ (cf. Eq. 9.93). Comparing Eqs. 10.49 and 10.51, we see that physical/chemical exchange and Heisenberg spin exchange are kinetically equivalent, with the following identities: $\tau_i^{-1} = f_k\tau_{HE}^{-1}$ and $\tau_k^{-1} = f_i\tau_{HE}^{-1}$, where the fractional population of state i is $f_i = N_i/\left(N_i + N_k\right)$ and $f_i + f_k = 1$.

The two-site spin exchange situation is applicable specifically to double-labelling experiments (Snel and Marsh 1994; Páli et al. 1999). Commonly, however, Heisenberg exchange takes place between more than two distinguishable spin-label species, in concentration-dependent single-labelling experiments. Depending on spin-label concentration, the relaxation enhancement then can be considerably more efficient than in the two-site case. We return to this more usual single-labelling case for Heisenberg exchange after first treating two-site exchange.

10.12 RELAXATION BY SLOW TWO-SITE EXCHANGE

Consider exchange between two sites b and f where spin transitions are induced at rate $2W$ by irradiating with the B_1-field at site b. The steady-state condition for the spin population difference at site b, including spin–lattice relaxation and exchange, is:

$$\frac{dn_b}{dt} = -\frac{n_b - n_b^o}{T_{1,b}^o} - \tau_b^{-1}n_b + \tau_f^{-1}n_f - 2Wn_b = 0 \tag{10.52}$$

where n_b^o is the spin population difference at Boltzmann equilibrium, and $T_{1,b}^o$ is the spin–lattice relaxation time in the absence of exchange. A corresponding steady-state condition applies to the spins at site f that are not irradiated (i.e., with $W=0$):

$$\frac{dn_f}{dt} = -\frac{n_f - n_f^o}{T_{1,f}^o} - \tau_f^{-1}n_f + \tau_b^{-1}n_b = 0 \tag{10.53}$$

Solution of rate equations 10.52 and 10.53, together with Eq. 10.48 (which holds also for population differences at Boltzmann equilibrium), yields the standard expression for saturation of the spin system at site b. This is Eq. 10.1 with $b \equiv p$. The effective spin–lattice relaxation time in the presence of exchange, i.e., $T_{1,b}^{eff}$, is thus defined by (Horváth et al. 1993):

$$\frac{T_{1,b}^{o}}{T_{1,b}^{eff}} = 1 + \frac{(1-f_b)T_{1,b}^{o}\tau_b^{-1}}{1-f_b+f_bT_{1,f}^{o}\tau_b^{-1}} \tag{10.54}$$

where f_b is the fractional population at site b. From this, we get the off-rate, τ_b^{-1}, for exchange (where $f_b\tau_b^{-1} = (1-f_b)\tau_f^{-1}$). The corresponding expression for Heisenberg spin-exchange between two distinguishable spin-labelled species is (Snel and Marsh 1994):

$$\frac{T_{1,b}^{o}}{T_{1,b}^{eff}} = 1 + \frac{(1-f_b)T_{1,b}^{o}\tau_{HE}^{-1}}{1+f_bT_{1,f}^{o}\tau_{HE}^{-1}} \tag{10.55}$$

which we get from the identities between the exchange rates in the two cases, viz., $\tau_b^{-1} = (1-f_b)\tau_{HE}^{-1}$ (see above).

10.13 CONCENTRATION DEPENDENCE OF HEISENBERG EXCHANGE BETWEEN SPIN LABELS

Generally, Heisenberg exchange occurs between many spin labels of the same type, which are distinguished either by their different nitrogen hyperfine states or by their different orientations with respect to the static magnetic field (see Figure 10.12). Thus, we must sum Eq. 10.51 over all states k with which the state i is exchanging. The steady-state rate equation for the population difference of the ith transition then becomes (Marsh 1992):

$$\frac{dn_i}{dt} = -\frac{n_i - n_i^o}{T_1^o} - k_{HE}(Nn_i - N_in) - 2Wn_p\delta_{i,p} = 0 \tag{10.56}$$

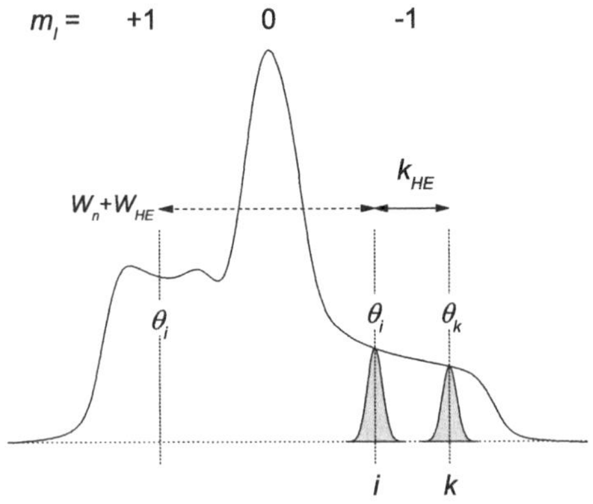

FIGURE 10.12 Heisenberg spin exchange within powder absorption spectrum of ^{14}N-nitroxide. Spin exchange with rate constant k_{HE} between nitroxides at orientations θ_i and θ_k within a single hyperfine manifold m_I, and at rate W_{HE} between manifolds. Additionally, nuclear relaxation between manifolds at rate W_n, with fixed θ_i.

where $N = \sum_k N_k$ and $n = \sum_k n_k$. Summation over all k is allowed, because the term with $k = i$ cancels (see Eq. 10.51). We assume that transitions are induced by the B_1-field in the pth spin system, i.e., $\delta_{i,p} = 1$ for $i = p$ and is zero otherwise. Summing Eq. 10.56 over all i yields the further condition:

$$-\frac{n - \sum n_i^o}{T_1^o} - 2Wn_p = 0 \tag{10.57}$$

Solving Eq. 10.56 (for $i = p$) and Eq. 10.57, together with the relation $\sum n_i^o = (n_p^o/N_p) \times N$, yields the standard expression for saturation of the pth spin system (see Eq. 10.1). The effective spin–lattice relaxation time then becomes (Marsh 1992):

$$T_{1,p}^{eff} = T_1^o\frac{1+Z_pT_1^o\tau_{HE}^{-1}}{1+T_1^o\tau_{HE}^{-1}} \tag{10.58}$$

Here, the exchange frequency is the pseudo-first-order rate constant: $\tau_{HE}^{-1} = k_{HE}N$, which gives the dependence on spin-label concentration, N; and $Z_p = N_p/N$ is the fractional population (or degeneracy) of the transition being saturated. Note that with $Z_p = \frac{1}{3}$, Eq. 10.58 becomes identical to Eq. 10.6 for the $m_I = 0$ manifold of a ^{14}N-nitroxide, when we neglect nuclear relaxation (i.e., $b = 0$). The CW-ELDOR reduction factor that corresponds to Eq. 10.58 is given by Eq. A10.19 in Appendix A10.2 at the end of the chapter.

Figure 10.13 shows the dependence of the reduction in effective spin–lattice relaxation time on Heisenberg spin-exchange frequency that we predict from Eq. 10.58. When the exchange frequency is low $(T_1^o\tau_{HE}^{-1} << 1)$, we get a linear dependence. When the degree of degeneracy is high $(Z_p \approx 0)$, as in a powder pattern, a large number of states is available for redistribution of

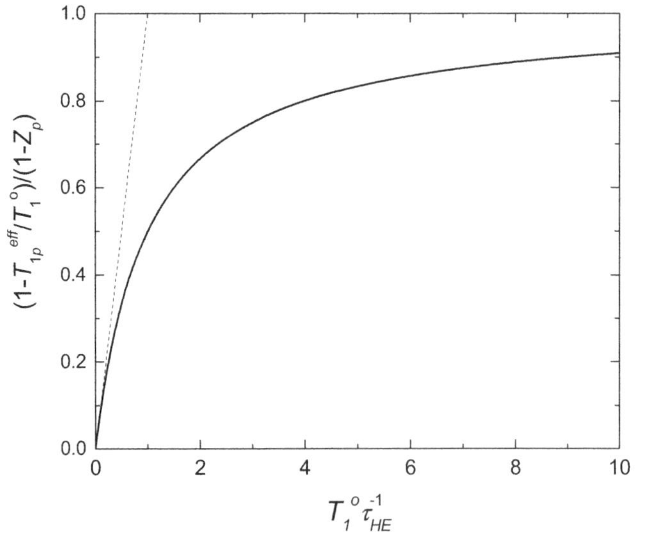

FIGURE 10.13 Dependence of normalized reduction in effective spin–lattice relaxation time $1-T_{1,p}^{eff}/T_1^o$ on scaled Heisenberg spin-exchange frequency $T_1^o\tau_{HE}^{-1}$ (Eq. 10.58). y-axis normalized to redistribution factor $1-Z_p$. At high exchange rates, normalized reduction in $T_{1,p}^{eff}$ reaches limiting value of 1.

saturation. Heisenberg exchange then has the same effect as a true relaxation enhancement, rather than cross-relaxation. The effective relaxation rate becomes simply:

$$\frac{1}{T_1^{eff}} = \frac{1}{T_1^o} + \tau_{HE}^{-1} \tag{10.59}$$

We find this to apply often in practice, for low exchange rates. On the other hand, at high exchange frequencies, T_1^{eff} tends to a limiting value of $Z_p T_1^o$ that represents the maximum degree of redistribution of saturation throughout the various distinct spin label states. Beyond this limit, T_2-relaxation becomes sensitive to spin exchange, so then we can determine exchange frequencies from linewidths and lineshapes of the conventional EPR spectra (see Sections 9.10 and 9.14–9.16 in Chapter 9).

We can combine the approach given by the rate equations of Eqs. 10.2 and 10.3 with that of Eqs. 10.56 and 10.57 to give effective spin–lattice relaxation times for the powder pattern of a ^{14}N-nitroxide, including Heisenberg spin exchange (see Figure 10.12). The effective spin–lattice relaxation times for the $m_I = 0, \pm 1$ hyperfine lines then finally become (Marsh 1992):

$$T_{1,p}^{eff}(0) = T_1^o \frac{1 + Z_p T_1^o \tau_{HE}^{-1} + b\left(1 + 3Z_p T_1^o \tau_{HE}^{-1}\right)/\left(1 + T_1^o \tau_{HE}^{-1}\right)}{1 + 3b + T_1^o \tau_{HE}^{-1}} \tag{10.60}$$

$$T_{1,p}^{eff}(\pm 1) = T_{1,p}^{eff}(0) + \frac{T_1^o b}{\left(1 + b + T_1^o \tau_{HE}^{-1}\right)\left(1 + 3b + T_1^o \tau_{HE}^{-1}\right)} \tag{10.61}$$

where again $b \equiv W_n/(2W_e)$ and $Z_p = N_p/N$. A derivation for ^{15}N-nitroxides is given in Appendix A10.1 at the end of the chapter (see Eqs. A10.8–A10.10). CW-ELDOR reduction factors that correspond to Eqs. 10.60 and 10.61 are given by Eqs. A10.20–A10.24 in Appendix A10.2 at the end of the chapter.

10.14 CONCLUDING SUMMARY

1. *Effective T_1-relaxation times in progressive saturation*: Effective electron spin–lattice relaxation times $T_{1,p}^{eff}$ in progressive-saturation CW-EPR are defined from the dependence of the population difference of the transition p irradiated on the microwave power, P:

$$n_p = \frac{n_p^{o,eff}}{1 + 2WT_{1,p}^{eff}} \tag{10.62}$$

where $W\left(\propto \sqrt{P}\right)$ is the rate at which the microwave B_1-field induces transitions. (Only when we neglect cross relaxation is $n_p^{o,eff}$ equal to n_p^o, the intrinsic population difference for a two-level system at Boltzmann equilibrium.)

2. *Nuclear relaxation and effective T_{1e}*: The main reason for departures of T_{1,m_I}^{eff} from the intrinsic value $T_{1e}^o = 1/(2W_e)$ is nuclear relaxation with rate W_n. For ^{15}N-nitroxides, i.e., $I = \frac{1}{2}$:

$$T_{1,\pm}^{eff} = \frac{1 + \frac{1}{2}b}{1 + b}\frac{1}{2W_e} \tag{10.63}$$

for $m_I = \pm\frac{1}{2}$, where $b = W_n/W_e$ is the normalized nuclear relaxation rate. For ^{14}N-nitroxides, i.e., $I = 1$:

$$T_{1,0}^{eff} = \frac{1 + b}{1 + 3b}\frac{1}{2W_e} \tag{10.64}$$

$$T_{1,\pm 1}^{eff} = \frac{1 + 3b + b^2}{(1 + b)(1 + 3b)}\frac{1}{2W_e} \tag{10.65}$$

for $m_I = 0, \pm 1$, where $b = W_n/(2W_e)$ is the normalized rate in this case.

3. *Cross relaxation and m_I-dependence of effective T_{1e}*: Effective spin–lattice relaxation times T_{1,m_I}^{eff} also contain significant contributions from cross relaxation (c), and from the m_I-dependence of the intrinsic spin–lattice relaxation rate (d_1 and d). For ^{15}N-nitroxides, i.e., $I = \frac{1}{2}$:

$$T_{1,\pm}^{eff} = \frac{1}{2\bar{W}_e}\frac{1 \mp d + \frac{1}{2}(b + c)}{(1 + b)(1 + c) - d^2} \tag{10.66}$$

where $2\bar{W}_e = W_{e,+} + W_{e,-}$, with $b \equiv W_n/\bar{W}_e$, $c \equiv W_x/\bar{W}_e$ and $d \equiv \delta/\bar{W}_e\ (2\delta = W_{e,+} - W_{e,-})$. For ^{14}N-nitroxides, i.e., $I = 1$:

$$T_{1,0}^{eff} = \frac{1}{2W_{e,0}}\frac{1 + d_1 + b + c}{1 + d_1 + (3 + 2d_1)(b + c) + 8bc} \tag{10.67}$$

$$T_{1,\pm 1}^{eff} = \frac{1}{2W_{e,0}(1 + d_1 + b + c)} \times \left(1 + \frac{(b - c)^2 \mp (1 + 2b + 2c)d}{1 + d_1 + (3 + 2d_1)(b + c) + 8bc}\right) \tag{10.68}$$

where $W_{e,0} = \bar{W}_e - \frac{2}{3}\delta_1$, with $b \equiv W_n/2W_{e,0}$, $c \equiv W_x/2W_{e,0}$, $d_1 \equiv \delta_1/W_{e,0}$ and $d \equiv \delta/W_{e,0}\ (2\delta = W_{e,+1} - W_{e,-1})$. The contribution of cross relaxation to T_1^{eff} is less important than that to the recovery rate in SR-EPR (see Chapter 12), because $c << b$ except for very short correlation times.

4. *Relaxation enhancement by paramagnetic species*: Paramagnetic relaxation enhancement at rate W_{RL} simply adds to the intrinsic spin–lattice relaxation rate W_e. Neglecting cross relaxation and the m_I-dependence of W_e, effective spin–lattice relaxation times for ^{14}N-nitroxides are:

$$T_{1,0}^{eff}(b''') = \frac{1 + b''' + b}{1 + b''' + 3b} \cdot \frac{1}{2W_e(1 + b''')} \tag{10.69}$$

$$T_{1,\pm1}^{eff}(b''') = \frac{(1+b''')^2 + 3(1+b''')b + b^2}{(1+b'''+b)(1+b'''+3b)} \times \frac{1}{2W_e(1+b''')} \tag{10.70}$$

where $b''' \equiv W_{RL}/W_e$ is the normalized rate of relaxation enhancement.

5. *Relaxation enhancement by slow two-site exchange*:
   Effective spin–lattice relaxation time $T_{1,b}^{eff}$ of spin-label b undergoing slow two-site exchange ($b \leftrightarrow f$) with spin label f is:

$$\frac{T_{1,b}^o}{T_{1,b}^{eff}} = 1 + \frac{(1-f_b)T_{1,b}^o \tau_b^{-1}}{1 - f_b + f_b T_{1,f}^o \tau_b^{-1}} \tag{10.71}$$

   where $\tau_b^{-1} = \tau_f^{-1}(1-f_b)/f_b$ is the rate of leaving site b and $T_{1,b}^o$ is the effective relaxation time in the absence of exchange.

6. *Relaxation enhancement by Heisenberg spin-exchange*:
   Heisenberg spin exchange between ^{14}N-nitroxides is less restrictive for the $m_I = \pm1$ hyperfine manifolds than is nuclear relaxation. Neglecting cross relaxation and the m_I-dependence of W_e, the effective relaxation times become:

$$T_{1,0}^{eff}(b'') = \frac{1+b+b''}{1+3b+3b''} \cdot \frac{1}{2W_e} \tag{10.72}$$

$$T_{1,\pm1}^{eff}(b'') = T_{1,0}^{eff}(b'') + \frac{b}{(1+b+3b'')(1+3b+3b'')} \cdot \frac{1}{2W_e} \tag{10.73}$$

   where $b'' \equiv \tau_{HE}^{-1}/(6W_e) = k_{HE}[\mathrm{SL}]/(6W_e)$ is the normalized Heisenberg exchange rate.

   For Heisenberg exchange between spin labels at different orientations in a powder pattern, the effective spin–lattice relaxation time when irradiating at position p is:

$$T_{1,p}^{eff} = T_1^o \frac{1 + Z_p T_1^o \tau_{HE}^{-1}}{1 + T_1^o \tau_{HE}^{-1}} \tag{10.74}$$

   where $Z_p = N_p/N$ is the fractional population (or degeneracy) of the transition being saturated, and $T_1^o \equiv 1/(2W_e)$ is the effective relaxation time in the absence of spin exchange. Incorporating also nuclear relaxation for a ^{14}N-nitroxide, the effective relaxation times become:

$$\frac{T_{1,p}^{eff}(0)}{T_1^o} = \frac{1 + 3Z_p b'' + b(1+9Z_p b'')/(1+3b'')}{1+3b+3b''} \tag{10.75}$$

$$T_{1,p}^{eff}(\pm1) = T_{1,p}^{eff}(0) + \frac{bT_1^o}{(1+b+3b'')(1+3b+3b'')} \tag{10.76}$$

   where $b \equiv W_n/(2W_e)$ and $b'' \equiv \tau_{HE}^{-1}/(6W_e)$.

7. *CW-ELDOR reduction factors*:
   Reduction factors in CW-ELDOR are defined by the population differences $n_{obs}(\mathrm{on})$ and $n_{obs}(\mathrm{off})$ at the transition observed, when microwaves pumping at position p are switched on and off, respectively:

$$R^{p,obs} = \frac{n_{obs}(\mathrm{off}) - n_{obs}(\mathrm{on})}{n_{obs}(\mathrm{off})} = 1 - \frac{n_{obs}(\mathrm{on})}{n_{obs}(\mathrm{off})} \tag{10.77}$$

8. *CW-ELDOR and nuclear relaxation*:
   CW-ELDOR reduction factors, extrapolated to infinite pumping power, that correspond to the situations given for effective relaxation times of ^{14}N-nitroxides under point 2 are:

$$R_\infty^{0,\pm1} = b/(1+b) \tag{10.78}$$

$$R_\infty^{\pm1,0} = \frac{b(1+b)}{1+3b+b^2} \tag{10.79}$$

$$R_\infty^{\pm1,\mp1} = \frac{b^2}{1+3b+b^2} \tag{10.80}$$

9. *CW-ELDOR and Heisenberg spin-exchange*:
   CW-ELDOR reductions corresponding to Heisenberg exchange between hyperfine manifolds of a ^{14}N-nitroxide, as under point 6 are:

$$\frac{1}{R_\infty^{0,\pm1}(b'')} = 1 + \frac{1+3b''}{b + b''(1+3b''+3b)} \tag{10.81}$$

$$\frac{1}{R_\infty^{\pm1,0}(b'')} = 1 + \frac{(1+3b'')(1+3b''+2b)}{(b + b''(1+3b''+3b))(1+3b''+b)} \tag{10.82}$$

$$\frac{1}{R_\infty^{\pm1,\mp1}(b'')} = 1 + \frac{(1+3b'')(1+3b''+3b)}{b^2 + b''(1+3b''+3b)(1+3b''+b)} \tag{10.83}$$

For Heisenberg exchange between spin labels at different orientations in a powder pattern:

$$\frac{1}{R_\infty^{p,obs}} = 1 + \frac{1}{Z_p T_1^o \tau_{HE}^{-1}} \tag{10.84}$$

which corresponds to the same situation as under point 6.

Expressions for the effective electron spin–lattice relaxation times T_{1,m_I}^{eff} in progressive-saturation EPR are summarized in Table 10.6. Corresponding expressions for the

TABLE 10.6 Effective electron spin–lattice relaxation times T_{1,m_I}^{eff} for ^{15}N- and ^{14}N-nitroxides

	^{15}N-*nitroxide*[a]	^{14}N-*nitroxide*[b]
nuclear relaxation	$b = W_n/\bar{W}_e$	$b = W_n/(2W_{e,0})$
	$T_{1,\pm}^{eff} = \frac{1 \mp d + \frac{1}{2}(b+c)}{(1+b)(1+c)-d^2}\frac{1}{2\bar{W}_e}$	$T_{1,0}^{eff} = \frac{1+d_1+b+c}{1+d_1+(3+2d_1)(b+c)+8bc}\frac{1}{2W_{e,0}}$
	–	$T_{1,\pm1}^{eff} = \left(1+\frac{(b-c)^2 \mp (1+2b+2c)d}{1+d_1+(3+2d_1)(b+c)+8bc}\right)\frac{1}{2W_{e,0}(1+d_1+b+c)}$
+Heisenberg exchange[c]	$b'' = \tau_{HE}^{-1}/(2\bar{W}_e)$	$b'' = \tau_{HE}^{-1}/(6W_e)$
	$T_{1,\pm}^{eff} = \frac{1 \mp d + \frac{1}{2}(b+b''+c)}{(1+b+b'')(1+c)-d^2}\frac{1}{2\bar{W}_e}$	$T_{1,0}^{eff} = \frac{1+b+b''}{1+3b+3b''}\frac{1}{2W_e}$
	–	$T_{1,\pm1}^{eff} = T_{1,0}^{eff} + \frac{b/(2W_e)}{(1+b+3b'')(1+3b+3b'')}$
exchange multiple sites[d]	$\frac{T_{1,p}^{eff}}{T_1^o} = \frac{1+Z_p b''}{1+b''}$	$\frac{T_{1,p}^{eff}}{T_1^o} = \frac{1+3Z_p b''}{1+3b''}$
+nuclear relaxation	$\frac{T_{1,p}^{eff}}{T_1^o} = \left(\frac{1+\frac{1}{2}b+b''}{1+b+b''}+Z_p b''\right)\frac{1}{1+b''}$	$\frac{T_{1,p}^{eff}(0)}{T_1^o} = \frac{1+3Z_p b''+b(1+9Z_p b'')/(1+3b'')}{1+3b+3b''}$
	–	$T_{1,p}^{eff}(\pm 1) = T_{1,p}^{eff}(0) + \frac{bT_1^o}{(1+b+3b'')(1+3b+3b'')}$

[a] $c \equiv W_x/\bar{W}_e$ and $d \equiv \delta/\bar{W}_e$, where $2\bar{W}_e = W_{e,+} + W_{e,-}$ and $2\delta = W_{e,+} - W_{e,-}$.

[b] $c \equiv W_x/2W_{e,0}$, $d_1 \equiv \delta_1/W_{e,0}$ and $d \equiv \delta/W_{e,0}$, where $W_{e,\pm1} = W_{e,0} + \delta_1 \pm \delta$.

[c] τ_{HE}^{-1} ($\equiv k_{HE}[SL]$) is the pseudo-first order rate constant for Heisenberg spin exchange. N.B. cross relaxation and m_I-dependent W_e are included for ^{15}N but not for ^{14}N.

[d] Z_p is the fractional spin population (or degeneracy) at site p which is irradiated; $T_1^o \equiv 1/(2W_e)$ is the intrinsic electron spin–lattice relaxation time in the absence of Heisenberg exchange.

CW-ELDOR reduction factors $R_\infty^{p,obs}$ are listed in Table 10.7. For ELDOR reduction factors that include cross relaxation and m_I-dependent spin–lattice relaxation, see Appendices A10.1–A10.3. See Figures 10.1, 10.11, A10.1 and A10.4 for energy-level diagrams with the corresponding transitions.

APPENDIX A10.1: EFFECTIVE SPIN–LATTICE RELAXATION TIMES AND ELDOR REDUCTION FACTORS FOR ^{15}N-NITROXIDES

Figure A10.1 shows the energy level scheme for a $S = \frac{1}{2}, I = \frac{1}{2}$ spin system. We include different transition probabilities $W_{e,\pm} \left(\equiv \bar{W}_e \pm \delta\right)$ for electron spin–lattice relaxation in the $m_I = \pm\frac{1}{2}$ hyperfine manifolds, and cross relaxation with probabilities per unit time W_{x1} and W_{x2}, for transitions with $\Delta m_I = \pm 1$, or $\Delta m_I = \mp 1$, respectively. This important modification was not included for ^{14}N-nitroxides in the main text but will be given later in Appendix 10.3 below.

The total spin population $N_\pm$ is the same in either hyperfine state, because their energy difference ha_o is very small compared with $k_B T$. Because $N_+ = N_-$, we get rate equations for the population differences $n_\pm$ immediately from either of the rate equations for the absolute spin populations, $N_\pm \pm n_\pm$. The steady-state rate equations for the spin population differences in the $m_I = \pm\frac{1}{2}$ hyperfine levels of a ^{15}N-nitroxide are (cf. Eqs. 10.2 and 10.3 for ^{14}N):

$$\frac{dn_\pm}{dt} = -2W_{e,\pm}\left(n_\pm - n_\pm^o\right) - W_n\left(n_\pm - n_\mp\right) - W_x\left(n_\pm + n_\mp\right) - 2Wn_\pm\delta_{m_I,p} = 0 \qquad \text{(A10.1)}$$

where $n_+^o = n_-^o \equiv n^o$ is the population difference at Boltzmann equilibrium, and the transition irradiated is that for which $m_I = p$. The nuclear spin–lattice relaxation rate W_n is defined by Eq. 5.20 of Chapter 5, with $I = \frac{1}{2}$ for a ^{15}N-nitroxide. Cross relaxation enters as the mean transition probability: $W_x = \frac{1}{2}\left(W_{x1} + W_{x2}\right)$, as found in Section 12.4 of Chapter 12 (see Eq. 12.5).

Solution of the two simultaneous Eqs. A10.1 yields the standard expression for the dependence of the steady-state population difference on W:

$$n_p = \frac{n_p^{o,eff}}{1 + 2WT_{1,p}^{eff}} \qquad \text{(A10.2)}$$

This expression is very similar to Eq. 10.1, but $n_p^{o,eff}$ no longer corresponds simply to n_p^o, the equilibrium population difference between $M_{S,p} = \pm\frac{1}{2}$ states in a two-level system. This is because longitudinal

TABLE 10.7 CW-ELDOR reduction factors $R_{\infty}^{p,obs}$ for ^{15}N- and ^{14}N-nitroxides[a]

	^{15}N-*nitroxide*	^{14}N-*nitroxide*
nuclear relaxation	$b = W_n/W_e$	$b = W_n/(2W_e)$
	$\frac{1}{R_{\infty}} = 1 + \frac{2}{b}$	$\frac{1}{R_{\infty}^{0,\pm1}} = 1 + \frac{1}{b}$
	–	$\frac{1}{R_{\infty}^{\pm1,0}} = 1 + \frac{1+2b}{b(1+b)}$
	–	$\frac{1}{R_{\infty}^{\pm1,\mp1}} = 1 + \frac{1+3b}{b^2}$
+Heisenberg exchange[b]	$b'' = \tau_{HE}^{-1}/(2W_e)$	$b'' = \tau_{HE}^{-1}/(6W_e)$
	$\frac{1}{R_{\infty}} = 1 + \frac{2}{b+b''}$	$\frac{1}{R_{\infty}^{0,\pm1}} = 1 + \frac{1+3b''}{b+b''(1+3b''+3b)}$
	–	$\frac{1}{R_{\infty}^{\pm1,0}} = 1 + \frac{(1+3b'')(1+3b''+2b)}{(b+b''(1+3b''+3b))(1+3b''+b)}$
	–	$\frac{1}{R_{\infty}^{\pm1,\mp1}} = 1 + \frac{(1+3b'')(1+3b''+3b)}{b^2+b''(1+3b''+3b)(1+3b''+b)}$
exchange multiple sites[c]	$\frac{1}{R_{\infty}^{p}} = 1 + \frac{1}{Z_p b''}$	$\frac{1}{R_{\infty}^{p}} = 1 + \frac{1}{3Z_p b''}$
+nuclear relaxation	$\frac{1}{R_{\infty}^{p,obs}} = 1 + \frac{1}{Z_p b''}\left(\frac{1+b''+\frac{1}{2}b}{1+b''+b}\right)$	$\frac{1}{R_{\infty}^{0,\pm1}} = 1 + \frac{1+3b''+b}{3Z_p b''(1+3b''+3b)}$
	–	$\frac{1}{R_{\infty}^{\pm1,0}} = \frac{1}{R_{\infty}^{\pm1,\mp1}} = \frac{1}{R_{\infty}^{0,\pm1}} + \frac{b(1+3b'')}{3Z_p b''(1+3b''+3b)(1+3b''+b)}$

[a] *p* is the pumping position and *obs* the observing position.
[b] τ_{HE}^{-1} ($\equiv k_{HE}[SL]$) is the pseudo-first order rate constant for Heisenberg spin exchange.
[c] Z_p is the fractional spin population (or degeneracy) at site *p* which is pumped; $T_1^o \equiv 1/(2W_e)$ is the electron spin–lattice relaxation time in the absence of Heisenberg spin exchange.

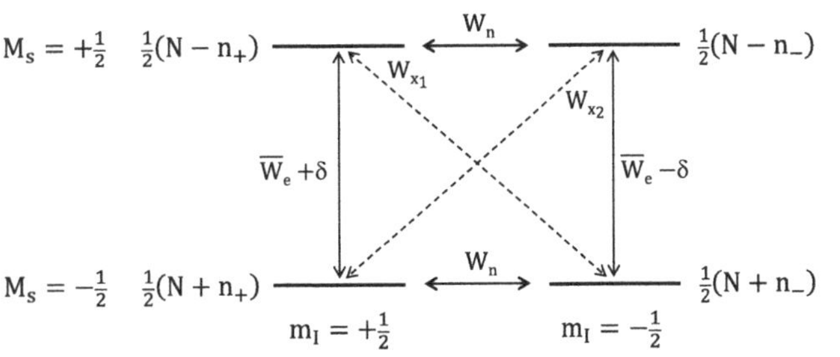

FIGURE A10.1 Energy levels, spin populations and transitions for ^{15}N-nitroxide spin label: $S = \frac{1}{2}$, $I = \frac{1}{2}$. Total spin population *N* in each hyperfine state. *Solid arrows*: spin–lattice relaxation pathways with transition probabilities per unit time $W_e \pm \delta$ for electron transitions, W_n for nuclear transitions. *Dashed arrows*: cross relaxation (i.e., simultaneous electron and nuclear transitions) with transition probabilities W_{x_1} and W_{x_2} for $\Delta M_S = \pm1$, $\Delta m_I = \pm1$ and $\Delta M_S = \pm1$, $\Delta m_I = \mp1$.

magnetizations of the electron and nuclear spins are now coupled (Marsh 2016a). The effective electron spin–lattice relaxation times T_{1,m_I}^{eff} that govern the saturation behaviour of the $m_I = \pm\frac{1}{2}$ hyperfine manifolds according to Eq. A10.2 are then (cf. Eqs. 10.4, 10.5):

$$T_{1,\pm}^{eff} = \frac{1}{2\overline{W}_e}\frac{1 \mp d + \frac{1}{2}(b+c)}{(1+b)(1+c)-d^2} \tag{A10.3}$$

where $b \equiv W_n/\overline{W}_e \left(= T_{1e}/T_{1n}\right)$, $c \equiv W_x/\overline{W}_e$ and $d \equiv \delta/\overline{W}_e$. This definition of *b* differs from that for ^{14}N-nitroxides in Section 10.2 but is consistent with the usage for CW-ELDOR (Hyde et al. 1968). Note that *b* is given by $b \times 2W_e = j^{AA}(\omega_a)$ for both ^{15}N- and ^{14}N-nitroxides (see Eq. 5.20), but $W_e = \overline{W}_e$ (the mean value) and $W_e = W_{e,0}$ (the $m_I =$ 0 value) in the two cases, respectively (see Appendix A10.3, and Sections 12.4, 12.6). See Marsh (2016b) for further discussion. For $c = 0 = d$, Eq. A10.3 reduces to the standard expression for a ^{15}N-nitroxide, including only nuclear relaxation: $2\overline{W}_e T_{1,\pm}^{eff} = \left(1 + \frac{1}{2}b\right)/(1+b)$ (Hyde et al. 1968, Marsh 2016b).

The effective population difference at Boltzmann equilibrium that appears in Eq. A10.2 becomes:

$$n_{\pm}^{o,eff} = \frac{1 + b \pm cd - d^2}{(1+b)(1+c) - d^2}n^o \tag{A10.4}$$

for the system coupled by cross relaxation and having different intrinsic spin–lattice relaxation rates. Without cross relaxation ($c = 0$), this reduces simply to $n_{\pm}^{o,eff} = n^o$, as for a two-level system.

The rate constants for nuclear relaxation W_n, cross relaxation W_x and electron spin–lattice relaxation $\overline{W}_e \pm \delta$, all depend on rotational dynamics. However, of the normalized rates *b*, *c* and *d*, only *b* depends on dynamics, because W_x and δ are determined by the same spectral density as $\overline{W}_e$. The upper panel of Figure A10.2

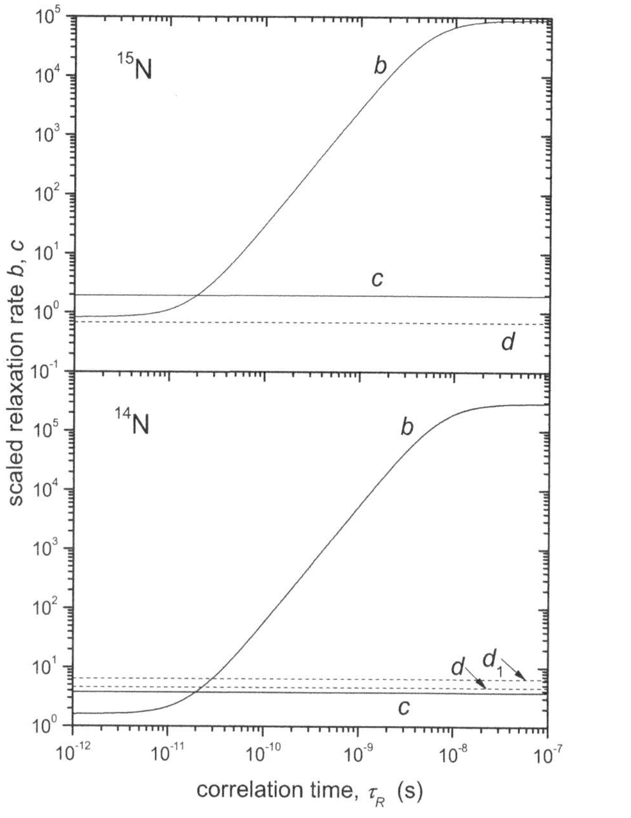

FIGURE A10.2 Normalized relaxation rates as function of isotropic rotational correlation time, τ_R. *Top*: ^{15}N-nitroxide, $b = W_n/W_e$, $c = W_x/W_e$, $d = \delta/W_e$, where $W_{e,\pm} \equiv W_e \pm \delta$. Spin-Hamiltonian parameters for ^{15}N-TEMPOL from Table A2.4. *Bottom*: ^{14}N-nitroxide, $b = W_n/2W_e$, $c = W_x/2W_e$, $d_1 = \delta_1/W_e$, $d = \delta/W_e$, where $W_{e,0} \equiv W_e$ and $W_{e,\pm 1} \equiv W_e + \delta_1 \pm \delta$. Spin-Hamiltonian parameters for ^{14}N-TEMPOL from Percival et al. (1975). EPR frequency 9.3 GHz.

shows the dependence of the normalized rates on correlation time for isotropic rotational diffusion of a ^{15}N-nitroxide. This plot is obtained from Eqs. 5.19 to 5.22 of Chapter 5, together with Eqs. 5.26, 5.30, 5.32 and 5.40, 5.41. For correlation times longer than 5×10^{-11} s, nuclear relaxation is the most important contributor by far. At correlation times shorter than this, cross relaxation and m_I-dependent contributions to the intrinsic electron spin–lattice relaxation rate are also significant. These have the constant relative rates: $c = 1.94$ and $d = 0.68$, independent of the correlation time for isotropic rotation.

The rate of change in population difference of the $m_I = \pm\frac{1}{2}$ states that results from Heisenberg spin exchange is (see, e.g., Figures 10.1 and 10.11):

$$\frac{dn_\pm}{dt} = -\tfrac{1}{2}k_{HE}\left((N+n_\pm)(N-n_\mp) - (N-n_\pm)(N+n_\mp)\right) = -k_{HE}N\left(n_\pm - n_\mp\right) \qquad \text{(A10.5)}$$

where k_{HE} is the second-order rate constant for spin exchange. Comparing with Eq. A10.1, we see that Heisenberg exchange has the same effect as nitrogen nuclear relaxation for the two-level system of ^{15}N-nitroxides. It therefore simply adds to the nuclear relaxation rate. This is unlike the situation for ^{14}N-nitroxides. Including nuclear relaxation, cross relaxation and Heisenberg exchange, the effective electron spin–lattice relaxation times for ^{15}N-nitroxides then become (cf. Eq. A10.3):

$$T_{1,\pm}^{eff} = \frac{1}{2\bar{W}_e}\frac{1 \mp d + \frac{1}{2}(b + b'' + c)}{(1 + b + b'')(1 + c) - d^2} \qquad \text{(A10.6)}$$

where $b'' \equiv k_{HE}N/\bar{W}_e = \tau_{HE}^{-1}/(2\bar{W}_e)$ is the normalized Heisenberg exchange rate. For $c = 0 = d$, Eq. A10.6 reduces to the standard expression for a ^{15}N-nitroxide, including only nuclear relaxation: $2\bar{W}_e T_{1,\pm}^{eff} = \left(1 + \frac{1}{2}b + \frac{1}{2}b''\right)/\left(1 + b + b''\right)$ (Hyde et al. 1968; Marsh 2016b).

In the presence of a paramagnetic relaxant, the effective spin–lattice relaxation times of a ^{15}N-nitroxide become correspondingly (cf. Eqs. 10.6 and 10.7):

$$T_{1,\pm}^{eff} = \frac{1}{2\bar{W}_e(1 + b''')} \cdot \frac{1 + b''' \mp d + \frac{1}{2}(b + b'' + c)}{(1 + b''' + b + b'')(1 + c) - d^2} \qquad \text{(A10.7)}$$

where $b''' \equiv W_{RL}/\bar{W}_e$ is the normalized paramagnetic relaxation enhancement rate that simply adds to the intrinsic spin–lattice relaxation rate (see Section 10.5). For $c = 0 = d$, Eq. A10.7 reduces to: $2\bar{W}_e T_{1,\pm}^{eff} = \left(1 + b''' + \frac{1}{2}b + \frac{1}{2}b''\right)/\left[\left(1 + b''' + b + b''\right)\left(1 + b'''\right)\right]$.

The rate equations for the population differences at position i within a powder pattern of a ^{15}N-nitroxide, including Heisenberg spin exchange, are (cf. Eq. 10.56):

$$\frac{dn_{i,\pm}}{dt} = -\frac{n_{i,\pm} - n_{i,\pm}^o}{T_1^o} - W_n\left(n_{i,\pm} - n_{i,\mp}\right) - k_{HE}\left(Nn_{i,\pm} - N_{i,\pm}n\right) - 2Wn_{i,\pm}\delta_{i,p}\delta_{m_I,\pm} = 0 \qquad \text{(A10.8)}$$

where $T_1^o \equiv 1/(2W_e)$ is the electron spin–lattice relaxation time in the absence of exchange, and p is the position irradiated. Summing Eq. A10.8 over all i, and both values of m_I, we get (cf. Eq.10.57):

$$-\frac{n - (N/N_p)n_p^o}{T_1^o} - 2W\delta_{i,p}\left(n_{i,+}\delta_{m_I,+} + n_{i,-}\delta_{m_I,-}\right) = 0 \qquad \text{(A10.9)}$$

where $n = \sum_k \left(n_{k,+} + n_{k,-}\right)$ and $N = \sum_k \left(N_{k,+} + N_{k,-}\right)$. The effective spin–lattice relaxation times for a ^{15}N-nitroxide powder pattern, including Heisenberg spin exchange, then become (cf. Eqs. 10.60 and 10.61 for ^{14}N):

$$T_{1,p}^{eff} = T_1^o\left(1 + Z_p T_1^o \tau_{HE}^{-1} - \frac{\frac{1}{2}b}{1 + b + T_1^o\tau_{HE}^{-1}}\right)\frac{1}{1 + T_1^o\tau_{HE}^{-1}} \qquad \text{(A10.10)}$$

where $Z_p = N_p/N$ is the fractional spin population at the position p irradiated, and $\tau_{HE}^{-1} = k_{HE}N$ is the pseudo-first order rate constant for spin exchange. For $\tau_{HE}^{-1} = 0$, Eq. A10.10 reduces to Eq. A.10.3 with $c = 0 = d$; and for $b = 0$, it reduces to Eq. 10.58.

CW-ELDOR reduction factors are relatively straightforward for ^{15}N-nitroxides. One of the two $m_I = \pm\frac{1}{2}$ hyperfine manifolds is pumped (p), and the other is observed (obs). From Eq. A10.1, we find that the equivalent of Eq. 10.11 for the dependence on pumping power W is given by (Marsh 2016a):

$$\frac{1}{R} = \left(1 + \frac{2W_{e,obs}n^o}{(W_n - W_x)n_p^{o,eff}}\right)\left(1 + \frac{1}{2WT_{1,p}^{eff}}\right) \quad \text{(A10.11)}$$

with the appropriate value of $T_{1,p}^{eff}$ from Eq. A10.3. The reduction factor extrapolated to infinite pump power, for ^{15}N-nuclear relaxation, then becomes:

$$R_\infty^\pm = \frac{b - c}{2(1 \pm d)\left(n^o/n_p^{o,eff}\right) + b - c} \quad \text{(A10.12)}$$

when observing the $m_I = \pm\frac{1}{2}$ hyperfine manifolds, respectively, where $b = W_n/W_e$, $c \equiv W_x/W_e$ and $d \equiv \delta/W_e$. The factor $n^o/n_p^{o,eff}$ is given by Eq. A10.4. For $c = 0 = d$, we get $n^o/n_p^{o,eff} = 1$ and Eq. A10.12 reduces to the standard expression for a ^{15}N-nitroxide, including only nuclear relaxation: $R_\infty = b/(2 + b)$ (Hyde et al. 1968, Marsh 2016b). This corresponds to Eqs. 10.12–10.14 for ^{14}N-nitroxides.

Including Heisenberg spin exchange in Eq. A10.12, we get:

$$R_\infty^\pm = \frac{b + b'' - c}{2(1 \pm d)\left(n^o/n_p^{o,eff}\right) + b + b'' - c} \quad \text{(A10.13)}$$

where $b'' = \tau_{HE}^{-1}/(2W_e)$. For $I = \frac{1}{2}$, Heisenberg exchange is formally equivalent to nuclear relaxation (see Figure A10.1), and b'' simply adds to b. Thus Eq. A10.4 becomes: $n^o/n_\pm^{o,eff} = \left[(1 + b + b'')(1 + c) - d^2\right]/\left[1 + b + b'' \pm cd - d^2\right]$ to be substituted in Eq. A10.13. For $c = 0 = d$, Eq. A10.13 reduces simply to $R_\infty = (b + b'')/(2 + b + b'')$.

The ELDOR reduction factor that corresponds to the effective relaxation time given by Eq. A10.10 for a ^{15}N-nitroxide is:

$$\left(R_\infty^{p,obs}\right)^{-1} = 1 + \frac{1 + \frac{1}{2}b + T_1^o\tau_{HE}^{-1}}{Z_pT_1^o\tau_{HE}^{-1}\left(1 + b + T_1^o\tau_{HE}^{-1}\right)} \quad \text{(A10.14)}$$

where Z_p is the fractional spin population at the position of the pump, p. Observation in the complementary manifold (or even in the same manifold) is at a general position ω_{obs} that does not correspond to the nitroxide orientation θ_p at the pump position ω_p (cf. Figure A10.3). For $b = 0$, Eq. A10.14 reduces to $R_\infty^{-1} = 1 + 1/\left(Z_pT_1^o\tau_{HE}^{-1}\right)$, and for $\tau_{HE}^{-1} = 0$, it reduces to $R_\infty = 0$ because there is then no pathway to the observing position. For observation at the unique orientation θ_p in the complementary manifold, we get:

$$R_\infty^{p+,p-} = \frac{1}{1 + \frac{1}{2}b + T_1^o\tau_{HE}^{-1}} \times \left(\frac{1}{2}b + \frac{1 + T_1^o\tau_{HE}^{-1}}{1 + \left(Z_pT_1^o\tau_{HE}^{-1}\right)^{-1}\left(1 + T_1^o\tau_{HE}^{-1} + \frac{1}{2}b\right)/\left(1 + T_1^o\tau_{HE}^{-1} + b\right)}\right) \quad \text{(A10.15)}$$

When $\tau_{HE}^{-1} = 0$, this reduces to Eq. A10.12 with $c = 0 = d$, $n^o/n_p^{o,eff} = 1$, and for $Z_p = \frac{1}{2}$ it reduces to Eq. A10.13 with the same restrictions.

APPENDIX A10.2: FURTHER CW-ELDOR REDUCTION FACTORS FOR ^{14}N-NITROXIDES

We introduce Heisenberg spin exchange, in addition to nitrogen spin–lattice relaxation, as done for the effective T_1 s in Eqs. 10.6 and 10.7 of Section 10.2. The CW-ELDOR reductions corresponding to Eqs. 10.12–10.14 of Section 10.3 for a ^{14}N-nitroxide then become (Marsh 1992; Hyde et al. 1968):

$$\left(R_\infty^{0,\pm1}\right)^{-1} = 1 + \frac{1 + 3b''}{b + b''(1 + 3b'' + 3b)} \quad \text{(A10.16)}$$

$$\left(R_\infty^{\pm1,0}\right)^{-1} = 1 + \frac{(1 + 3b'')(1 + 3b'' + 2b)}{\left(b + b''(1 + 3b'' + 3b)\right)(1 + 3b'' + b)} \quad \text{(A10.17)}$$

$$\left(R_\infty^{\pm1,\mp1}\right)^{-1} = 1 + \frac{(1 + 3b'')(1 + 3b'' + 3b)}{b^2 + b''(1 + 3b'' + 3b)(1 + 3b'' + b)} \quad \text{(A10.18)}$$

where $b \equiv W_n/(2W_e)$, and $b'' \equiv \tau_{HE}^{-1}/(6W_e)$ $\left(= \frac{1}{3}T_1^o\tau_{HE}^{-1}\right)$. For $R_\infty^{p,obs}$, the pump position is at $m_I = p$, and the observing position is at $m_I = obs$, in a three-line motionally narrowed spectrum.

For Heisenberg spin exchange between the pump position p and many other sites, the CW-ELDOR reduction factor that corresponds to Eq. 10.58 for the effective T_1 in Section 10.13 is (Marsh 1992):

$$\left(R_\infty^{p,obs}\right)^{-1} = 1 + \frac{1}{Z_pT_1^o\tau_{HE}^{-1}} \quad \text{(A10.19)}$$

where Z_p is the fractional spin population at the position p of the microwave pump, T_1^o is the spin–lattice relaxation time in the absence of exchange, and $\tau_{HE}^{-1} = k_{HE}N$ is the pseudo-first-order rate constant for spin exchange. Equation A10.19 applies to any observing position, *obs*, that undergoes Heisenberg exchange with the pump position. Including nuclear relaxation for a ^{14}N-nitroxide, the reduction factors that correspond to the effective T_1 s of Eqs. 10.60 and 10.61 then become (Marsh 2016b):

$$\left(R_{\infty}^{0,\pm1}\right)^{-1}=1+\frac{1+T_1^o\tau_{HE}^{-1}+b}{Z_pT_1^o\tau_{HE}^{-1}\left(1+T_1^o\tau_{HE}^{-1}+3b\right)} \quad \text{(A10.20)}$$

$$\left(R_{\infty}^{\pm1,0}\right)^{-1}=\left(R_{\infty}^{\pm1,\mp1}\right)^{-1}=\left(R_{\infty}^{0,\pm1}\right)^{-1}+\frac{b\left(1+T_1^o\tau_{HE}^{-1}\right)}{Z_pT_1^o\tau_{HE}^{-1}\left(1+T_1^o\tau_{HE}^{-1}+3b\right)\left(1+T_1^o\tau_{HE}^{-1}+b\right)} \quad \text{(A10.21)}$$

Equations A10.20 and A10.21 apply to observation at any point within the specified observing manifold, except that which corresponds to the nitroxide orientation θ_p at the pump position (see Figure A10.3). As in the case of ^{15}N-nitroxides, Eqs. A10.20 and A10.21 both reduce to Eq. A10.19 when $b=0$, and for $\tau_{HE}^{-1}=0$ they reduce to $R_{\infty}=0$ because then there is no pathway to the observing position. Observation at the unique position that corresponds to nitroxide orientation θ_p results in the following reduction factors (Marsh 1992):

$$\left(R_{\infty}^{0,\pm1}\right)^{-1}=1+\frac{1+T_1^o\tau_{HE}^{-1}}{b+Z_pT_1^o\tau_{HE}^{-1}\left(1+T_1^o\tau_{HE}^{-1}+3b\right)} \quad \text{(A10.22)}$$

$$\left(R_{\infty}^{\pm1,0}\right)^{-1}=1+\frac{\left(1+T_1^o\tau_{HE}^{-1}\right)\left(1+T_1^o\tau_{HE}^{-1}+2b\right)}{\left(b+Z_pT_1^o\tau_{HE}^{-1}\left(1+T_1^o\tau_{HE}^{-1}+3b\right)\right)\left(1+T_1^o\tau_{HE}^{-1}+b\right)} \quad \text{(A10.23)}$$

$$\left(R_{\infty}^{\pm1,\mp1}\right)^{-1}=1+\frac{\left(1+T_1^o\tau_{HE}^{-1}\right)\left(1+T_1^o\tau_{HE}^{-1}+3b\right)}{b^2+Z_pT_1^o\tau_{HE}^{-1}\left(1+T_1^o\tau_{HE}^{-1}+3b\right)\left(1+T_1^o\tau_{HE}^{-1}+b\right)} \quad \text{(A10.24)}$$

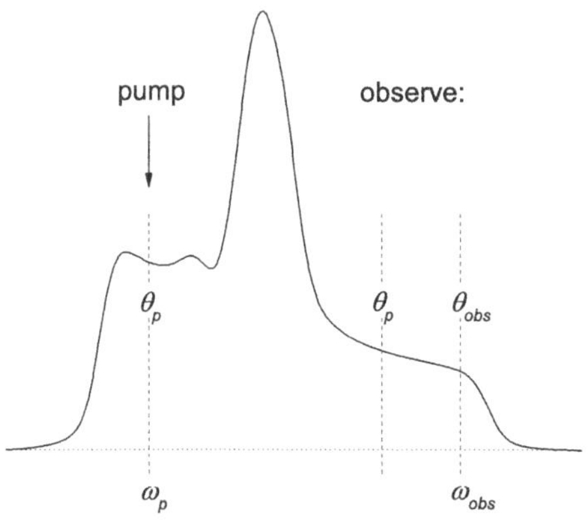

FIGURE A10.3 CW-ELDOR in absorption powder pattern of ^{14}N-nitroxide. Pump position ω_p corresponds to nitroxide orientation θ_p in one particular ^{14}N-hyperfine manifold. Observe position ω_{obs} corresponds to general orientation θ_{obs} in a different manifold (or same manifold). Special case: observe in different manifold but at specific position $\theta_{obs}=\theta_p$.

For $b=0$, Eqs. A10.22–A10.24 all reduce to Eq. A10.19. For $Z_p=\frac{1}{3}$, Eqs. A10.22–A10.24 reduce to Eqs. A10.16–A10.18.

APPENDIX A10.3: CROSS RELAXATION AND m_I-DEPENDENT INTRINSIC SPIN–LATTICE RELAXATION RATES FOR ^{14}N-NITROXIDES

We now incorporate cross relaxation and different electron spin–lattice transition probabilities W_{e,m_I} into Eqs. 10.2 and 10.3 of Section 10.2. The steady-state rate equations for the spin population differences in the $m_I=0,\pm1$ hyperfine levels then become (see Figure A10.4 and cf. Eq. A10.1):

$$\frac{dn_0}{dt}=-2W_{e,0}\left(n_0-n_0^o\right)-W_n\left(2n_0-n_{+1}-n_{-1}\right)$$
$$-W_x\left(2n_0+n_{+1}+n_{-1}\right)-2Wn_0\delta_{mI,p}=0 \quad \text{(A10.25)}$$

$$\frac{dn_{\pm1}}{dt}=-2W_{e,\pm1}\left(n_{\pm1}-n_{\pm1}^o\right)-W_n\left(n_{\pm1}-n_0\right)-W_x\left(n_{\pm}+n_0\right)$$
$$-2Wn_{\pm1}\delta_{mI,p}=0 \quad \text{(A10.26)}$$

where $n_0^o=n_{+1}^o=n_{-1}^o\equiv n^o$, and $W_x=\frac{1}{2}\left(W_{x1}+W_{x2}\right)$ (cf. Eqs. 12.5, 12.6 in Chapter 12).

When irradiating the $m_I=0$ transition, solution of Eqs. A10.25 and A10.26 gives a steady-state spin population that conforms to Eq. A10.2 with $p\equiv0$. The effective spin–lattice relaxation time is then (Marsh 2016a):

$$T_{1,0}^{eff}=\frac{1}{2W_{e,0}}\frac{1+d_1+b+c}{1+d_1+(3+2d_1)(b+c)+8bc} \quad \text{(A10.27)}$$

where $b\equiv W_n/2W_{e,0}$, $c\equiv W_x/2W_{e,0}$ and $d_1\equiv\delta_1/W_{e,0}$, when $W_{e,\pm1}\equiv W_{e,0}+\delta_1\pm\delta$. Equation A10.27 is correct to first order in the difference 2δ between the spin–lattice relaxation rates of the $m_I=\pm1$ transitions. The j^{gA} term in Eq. 5.19 for the

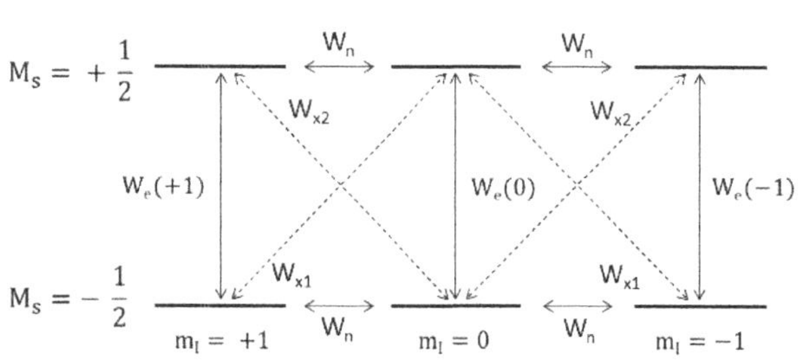

FIGURE A10.4 Energy levels and transitions for ^{14}N-nitroxide $\left(S=\frac{1}{2},\ I=1\right)$. Transition probabilities per unit time for leaving a particular state are: $W_e\left(m_I\right)$ for electron transitions (Eq. 5.19), W_n for nuclear transitions (Eq. 5.20) and W_{x1}, W_{x2} for cross relaxation (Eqs. 5.21, 5.22). $W_e(0)\equiv W_e-\frac{2}{3}\delta_1$, $W_e(+1)-W_e(-1)\equiv\delta$.

transition probability contributes to the effective T_1-relaxation time only in second order. For $c = 0 = d_1$, Eq. A10.27 reduces to Eq. 10.4, which includes only nuclear relaxation. The effective population difference at Boltzmann equilibrium in Eq. A10.2 is given by:

$$n_0^{o,eff} = 2W_{e,0}T_{1,0}^{eff}\left(1+\frac{2(b-c)}{1+d_1+b+c}\right)n^o \tag{A10.28}$$

for the coupled system. This reduces to $n_0^{o,eff} = n^o$ when $c = 0 = d_1$, as in Section 10.2.

When we irradiate the $m_I = \pm 1$ transitions, solution of Eqs. A10.25 and A10.26 together with Eq. A10.2 yields an effective spin–lattice relaxation time that is given by:

$$T_{1,\pm 1}^{eff} = \frac{1}{2W_{e,0}(1+d_1+b+c)}\left(1+\frac{(b-c)^2 \mp (1+2b+2c)d}{1+d_1+(3+2d_1)(b+c)+8bc}\right) \tag{A10.29}$$

where $d \equiv \delta/W_{e,0}$. Again, Eq. A10.29 is correct to first order in δ. With $c = 0$, $d_1 = 0$ and $d = 0$, Eq. A10.29 reduces to Eq. 10.5, which includes only nuclear relaxation. Correspondingly, the effective population difference at Boltzmann equilibrium is given by:

$$n_{\pm 1}^{o,eff} = 2W_{e,0}T_{1,\pm 1}^{eff} \times \left(1+d_1 \pm d+\frac{(b-c)\left(1+2b+(d_1 \mp d)(1+b-c)\right)}{1+3(b+c)+(b+c)^2+4bc+(d_1 \mp d)(1+2b+2c)}\right)n^o \tag{A10.30}$$

which reduces to $n_{\pm 1}^{o,eff} = n^o$ when $c = 0$, $d_1 = 0$ and $d = 0$.

For CW-ELDOR, assume that we pump the $m_I = 0$ hyperfine manifold and observe at the $m_I = \pm 1$ manifold. From Eqs. A10.2 and A10.26, we find that the ELDOR reduction factor again is given by Eq. A10.11, with $p \equiv 0$. The reduction factor extrapolated to infinite microwave power $R_\infty^{p,obs}$ then becomes:

$$R_\infty^{0,\pm 1} = \frac{b-c}{(1+d_1 \pm d)\left(n^o/n_0^{o,eff}\right)+b-c} \tag{A10.31}$$

With $c = 0$, $d_1 = 0$ and $d = 0$, Eq. A10.31 reduces to Eq. 10.12, which includes only nuclear relaxation. When pumping the $m_I = \pm 1$ transition, the reduction factor at infinite power is given by:

$$\frac{1}{R_\infty^{\pm 1,0}} = 1+\frac{n^o}{n_{\pm 1}^{o,eff}}\left(\frac{1}{b-c}+\frac{1+d_1 \mp d}{1+d_1 \pm d+b+c}\right) \tag{A10.32}$$

if observing the $m_I = 0$ transition and by:

$$\frac{1}{R_\infty^{\pm 1,\mp 1}} = 1+\frac{n^o/n_{\pm 1}^{o,eff}}{b-c}\left(1+\frac{(1+2d_1 \mp 2d)(1+2b+2c)}{b-c}\right) \tag{A10.33}$$

if observing the $m_I = \mp 1$ transition. With $c = 0$, $d_1 = 0$ and $d = 0$, Eqs. A10.32 and A10.33 reduce to Eqs. 10.13 and 10.14, respectively, which include only nuclear relaxation.

The lower panel of Figure A10.2 in Appendix A10.1 shows the dependence of the normalized relaxation rates b, c, d_1 and d, used in this appendix, on correlation time for isotropic rotational diffusion of ^{14}N-nitroxides. Nuclear relaxation dominates for correlation times longer than 5×10^{-11} s. At shorter correlation times, the constant relative rates for cross relaxation and m_I-dependent contributions to the intrinsic electron spin–lattice relaxation become more significant. Their values are: $c = 3.72$, $d_1 = 6.38$ and $d = 4.52$, independent of the correlation time for isotropic diffusion.

APPENDIX A10.4: HEISENBERG-EXCHANGE RATE CONSTANTS FOR PARAMAGNETIC RELAXANTS

TABLE A10.1 Spin-exchange rate constants τ_{HE}^{-1}/c_R for ferricyanide $K_3Fe(CN)_6$, bis-diphenyl chromium $Cr(C_6H_5\text{-}C_6H_5)_2I$, or $NiCl_2$, and different nitroxides in water or aqueous buffer[a]

	$\tau_{HE}^{-1}/c_R \times 10^{-9}\,(\mathrm{M^{-1}\,s^{-1}})$		
nitroxide[b]	$Fe(CN)_6^{3-}$	$Cr(C_6H_5-C_6H_5)_2^+$	Ni^{2+}
TEMPONE	1.64 ± 0.02		1.86 ± 0.03
	1.22	2.84	
TEMPOL	1.48 ± 0.07		1.48 ± 0.07
TEMPO-OOCMe	0.91	2.45	
TEMPO-OOCCH$_2$I	0.77	2.12	
TEMPO-OCH$_2$CH$_2$CN	1.08	2.48	
TEMPENE-COOMe	0.88	2.50	
TEMPO-NH$_2$	0.98	2.73	
PROXYL-NH$_2$	0.91	2.45	
TEMPO-NH$_3^+$	3.50	2.00	0.90 ± 0.03
PROXYL-NH$_3^+$	1.68	2.00	
PYCOO$^-$	0.18	3.15	
TEMPENE-COO$^-$	0.23	2.90	
TEMPENE-CH$_2$COO$^-$	0.28	2.73	
TEMPO-O(CH$_2$)$_2$COO$^-$	0.35	2.27	
TEMPO-SO$_4^-$	0.25 ± 0.03		
TEMPO-PO$_4^{2-}$	0.063 ± 0.004		

[a] Determined from exchange broadening: $\tau_{HE}^{-1} = (3\sqrt{3}/4)\gamma_e \Delta\Delta B_{pp}$ for ^{14}N-nitroxides ($I = 1$), where $\Delta\Delta B_{pp}$ is increase in first-derivative peak-to-peak linewidth and c_R molar concentration of relaxant. Values without uncertainty range are in 60 mM phosphate buffer at ambient temperatures (Grebenshchikov et al. 1972b); those with uncertainty range are in water, pH 4–6, at 25±1°C (Keith et al. 1977); except data for unprotonated TEMPO-NH$_2$ and PROXYL-NH$_2$ are from Molin et al. (1980).

[b] Nitroxides: TEMPONE, 1-oxyl-2,2,6,6-tetramethylpiperidin-4-yl oxide; TEMPOL, 1-oxyl-2,2,6,6-tetramethylpiperidin-4-ol; TEMPO-NH$_2$, 1-oxyl-2,2,6,6-tetramethylpiperidin-4-ylamine; PROXYL-NH$_2$, 1-oxyl-2,2,5,5-tetramethylpyrrolidin-3-ylamine; TEMPENE-COOMe, 1-oxyl-2,2,6,6-tetramethylpiperid-3-en-4-ylcarboxymethyl ester; TEMPO-OOCMe, 1-oxyl-2,2,6,6-tetramethylpiperidin-4-yl acetate; TEMPO-OOCCH$_2$I, 1-oxyl-2,2,6,6-tetramethylpiperidin-4-yl iodoacetate; TEMPO-OCH$_2$CH$_2$CN, 1-oxyl-2,2,6,6-tetramethylpiperidin-4-yloxyethyl cyanide; PYCOO$^-$, 1-oxyl-2,2,5,5-tetramethylpyrrolin-3-yl carboxylate; TEMPENE-COO$^-$, 1-oxyl-2,2,6,6-tetramethylpiperid-3-en-4-yl carboxylate; TEMPENE-CH$_2$COO$^-$, 1-oxyl-2,2,6,6-tetramethylpiperid-3-en-4-ylmethyl carboxylate; TEMPO-O(CH$_2$)$_2$COO$^-$, 1-oxyl-2,2,6,6-tetramethylpiperidin-4-yloxyethyl carboxylate; TEMPO-SO$_4$, 1-oxyl-2,2,6,6-tetramethylpiperidin-4-yl sulphate; TEMPO-PO$_4$, 1-oxyl-2,2,6,6-tetramethylpiperidin-4-yl phosphate.

TABLE A10.2 Spin-exchange rate constants τ_{HE}^{-1}/c_R for PADS $(SO_3^-)_2NO$ (Pearson and Buch 1962), TEMPONE (Keith et al. 1977; Anisimov et al. 1971; Skubnevskaya and Molin 1967), TEMPO and TEMPO-NH_3^+; (Anisimov et al. 1971; Molin et al. 1980) with aqueous paramagnetic ions or their complexes[a]

	$\tau_{HE}^{-1}/c_R \times 10^{-9}\ (M^{-1}\ s^{-1})$			
relaxant	$PADS^{2-}$	TEMPONE	TEMPO	TEMPO-NH_3^+
$Cr(H_2O)_6^{3+}$	15.4	1.24 ± 0.07	1.2	0.65
$Co(H_2O)_6^{2+}$	14.0	0.81 ± 0.07		
$Ni(H_2O)_6^{2+}$	14.3	1.86 ± 0.03		
$Cu(H_2O)_6^{2+}$	14.3	1.4 2.66 ± 0.02 2.20 ± 0.02	2.9	1.2
$Mn(H_2O)_6^{2+}$		2.07 ± 0.07	2.0	1.1
$Fe(H_2O)_6^{2+}$		0.9		
$VO(H_2O)_5^{2+}$		1.2		
$Ce(H_2O)_n^{3+}$	2.4			
$Pr(H_2O)_n^{3+}$	1.2			
$Nd(H_2O)_n^{3+}$	1.5			
$Gd(H_2O)_n^{3+}$	15.4			
$Cr(en)_3^{3+}$	15.4			
$Cr(NH_3)_5Cl^{2+}$	14.5			
$Cr(en)_2(NCS)_2^+$	8.8			
$Cr(en)(C_2O_4)_2^-$	2.3			
$Cr(CN)_6^{3-}$	0.17			
$Cu(en)_2(H_2O)_2^{2+}$		2.59 ± 0.07	2.9	1.2
Cu-tris		1.43 ± 0.02		
Cu-EDTA		2.09 ± 0.04		
$Ni(en)_2(H_2O)_2^{2+}$		2.10 ± 0.07		
Ni-tris		0.97 ± 0.01		
Ni-EDTA		0.74 ± 0.01		
$Fe(CN)_6^{3-}$		1.64 ± 0.02		

[a] Determined from exchange broadening: $\tau_{HE}^{-1} = (3\sqrt{3}/4)\gamma_e \Delta\Delta B_{pp}$ for ^{14}N-nitroxides ($I = 1$), where $\Delta\Delta B_{pp}$ is increase in first-derivative peak-to-peak linewidth, and c_R molar concentration of relaxant.
Ligand: en, $NH_2CH_2CH_2NH_2$.
Nitroxides: PADS, peroxylamine disulphonate $(SO_3^-)_2NO$; TEMPONE, 1-oxyl-2,2,6,6-tetramethylpiperidin-4-yl oxide; TEMPO, 1-oxyl-2,2,6,6-tetramethylpiperidine; TEMPO-NH_3^+, 1-oxyl-2,2,6,6-tetramethylpiperidin-4-ylamine.

TABLE A10.3 Spin-exchange rate constants τ_{HE}^{-1}/c_R for 10-DOXYL phosphatidylcholine[a] with paramagnetic-ion salts in methanol (or methanol–water) (Livshits et al. 2001)

relaxant	$\tau_{HE}^{-1}/c_R \times 10^{-9}\ (M^{-1}\ s^{-1})$[b]
$Ni(ClO_4)_2$	1.3 ± 0.1
$NiCl_2$	1.1 ± 0.1
$NiSO_4$	0.67 ± 0.1
$Cu(ClO_4)_2$	1.9 ± 0.1
$CuCl_2$	1.8 ± 0.1
$CoCl_2$	0.13 ± 0.1

(Continued)

TABLE A10.3 (*Continued*) Spin-exchange rate constants τ_{HE}^{-1}/c_R for 10-DOXYL phosphatidylcholine[a] with paramagnetic-ion salts in methanol (or methanol–water) (Livshits et al. 2001)

relaxant	$\tau_{HE}^{-1}/c_R \times 10^{-9}\ (M^{-1}\ s^{-1})$[b]
$DyCl_3$	0.05
$MnCl_2$	1.6 ± 0.1
$MnSO_4$ [c]	0.77 ± 0.2
$K_3Fe(CN)_6$ [c]	0.072 ± 0.02

[a] 10-DOXYL phosphatidylcholine: 1-acyl-2-(10-(3-oxyl-4,4-dimethyloxazolidin-2-yl) stearoyl)-*sn*-glycero-3-phosphocholine.

[b] From exchange broadening: $\tau_{HE}^{-1} = (3\sqrt{3}/4)\gamma_e(\Delta\Delta B_{pp} + 0.0783\ \mathrm{mT})$, where $\Delta\Delta B_{pp}$ is increase in first-derivative peak-to-peak linewidth (plus correction for inhomogeneous broadening in DOXYL labels) and c_R molar concentration of relaxant. Diffusion-controlled rate constant for bimolecular collisions at 20°C is: $k_{2,diff} = \frac{8}{3} \cdot 10^3 N_A k_B T/\eta = 11.10$ and $3.58 \times 10^9\ M^{-1}\ s^{-1}$ in methanol and 1:1 v/v methanol-water, respectively.

[c] In 1:1 v/v methanol-water.

REFERENCES

Abragam, A. 1961. *The Principles of Nuclear Magnetism*, Oxford: Oxford University Press.

Abragam, A. and Bleaney, B. 1970. *Electron Paramagnetic Resonance of Transition Ions*, Oxford: Oxford University Press.

Alsaadi, B.M., Rossotti, F.J.C., and Williams, R.J.P. 1980. Electron relaxation rates of lanthanide aquo-cations. *J. Chem. Soc.-Dalton Trans.* 11:2147–2150.

Amdur, I. and Hammes, G.G. 1966. *Chemical Kinetics. Principles and Selected Topics*, New York: McGraw-Hill Book Company.

Anisimov, O.A., Nikitaev, A.T., Zamaraev, K.I., and Molin, Yu.N. 1971. Separation of exchange and dipole-dipole broadening on the basis of viscosity changes in ESR spectra (translated from Russian). *Theor. Expmtl. Chem.* 7:556–559.

Banci, L., Bertini, I., Briganti, F., and Luchinat, C. 1986. The electron nucleus dipolar coupling in slow rotating systems. 4. The effect of zero-field splitting and hyperfine coupling when S=5/2 and I=5/2. *J. Magn. Reson.* 66:58–65.

Banci, L., Bertini, I., and Luchinat, C. 1991. *Nuclear and Electron Relaxation. The Magnetic Nucleus-Unpaired Electron Coupling in Solution*, Weinheim New York: VCH.

Bertini, I., Briganti, I., Xia, Z.C., and Luchinat, C. 1993. Nuclear magnetic relaxation dispersion studies of hexaaquo Mn(II) ions in water glycerol mixtures. *J. Magn. Reson. Series A* 101:198–201.

Bertini, I. and Luchinat, C. 1986. *NMR of Paramagnetic Molecules in Biological Systems*, Menlo Park, CA: Benjamin/Cummings.

Bloembergen, N. 1949. On the interaction of nuclear spins in a crystalline lattice. *Physica* 15:386–426.

Chmelnik, A.M. and Fiat, D. 1967. Magnetic resonance studies of ion solvation. Hydration of cobaltous ion. *J. Chem. Phys.* 47:3986–3990.

Connick, R.E. and Fiat, D. 1966. Oxygen-17 nuclear magnetic resonance study of hydration shell of nickelous ion. *J. Chem. Phys.* 44:4103–4107.

Fiat, D., Luz, Z., and Silver, B.L. 1968. Solvation of Co(II) in methanol and water enriched in oxygen-17. *J. Chem. Phys.* 49:1376–1379.

Friedman, H.L., Holz, M., and Hertz, H.G. 1979. EPR relaxations of aqueous Ni^{2+} ion. *J. Chem. Phys.* 70:3369–3383.

Goldman, S.A., Bruno, G.V., and Freed, J.H. 1973. ESR studies of anisotropic rotational reorientation and slow tumbling in liquid and frozen media. 2. Saturation and nonsecular effects. *J. Chem. Phys.* 59:3071–3091.

Grebenshchikov, Yu.B., Ponomarev, G.V., Yevstigneyeva, R.P., and Likhtenshtein, G.I. 1972a. Spin relaxation of stable 2,2,6,6-tetramethyl-4-oxy-piperidine-1-oxyl radical on interaction in solutions with the porphyrin complexes with VO(II), Fe(III) Ni(II), Cu(II) and haemoglobin (translated from Russian). *Biophysics* 956–959.

Grebenshchikov, Yu.B., Likhtenshtein, G.I., Ivanov, V.P., and Rozantsev, E.G. 1972b. Investigation of electrostatic charges in proteins by the paramagnetic probe method (translated from Russian). *Mol. Biol.* 6: 400–406.

Hausser, R. and Noack, F. 1964. Kernmagnetische Relaxation und Korrelation in Zwei-Spin-Systemen. *Zeitschrift fur Physik* 182:93–98.

Horváth, L.I., Brophy, P.J., and Marsh, D. 1993. Exchange rates at the lipid-protein interface of the myelin proteolipid protein determined by saturation transfer electron spin resonance and continuous wave saturation studies. *Biophys. J.* 64:622–631.

Hwang, J.S., Mason, R.P., Hwang, L.P., and Freed, J.H. 1975. Electron spin resonance studies of anisotropic rotational reorientation and slow tumbling in liquid and frozen media. 3. Perdeuterated 2,2,6,6-tetramethyl-4-piperidone *N*-oxide and analysis of fluctuating torques. *J. Phys. Chem.* 79:489–511.

Hyde, J.S., Chien, J.C.W., and Freed, J.H. 1968. Electron-electron double resonance of free radicals in solution. *J. Chem. Phys.* 48:4211–4226.

Hyde, J.S. and Dalton, L.R. 1979. Saturation-transfer spectroscopy. In *Spin-Labeling II. Theory and Applications*, ed. Berliner, L.J., 1–70. New York: Academic Press.

Hyde, J.S. and Subczynski, W.K. 1984. Simulation of electron spin resonance spectra of the oxygen-sensitive spin label probe CTPO. *J. Magn. Reson.* 56:125–130.

Hyde, J.S. and Subczynski, W.K. 1989. Spin-label oximetry. In *Biological Magnetic Resonance*, Vol. 8, Spin Labeling. Theory and Applications, eds. Berliner, L.J. and Reuben, J., 399–425. New York and London: Plenum Press.

Keith, A.D., Snipes, W., Mehlhorn, R.J., and Gunter, T. 1977. Factors restricting diffusion of water-soluble spin labels. *Biophys. J.* 19:205–218.

Kooser, R.G., Volland, W.V., and Freed, J.H. 1969. ESR relaxation studies on orbitally degenerate free radicals. I. Benzene anion and tropenyl. *J. Chem. Phys.* 50:5243–5247.

Kulikov, A.V. and Likhtenstein, G.I. 1977. Use of spin relaxation phenomena in investigation of structure of model and biological systems by method of spin labels. *Adv. Mol. Relax. Processes* 10:47–79.

Leigh Jr., J.S. 1970. ESR rigid-lattice line shape in a system of two interacting spins. *J. Chem. Phys.* 52: 2608–2612.

Lewis, W.B., Alei Jr., M., and Morgan, L.O. 1966. Magnetic-resonance studies on copper(II) complex ions in solution. II. Oxygen-17 NMR and copper(II) EPR in aqueous solutions of $Cu(en)(H_2O)_4^{2+}$ and $Cu(en)_2(H_2O)_2^{2+}$. *J. Chem. Phys.* 45: 4003–4013.

Lindner, U. 1965. Protonenrelaxation in paramagnetischen Lösungen unter Berücksichtigung der Nullfeldaufspaltung. *Annalen der Physik* B 16:319–335.

Livshits, V.A., Dzikovski, B.G., and Marsh, D. 2001. Mechanism of relaxation enhancement of spin labels in membranes by paramagnetic ion salts: dependence on 3*d* and 4*f* ions and on the anions. *J. Magn. Reson.* 148:221–237.

Livshits, V.A., Páli, T., and Marsh, D. 1998. Relaxation time determinations by progressive saturation EPR: Effects of molecular motion and Zeeman modulation for spin labels. *J. Magn. Reson.* 133:79–91.

Marsh, D. 1992. Influence of nuclear relaxation on the measurement of exchange frequencies in CW saturation EPR studies. *J. Magn. Reson.* 99:332–337.

Marsh, D. 1995a. Paramagnetic relaxation enhancement in continuous wave saturation EPR experiments with nuclear relaxation. *Spectrochim. Acta A* 51:L1–L6.

Marsh, D. 1995b. Erratum to "Paramagnetic relaxation enhancement in continuous wave saturation EPR experiments with nuclear relaxation" [Spectrochim. Acta Part A, 51 (1995) L1-L6]. *Spectrochim. Acta A* 51:1453.

Marsh, D. 2013. *Handbook of Lipid Bilayers*, Boca Raton, FL: CRC Press.

Marsh, D. 2016a. Cross relaxation in nitroxide spin labels. *J. Magn. Reson.* 272:172–180.

Marsh, D. 2016b. Nuclear spin-lattice relaxation in nitroxide spin-label EPR. *J. Magn. Reson.* 272:166–171.

Marsh, D., Páli, T., and Horváth, L.I. 1998. Progressive saturation and saturation transfer EPR for measuring exchange processes and proximity relations in membranes. In *Biological Magnetic Resonance*, Vol. 14, Spin Labeling. The Next Millenium, ed. Berliner, L.J., 23–82. New York: Plenum Press.

McGarvey, B.R. 1957. Linewidths in the paramagnetic resonance of transition ions in solution. *J. Phys. Chem.* 61:1232–1237.

McLaughlin, A., Gratwohl, G., and McLaughlin, S. 1978. The adsorption of divalent cations to phosphatidylcholine bilayer membranes. *Biochim. Biophys. Acta* 513:338–357.

Molin, Y.N., Salikov, K.M., and Zamaraev, K.I. 1980. *Spin Exchange. Principles and Applications in Chemistry and Biology*, Berlin: Springer Verlag.

Páli, T., Bartucci, R., Horváth, L.I., and Marsh, D. 1992. Distance measurements using paramagnetic ion-induced relaxation in the saturation transfer electron spin resonance of spin-labeled biomolecules. Application to phospholipid bilayers and interdigitated gel phases. *Biophys. J.* 61:1595–1602.

Páli, T., Finbow, M.E., and Marsh, D. 1999. Membrane assembly of the 16-kDa proteolipid channel from *Nephrops norvegicus* studied by relaxation enhancements in spin-label ESR. *Biochemistry* 38:14311–14319.

Pearson, R.G. and Buch, T. 1962. Electron paramagnetic relaxation in ionic solutions. *J. Chem. Phys.* 36:1277–1282.

Percival, P.W., Hyde, J.S., Dalton, L.A., and Dalton, L.R. 1975. Molecular and applied modulation effects in electron-electron double resonance. 5. Passage effects in high-resolution frequency and field swept ELDOR. *J. Chem. Phys.* 62:4332–4342.

Plachy, W. and Kivelson, D. 1967. Spin exchange in solutions of di-tertiary-butyl nitroxide. *J. Chem. Phys.* 47:3312–3318.

Robinson, R.A. and Stokes, R.H. 1955. *Electrolyte Solutions*, London: Butterworths.

Salikov, K.M., Doctorov, A.B., and Molin, Yu.N. 1971a. Exchange broadening of ESR lines for solutions of free radicals and transition metal complexes. *J. Magn. Reson.* 5:189–205.

Salikhov, K.M., Nikitaev, A.T., Senyukova, G.A., and Zamaraev, K.I. 1971b. Exchange broadening of ESR lines in dilute solutions containing two types of paramagnetic particles (translated from Russian). *Theoret. Expmtl. Chem.* 7:503–507.

Skubnevskaya, G.I. and Molin, Yu.N. 1967. Exchange interactions of aquo-complexes of ions of the iron group with paramagnetic particles in solutions (translated from Russian). *Kinet. Catal.* 8:1012–1016.

Skubnevskaya, G.I., Salikhov, K.M., Smirnova, L.M., and Molin, Yu.N. 1970. Effect of electron spin relaxation on exchange broadening of EPR lines in dilute solutions (translated from Russian). *Kinet. Catal.* 11:733–737.

Skubnevskaya, G.I., Zaev, E.E., Zusman, R.I., and Molin, Y.N. 1966. Influence of complex formation on the effectiveness of electron exchange interactions in solutions (translated from Russian). *Doklady Phys. Chem.* 170:603–606.

Slichter, C.P. 1978. *Principles of Magnetic Resonance*, Berlin-Heidelberg-New York: Springer-Verlag.

Snel, M.M.E. and Marsh, D. 1994. Membrane location of apocytochrome *c* and cytochrome *c* determined from lipid-protein spin exchange interactions by continuous wave saturation electron spin resonance. *Biophys. J.* 67:737–745.

Solomon, I. 1955. Relaxation processes in a system of 2 spins. *Phys. Rev.* 99:559–565.

Stephens, E.M. and Grisham, C.M. 1979. Li-7 nuclear magnetic resonance, water proton nuclear magnetic resonance and gadolinium electron paramagnetic resonance studies of the sarcoplasmic reticulum calcium-ion transport adenosine triphosphatase. *Biochemistry* 18:4876–4885.

Subczynski, W. and Hyde, J.S. 1981. The diffusion concentration product of oxygen in lipid bilayers using the spin-label T_1 method. *Biochim. Biophys. Acta* 643:283–291.

Subczynski, W.K. and Hyde, J.S. 1984. Diffusion of oxygen in water and hydrocarbons using an electron spin resonance spin label technique. *Biophys. J.* 45:743–748.

Swift, T.J. and Connick, R.E. 1962. NMR relaxation mechanisms of O-17 in aqueous solutions of paramagnetic cations and lifetime of water molecules in first coordination sphere. *J. Chem. Phys.* 37:307–320.

Teng, C.L., Hong, H., Kühne, S., and Bryant, R.G. 2001. Molecular oxygen spin-lattice relaxation in solutions measured by proton magnetic relaxation dispersion. *J. Magn. Reson.* 148:31–34.

Waite, T.R. 1958. General theory of bimolecular reaction rates in solids and liquids. *J. Chem. Phys.* 28:103–106.

Windrem, D.A. and Plachy, W.Z. 1980. The diffusion-solubility of oxygen in lipid bilayers. *Biochim. Biophys. Acta* 600:655–665.

Yin, J.J. and Hyde, J.S. 1987. Application of rate equations to ELDOR and saturation recovery experiments on N-14-N-15 spin label pairs. *J. Magn. Reson.* 74:82–93.

Zitserman, V.Yu. 1971. On Heisenberg exchange of high spins. *Mol. Phys.* 20: 1005–1011.

Nonlinear and Saturation-Transfer EPR

11

11.1 INTRODUCTION

Nonlinear EPR refers to CW-EPR that is performed at higher microwave powers, such that the spin system is partially saturated. The amplitude of the EPR signal then is no longer linearly proportional to the radiation B_1-field, i.e., to the square root of the microwave power. Measurements made under saturating powers are important because they are sensitive to electron spin–lattice relaxation. As we saw in Chapter 10, this allows us to study both much slower motions, and considerably weaker magnetic interactions between spin labels, than is possible simply from linewidths and lineshapes of unsaturated nitroxide EPR spectra.

Progressive-saturation EPR, where we monitor the signal amplitude or intensity as a function of microwave power, is the most common form of nonlinear EPR. Saturation broadening of the EPR lineshape accompanies true saturation of the integrated intensity. Both affect the saturation behaviour of the EPR amplitude, which depends also on the degree of inhomogeneous broadening and on rotational motion of the nitroxide spin label.

In a second class of nonlinear experiment, we detect the EPR spectrum 90°-out-of-phase with respect to the field modulation used in CW-spectrometers. Only at moderately high-microwave powers, where the EPR signal no longer depends linearly on the B_1-field, does the out-of-phase signal have appreciable intensity. The strength of the nonlinear signal therefore increases with increasing spin–lattice relaxation time. Correspondingly, the out-of-phase lineshape depends on rotational motions taking place on the timescale of spin–lattice relaxation. We call this saturation-transfer (ST) EPR, because saturation is alleviated preferentially in those parts of the powder spectrum where the resonance position depends most strongly on orientation of the nitroxide to the static B_o-field direction.

We start the chapter with progressive-saturation EPR by introducing the microwave B_1-field into the Bloch equations. This gives us the response of the EPR amplitudes and linewidths to increasing microwave power. Next, we introduce the modulation field, which gives us the various nonlinear out-of-phase EPR displays. Most important are the first-harmonic absorption spectrum detected 90°-out-of-phase (V_1'-EPR), and the second-harmonic absorption spectrum detected 90°-out-of-phase (V_2'-EPR). V_1'-EPR spectra are sensitive primarily to spin–lattice relaxation and not to ultraslow rotational motion. V_2'-EPR spectra are the classical ST-EPR display, whose lineshapes are exquisitely sensitive to very slow rotational motion. A less used, alternative ST-EPR display is the first-harmonic dispersion spectrum detected 90°-out-of-phase (U_1'-EPR), which also is sensitive to very slow rotation.

For reference, we list the different spectral displays used in nonlinear CW-EPR in Table 11.1.

11.2 PROGRESSIVE SATURATION: BLOCH EQUATIONS

We described the Bloch equations for the transverse components, u and v, of the magnetization in the rotating frame already in Section 6.2 of Chapter 6. These specify the unsaturated lineshape, which depends on the transverse relaxation time T_2 (see Eqs. 6.5 and 6.6). To allow for saturation, we must add the third Bloch equation, that for the M_z-component of the magnetization, which depends on the spin–lattice relaxation time T_1 (Carrington and McLachlan 1969):

$$\frac{du}{dt} = (\omega_0 - \omega)v - \frac{u}{T_2^o} \quad (11.1)$$

$$\frac{dv}{dt} = -(\omega_0 - \omega)u + \gamma_e B_1 M_z - \frac{v}{T_2^o} \quad (11.2)$$

$$\frac{dM_z}{dt} = -\gamma_e B_1 v - \frac{(M_z - M_o)}{T_1} \quad (11.3)$$

TABLE 11.1 Notation for linear and nonlinear continuous-wave EPR displays with field modulation

mode	*harmonic*	*in-phase*[a]	*out-of-phase*[b]
absorption	integral ($n = -1$)	S	-
	zeroth ($n = 0$)	V_0	-
	first ($n = 1$)	V_1	V_1'
	second ($n = 2$)	V_2	V_2'
dispersion	first ($n = 1$)	U_1	U_1'

[a] Detection in-phase with the field modulation (normal display).

[b] Detection 90° out-of-phase with respect to the field modulation (saturation-transfer display). Zero intensity at subsaturating microwave powers.

where M_o is the equilibrium value of M_z in the absence of saturation and γ_e is the electron gyromagnetic ratio. The longitudinal relaxation term in Eq. 11.3 follows directly from Eq. 1.17 for the population difference between $M_S = \pm\frac{1}{2}$ oriented spins that we introduced in Chapter 1. Written in matrix form, the Bloch equations 11.1–11.3 for the magnetization vector $\widehat{\mathbf{M}} \equiv (u, v, M_z)$ in the rotating frame become (Livshits et al. 1998a):

$$\frac{d\widehat{\mathbf{M}}}{dt} = -\mathbf{A}\cdot\widehat{\mathbf{M}} + \mathbf{M}_o \tag{11.4}$$

where $\mathbf{M}_o \equiv M_o(0,0,T_1^{-1})$ and the Bloch matrix is defined by:

$$\mathbf{A} \equiv \begin{pmatrix} (T_2^o)^{-1} & \Delta\omega & 0 \\ -\Delta\omega & (T_2^o)^{-1} & -\gamma_e B_1 \\ 0 & \gamma_e B_1 & T_1^{-1} \end{pmatrix} \tag{11.5}$$

with $\Delta\omega \equiv \omega - \omega_o$ for the offset from the angular resonance frequency ω_o. In the steady-state where $d\widehat{\mathbf{M}}/dt = 0$, solution of Eq. 11.4 is straightforward: $\widehat{\mathbf{M}} = \mathbf{A}^{-1}\cdot\mathbf{M}_o$. This is simply the determinantal solution for linear simultaneous equations 11.1–11.3 that is given by Cramer's rule. We introduce the matrix formalism here, because we need it later in Section 11.4 and following sections, when dealing with the modulation-coupled Bloch equations.

The steady-state solutions for the transverse magnetisation components, u and v, that give the EPR lineshapes under saturation are:

$$u(\omega) = M_o \frac{\gamma_e B_1 (T_2^o)^2 (\omega_o - \omega)}{1 + \gamma_e^2 B_1^2 T_1 T_2^o + (T_2^o)^2 (\omega_o - \omega)^2} \tag{11.6}$$

$$v(\omega) = M_o \frac{\gamma_e B_1 T_2^o}{1 + \gamma_e^2 B_1^2 T_1 T_2^o + (T_2^o)^2 (\omega_o - \omega)^2} \tag{11.7}$$

These reduce to Eqs. 6.9 and 6.10 for weak microwave fields when $\gamma_e^2 B_1^2 T_1 T_2^o << 1$. We therefore define a saturation parameter:

$$\sigma = \gamma_e^2 T_1 T_2^o B_1^2 \equiv P\cdot B_1^2 \tag{11.8}$$

where T_2^o is the intrinsic transverse relaxation time without saturation, and $P \equiv \gamma_e^2 T_1 T_2^o$. The saturation parameter is directly proportional to the microwave radiation power: B_1^2.

Equation 11.6 gives us the EPR dispersion lineshape, and Eq. 11.7 gives the usual absorption lineshape. Comparing Eq. 11.7 with Eq. 6.10, we see that the EPR absorption line is still Lorentzian but with saturation broadening given by:

$$\frac{1}{T_2} = \frac{1}{T_2^o}\sqrt{1 + P\cdot B_1^2} \tag{11.9}$$

The central amplitude of the absorption, $v(\omega_o)$, is reduced in the ratio T_2/T_2^o by this broadening, in addition to the intrinsic decrease in EPR intensity from saturation. From Eq. 11.7, the combined effect is given by:

$$v(\omega_o) = (v_o T_2^o)\frac{B_1}{1 + P\cdot B_1^2} \tag{11.10}$$

where $v_o T_2^o B_1$ is the amplitude in the absence of saturation, with $v_o \equiv \gamma_e M_o$. The intrinsic saturation behaviour, i.e., that of the integrated EPR absorption intensity: $S = \int d\omega\cdot v(\omega)$, is given by integrating Eq. 11.7:

$$S = S_o \frac{B_1}{\sqrt{1 + P\cdot B_1^2}} \tag{11.11}$$

where $S_o B_1$ is the integrated intensity in the absence of saturation, which like the unsaturated amplitude is directly proportional to B_1. We see immediately from Eqs. 11.9 and 11.10 that the reduced degree of saturation of the integrated intensity, relative to $v(\omega_o)$, equals that for the amplitude multiplied by the linewidth.

Figure 11.1 compares progressive-saturation curves of the EPR absorption amplitude, V_0, and integrated intensity, S, for increasing amplitude of the B_1-field (or square root of the microwave power). Initially there is no saturation, and all curves increase linearly with increasing B_1 as indicated by the dashed line in Figure 11.1. The integrated intensity increases monotonically with further increase in B_1, finally reaching a maximum saturating value at high microwave powers. On the other hand, the absorption amplitude displays a biphasic dependence on B_1. The V_0-signal passes through a maximum, as saturation broadening begins to dominate and then decreases steadily to zero. Figure 11.1 also shows the saturation curve for the peak-to-peak amplitude of the normal first-derivative EPR absorption, V_1, which

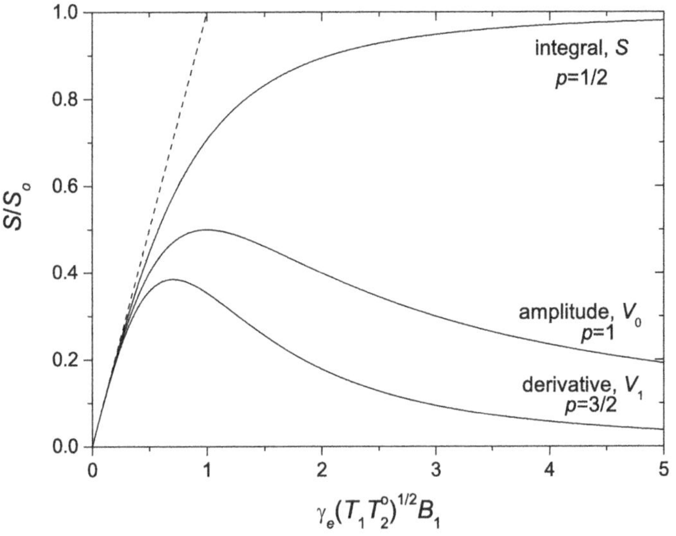

FIGURE 11.1 Saturation curves for integrated intensity S (Eq. 11.11), amplitude V_0 (Eq. 11.10) and first-derivative V_1 (Eq. 11.12 with $n = 1$) of EPR absorption as function of scaled microwave radiation-field intensity: $\sqrt{\sigma} \propto B_1$. Curves normalized to give same initial, linear dependence for unsaturated absorption spectrum (dashed line).

is measured at resonance offsets: $\Delta\omega = \pm\left(\sqrt{3}T_2^o\right)^{-1}\sqrt{1+P\cdot B_1^2}$ (cf. Eq. 1.24 in Chapter 1). Relative to the absorption amplitude, V_0, the derivative saturates more readily: the maximum occurs at a lower B_1 and the subsequent decay to zero is more rapid. This is because saturation broadening is the same, given by Eq. 11.9 for both amplitude and derivative, and therefore its effect is stronger for the derivative display. Figure 11.2 illustrates the saturation behaviour of the first-derivative Lorentzian absorption lineshape according to Eq. 11.7. We see clearly a progressive broadening of the peak-to-peak linewidth with increasing saturation parameter $\sigma \propto B_1^2$. The spectral amplitude is normalized to B_1 and decreases because of both intrinsic loss in intensity and saturation broadening.

Quite generally, we get the saturation curve for the amplitude of the nth derivative of the absorption by successive differentiation of Eq. 11.7:

$$\frac{d^n v}{d\omega^n} = v_{o,n}\left(T_2^o\right)^{n+1}\frac{B_1}{\left(1+P\cdot B_1^2\right)^{n/2+1}} \tag{11.12}$$

Correspondingly, the position of the maximum in the saturation curve is given by:

$$\left(B_1^2\right)_{max} = \frac{1}{P}\left(\frac{1}{n+1}\right) \tag{11.13}$$

which decreases systematically towards lower microwave powers for the higher derivatives. With $n=0$ and $n=-1$, Eq. 11.12 applies also to progressive saturation of the absorption amplitude and integrated intensity, respectively (cf. Eqs. 11.10 and 11.11). For normal first-derivative EPR (Figure 11.2), the exponent in the denominator $\left(1+P\cdot B_1^2\right)^p$ of the saturation curve is $p=3/2$, as compared with $p=1$ and $p=1/2$ for the absorption amplitude and integral, respectively, assuming purely homogeneous broadening.

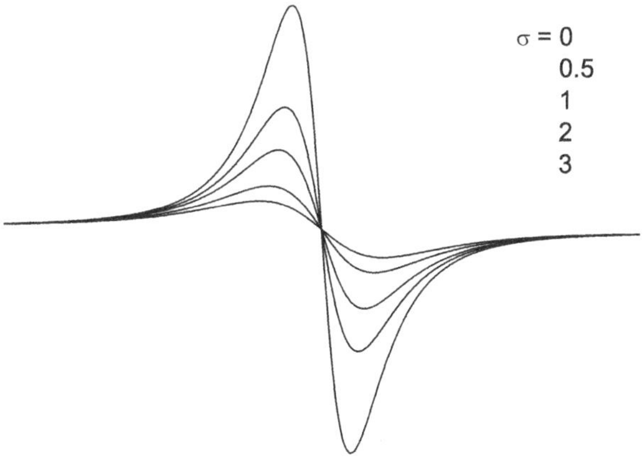

FIGURE 11.2 Saturation of single homogeneous EPR line for increasing values of saturation parameter $\sigma = \gamma_e^2 T_1 T_2^o B_1^2$, where B_1 is microwave magnetic field strength. First-derivative spectral amplitude of EPR absorption (Eq. 11.7) is divided by B_1 to show intrinsic decrease in intensity and effect of saturation broadening. Largest amplitude for $\sigma = 0$, and smallest for $\sigma = 3$.

11.3 PROGRESSIVE SATURATION: INHOMOGENEOUS BROADENING

As we saw in Section 11.2, saturation curves of homogeneously broadened lines differ between the different EPR displays (total intensity, absorption amplitude, first derivative, etc.) because of saturation broadening. Inhomogeneous broadening makes saturation broadening of the individual spin packets less significant. Therefore, the saturation behaviour of the normal first-derivative EPR absorption spectrum depends strongly on the degree of inhomogeneous broadening. In the limit of completely inhomogeneous broadening, the first derivative will saturate as does the integrated intensity of the absorption, i.e., according to Eq. 11.11.

We can allow for inhomogeneous broadening in the first-derivative spectrum by using an empirical expression for the saturation curve (Castner, Jr. 1959):

$$V_1 = V_{1,o}\frac{B_1}{\left(1+P\cdot B_1^2\right)^p} \tag{11.14}$$

where $P \equiv \gamma_e^2 (T_1T_2)^{eff}$ and the exponent p are parameters to be fitted (cf. Eqs. 11.10 and 11.11). The exponent should lie within the range from $p=1/2$ for entirely inhomogeneous broadening to $p=3/2$ for entirely homogeneous broadening. The saturation parameter, P, specifies the effective T_1T_2 relaxation-time product (cf. Eq. 11.8), which is the experimental quantity that we extract from progressive-saturation studies. Instead of fitting the complete saturation curve, we can estimate the $(T_1T_2)^{eff}$ product from the half-saturation power, i.e., the microwave power at which $V_1/(V_{1,o}B_1) = 1/2$, that we get by extrapolating the linear region of the saturation curve (the dashed line in Figure 11.1). This demands good data at low powers. From Eq. 11.14, the microwave magnetic field intensity at half saturation is:

$$\left(B_1^2\right)_{1/2} = \left(\frac{2^{1/p}-1}{\gamma_e^2}\right)\frac{1}{(T_1T_2)^{eff}} \tag{11.15}$$

which needs an informed guess for the value of p.

A more rigorous approach to inhomogeneous broadening is to use the integrated absorption intensity, S, instead of the first-derivative amplitude V_1 (Páli et al. 1993). This eliminates the effects of saturation broadening, and hence of inhomogeneous broadening too. We can show this quite generally. Individual spin packets saturate independently in a homogeneous way, and the EPR absorption amplitude is (Portis 1953; and see Eq. 1.21 in Chapter 1):

$$v(\omega-\omega', B_1) = \frac{v_o B_1 g(\omega-\omega')}{1+\pi\gamma_e^2 B_1^2 g(\omega-\omega')T_1} \tag{11.16}$$

where $g(\omega-\omega')$ is the lineshape for a spin packet centred about ω'. The spin packets are distributed according to the inhomogeneous envelope function $h(\omega'-\omega_o)$, which is centred about ω_o (see Figure 11.3). We get the EPR absorption lineshape by

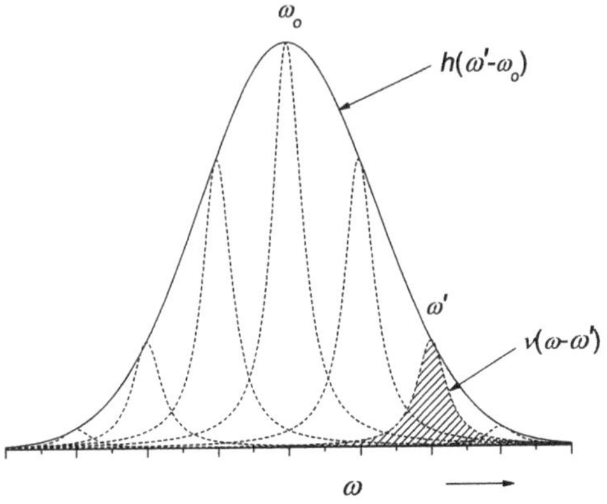

FIGURE 11.3 Inhomogeneous broadening of EPR absorption lineshape. Partially saturated homogeneous spin packets $v(\omega-\omega')$, each centred about ω', are distributed according to inhomogeneous envelope function $h(\omega'-\omega_o)$ that is centred about ω_o.

convoluting the homogeneous lineshape with the inhomogeneous envelope function (see Section 2.6, and Fajer et al. 1992):

$$V_{0,inh}(\omega,B_1)=\int_{-\infty}^{\infty}\mathrm{d}\omega'\cdot h(\omega'-\omega_o)v(\omega-\omega',B_1) \quad (11.17)$$

The total integrated intensity of the powder absorption spectrum (or equivalently the double integral of the usual first-derivative spectrum) is then:

$$S_{inh}(B_1)\equiv\int_{-\infty}^{\infty}\mathrm{d}\omega\cdot V_{0,inh}(\omega,B_1)$$

$$=\int_{-\infty}^{\infty}\mathrm{d}\omega'\cdot h(\omega'-\omega_o)\int_{-\infty}^{\infty}\mathrm{d}\omega\cdot v(\omega-\omega',B_1) \quad (11.18)$$

A simple change of variables shows that the two integrals are separable. The integral over the inhomogeneous envelope function reduces to the normalization condition: $\int\mathrm{d}\omega'\cdot h(\omega')=1$, independent of ω_o. Therefore, the integral over the entire absorption lineshape saturates as does the integrated intensity of an individual spin packet, viz. $\int\mathrm{d}\omega\cdot v(\omega,B_1)$, independent of ω'. For Lorentzian spin packets, this is simply the saturation curve given by Eq. 11.11.

We now see the advantage of progressive-saturation experiments that use the integrated absorption intensity S, instead of the first derivative V_1. The exponent in the saturation curve is rigorously $p=1/2$, irrespective of the extent of inhomogeneous broadening. It is no longer a fitting parameter as in Eq. 11.14 for the first derivative. Although changes in integral intensity with saturation are less than those in spectral line heights (see Figure 11.1), the precision in determining the saturation parameters is increased by fitting the exact analytical expression (with $p=1/2$) for the saturation curve. A further advantage of using the integrated intensity is that it is directly additive in multicomponent systems, even at saturating powers, which is not the case for the line heights (Páli et al. 1993).

Extending the above treatment shows that it applies equally to saturation of the integrated intensities of powder EPR spectra (see Páli et al. 1993). Pursuing this further, we turn to saturation of the first-derivative amplitudes at the characteristic turning points in the powder lineshape. For simplicity we assume axial anisotropy, the equivalent of Eq. 11.17 then becomes, for a randomly oriented powder sample (Fajer et al. 1992):

$$V_{0,inh}(\omega,B_1)=\int_{\omega_\perp}^{\omega_\parallel}\mathrm{d}\omega_o\cdot I(\omega_o)\int_{-\infty}^{\infty}\mathrm{d}\omega'\cdot h(\omega'-\omega_o)v(\omega-\omega',B_1) \quad (11.19)$$

where $I(\omega_o)$ is the powder distribution of inhomogeneously broadened lines, i.e., for a random orientation of nitroxide axes to the static magnetic field, B_o. This distribution extends from $\omega_o=\omega_\perp$ to $\omega_o=\omega_\parallel$ for B_o oriented perpendicular and parallel, respectively, to the nitroxide z-axis. For narrow spin packets, we approximate the inhomogeneous envelope function by $h(\omega-\omega_o)$, which allows separation of variables. The first-derivative powder spectrum then becomes:

$$\frac{\mathrm{d}V_{0,inh}(B_1)}{\mathrm{d}\omega}=\int_{\omega_\perp}^{\omega_\parallel}\mathrm{d}\omega_o\cdot I(\omega_o)\left(\frac{\mathrm{d}h(\omega-\omega_o)}{\mathrm{d}\omega}\int_{-\infty}^{\infty}\mathrm{d}\omega'\cdot v(\omega-\omega',B_1)\right.$$
$$\left.+h(\omega-\omega_o)\int_{-\infty}^{\infty}\mathrm{d}\omega'\frac{\mathrm{d}v(\omega-\omega',B_1)}{\mathrm{d}\omega}\right) \quad (11.20)$$

Changing variables from ω and ω' to $\omega-\omega'$ in the two integrals over ω' shows that the first is the integrated intensity of an individual spin packet, and the second vanishes because it is the integral over the corresponding first-derivative lineshape.

The powder lineshape distribution $I(\omega_o)$ is stationary in the vicinity of the outer extrema, where $\omega_o=\omega_\parallel$ for 9-GHz spectra. We therefore assume a constant value, $I(\omega_\parallel)$, in the integral over ω_o in Eq. 11.20. Changing variables from ω and ω_o to $\omega-\omega_o$ then yields the first-derivative lineshape around the $\omega_\parallel$-extremum (Fajer et al. 1992):

$$\left.\frac{\mathrm{d}V_{0,inh}(B_1)}{\mathrm{d}\omega}\right|_{\omega_\parallel}=I(\omega_\parallel)h(\omega-\omega_\parallel)\int_{-\infty}^{\infty}\mathrm{d}(\omega-\omega')\cdot v(\omega-\omega',B_1) \quad (11.21)$$

The first-derivative powder spectrum at the parallel turning point therefore has the inhomogeneous absorption lineshape (cf. Eq. 2.22 in Section 2.4) and saturates in the same way as does the integrated absorption of a single spin packet. For Lorentzian spin packets (and strong inhomogeneous broadening), the saturation curve for the line heights of the outer peaks in the powder spectrum is therefore given by Eq. 11.11, i.e., with an exponent $p = 1/2$.

On the other hand, if inhomogeneous broadening is very weak, the powder lineshape is given by Eq. 11.19 with $h(\omega' - \omega_o) = \delta(\omega' - \omega_o)$. Instead of Eq. 11.21, the lineshape at the $\omega_{\parallel}$-extremum with only homogeneous broadening is:

$$\left.\frac{\mathrm{d}V_{0,\mathrm{hom}}(B_1)}{\mathrm{d}\omega}\right|_{\omega_{\parallel}} = I(\omega_{\parallel})v(\omega - \omega', B_1) \tag{11.22}$$

The homogeneously broadened first-derivative powder pattern at the parallel turning point therefore has the homogeneous spin-packet absorption lineshape and also saturates in the same way as does the spin-packet absorption. This agrees with the considerations of powder lineshapes given in Section 2.4. For Lorentzian spin-packet lineshapes, saturation of the EPR amplitudes in the outer wings is then given by Eq. 11.10, i.e., with an exponent $p = 1$.

11.4 FIELD MODULATION: MODULATION-COUPLED BLOCH EQUATIONS

Solutions of the Bloch equations given in Section 11.2 correspond to true slow-passage conditions. In practice, we make CW-saturation measurements with conventional field modulation $\approx B_{\mathrm{mod}} \cos\omega_{\mathrm{mod}} t$, and phase-sensitive detection. If the spin–lattice relaxation time is long, the passage conditions depend on the modulation frequency $\omega_{\mathrm{mod}}/2\pi$.

Including field modulation, the resonance offset frequency $\Delta\omega$ that appears in the Bloch matrix $\mathbf{A}$ becomes (see Eq. 11.5):

$$\Delta\omega = \omega - \omega_o + \gamma_e B_{\mathrm{mod}} \cos\omega_{\mathrm{mod}} t \tag{11.23}$$

where B_{mod} is half the peak-to-peak field modulation amplitude. For a steady-state solution of the Bloch equations (i.e., of Eq. 11.4), we expand the magnetization components (u, v, M_z) in Fourier harmonics of ω_{mod} (Halbach 1954):

$$u = \sum_{n=-\infty}^{\infty} u_n \exp(-in\omega_{\mathrm{mod}} t); \quad v = \sum_{n=-\infty}^{\infty} v_n \exp(-in\omega_{\mathrm{mod}} t);$$

$$M_z = \sum_{n=-\infty}^{\infty} M_{z,n} \exp(-in\omega_{\mathrm{mod}} t) \tag{11.24}$$

where $u_n, v_n, M_{z,n}$ are the complex Fourier amplitudes. These correspond to the absorption and dispersion EPR signals (v_n and u_n) detected at different harmonics $n \times \omega_{\mathrm{mod}}$ of the modulation frequency. The conventional CW-EPR signal is the first harmonic (i.e., $n = 1$). Because the modulation varies as $\cos\omega_{\mathrm{mod}} t$, the real part of the Fourier amplitude gives the EPR signal that we detect in phase with the field modulation. This is the usual EPR spectrum (i.e., V_1). The imaginary part of the Fourier amplitude is the EPR signal that lags 90° out-of-phase with respect to the field modulation. This is the nonlinear rapid-passage signal, of which the second-harmonic saturation transfer EPR spectrum V_2' is best known. The out-of-phase signal has appreciable intensity only in the presence of saturation; we use this property for setting the phase in nonlinear EPR experiments.

The modulation term in Eq. 11.23 couples the different Fourier harmonics of Eqs. 11.24. Substituting Eqs. 11.24 into Eq. 11.4 gives an infinite system of coupled equations for the Fourier amplitudes $\widehat{\mathbf{M}}_n \equiv (u_n, v_n, M_{z,n})$ of the spin magnetization (cf. Livshits et al. 1998a):

$$\mathbf{A}_0 \cdot \widehat{\mathbf{M}}_0 = \mathbf{M}_o + \tfrac{1}{2}\boldsymbol{\gamma} \cdot B_{\mathrm{mod}}\left(\widehat{\mathbf{M}}_1 + \widehat{\mathbf{M}}_{-1}\right) \tag{11.25}$$

$$\mathbf{A}_n \cdot \widehat{\mathbf{M}}_n = \tfrac{1}{2}\boldsymbol{\gamma} \cdot B_{\mathrm{mod}}\left(\widehat{\mathbf{M}}_{n+1} + \widehat{\mathbf{M}}_{n-1}\right) \tag{11.26}$$

$$\text{with } \mathbf{A}_n \equiv \begin{pmatrix} (T_2^o)^{-1} + \tau_l^{-1} - in\omega_{\mathrm{mod}} & \omega - \omega_o & 0 \\ -(\omega - \omega_o) & (T_2^o)^{-1} + \tau_l^{-1} - in\omega_{\mathrm{mod}} & -\gamma_e B_1 \\ 0 & \gamma_e B_1 & T_1^{-1} + \tau_l^{-1} - in\omega_{\mathrm{mod}} \end{pmatrix} \tag{11.27}$$

$$\text{and } \boldsymbol{\gamma} \equiv \begin{pmatrix} 0 & -\gamma_e & 0 \\ \gamma_e & 0 & 0 \\ 0 & 0 & 0 \end{pmatrix} \tag{11.28}$$

The term on the right in Eqs. 11.25 and 11.26 gives the coupling of the Fourier amplitudes by the modulation field B_{mod} explicitly. Note that Eq. 11.25 for the zeroth harmonic ($n = 0$) is a special case, because it contains the equilibrium magnetization vector $\mathbf{M}_o$ (cf. Eq. 11.4).

For small-modulation amplitudes $\left(\gamma_e B_{\mathrm{mod}} < 1/T_2^o\right)$, we get the Fourier amplitudes u_n, v_n, $M_{z,n}$ of the nth-harmonic EPR signal by expanding in a power series of the modulation amplitude (Halbach 1954):

$$u_n = \sum_{v=0}^{\infty} u_{n,v}\left(\gamma_e B_{\text{mod}}\right)^v; \quad v_n = \sum_{v=0}^{\infty} v_{n,v}\left(\gamma_e B_{\text{mod}}\right)^v; \qquad (11.29)$$

$$M_{z,n} = \sum_{v=0}^{\infty} M_{z,n,v}\left(\gamma_e B_{\text{mod}}\right)^v$$

The coefficients $u_{n,v}$, $v_{n,v}$, $M_{z,n,v}$ with $v < |n|$ disappear because of the symmetry properties of the Bloch equations. To lowest order in modulation amplitude (i.e., $v = n$), the nth-harmonic signal therefore depends on B_{mod}^n.

From Eqs. 11.25 and 11.26, the zero-order coefficients $\widehat{\mathbf{M}}_{0,0} \equiv \left(u_{0,0}, v_{0,0}, M_{z,0,0}\right)$, and the vector coefficients $\widehat{\mathbf{M}}_{n,n} \equiv \left(u_{n,n}, v_{n,n}, M_{z,n,n}\right)$ defining harmonic components that depend on the nth power of B_{mod}, become:

$$\mathbf{A}_0 \cdot \widehat{\mathbf{M}}_{0,0} = \mathbf{M}_o \qquad (11.30)$$

$$\mathbf{A}_n \cdot \widehat{\mathbf{M}}_{n,n} = \left(\boldsymbol{\gamma}/\gamma_e\right) \cdot \widehat{\mathbf{M}}_{n-1,n-1} \qquad (11.31)$$

Equation 11.30 for the zeroth harmonic is equivalent to Eq. 11.4 in the steady state and yields the usual slow-passage solutions of the Bloch equations, i.e., Eqs. 11.6 and 11.7 for $u_{0,0}$ and $v_{0,0}$, respectively (see Section 11.2). We then get the first-harmonic solution $\widehat{\mathbf{M}}_{1,1}$ by substituting $\widehat{\mathbf{M}}_{0,0}$ in Eq. 11.31 with $n = 1$. The corresponding solution for the second-harmonic $\widehat{\mathbf{M}}_{2,2}$ comes from substituting $\widehat{\mathbf{M}}_{1,1}$ in Eq. 11.31 with $n = 2$, and so on.

To lowest order in modulation amplitude, the Fourier coefficients of the nth harmonic are $u_n \approx u_{n,n}B_{\text{mod}}^n$ and $v_n \approx v_{n,n}B_{\text{mod}}^n$. Thus the conventional in-phase EPR spectrum, and the first-harmonic dispersion and second-harmonic absorption out-of-phase ST-EPR spectra, are given by:

$$V_1(\omega) \approx \text{Re}\left\{v_{1,1}(\omega)\right\} \times B_{\text{mod}} \qquad (11.32)$$

$$U_1'(\omega) \approx \text{Im}\left\{u_{1,1}(\omega)\right\} \times B_{\text{mod}} \qquad (11.33)$$

$$V_2'(\omega) \approx \text{Im}\left\{v_{2,2}(\omega)\right\} \times B_{\text{mod}}^2 \qquad (11.34)$$

respectively, where the prime denotes out-of-phase detection (see Table 11.1). By restricting to the lowest terms in the expansions of Eq. 11.29, these approximations are valid for low modulation amplitudes.

Spectral simulations for progressive-saturation experiments show that, with $T_1 = 5\,\mu$s and a spin packet width $\Delta B_o = 0.1$ mT, the saturation factor is approximately halved at a modulation frequency of 100 kHz, relative to 10-kHz modulation (Livshits et al. 1998a). At 10 kHz, the saturation curve already is close to that for slow passage. For T_1 shorter than 1–2 μs, the dependence of the saturation curves on modulation frequency is also small.

We shall return to this section, particularly Eqs. 11.33 and 11.34, later when treating nonlinear and ST-EPR displays in more detail. Our immediate concern is the effects of spin-label motion on progressive-saturation measurements. For this, we already need solution of the Bloch equations with field modulation.

11.5 PROGRESSIVE SATURATION: SUDDEN-JUMP ROTATIONAL MOBILITY

We now introduce rotation of the spin-label molecule into the Bloch equations (i.e., Eq. 11.4) that include the microwave and Zeeman modulation fields. To do this, we choose the random sudden-jump model, described already in Section 6.4 of Chapter 6. When the static magnetic field B_o is oriented at polar angle $\Omega \equiv (\theta, \phi)$ to the nitroxide axes, the time dependence of the spin magnetization vector $\widehat{\mathbf{M}} \equiv \left(u, v, M_z\right)$ in the rotating frame is (Livshits et al. 1998a; and cf. Eq. 6.29):

$$\frac{\mathrm{d}\widehat{\mathbf{M}}}{\mathrm{d}t} + \left(\mathbf{A} + \tau_l^{-1}\mathbf{1}\right) \cdot \widehat{\mathbf{M}} = \mathbf{M}_o + \tau_l^{-1}P_o \int \widehat{\mathbf{M}}\,\mathrm{d}\Omega \qquad (11.35)$$

where τ_l^{-1} is the frequency of isotropic rotational jumps and $P_o \equiv P_o(\Omega)$ is the spatial angular degeneracy for orientation Ω. The term containing τ_l^{-1} on the left-hand side of Eq. 11.35 is the rate of transfer of spin magnetization from Ω to other orientations (**1** is the identity matrix) and that on the right-hand side is the rate of transfer to Ω from all other orientations. The Bloch matrix **A** is given by Eq. 11.5 of Section 11.2, together with Eq. 11.23 for the resonance offset. But now the resonance frequency is angular dependent: $\omega_o \equiv \omega_{m_I}(\theta, \phi)$ given by Eq. 6.33 (see Chapters 2 and 6).

We first introduce Fourier expansion of the magnetization components (i.e., Eqs. 11.24) into Eq. 11.35. In the presence of molecular motion, Eqs. 11.25 and 11.26 for the coupled Fourier amplitudes thus become:

$$\mathbf{A}_0 \cdot \widehat{\mathbf{M}}_0 = \mathbf{M}_o + \tau_l^{-1}P_o \int \widehat{\mathbf{M}}_0\,\mathrm{d}\Omega + \tfrac{1}{2}\boldsymbol{\gamma} \cdot B_{\text{mod}}\left(\widehat{\mathbf{M}}_1 + \widehat{\mathbf{M}}_{-1}\right) \qquad (11.36)$$

$$\mathbf{A}_n \cdot \widehat{\mathbf{M}}_n = \tau_l^{-1}P_o \int \widehat{\mathbf{M}}_n\,d\Omega + \tfrac{1}{2}\boldsymbol{\gamma} \cdot B_{\text{mod}}\left(\widehat{\mathbf{M}}_{n+1} + \widehat{\mathbf{M}}_{n-1}\right) \qquad (11.37)$$

where matrices $\mathbf{A}_n$ and $\boldsymbol{\gamma}$ are given respectively by Eq. 11.27 with $\omega_o \equiv \omega_{m_I}(\theta, \phi)$, and by Eq. 11.28. Next we substitute the power series in modulation amplitude (i.e., Eqs. 11.29 for $v = 0$ and 1) into Eqs. 11.36 and 11.37. The zero-order coefficients $\widehat{\mathbf{M}}_{0,0} \equiv \left(u_{0,0}, v_{0,0}, M_{z,0,0}\right)$, and vector coefficients $\widehat{\mathbf{M}}_{1,1} \equiv \left(u_{1,1}, v_{1,1}, M_{z,1,1}\right)$ for magnetization components linear in B_{mod}, are then:

$$\mathbf{A}_0 \cdot \widehat{\mathbf{M}}_{0,0} = \mathbf{M}_o + \tau_l^{-1}P_o \int \widehat{\mathbf{M}}_{0,0}\,\mathrm{d}\Omega \qquad (11.38)$$

$$\mathbf{A}_1 \cdot \widehat{\mathbf{M}}_{1,1} = \tau_l^{-1}P_o \int \widehat{\mathbf{M}}_{1,1}\,\mathrm{d}\Omega + \left(\boldsymbol{\gamma}/\gamma_e\right) \cdot \widehat{\mathbf{M}}_{0,0} \qquad (11.39)$$

which correspond to Eqs. 11.30 and 11.31 in the absence of motion.

Solution of integral Eqs. 11.38 and 11.39 is relatively straightforward because the vector integrals $\mathbf{I}_0 \equiv \int \widehat{\mathbf{M}}_{0,0}\,\mathrm{d}\Omega$ and $\mathbf{I}_1 \equiv \int \widehat{\mathbf{M}}_{1,1}\,\mathrm{d}\Omega$ do not depend on the angular variables, $\Omega \equiv (\theta, \phi)$.

Thus, solving Eq. 11.38 for $\widehat{\mathbf{M}}_{0,0}$ and integrating over Ω, we get (cf. Eq. 6.31):

$$\mathbf{I}_0 = \left(\mathbf{1} - \tau_l^{-1} P_o \mathbf{J}_0\right)^{-1} \cdot \mathbf{J}_0 \cdot \mathbf{M}_o \tag{11.40}$$

where $\mathbf{J}_0 \equiv \int \mathbf{A}_0^{-1}\, \mathrm{d}\Omega$. Substituting Eq. 11.40 into Eq. 11.38 then gives the solution:

$$\widehat{\mathbf{M}}_{0,0} = \mathbf{A}_0^{-1} \cdot \left[\mathbf{M}_o + \tau_l^{-1} P_o \left(\mathbf{1} - \tau_l^{-1} P_o \mathbf{J}_0\right)^{-1} \cdot \mathbf{J}_0 \cdot \mathbf{M}_o\right] \tag{11.41}$$

Similarly, solving Eq. 11.39 for $\widehat{\mathbf{M}}_{1,1}$ and integrating over Ω, we get:

$$\mathbf{I}_1 = \left(\mathbf{1} - \tau_l^{-1} P_o \mathbf{J}_1\right)^{-1} \cdot \mathbf{J}_{10} \tag{11.42}$$

where $\mathbf{J}_1 = \int \mathbf{A}_1^{-1}\, \mathrm{d}\Omega$ and $\mathbf{J}_{10} = \int \mathbf{A}_1^{-1} \cdot \gamma \cdot \widehat{\mathbf{M}}_{0,0}\, \mathrm{d}\Omega$. The solution for $\widehat{\mathbf{M}}_{0,0}$ in Eq. 11.41 is then used to get $\mathbf{J}_{10}$.

Finally, we get the lineshape of the first-harmonic in-phase absorption spectrum by taking the real part of the solution:

$$V_1(\omega) = \mathrm{Re}\left\{V_{1,1}(\omega)\right\} \tag{11.43}$$

where $\mathbf{I}_1 \equiv \left(U_{1,1}, V_{1,1}, M_{z,1,1}\right)$ comes from Eq. 11.42. We obtain this result in closed form and only need to evaluate the integrals numerically, for each value of the static field B_o.

As explained in Section 6.4 of Chapter 6, using the full anisotropic angular resonance frequency $\omega_{m_I}(\theta,\phi)$ from Eq. 6.33 in Eq. 11.35 is an adiabatic approximation that is not valid in the fast motional regime. For fast rotation, we must use the isotropically averaged value $\langle \omega_{m_I} \rangle$, omit all terms containing τ_l^{-1}, and instead add the $T_2^{-1}(m_I)$ contribution from Eqs. 5.58–5.62 to the Bloch matrix in Eq. 11.5 (and Eq. 11.27). Nonetheless, power saturation curves for the double-integrated intensity found thus from time-dependent perturbation theory are almost identical with those that we get using the adiabatic approach, despite significant differences in spectral lineshapes (Livshits et al. 1998a).

Figure 11.4 shows power-saturation curves for the integrated intensity of CW-EPR absorption spectra simulated for different rates of jump reorientation τ_l^{-1}. As expected, we find sensitivity to rotational motion that corresponds to spectral diffusion of saturation in the slow-motional regime and to the conventional rotational contribution to transverse relaxation in the faster motional regime. Nonetheless, all saturation curves for the integrated intensity correspond quite closely to the form predicted by Eq. 11.11, for true slow-passage conditions without molecular reorientation. The single fitting parameter for the simulated saturation curves, $P \equiv \gamma_e^2 (T_1T_2)^{eff}$, reveals that the effective T_1T_2 -relaxation rate depends linearly on the spin–lattice relaxation rate (Livshits et al. 1998a):

$$\frac{1}{(T_1T_2)^{eff}} = \frac{1}{T_2^{eff}}\left(\frac{1}{T_1}\right) + a \tag{11.44}$$

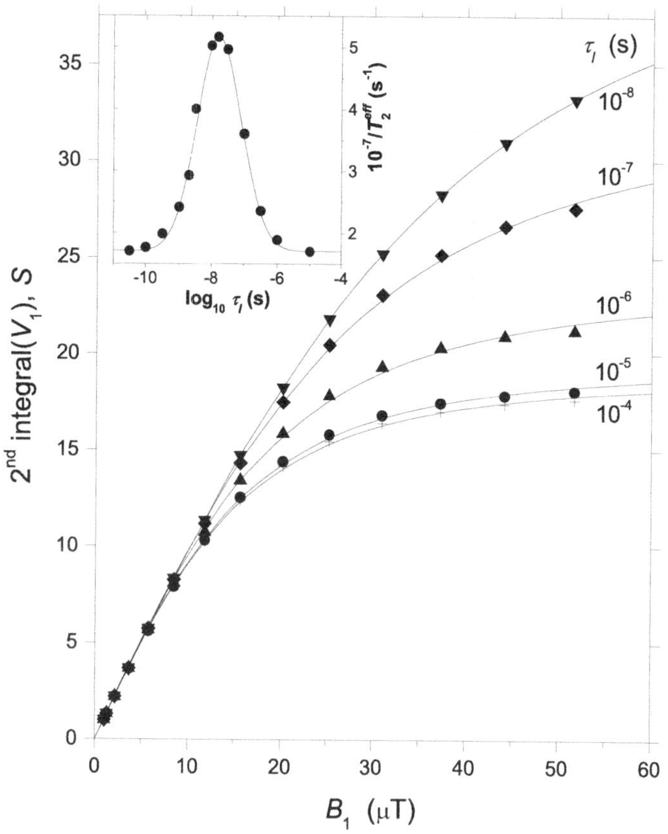

FIGURE 11.4 Saturation curves for double-integrated intensity of V_1-absorption signal for different lifetimes τ_l between sudden angular jumps, fixed spin–lattice relaxation time $T_1^o = 1.8$ μs and modulation frequency $\omega_{mod}/2\pi = 100$ kHz. Solid lines: fits of Eq. 11.11. *Inset*: dependence of effective T_2-relaxation rate from saturation factor (Eq. 11.44) on τ_l; intrinsic $T_2^o = 57$ ns; solid line: fit by Eq. 11.45 (see Livshits et al. 1998a).

where a is a small intercept. As shown in the insert to Figure 11.4, the effective transverse relaxation rate $1/T_2^{eff}$ depends strongly on the rotational reorientation time, with a characteristic maximum in the region of $\tau_l \approx 10^{-7} - 10^{-8}$ s that corresponds to the frequency equivalent of the ^{14}N-hyperfine anisotropy. Numerically, we can represent this by a log-normal distribution:

$$\frac{1}{T_2^{eff}} \approx \frac{1}{T_2^o} + \frac{3.3 \cdot 10^7\ \mathrm{s}^{-1}}{w} \exp\left(\frac{-\left(\log_{10} \tau_l + 7.78\right)^2}{w^2}\right) \tag{11.45}$$

where τ_l is in seconds, and $w \approx 0.93 - 0.97$ depends only weakly on the intrinsic linewidth: $1/T_2^o = 1.75 \times 10^7\ \mathrm{s}^{-1}$ (Livshits et al. 1998a). Correspondingly, calibration for the intercept correction, a, is parameterized by:

$$a/\left(10^{12}\mathrm{s}^2\right) \approx 8.9 \exp\left(-\left(\log_{10} \tau_l + 6.9\right)^2/0.7\right) + 6.1 \exp\left(-\left(\log_{10} \tau_l + 8.37\right)^2/0.41\right) + 1.6 \tag{11.46}$$

with the same conditions as for Eq. 11.45.

At the extremes for very fast $(\tau_l < 10^{-10}\text{ s})$ or very slow $(\tau_l > 10^{-6}\text{ s})$ motion, the saturation properties are insensitive to motional rates. The effective T_2 is then equal to the intrinsic T_2^o, and the intercept a is close to zero. Around the maximum in $1/T_2^{eff}$, on the other hand, motional broadening far exceeds the intrinsic linewidth and does not depend much on reorientation time for $3.10^{-7}\text{ s} \le \tau_l \le 10^{-8}\text{ s}$. Under these circumstances, saturation properties are insensitive both to intrinsic line broadening and to motional rates. We can then obtain T_1 directly from progressive-saturation measurements, without knowing T_2^o or τ_l precisely. Calibrations to determine spin–lattice relaxation times from saturation measurements in the region of extreme motional broadening, and those to determine the T_1T_2-relaxation time product for faster or slower motions outside this regime, are given in Appendix A11.1 at the end of this chapter.

11.6 PROGRESSIVE SATURATION: BROWNIAN ROTATIONAL DIFFUSION

We turn now to the influence of Brownian rotational diffusion by the spin label on progressive-saturation CW-EPR (Haas et al. 1993). This complements the previous section, where we used the random sudden-jump model. Also, we concentrate here on saturation curves for the first-derivative amplitude of the low-field ($m_I = +1$) line of a ^{14}N-nitroxide (at 9 GHz), instead of those for the integrated intensity. In practice, the amplitude of the central ($m_I = 0$) line is that used most often, to get better signal-to-noise. However, unlike the low-field turning point, this region is complicated by overlapping components in the slow-motion regime. For instance, effective values of the fitting exponent in Eq. 11.14 that lie outside the range for Lorentzian spin packets (viz. $p > 3/2$) can arise when using the central peak. Consequently, as seen already in Section 11.3, interpretation of the low-field amplitude is more straightforward.

To treat slow Brownian rotational diffusion, we use simulations with the stochastic-Liouville equation, as described in Section 6.7 of Chapter 6. We introduce saturation into Eq. 6.54 by including a Redfield-type term $\Gamma_R\chi(\Omega,t)$ that contains the intrinsic T_1-relaxation time (and T_2^o). The equation of motion for the reduced spin density-matrix $\chi(\Omega,t)$ then becomes:

$$\frac{\partial\chi(\Omega,t)}{\partial t} + \frac{i}{\hbar}\left[\mathcal{H}(\Omega),\chi(\Omega,t)\right] + \Gamma_\Omega\chi(\Omega,t) + \Gamma_R\chi(\Omega,t) = \frac{i}{\hbar}\left[\mathcal{H}(\Omega),\sigma_o(\Omega)\right] \tag{11.47}$$

where $\Gamma_R\chi(\Omega,t)$ contains all relaxation processes not involving rotational diffusion. Additionally, we must retain the time-dependent term $\varepsilon(t)$ that contains the microwave radiation field, when expanding the Hamiltonian $\mathcal{H}(\Omega)$ according to Eq. 6.55. From Section 6.7, Eq. 11.47 then becomes (cf. Eq. 6.58):

$$\frac{\partial\chi(\Omega,t)}{\partial t} + i\omega_o\left[S_z,\chi(\Omega,t)\right] + \frac{i}{\hbar}\left[\mathcal{H}_1(\Omega),\chi(\Omega,t)\right] + \frac{i}{\hbar}\left[\varepsilon(t),\chi(\Omega,t)\right] + \Gamma_\Omega\chi(\Omega,t) + \Gamma_R\chi(\Omega,t) = \frac{i\gamma_e B_1\hbar\omega_o}{2N_Z k_B T}S_-\exp(i\omega t) \tag{11.48}$$

If passage conditions are significant, $\varepsilon(t)$ must also include the field modulation at frequency ω_{mod} (see Section 11.15, later). Solution for the EPR lineshapes under saturation again uses the eigenfunction-expansion method of Section 6.7 (see Eq. 6.63); this is described by Dalton et al. (1976) and Robinson et al. (1985).

Simulations with the stochastic Liouville equation for isotropic Brownian rotational diffusion show that saturation curves for the positive low-field amplitude are well fitted by Eq. 11.14 for all correlation times, τ_R, of the spin label (Haas et al. 1993). Table 11.2 lists values of the exponent p that are obtained by nonlinear least-squares fitting. For very fast motion $(\tau_R = 10^{-11}\text{ s})$, the exponent is $p = 1.49 \approx 3/2$, as expected for a homogeneously broadened first-derivative lineshape. (Inhomogeneous broadening is not included in the simulations.) For very slow motion $(\tau_R = 10^{-5}\text{ s})$, the exponent $p = 1.1-1.2$ approaches the value of unity for a homogeneously broadened absorption lineshape. This is expected for the low-field turning point in a powder spectrum (see Section 11.3). However, lineshapes for slow-motion spectra with $\tau_R \approx 10^{-7}$ s already approach the rigid limit, but the exponent $p = 1.35-1.4$ is closer to $p = 3/2$ than to unity. Because spin–lattice relaxation is much slower than transverse relaxation, these spin packets still saturate predominantly as the homogeneous first harmonic, although the lineshape itself is barely sensitive to motion.

Simulations for different values of the spin–lattice relaxation time (Haas et al. 1993) show that the saturation parameter

TABLE 11.2 Exponent p for progressive saturation of the low-field ($m_I = +1$) amplitude in the first-derivative 9-GHz EPR spectrum (Eq. 11.14) and calibration parameter T_2^{eff} in Eq. 11.44 (with $a = 0$) for T_1 in the range 0.25–5 μs[a]

τ_R (s)	$T_2^o \times 10^9$ (s)[b]	p	$T_2^{eff} \times 10^9$ (s)
1×10^{-11}	30	1.465	29.9
	100	1.485	100.6 ± 0.3
1×10^{-9}	30	1.30–1.35	23.7 ± 0.1
	100	1.447	42.2
1×10^{-8}	30	1.21–1.22	6.88
	100	1.224	7.45
3×10^{-8}	30	1.31–1.33	7.0 ± 0.1
	100	1.33–1.42	8.9 ± 0.2
1×10^{-7}	30	1.375–1.384	6.32 ± 0.03
	100	1.418	8.05 ± 0.09
1×10^{-5}	30	1.137	22.09
	100	1.224	42.0

From stochastic–Liouville simulations for isotropic Brownian rotation with correlation time τ_R (Haas et al. 1993).

[a] 5–15 μs for $\tau_R = 1\times10^{-7}$ s.

[b] Values of T_2^o correspond to intrinsic linewidths: $\Delta B_{1/2}^o \equiv 1/(\gamma_e T_2^o)$ of 0.19 and 0.057 mT.

$P \equiv \gamma_e^2 (T_1 T_2)^{eff}$ depends on T_1 according to Eq. 11.44, just as for jump diffusion. In this case, however, the intercept a is almost zero. Table 11.2 lists values of the fitting parameter T_2^{eff} for different rotational correlation times. Only for very fast motion $\left(\tau_R = 10^{-11}\ \text{s}\right)$ is the effective T_2^{eff} equal to the intrinsic T_2^o. In all other cases, it is considerably shorter.

Figure 11.5 shows the dependence of the effective transverse relaxation rate $1/T_2^{eff}$ on rotational correlation time for Brownian diffusion. The rate exhibits a pronounced maximum for correlation times in the region of $\tau_R = 10^{-8}$–10^{-7} s, as for the jump model in Figure 11.4. Haas et al. (1993) modify the spectral densities of motional narrowing theory given in Chapter 5 to model the motional dependence of the effective rate:

$$\frac{1}{T_2^{eff}} = \frac{1}{T_2^o} + k\frac{\Delta\omega_{eff}^2 \tau_R}{1+\left(\Delta\omega_{eff}\tau_R\right)^\beta} \tag{11.49}$$

where β is an empirical exponent, $\Delta\omega_{eff}$ is the extent of spectral anisotropy of the low-field $m_I = +1$ hyperfine manifold and k is a scaling factor that represents contributions of several spectral densities. The Lorentzian spectral densities of motional-narrowing theory correspond to $\beta = 2$ in Eq. 11.49. Application to longer correlation times requires empirical extension beyond the limits valid for time-dependent perturbation theory. The physical idea behind this is that, because $T_1 >> T_2$ for spin labels, saturation is alleviated by spectral diffusion, which arises from slow rotation that takes place within the spin–lattice relaxation time. This is reflected by changes in T_2^{eff}. Spectral diffusion of saturation (i.e., saturation transfer) results in a distribution of relaxation rates, which is represented by the Cole–Davidson spectral density in Eq. 11.49 (where $1 < \beta < 2$). We use values of $\beta \approx 1.3$ from Haas et al. (1993) in Figure 11.5. Data simulated with the jump model of the previous section do not evidence such a pronounced asymmetry (see Figure 11.4). The dashed line in Figure 11.5 shows a fit of Eq. 11.45 to the peak, but with the same value of the width w that we used for the jump model.

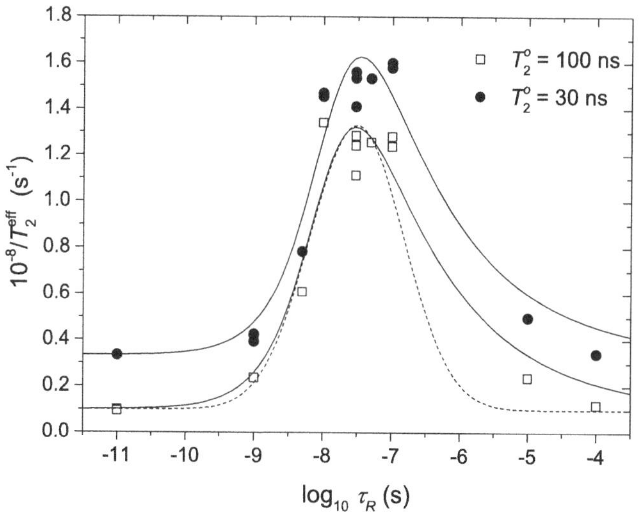

FIGURE 11.5 Dependence of effective T_2^{eff}-relaxation rate (from saturation parameter $P = \gamma_e^2 T_1 T_2^{eff}$ of low-field first-derivative EPR amplitude for ^{14}N-nitroxide at 9 GHz) on correlation time τ_R for Brownian rotation. T_2^o: intrinsic T_2-relaxation time in stochastic-Liouville simulations. *Solid lines*: Eq. 11.49 with $\beta = 1.33 - 1.34$ (Haas et al. 1993). *Dashed line*: Eq. 11.45 with $w = 0.97$, and adjusted amplitude and centre.

11.7 NONLINEAR DISPLAYS: OUT-OF-PHASE SPECTRA (ST-EPR)

In the linear region, at low microwave fields (indicated by the dashed line in Figure 11.1), there is no saturation of the EPR signal. Because T_2-relaxation rates exceed the modulation frequency ($\omega_{\text{mod}}/2\pi \approx 10^5\ \text{s}^{-1}$, or less), the EPR response then remains in phase with the static field modulation. Setting the phase-sensitive detector 90° out-of-phase with respect to the field modulation therefore gives a null signal, in the absence of saturation. Under partial saturation – optimally, close to the maximum amplitude in the saturation curve (see Figure 11.1) – we obtain rapid-passage conditions when the modulation frequency is comparable to the spin–lattice relaxation rate (Portis 1955; Weger 1960). The EPR response then lags behind the field modulation, and we obtain an out-of-phase signal with a non-vanishing intensity that depends on T_1 and the modulation frequency. Any process that alleviates saturation reduces the intensity of the nonlinear out-of-phase signal. In particular, rotational diffusion at a rate comparable to T_1-relaxation transfers saturation to a different part of the spectrum. The rate of transfer (or spectral diffusion) depends on spectral position, and the lineshape therefore depends on rotational rate. See Marsh et al. (2005) for an introduction to the development of saturation-transfer (ST) spectroscopy.

Out-of-phase nonlinear displays were used originally in CW-NMR to determine spin–lattice relaxation times (see Halbach 1954). Their introduction in spin-label EPR was aimed at finding spectral displays sensitive to very slow rotational motion that takes place on the timescale of T_1 (Hyde and Dalton 1972). Figure 11.6 shows the various in-phase and out-of-phase displays; these spectra are simulated by using the jump model as described in Section 11.5. Both the first-harmonic dispersion (U_1') and second-harmonic absorption (V_2') out-of-phase EPR signals are sensitive to microsecond rotations (see Table 11.1 for nomenclature). In practice, the second-harmonic V_2'-spectrum is the standard saturation-transfer EPR display, chosen for its large changes in lineshape with motional rate (Hyde and Thomas 1973; Thomas et al. 1976).

First-harmonic dispersion out-of-phase U_1'-lineshapes are not richly detailed, as typically are those in V_2'-STEPR. For the rigid limit, they correspond simply to the pure zeroth-harmonic absorption spectrum, V_0. Motionally-dependent changes in lineshape are therefore quantified by the $V_0 - U_1'$ difference spectrum (Fajer and Marsh 1983a). We obtain the V_0-lineshape by integration of the normal unsaturated in-phase absorption spectrum, V_1. This integrated reference spectrum offers a general approach to detecting saturation transfer in complex or multi-component systems by using U_1' signals.

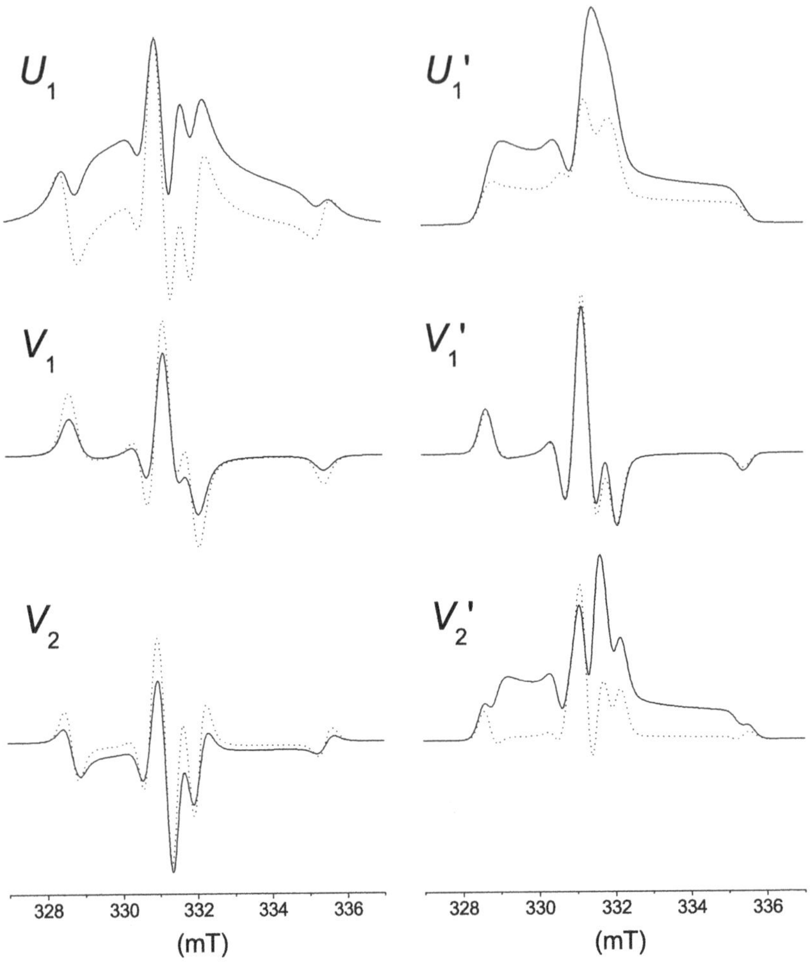

FIGURE 11.6 Out-of-phase (right column) and in-phase (left column) displays of 9-GHz ^{14}N-nitroxide EPR spectra under saturation, with 100-kHz field modulation (or detection). *Top row*: first-harmonic dispersion, U_1 and U_1'; *middle row*: first-harmonic absorption, V_1 and V_1'; *bottom row*: second-harmonic absorption, V_2 and V_2'. *Solid line*: no motion; *dashed line*: jump reorientation time $\tau_j = 10^{-6}$ s. (See Marsh et al. 2005.)

On the other hand, the out-of-phase first-harmonic absorption spectrum (V_1') is unaffected by ultraslow rotation (see Figure 11.6), and is sensitive only to the intrinsic T_1. Its lineshape is similar to that of the conventional in-phase EPR spectrum. We can use such pure T_1-sensitive displays for detecting spin–lattice relaxation enhancements discussed in Chapter 10. Other nonlinear displays may also be used to determine relaxation enhancements, e.g., V_2'-STEPR, but then we must correct for motional contributions, if these also change. The hallmark of a simple T_1-relaxation enhancement is that the intensity of the out-of-phase spectrum decreases, but the lineshape does not change appreciably.

11.8 T_1-SENSITIVE NONLINEAR EPR DISPLAYS

Here, we consider the dependence of the nonlinear out-of-phase spectra on the relaxation times, in the absence of rotational motion. This gives the necessary background for their use to study spin–lattice relaxation enhancements.

Solutions of the Bloch equations that explicitly include low Zeeman modulation fields (see Section 11.4) yield the different out-of-phase dispersion (U_n') and absorption (V_n') displays for a single homogeneously broadened line. We use iterative solutions of Eq. 11.31 for the vector coefficients $\widehat{\mathbf{M}}_{n,n} \equiv (u_{n,n}, v_{n,n}, M_{z,n,n})$:

$$\widehat{\mathbf{M}}_{n,n} = \mathbf{A}_n^{-1} \cdot (\gamma/\gamma_e) \cdot \widehat{\mathbf{M}}_{n-1,n-1} \tag{11.50}$$

and relate these to the zeroth-harmonic absorption amplitude $v_{0,0}$ by using Eq. 11.30. Explicitly, for the transverse components corresponding to dispersion and absorption, we get the following recursion relations (Páli et al. 1996):

$$u_{n,n} = \frac{-\left[1 - in\omega_{\text{mod}}T_2 + \sigma/(1 - in\omega_{\text{mod}}T_1)\right]v_{n-1,n-1} - \Delta\omega T_2 u_{n-1,n-1}}{(\Delta\omega)^2 T_2^2 + (1 - in\omega_{\text{mod}}T_2)\left[1 - in\omega_{\text{mod}}T_2 + \sigma/(1 - in\omega_{\text{mod}}T_1)\right]} \tag{11.51}$$

$$v_{n,n} = \frac{(1 - in\omega_{\text{mod}}T_2)u_{n-1,n-1} - \Delta\omega T_2 v_{n-1,n-1}}{(\Delta\omega)^2 T_2^2 + (1 - in\omega_{\text{mod}}T_2)\left[1 - in\omega_{\text{mod}}T_2 + \sigma/(1 - in\omega_{\text{mod}}T_1)\right]} \tag{11.52}$$

where $\sigma \equiv \gamma_e^2 B_1^2 T_1 T_2$ is the usual slow-passage saturation parameter (see Eq. 11.8). Like Eq. 11.50, Eqs. 11.51 and 11.52 do not apply to the zeroth harmonic (i.e., when n is equal to zero).

At the centre of the resonance $(\Delta\omega \equiv \omega - \omega_o = 0)$, the complex Fourier amplitude of the first-harmonic dispersion is simply (Halbach 1954):

$$u_{1,1}(0) = \frac{1}{1 - i\omega_{\text{mod}} T_2} v_{0,0}(0) \tag{11.53}$$

where $v_{0,0}(0) = \gamma_e B_1 T_2/(1+\sigma)$ is the absorption amplitude at $\Delta\omega = 0$. The complex Fourier amplitude of the first-harmonic absorption, at resonance offset $\Delta\omega$, is correspondingly (Halbach 1954):

$$v_{1,1}(\Delta\omega) = \frac{\Delta\omega T_2 (2 - i\omega_{\text{mod}} T_2)}{(\Delta\omega T_2)^2 + (1 - i\omega_{\text{mod}} T_2)\left[1 - i\omega_{\text{mod}} T_2 + \sigma/(1 - i\omega_{\text{mod}} T_1)\right]} \times v_{0,0}(\Delta\omega) \tag{11.54}$$

where $v_{0,0}(\Delta\omega) = \gamma_e B_1 T_2 / \left(1 + \sigma + \Delta\omega^2 T_2^2\right)$ is the absorption amplitude at $\Delta\omega$. At the centre of the second-harmonic absorption, the complex Fourier amplitude is (Páli et al. 1996):

$$v_{2,2}(0) = \frac{1}{(1 - i\omega_{\text{mod}} T_2)\left[1 - 2i\omega_{\text{mod}} T_2 + \sigma/(1 - 2i\omega_{\text{mod}} T_1)\right]} v_{0,0}(0) \tag{11.55}$$

We then get the nonlinear out-of-phase EPR signals from the imaginary parts of the complex Fourier amplitudes, as in Eqs. 11.33 and 11.34 of Section 11.4.

We retain all saturation factors but use approximations introduced in the following Section 11.9. From Eq. 11.53, the amplitude of the out-of-phase first-harmonic dispersion spectrum at the centre of the resonance is:

$$U_1'(0) = \frac{1}{2}\gamma_e B_{\text{mod}} \omega_{\text{mod}} T_2^2 \left(\frac{\gamma_e B_1 T_2}{1+\sigma}\right) \tag{11.56}$$

The sensitivity of the $U_1'(0)$ amplitude to T_1 therefore does not extend beyond that of the conventional saturation factor $(1+\sigma)^{-1}$. This follows immediately from $v_{0,0}(0)$ in Eq. 11.53. From Eq. 11.54, the amplitude of the out-of-phase first-harmonic absorption spectrum measured at a distance from the resonance position equal to the zeroth-harmonic linewidth $\left(\Delta\omega_{1/2} = 1/T_2\right)$ is (Livshits et al. 1998b):

$$V_1'\left(\Delta\omega_{1/2}\right) = \gamma_e B_{\text{mod}} \omega_{\text{mod}} T_2^2 \frac{1 - \sigma\left(T_1/T_2\right)/\left(1 + \omega_{\text{mod}}^2 T_1^2\right)}{\left[2 + \sigma/\left(1 + \omega_{\text{mod}}^2 T_1^2\right)\right]^2} \times V_0\left(\Delta\omega_{1/2}\right) \tag{11.57}$$

where $V_0\left(\Delta\omega_{1/2}\right) = v_{0,0}\left(\Delta\omega_{1/2}\right)$ for the zeroth harmonic. Note that the amplitude at the centre of the line is zero (see Figure 11.6).

This out-of-phase signal displays a considerably greater sensitivity to T_1-relaxation than does U_1' or the conventional in-phase absorption spectrum V_0. From Eq. 11.55, the central amplitude of the second-harmonic out-of-phase absorption spectrum is (Marsh et al. 1997):

$$V_2'(0) = \frac{1}{4}\gamma_e^2 B_{\text{mod}}^2 \omega_{\text{mod}} T_2^3 \frac{3 - 2\sigma\left(T_1/T_2\right)/\left(1 + 4\omega_{\text{mod}}^2 T_1^2\right)}{1 + 2\sigma/\left(1 + 4\omega_{\text{mod}}^2 T_1^2\right)} \times V_0(0) \tag{11.58}$$

This spectral display corresponds to standard second-harmonic saturation-transfer spectroscopy. Again, as for the first-harmonic, the second-harmonic out-of-phase absorption signal possesses additional sensitivity to T_1, beyond that simply from the saturation factor. Both first- and second-harmonic nonlinear absorption EPR spectra are more sensitive to T_1-relaxation than are conventional progressive-saturation experiments performed on in-phase V_1-spectra. As we see from Figure 11.7, the first-harmonic out-of-phase V_1'-spectrum has advantages over the second-harmonic out-of-phase V_2'-spectrum, for determining

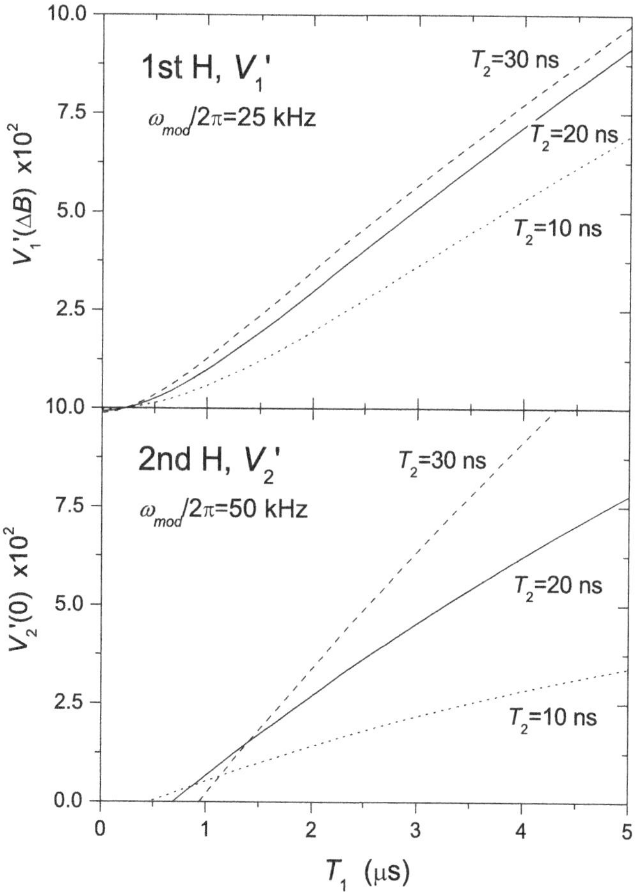

FIGURE 11.7 Dependence of out-of-phase EPR absorption amplitude on T_1-relaxation time for: first-harmonic (top, Eq. 11.57), and second-harmonic (bottom, Eq. 11.58), with values of T_2 indicated. Amplitudes normalized to in-phase absorption signal, by using factor $V_0(\Delta\omega) \times \gamma_e B_{\text{mod}} T_2$. B_1 = 25 μT; B_{mod} = 0.5 mT for V_2' (Marsh et al. 1997).

T_1-relaxation enhancements. The V_1'-spectrum depends far less on T_2-relaxation than does the V_2'-spectrum. As seen in Figure 11.6, it is also insensitive to molecular motion. This makes it an almost pure T_1-display and the nonlinear display of choice for studies of spin–lattice relaxation enhancement.

The V_2'-spectrum depends strongly on microsecond motion (see Figure 11.6), and its principal application is detecting ultraslow rotation in saturation-transfer spectroscopy. We can use it to determine spin–lattice relaxation enhancements when these produce only negligible change in T_2 and rotational rates remain unchanged. Before going on to saturation-transfer EPR proper, we consider first using V_1'- and V_2'-spectra to determine T_1 -relaxation enhancements of nitroxides.

11.9 FIRST-HARMONIC, OUT-OF-PHASE ABSORPTION V_1'-EPR

As we have seen, the first-harmonic, out-of-phase absorption spectrum (V_1') is sensitive to T_1 but relatively insensitive to both molecular motion (Hyde and Thomas 1973) and T_2-relaxation (Livshits et al. 1998b). It is helpful to use out-of-phase/in-phase absorption intensity ratios because these do not depend on the amount of sample and various instrumental parameters.

From Eq. 11.54 for the first-harmonic Fourier coefficient, the out-of-phase $V_{1,1}' \equiv \mathrm{Im}\{v_{1,1}\}$ and in-phase $V_{1,1} \equiv \mathrm{Re}\{v_{1,1}\}$ amplitudes for a single homogeneous line become (Livshits et al. 1998b):

$$V_{1,1}'(\Delta\omega) = \omega_{\mathrm{mod}}\Delta\omega T_2^2 \frac{(\Delta\omega)^2T_2^2 + A + 2B}{\left((\Delta\omega)^2T_2^2 + A\right)^2 + B^2\omega_{\mathrm{mod}}^2T_2^2} v_{0,0}(\Delta\omega) \tag{11.59}$$

$$V_{1,1}(\Delta\omega) = \Delta\omega T_2 \frac{2\left((\Delta\omega)^2T_2^2 + A\right)^2 - B\omega_{\mathrm{mod}}^2T_2^2}{\left((\Delta\omega)^2T_2^2 + A\right)^2 + B^2\omega_{\mathrm{mod}}^2T_2^2} v_{0,0}(\Delta\omega) \tag{11.60}$$

$$\text{with } A \equiv 1 - \omega_{\mathrm{mod}}^2T_2^2 + \sigma\frac{1+\omega_{\mathrm{mod}}^2T_1T_2}{1+\omega_{\mathrm{mod}}^2T_1^2} \approx 1 + \frac{\sigma}{1+\omega_{\mathrm{mod}}^2T_1^2} \tag{11.61}$$

$$B \equiv -2 + \sigma\left(T_1/T_2\right)\frac{1-T_2/T_1}{1+\omega_{\mathrm{mod}}^2T_1^2} \approx -2 + \frac{\sigma T_1/T_2}{1+\omega_{\mathrm{mod}}^2T_1^2} \tag{11.62}$$

The rightmost expressions in Eqs. 11.61 and 11.62 hold when $T_1 >> T_2$, $\omega_{\mathrm{mod}}^2T_2^2 << 1$ and $\omega_{\mathrm{mod}}^2T_1T_2 << 1$, which are justified by Livshits et al. (1998b). From Eqs. 11.59 and 11.60, the out-of-phase to in-phase first-harmonic amplitude ratio becomes:

$$\rho_1'(\Delta\omega) \equiv \frac{V_{1,1}'(\Delta\omega)}{V_{1,1}(\Delta\omega)} = -\omega_{\mathrm{mod}}T_2\frac{(\Delta\omega)^2T_2^2 + A + 2B}{2\left((\Delta\omega)^2T_2^2 + A\right) - B\omega_{\mathrm{mod}}^2T_2^2} \tag{11.63}$$

Assuming that $A >> B\omega_{\mathrm{mod}}^2T_2^2$ and substituting the approximate values from Eqs. 11.61 and 11.62, we get:

$$\rho_1'(\Delta\omega) \approx -\left(\frac{\omega_{\mathrm{mod}}T_1}{1+\omega_{\mathrm{mod}}^2T_1^2}\right)\frac{\sigma}{1+(\Delta\omega)^2T_2^2 + \sigma/\left(1+\omega_{\mathrm{mod}}^2T_1^2\right)} - \frac{\omega_{\mathrm{mod}}T_2}{2} \tag{11.64}$$

where $\sigma \equiv \gamma_e^2B_1^2T_1T_2$, and the final T_1-independent term is small because $T_1 >> T_2$. This result holds for $\omega_{\mathrm{mod}}T_1 < 1$, and for $\omega_{\mathrm{mod}}T_1 > 1$ with weak saturation (i.e., $\sigma \leq 1$) (Livshits et al. 1998b).

Figure 11.8 shows how the out-of-phase/in-phase first-harmonic amplitude ratio depends on B_1-field strength. Amplitude ratios for different values of T_1 are calculated by using Eq. 11.63, without approximation. The ratio increases strongly with increasing B_1, such that the out-of-phase amplitude becomes comparable with that of the in-phase signal at high B_1, if T_1 is long. In addition, high B_1-field strengths give better discrimination between different values of T_1.

As with progressive saturation measurements, there are advantages to using double-integrated intensities of the out-of-phase and in-phase first-harmonic signals. For the approximations in Eqs. 11.61 and 11.62, and conditions given following Eq. 11.64, we find that $A^2 >> B^2\omega_{\mathrm{mod}}^2T_2^2$. Then we can integrate the first-harmonic amplitudes analytically (Livshits et al. 1998b). The out-of-phase/in-phase ratio for integrated first-harmonic intensities becomes:

$$\rho_1' \equiv \frac{\iint V_1'(\omega)\cdot \mathrm{d}^2\omega}{\iint V_1(\omega)\cdot \mathrm{d}^2\omega} \approx -\frac{\omega_{\mathrm{mod}}T_2}{2}\left(\frac{B}{A+\sqrt{A(1+\sigma)}}+1\right) \tag{11.65}$$

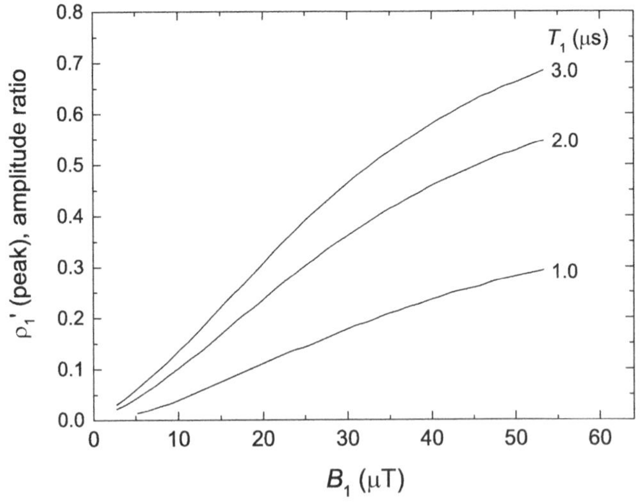

FIGURE 11.8 Dependence of out-of-phase/in-phase ratio of peak amplitudes ρ_1' on B_1-field strength, for first-harmonic absorption EPR spectra. Equation 11.63 without approximation, for values of T_1 indicated. Modulation frequency, $\omega_{\mathrm{mod}}/2\pi$ = 100 kHz; intrinsic homogeneous linewidth, $\Delta B_{1/2}^o \left(=1/\gamma_e T_2\right)$ = 0.2 mT (Livshits et al. 1998b).

For moderate saturation $(B_1 \geq 20\,\mu\text{T}; T_1 \geq 0.5\,\mu s)$, Eq. 11.62 approximates to $B \approx \sigma(T_1/T_2)/(1+\omega_{mod}^2 T_1^2)$. Along with the approximate Eq. 11.61, Eq. 11.65 for double-integrated intensities then becomes:

$$\rho_1' \approx -\frac{\omega_{mod} T_1}{2} \times \left(\frac{\sigma}{1+\sigma+\omega_{mod}^2 T_1^2 + \sqrt{(1+\sigma)(1+\omega_{mod}^2 T_1^2)(1+\sigma+\omega_{mod}^2 T_1^2)}} + \frac{T_2}{T_1} \right) \tag{11.66}$$

For low modulation frequencies such that $\omega_{mod}^2 T_1^2 << 1$, this reduces further to:

$$\rho_1' \approx -\frac{\omega_{mod} T_1}{2} \left(\frac{\sigma}{2(1+\sigma)} + \frac{T_2}{T_1} \right) \tag{11.67}$$

where $T_2/T_1 << 1$. These two equations display a very direct dependence on T_1.

We can explore the effects of molecular motion on the first-harmonic out-of-phase/in-phase ratios by using the jump model described in Section 11.5. The in-phase spectrum is given by Eq. 11.43, and the corresponding out-of-phase spectrum by $V_1'(\omega) = \text{Im}\{V_{1,1}(\omega)\}$.

Figure 11.9 shows the dependence of out-of-phase/in-phase integrated intensity ratio, ρ_1', on spin–lattice relaxation time that we obtain from spectral simulations for different jump reorientation times τ_j. The dependence on molecular motion is relatively small, throughout the entire range of reorientation times: $\tau_j = 10^{-10}$–10^{-4} s, especially for a Zeeman modulation frequency of 25 kHz. The dependence on intrinsic T_2^o also is similarly slight (Livshits et al. 1998b; Livshits and Marsh 2000).

As seen in Figure 11.8, we get best sensitivity to T_1 at relatively high-microwave magnetic field intensities, $B_1 \geq 30$ μT. The data in Figure 11.9 are calculated for $B_1 = 50$ μT. We can fit the dependence of out-of-phase/in-phase ratio on T_1 with a semi-empirical expression based on Eq. 11.66:

$$\rho_1' \equiv \frac{\iint V_1'(B)\cdot d^2 B}{\iint V_1(B)\cdot d^2 B} = \frac{a_1 T_1^m}{1+b_1 T_1^m} + \rho_1'^{o} \tag{11.68}$$

where, a_1, b_1, $\rho_1'^{o}$ and exponent m are fitting parameters. Calibration values are listed in Appendix A11.2 at the end of the chapter. A similar approach applies also to the amplitude ratio $\rho_1'(\Delta\omega)$, based on Eq. 11.64 (see Appendix A11.2).

The most striking feature of Figure 11.9 is the dependence on Zeeman modulation frequency. For $\omega_{mod}/2\pi$ = 100 kHz, the out-of-phase/in-phase ratio is most sensitive to short spin–lattice relaxation times, $T_1 \leq 2.5$ μs. The lower modulation frequency of 25 kHz extends this range of sensitivity to the T_1 = 2.5–5 μs range, for which the out-of-phase intensities become comparable to those at 100-kHz modulation frequency. Thus, the sensitivity of the first-harmonic out-of-phase display can be tuned to different ranges of spin–lattice relaxation time by changing the modulation frequency. Practical examples are given by Livshits et al. (1998b).

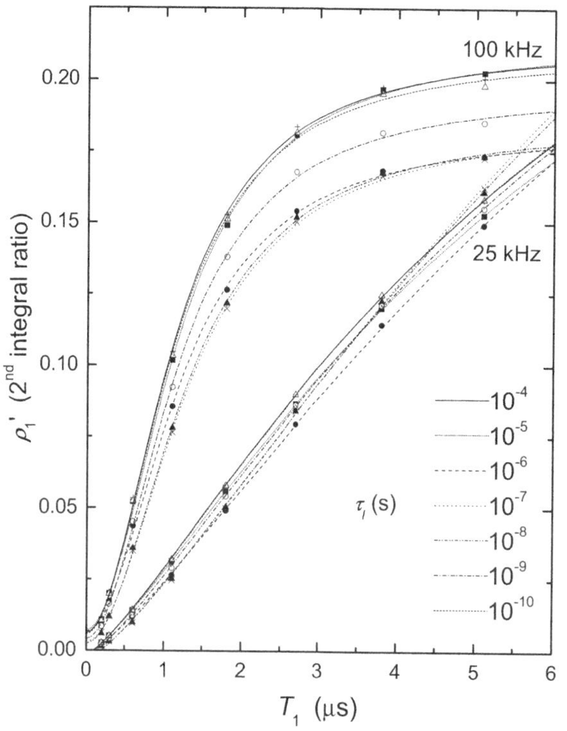

FIGURE 11.9 Dependence of out-of-phase/in-phase double-integrated intensity ratio ρ_1', for first-harmonic absorption spectra, on spin–lattice relaxation time T_1. Jump model from Section 11.5 with angular reorientation times τ_j shown. Zeeman modulation frequencies: $\omega_{mod}/2\pi$ = 100 kHz and 25 kHz; microwave field B_1 = 50 μT. *Lines*: calibration fits with Eq. 11.68. (Livshits and Marsh 2000.)

11.10 SECOND-HARMONIC, OUT-OF-PHASE ABSORPTION V_2'-EPR INTENSITIES

Although the second-harmonic out-of-phase absorption spectrum is less optimal for estimating T_1-relaxation times than is the first-harmonic equivalent, it is the classical ST-EPR display. In consequence, we have used it extensively to measure spin–lattice relaxation enhancements (see, e.g., Marsh et al. 1998).

The normalized integrated intensity of the V_2'-STEPR spectrum is defined by:

$$I_{ST} \equiv \frac{\int V_2'(B)\cdot dB}{\iint V_1(B)\cdot d^2 B} \tag{11.69}$$

We use the first integral of the V_2'-spectrum because it lies wholly above the baseline, in the absence of motion (see Figure 11.6).

The second integral of the conventional V_1-spectrum, recorded at the same B_1-field, normalizes the ST-EPR intensity to sample concentration. Conventionally, both signals are normalized to gain of 10× and a modulation amplitude $B_{mod} = 0.1$ mT (Horváth and Marsh 1983). The latter is inadequate for the V_2'-spectrum because its amplitude depends on $B_{mod}{}^2$: for consistency, we therefore always must use the fixed standard value of $B_{mod} = 0.5$ mT for recording the V_2'-spectrum (cf. Hemminga et al. 1984).

Figure 11.10 illustrates sensitivity of V_2'-intensities to relaxation enhancement induced by spin–spin interactions (i.e., cross relaxation) for a spin-labelled lipid in gel-phase bilayer membranes. Theoretical dependences of the integrated second-harmonic V_2'-intensity on spin–lattice relaxation time from simulation with the modulation-coupled Bloch equations are shown in the inset to the figure. The V_2'-intensity ratio, $\rho_2' \propto \int \mathrm{Im}(v_{22})\cdot \mathrm{d}B \Big/ \iint \mathrm{Re}(v_{11})\cdot \mathrm{d}^2B$, changes strongly with spin–lattice relaxation time in the range T_1 = 1–5 μs but also depends significantly on the intrinsic T_2. From the correlation with effective T_1T_2-relaxation time product obtained in progressive-saturation experiments (see Figure 11.10), we find that a consistent value of T_2^o = 8 ns at 25°C (8.5 ns at 15°C and 6.5 ns at 35°C) describes the experimental correlation between normalized ST-EPR intensities, $I_{ST,}$ and progressive-saturation results. This is established by no-motion simulations as in Sections 11.4 and 11.8. From linear regression, the numerical calibration for dependence on T_1 becomes (Páli et al. 1996):

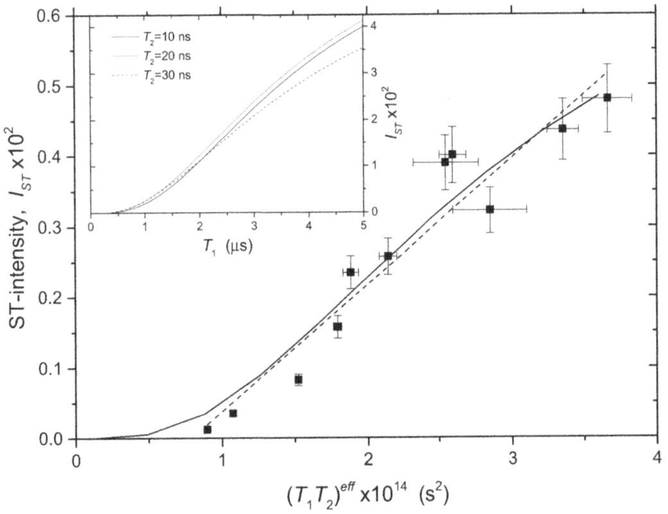

FIGURE 11.10 Correlation of normalized integrated intensity I_{ST} for second-harmonic, out-of-phase absorption V_2'-spectra with effective T_1T_2-relaxation time product from progressive saturation of double-integrated conventional V_1-EPR spectra. Spin-labelled phosphatidylcholine (5-DOXYL-PC), at increasing concentrations in gel-phase (T = 25°C) dipalmitoyl phosphatidylcholine bilayers. *Solid line*: dependence on T_1 from spectral simulations with modulation-coupled Bloch equations for T_2^o = 7.8 ± 1.1 ns and no motion; *dashed line*: linear regression. *Inset*: dependence of integrated V_2'-intensity on T_1 from no-motion spectral simulations (Section 11.4) for different values of T_2. (Páli et al. 1996).

$$I_{ST} \times 10^3 \approx 0.17 \times T_1(\mu s) T_2^{eff}(ns) - 1.3 \tag{11.70}$$

where $T_2^{eff} \approx 8.5$ ns at both 15°C and 25°C, which correspond essentially to a "no motion" situation.

11.11 EXAMPLE OF VERY SLOW TWO-SITE EXCHANGE IN LIPID–PROTEIN INTERACTIONS

Here, we give an example of the advantages that integrated intensities bring, both in ST-EPR and in progressive-saturation EPR. Consider the exchange equilibrium of a spin-labelled lipid **L*** with an unlabelled lipid **L** at the lipid–protein interface of a membrane-spanning protein **P**:

$$\mathbf{P}\cdot\mathbf{L}_b + \mathbf{L}_f^* \leftrightarrow \mathbf{P}\cdot\mathbf{L}_b^* + \mathbf{L}_f$$

where subscripts f and b indicate locations in the bulk lipid and at the lipid–protein interface, respectively. This is the two-site exchange situation that we treated in Section 10.12. The fractional population of **L*** at the lipid–protein interface is f_b and that in the bulk lipid is $1 - f_b$.

If exchange is slow on the T_1-timescale, progressive saturation of the integrated intensity is described by a simple weighted sum (cf. Eq. 11.11):

$$S = S_o B_1 \left(\frac{f_b}{\sqrt{1+\sigma_{o,b}}} + \frac{1-f_b}{\sqrt{1+\sigma_{o,f}}} \right) \tag{11.71}$$

where the saturation parameter is $\sigma_{o,b} = \gamma_e^2 B_1^2 T_{1,b}^o T_{2,b}^o$ with superscript "o" on T_1 and T_2 indicating no exchange, and similarly for $\sigma_{o,f}$ (see Eq. 11.8). Because we are using integrated intensities, the exponent in the denominators is unambiguous; also we need no additional scaling in the weighting factors to allow for different linewidths. Figure 11.11 illustrates application of Eq. 11.71 to detecting exchange between different lipid environments in membranes.

The figure shows saturation curves for a spin-labelled lipid in a myelin proteolipid protein/phosphatidylcholine recombinant membrane (P.L), and in corresponding lipid-alone (L) and protein-alone (P) environments. Spin labels associated with delipidated protein (P) saturate more readily than in bulk lipid (L) because of chain-mobility differences on the T_1-timescale. The left panel gives data for the gel phase, below the lipid chain-melting temperature T_m, where we do not expect exchange between the two membrane environments. The right panel gives data for the fluid phase, above T_m, where exchange on the T_1-timescale is possible. Spin labels saturate less readily in the fluid phase than in the gel phase because of higher chain mobility, but the striking difference is in exchange rate. We get the fractional population $f_b = 0.4$ at the lipid–protein interface from subtraction with

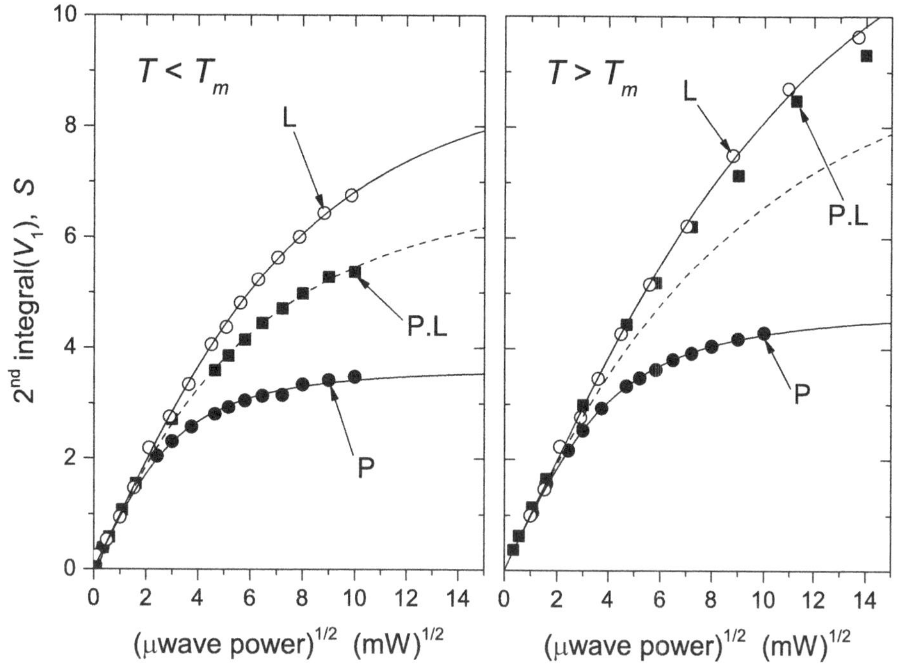

FIGURE 11.11 Progressive-saturation curves for integrated intensities *S* of EPR absorption spectra from spin-labelled phosphatidylcholine (14-DOXYL-PC) in dimyristoyl phosphatidylcholine membranes alone (open circles, L), associated with delipidated myelin proteolipid protein (solid circles, P), and in myelin proteolipid protein/dimyristoyl phosphatidylcholine membranes of lipid/protein ratio 24:1 mol/mol (solid squares, P.L). *Left panel*: samples in the lipid gel phase at 4°C; *right panel*: samples in the lipid fluid phase at 30°C. *Solid lines*: fits for single components with Eq. 11.11; *dashed lines*: predictions for lipid/protein membranes with no exchange (Eq. 11.71; $f_b = 0.4$). Data from Horváth et al. (1993a).

unsaturated spectral components in the fluid membrane phase, where lineshapes evidence clear differences in chain mobility between bulk lipid and the lipid-protein boundary (see Marsh and Horváth 1998; Marsh 2018). Saturation curves predicted by Eq. 11.71 with this value of f_b and saturation parameters $\sigma_{o,b}, \sigma_{o,f}$ derived from the single components are given by dashed lines in Figure 11.11. Spin labels in the gel-phase lipid–protein membrane (left panel) saturate just as predicted by linear additivity. Those in the fluid phase, however, saturate less readily than predicted by additivity: at a rate closer to that in bulk fluid lipids. Alleviation of saturation arises from exchange of lipids at the protein interface with those in the bulk, at a rate comparable to that of spin–lattice relaxation. This is consistent with our resolving two-component spectra, which indicates that exchange is slow on the conventional EPR timescale (Horváth et al. 1988a; Horváth et al. 1988b).

In the presence of exchange, Eq. 11.71 becomes:

$$S = S_o B_1 \left(\frac{f_b}{\sqrt{1+\sigma_{o,b}\left(T_{1,b}^{eff}/T_{1,b}^{o}\right)}} + \frac{1-f_b}{\sqrt{1+\sigma_{o,f}\left(T_{1,f}^{eff}/T_{1,f}^{o}\right)}} \right) \tag{11.72}$$

where the relaxation enhancements $T_{1,b}^{eff}/T_{1,b}^{o}$ are related to the exchange rate τ_b^{-1} by Eq. 10.54, and correspondingly for $T_{1,f}^{eff}/T_{1,f}^{o}$. In practice, it is not always easy to distinguish two components unambiguously in a saturation curve. Criteria for doing this are discussed by Páli et al. (1993). Therefore we turn to ST-EPR for determining the quantitative exchange rate.

Normalized ST-EPR integrals I_{ST} are additive in two-component systems without exchange. The equivalent of Eq. 11.71 is simply:

$$I_{ST} = f_b I_{ST,b}^{o} + (1-f_b) I_{ST,f}^{o} \tag{11.73}$$

where $I_{ST,b}^{o}$ and $I_{ST,f}^{o}$ are the values of I_{ST} for the protein-alone and lipid-alone samples, respectively. If we assume that I_{ST} is approximately proportional to T_1 (see Figure 11.10), the ST-EPR integral in the presence of exchange becomes simply:

$$I_{ST} = f_b I_{ST,b}^{o} \frac{T_{1,b}^{eff}}{T_{1,b}^{o}} + (1-f_b) I_{ST,f}^{o} \frac{T_{1,f}^{eff}}{T_{1,f}^{o}} \tag{11.74}$$

where the relaxation enhancements are given by Eq. 10.54 and equivalent, as for Eq. 11.72. Figure 11.12 shows the dependence of I_{ST} on f_b for the same myelin proteolipid membrane as in Figure 11.11. The lipid/protein ratio is fixed and we vary f_b, which we determine from unsaturated two-component spectra with conventional subtraction techniques, by choosing spin-labelled lipids with different selectivities for the protein. For each labelled species, the ST-intensity is greater in the gel phase than in the fluid phase because T_1 is longer, and I_{ST} depends linearly on f_b consistent with Eq. 11.73, indicating no exchange for $T < T_m$. In the fluid phase, however, the dependence of I_{ST} on f_b is markedly nonlinear and lies below the linear prediction given by

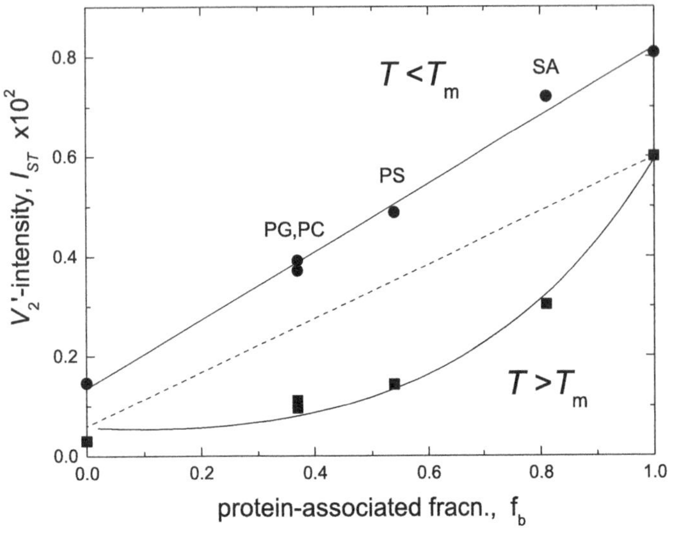

FIGURE 11.12 Dependence of normalized integrated V_2'-intensity I_{ST} on fraction f_b of different spin-labelled lipid species (SA, PS, PG, PC) associated with myelin proteolipid protein in dimyristoyl phosphatidylcholine membranes of fixed lipid/protein ratio 37:1 mol/mol. Samples in gel phase at 4°C (circles) and in fluid phase at 30°C (squares). *Straight solid line*: linear regression to circles; *curved solid line*: least-squares fit of Eq. 11.74 (with Eq. 10.54 and equivalent) to squares. *Dashed line*: dependence for no exchange at 30°C. Data from Horváth et al. (1993a).

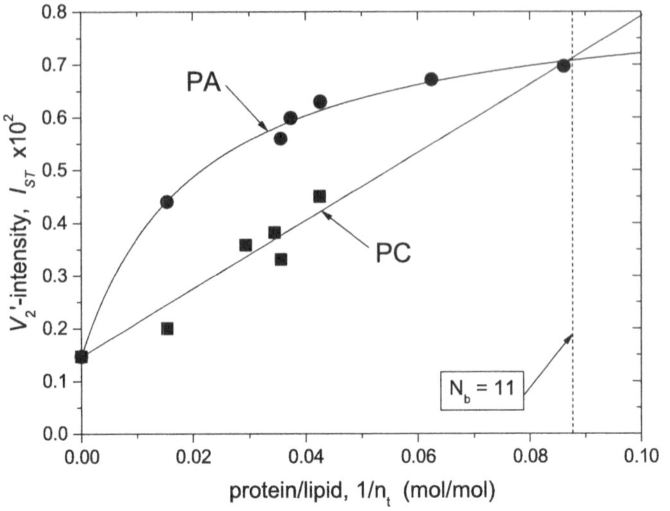

FIGURE 11.13 Dependence on protein/lipid ratio $1/n_t$ of V_2'-intensities I_{ST} from 14-DOXYL-PC (squares) and 14-DOXYL-PA (circles) in recombinant membranes of myelin proteolipid protein with dimyristoyl phosphatidylcholine in the gel phase at 0°C. *Straight solid line*: linear regression to squares; *curved solid line*: least-squares fit of Eq. 11.73 (with Eq. 11.75) to circles. *Dashed vertical line*: lipid/protein stoichiometry $N_b = 11$ mol/mol. Data from Horváth et al. (1993b).

the dashed line. Saturation is alleviated by exchange on and off the protein at a rate that affects spin–lattice relaxation. The solid curved line in Figure 11.12 is a least-squares fit of Eq. 11.74 combined with Eq. 10.54 and its equivalent for $T_{1,f}^{eff}/T_{1,f}^{o}$. This yields a normalized on-rate constant $T_{1,b}^{o}\tau_f^{-1} = 0.29$ for fixed lipid/protein ratio. Using the relation for detailed balance: $f_b\tau_b^{-1} = (1-f_b)\tau_f^{-1}$, the intrinsic off-rates of lipids with different selectivities for the protein are: $\tau_b^{-1} = 0.7, 0.9, 2.5$ and 4.9×10^6 s^{-1} for stearic acid, phosphatidic acid, phosphatidylserine and PC, respectively. These values are based on $T_{1,b}^{o} = 1$ µs for the intrinsic spin–lattice relaxation time (Horváth et al. 1993a).

Finally, Figure 11.13 shows how we can use ST-EPR to study the selectivity of lipid–protein interactions in the gel phase. Here, conventional lineshapes in both environments are rather similar, which makes quantifying relative populations difficult. We use Eq. 11.73, because exchange does not contribute, and vary f_b by changing the total lipid/protein mole ratio, n_t. Spin-labelled phosphatidylcholine (PC) shows no selectivity for the protein because I_{ST} increases linearly with n_t, which implies that $f_b = N_b/n_t$ where N_b is number of lipids per protein at the boundary interface. From gradient and intercept for PC in Figure 11.13, we find that $N_b = 11.0$ mol/mol. Spin-labelled phosphatidic acid (PA), on the other hand, has higher values of I_{ST} than PC at the same lipid/protein ratio. Also, the dependence on n_t is very nonlinear. When K_r^{PA} is the association constant of spin-labelled PA with the protein, relative to the background host lipid, the equation for equilibrium association gives (Marsh 1989):

$$f_b = \frac{K_r^{PA}}{n_t/N_b + K_r^{PA} - 1} \tag{11.75}$$

when the spin-labelled lipid is at probe concentrations. The solid curved line in Figure 11.13 is a nonlinear fit of Eq. 11.73 combined with Eq. 11.75. This yields $K_r^{PA} = 5.2$ relative to PC, and a consistent value of $N_b = 11.4$ mol/mol. These results are similar to those obtained for lipid-protein interactions in the fluid phase by conventional spectral subtraction (Brophy et al. 1984; Horváth et al. 1990a).

11.12 SATURATION TRANSFER EPR: ULTRASLOW ROTATIONAL MOTION

Saturation-transfer EPR (ST-EPR) is the original nonlinear EPR method that was designed to study very slow rotational diffusion of spin labels (Hyde 1978). It applies to correlation times in the microsecond regime or longer. The underlying idea is that saturation of a given spin packet in an orientationally dispersed powder spectrum (see Figure 11.14) is alleviated by rotation of the spin label on a timescale comparable to that of spin–lattice (T_1) relaxation. Transfer of saturation between different spin-label orientations is greatest in those regions of the spectrum where the change in resonance position with angular orientation $\partial\omega/\partial\theta$ is at a maximum. As we saw in Figure 11.6, this ω-dependent spectral diffusion makes the out-of-phase EPR lineshape sensitive to rotational correlation times in the sub-millisecond regime.

We illustrate the principles involved with a simple semi-analytical model. From Section 11.10, we know that the intensity

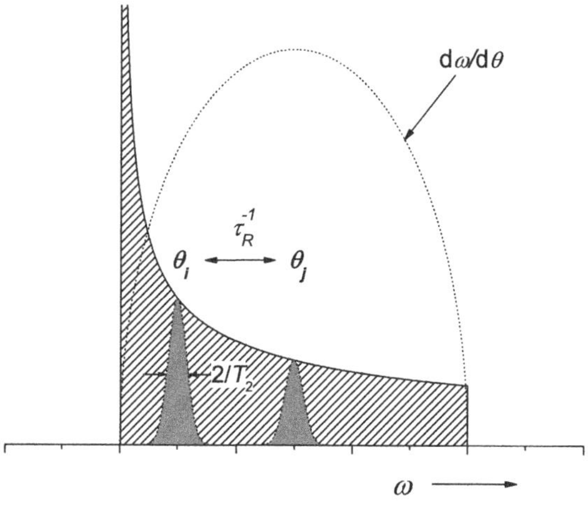

FIGURE 11.14 Saturation transfer between orientationally-selected spin packets (nitroxides at angles θ_i and θ_j in axial powder spectrum), induced by slow rotational diffusion with rate τ_R^{-1}. *Dotted line*: rate of change in resonance position ω with spin-label orientation, $\partial\omega/\partial\theta$. *Heavy line*: axial absorption powder lineshape (see Figure 2.6), approximating high-field manifold of ^{14}N-nitroxide at 9 GHz.

of the out-of-phase ST-EPR signal, relative to the conventional in-phase EPR, is roughly proportional to the effective spin–lattice relaxation time T_1^{eff} (see Figure 11.10 and Eq. 11.70). Therefore, we approximate the ST-EPR intensity at spectral position ω by (Marsh and Horváth 1992b):

$$I(\omega) = \frac{I_o(\omega)}{T_1^o} T_1^{eff}(\omega) \tag{11.76}$$

where $I_o(\omega)$ is the intensity, and T_1^o the spin–lattice relaxation time in the absence of spin-label rotation. Spectral diffusion of saturation throughout the powder lineshape by rotational diffusion is formally equivalent to the redistribution of saturation by exchange processes that we treated in Sections 10.11–10.13 of Chapter 10. For a given spin packet at resonance position ω, the effective spin–lattice relaxation time, $T_1^{eff}(\omega)$, depends on spectral diffusion rate, $\tau_{sd}(\omega)^{-1}$, according to Eq. 10.58 for spin exchange. The approximate ST-EPR lineshape in Eq. 11.76 therefore becomes:

$$I(\omega) = I_o(\omega)\cdot\frac{1+\left[I_o(\omega)/T_2\right]T_1^o\tau_{sd}(\omega)^{-1}}{1+T_1^o\tau_{sd}(\omega)^{-1}} \tag{11.77}$$

where $I_o(\omega)/T_2$ is the fractional population for a spectral segment of width $\propto 1/T_2$ (i.e., Z_p in Eq. 10.58).

To alleviate saturation, the resonance position must change at least by the spin-packet width $1/T_2$ (cf. Figure 11.14). The change $\Delta\theta_{sd}$ in spin-label orientation needed to achieve this is $(\partial\omega/\partial\theta)^{-1}T_2^{-1}$. For Brownian rotation, the corresponding diffusion time τ_{sd} is given by the Einstein relation: $\left\langle(\Delta\theta_{sd})^2\right\rangle = 2D_R\tau_{sd}$ (see Eq. 7.5 in Section 7.2). The spectral diffusion rate at resonance position ω then becomes (cf. Fajer et al. 1986):

$$\tau_{sd}(\omega)^{-1} = \frac{1}{3}\left(\frac{\partial\omega}{\partial\theta}\right)^2 T_2^2\tau_R^{-1} \tag{11.78}$$

where $\tau_R \equiv 1/(6D_R)$ is the rotational correlation time.

For simplicity, we use the high-field approximation for axial dependence of the resonance position on orientation θ of the spin label to the static field (cf. Eq. 2.11):

$$\omega = (\omega_{\parallel} - \omega_{\perp})\cos^2\theta + \omega_{\perp} \tag{11.79}$$

where $\omega_{\parallel}$ and $\omega_{\perp}$ are the resonance line positions corresponding to the magnetic field oriented parallel or perpendicular to the principal axis, i.e., $\theta = 0°$ and $\theta = 90°$, respectively. Equation 11.79 contains the essential features of anisotropic powder patterns (cf. Marsh 1990). The resulting rate of change of the resonance position with angle is:

$$\frac{\partial\omega}{\partial\theta} = -2\left[(\omega_{\parallel}-\omega)(\omega-\omega_{\perp})\right]^{1/2} \tag{11.80}$$

This is shown by the dotted line in Figure 11.14. From Eq. 2.15 of Section 2.4, the normalized lineshape in the absence of rotational diffusion is:

$$I_o(\omega) = \frac{1}{2}\left[(\omega_{\parallel}-\omega_{\perp})(\omega-\omega_{\perp})\right]^{-1/2} \tag{11.81}$$

which is valid for the range $\omega_{\perp} < \omega < \omega_{\parallel}$.

Figure 11.15 shows model lineshapes predicted from Eqs. 11.77–11.81 for various values of rotational correlation time, τ_R. They reproduce the well-known sensitivity of ST-EPR lineshapes to rotational diffusion at rates comparable to that of spin–lattice relaxation (cf. Thomas et al. 1976). The line height at an intermediate spectral position P', relative to that at the invariant turning point P, decreases progressively with decreasing correlation time (cf. Figure 11.6). The inset to Figure 11.15 shows that the relative intensity at a position 1/3 of the way in from the $\omega_{\parallel}$-turning point depends on τ_R like the experimental diagnostic ST-EPR line-height ratios L''/L, C'/C and H''/H, which we define in the next section.

11.13 SATURATION-TRANSFER EPR: V_2'-LINESHAPES

Out-of-phase second-harmonic spectra have the appearance of a second-harmonic in-phase lineshape superimposed on that of the absorption. Figure 11.16 shows typical second-harmonic out-of-phase absorption V_2'-lineshapes. These are calculated by using the jump model for rotational motion described in Section 11.5, where the out-of-phase lineshape is given by $V_2'(\omega) = \mathrm{Im}\{v_{2,2}(\omega)\}$. The characteristic preferential loss of intensity at intermediate field positions, relative to those corresponding to stationary turning points, is evident with decreasing

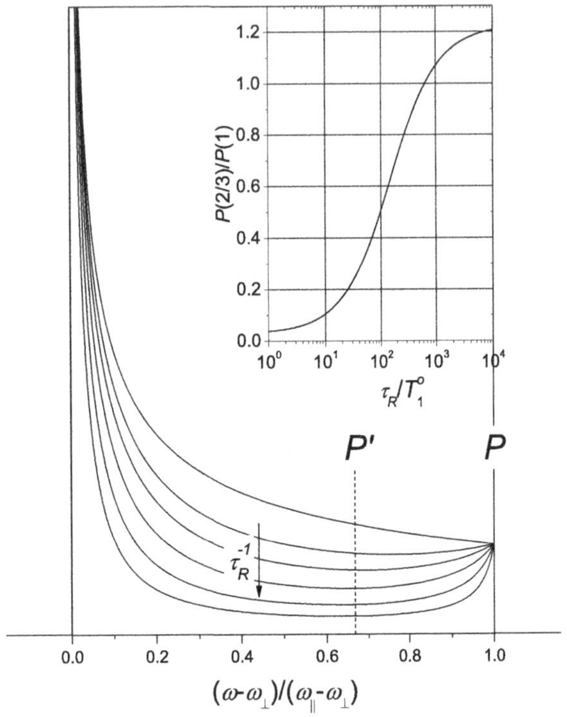

FIGURE 11.15 Model axial ST-EPR powder absorption lineshapes (heavy lines) from Eq. 11.77 (with Eqs. 11.78, 11.80, 11.81), for increasing rates of rotational diffusion. *Top to bottom*: $T_1^o\tau_R^{-1}$ = 0, 0.0025, 0.005, 0.01, 0.02 and 0.04; $T_2^{-1} = (\omega_\parallel - \omega_\perp)/25$. *Inset*: diagnostic ST-EPR line-height ratio at point *P'*, relative to turning point *P* at right-hand extreme of spectrum. Data from Marsh and Horváth (1992a).

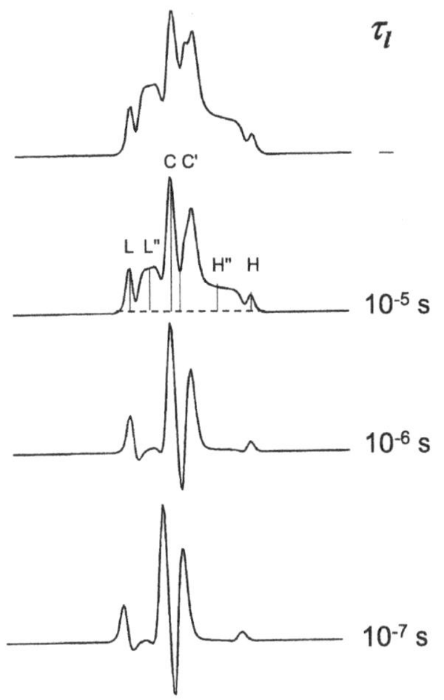

FIGURE 11.16 Second-harmonic, 90°-out-of-phase absorption ST-EPR spectra (V_2'-display) for decreasing reorientation times, τ_l. Simulations for ^{14}N-nitroxide at 9 GHz with random sudden-jump diffusion (Section 11.5). Spectra normalized to maximum line height and do not reflect decreasing absolute intensity with decreasing τ_l. Positions defining diagnostic line-height ratios, *L″/L*, *C′/C* and *H″/H*, shown for $\tau_l = 10^{-5}$ s. (See Marsh et al. 2005.)

angular reorientation time τ_l (cf. Figure 11.15). This behaviour of the lineshape defines the diagnostic line-height ratios introduced by Thomas et al. (1976). *L″*, *C′* and *H″* shown in Figure 11.16 are line heights at the middle of the intermediate regions in the powder pattern, for the low-field, central and high-field ^{14}N-hyperfine manifolds, respectively. *L*, *C* and *H* are line heights at the corresponding stationary turning points, in the 9-GHz ST-EPR spectrum.

Experimental calibration of the dependence of ST-EPR lineshapes on rotational mobility plays an important role in saturation-transfer spectroscopy. Figure 11.17 shows the dependence of the diagnostic line-height ratios on rotational correlation time for spin-labelled haemoglobin in glycerol-water mixtures of different viscosities, η. The correlation time for isotropic rotational diffusion is determined from the Debye equation: $\tau_R = \eta V_{mol}/(k_B T)$, where V_{mol} is the molecular volume (see Section 7.3). Maximum sensitivity occurs in the region of τ_R = 8, 70 and 200 µs for *C′/C*, *L″/L* and *H″/H*, respectively. This correlates with the extent of spectral anisotropy $(\omega_\parallel - \omega_\perp)$ in the three hyperfine manifolds (cf. Figure 11.15). As we saw in previous sections, the out-of-phase intensities depend on our choice of microwave B_1-field, and on the modulation field B_{mod} and

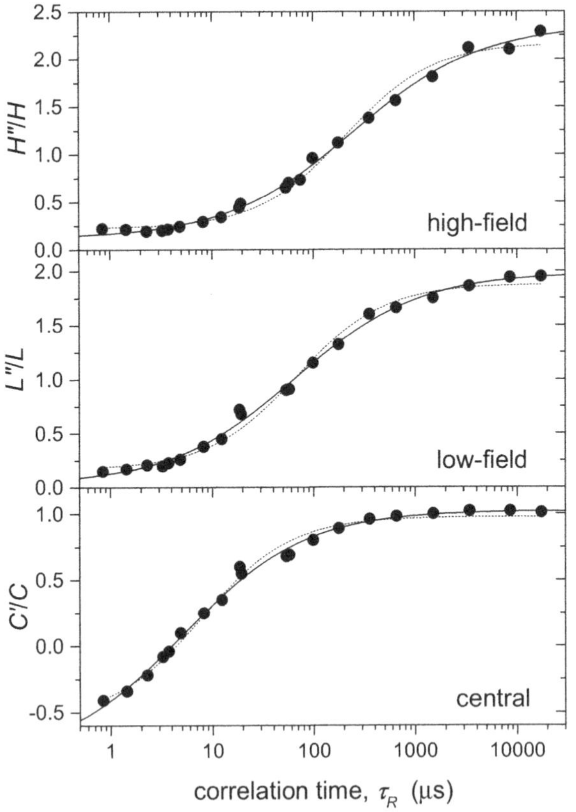

FIGURE 11.17 Second-harmonic V_2'-STEPR rotational correlation-time calibrations for low-field, central and high-field diagnostic 9-GHz line-height ratios (*L″/L*, *C′/C*, *H″/H*) from ^{14}N spin-labelled haemoglobin (Horváth and Marsh 1983). Fitting parameters in the logistic Eq. A11.6, for line-height ratios (solid lines) and normalized integrals (I_{ST}) given in Table A11.8 of Appendix A11.3. Dashed lines: fits of semi-empirical Eq. 11.83 (see Marsh and Horváth 1992a).

TABLE 11.3 Standard experimental conditions for V_2'-STEPR spectra at 9 GHz (Hemminga et al. 1984; Thomas et al. 1976)[a]

sample:		*microwaves:*		*modulation:*		*detection:*	
[SL] (μM)	*length* (mm)	$\omega_e/2\pi$ (GHz)	$B_1(0)$ (μT)	$\omega_{mod}/2\pi$ (kHz)	$B_{mod}(0)$ (mT)	$\omega_{ref}/2\pi$ (kHz)	*phase*
≥30	5	9	25	50	0.5	100	90°
≤30	>20	9	25	50	1.0	100	90°

[a] A standard rectangular cavity resonator is assumed, and the sample is contained in a capillary of 1-mm inner diameter, and centred in the cavity ($z = 0$) (see Fajer and Marsh 1982).

frequency $\omega_{mod}/2\pi$. This is also true of the lineshapes. Table 11.3 gives standard experimental conditions for recording 9-GHz ST-EPR spectra (Hemminga et al. 1984).

We can use the analysis given in Section 11.12 to parameterize the ST-EPR correlation-time calibrations. Figure 11.15 and Eq. 11.77 suggest that spectral line-height ratios and integrated intensities, R, have the following dependence on τ_R:

$$R = R_o \frac{1 + a/\tau_R}{1 + b/\tau_R} \tag{11.82}$$

where R_o is the value of R in the absence of rotational diffusion; and a, b are constants to be fitted that depend only on intrinsic spectral parameters. The ratio a/b is effectively related to the orientational degeneracy parameter $\left(Z(\omega_\theta) \approx \sin\theta\right)$ at ω_θ that corresponds to the diagnostic spectral position P'. As we see from the dashed lines in Figure 11.17, Eq. 11.82 describes the dependence of the diagnostic ST-EPR line-height ratios and intensities on rotational correlation time rather well (Marsh and Horváth 1992a). Therefore, we use it to give the following simple expression for the correlation-time calibrations of the experimental ST-EPR spectra:

$$\tau_R = \frac{k}{R_o - R} - b \tag{11.83}$$

where values of the calibration constants, k, R_o and b, for the different diagnostic spectral parameters are given in Table 11.4.

TABLE 11.4 Second-harmonic 9-GHz V_2'-STEPR rotational correlation-time calibrations from ^{14}N spin-labelled haemoglobin (Horváth and Marsh 1983)

R	*range fitted*	R_o	k (μs)	b (μs)
L''/L[a]	0.2, 2.0	1.825	105.6	63.8
H''/H	0.2, 2.0	2.17	407	210
C'/C	0.2, 1.0	1.01	21.3	21.1
	−0.4, 1.0	0.976	11.9	7.82
I_{ST} (total)	0.15, 1.0	1.07×10^{-2}	0.400	43.6
I_{ST} ($m_I = -1$)[b]	0.06, 0.6	6.97×10^{-4}	2.85×10^{-2}	45.0

Fitting parameters for diagnostic spectral line-height ratios (L''/L, H''/H, C'/C), and normalized integrals (I_{ST}), in Eq. 11.83.
[a] Values of L'' correspond to the maximum in the low-field diagnostic region. For values corresponding to the position 1/3 of the way in from the low-field turning point (L), as given originally in Horváth and Marsh (1983), see Marsh (1992).
[b] Integrated intensity for the high-field manifold alone; used to detect anisotropic rotational diffusion (Horváth and Marsh 1983).

The solid lines in Figure 11.17 are from a totally empirical fit with the logistic equation, which contains one more adjustable parameter than Eq. 11.83. This description is essentially quantitative: fitting parameters are listed for practical reference in Table A11.8 of Appendix A11.3 at the end of the chapter.

From the conventional-EPR powder lineshapes in Figure 2.8 of Chapter 2, we see that ST-EPR spectra at higher microwave frequencies can explore different modes of ultraslow anisotropic rotation (cf. Fajer and Marsh 1983b). Alternatively, they should provide consistency checks for isotropic rotation. The larger spectral anisotropy at high-field/frequency also increases sensitivity of saturation transfer to very slow rotation (see Eq. 11.80 and Figure 11.14). Calibrations for different line-height ratios in 35-GHz ST-EPR spectra are given by Johnson and Hyde (1981). Calibrations for high-field (94 GHz) ST-EPR of a ^{15}N-nitroxide are given by Song et al. (2010). Note that we expect high-field EPR to drive any slow segmental motion into the rigid limit (cf. Section 8.15), which should help in promoting rigid label attachment that we need for ST-EPR.

11.14 ROTATIONAL DIFFUSION OF MEMBRANE PROTEINS

An important application of ST-EPR is to study rotational motion of spin-labelled membrane-spanning proteins. We treated rotational diffusion in isotropic media in Section 7.3 of Chapter 7. Here we give modifications needed in the highly anisotropic environment of the membrane. An integral protein primarily undergoes uniaxial rotation about the transmembrane diffusion axis (see Figure 11.18). The diffusion coefficient for uniaxial rotation is related to the corresponding frictional coefficient $f_{R\parallel}$ by the Einstein relation: $D_{R\parallel} = k_B T / f_{R\parallel}$ (see Eq. 7.9).

Because the intramembrane viscosity, $\eta_m \sim 0.2$–0.5 Pa s (Cherry and Godfrey 1981), is much greater than that of the external aqueous viscosity ($\eta_o \sim 1$ mPa s), the dominant frictional torque acts solely on the intramembrane sections of the protein. For a circular cylindrical transmembrane protein, the friction coefficient for uniaxial rotation is (Saffman 1976):

$$f^o_{R\parallel,m} = 4\eta_m V_m = 4\eta_m A_m h_m \tag{11.84}$$

(cf. Eq. 7.10), where A_m is the cross-sectional area of the cylinder and h_m is the membrane thickness (see Figure 11.18).

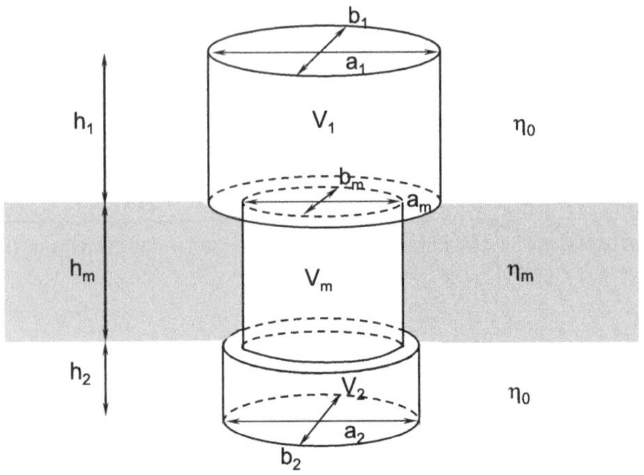

FIGURE 11.18 Hydrodynamic model for rotational diffusion of transmembrane protein with central cylindrical section (volume V_m) embedded in membrane of viscosity η_m, and two extramembranous sections (total volume $V_o = V_1 + V_2$) in aqueous medium of viscosity η_o. Heights, h_i, and cross-sectional dimensions, a_i and b_i, of sections are indicated. Uniaxial rotational diffusion occurs about cylinder axis.

This direct dependence on the cross-sectional area (A_m) of the rotating species lets us determine the oligomerisation state of a membrane-spanning protein. If a cylinder lacks circular symmetry, the friction coefficient is related to a shape factor $F_{R\parallel}$ by:

$$f_{R\parallel} = f^o_{R\parallel} F_{R\parallel} \tag{11.85}$$

cf. Eq. 7.11. For elliptical cross-sections, the shape factor is (Jähnig 1986):

$$F_{R\parallel} = \frac{1}{2}\left(\frac{a_m}{b_m} + \frac{b_m}{a_m}\right) \tag{11.86}$$

where a_m and b_m are the semi-axes of the ellipse, and $F_{R\parallel} = 1$ for a circular cylinder $(a_m = b_m)$. The correlation time for uniaxial rotation $\tau_{R\parallel} \equiv 1/(6D_{R\parallel})$ then becomes:

$$\tau_{R\parallel,m} = \frac{2\eta_m h_m A_m}{3k_BT} \cdot F_{R\parallel,m} = \frac{\pi\eta_m h_m}{3k_BT}\left(a_m^2 + b_m^2\right) \tag{11.87}$$

which gives the explicit dependence on the intramembranous section of the protein, when external viscosity can be neglected.

If we increase the external viscosity, e.g., by addition of glycerol or sucrose, the rotational correlation time then depends also on the dimensions of the extramembranous sections of the protein. Using Eq. 7.6, we sum the individual contributions to the overall frictional coefficient:

$$f_{R\parallel} = \sum_i f^o_{R\parallel,i} F_{R\parallel,i} \tag{11.88}$$

because the contributions of the separate sections, i, of the protein to the net frictional torque are additive (see Figure 11.18 and Eq. 7.11). The dependence of the rotational correlation time on external viscosity, η_o, then becomes:

$$\tau_{R\parallel} = \tau_{R\parallel,m} + \frac{2\eta_o}{3k_BT}\sum_{j\neq m} V_j F_{R\parallel,j} \tag{11.89}$$

where summation on the right extends only over the extramembrane sections of the protein with volumes V_j, and $\tau_{R\parallel,m}$ is given by Eq. 11.87. Application to size determination of the intramembrane and extramembrane portions the ion-transporting enzyme Na,K-ATPase is given by Esmann et al. (1994).

An important practical consideration is the extent to which we can apply correlation time calibrations obtained with isotropically rotating species (see Section 11.13) to systems undergoing anisotropic rotation. In 9-GHz spectra, the outer line-height ratios, L''/L and H''/H, are sensitive only to rotation of the nitroxide z-axis, via modulation of the ^{14}N-hyperfine interaction. Then we need to know the orientation, θ, of the nitroxide z-axis to the rotational diffusion axis. For axially anisotropic rotational diffusion, Robinson and Dalton (1980) showed by simulation of U_1' ST-EPR spectra that we can still use isotropic calibration spectra. The effective values of $\tau_R^{eff}(m_I)$ that we then get from the L''/L and H''/H ratios are related approximately to the true correlation times by:

$$\tau_R^{eff}(\pm 1) = \frac{2\tau_{R\parallel}\tau_{R\perp}}{\tau_{R\perp}\sin^2\theta + \tau_{R\parallel}\left(1+\cos^2\theta\right)} \tag{11.90}$$

where $\tau_{R\parallel}$, $\tau_{R\perp}$ are correlation times for rotation about and perpendicular to the principal rotational axis ($\tau_{R\parallel} << \tau_{R\perp}$), respectively. For a transmembrane peptide or protein, we can assume that $\tau_{R\perp}$ is so long that $\tau_R^{eff}(\pm 1)$ is determined solely by $\tau_{R\parallel}$. This rotational correlation time is then given by $\tau_{R\parallel} = \frac{1}{2}\tau_R^{eff}(\pm 1)\sin^2\theta$, which has a maximum value: $\tau_{R\parallel}^{\max} = \frac{1}{2}\tau_R^{eff}(\pm 1)$. The latter also corresponds to statistically the most likely orientation $\theta = 90°$ of the nitroxide axes relative to the rotation axis (see Figure 2.5).

The central line-height ratio, C'/C, is sensitive to rotation of all three nitroxide axes, via modulation of the g-value anisotropy. Simulations of V_2' ST-EPR spectra for uniaxial rotational diffusion indicate that the C'/C ratio is relatively insensitive to the orientation, θ, of the spin-label axes (Beth and Hustedt 2005). For the extreme $\theta = 0°$ orientation, uniaxial rotation involves only interchange of the x- and y-nitroxide axes, whereas for the opposite extreme of $\theta = 90°$, z-interchanges with the x-/y-axes. Nevertheless, the C'/C ratios are comparable (Beth and Robinson 1989), although the C' position shifts from the lower to the upper half of the central manifold (at 9 GHz). Therefore, we can use C'/C to get an upper estimate for $\tau_{R\parallel}$, in the case of uniaxial rotation. In general, different values of effective correlation time deduced from C'/C and L''/L (or H''/H) using isotropic calibrations clearly indicate that the rotational diffusion is anisotropic (Fajer and Marsh 1983b; Marsh 1980). However, a practical limitation is that the central region of the ST-EPR spectrum becomes obscured by any residual mobile spin-label species.

Table 11.5 lists some ST-EPR results obtained with different spin-labelled transmembrane proteins. Maximum values for the uniaxial correlation time $\tau_{R\parallel}^{\max}(\pm 1)$ are given, together with (where appropriate) the effective correlation time deduced from the central region of the spectrum $\tau_R^{eff}(0)$. These are compared with values $\tau^o_{R\parallel}$ predicted from the protein size by using Eq. 11.87.

TABLE 11.5 Maximal values for uniaxial rotational correlation times, $\tau_{R\parallel}^{max}$, deduced from *L″/L* or *H″/H* ST-EPR line-height ratios of membrane-spanning proteins[a]

protein	$a_m \times b_m$ (nm)	$\tau_{R\parallel}^{o}$ (μs)	$\tau_{R\parallel}^{max}(\pm 1)$ (μs)	$\tau_R^{eff}(0)$ (μs)	*Ref.*[b]
Na,K-ATPase	3.25 × 3.75	7.2–14.3	25	-	1, 2
ADP–ATP carrier	1.5 × 1.5	1.3–2.5	2[c]	10[c]	3
cytochrome oxidase	4.5 × 7.5[d]	22–43	25	-	4
Ca-ATPase	2.5 × 2.5	3.4–7.7	25–40	(2–10)	5
M13 coat protein			10		6
Rhodopsin	2.25 × 1.85	2.2–4.5	3.5		7
GalP	3.1 × 1.5[e]	3.1–6.2	17	5	8
V-ATPase	3.4 × 3.4[f]	7.1–14.2	60–75	95	9, 10
alamethicin	0.5 × 0.5	0.14–0.28	<0.1		11

[a] Values of $\tau_R^{eff}(0)$ are deduced from *C′/C*. a_m and b_m: intramembranous cross-sectional semi-axes of the monomer, unless noted otherwise (see Marsh 2007 for references). $\tau_{R\parallel}^{o}$: theoretical value of uniaxial rotational correlation time (Eq. 11.87 with $\eta = 0.25$–0.5 Pa s, $h_m = 4.5$ nm).
[b] References: 1. Esmann et al. (1989); 2. Esmann et al. (1987); 3. Horváth et al. (1989); 4. Fajer et al. (1989); 5. Horváth et al. (1990b); 6. de Jongh et al. (1990); 7. Ryba and Marsh (1992); 8. Marsh and Henderson (2001); 9. Páli et al. (2004); 10. Páli et al. (1999); 11. Marsh et al. (2007).
[c] Value of $\tau_{R\parallel}^{max}$ deduced from high-field ST-EPR integral, and of $\tau_R^{eff}(0)$ from total integral, I_{ST}.
[d] For dimer.
[e] For Lac permease monomer.
[f] For hexamer.

In one notable example, both orientation and rotational correlation time of a maleimide affinity spin label attached to the anion exchange transporter in red blood cells were determined from Bloch-/transition-matrix simulations of ST-EPR spectra at 3 modulation frequencies (Hustedt and Beth 1995) and 2 microwave frequencies (Beth and Hustedt 2005). These studies required using an isotopically labelled [^{15}N,^{2}H$_{13}$]-dihydrostilbene disulphonate maleimide to improve spectral resolution. An analogous study of spin-labelled epidermal growth factor bound to its receptor also used multiple modulation frequencies and ST-EPR simulations (Stein et al. 2002). The EGF-receptor spans the membrane with a single α-helix, and the ST-EPR results were consistent with coexisting monomers (cf. $\tau_{R\parallel}^{o}$ for alamethicin in Table 11.5) and dimers. In the two sections following, we describe briefly ST-EPR simulation methods for Brownian rotational diffusion.

11.15 ST-EPR SIMULATIONS WITH THE BLOCH EQUATIONS

In Section 6.5 of Chapter 6, we used the transverse Bloch equations coupled by rotational diffusion to simulate unsaturated in-phase EPR lineshapes for nitroxides undergoing slow rotational motion (McCalley et al. 1972). We can use a similar technique to calculate the out-of-phase lineshapes under saturation (Thomas and McConnell 1974). To do this, we must include the Bloch equation for the longitudinal magnetisation (Eq. 11.3) and also introduce field modulation (Eqs. 11.23, 11.24), as in Sections 11.2 and 11.4, respectively.

As explained in Section 6.5, Brownian rotational diffusion is modelled by transitions between closely-spaced angular zones. For the kth zone, the probabilities per unit time $W_{k,k\pm1}$ of transition to the adjacent ($k\pm1$)th zones are given by Eqs. 6.38–6.40 and similarly for other zones. Instead of the steady-state Eqs. 6.42 and 6.43 in the rotating frame, we now get three coupled relations for the components u_{nk}, v_{nk}, $M_{z,nk}$ of the complex magnetization from the kth angular zone that oscillate at the nth harmonic of the modulation frequency ω_{mod} (Thomas and McConnell 1974):

$$\left(\frac{1}{T_2} + W_{k,k+1} + W_{k,k-1} - in\omega_{mod}\right)u_{nk} - (\omega_k - \omega)v_{nk} - W_{k+1,k}u_{n(k+1)} - W_{k-1,k}u_{n(k-1)} - \frac{1}{2}\gamma_e B_{mod}\left(v_{(n-1)k} + v_{(n+1)k}\right) = 0 \quad (11.91)$$

$$\left(\frac{1}{T_2} + W_{k,k+1} + W_{k,k-1} - in\omega_{mod}\right)v_{nk} + (\omega_k - \omega)u_{nk} - W_{k+1,k}v_{n(k+1)} - W_{k-1,k}v_{n(k-1)} + \frac{1}{2}\gamma_e B_{mod}\left(u_{(n-1)k} + u_{(n+1)k}\right) = \gamma_e B_1 M_{z,nk} \quad (11.92)$$

$$\begin{aligned}\left(\frac{1}{T_1} + W_{k,k+1} + W_{k,k-1} - in\omega_{mod}\right)M_{z,nk} - \gamma_e B_1 v_{nk} - W_{k+1,k}M_{z,n(k+1)} - W_{k-1,k}M_{z,n(k-1)} &= \left(1 + (\omega_k - \omega)/\omega_e\right)M_{o,k}/T_1 \quad (n = 0)\\ &= -\frac{1}{2}\left(\gamma_e B_{mod}/\omega_e\right)M_{o,k}/T_1 \quad (|n| = 1)\\ &= 0 \quad (|n| \geq 1)\end{aligned} \quad (11.93)$$

where ω_e is the fixed microwave angular frequency, the angular resonance frequency for the kth angular zone is $\omega_k \equiv \gamma_e B_{res}(\theta_k)$, and the spectral scan parameter is $\omega \equiv \gamma_e B$. $M_{o,k}$ is the equilibrium z-magnetization of the kth zone as in Section 6.5, and B_{mod} again is half the peak-to-peak field modulation amplitude.

The system of coupled equations, involving now both k and n, is solved numerically for each value of the spectral sweep ω (i.e., B). This yields the complex Fourier coefficients v_{nk} (for absorption) and u_{nk} (for dispersion), where the real part gives the in-phase signal and the imaginary part gives that 90° out-of-phase (cf. Eqs. 11.32–11.34). Summing v_{nk} and u_{nk} over all orientations k gives the amplitude of the nth harmonic spectrum at a given field point. For simplicity, an effective $T_2 = T_2^* = 2.4 \times 10^{-8}$ s is used to approximate intrinsic line broadening. Then, using at least 15 to 18 angular zones (k) and including 3 or 4 harmonics (n) is sufficient for convergence, up to the second harmonic V_2'-EPR spectra (Thomas and McConnell 1974). For narrower linewidths, up to 90 angular zones and five harmonics may be required (Hyde 1978). Note that restricting to the lowest power of the modulation amplitude, $v = n$ in Eq. 11.29, does not guarantee convergence. This is because ST-EPR spectra are recorded at higher modulation amplitudes than used for conventional EPR (see Table 11.3).

Simulations using the diffusion- and modulation-coupled Bloch equations account satisfactorily for correlation time calibrations of line-height ratios L''/L and H''/H from V_2'-STEPR spectra of spin-labelled haemoglobin (Thomas et al. 1976). To account for the C'/C ratios would need us to relax the axial approximation that was used in Section 6.5. Application of this model to uniaxial rotation for membrane proteins, with a fixed orientation of the nitroxide z-axis to the rotation axis, is described by Hustedt and Beth (1995).

11.16 STOCHASTIC-LIOUVILLE ST-EPR SIMULATIONS

To treat ST-EPR using the stochastic-Liouville equation, we must include field modulation in the time-dependent term $\varepsilon(t)$ of Eq. 11.48, as described in Section 11.6. From Eqs. 6.57 and 11.23, the combined effect of microwave and modulation fields becomes:

$$\begin{aligned}\varepsilon(t) &= \tfrac{1}{2}\gamma_e B_1\left(S_- \exp(i\omega t) + S_+ \exp(-i\omega t)\right)\\ &\quad + \tfrac{1}{2}\gamma_e B_{\mathrm{mod}} S_z\left(\exp(i\omega_{\mathrm{mod}}t) + \exp(-i\omega_{\mathrm{mod}}t)\right)\end{aligned} \quad (11.94)$$

where $\pm B_{\mathrm{mod}}$ is the amplitude, and ω_{mod} the angular frequency, of the Zeeman modulation. Solutions of Eq. 11.48 are then expanded in harmonics n of the modulation frequency (cf. Eq. 11.24):

$$\chi(\Omega,t) = \exp(i\omega t)\sum_n Z^{(\pm n)}(\Omega,\omega)\exp(\pm in\omega_{\mathrm{mod}}t) \quad (11.95)$$

where $Z^{(n)}(\Omega,\omega) \equiv Z'^{(n)} + iZ''^{(n)}$ are harmonic components of the reduced spin density-matrix in the rotating frame (cf. Eq. 6.59).

By comparing signs in Eqs 11.94 and 11.95, we see that EPR signals in phase with the Zeeman modulation (i.e., $\approx \cos\omega_{\mathrm{mod}}t$) are given by $Z^{(n)} + Z^{(-n)}$, and those 90°–out-of-phase ($\approx \sin\omega_{\mathrm{mod}}t$) by $Z^{(n)} - Z^{(-n)}$. The in-phase and 90°–out-of-phase nth harmonic dispersion signals are therefore given by:

$$U_n = Z_{+-}'^{(n)} + Z_{+-}'^{(-n)} \quad (11.96)$$

$$U_n' = Z_{+-}''^{(n)} - Z_{+-}''^{(-n)} \quad (11.97)$$

respectively. Correspondingly, the in-phase and 90°–out-of-phase nth harmonic absorption signals are:

$$V_n = Z_{+-}''^{(n)} + Z_{+-}''^{(-n)} \quad (11.98)$$

$$V_n' = Z_{+-}'^{(n)} - Z_{+-}'^{(-n)} \quad (11.99)$$

respectively. The real part, $Z'^{(n)}$, and imaginary part, $Z''^{(n)}$, of $Z^{(n)}(\Omega,\omega)$ give the in-phase dispersion and absorption signals, respectively (cf. Section 6.7), and vice-versa for signals 90° out-of-phase.

Nitrogen hyperfine coupling is ignored in Eqs. 11.95–11.99, in particular the important pseudosecular terms. It is included explicitly by introducing the $\chi_{-+}^{m_I m_I'}$ elements of the reduced spin density-matrix (and correspondingly for $Z^{(n)}$), as described in Sections 6.6 and 6.9 of Chapter 6. We solve by using the eigenfunction expansion method of Section 6.7 (see Eqs. 6.63–6.65). Solutions for the ST-EPR lineshapes, Eqs. 11.97 and 11.99, are described in Dalton et al. (1976), Dalton and Dalton (1979) and Robinson et al. (1985).

11.17 CONCLUDING SUMMARY

1. *Saturation and T_1-timescale:*
   T_1-sensitive measurements, performed at partially saturating radiation power, allow us to study rotational motion on the microsecond timescale or longer, where EPR lineshapes are no longer sensitive. We also can determine much weaker relaxation enhancements than is possible from linewidths (cf. Chapter 10).
2. *Progressive saturation and T_1T_2-product*:
   Progressive-saturation CW-EPR, with increasing microwave power, is the most common nonlinear EPR experiment. The dependence of the EPR signal S on microwave field strength B_1 is:

$$S = S_o \frac{B_1}{\left(1 + P\cdot B_1^2\right)^p} \quad (11.100)$$

where $P = \gamma_e^2 (T_1T_2)^{eff}$. For homogeneously broadened (i.e., Lorentzian) lines, exponent p is 1/2, 1 and 3/2 for the integrated intensity, absorption and first derivative, respectively. The different values of p reflect the contribution of saturation broadening to absorption and derivative amplitudes. For completely inhomogeneous broadening, $p = \tfrac{1}{2}$ in all cases. When inhomogeneous broadening is partial, we treat p as a fitting parameter. In general, the corresponding fitted value of the saturation parameter P gives:

$$1/(T_1T_2)^{eff} \approx \left(1/T_2^{eff}\right) \times 1/T_1 + const \quad (11.101)$$

where *const* is small, hence establishing connection with the spin–lattice relaxation rate (see Tables A11.1, A11.2 and Eq. 11.49). Progressive saturation of the integrated intensity is insensitive to inhomogeneous broadening; exponent is rigorously $p = \frac{1}{2}$.

3. *Nonlinear out-of-phase EPR and T_1-measurement*:
EPR signals detected 90°-out-of-phase with the field modulation have zero intensity in the absence of saturation, but non-vanishing intensities under saturation that are approximately proportional to T_1 modulated by a saturation factor. The out-of-phase first-harmonic absorption signal V_1' is a pure T_1-display, which is insensitive to motion and to T_2^o. The out-of-phase/in-phase ratio has the semi-empirical T_1-dependence:

$$\rho_1' = \frac{a_1 T_1^m}{1 + b_1 T_1^m} + \rho_1'^o \tag{11.102}$$

where calibration parameters a_1, b_1, $\rho_1'^o$ and exponent m are given for double-integrated intensities and peak amplitudes in Tables A11.3–A11.7. Out-of-phase/in-phase ratios with $\omega_{mod}/2\pi = 25$ kHz are sensitive to longer T_1 than are those at $\omega_{mod}/2\pi = 100$ kHz.

4. *Saturation-transfer EPR and ultraslow motion*:
Second-harmonic, 90°-out-of-phase, absorption V_2'-STEPR spectra are those with lineshapes most sensitive to rotational diffusion on the μs-timescale of T_1-relaxation. For ^{14}N-nitroxides at 9 GHz, ratios R of line heights (L'', C', H'') at intermediate spectral regions, where $\partial\omega/\partial\theta$ and hence saturation transfer are maximum, to those (L, C, H) at the stationary turning points where $\partial\omega/\partial\theta = 0$, depend on the rotational correlation time:

$$\tau_R = k/(R_o - R) - b \tag{11.103}$$

Calibration constants k, R_o and b for $R \equiv L''/L, C'/C, H''/H$ are listed in Table 11.4.

5. *ST-EPR and membrane proteins*:
To analyse anisotropic rotational diffusion, in particular for integral transmembrane proteins, we need the orientation θ of the nitroxide z-axis to that of the principal rotation axis (||). Values of $\tau_R^{eff}(m_I)$ from L''/L and H''/H ratios are approximately:

$$\tau_R^{eff}(\pm 1) = \frac{2\tau_{R\parallel}\tau_{R\perp}}{\tau_{R\perp}\sin^2\theta + \tau_{R\parallel}\left(1 + \cos^2\theta\right)} \tag{11.104}$$

When $\tau_{R\parallel} << \tau_{R\perp}$, as for a transmembrane peptide or protein, the principal correlation time is then $\tau_{R\parallel} = \frac{1}{2}\tau_R^{eff}(\pm 1)\sin^2\theta$. More precisely, multiple modulation (and microwave) frequencies are combined with detailed spectral simulation.

APPENDICES: CALIBRATIONS OF NONLINEAR EPR SPECTRA FOR T_1, AND FOR τ_R

APPENDIX A11.1: CALIBRATIONS FOR T_1-MEASUREMENTS IN *PROGRESSIVE-SATURATION* CW-EPR WITH MOLECULAR MOTION

The dependence of the double-integrated intensity, S, of the conventional first-derivative V_1-EPR spectrum on microwave magnetic field B_1 is fitted by (cf. Eq. 11.11):

$$S(B_1) \equiv \iint d^2B \cdot V_1(B) = S_o' \frac{(\gamma_e B_1)}{\sqrt{1 + (T_1T_2)^{eff}(\gamma_e B_1)^2}} \tag{A11.1}$$

where S_o' is a scaling factor, and $(T_1T_2)^{eff}$ is the fitting parameter that depends on T_1.

For rotational reorientation times in the fast regime (i.e., motional narrowing, $\tau_l \le 10^{-9}$ s), and in the slow regime ($\tau_l \ge 10^{-7}$ s), the effective $(T_1T_2)^{eff}$ relaxation-time product is related to the spin–lattice relaxation time by (Livshits et al. 1998a):

$$\frac{1}{(T_1T_2)^{eff}} = \frac{1}{T_2^{eff}}\left(\frac{1}{T_1}\right) + a \tag{A11.2}$$

where a and T_2^{eff} are fitting parameters that depend on the reorientation time τ_l and the intrinsic transverse relaxation time T_2^o. Values of these calibration constants are listed in Table A11.1. Relaxation times are in seconds.

In the extreme motional broadening region (10^{-9} s $\le \tau_l \le 10^{-7}$ s), the effective $(T_1T_2)^{eff}$ relaxation-time product depends relatively weakly on intrinsic T_2^o and reorientation time τ_l. It is then related to the spin–lattice relaxation time by Eq. A11.2 with the calibration parameters a and T_2^{eff} given in Table A11.2 (and cf. Eqs. 11.45, 11.46). Again, relaxation times are in seconds.

TABLE A11.1 Calibration of fitting parameters a and T_2^{eff} (Eq. A11.2), for progressive-saturation EPR of integral intensity S, (T_1-range: 0.2–5.1 μs) Spin-label jump reorientation times: 3.10^{-11} s $\leq \tau_l \leq 10^{-9}$ s and 10^{-7} s $\leq \tau_l \leq 10^{-5}$ s, i.e., outside range of extreme motional broadening. (Livshits et al. 1998a)

τ_l (s)	$T_2^o \times 10^9$ (s)[a]	$a \times 10^{-12}$ (s^{-2})	$T_2^{eff} \times 10^9$ (s)
3.16×10^{-11}	57	1.8	56.8
	114	0.9	110.3
1.0×10^{-10}	57	1.9	54.1
	114	1.1	97.1
3.16×10^{-10}	57	2.5	46.9
	114	1.6	77.3
1.0×10^{-9}	57	4.3	35.5
	114	3.7	51.6
1.0×10^{-7}	18.9	17.2	14.2
	57	12.6	28.0
	114	9.0	38.4
1.67×10^{-7}	18.9	16.4	15.5
	57	11.5	34.0
3.16×10^{-7}	18.9	16	17.2
	57	10	42.4
1.0×10^{-6}	20.3	11	20.3
	57	3.7	53.1
1.0×10^{-5}	18.9	6	20.4
	57	1	58.5

EPR frequency 9 GHz; field-modulation frequency $\omega_{mod}/2\pi = 100$ kHz.

[a] Values of T_2^o correspond to intrinsic linewidths: $\Delta B_{1/2}^o \equiv 1/(\gamma_e T_2^o) =$ 0.1–0.05 mT, 0.3–0.1 mT, and 0.28–0.10 mT.

APPENDIX A11.2: CALIBRATIONS FOR T_1-MEASUREMENTS IN *FIRST-HARMONIC OUT-OF-PHASE* V_1'-EPR WITH MOLECULAR MOTION

Calibrations of the out-of-phase/in-phase ratios (V_1'/V_1) are given for the integrated spectral intensities of the first-harmonic absorption:

$$\rho_1' \equiv \frac{\iint V'(B)\cdot \mathrm{d}^2 B_1}{\iint V_1(B)\cdot \mathrm{d}^2 B} \tag{A11.3}$$

and for the low-field $(m_I = +1)$ spectral amplitudes:

$$\rho_1'(m_I) \equiv \frac{V_1'(m_I)}{V_1(m_I)} \tag{A11.4}$$

of a ^{14}N-nitroxide at 9 GHz.

TABLE A11.2 Calibration of fitting parameters a and T_2^{eff} (Eq. A11.2), for progressive-saturation EPR of integral intensity S (T_1-range: 0.2–6 μs). Spin-label jump reorientation times 10^{-9} s $\leq \tau_l \leq 10^{-7}$ s span range of extreme motional broadening (Livshits et al. 1998a)

τ_l (s)	$T_2^o \times 10^9$ (s)[a]	$a \times 10^{-12}$ (s^{-2})	$T_2^{eff} \times 10^9$ (s)
1.0×10^{-9}	57	4.34	35.4
	114	3.72	51.2
3.16×10^{-9}	38	7.13	25.0
	114	6.82	32.2
1.0×10^{-8}	57	7.75	19.9
	114	7.44	24.1
3.16×10^{-8}	38	9.00	17.4
	114	6.82	23.5
1.0×10^{-7}	38	13.6	22.2
	114	9.92	38.9

EPR frequency: 9 GHz; field-modulation frequency: $\omega_{mod}/2\pi = 100$ kHz.

[a] Values of T_2^o correspond to intrinsic linewidths: $\Delta B_{1/2}^o \equiv 1/(\gamma_e T_2^o) =$ 0.1–0.05 mT and 0.15–0.05 mT.

For all calibrations, the (V_1'/V_1) ratio depends on T_1 according to the following semi-empirical expression (Livshits et al. 1998b; Livshits and Marsh 2000):

$$\rho_1' = \frac{a_1 T_1^m}{1 + b_1 T_1^m} + \rho_1'^o \tag{A11.5}$$

where $\rho_1'^o$ is a small intercept (cf. Eqs. 11.64 and 11.66, 11.67). The parameters a_1 and b_1 depend quite strongly on B_1 and the Zeeman modulation frequency, ω_{mod}, but less strongly on T_2^o and the spin-label reorientation time, τ_l. The exponent m in Eq. A11.5 depends on the modulation frequency, and also on whether spectral amplitudes or integrated intensities are measured.

Table A11.3 gives calibrations for the "no-motion" situation, which applies to both quasi-rigid limit and extreme motional narrowing spectra (Livshits et al. 1998b). These calibrations are for a single homogeneously broadened line. Tables A11.4–A11.7 give corresponding calibrations that specifically take account of molecular motion by using the random sudden-jump model (Livshits and Marsh 2000). These calibrations come from spectral simulations for a ^{14}N-nitroxide at a microwave frequency of 9 GHz. Tables A11.4 and A11.5 give calibrations for the ratio of double-integrated intensities (Eq. A11.3). Tables A11.6 and A11.7 give calibrations for the amplitude of the low-field $(m_I = +1)$ maximum amplitudes (Eq. A11.4).

Calibrations are given not only for the standard modulation frequency $\omega_{mod}/2\pi = 100$ kHz (Tables A11.4 and A11.6) but also for a modulation frequency of 25 kHz (Tables A11.5 and A11.7). Calibrations for $\omega_{mod}/2\pi = 25$ kHz show enhanced sensitivity to longer T_1 relaxation times (see Figure 11.9).

A corresponding treatment for anisotropic rotational motion is given by Livshits et al. (2003). In this case, the angular amplitude of motion (or orientational order parameter) must be obtained from spectral simulations, in addition to the rotational frequency/correlation time.

TABLE A11.3 Calibration parameters ($\rho_1'^o$, a_1, b_1, m) for T_1-dependence of first-harmonic absorption out-of-phase/in-phase ratio (Eq. A11.5), for double-integral V_1'-intensities (Eq. A11.3) and V_1'-amplitudes (Eq. A11.4).[a] Parameters for rigid-limit or extreme-narrowing spectra. (Livshits et al. 1998b)

$\omega_{mod}/2\pi$ (kHz)	B_1 (μT)	$\Delta B^o_{1/2}$ (mT)	$\rho_1'^o$	a_1	b_1	m
integrals:						
100	50	0.2	0.0067	0.1562	0.754	2.0
100	40	0.2	0.0041	0.1265	0.680	2.0
100	30	0.2	0.0019	0.0894	0.590	2.0
25	50	0.2	0.0026	0.0347	0.0967	1.3
25	40	0.2	0.0035	0.0315	0.0897	1.3
25	30	0.2	0.0045	0.0263	0.0760	1.3
amplitudes:						
100	50	0.1	−0.0160	0.553	0.538	1.6
100	50	0.15	−0.0170	0.496	0.507	1.6
100	50	0.2	−0.0177	0.489	0.483	1.6
100	50	0.25	−0.0179	0.410	0.464	1.6
25	50	0.1	−0.0037	0.1055	0.0787	1.3
25	50	0.15	−0.0064	0.1014	0.0753	1.3
25	50	0.2	−0.0086	0.0975	0.0720	1.3
25	50	0.25	−0.0100	0.0938	0.0690	1.3

Modulation frequency ω_{mod}, microwave field intensity B_1, and intrinsic linewidth $\Delta B^o_{1/2}\left(=1/\gamma_e T_2^o\right)$.

[a] Parameters a_1 and b_1 are for T_1 in μs.

TABLE A11.4 Calibration parameters ($\rho_1'^o$, a_1, b_1, $m = 2$ in Eq. A11.5) for T_1-dependence of first-harmonic absorption out-of-phase/in-phase integrated V_1'-intensity ratio (Eq. A11.3). Modulation frequency = 100 kHz; T_1-range = 0.2–5.1 μs; and jump reorientation times τ_l listed. (Livshits and Marsh 2000)[a]

τ_l (s)	$\rho_1'^o$	a_1	b_1
10^{-4}	0.0134–0.0037	0.202–0.133	0.91–0.69
10^{-5}	0.0146–0.0044	0.180–0.123	0.81–0.64
3.16×10^{-6}	0.0147–0.0047	0.162–0.114	0.78–0.61
10^{-6}	0.0129–0.0042	0.142–0.104	0.78–0.62
3.16×10^{-7}	0.0096–0.0031	0.115–0.094	0.63–0.57
10^{-7}	0.0053–0.0022	0.113–0.093	0.56–0.55
3.16×10^{-8}	0.0037–0.0016	0.116–0.094	0.57–0.55
10^{-8}	0.0040–0.0017	0.120–0.097	0.60–0.57
3.16×10^{-9}	0.0053–0.0022	0.134–0.105	0.68–0.60
10^{-9}	0.0076–0.0026	0.164–0.118	0.80–0.65
3.16×10^{-10}	0.0102–0.0031	0.190–0.128	0.88–0.68
10^{-10}	0.0120–0.0034	0.200–0.133	0.91–0.69
3.16×10^{-11}	0.0127–0.0035	0.204–0.134	0.92–0.70

[a] Range of parameters given for intrinsic linewidth $\Delta B^o_{1/2}\left(=1/\gamma_e T_2^o\right)$, in range 0.1–0.3 mT $\left(T_2^o = 18.9\text{–}57\ \text{ns}\right)$. Microwave field amplitude, $B_1 = 50\ \mu\text{T}$. Values of a_1 and b_1 are for T_1 in μs.

TABLE A11.5 Calibration parameters ($\rho_1'^o$, a_1, b_1, $m = 1.3$ in Eq. A11.5) for T_1-dependence of first-harmonic absorption out-of-phase/in-phase integrated V_1'-intensity ratio (Eq. A11.3). Modulation frequency = 25 kHz; T_1-range = 0.2–5.1 μs; and jump reorientation times τ_l listed. (Livshits and Marsh 2000)[a]

τ_l (s)	$\rho_1'^o$	a_1	b_1
10^{-4}	0.0007 to −0.0025	0.0361–0.0303	0.1051–0.0724
10^{-5}	0.0010 to −0.0023	0.0344–0.0291	0.1041–0.0701
3.16×10^{-6}	0.0013 to −0.0021	0.0310–0.0273	0.0928–0.0632
10^{-6}	0.0008 to −0.0023	0.0273–0.0254	0.0635–0.050
3.16×10^{-7}	−0.0013 to −0.0030	0.0278–0.0254	0.0486–0.0430
10^{-7}	−0.0030 to −0.0034	0.0300–0.0260	0.0490–0.0429
3.16×10^{-8}	−0.0035 to −0.0037	0.0309–0.0263	0.0510–0.0438
10^{-8}	−0.0032 to −0.0036	0.0305–0.0264	0.0542–0.0460
3.16×10^{-9}	−0.0023 to −0.0033	0.0309–0.0269	0.0664–0.0529
10^{-9}	−0.0012 to −0.0030	0.0334–0.0285	0.0868–0.0633
3.16×10^{-10}	−0.0002 to −0.0028	0.0355–0.0294	0.0990–0.0673
10^{-10}	−0.0003 to −0.0027	0.0363–0.0302	0.1033–0.0718
3.16×10^{-11}	−0.0005 to −0.0026	0.0366–0.0305	0.1047–0.0735

[a] Range of parameters given for intrinsic linewidth, $\Delta B^o_{1/2}\left(=1/\gamma_e T_2^o\right)$, in range 0.1–0.3 mT $\left(T_2^o = 18.9\text{–}57\ \text{ns}\right)$. Microwave field amplitude, $B_1 = 50\ \mu\text{T}$. Values of a_1 and b_1 are for T_1 in μs.

TABLE A11.6 Calibration parameters [$\rho_1^{\prime o}(m_I=+1)$, $a_1(m_I=+1)$, $b_1(m_I=+1)$, $m_I=1.6$ in Eq. A11.5] for T_1-dependence of out-of-phase/in-phase ratio of first-harmonic $m_I=+1$ V_1'-amplitudes (Eq. A11.4). Modulation frequency = 100 kHz; T_1-range = 0.2–5.1 μs; and jump reorientation times τ_l listed. (Livshits and Marsh 2000)[a]

τ_l(s)	$\rho_1^{\prime o}(m_I=+1)$	$a_1(m_I=+1)$	$b_1(m_I=+1)$
10^{-4}	−0.0060 to −0.0116	0.352–0.260	0.629–0.560
10^{-5}	0.0007 to −0.0065	0.337–0.250	0.623–0.555
3.16×10^{-6}	0.0007 to −0.0080	0.269–0.216	0.544–0.487
10^{-6}	−0.0001 to −0.0076	0.219–0.190	0.555–0.485
3.16×10^{-7}	−0.0060 to −0.0095	0.175–0.171	0.473–0.452
10^{-7}	−0.0120 to −0.0110	0.174–0.169	0.436–0.437
3.16×10^{-8}	−0.0134 to −0.0108	0.188–0.177	0.461–0.451
10^{-8}	−0.0126 to −0.0120	0.251–0.218	0.528–0.498
3.16×10^{-9}	−0.0173 to −0.0156	0.370–0.285	0.494–0.481
10^{-9}	−0.0169 to −0.0171	0.440–0.330	0.500–0.480
3.16×10^{-10}	−0.0151 to −0.0175	0.464–0.352	0.507–0.483
10^{-10}	−0.0141 to −0.0176	0.470–0.360	0.508–0.485
3.16×10^{-11}	−0.0139 to −0.0176	0.470–0.361	0.508–0.486

[a] Range of parameters for intrinsic linewidth, $\Delta B^o_{1/2}\left(=1/\gamma_e T_2^o\right)$, range 0.1–0.3 mT $\left(T_2^o=18.9\text{–}57\text{ ns}\right)$; peak-to-peak inhomogeneous linewidth = 0.14 mT. Microwave field amplitude, $B_1=50\,\mu$T. Values of $a_1(m_I=+1)$ and $b_1(m_I=+1)$ are for T_1 in μs.

TABLE A11.7 Calibration parameters [$\rho_1^{\prime o}(m_I=+1)$, $a_1(m_I=+1)$, $b_1(m_I=+1)$, $m=1.3$ in Eq. A11.5] for T_1-dependence of out-of-phase/in-phase ratio of first-harmonic $m_I=+1$ V_1'-amplitudes (Eq. A11.4). Modulation frequency = 25 kHz; T_1-range = 0.2–5.1 μs; and jump reorientation times τ_l listed. (Livshits and Marsh 2000)[a]

τ_l (s)	$\rho_1^{\prime o}(m_I=+1)$	$a_1(m_I=+1)$	$b_1(m_I=+1)$
10^{-4}	−0.0027 to −0.0055	0.0677–0.0568	0.090–0.073
10^{-5}	−0.0017 to −0.0053	0.0620–0.0540	0.087–0.063
3.16×10^{-6}	−0.0007 to −0.0050	0.0530–0.0495	0.072–0.051
10^{-6}	−0.0016 to −0.0057	0.0440–0.0458	0.040–0.036
3.16×10^{-7}	−0.0059 to −0.0074	0.0439–0.0456	0.026–0.028
10^{-7}	−0.0084 to −0.0082	0.0471–0.0465	0.027–0.029
3.16×10^{-8}	−0.0085 to −0.0075	0.0491–0.0473	0.033–0.033
10^{-8}	−0.0074 to −0.0077	0.0586–0.0537	0.056–0.050
3.16×10^{-9}	−0.0092 to −0.0090	0.0846–0.0690	0.075–0.066
10^{-9}	−0.0081 to −0.0093	0.0948–0.0780	0.079–0.071
3.16×10^{-10}	−0.0071 to −0.0093	0.0972–0.0814	0.079–0.073
10^{-10}	−0.0066 to −0.0092	0.0978–0.0826	0.079–0.074
3.16×10^{-11}	−0.0065 to −0.0092	0.0978–0.0829	0.078–0.074

[a] Range of parameters for intrinsic linewidth $\Delta B^o_{1/2}\left(=1/\gamma_e T_2^o\right)$ in range 0.1–0.3 mT $\left(T_2^o=18.9\text{–}57\text{ ns}\right)$; peak-to-peak inhomogeneous linewidth = 0.14 mT. Microwave field amplitude, $B_1=50\,\mu$T. Values of $a_1(m_I=+1)$ and $b_1(m_I=+1)$ are for T_1 in μs.

TABLE A11.8 Second-harmonic V_2'-STEPR rotational correlation-time calibrations from ^{14}N spin-labelled haemoglobin at 9 GHz (Horváth and Marsh 1983). Fitting parameters for spectral line-height ratios (*L''/L*, *H''/H*, *C'/C*) and normalized integrals (I_{ST}) in Eq. A11.6

R	range fitted	R_o	r	p	τ_o (μs)
L''/L[a]	0.2, 2.0	1.983	0.986	0.704	64.8
H''/H	0.2, 2.0	2.360	0.949	0.690	253
C'/C	−0.4, 1.0	1.023	1.830	0.728	5.11
I_{ST} (total)	0.15, 1.0	1.07×10^{-2}	0.400		43.6
I_{ST} ($m_I = -1$)[b]	0.06, 0.6	6.97×10^{-4}	2.85×10^{-2}		45.0

[a] Values of *L''* are for maximum in low-field diagnostic region.
[b] Integrated intensity for high-field manifold alone; used to detect anisotropic rotational diffusion (Horváth and Marsh 1983).

APPENDIX A11.3: CALIBRATIONS FOR τ_R IN *SECOND-HARMONIC ABSORPTION, OUT-OF-PHASE* V_2'-EPR

Diagnostic line-height ratios *H''/H*, *L''/L* and *C'/C* (see Figure 11.16) and integrated intensities I_{ST} of V_2'-STEPR spectra from spin-labelled haemoglobin are fitted, as a function of rotational correlation time τ_R, to the logistic equation:

$$R = R_o\left(1 - \frac{r}{1+\left(\tau_R/\tau_o\right)^p}\right) \qquad \text{(A11.6)}$$

where $R \equiv H''/H, L''/L, C'/C$ or I_{ST}. The fitting parameters R_o, τ_o, r and p are listed for the different *R*-functions in Table A11.8. Fits are shown by the solid lines in Figure 11.17.

REFERENCES

Beth, A.H. and Hustedt, E.J. 2005. Saturation transfer EPR. In *Biological Magnetic Resonance*, Vol. 24, eds. Eaton, G.R., Eaton, S.S., and Berliner, L.J., 369–407. New York: Kluwer Publishing.

Beth, A.H. and Robinson, B.H. 1989. Nitrogen-15 and deuterium substituted spin labels for studies of very slow rotational motion. In *Biological Magnetic Resonance*, Vol. 8, Spin Labeling. Theory and Applications. eds. Berliner, L.J. and Reuben, J., 179–253. New York: Plenum Publishing Corp.

Brophy, P.J., Horváth, L.I., and Marsh, D. 1984. Stoichiometry and specificity of lipid-protein interaction with myelin proteolipid protein studied by spin-label electron spin resonance. *Biochemistry* 23:860–865.

Carrington, A. and McLachlan, A.D. 1969. *Introduction to Magnetic Resonance with Applications to Chemistry and Chemical Physics*, New York: Harper and Row.

Castner, T.G., Jr. 1959. Saturation of the paramagnetic resonance of a V center. *Phys. Rev.* 115:1506–1515.

Cherry, R.J. and Godfrey, R.E. 1981. Anisotropic rotation of bacteriorhodopsin in lipid membranes. *Biophys. J.* 36:257–276.

Dalton, L.A. and Dalton, L.R. 1979. Modulation effects in multiple electron resonance spectroscopy. In *Multiple Electron Resonance Spectroscopy*, eds. Dorio, M.M. and Freed, J.H., 169–228. New York: Plenum.

Dalton, L.R., Robinson, B.H., Dalton, L.A., and Coffey, P. 1976. Saturation transfer spectroscopy. *Adv. Magn. Reson.* 8:149–259.

de Jongh, H.H.J., Hemminga, M.A., and Marsh, D. 1990. ESR of spin-labeled bacteriophage M13 coat protein in mixed phospholipid bilayers. *Biochim. Biophys. Acta* 1024:82–88.

Esmann, M., Horváth, L.I., and Marsh, D. 1987. Saturation-transfer electron spin resonance studies on the mobility of spin-labeled sodium and potassium ion activated adenosinetriphosphatase in membranes from *Squalus acanthias*. *Biochemistry* 26:8675–8683.

Esmann, M., Hankovszky, H.O., Hideg, K., and Marsh, D. 1989. A novel spin-label for study of membrane protein rotational diffusion using saturation transfer electron spin resonance. Application to selectively labelled class I and class II -SH groups of the shark rectal gland Na^+/K^+-ATPase. *Biochim. Biophys. Acta* 978:209–215.

Esmann, M., Hideg, K., and Marsh, D. 1994. Influence of poly(ethylene glycol) and aqueous viscosity on the rotational diffusion of membranous Na,K-ATPase. *Biochemistry* 33:3693–3697.

Fajer, P. and Marsh, D. 1982. Microwave and modulation field inhomogeneities and the effect of cavity Q in saturation transfer ESR spectra. Dependence on sample size. *J. Magn. Reson.* 49:212–224.

Fajer, P. and Marsh, D. 1983a. Analysis of dispersion mode saturation-transfer ESR spectra. Application to model membranes. *J. Magn. Reson.* 55:205–215.

Fajer, P. and Marsh, D. 1983b. Sensitivity of saturation transfer ESR spectra to anisotropic rotation. Application to membrane systems. *J. Magn. Reson.* 51:446–459.

Fajer, P., Thomas, D.D., Feix, J.B., and Hyde, J.S. 1986. Measurement of rotational molecular motion by time-resolved saturation transfer electron paramagnetic resonance. *Biophys. J.* 50:1195–1202.

Fajer, P., Knowles, P.F., and Marsh, D. 1989. Rotational motion of yeast cytochrome oxidase in phosphatidylcholine complexes studied by saturation-transfer electron spin resonance. *Biochemistry* 28:5634–5643.

Fajer, P., Watts, A., and Marsh, D. 1992. Saturation transfer, continuous wave saturation, and saturation recovery electron spin resonance studies of chain-spin labeled phosphatidylcholines in the low temperature phases of dipalmitoyl phosphatidylcholine bilayers. Effects of rotational dynamics and spin-spin interactions. *Biophys. J.* 61:879–891.

Haas, D.A., Mailer, C., and Robinson, B.H. 1993. Using nitroxide spin labels. How to obtain T_{1e} from continuous wave electron paramagnetic resonance spectra at all rotational rates. *Biophys. J.* 64:594–604.

Halbach, K. 1954. Über eine neue Methode zur Messung von Relaxationszeiten und über den Spin von Cr^{53}. *Helv. Phys. Acta* 27:259–282.

Hemminga, M.A., De Jager, P.A., Marsh, D., and Fajer, P. 1984. Standard conditions for the measurement of saturation transfer ESR spectra. *J. Magn. Reson.* 59:160–163.

Horváth, L.I. and Marsh, D. 1983. Analysis of multicomponent saturation transfer ESR spectra using the integral method: application to membrane systems. *J. Magn. Reson.* 54:363–373.

Horváth, L.I., Brophy, P.J., and Marsh, D. 1988a. Exchange rates at the lipid-protein interface of myelin proteolipid protein studied by spin-label electron spin resonance. *Biochemistry* 27:46–52.

Horváth, L.I., Brophy, P.J., and Marsh, D. 1988b. Influence of lipid headgroup on the specificity and exchange dynamics in lipid-protein interactions. A spin label study of myelin proteolipid apoprotein-phospholipid complexes. *Biochemistry* 27:5296–5304.

Horváth, L.I., Munding, A., Beyer, K., Klingenberg, M., and Marsh, D. 1989. Rotational diffusion of mitochondrial ADP/ATP carrier studied by saturation-transfer electron spin resonance. *Biochemistry* 28:407–414.

Horváth, L.I., Brophy, P.J., and Marsh, D. 1990a. Influence of polar residue deletions on lipid-protein interactions with the myelin proteolipid protein. Spin-label ESR studies with DM-20/lipid recombinants. *Biochemistry* 29:2635–2638.

Horváth, L.I., Dux, L., Hankovszky, H.O., Hideg, K., and Marsh, D. 1990b. Saturation transfer electron spin resonance of Ca^{2+}-ATPase covalently spin-labeled with β-substituted vinyl ketone- and maleimide-nitroxide derivatives. Effects of segmental motion and labeling levels. *Biophys. J.* 58:231–241.

Horváth, L.I., Brophy, P.J., and Marsh, D. 1993a. Exchange rates at the lipid-protein interface of the myelin proteolipid protein determined by saturation transfer electron spin resonance and continuous wave saturation studies. *Biophys. J.* 64:622–631.

Horváth, L.I., Brophy, P.J., and Marsh, D. 1993b. Spin label saturation transfer EPR determinations of the stoichiometry and selectivity of lipid-protein interactions in the gel phase. *Biochim. Biophys. Acta* 1147:277–280.

Hustedt, E.J. and Beth, A.H. 1995. Analysis of saturation transfer electron paramagnetic resonance spectra of a spin-labeled integral membrane protein, band 3, in terms of the uniaxial rotational diffusion model. *Biophys. J.* 69:1409–1423.

Hyde, J.S. 1978. Saturation-transfer spectroscopy. In *Methods in Enzymology*, 49G, eds. Hirs, C.H.W. and Timasheff, S.N., 480–511. New York: Academic Press.

Hyde, J.S. and Dalton, L. 1972. Very slowly tumbling spin labels: adiabatic rapid passage. *Chem. Phys. Lett.* 16:568–572.

Hyde, J.S. and Thomas, D.D. 1973. New EPR methods for the study of very slow motion: application to spin-labeled hemoglobin. *Ann. N. Y. Acad. Sci.* 222:680–692.

Jähnig, F. 1986. The shape of a membrane protein derived from rotational diffusion. *Eur. Biophys. J.* 14:63–64.

Johnson, M.E. and Hyde, J.S. 1981. 35-GHz (Q-band) saturation transfer electron paramagnetic resonance studies of rotational diffusion. *Biochemistry* 20:2875–2880.

Livshits, V.A. and Marsh, D. 2000. Spin relaxation measurements using first-harmonic out-of-phase absorption EPR signals: rotational motion effects. *J. Magn. Reson.* 145:84–94.

Livshits, V.A., Páli, T., and Marsh, D. 1998a. Relaxation time determinations by progressive saturation EPR: Effects of molecular motion and Zeeman modulation for spin labels. *J. Magn. Reson.* 133:79–91.

Livshits, V.A., Páli, T., and Marsh, D. 1998b. Spin relaxation measurements using first-harmonic out-of-phase absorption EPR signals. *J. Magn. Reson.* 134:113–123.

Livshits, V.A., Dzikovski, B.G., and Marsh, D. 2003. Anisotropic motion effects in CW non-linear EPR spectra: relaxation enhancement of lipid spin labels. *J. Magn. Reson.* 162:429–442.

Marsh, D. 1980. Molecular motion in phospholipid bilayers in the gel phase: Long axis rotation. *Biochemistry* 19:1632–1637.

Marsh, D. 1989. Experimental methods in spin-label spectral analysis. In *Biological Magnetic Resonance*, 8, eds. Berliner, L.J. and Reuben, J., 255–303. New York: Plenum Publishing Corp.

Marsh, D. 1990. Sensitivity analysis of magnetic resonance spectra from unoriented samples. *J. Magn. Reson.* 87:357–362.

Marsh, D. 1992. Exchange and dipolar spin-spin interactions and rotational diffusion in saturation transfer EPR spectroscopy. *Appl. Magn. Reson.* 3:53–65.

Marsh, D. 2007. Saturation transfer EPR studies of slow rotational motion in membranes. *Appl. Magn. Reson* 31:387–410.

Marsh, D. 2018. Distinct populations in spin-label EPR spectra from nitroxides. *J. Phys. Chem. B* 122:6129–6133.

Marsh, D. and Henderson, P.J.F. 2001. Specific spin labelling of the sugar-H^+ symporter, GalP, in cell membranes of *Escherichia coli*: site mobility and overall rotational diffusion of the protein. *Biochim. Biophys. Acta* 1510:464–473.

Marsh, D. and Horváth, L.I. 1992a. A simple analytical treatment of the sensitivity of saturation transfer EPR spectra to slow rotational diffusion. *J. Magn. Reson.* 99:323–331.

Marsh, D. and Horváth, L.I. 1992b. Influence of Heisenberg spin exchange on conventional and phase-quadrature EPR lineshapes and intensities under saturation. *J. Magn. Reson.* 97:13–26.

Marsh, D. and Horváth, L.I. 1998. Structure, dynamics and composition of the lipid-protein interface. Perspectives from spin-labelling. *Biochim. Biophys. Acta* 1376:267–296.

Marsh, D., Livshits, V.A., and Páli, T. 1997. Non-linear, continuous-wave EPR spectroscopy and spin-lattice relaxation: spin-label EPR methods for structure and dynamics. *J. Chem. Soc., Perkin Trans.* 2:2545–2548.

Marsh, D., Páli, T., and Horváth, L.I. 1998. Progressive saturation and saturation transfer EPR for measuring exchange processes and proximity relations in membranes. In *Biological Magnetic Resonance*, Vol. 14, Spin Labeling. The Next Millenium. ed. Berliner, L.J., 23–82. New York: Plenum Press.

Marsh, D., Horváth, L.I., Páli, T., and Livshits, V.A. 2005. Saturation transfer spectroscopy of biological membranes. In *Biological Magnetic Resonance*, Vol. 24. eds. Eaton, G.R., Eaton, S.S., and Berliner, L.J., 309–368. New York: Kluwer Publishing.

Marsh, D., Jost, M., Peggion, C., and Toniolo, C. 2007. Lipid chain-length dependence for incorporation of alamethicin in membranes: electron paramagnetic resonance studies on TOAC-spin labelled analogs. *Biophys. J.* 92:4002–4011.

McCalley, R.C., Shimshick, E.J., and McConnell, H.M. 1972. The effect of slow rotational motion on paramagnetic resonance spectra. *Chem. Phys. Lett.* 13:115–119.

Páli, T., Horváth, L.I., and Marsh, D. 1993. Continuous-wave saturation of two-component, inhomogeneously broadened, anisotropic EPR spectra. *J. Magn. Reson. A* 101:215–219.

Páli, T., Livshits, V.A., and Marsh, D. 1996. Dependence of saturation-transfer EPR intensities on spin-lattice relaxation. *J. Magn. Reson. B* 113:151–159.

Páli, T., Finbow, M.E., and Marsh, D. 1999. Membrane assembly of the 16-kDa proteolipid channel from *Nephrops norvegicus* studied by relaxation enhancements in spin-label ESR. *Biochemistry* 38:14311–14319.

Páli, T., Whyteside, G., and Dixon, N. et al. 2004. Interaction of inhibitors of the vacuolar H^+-ATPase with the transmembrane V_0-sector. *Biochemistry* 43:12297–12305.

Portis, A.M. 1953. Electronic structure of F centers: saturation of the electron spin resonance. *Phys. Rev.* 91:1071–1078.

Portis, A.M. 1955. Rapid passage in electron spin resonance. *Phys. Rev.* 100:1219–1224.

Robinson, B., Thomann, H., Beth, A., Fajer, P., and Dalton, L. 1985. *EPR and Advanced EPR Studies of Biological Systems*, Boca Raton, FL: CRC Press Inc.

Robinson, B.H. and Dalton, L.R. 1980. Anisotropic rotational diffusion studied by passage saturation transfer electron paramagnetic resonance. *J. Chem. Phys.* 72:1312–1324.

Ryba, N.J.P. and Marsh, D. 1992. Protein rotational diffusion and lipid/protein interactions in recombinants of bovine rhodopsin with saturated diacylphosphatidylcholines of different chain lengths studied by conventional and saturation transfer electron spin resonance. *Biochemistry* 31:7511–7518.

Saffman, P.G. 1976. Brownian motion in thin sheets of viscous fluid. *J. Fluid Mech.* 73:593–602.

Song, L., Larion, M., Chamoun, J., Bonora, M., and Fajer, P.G. 2010. Distance and dynamics determination by W-band DEER and W-band ST-EPR. *Eur. Biophys. J.* 39:711–719.

Stein, R.A., Hustedt, E.J., Staros, J.V., and Beth, A.H. 2002 Rotational dynamics of the epidermal growth factor receptor. *Biochemistry* 41:1957–1964.

Thomas, D.D., Dalton, L.R., and Hyde, J.S. 1976. Rotational diffusion studied by passage saturation transfer electron paramagnetic resonance. *J. Chem. Phys.* 65:3006–3024.

Thomas, D.D. and McConnell, H.M. 1974. Calculation of paramagnetic resonance spectra sensitive to very slow rotational motion. *Chem. Phys. Lett.* 25:470–475.

Weger, M. 1960. Passage effects in paramagnetic resonance experiments. *Bell Syst. Tech. J.* 39:1013–1112.

Saturation-Recovery EPR and ELDOR

12

12.1 INTRODUCTION

In a saturation-recovery experiment, we apply a single intense pulse of microwave radiation that saturates the EPR transition irradiated, reducing the z-magnetization to zero. The spin system then relaxes back to thermal equilibrium by T_1-processes. We detect return to equilibrium by monitoring the transverse magnetization with weak continuous-wave irradiation, as in conventional CW-EPR but without field modulation (see Figure 12.1).

If we observe the same transition as that saturated, we have saturation-recovery EPR (SR-EPR), which reports primarily on the electron spin–lattice relaxation rate $2W_e = 1/T_{1e}$. If we observe at a frequency other than that saturated, we have saturation-recovery electron–electron double resonance (SR-ELDOR). This depends strongly on the rate of spectral diffusion that transfers saturation from the pump to the observing frequency. One source of spectral diffusion is nuclear spin–lattice relaxation, which takes place between different nitrogen hyperfine manifolds of the nitroxide at rate W_n (see Figure 12.2). Another contributor is slow rotational diffusion that transfers saturation between different orientations within a single manifold of a powder pattern (see Section 11.12 in Chapter 11). This takes place at rate $W_R \approx 1/\tau_R$, where τ_R is the rotational correlation time. At higher spin-label concentrations, or in the presence of a fast-relaxing paramagnetic species, Heisenberg spin exchange also contributes to spectral diffusion (cf. Sections 10.6, 10.7 and 10.11, 10.13 in Chapter 10).

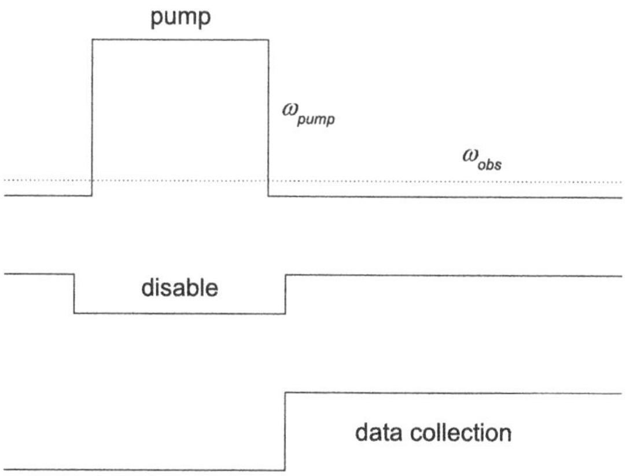

FIGURE 12.1 Timing diagram for saturation recovery. *Top*: saturating pulse at microwave frequency ω_{pump} (solid line); weak CW observing microwave radiation of frequency ω_{obs} (dotted line). *Middle*: defence pulse that disables microwave detection channel to prevent overload by pump pulse. *Bottom*: data collection begins at dead-time delay (≈100 ns) after end of pump pulse. Pump-pulse length: typically 1–4 μs for long pulses; 0.1–0.5 μs for short pulses.

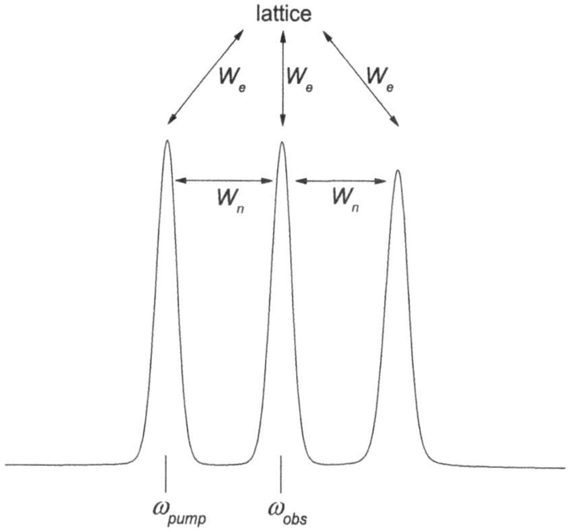

FIGURE 12.2 Electron spin–lattice relaxation, W_e, and spectral diffusion by nuclear spin–lattice relaxation, W_n, for ^{14}N-nitroxides. "lattice" represents the surrounding environment that establishes thermal equilibrium of the spin system. One hyperfine line is strongly irradiated at pump frequency ω_{pump}, and observed at the same frequency for SR-EPR, or at the frequency of a different hyperfine line ω_{obs} for SR-ELDOR.

To give optimum sensitivity to the spectral diffusion processes, the saturating pump pulse should be short relative to the timescale of spectral diffusion. We call this *short*-pulse SR-EPR. On the other hand, to concentrate as far as possible on electron spin–lattice relaxation (and cross relaxation), the pump pulse should be long enough to establish a steady state between the different states coupled by spectral diffusion, before the pulse ends. We call this *long*-pulse SR-EPR.

12.2 SATURATION-RECOVERY DETECTION

As already mentioned, we do not detect the z-magnetization $M_z(t)$ of the electron spin directly in SR-EPR but instead use weak observing radiation to monitor absorption by the transverse magnetization. Percival and Hyde (1975) use the Bloch equations to predict the dependence of the detected signal $M_y(t)$ on the weak observing microwave field B_1:

$$M_y(t) = M_y(0)\exp\left(-\frac{t}{T_{2e}}\right)$$
$$+\gamma_e B_1 T_{2e}\left(\frac{M_z(0)-M_z(\infty)}{1+\gamma_e^2 B_1^2 T_{1e}T_{2e}}\right)\exp\left(-\frac{t}{T_{1e}}-\gamma_e^2 B_1^2 T_{2e}t\right)$$
$$+\frac{M_z(\infty)\gamma_e B_1 T_{2e}}{1+\gamma_e^2 B_1^2 T_{1e}T_{2e}} - M_z(0)\gamma_e B_1 T_{2e}\exp\left(-\frac{t}{T_{2e}}\right) \quad (12.1)$$

where T_{1e} and T_{2e} are the longitudinal and transverse relaxation times, respectively, of the electron spin magnetization and $T_{1e} >> T_{2e}$. The second term on the right is the saturation-recovery signal $SR(t)$ that we require. The third term is simply the usual steady-state CW EPR signal, which does not depend on time. The first term is the straightforward free-induction decay, which together with the last term decreases exponentially at the T_{2e}-relaxation rate that is much faster than T_{1e}-relaxation. The last term is eliminated if the pump power is sufficiently high to ensure complete saturation: $M_z(0) = 0$. This also maximizes the saturation-recovery amplitude. The $SR(t)$ signal improves with higher observing microwave field B_1, but this should not be so high as to shorten the recovery time by the second term in the saturation-recovery exponential.

12.3 NUCLEAR RELAXATION IN ^{15}N- AND ^{14}N-NITROXIDES

Before proceeding with analysis of saturation recovery, we address the issue of how to define apparent nitrogen spin–lattice relaxation times in nitroxide EPR. Given that we observe only EPR transitions, the parameter directly involved is the transition probability per unit time for nuclear spin flips, W_n. We seek a useful convention to associate this fundamental rate with a characteristic time. The simplest is to take the spin–lattice relaxation time of a nitrogen nucleus, in the absence of an unpaired electron spin (Marsh 2016b).

Figure 12.3 gives the relaxation pathways for ^{15}N- and ^{14}N-nuclei, in the absence of coupling to an unpaired electron spin. The vertical transitions in Figure 12.3 correspond to the horizontal transitions given in the energy-level scheme for a ^{14}N-nitroxide in Figure 10.1 of Chapter 10. For $I = \frac{1}{2}$, the two-level system has only one relaxation pathway, $\Delta m_I = \pm 1$, with transition probability per unit time W_n (see Figure 12.3, left). As we know from Section 1.6, the spin–lattice relaxation time for this case is simply related to the forward and reverse transition rates: $T_{1n} = 1/(W_\uparrow + W_\downarrow) = 1/(2W_n)$ (see Eq. 1.17). For $I = 1$, however, the three-level system has several relaxation pathways

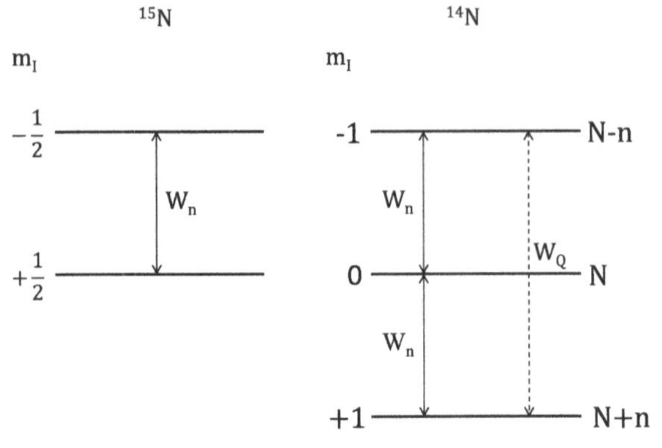

FIGURE 12.3 Energy-level scheme and relaxation pathways for ^{15}N-nuclei, $I = \frac{1}{2}$ (left); and ^{14}N-nuclei, $I = 1$ (right). Transition probabilities per unit time: W_n, $\Delta m_I = \pm 1$ transitions; W_Q, $\Delta m_I = \pm 2$ transitions. Relative spin populations for ^{14}N are given at the extreme right.

(see Figure 12.3, right). We assign the rate W_n to transitions with $\Delta m_I = \pm 1$, as in the two-level case. But we assign a different rate W_Q to the transition with $\Delta m_I = \pm 2$, because this can be induced only by nuclear quadrupole interactions.

In Figure 12.3, we assume a common spin temperature, T_S, to assign populations $N+n$, N and $N-n$ to the three hyperfine levels of the ^{14}N-nucleus (see, e.g., Carrington and McLachlan 1969). These fulfill the requirement that the ratio of populations between levels with the same energy separations are equal, given by a quasi-Boltzmann factor, e.g., $\exp(-a_o\Delta m_I/k_B T_S)$. For our case: $(N-n)/N = 1 - n/N \approx N/(N+n)$, because n is small. Summing the different relaxation pathways, we find that $dN/dt = 0$ for the $m_I = 0$ level. For either of the $m_I = \pm 1$ levels, the rate equation for the population difference then becomes:

$$\frac{dn}{dt} = -(W_n + 2W_Q)n + \text{constant} \quad (12.2)$$

We need the constant term on the right of Eq. 12.2 to maintain detailed balance for the up and down transitions: $W_\uparrow N_- = W_\downarrow N_+$, for each of the W_i s. In particular, the Boltzmann population difference at equilibrium is $n = n_o$, when $dn/dt = 0$ (see Solomon 1955). The rate equation for the population difference n that corresponds to an NMR transition $(\Delta m_I = \pm 1)$ thus becomes:

$$\frac{dn}{dt} = -(W_n + 2W_Q)(n - n_o) \quad (12.3)$$

The population difference n directly determines the total nuclear z-magnetization of the $I = 1$ spin system: $\langle I_z \rangle = \sum_m f_m \langle m|I_z|m\rangle = \frac{2}{3}n/N$, where f_m is the fractional population of the level with $m_I = m$. Therefore, Eq. 12.3 is the standard rate equation for spin–lattice relaxation, and we see that the relaxation time for a ^{14}N-nucleus alone is:

$$\frac{1}{T_{1n}} = W_n + 2W_Q \quad (12.4)$$

If as usual we neglect nitrogen quadrupolar relaxation (cf. Section 10.2), the spin–lattice relaxation time becomes $T_{1n} = 1/W_n$ for a ^{14}N-nucleus, as opposed to $T_{1n} = 1/(2W_n)$ for the corresponding transition $(\Delta m_I = \pm 1)$ of a ^{15}N-nucleus. Both are then determined primarily by the END interaction.

This is the simplest rational way to relate the transition probability per unit time W_n of a nitroxide to some characteristic relaxation time T_{1n}. In general, T_{1n} so defined is not the nuclear spin–lattice relaxation time that we would measure for a nitroxide, in an NMR experiment. The latter is more complicated, because of coupling to the electron spin. It only reduces to this form in the limit of either fast $(W_n >> W_e)$ or slow $(W_n << W_e)$ nuclear relaxation (Marsh 2016b).

Table 12.1 lists the electron and nuclear spin–lattice relaxation times for those ^{14}N- and ^{15}N-nitroxides where we have data on nuclear relaxation from ELDOR measurements. Note that the values of electron spin–lattice relaxation time contain contributions from cross relaxation, because we get them from long-pulse SR-EPR (see below in Sections 12.4 and 12.6). In principle, values for the nuclear relaxation time correspondingly contain contributions from electron relaxation (and cross relaxation), but they approximate closely to T_{1n}, because nuclear relaxation is considerably faster. We use $T_{1n} = 1/(2W_n)$ for a ^{15}N-nitroxide, and $T_{1n} = 1/W_n$ for a ^{14}N-nitroxide, where W_n is the true transition probability given by Eq. 5.20 of Section 5.4. Values for ^{15}N-nitroxides remain as given in the original literature, but values for ^{14}N-nitroxides from saturation-recovery measurements are redefined and are marked by asterisks. Values of the normalized nuclear relaxation rate b are also included in Table 12.1. The definition of b for ^{14}N-nitroxides differs from the original literature, as explained in Section 10.2, but the numerical values of b stay unchanged. Note that the ratio of ^{14}N to ^{15}N relaxation rates for PYCONH$_2$ is $T_{1n}(^{15}\text{N})/T_{1n}(^{14}\text{N}) = 0.46 \pm 0.03$ from Table 12.1, which is reasonably close to the value of $\left(\gamma_n(^{14}\text{N})/\gamma_n(^{15}\text{N})\right)^2 = 0.508$ predicted for an END mechanism (see Section 5.6).

TABLE 12.1 Electron (T_{1e}^{eff}) and nitrogen nuclear (T_{1n}) spin–lattice relaxation times for nitroxides with rotational correlation time τ_R. Deduced from SR-EPR and SR-ELDOR at EPR frequency ν_e

nitroxide[a]	ν_e (GHz)	τ_R (s)	T_{1e}^{eff} (μs)[b]	T_{1n} (μs)[c]	b[d]	*Ref.*[e]
^{15}N-TEMPOL	9	1.6×10^{-7}	19.0	0.225	84.7	1
	9	3.0×10^{-7}	20.0	0.342	58.3	1
	9	1.0×10^{-6}	25.2	1.05	24.0	1
	9	5.0×10^{-6}	25.9	3.67	7.06	1
	9	9.3×10^{-6}	26.1	4.43	5.89	1
PYCONH$_2$	9	5×10^{-11}	0.55 ± 0.05	2.40 ± 0.2*	0.23	2
	9	4.3×10^{-8}	1.7 ± 0.1	1.4 ± 0.4*	1.2	2
	94		1.25 ± 0.04	7.0 ± 0.2*	0.18	3
^{15}N-PYCONH$_2$	9	5.9×10^{-6}	26.7	2.66	10.1	1
	94		1.48 ± 0.04	3.23 ± 0.1	0.46	3
TEMPONE	9	2.1×10^{-10}	2.25	0.64	3.5	4,5
	94		0.85 ± 0.03			3
^{15}N-TEMPO-Mal/Hb	9	4×10^{-6}	10.7	0.50	21.2	1
	9	4×10^{-6}	15.3	0.36	42.4	1
16-DOXYL-SA/(14:0)$_2$PC	9	2.3×10^{-10}	1.2	1.0	1.18	6
	9	9×10^{-10}	2.25	0.16	13.8	6
	9	(54°C)	1.72 ± 0.05	1.6 ± 0.1*	1.1	7
16-DOXYL-SA/(18:1)$_2$PC	9	(54°C)	1.7 ± 0.1	1.44 ± 0.02*	1.15	8
3-DOXYL-cholestane/(14:0)$_2$PC	9	(27°C)	4.23 ± 0.08	0.37 ± 0.01	11.3	9
	9	(37°C)	3.26 ± 0.08	0.86 ± 0.02	3.8	9
^{15}N-3-DOXYL-cholestane/(14:0)$_2$PC	9	(27°C)	4.62 ± 0.08			9
	9	(37°C)	3.64 ± 0.07			9

[a] Abbreviations: TEMPOL, 1-oxyl-2,2,6,6-tetramethylpiperidin-4-ol; PYCONH$_2$, 1-oxyl-2,2,5,5-tetramethylpyrrolin-3-ylamide; TEMPONE, 1-oxyl-2,2,6,6-tetramethylpiperdin-4-yl oxide; TEMPO-Mal, *N*-(1′-oxyl-2′,2′,6′,6′-tetramethylpiperid-4′-yl)maleimide; 16-DOXYL SA, 16-(3′-oxyl-4′,4′-dimethyloxazolidin-2′-yl)stearic acid; 3-DOXYL cholestane, 3-(4′,4′-dimethyloxazolidinyl-3′-oxyl-2′-yl)-5α-cholestane. Naturally occurring ^{14}N-isotope, unless indicated otherwise. Hb, haemoglobin; (14:0)$_2$PC, dimyristoyl phosphatidylcholine; (18:1)$_2$PC, dioleoyl phosphatidylcholine.

[b] $1/T_{1e}^{eff} \equiv 1/T_{SR,e} = 2\bar{W}_e + 2\bar{W}_x$, which includes cross relaxation, and the rates W are averaged over all N-hyperfine manifolds m_I.

[c] $T_{1n} \equiv 1/(2W_n)$ for ^{15}N, but $T_{1n} \equiv 1/W_n$ for ^{14}N (cf. Eq. 12.4). Values redefined from original publication are asterisked.

[d] $b \equiv W_n/W_e$ for ^{15}N, and $b \equiv W_n/(2W_e)$ for ^{14}N. (Note that $b = T_{1e}/T_{1n}$ for both.)

[e] References: 1. Haas et al. (1993); 2. Hyde et al. (1984); 3. Froncisz et al. (2008); 4. Percival and Hyde (1976); 5. Dalton et al. (1976); 6. Popp and Hyde (1982); 7. Yin and Hyde (1987b); 8. Yin et al. (1987); 9. Yin et al. (1988).

12.4 RATE EQUATIONS FOR SR-EPR AND SR-ELDOR OF ^{15}N-NITROXIDES

The observable in saturation-recovery experiments is the departure of the population difference between the $M_S = \pm\frac{1}{2}$ electron spin levels from its equilibrium value. As we have seen in Chapter 5 (Sections 5.4–5.6), in addition to nuclear relaxation, it is important to include cross relaxation and also to allow for different intrinsic spin–lattice relaxation rates in the different hyperfine states, m_I. This complicates matters, and for simplicity, we start with a ^{15}N-nitroxide. Figure 12.4 shows the energy levels, populations and transitions for this $S = \frac{1}{2}$, $I = \frac{1}{2}$ spin system. Electron spin–lattice transitions ($\Delta M_S = \pm 1$) take place with rate constants $W_{e,\pm} \equiv \bar{W}_e \pm \delta$, for the $m_I = \pm\frac{1}{2}$ hyperfine manifolds, respectively, where $\delta = 2j^{gA}(\omega_e)B_o$ from Eq. 5.19. Nuclear spin–lattice transitions ($\Delta m_I = \pm 1$) take place with rate constant W_n, and cross relaxation ($\Delta M_S = \pm 1$ and $\Delta m_I = \pm 1$, or $\Delta m_I = \mp 1$) with rate constants W_{x_1} and W_{x_2}, respectively.

Referring to Figure 12.4, we see that cross relaxation changes the spin population $\frac{1}{2}(N - n_+)$ at a rate: $\frac{1}{2}d(N - n_+)/dt = -\frac{1}{2}W_{x_1}(n_+ + n_-)$, and correspondingly for $\frac{1}{2}(N + n_+)$. The rate of change in the population difference n_+ that is contributed by cross relaxation is therefore:

$$\frac{dn_+}{dt} = \frac{1}{2}\left(\frac{d(N + n_+)}{dt} - \frac{d(N - n_+)}{dt}\right) = -\frac{1}{2}(W_{x_1} + W_{x_2})(n_+ + n_-) \tag{12.5}$$

and similarly for n_-. Immediately after the pump pulse, the complete rate equations for electron-spin population differences $n_\pm$ in the $m_I = \pm\frac{1}{2}$ hyperfine manifolds thus become (see Figure 12.4):

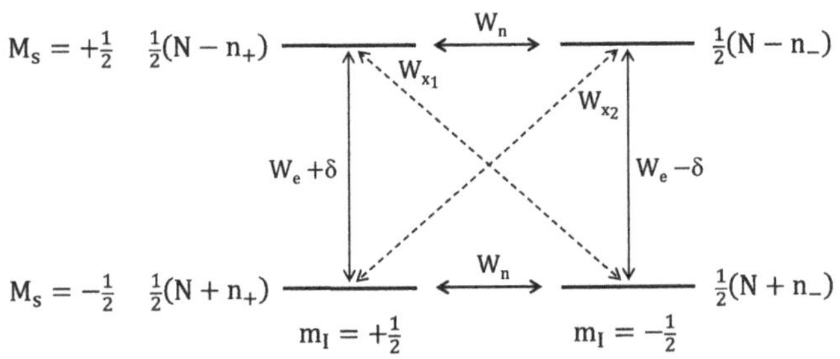

FIGURE 12.4 Energy levels, spin populations and transitions for ^{15}N-nitroxide spin label $\left(S = \frac{1}{2}, I = \frac{1}{2}\right)$ in a magnetic field. Spin states labelled by their electron and nuclear magnetic quantum numbers: M_S and m_I, respectively. Total spin population N is the same in both hyperfine states. *Solid arrows*: spin–lattice relaxation pathways with transition probabilities per unit time $W_e \pm \delta$ for electron-spin transitions and W_n for nuclear spin transitions. *Dashed arrows*: cross relaxation, i.e., simultaneous electron and nuclear transitions with transition probabilities W_{x_1} and W_{x_2} for $\Delta M_S = \pm 1$, $\Delta m_I = \pm 1$ and $\Delta M_S = \pm 1$, $\Delta m_I = \mp 1$, respectively.

$$\frac{dn_\pm}{dt} = -2W_{e,\pm}n_\pm - W_n(n_\pm - n_\mp) - W_x(n_\pm + n_\mp) + const. \tag{12.6}$$

where the cross-relaxation rate is $W_x = \frac{1}{2}(W_{x_1} + W_{x_2})$, and *const* is a constant term that we need to ensure the correct population differences at thermal equilibrium (see Solomon 1955). Correspondingly, the rate equations for departures, $\Delta n_\pm = n_\pm - n_\pm^o$, of the population differences from Boltzmann equilibrium become:

$$\frac{d\Delta n_\pm}{dt} = -2W_{e,\pm}\Delta n_\pm - W_n(\Delta n_\pm - \Delta n_\mp) - W_x(\Delta n_\pm + \Delta n_\mp) \tag{12.7}$$

where $n_+^o = n_-^o$ at equilibrium, in the high-temperature approximation.

Particular solutions of Eqs. 12.7 are exponentials of the form $\Delta n_\pm = C_\pm \exp(-\lambda t)$. By substituting in Eq. 12.7, we get linear simultaneous equations for the coefficients $C_\pm$:

$$-(2W_{e,+} + W_n + W_x - \lambda)C_+ + (W_n - W_x)C_- = 0 \tag{12.8}$$

$$(W_n - W_x)C_+ - (2W_{e,-} + W_n + W_x - \lambda)C_- = 0 \tag{12.9}$$

The condition for a non-trivial solution of Eqs. 12.8 and 12.9 comes from the secular equation:

$$\begin{vmatrix} -(2W_{e,+} + W_n + W_x - \lambda) & W_n - W_x \\ W_n - W_x & -(2W_{e,-} + W_n + W_x - \lambda) \end{vmatrix} = 0 \tag{12.10}$$

– see also Appendix G.2 and G.4, for treatment of secular equations. Solution of the resulting quadratic equation gives the two rate constants for saturation recovery:

$$\lambda_{1,2} = 2\bar{W}_e + W_n + W_x \mp \sqrt{(W_n - W_x)^2 + 4\delta^2} \tag{12.11}$$

where $\delta = \frac{1}{2}(W_{e,+} - W_{e,-})$ and $\bar{W}_e = \frac{1}{2}(W_{e,+} + W_{e,-})$, as defined above. This is the result that Robinson et al. (1999) obtained by using a rather more involved approach.

From Section 5.6, we know that $W_n \gg W_x, \delta$, for all except extremely short correlation times. Thus we can replace Eq. 12.11, correct to second order, by:

$$\lambda_1 = 2\bar{W}_e + 2W_x - \frac{2\delta^2}{W_n - W_x} \tag{12.12}$$

$$\lambda_2 = 2\bar{W}_e + 2W_n + \frac{2\delta^2}{W_n - W_x} \tag{12.13}$$

From Eq. 12.12, we see that the first rate constant is dominated by electron spin–lattice relaxation combined with cross relaxation. Correspondingly, we see from Eq. 12.13 that the second rate constant is dominated by nuclear spin–lattice relaxation (but combined with electron spin–lattice relaxation). This arises because nuclear relaxation is much faster than the difference δ

in electron relaxation rates. To first order, the effective rate constants that we deduce from saturation recovery of ^{15}N-nitroxides are therefore:

$$1/T_{SR,e} \equiv 2\bar{W}_e + 2W_x = 2\bar{W}_e(1+c) \tag{12.14}$$

$$1/T_{SR,n} \equiv 2W_n + 2\bar{W}_e = 2W_n\left(1+b^{-1}\right) \tag{12.15}$$

where $c \equiv W_x/\bar{W}_e$ and $b \equiv W_n/\bar{W}_e$ (Marsh 2016b). Neither of these correspond to the true spin–lattice relaxation times for a simple two-level system, which are defined by $T_{1e}^o \equiv 1/(2W_e)$ and $T_{1n}^o \equiv 1/(2W_n)$, respectively (see Section 12.3). In particular, Eq. 12.14 contains an important contribution from cross relaxation. Also, we are able to determine the nuclear relaxation rate W_n only when $W_n >> \bar{W}_e$, because combining measurements of $1/T_{SR,n}$ and $1/T_{SR,e}$ still leaves a first-order contribution from cross relaxation.

The general solution of the rate equation for Δn_+ (Eq. 12.7) is:

$$\Delta n_+(t) = A_1 \exp(-\lambda_1 t) + A_2 \exp(-\lambda_2 t) \tag{12.16}$$

where the amplitudes A_1 and A_2 (but not the rate constants λ_1 and λ_2) depend upon the initial conditions. The corresponding general solution for Δn_- has amplitudes that are related to those of Δn_+ by the ratio C_-/C_+ that we get from Eq. 12.8 (or Eq. 12.9):

$$\Delta n_-(t) = \frac{2W_{e,+} + W_n + W_x - \lambda_1}{W_n - W_x} A_1 \exp(-\lambda_1 t) + \frac{2W_{e,+} + W_n + W_x - \lambda_2}{W_n - W_x} A_2 \exp(-\lambda_2 t) \tag{12.17}$$

For a *short* pulse applied to the $m_I = +\frac{1}{2}$ hyperfine manifold, we have the initial condition $\Delta n_-(0) = 0$, which from Eq. 12.17 gives:

$$A_2 = -\frac{2W_{e,+} + W_n + W_x - \lambda_1}{2W_{e,+} + W_n + W_x - \lambda_2} A_1 \tag{12.18}$$

Substituting in Eq. 12.16, the short-pulse SR-EPR signal becomes:

$$\Delta n_+(t) = A_1\left(\exp\left(-2\left(\bar{W}_e + W_x\right)t\right) + \frac{1+2\delta/(W_n - W_x)}{1-2\delta/(W_n - W_x)} \exp\left(-2\left(W_n + \bar{W}_e\right)t\right)\right) \tag{12.19}$$

and from Eq. 12.17, the short-pulse SR-ELDOR signal is:

$$\Delta n_-(t) = \left(1 + \frac{2\delta}{W_n - W_x}\right) A_1 \times\left(\exp\left(-2\left(\bar{W}_e + W_x\right)t\right) - \exp\left(-2\left(W_n + \bar{W}_e\right)t\right)\right) \tag{12.20}$$

where we use first-order approximations (Eqs. 12.14 and 12.15) for the rate constants. As is well known, SR-ELDOR discriminates better between electron and nuclear relaxation than does SR-EPR, because its recovery is biphasic (compare Eq. 12.20 with Eq. 12.19).

For a *long* pulse applied to the $m_I = +\frac{1}{2}$ hyperfine manifold, on the other hand, we have the initial condition $d\Delta n_-(0)/dt = 0$, which gives:

$$A_2 = -\frac{2W_{e,+} + W_n + W_x - \lambda_1}{2W_{e,+} + W_n + W_x - \lambda_2}\left(\frac{\lambda_1}{\lambda_2}\right) A_1 \tag{12.21}$$

Because $\lambda_1/\lambda_2 \approx \bar{W}_e/W_n << 1$ (see Eqs. 12.12 and 12.13), a long pulse reduces the amplitude A_2 of the second term in the SR-EPR signal, relative to a short pulse. This suppresses the recovery that depends on nuclear relaxation, allowing easier measurement of the combined electron and cross-relaxation rate $\lambda_1 = 2\bar{W}_e + 2W_x$ in SR-EPR (cf. Eq. 12.19). Irrespective of which hyperfine line we pump or observe in a ^{15}N-nitroxide, the recovery rate constant depends on the mean electron spin–lattice relaxation rate: $2\bar{W}_e = W_{e,+} + W_{e,-}$ and on the mean cross-relaxation rate $2\bar{W}_x \equiv 2W_x = W_{x1} + W_{x2}$ but not on the rates individually.

The saturation-recovery curve that we observe experimentally is given by:

$$SR(t) = SR(\infty)\left(1 - \frac{\Delta n_{m_I}(t)}{\Delta n_{m_I}(0)}\right) \tag{12.22}$$

where $SR(\infty)$ is the final signal obtained at long times when equilibrium is re-established. Figure 12.5 shows representative SR-EPR (top panel) and SR-ELDOR (bottom panel) recovery signals that we deduce from Eqs. 12.19 and 12.20 for a short pulse pumping the $m_I = +\frac{1}{2}$ transition of a ^{15}N-nitroxide. For SR-EPR, we observe the same transition as that pumped (i.e., $m_I = +\frac{1}{2}$), and for SR-ELDOR, we observe the $m_I = -\frac{1}{2}$ transition. The ratio of nuclear to electron spin–lattice relaxation rates is given by $b \equiv W_n/\bar{W}_e = 10$ in this example. Additionally, the relative cross-relaxation rate, and the difference 2δ in electron spin–lattice rate of the $m_I = \pm\frac{1}{2}$ manifolds, have the constant values $c \equiv W_x/\bar{W}_e = 2$ and $d \equiv \delta/\bar{W}_e = 0.7$, respectively, predicted for an EPR frequency of 9.3 GHz (see Figure A10.2 in Appendix A10.1). The SR-EPR recovery (solid line) is monotonic but, in this case, clearly not mono-exponential. Component exponentials are shown in the figure by dashed lines. In contrast, the SR-ELDOR recovery curve is strongly biphasic. This corresponds first to transfer of saturation from the pumped transition to the observed transition by nuclear relaxation, which then is followed by slower relaxation of the electron spin. SR-ELDOR therefore distinguishes far more decisively between nuclear relaxation (or spectral diffusion processes in general) and electron relaxation than does SR-EPR. The dotted line in the top panel of Figure 12.5 is the SR-EPR recovery curve predicted from Eq. 12.21 for a long pumping pulse. As expected, this approaches more closely to a single exponential than does the short-pulse recovery that is given by the solid line. The amplitude of the exponential that contains nuclear relaxation is reduced by a factor $(1+b)/(1+c) = \frac{11}{3}$, in the long-pulse recovery (see Eq. 12.21).

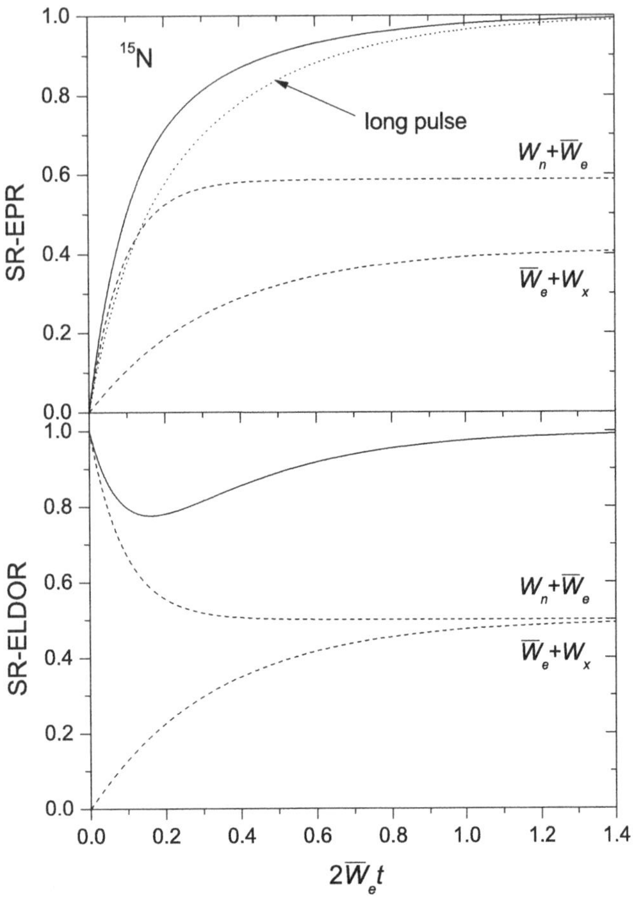

FIGURE 12.5 Short-pulse SR-EPR and SR-ELDOR recoveries (Eqs. 12.19, 12.20, 12.22) for ^{15}N-nitroxide. $m_I = +\frac{1}{2}$ transition is pumped, and also observed for SR-EPR (top panel); $m_I = -\frac{1}{2}$ manifold is observed for SR-ELDOR (bottom panel). *Solid lines*: recovery curves; *dashed lines*: constituent exponential components. Ratio of nuclear to electron spin–lattice relaxation rates: $W_n/\overline{W}_e = 10$. Relative cross-relaxation rate and difference in electron spin–lattice rates: $W_x/\overline{W}_e = 2$ and $2\delta/\overline{W}_e = 1.4$ (see top panel of Figure A10.2 in Chapter 10). Recovery curves normalized to unity at equilibrium. *Dotted line*, top panel: SR-EPR recovery after long pulse (Eq. 12.21).

12.5 ROTATIONAL DYNAMICS AND SR-EPR, SR-ELDOR OF ^{15}N-NITROXIDES

As described by the relaxation mechanisms given in Chapter 5, we expect spin–lattice relaxation rates from saturation-recovery experiments to depend on molecular dynamics. Heavy solid lines in Figure 12.6 show the recovery rate constants $1/T_{SR}$ as a function of rotational correlation time τ_R, for a ^{15}N-nitroxide. These are predicted from Eqs. 12.14 and 12.15, for isotropic rotational diffusion. Dashed lines show the dependences on τ_R of the component rates $2W_e$, $2W_x$, $2W_{e,SJ}$ and $2W_n$. We get these from Eqs. 5.19–5.22 and 5.41 of Chapter 5, together with the spectral densities of Eqs. 5.26, 5.30, 5.32 and 5.40. Note that here $\overline{W}_e \equiv W_e + W_{e,SJ}$, where W_e is the contribution to $\frac{1}{2}(W_{e,+} + W_{e,-})$ from g-tensor and A-tensor anisotropy. Spin-rotation coupling $2W_{e,SJ}$ is particularly important at short correlation times (see Eqs. 5.40 and 5.41). For $\tau_R < 10^{-11}$ s, spin-rotation makes the greatest contribution to the rate constant $1/T_{SR,e}$ that we get from SR-EPR. On the other hand, nuclear relaxation $2W_n$ makes an overwhelming contribution to the rate constant $1/T_{SR,n}$ that we derive primarily from SR-ELDOR. For $\tau_R > 10^{-10}$ s, nuclear relaxation is practically the sole contributor to $1/T_{SR,n}$. Only for rotation faster than this does the $2\overline{W}_e$ contribution become appreciable.

A significant feature of Figure 12.6 is that cross relaxation makes an important contribution to the recovery rate in SR-EPR. Indeed, of the mechanisms shown, it is the major contributor to $1/T_{SR,e}$ for correlation times longer than 2×10^{-11} s. This also is implicit in the relaxation-matrix treatment of Robinson et al. (1999), where cross relaxation is subsumed in the total contribution from electron-nuclear dipolar interaction (END) – excluding, of course, pseudo-secular terms. Including cross relaxation, the contribution that pure END makes to the SR-EPR relaxation rate for a ^{15}N-nitroxide is (see Eqs. 5.19, 5.21, 5.22 and 5.26):

$$\lambda_1(END) = \frac{10}{3} j^{AA}(\omega_e) = \frac{1}{6} \sum_{i=x,y,z} (A_{ii} - a_o)^2 \times \frac{\tau_R}{1+\omega_e^2\tau_R^2} \qquad (12.23)$$

This agrees with expressions given by Robinson et al. (1994), and Mailer et al. (2005). For a ^{15}N-nitroxide, cross relaxation makes a greater contribution to SR-EPR than does the pure END part of the electron spin–lattice relaxation $2\overline{W}_e(END) = j^{AA}(\omega_e)$. The exact factor is thus 7/3 times.

Figure 12.7 gives the dependence of the saturation-recovery rate constants $1/T_{SR,e}$ and $1/T_{SR,n}$ on isotropic rotational correlation time for ^{15}N-TEMPOL (1-oxyl-2,2,6,6-tetramethylpiperidin-4-ol) in aqueous glycerol. We compare the experimental data with theoretical predictions of contributions to the recovery rates that are based on calculations like those in Figure 12.6. The $1/T_{SR,n}$ rate constant (circles) shows a pronounced maximum characteristic of a single Lorentzian spectral density. In this case, the spectral density is that for END. The maximum occurs close to the rotational correlation time for which $\omega_a\tau_R = 1$, i.e., $\tau_R \approx 2/a_o^N = 5.6\times10^{-9}$ s. Over a wide range of correlation times, ^{15}N-nuclear relaxation $2W_n$ (see Eq. 5.42) accounts almost quantitatively for the measured rates. Earlier SR-ELDOR measurements on ^{14}N-PYCONH$_2$ (1-oxyl-2,2,5,5-tetramethylpyrrolin-3-ylamide) in *sec*-butyl-benzene failed to detect a pronounced maximum in rate, apparently because of bandwidth limitations (Hyde et al. 1984; Mailer et al. 1992).

The $1/T_{SR,e}$ rate constant (squares) is given to first order by Eq. 12.14 and contains several contributions to the electron relaxation, as specified in Sections 5.5 and 5.6 of Chapter 5. Dashed lines in Figure 12.7 give the predicted dependences on rotational correlation time for the different contributions. In addition, a substantial contribution comes from spin exchange with molecular oxygen (see Chapter 9). Robinson et al. (1994) estimate that oxygen contributes a rate constant $2W_{O_2} = 3.3\times10^{-5}/\tau_R$, by

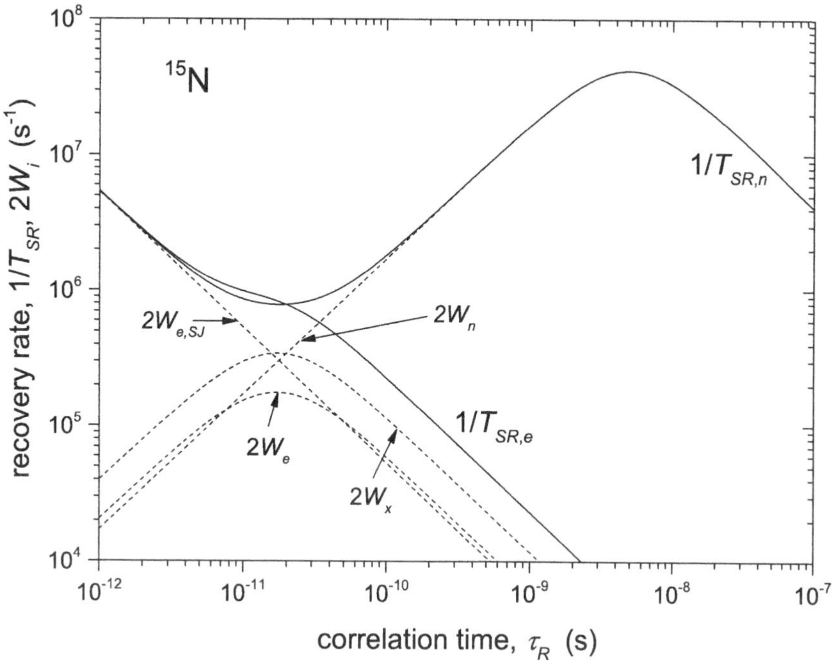

FIGURE 12.6 Saturation-recovery rate constants $1/T_{SR,e}$ and $1/T_{SR,n}$ for a ^{15}N-nitroxide spin label (solid lines, Eqs. 12.14, 12.15) and their component contributions, $2W_e$, $2W_x$, $2W_{e,SJ}$ and $2W_n$ (dashed lines). Rate constants as function of isotropic rotational correlation time τ_R (Eqs. 5.19–5.22 together with Eqs. 5.26, 5.30, 5.32, 5.40, 5.41). Rates $2W_e = W_{e,+} + W_{e,-}$, $2W_x = W_{x_1} + W_{x_2}$ are means for corresponding transitions (see Figure 12.4), with W_e contribution from *g*- and *A*-tensor anisotropy. Spin-Hamiltonian parameters for ^{15}N-TEMPOL from Table A2.4; EPR frequency 9.3 GHz.

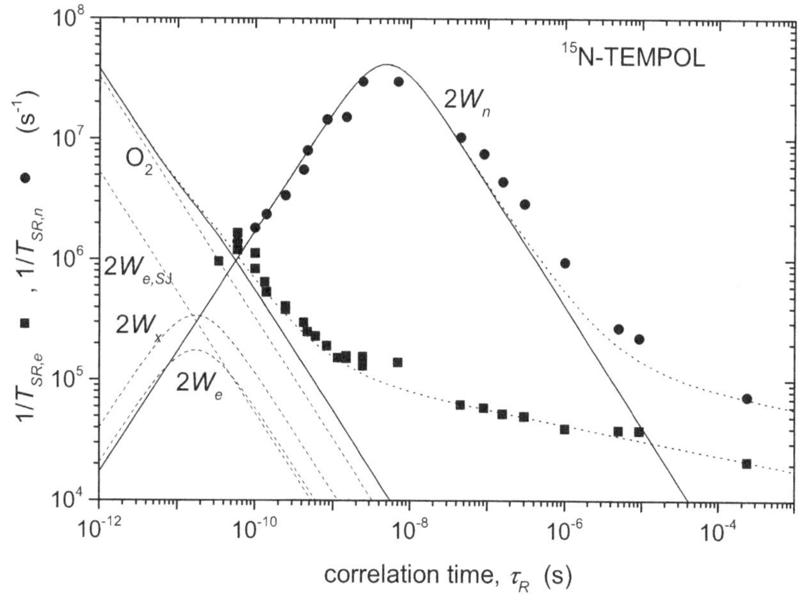

FIGURE 12.7 Dependence of "electron" (squares) and "nuclear" (circles) saturation-recovery rates, $1/T_{SR,e}$ and $1/T_{SR,n}$ (Eqs. 12.14, 12.15) for ^{15}N-TEMPOL in glycerol–water on correlation time τ_R for isotropic rotation. *Dashed lines*: contributions to $1/T_{SR,e}$ from spin–lattice relaxation by modulation of *g*- and *A*-tensor anisotropy ($2W_e$), cross relaxation ($2W_x$), spin-rotation interaction ($2W_{e,SJ}$) and Heisenberg exchange with oxygen (O_2). *Solid lines*: total contribution for $1/T_{SR,e}$ and contribution from nuclear relaxation ($2W_n$) for $1/T_{SR,n}$. *Dotted lines*: solid lines plus spectral diffusion (dashed lines from Figure 12.8). Experimental data from Robinson et al. (1994); EPR frequency 9.3 GHz.

comparing air-saturated and deoxygenated samples. Note that the collision rate with molecular oxygen depends inversely on viscosity and hence on τ_R. The solid line in Figure 12.7 is the sum of all these contributions to $1/T_{SR,e}$. We see that the experimental rates clearly are faster than those predicted, especially at long correlation times. Figure 12.8 gives the residual rates $\Delta\left(1/T_{SR,e}\right)$ and $\Delta\left(1/T_{SR,n}\right)$ after subtraction of predictions for the mechanisms considered above. The residual rate $\Delta\left(1/T_{SR,e}\right)$ has only

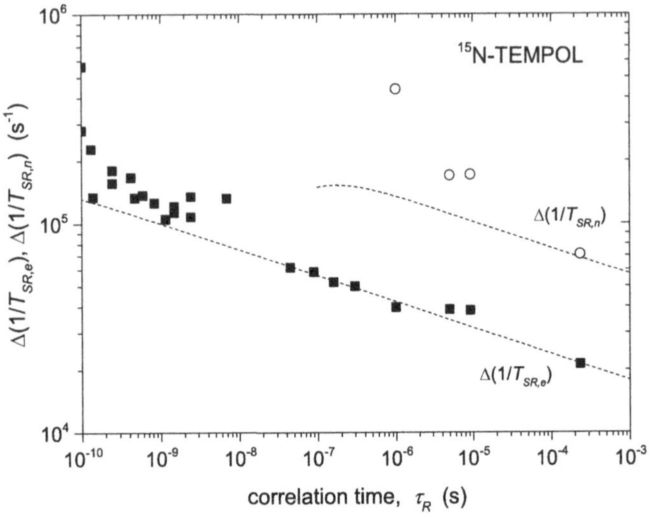

FIGURE 12.8 Differences $\Delta(1/T_{SR,e})$ and $\Delta(1/T_{SR,n})$ in experimental saturation-recovery rates (squares and circles) from theoretical predictions (solid lines in Figure 12.7) for ^{15}N-TEMPOL, as function of isotropic rotational correlation time τ_R. Dashed lines: predictions for spectral-diffusion mechanism (Eq. 12.14) with adjustable amplitude (see text and Robinson et al. 1994; Mailer et al. 2005).

a weak dependence on rotational correlation time. The straight line in the figure is given by:

$$\Delta(1/T_{SR,e}) = R_{sd}\left(\frac{\tau_D}{1+(\omega_e\tau_D)^{3/2}}\right)^{1/4} \approx R_{sd}(\omega_e^3\tau_R)^{-1/8} \quad (12.24)$$

where R_{sd} is an adjustable constant. The first equality in Eq. 12.24 is the dependence predicted for a spin-diffusion mechanism that involves solvent protons (Robinson et al. 1994; De Gennes 1958). Here, the translational diffusion time is $\tau_D = \varepsilon\tau_R$ with $\varepsilon = 0.5$–3.0, and the relaxation rate has a maximum at τ_R given by $(\varepsilon\omega_e\tau_R)^{3/2} = 2$. A $\tau_R^{-1/8}$ power-law dependence as in Eq. 12.24 was found also for saturation recovery of ^{14}N-nitroxide labelled haemoglobin (Fajer et al. 1986). The dashed prediction for $\Delta(1/T_{SR,n})$ in Figure 12.8 is Eq. 12.24 with ω_n replacing ω_e but retaining the value of R_{sd} used for fitting $\Delta(1/T_{SR,e})$.

In spite of the proposed spin diffusion, deuterating the solvent has little effect on nitroxide spin–lattice relaxation times (Owenius et al. 2004; Eaton and Eaton 2000; Biller et al. 2011). Therefore, proton spins are not involved, and Eq. 12.24 with $\tau_D = \tau_R$ is best regarded as a phenomenological spectral/spin diffusion mechanism of unclear origin. Deuterating the spin label also has only small effect, if any, on nitroxide spin–lattice relaxation times (Sato et al. 2008; Owenius et al. 2004; Percival and Hyde 1976). This excludes methyl-group rotation as a likely mechanism for the residual relaxation rate. Note that relaxation by methyl-group rotation is predicted to have an effective $\tau_R^{-1/2}$ dependence (Mailer et al. 2005; Owenius et al. 2004).

12.6 RATE EQUATIONS FOR SR-EPR AND SR-ELDOR OF ^{14}N-NITROXIDES

We now turn to the more common, but more complicated, case of ^{14}N-nitroxides. Figure 12.9 shows the energy levels and transitions for this $S = \frac{1}{2}$, $I = 1$ spin system. Electron spin–lattice transitions ($\Delta M_S = \pm 1$) take place with probabilities per unit time of $W_{e,0}$ and $W_{e,\pm1} \equiv W_{e,0} + \delta_{\pm1}$ for the $m_I = 0, \pm 1$ hyperfine manifolds, respectively. The rate equations for departures, $\Delta n_0, \Delta n_{\pm1}$, of the population differences in the $m_I = 0, \pm 1$ hyperfine levels from Boltzmann equilibrium are:

$$\frac{d\Delta n_0}{dt} = -(2W_{e,0} + 2W_n + 2W_x)\Delta n_0 + (W_n - W_x)(\Delta n_{+1} + \Delta n_{-1}) \quad (12.25)$$

$$\frac{d\Delta n_{\pm1}}{dt} = -(2W_{e,\pm1} + W_n + W_x)\Delta n_{\pm1} + (W_n - W_x)\Delta n_0 \quad (12.26)$$

immediately after the pump pulse. On achieving equilibrium, we have $\Delta n_0 = \Delta n_{+1} = \Delta n_{-1} = 0$. As for ^{15}N-nitroxides, we define: $W_x = \frac{1}{2}(W_{x1} + W_{x2})$, but here W_x is no longer the cross-relaxation rate averaged over all hyperfine states.

Particular solutions of Eqs. 12.25, 12.26 take the form $\Delta n_{m_I} = C_{m_I}\exp(-\lambda t)$, where we get the coefficients C_{m_I} from the simultaneous equations:

$$-(2W_{e,0} + 2W_n + 2W_x - \lambda)C_0 + (W_n - W_x)(C_{+1} + C_{-1}) = 0 \quad (12.27)$$

$$(W_n - W_x)C_0 - (2W_{e,+1} + W_n + W_x - \lambda)C_{+1} = 0 \quad (12.28)$$

$$(W_n - W_x)C_0 - (2W_{e,-1} + W_n + W_x - \lambda)C_{-1} = 0 \quad (12.29)$$

The condition for a non-trivial solution is:

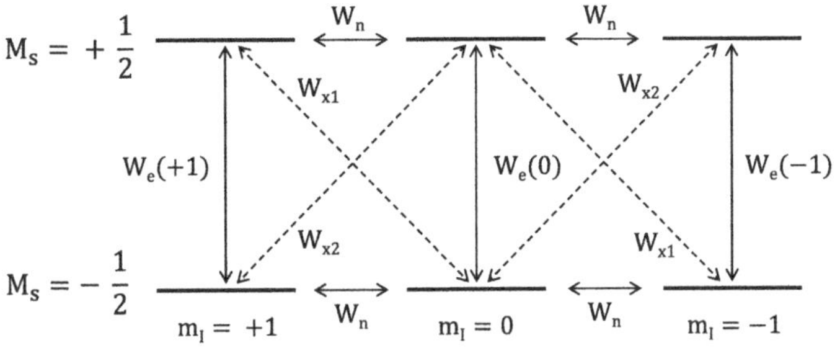

FIGURE 12.9 Energy levels and transitions for ^{14}N-nitroxide $(S = \frac{1}{2}, I = 1)$. Transition probabilities per unit time for leaving a particular state: $W_e(m_I)$ for electron transitions (Eq. 5.19); W_n for nuclear transitions (Eq. 5.20); W_{x1}, W_{x2} for cross relaxation (Eqs. 5.21, 5.22).

$$\begin{vmatrix} -(2(W_n + W_{e,0} + W_x) - \lambda) & W_n - W_x & W_n - W_x \\ W_n - W_x & -(W_n + 2W_{e,+1} + W_x - \lambda) & 0 \\ W_n - W_x & 0 & -(W_n + 2W_{e,-1} + W_x - \lambda) \end{vmatrix} = 0 \tag{12.30}$$

Solution of Eq. 12.30 simplifies considerably if we assume that $W_{e,+1} = W_{e,-1}$, which amounts to neglecting the cross-term $j^{gA}(\omega_e)$ in Eq. 5.19. This immediately yields one of the rate constants: $\lambda_2 = 2W_{e,\pm 1} + W_n + W_x$, and the other two come from the quadratic solution:

$$\lambda_{1,3} = 2W_{e,0} + \delta_1 + \tfrac{3}{2}(W_n + W_x) \mp \tfrac{1}{2}\sqrt{9W_n^2 - (14W_x + 4\delta_1)W_n + 9W_x^2 + 4\delta_1^2 - 4W_x\delta_1} \tag{12.31}$$

where $\delta_1 \equiv \delta_{+1} = \delta_{-1}$, and from Eq. 5.19: $\delta_1 = 2j^{AA}(\omega_e)$. (Note that $\delta_{\pm 1} = W_{e,\pm 1} - W_e \equiv W_{e,\pm 1} - W_{e,0}$, from the definitions given before Eqs. 12.25 and 12.26.) Using the approximation $W_n >> W_x, \delta_1$, which holds except in the limit of extreme motional narrowing, we find that to first order, the three rate constants for saturation recovery become:

$$\lambda_1 = 2(W_{e,0} + \tfrac{2}{3}\delta_1) + \tfrac{8}{3}W_x \tag{12.32}$$

$$\lambda_2 = 2(W_{e,0} + \delta_1) + W_n + W_x \tag{12.33}$$

$$\lambda_3 = 2(W_{e,0} + \tfrac{1}{3}\delta_1) + 3W_n + \tfrac{1}{3}W_x \tag{12.34}$$

For $W_x = 0 = \delta_1$, Eqs. 12.32–12.34 reduce to the rate constants obtained by Yin et al. (1987) who neglect cross relaxation and assume equal intrinsic spin–lattice relaxation rates for all three hyperfine manifolds.

Note that, as for ^{15}N-nitroxides, the SR rate constants do not depend on which hyperfine line we pump or observe. This contrasts with the well-known dependence of nitroxide linewidths on hyperfine manifold and predictions for the electron spin–lattice relaxation rate alone given by Eq. 5.41 in Chapter 5. For ^{14}N-nitroxides, however, we cannot assume a model with equal *effective* electron spin–lattice relaxation rates in each hyperfine state, because δ_1 makes different first-order contributions to Eqs. 12.32–12.34. The first rate constant $\lambda_1\ (= 2\bar{W}_e + 2\bar{W}_x)$ depends directly on the mean spin–lattice relaxation rate $\bar{W}_e = \tfrac{1}{3}(W_{e,-1} + W_{e,0} + W_{e,+1})$ and cross-relaxation rate averaged over the three hyperfine manifolds $\bar{W}_x = \tfrac{2}{3}(W_{x1} + W_{x2})$, but the two other rate constants do not.

The general solution of the rate equation for $\Delta n_{\pm 1}$ is (cf. Yin and Hyde 1987b):

$$\Delta n_{\pm 1}(t) = A_1\exp(-\lambda_1 t) \pm A_2\exp(-\lambda_2 t) + A_3\exp(-\lambda_3 t) \tag{12.35}$$

where the amplitudes A_1, A_2 and A_3 depend upon the initial conditions. The corresponding general solution for Δn_0 has amplitudes that are related to those of $\Delta n_{\pm 1}$ by the ratio $C_0/C_{\pm 1}$ that we get from Eq. 12.28 (or Eq. 12.29):

$$\Delta n_0(t) = \frac{\lambda_2 - \lambda_1}{W_n - W_x}A_1\exp(-\lambda_1 t) + \frac{\lambda_2 - \lambda_3}{W_n - W_x}A_3\exp(-\lambda_3 t) \tag{12.36}$$

where we have substituted λ_2 from Eq. 12.33.

For a *short* pulse applied to the $m_I = \pm 1$ hyperfine manifold, we have the initial conditions $\Delta n_{\mp 1}(0) = 0 = \Delta n_0(0)$, which from Eqs. 12.35 and 12.36 gives:

$$A_2 = \pm\frac{\lambda_1 - \lambda_3}{\lambda_2 - \lambda_3}A_1 \tag{12.37}$$

$$A_3 = \frac{\lambda_1 - \lambda_2}{\lambda_2 - \lambda_3}A_1 \tag{12.38}$$

Substituting in Eq. 12.35, the short-pulse SR-EPR signal becomes:

$$\Delta n_{\pm 1}(t) = A_1\left(\exp(-\lambda_1 t) \pm \frac{\lambda_3 - \lambda_1}{\lambda_3 - \lambda_2}\exp(-\lambda_2 t) + \frac{\lambda_2 - \lambda_1}{\lambda_3 - \lambda_2}\exp(-\lambda_3 t)\right) \tag{12.39}$$

and from Eq. 12.36, the short-pulse SR-ELDOR signal is:

$$\Delta n_0(t) = \left(\frac{\lambda_2 - \lambda_1}{W_n - W_x}\right)A_1\left[\exp(-\lambda_1 t) - \exp(-\lambda_3 t)\right] \tag{12.40}$$

For a *long* pulse applied to the $m_I = \pm 1$ hyperfine manifold, on the other hand, we have the initial conditions $d\Delta n_{\mp 1}(0)/dt = 0 = d\Delta n_0(0)/dt$, which from Eqs. 12.35 and 12.36 gives:

$$A_2 = \pm\frac{\lambda_1 - \lambda_3}{\lambda_2 - \lambda_3}\left(\frac{\lambda_1}{\lambda_2}\right)A_1 \tag{12.41}$$

$$A_3 = \frac{\lambda_1 - \lambda_2}{\lambda_2 - \lambda_3}\left(\frac{\lambda_1}{\lambda_3}\right)A_1 \tag{12.42}$$

Because $\lambda_1/\lambda_2 \approx 2W_{e,0}/W_n$ and $\lambda_1/\lambda_3 \approx 2W_{e,0}/3W_n$ (see Eqs. 12.32–12.34) where $W_n >> W_{e,0}$, a long pulse reduces the amplitudes A_2 and A_3 in the SR-EPR signal, relative to those for a short pulse (cf. Eqs. 12.37 and 12.38).

For a *short* pulse applied to the $m_I = 0$ hyperfine manifold, we have the initial conditions $\Delta n_{\pm 1}(0) = 0$, which gives: $A_2 = 0$

and $A_3 = -A_1$. Substituting in Eq. 12.36, the short-pulse SR-EPR signal becomes:

$$\Delta n_0(t) = \frac{A_1}{W_n - W_x}\left[(\lambda_2 - \lambda_1)\exp(-\lambda_1 t) - (\lambda_2 - \lambda_3)\exp(-\lambda_3 t)\right] \tag{12.43}$$

and from Eq. 12.35, the short-pulse SR-ELDOR signal is:

$$\Delta n_{\pm 1}(t) = A_1\left[\exp(-\lambda_1 t) - \exp(-\lambda_3 t)\right] \tag{12.44}$$

For a *long* pulse applied to the $m_I = 0$ hyperfine manifold, we have the initial conditions $d\Delta n_{\pm 1}(0)/dt = 0$, which gives: $A_2 = 0$ and $A_3 = -(\lambda_1/\lambda_3)A_1$. Again, a long pulse reduces the amplitude of the λ_3-component in the SR-EPR signal, relative to that for a short pulse (cf. Eq. 12.43). As expected, Eqs. 12.39, 12.40, 12.43 and 12.44 with $W_x = 0 = \delta_1$, reproduce the recovery amplitudes for short pulses that Yin et al. (1987) obtain by neglecting cross relaxation and assuming equal spin–lattice relaxation rates for all three hyperfine manifolds (see Table 12.2).

Figure 12.10 shows representative SR-EPR (top panel) and SR-ELDOR (bottom panel) recovery signals that we deduce from Eq. 12.35 for a short pulse pumping the $m_I = +1$ transition of a ^{14}N-nitroxide. For SR-EPR, we observe the $m_I = +1$ transition, and for the SR-ELDOR case shown in Figure 12.10, we observe the $m_I = -1$ transition. Relative amplitudes therefore come from Eqs. 12.37 and 12.38. The ratio of nuclear to electron spin–lattice relaxation rates is given by $b \equiv W_n/2W_{e,0} = 5$ in this example, and for simplicity we assume that $c \equiv W_x/W_{e,0} = 0$ and $d_1 \equiv \delta_1/W_{e,0} = 0$ (also $d \equiv \delta/W_{e,0} = 0$). Note that in the figure: $W_e \equiv W_{e,+1} = W_{e,0} = W_{e,-1}$. As for ^{15}N-nitroxides in Section 12.4, the SR-EPR recovery (solid line) is monotonic but clearly not mono-exponential. Correspondingly, the SR-ELDOR recovery curve is again strongly biphasic. The dotted line in the top panel of Figure 12.10 is the SR-EPR recovery curve predicted from Eqs. 12.41 and 12.42 for a long pumping pulse. Again, this approaches more closely to a single exponential than does the short-pulse recovery that is given by the solid line. The amplitudes of the exponentials that contain nuclear relaxation are reduced by factors of $1 + b = 6$ and $1 + 3b = 16$, respectively, in long-pulse recoveries (see Eqs. 12.41 and 12.42).

TABLE 12.2 Relative amplitudes of exponential decays $\exp(-2W_e t)$, $\exp(-(2W_e + W_n)t)$ and $\exp(-(2W_e + 3W_n)t)$ in *short*-pulse SR-EPR and SR-ELDOR of a ^{14}N-nitroxide (from Eqs. 12.39, 12.40, 12.43 and 12.44, with $W_x = 0 = \delta_1$)[a]

pump	*observe*	$2W_e$	$2W_e + W_n$	$2W_e + 3W_n$
±1	±1	1	$\frac{3}{2}$	$\frac{1}{2}$
±1	0	1	0	−1
±1	∓1	1	$-\frac{3}{2}$	$\frac{1}{2}$
0	0	1	0	2
0	±1	1	0	−1

[a] $W_e \equiv W_{e,0} = W_{e,+1} = W_{e,-1}$.
For a *long* pulse, amplitudes in the $2W_e + W_n$ column must be divided by $(1+b)$ and those in the $2W_e + 3W_n$ column by $(1+3b)$, where $b = W_n/(2W_e)$ (see Eqs. 12.41 and 12.42, with $W_x = 0 = \delta_1$).

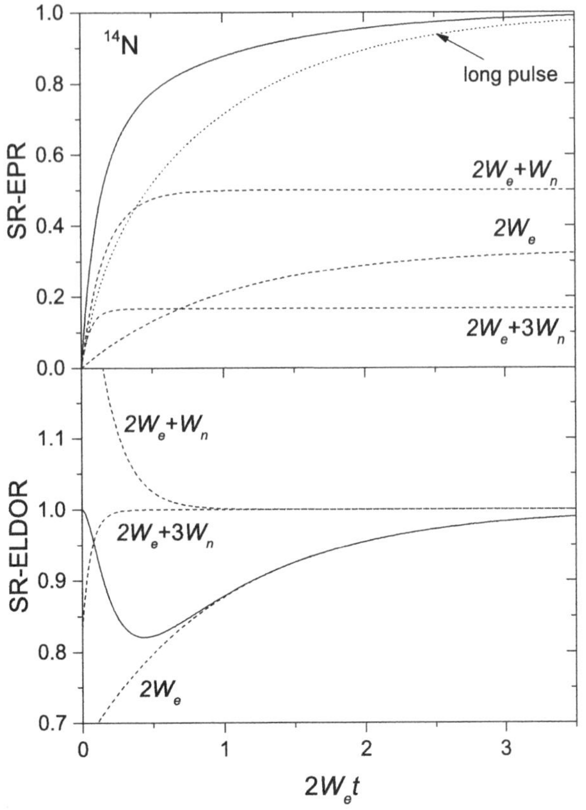

FIGURE 12.10 Short-pulse SR-EPR and SR-ELDOR recoveries (Eq. 12.35) for ^{14}N-nitroxide. $m_I = +1$ transition is pumped, and also observed for SR-EPR (top panel); $m_I = -1$ manifold is observed for SR-ELDOR (bottom panel). *Solid lines*: recovery curves; *dashed lines*: constituent exponential components (rate constants from Eqs. 12.32–12.34). *Dotted line*, top panel: SR-EPR recovery after long pulse. Ratio of nuclear to electron spin–lattice relaxation rates: $W_n/W_e = 10$; and $W_x = 0 = \delta_1 = \delta$ (i.e., $W_{e,+1} = W_{e,-1} = W_{e,0} \equiv W_e$). Recovery curves normalized to unity at equilibrium; bottom panel: $2W_e + W_n$ recovery curve displaced upward by +1.

12.7 ROTATIONAL DYNAMICS AND SR-EPR, SR-ELDOR OF ^{14}N-NITROXIDES

We now look at the dependence on rotational dynamics for the most common case in spin labelling, that of ^{14}N-nitroxides. Heavy solid lines in Figure 12.11 show dependences of the saturation-recovery rate constants for a ^{14}N-nitroxide (see Eqs. 12.32–12.34) on isotropic rotational correlation time. Component contributions to the different rates (dashed and light solid lines) are given by Eqs. 5.19–5.22 and Eq. 5.41 of Chapter 5, together with the spectral densities of Eqs. 5.26, 5.30, 5.32 and 5.40. (Note that W_{e,m_I} are the electron spin–lattice relaxation rates from modulation of g- and A-tensor anisotropy, without spin-rotation.) We see that cross relaxation makes an important contribution to the recovery rate constant λ_1 in long-pulse SR-EPR. From Eqs. 5.19, 5.21, 5.22,

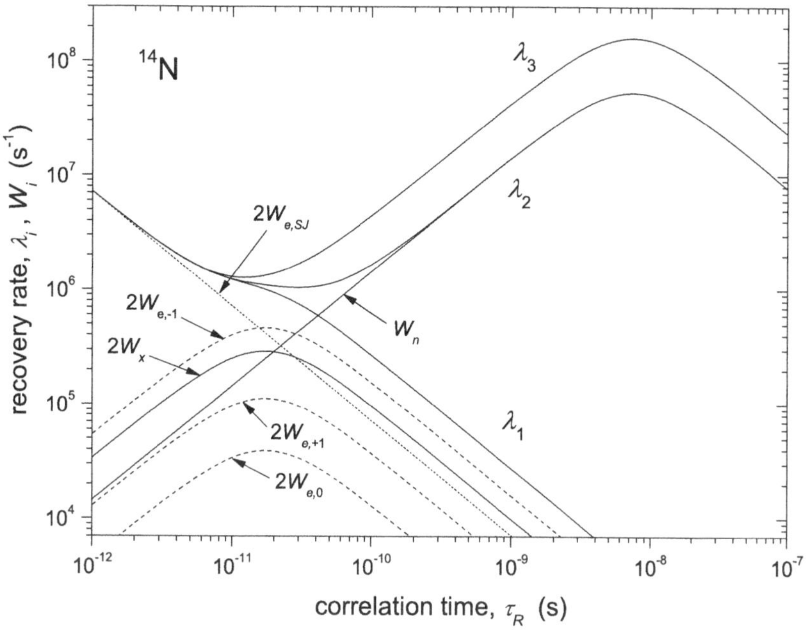

FIGURE 12.11 Saturation-recovery rate constants λ_1, λ_2 and λ_3 for ^{14}N-nitroxide spin label (solid lines, Eqs. 12.32–12.34), and component contributions $2W_{e,m_I}$, $2W_x$, $2W_{e,SJ}$, W_n (dashed lines), with W_{e,m_I} from modulation of *g*- and *A*-tensor anisotropies. Rate constants given as function of isotropic rotational correlation time τ_R (Eqs. 5.19–5.22, together with Eqs. 5.26, 5.30, 5.32, 5.40, 5.41). Cross-relaxation rate $2W_x = W_{x_1} + W_{x_2}$: mean for two coupled transitions (see Figure 12.9). Spin-Hamiltonian parameters for ^{14}N-TEMPOL from Percival et al. (1975); EPR frequency 9.3 GHz.

5.26 and 12.32, the contribution of pure END to the SR-EPR relaxation rate of a ^{14}N-nitroxide is:

$$\lambda_1(END) = \frac{4}{3}(2W_x + \delta_1) = \frac{80}{9} j^{AA}(\omega_e)$$

$$= \frac{4}{9}\sum_{i=x,y,z} (A_{ii} - a_o)^2 \times \frac{\tau_R}{1+\omega_e^2\tau_R^2} \quad (12.45)$$

This agrees with the expression given by Mailer et al. (2005), who also assume that $\delta_{+1} = \delta_{-1}$. For ^{14}N-nitroxides, the contribution of cross relaxation is again greater than the pure END part of the electron spin–lattice relaxation: $2\overline{W}_e(END) = \frac{4}{3}\delta_1 = \frac{8}{3} j^{AA}(\omega_e)$. As for ^{15}N-nitroxides, the exact factor is 7/3 times.

On the other hand, nuclear relaxation is the most significant contributor to the recovery rate constants λ_2 and λ_3 that are important in short-pulse SR-ELDOR. For correlation times longer than 2×10^{-10} s, nuclear relaxation is almost the sole contributor. To a good approximation then $\lambda_3 = 3\lambda_2$ (see Eqs. 12.33 and 12.34). Cross relaxation contributes relatively little to λ_2 and λ_3. This means that we cannot use λ_1 to correct for the contributions to λ_2 and λ_3 from electron relaxation, which become non-vanishing at shorter correlation times.

Figure 12.12 shows the dependence on rotational correlation time of the SR-EPR recovery rates for ^{14}N-TEMPOL in *sec*-butyl benzene. The rate constant measured by using long pulses is essentially λ_1 (see, e.g., Eqs. 12.41 and 12.42). Comparing the squares, circles and triangles ($m_I = 0,+1,-1$, respectively) in Figure 12.12, we see that there is no significant dependence of the rate constant on the hyperfine manifold pumped and observed. This agrees with our analysis in Section 12.6, see Eqs. 12.35 and 12.36. The experimental rate constants are compared with predictions (lines) for the standard contributions that were given in Figure 12.11. As for the case of ^{15}N-TEMPOL in Figure 12.7, the experimental recovery rates exceed the sum

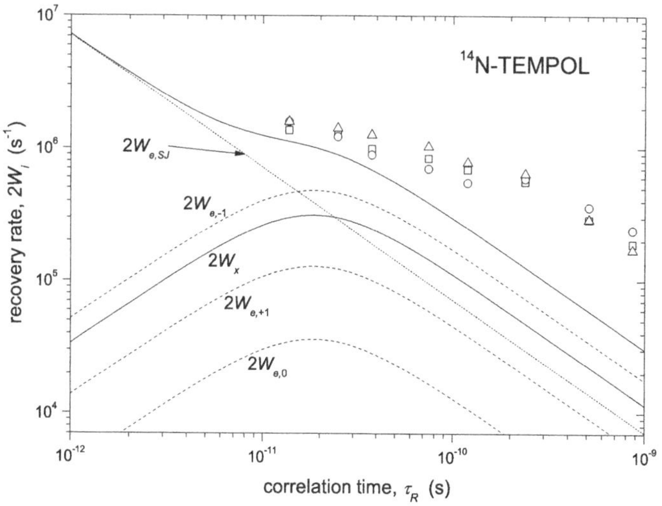

FIGURE 12.12 Contributions $2W_i$ (cf. Figure 12.11) to long-pulse SR-EPR recovery rate (Eq. 12.32, heavy solid line) of ^{14}N-TEMPOL in *sec*-butyl-benzene, as function of isotropic rotational correlation time τ_R (Eqs. 5.41, 5.43, 5.44). Experimental data for $m_I = 0$ (squares), $m_I = +1$ (circles) and $m_I = -1$ (triangles) hyperfine lines from Percival and Hyde (1976). Spin-Hamiltonian parameters from Percival et al. (1975); EPR frequency 8.62 GHz.

of the theoretical predictions (heavy solid line), particularly at long correlation times. The residual rate displays a rather weak dependence on rotational correlation time. It remains essentially constant at $\approx 4\times10^5$ s^{-1} up to $\tau_R \approx 2\times10^{-10}$ s but then decreases according to $\tau_R^{-0.8}$ as does also the total rate in this region. Note that rotational correlation times shorter than $\approx 5\times10^{-11}$ s are underestimated if non-secular contributions to the line widths are neglected (see Section 7.8), which would lead to overestimates of the residual SR-rate.

Figure 12.13 shows the dependence of the long-pulse SR rate constant on rotational correlation time for TEMPOL in glycerol–water mixtures at 20°C. Saturation-recovery measurements are at three different microwave frequencies: 1.9 GHz (L-band), 3.1 GHz (S-band) and 9.22 GHz (X-band). The dependence on τ_R predicted for the standard liquid-state contributions to the recovery rate is given for each microwave frequency. The dependence on EPR frequency arises both from contributions to the relaxation rate from Zeeman anisotropy (see Eq. 5.19) and from the spectral densities (see Eq. 5.13). In each case, the solid line represents the sum of the contributions from cross relaxation, modulation of *g*-value and hyperfine anisotropy, and spin-rotation interaction. With the exception of L-band, the experimental rates exceed the predictions for all correlation times measured. For longer correlation times, the experimental rates at L-band also exceed the predictions. We can describe the experimental rates reasonably well by adding a residual rate with a weak power-law dependence on correlation time. This results in the dotted

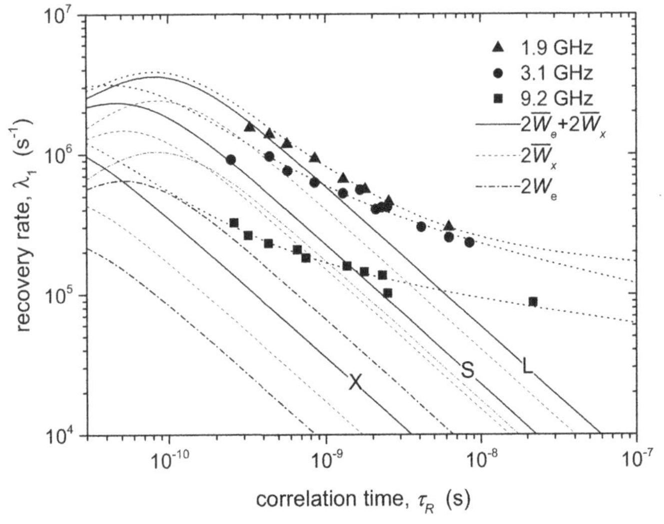

FIGURE 12.13 Saturation-recovery EPR rate constant $\lambda_1 = 2\overline{W}_e + 2\overline{W}_x$ (Eq. 12.32) of perdeuterated ^{14}N-TEMPOL in aqueous glycerol at X-band (9.22 GHz, squares), S-band (3.1 GHz, circles) and L-band (1.9 GHz, triangles) EPR frequencies, as function of rotational correlation time at 20°C. Experimental data from Owenius et al. (2004). Contributions from cross relaxation $2\overline{W}_x$ (dashed lines), from modulation of *g*- and *A*-tensor anisotropy $2\overline{W}_e$ (dashed-and-dotted lines), and their sum plus spin-rotation (solid lines), for each frequency. Residual relaxation rate fitted to power law $\propto \tau_R^m$ (dotted lines for total rate): $m = -0.17\pm0.03, -0.25\pm0.04, -0.08\pm0.04$ for X-, S- and L-band, respectively.

lines shown in Figure 12.13; the exponents required are $\approx \tau_R^{-0.2}$ for X-band and S-band, and $\approx \tau_R^{-0.1}$ for L-band.

Complications arise at high fields because the frequency dependence of the recovery rate becomes anomalous. For standard relaxation mechanisms, we expect the SR-rate to reach a constant value as the EPR frequency increases and modulation of the Zeeman anisotropy comes to dominate over that of END. Instead, the experimental recovery rate at an EPR frequency of 94 GHz increases from that at 35 GHz, which is not explained by current mechanisms (Froncisz et al. 2008; Subczynski et al. 2011).

12.8 SOLID-STATE (VIBRATIONAL) CONTRIBUTIONS AND GLASSY SOLVENTS

Given that the usual liquid-state relaxation mechanisms of Chapter 5 do not account fully for the observed saturation-recovery rates, we need to ask whether vibrational mechanisms that operate in the solid state may contribute. Of these, Raman and local-mode processes are most likely to contribute at temperatures of 100 K and above (Du et al. 1995; Sato et al. 2007; Sato et al. 2008). The Raman mechanism is a two-phonon process that depends on T^2 in the high-temperature regime, above the Debye temperature (Orton 1968). The local mode is characterized by a single vibrational frequency, which determines its exponential temperature dependence (Feldman et al. 1965). Contribution of the two mechanisms to the spin–lattice relaxation rate is given by (cf. Thomas et al. 1976):

$$\frac{1}{T_1^{solid}} = C_{Ram}T^2 + C_{loc}\exp\left(-\Delta_{loc}/T\right) \qquad (12.46)$$

where C_{Ram} and C_{loc} are constants, and Δ_{loc} is the energy of the local vibration in degrees Kelvin. Extensive measurements on various nitroxides in high-temperature glassy solvents demonstrate the generality of Eq. 12.46, and universality when a scaled temperature $T' = TV_{mol}^{\gamma}$, where V_{mol} is the molecular volume of the spin label and γ is a solvent-specific exponent, is used (Sato et al. 2007).

Figure 12.14 shows the temperature dependence for the SR rate constant of ^{14}N-TEMPONE in glycerol (circles) and in a 1:1 v/v mixture of glycerol and water (squares). Transition to a glassy state takes place in the region of 175 K and 186 K for glycerol–water and glycerol, respectively. Below 220 K, the recovery rate constants are identical in both solvents. The solid line shows the result of fitting Eq. 12.46 in the low-temperature glassy region (Sato et al. 2008). In the much more viscous glycerol solvent, the recovery rate constant does not depart from this dependence until approximately 260 K. In glycerol–water, however, rotational contributions to the SR rate that are characteristic of the liquid state begin shortly above 220 K and increase

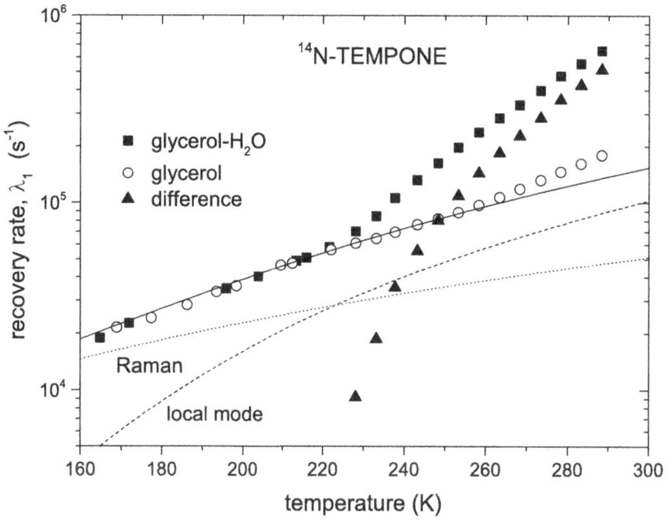

FIGURE 12.14 Temperature dependence of long-pulse SR-EPR rate constant λ_1 for ^{14}N-TEMPONE (1-oxyl-2,2,6,6-tetramethylpiperidin-4-yl oxide) in pure glycerol (circles) and in glycerol–water 1:1 v/v (squares). *Dotted line*: contribution from Raman processes $\left(C_{Ram} = 0.57\ \mathrm{s^{-1}\ K^{-2}}\right)$; *dashed line*: from local mode ($C_{loc} = 3.7 \times 10^6\ \mathrm{s^{-1}}$, $\Delta_{loc} = 1{,}090$ K) – see Eq. 12.46. Data from Sato et al. (2008). *Triangles*: rotational contribution to relaxation in glycerol-water, after subtracting Raman and local-mode contributions (solid line).

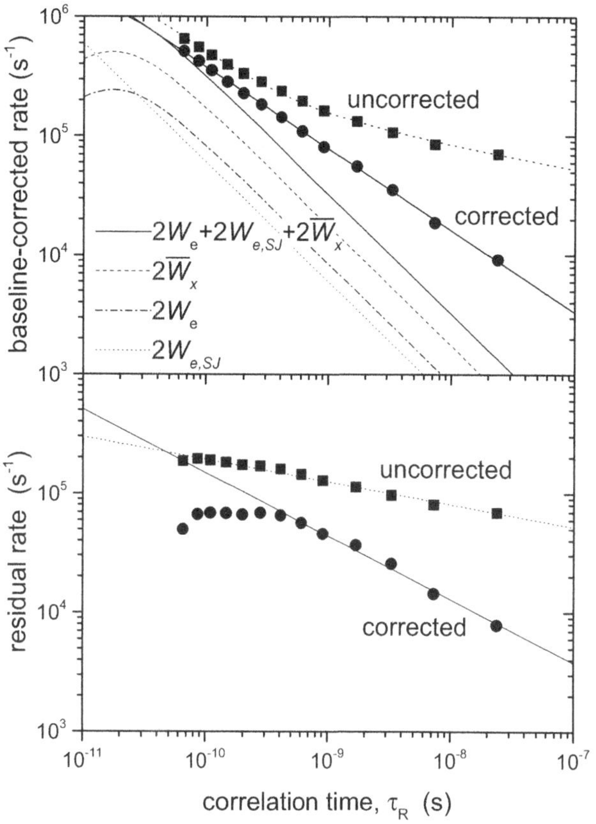

FIGURE 12.15 Dependence on rotational correlation time of SR-EPR rate constant for ^{14}N-TEMPONE in glycerol–water 1:1 v/v at different temperatures above the glass transition. *Top panel*: rate corrected for extrapolated low-temperature mechanisms as in Figure 12.14 (circles); uncorrected rate (squares); contributions from rotational dynamics: cross relaxation $2\overline{W}_x$ (dashed lines), modulation of *g*- and *A*-tensor anisotropy $2W_e$ (dashed-and-dotted line), spin-rotation $2W_{e,SJ}$ (dotted line) and their sum (solid line). *Bottom panel*: residual relaxation rate after subtracting total theoretical contribution (upper panel) from baseline-corrected (circles) and uncorrected (squares) rates. Experimental data from Sato et al. (2008).

substantially with increasing temperature. The triangles in Figure 12.14 represent the rotational contribution that we obtain by subtracting the contribution from Raman and local-mode processes. Given that the glass is a supercooled state, and that there are no obvious discontinuities in temperature dependence at the glass transitions, it seems reasonable to make this extrapolation into the high-temperature region.

Figure 12.15 shows the dependence of the SR-EPR rate constant λ_1 for ^{14}N-TEMPONE on rotational correlation time in glycerol–water. The baseline-corrected rate constant (circles) that is given by the triangles in Figure 12.14 is compared with the uncorrected rate constant (squares). Even after subtracting the contributions from Raman and local-mode processes, we see from the upper panel in Figure 12.15 that the measured rate constant still exceeds predictions for the standard liquid-state mechanisms. The lower panel in Figure 12.15 shows the residual rate constant that we obtain after subtracting the contributions from cross relaxation, *g*-tensor and *A*-tensor anisotropy, and spin-rotation. Of course, this residual rate is less than without correction for extrapolated vibrational contributions. But it still corresponds to an almost constant rate of $\approx 7 \times 10^4\ \mathrm{s^{-1}}$ for correlation times up to $\tau_R \approx 4 \times 10^{-10}$ s. Beyond this, the residual rate decreases according to $\tau_R^{-0.53}$, as opposed to $\tau_R^{-0.19}$ for the residual rate without vibrational contributions subtracted. Clearly, if quasi-solid vibrational contributions are important, and extensive measurements in glasses suggest that they are (Sato et al. 2007), exponents in the τ_R-dependence of the residual rates at long correlation times may need to be reassessed.

12.9 SPIN–LATTICE RELAXATION ENHANCEMENTS IN SR-EPR

We treated enhancement of spin–lattice relaxation by fast-relaxing paramagnetic species such as molecular oxygen and paramagnetic metal ions in Sections 10.5–10.10 of Chapter 10. In the context of CW-saturation, we saw there that the fast-relaxing species constitutes a true relaxation sink (see Figure 10.4). The electron spin–lattice relaxation rate then simply becomes:

$$W_e = W_e^o + W_{RL} \tag{12.47}$$

where W_{RL} is the rate of the additional paramagnetic spin–lattice relaxation process. We therefore incorporate Eq. 12.47

in Eqs. 12.35 and 12.36 for a ^{14}N-nitroxide, giving (cf. Yin and Hyde 1987b):

$$\Delta n_{\pm 1}(t) = A_1 \exp\left(-\left(\lambda_1 + 2W_{RL}\right)t\right) \pm A_2 \exp\left(-\left(\lambda_2 + 2W_{RL}\right)t\right) + A_3 \exp\left(-\left(\lambda_3 + 2W_{RL}\right)t\right) \tag{12.48}$$

$$\Delta n_0(t) = \frac{\lambda_2 - \lambda_1}{W_n - W_x} A_1 \exp\left(-\left(\lambda_1 + 2W_{RL}\right)t\right) + \frac{\lambda_2 - \lambda_3}{W_n - W_x} A_3 \exp\left(-\left(\lambda_3 + 2W_{RL}\right)t\right) \tag{12.49}$$

where λ_1, λ_2 and λ_3 are as given by Eqs. 12.32–12.34, and the pre-exponential coefficients A_1, A_2 and A_3 remain unchanged. Measuring paramagnetic spin–lattice relaxation enhancements by SR-EPR therefore amounts to determining $1/T_{1e}$ with long-pulse techniques.

When enhancement occurs by Heisenberg exchange, SR-EPR experiments at room temperature give accessibilities of paramagnetic relaxants to the spin label. Analysis is just the same as in CW-saturation studies (see Section 10.6). On the other hand, when relaxation enhancement comes from static dipole–dipole interactions, we can get distance information from SR-EPR experiments, not only on frozen samples but also on fluid samples (Yang et al. 2014). For frozen samples, the dipolar enhancement in relaxation rate is given by Eqs. 10.29–10.33, and we must integrate over the powder distribution of interspin orientations as explained already in Section 10.9. For liquid samples, we use the classic Solomon–Bloembergen equation for dipolar spin–lattice relaxation induced by rotational diffusion, when spin label and relaxant are fixed in the rotating species, i.e., Eqs. 10.27 and 10.28. In this case, we need spin labels that remain strongly immobilized at ambient temperatures, so that T_{1e} in the absence of relaxant is still long (Yang et al. 2015; and see Table 12.1). We discussed the choice of paramagnetic metal ions that maximizes relaxivity already in Section 10.9 (see also Table 10.5).

12.10 SLOW TWO-SITE EXCHANGE AND SR-EPR

We treated kinetics of exchange between two physical environments or between two chemical conformations, in Section 10.11 of Chapter 10. The two-site model applies to SR-EPR if nitrogen nuclear relaxation is sufficiently fast ($W_n >> W_e$) that spin–lattice relaxation is effectively averaged between the hyperfine manifolds. Each site then relaxes as a single species in the absence of exchange.

Let τ_b^{-1} and τ_f^{-1} be the rates of transfer from site b to site f, and vice-versa, respectively (see Figure 12.16). In general, b and f correspond to spin labels with different spin–lattice relaxation rates, $W_{e,b}$ and $W_{e,f}$, respectively. We assume arbitrarily that site b corresponds to that of lower mobility, and hence $W_{e,b} < W_{e,f}$. The rate of change in population difference n_b between the $M_{S,b} = \pm\frac{1}{2}$ spin states at site b that comes from exchange with site f is (see Eq. 10.49):

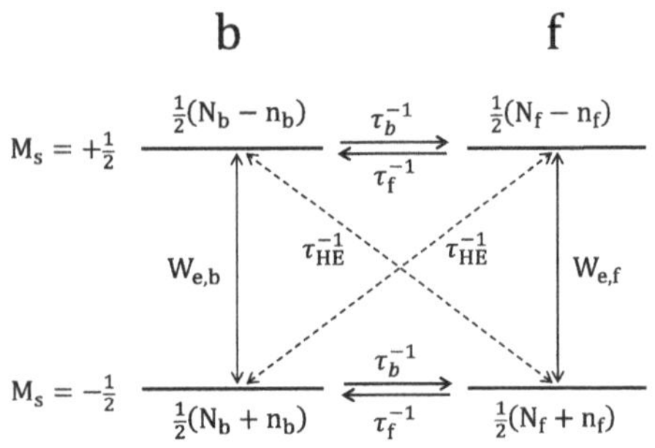

FIGURE 12.16 Energy-level scheme for $S = \frac{1}{2}$ spin label undergoing physical (or chemical) exchange between environments b and f, at transfer rates τ_b^{-1} and τ_f^{-1}. Populations of states are indicated; n_b, n_f: population differences between $M_S = \pm\frac{1}{2}$ levels. Fractional total populations in the two states/environments: $f_b = N_b/(N_b + N_f)$, $f_f = 1 - f_b$. Spin–lattice relaxation rates $W_{e,b}$, $W_{e,f}$ differ but are uniformly averaged by fast internal spectral diffusion within each state. Dashed lines: Heisenberg spin exchange between the two species at rate τ_{HE}^{-1}.

$$\frac{dn_b}{dt} = -\left(\tau_b^{-1} n_b - \tau_f^{-1} n_f\right) \tag{12.50}$$

and similarly for n_f. As usual, we need rate equations for deviations, $\Delta n_i = n_i - n_i^o$, of the population differences from their values n_i^o at Boltzmann equilibrium. These become (cf. Eqs. 10.52 and 10.53):

$$\frac{d\Delta n_b}{dt} = -2W_{e,b}\Delta n_b - \tau_b^{-1}\Delta n_b + \tau_f^{-1}\Delta n_f \tag{12.51}$$

$$\frac{d\Delta n_f}{dt} = -2W_{e,f}\Delta n_f - \tau_f^{-1}\Delta n_f + \tau_b^{-1}\Delta n_b \tag{12.52}$$

where we use the condition, $\tau_b^{-1} n_b^o = \tau_f^{-1} n_f^o$, for detailed balance at Boltzmann and exchange equilibrium (cf. Eq. 10.48).

Particular solutions of the coupled differential equations are single exponentials $\Delta n_i = C_i \exp(-\lambda t)$. From Eqs. 12.51 and 12.52, we then get equations for the coefficients C_i:

$$-\left(2W_{e,b} + \tau_b^{-1} - \lambda\right)C_b + \tau_f^{-1} C_f = 0 \tag{12.53}$$

$$\tau_b^{-1} C_b - \left(2W_{e,f} + \tau_f^{-1} - \lambda\right)C_f = 0 \tag{12.54}$$

The condition for a non-trivial solution is:

$$\begin{vmatrix} -(2W_{e,b} + \tau_b^{-1} - \lambda) & \tau_f^{-1} \\ \tau_b^{-1} & -(2W_{e,f} + \tau_f^{-1} - \lambda) \end{vmatrix} = 0 \tag{12.55}$$

from which we get the rate constants for recovery (see also Bridges et al. 2010; Kawasaki et al. 2001):

$$\lambda_{1,2} = W_{e,b} + W_{e,f} + \tfrac{1}{2}\left(\tau_b^{-1} + \tau_f^{-1}\right)$$
$$\mp\sqrt{\left(W_{e,b} - W_{e,f}\right)^2 + \left(W_{e,b} - W_{e,f}\right)\left(\tau_b^{-1} - \tau_f^{-1}\right) + \tfrac{1}{4}\left(\tau_b^{-1} + \tau_f^{-1}\right)^2} \tag{12.56}$$

The general solution for Δn_b becomes:

$$\Delta n_b(t) = A_1 \exp(-\lambda_1 t) + A_2 \exp(-\lambda_2 t) \tag{12.57}$$

where the amplitudes A_1 and A_2 depend upon the initial conditions. We note in passing that only the amplitudes, and not the rate constants λ_1 and λ_2, depend upon spectral overlap between the two components (Yin et al. 1988). The general solution for Δn_f has amplitudes that are related to those of Δn_b by the ratio C_f/C_b that we get from Eq. 12.53 (or Eq. 12.54):

$$\Delta n_f(t) = \frac{2W_{e,b} + \tau_b^{-1} - \lambda_1}{\tau_f^{-1}} A_1 \exp(-\lambda_1 t) + \frac{2W_{e,b} + \tau_b^{-1} - \lambda_2}{\tau_f^{-1}} A_2 \exp(-\lambda_2 t) \tag{12.58}$$

For a *short* pulse applied at site b, we have the initial condition $\Delta n_f(0) = 0$, which from Eq. 12.58 gives:

$$A_2 = -\frac{2W_{e,b} + \tau_b^{-1} - \lambda_1}{2W_{e,b} + \tau_b^{-1} - \lambda_2} A_1 \tag{12.59}$$

Substituting in Eq. 12.57, the short-pulse SR-EPR signal becomes:

$$\Delta n_b(t) = A_1\left(\exp(-\lambda_1 t) - \frac{2W_{e,b} + \tau_b^{-1} - \lambda_1}{2W_{e,b} + \tau_b^{-1} - \lambda_2}\exp(-\lambda_2 t)\right) \tag{12.60}$$

and from Eq. 12.58, the short-pulse SR-ELDOR signal is:

$$\Delta n_f(t) = \frac{2W_{e,b} + \tau_b^{-1} - \lambda_1}{\tau_f^{-1}} A_1\left(\exp(-\lambda_1 t) - \exp(-\lambda_2 t)\right) \tag{12.61}$$

For a *long* pulse applied at site b, on the other hand, we have the initial condition $d\Delta n_f(0)/dt = 0$, which from Eq. 12.58 gives:

$$A_2 = -\frac{2W_{e,b} + \tau_b^{-1} - \lambda_1}{2W_{e,b} + \tau_b^{-1} - \lambda_2}\left(\frac{\lambda_1}{\lambda_2}\right) A_1 \tag{12.62}$$

Because $\lambda_1 < \lambda_2$ (see Eq. 12.56), a long pulse reduces the amplitude A_2 of the second term in the SR-EPR signal, relative to a short pulse. But this term decays according to λ_2, which is more sensitive to the exchange rate constants $\tau_b^{-1} + \tau_f^{-1}$ (see Eq. 12.56). Therefore, short pulses are preferred to long pulses for measuring exchange rates with SR-EPR.

Note that generally the exchange rate that we measure by saturation recovery is the composite quantity $\tau_b^{-1} + \tau_f^{-1}$. However, the forward and reverse rate constants are related by material balance:

$$f_b \tau_b^{-1} = \left(1 - f_b\right)\tau_f^{-1} \tag{12.63}$$

where f_b is the fractional population of the spin label in state b. We can obtain values for f_b from spectral subtractions with the conventional (unsaturated) CW-EPR spectra or possibly from lineshape simulations. Note that the partition coefficient (in the case of two environments), or equilibrium constant (in the case of two conformations), is given by $K_b = f_b/\left(1 - f_b\right)$.

The additional relation provided by Eq. 12.63 also lets us investigate some limiting cases for short-pulse SR-EPR. For slow exchange when $\tau_b^{-1} + \tau_f^{-1} \ll \left|W_{e,b} - W_{e,f}\right|$, the rate constant with appreciable amplitude is $\lambda_2 \approx 2W_{e,b} + \left(1 - f_b\right)\left(\tau_b^{-1} + \tau_f^{-1}\right)$. For fast exchange when $\tau_b^{-1} + \tau_f^{-1} \gg \left|W_{e,b} - W_{e,f}\right|$, both components have significant amplitudes: $A_2/A_1 \approx \left(1 - f_b\right)/f_b$, and the rate constants are: $\lambda_2 \approx 2\left(1 - f_b\right)W_{e,b} + 2f_b W_{e,f} + \tau_b^{-1} + \tau_f^{-1}$ and $\lambda_1 \approx 2f_b W_{e,b} + 2\left(1 - f_b\right)W_{e,f}$. We get best sensitivity in the intermediate case $\tau_b^{-1} + \tau_f^{-1} \approx \left|W_{e,b} - W_{e,f}\right|$, because then both rate constants λ_1 and λ_2 depend on the exchange process.

Determining absolute exchange rates from SR-EPR is problematic, if the spin–lattice relaxation rate of at least one of the components is not known independently. In this case, it helps to determine λ_1 and λ_2 at different concentrations of paramagnetic relaxant, e.g., oxygen or Ni^{2+}, and also to vary the spin-label concentration (Kawasaki et al. 2001; Bridges et al. 2010). Note that the possibility of enhancement by paramagnetic relaxants is included already here in the separate rates $W_{e,b}$ and $W_{e,f}$. Both then contain terms linear in relaxant concentration, with individual rate constants for enhancement $k_{RL,b}$ and $k_{RL,f}$.

12.11 HEISENBERG SPIN EXCHANGE IN SR-EPR

In principle, Heisenberg spin exchange is a cross-relaxation process (see Figure 12.16). However, we saw in Section 10.11 that we can express the rate equation in a form that is equivalent to that for physical or chemical exchange between different sites. The rate at which the spin population difference at site b changes on spin exchange with site f is (Marsh 1992):

$$\frac{dn_b}{dt} = -k_{HE}\left(N_f n_b - N_b n_f\right) \tag{12.64}$$

where N_b and N_f are the total numbers of spins at sites b and f, respectively (see Eq. 10.51). The pseudo-first order rate constant for spin exchange (or exchange frequency) that we define in Section 9.10 of Chapter 9 is thus: $\tau_{HE}^{-1} = k_{HE}\left(N_b + N_f\right)$ (cf. Eq. 9.93). Comparing Eqs. 12.50 and 12.64, we see that physical/chemical exchange and Heisenberg spin exchange are kinetically equivalent, with the following identities: $\tau_b^{-1} = \left(1 - f_b\right)\tau_{HE}^{-1}$ and $\tau_f^{-1} = f_b \tau_{HE}^{-1}$, where the fractional population of state b is $f_b = N_b/\left(N_b + N_f\right)$.

SR-EPR recovery curves for Heisenberg spin exchange between sites b and f are given, therefore, by the same expressions as those of the previous section. We must simply substitute the above identities in Eqs. 12.56–12.62 (see also Yin et al. 1987). In particular, the rate constants for recovery λ_1 and λ_2 are given by substituting $\tau_b^{-1} + \tau_f^{-1} = \tau_{HE}^{-1}$ and $\tau_b^{-1} - \tau_f^{-1} = (1 - 2f_b)\tau_{HE}^{-1}$ in Eq. 12.56. If physical exchange and Heisenberg exchange are present simultaneously (see Figure 12.16), we must add the corresponding spin-exchange terms to Eqs. 12.56–12.62. The recovery rate constants from Eq. 12.56 then become (see also Kawasaki et al. 2001):

$$\lambda_{1,2} = W_{e,b} + W_{e,f} + \tfrac{1}{2}\left(\tau_b^{-1} + \tau_f^{-1} + \tau_{HE}^{-1}\right) \mp \sqrt{\left(W_{e,b} - W_{e,f}\right)^2 + \left(W_{e,b} - W_{e,f}\right)\left(1 - 2f_b\right)\left(\tau_b^{-1} + \tau_{HE}^{-1}\right) + \tfrac{1}{4}\left(\tau_b^{-1} + \tau_f^{-1} + \tau_{HE}^{-1}\right)^2} \tag{12.65}$$

where we use also the identity: $\tau_b^{-1} - \tau_f^{-1} = \left(1 - 2f_b\right)\tau_b^{-1}$ from Eq. 12.63.

We turn now to the case of Heisenberg exchange between identical ^{14}N-nitroxides, i.e., $W_{e,b} = W_{e,f}$. For the two-site model, we assumed that spin–lattice relaxation of the three hyperfine manifolds is short-circuited by fast nitrogen nuclear relaxation. From Eq. 12.65, we see that the recovery rate constants are then given by simply:

$$\lambda_1 = 2W_e \tag{12.66}$$

$$\lambda_2 = 2W_e + \tau_b^{-1} + \tau_f^{-1} + \tau_{HE}^{-1} \tag{12.67}$$

with relative amplitudes $A_2/A_1 = \left(1 - f_b\right)/f_b$, for short-pulse SR-EPR.

We can use Figure 12.17 for a ^{14}N-nitroxide to analyse what happens when the averaging condition $W_n \gg W_e$ no longer holds. Relaxation pathways for Heisenberg exchange at rate $W_{HE} \equiv \tau_{HE}^{-1}/(2I + 1)$ are represented by the two-site exchange equivalents in Figure 12.17. The figure includes, besides nuclear relaxation, also cross relaxation and differences in intrinsic electron spin–lattice relaxation rates (cf. Section 12.6 and Figure 12.9). Incorporating spin exchange between the three hyperfine manifolds, the rate equations for the departures, $\Delta n_{m_I} = n_{m_I} - n_{m_I}^o$, of the spin population differences from Boltzmann equilibrium become (cf. Eqs. 12.25 and 12.26):

$$\frac{d\Delta n_{\pm 1}}{dt} = -\left(2W_{e,0} + 2\delta_{\pm 1} + W_n + W_x + 2W_{HE}\right)\Delta n_{\pm 1} + \left(W_n - W_x + W_{HE}\right)\Delta n_0 + W_{HE}\Delta n_{\mp 1} \tag{12.68}$$

$$\frac{d\Delta n_0}{dt} = -\left(2W_{e,0} + 2W_n + 2W_x + 2W_{HE}\right)\Delta n_0 + \left(W_n - W_x + W_{HE}\right)\left(\Delta n_{+1} + \Delta n_{-1}\right) \tag{12.69}$$

where $n_0^o = n_{+1}^o = n_{-1}^o$ at equilibrium, and $W_x = \tfrac{1}{2}\left(W_{x_1} + W_{x_2}\right)$. The third term on the right in Eq. 12.68 represents the additional pathway for exchange that is forbidden by the selection rules for nuclear relaxation.

Solution of the differential equations, Eqs. 12.68 and 12.69, follows that given for the case without exchange in Section 12.6. For simplicity, we assume again that $\delta_{+1} = \delta_{-1} \equiv \delta_1$. The secular equation then immediately gives one of the rate constants: $\lambda_2 = 2W_{e,0} + 2\delta_1 + W_n + W_x + 3W_{HE}$. The other two come from the quadratic solution (cf. Eq. 12.31):

$$\lambda_{1,3} = 2W_{e,0} + \delta_1 + \tfrac{3}{2}\left(W_n + W_x + W_{HE}\right) \mp \tfrac{1}{2}\sqrt{9\left(W_n + W_{HE}\right)^2 - \left(14W_x + 4\delta_1\right)\left(W_n + W_{HE}\right) + 9W_x^2 + 4\delta_1^2 - 4W_x\delta_1} \tag{12.70}$$

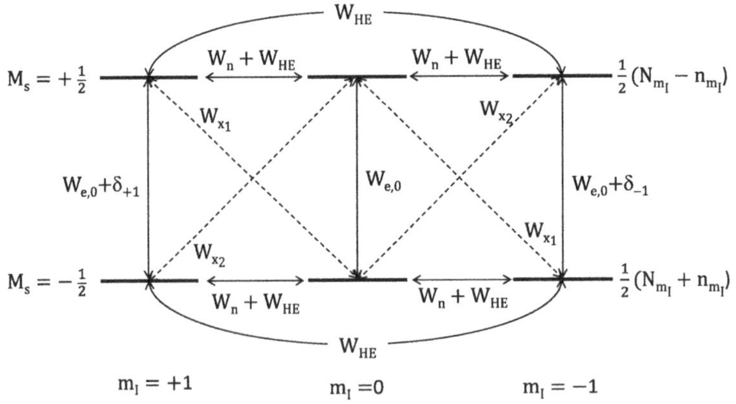

FIGURE 12.17 Energy-level scheme for $S = \tfrac{1}{2}$, $I = 1$ ^{14}N-nitroxide spin label with Heisenberg spin exchange. Populations shown on the right; n_{m_I}: population difference between $M_S = \pm\tfrac{1}{2}$ levels. *Straight solid arrows*: spin–lattice relaxation pathways with rates $W_{e,0} + \delta_{m_I}$ (electron transitions), W_n (nuclear transitions). *Dashed arrows*: cross-relaxation pathways with rates W_{x_1}, W_{x_2}. Spin exchange at rate W_{HE} is represented by saturation transfer pathways based on equivalence with two-site exchange (cf. Figure 12.16, and see text).

Using the approximation $W_n >> W_x, \delta_1$, as in Section 12.6, the recovery rate constants become to first order:

$$\lambda_1 = 2\left(W_{e,0} + \tfrac{2}{3}\delta_1\right) + \tfrac{8}{3}W_x \quad (12.71)$$

$$\lambda_2 = 2\left(W_{e,0} + \delta_1\right) + W_n + W_x + 3W_{HE} \quad (12.72)$$

$$\lambda_3 = 2\left(W_{e,0} + \tfrac{1}{3}\delta_1\right) + 3W_n + \tfrac{1}{3}W_x + 3W_{HE} \quad (12.73)$$

where $W_{HE} = \frac{1}{3}\tau_{HE}^{-1} \equiv \frac{1}{3}k_{HE}[\mathrm{SL}]$. Here k_{HE} is related to the second-order rate constant k_{coll} for bimolecular collisions between ^{14}N-nitroxides by: $k_{HE} = p_{ex}k_{coll}$ (see Sections 9.10, 9.14). When $W_x = 0 = \delta_1$, Eqs. 12.71–12.73 reduce to the rate constants given by Yin and Hyde (1987b) and Yin et al. (1987), which neglect cross relaxation and differences in intrinsic electron spin–lattice relaxation rate. When $W_{HE} = 0$, they reduce to Eqs. 12.32–12.34, as we expect in the absence of Heisenberg exchange.

As in Section 12.6, the general solution of the rate equation for $\Delta n_{\pm 1}$ is:

$$\Delta n_{\pm 1}(t) = A_1 \exp(-\lambda_1 t) \pm A_2 \exp(-\lambda_2 t) + A_3 \exp(-\lambda_3 t) \quad (12.74)$$

where the amplitudes A_1, A_2 and A_3 depend upon the initial conditions. The corresponding general solution for Δn_0 has amplitudes that are related to those of $\Delta n_{\pm 1}$ by the ratio $C_0/C_{\pm 1}$ that we get from the secular equation:

$$\Delta n_0(t) = \frac{\lambda_2 - \lambda_1 - 2W_{HE}}{W_n - W_x + W_{HE}} A_1 \exp(-\lambda_1 t)$$
$$- \frac{2W_{HE}}{W_n - W_x + W_{HE}} A_2 \exp(-\lambda_2 t) - \frac{\lambda_3 - \lambda_2 + 2W_{HE}}{W_n - W_x + W_{HE}} A_3 \exp(-\lambda_3 t) \quad (12.75)$$

where we have substituted λ_2 from Eq. 12.72. We then get the amplitudes for short-pulse SR-EPR and SR-ELDOR, and for long-pulse SR-EPR, by using the initial conditions given in Section 12.6 for the situation without Heisenberg exchange. If $W_{HE} << W_n$, the amplitudes are the same as those given in Section 12.6, and with the additional approximation $W_n >> W_x, \delta_1$ they correspond to those given in Table 12.2.

The situation is simpler for ^{15}N-nitroxides because the nuclear spin is $I = \frac{1}{2}$, which gives only two hyperfine lines. Heisenberg exchange then does not introduce any new pathways that are not allowed for nuclear relaxation. In Figure 12.4, we therefore simply replace W_n by $W_n + W_{HE}$ and likewise in Eqs. 12.7 and 12.11. For all except very short rotational correlation times: $W_n >> W_x, \delta$, and the recovery rate constants for ^{15}N-nitroxides become (cf. Eqs. 12.14 and 12.15):

$$\lambda_1 = 2\bar{W}_e + 2W_x \quad (12.76)$$

$$\lambda_2 = 2W_n + 2W_{HE} + 2\bar{W}_e \quad (12.77)$$

To within the same approximation, general solutions for the departures of the population differences from Boltzmann equilibrium are (cf. Eqs. 12.16 and 12.17):

$$\Delta n_{\pm}(t) = A_1 \exp\left(-2\left(\bar{W}_e + W_x\right)t\right) \pm A_2 \exp\left(-2\left(\bar{W}_e + W_n + W_{HE}\right)t\right) \quad (12.78)$$

For a *short* pulse applied to the $m_I = \pm\frac{1}{2}$ hyperfine manifold, we have the initial condition $\Delta n_{\mp}(0) = 0$ and the relative amplitudes become simply $A_2/A_1 = 1$. For a *long* pulse, on the other hand, we have the initial condition $\mathrm{d}\Delta n_{\mp}(0)/\mathrm{d}t = 0$ and get: $A_2/A_1 \cong 1/(1 + b + b'')$, where $b = W_n/\bar{W}_e$ and $b'' = W_{HE}/\bar{W}_e$ $\left(= \tau_{HE}^{-1}/\left(2\bar{W}_e\right)\right)$. This reduces the amplitude of the λ_2-component that is sensitive to Heisenberg exchange and nuclear relaxation.

12.12 ^{14}N-^{15}N NITROXIDE PAIRS AND HEISENBERG EXCHANGE

Because of their different nuclear spins ^{14}N- $(I = 1)$ and ^{15}N- $\left(I = \frac{1}{2}\right)$ nitroxides have distinguishable EPR spectra and can be used for labelling either different molecular species or the same molecular species. This lets us study Heisenberg spin exchange not only between different species but also between identical molecules. The latter case provides a very specific way to study bimolecular collision rates at low concentrations by SR-EPR, without uncertainties from nuclear relaxation (Feix et al. 1984). Spectral diffusion between different nuclear species is possible only by Heisenberg exchange. This principle was used first in CW-ELDOR (Stetter et al. 1976).

Yin and Hyde (1987a) give a rather complete analysis of SR-EPR and SR-ELDOR for ^{14}N-nitroxide (spin I) and ^{15}N-nitroxide (spin II) pairs. The energy level scheme and relaxation paths are shown in Figure 12.18. Electron relaxation rates and nuclear transfer rates for the two species are: $2W_{e,I}$, $2W_{e,II}$ and $W_{n,I}$, $W_{n,II}$, respectively. Here, we define nuclear relaxation rates as in Section 5.4: $W_{n,I} \equiv j^{AA}(\omega_a)$ for ^{14}N-nitroxides, and $W_{n,II} \equiv \frac{1}{2}j^{AA}(\omega_a)$ for ^{15}N-nitroxides (see Eq. 5.20). In the presence of a paramagnetic relaxant, the electron relaxation rates contain a term linear in the relaxant concentration, [R]:

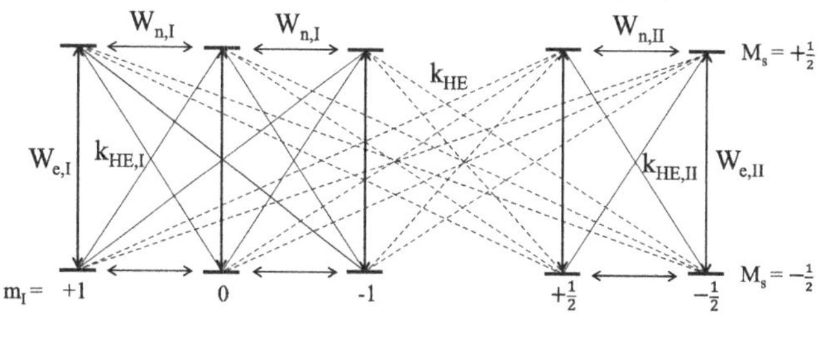

FIGURE 12.18 Energy levels $(M_S = \pm\frac{1}{2})$ and transitions for ^{14}N–^{15}N pair system. Species I (^{14}N): $m_I = +1, 0, -1$; species II (^{15}N): $m_I = \pm\frac{1}{2}$. *Solid lines*: intraspecies relaxations (W_e, W_n) including spin exchange (W_{HE}, Eqs. 12.81, 12.82). *Dashed lines*: interspecies spin-exchange pathways with rates W_{I-II} (^{14}N-nitroxide with population of ^{15}N-nitroxides, Eq. 12.83) and W_{II-I} (vice-versa, Eq. 12.84).

$$W_{e,I} = W_{e,I}^{o} + k_{RL,I}[\mathrm{R}] \tag{12.79}$$

$$W_{e,II} = W_{e,II}^{o} + k_{RL,II}[\mathrm{R}] \tag{12.80}$$

where $W_{e,I}^{o}$ is the spin–lattice relaxation rate of species I in the absence of relaxant and $k_{RL,I}$ is the rate constant for paramagnetic relaxation enhancement of species I and similarly for species II (see Section 10.6). Intraspecies Heisenberg exchange rates are:

$$W_{HE,I} = \tfrac{1}{3} k_{HE,I}[\mathrm{I}] \tag{12.81}$$

$$W_{HE,II} = \tfrac{1}{2} k_{HE,II}[\mathrm{II}] \tag{12.82}$$

where $k_{HE,I}$ is the rate constant for Heisenberg exchange within species I, [I] is the molar concentration of species I and similarly for species II. The statistical factors, $1/(2I+1)$, in Eqs. 12.81 and 12.82 are those appropriate to ^{14}N- and ^{15}N-nitroxides, respectively. Interspecies exchange rates are:

$$W_{I-II} = \tfrac{1}{2} k_{HE}[\mathrm{II}] \tag{12.83}$$

$$W_{II-I} = \tfrac{1}{3} k_{HE}[\mathrm{I}] \tag{12.84}$$

where k_{HE} is the rate constant for Heisenberg exchange between species I and species II.

The time dependence of the saturation-recovery observable Δn contains five exponential components and has the general form:

$$\Delta n(t) = A_1 \exp(-\lambda_1 t) + A_2 \exp(-\lambda_2 t) + A_3 \exp(-\lambda_3 t) + A_4 \exp(-\lambda_4 t) + A_5 \exp(-\lambda_5 t) \tag{12.85}$$

where, as always, the pre-exponential coefficients A_1 to A_5 depend on initial conditions immediately after the pump pulse. The rate constants for recovery are given by (Yin and Hyde 1987a):

$$\lambda_1 = 2W_{e,I} + 3W_{n,I} + 3W_{HE,I} + 2W_{I-II} \tag{12.86}$$

$$\lambda_2 = 2W_{e,I} + W_{n,I} + 3W_{HE,I} + 2W_{I-II} \tag{12.87}$$

$$\lambda_3 = 2W_{e,II} + 2W_{n,II} + 2W_{HE,II} + 3W_{II-I} \tag{12.88}$$

TABLE 12.3 Relative amplitudes A_1 to A_5 (see Eq. 12.85) of component exponential recoveries with rate constants λ_1 to λ_5 (see Eqs. 12.86–12.89), for *short*-pulse SR-EPR and SR-ELDOR of exchange-coupled pairs of ^{14}N- and ^{15}N-nitroxides (species I and II, respectively). Results from Yin and Hyde (1987a)

m_I (observe)	A_1	A_2	A_3	A_4	A_5
pump $m_I = \pm 1$					
+1	1	±3	0	$1+E/D$	$1-E/D$
0	−2	0	0	$1+E/D$	$1-E/D$
−1	1	∓3	0	$1+E/D$	$1-E/D$
$+\frac{1}{2}$	0	0	0	$-\frac{3}{2}W_{I-II}/D$	$\frac{3}{2}W_{I-II}/D$
$-\frac{1}{2}$	0	0	0	$-\frac{3}{2}W_{I-II}/D$	$\frac{3}{2}W_{I-II}/D$
pump $m_I = 0$					
+1	−1	0	0	$\frac{1}{2}(1+E/D)$	$\frac{1}{2}(1-E/D)$
0	2	0	0	$\frac{1}{2}(1+E/D)$	$\frac{1}{2}(1-E/D)$
−1	−1	0	0	$\frac{1}{2}(1+E/D)$	$\frac{1}{2}(1-E/D)$
$+\frac{1}{2}$	0	0	0	$-3W_{I-II}/D$	$3W_{I-II}/D$
$-\frac{1}{2}$	0	0	0	$-3W_{I-II}/D$	$3W_{I-II}/D$
pump $m_I = \pm\frac{1}{2}$					
+1	0	0	0	$-W_{II-I}/D$	W_{II-I}/D
0	0	0	0	$-W_{II-I}/D$	W_{II-I}/D
−1	0	0	0	$-W_{II-I}/D$	W_{II-I}/D
$+\frac{1}{2}$	0	0	±1	$\frac{1}{2}(1-E/D)$	$\frac{1}{2}(1+E/D)$
$-\frac{1}{2}$	0	0	∓1	$\frac{1}{2}(1-E/D)$	$\frac{1}{2}(1+E/D)$

$D = \sqrt{(W_{e,I} - W_{e,II})^2 + (W_{e,I} - W_{e,II})(2W_{I-II} - 3W_{II-I}) + \frac{1}{4}(2W_{I-II} + 3W_{II-I})^2}$

$E = (W_{e,I} - W_{e,II}) + \frac{1}{2}(W_{I-II} - W_{II-I})$

12.13 SLOW ROTATIONAL DIFFUSION FOR SR-ELDOR AND SR-EPR

Fast rotational diffusion affects the intrinsic spin–lattice relaxation rate W_e, as we have seen in Chapter 5 and Sections 12.5, 12.7.

$$\lambda_{4,5} = W_{e,I} + W_{e,II} + \tfrac{1}{2}(2W_{I-II} + 3W_{II-I}) \pm \sqrt{(W_{e,I} - W_{e,II})^2 + (W_{e,I} - W_{e,II})(2W_{I-II} - 3W_{II-I}) + \tfrac{1}{4}(2W_{I-II} + 3W_{II-I})^2} \tag{12.89}$$

We readily see a correspondence between Eqs. 12.86, 12.87 and Eqs. 12.73, 12.72 for a ^{14}N-nitroxide, and between Eqs. 12.89 and 12.56 for two-site exchange. Equation 12.88 corresponds to Eq. 12.77 for a ^{15}N-nitroxide. Relative amplitudes of the five components for short-pulse SR-EPR and SR-ELDOR are given in Table 12.3. The possibility of differential spin–lattice relaxation enhancement by added paramagnetic relaxants is included in $W_{e,I}$ and $W_{e,II}$. Both contain terms linear in relaxant concentration, with individual rate constants $k_{RL,I}$ and $k_{RL,II}$ (see Eqs. 12.79, 12.80).

On the other hand, slow rotational diffusion transfers saturation from one part of the powder pattern to another, as illustrated in Figure 12.19. This is the basis of ST-EPR spectroscopy for studying ultraslow motion, as discussed in Section 11.12 of Chapter 11. Saturation transfer EPR is a CW technique. SR-ELDOR, however, offers the possibility of detecting saturation transfer by rotational diffusion directly, in a time-resolved fashion.

The simplest approach is to model rotational diffusion as two-site exchange, which we treated already in Section 12.10.

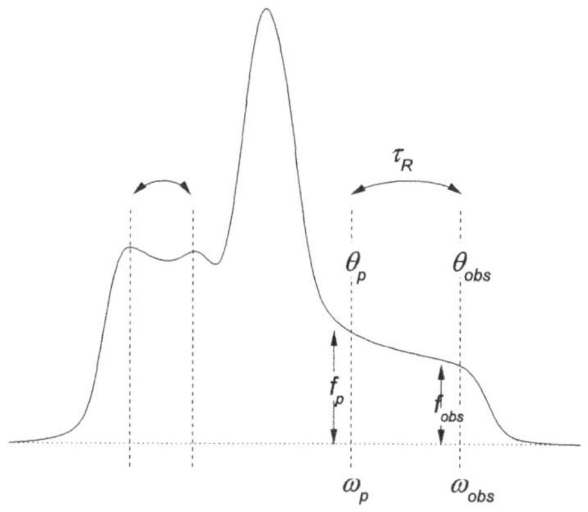

FIGURE 12.19 Spectral diffusion by slow rotational diffusion at rate $1/\tau_R$ in ^{14}N-nitroxide absorption powder pattern. Pumping pulse at ω_p corresponds to nitroxides oriented at angle θ_p to the static field. SR-ELDOR observing pulse at ω_{obs} corresponds to nitroxides oriented at angle θ_{obs}.

The two sites are defined by the pump and observe positions ω_p and ω_{obs} in the powder pattern (see Figure 12.19), which correspond to different angular orientations of the nitroxide, θ_p and θ_{obs}, relative to the static magnetic field. The recovery rate constants are then given simply by (see Eqs. 12.66 and 12.67):

$$\lambda_1 = 2W_e \tag{12.90}$$

$$\lambda_2 = 2W_e + \tau_p^{-1} + \tau_{obs}^{-1} \tag{12.91}$$

where τ_p^{-1} and τ_{obs}^{-1} are the rates at which saturation is transferred from the pump and observe positions, respectively, both of which are related directly to the rate of rotational diffusion τ_R^{-1} (Sugano et al. 1987). From Eq. 12.61, the short-pulse SR-ELDOR signal is then:

$$\Delta n_{obs}(t) \propto \exp(-2W_e t) - \exp\left(-(2W_e + \tau_p^{-1} + \tau_{obs}^{-1})t\right) \tag{12.92}$$

If f_p and f_{obs} are the spin populations at the pump and observe positions, respectively, then $\tau_p^{-1} \approx f_p \tau_R^{-1}$ and $\tau_{obs}^{-1} \approx f_{obs} \tau_R^{-1}$ (see Figure 12.19).

The two-site exchange model assumes nitrogen nuclear relaxation is sufficiently fast that each site relaxes as does a single species without hyperfine structure (see Section 12.10). We now include nuclear relaxation and rotational diffusion in the rate equation for a ^{15}N-nitroxide oriented at angle θ to the static field. To simplify, we neglect cross relaxation and differences in W_e between hyperfine states. These can be incorporated easily (see Section 12.4) but do not affect the way that we determine correlation times. For departures of spin population differences from Boltzmann equilibrium $\Delta n_\pm(\theta) = n_\pm - n_\pm^o$, in the $m_I = \pm\frac{1}{2}$ hyperfine manifolds, we get (cf. Eq. 12.7):

$$\frac{d\Delta n_\pm(\theta)}{dt} = -2W_e\Delta n_\pm - W_n\left(\Delta n_\pm - \Delta n_\mp\right) - D_R\nabla_\theta^2\Delta n_\pm \tag{12.93}$$

where D_R is the rotational diffusion coefficient, ∇_θ^2 is the angular Laplacian operator and $n_+^o = n_-^o$. Solution of the rotational diffusion equation is well known; it is a linear expansion in spherical harmonics (see Section 7.5). From Eq. 7.21 to lowest order, the conditional probability that a nitroxide with orientation θ_p at time $t=0$ has orientation θ_{obs} later at time t is: $P(\theta_p;\theta_{obs},t) = 1 + P_2(\cos\theta_p)P_2(\cos\theta_{obs})\exp(-t/\tau_R)$. Here, $P_2(\cos\theta)$ is a second-order Legendre polynomial and the isotropic rotational correlation time is defined by Eq. 7.22: $\tau_R \equiv 1/(6D_R)$. Making a separation of variables, an approximate solution to Eq. 12.93 for short-pulse SR-ELDOR becomes (Haas et al. 1993):

$$\Delta n_\pm(\theta_p;\theta_{obs},t) \propto \left[\exp(-2W_e t) \pm \exp\left(-2(W_e + W_n)t\right)\right]$$
$$\times\left(1 + P_2(\cos\theta_p)P_2(\cos\theta_{obs})\exp(-t/\tau_R)\right) \tag{12.94}$$

where we take the angle-independent part of the solution from Eqs. 12.19 and 12.20, with the simplification $W_x = 0 = \delta$.

Equation 12.94 is for a ^{15}N-nitroxide, and gives an SR-ELDOR recovery signal that contains four separate exponential components. Generalizing this approach to a ^{14}N-nitroxide, we get six components. The first three recovery rate constants λ_1 to λ_3 are given by Eqs. 12.32–12.34 with $W_x = 0 = \delta_1$. The three remaining rate constants are given by adding the rotational contribution ($W_R \equiv 1/\tau_R$) to λ_1 to λ_3, as in Eq. 12.94:

$$\lambda_4 = 2W_e + \tau_R^{-1} \tag{12.95}$$

$$\lambda_5 = 2W_e + W_n + \tau_R^{-1} \tag{12.96}$$

$$\lambda_6 = 2W_e + 3W_n + \tau_R^{-1} \tag{12.97}$$

where again we use the simplification $W_x = 0 = \delta_1$. We know from Section 12.6 that the λ_2 and λ_5 components vanish when we observe the central $m_I = 0$ hyperfine line in either SR-EPR or SR-ELDOR (see Eq. 12.36). For this observing position, the number of exponential components in the recovery of a ^{14}N-nitroxide is reduced to four.

Combining SR-EPR and SR-ELDOR, with various pump and observe positions lets us determine T_{1e}, T_{1n} and correlation time τ_R for slow rotational motion directly. Haas et al. (1993) demonstrate this experimentally for a ^{15}N-nitroxide. Extension to include cross relaxation and differences in W_e in Eqs. 12.93–12.97 is straightforward, as seen in Section 12.6. It affects the determination of T_{1e} and T_{1n} but not of the rotational correlation times.

12.14 SATURATION-RECOVERY EPR AND MOLECULAR ORDERING

We treated the effects of rotational diffusion on saturation-recovery rates in Sections 12.5 and 12.7 but without allowing for molecular ordering which limits the rotational amplitudes. In Section 8.8 of Chapter 8, we showed how molecular order influences spin–lattice relaxation and cross-relaxation rates. An important feature is that relaxation rates depend on orientation of the static magnetic field to the director axis of ordering.

We get the direct contributions of spin–lattice relaxation and cross relaxation to the saturation-recovery rate ($1/T_{SR,e} = 2\bar{W}_e + 2\bar{W}_x$ for long pulses) from Eqs. 8.61, 8.65 and 8.66. Substituting from Eqs. 8.49–8.51 (or equivalents) for the angular averages, the SR-EPR rate for a ^{15}N-nitroxide $\left(I = \frac{1}{2}\right)$ then becomes (cf. Eq. 12.14):

$$1/T_{SR,e} = \left(\tfrac{5}{36}\left((\Delta A)^2 + 3(\delta A)^2\right) + \left[\left((\Delta g)^2 + 3(\delta g)^2\right)\frac{\beta_e^2}{\hbar^2}B_o^2 - \tfrac{1}{12}\left((\Delta A)^2 + 3(\delta A)^2\right)\right]\langle \cos^2\theta \rangle - \left((\Delta g)^2 + 3(\delta g)^2\right)\frac{\beta_e^2}{\hbar^2}B_o^2\langle \cos^4\theta \rangle \right) \times \frac{\tau_c}{1+\omega_e^2\tau_c^2} \quad (12.98)$$

and for a ^{14}N-nitroxide ($I = 1$) becomes (cf. Eq. 12.32):

$$1/T_{SR,e} = \left(\tfrac{10}{27}\left((\Delta A)^2 + 3(\delta A)^2\right) + \left[\left((\Delta g)^2 + 3(\delta g)^2\right)\frac{\beta_e^2}{\hbar^2}B_o^2 - \tfrac{2}{9}\left((\Delta A)^2 + 3(\delta A)^2\right)\right]\langle \cos^2\theta \rangle - \left((\Delta g)^2 + 3(\delta g)^2\right)\frac{\beta_e^2}{\hbar^2}B_o^2\langle \cos^4\theta \rangle \right) \times \frac{\tau_c}{1+\omega_e^2\tau_c^2} \quad (12.99)$$

where $\theta(t)$ is the instantaneous angle that the nitroxide z-axis makes to the static field B_o, and τ_c is the rotational correlation time for restricted rotation of the nitroxide z-axis. Angular brackets indicate a time (or ensemble) average. Using Abragam's spin-operator method (see Appendix O), the general result for arbitrary nuclear spin I is (Marsh 2018):

$$1/T_{SR,e} = \left(\tfrac{5}{27}I(I+1)\left((\Delta A)^2 + 3(\delta A)^2\right) + \left[\left((\Delta g)^2 + 3(\delta g)^2\right)\frac{\beta_e^2}{\hbar^2}B_o^2 - \tfrac{1}{9}I(I+1)\left((\Delta A)^2 + 3(\delta A)^2\right)\right]\langle \cos^2\theta \rangle - \left((\Delta g)^2 + 3(\delta g)^2\right)\frac{\beta_e^2}{\hbar^2}B_o^2\langle \cos^4\theta \rangle \right) \times \frac{\tau_c}{1+\omega_e^2\tau_c^2} \quad (12.100)$$

For isotropic rotation, where $\langle \cos^2\theta \rangle = \frac{1}{3}$ and $\langle \cos^4\theta \rangle = \frac{1}{5}$, Eq. 12.100 corresponds with the result obtained for axial tensors in Marsh (2017) (see also Mailer et al. 2005). Note that the quantities calculated as W_{x1} and W_{x2} in Marsh (2017) are sums of the cross-relaxation rates for a given value of m_I, and not the single-transition probabilities as defined by Eqs. 8.65 and 8.66 in Section 8.8.

An interesting feature of Eqs. 12.98–12.100 is that the $\langle \cos^4\theta \rangle$ term is contributed only by the field-dependent Zeeman anisotropy. Thus, at low operating frequencies (below 9 GHz), we expect the long-pulse recovery rate to depend almost linearly on the S_{zz}-order parameter.

To relate the recovery rate $2W_{SR,e} = 1/T_{SR,e}$ to ordering of the nitroxide z-axis relative to the director for molecular order, we must relate averages over θ in Eqs. 12.98–12.100 to those over θ_z, the tilt angle of the nitroxide z-axis to the director. We do this via the fixed angle γ of the static field to the director, by using Eqs. 8.67 and 8.68 from Section 8.8. For the magnetic field parallel to the director ($\gamma = 0$), we have simply: $\theta = \theta_z$. For the magnetic field perpendicular to the director ($\gamma = 90°$), on the other hand, we get: $\langle \cos^2\theta \rangle = \frac{1}{2}\left(1 - \langle \cos^2\theta_z \rangle\right)$ and $\langle \cos^4\theta \rangle = \frac{3}{8}\left(\langle \cos^4\theta_z \rangle - 2\langle \cos^2\theta_z \rangle + 1\right)$.

Figure 12.20 shows the dependence of saturation-recovery rate at 9.4 GHz on conventional order parameter $S_{zz} \equiv \frac{1}{2}\left(3\langle \cos^2\theta_z \rangle - 1\right)$, for the two extreme orientations of the magnetic field. We use the three different motional models introduced in Section 8.8 to relate the $\langle \cos^4\theta_z \rangle$ average to S_{zz}

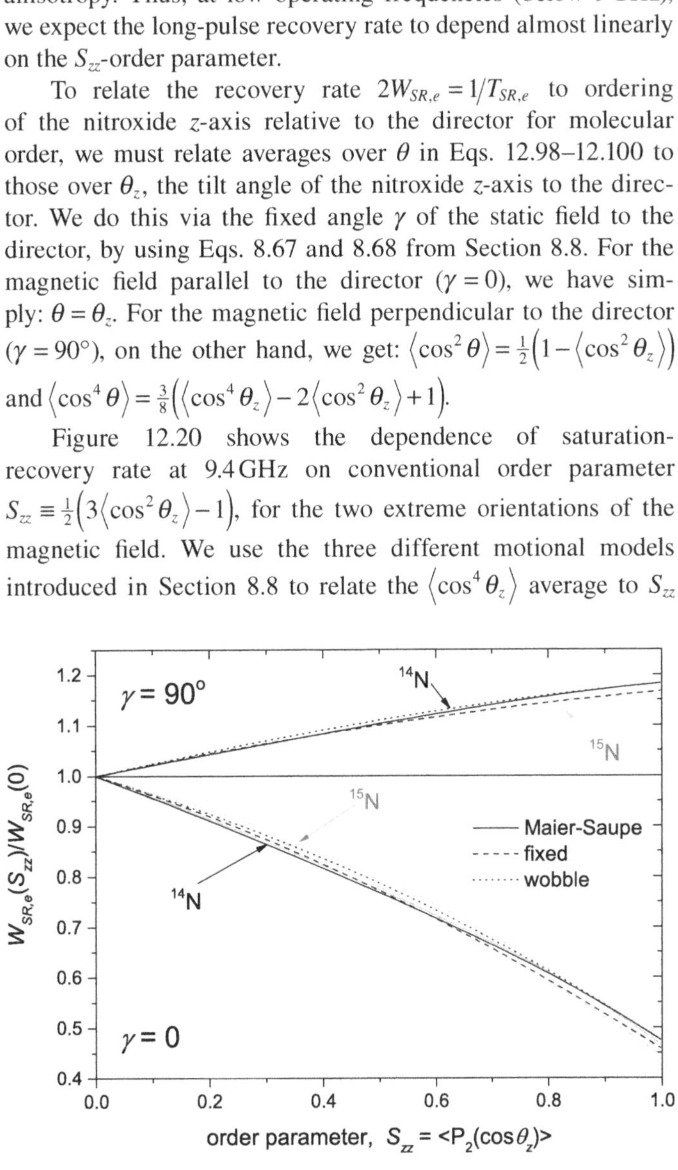

FIGURE 12.20 Saturation-recovery rate $W_{SR,e} = 1/(2T_{SR,e})$ (Eq. 12.100) for long pulses at 9.4 GHz, as function of order parameter $S_{zz} \equiv \langle P_2(\cos\theta_z)\rangle$, with angular averages determined by Maier–Saupe orientation potential (solid line), or corresponding to fixed θ_z (dashed line) or random walk within a cone (dotted line). Recovery rates $W_{SR,e}(S_{zz})$, with static magnetic field parallel (*bottom*; $\gamma = 0$), perpendicular (*top*; $\gamma = 90°$ in Eqs. 8.67, 8.68) to the director, are normalized to values $W_{SR,e}(0)$ for $S_{zz} = 0$: ^{14}N-nitroxides (black lines, Eq. 12.99) and ^{15}N-nitroxides (grey lines, Eq. 12.98). ^{14}N-DOXYL spin-Hamiltonian tensors from Jost et al. (1971), Smirnova et al. (1995); and corrected by nuclear gyromagnetic ratio for ^{15}N-DOXYL.

(cf. Eqs. 8.62–8.64). From the lower part of Figure 12.20 (i.e., $\gamma = 0$), we see that uniaxial molecular ordering can make up to a factor of two difference in saturation-recovery rates at a microwave frequency of 9.4 GHz, for a fixed correlation time. The rate of saturation recovery decreases with increasing order parameter, when the magnetic field is parallel to the director. This is because restricting the amplitude of angular fluctuation reduces its effectiveness as a relaxation process. Differences in normalized recovery rates between ^{14}N- and ^{15}N-nitroxides are relatively small at an EPR frequency of 9.4 GHz. Also, the various motional models do not change the dependence of the rates on order parameter greatly.

The upper part of Figure 12.20 shows the dependence of saturation-recovery rate on order parameter when the magnetic field is perpendicular to the director (i.e., $\gamma = 90°$). Most notably, the recovery rate now increases with increasing order parameter. The changes in rate are not as large, however, as when the magnetic field is parallel to the director. Again, differences in the normalized recovery rates of ^{14}N- and ^{15}N-nitroxides are not large at 9.4 GHz, and differences between the three motional models are not much greater than for the parallel field orientation.

If we make saturation-recovery measurements on the absorption maximum of a 9-GHz powder pattern, as is commonly the case, contributions will come mainly from field orientations perpendicular to the director axis, in the $m_I = 0$ manifold. Positions in the powder pattern are then specified by $\langle g_\perp \rangle$, which comes from Eq. 8.8 or the equivalent of Eq. 8.6 for axial symmetry (see Section 8.3). We then expect the dependence of recovery rate constant on order parameter contributed by modulation of the spin-Hamiltonian anisotropies to be approximated by the upper part of Figure 12.20. This implies that rotational correlation times will be more significant than order parameters in determining saturation-recovery rates, as is often tacitly assumed (e.g., Mailer et al. 2005; Mainali et al. 2011a). Note that at high field, corresponding to 94-GHz EPR, the strongest absorption peak in the powder pattern occurs for the magnetic field directed perpendicular to the director axis, i.e., again the $\gamma = 90°$ orientation. The situation at other EPR frequencies is discussed by Marsh (2018).

For chain-labelled lipids in membranes, Mainali et al. (2011a) find that the dependence of the saturation-recovery rate at 9 GHz on nitroxide position n down the chain tracks with the rotational diffusion rate $D_{R\perp}$ and is opposite in direction to that of the order parameter S_{zz}. For a series of other membrane systems at 9 GHz, Mainali et al. (2011b; 2012; 2013a) also find opposite dependences of $1/T_{SR,e}$ and S_{zz} on n in DOXYL-labelled chains. These latter findings are consistent with Figure 12.20 for the $\gamma = 90°$ orientation.

As illustration, Table 12.4 lists saturation-recovery rates at 9 GHz of chain-labelled phosphatidylcholines in fluid bilayer membranes of dimyristoyl phosphatidylcholine containing 0 mol% or 50 mol% cholesterol. The spin-label order parameter S_{zz} decreases with increasing position n down the hydrocarbon chain and increases on adding cholesterol. Recovery rates are normalized to the reduced spectral density $j(\omega_e) = \tau_c / (1 + \omega_e^2 \tau_c^2)$, where $\tau_c = 1/6D_{R\perp}$ is deduced from the rotational diffusion coefficient $D_{R\perp}$ that is obtained by simulating the spectral lineshapes (Mainali et al. 2011a). This normalized rate should reflect directly the dependence on order parameter given in the upper part of Figure 12.20 (cf. Eqs. 12.99 and 12.100 with substitutions for $\gamma = 90°$ in angular averages). We see that $(1/T_{SR,e})/j(\omega_e)$ remains approximately constant, or increases slightly with increasing order parameter (i.e., with decreasing n), as predicted in Figure 12.20. However, we saw in Section 8.14 that spectra of chain-labelled lipids contain motional components in both fast and slow regimes of 9-GHz nitroxide EPR. Significantly, the single correlation times from lineshape simulations used in Table 12.4 are considerably longer than those used to interpret multifrequency saturation-recovery experiments in the absence of lineshape analysis (Mailer et al. 2005).

TABLE 12.4 Rate constants $1/T_{SR,e}$ for saturation recovery at 9.2 GHz of *n*-DOXYL phosphatidylcholine spin labels in bilayer membranes of dimyristoyl phosphatidylcholine with and without 50 mol% cholesterol, at 27°C[a]

	$(1/T_{SR,e})/j(\omega_e)\ (\times 10^{-18}\ s^{-2})$	
n (C-atom)	*0 mol% cholesterol*	*50 mol% cholesterol*
5	2.49	3.56
7	2.45	3.85
10	2.50	2.27
12	2.30	2.21
14	2.34	2.14
16	1.80	1.62

Data from Mainali et al. (2011a).
[a] Recovery rates normalized to reduced spectral density: $j(\omega_e) = \tau_c / (1 + \omega_e^2 \tau_c^2)$, where $\tau_c = 1/(6D_{R\perp})$ comes from lineshape simulations.

At high field, corresponding to 94-GHz EPR, the $\langle \cos^4 \theta \rangle$ angular average makes a much larger contribution, relative to that from $\langle \cos^2 \theta \rangle$, than at 9 GHz (see Eq. 12.100). For the $\gamma = 90°$ orientation at the absorption maximum, we then predict that the recovery rate barely increases, or slowly decreases, with increasing order parameter, before decreasing strongly at higher S_{zz}-values (Marsh 2018). In contrast, the normalized experimental recovery rate $(1/T_{SR,e})/j(\omega_e)$ at 94 GHz remains approximately constant or increases with increasing order parameter (Mainali et al. 2013b), similar to the situation at 9 GHz. See Marsh (2018) for further discussion.

12.15 CONCLUDING SUMMARY

1. *SR-EPR is independent of hyperfine state:*
   Rate constants λ_i in saturation-recovery EPR of ^{14}N- and ^{15}N-nitroxides do not depend on which hyperfine manifold we saturate, even when intrinsic electron spin–lattice relaxation rates are m_I-dependent.

2. *Long-pulse SR-EPR:*
Recovery rate constants in long-pulse SR-EPR contain major contributions from cross relaxation in addition to intrinsic electron spin–lattice relaxation. For ^{15}N-nitroxides, i.e., $I = \frac{1}{2}$:

$$\lambda_1 = 2\bar{W}_e + 2W_x \tag{12.101}$$

where $\bar{W}_e = \frac{1}{2}\left(W_{e,+} + W_{e,-}\right)$ and $W_x = \frac{1}{2}\left(W_{x,1} + W_{x,2}\right)$ are the mean spin–lattice relaxation and cross-relaxation rates, respectively. For ^{14}N-nitroxides, i.e., $I = 1$:

$$\lambda_1 = 2\bar{W}_e + \tfrac{8}{3}W_x \tag{12.102}$$

where the mean electron spin–lattice relaxation rate is $\bar{W}_e = \frac{1}{3}\left(W_{e,-1} + W_{e,0} + W_{e,+1}\right)$ and again $W_x = \frac{1}{2}\left(W_{x,1} + W_{x,2}\right)$. For $I = 1$, the actual mean cross relaxation rate is $\bar{W}_x = \frac{2}{3}\left(W_{x,1} + W_{x,2}\right)$, and thus $\lambda_1 = 2\bar{W}_e + 2\bar{W}_x$ as for the $I = \frac{1}{2}$ case.

3. *Short-pulse SR-ELDOR:*
The additional recovery rate constants that we get (optimally) from short-pulse SR-ELDOR are dominated mostly by nuclear relaxation (except for very short correlation times). For ^{15}N-nitroxides, i.e., $I = \frac{1}{2}$:

$$\lambda_2 = 2\bar{W}_e + 2W_n \tag{12.103}$$

For ^{14}N-nitroxides, i.e., $I = 1$:

$$\lambda_2 = 2\bar{W}_e + \tfrac{2}{3}\delta_1 + W_n + W_x \tag{12.104}$$

$$\lambda_3 = 2\bar{W}_e - \tfrac{2}{3}\delta_1 + 3W_n + \tfrac{1}{3}W_x \tag{12.105}$$

where $\delta_1 = \frac{1}{2}\left(W_{e,-1} + W_{e,+1}\right) - W_{e,0}$. In either case, we cannot express λ_2 or λ_3 directly in terms of the combination of spin–lattice and cross relaxation that we get from λ_1 with long-pulse SR-EPR (cf. point 2).

4. *Electron spin–lattice relaxation and rotational dynamics:*
From the dependence on rotational correlation time τ_R, the electron spin–lattice relaxation rate (i.e., λ_1 from long-pulse SR-EPR) is determined by modulation of Zeeman and hyperfine (END) anisotropies (maximum at $\omega_e\tau_R = 1$). At short τ_R, modulation comes from spin-rotation interaction $\left(W_{e,SJ} \propto 1/\tau_R\right)$; at long τ_R, the relaxation rate is dominated by a residual contribution with weak dependence on τ_R.

5. *Nuclear spin–lattice relaxation and rotational dynamics:*
Nuclear relaxation rate W_n (best obtained from short-pulse SR-ELDOR) is determined primarily by rotational modulation of the END interaction (maximum at $\omega_a\tau_R = 1$, where $\omega_a = \frac{1}{2}a_o$), with additional contributions at long τ_R.

TABLE 12.5 Rate constants λ_i for SR-EPR and SR-ELDOR

	^{15}N-*nitroxide*	^{14}N-*nitroxide*	*Ref.*[a]
nuclear relaxation[b]	$T_{1n} = 1/(2W_n)$	$T_{1n} = 1/W_n$	
	$\lambda_1^{(o)} = 2\bar{W}_e + 2W_x$	$\lambda_1^{(o)} = 2\bar{W}_e + \frac{8}{3}W_x$	1,2
	$\lambda_2^{(o)} = 2\bar{W}_e + 2W_n$	$\lambda_2^{(o)} = 2\left(\bar{W}_e + \frac{1}{3}\delta_1\right) + W_x + W_n$	
	–	$\lambda_3^{(o)} = 2\left(\bar{W}_e + \frac{1}{3}\delta_1\right) + \frac{1}{3}W_x + 3W_n$	
+Heisenberg exchange[c]	$\tau_{HE}^{-1} = k_{HE}[\mathrm{SL}]$	$\tau_{HE}^{-1} = k_{HE}[\mathrm{SL}]$	
	$\lambda_1 = \lambda_1^{(o)}$	$\lambda_1 = \lambda_1^{(o)}$	1,3
	$\lambda_2 = \lambda_2^{(o)} + \tau_{HE}^{-1}$	$\lambda_2 = \lambda_2^{(o)} + \tau_{HE}^{-1}$	
	–	$\lambda_3 = \lambda_3^{(o)} + \tau_{HE}^{-1}$	
2-site exchange[d]	$f_b\tau_b^{-1} = \left(1 - f_b\right)\tau_f^{-1}$		
	$\lambda_{1,2} = W_{e,b} + W_{e,f} + \frac{1}{2}\left(\tau_b^{-1} + \tau_f^{-1}\right) \mp \sqrt{\left(W_{e,b} - W_{e,f}\right)^2 + \left(W_{e,b} - W_{e,f}\right)\left(\tau_b^{-1} - \tau_f^{-1}\right) + \frac{1}{4}\left(\tau_b^{-1} + \tau_f^{-1}\right)^2}$		4
+Heisenberg exchange	$\lambda_{1,2} = W_{e,b} + W_{e,f} + \frac{1}{2}\left(\tau_b^{-1} + \tau_f^{-1} + \tau_{HE}^{-1}\right) \mp \sqrt{\left(W_{e,b} - W_{e,f}\right)^2 + \left(W_{e,b} - W_{e,f}\right)\left(1 - 2f_b\right)\left(\tau_b^{-1} + \tau_{HE}^{-1}\right) + \frac{1}{4}\left(\tau_b^{-1} + \tau_f^{-1} + \tau_{HE}^{-1}\right)^2}$		5
rotation[e]	$\tau_R = 1/(6D_R)$		
	$\lambda_3 = \lambda_1^{(o)} + \tau_R^{-1}$	$\lambda_4 = \lambda_1^{(o)} + \tau_R^{-1}$	1,6
	$\lambda_4 = \lambda_2^{(o)} + \tau_R^{-1}$	$\lambda_5 = \lambda_2^{(o)} + \tau_R^{-1}$	
	–	$\lambda_6 = \lambda_3^{(o)} + \tau_R^{-1}$	

[a] References: 1. Marsh (2016a); 2. Hyde et al. (1984); 3. Yin et al. (1987); 4. Bridges et al. (2010); 5. Kawasaki et al. (2001); 6. Haas et al. (1993).
[b] Electron spin–lattice relaxation: $\bar{W}_e = \frac{1}{2}\left(W_{e,-} + W_{e,+}\right)$ for ^{15}N-nitroxides, and $\bar{W}_e = \frac{1}{3}\left(W_{e,-1} + W_{e,0} + W_{e,+1}\right)$ with $\delta_1 = \frac{1}{2}\left(W_{e,-1} + W_{e,+1}\right) - W_{e,0}$ for ^{14}N-nitroxides. Cross relaxation: $W_x = \frac{1}{2}\left(W_{x_1} + W_{x_2}\right)$.
[c] k_{HE} is the second-order rate constant for Heisenberg spin exchange, and [SL] is the nitroxide concentration.
[d] τ_b^{-1} and τ_f^{-1} are the rate constants for leaving sites *b* and *f*, respectively, and $f_b = \left(1 - f_f\right)$ is the fractional spin population at site *b*.
[e] Rate constants not listed are identical to $\lambda_1^{(o)}$, $\lambda_1^{(o)}$ and $\lambda_3^{(o)}$. τ_R is the correlation time and D_R the diffusion coefficient for isotropic rotation.

6. *Paramagnetic relaxation enhancement:*
   Long-pulse SR-EPR determines paramagnetic spin–lattice relaxation enhancements directly: $\lambda_1 = \lambda_1^o + k_{RL}[\text{R}]$, where [R] is the relaxant concentration (cf. Section 10.5).
7. *Two-site exchange:*
   Short-pulse (ideally) SR-EPR detects two-site exchange ($b \leftrightarrow f$) directly:

$$\lambda_{1,2} = W_{e,b} + W_{e,f} + \tfrac{1}{2}\left(\tau_b^{-1} + \tau_f^{-1}\right)$$
$$\mp\sqrt{\left(W_{e,b} - W_{e,f}\right)^2 + \left(W_{e,b} - W_{e,f}\right)\left(\tau_b^{-1} - \tau_f^{-1}\right) + \tfrac{1}{4}\left(\tau_b^{-1} + \tau_f^{-1}\right)^2}$$

   where $\tau_b^{-1} = \tau_f^{-1}\left(1 - f_b\right)/f_b$ is the rate of leaving site b. For slow exchange, the component with $\lambda_2 \approx 2W_{e,b} + \left(1 - f_b\right)\left(\tau_b^{-1} + \tau_f^{-1}\right)$ dominates, where f_b is the fractional population of b. For fast exchange, both components are significant: $\lambda_1 \approx 2\overline{W}_e$ and $\lambda_2 \approx 2\left(W_{e,b} + W_{e,f} - \overline{W}_e\right) + \tau_b^{-1} + \tau_f^{-1}$, where $\overline{W}_e = f_b W_{e,b} + \left(1 - f_b\right)W_{e,f}$.
8. *Heisenberg spin exchange:*
   Short-pulse SR-EPR detects Heisenberg exchange between spin labels. The Heisenberg spin-exchange rate $\tau_{HE}^{-1} = k_{HE}[\text{SL}]$ simply adds to the (short-pulse) rate constants λ_2 and λ_3 given under point 3, but the λ_1 (long-pulse) rate constant given under point 2 is unaffected.
9. *Spectral diffusion by slow rotation:*
   Slow rotational diffusion contributes to spectral diffusion detected by SR-ELDOR (and SR-EPR). In addition to the recovery components with rate constants λ_1, λ_2 and λ_3 given under points 2 and 3, we get further recovery components with these rate constants increased exactly by the rotational diffusion rate τ_R^{-1} (e.g., $\lambda_4 = \lambda_1 + \tau_R^{-1}$).

Expressions for the various recovery rate constants of primary importance are summarized in Table 12.5. See Figures. 12.4, 12.9 and 12.16, 12.17 for energy-level diagrams with the corresponding transitions.

REFERENCES

Biller, J.R., Meyer, V., Elajaili, H. et al. 2011. Relaxation time and line widths of isotopically substituted nitroxides in aqueous solution at X-band. *J. Magn. Reson.* 212:370–377.

Bridges, M.D., Hideg, K., and Hubbell, W.L. 2010. Resolving conformational and rotameric exchange in spin-labeled proteins using saturation recovery EPR. *Appl. Magn. Reson.* 37:363–390.

Carrington, A. and McLachlan, A.D. 1969. *Introduction to Magnetic Resonance. With Applications to Chemistry and Chemical Physics*, New York: Harper and Row.

Dalton, L.R., Robinson, B.H., Dalton, L.A., and Coffey, P. 1976. Saturation transfer spectroscopy. *Adv. Magn. Reson.* 8:149–259.

De Gennes, P.G. 1958. Sur la relaxation nucléaire dans les cristaux ioniques. *J. Phys. Chem. Solids* 7:345–350.

Du, J.L., Eaton, G.R., and Eaton, S.S. 1995. Temperature, orientation, and solvent dependence of electron spin-lattice relaxation rates for nitroxyl radicals in glassy solvents and doped solids. *J. Magn. Reson. A* 115:213–221.

Eaton, S.S. and Eaton, G.R. 2000. Relaxation times of organic radicals and transition metal ions. In *Biological Magnetic Resonance Vol. 19. Distance Measurements in Biological Systems*, eds. Berliner, L.J., Eaton, S.S., and Eaton, G.R., 29–154. New York: Springer.

Fajer, P., Thomas, D.D., Feix, J.B., and Hyde, J.S. 1986. Measurement of rotational molecular motion by time-resolved saturation transfer electron paramagnetic resonance. *Biophys. J.* 50:1195–1202.

Feix, J.B., Popp, C.A., Venkataramu, S.D. et al. 1984. An electron-electron double resonance study of interactions between ^{14}N stearic and ^{15}N stearic acid spin label pairs - Lateral diffusion and vertical fluctuations in dimyristoylphosphatidylcholine. *Biochemistry* 23:2293–2299.

Feldman, D.W., Castle, J.G., and Murphy, J. 1965. Spin relaxation of atomic hydrogen in CaF_2 - evidence for local modes. *Phys. Rev.* 138:A1208–A1216.

Froncisz, W., Camenisch, T.G., Ratke, J.J. et al. 2008. Saturation recovery EPR and ELDOR at W-band for spin labels. *J. Magn. Reson.* 193:297–304.

Haas, D.A., Sugano, T., Mailer, C., and Robinson, B.H. 1993. Motion in nitroxide spin labels - direct measurement of rotational correlation times by pulsed electron double resonance. *J. Phys. Chem.* 97:2914–2921.

Hyde, J.S., Froncisz, W., and Mottley, C. 1984. Pulsed ELDOR measurement of nitrogen T_1 in spin labels. *Chem. Phys. Lett.* 110:621–625.

Jost, P.C., Libertini, L.J., Hebert, V.C., and Griffith, O.H. 1971. Lipid spin labels in lecithin multilayers. A study of motion along fatty acid chains. *J. Mol. Biol.* 59:77–98.

Kawasaki, K., Yin, J.J., Subczynski, W.K., Hyde, J.S., and Kusumi, A. 2001. Pulse EPR detection of lipid exchange between protein-rich raft and bulk domains in the membrane: methodology development and its application to studies of influenza viral membrane. *Biophys. J.* 80:738–748.

Mailer, C., Robinson, B.H., and Haas, D.A. 1992. New developments in pulsed electron paramagnetic resonance relaxation mechanisms of nitroxide spin labels. *Bull. Magn. Reson.* 14:30–34.

Mailer, C., Nielsen, R.D., and Robinson, B.H. 2005. Explanation of spin-lattice relaxation rates of spin labels obtained with multifrequency saturation recovery EPR. *J. Phys. Chem. A* 109:4049–4061.

Mainali, L., Feix, J.B., Hyde, J.S., and Subczynski, W. 2011a. Membrane fluidity profiles as deduced by saturation recovery EPR measurements of spin-lattice relaxation times of spin labels. *J. Magn. Reson.* 212:418–425.

Mainali, L., Raguz, M., Camenisch, T.G., Hyde, J.S., and Subczynski, W.K. 2011b. Spin-label saturation recovery EPR at W-band: applications to eye lens lipid membranes. *J. Magn. Reson.* 212:86–94.

Mainali, L., Raguz, M., and Subczynski, W. 2012. Phases and domains in sphingomyelin-cholesterol membranes: structure and properties using EPR spin-labeling methods. *Eur. Biophys. J.* 41:147–159.

Mainali, L., Raguz, M., O'Brien, W.J., and Subczynski, W. 2013a. Properties of membranes derived from the total lipids extracted from the human lens cortex and nucleus. *Biochim. Biophys. Acta* 1828:1432–1440.

Mainali, L., Hyde, J.S., and Subczynski, W.K. 2013b. Using spin-label W-band EPR to study membrane fluidity profiles in samples of small volume, *J. Magn. Reson.* 226:35–44.

Marsh, D. 1992. Influence of nuclear relaxation on the measurement of exchange frequencies in CW saturation EPR studies. *J. Magn. Reson.* 99:332–337.

Marsh, D. 2016a. Cross relaxation in nitroxide spin labels. *J. Magn. Reson.* 272:172–180.

Marsh, D. 2016b. Nuclear spin-lattice relaxation in nitroxide spin-label EPR. *J. Magn. Reson.* 272:166–171.

Marsh, D. 2017. Coherence transfer and electron T_1-, T_2-relaxation in nitroxide spin labels. *J. Magn. Reson.* 277:86–94.

Marsh, D. 2018. Molecular order and T_1-relaxation, cross-relaxation in nitroxide spin labels. *J. Magn. Reson.* 290:38–45.

Orton, J.W. 1968. *Electron Paramagnetic Resonance*, London: Iliffe.

Owenius, R., Terry, G.E., Williams, M.J., Eaton, S.S., and Eaton, G.R. 2004. Frequency dependence of electron spin relaxation of nitroxyl radicals in fluid solutions. *J. Phys. Chem. B* 108:9475–9481.

Percival, P.W. and Hyde, J.S. 1975. Pulsed EPR spectrometer, II. *Rev. Sci. Instrum.* 46:1522–1529.

Percival, P.W. and Hyde, J.S. 1976. Saturation-recovery measurements of spin-lattice relaxation-times of some nitroxides in solution. *J. Magn. Reson.* 23:249–257.

Percival, P.W., Hyde, J.S., Dalton, L.A., and Dalton, L.R. 1975. Molecular and applied modulation effects in electron-electron double resonance. 5. Passage effects in high-resolution frequency and field swept ELDOR. *J. Chem. Phys.* 62:4332–4342.

Popp, C.A. and Hyde, J.S. 1982. Electron-electron double resonance and saturation recovery studies of nitroxide electron and nuclear spin-lattice relaxation times and Heisenberg exchange rates: lateral diffusion in dimyristoylphosphatidylcholine. *Proc. Natl. Acad. Sci. USA* 79:2559–2563.

Robinson, B.H., Haas, D.A., and Mailer, C. 1994. Molecular dynamics in liquids: spin-lattice relaxation of nitroxide spin labels. *Science* 263:490–493.

Robinson, B.H., Reese, A.W., Gibbons, E., and Mailer, C. 1999. A unified description of the spin-spin and spin-lattice relaxation rates applied to nitroxide spin labels in viscous liquids. *J. Phys. Chem. B* 103:5881–5894.

Sato, H., Kathirvelu, V., Fielding, A. et al. 2007. Impact of molecular size on electron spin relaxation rates of nitroxyl radicals in glassy solvents between 100 and 300 K. *Mol. Phys.* 105:2137–2151.

Sato, H., Bottle, S.E., Blinco, J.P. et al. 2008. Electron spin-lattice relaxation of nitroxyl radicals in temperature ranges that span glassy solutions to low-viscosity liquids. *J. Magn. Reson.* 191:66–77.

Smirnova, T.I., Smirnov, A.I., Clarkson, R.B., and Belford, R.L. 1995. W-Band (95 GHz) EPR spectroscopy of nitroxide radicals with complex proton hyperfine structure: fast motion. *J. Phys. Chem.* 99:9008–9016.

Solomon, I. 1955. Relaxation processes in a system of 2 spins. *Phys. Rev.* 99:559–565.

Stetter, E., Vieth, H.M., and Hausser, K.H. 1976. ELDOR studies of nitroxide radicals - discrimination between rotational and translational correlation times in liquids. *J. Magn. Reson.* 23:493–504.

Subczynski W., Mainali, L., Camenisch, T.G., Froncisz, W., and Hyde, J.S. 2011. Spin-label oximetry at Q- and W-band. *J. Magn. Reson.* 209:142–148.

Sugano, T., Mailer, C., and Robinson, B.H. 1987. Direct detection of very slow 2-jump processes by saturation recovery electron paramagnetic resonance spectroscopy. *J. Chem. Phys.* 87:2478–2488.

Thomas, D.D., Dalton, L.R., and Hyde, J.S. 1976. Rotational diffusion studied by passage saturation transfer electron paramagnetic resonance. *J. Chem. Phys.* 65:3006–3024.

Yang, Z.Y., Jimenez-Oses, G., Lopez, C.J. et al. 2014. Long-range distance measurements in proteins at physiological temperatures using saturation recovery EPR spectroscopy. *J. Am. Chem. Soc.* 136:15356–15365.

Yang, Z., Bridges, M., Lerch, M.T., Altenbach, C., and Hubbell, W.L. 2015. Saturation recovery EPR and nitroxide spin labeling for exploring structure and dynamics in proteins. *Methods Enzymol.* 564:3–27.

Yin, J.J. and Hyde, J.S. 1987a. Application of rate equations to ELDOR and saturation recovery experiments on N-14-N-15 spin label pairs. *J. Magn. Reson.* 74:82–93.

Yin, J.J. and Hyde, J.S. 1987b. Spin-label saturation recovery electron spin resonance measurements of oxygen transport in membranes. *Z. Phys. Chem.* 153:57–65.

Yin, J.J., Pasenkiewicz-Gierula, M., and Hyde, J.S. 1987. Lateral diffusion of lipids in membranes by pulse saturation recovery electron spin resonance. *Proc. Natl. Acad. Sci. USA* 84:964–968.

Yin, J.J., Feix, J.B., and Hyde, J.S. 1988. Solution of the nitroxide spin label spectral overlap problem using pulse electron spin resonance. *Biophys. J.* 53:525–531.

Spin-Echo EPR

13

13.1 INTRODUCTION

Pulse-EPR allows direct time-dependent measurement of spin-label dynamics. A single 90°-pulse of microwave radiation tips the electron spin magnetization into the x, y-plane, perpendicular to the static magnetic field. The resulting EPR-detected transverse magnetization decays according to the effective T_2-relaxation time. Because spin-label electron T_2's are rather short, we lose a substantial part of the free-induction decay (FID) following the single pulse, in the dead time of the receiver. Therefore, we detect transverse relaxation instead by spin echoes or other multiple-pulse techniques. In fact, most one-dimensional pulse-EPR with spin labels is based on spin echoes, e.g., echo-detected spectra with magnetic field sweep or echo-detected inversion recovery determinations of T_1.

In this chapter, we deal first with the two-pulse primary echo (see Figure 13.1 top), where the dependence of the amplitude on echo delay time is a direct measure of T_2 or the homogeneous linewidth. Then we go on to partially relaxed echo-detected spectra, including analysis of librational motion. Finally, we consider the three-pulse stimulated echo (see Figure 13.1 bottom), which is sensitive to slower motions than is the primary echo. Measurements are confined to the low-temperature regime, because spin-label T_2-relaxation is mostly too fast to produce detectable echo intensities at ambient temperatures. The echo intensity is recovered at high temperatures in the extreme motional narrowing regime, well beyond the T_2 minimum, but this is a region readily studied by CW-EPR (see Chapters 5 and 7).

13.2 MICROWAVE PULSES IN THE VECTOR MODEL

We can understand the basic principles of pulse EPR from the classical vector model for the electron spin magnetization, $\mathbf{M}$. We begin with the equilibrium situation, in the absence of microwave radiation. The net magnetization from the ensemble of electron spins precesses at the Larmor frequency ω_o around the static magnetic field $\mathbf{B}_o$, which is directed along the z-axis (see Section 6.2 in Chapter 6). The transverse magnetization is therefore zero; the only non-vanishing resultant component of magnetization at equilibrium, $\mathbf{M}_o$, is directed along the static magnetic field. This

is the situation immediately before the first pulse in any sequence (see Figure 13.1).

We now apply a pulse of circularly polarized microwave radiation, where the $\mathbf{B}_1$-field rotates in the x, y-plane with components $B_{1,x} = B_1 \cos \omega t$ and $B_{1,y} = B_1 \sin \omega t$. Viewed in the axis frame that rotates about $\mathbf{B}_o$ at angular frequency ω, this $\widehat{\mathbf{B}}_1$-field remains fixed along the x-axis. If the radiation is at the resonance frequency $\omega = \omega_o$, the equations of motion in the rotating frame are (see Eqs. 6.4–6.6 in Chapter 6):

$$\frac{\mathrm{d}\widehat{M}_x}{\mathrm{d}t} = 0 \tag{13.1}$$

$$\frac{\mathrm{d}\widehat{M}_y}{\mathrm{d}t} = -\omega_1 \widehat{M}_z \tag{13.2}$$

$$\frac{\mathrm{d}\widetilde{M}_z}{\mathrm{d}t} = \omega_1 \widetilde{M}_y \tag{13.3}$$

where $\omega_1 \equiv \gamma_e B_1$, and we ignore spin relaxation for the moment. Solution of Eqs. 13.2 and 13.3, for the initial conditions before the pulse ($\widehat{M}_z = M_o$, $\widehat{M}_x = \widehat{M}_y = 0$), gives:

$$\widehat{M}_y = -M_o \sin \omega_1 t_p \tag{13.4}$$

$$\widehat{M}_z = M_o \cos \omega_1 t_p \tag{13.5}$$

where t_p is the duration of the microwave pulse. As expected, the magnetization precesses about B_1 (i.e., the x-axis) because, in the rotating frame, the effective field along the z-axis ($\widehat{B}_z = B_o - \omega_o/\gamma_e$, see Eq. 6.4) vanishes at resonance. For $\omega_1 t_p = \pi/2$, Eqs. 13.4 and 13.5 become: $\widehat{M}_y = -M_o$ and $\widehat{M}_z = 0$. The static magnetization M_o is rotated from the z-axis to along the $-y$-axis by this 90°-pulse along the x-axis. Similarly, for $\omega_1 t_p = \pi$, the resulting 180°-pulse

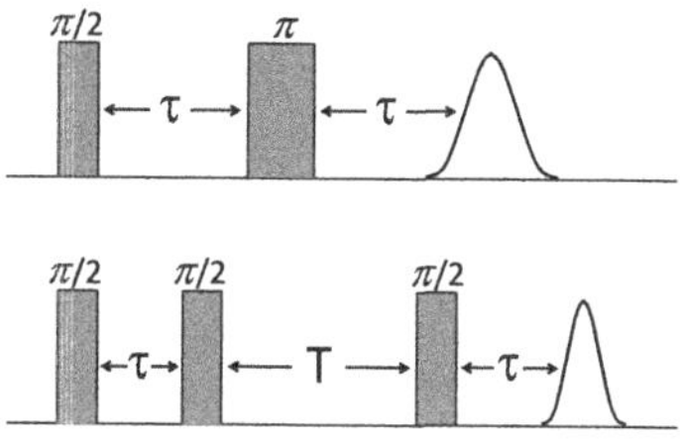

FIGURE 13.1 Two-pulse primary spin-echo sequence (upper), and three-pulse stimulated echo sequence (lower). Shaded areas indicate times when microwave radiation is switched on.

reverses (i.e., inverts) the direction of the magnetization. In general, the applied pulse rotates the magnetization about the x-axis by an angle $\beta_p = \omega_1 t_p \equiv \gamma_e B_1 t_p$ that depends on both the microwave field strength and the pulse length.

We detect the transverse magnetization that is created by a 90°-pulse as an EPR signal: the free induction decay (FID). If we include T_2-relaxation and a resonance offset $\Delta\omega_o = \omega - \omega_o$, the Bloch equations (Eqs. 13.1 and 13.2) after a 90°-pulse along x become:

$$\frac{d\hat{M}_x}{dt} = -\Delta\omega_o \hat{M}_y - \frac{\hat{M}_x}{T_2} \tag{13.6}$$

$$\frac{d\hat{M}_y}{dt} = \Delta\omega_o \hat{M}_x - \frac{\hat{M}_y}{T_2} \tag{13.7}$$

– see Eqs. 6.5 and 6.6 with $M_z = 0$ in Chapter 6. Solution of Eqs. 13.5–13.7 for evolution of the transverse magnetization at time t after the pulse gives:

$$\hat{M}_x = M_o \exp\left(-\frac{t}{T_2}\right) \sin \Delta\omega_o t \tag{13.8}$$

$$\hat{M}_y = -M_o \exp\left(-\frac{t}{T_2}\right) \cos \Delta\omega_o t \tag{13.9}$$

with the initial conditions: $\hat{M}_y = -M_o$ and $\hat{M}_x = 0$ at $t = 0$. For a homogeneously broadened spectral line, the FID is exponential in time with T_2 as the decay time and a sinusoidal modulation at the offset frequency $\Delta\omega_o$. Fourier transformation of the FID gives a Lorentzian lineshape with half-width at half-height of $1/T_2$ (see Table T.1 in Appendix T). For an inhomogeneously broadened spectral line, the decay of the FID is more rapid, determined by an effective T_2^* that characterizes the distribution of resonance offsets $\Delta\omega_o$ over the inhomogeneous lineshape.

For pulse angles less than 90°, the amplitude of the FID is determined by the projection of the magnetization on the x, y-plane. From Eq. 13.4, we see that the FID-amplitude is then scaled by the factor: $p_{\pi/2} = \sin\beta_p$. As mentioned already in the introduction, FIDs from spin labels decay too rapidly to be of much practical use, except in the extreme motional-narrowing region.

13.3 FINITE PULSE POWER AND WIDTH

An important feature of pulse-EPR with spin labels is that a nitroxide powder spectrum is normally broader than the frequency range ~ $\pm\omega_1$ that can be covered by the microwave pulse. The applied pulse therefore excites spins from only a relatively small range of nitroxide orientations within the powder-pattern lineshape. The frequency distribution of spins that we can excite is given by Fourier synthesis of the pulse of sinusoidal radiation. However, only in the region of linear response, i.e., low pulse angles, does Fourier transformation of the pulse shape describe the excitation spectrum properly (Ernst et al. 1987; Schweiger and Jeschke 2001).

We noted already in Section 6.2 of Chapter 6 that the effective field B_{eff} in the rotating frame is inclined at an angle θ_{eff} to the static field direction (see Figure 6.2). Only exactly on resonance, $\Delta\omega_o = 0$, does B_{eff} lie along the B_1-radiation field. During a microwave pulse, the magnetization precesses around the effective field at angular frequency $\omega_{eff} = \gamma_e B_{eff}$ that is given by:

$$\omega_{eff} = \sqrt{(\Delta\omega_o)^2 + \omega_1^2} \tag{13.10}$$

As we see from Figure 13.2, the magnetization **M** starts from its equilibrium value M_o along the z-axis at the beginning of the pulse and precesses in a cone of angle θ_{eff} about the ω_{eff}-direction, where

$$\sin\theta_{eff} = \omega_1/\omega_{eff} \tag{13.11}$$

After time t_p, at the end of the pulse, the magnetization has precessed by azimuthal angle $\omega_{eff} t_p$ about the ω_{eff}-direction.

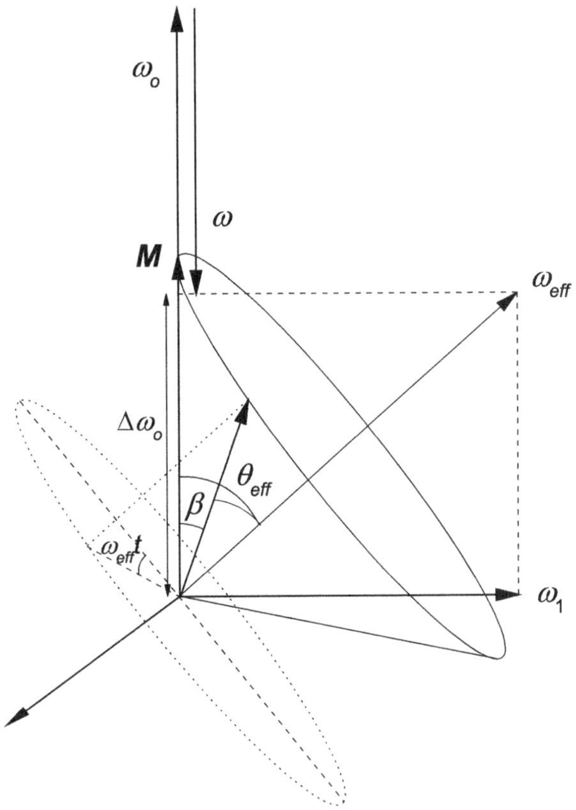

FIGURE 13.2 Effective field $\omega_{eff} = \gamma_e B_{eff}$ in the rotating frame, with transverse radiation field $\omega_1 = \gamma_e B_1$. Fields expressed as angular frequencies: ω, angular frequency of microwave radiation; $\omega_o = \gamma_e B_o$, Larmor frequency; $\Delta\omega_o = \omega_o - \omega$, resonance offset. Spin magnetization **M** precesses about ω_{eff}-direction in cone of half-angle θ_{eff}, at angular rate ω_{eff}.

From the geometry in Figure 13.2, we see that the magnetization is then inclined at a turning angle β to the z-axis, which is given by:

$$\cos\beta = \cos^2\theta_{eff} + \sin^2\theta_{eff}\cos\omega_{eff}t_p \tag{13.12}$$

This immediately gives us the z-magnetization $M_z = M_o\cos\beta$ that remains after the pulse. Correspondingly, the fraction of z-magnetization that has been reduced to zero is $1-\cos\beta$. The fraction of individual spins p_π that are flipped by the pulse is therefore:

$$p_\pi\left(+\tfrac{1}{2}\rightarrow-\tfrac{1}{2}\right) = \tfrac{1}{2}(1-\cos\beta) = \sin^2\theta_{eff}\sin^2\tfrac{1}{2}\omega_{eff}t_p \tag{13.13}$$

because each spin flipped changes its z-projection by $\Delta M_S = -1$. This result comes originally from Rabi (Rabi et al. 1938). It is important for us because it gives the excitation spectrum $p_\pi(\omega_{eff})$ when we invert the magnetization by using a nominally 180°-pulse.

The total transverse magnetization is given by: $\sqrt{M_y^2 + M_x^2} = M_o\sin\beta$. This is what we would get for the absolute-value spectrum, if we use quadrature phase detection. However, single-channel detection is more usual for nitroxide spin labels. Thus, we need expressions for the transverse magnetization components separately, because they are phase-shifted by 90° relative to one another (compare Eqs. 13.8 and 13.9). During a pulse along the x-axis, the magnetization precesses about the ω_{eff}-direction, which generates both x- and y-components of transverse magnetization (see Figure 13.2). Unlike the special situation on resonance where $\Delta\omega_o = 0$ and $\omega_{eff} \equiv \omega_1$, the full transverse component $M_o\sin\beta$ then no longer lies along the y-axis.

From Figure 13.2, we see that calculating the transverse components is best done by rotating the initial magnetization from the z-axis to the ω_{eff}-direction, rotating about this direction by the angle $\omega_{eff}t_p$ and then rotating back again. Starting with magnetization along the z-axis, the sequence for applying the rotation matrices is: $R^{-1}(\theta_{eff},y)R(\omega_{eff}t_p,z)R(\theta_{eff},y)$, where $R(\theta_{eff},y)$ rotates by angle θ_{eff} about the current y-axis, etc. (see Appendix I). This transformation yields all three magnetization components simultaneously (see Ernst et al. 1987). For a pulse along the x-axis, the y-component of the transverse magnetization is then given by:

$$M_y = -M_o\sin\theta_{eff}\sin\omega_{eff}t_p \tag{13.14}$$

Correspondingly, the x-component becomes:

$$M_x = M_o\sin\theta_{eff}\cos\theta_{eff}\left(1-\cos\omega_{eff}t_p\right) \tag{13.15}$$

which also can be derived from the expression for the total transverse magnetization, together with Eq. 13.14. The z-component agrees, of course, with Eq. 13.12.

Figure 13.3 shows the excitation profiles for M_y and M_x that result from a nominal 90°-pulse (i.e., for $\omega_1 t_p = \pi/2$). We recognize the 90°-phase shift between the two, as a function of offset frequency. Beneath them is the profile for the absolute magnitude of the transverse magnetization (solid line). We compare this with the absolute-value spectrum obtained from Fourier transforming a rectangular pulse of duration t_p (see Table T.1 in Appendix T):

$$P(\Delta\omega_o) = \left|\frac{\sin\left(\frac{1}{2}\Delta\omega_o t_p\right)}{\frac{1}{2}\Delta\omega_o}\right| \tag{13.16}$$

which is given by the dashed line. There are similarities, but the absolute magnitude profile remains constant over a much wider range of offsets than does the Fourier spectrum of the pulse shape. In angular frequency units, the half-width at half-height from the pulse shape is $\Delta\Delta\omega_{1/2} = 3.791/t_p$, as opposed to $\Delta\Delta\omega_{1/2} = 4.298/t_p$ for the absolute-magnitude excitation spectrum. The nth null point occurs close to $\Delta\omega_n = \pm 2\pi n/t_p$ for both profiles. The bottom panel of Figure 13.3 shows the profile for inversion of the z-magnetization by a nominally 180°–pulse ($\omega_1 t_p = \pi$). This has a width $\Delta\Delta\omega_{1/2} = 2.508/t_p$ intermediate between that for the total transverse magnetization and that of M_y ($\Delta\Delta\omega_{1/2} = 1.698/t_p$). The effects that finite pulse widths have on excitation profiles are treated in Section 13.15 and in Section 15.10 of Chapter 15.

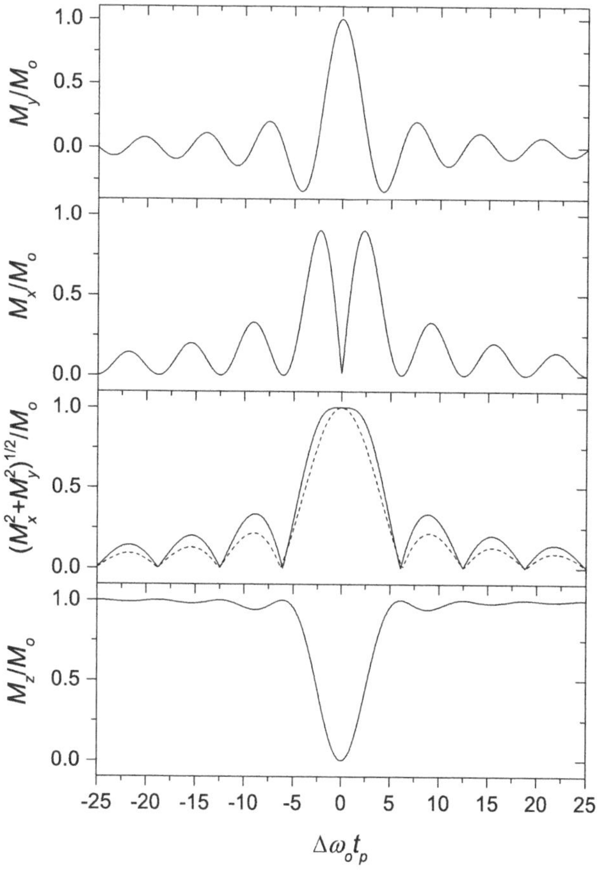

FIGURE 13.3 Excitation profiles for x-axis pulse with 90°-turning angle on resonance ($\omega_1 t_p = \pi/2$). *Top to bottom:* y-axis magnetization in rotating frame (Eq. 13.14); x-axis magnetization (Eq. 13.15); total transverse magnetization, $M_o\sin\beta$ with Eq. 13.12 (dashed line: Fourier transform of rectangular pulse, Eq. 13.16); z-axis magnetization $M_o\cos\beta$ (Eq. 13.12).

13.4 VECTOR MODEL FOR PRIMARY SPIN ECHO

The primary echo sequence is given by $(\pi/2)_x$–τ–$(\pi)_x$–τ–*echo* (see Figure 13.1), where both pulses are applied along the x-axis in the rotating frame. Figure 13.4 shows the behaviour of the macroscopic magnetization at various stages in a two-pulse primary echo sequence, as viewed in the rotating frame. The first pulse is directed along the x-axis and fulfills the condition $\omega_1 t_p = \pi/2$. It tips the magnetization from its equilibrium direction along the z-axis to along the $-y$-axis, as described in the previous section (see first panel in Figure 13.4). After the first pulse is switched off, the magnetization remains in the x, y-plane and precesses about the static $\mathbf{B}_o$-field. However, the magnetization does not necessarily remain fixed wholly along the $-y$-axis of the rotating frame during this period of free precession. Inhomogeneous spectral broadening corresponds to a spread of resonance frequencies (i.e., of Larmor precession frequencies ω_o), which causes the transverse magnetization progressively to fan out away from the $-y$-axis (see second panel in Figure 13.4).

The first period of free precession, or evolution, is terminated after time τ by the 180°-pulse in the primary echo sequence. This second pulse is also directed along the x-axis in the rotating frame and satisfies the condition $\omega_1 t_p = \pi$. It therefore reverses the direction of each individual y-component in the dephasing transverse magnetization, without affecting the x-components. Correspondingly, the sense in which the magnetization defocuses is also reversed, because the direction of precession of the individual spin packets remains unchanged by the pulse (see third panel in Figure 13.4). After a further evolution time τ, all the transverse magnetization components therefore refocus to align along the $+y$-axis. This resultant transverse magnetization is detected as the spin echo (see fourth panel in Figure 13.4).

Immediately before the 180°-pulse, the transverse magnetization is given by Eqs. 13.8 and 13.9 with $t = \tau$. Applying the 180°-pulse along x leaves $\hat{M}_x$ unchanged and reverses the sign of $\hat{M}_y$. Evolution of the transverse magnetization in the subsequent period of free precession is described by the Bloch equations, Eqs. 13.6 and 13.7, with initial conditions given by Eq. 13.8 and Eq. 13.9 after the change of sign. At time t after the 180°-pulse, the required solutions are:

$$\hat{M}_x = M_o \exp\left(-\frac{\tau + t}{T_2}\right) \sin \Delta\omega_o(\tau - t) \tag{13.17}$$

$$\hat{M}_y = M_o \exp\left(-\frac{\tau + t}{T_2}\right) \cos \Delta\omega_o(\tau - t) \tag{13.18}$$

Thus for $t = \tau$, we get $\hat{M}_x = 0$ and the refocused y-magnetization becomes:

$$\hat{M}_y(2\tau) = M_o \exp\left(-\frac{2\tau}{T_2}\right) \tag{13.19}$$

which is independent of $\Delta\omega_o$, as it must be for the echo signal. Thus, dephasing of the transverse magnetization is reversed exactly, and the primary echo decays with the intrinsic T_2-relaxation time, which corresponds to the homogeneous linewidth. All effects of inhomogeneous broadening are cancelled in the refocused echo.

Again, as for a single pulse, only a small range of frequencies within the nitroxide powder lineshape is excited by the spin-echo sequence. We obtain an echo-detected spectrum by recording the integrated echo intensity at fixed delay time τ, whilst slowly sweeping the static magnetic field B. Short τ-delays yield the absorption powder spectrum, whereas progressively longer τ-delays yield partially relaxed absorption spectra. These contain more information about the relaxation mechanism than do T_{2M}-measurements at a single magnetic field. Soft pulses with weak B_1-field and long t_p improve the spectral resolution but can complicate the relaxation analysis.

We also obtain spin echoes when the first pulse is not a 90°-pulse and the second pulse is not a 180°-pulse. A general echo sequence is β_1–τ–β_2–τ–*echo*, where β_1 and β_2 are the flip angles of the first and second pulses, respectively. We know the result for the first pulse from Eq. 13.4 in Section 13.2, if the resonance condition $\omega = \omega_o$ is met. Then the y-magnetization is scaled by the factor $\sin\beta_1$ (see also Eq. P.10 in Appendix P). For the second pulse, Eq. 13.13 in Section 13.3 gives us the fraction of magnetization that is reversed in direction by the action of a nominally 180°-pulse. Only this fraction gives rise to an echo. When we are on resonance, $\sin\theta_{eff} = 1$ and $\omega_{eff} = \omega_1$, and the echo amplitude is therefore scaled by the factor $\sin^2\beta_2/2$. Thus, the

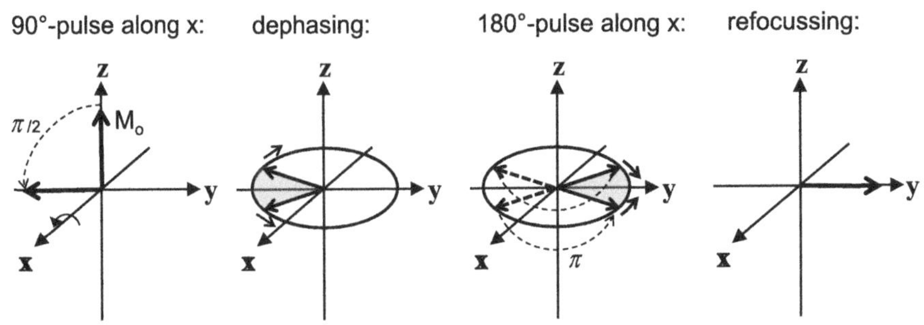

FIGURE 13.4 Spin magnetization vector in $\pi/2$–τ–π–τ–*echo* spin-echo pulse sequence, viewed in axis-frame rotating at Larmor frequency about static magnetic field z-direction. *From left to right*: $\pi/2$-pulse; free-precession period τ; π-pulse; second free-precession period τ.

echo amplitude for general flip angles of both pulses is multiplied by the factor $\sin\beta_1 \times \sin^2(\beta_2/2)$, which has a maximum value of unity when $\beta_1 = \pi/2$ and $\beta_2 = \pi$.

The above considerations are for narrow lines. For broad lines, we must combine the results of Section 13.3 for the excitation profiles of the two pulses. These are centred about a particular angular offset frequency $\Delta\omega_o$ within the nitroxide EPR powder spectrum. From Eqs. 13.13–13.15, the echo signal for the sequence $\beta_1 t_{p1}-\tau-\beta_2 t_{p2}-(\tau+t)$ becomes (Bloom 1955):

$$E(2\tau) = \sin^3\theta_{eff}\sin\omega_{eff}t_{p1}\sin^2\frac{\omega_{eff}t_{p2}}{2}\cos\Delta\omega_o t$$
$$-\sin 2\theta_{eff}\sin^2\frac{\omega_{eff}t_{p1}}{2}\sin^2\frac{\omega_{eff}t_{p2}}{2}\sin\Delta\omega_o t \qquad (13.20)$$

This result is obtained by straightforward multiplication of the two profiles and taking into account the phase relation of the M_y and M_x components.

13.5 PRODUCT SPIN-OPERATOR METHOD FOR PRIMARY SPIN ECHO

It is instructive to complement the classical approach of the previous section with a quantum-mechanical treatment by the product spin-operator method described in Appendix P. The ideal primary-echo sequence is dealt with already as an example in Section P.2 of that Appendix. We extend this here by considering a general second pulse, whilst still assuming that the first pulse is a 90°-pulse: $\beta_1 = \pi/2$. Evolution of magnetization in the rotating frame is represented solely by spin operators. Before the first pulse, the equilibrium magnetization is S_z, which is rotated to the negative y-axis by a $\pi/2$-pulse along the x-axis. Under subsequent free precession by an angle $\Delta\omega_o\tau$ about the z-axis, the resulting y-magnetization transforms according to Eq. P.6 in Appendix P:

$$S_z \xrightarrow{\pi/2 S_x} -S_y \xrightarrow{\Delta\omega_o\tau S_z} S_x\sin\Delta\omega_o\tau - S_y\cos\Delta\omega_o\tau \qquad (13.21)$$

where $\Delta\omega_o$ is the offset in angular resonance frequency. Rotation by the β_2-pulse along the x-axis then transforms this further according to Eq. P.4:

$$S_x\sin\Delta\omega_o\tau - S_y\cos\Delta\omega_o\tau \xrightarrow{\beta_2 S_x}$$
$$S_x\sin\Delta\omega_o\tau - \left(S_y\cos\beta_2 - S_z\sin\beta_2\right)\cos\Delta\omega_o\tau \qquad (13.22)$$

During the final period of free precession, only evolution of the y-magnetization interests us, because this produces the echo. Rotation by an angle $\Delta\omega_o t$ about the z-axis then gives (see Eq. P.6):

$$S_x\sin\Delta\omega_o\tau - S_y\cos\beta_2\cos\Delta\omega_o\tau \xrightarrow{\Delta\omega_o t S_z}$$
$$S_y\left(\sin\Delta\omega_o\tau\sin\Delta\omega_o t - \cos\beta_2\cos\Delta\omega_o\tau\cos\Delta\omega_o t\right) \qquad (13.23)$$

For the echo, we need the terms containing both t and τ that become independent of $\Delta\omega_o$ on refocusing at $t=\tau$. Rewriting terms containing $\Delta\omega_o$ in Eq. 13.23, the average value of this part of the y-magnetization is:

$$\overline{\langle S_y(\tau,t)\rangle} = \tfrac{1}{2}\left(\left[\cos\Delta\omega_o(\tau-t) - \cos\Delta\omega_o(\tau+t)\right]\right.$$
$$\left.-\cos\beta_2\left[\cos\Delta\omega_o(\tau-t) + \cos\Delta\omega_o(\tau+t)\right]\right) \qquad (13.24)$$

The refocused echo signal at $t=\tau$ therefore becomes:

$$\overline{\langle S_y(2\tau)\rangle} = \tfrac{1}{2}\left(1-\cos\beta_2\right) = \sin^2\frac{\beta_2}{2} \qquad (13.25)$$

where we discard the terms from Eq. 13.24 that contain $\Delta\omega_o 2\tau$ because they correspond to the remanent FID at time 2τ after the first pulse. Equation 13.25 is the echo amplitude for a general flip angle of the second pulse and agrees with the vector-model result obtained in the previous section.

13.6 DENSITY MATRIX, PULSES AND TRANSVERSE RELAXATION

We can best treat relaxation mechanisms in pulse-EPR by using the density-matrix formalism (see Appendix N). The transverse magnetization that generates the spin echo is proportional to the ensemble expectation value $\overline{\langle S_+\rangle}$, which is determined by the off-diagonal elements of the spin density-matrix (Eq. N.8 in Appendix N):

$$\overline{\langle S_+\rangle} = \sigma_{-+} \qquad (13.26)$$

just as in CW-EPR – see also Section 6.6 in Chapter 6. In turn, the time evolution of the spin density-matrix, under the influence of spin Hamiltonian $\mathcal{H}$, is given by solution of the quantum-mechanical Liouville equation:

$$\sigma(t) = \exp\left(-i\mathcal{H}t/\hbar\right)\sigma(0)\exp\left(i\mathcal{H}t/\hbar\right) \qquad (13.27)$$

where $\sigma(0)$ is the initial value of the spin density-matrix – see Eq. N.19 in Appendix N.

If the microwave pulses are sufficiently intense, the Hamiltonian in the rotating frame is dominated by the resonant microwave field B_1 that is directed along the x-axis: $\mathcal{H} \approx g\beta_e B_1 S_x \equiv \hbar\omega_1 S_x$, which remains constant during the pulse. Equation 13.27 then corresponds to rotation of $\sigma(0)$ through an angle $\beta_p = \omega_1 t_p$ about the x-axis, where t_p is the pulse duration. We see this readily from Appendix R.1, where the operator for rotation by angle β_p about the x-axis is given by $R(\beta_p,x)=\exp(-i\beta_p S_x)$, together with Eq. B.25 of Appendix B that gives the unitary transformation of $\sigma(0)$ by rotation R as: $\sigma'(0)=R\sigma(0)R^{-1}$. If the spin-echo pulses with $\beta_p=\pi/2$ and π are sufficiently short and intense, we can assume that their only effect is to rotate the magnetization.

We therefore concentrate on free precession in the periods when the microwaves are switched off (see Mims 1968). The spin density-matrix then evolves under the influence of the time-independent Hamiltonian $\mathcal{H}_o$ in Eq. 13.27. We retain only the secular terms in the Hamiltonian (i.e., $\mathcal{H}_o \approx \hbar\omega_o S_z$) with eigenfunctions $|M_S=\pm\frac{1}{2}\rangle$, because the Zeeman interaction is by far the largest contribution. In this representation, the time evolution of the off-diagonal element of the spin density-matrix is given by Eq. N.20 of Appendix N:

$$\sigma_{-+}(t)=\exp(-i\omega_o t)\sigma_{-+}(0) \tag{13.28}$$

where $\omega_o\ \left(\equiv(E_{+1/2}-E_{-1/2})/\hbar\right)$ is the angular resonance frequency. This describes the evolution after the first pulse, in the time interval from $t=0$ to $t=\tau$. The second pulse reverses the phase of the transverse magnetization, which transforms $\sigma_{-+}(\tau)$ into its complex conjugate $\sigma^*_{-+}(\tau)$, and the subsequent evolution is given by:

$$\begin{aligned}\sigma_{-+}(t)&=\exp(-i\omega_o(t-\tau))\exp(i\omega_o\tau)\sigma^*_{-+}(0)\\&=\exp(-i\omega_o(t-2\tau))\sigma^*_{-+}(0)\end{aligned} \tag{13.29}$$

Thus, we get the echo signal, $\sigma^*_{-+}(0)$, at time $t=2\tau$. So far, however, this analysis neglects transverse relaxation that occurs during the time of the echo sequence.

Because of molecular motion (or flips of other adjacent spins), the angular resonance frequency ω_o does not remain constant but instead fluctuates by amounts $\Delta\omega_o(t')$ during the periods of free precession. This causes loss of phase coherence for the precessing spins. In Eq. 13.28, we then replace ω_o by $\omega_o+\Delta\omega_o(t')$ and must integrate over the time dependence of the $\omega_o t$ terms. The appropriate integral is: $\int_0^\tau \Delta\omega_o(t)\,\mathrm{d}t-\int_\tau^{2\tau}\Delta\omega_o(t)\,\mathrm{d}t\equiv\int_0^{2\tau}s(t)\Delta\omega_o(t)\,\mathrm{d}t$, where $s(t)$ defines a switching function that brings about phase reversal by the second pulse at $t=\tau$: $s(t)=1$ when $t<\tau$ and $s(t)=-1$ when $t>\tau$. Additionally, we must take a time (or ensemble) average over the fluctuations $\Delta\omega_o(t')$. From Eqs. 13.28 and 13.29, the required spin density-matrix element then becomes:

$$\sigma_{-+}(2\tau)=\sigma^*_{-+}(0)\left\langle\exp\left(-i\int_0^{2\tau}s(t)\Delta\omega_o(t)\cdot\mathrm{d}t\right)\right\rangle_t \tag{13.30}$$

If we assume that the fluctuation in resonance frequency is a Gaussian random function, the integral $X(\tau)$ in Eq. 13.30 is distributed according to a normal probability curve. The average in Eq. 13.30 is then given by (see Table T.1 in Appendix T):

$$\begin{aligned}\langle\exp(-iX)\rangle&=\frac{1}{\sqrt{2\pi\langle X^2\rangle}}\int_{-\infty}^{\infty}\exp\left(\frac{-X^2}{2\langle X^2\rangle}\right)\exp(-iX)\,\mathrm{d}X\\&=\exp\left(-\tfrac{1}{2}\langle X^2\rangle\right)\end{aligned} \tag{13.31}$$

where the second moment is given by: $\langle X^2\rangle=\int\int s(t')s(t)\langle\Delta\omega_o(t')\Delta\omega_o(t)\rangle\mathrm{d}t\mathrm{d}t'$ and the autocorrelation function $\langle\Delta\omega_o(t')\Delta\omega_o(t)\rangle$ is non-vanishing (Anderson 1954). The resulting dependence of the off-diagonal density matrix element on pulse separation τ is therefore:

$$\sigma_{-+}(2\tau)=\sigma^*_{-+}(0)\exp\left(-\frac{1}{2}\int_0^{2\tau}\mathrm{d}t'\int_0^{t'}s(t')s(t)\langle\Delta\omega_o(t')\Delta\omega_o(t)\rangle\mathrm{d}t\right) \tag{13.32}$$

Equation 13.32 describes the dependence of the echo decay on the fluctuating local fields experienced by the electron spins. It corresponds to the Gauss–Markov model of Mims (Mims 1968).

13.7 PRIMARY-ECHO DECAY AND ROTATIONAL DYNAMICS (PHASE-MEMORY TIME)

The echo-decay envelope, $E(2\tau)$, depends on fluctuations (or a distribution) in the local fields that are experienced by the spins under observation. We express the resulting fluctuations about the mean resonance frequency ω_o by the autocorrelation function, $\langle\Delta\omega_o(t')\Delta\omega_o(t)\rangle$, and the dependence of the echo decay envelope on τ becomes (Klauder and Anderson 1962; Mims 1968):

$$E(2\tau)=\exp\left(-\frac{1}{2}\int_0^{2\tau}\mathrm{d}t'\int_0^{t'}s(t')s(t)\langle\Delta\omega_o(t')\Delta\omega_o(t)\rangle\mathrm{d}t\right) \tag{13.33}$$

which derives directly from Eq. 13.32 for the transverse magnetization. Assuming a Gauss–Markov process implies exponential correlation functions, as for Brownian rotational diffusion – see Section 7.5:

$$\langle\Delta\omega_o(t')\Delta\omega_o(t)\rangle=\langle\Delta\omega_o(0)^2\rangle\exp\left(-(t'-t)/\tau_c\right) \tag{13.34}$$

where τ_c is the correlation time and $\langle\Delta\omega_o^2\rangle\equiv\langle\Delta\omega_o(0)^2\rangle$ is the second moment of the fluctuations in angular resonance frequency. Expanding the integrals in Eq. 13.33 according to the $s(t')s(t)$ product of switching functions, we get:

$$\int_0^{2\tau} dt' \int_0^{t'} s(t')s(t)G(t',t)\cdot dt = \int_0^{\tau} dt' \int_0^{t'} G(t',t)\cdot dt - \int_{\tau}^{2\tau} dt' \int_0^{\tau} G(t',t)\cdot dt$$
$$+ \int_{\tau}^{2\tau} dt' \int_{\tau}^{t'} G(t',t)\cdot dt \qquad (13.35)$$

where $G(t',t) \equiv \langle \Delta\omega_o(t')\Delta\omega_o(t)\rangle$. From Eqs. 13.34 and 13.35, the echo envelope then becomes (see also Klauder and Anderson 1962; Mims 1968; Brown 1974):

$$E(2\tau) = \exp\left(-\langle \Delta\omega_o^2\rangle \tau_c^2 C(\tau/\tau_c)\right) \qquad (13.36)$$

where

$$C(t) \equiv 2t - 3 + 4e^{-t} - e^{-2t} \qquad (13.37)$$

Figure 13.5 shows the dependence of this function on the interpulse spacing, τ. The limiting dependence for fast motion (long echo delays) is:

$$C(\tau >> \tau_c) = 2\tau/\tau_c - 3 \qquad (13.38)$$

and for slow motion (short echo delays) is:

$$C(\tau << \tau_c) = \tfrac{2}{3}(\tau/\tau_c)^3 \qquad (13.39)$$

as we see in Figure 13.5.

For fast motion, i.e., short correlation times: $\tau_c << \tau$, the echo decay is therefore mono-exponential (see Eqs. 13.36 and 13.38):

$$E(2\tau) = E(0)\exp(-2\tau/T_{2M}) \qquad (13.40)$$

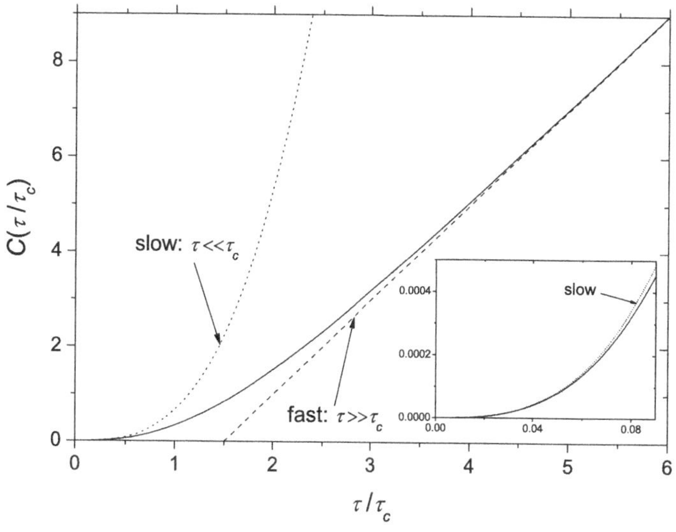

FIGURE 13.5 Dependence of decay function $C(\tau/\tau_c)$ (Eq. 13.37) on interpulse spacing τ in primary spin-echo sequence; correlation time τ_c. Limiting dependences: $\tau >> \tau_c$ (dashed line, Eq. 13.38); $\tau << \tau_c$ (dotted line, Eq. 13.39). Echo decay envelope $E(2\tau)$ given by Eq. 13.36: $\ln(1/E(2\tau)) = \langle \Delta\omega_o^2\rangle \tau_c^2 C(\tau/\tau_c)$. *Inset*: expanded version of the slow-motion regime.

where the phase-memory time T_{2M} is inversely proportional to the correlation time:

$$\frac{1}{T_{2M}} = \langle \Delta\omega_o^2\rangle \tau_c \qquad (13.41)$$

which we readily recognize as the secular contribution to the T_2-relaxation rate (see Eq. 5.49 in Chapter 5). We give an experimental illustration in Figure 13.6. In the motional-narrowing regime, the echo decay therefore becomes slower for faster motion (at constant $\langle \Delta\omega_o^2\rangle$). On the other hand for slow motion, i.e., long correlation times: $\tau_c >> \tau$, the echo decay is non-exponential, which

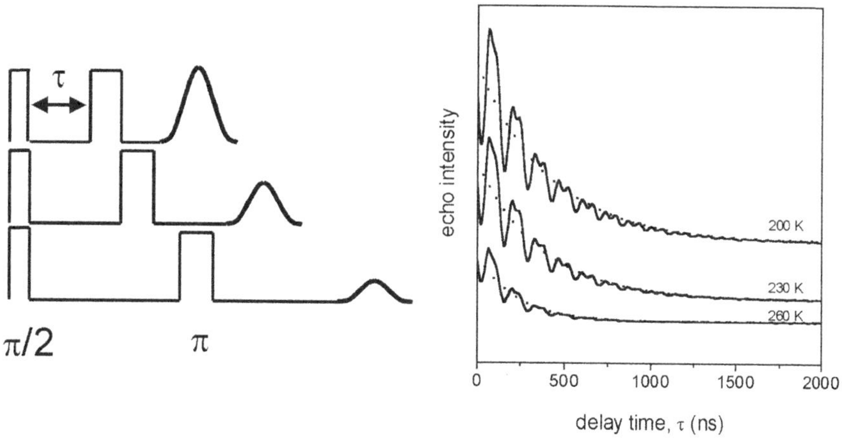

FIGURE 13.6 *Left*: two-pulse spin echo sequence with increasing echo-delay time τ. *Right*: echo-decay curves for spin-labelled lipid (14-DOXYL-PC, see Figure 1.1) in dipalmitoyl phosphatidylcholine membranes at temperatures indicated. Dotted lines: exponential fits (superimposed on modulation from proton hyperfine interactions), giving phase-memory time T_{2M} (Bartucci et al. 2003).

we can loosely interpret as a distribution of phase-memory times (cf. Guzzi et al. 2014). Additionally, the echo decay rate decreases for slower motion, when in the slow-motion regime.

13.8 ROTATIONAL DIFFUSION AND PRIMARY-ECHO DECAY

We dealt with Brownian rotational diffusion in Section 7.5 of Chapter 7. The relevant correlation functions are exponential, and the rotational correlation time is given by $\tau_R = 1/(6D_R)$ for isotropic diffusion, where D_R is the rotational diffusion coefficient. Thus, the general results that we obtained in the previous section apply specifically to diffusive rotational motion. As for the analysis of electron spin relaxation in Chapter 5, the principal contributions to fluctuations of the local field come from the angular anisotropy of the nitrogen hyperfine tensor and of the nitroxide g-tensor. This is modulated by rotational diffusion.

The hyperfine tensor determines the local field at the electron spin that arises from the nitrogen nuclear moment. For an axially symmetric tensor, expressed in angular frequency units, the anisotropy in angular resonance frequency $\Delta\omega_o(\theta)$ is given from Eq. 5.15 in Chapter 5 by:

$$\Delta\omega_o(\theta) = \tfrac{1}{3}(A_{\parallel} - A_{\perp})(3\cos^2\theta - 1)m_I \tag{13.42}$$

where θ is the angle of the static magnetic field B_o to the principal (||)-axis of the hyperfine tensor, and m_I is the nuclear magnetic quantum number. A similar expression holds for the contribution to the local field from the g-value anisotropy:

$$\Delta\omega_o(\theta) = \tfrac{1}{3}(g_{\parallel} - g_{\perp})\beta_e B_o/\hbar \times (3\cos^2\theta - 1) \tag{13.43}$$

– see Eq. 5.15. Note that, for simplicity, we retain only secular terms in the time-dependent spin Hamiltonian.

Consider first fast rotational diffusion (i.e., $\tau_R << \tau$), where there are many small reorientations in time 2τ and we can replace the time average by a spatial average. The second moment of the fluctuations in angular resonance frequency comes from averaging $\Delta\omega_o(\theta)^2$ over the angle θ, according to Eq. 5.16. The phase-memory time for the m_I-hyperfine manifold is then given from Eq. 13.41 by:

$$\frac{1}{T_{2M}} = \langle \Delta\omega_o^2 \rangle \tau_R$$
$$= \frac{4}{45}\left((\Delta A)^2 m_I^2 + 2\Delta A \Delta g \frac{\beta_e}{\hbar} B_o m_I + \left(\frac{\Delta g \beta_e}{\hbar} \right)^2 B_o^2 \right)\tau_R \tag{13.44}$$

where $\Delta A = A_{\parallel} - A_{\perp}$ and $\Delta g = g_{\parallel} - g_{\perp}$. As anticipated, Eq. 13.44 bears a direct relation to the well-known dependence of nitroxide linewidths (or T_2-relaxation rates) on nitrogen hyperfine state, for fast motion. Comparison with Eqs. 5.63–5.65 for the motional narrowing regime in Chapter 5 shows that the phase-memory time is identical to the intrinsic T_2-relaxation time, at the same level of approximation.

On the other hand, for slow rotational diffusion (i.e., $\tau_R >> \tau$), there are only a few small orientational steps during the echo delay time. We can then replace the time average $\langle \Delta\omega_o^2 \rangle$ by the instantaneous value $\Delta\omega_o(\theta)^2$ from Eqs. 13.42 and 13.43, i.e., the adiabatic approximation. For non-selective pulses, we must integrate the angular-dependent expression for $E(2\tau)$ over all θ for each value of τ (Brown 1974). From Eqs. 13.36 and 13.39, the normalized echo envelope therefore becomes:

$$E(2\tau) = \int_0^{\pi/2} \exp\left[-\frac{2}{27}\left(\Delta A \cdot m_I + \Delta g \frac{\beta_e}{\hbar} B_o \right)^2 \frac{\tau^3}{\tau_R} (3\cos^2\theta - 1)^2 \right] \times \sin\theta \cdot d\theta \tag{13.45}$$

Figure 13.7 shows the results of numerical integration of Eq. 13.45 for various values of the rotational correlation time τ_R. Decay profiles are given for the central hyperfine manifold, $\Delta m_I = 0$, of a ^{14}N-nitroxide at 9 GHz. Non-selective excitation is practicable only for this manifold; excitation within the $\Delta m_I = \pm 1$ manifolds depends on field position and is determined by convolution of the excitation spectrum with the powder lineshape. Decay rates deduced from Eq. 13.45 are very rapid, unless τ_R is very long, i.e., approaching the microsecond regime. As seen from Figure 13.7, the echo decay is non-exponential for slow motion or short delay times. For very short τ, expansion of the exponential in Eq. 13.45 shows that the initial decay has the form $1 - \text{const.} \times \tau^3$, which results from Eq. 13.39. For longer delay times, we can approximate the echo decay by multiple exponentials or a stretched exponential. As for CW-EPR, we obtain a

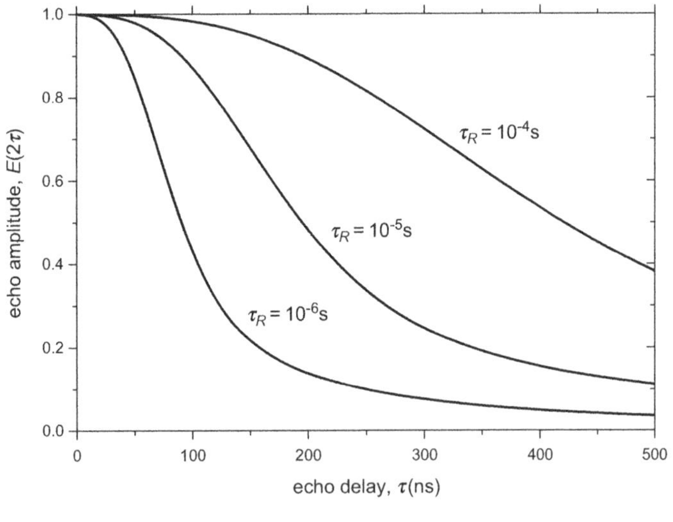

FIGURE 13.7 Dependence of two-pulse echo amplitude $E(2\tau)$ on interpulse spacing τ (Eq. 13.45), for rotational correlation times $\tau_R >> \tau$ in slow-motion regime, with complete excitation of $m_I = 0$ ^{14}N-hyperfine manifold. Axial symmetry assumed with g-tensor anisotropy: $g_{\parallel} - g_{\perp} = -0.0053$, and $B_o = 0.335$ T.

more detailed analysis of spin-echo decay in the slow-motion regime by using the stochastic Liouville equation (cf. Sections 6.7–6.10 in Chapter 6).

13.9 STOCHASTIC-LIOUVILLE SIMULATIONS OF PRIMARY-ECHO DECAY

Slow-motion simulations require inclusion of the pseudo-secular hyperfine contributions (see, e.g., Eq. 6.77 in Section 6.9). We ignored these in the previous sections, which simply used a scalar angular resonance frequency. A general approach using the stochastic Liouville equation (Schwartz et al. 1982) closely parallels slow-motion simulations of CW-EPR lineshapes (see Sections 6.7–6.10 in Chapter 6), if we assume short intense pulses whose only effect is to rotate the magnetization.

Again, we are interested in the time evolution of the stochastic Liouville equation in the periods of free precession, when the microwaves are switched off. Our methods are those of Chapter 6, except that there we needed the frequency-dependent solution under steady-state continuous irradiation. Here we want the time-dependent solution, at fixed frequency. The time-independent spin Hamiltonian is (cf. Eqs. 6.55 and 6.56):

$$\mathcal{H} = \mathcal{H}_o + \mathcal{H}_1(\Omega) = \hbar\omega_o S_z + \mathcal{H}_1(\Omega) \tag{13.46}$$

where Ω represents the angular dependence, and ω_o is the Larmor frequency. The second equality uses the high-field approximation. The time dependence of the reduced spin density-matrix $\chi(\Omega,t)$ during periods of free precession is given by (see Eq. 6.58):

$$\frac{\partial \chi(\Omega,t)}{\partial t} + i\omega_o[S_z, \chi(\Omega,t)] + \frac{i}{\hbar}[\mathcal{H}_1(\Omega), \chi(\Omega,t)] + \Gamma_\Omega \chi(\Omega,t) = 0 \tag{13.47}$$

where Γ_Ω is the Markov operator that describes rotational diffusion. Taking the $\langle M_S = -\tfrac{1}{2}, m_I' | \ldots | M_S = +\tfrac{1}{2}, m_I \rangle$ matrix elements for the allowed and nuclear-forbidden transitions then yields (cf. Eq. 6.60):

$$\frac{\partial \chi_{-+}}{\partial t} = i\omega_o \chi_{-+} - \frac{i}{\hbar}[\mathcal{H}_1(\Omega), \chi(\Omega,t)]_{-+} - (\Gamma_\Omega \chi(\Omega,t))_{-+} \tag{13.48}$$

where $\chi_{-+} \equiv \langle -\tfrac{1}{2}, m_I' | \chi(\Omega,t) | +\tfrac{1}{2}, m_I \rangle$ are the off-diagonal matrix elements needed to obtain the echo intensity (cf. Eq. 6.64). To obtain ensemble averages over the spin-label orientations Ω, we expand the χ_{-+} in a complete orthogonal set of eigenfunctions, $G_{m'}(\Omega)$, of the rotational diffusion operator Γ_Ω (cf. Eq. 6.63):

$$\chi_{-+}(\Omega,t) = \sum_{m'} (C_{m'}(t))_{-+} G_{m'}(\Omega) \tag{13.49}$$

where the matrix of expansion coefficients, $C_{m'}$, is an operator in spin space that is time dependent but independent of the orientation Ω. In terms of ensemble averages, Eq. 13.48 then becomes (cf. Eq. 6.66):

$$\frac{\partial \bar{\chi}_{-+}}{\partial t} = i\omega_o \bar{\chi}_{-+} - \frac{i}{\hbar}\sum_{m'} \int [\mathcal{H}_1(\Omega), C_{m'}]_{-+} G_{m'}(\Omega) p_o(\Omega) \cdot \mathrm{d}\Omega \tag{13.50}$$

where $\bar{\chi}_{-+} \equiv \langle -\tfrac{1}{2}, m_I' | \bar{\chi}(t) | +\tfrac{1}{2}, m_I \rangle$ is the average over Ω and $p_o(\Omega)$ is the normalized distribution of spin-label orientations. Finally, pre-multiplying by $G_m^*(\Omega)$ and integrating over Ω to exploit the orthogonality of the eigenfunctions, the time dependence of the $C_m(t)$ coefficients becomes (cf. Eq. 6.67):

$$\frac{\partial (C_m)_{-+}}{\partial t} = -(\tau_m^{-1} - i\omega_o)(C_m)_{-+} - \frac{i}{\hbar}\sum_{m'} \int G_m^*(\Omega)[\mathcal{H}_1(\Omega), C_{m'}]_{-+} G_{m'}(\Omega) \cdot \mathrm{d}\Omega \tag{13.51}$$

where τ_m^{-1} are the eigenvalues of the rotational diffusion equation.

Eq. 13.51 is an infinite series of coupled equations, which we solve by truncating at some upper value of $m' \equiv m_{\max}$, as described in Sections 6.8 and 6.10 of Chapter 6. Expressed in matrix form (cf. Section 6.10), Eq. 13.51 becomes (Schwartz et al. 1982):

$$\frac{\partial \mathbf{C}(t)}{\partial t} = -\mathbf{B} \cdot \mathbf{C}(t) \tag{13.52}$$

where $\mathbf{C}(t)$ is a column vector containing the values $(C_m)_{-+}$ as elements, and $\mathbf{B}$ is a square matrix with diagonal elements $\tau_m^{-1} - i\omega_o$ and off-diagonal elements $(i/\hbar)\int G_m^*(\Omega)[\mathcal{H}_1(\Omega), C_{m'}]_{-+} G_{m'}(\Omega) \cdot \mathrm{d}\Omega$. Values for specific anisotropic spin Hamiltonians, $\mathcal{H}_1(\Omega)$, are given in Sections 6.8–6.10. Eq. 13.52 is the time-dependent analogue of the frequency-dependent Eq. 6.110 for the spectral lineshape in Section 6.10. A formal solution to Eq. 13.52 is given by:

$$\mathbf{C}(t) = \exp(-\mathbf{B}t) \cdot \mathbf{C}(0) \tag{13.53}$$

where $\mathbf{C}(0)$ is the value at $t = 0$, immediately after the first (i.e., 90°) pulse. At the end of the free precession period just before the second pulse, we therefore have: $\mathbf{C}(\tau) = \exp(-\mathbf{B}\tau)\mathbf{C}(0)$. The second (i.e., 180°) pulse transforms $\mathbf{C}(t)$ to the complex conjugate $\mathbf{C}^*(t)$. Therefore, at the end of the 90°–τ–180°–τ sequence, we get:

$$\mathbf{C}(2\tau) = \exp(-\mathbf{B}\tau)\exp(-\mathbf{B}^*\tau)\mathbf{C}^*(0) \tag{13.54}$$

The echo envelope is then given by (Schwartz et al. 1982):

$$E(2\tau) = \mathrm{Re}\left[\mathbf{C}^{\mathrm{T}}(0)\exp(-\mathbf{B}\tau)\exp(-\mathbf{B}^*\tau)\mathbf{C}^*(0)\right] \tag{13.55}$$

where superscript T indicates the transpose: $[C_{ij}]^{\mathrm{T}} = [C_{ji}]$. To evaluate Eq. 13.55, we diagonalize $\mathbf{B}$ with the orthogonal transformation (see Appendix B.4):

$$\mathbf{O}^{-1}\mathbf{B}\mathbf{O} = \mathbf{\Lambda} \tag{13.56}$$

where $\boldsymbol{\Lambda}$ is a diagonal matrix of eigenvalues Λ_j, and the transformation matrix $\mathbf{O}$ is orthogonal: $\mathbf{O}^{-1} = \mathbf{O}^{\mathrm{T}}$. The echo envelope then becomes:

$$E(2\tau) = \mathrm{Re}\left[\mathbf{C}^{\mathrm{T}}(0)\mathbf{O}\exp(-\boldsymbol{\Lambda}\tau)\mathbf{O}^{\mathrm{T}}\mathbf{O}^{*}\exp\left(-\boldsymbol{\Lambda}^{*}\tau\right)\mathbf{O}^{\mathrm{T}*}\mathbf{C}^{*}(0)\right]$$
$$\equiv \mathrm{Re}\sum_{i,j,k,l,m} C(0)_k O_{kl}O_{il}O_{ij}^{*}O_{mj}^{*}C^{*}(0)_m \exp\left(-\left(\Lambda_l + \Lambda_j^{*}\right)\tau\right) \quad (13.57)$$

where the identity on the right is the expansion in matrix elements O_{kl}, et seq.

In general, therefore, the echo decay is not a single exponential but a sum of exponentials:

$$E(2\tau) = \mathrm{Re}\sum_{j,l} a_{jl}\exp\left(-\left(\Lambda_l + \Lambda_j^{*}\right)\tau\right) \quad (13.58)$$

This result agrees with the simpler analysis given in Section 13.7 that predicts a stretched exponential decay for slow motions, which is characteristic of a distribution of single exponentials.

For a single-line spectrum, the C-vector in Eq. 13.57 has just one non-zero matrix element: $C(0)_k \propto i\delta_{k,1}$, immediately after the first pulse. For a ^{14}N-nitroxide, which has three $\left(m_I = 0, \pm 1\right)$ hyperfine lines, $\mathbf{C}(0)$ is again purely imaginary but with three equal non-zero matrix elements. The expansion coefficients in Eq. 13.58 are then given by:

$$a_{j,l} = \sum_{k=1}^{3}\sum_{i} O_{kl}O_{kj}^{*}O_{il}O_{ij}^{*} \quad (13.59)$$

where the first summation is over the three hyperfine states. Also, we can restrict the summation in Eq. 13.58 to those hyperfine components that are strongly irradiated by the pulses. This means that we retain only those eigenvalues Λ_l that satisfy the condition: $\left|\omega_o - \mathrm{Im}\,\Lambda_l\right| \le \gamma_e B_1$, where B_1 is the magnetic-field strength of the microwave pulse.

13.10 SLOW-MOTION SIMULATIONS OF PRIMARY-ECHO DECAY

Stochastic-Liouville simulations of the echo envelopes, as described in the previous section, reveal the expected departures from single-exponential decay for long rotational correlation times (Schwartz et al. 1982). At short echo delays, an $\exp\left(-c\tau^3\right)$ dependence is found, in accordance with the predictions of Eq. 13.39 together with Eq. 13.36. With increasing delay times, this gradually goes over to a single-exponential tail at long delays. Qualitatively, this is in agreement with the predictions of the simple model embodied in Eqs. 13.36–13.39. The exponential tail of the slow-motion simulations, at long echo delays, τ:

$$E(2\tau) \propto \exp\left(-2\tau/T_{2M}^{\infty}\right) \quad (13.60)$$

corresponds to the smallest eigenvalue, Λ_1, of the $\mathbf{B}$ matrix in Eq. 13.58, i.e., $T_{2M}^{\infty} = 1/\mathrm{Re}\,\Lambda_1$. In practice, the asymptotic phase-memory time T_{2M}^{∞} corresponds experimentally to obtaining a single-exponential fit by excluding initial points that deviate from the fit. This procedure selects out the slowest-decaying contribution to the overall slow-motion echo envelope. Instrumental limitations, in any case, make it difficult to determine the whole of the initial echo decay.

Results of slow-motion simulations for a ^{14}N-nitroxide at 9-GHz EPR frequency are shown in Figure 13.8. Simulations are for isotropic Brownian rotational diffusion, for collision-limited free diffusion and for jump diffusion. We discussed these three slow-motional models in Section 7.12 of Chapter 7. For Brownian rotational diffusion, the isotropic rotational correlation time is given by: $\tau_R = 1/(6D_R)$, where D_R is the rotational diffusion coefficient. For collision-limited free diffusion: $\tau_R = \sqrt{1+6D_R\tau_J}/(6D_R)$, where τ_J is the time of inertial diffusion between instantaneous reorientations and we assume that $D_R\tau_J = 1$, i.e., $\tau_R = \sqrt{7}/(6D_R)$. For jump diffusion: $\tau_R = (1+6D_R\tau_l)/(6D_R)$, where τ_l is the lifetime in a fixed orientation before jumping instantaneously to a new orientation and we assume that $D_R\tau_l = 1$, i.e., $\tau_R = 7/(6D_R)$. The asymptotic phase-memory time T_{2M}^{∞} that we get from the long-time tails of simulated echo decays, is given as a function of τ_R in the log–log plot of Figure 13.8. We find a power-law dependence for each of the three slow-motional models. This is summarized by the following empirical calibrations (Schwartz et al. 1982):

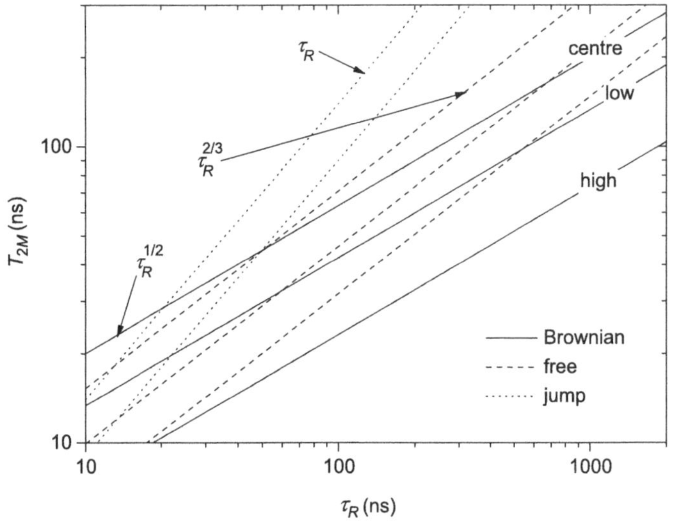

FIGURE 13.8 Slow-motion calibrations of echo phase-memory time T_{2M}, for ^{14}N-nitroxide undergoing Brownian diffusion (solid lines), free diffusion (dashed lines) or jump diffusion (dotted lines), with effective isotropic rotational correlation time τ_R. *Highest to lowest lines* (for each diffusion model): echo decay recorded at central maximum, low- and high-field extrema (as for Brownian diffusion). Calibrations (Eqs. 13.61–13.63) from stochastic-Liouville simulations (Schwartz et al. 1982) with axial hyperfine and g-tensor anisotropies: $\Delta A = 4.96\times10^{8}$ rad s^{-1} (2.80 mT) and $\Delta g = -0.0053$; and $B_o = 0.335$ T.

$$T_{2M}^{\infty} = C_1\left(\left|\Delta\mathcal{H}_1\right|\tau_R\right)^{-1/2} \times \tau_R, \text{ for Brownian diffusion} \quad (13.61)$$

$$T_{2M}^{\infty} = C_2\left(\left|\Delta\mathcal{H}_1\right|\tau_R\right)^{-1/3} \times \tau_R, \text{ for free diffusion} \quad (13.62)$$

$$T_{2M}^{\infty} = C_3 \times \tau_R, \text{ for jump diffusion} \quad (13.63)$$

where C_1, C_2, C_3 are dimensionless calibration constants, and $\left|\Delta\mathcal{H}_1\right| \equiv \frac{2}{3}\left(\Delta A + \Delta g \beta_e B_o/\hbar\right)$ with $\Delta A \equiv A_{zz} - \frac{1}{2}\left(A_{xx} + A_{yy}\right)$ in angular frequency units and $\Delta g \equiv g_{zz} - \frac{1}{2}\left(g_{xx} + g_{yy}\right)$. Values of the calibration constants are: $C_1 \cong 2.0, 3.0, 1.1$, $C_2 \cong 1.3, 2.0, 0.9$ and $C_3 \cong 0.9, 1.4, 0.9$ for echoes detected at the low-field extremum, central region and high-field extremum, respectively, of the nitroxide spectrum with a centre field of $B_o \equiv 0.335$ T.

For jump diffusion, the dependence of T_{2M}^{∞} on τ_R is linear, as we expect for lifetime broadening according to the uncertainty principle. Also, there is relatively little dependence on spectral position; in particular, the values of T_{2M}^{∞} determined at the low- and high-field extrema are identical. Again, we expect this for the jump model of slow diffusion because the jumps are large and therefore independent of the initial or final orientations. For Brownian diffusion, the dependence $T_{2M}^{\infty} \propto \tau_R^{1/2}$ is weaker than that for jump diffusion, because reorientation takes place in very small steps. Kivelson and Lee (1982) assume that orientational fluctuations about the parallel ($\theta = 0$) extrema are limited by the T_2-relaxation time and deduce that this causes a $\tau_R^{1/2}$ -dependent line broadening in the slow-motion regime. We follow this line of argument here. In Section 2.4, we found that the angular dependence of the resonance frequency close to $\theta = 0$ is approximated by $\omega_o \cong \omega_\parallel \cos\theta \cong \omega_\parallel\left(1 - \frac{1}{2}\theta^2\right)$. Averaging over the characteristic phase-memory time T_{2M}, fluctuations in ω_o then become $\left\langle\Delta\omega_o\right\rangle \cong \frac{1}{2}\omega_\parallel\left\langle\theta^2\right\rangle = \omega_\parallel D_R T_{2M}$ from Einstein's relation for Brownian diffusion, where D_R is the rotational diffusion coefficient (see Eq. 7.5 in Section 7.2). For the motional regime where $\left\langle\Delta\omega_o\right\rangle T_{2M} \approx 1$ (Kivelson and Lee 1982), T_{2M} is inversely proportional to $\left\langle\Delta\omega_o\right\rangle$, and hence $T_{2M} \propto 1/D_R^{1/2} \propto \tau_R^{1/2}$ for slow Brownian diffusion. For collision-limited free diffusion, the $\tau_R^{2/3}$ dependence of the phase-memory time is intermediate between those of the two other extreme models for slow motion.

Although we can use Eqs. 13.61–13.63 as calibrations to obtain correlation times, they are approximations. In particular, detailed simulations reveal some differences in exponent between the low-, central- and high-field regions of the spectrum (Schwartz et al. 1982). An empirical correlation-time calibration then applies to each specific spectral region, $m \equiv$ low, central, high:

$$\tau_R(\text{ns}) = a_m \times \left(T_{2M}(\text{ns})\right)^{b_m} \quad (13.64)$$

where rotational correlation time and phase-memory time are in nanoseconds, and a_m and b_m are calibration constants. We list values of these calibration constants for practical application to ^{14}N-nitroxides in Table 13.1. This parallels the empirical slow-motional calibrations that we used for linewidths and hyperfine splittings of the slow-motion spectra from CW-EPR, in Section 7.14 of Chapter 7.

TABLE 13.1 Slow-motion rotational correlation time (τ_R) calibrations for spin-echo phase memory times T_{2M}, according to $\tau_R(\text{ns}) = a_m \times \left(T_{2M}(\text{ns})\right)^{b_m}$, from stochastic-Liouville simulations in Schwartz et al. (1982)

model	*position*	a_m	b_m
Brownian	centre	0.072	1.75
	low-field	0.095	1.85
	high-field	0.172	2.00
free	centre	0.160	1.50
	low-field	0.284	1.51
	high-field	0.366	1.61
jump	centre	0.790	1.03
	low- and high-field	1.234	0.99

Note: τ_R and T_{2M} are in nanoseconds.

13.11 PRIMARY-ECHO DECAY IN SUDDEN-JUMP MODEL

Later in the chapter (in Section 13.19), we use the model of Zhidomirov and Salikhov (1969) for uncorrelated sudden jumps to describe the effect of angular jump dynamics on three-pulse stimulated echoes. Putting the second delay (T) of the stimulated echo equal to zero immediately gives the results for the primary echo. Successive jumps shift the angular resonance frequency by $\pm\Delta\omega_{m_I}$ at a rate $1/\tau_c$. From Eq. 13.133 in Section 13.19, we find that the two-pulse echo decay is then (cf. Milov et al. 1973a; Erilov et al. 2004a):

$$E(2\tau) = \exp\left(-2\tau/\tau_c\right)\left[1 + \frac{1}{R_{m_I}\tau_c}\sinh 2R_{m_I}\tau + \frac{2}{R_{m_I}^2\tau_c^2}\sinh^2 R_{m_I}\tau\right] \quad (13.65)$$

where $R_{m_I}^2 \equiv \left(1/\tau_c\right)^2 - \Delta\omega_{m_I}^2$. When $R_{m_I}^2 < 0$, i.e., R_{m_I} is imaginary, we express the hyperbolic functions as normal trignometric functions, and the primary-echo decay becomes (see Eq. 13.134 given later):

$$E(2\tau) = \exp\left(-2\tau/\tau_c\right)\left[1 + \frac{1}{\left|R_{m_I}\right|\tau_c}\sin 2\left|R_{m_I}\right|\tau + \frac{2}{\left|R_{m_I}^2\right|\tau_c^2}\sin^2\left|R_{m_I}\right|\tau\right] \quad (13.66)$$

for the condition $\Delta\omega_{m_I}^2\tau_c^2 > 1$.

For fast motion (i.e., $\Delta\omega_{m_I}^2\tau_c^2 << 1$ and thus $R_{m_I} \to 1/\tau_c$), Eq. 13.65 becomes:

$$E(2\tau) \approx \exp\left(-2\tau/\tau_c\right)\left[1 + \sinh 2R_{m_I}\tau + \sinh^2 R_{m_I}\tau\right]$$
$$= \exp\left(-2\tau/\tau_c\right)\exp\left(2R_{m_I}\tau\right) \quad (13.67)$$

Expanding $R_{m_I} - 1/\tau_c$ to lowest order, we therefore get exponential decay of the primary echo:

$$E(2\tau) \approx \exp\left(-\Delta\omega_{m_I}^2 \tau_c \tau\right) \tag{13.68}$$

with phase-memory decay rate: $1/T_{2M} = \frac{1}{2}\Delta\omega_{m_I}^2 \tau_c$, for rapid jumps. This is what we found already for fast Brownian diffusion (Eqs. 13.36 and 13.38) and conforms to the more general result that fast-motion decays are independent of mechanism (see, e.g., Slichter 1990).

For slow motion (i.e., $\Delta\omega_{m_I}^2 \tau_c^2 >> 1$ and thus $R_{m_I} \to i\Delta\omega_{m_I}$), the primary-echo decay becomes simply:

$$E(2\tau) = \exp\left(-\frac{2\tau}{\tau_c}\right) \tag{13.69}$$

when we neglect terms in $1/\left(\Delta\omega_{m_I}\tau_c\right)$ and $1/\left(\Delta\omega_{m_I}\tau_c\right)^2$ compared to unity in Eq. 13.66. For slow jump rates, the two-pulse echo decay is therefore also monoexponential. The phase-memory time is then equal to the lifetime: $T_{2M} = \tau_c$, in accord with Eq. 13.63 for slow jump diffusion in Section 13.10.

Figure 13.9 shows the primary-echo decay profiles that we predict from Eq. 13.65 for different values of the time τ_c between jumps. We find non-exponential decays of this shape for the nitroxide TEMPONE in glasses at low temperature, where electron spin dephasing is dominated by nuclear spin flips in the solvent (Zecevic et al. 1998). Fits of Eq. 13.65 to the decay curves give the proton flip rate (see Table 13.3 given later). For slow jumps, where $\Delta\omega_{m_I}\tau_c > 1$, the echo decay rate becomes more rapid than indicated by the solid lines in Figure 13.9, and the phase-memory time approaches the jump time, as in Eq. 13.69 (see dashed line).

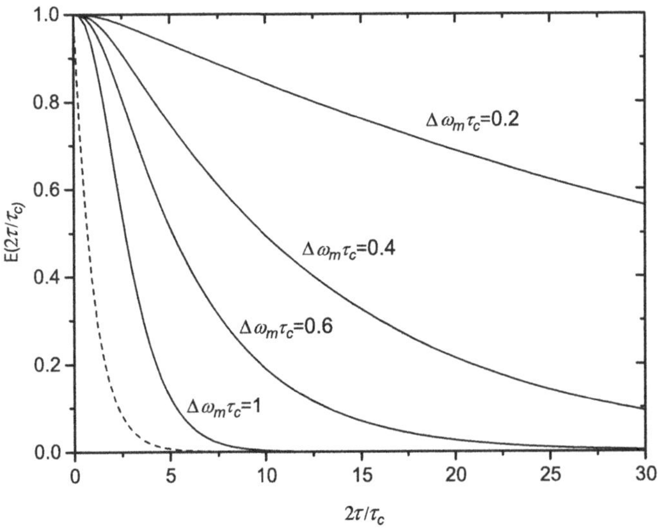

FIGURE 13.9 Dependence of two-pulse echo amplitude, $E(2\tau)$, on interpulse spacing, τ, for uncorrelated sudden-jump model (Eq. 13.65). *Solid lines*: echo decays for different jump times τ_c, scaled by jump length $\Delta\omega_{m_I}$. *Dashed line*: single-exponential decay (Eq. 13.69) for slow jump rates. Echo delay, 2τ, is normalized to τ_c.

13.12 EXPERIMENTAL PHASE-MEMORY TIMES

Figure 13.10 gives phase-memory times determined in the central region of the spectrum from ^{14}N-TEMPONE in 85% aqueous glycerol (solid symbols), over a wide range of temperatures. For comparison, we include corresponding data for peroxylamine disulphonate (PADS), a nitroxide that contains no H-atoms, as open symbols in Figure 13.10. We see that results for PADS are displaced along the $1/T$-axis, relative to TEMPONE, because of its smaller diffusion radius. The rotational correlation times have an Arrhenius-law temperature dependence: $\tau_R = \tau_{R,0}\exp\left(E_a/RT\right)$, with $E_a \approx 63$ kJ mol^{-1} and $\tau_{R,0} \approx 2.3 \times 10^{-21}$ s^{-1} found from CW-EPR spectra for TEMPONE (Hwang et al. 1975, and see Sections 7.8 and 7.9 in Chapter 7). Accordingly, we present Figure 13.10 as an Arrhenius plot: $\log_{10} T_{2M}$ against $1/T$.

At high temperatures, the phase-memory time increases with increasing temperature (shorter correlation times) and is inversely proportional to τ_R as predicted by Eq. 13.44 for the fast motional regime. As also expected from Eq. 13.44, the phase-memory times of TEMPONE in this regime are very close to the T_2-relaxation times that are determined from the central linewidths of the motionally narrowed CW-EPR spectra (see Hwang et al. 1975, and Section 7.8 in Chapter 7).

At lower temperatures, in the slow-motion regime, the phase-memory times increase with decreasing temperature (longer correlation times) but with a weaker power-law dependence on correlation time that is closer to the prediction for Brownian diffusion

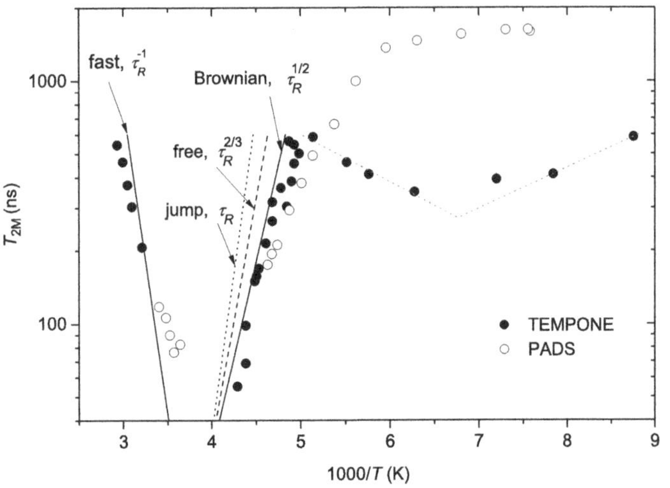

FIGURE 13.10 Temperature dependence of phase-memory times, T_{2M}, from spin-echo decays of 2,2,6,6-tetramethyl-4-oxo-piperidine-1-oxyl (TEMPONE, solid circles) and peroxylamine disulphonate (PADS, open circles) in 85% aqueous glycerol. Data from Stillman et al. (1980) and Brown (1974), given as Arrhenius plot: $\log T_{2M}$ vs. $1/T$. Prediction for motional narrowing (Eq. 13.41): solid line at high temperatures. Predictions for slow-motion regime at lower temperatures (Schwartz et al. 1982): Brownian diffusion (Eq. 13.61, solid line); collision-limited free diffusion (Eq. 13.62, dashed line); jump diffusion (Eq. 13.63, dotted line).

(viz., $\tau_R^{1/2}$) than those for either free or jump diffusion. The T_{2M} -data for TEMPONE in this regime are also consistent with later measurements on the same system that used partially relaxed echo-detected spectra (Millhauser and Freed 1984). However, the slow-motion CW-EPR lineshapes can be fit better with a jump or collision-limited free diffusion model than by Brownian diffusion (Hwang et al. 1975, and see Section 7.12 in Chapter 7).

By interpolation, the phase-memory time reaches a deep minimum at ca. 266 K, between the fast- and slow-motional regimes. At this temperature, the correlation time is predicted to be $\tau_R \cong 5$ ns (Hwang et al. 1975). Because the frequency equivalent of the g-value anisotropy of the central ($m_I = 0$) hyperfine line is: $\Delta\omega_o = (g_{xx} - g_{zz})\beta_e B_o/\hbar \cong 1.9\times10^{-8}$ rad s^{-1} for $B_o = 0.335$ T, we find that $\Delta\omega_o \times \tau_R \approx 1$ at the minimum in T_{2M}.

At yet lower temperatures, in the glassy state, the phase-memory time of PADS increases to an almost temperature-independent value. TEMPONE, however, has shorter values of T_{2M} than PADS that exhibit a broad minimum at ca. 150 K. This enhanced relaxation of TEMPONE, in the glassy regime, is caused by fluctuations in dipolar couplings with protons, which are absent in PADS. Methyl-group rotation is the only dynamic process of large amplitude in glasses at low temperature (Tsvetkov and Dzuba 1990). For instance, Dzuba et al. (1984) observe a similar T_{2M}-minimum for imidazoline nitroxides in di-n-butyl phthalate at ca. 130 K and suggest that the jump frequency for methyl-group rotation corresponds to the proton dipolar interaction at these temperatures. Also in 50% aqueous glycerol, TEMPONE displays a T_{2M}-minimum attributable to methyl rotation at ca. 150 K (Nakagawa et al. 1992). On the fast-motion side of the minimum, the dephasing rate $1/T_{2M}$ is given by Eq. 13.41 and is directly proportional to the correlation time for methyl rotation. On the slow-motion side of the minimum, the dephasing time T_{2M} is given by Eq. 13.63 for jump diffusion and is approximately equal to the time between jumps, τ_J. The activation energies for methyl-group jumps in TEMPONE are: $E_a \approx 3.5$ kJ mol^{-1}, on both sides of the minimum in Figure 13.10, and the root-mean-square value of the splitting being averaged is: $\sqrt{\langle\Delta\omega_o^2\rangle} \approx 3$ MHz. Rather higher values are found for TEMPONE in 50% glycerol–water: $E_a \approx 8.7$ kJ mol^{-1} and $\sqrt{\langle\Delta\omega_o^2\rangle} \approx 8.8$ MHz but overall are typical of methyl-group rotation (Nakagawa et al. 1992).

13.13 PRIMARY ECHO-DETECTED SPECTRA AND LIBRATIONAL DYNAMICS

In the slow-motion regime, the phase-memory times that we predict for isotropic rotational diffusion are rather long. For glassy states, rapid, low-amplitude torsional librations give more effective phase relaxation than does slow isotropic rotation (Dzuba 1996). Because librations are of strictly limited angular amplitude, they also produce stronger changes in the echo-detected lineshapes than those arising from overall rotation.

The contribution of rapid libration to T_{2M} is given by Eq. 13.41 for fast motion, and the two-pulse echo decay is mono-exponential as in Eq. 13.40. Assuming axial anisotropy, the fluctuations in angular resonance frequency are given by Eq. 13.42 or 13.43, which we can abbreviate to: $\omega_o(\theta) = (\omega_\parallel - \omega_\perp)\cos^2\theta + \omega_\perp$. For small librations $\Delta\theta$ around the non-unique perpendicular axes, an acceptable approximation is: $\Delta\omega_o \approx (\partial\omega_o/\partial\theta)\Delta\theta$. The angular dependent phase-memory time then becomes:

$$\frac{1}{T_{2M}(\theta)} = \langle\Delta\omega_o(\theta)^2\rangle\tau_R = (\omega_\parallel - \omega_\perp)^2\langle\Delta\theta^2\rangle\tau_R \times \sin^2 2\theta \qquad (13.70)$$

where $\langle\Delta\theta^2\rangle$ is the mean-square librational amplitude. When considering ED-lineshapes, it helps us to define a dimensionless scan parameter: $\delta \equiv (\omega_o - \omega_\perp)/(\omega_\parallel - \omega_\perp)$. From Eq. 2.15 in Chapter 2, the absorption powder lineshape is then given by: $E(\delta) = \sin\theta/(\partial\delta/\partial\theta) = \frac{1}{2}\delta^{-1/2}$ (see Bartucci et al. 2003, and cf. Figure 2.6), and the spectrally resolved phase-memory time becomes:

$$\frac{1}{T_{2M}(\delta)} = (\omega_\parallel - \omega_\perp)^2\langle\Delta\theta^2\rangle\tau_R \times \delta(1-\delta) \qquad (13.71)$$

From Eq. 13.40, the ED-lineshape for echo delay τ is then given by:

$$E(2\tau,\delta) = E(0,\delta)\exp\left(-2\tau/T_{2M}(\delta)\right) \qquad (13.72)$$

where $E(0,\delta) \equiv E(\delta)$ is the non-relaxed lineshape at $\tau = 0$, and we neglect T_2-relaxation other than by librational motion. Figure 13.11 shows model ED-lineshapes that we predict from Eqs. 13.71 and 13.72 for progressively increasing echo delay times, τ. We find greatest dependence on τ for intermediate spectral positions δ between the $\delta_\perp = 0$ discontinuity and the $\delta_\parallel = 1$ shoulder of the powder pattern. The maximum reduction in ED-intensity corresponds to the region of the spectrum where the resonance frequency changes most with orientation θ of the magnetic field. We see this from the values of $\partial\omega_o/\partial\theta$ given in the insert to Figure 13.11. These model calculations reproduce the qualitative features of the τ-dependence in the high-field region of ED-spectra from nitroxide spin labels quite well (see, e.g., Dzuba 1996; Bartucci et al. 2003).

Specifically for a ^{14}N-nitroxide, the angular resonance frequency (i.e., spectral line position) is given by:

$$\omega_{m_I}(\theta,\phi) = g(\theta,\phi)\frac{\beta_e}{\hbar}B + m_I A(\theta,\phi), \qquad (13.73)$$

where the static magnetic field B makes polar angles (θ, ϕ) with the nitroxide x, y, z-axes and the hyperfine constant is in angular frequency units. The angular-dependent g-value, $g(\theta,\phi)$, and hyperfine coupling constant, $A(\theta,\phi)$, are given by Eqs. 2.11 and 2.12, respectively, of Chapter 2. In terms of polar coordinates, the shift in resonance frequency caused by angular libration is given by:

$$\Delta\omega_{m_I}(\theta,\phi) = \frac{\partial\omega_{mI}}{\partial\theta}\Delta\theta + \frac{1}{\sin\theta}\frac{\partial\omega_{mI}}{\partial\phi}\Delta\phi \qquad (13.74)$$

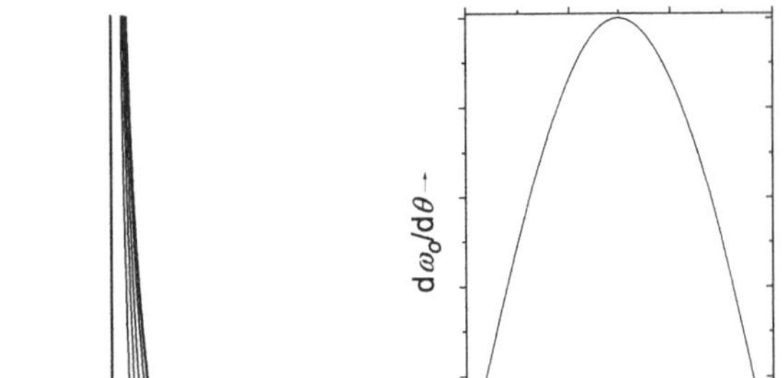

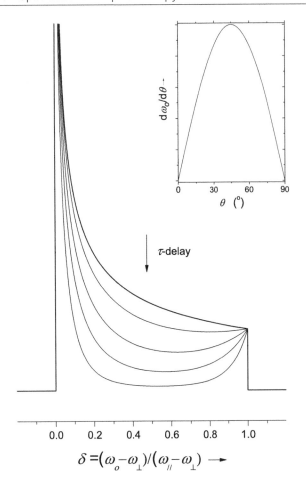

FIGURE 13.11 Model ED-lineshapes for axially anisotropic spectrum (Eqs. 13.71, 13.72). Interpulse spacing of the primary-echo sequence increases from *top to bottom*: τ = (0, 1, 3, 6, 12) ÷ $\left((\omega_{\parallel}-\omega_{\perp})^2\langle\Delta\theta^2\rangle\tau_R\right)$. $\delta = 0$ corresponds to angular resonance frequency $\omega_o = \omega_\perp$ and $\delta = 1$ to $\omega_o = \omega_\parallel$. *Insert*: angular dependence of $\partial\omega_o/\partial\theta \propto \sin 2\theta$ (cf. Eq. 13.70).

For rotations of small amplitude $\pm\alpha$ around an axis that makes polar angles (Θ,Φ) with the nitroxide axes (see Figure 13.12), the polar increments $\Delta\theta$ and $\Delta\phi$ are given by (Dzuba et al. 1995):

$$\Delta\theta = \sin\Theta\sin(\phi-\Phi)\times\alpha \quad (13.75)$$

$$\Delta\phi = (\cos\theta\sin\Theta\cos(\phi-\Phi) - \sin\theta\cos\Theta)\times\alpha \quad (13.76)$$

If the α-libration is around the nitroxide x-axis (i.e., $\Theta = 90°$, $\Phi = 0°$), then $\Delta\theta = \alpha\sin\phi$ and $\Delta\phi = \alpha\cos\theta\cos\phi$, and the shift in angular resonance frequency is given by (Kirilina et al. 2001):

$$\Delta\omega_{m_I}(\theta,\phi) = \alpha\left\{\frac{\beta_e}{\hbar}B\left(g_{yy}-g_{zz}\right) + \frac{m_I\left(A_{yy}^2 - A_{zz}^2\right)}{A(\theta,\phi)}\right\}\sin\theta\cos\theta\sin\phi \quad (13.77)$$

Similarly, for torsional rotations around the nitroxide y- or z-axes, the shifts in resonance frequency are (Erilov et al. 2004b):

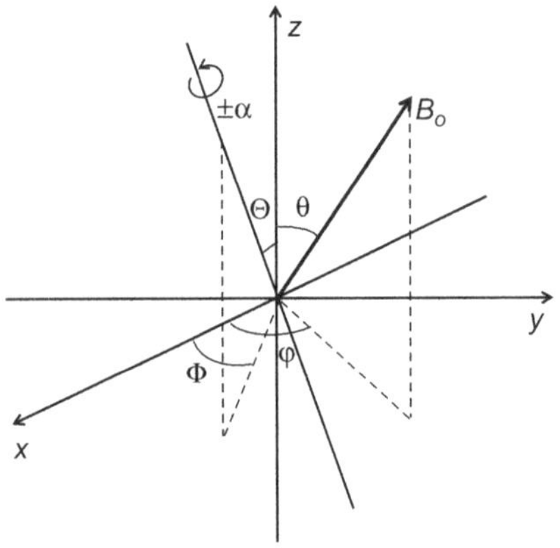

FIGURE 13.12 Orientation (Θ,Φ) of axis for librational torsions with amplitude $\pm\alpha$. $B_o(\theta,\phi)$: static magnetic-field direction; x,y,z: principal axes of nitroxide A- and g-tensors.

$$\Delta\omega_{m_I}(\theta,\phi) = \alpha\left\{\frac{\beta_e}{\hbar}B\left(g_{zz}-g_{xx}\right) + \frac{m_I\left(A_{zz}^2 - A_{xx}^2\right)}{A(\theta,\phi)}\right\}\sin\theta\cos\theta\cos\phi \quad (13.78)$$

$$\Delta\omega_{m_I}(\theta,\phi) = \alpha\left\{\frac{\beta_e}{\hbar}B\left(g_{xx}-g_{yy}\right) + \frac{m_I\left(A_{xx}^2 - A_{yy}^2\right)}{A(\theta,\phi)}\right\}\sin^2\theta\sin\phi\cos\phi \quad (13.79)$$

because $\Delta\theta = -\alpha\cos\phi$, $\Delta\phi = \alpha\cos\theta\sin\phi$ and $\Delta\theta = 0$, $\Delta\phi = -\alpha\sin\theta$, respectively. Note that in each case: $A(\theta,\phi) = \sqrt{\left(A_{xx}^2\cos^2\phi + A_{yy}^2\sin^2\phi\right)\sin^2\theta + A_{zz}^2\cos^2\theta}$, – see Eq. 2.12. So-called "isotropic" libration is modelled by assuming simultaneous, independent torsional librations, each of amplitude α, around the three perpendicular x-, y- and z-axes of the nitroxide (Erilov et al. 2004a). The net relaxation is then given by the product of the three independent relaxations induced by each librational mode.

We get the echo-detected absorption lineshape by summing the echo amplitude over all orientations θ,ϕ of the magnetic field, and over all three $m_I = 0, \pm1$ ^{14}N-hyperfine manifolds:

$$E(2\tau,B) = \sum_{m_I}\iint g\left(B - \omega_{m_I}(\theta,\phi)/\gamma_e\right) \times \exp\left(-2\tau/T_{2M}(\theta,\phi)\right)\sin\theta\, d\theta\, d\phi \quad (13.80)$$

where $g(\Delta B)$ is the lineshape of an individual inhomogeneously broadened spin packet. The angular-dependent rate of spin dephasing in Eq. 13.80 comes from Eqs. 13.77 to 13.79 by using Eq. 13.41:

$$1/T_{2M}(\theta,\phi) = \left\langle\Delta\omega_{m_I}^2(\theta,\phi)\right\rangle\tau_R \quad (13.81)$$

and is directly proportional to the product of the mean-square torsional amplitude and the librational correlation time, $\left\langle\alpha^2\right\rangle\tau_R$ (cf. Eq. 13.71).

Figure 13.13 shows simulated echo-detected EPR spectra of a nitroxide spin label that is undergoing librational motion according to the "isotropic" model. The spectra are normalized to the same maximum line height, so as to display only the field-dependent part of the relaxation. For a typical value of librational amplitude-correlation time product: $\langle\alpha^2\rangle\tau_R = 5\times10^{-12}$ rad^2 s, pulse delay times used in the simulations are τ = 0, 100, 200, 400, 600, 800 and 1000 ns. With increasing τ, amplitudes in intermediate spectral regions at low- and high-field decrease systematically, relative to the corresponding outer peaks (or shoulders), just as in the model calculations of Figure 13.11. We can use the ratio, P''/P, of line heights in the intermediate regions (P'') to those at the outer turning points (P) to determine the rate of anisotropic relaxation. Figure 13.14 shows that the dependence of the low-field diagnostic line height ratio, L''/L, on τ is close to exponential, as found experimentally (Bartucci et al. 2003). Near-exponential relaxation is not expected automatically, because net intensity at a fixed field position, i.e., $E(2\tau, B)$, arises from molecules with different polar orientations, not just a single θ, ϕ-orientation (see Eq. 13.80). Slopes of the logarithmic plots in Figure 13.14 are directly proportional to the amplitude-correlation time product: $d\ln(L''/L)/d\tau = (1.13\times10^{17}\,\text{rad}^{-2}\,\text{s}^{-2}) \times \langle\alpha^2\rangle\tau_R$, which gives us a useful means of calibration (see Table 13.2).

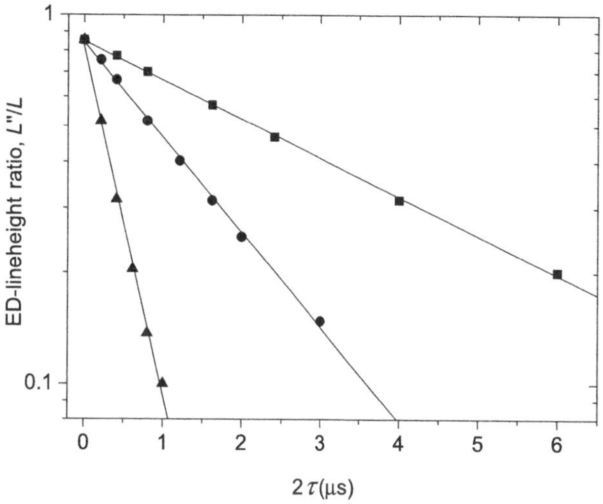

FIGURE 13.14 Dependence of low-field diagnostic line-height ratio L''/L on interpulse spacing τ for ED-spectra simulated with "isotropic" librational model and amplitude-correlation time product: $\langle\alpha^2\rangle\tau_R = 2\times10^{-12}$ rad^{2}s (squares), 5×10^{-12} rad^{2}s (circles) and 2×10^{-11} rad^{2}s (triangles). *y*-axis is logarithmic. Data from Erilov et al. (2004b).

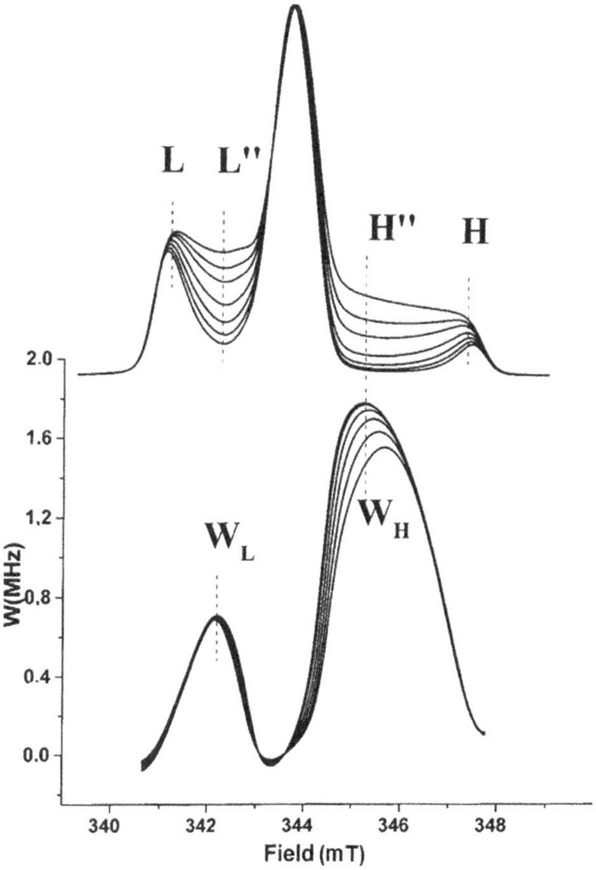

FIGURE 13.13 *Upper panel*: echo-detected spectra simulated with "isotropic" librational model, for interpulse spacings: $\tau\times\langle\alpha^2\rangle\tau_R = 0$, 0.5, 1, 2, 3, 4 and 5 ×10^{-18}rad^{2}s^2 (top to bottom). Spin-Hamiltonian parameters: (g_{xx}, g_{yy}, g_{zz}) = (2.0089, 2.0059, 2.0024), (A_{xx}, A_{yy}, A_{zz}) = (7.4, 6.3, 60.9) × 10^{7}rad s^{-1} ≡ (0.42, 0.36, 3.46) mT. *Lower panel*: anisotropic part of relaxation rate W (Eq. 13.82) for pairs of ED-spectra with interpulse separations: $\tau_1 = 0$ and $\tau_2\times\langle\alpha^2\rangle\tau_R = 0$, 0.5, 1, 2, 3, 4 and 5 × 10^{-18}rad^{2}s^2. Data from Erilov et al. (2004a).

The lower panel of Figure 13.13 shows an alternative way to analyse partially relaxed ED-lineshapes. The anisotropic lineshapes are replotted as the ratio of ED-spectra recorded at two different values, τ_1 and τ_2, of interpulse delay according to:

$$W(B,\tau_1,\tau_2) = \ln\left[\frac{E(2\tau_1,B)}{E(2\tau_2,B)}\right]\cdot\frac{1}{2(\tau_2-\tau_1)} \tag{13.82}$$

where $W(B,\tau_1,\tau_2)$ represents the anisotropic relaxation rate at field position B, averaged over the time interval from τ_1 to τ_2. The near coincidence of W-relaxation curves derived from different pairs of τ-values implies that the anisotropic relaxation is almost exponential: especially on the low-field side, just as we find from L''/L line height ratios in Figure 13.14. We can use the approximately invariant values, W_L and W_H, of the low-field and high-field peaks in W-relaxation to characterize librational dynamics. For example, $W_L = (1.41\times10^{17}\,\text{rad}^{-2}\text{s}^{-2}) \times \langle\alpha^2\rangle\tau_R$ for the low-field peak. A complete set of calibration constants for the W-relaxation parameters is included in Table 13.2.

ED-spectra of spin-labelled lipids in the glassy state of bilayer membranes (Erilov et al. 2004a; Erilov et al. 2004b), and of peroxylamine disulphonate (PADS) similarly in a glycerol glass (Kirilina et al. 2005), are better described by the "isotropic" librational model than by libration about a single nitroxide axis. At 9-GHz EPR frequency, we can readily distinguish libration about the nitroxide z-axis from that about the x- or y-axis. However, we need higher EPR frequencies, which exploit the non-axiality of the g-tensor, to distinguish clearly between x- and y-axis libration (Kirilina et al. 2005).

TABLE 13.2 Multiplicative calibration factors (in rad^{-2}s^{-2}) relating τ-dependence of diagnostic line-height ratios ln(L''/L) and ln(H''/H), and averaged W-relaxation rates W_L and W_H (Eq.13.82), from echo-detected spectra, to amplitude-correlation time product $\langle\alpha^2\rangle\tau_R$ of "isotropic" libration (Erilov et al. 2004b)

parameter	5-DOXYL-PC	14-DOXYL-PC
dln(L''/L)/dτ	$1.13 \pm 0.01 \times 10^{17}$	$9.30 \pm 0.15 \times 10^{16}$
dln(H''/H))/dτ	$2.3 \pm 0.1 \times 10^{17}$	$2.2 \pm 0.1 \times 10^{17}$
W_L	$1.41 \pm 0.02 \times 10^{17}$	$1.05 \pm 0.02 \times 10^{17}$
W_H	$3.3 \pm 0.1 \times 10^{17}$	$2.9 \pm 0.1 \times 10^{17}$

From spectral simulations with spin-Hamiltonian parameters and linewidths appropriate to 5-DOXYL-PC, and 14-DOXYL-PC phosphatidylcholine (see Figure 1.1 for structures). For example, for 5-DOXYL-PC: $d\ln(L''/L)/d\tau = (1.13 \pm 0.01 \times 10^{17}\ \mathrm{rad^{-2}\ s^{-2}}) \times \langle\alpha^2\rangle\tau_R$, and $W_L(\mathrm{s^{-1}}) = (1.41 \pm 0.02 \times 10^{17}\ \mathrm{rad^{-2}\ s^{-2}}) \times \langle\alpha^2\rangle\tau_R$, with τ and τ_R in sec.

We can estimate the mean-square librational amplitude $\langle\alpha^2\rangle$ directly from the outer hyperfine splitting in the CW-EPR spectrum. For fast small-amplitude librations about the x-axis, the partially averaged hyperfine splitting $\langle A_{zz}\rangle$ is given by (see Eq. 8.3):

$$\langle A_{zz}\rangle = A_{zz} - (A_{zz} - A_{yy})\langle\sin^2\alpha\rangle \approx A_{zz} - (A_{zz} - A_{yy})\langle\alpha^2\rangle \tag{13.83}$$

where the angular brackets indicate motional averaging of the hyperfine tensor with principal elements: A_{xx}, A_{yy} and A_{zz}. Combining with the value of $\langle\alpha^2\rangle\tau_R$ from the ED-spectra, we then get the librational correlation time τ_R. Typical values in glass-like states are: $\sqrt{\langle\alpha^2\rangle} \approx 0°\text{–}15°$ and $\tau_R \approx 0.1\text{–}2$ ns (Guzzi et al., 2009; Bartucci et al. 2008; Bartucci et al. 2006). These readily fulfill the small-amplitude requirement, and the condition for motional narrowing: $(\Delta\omega)^2\tau_R^2 = (\Delta A)^2\langle\alpha^2\rangle\tau_R^2 << 1$.

13.14 SPIN–SPIN INTERACTIONS AND SPECTRAL DIFFUSION

Historically, dipolar interactions between the electron spins were among the first to be considered in discussing phase-memory times because early electron spin-echo experiments were performed on solids. Correspondingly, in dilute electron-spin systems, electron–nuclear dipolar interactions are the main contributor to echo dephasing at low temperatures, where spin diffusion of solvent protons is the dominant mechanism (see Brown 1979; Milov et al. 1973a). We treat electron spin–spin interactions here, but the treatment of electron-nuclear spin–spin interactions is exactly equivalent.

Electron spins excited by the microwave pulse we call A-spins, and spins within the inhomogeneously broadened lineshape that are not excited by the pulse we call B-spins. The local dipolar field at the A-spins is modulated by spin–lattice relaxation, or by mutual spin flips, of the surrounding B spins. For a particular A-spin, the dipolar shift in Larmor frequency $\Delta\omega_{dd}$ is a sum over contributions from the surrounding B-spins. The characteristic time τ_c for fluctuations in $\Delta\omega_{dd}$ of an A-spin is determined by the T_1- or T_2-relaxation time of the B-spins, whichever is the more effective spin-flip mechanism (Salikov and Tsvetkov 1979). The echo-decay envelope then results from averaging over the various B-spin environments that are experienced by the different A-spins.

Although from the same species, the A-spins and B-spins are magnetically unlike whenever their mutual dipolar interaction is much smaller than the overall width of the nitroxide powder pattern (see Section 9.4 in Chapter 9, and Marsh 2014). The dipolar Hamiltonian for two unlike spins contains only the $s_{1z}s'_{2z}$ secular terms and the shift in Larmor frequency $\Delta\omega_{d,j}$ of an A-spin by the jth B-spin is (see Eq. 9.19 in Chapter 9):

$$\Delta\omega_{d,j} = \frac{\mu_o}{4\pi}\frac{g^2\beta_e^2}{\hbar}\frac{(1-3\cos^2\theta_j)}{r_j^3}s'_{jz} \equiv \Delta\omega_{d,j}^o \times s'_{jz} \tag{13.84}$$

where r_j is the distance between the A-spin and the B-spin, θ_j is the angle between the interspin vector and the static B_o field, and s'_{jz} is the z-component of the B-spin. Equation 13.84 specifies the local magnetic field, $\Delta\omega_{d,j}/\gamma_e$, from the jth B-spin. Flipping the B-spin causes an abrupt change $\pm\Delta\omega_{d,j}^o$ in Larmor frequency of the A-spin, because $s'_{jz} = \pm\frac{1}{2}$ for each B-spin. For relatively dilute systems, we can model this by uncorrelated sudden jumps (Zhidomirov and Salikhov 1969; Hu and Hartmann 1974) but not as a Gauss–Markov process which inherently involves small steps (as in rotational diffusion – cf. Sections 13.6–13.8). However, because of the central-limit theorem, the probability density of a random quantity tends to Gaussian at long times (i.e., for fast motion), irrespective of mechanism, and then we can also use the Gaussian–Markov model (Milov et al. 1973b).

13.14.1 Jump Model for Short τ

We begin with the sudden-jump theory at short delay times (Hu and Hartmann 1974). The time dependence of the dipolar shift in resonance frequency by the jth B-spin is:

$$\Delta\omega_{d,j}(t) = \Delta\omega_{d,j}^o h(t) \tag{13.85}$$

where $h(t) = \pm 1$ is the switching function, which changes sign every time the spin flips. From Eq. 13.30, the echo decay envelope for a single B-spin configuration is then given by:

$$E(2\tau) = \left\langle \exp\left(-i\int_0^{2\tau} s(t)\Delta\omega_{dd}(t)\,\mathrm{d}t\right)\right\rangle_t = \exp\left(-i\sum_j \Delta\omega_{d,j}^o\left\langle\int_0^{2\tau} s(t)h(t)\,\mathrm{d}t\right\rangle_t\right) \tag{13.86}$$

where the average is over all histories of B-spin flips, which we shall perform later. To average Eq. 13.86 over all possible configurations of B-spins, we use a method from the statistical theory of dipolar broadening (see Section 15.9 in Chapter 15). The probability that the jth spin is situated at position (r_j, θ_j) is just dV_j/V, where $dV_j = 2\pi r_j^2 \sin\theta_j dr_j d\theta_j$ and V is the total volume. For N B-spins in total, the summed echo envelope is then given from Eq. 13.86 by:

$$E(2\tau) = \int_{V^N} \exp\left(-i\left(\Delta\omega_{d,1}^o + \cdots + \Delta\omega_{d,N}^o\right) \left\langle \int_0^{2\tau} s(t)h(t)\,dt \right\rangle_t \right) \cdot \frac{dV_1}{V} \cdots \frac{dV_N}{V} \tag{13.87}$$

where the integration is over each of the volume elements dV_j. We can express the N equivalent definite integrals as the product:

$$E(2\tau) = \left[\frac{1}{V} \int_V \exp\left(-i\Delta\omega_d^o \left\langle \int_0^{2\tau} s(t)h(t)\,dt \right\rangle_t \right) \cdot dV \right]^N \tag{13.88}$$

where we now omit the superfluous subscript j. Using the methods of Section 15.9 to avoid divergences as N and V are allowed to approach infinity, we get the standard statistical-mechanical replacement:

$$E(2\tau) = \left(1 - V'(\tau)/V\right)^N = \exp(-n_B V'(\tau)) \tag{13.89}$$

where $n_B = N/V$ is the number of B-spins per unit volume. From Eqs. 13.84 and 13.87, V' is defined by:

$$V'(\tau) \equiv 2\pi \int_0^{\pi} \sin\theta \cdot d\theta \times \int_0^{\infty} \left[1 - \exp\left(-i \left(\frac{\mu_o}{4\pi} \right) \frac{g^2\beta_e^2}{\hbar} \frac{(1 - 3\cos^2\theta)}{r^3} \left\langle \int_0^{2\tau} s(t)h(t)\,dt \right\rangle_t \right) \right] r^2\,dr \tag{13.90}$$

We evaluate the integrals in Eq. 13.90 as described in Section 15.9 by substituting $x = 1/r^3$, using $\int_{-\infty}^{\infty} (1 - \exp(-ibx))/x^2 \cdot dx = \pi|b|$ and $\int_0^{\pi} \left|1 - 3\cos^2\theta\right| \sin\theta \cdot d\theta = 8/(3\sqrt{3})$, which gives:

$$V'(\tau) = \frac{8\pi^2}{9\sqrt{3}} \left(\frac{\mu_o}{4\pi} \right) \frac{g^2\beta_e^2}{\hbar} \left\langle \left| \int_0^{2\tau} s(t)h(t)\,dt \right| \right\rangle_t \tag{13.91}$$

Substituting Eq. 13.91 in Eq. 13.89, the electron spin-echo decay envelope then becomes:

$$E(2\tau) = \exp\left(-\Delta\omega_{1/2} \left\langle \left| \int_0^{2\tau} s(t)h(t)\,dt \right| \right\rangle_t \right) \tag{13.92}$$

where the dipolar half-width at half-height is given from Eq. 15.31 in Chapter 15 by:

$$\Delta\omega_{1/2} = \frac{4\pi^2}{9\sqrt{3}} \left(\frac{\mu_o}{4\pi} \right) \frac{g^2\beta_e^2}{\hbar} n_{SL} \tag{13.93}$$

and is directly proportional to the spin concentration, n_{SL}.

Now we must evaluate the averages $\langle \ldots \rangle_t$ over the different B spin-flip histories (Hu and Hartmann 1974). The probability that the B-spin does not flip in time t is $\exp(-t/\tau_S)$, where the spin-flip rate is $1/\tau_S$. The probability that the spin flips just once in time t and that the flip occurs in the interval t_1 to $t_1 + dt_1$ is $(dt_1/\tau_S) \times \exp(-t/\tau_S)$, and that it flips just twice is $(dt_2/\tau_S)(dt_1/\tau_S) \times \exp(-t/\tau_S)$ and so on. We have a Poisson process, where the probability that there will be a total of m B-spin flips in time t is:

$$W_m(t) = w_m(t) \int_0^t dt_1 \int_{t_1}^t dt_2 \cdots \int_{t_{m-1}}^t dt_m = \frac{1}{m!} \left(\frac{t}{\tau_S} \right)^m \exp\left(-\frac{t}{\tau_S} \right) \tag{13.94}$$

where $w_m(t) \equiv (1/\tau_S)^m \exp(-t/\tau_S)$. From the definition of $s(t)$, the contribution of the $m = 0$ term to the average in Eq. 13.92 vanishes:

$$w_0(2\tau) \left| \int_0^{2\tau} s(t)h_0(t) \cdot dt \right| = w_0(2\tau) \left| \int_0^{\tau} dt - \int_{\tau}^{2\tau} dt \right| = 0 \tag{13.95}$$

where $h_0(t) \equiv 1$ because there are no spin flips. Similarly, the contribution from histories with one flip only is (cf. Eq. 13.94):

$$w_1(2\tau) \int_0^{2\tau} dt_1 \left| \int_0^{2\tau} s(t)h_1(t)\,dt \right| = w_1(2\tau) \int_0^{2\tau} dt_1 \left| \int_0^{\tau} h_1(t)\,dt - \int_{\tau}^{2\tau} h_1(t)\,dt \right| \tag{13.96}$$

where $h_1(t)$ changes from +1 to −1 at time $t = t_1$. Thus, the first integral in Eq. 13.96 must be evaluated separately, depending on whether $t_1 \le \tau$ or $t_1 \ge \tau$, respectively:

$$\left| \int_0^{\tau} h_1(t)\,dt - \int_{\tau}^{2\tau} h_1(t)\,dt \right| = \left| \left(\int_0^{t_1} dt - \int_{t_1}^{\tau} dt \right) + \int_{\tau}^{2\tau} dt \right| = 2t_1 \tag{13.97a}$$

$$\left| \int_0^{\tau} h_1(t)\,dt - \int_{\tau}^{2\tau} h_1(t)\,dt \right| = \left| \int_0^{\tau} dt - \left(\int_{\tau}^{t_1} dt - \int_{t_1}^{2\tau} dt \right) \right| = 2(2\tau - t_1) \tag{13.97b}$$

Substituting Eqs. 13.97a,b in Eq. 13.96, the contribution from single-flip histories then becomes:

$$2w_1(2\tau)\left|\int_0^{\tau} t_1\,dt_1 + \int_{\tau}^{2\tau}(2\tau - t_1)dt_1\right| = 2\left(\frac{\tau^2}{\tau_S}\right)\exp\left(-\frac{2\tau}{\tau_S}\right) \quad (13.98)$$

This is therefore the leading contribution for short delay times, $\tau << \tau_S$, and from Eq. 13.92 the echo decay envelope then becomes:

$$E(2\tau) = \exp\left(-2\Delta\omega_{1/2}\tau^2/\tau_S\right) \quad (13.99)$$

when $\exp\left(-2\tau/\tau_S\right) \approx 1$. Such a τ^2-dependence at short times was predicted first by Klauder and Anderson (1962) for a Lorentz–Markov model that relates directly to the Lorentzian dipolar line broadening predicted by the statistical theory (see Section 15.9). The Lorentz–Markov model is described in Appendix A13 at the end of the chapter.

A τ^2-dependence is observed experimentally for electron spin–spin interactions of a nitroxide in frozen solution (Salikov and Tsvetkov 1979, and see Section 13.15). It is found also, at short times, for the analogous case of transverse relaxation induced by proton spin flips in glassy solvents that do not contain (rotating) methyl groups (Zecevic et al. 1998). Then the effective flip frequency is directly proportional to the proton concentration in the solvent, as expected from Eq. 13.93 for the corresponding electron-spin situation. Table 13.3 gives results for TEMPONE at 40 K in glasses of solvents that contain no methyl groups. The mean first-order rate constant for mutual proton spin flips is $k_{flip} = 30 \pm 5\ \text{Hz M}^{-1}$, which corresponds to a frequency of 3 kHz for a typical proton concentration of ≈100 M. These flip rates come from fitting the echo decay curves with Eq. 13.65 for the strong-jump model given in Section 13.11 (see Zecevic et al. 1998). Note that deuterating the solvent can increase the phase-memory time at low temperatures dramatically (Lindgren et al. 1997), as can deuteration of spin-labelled proteins (Ward et al. 2010). El Mkami et al. (2014) find that much of the effect is due to protons within 2.5 nm of the nitroxide.

TABLE 13.3 Stretched-exponential fits to two-pulse echo decays of 1-oxyl-2,2,6,6-tetramethylpiperidin-4-yl oxide (TEMPONE) at 40 K in protonated solvents (Zecevic et al. 1998)[a]

solvent	[H] (M)[b]	T_{2M} (μs)	β (*exponent*)	k_{flip} ($M^{-1}s^{-1}$)[c]
o-terphenyl (OTP)	49	8.5	2.5	38
19:1 OTP-decalin	64	7.9	2.5	28
3:1 OTP-decalin	75	6.7	2.3	28
1:1 OTP-decalin	89	6.3	2.2	25
1:1 H_2O-glycerol	110	4.6	2.3	33
glycerol	110	5.2	2.4	25
triethanolamine	113	4.7	2.2	30
decalin	120	4.2	2.1	33

[a] Echo amplitude: $E(2\tau) = E(0)\exp\left(-\left(2\tau/T_{2M}\right)^{\beta}\right)$.
[b] Total proton concentration in solvent (mole/litre).
[c] Proton spin-flip rate = $k_{flip}[\text{H}]$.

13.14.2 Sudden Jump Model: General Case

Hu and Hartmann (1974) give a general analytical solution for the $\langle\cdots\rangle$ average in Eq. 13.92. This is based on the method used to obtain the contribution from a single spin flip in Eq. 13.98. Accordingly, we write the decay envelope of the primary echo as:

$$E(2\tau) = \exp\left(-2\Delta\omega_{1/2}\tau K\left(2\tau/\tau_S\right)\right) \quad (13.100)$$

The universal function $K(z)$ in Eq. 13.100 is defined by:

$$K(z) \equiv \exp(-z)\left[I_1(z) + \tfrac{1}{2}\left(I_1(z)L_o(z) - L_1(z)I_o(z)\right)\right] \quad (13.101)$$

where $I_n(z)$ and $L_n(z)$ are the modified Bessel functions and Struve functions, respectively, of order n. Figure 13.15 plots the $K(z)$ function, together with asymptotic forms for small and large z, which are given by dashed lines. We list numerical values of $K(z)$ in Table 13.4. The z-dependence of $K(z)$ is biphasic, with a maximum value: $K(z) \approx \frac{1}{2}\pi^{-1/3}$ at $z \approx 2$. For small z: $K(z) \approx \frac{1}{2}z$, and for large z: $K(z) \approx \sqrt{2/\pi} \times z^{-1/2}$.

Figure 13.16 shows the echo decay curves predicted by Eqs. 13.100 and 13.101, for different values of the normalized spin-flip time $\Delta\omega_{1/2}\tau_S$. Asymptotic τ-dependences are shown by dotted lines. That for short delay times $(\tau << \tau_S)$ is given for the longest value of τ_S, and that for long delays $(\tau >> \tau_S)$ is given for the shortest value of τ_S. This illustrates the way in which the form of the echo decay goes over from a τ^2-dependence at slow spin-flip rates (cf. Eq. 13.99) to a $\tau^{1/2}$-dependence at fast flip rates. The dashed straight line in Figure 13.16 is the locus of the intersection of the two asymptotic forms for different values of τ_S (see Hu and Hartmann 1974). It forms a tangent to each of the complete decay curves, all of which lie above this line that therefore specifies the lower limit for the echo amplitude.

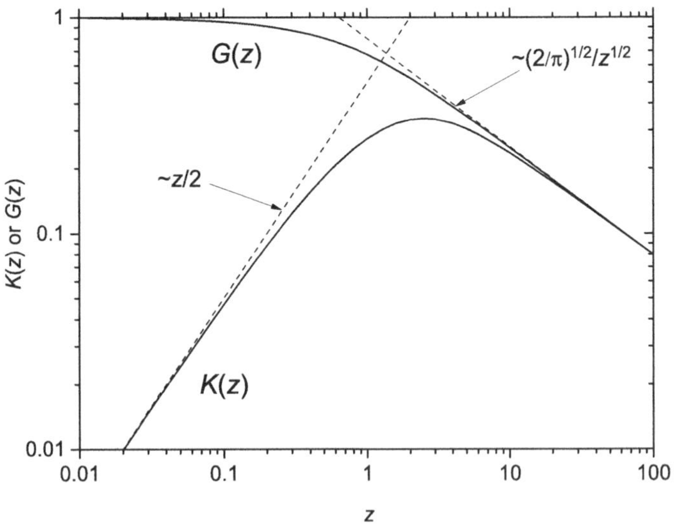

FIGURE 13.15 Generalized functions $K(z) \equiv \exp(-z)\left[I_1(z) + \frac{1}{2}\left(I_1(z)L_o(z) - L_1(z)I_o(z)\right)\right]$ and $G(z) \equiv \exp(-z)\left[I_o(z) + I_1(z)\right]$ that determine decay rates of primary and stimulated echoes (Eqs. 13.100 and 13.143), for uncorrelated sudden-jump model of dipolar relaxation (see Hu and Hartmann 1974).

TABLE 13.4 Generalized functions $K(z)$ and $G(z)$ (Eqs. 13.101 and 13.144) determining decay rates of primary and stimulated echoes, respectively, for uncorrelated sudden-jump model of dipolar relaxation (Hu and Hartmann 1974)

z	$K(z)$	$G(z)$	z	$K(z)$	$G(z)$
0.04	0.0195	0.98	1.3	0.304	0.618
0.05	0.0242	0.976	1.6	0.323	0.572
0.06	0.0288	0.971	2	0.336	0.524
0.08	0.0379	0.962	2.5	0.34	0.477
0.1	0.0468	0.952	3	0.338	0.44
0.13	0.0597	0.939	4	0.323	0.386
0.16	0.072	0.926	5	0.305	0.348
0.2	0.0878	0.909	6	0.287	0.319
0.25	0.1063	0.888	8	0.258	0.278
0.3	0.1236	0.87	10	0.236	0.249
0.4	0.155	0.834	13	0.21	0.219
0.5	0.182	0.801	16	0.191	0.198
0.6	0.206	0.772	20	0.173	0.177
0.8	0.245	0.719	25	0.156	0.159
1	0.274	0.674	30	0.143	0.145

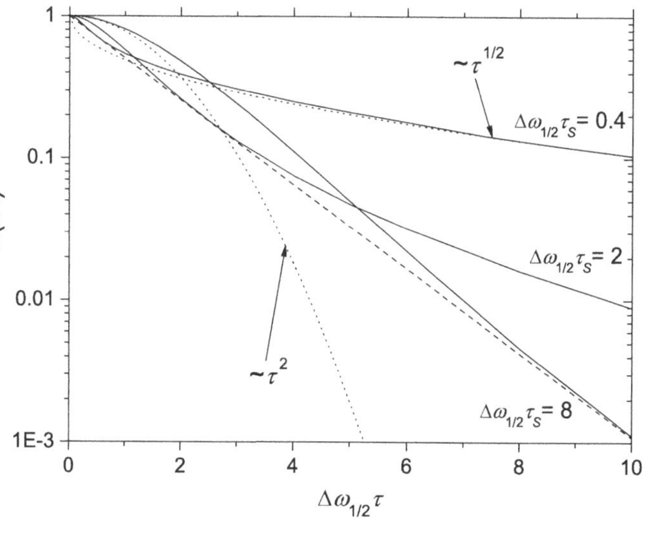

FIGURE 13.16 Dependence of primary-echo amplitude $E(2\tau)$ on interpulse spacing τ (Eq. 13.100), for dipolar relaxation by uncorrelated sudden-jumps (Hu and Hartmann 1974); τ_S, time between spin flips. *Dotted lines*: limiting dependence for $\tau \ll \tau_S$ (Eq. 13.99) and $\tau \gg \tau_S$ (Eq. 13.106). *Dashed straight line*: exponential τ-dependence $E(2\tau) = \exp\left(-2\pi^{-1/3}\Delta\omega_{1/2}\tau\right)$. *y*-axis is logarithmic; *x*-axis in units of $\Delta\omega_{1/2}$.

13.14.3 Gauss–Markov Model: Long τ

In view of the central-limit theorem from probability theory, it is interesting to compare the asymptotic $\tau^{1/2}$-dependence shown for the jump model in Figure 13.16 with the predictions of the Gauss–Markov model at long delays $\tau \gg \tau_S$. For a fixed configuration of B-spins, the contribution to the echo decay envelope from a particular A-spin is given by Eqs. 13.36 and 13.38 of Section 13.7. We therefore need the second moment $\left\langle \Delta\omega_{dd}^2 \right\rangle$ for the shift in Larmor frequency of the A-spin. This is the sum of the individual second-moment contributions from each B-spin (see Appendix U), which from Eq. 13.84 becomes:

$$\left\langle \Delta\omega_{dd}^2 \right\rangle = \left\langle \sum_j \Delta\omega_{d,j}^2 \right\rangle = \tfrac{1}{4}\sum_j \left(\Delta\omega_{d,j}^o\right)^2 \tag{13.102}$$

Here, the average that we indicate by angular brackets corresponds to the mean-square amplitude for the time dependence of the local field from a fixed configuration of the surrounding B-spins, i.e., $\left\langle s_{jz}'^2 \right\rangle = \frac{1}{4}$ for all j (see Eq. 13.34).

We sum over the different B-spin configurations in the same way as for the jump model (Mims 1968). From Eqs. 13.36, 13.38 and 13.102, the equivalent of Eq. 13.88 for the echo decay envelope is:

$$E(2\tau) = \left[\frac{1}{V}\int_V \exp\left(-\tfrac{1}{2}\left(\Delta\omega_d^o\right)^2 \tau_S\tau\right)\cdot \mathrm{d}V\right]^N \tag{13.103}$$

Again we express the integral as in Eq. 13.89. From Eqs. 13.84 and 13.103, V' for this case is defined by:

$$V'(\tau) \equiv 2\pi\int_0^{\pi} \sin\theta\cdot \mathrm{d}\theta \times \int_0^{\infty}\left[1-\exp\left(-\left(\frac{\mu_o}{4\pi}\right)^2 \frac{g^4\beta_e^4}{2\hbar^2}\frac{\left(1-3\cos^2\theta\right)^2}{r^6}\tau_S\tau\right)\right] r^2\,dr \tag{13.104}$$

We evaluate the integral over r in Eq. 13.104 by substituting $x = r^3/a$, giving $\frac{1}{3}aI_1$ where $I_1 \equiv \int_0^{\infty}\left(1-\exp\left(-1/x^2\right)\right)\cdot \mathrm{d}x = \sqrt{\pi}$ and $a \equiv \left(\mu_o/4\pi\right)\left(g^2\beta_e^2/\hbar\right)\sqrt{\tau_S\tau/2}\left|1-3\cos^2\theta\right|$. The integral over θ is then:

$$V'(\tau) = \frac{\sqrt{2}\pi^{3/2}}{3}\left(\frac{\mu_o}{4\pi}\right)\frac{g^2\beta_e^2}{\hbar}\sqrt{\tau_S\tau}\int_0^{\pi}\left|1-3\cos^2\theta\right|\sin\theta\cdot \mathrm{d}\theta$$

$$= \frac{8\sqrt{2}\pi^{3/2}}{9\sqrt{3}}\left(\frac{\mu_o}{4\pi}\right)\frac{g^2\beta_e^2}{\hbar}\sqrt{\tau_S\tau} \tag{13.105}$$

Substituting Eq. 13.105 in Eq. 13.89, the electron spin-echo decay envelope for long delays, $\tau \gg \tau_S$, finally becomes:

$$E(2\tau) = \exp\left(-2\sqrt{2/\pi}\,\Delta\omega_{1/2}\sqrt{\tau_S\tau}\right) \tag{13.106}$$

where the dipolar half-width $\Delta\omega_{1/2}$ is again given by Eq. 13.93. This result for the Gauss–Markov model was obtained originally by Mims (1968). In Figure 13.16, we see that it is identical to the asymptotic form of Eqs. 13.100 and 13.101 for the jump model (cf. Hu and Hartmann 1974), as is required by the central-limit theorem from probability theory.

Note that the Gauss–Markov model predicts a $\tau^{3/2}$-dependence for short echo delays (Mims 1968). As we might anticipate, this does not agree with Eq. 13.99 for the jump model. Nor is it found experimentally for spin–spin interactions of nitroxides (Milov et al. 1973b).

13.15 SPIN–SPIN INTERACTIONS AND INSTANTANEOUS DIFFUSION

An important aspect of spin–spin interactions in echo EPR is so-called instantaneous diffusion, which is induced by direct action of the inverting π-pulse. For ED-spectra of ^{14}N-nitroxides, instantaneous diffusion is characterized by preferential reduction in intensity of the central manifold, to an extent that depends on the microwave power. We illustrate this effect on spin-label ED-lineshapes from a membrane-associated peptide in Figure 13.17. The bottom panel shows that the amplitude of the central peak is reduced, relative to that of the outer wings, in spectra from pore-forming aggregates of the spin-labelled peptide. ED-spectra from superficially associated peptide at lower concentration, which are shown in the upper panel, serve as a control.

The A-spins are those that are excited by the microwave pulse. A π-pulse reverses the direction of the local dipolar field created by the A-spins, which instantaneously shifts their resonance frequency by an amount $\Delta\omega_{dd}^{o}$ (see Section 13.14). This process has diffusive character because the distribution of local fields from the different A-spins that surround a given A-spin is random. In addition, because of the non-vanishing width of the pulse, not all spins of the A-population are inverted by the nominal π-pulse (cf. Figure 13.3).

Clearly, the extent of instantaneous diffusion depends on the probability $p_{\beta,j}$ that spin j is flipped by the second pulse (see Klauder and Anderson 1962). For a particular A-spin, the contribution to the echo envelope from instantaneous diffusion is therefore (cf. Eq. 13.86):

$$E(2\tau) \sim \exp\left(-i\sum_{j} 2\Delta\omega_{d,j}^{o}\left\langle p_{\beta,j}\right\rangle_{g}\tau\right) \quad (13.107)$$

where $\Delta\omega_{d,j}^{o}$ is defined by the equivalent of Eq. 13.84 for A-spin pairs. The summation in Eq. 13.107 is over one particular configuration of surrounding A-spins, and the average is over the inhomogeneous lineshape $g(\Delta\omega_o)$ (Raitsimring et al. 1974). We average Eq. 13.107 over all A-spins exactly as done in Section 13.14 (see Eq. 13.88):

$$E(2\tau) \sim \left[\frac{1}{V}\int_{V} \exp\left(-i2\Delta\omega_{d}^{o}\left\langle p_{\beta}\right\rangle_{g}\tau\right)\cdot dV\right]^{N} \quad (13.108)$$

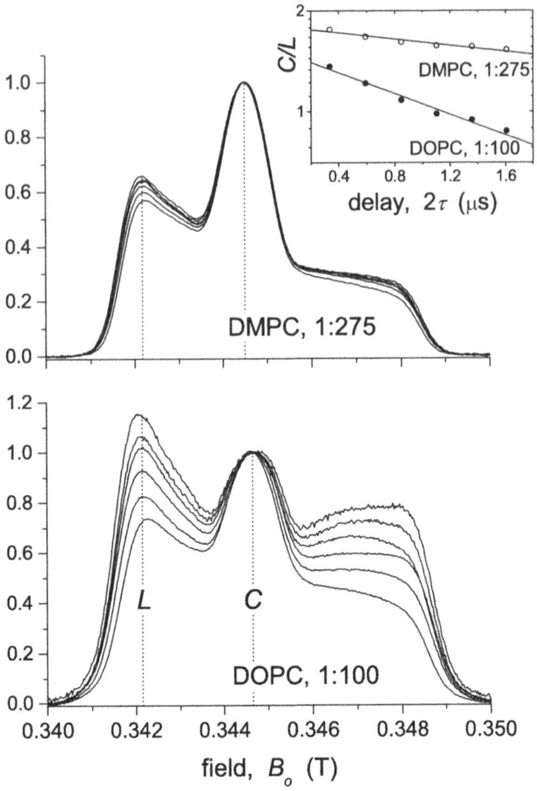

FIGURE 13.17 Two-pulse echo-detected EPR spectra of channel-forming peptide alamethicin (Alm) with first amino acid substituted by TOAC ^{14}N-nitroxide. $T = 77$ K. *Top*: TOAC[1]-Alm superficially associated with dimyristoyl phosphatidylcholine (DMPC) membranes at peptide/lipid ratio 1:275 mol/mol. *Bottom*: TOAC[1]-Alm fully integrated in dioleoyl phosphatidylcholine (DOPC) membranes at peptide/lipid ratio 1:100 mol/mol. Spectra normalized to central line height *C* and recorded with increasing interpulse spacing τ (bottom to top in each set). *Inset*: height-ratio *C/L* of central line ($m_I = 0$) to low-field shoulder ($m_I = +1$) as function of echo delay time 2τ. *y*-axis logarithmic. Data from Bartucci et al. (2009).

Again we express the integral by Eq. 13.89 where now:

$$V'(\tau) \equiv 2\pi\int_{0}^{\pi} \sin\theta\cdot d\theta \times \int_{0}^{\infty}\left[1-\exp\left(-i\left(\frac{\mu_o}{4\pi}\right)\frac{g^2\beta_e^2}{\hbar}\frac{\left(1-3\cos^2\theta\right)}{r^3}\left\langle p_{\beta}\right\rangle_{g}\right)\right]r^2\,dr \quad (13.109)$$

which we again evaluate as for Eq. 13.90. Thus, the electron spin-echo decay envelope for instantaneous diffusion becomes:

$$E(2\tau) \sim \exp\left(-2\Delta\omega_{1/2}\left\langle p_{\beta}\right\rangle_{g}\tau\right) \quad (13.110)$$

where the dipolar half-width $\Delta\omega_{1/2}$ is given by Eq. 13.93 and is directly proportional to the concentration of spin labels.

Now we see why instantaneous diffusion has such a pronounced effect on the lineshape of echo-detected spectra.

A larger fraction of spins $\langle p_\beta \rangle_{m_I}$ is excited in the central $m_I = 0$ manifold than in the outer wings ($m_I = \pm 1$), because the spread of resonance frequencies is narrower in this part of the powder pattern (see, e.g., Figure 9.5 in Chapter 9). From Eq. 13.110, this implies a more rapid decay of intensity in the centre of the ED-spectrum than in the wings, just as we see in Figure 13.17. Eq. 13.110 also predicts that the change in lineshape depends on τ according to a simple exponential decay. This is just what we find for the *C*/*L* amplitude ratio of the ED-spectra in the inset to Figure 13.17, where the *y*-axis is logarithmic.

From Sections 13.4 and 13.5, we know that the probability of a spin being flipped by the second pulse of rotation angle β_2 is given by Eq. 13.25. Thus, the average in Eq. 13.110 becomes:

$$\langle p_\beta \rangle_g = \left\langle \sin^2 \frac{\beta_2}{2} \right\rangle_g = \int_{-\infty}^{\infty} g(\Delta\omega_o) \sin^2 \frac{\beta_2(\Delta\omega_o)}{2} \cdot \mathrm{d}\Delta\omega_o \qquad (13.111)$$

where integration is over the normalized inhomogeneous lineshape $g(\Delta\omega_o)$. We also know from Eq. 13.4 of Section 13.2 that the echo amplitude is scaled by a factor $\sin\beta_1$ for a first pulse of angle β_1. Incorporating these results in Eq. 13.110, the echo decay envelope becomes (Raitsimring et al. 1974):

$$E(2\tau) = \left\langle \sin\beta_1 \sin^2 \frac{\beta_2}{2} \right\rangle_g \exp\left(-2\Delta\omega_{1/2} \left\langle \sin^2 \frac{\beta_2}{2} \right\rangle_g \tau \right) \qquad (13.112)$$

for the instantaneous diffusion mechanism.

For a second pulse of non-vanishing length, t_p, the value of $\sin(\beta_2/2)$ is given by Eq. 13.13 of Section 13.3. Integrating over the lineshape $g(\Delta\omega_o)$ as in Eq. 13.111, the average flip probability $\langle p_\beta \rangle_g$ that appears in Eqs. 13.110 and 13.112 becomes (Erilov et al. 2004a):

$$\left\langle \sin^2 \frac{\beta_2}{2} \right\rangle_g = \int_{-\infty}^{\infty} g(\Delta\omega_o) \frac{\omega_1^2}{(\Delta\omega_o)^2 + \omega_1^2} \times \sin^2\left(\frac{t_p}{2} \sqrt{(\Delta\omega_o)^2 + \omega_1^2} \right) \mathrm{d}\Delta\omega_o \qquad (13.113)$$

where $\omega_1 \equiv \gamma_e B_1$ for a second pulse of microwave field amplitude B_1. Equation 13.113 gives the explicit dependence on microwave power (via the ω_1^2-terms), which is a defining feature of instantaneous diffusion.

Raitsimring et al. (1974) studied the concentration dependence of the two-pulse echo decay from the nitroxide TEMPONE in frozen methanol solution at 77 K. Rotational and librational motion does not contribute to the echo decay at this low temperature. The experimental decay law is found to be:

$$E(2\tau) = E(0) \exp\left(-2b_{id}\tau - m\tau^2 \right) \qquad (13.114)$$

where the first term corresponds to the contribution from instantaneous diffusion (cf. Eq. 13.112), and the second term to that from dipolar-induced spectral diffusion (cf. Eq. 13.99). Figure 13.18 shows the concentration dependence of the experimental decay parameters b_{id} and m. The phase-memory parameter depends linearly on the spin-label concentration: $b_{id} = 5.5 \times 10^{-13} n_{SL}\ \mathrm{cm^3\ s^{-1}}$, as we expect for instantaneous diffusion from Eqs. 13.112 and 13.93. The spectral-diffusion parameter, on the other hand, depends quadratically on the spin-label concentration: $m = 4 \times 10^{-27} n_{SL}^2\ \mathrm{cm^6\ s^{-2}}$, which implies that the spin-flip rate $1/\tau_S$ in Eq. 13.99 is directly proportional to the spin-label concentration. We expect this if spectral diffusion takes place via a T_2-mechanism of mutual spin flips, i.e., by spin diffusion that depends on the concentration of neighbouring spins.

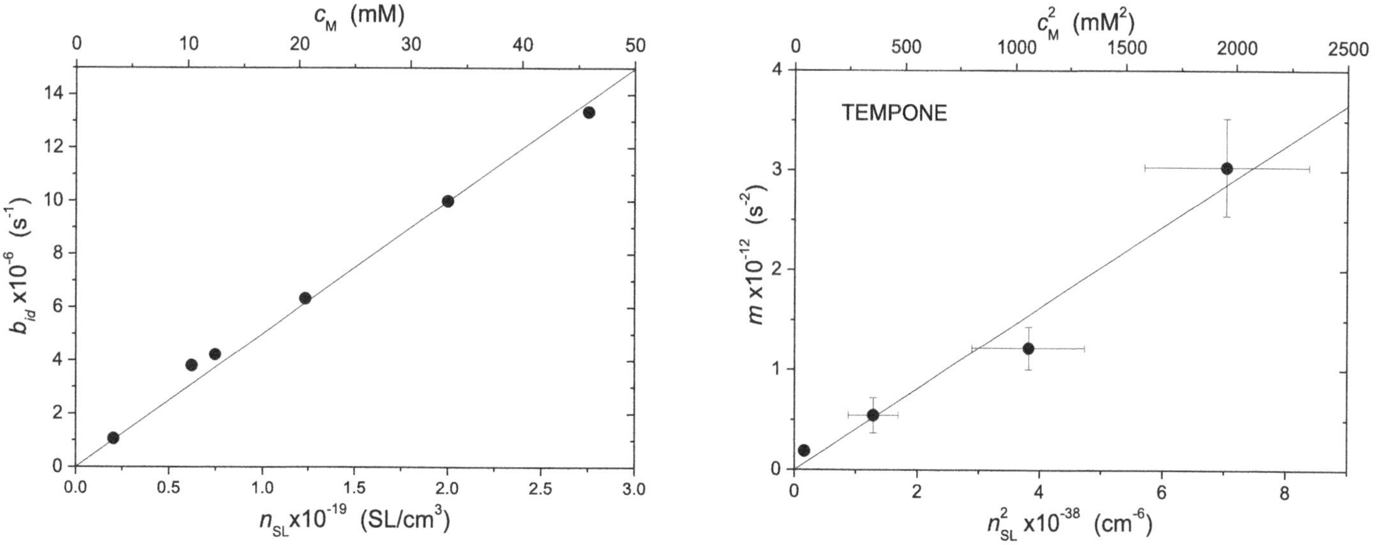

FIGURE 13.18 Concentration dependence of two-pulse echo decay constants for 2,2,6,6-tetramethyl-4-oxo-piperidine-*N*-oxyl (TEMPONE) solution in methanol at 77 K. *Left*: phase relaxation rate, b_{id}; *right*: spectral diffusion parameter, m (see Eq. 13.114). Data from Raitsimring et al. (1974).

13.16 VECTOR MODEL FOR STIMULATED ECHOES

The stimulated-echo sequence is given by $(\pi/2)_x-\tau-(\pi/2)_x-T-(\pi/2)_x-\tau-echo$ (see Figure 13.1), where all three π/2-pulses are applied along the x-axis in the rotating frame. In Fourier-transform EPR (see Chapter 14), the first delay (τ) is called the evolution period, the second delay (T) is the mixing period and the final delay (τ) is the detection period. Spin-magnetization is stored along the z-direction by the action of the second π/2-pulse. The rate of decay in the following mixing period, T, is therefore determined by the spin–lattice relaxation time T_1, which is much longer than T_2 or the phase-memory time. This extends the time range of the echo sequence, which gives sensitivity to slower motions and also improves resolution of ESEEM spectra (see Chapter 14).

Figure 13.19 shows the behaviour of the macroscopic magnetization at various stages in a three-pulse stimulated-echo sequence, starting with the first period of free precession and viewed in the rotating frame. The first 90°-pulse and subsequent evolution period are exactly as in the two-pulse primary echo sequence (see Figure 13.4). Unlike for the primary echo, however, the evolution period is terminated by another 90°-pulse along x. This second pulse tips the dephasing transverse magnetization from the x–y plane into the x–z plane (see second panel in Figure 13.19). The off-axis components then precess about the z-axis during the mixing period T that follows the second pulse (see third panel in Figure 13.19). This results in dephasing of their transverse components. The final 90°-pulse along x tips the remaining x–z plane projections of the magnetization back into the x–y plane. This is illustrated in the fourth panel of Figure 13.19, which shows that not the whole of the original dephasing magnetization given in the first panel can be refocused into the stimulated echo.

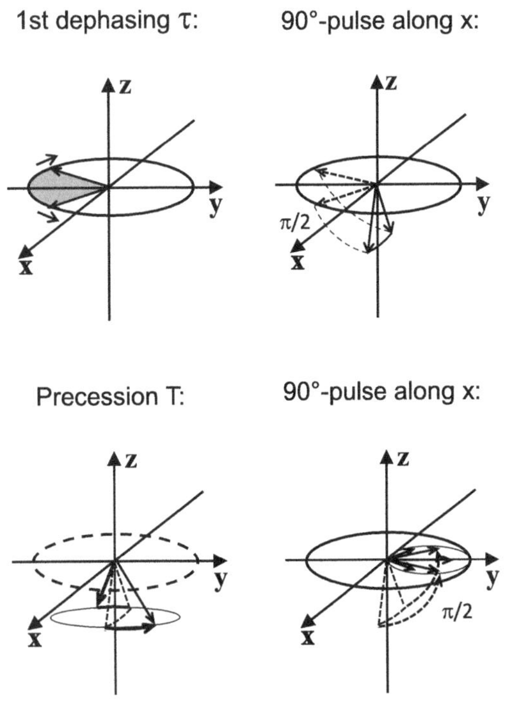

FIGURE 13.19 Spin-magnetization vector in $\pi/2-\tau-\pi/2-T-\pi/2-\tau-echo$ stimulated-echo pulse sequence, viewed in axis-frame that rotates at Larmor frequency about static magnetic-field z-direction. *From left to right*: free precession period τ after first pulse; second π/2-pulse; evolution period T; third π/2-pulse.

We can follow the pulse sequence quantitatively by using the spin-operator formalism that we introduced for the primary echo in Section 13.5 and describe in Appendix P. The spin operator after the second pulse along the x-axis with rotation angle $\beta_2 = \pi/2$ is given from Eq. 13.22 by:

$$\sigma(\tau) = S_x \sin\Delta\omega_o\tau + S_z \cos\Delta\omega_o\tau \tag{13.115}$$

for a spin packet with offset $\Delta\omega_o$ in angular resonance frequency (see Section 13.5). The S_z-component of the magnetization has an oscillatory dependence on the first precession period τ, and hence is known as a polarization pattern or grating (Mims 1972; Schweiger and Jeschke 2001). After storage along z for time T, the third 90°-pulse along x rotates this polarization grating into the x–y plane, where it then precesses further about the z-axis:

$$S_z \cos\Delta\omega_o\tau \xrightarrow{\pi/2\,S_x} S_y \cos\Delta\omega_o\tau \xrightarrow{\Delta\omega_o t S_z}$$

$$\left(S_y \cos\Delta\omega_o t + S_x \sin\Delta\omega_o t\right)\cos\Delta\omega_o\tau \tag{13.116}$$

At time t after the third pulse, the average value of the S_y-magnetization, which produces the echo, is therefore given by:

$$\overline{\langle S_y(\tau,T,t)\rangle} = \tfrac{1}{2}\left(\cos\Delta\omega_o(t-\tau) + \cos\Delta\omega_o(t+\tau)\right)S_y \tag{13.117}$$

where we have expanded the $S_y \cos\Delta\omega_o t \cos\Delta\omega_o\tau$ term from Eq. 13.116. For the echo, we need the part that becomes independent of $\Delta\omega_o$ on refocusing. The refocused echo signal at time $t=\tau$ therefore becomes: $\overline{\langle S_y(2\tau+T)\rangle} = \tfrac{1}{2}$, where we discard the term from Eq. 13.117 that contains $\Delta\omega_o 2\tau$ because it is not refocused and corresponds to residual FID. Comparing with Eq. 13.25 for $\beta_2 = \pi$, we see that the amplitude of the stimulated echo is half that of the primary echo, because not all of the magnetization is refocused.

For pulse angles β_p less than 90°, the amplitude of the stimulated echo is determined by the projection of the magnetization on the x, y-plane, or x, z-plane. For each pulse, the echo amplitude is scaled by a factor $\sin\beta_p$. The refocused stimulated-echo signal at time $t=\tau$ then becomes (Mims 1972):

$$\overline{\langle S_y(2\tau+T)\rangle} = \tfrac{1}{2}\sin\beta_1 \sin\beta_2 \sin\beta_3 \tag{13.118}$$

where β_1, β_2 and β_3 are the turning angles of the three pulses.

So far, we have ignored T_1- and T_2-relaxation during the stimulated-echo sequence. From Eqs. 13.8 and 13.9 for evolution of the transverse magnetization in Section 13.2, we see that the echo amplitude is reduced by a factor $\exp(-\tau/T_2)$ in each of the free precession periods τ. Correspondingly, the echo amplitude is reduced by a factor $\approx \exp(-T/T_1)$ during the mixing period T in which the polarization grating is stored along the z-direction (see second panel in Figure 13.19). The decay of the stimulated-echo envelope is therefore scaled by both these relaxation factors:

$$E(2\tau+T)\sim\exp\left(-\frac{2\tau}{T_2}-\frac{T}{T_1}\right) \tag{13.119}$$

which can be compared with Eq. 13.19 for the primary echo. A simple case of sensitivity to dynamics comes from applying the strong-jump model to Eq. 13.119. Strong jump shifts the resonance frequency to a position in the powder lineshape that lies outside the excitation bandwidth of the pulse. Both relaxation rates are then limited by the lifetime, τ_c, between jumps. For example: $1/T_2 = 1/T_2^o + 1/\tau_c$ in the two-site exchange model of Chapter 6, when the exchange rate is slow (see Eq. 6.20). The corresponding result for T_1 is $1/T_1 = 1/T_1^o + 1/\tau_c$ (McConnell 1958). The stimulated-echo decay therefore becomes:

$$E(2\tau+T)\sim\exp\left(-\frac{2\tau}{T_2^o}-\frac{T}{T_1^o}\right)\exp\left(-\frac{2\tau+T}{\tau_c}\right) \tag{13.120}$$

and the dynamic contribution to the stimulated-echo decay time T_{SE} is equal to the exchange lifetime τ_c (assuming $\tau_c << T_1^o$). Equation 13.120 emphasizes that the intrinsic spin–lattice relaxation time T_1^o also contributes directly to the stimulated-echo decay. We must correct for this, by using T_1-measurements from echo-detected saturation recovery or inversion recovery, to isolate the τ_c-contribution from slow molecular motion (Leporini et al. 2003; Erilov et al. 2004a). Routinely, the stimulated-echo decay is divided by the normalized extent of T_1-relaxation.

13.17 UNWANTED ECHOES AND PHASE CYCLING

In Section 13.16, we concentrated on generation of the stimulated echo at time τ after the third pulse. Other components of the magnetization were ignored. Clearly, the situation is more complicated than this. From Sections 13.2 and 13.4, we know that single pulses in general produce FIDs, and pairs of pulses correspondingly generate primary echoes, even if the flip angles differ from the optimal ideal values. Depending on the values of T and τ, these unwanted echoes and FIDs might overlap with the stimulated echo (see Figure 13.20). We remove them by cycling the phase of the pulses and of the detection.

Reversing the phase of a pulse along the positive x direction produces a pulse in the $-x$ direction, which reverses the sense of rotation of the spin magnetization by the pulse. This therefore reverses the direction of the transverse magnetization at the end of the pulse sequence, producing a negative echo. Subtracting this result from that of the original $+x$ pulse, then reinforces the echo. Applying this principle to the stimulated-echo sequence, we reverse the phase of the first pulse; then we subtract, i.e., reverse the detection phase for the y-magnetization from $+y$ to $-y$. The stimulated echo is reinforced by this operation. However, the

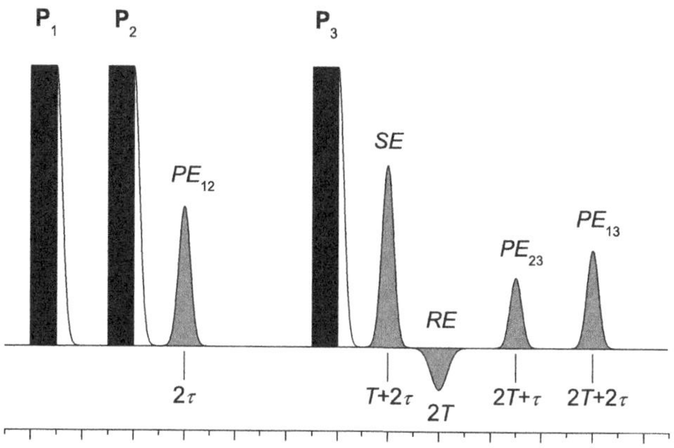

FIGURE 13.20 Echoes generated by stimulated-echo sequence $(\pi/2)_1-\tau-(\pi/2)_2-T-(\pi/2)_3-\tau-SE$, where all pulses are along x-axis. PE_{ij}: primary echo generated by pulses $\mathbf{P}_i$ and $\mathbf{P}_j$; RE: refocused PE_{12}-echo; FIDs: solid lines (without infill) immediately following each pulse.

primary echo PE_{23} produced by the second and third pulses, and the FIDs that follow the second and third pulses are not affected by the first pulse and consequently are cancelled out by the subtraction procedure.

Applying the same phase cycle to the second pulse removes the primary echo PE_{13} that is produced by the first and third pulses, as well as the FID from the first pulse, but also subtracts again the FID from the third pulse. We complete the four-step phase cycle by reversing the phase of both first and second pulses and adding the result (see Table 13.5). This compensates the extra subtraction of the FID from the third pulse. The primary echo PE_{12} from the first and second pulses is removed because the second pulse acts as a 180°-pulse for this echo, and the sign of the echo effectively changes three times and is negative. The PE_{12} echo is refocused by the third pulse, whose phase is not changed, acting as a 180°-pulse. Hence, the sign of the refocused echo RE is changed like that of PE_{12}, and both are removed by addition to the result of the previous phase cycle.

We illustrate the effect of the final step in the phase cycle in Figure 13.21 by the so-called coherence-transfer pathways. The coherence order, p, follows the changes in transverse magnetization (or coherence) that are needed to produce a particular echo (see Schweiger and Jeschke 2001). We have $p = 0$ (z-magnetization) for thermal equilibrium at the start of a pulse sequence, $|p| = 1$ (y-magnetization) for detection and opposite signs of p for dephasing and refocussing. Changes in coherence order are: $\Delta p = \pm 1$ for x-pulses that act in echo generation as 90°-pulses, $\Delta p = \pm 2$ for x-pulses that act on y-magnetization as 180°-pulses and $\Delta p = 0$ for x-pulses that are not needed to generate the echo (or act on z-magnetization as 180°-pulses). Reversing the phase of a pulse changes the sign of Δp for that pulse. We need a net $\Delta p = \pm 1$ at the end of the sequence to produce detectable coherence.

As shown in Figure 13.21, the stimulated-echo sequence SE has $\Delta p = 1$ at each of the three pulses. Reversing the phase of the first two pulses therefore leaves the sign of the stimulated echo unchanged. The primary echo PE_{12} has $\Delta p = 1$ at the first pulse,

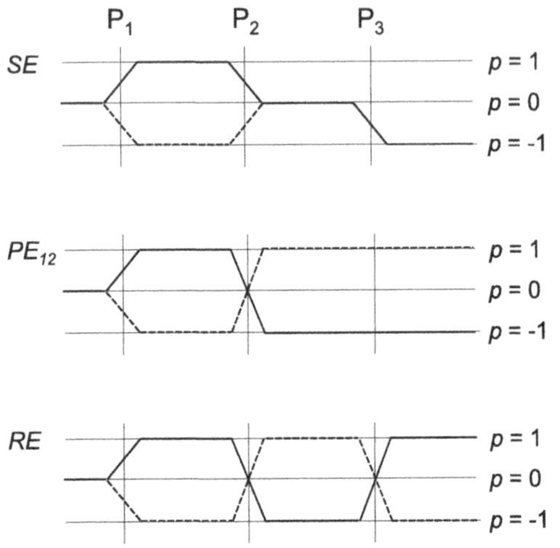

FIGURE 13.21 Coherence-transfer pathways p for stimulated-echo sequence $\mathbf{P}_1$–τ–$\mathbf{P}_2$–T–$\mathbf{P}_3$–τ–*echo*. *SE*: stimulated echo; PE_{12}: intermediate primary echo (at time 2τ); *RE*: refocused PE_{12}-echo (at time $2T$). Pathways for PE_{23} and PE_{13} primary echoes analogous to that for PE_{12}.

and $\Delta p = -2$ at the second pulse. Reversing the phase of these two pulses thus changes the sign of the resulting primary echo. The refocused echo *RE* is produced by a change in coherence order $\Delta p = 2$ at the third pulse, which results in $p = 1$ at the time of detection. As seen from Figure 13.20, the refocused echo has the opposite sign to the original PE_{12} echo, which is detected with $p = -1$ according to the scheme of Figure 13.21. Reversing the phase of the first two pulses does not change the relation of *RE* to PE_{12}, and thus similarly changes the sign of the refocused echo.

The scheme given in Table 13.5 is a complete phase cycle for the stimulated-echo sequence. Each step preserves the stimulated echo, all unwanted echoes and FIDs are eliminated, and no additional artefacts are introduced. We see that PE_{13}, PE_{12}, *RE* and the FID after the first pulse are unchanged by the first two steps, which we use to eliminate PE_{23} and the FIDs after the second and third pulses. Correspondingly, PE_{23} and the FIDs after the second and third pulses are unchanged by the final two steps, which we use to eliminate PE_{13}, PE_{12}, *RE* and the FID after the first pulse. This is a considerable achievement that becomes increasingly difficult as we increase the number of steps in the phase cycle. It helps to restrict the number of pulses that are cycled: here, for example, the phase of the third pulse stays unchanged.

TABLE 13.5 Four-step phase cycle for stimulated-echo pulse sequence, $(\pi/2)_x$–τ–$(\pi/2)_x$–T–$(\pi/2)_x$–τ–$(detection)_y$, giving phases of each pulse ($\pm x$) and of the detection ($\pm y$). (Fauth et al. 1986)

$\mathbf{P}_1$ $(\pi/2)_x$	$\mathbf{P}_2$ $(\pi/2)_x$	$\mathbf{P}_3$ $(\pi/2)_x$	*detection*
x	x	x	y
$-x$	x	x	$-y$
x	$-x$	x	$-y$
$-x$	$-x$	x	y

A simpler way of removing unwanted echoes is to reverse the phase of the first two pulses in alternate spin-echo sequences, and add the result. This alternately reverses the sign of the two-pulse echoes but leaves the sign of the three-pulse echo unchanged (Mims 1984). However, the FID from the final pulse also remains.

13.18 STIMULATED ECHO AND ROTATIONAL DYNAMICS

We can treat the sensitivity of the stimulated-echo decay to dynamics in the same way as for the primary echo in Section 13.7. Only the free-precession periods τ, and not the storage period T, interest us for generation of the stimulated echo. From the equivalent of Eq. 13.30 for the primary echo, the decay envelope of the stimulated echo is given by (Klauder and Anderson 1962):

$$E(2\tau+T) = \left\langle \exp\left(-i\int_0^{\tau}\Delta\omega_o(t)\,\mathrm{d}t\right)\exp\left(i\int_{\tau+T}^{2\tau+T}\Delta\omega_o(t)\,\mathrm{d}t\right)\right\rangle_t$$

$$\equiv \left\langle \exp\left(-i\int_0^{2\tau+T} s(t)\Delta\omega_o(t)\,\mathrm{d}t\right)\right\rangle_t \qquad (13.121)$$

where we now define the switching function by: $s(t) = 1$ when $t < \tau$, $s(t) = 0$ when $\tau < t < \tau + T$, and $s(t) = -1$ when $t > \tau + T$. For fluctuations in resonance frequency induced by small diffusive steps, the echo decay envelope is given by the equivalent of Eq. 13.33 for the primary echo:

$$E(2\tau+T) = \exp\left(-\frac{1}{2}\int_{\tau+T}^{2\tau+T}\mathrm{d}t'\int_0^{t'} s(t')s(t)G(t',t)\cdot\mathrm{d}t\right) \qquad (13.122)$$

where the autocorrelation function $G(t',t) \equiv \langle\Delta\omega_o(t')\Delta\omega_o(t)\rangle$ is exponential and given by Eq. 13.34. Expanding the integrals in Eq. 13.122 according to the $s(t')s(t)$ product of switching functions, we get:

$$\int_{\tau+T}^{2\tau+T}\mathrm{d}t'\int_0^{t'} s(t')s(t)G(t',t)\cdot\mathrm{d}t = \int_0^{\tau}\mathrm{d}t'\int_0^{t'}G(t',t)\cdot\mathrm{d}t$$

$$-\int_{\tau+T}^{2\tau+T}\mathrm{d}t'\int_0^{\tau}G(t',t)\cdot\mathrm{d}t + \int_{\tau+T}^{2\tau+T}\mathrm{d}t'\int_{\tau+T}^{t'}G(t',t)\cdot\mathrm{d}t \qquad (13.123)$$

Evaluating the integrals with Eq. 13.34 and substituting in Eq. 13.122, the stimulated-echo decay becomes the equivalent of Eq. 13.36 for the primary echo:

$$E(2\tau+T) = \exp\left(-\langle\Delta\omega_o^2\rangle\tau_c^2 C\left(\tau/\tau_c, T/\tau_c\right)\right) \qquad (13.124)$$

where τ_c is the correlation time, and $\langle\Delta\omega_o^2\rangle \equiv \langle\Delta\omega_o(0)^2\rangle$ is the second moment of the fluctuations in angular resonance frequency. The relaxation function in Eq. 13.124 is defined by:

$$C(t_1,t_2) \equiv 2\left(t_1 - 1 + e^{-t_1}\right) - \left(1 - 2e^{-t_1} + e^{-2t_1}\right)e^{-t_2} \quad (13.125)$$

which reduces to Eq. 13.37 when $t_2 = 0$. The limiting dependence for fast motion (long echo delays) is:

$$C(\tau >> \tau_c) = 2\tau/\tau_c - 3 \quad (13.126)$$

which does not depend on T and is identical with Eq. 13.38 for the primary echo. The stimulated-echo decay time T_{SE} is therefore not sensitive to fast motion. For slow motion (short echo delays), however, we get:

$$C(\tau << \tau_c) = \tfrac{2}{3}\left(\tau/\tau_c\right)^3 + \tau^2 T/\tau_c^3 \quad (13.127)$$

which depends directly on the stimulated-echo delay time, T. Equation 13.119, together with Eq. 13.127, therefore predicts a simple exponential dependence of the stimulated-echo intensity on T at constant τ, because $T >> \tau$. The stimulated-echo decay rate depends quadratically, however, on the fixed value of τ: $1/T_{SE} = \langle\Delta\omega_o^2\rangle\tau^2/\tau_c$. Exponential decay of the stimulated echo is found experimentally in the microsecond range for slow motion of spin-labelled lipids, after correcting for intrinsic spin–lattice relaxation by using inversion-recovery data (Erilov et al. 2004a).

13.19 STIMULATED-ECHO DECAY IN SUDDEN-JUMP MODEL

In Section 13.14, we used a model of uncorrelated sudden jumps to interpret the primary echo decay that arises from spin–spin interactions. Here, we use this model from Zhidomirov and Salikhov (1969) to describe the effect of angular jump dynamics on the stimulated echo. We restrict ourselves to small angular jumps, because these are most appropriate for motions in the ultra-slow regime covered by the stimulated-echo sequence (Leporini et al. 2003; Isaev and Dzuba 2008). Note that, in their treatment of spin–spin interactions, Hu and Hartmann (1974) first average over the distributions of A and B spins, before averaging over the time dependence of the spin flips (see Section 13.14). Zhidomirov and Salikhov (1969), on the other hand, first treat the time-average over the spin-flip histories. Therefore, we can adopt this latter model for sudden jumps in molecular orientation.

By analogy with Eq. 13.85 for the spin–spin case, the shift in angular resonance frequency arising from the angular jumps is given by:

$$\Delta\omega_{m_I}(\theta,\phi,t) = \Delta\omega_{m_I}(\theta,\phi)h(t) \quad (13.128)$$

where $h(t) = \pm 1$ is the switching function, which changes sign at each jump. The shift $\Delta\omega_{m_I}(\theta,\phi)$ for a single angular jump, $\pm\alpha$, is given by Eqs. 13.74, 13.77–13.79 in Section 13.13 on angular libration. The polar angles (θ, ϕ) define the orientation of the nitroxide x, y, z-axes, relative to the static magnetic field B_o, for the spin label of interest. To get the contribution to the stimulated echo from nitroxides at this orientation, we substitute Eq. 13.128 into Eq. 13.121:

$$E(2\tau + T) = \left\langle \exp\left(-i\Delta\omega_{m_I}(\theta,\phi)\left|\int_0^{2\tau+T} s(t)h(t)\,\mathrm{d}t\right|\right)\right\rangle_t \quad (13.129)$$

The two-site jump is a Poisson random process (see Section 13.14) and $h(t) = (-1)^{m(t)}$, where $m(t)$ is the number of jumps in time t. Substituting in Eq. 13.129 and taking the real part as the solution for the echo amplitude, we get:

$$E(2\tau + T) = \left\langle \cos\left(\Delta\omega_{m_I}(\theta,\phi)\left|\int_0^{2\tau+T} s(t)(-1)^{m(t)}\,\mathrm{d}t\right|\right)\right\rangle_t \quad (13.130)$$

Expanding the cosine in a series, the average then becomes (Zhidomirov and Salikhov 1969):

$$E(2\tau + T) = \sum_k (-1)^k \frac{(\Delta\omega_{m_I})^{2k}}{(2k)!}$$

$$\times \int_0^{2\tau+T} \mathrm{d}t_{2k} \int_0^{t_{2k}} \mathrm{d}t_{2k-1} \cdots \int_0^{t_2} \mathrm{d}t_1 s(t_{2k}) \cdots s(t_1)\left\langle(-1)^M\right\rangle \quad (13.131)$$

where $\Delta\omega_{m_I} \equiv \Delta\omega_{m_I}(\theta,\phi)$, and $M = \sum m(t_i)$. Averaging over the Poisson distribution, which is given by Eq. 13.94, we obtain a product of exponentials with alternating signs:

$$\left\langle(-1)^M\right\rangle = \exp\left(-2\left(t_{2k} - t_{2k-1} + \cdots + t_2 - t_1\right)/\tau_c\right) \quad (13.132)$$

where $1/\tau_c$ is the jump rate.

Performing the integrals in Eq. 13.131 by using the Laplace transform, Zhidomirov and Salikhov (1969) finally obtain the following result for the stimulated-echo signal (see Erilov et al. 2004a):

$$E(2\tau + T) = \exp\left(-2\tau/\tau_c\right)\left[1 + \frac{1}{R_{m_I}\tau_c}\sinh 2R_{m_I}\tau\right.$$

$$\left. + \frac{2}{R_{m_I}^2\tau_c^2}\sinh^2 R_{m_I}\tau - \left(1 - \exp\left(-2T/\tau_c\right)\right)\frac{\Delta\omega_{m_I}^2}{R_{m_I}^2}\sinh^2 R_{m_I}\tau\right] \quad (13.133)$$

where $R_{m_I}^2 \equiv \left(1/\tau_c\right)^2 - \Delta\omega_{m_I}^2$. When $R_{m_I}^2 < 0$, i.e., R_{m_I} is imaginary, we can express the hyperbolic functions more conveniently as normal trignometric functions. Then Eq. 13.133 becomes:

$$E(2\tau+T)=\exp\left(-2\tau/\tau_c\right)\left[1+\frac{1}{|R_{m_I}|\tau_c}\sin 2|R_{m_I}|\tau\right.$$
$$\left.+\frac{2}{|R_{m_I}^2|\tau_c^2}\sin^2|R_{m_I}|\tau-\left(1-\exp\left(-2T/\tau_c\right)\right)\frac{\Delta\omega_{m_I}^2}{|R_{m_I}^2|}\sin^2|R_{m_I}|\tau\right] \tag{13.134}$$

for the condition $\Delta\omega_{m_I}^2\tau_c^2>1$.

We get the primary-echo amplitude $E(2\tau)$ by omitting the last term on the right in Eq. 13.133, because it vanishes for $T=0$. For fast motion (i.e., when $\Delta\omega_{m_I}^2\tau_c^2<<1$ and thus $R_{m_I}=1/\tau_c$) we can neglect this term, showing again that the stimulated-echo decay time T_{SE} is insensitive to fast motion (cf. Eq. 13.126). Expanding the remaining terms to the lowest non-vanishing order in $\Delta\omega_{m_I}$, we find that the stimulated-echo decay becomes exponential:

$$E(2\tau+T)\approx\exp\left(-\Delta\omega_{m_I}^2\tau_c\tau\right) \tag{13.135}$$

and is identical to the primary-echo decay. This is also the decay law found from Eqs. 13.124 and 13.126 for Brownian diffusion, which confirms that fast-motion results are independent of mechanism, as required by the central-limit theorem of probability theory.

For the opposite situation of slow motion, (i.e., when $\Delta\omega_{m_I}^2\tau_c^2>>1$ and thus $R_{m_I}=i\Delta\omega_{m_I}$), the stimulated-echo amplitude becomes (see Eq. 13.134):

$$E(2\tau+T)=\cos^2\left(\Delta\omega_{m_I}\tau\right)\times\exp\left(-\frac{2\tau}{\tau_c}\right)$$
$$+\sin^2\left(\Delta\omega_{m_I}\tau\right)\exp\left(-\frac{2(\tau+T)}{\tau_c}\right) \tag{13.136}$$

where we neglect terms in $1/\left(\Delta\omega_{m_I}\tau_c\right)$ and $1/\left(\Delta\omega_{m_I}\tau_c\right)^2$, compared with unity. Thus, the primary echo-detected spectrum (where $T=0$) is insensitive to slow motion, because the echo amplitude then no longer depends on $\Delta\omega_{m_I}(\theta,\phi)$. The stimulated-echo decay is monoexponential for slow jumps, with a decay time determined by the jump time: $T_{SE}=\tau_c/2$ and independent of the value of τ (cf. Eq. 13.120). By comparing lineshapes detected from two-pulse and three-pulse echo experiments, we therefore can learn about the timescale of motion and its distribution between fast and slow components.

13.20 STIMULATED ECHO AND SPIN–SPIN INTERACTION

We can extend the sudden-jump model, used in Section 13.14 to describe the effects of dipolar spin–spin interactions on the primary-echo decay, immediately to the stimulated echo (Hu and Hartmann 1974). Equation 13.92 for the primary spin-echo decay envelope then becomes:

$$E(2\tau+T)=\exp\left(-\Delta\omega_{1/2}\left\langle\left|\int_0^{2\tau+T}s(t)h(t)\mathrm{d}t\right|\right\rangle_t\right) \tag{13.137}$$

where the dipolar half-width $\Delta\omega_{1/2}$ is again given by Eq. 13.93. We take the definition of the switching function $s(t)$ for the stimulated echo from Section 13.18. The contribution to the average in Eq. 13.137 for no spin flips ($m=0$) vanishes:

$$w_0(2\tau+T)\left|\int_0^{2\tau+T}s(t)\cdot\mathrm{d}t\right|=w_0(2\tau+T)\left|\int_0^{\tau}\mathrm{d}t-\int_{\tau+T}^{2\tau+T}\mathrm{d}t\right|=0 \tag{13.138}$$

as for the primary echo (cf. Eq. 13.95). The contribution from histories with one spin flip only is (cf. Eq. 13.96):

$$w_1(2\tau+T)\int_0^{2\tau+T}\mathrm{d}t_1\left|\int_0^{2\tau+T}s(t)h_1(t)\mathrm{d}t\right|$$
$$=w_1(2\tau+T)\int_0^{2\tau+T}\mathrm{d}t_1\left|\int_0^{\tau}h_1(t)\mathrm{d}t-\int_{\tau+T}^{2\tau+T}h_1(t)\mathrm{d}t\right| \tag{13.139}$$

where $h_1(t)$ changes from +1 to −1 at time $t=t_1$. Thus, we must evaluate the first integral in Eq. 13.139 separately, depending on whether $t_1\le\tau$, $\tau\le t_1\le\tau+T$ or $t_1\ge\tau+T$, respectively:

$$\left|\int_0^{\tau}h_1(t)\mathrm{d}t-\int_{\tau+T}^{2\tau+T}h_1(t)\mathrm{d}t\right|=\left|\left(\int_0^{t_1}\mathrm{d}t-\int_{t_1}^{\tau}\mathrm{d}t\right)+\int_{\tau+T}^{2\tau+T}\mathrm{d}t\right|=2t_1 \tag{13.140a}$$

$$\left|\int_0^{\tau}h_1(t)\mathrm{d}t-\int_{\tau+T}^{2\tau+T}h_1(t)\mathrm{d}t\right|=\left|\int_0^{\tau}\mathrm{d}t+\int_{\tau+T}^{2\tau+T}\mathrm{d}t\right|=2\tau \tag{13.140b}$$

$$\left|\int_0^{\tau}h_1(t)\mathrm{d}t-\int_{\tau+T}^{2\tau+T}h_1(t)\mathrm{d}t\right|=\left|\int_0^{\tau}\mathrm{d}t-\left(\int_{\tau+T}^{t_1}\mathrm{d}t-\int_{t_1}^{2\tau+T}\mathrm{d}t\right)\right|$$
$$=2(2\tau+T-t_1) \tag{13.140c}$$

Substituting Eqs. 13.140a,b,c into Eq. 13.139, the contribution from single-flip histories then becomes:

$$2w_1(2\tau+T)\left|\int_0^{\tau}t_1\,\mathrm{d}t_1+\int_{\tau}^{\tau+T}\tau\cdot\mathrm{d}t_1+\int_{\tau+T}^{2\tau+T}(2\tau+T-t_1)\mathrm{d}t_1\right|$$
$$=2\left(\frac{\tau^2+\tau T}{\tau_S}\right)\exp\left(-\frac{2\tau+T}{\tau_S}\right) \tag{13.141}$$

where $w_1(2\tau+T)$ comes from the definition following Eq. 13.94 in Section 13.14, and $1/\tau_S$ is the spin-flip rate. Equation 13.141 is the leading contribution for short delay times or slow spin flips, $T<<\tau_S$, and from Eq. 13.137 the stimulated-echo decay envelope then becomes (Hu and Hartmann 1974):

$$E(2\tau+T)=\exp\left(-2\Delta\omega_{1/2}\left(\tau^2+\tau T\right)/\tau_S\right) \quad (13.142)$$

when $\exp\left(-(2\tau+T)/\tau_S\right)\approx 1$. This again predicts an exponential decay of the stimulated echo for constant τ, because $T >> \tau$. The stimulated-echo decay rate depends on the fixed value of τ: $1/T_{SE}=2\Delta\omega_{1/2}\tau/\tau_S$, and in a way different from that for Brownian diffusion (cf. Eq. 13.127).

Hu and Hartmann (1974) give a general analytical solution for the $\langle\ldots\rangle_t$ average in Eq. 13.137. The decay envelope of the stimulated echo is written as:

$$E(2\tau+T)=\exp\left(-2\Delta\omega_{1/2}\tau K\left(2\tau/\tau_S\right)\right)$$
$$\times\exp\left[-\Delta\omega_{1/2}\tau\left(G\left(2\tau/\tau_S\right)-K\left(2\tau/\tau_S\right)\right)\times\left(1-e^{-T/\tau_S}\right)\right] \quad (13.143)$$

where the first exponential term is the two-pulse echo decay function $E(2\tau)$ – see Eq. 13.100. The universal function $K(z)$ is given for the primary echo by Eq. 13.101; and $G(z)$, which characterizes the free-induction decay (FID), is defined by:

$$G(z)\equiv\exp(-z)\left[I_o(z)+I_1(z)\right] \quad (13.144)$$

where $I_n(z)$ are the modified Bessel functions of order n. Figure 13.15, which is given already in Section 13.14 above, plots the $G(z)$ and $K(z)$ functions, together with the asymptotic forms for small and large z. We include numerical values for $G(z)$ in Table 13.4, given likewise in Section 13.14. For $z << 1$, we have $G(z)=1-\frac{1}{2}z$ and $K(z)=\frac{1}{2}z$, so that together with $T << \tau_S$, Eq. 13.143 reduces to Eq. 13.142 derived for short times and slow spin flips. For $z >> 1$, we have $G(z)=K(z)$ as seen in Figure 13.15; the stimulated-echo decay is then independent of the T-delay and becomes identical to the primary-echo decay for long times and fast spin flips (see Eq. 13.106).

Figure 13.22 shows how the amplitude of the stimulated echo depends on the second delay time, T, for different fixed values of the parameter $\Delta\omega_{1/2}\tau\left[G\left(2\tau/\tau_S\right)-K\left(2\tau/\tau_S\right)\right]$ appearing in Eq. 13.143. The y-axis is logarithmic. These curves confirm that the echo-decay profiles become independent of the second delay, when T is long compared with the time between spin flips.

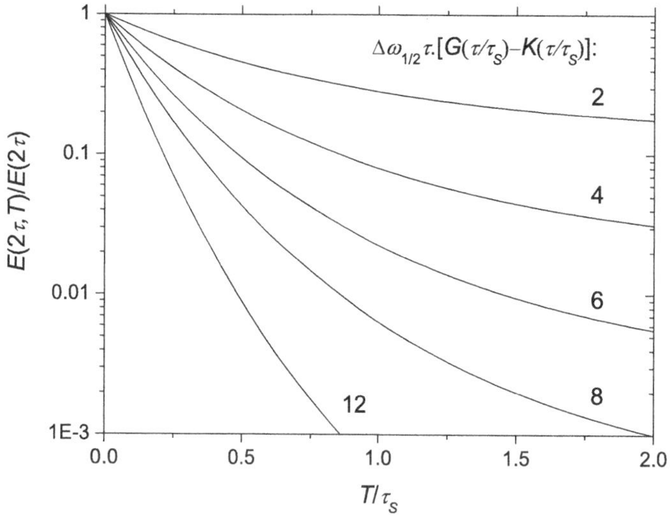

FIGURE 13.22 Dependence of stimulated-echo amplitude $E(2\tau+T)$ on second interpulse spacing T (Eq. 13.143), for dipolar relaxation by uncorrelated sudden jumps (Hu and Hartmann 1974); τ_S, time between spin flips. Decay curves for fixed values of $\Delta\omega_{1/2}\tau\left[G\left(2\tau/\tau_S\right)-K\left(2\tau/\tau_S\right)\right]$ (see Eqs. 13.101 and 13.144), where $\tau(>> \tau_S)$ is first interpulse spacing. Logarithmic y-axis normalized to amplitude $E(2\tau)$ of primary echo; x-axis normalized to τ_S.

13.21 EXPERIMENTAL STIMULATED-ECHO DECAYS

Figure 13.23 shows the stimulated-echo (SE) decay rates, $1/T_{SE}$, for orientation-selective measurements in the high-field spectra from a spin-labelled ionic polymer. Decay rates are recorded at the canonical g_{xx}-, g_{yy}- and g_{zz}-turning points in the 94-GHz echo-detected spectra. Decay rates are twice as fast in the x- and y-regions as in the z-region, because the latter is sensitive only to fluctuations in the polar angle θ, whereas the x- and y-regions are sensitive to fluctuations in both θ and ϕ (Leporini et al. 2003). The SE-decay rates are not strongly dependent on the first delay τ (see Figure 13.23), which is inconsistent with Eq. 13.127 for Brownian diffusion but tends to favour a jump model (see Eq. 13.136). Leporini et al. (2003) conclude that the waiting time τ_ϕ between jumps in ϕ determines the SE-decay rate in the x-, y-regions, i.e., $\tau_\phi\approx T_{SE\,x,y}$, where the decay time is $T_{SE\,x,y}\approx 18\,\mu s$ at temperatures below 260 K.

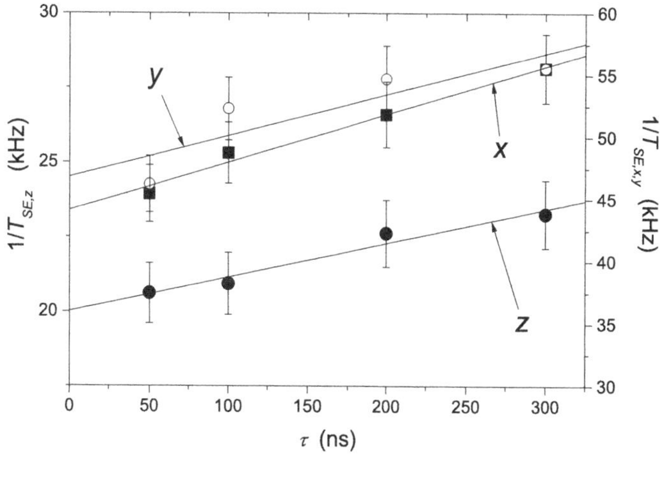

FIGURE 13.23 Stimulated-echo decay rate, $1/T_{SE}$, as function of first delay time τ, measured at x-, y- or z-turning points in 94-GHz spectrum of 4-carboxy-TEMPO K+ salt in end-group sulphonated *poly*(isoprene) at 230 K. Decay rate determined from exponential dependence of echo intensity on second delay time T (cf. Eqs. 13.127 and 13.136). Data from Leporini et al. (2003).

13.22 CONCLUDING SUMMARY

1. *Spin echoes for short T_2:*
In pulse EPR, electron spin echoes refocus the transverse magnetization after a suitable time delay. This overcomes the deadtime problem associated with the short T_2-relaxation times of nitroxide spin labels.
2. *Primary echo and fast motion (phase-memory time):*
The two-pulse primary echo $(\pi/2)_x$–τ–$(\pi)_x$–τ–*echo* amplitude decays exponentially for fast motion (i.e., short correlation times $\tau_c \ll \tau$):

$$E(2\tau) = E(0)\exp\left(-2\tau/T_{2M}\right) \tag{13.145}$$

The phase-memory time T_{2M} is given by:

$$1/T_{2M} = \left\langle \Delta\omega_o^2 \right\rangle \tau_c \tag{13.146}$$

where $\left\langle \Delta\omega_o^2 \right\rangle$ is the second moment of the fluctuations in angular resonance frequency, just as for Lorentzian linewidths.
3. *Primary echo and slow motion (Brownian, free and jump diffusion):*
For slow Brownian motion (i.e., long correlation times $\tau_c \gg \tau$), or with short delays, the primary-echo decay is non-exponential:

$$E(2\tau) = E(0)\exp\left(-\left\langle \Delta\omega_o^2 \right\rangle 2\tau^3/3\tau_c\right) \tag{13.147}$$

We must integrate Eq. 13.147 with angular-dependent $\left\langle \Delta\omega_o(\theta)^2 \right\rangle$ over all θ, for slow motion (see Eq. 13.45). The exponential tail at long delays characterizes the longest decay time, where $T_{2M}^{\infty} \propto \tau_R^{1/2}, \propto \tau_R^{2/3}$ and $\propto \tau_R$ for Brownian diffusion, collision-limited free diffusion and jump diffusion, respectively. Table 13.1 gives more precise calibration parameters. The general solution, which corresponds to Eq. 13.147 for slow motion, is given by Eqs. 13.36 and 13.37 for Brownian diffusion with unrestricted τ/τ_c.
4. *Echo-detected spectra and librational motion:*
Field-swept spectra detected by the primary echo have anisotropic lineshapes that are sensitive to rapid, low-amplitude librations. For dimensionless scan parameter $\delta = (\omega - \omega_\perp)/(\omega_\parallel - \omega_\perp)$, the dependence of the phase-memory time is modelled by:

$$1/T_{2M}(\delta) \approx (\omega_\parallel - \omega_\perp)^2 \left\langle \alpha^2 \right\rangle \tau_R \times \delta(1-\delta) \tag{13.148}$$

where $\left\langle \alpha^2 \right\rangle$ is the mean-square angular amplitude of libration. We get the amplitude-correlation time product $\left\langle \alpha^2 \right\rangle \tau_R$ from partially relaxed echo-detected spectra, as a function of the echo delay-time τ (see Table 13.2). The libration amplitude $\left\langle \alpha^2 \right\rangle$ comes independently from the small reductions in outer hyperfine splitting $\langle 2A_{zz} \rangle$.
5. *Primary echoes and dipolar spin–spin interaction:*
Primary-echo decays are affected by dipolar spin–spin interactions. For dilute electron spins, phase-memory times are limited by spin flips of surrounding nuclei. For delay times short compared with mean time τ_S between spin flips ($\tau \ll \tau_S$):

$$E(2\tau) = \exp\left(-2\Delta\omega_{1/2}\tau^2/\tau_S\right) \tag{13.149}$$

where $\Delta\omega_{1/2}$ defined by Eq. 13.93 is the dipolar half-width, proportional to the concentration of spin-flipping species. For long delay times ($\tau \gg \tau_S$):

$$E(2\tau) = \exp\left(-2\sqrt{2/\pi}\,\Delta\omega_{1/2}\sqrt{\tau_S \tau}\right) \tag{13.150}$$

which, by virtue of the central-limit theorem, is also the result predicted with a Gauss–Markov model. The general solution is given by Eq. 13.100 and Table 13.4 for unrestricted τ/τ_S.
6. *Primary echoes and instantaneous diffusion:*
Instantaneous diffusion comes from flipping of surrounding electron spins by the second pulse in the primary-echo sequence. The echo decay becomes:

$$E(2\tau) \sim \exp\left(-2\Delta\omega_{1/2}\left\langle p_\beta \right\rangle_g \tau\right) \tag{13.151}$$

where $\left\langle p_\beta \right\rangle_g$ is the probability that a spin is flipped by the second pulse, averaged over the inhomogeneous lineshape $g(\Delta\omega_o)$. This depends on the microwave power B_1^2 of the second pulse:

$$\left\langle p_\beta \right\rangle_g = \int_{-\infty}^{\infty} g(\Delta\omega_o) \frac{\gamma_e^2 B_1^2}{(\Delta\omega_o)^2 + \gamma_e^2 B_1^2} \times \sin^2\left(\frac{t_p}{2}\sqrt{(\Delta\omega_o)^2 + \gamma_e^2 B_1^2}\right) \mathrm{d}\Delta\omega_o \tag{13.152}$$

where t_p is the pulse length. Instantaneous diffusion preferentially suppresses intensity in the central line of an echo-detected ^{14}N-nitroxide spectrum, increasingly with increasing microwave power.
7. *Stimulated, three-pulse echoes:*
Stimulated echoes $(\pi/2)_x$–τ–$(\pi/2)_x$–T–$(\pi/2)_x$–τ–*echo* are sensitive to ultraslow motions on the T_1-timescale, when the second delay T is incremented ($T \gg \tau$). The stimulated-echo decay is exponential for slow two-site exchange or strong jumps:

$$E(2\tau + T) \sim \exp\left(-2\tau\left(\frac{1}{T_2^o} + \frac{1}{\tau_c}\right)\right)\exp\left(-T\left(\frac{1}{T_1^o} + \frac{1}{\tau_c}\right)\right) \tag{13.153}$$

The stimulated-echo decay rate is $1/T_{SE} = 1/T_1^o + 1/\tau_c$ (assuming $\tau_c << T_1^o$). We must correct the decay by using T_1^o-measurements from saturation- or inversion-recovery to get the dynamic contribution, i.e., τ_c.

8. *Stimulated echoes and slow motion:*
For short correlation times or long delay times ($T >> \tau_c$), the stimulated-echo decay does not depend on T, i.e., is insensitive to fast motion. For slow Brownian diffusion or with short delay times ($T << \tau_c$), the stimulated-echo decay is exponential:

$$E(2\tau + T) \approx \exp\left(-\left\langle \Delta\omega_o^2 \right\rangle \tau^2 T/\tau_c\right) \tag{13.154}$$

after correction for T_1-relaxation, where $\left\langle \Delta\omega_o^2 \right\rangle$ is the second moment of fluctuations in angular resonance frequency. Again we must integrate over all angles for slow motion. The stimulated echo decay rate then depends quadratically on fixed delay τ: $1/T_{SE} = \left\langle \Delta\omega_o^2 \right\rangle \tau^2/\tau_c$. The general solution, which corresponds to Eq. 13.154, is given by Eqs. 13.124 and 13.125 for Brownian diffusion with unrestricted T/τ_c.

9. *Stimulated echoes and dipolar spin–spin interaction:*
For fast spin flips or long delay times ($T >> \tau_S$), the effect of dipolar interactions on the stimulated echo is independent of T-delay. For slow spin flips or short delays ($T << \tau_S$), the stimulated-echo decay is exponential:

$$E(2\tau + T) \approx \exp\left(-2\Delta\omega_{1/2}\tau T/\tau_S\right) \tag{13.155}$$

after correction for T_1-relaxation, where $\Delta\omega_{1/2}$ is the dipolar half-width. The stimulated-echo decay rate then depends linearly on fixed delay τ: $1/T_{SE} = 2\Delta\omega_{1/2}\tau/\tau_S$. The general solution for unrestricted τ/τ_S is given by Eq. 13.143 and Table 13.4.

APPENDIX A13: LORENTZIAN SPECTRAL DIFFUSION

A13.1 Conditional Probabilities and Spectral Diffusion

Where appropriate, we compare Gaussian spectral diffusion (already dealt with in the main body of the chapter) with equivalent results for Lorentzian spectral diffusion.

Klauder and Anderson (1962) describe an approximate treatment (valid for short delay times), where the frequency Fourier-transform of the conditional probability, $P(t;\Delta\omega) = P(\omega_o;\omega,t)$, is given by the general expression:

$$p(t;\tau) \equiv \mathcal{F}^{-1}\{P(t;\Delta\omega)\} = \exp\left(-f(\tau)t\right) \tag{A13.1}$$

cf. Eq. T.2 in Appendix T. For Gaussian spectral diffusion, the functional dependence on τ is:

$$f(\tau) = k\tau^2 \tag{A13.2}$$

and for Lorentzian spectral diffusion is defined by:

$$f(\tau) = m|\tau| \tag{A13.3}$$

where k and m are constant coefficients determining the intensity of spectral diffusion.

Specifically, from back Fourier-transform of Eq. A13.1, the conditional probability that corresponds to Eq. A13.3 is (see Table T.1 in Appendix T):

$$P(t;\Delta\omega) \equiv \mathcal{F}\{p(t;\tau)\} = \frac{2mt}{(\Delta\omega)^2 + (mt)^2} \tag{A13.4}$$

which is a Lorentzian frequency distribution. Eq. A13.4 describes spectral diffusion with a Lorentzian half-width, $\Delta\omega_{1/2,L} = mt$, that increases linearly with time. We call it the diffusion kernel. The conditional probability is not stationary because the right-hand side of Eq. A13.4 tends to zero, instead of to a limiting distribution, at long times. Therefore, this simplified treatment applies only for short times.

Correspondingly, Eq. A13.2 leads to a Gaussian frequency distribution (see Table T.1) with a second moment, $\left\langle \Delta\omega^2 \right\rangle_G = 2kt$, that increases linearly with time (cf. Eq. 13.31 for a Gaussian random variable in Section 13.6).

A13.2 Echo Decays

We return to Eq. 13.30 in Section 13.6, but consider only the time dependence of the echo decay, ignoring the absolute amplitude. Klauder and Anderson (1962) show that, after averaging over the equilibrium frequency distribution, the echo decay specified by Eq. A13.1 becomes:

$$E(t) = \exp\left(-\int_0^t f\left(\int_0^{t'} s(t'')\cdot dt''\right)dt'\right) \tag{A13.5}$$

at delay time t. Expanding the integrals according to the switching function $s(t'')$, the two-pulse echo decay is then given by:

$$E(2\tau) = \exp\left(-2\int_0^{\tau} f(t')\cdot dt'\right) \tag{A13.6}$$

For the stimulated echo (see Section 13.16), a sinusoidal magnetization pattern that depends on the integral over the frequency distribution $P(T;\Delta\omega)$ is stored along the z-direction during the mixing time T. This S_z-magnetization pattern is periodic in $\Delta\omega\tau$ (see Eq. 13.115), and therefore decays as the Fourier transform $p(T;\tau)$ of the conditional probability, i.e., diffusion kernel (Mims

1968). From Eq. A13.1, the decay of the three-pulse echo thus becomes simply:

$$E(2\tau+T)=E(2\tau)p(T;\tau)=E(2\tau)\exp(-f(\tau)T) \quad \text{(A13.7)}$$

where τ is the first delay and T is the second delay in the three-pulse echo sequence (see Figure 13.1). Such a simple separation is not possible, however, in the case of longer τ-delays (Klauder and Anderson 1962).

For Gaussian spectral diffusion, we get the decays of the primary and stimulated echoes from Eqs. A13.2 and A13.6, A13.7:

$$E(2\tau)=\exp\left(-2k\tau^3/3\right) \quad \text{(A13.8)}$$

$$E(2\tau+T)=E(2\tau)\exp\left(-k\tau^2T\right) \quad \text{(A13.9)}$$

which agree with Eqs. 13.36, 13.39 and 13.124, 13.127, in Sections 13.7 and 13.18, respectively, for short delay times. From this we find that $k=\langle\Delta\omega(0)^2\rangle/\tau_c$, in the case of Brownian rotational diffusion.

For Lorentzian spectral diffusion, on the other hand, we get the following echo decays from Eqs. A13.3 and A13.6, A13.7:

$$E(2\tau)=\exp\left(-m\tau^2\right) \quad \text{(A13.10)}$$

$$E(2\tau+T)=E(2\tau)\exp(-m\tau T) \quad \text{(A13.11)}$$

As anticipated from the statistical theory of dipolar broadening, these results agree with Eqs. 13.99 and 13.142, which apply at short delay times for relaxation induced by a random distribution of dilute B-spins (see Sections 13.14 and 13.20, respectively). Flipping B-spins at a given rate $1/\tau_S$ is the same as introducing an equivalent number of randomly placed static B-spins. From the statistical theory presented in Section 15.9 of Chapter 15, we know that this causes (truncated) Lorentzian broadening with half-width at half-height $\Delta\omega_{1/2}$ given by Eq. 13.93. The resulting Lorentzian spectral diffusion is then characterized by the coefficient $m=2\Delta\omega_{1/2}/\tau_S$. As we see in the next section, however, this analogy does not apply at longer times when the B-spins undergo multiple flips (see Mims 1968).

A13.3 Stationary Probabilities and Spectral Diffusion

Klauder and Anderson (1962) give an exact treatment (valid at all times) for a stationary conditional probability. To avoid divergences at long times, we replace Eq. A13.1 by:

$$\begin{aligned} p(t;\tau) &= \exp\left(-f(\tau)\int_0^t \tau_c\exp\left(-t'/\tau_c\right)\cdot dt'\right) \\ &= \exp\left(-f(\tau)\tau_c\left[1-\exp\left(-t/\tau_c\right)\right]\right) \end{aligned} \quad \text{(A13.12)}$$

where τ_c is some characteristic decay time. At short times, Eq. A13.12 reduces to Eq. A13.1. From Eqs. A13.3 and A13.12, the conditional probability describing Lorentzian spectral diffusion then becomes (cf. Mims 1968):

$$P(t;\Delta\omega)=\frac{2m\tau_c\left[1-\exp\left(-t/\tau_c\right)\right]}{(\Delta\omega)^2+m^2\tau_c^2\left[1-\exp\left(-t/\tau_c\right)\right]^2} \quad \text{(A13.13)}$$

instead of the non-stationary value given by Eq. A13.4. (Note that the factor of two scaling $R\equiv 1/\tau_c$ in Mims 1968 is unnecessary: see Hu and Hartmann 1974.) The time-dependent Lorentzian half-width is now: $\Delta\omega_{1/2,L}=m\tau_c\left[1-\exp\left(-t/\tau_c\right)\right]$, which tends to a limiting value $\Delta\omega_{1/2,L}=m\tau_c$ at long times.

After averaging over the equilibrium frequency distribution, Klauder and Anderson (1962) show that the echo decay specified by Eq. A13.12 becomes:

$$\begin{aligned} E(t) = \exp\Bigg(&-\int_0^t f\left(-e^{-t'/\tau_c}\int_0^{t'} e^{-t''/\tau_c}s(t'')\cdot dt''\right)dt' \\ &-\int_0^\infty f\left(-e^{-t'/\tau_c}\int_0^{t'} e^{-t''/\tau_c}s(t'')\cdot dt''\right)dt'\Bigg) \end{aligned} \quad \text{(A13.14)}$$

instead of Eq. A13.5. For Gaussian spectral diffusion (i.e., Eq. A13.2), this reduces to (Klauder and Anderson 1962):

$$E(t)=\exp\left(-k\tau_c\int_0^t dt'\int_0^{t'} s(t'')s(t')\exp\left(-(t'-t'')/\tau_c\right)\cdot dt''\right) \quad \text{(A13.15)}$$

which essentially is equivalent to Eqs. 13.33 and 13.34 in Section 13.7. Together with Eq. 13.35, where $G(t',t'')\equiv\exp\left(-(t'-t'')/\tau_c\right)$, the primary echo decay for Gaussian spectral diffusion then becomes:

$$E(2\tau)=\exp\left[-k\tau_c^3\left(2\tau/\tau_c-3+4e^{-\tau/\tau_c}-e^{-2\tau/\tau_c}\right)\right] \quad \text{(A13.16)}$$

which is identical with Eqs. 13.36 and 13.37 for Brownian diffusion, where again $k=\langle\Delta\omega(0)^2\rangle/\tau_c$. The short-time $(\tau<<\tau_c)$ behavior of Eq. A13.16 reduces to Eq. A13.8. For the three-pulse echo, we get the decay profile from Eqs. 13.124 and 13.125 of Section 13.18 for Brownian diffusion, together with Eq. A13.16:

$$E(2\tau+T)=E(2\tau)\exp\left[-k\tau_c^3\left(1-2e^{-\tau/\tau_c}+e^{-2\tau/\tau_c}\right)\left(1-e^{-T/\tau_c}\right)\right] \quad \text{(A13.17)}$$

which reduces to $E(2\tau+T)=E(2\tau)$ for $T=0$. By definition, short- and long-time limiting behaviours are those of Eqs. 13.124 and 13.125. In particular, Eq. A13.17 reduces to Eq. A13.9 for $T,\tau<<\tau_c$. In the limit of short τ, the term multiplying $E(2\tau)$ in Eq. A13.17 becomes equal to $p(T;\tau)$, the Fourier transform of the Gaussian diffusion kernel (see Eqs. A13.2 and A13.12). This agrees with our reasoning given in Section A13.2 (cf. Eq. A13.7).

In the case of Lorentzian spectral diffusion, the general expression for the primary echo decay that corresponds to Eq. A13.12 is (Klauder and Anderson 1962):

$$E(2\tau) = \exp\left[-m\tau_c^2\left(\tau/\tau_c - \ln\left(2 - e^{-\tau/\tau_c}\right)\right)\right] \qquad (A13.18)$$

which differs in detail from Eq. 13.100 of Section 13.14 for the statistical approach to a dilute distribution of interacting spins (see Figure A13.1). The short-time $(\tau << \tau_c)$ behaviour of Eq. A13.18 reduces to Eq. A13.10, as it must (see dotted line in Figure A13.1). At long times $(\tau >> \tau_c)$, we get a simple exponential decay: $E(2\tau) \approx \exp(-m\tau_c\tau)$ (see dashed line in Figure A13.1). Eq. A13.16 for Gaussian spectral diffusion has the same exponential τ-dependence at long times, in accord with the central-limit theorem for probability. The decay of the three-pulse echo under Lorentzian spectral diffusion is given from Eqs. A13.3 and A13.12 by (cf. Mims 1968):

$$E(2\tau + T) = E(2\tau)\exp\left[-m\tau_c\tau\left(1 - e^{-T/\tau_c}\right)\right] \qquad (A13.19)$$

where we assume that the factor multiplying $E(2\tau)$ is equal to the Fourier transform of the Lorentzian diffusion kernel (see Eq. A13.7). This limits us to short values of the first delay $(\tau << \tau_c)$. Eq. A13.19 reduces to Eq. A13.11 at short times $(T << \tau_c)$. The behaviour at long times or for fast motion $(T >> \tau_c)$ reduces to: $E(2\tau + T) \approx \exp(-2m\tau_c\tau)$, which is independent of the T-delay, as is Eq. A13.17 for Gaussian spectral diffusion at long times.

Equation A13.19 is similar in form, but not identical, to Eq. 13.143 of Section 13.20 for dipolar relaxation by a statistical distribution of dilute spins. The latter has a more complex dependence of the T-dependent term on τ. Mims (1968) gives a critical comparison of Lorentzian spectral diffusion with the sudden-jump model of interacting spins (see also Zhidomirov

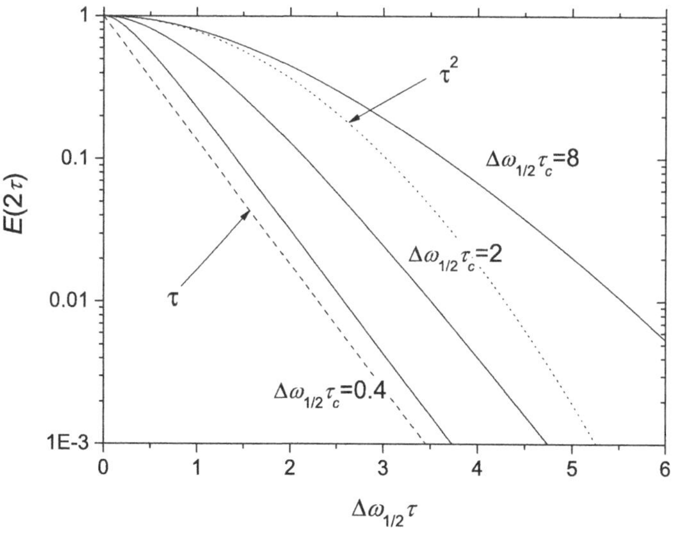

FIGURE A13.1 Two-pulse echo decay curves predicted by Eq. A13.18 for Lorentzian spectral diffusion with half-width at half-height $\Delta\omega_{1/2}$ and rate $1/\tau_c$, where $m = 2\Delta\omega_{1/2}/\tau_c$. *Dotted line*: prediction for short decay times (Eq. A13.10 with $\Delta\omega_{1/2}\tau_c = 8$); *dashed line*: single-exponential decay predicted for long decay times.

and Salikhov 1969). At long times, Lorentzian spectral diffusion leads to simple exponential decay, instead of the stretched exponential with exponent $\beta = \frac{1}{2}$ from the sudden-jump model (cf. Section 13.14 and Eq. 13.106). Also, Lorentzian spectral diffusion does not give rise to exchange narrowing.

REFERENCES

Anderson, P.W. 1954. A mathematical model for the narrowing of spectral lines by exchange or motion. *J. Phys. Soc. Japan* 9:316–339.

Bartucci, R., Guzzi, R., Marsh, D., and Sportelli, L. 2003. Chain dynamics in the low-temperature phases of lipid membranes by electron spin-echo spectroscopy. *J. Magn. Reson.* 162:371–379.

Bartucci, R., Erilov, D.A., Guzzi, R. et al. 2006. Time-resolved electron spin resonance studies of spin-labelled lipids in membranes. *Chem. Phys. Lipids* 141:142–157.

Bartucci, R., Guzzi, R., De Zotti, M. et al. 2008. Backbone dynamics of alamethicin bound to lipid membranes: spin-echo EPR of TOAC spin labels. *Biophys. J.* 94:2698–2705.

Bartucci, R., Guzzi, R., Sportelli, L., and Marsh, D. 2009. Intramembrane water associated with TOAC spin-labelled alamethicin: electron spin-echo envelope modulation by D_2O. *Biophys. J.* 96:997–1007.

Bloom, A.L. 1955. Nuclear induction in inhomogeneous fields. *Phys. Rev.* 98:1105–1111.

Brown, I.M. 1974. Electron spin echo envelope decays and molecular motion - rotational and translational diffusion. *J. Chem. Phys.* 60:4930–4938.

Brown, I.M. 1979. Electron spin-echo studies of relaxation processes in molecular solids. In *Time Domain Electron Spin Resonance*, eds. Kevan, L. and Schwartz, R.N., 195–277. New York: Wiley.

Dzuba, S.A. 1996. Librational motion of guest spin probe molecules in glassy media. *Phys. Lett. A* 213:77–84.

Dzuba, S.A., Maryasov, A.G., Salikov, K.M., and Tsvetkov, Y.D. 1984. Superslow rotations of nitroxide radicals studied by pulse EPR spectroscopy. *J. Magn. Reson.* 58:95–117.

Dzuba, S.A., Watari, H., Shimoyama, Y. et al. 1995. Molecular motion of the cholestane spin label in a multibilayer in the gel phase studied using echo-detected EPR. *J. Magn. Reson. A* 115:80–86.

El Mkami, H., Ward, R., Bowman, A., Owen-Hughes, T., and Norman, D.G. 2014. The spatial effect of protein deuteration on nitroxide spin-label relaxation: implications for EPR distance measurements. *J. Magn. Reson.* 248:36–41.

Erilov, D.A., Bartucci, R., Guzzi, R. et al. 2004a. Echo-detected electron paramagnetic resonance spectra of spin-labeled lipids in membrane model systems. *J. Phys. Chem. B* 108:4501–4507.

Erilov, D.A., Bartucci, R., Guzzi, R. et al. 2004b. Librational motion of spin-labeled lipids in high-cholesterol containing membranes from echo-detected EPR spectra. *Biophys. J.* 87:3873–3881.

Ernst, R.R., Bodenhausen, G., and Wokaun, A. 1987. *Principles of NMR in One and Two Dimensions*, Oxford: Clarendon Press.

Fauth, J.M., Schweiger, A., Braunschweiler, L., Forrer, J., and Ernst, R.R. 1986. Elimination of unwanted echoes and reduction of dead time in 3-pulse electron spin-echo spectroscopy. *J. Magn. Reson.* 66:74–85.

Guzzi, R., Bartucci, R., Sportelli, L., Esmann, M., and Marsh, D. 2009. Conformational heterogeneity and spin-labelled -SH groups: Pulsed EPR of Na,K-ATPase. *Biochemistry* 48:8343–8354.

Guzzi, R., Bartucci, R., and Marsh, D. 2014. Heterogeneity of protein substates visualized by spin-label EPR. *Biophys. J.* 106:716–722.

Hu, P. and Hartmann, S.R. 1974. Theory of spectral diffusion decay using an uncorrelated sudden-jump model. *Phys. Rev B* 9:1–13.

Hwang, J.S., Mason, R.P., Hwang, L.P., and Freed, J.H. 1975. Electron spin resonance studies of anisotropic rotational reorientation and slow tumbling in liquid and frozen media. 3. Perdeuterated 2,2,6,6-tetramethyl-4-piperidone *N*-oxide and analysis of fluctuating torques. *J. Phys. Chem.* 79:489–511.

Isaev, N.P. and Dzuba, S.A. 2008. Fast stochastic librations and slow rotations of spin labeled stearic acids in a model phospholipid bilayer at cryogenic temperatures. *J. Phys. Chem. B* 112:13285–13291.

Kirilina, E.P., Dzuba, S.A., Maryasov, A.G., and Tsvetkov, Y.D. 2001. Librational dynamics of nitroxide molecules in a molecular glass studied by echo-detected EPR. *Appl. Magn. Reson* 21:203–221.

Kirilina, E.P., Prisner, T.F., Bennati, M. et al. 2005. Molecular dynamics of nitroxides in glasses as studied by multi-frequency EPR. *Mag. Reson. Chem.* 43:S119–S129.

Kivelson, D. and Lee, S. 1982. Theory of electron spin resonance parallel-edge lines of slowly tumbling molecules. *J. Chem. Phys.* 76:5746–5754.

Klauder, J.R. and Anderson, P.W. 1962. Spectral diffusion decay in spin resonance experiments. *Phys. Rev.* 125:912–932.

Leporini, D., Schadler, V., Wiesner, U., Spiess, H.W., and Jeschke, G. 2003. Electron spin relaxation due to small-angle motion: theory for the canonical orientations and application to hierarchic cage dynamics in ionomers. *J. Chem. Phys.* 119:11829–11846.

Lindgren, M., Eaton, G.R., Eaton, S.S. et al. 1997. Electron spin echo decay as a probe of aminoxyl environment in spin-labeled mutants of human carbonic anhydrase II. *J. Chem. Soc., Perkin Trans.* 2:2549–2554.

Marsh, D. 2014. Intermediate dipolar distances from spin labels. *J. Magn. Reson.* 238:77–81.

McConnell, H.M. 1958. Reaction rates by nuclear magnetic resonance. *J. Chem. Phys.* 28:430–431.

Millhauser, G.L. and Freed, J.H. 1984. Two dimensional electron spin echo spectroscopy and slow motions. *J. Chem. Phys.* 81:37–48.

Milov, A.D., Salikhov, K.M., and Tsvetkov, Y.D. 1973a. Phase relaxation of hydrogen atoms stabilized in an amorphous matrix. *Sov. Phys. Solid State* 15:802–806.

Milov, A.D., Salikhov, K.M., and Tsvetkov, Y.D. 1973b. Effect of spin dipole-dipole interaction on phase relaxation in magnetically dilute solid bodies. *Sov. Phys. JETP* 36:1229–1232.

Mims, W.B. 1968. Phase memory in electron spin echoes, lattice relaxation effects in $CaWO_4$:Er, Ce, Mn. *Phys. Rev.* 168:370–389.

Mims, W.B. 1972. Electron spin echoes. In *Electron Paramagnetic Resonance*, ed. Geschwind, S., 263–351. New York: Plenum.

Mims, W.B. 1984. Elimination of the dead-time artifact in electron spin echo evelope spectra. *J. Magn. Reson.* 59:291–306.

Nakagawa, K., Candelaria, M.B., Chik, W.W.C., Eaton, S.S., and Eaton, G.R. 1992. Electron spin relaxation times of chromium (V). *J. Magn. Reson.* 98:81–91.

Rabi, I.I., Zacharias, J.R., Millman, S., and Kusch, P. 1938. A new method of measuring nuclear magnetic moment. *Phys. Rev.* 53:318–318.

Raitsimring, A.M., Salikhov, K.M., Umanskii, B.A., and Tsvetkov, Y.D. 1974. Instantaneous diffusion in the electron spin echo of paramagnetic centers stabilized in a solid host. *Sov. Phys. Solid State* 16:492–497.

Salikov, K.M. and Tsvetkov, Y.D. 1979. Electron spin-echo studies of spin-spin interactions in solids. In *Time Domain Electron Spin Resonance*, eds. Kevan, L. and Schwartz, R.N., 231–278. New York: Wiley-Interscience.

Schwartz, L.J., Stillman, A.E., and Freed, J.H. 1982. Analysis of electron spin echoes by spectral representation of the stochastic Liouville equation. *J. Chem. Phys.* 77:5410–5425.

Schweiger, A. and Jeschke, G. 2001. *Principles of Pulse Electron Paramagnetic Resonance*, Oxford: Oxford University Press.

Slichter, C.P. 1990. *Principles of Magnetic Resonance*, New York: Springer-Verlag.

Stillman, A.E., Schwartz, L.J., and Freed, J.H. 1980. Direct determination of rotational correlation time by electron spin echoes. *J. Chem. Phys.* 73:3502–3503.

Tsvetkov, Y.D. and Dzuba, S.A. 1990. Pulsed ESR and molecular motions. *Appl. Magn. Reson.* 1:179–194.

Ward, R., Bowman, A., Sozudogru, E. et al. 2010. EPR distance measurements in deuterated proteins. *J. Magn. Reson.* 207:164–167.

Zecevic, A., Eaton, G.R., Eaton, S.S., and Lindgren, M. 1998. Dephasing of electron spin echos for nitroxyl radicals in glassy solvents by non-methyl and methyl protons. *Mol. Phys.* 95:1255–1263.

Zhidomirov, G.M. and Salikhov, K.M. 1969. Contribution to theory of spectral diffusion in magnetically diluted solids. *Sov. Phys. JETP* 29:1037–1040.

ESEEM and ENDOR: Hyperfine Spectroscopy

14

14.1 INTRODUCTION

There are two main EPR methods for determining unresolved hyperfine couplings of nitroxide spin labels. One is a pulse technique, electron-spin-echo envelope modulation (ESEEM), and the other is a double-resonance technique, electron–nuclear double resonance (ENDOR) that has both CW and pulse variants.

ESEEM spectroscopy exploits periodic oscillations of the spin echo decay that occur at the nuclear hyperfine frequencies, following narrow pulses (see Figure 14.1). It is best suited to determining weak hyperfine interactions, especially from low-γ_n nuclei, e.g., for deuterium nuclear frequencies or ^{14}N-quadrupole splittings. The three-pulse echo offers better ESEEM resolution than does the two-pulse echo, because the extended decay period provides more oscillation cycles. Also, three-pulse ESEEM does not involve the sum and difference frequencies that appear in two-pulse ESEEM. Spin-label applications include detection of exposure to water (D_2O) from ^{2}H-ESEEM intensities and investigation of environmental polarity from ^{14}N-quadrupole couplings. Dikanov and Tsvetkov (1992) have published a monograph on ESEEM spectroscopy, and a detailed treatment is also given in the textbook on pulse EPR by Schweiger and Jeschke (2001).

ENDOR involves EPR detection of NMR transitions induced by radiofrequency radiation, which alleviate saturation of the corresponding EPR line. It is better suited to stronger hyperfine interactions from high-γ_n nuclei, principally protons and also fluorine or phosphorus. The main application is in structural studies, which exploit the distance dependence of dipolar hyperfine interactions. Several such studies use CW-ENDOR (Makinen et al. 1998). Pulse-ENDOR has the advantage that it depends less critically on relaxation times (and hence temperature) than does CW-ENDOR. There are two main pulse variants: Davies-ENDOR and Mims-ENDOR that are suited to medium to large hyperfine couplings and to small hyperfine couplings, respectively. An alternative to pulse-ENDOR that is applicable to nitroxides, and particularly also to determining ^{14}N-quadrupole couplings, is ELDOR-detected NMR. This involves pulses at a second microwave frequency, instead of the radiofrequency pulses in ENDOR; it relies on exciting "forbidden" transitions, as does ESEEM.

Note that, because we are dealing throughout with nuclear frequencies, all hyperfine constants in this chapter are in angular frequency units. They are related to the hyperfine constants in energy units by multiplying by $\hbar$.

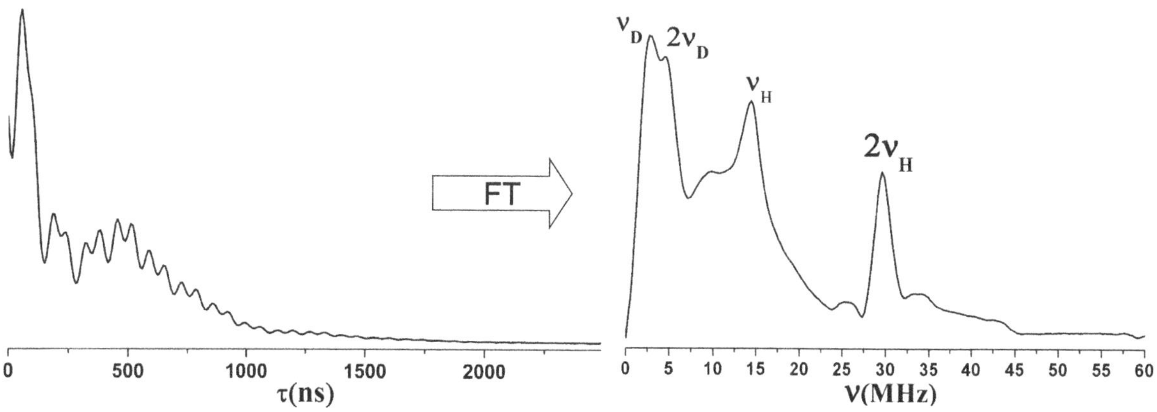

FIGURE 14.1 Electron-spin-echo-envelope-modulation spectroscopy. *Left*: decay of two-pulse electron-spin-echo amplitude with pulse delay, τ, for spin-labelled lipids in membranes dispersed in D_2O. Rapid oscillations in echo amplitude are produced by weak hyperfine interactions with matrix protons. Slow modulation superimposed on the rapid oscillations is from deuterium hyperfine interactions with D_2O. *Right*: Fourier transform of the echo decay produces the ESEEM spectrum. Peaks at ν_D (=2.7 MHz) and $2\nu_D$ are from ^{2}H-hyperfine interactions with D_2O; peaks at ν_H (=14.4 MHz) and $2\nu_H$ are from proton hyperfine interactions. (Bartucci et al. 2003).

14.2 ECHO ENVELOPE MODULATION

Pulse EPR and spin echoes are described already in Chapter 13. Consider for simplicity a two-pulse echo sequence (see Figure 13.1). We can describe the ESEEM effect in terms of the mutual hyperfine fields experienced by the electron and the nucleus. The nuclear spin magnetization precesses about the effective field produced by the electron spin, which generally is stronger than the static laboratory magnetic field (see Section 2.2 in Chapter 2).

In a spin-echo sequence, the initial 90°-pulse tilts the electron spin magnetization into the *x,y*-plane (see Figure 13.4). The nuclear magnetization then precesses about the new direction of the effective field established by the electron magnetization (cf. Figure 2.3). This causes the electron to experience a different time-varying field from the nucleus. Only if the echo delay time τ corresponds to an exact multiple of the nuclear precession frequency, is the nuclear field at the time of the echo exactly the same as it is at the time of the 180°-pulse in Figure 13.4. Otherwise, not all of the electron spin magnetization is refocused at time 2τ. The echo amplitude is then decreased with a periodic dependence on τ that is determined by the nuclear frequency. This is the origin of the modulation of the electron spin echo decay envelope.

14.3 TWO-PULSE ESEEM FOR $I = \frac{1}{2}$ NUCLEI

Two-pulse ESEEM (see Figure 13.1) is more straightforward to analyse than is three-pulse ESEEM, which we consider later. The simplest example is interaction of the $S = \frac{1}{2}$ spin label with a single $I = \frac{1}{2}$ nucleus, e.g., a proton. We use the product-operator formalism that we introduced in Chapter 13 and describe in Appendix P.

If the spin Hamiltonian contains only secular terms, the hyperfine interaction AI_zS_z behaves exactly as an offset in Larmor frequency $\Delta\omega_oS_z$ and is refocused by the echo (see Section 13.4). To obtain the ESEEM effect, we therefore must include the pseudosecular hyperfine terms BI_xS_z in the spin Hamiltonian:

$$\mathcal{H}_o = \hbar\left(\Delta\omega_oS_z + \omega_I I_z + AI_zS_z + BI_xS_z\right) \quad (14.1)$$

where ω_I is the nuclear Larmor angular frequency, and the hyperfine constants A and B are in angular frequency units. Note that $\Delta\omega_o$, A and B depend on the orientation θ of the static magnetic field. For an axial hyperfine interaction with principal elements $A_\parallel$ and $A_\perp$, we have from Eq. B.29 in Appendix B:

$$A = A_\parallel\cos^2\theta + A_\perp\sin^2\theta \quad (14.2)$$

$$B = \left(A_\parallel - A_\perp\right)\sin\theta\cos\theta \quad (14.3)$$

For convenience, we choose $\mathbf{B}_o$ to lie in the x, z-plane, but for an axial hyperfine tensor, this is readily generalized. We deal with the dependence on θ later, in Section 14.8 on ESEEM powder spectra from disordered systems.

We treat the effects of a single 90°-pulse on this $S = \frac{1}{2}, I = \frac{1}{2}$ spin system in Appendix P.4, which should be consulted for what follows. Diagonalizing $\mathcal{H}_o$ transforms Eq. 14.1 to (see Eqs. P.35–P.37):

$$\mathcal{H}_o^d = \hbar\left(\Delta\omega_oS_z + \tfrac{1}{2}\omega_+ I'_z + \omega_- I'_zS_z\right) \quad (14.4)$$

where:

$$\tfrac{1}{2}(\omega_+ + \omega_-) = \sqrt{\left(\omega_I + \tfrac{1}{2}A\right)^2 + \tfrac{1}{4}B^2} \equiv \omega_\alpha \quad (14.5)$$

$$\tfrac{1}{2}(\omega_+ - \omega_-) = \sqrt{\left(\omega_I - \tfrac{1}{2}A\right)^2 + \tfrac{1}{4}B^2} \equiv \omega_\beta \quad (14.6)$$

Here, ω_α and ω_β are angular frequencies for nuclear transitions within the $M_S = +\frac{1}{2}$ and $M_S = -\frac{1}{2}$ electron spin states, respectively (see Figure 14.2). This diagonalization corresponds to the following product-operator transformation (see Eq. P.38):

$$\omega_I + AI_zS_z + BI_xS_z \xrightarrow{\xi I_y + 2\eta I_yS_z} \tfrac{1}{2}\omega_+ I_z + \omega_- I_zS_z \quad (14.7)$$

where:

$$\eta = \eta_\alpha - \eta_\beta \quad (14.8)$$

$$\tan 2\eta_{\alpha,\beta} = \frac{\mp B}{2\omega_I \pm A} \quad (14.9)$$

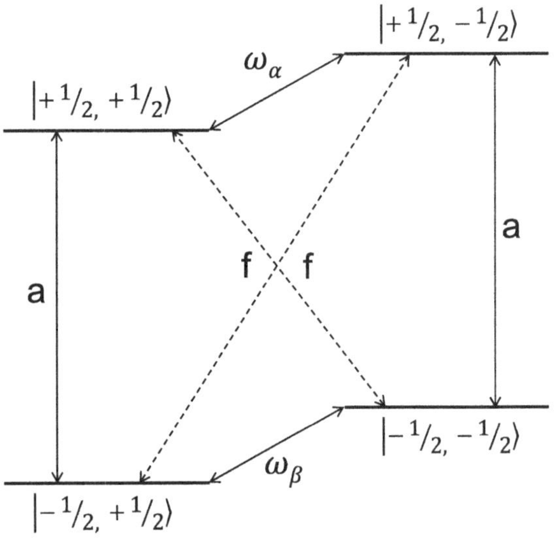

FIGURE 14.2 Energy-level diagram for $S = \frac{1}{2}$, $I = \frac{1}{2}$ with $\frac{1}{2}\left|A_{hf}\right| < \left|\omega_I\right|$, showing allowed (*a*) and forbidden (*f*) electron-spin transitions, and pure nuclear transitions ω_α and ω_β. Spin states are specified as $\left|M_S, m_I\right\rangle$.

which follow from Eqs. P.33 and P.34, with the upper signs applying to η_α and the lower to η_β. As seen in Section P.4 of Appendix P, we can ignore the ξI_y term in this transformation, because it does not affect the transverse electron magnetization S_y that generates the echo. We need retain only $2\eta I_y S_z$ of the product-operator transformation.

Now we can write the $(\pi/2)_x - \tau - (\pi)_x - \tau - echo$ sequence as a series of product-operator transformations (see Appendix P), starting from the equilibrium spin density matrix, $\sigma(0)$:

$$\sigma(0)\xrightarrow{\frac{1}{2}\pi S_x}\sigma_1(0)\xrightarrow{2\eta I_y S_z}$$
$$\sigma_1'(0)\xrightarrow{\frac{1}{2}\omega_-\tau I_z+\omega_+\tau I_z S_z}\sigma'(\tau)\xrightarrow{-2\eta I_y S_z}\sigma(\tau)$$
$$\sigma(\tau)\xrightarrow{\pi S_x}\sigma_2(\tau)\xrightarrow{2\eta I_y S_z}$$
$$\sigma_2'(\tau)\xrightarrow{\frac{1}{2}\omega_-\tau I_z+\omega_+\tau I_z S_z}\sigma'(2\tau)\xrightarrow{-2\eta I_y S_z}\sigma(2\tau) \tag{14.10}$$

where subscripts on σ indicate the action of a pulse: $\sigma(\tau)$ in the third row starts at the second pulse. Primes indicate that σ is transformed to the basis where the Hamiltonian is diagonal, and the corresponding negative product operator transforms σ' back to the Cartesian laboratory system. The echo amplitude $\overline{\langle S_y\rangle}$ is given by the coefficient of S_y in $\sigma(2\tau)$. Therefore, we need track only the S_y and $I_y S_x$ terms of the density matrix through the series of transformations in Eq. 14.10 (see Section P.3 in Appendix P). Specifically, we need the following (truncated) transformations (see Eqs. P.18, P.26, P.27 and Tables P.1, P.2):

$$S_y\xrightarrow{2\eta I_y S_z}S_y\cos\eta-2I_y S_x\sin\eta \tag{14.11}$$

$$I_y S_x\xrightarrow{2\eta I_y S_z}I_y S_x\cos\eta+\tfrac{1}{2}S_y\sin\eta \tag{14.12}$$

$$I_y S_x\xrightarrow{\frac{1}{2}\phi I_z}I_y S_x\cos\tfrac{1}{2}\phi \tag{14.13}$$

$$S_y\xrightarrow{\phi I_z S_z}S_y\cos\tfrac{1}{2}\phi \tag{14.14}$$

$$I_y S_x\xrightarrow{\phi I_z S_z}I_y S_x \tag{14.15}$$

where $\phi\equiv\omega_\pm\tau$.

First applying the 90°-pulse of Eq. 14.10 and then transforming to the frame where $\mathcal{H}_o$ is diagonal, we get (see Eq. P.39):

$$\sigma(0)=S_z\xrightarrow{(\pi/2)S_x}-S_y\xrightarrow{2\eta I_y S_z}$$
$$-S_y\cos\eta+2I_y S_x\sin\eta=\sigma_1'(0) \tag{14.16}$$

Subsequent evolution under $\mathcal{H}_o^d$ for time τ is given by (see Eqs. P.40, P.41):

$$\sigma_1'(0)\xrightarrow{\frac{1}{2}\omega_+\tau I_z+\omega_-\tau I_z S_z}$$
$$-S_y\cos\eta\cos\tfrac{1}{2}\omega_-\tau+2I_y S_x\sin\eta\cos\tfrac{1}{2}\omega_+\tau=\sigma'(\tau) \tag{14.17}$$

Transforming back to the laboratory frame, we then get (see Eq. P.42):

$$\sigma'(\tau)\xrightarrow{-2\eta I_y S_z}-S_y\left(\cos^2\eta\cos\tfrac{1}{2}\omega_-\tau+\sin^2\eta\cos\tfrac{1}{2}\omega_+\tau\right)$$
$$-I_y S_x\sin 2\eta\left(\cos\tfrac{1}{2}\omega_-\tau-\cos\tfrac{1}{2}\omega_+\tau\right)=\sigma(\tau) \tag{14.18}$$

Applying the refocusing 180°-pulse simply changes the sign of the S_y-component in $\sigma(\tau)$. For the following detection period, we must again transform to the diagonal frame with the $2\eta I_y S_x$ operator, again perform the evolution under $\mathcal{H}_o^d\tau$, and again transform back to the laboratory frame with the $-2\eta I_y S_x$ operator. We do this by using Eqs. 14.11–14.15, as in Eqs. 14.16–14.18. From the coefficient of S_y in the resulting expression for $\sigma(2\tau)$, we get (Schweiger and Jeschke 2001):

$$\overline{\langle S_y(2\tau)\rangle}=\cos 2\eta\left(\cos^2\eta\cos^2\frac{\omega_-\tau}{2}-\sin^2\eta\cos^2\frac{\omega_+\tau}{2}\right)$$
$$+\cos^2\eta\sin^2\frac{\omega_-\tau}{2}+\sin^2\eta\sin^2\frac{\omega_+\tau}{2}$$
$$+\sin^2 2\eta\cos\frac{\omega_-\tau}{2}\cos\frac{\omega_+\tau}{2} \tag{14.19}$$

for the echo amplitude. After trigonometric rearrangements in Eq. 14.19, this becomes (cf. Mims 1972):

$$E_{2p}(2\tau)=1-\frac{k}{2}(1-\cos\omega_\alpha\tau)(1-\cos\omega_\beta\tau) \tag{14.20}$$

where we use Eqs. 14.5 and 14.6 to introduce the individual nuclear frequencies ω_α and ω_β. The dimensionless modulation-depth parameter, k, is given by:

$$k=\sin^2 2\eta=\left(\frac{B\omega_I}{\omega_\alpha\omega_\beta}\right)^2 \tag{14.21}$$

where we use Eqs. 14.5, 14.6, 14.8 and 14.9 for the second identity. Note that from Eqs. H.20 and H.21 in Appendix H, $k=4I_a I_f$ so both allowed and forbidden transitions (see Figure 14.2) must have non-vanishing intensities I_a and I_f for us to get ESEEM. From the treatment given in Section 13.4 of Chapter 13, we see that for general pulse flip angles β_1 and β_2, the modulation depth k is multiplied by the factor $\sin\beta_1\times\sin^2(\beta_2/2)$. Also, we can account for relaxation by multiplying the echo amplitude $E_{2p}(2\tau)$ by the factor $\exp(-2\tau/T_{2M})$, where T_{2M} is the phase-memory time of the electron spin.

Equation 14.20 shows that the two-pulse ESEEM spectrum contains not only the basic nuclear frequencies ω_α and ω_β (cf. Figure 14.2) but also the combination frequencies $\omega_- = \omega_\alpha - \omega_\beta$ and $\omega_+ = \omega_\alpha + \omega_\beta$. The latter components come from the cross-term in Eq. 14.20: $\cos\omega_\alpha\tau\cos\omega_\beta\tau = \frac{1}{2}(\cos\omega_-\tau + \cos\omega_+\tau)$ and appear with half the intensity and opposite phase to the basic nuclear components. The nuclear frequencies and hyperfine constants are all directly proportional to the nuclear gyromagnetic ratio, γ_n (see Sections 3.3 and 3.4 in Chapter 3). Therefore, the modulation depth k is independent of γ_n. This suits low-γ_n nuclei to ESEEM as opposed to ENDOR, because nuclear transition probabilities for the latter are proportional to γ_n^2. For very weak hyperfine couplings, i.e., $A, B << \omega_I$, the basic nuclear frequencies are close to the Larmor frequency: $\omega_\alpha \approx \omega_\beta \approx \omega_I$, and the combination frequency is twice this: $\omega_+ \approx 2\omega_I$ with $\omega_- \approx 0$ (see Figure 14.1). The modulation depth is then: $k \approx (B/\omega_I)^2 \propto 1/B_0^2$, which decreases with the square of the static magnetic field, B_o. For very strong hyperfine couplings, i.e., $A, B >> \omega_I$, the basic nuclear frequencies are again equal: $\omega_\alpha \approx \omega_\beta \approx \frac{1}{2}\sqrt{A^2+B^2}$ and ω_+ is twice this value with $\omega_- \approx 0$. The modulation depth is then: $k \approx 16B^2\omega_I^2/(A^2+B^2)^2 \propto B_0^2$, which increases with the square of the static magnetic field strength.

For weak hyperfine couplings, i.e., $A, B < \omega_I$, the basic nuclear frequencies are split about the Larmor frequency (see Eqs. 14.5 and 14.6):

$$\omega_{\alpha,\beta} \approx \omega_I \pm \tfrac{1}{2}A \tag{14.22}$$

where the upper sign applies to ω_α and the lower to ω_β. The combination frequencies are then: $\omega_+ \approx 2\omega_I$ and $\omega_- \approx A$. This is often the situation with spin labels, when we are concerned with deuterium ESEEM or with ESEEM from unresolved proton hyperfine interactions. For strong hyperfine couplings, i.e., $A, B > \omega_I$, the basic nuclear frequencies are split about the hyperfine frequency, A_{eff} (see Eqs. 14.5 and 14.6):

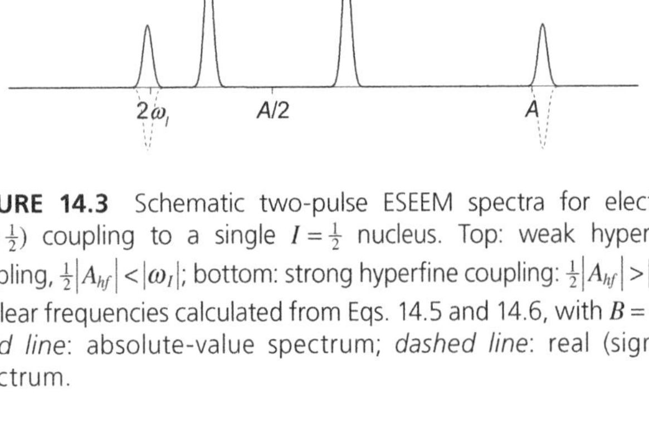

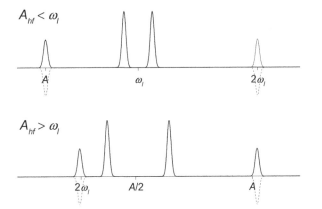

FIGURE 14.3 Schematic two-pulse ESEEM spectra for electron ($S = \frac{1}{2}$) coupling to a single $I = \frac{1}{2}$ nucleus. Top: weak hyperfine coupling, $\frac{1}{2}|A_{hf}| < |\omega_I|$; bottom: strong hyperfine coupling: $\frac{1}{2}|A_{hf}| > |\omega_I|$. Nuclear frequencies calculated from Eqs. 14.5 and 14.6, with $B = \frac{1}{5}A$. *Solid line*: absolute-value spectrum; *dashed line*: real (signed) spectrum.

$$\omega_{\alpha,\beta} \approx \tfrac{1}{2}\sqrt{A^2+B^2} \pm \frac{A\omega_I}{\sqrt{A^2+B^2}} \equiv \tfrac{1}{2}A_{eff} \pm (A/A_{eff})\omega_I \tag{14.23}$$

The combination frequencies are then: $\omega_+ \approx A_{eff}$ and $\omega_- \approx 2\omega_I A/A_{eff}$, where $A_{eff} \equiv \sqrt{A^2+B^2}$. This is the case for nitrogen ESEEM of spin labels, but there we are interested primarily in small ^{14}N-quadrupole splittings within a single manifold (see later). Figure 14.3 illustrates the two-pulse ESEEM spectra for the two extreme cases. Solid lines correspond to the absolute-value spectra that we normally use in spin-label ESEEM (cf. Figure 14.1).

14.4 TWO-PULSE ESEEM FOR $I = 1$ NUCLEI

The spin Hamiltonian for nuclei with $I > \frac{1}{2}$ contains an extra term that arises from the nuclear quadrupole interaction (see Section 3.10):

$$\mathcal{H}_o = \hbar\left(\Delta\omega_o S_z + \omega_I I_z + AI_z S_z + BI_x S_z + PI_z^2\right) \tag{14.24}$$

where we include just the diagonal part of the quadrupole interaction PI_z^2, because the remainder contributes only in second order. Note that P depends on the orientation of the static magnetic field, B_o (as do also A and B). For an axial quadrupole interaction: $P = \omega_Q(3\cos^2\theta_Q - 1)$, where θ_Q is the angle between the principal axis of the quadrupole tensor and the magnetic field, and $\omega_Q = 3e^2qQ/8I(2I-1) \equiv \frac{3}{4}P_{zz}$ (see Eq. B.32 in Appendix B, and Section 3.10). Diagonalization of the Hamiltonian proceeds exactly as in the previous section, and we treat the quadrupole interaction with first-order perturbation theory.

If we neglect the quadrupole interaction (i.e., $P = 0$), the echo amplitude for $I = 1$ becomes (Mims 1972):

$$E_{2p}(2\tau) = 1 - \frac{4k}{3}(1-\cos\omega_\alpha\tau)(1-\cos\omega_\beta\tau) + \frac{4k^2}{3}(1-\cos\omega_\alpha\tau)^2(1-\cos\omega_\beta\tau)^2 \tag{14.25}$$

where again ω_α and ω_β are nuclear transition frequencies within the $M_S = +\frac{1}{2}$ and $M_S = -\frac{1}{2}$ electron spin states, respectively, and k is given by Eq. 14.21. If B is small, we may neglect the term in k^2 and the modulation pattern is the same as for the $I = \frac{1}{2}$ case. In general, if the term linear in k dominates, the modulation depth is increased by a factor $\frac{4}{3}I(I+1)$ relative to the value for $I = \frac{1}{2}$. Thus ESEEM intensities of deuterium are 8/3 times those of protons, when we ignore broadening by the quadrupole interaction. If the k^2 term is appreciable, higher combination frequencies appear in the two-pulse ESEEM spectra for $I = 1$ nuclei.

Group-theoretical arguments show that we can write the two-pulse echo amplitude for a general nuclear spin I as (Dikanov et al. 1981):

$$E_{2p,I}(2\tau) = \frac{1}{2I+1}\sum_{m_I=-I}^{+I}\exp(im_I 2\xi) = \frac{1}{2I+1}\frac{\sin((2I+1)\xi)}{\sin\xi} \tag{14.26}$$

where the first factor is a normalization, and the second factor is the character $\chi^{(I)}(2\xi)$ for rotation 2ξ in the full rotation group (see, e.g., Heine 1960). We can express Eq. 14.26 for general I in terms of the result for $I = \frac{1}{2}$, because:

$$E_{2p,1/2}(2\tau) = \cos\xi \tag{14.27}$$

where $E_{2p,1/2}(2\tau)$ is given explicitly by Eq. 14.20, which then leads eventually to Eq. 14.25 for $I = 1$.

For weak but non-vanishing quadrupole interaction, we may use perturbation theory to determine the effect on the ESEEM frequencies without worrying about the amplitudes (Shubin and Dikanov 1983). For $I = 1$, there are then four distinct nuclear transition frequencies defined by $\Delta M_S = 0, \Delta m_I = \pm 1$ (see Figure 14.4). Note that the quadrupole interaction does not contribute to the conventional EPR spectrum $(\Delta M_S = 1, \Delta m_I = 0)$ because it shifts all energy levels by the same amount. To first order (i.e., neglecting B), the nuclear frequencies within the α and β (i.e., $M_S = \pm\frac{1}{2}$) electron spin manifolds, $\omega_{\alpha,\Delta m_I}$ and $\omega_{\beta,\Delta m_I}$, that we get from Eq. 14.24 are:

$$\omega_{\alpha,\pm 1} = \left|\omega_I + \tfrac{1}{2}A \pm P\right| \tag{14.28}$$

$$\omega_{\beta,\pm 1} = \left|\omega_I - \tfrac{1}{2}A \pm P\right| \tag{14.29}$$

respectively, where P and A are in angular frequency units. The absolute values allow for cases where P or A is the larger term. The transition frequencies in Eqs. 14.28 and 14.29 correspond to Eq. 14.22 when we neglect quadrupole interactions. At the same level of approximation as in Eqs. 14.28 and 14.29, the two-pulse echo amplitude for weakly coupled $I = 1$ nuclei becomes (Shubin and Dikanov 1983):

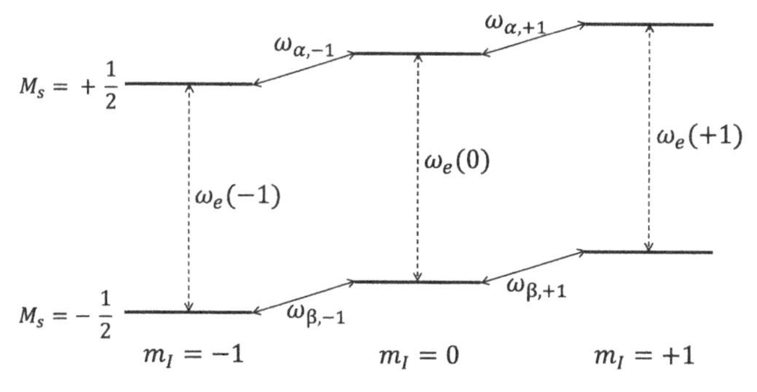

FIGURE 14.4 Energy levels and nuclear transitions for $S = \frac{1}{2}$, $I = 1$. Four nuclear transitions, $\omega_{\alpha,\pm 1}$ and $\omega_{\beta,\pm 1}$, are distinct for non-vanishing quadrupole interaction (see Eqs. 14.28, 14.29). EPR transitions $\omega_e(m_I)$ do not depend on the quadrupole interaction.

$$E_{2p}(2\tau) = 1 - \frac{4}{3}\frac{B^2}{\omega_I^2}\left[1 - 2\cos\omega_I\tau\cos\tfrac{1}{2}A\tau\cos P\tau + \cos 2\omega_I\tau\cos 2P\tau + \cos A\tau\right] \tag{14.30}$$

For $P = 0$, this reduces to the first term in Eq. 14.25 with the nuclear frequencies approximated by Eq. 14.22. The effect of quadrupole interaction is to modulate the components having frequencies ω_I and $2\omega_I$. The ESEEM peaks are split by the quadrupole interaction, just as they are by the hyperfine interaction in the upper spectrum of Figure 14.3, except that the peak at $2\omega_I$ is now also split (by an amount $2P$).

14.5 THREE-PULSE ESEEM

In three-pulse ESEEM, the delay τ between the first two pulses remains fixed, whilst the delay T between the second and third pulses is incremented (see Figure 13.1). The echo decay as a function of T depends on the T_1-relaxation time, and thus persists much longer than the two-pulse echo decay that depends on the shorter phase-memory time (see Section 13.16 in Chapter 13). We then detect echo modulation over a much extended time period, with consequent increase in resolution of the ESEEM spectrum after Fourier transformation.

The electron spins are in a definite state, either α or β $(M_S = \pm\frac{1}{2})$, during the evolution period T when electron magnetization is stored along the z-direction. Therefore, only the basic nuclear frequencies appear in the three-pulse ESEEM spectrum. The amplitude of the three-pulse echo is the sum of contributions from the two states:

$$E_{3p}(2\tau + T) = \frac{1}{2}\left[E_{3p}^{\alpha}(2\tau + T) + E_{3p}^{\beta}(2\tau + T)\right] \tag{14.31}$$

For $I = \frac{1}{2}$, the two contributions are (cf. Eq 14.20):

$$E_{3p}^{\alpha}(2\tau + T) = 1 - \frac{k}{2}(1 - \cos\omega_\beta\tau)(1 - \cos\omega_\alpha(\tau + T)) \tag{14.32}$$

$$E_{3p}^{\beta}(2\tau + T) = 1 - \frac{k}{2}(1 - \cos\omega_\alpha\tau)(1 - \cos\omega_\beta(\tau + T)) \tag{14.33}$$

where k is given by Eq. 14.21. The echo amplitude in the three-pulse ESEEM experiment for $I = \frac{1}{2}$ then becomes (Mims 1972):

$$E_{3p}(2\tau + T) = 1 - \frac{k}{4}\left[(1 - \cos\omega_\beta\tau)(1 - \cos\omega_\alpha(\tau + T)) + (1 - \cos\omega_\alpha\tau)(1 - \cos\omega_\beta(\tau + T))\right] \tag{14.34}$$

We see explicitly from the dependence on the incremented time T that the three-pulse ESEEM spectrum contains only the basic nuclear frequencies ω_α and ω_β, without the combination frequencies that appear in two-pulse ESEEM.

The modulation depth depends on the nuclear frequency, via the terms that contain only the fixed τ-delay. We can optimize for a particular nucleus by tuning the fixed time delay to give $\cos\omega_I\tau = -1$, i.e., $\tau = 2\pi(n+\frac{1}{2})/\omega_I$, where $n = 0,1,2,...$ and ω_I is the angular nuclear Larmor frequency. Correspondingly, the three-pulse ESEEM spectrum has blind spots for nuclear frequencies where $\cos\omega_I\tau = +1$, i.e., with fixed delay $\tau = 2\pi n/\omega_I$. We can use this latter condition to suppress unwanted ESEEM components from weakly coupled matrix protons. For instance, $\tau \approx 204\,\text{ns}$ gives a blind spot for protons $(\omega_H/2\pi \approx 14.6\,\text{MHz}, n=3)$ but is optimized for hydrogen-bonded deuterons $(\omega_D/2\pi \approx 2.45\,\text{MHz}, n=0)$, at an EPR frequency of 9.4 GHz $(B_o \approx 0.34\,\text{T})$. On the other hand, $\tau \approx 168\,\text{ns}$ maximizes the deuterium and proton modulations simultaneously at this EPR frequency (Erilov et al. 2005).

For modulation by a nucleus with $I = 1$, we can obtain the amplitude of the three-pulse echo from Eq. 14.34 for the $I = \frac{1}{2}$ case by the same method that we used for the two-pulse echo in Section 14.4. Equations 14.26 and 14.27 apply equally well for the $E_{3p}^{\alpha}(2\tau+T)$ and $E_{3p}^{\beta}(2\tau+T)$ contributions to the three-pulse echo. Eliminating the rotation angles ξ_α and ξ_β, the result for $I = 1$ then becomes:

$$\begin{aligned} E_{3p}(2\tau+T) = 1 - \frac{2k}{3}\Big[&(1-\cos\omega_\beta\tau)(1-\cos\omega_\alpha(\tau+T)) \\ &+(1-\cos\omega_\alpha\tau)(1-\cos\omega_\beta(\tau+T))\Big] \\ &+\frac{k^2}{6}\Big[(1-\cos\omega_\beta\tau)^2(1-\cos\omega_\alpha(\tau+T))^2 \\ &+(1-\cos\omega_\alpha\tau)^2(1-\cos\omega_\beta(\tau+T))^2\Big] \end{aligned} \tag{14.35}$$

when we neglect quadrupole interactions. Again for small k, the ESEEM pattern is similar to that for $I = \frac{1}{2}$, with the modulation depth increased by a factor $\frac{4}{3}I(I+1) = 8/3$. To the same level of approximation as we used in treating quadrupole interactions for the two-pulse echo (cf. Eqs. 14.28 and 14.29), the three-pulse echo amplitude for $I = 1$ with weak quadrupole interaction and hyperfine coupling is given by (Shubin and Dikanov 1983):

$$\begin{aligned} E_{3p}(2\tau+T) = 1 - \frac{4}{3}\frac{B^2}{\omega_I^2}\Big[&1-\cos\omega_I\tau\cos\tfrac{1}{2}A\tau\cos P\tau \\ &-\cos\omega_I(\tau+T)\cos\tfrac{1}{2}A(\tau+T)\cos P(\tau+T) \\ &+\tfrac{1}{2}\cos\omega_I(2\tau+T)\cos\tfrac{1}{2}AT\cos P(2\tau+T) \\ &+\tfrac{1}{2}\cos\omega_I T\cos\tfrac{1}{2}A(2\tau+T)\cos PT\Big] \end{aligned} \tag{14.36}$$

where we are interested in the terms that depend on the incremented time delay T. Choosing the fixed time delay to satisfy the condition $\cos\omega_I\tau = -1$ (which optimizes the modulation), the expression for the three-pulse echo amplitude then becomes:

$$\begin{aligned} E_{3p}(2\tau+T) = 1 - \frac{4}{3}\frac{B^2}{\omega_I^2}\Big\{&1+\cos\tfrac{1}{2}A\tau\cos P\tau+\cos\omega_I T \\ &\times\Big[\cos\tfrac{1}{2}A(\tau+T)\cos P(\tau+T)+\tfrac{1}{2}\cos\tfrac{1}{2}AT\cos P(2\tau+T) \\ &+\tfrac{1}{2}\cos\tfrac{1}{2}A(2\tau+T)\cos PT\Big]\Big\} \end{aligned} \tag{14.37}$$

Thus, the quadrupole interaction imposes an additional modulation on the component with frequency ω_I. As for the two-pulse case, this leads to quadrupole splitting of the ESEEM spectrum, which is more likely to be resolved in the three-pulse case with its extended time axis. For remote deuterium nuclei, the quadrupole interaction exceeds the dipolar hyperfine interaction with the nitroxide and a doublet splitting is observed (see Figure 14.5 given later).

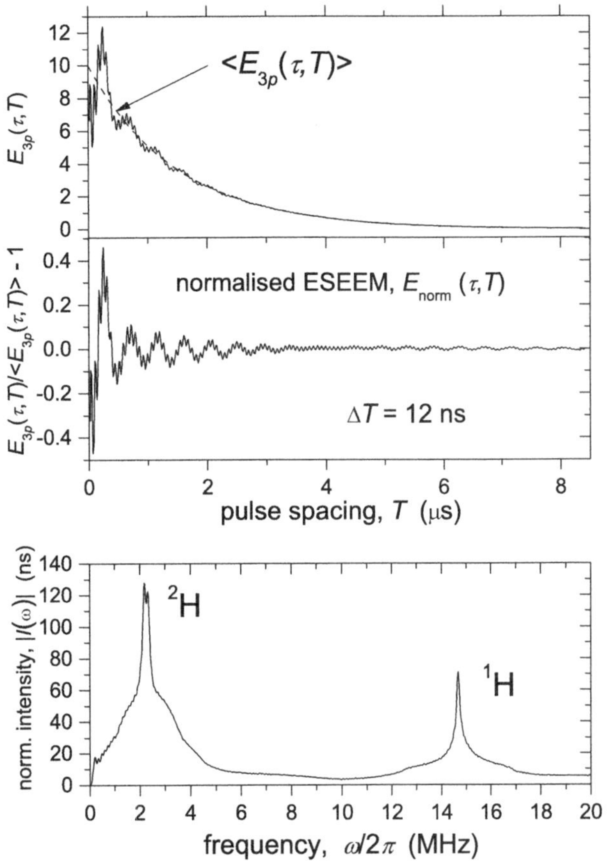

FIGURE 14.5 Normalized ESEEM amplitudes. *Top*: Spin-echo amplitude, $E_{3p}(\tau,T)$, decay as function of second delay time, T, in a three-pulse echo sequence (solid line). Average echo decay, $\langle E_{3p}(\tau,T)\rangle$, obtained by fitting with a biexponential function (dashed line). *Middle*: Normalized ESEEM signal (Eq. 14.38). *Bottom*: Calibrated absolute-value ESEEM spectrum (cf. Eq. 14.39) from real and imaginary components of the Fourier transform of $E_{norm}(\tau,T)$ in the middle panel. Fixed τ-delay is chosen to maximize deuterium (^{2}H) and proton (^{1}H) modulations simultaneously, for spin-labelled membranes dispersed in D_2O (Guzzi et al. 2009).

14.6 STANDARD INTENSITIES FOR 3-PULSE ESEEM

To compare intensities between different ESEEM spectra, we must allow for different increments, ΔT, in the T-delay time that we might use in recording the 3-pulse ESEEM signals. Discrete Fourier-transform software supplied with pulse-EPR spectrometers does not necessarily do this. Data-processing protocols that we give below result in machine-independent spectral densities, with dimensions of time (Bartucci et al. 2009). We may use these quite generally for comparing different systems.

Figure 14.5 shows how to process the time-dependent amplitudes, $E_{3p}(\tau,T)$, of three-pulse echoes to obtain standardized ESEEM intensities, $E_{norm}(\tau,T)$. First, we fit the T-dependence with a biexponential function to get the average experimental echo decay, $\langle E_{3p}(\tau,T)\rangle$ (see Figure 14.5 top). Then we subtract the average decay from $E_{3p}(\tau,T)$ and normalize the resulting difference to the average value (Erilov et al. 2005):

$$E_{norm}(\tau,T) = \frac{E_{3p}(\tau,T)}{\langle E_{3p}(\tau,T)\rangle} - 1 \tag{14.38}$$

as shown in the middle panel of Figure 14.5. We evaluate the complex Fourier transform (FT) of this normalized time-domain signal numerically (Bartucci et al. 2009):

$$I(\omega_k) = \Delta T \sum_{j=0}^{N} E_{norm}(\tau, T_o + j\Delta T)\exp\left(-i\omega_k(\tau + T_o + j\Delta T)\right) \tag{14.39}$$

where $\omega_k = 2\pi k/(N\Delta T)$ with $k = -N/2$ to $+N/2$. This calculation explicitly includes the factor ΔT for the time increment. We then use the real and imaginary Fourier coefficients in Eq. 14.39 to obtain the absolute-value ESEEM spectrum, which is given in the bottom panel of Figure 14.5. We use the absolute-value FT-spectrum because it eliminates blind spots that arise from the $(1-\cos\omega\tau)$ factors in the modulation depth for broad lines (Fauth et al. 1989).

14.7 WATER-PENETRATION PROFILES AND H-BONDING IN ^{2}H-ESEEM

Frequently, we use deuterium ESEEM intensities to determine the accessibility of deuterated species (e.g., water) to a spin label or the proximity of spin labels to a deuterated group (e.g., amino-acid side chain). Penetration of D_2O into lipid-bilayer biomembranes provides an instructive example. We monitor the D_2O profile with phospholipid probes (n-DOXYL-PC) spin labelled at specific positions, n, down their hydrocarbon chains (see Figure 1.1). Typically, D_2O-ESEEM spectra consist of a broad and a narrow component, as seen in the bottom panel of Figure 14.5. The broad component arises from D_2O molecules directly H-bonded to the spin label, and the narrow component originates from D_2O molecules distributed at locations more remote from the spin label. The identity of the H-bonded component is established by quantum-chemical calculations, whereas the narrow component is determined by classical dipolar interactions that depend on the density distribution of surrounding nuclear moments (Erilov et al. 2005).

Figure 14.6 shows transmembrane profiles of D_2O penetration for phosphatidylcholine bilayers, with and without 50 mol% cholesterol, that we get from three-pulse ^{2}H-ESEEM spectroscopy. The amplitude, I_{tot}, of the total D_2O signal decreases with depth into the membrane, with a sharp change at an intermediate chain position. This is well described by the Boltzmann sigmoid (Eq. 4.44), which we introduced in Chapter 4 for transmembrane polarity profiles. The ^{2}H-ESEEM measurements given in Figure 14.6 for frozen membranes accord with CW-EPR measurements of polarity in fluid membranes presented in Section 4.6: cholesterol increases water concentration in the outer regions of the membrane ($n<8$), by increasing the separation of the phospholipid head-groups (cf. Figure 4.16). Direct observation of intramembrane water by ^{2}H-ESEEM spectroscopy therefore fully supports interpretations of transmembrane polarity profiles that we arrived at by more indirect spin-label EPR methods in Chapter 4. The pattern of water permeation is controlled by the hydrophobic membrane barrier, which is reflected inversely in the polarity profile.

Although the overall intensity I_{tot} of the D_2O-ESEEM spectrum varies with n, the ratio I_{narrow}/I_{broad} of the sharp to the broad component remains constant in a given membrane, as we

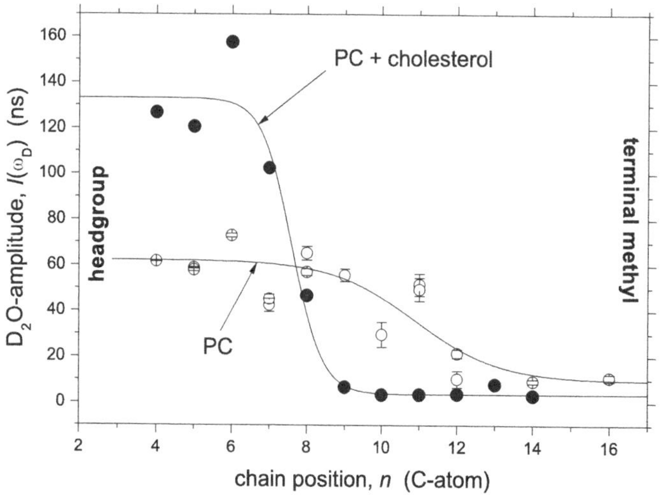

FIGURE 14.6 Water-penetration profile in lipid membranes. Dependence on spin-label position, n, of ^{2}H-ESEEM spectral amplitudes from chain-labelled phosphatidylcholines (n-DOXYL-PC, see Figure 1.1) in dipalmitoyl phosphatidylcholine (PC) bilayers, with (solid circles) and without (open circles) 50 mol% cholesterol, dispersed in D_2O (Erilov et al. 2005). Solid lines are Boltzmann sigmoidal fits (see Eq. 4.44). Amplitudes normalized as described in Section 14.6.

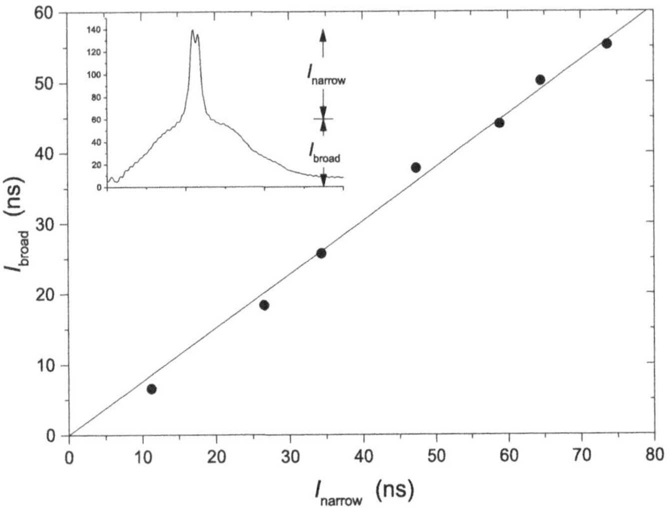

FIGURE 14.7 Linear dependence of intensity of broad component, I_{broad}, on that of narrow component, I_{narrow}, in D_2O-ESEEM spectra of spin-labelled stearic acid (*n*-DOXYL-SA, see Figure 1.1) in Na,K-ATPase membranes. Broad component is from D_2O directly H-bonded to the nitroxide and narrow component from slightly more remote ^{2}H-nuclei. (Bartucci et al. 2014).

see from Figure 14.7. This correlation between the ^{2}H-ESEEM amplitudes of the H-bonded and free spectral components suggests that the H-bonding equilibrium is frozen in (Erilov et al. 2005). Successive association of one and two water molecules (W) with the N–O radical (R) is given by local equilibria at the membrane depth of the spin label:

$$R + W \xleftrightarrow{K_1} W_1R$$

$$W_1R + W \xleftrightarrow{K_2} W_2R$$

where K_1 and K_2 are association constants for binding the first and second water molecules, respectively. If binding the second water molecule is not strongly influenced by binding the first, K_1 and K_2 are related to the intrinsic binding constant, K, for binding to an isolated single site, simply by statistical factors. Applying the law of mass action then yields:

$$\frac{I_{broad}}{I_o} = \frac{2K[W]}{1+K[W]} \tag{14.40}$$

where $I_o \approx 115$ ns is the ^{2}H-ESEEM intensity for a single D_2O molecule bound (permanently) to the nitroxide. This calibration value is determined from calculations with density functional theory (DFT) (Erilov et al. 2005).

Equation 14.40 gives the dependence of the amplitude of the broad line, I_{broad}, in the deuterium ESEEM spectrum on the local water concentration, in terms of the composite quantity K[W]. The ratio of populations of double to singly H-bonded nitroxides is: $[W_2R]/[W_1R] = K[W]/2$, and the fraction of spin labels with a single water molecule bound becomes:

$$f_{1W} = 2/(1/K[W] + 2 + K[W]) \tag{14.41}$$

At the C-4 position of the lipid chains in Figure 14.6, the fraction that is singly H-bonded is $f_{1W} \approx 0.36$ and 0.20, for membranes with and without cholesterol, respectively. Corresponding fractions for two water molecules bound are: $f_{2W} = \frac{1}{2}\left(I_{broad}/I_o - f_{1W}\right) \approx 0.06$ and 0.01. These values fall progressively to zero towards the membrane mid-plane, as we see from Figure 14.6. These ESEEM results reveal heterogeneity in the number of water molecules that are H-bonded to the nitroxide, for spin labels located at the head-group end of the lipid chains. This causes inhomogeneous broadening (*g*-strain) of the polarity-sensitive g_{xx}-feature in high-field EPR spectra from spin labels in this region, relative to those located close to the middle of the membrane, as we saw in Section 4.4.4 of Chapter 4.

14.8 ESEEM IN DISORDERED SAMPLES: POWDER SPECTRA

We assume weak hyperfine coupling, where the ESEEM frequencies are given by Eq. 14.22, because this is a frequent situation for spin labels. From Eq. 14.34, the three-pulse echo amplitude for $I = \frac{1}{2}$ then becomes:

$$E_{3p}(2\tau + T) = 1 - \frac{B^2}{\omega_I^{\,2}}(1 - \cos\omega_I\tau)$$
$$\times\left[1 - \cos\omega_I(\tau + T)\cos\tfrac{1}{2}A(\tau + T)\right] \tag{14.42}$$

where we use the approximation $\omega_\alpha \cong \omega_\beta \cong \omega_I$ for all terms that do not contain the incremented time delay *T*. Note that Eq. 14.42 is also the leading contribution for $I = 1$ because *k* is small (cf. Eq. 14.35).

For weak hyperfine interactions, we can neglect the isotropic contact interaction in comparison with the dipolar terms. This holds for remote nuclei such as deuterium or protons. We then have the dipolar hyperfine elements $A_\parallel = 2T_\perp^d$ and $A_\perp = -T_\perp^d$, which yield (see Eqs. 14.2 and 14.3):

$$A = T_\perp^d\left(3\cos^2\theta - 1\right) \tag{14.43}$$

$$B = 3T_\perp^d \sin\theta\cos\theta \tag{14.44}$$

where θ is the angle that the electron–nuclear vector **r** makes with the magnetic field direction. The strength of the dipolar hyperfine interaction is defined by:

$$T_\perp^d = \left(\mu_o/4\pi\right)g_e\beta_e g_n\beta_n/\left(\hbar r^3\right) \tag{14.45}$$

in angular frequency units (cf. Section 3.4 in Chapter 3). For a disordered sample, we must average Eq. 14.42 over all angles θ with respect to the static field. The angular average that we need for B^2 is $\langle\sin^2\theta\cos^2\theta\rangle \equiv \int_0^\pi \sin^2\theta\cos^2\theta\sin\theta\cdot d\theta \Big/ \int_0^\pi \sin\theta\cdot d\theta = \frac{2}{15}$.

The three-pulse echo amplitude of a powder sample therefore becomes:

$$\langle E_{3p}(2\tau+T)\rangle = 1-\frac{6}{5}\left(\frac{T_\perp^d}{\omega_I}\right)^2(1-\cos\omega_I\tau)$$
$$\times\left[1-\cos\omega_I(\tau+T)F\left(T_\perp^d(\tau+T)\right)\right] \quad (14.46)$$

where the function $F\left(T_\perp^d t\right)$ defines the integral:

$$F\left(T_\perp^d t\right)=\frac{15}{4}\int_0^\pi \sin^2\theta\cos^2\theta\cos\left(\tfrac{1}{2}T_\perp^d t\left(3\cos^2\theta-1\right)\right)\sin\theta\cdot d\theta \quad (14.47)$$

which is normalized to unity at the origin (Yudanov et al. 1976). We show results from numerical integration of Eq. 14.47 in Figure 14.8. As a function of the time delay T, the three-pulse ESEEM amplitude is modulated by a single frequency ω_I according to Eq. 14.46. To maximize the amplitude term $(1-\cos\omega_I\tau)F\left(T_\perp^d(\tau+T)\right)$, we adjust the fixed time delay τ close to π/ω_I, for a weakly coupled nucleus with Larmor frequency ω_I in a powder sample. On increasing the incremented delay T, the modulation amplitude passes through zero at $T_\perp^d(\tau+T)=7.25$, which allows us to determine $T_\perp^d$ for more remote nuclei with small hyperfine couplings (Narayana et al. 1975). Note that for two-pulse echoes, which are modulated at two frequencies, $F\left(T_\perp^d\tau\right)$ scales the amplitude at ω_I but not that at $2\omega_I$. As the time increment τ increases, a single modulation at $2\omega_I$ therefore comes to dominate the two-pulse echo from a powder sample (Dikanov and Tsvetkov 1985).

In the case of non-vanishing quadrupole interaction, the terms containing $\cos Pt$ in Eq. 14.37 also depend on the magnetic field direction. For the axial case with $I=1$, the quadrupole coupling is given by (see Eq. B.32 in Appendix B):

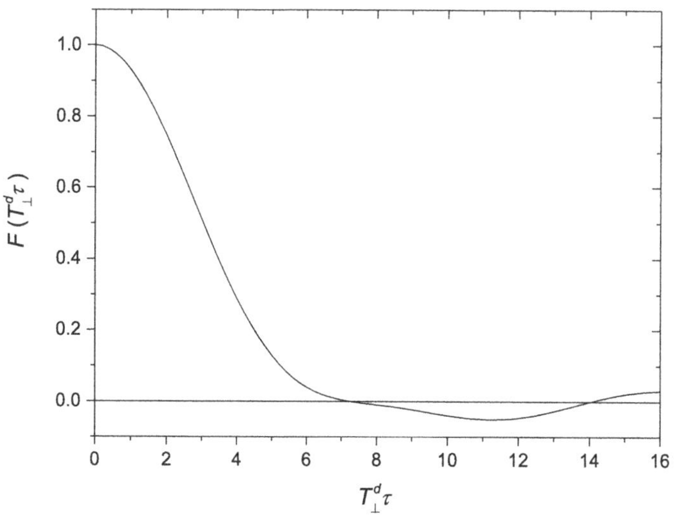

FIGURE 14.8 Integral $F\left(T_\perp^d\tau\right)$ (see Eq. 14.47) needed in Eq. 14.46 for powder averaged three-pulse echo intensity, as function of $T_\perp^d\tau$.

$$P=\frac{3e^2qQ}{8\hbar}\left(3\cos^2\theta_Q-1\right) \quad (14.48)$$

where Q is the quadrupole moment of the nucleus, q is the electric field gradient at the nucleus and θ_Q is the angle between the electric field gradient and the static magnetic field direction. Clearly, the quadrupole interaction increases the damping of the ω_I modulation beyond that by the hyperfine interaction. In general, the principal axes of the hyperfine tensor and quadrupole tensor do not coincide, i.e., θ_Q in Eq. 14.48 is different from θ. For deuterium nuclei, θ specifies the direction of the inter-nuclear vector, whereas θ_Q corresponds to the direction of the O–D bond in water or alcohols. Only for the ^{14}N-nucleus of the nitroxide itself do we expect collinear principal axes (Savitsky et al. 2008). Shubin and Dikanov (1983) consider the two-pulse echo of a disordered sample, where the principal qualitative effect of the quadrupole interaction is to damp the modulation at frequency $2\omega_I$, which otherwise is not damped.

14.9 ESEEM POWDER LINESHAPES

Methods for determining powder lineshapes are given in Section 2.4 of Chapter 2. However, we do not expect ESEEM powder spectra to exhibit discontinuities at the turning points that characterize the principal values of the hyperfine tensor as do EPR absorption spectra (and also ENDOR spectra). This is because of the angular dependence of the modulation depth parameter k. The latter is given from Eqs. 14.42 and 14.44, for weakly coupled nuclei, by:

$$k(\theta)=\frac{B^2}{\omega_I^2}=\left(\frac{3T_\perp^d}{\omega_I}\right)^2\sin^2\theta\cos^2\theta \quad (14.49)$$

where θ is the angle between the static magnetic field and the principal axis of the dipolar hyperfine tensor. Clearly, the modulation depth is zero for the magnetic field oriented along or perpendicular to this principal axis. It exhibits a single maximum at the intermediate orientation $\theta=45°$, where the hyperfine frequency $A(\theta)$ changes most rapidly (cf. Figure 13.11 inset). The result is a single hump-like feature for each region of angular anisotropy in the ESEEM powder pattern, from which we cannot easily extract tensor values.

However, these conclusions are strongly modified when we take into account the effects of spectrometer dead time on recording the echo amplitude (Astashkin et al. 1987). With increasing dead time, intensity is lost preferentially from those regions of the ESEEM powder pattern where spectral position changes fastest with field orientation. Sharp peaks then appear at the canonical turning points, which correspond to the principal hyperfine couplings. One example is the resolution of sharp doublet peaks corresponding to the perpendicular quadrupolar splitting in deuterium powder ESEEM (see, e.g., Figures 14.5 and 14.7). This enhanced resolution is an experimental artefact but one of considerable practical value.

These results have important implications for orientational selection in high-field ESEEM spectroscopy. According to simple considerations based on Eq. 14.49, we expect no nitrogen ESEEM signal when the field is set to the canonical x, y or z region of a high-field nitroxide spectrum (see, e.g., the 94-GHz spectrum in Figure 2.8). Nevertheless, the dead-time effect ensures that we can measure principal values of the hyperfine tensors in this way on powder samples. A notable example is measurement of the ^{14}N-quadrupole tensor for a nitroxide by 3-pulse ESEEM at 94 GHz (Savitsky et al. 2008; and see Table 4.9).

The latter work is a useful illustration of the matching of microwave frequency to hyperfine coupling in ESEEM-spectroscopy. At 9 GHz, the quadrupole interaction exceeds the nitrogen Zeeman interaction. At 94 GHz the reverse is true, and the quadrupole splitting is centred about the nitrogen Larmor frequency.

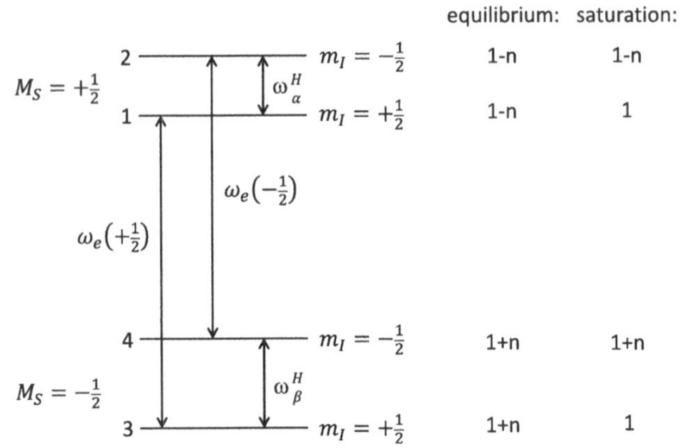

FIGURE 14.9 Energy-level diagram and proton-ENDOR transitions $\omega^{\mathrm{H}}_{\alpha,\beta}$ for $S=\frac{1}{2}, I=\frac{1}{2}$. Proton resonance frequency is greater than hyperfine coupling $\left(2\omega_{\mathrm{H}} > A^{\mathrm{H}}_{\parallel}\right)$. EPR transition $\omega_e(m_I)$ of a proton hyperfine line is saturated by fixed microwave radiation at fixed field, and monitored whilst scanning radiofrequency radiation through nuclear resonances that we detect by alleviation of EPR saturation. Electron-spin populations at thermal equilibrium, and for complete saturation of the $\omega_e\left(+\frac{1}{2}\right)$ transition, are indicated on the right. Numbering of the levels (1–4) is that of Schweiger and Jeschke (2001) (and Ernst et al. 1987).

14.10 ELECTRON–NUCLEAR DOUBLE RESONANCE (ENDOR)

ENDOR is basically a frequency-swept NMR experiment where the nuclear transitions are detected by their effect in alleviating saturation of the electron transitions. We thus combine the resolution of NMR spectroscopy with the sensitivity of EPR spectroscopy. The advantage for nitroxides is that we can determine hyperfine interactions with splittings that are too small to be resolved in the EPR spectrum. These are principally dipolar hyperfine interactions with remote protons, which we use to obtain structural information (Makinen et al. 1998). Also, the selection rules for ENDOR are much simpler than for hyperfine structure in EPR. Routinely, measurements are performed at low temperatures to prevent averaging of dipolar hyperfine couplings that we need for structural analysis. This also stabilizes conformations, traps enzymatic intermediates and ensures that the EPR spectrum is readily saturable.

14.11 CONTINUOUS-WAVE PROTON ENDOR

Figure 14.9 shows the energy level scheme and transitions for a $S=\frac{1}{2}, I=\frac{1}{2}$ system. This represents the interaction with a single proton for one of the nitrogen hyperfine lines of a nitroxide. Continuous microwave and radiofrequency radiation are applied simultaneously in CW-ENDOR. The static magnetic field and microwave frequency are fixed. High microwave power saturates the EPR transition (see Sections 11.2 and 11.3), which is monitored continuously as the radiofrequency is swept. Electron spin populations are equalized in the hyperfine state, e.g., $m_I=+\frac{1}{2}$, being saturated. The $\omega^{\mathrm{H}}_{\beta}$ NMR transition then repopulates the lower level $\left(M_S=-\frac{1}{2}\right)$ for this state, and correspondingly the $\omega^{\mathrm{H}}_{\alpha}$ NMR transition depopulates the higher level $\left(M_S=+\frac{1}{2}\right)$. Either of these redistributions alleviates saturation of the EPR transition, and consequently the two ENDOR transitions are detected.

Usually, proton hyperfine interactions in nitroxides are smaller than the proton Zeeman interaction. The ENDOR frequencies $\omega^{\mathrm{H}}_{\alpha,\beta}$ are then split about the nuclear Larmor frequency ω_{H} ($2\pi \times 14.5$ MHz at 0.34 T), as in Eq. 14.22:

$$\omega^{\mathrm{H}}_{\alpha,\beta} = \omega_{\mathrm{H}} \pm \tfrac{1}{2} A^{\mathrm{H}}(\theta) \qquad (14.50)$$

where $A^{\mathrm{H}}(\theta)$ is the proton hyperfine coupling when the static magnetic field B_o is inclined at angle θ to the vector $\mathbf{r}$ joining electron and proton spins. (For large hyperfine couplings, ENDOR frequencies are split by the Larmor frequency and centred about half the hyperfine coupling.) Using the magnetic dipole–dipole expression for the anisotropic part of the proton hyperfine coupling (cf. Eq. 3.20), the angle-dependent coupling is:

$$A^{\mathrm{H}}(\theta) = \left(\frac{\mu_o}{4\pi}\right)\frac{g_e\beta_e\gamma_{\mathrm{H}}}{r^3}\left(3\cos^2\theta - 1\right) + a_o^{\mathrm{H}} \qquad (14.51)$$

where γ_{H} is the proton gyromagnetic ratio and a_o^{H} is the isotropic hyperfine coupling constant. (Numerically $\left(\mu_o/4\pi\right)g_e\beta_e\left(\gamma_{\mathrm{H}}/2\pi\right) = 79.1\,\mathrm{kHz\,nm^3}$, for use with linear ^{1}H-ENDOR frequencies and r in nm.) Principal values of the hyperfine coupling, $A^{\mathrm{H}}_{\parallel}$ and $A^{\mathrm{H}}_{\perp}$, are given by Eq. 14.51 with $\theta = 0$ and $\theta = 90°$, respectively. Isotropic couplings are small and become negligible when $r > 0.5$ nm. The hyperfine tensor elements measured are then solely dipolar, so that $A^{\mathrm{H}}_{\perp} = -\frac{1}{2}A^{\mathrm{H}}_{\parallel}$. ENDOR is not sensitive to the sign of the hyperfine coupling but, whenever a_o^{H} is small, $A^{\mathrm{H}}_{\parallel}$ and $A^{\mathrm{H}}_{\perp}$ will have opposite signs.

Because unpaired electron spin density is distributed roughly equally between the nitrogen $\left(\rho^{\mathrm{N}}_{\pi}\right)$ and oxygen $\left(\rho^{\mathrm{O}}_{\pi}\right)$ atoms of the nitroxide, the dipolar hyperfine interaction in Eq. 14.51 consists of two parts:

$$A^{\mathrm{H},dip} = \left(\frac{\mu_o}{4\pi}\right) g_e \beta_e \gamma_{\mathrm{H}} \left(\frac{\rho_\pi^{\mathrm{O}}}{r_{\mathrm{HO}}^3}\left(3\cos^2\theta_{\mathrm{HO}} - 1\right) + \frac{\rho_\pi^{\mathrm{N}}}{r_{\mathrm{HN}}^3}\left(3\cos^2\theta_{\mathrm{HN}} - 1\right) \right) \tag{14.52}$$

where r_{HO} is the proton distance from the unpaired electron on the oxygen and θ_{HO} is the inclination of the $\mathbf{r}_{\mathrm{HO}}$ vector to the magnetic field. Similar definitions hold for the nitrogen atom. Proton-ENDOR measurements on the small nitroxide $\mathrm{PYCONH_2}$ (1-oxyl-2,2,5,5-tetramethylpyrrolin-3-ylamide) agree with the predictions of Eq. 14.52 when using crystal-structure data and EPR-derived values for spin densities ρ_π^{N} and ρ_π^{O} (Mustafi et al. 1991). They also confirm that the nitroxide unpaired spin can be represented by a point dipole situated at the effective vector position:

$$\mathbf{r}_e = \rho_\pi^{\mathrm{O}} \mathbf{r}_{\mathrm{HO}} + \rho_\pi^{\mathrm{N}} \mathbf{r}_{\mathrm{HN}} \tag{14.53}$$

along the N–O bond. This empirical relation is found to give a consistent description for all sets of ENDOR-measured distances (Makinen et al. 1998).

14.12 ORIENTATION SELECTION IN POWDER ENDOR

We learnt from Section 2.4 that EPR powder patterns consist of spectrally dispersed regions that correspond to different orientations of the magnetic field relative to the nitroxide axes. Indeed, the outer peaks in a 9-GHz powder pattern correspond uniquely to the magnetic field directed along the nitroxide z-axis, and at 94 GHz or above the x- and y-orientations are also resolved (see Figure 2.8). The principal axis for a proton hyperfine tensor is specified by the electron–nucleus interspin vector, which in general will not coincide with one of the nitroxide axes. Nevertheless, orientational selection by choosing specific field positions within the EPR powder pattern can simplify the ENDOR powder spectrum considerably. We illustrate the principle in Figure 14.10 for magnetic field parallel to the nitroxide z-axis (B_z) or in the nitroxide x,y-plane ($B_{x,y}$), and various positions of the proton (black balls). Protons lying on the nitroxide z-axis then give single-crystal like ENDOR spectra at the special field positions, with hyperfine couplings $A_\parallel^{\mathrm{H}}$ and $A_\perp^{\mathrm{H}}$ for B_z and $B_{x,y}$, respectively. Equatorial protons, lying in the nitroxide x,y-plane, are statistically more likely (cf. Figure 2.5). These will have single-crystal like ENDOR with hyperfine coupling $A_\perp^{\mathrm{H}}$ for B_z but a two-dimensional powder pattern between $A_\parallel^{\mathrm{H}}$ and $A_\perp^{\mathrm{H}}$ for orientation $B_{x,y}$. For 9-GHz spectra, the outer peaks of the EPR powder pattern give the unique B_z-orientation, but the central peak of a ^{14}N-nitroxide approximates more to an isotropic distribution of field directions B_x, B_y and B_z (see Figure 1.4).

Figure 14.11 shows ENDOR absorption spectra that we expect from orientation selection for an equatorial proton.

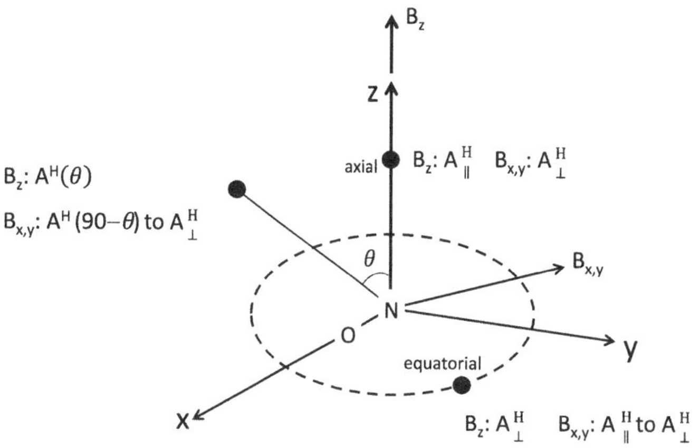

FIGURE 14.10 Orientation selection in proton-ENDOR spectra, depending on field-position in nitroxide powder EPR spectrum. *Axial proton* (lying along nitroxide *z*-axis): hyperfine coupling $A_\parallel^{\mathrm{H}}$ for field B_z parallel to *z*-axis; $A_\perp^{\mathrm{H}}$ for field $B_{x,y}$ in nitroxide *x,y*-plane. *Equatorial proton* (lying in nitroxide *x,y*-plane): hyperfine coupling $A_\perp^{\mathrm{H}}$ for field-orientation B_z; extends from $A_\parallel^{\mathrm{H}}$ to $A_\perp^{\mathrm{H}}$ for field-orientation $B_{x,y}$. *General proton location* (θ): hyperfine coupling $A^{\mathrm{H}}(\theta)$ (see Eq. 14.51) for field-orientation B_z; spans $A^{\mathrm{H}}(90-\theta)$ to $A_\perp^{\mathrm{H}}$ for field-orientation $B_{x,y}$.

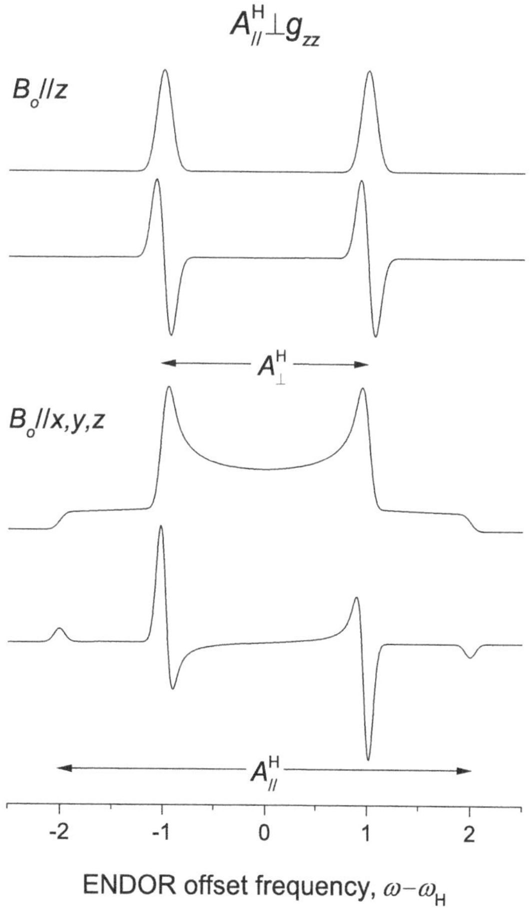

FIGURE 14.11 ENDOR spectra for proton in nitroxide *x,y*-plane, with magnetic field B_o parallel to nitroxide *z*-axis (top), and B_o distributed randomly (bottom). *Upper spectrum of each pair*: absorption; *lower*: first derivative.

Selecting the outer wings of the EPR spectrum ($B_o \parallel z$) gives a single-crystal like spectrum. Selecting the central peak of the EPR powder pattern ($B_o \parallel x, y, z$) may result in a spectrum with more intensity in the wings than the classic Pake pattern shown at the bottom of Figure 14.11 (see, e.g., Makinen et al. 1998). This would indicate some degree of orientational selection that introduces partial two-dimensional character to the powder pattern.

Choosing other positions in the EPR powder pattern resembles doing a single-crystal rotation experiment. Figure 14.12 shows schematically the variation of ENDOR powder-pattern peak splittings with field position for an equatorial proton, i.e., corresponding to the spectra in Figure 14.11. An experimental dependence on field position that corresponds to this situation is found for the H^{α} proton of spin-labelled methyl-L-alanate (Mustafi et al. 1990). For this particular location of the proton,

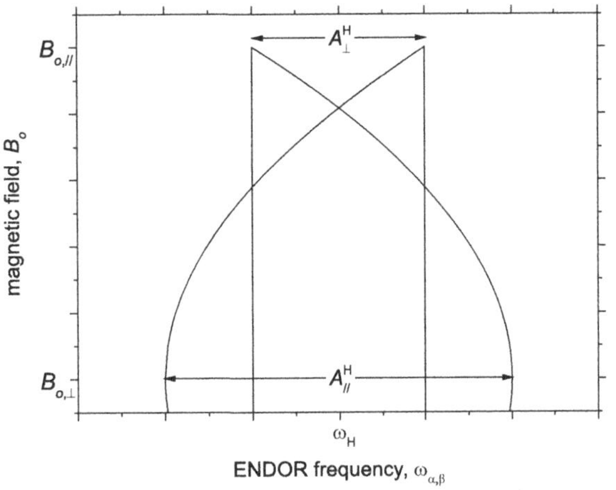

FIGURE 14.12 Schematic dependence of ENDOR frequencies $\omega_{\alpha,\beta}^{H}$, for a proton located in the nitroxide *x,y*-plane, on magnetic-field setting B_o within the EPR powder pattern. $B_{o,\perp}$ selects nitroxides with static field in their *x,y*-plane; $B_{o,\parallel}$ selects those with field along their *z*-axis. For ^{14}N-nitroxide at 9 GHz: $B_{o,\perp}$ corresponds to absorption maximum in the middle of the powder pattern; $B_{o,\parallel}$ to the outer turning points. $A_{\parallel}^{H}$: hyperfine coupling for protons with interspin vector **r** parallel to the static field; $A_{\perp}^{H}$: for protons with **r** perpendicular to the field.

there is no problem in determining both $A_{\perp}^{H}$ and $A_{\parallel}^{H}$. For a general location with polar angle θ, as indicated in Figure 14.10, we do not get $A_{\parallel}^{H}$ directly from either of the special field positions B_z and $B_{x,y}$. In this case, a systematic dependence as in Figure 14.12 can locate the field position corresponding to $A_{\parallel}^{H}$, which should give the largest splitting. Note that when we can neglect a_o^{H}, measuring $A_{\perp}^{H}$ alone is sufficient to give the dipolar splitting.

14.13 DISTANCES FROM ENDOR FREQUENCIES

For structural work we need to assign $A_{\perp}^{H}$ and $A_{\parallel}^{H}$ values to particular protons in the spin-labelled molecule. Dependences on magnetic field position as in Figure 14.12 can aid assignment, and occasionally a_o^{H} values, but in general we need deuteration at specific sites. Fluorine-ENDOR of specifically fluorinated derivatives is also helpful (Wells et al. 1990).

Table 14.1 lists the hyperfine couplings for protons, and a fluorine in the phenylalanine ring, of PYCOO-L-phenylalaninal adducts with methanol or Ser195 in α-chymotrypsin. This spin-labelled phenylalaninal (Phe-al) is a covalent active-site inhibitor of serine proteases. The hyperfine couplings are measured from ENDOR according to Eqs. 14.50 and 14.51. We know that $A_{\perp}^{H}$ (and $A_{\perp}^{F}$) must be negative; thus we can calculate the isotropic coupling a_o^{H} – except where $A_{\parallel}^{H}$ is unknown. Using Eq. 14.51, we then get the dipolar components $A_{\parallel}^{dip}$ and $A_{\perp}^{dip}$, and hence the electron–nucleus separations *r*. The three measured distances are very similar for tetrahedral methanol adduct and inhibitory enzyme adduct. They are compatible with both *R* and *S* configurations of the aldehyde carbonyl carbon. However, the $^{19}F^{\zeta}$ powder ENDOR spectrum with the central $B_{x,y,z}$ field setting for the enzyme adduct shows that the ^{19}F nucleus of the phenylalanine ring lies close to the nitroxide *x,y*-plane (Jiang et al. 1998). Consequently, only the *S* configuration of the aldehyde is possible for this transition-state inhibitor analogue.

TABLE 14.1 Hyperfine couplings and electron–nucleus distances *r* deduced from dipolar contributions, for adduct of *N*-(1-oxyl-2,2,5,5-tetramethylpyrrolinyl-3-carbonyl)-phenylalaninal (PYCOPhe-al) with methanol or with Ser195 of α-chymotrypsin. (Jiang et al. 1998)

nucleus, n	$A_{\parallel}^{(n)}$ (MHz)	$A_{\perp}^{(n)}$ (MHz)	$a_o^{(n)}$ (MHz)	$A_{\parallel}^{dip}$ (MHz)	$A_{\perp}^{dip}$ (MHz)	*r* (nm)
methanol adduct:						
H^N	1.193	0.577	0.013	1.180	−0.590	0.512 ± 0.004
H^{α}	0.542	0.262	0.006	0.536	−0.268	0.667 ± 0.010
H^{al}	0.402	0.227	−0.017	0.419	−0.210	0.724 ± 0.010
$^{19}F^{\zeta}$	0.245	0.135	−0.008	0.253	−0.127	0.838 ± 0.033
enzyme adduct:						
H^{α}	-	0.259	0[a]	0.518	−0.259	0.673 ± 0.02
H^{al}	-	0.223	0[a]	0.446	−0.223	0.710 ± 0.02
$^{19}F^{\zeta}$	0.229	0.110	0.003	0.226	−0.113	0.870 ± 0.04

[a] Assumed.

The conformation of the spin-labelled Phe-al inhibitor was obtained by molecular modelling subject to constraints provided by the ENDOR distances and the angle-dependent hyperfine coupling patterns for $B_{x,y,z}$ and B_z settings of the static field (Jiang et al. 1998). The peptide bond with the PYCOOH nitroxide is planar with dihedral $\omega = 175°$, as expected. Dihedral angles for the Phe side chain of the free inhibitor $\chi_1 = -62°$ and $\chi_2 = 90°$ (cf. Figure 16.1 for definition) are staggered as expected. In contrast, dihedral angle $\chi_2 \approx -10°$ of the enzyme-bound inhibitor is almost eclipsed, as found also in ^{1}H- and ^{19}F-ENDOR studies of the spin-labelled tryptophanyl acylenzyme intermediate (Wells et al. 1994). The high-energy nearly eclipsed conformation about C$^\beta$–C$^\gamma$ keeps the oxyanion binding site open for the substrate carbonyl to form a tetrahedral adduct, and favours subsequent dissociation of the carboxylic acid product. Torsional destabilization of bound substrate is likely a common feature driving formation of transition states in enzyme catalytic cycles.

Proton ENDOR has been applied similarly to defining the structure of a spin-labelled mixed-anhydride acylenzyme reaction intermediate in the catalytic site of carboxypeptidase A (Mustafi and Makinen 1994). Also, the conformation of spin-labelled penicillanic acid as a catalytically competent reaction intermediate in the active site of TEM-1 β-lactamase has been determined (Mustafi et al. 2001). For the latter enzyme, the conformation of the MTSSL side chain in site-directed spin labelling of the Glu240Cys mutant was defined by proton ENDOR (Mustafi et al. 2002; and see Table A16.2 in the Appendix to Chapter 16).

14.14 PULSE ENDOR

Because ENDOR experiments typically are performed at low temperature to preserve structurally sensitive dipolar hyperfine interactions, pulse ENDOR measurements offer an alternative. Although not necessarily favourable compared with CW-ENDOR in terms of sensitivity, pulse ENDOR depends less on relaxation times, thus the need to optimize temperature is not so severe. In the two pulse-sequences commonly used, a long selective radiofrequency pulse (whose frequency is swept) is applied during an evolution period T that is sensitive to T_1-relaxation (see Figure 14.13). Both sequences start with a preparation period where microwave pulse(s) create electron–nuclear two-spin order $2I_zS_z$. This is followed by a mixing period when the radiofrequency pulse is applied to change the nuclear polarization. Finally, in the detection period, the changes in nuclear polarization are recorded by an electron spin echo.

14.14.1 Davies ENDOR

Davies ENDOR is closer conceptually to CW-ENDOR, because it consists of an inversion recovery sequence with radiofrequency pulse spanning the fixed recovery period T (Davies 1974). The initial inverting microwave π-pulse (see Figure 14.13, top) is selective and burns a hole in the inhomogeneous lineshape.

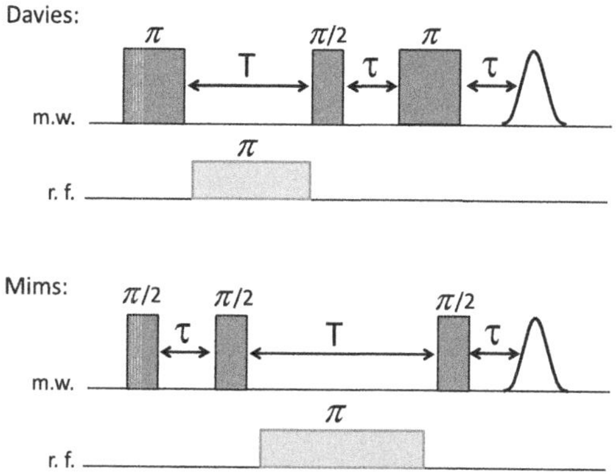

FIGURE 14.13 ENDOR pulse sequences. *Davies ENDOR* (upper panel): selective microwave π-pulse (m.w.) inverts the electron polarization for a particular transition; selective resonant radiofrequency π-pulse (r.f.) transfers nuclear polarization from one of the two states connected by the transition; primary echo (π/2–τ–π) reads out resulting change in electron polarization. *Mims ENDOR* (lower panel): selective radiofrequency pulse (r.f.), applied during evolution period T of a non-selective three-pulse echo sequence (m.w.), reduces stimulated-echo intensity when inducing nuclear transitions.

Refilling the hole by resonant nuclear spin flips is then monitored with a primary electron-echo sequence. Static field and all time delays remain fixed; only the radiofrequency is scanned.

To deal with selective pulses, we must use single-transition operators in the product-operator description. We describe these in Appendix P.5, which should be consulted for what follows. Figure P.1 of this appendix shows the energy-level scheme and transitions for a $S = \frac{1}{2}, I = \frac{1}{2}$ spin system, applicable to proton ENDOR. From Eq. P.50, we see that a π-pulse resonant with allowed EPR transition (1,3) creates two-spin order $\sigma_{prep} = -2I_zS_z$, in the preparation period. The subsequent radiofrequency x-pulse of effective angle β_2 is resonant with nuclear transition (1,2), where the single-transition operator is $I_x^{(12)} = I_xS^\alpha = I_x\left(\frac{1}{2}+S_z\right)$. We express σ_{prep} in terms of nuclear single-transition operators by using the relation: $I_z = \left(S^\alpha + S^\beta\right)I_z = I_z^{(12)} + I_z^{(34)}$, which is the polarization summed over the two transitions (cf. Eq. P.46 and see Table P.3). By applying transformations as in Eq. P.49 we get:

$$\sigma_{prep} = -2S_z\left(I_z^{(12)} + I_z^{(34)}\right)\xrightarrow{\beta_2 I_x^{(12)}}$$

$$-2S_z\left(I_z^{(12)}\cos\beta_2 - I_y^{(12)}\sin\beta_2 + I_z^{(34)}\right)$$

$$= -\left(1+\cos\beta_2\right)I_zS_z + \tfrac{1}{2}\left(1-\cos\beta_2\right)I_z + 2\sin\beta_2 I_y^{(12)}S_z = \sigma_{mix} \quad (14.54)$$

If $\beta_2 = \pi$ for a resonant pulse, then $\sigma_{mix} = I_z$ and nuclear polarization is transferred across the (1,2) transition, during the mixing period. If $\beta_2 = 0$, corresponding to a non-resonant pulse, then $\sigma_{mix} = -2I_zS_z$ and the nuclear polarization remains unchanged.

Detection with the primary-echo sequence uses x-pulses acting on the same (1,3) EPR transition as for the original preparation pulse. The only corresponding electron single-transition operator in σ_{mix} comes from expanding the term in $I_zS_z = \frac{1}{2}\left(I^\alpha - I^\beta\right)S_z = \frac{1}{2}\left(S_z^{(13)} - S_z^{(24)}\right)$ – see Eq. P.47 and

Table P.3. Applying the $\pi/2\text{–}\tau\text{–}\pi\text{–}\tau\text{–}echo$ sequence to just this term, we get:

$$\sigma_{mix} = -\tfrac{1}{2}(1+\cos\beta_2)S_z^{(13)}$$
$$\xrightarrow{(\pi/2)S_x^{(13)}} \xrightarrow{\mathcal{H}_o\tau} \xrightarrow{\pi S_x^{(13)}} \xrightarrow{\mathcal{H}_o\tau}$$
$$-\tfrac{1}{2}(1+\cos\beta_2)S_y^{(13)} = -\tfrac{1}{2}(1+\cos\beta_2)I^\alpha S_y = \sigma_{echo} \quad (14.55)$$

where we immediately discard the term in I_zS_y that comes from $I^\alpha = \frac{1}{2}+I_z$ because it does not contribute to the echo. Note also that $S_z^{(24)}$ does not contribute because this transition is not excited. For a non-resonant pulse, $\beta_2 = 0$ and $\sigma_{echo} = -\frac{1}{4}S_y$, and the echo intensity is maximum. For a resonant pulse, $\beta_2 = \pi$ and $\sigma_{echo} = 0$: the echo is reduced to zero. The transverse magnetization refocused by the echo is:

$$\overline{\langle S_y\rangle} = -\tfrac{1}{4}(1+\cos\beta_2) \quad (14.56)$$

which is the signal that we detect from one spin packet.

When inhomogeneous line broadening is much larger than the hyperfine splitting of individual spin packets ($\Gamma_{inh} >> A_{hf}$), the high-frequency hyperfine line of one spin packet superimposes on the low-frequency line of a second spin packet that is shifted by the hyperfine coupling A_{hf} (see Figure 14.14, bottom two traces). Both contribute equally to the hole burnt by the preparation π-pulse that is on resonance with the (1,3) transition of one spin packet and the

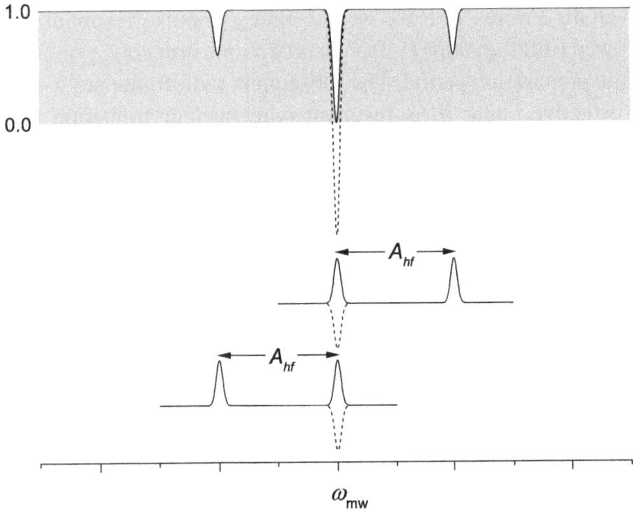

FIGURE 14.14 Hole burning in inhomogeneously broadened EPR line (shaded area with $\Gamma_{inh} \to \infty$) by Davies-ENDOR pulse sequence, for $S=\frac{1}{2}, I=\frac{1}{2}$. Two spin packets shifted by A_{hf} (lower traces) share a common line position. Microwave preparation π-pulse applied at this position ω_{mw} burns a central hole (dashed lines). Radiofrequency π-pulse ω_{rf} applied on resonance with low-frequency spin packet (bottom trace) burns a hole shifted down by A_{hf} from the central hole (heavy solid line), filling in part of the central hole. Similarly for high-frequency spin packet (middle trace): side hole is shifted up by A_{hf} and central hole further diminished. Amplitude of central hole is recorded as function of ω_{rf} (After Gemperle and Schweiger 1991).

(2,4) transition of the other. The radiofrequency π-pulse inverts the polarization of nuclear transition (1,2) or (2,3), depending with which it resonates. This decreases the depth of the central hole and induces two side holes that are separated from the central hole by A_{hf}. The ENDOR experiment consists of monitoring the intensity of the centre hole whilst sweeping the radiofrequency. Nuclear resonances correspond to increases in echo amplitude. The side holes must be sufficiently removed as not to overlap with the central hole. Therefore, Davies-ENDOR is a method for measuring relatively large unresolved hyperfine interactions, e.g., from directly hydrogen-bonded protons. For smaller hyperfine interactions, e.g., from hydrogen-bonded deuterium and more remote protons, Mims-ENDOR is more appropriate.

14.14.2 Example of Davies ENDOR

An interesting application of Davies ^{1}H-ENDOR is the high-field EPR study on hydrogen bonding of alcohols to a nitroxide by Cox et al. (2017). All protons are replaced by deuterium, except the alcohol –OH that is involved in hydrogen bonding. Samples are frozen glasses.

Figure 14.15 shows schematically the characteristic orientationally-selected ridges in the proton ENDOR powder pattern as a function of static-field position B_o in the 94-GHz EPR spectrum, where the ENDOR sequence is applied (cf. Figure 14.12). In this particular example, the proton lies in the x,y-plane of the nitroxide, as expected for hydrogen bonding with N–O. Expressed relative to a suitable reference, the static-field offset ΔB in the EPR spectrum is given by (cf. Eq. 2.23 in Section 2.4):

$$\Delta B(\theta,\phi) = \left(\Delta B_x\cos^2\phi + \Delta B_y\sin^2\phi\right)\sin^2\theta + \Delta B_z\cos^2\theta \quad (14.57)$$

where (θ,ϕ) are the polar angles of the static field in the nitroxide axis system. If we take the isotropic position $B_o = h\nu/g_o\beta_e$ as reference field, then $\Delta B_x = (g_{xx}-g_o)B_o/g_o$, $\Delta B_y = (g_{yy}-g_o)B_o/g_o$ and $\Delta B_z = (g_{zz}-g_o)B_o/g_o$. The nuclear–electron dipolar coupling depends on the angle Θ_D that the electron–nuclear vector makes with the static magnetic field. This is given by the spherical-harmonic addition theorem: $\cos\Theta_D = \cos\theta\cos\theta_D + \sin\theta\sin\theta_D\cos(\phi-\phi_D)$, where (θ_D,ϕ_D) are the polar angles of the electron-nuclear vector in the nitroxide axis system. For $\theta_D = 90°$, as in Figure 14.15, the nuclear dipolar frequency is given by (cf. Eq. 14.51):

$$\omega_D = T_\perp^d\left(3\cos^2\Theta_D - 1\right) = T_\perp^d\left(3\sin^2\theta\cos^2(\phi-\phi_D)-1\right) \quad (14.58)$$

where $T_\perp^d = (\mu_o/4\pi)g_e\beta_e\gamma_H/r_D^3$ for nuclear–electron separation r_D. The θ,ϕ-terms in Eq. 14.58 give the dependence of ω_D on field position ΔB according to Eq. 14.57.

We see from Figure 14.15 and Eq. 14.57 that there are two EPR regimes: between B_z and the x,y-region, where ΔB depends on θ; and between B_x and B_y where ΔB depends on ϕ. In the first region, $\Delta B(\theta)$ depends on $\sin^2\theta\,(=1-\cos^2\theta)$, so that ω_D and ΔB are linearly related, as we see from the two heavy (reflected)

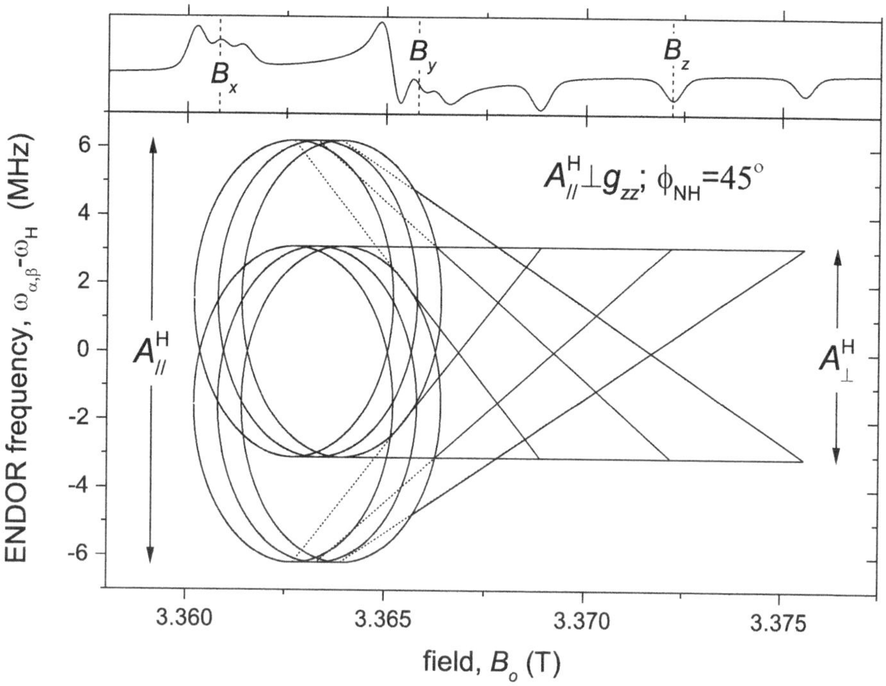

FIGURE 14.15 Ridges in powder [1]H-ENDOR spectrum for proton in nitroxide *x,y*-plane, as function of spectral position B_o in 94-GHz EPR spectrum of ^{14}N-nitroxide. Schematic diagram based on simulated data from Cox et al. (2017) and Linder et al. (1980). Interspin vector perpendicular to nitroxide *z*-axis ($\theta_D = 90°$) and 45° to nitroxide *x*-axis ($\phi_D = 45°$). Proton dipolar splitting depends linearly on field-offset in $B_z - B_y$ region and describes ellipse in $B_y - B_x$ region (see Eqs. 14.57, 14.58).

straight ridges in Figure 14.15. In the second region, however, the reflected ridges describe an ellipse that is limited by B_x and B_y and extends to the maximum dipolar splitting $A_{\parallel}^{H}$ (Linder et al. 1980). This is because the principal axis (∥) of the dipolar splitting lies in the *x,y*-plane but is aligned along neither of the nitroxide axes. In Figure 14.15, the dipolar axis is inclined at $\phi_D = 45°$ to the nitroxide *x*-axis. This corresponds to the field position where the dipolar splitting is maximum:

$$\Delta B\left(\omega_D^{\max}\right) = \left(\Delta B_x \cos^2\phi_D + \Delta B_y \sin^2\phi_D\right)\sin^2\theta_D + \Delta B_z \cos^2\theta_D \tag{14.59}$$

which allows us to determine ϕ_D when $\theta_D = 90°$ as in Figure 14.15. If the dipolar axis coincides with the nitroxide *x*- or *y*-axis, then the ellipse in Figure 14.15 collapses to a third (reflected) straight line. See Linder et al. (1980) for description of the different cases.

The schematic ridge plot in Figure 14.15 closely resembles Davies powder ^{1}H-ENDOR spectra of the solvent hydroxyl proton as a function of field position B_o in the 94-GHz EPR spectrum of ^{14}N-PYMeOH-d_{16} in CD_3OH (Cox et al. 2017). PYMeOH (1-oxy-2,2,5,5-tetramethylpyrrolin-3-yl methanol) is closely related to the MTSSL spin label that is used routinely in site-directed spin labelling (see Chapter 16). The three nitrogen hyperfine manifolds contribute separately and their intensity then superimposes. The Davies ENDOR spectrum therefore shows that the alcohol –OH, which participates in hydrogen bonding, lies in the nitroxide *x,y*-plane and makes an angle of $180° - \phi_D \approx 135°$ with the N–O bond (cf. Figure 4.13 for water). The H-bonding lone pair on the nitroxide oxygen therefore resides in a sp^2-hybrid and not a simple p_y-orbital. The experimental proton hyperfine splittings are $A_{\parallel}^{H} = 12.5\,\text{MHz}$ and $A_{\perp}^{H} = -5\,\text{MHz}$. This leads to a dipolar coupling: $A_{\parallel}^{dip} = 11.7\,\text{MHz}$ (and $a_o^{H} = 0.8\,\text{MHz}$). Contributions come from both the oxygen hydrogen-bond acceptor and the nitrogen according to Eq. 14.52.

Figure 14.16 shows the orientation of the proton hyperfine axes $\left(A_{xx}^{H}, A_{yy}^{H}, A_{zz}^{H}\right)$ relative to the nitroxide principal axes (*x,y,z*). Because the proton is close to the nitroxide, we must take into account the offsets of the 2pπ-orbitals from the nitroxide *x,y*-plane: $r_{2p,O}^{eff} = \left\langle r^{-3}\right\rangle_{2p,O}^{-1/3} = 0.0310\,\text{nm}$ and $r_{2p,N}^{eff} = 0.0363\,\text{nm}$

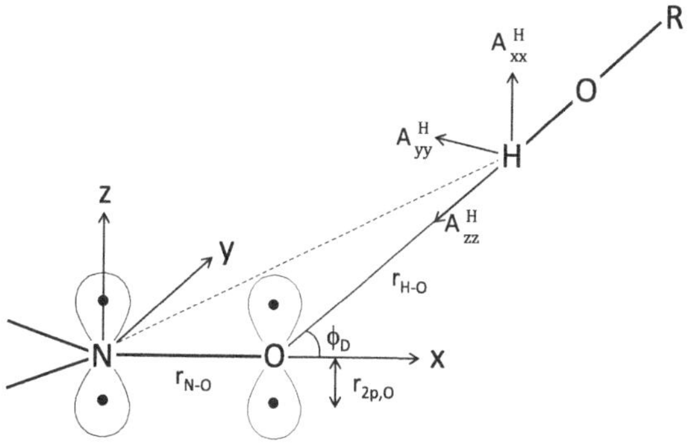

FIGURE 14.16 Orientation of $\left(A_{xx}^{H}, A_{yy}^{H}, A_{zz}^{H}\right)$ hyperfine tensor for hydrogen-bonding proton, relative to nitroxide *x,y,z*-axes. –OH bond lies in nitroxide *x,y*-plane (cf. Figure 4.13). Unpaired electron represented by point-dipoles (dots) located in nitrogen and oxygen 2*p*π-orbitals with *z*-offsets $r_{2p,N}$, $r_{2p,O}$. For –OD, principal ^{2}H-quadrupole tensor axis assumed parallel to A_{zz}^{H}.

(Morton 1964, and see Section 3.4). The distances that we require in Eq. 14.52 are then:

$$r_{HO} = \sqrt{r_{H-O}^2 + r_{2p,O}^2} \quad (14.60)$$

$$r_{HN} = \sqrt{r_{H-O}^2 + r_{N-O}^2 + 2r_{H-O}r_{N-O}\cos\phi_D + r_{2p,N}^2} \quad (14.61)$$

where r_{H-O}, r_{N-O} are the hydrogen-bond and N–O bond lengths, respectively. When the magnetic field is along the A_{zz}^{H}-direction (i.e. H–O), the corresponding angular variables are given by:

$$\cos\theta_{HO} = r_{H-O}/r_{HO} \quad (14.62)$$

$$\cos\theta_{HN} = \left(r_{H-O} + r_{N-O}\cos\phi_D\right)/r_{HN} \quad (14.63)$$

We now can determine the hydrogen bond length from $A_{\parallel}^{dip}$ by using Eqs. 14.52, 14.60–14.63. Taking $\rho_{\pi}^{O} \approx 0.45$ gives $r_{H-O} = 0.194\,\text{nm}$, which is slightly longer than values predicted for water by DFT calculations (see Table 4.7). As we shall see in Section 14.14.4, the discrepancy might arise from limitations of the simple point-dipole model.

When the magnetic field is along the A_{xx}^{H}-direction, instead of Eqs. 14.62 and 14.63, we have $\theta_{HO} = 90°$ and:

$$\cos\theta_{HN} = r_{2p,N}/r_{HN} \quad (14.64)$$

Correspondingly for A_{yy}^{H} we get:

$$\cos\theta_{HN} = \left(r_{N-O}/r_{HN}\right)\cos\phi_D \quad (14.65)$$

and again $\theta_{HO} = 90°$. As we anticipate from Figure 14.16, the dipolar contributions to A_{xx}^{H} and A_{yy}^{H} are no longer equal. The proton dipolar tensor is markedly non-axial: using Eqs. 14.64 and 14.65 with $r_{H-O} = 0.194\,\text{nm}$ deduced from A_{zz}^{dip} predicts $A_{xx}^{dip} = -6.2\,\text{MHz}$ and $A_{yy}^{dip} = -5.9\,\text{MHz}$. This non-axiality is not immediately apparent from the experimental ridge plot but can be accommodated by simulation (Cox et al. 2017).

14.14.3 Mims ENDOR

Mims ENDOR is based on the stimulated-echo sequence with non-selective pulses (Mims 1965). However, we neglect pseudosecular terms, which suppresses the ESEEM effect, because the radiofrequency pulse is applied during the longer delay time T. This comes after the first two $\pi/2$-pulses, which constitute the preparation sequence (see Figure 14.13). Detection by stimulated echo follows the third $\pi/2$-pulse. Using Table P.1 in Appendix P, the preparation sequence (see Figure 14.13) becomes:

$$\sigma_{eq} = S_z \xrightarrow{(\pi/2)S_x} -S_y \xrightarrow{\left(\Delta\omega_o S_z + aI_zS_z\right)\tau} \xrightarrow{(\pi/2)S_x} \sigma_{prep}$$
$$= -S_z\cos\tfrac{1}{2}a\tau\cos\Delta\omega_o\tau - 2I_zS_z\sin\tfrac{1}{2}a\tau\sin\Delta\omega_o\tau \quad (14.66)$$

where we omit terms other than those in S_z and I_zS_z because they do not contribute to the final echo. We now apply a selective radiofrequency x-pulse $\beta_2 I_x^{(12)}$ with flip angle β_2 to the (1,2) nuclear transition (see Figure P.1). This acts only on the $2I_zS_z = \left(S^{\alpha} - S^{\beta}\right)I_z = I_z^{(12)} - I_z^{(34)}$ term (that we expand using the equivalent of Eq. P.47 and Table P.3), which transforms according to (cf. Table P.1):

$$2I_zS_z \xrightarrow{\beta_2 I_x^{(12)}} I_z^{(12)}\cos\beta_2 - I_y^{(12)}\sin\beta_2 - I_z^{(34)} = I_zS_z\left(1+\cos\beta_2\right) \quad (14.67)$$

Again we retain only S_z and I_zS_z terms in the final result, which comes from expressions for single-transition operators in Table P.3. The spin density-operator after the mixing period therefore becomes:

$$\sigma_{mix} = -S_z\cos\tfrac{1}{2}a\tau\cos\Delta\omega_o\tau + I_zS_z\left(1+\cos\beta_2\right)\sin\tfrac{1}{2}a\tau\sin\Delta\omega_o\tau \quad (14.68)$$

Applying the third non-selective $\pi/2$-pulse to generate the stimulated echo after further evolution time τ, we get:

$$\sigma_{mix} \xrightarrow{(\pi/2)S_x} \xrightarrow{\left(\Delta\omega_o S_z + aI_zS_z\right)\tau} \sigma_{echo} =$$
$$-S_y\left[\cos^2\tfrac{1}{2}a\tau\cos^2\Delta\omega_o\tau + \tfrac{1}{2}\left(1+\cos\beta_2\right)\sin^2\tfrac{1}{2}a\tau\sin^2\Delta\omega_o\tau\right] \quad (14.69)$$

retaining only the term in S_y because this produces the echo from x-pulses. Applying a resonant radiofrequency pulse with $\beta_2 = \pi$ therefore reduces the echo intensity, relative to a non-resonant pulse where effectively $\beta_2 = 0$. To allow for inhomogeneous broadening we must average σ_{echo} over all offset frequencies $\Delta\omega_o$ giving:

$$\overline{\langle S_y \rangle} = -\tfrac{1}{4}(1+\cos a\tau) \quad (14.70)$$

for $\beta_2 = \pi$. Depending on the hyperfine coupling a, Mims ENDOR has blind spots when the delay time τ of the stimulated-echo sequence is an integral multiple of $2\pi/a$. To avoid missing any ENDOR lines, we must repeat the sequence with different τ-values.

14.14.4 Example of Mims ENDOR

An interesting example of Mims ^{2}H-ENDOR is a high-field study of 2-isopropanol-d_1 hydrogen bonding to a DOXYL nitroxide by Smirnova et al. (2007). Only the alcohol hydroxyl, i.e., that taking part in hydrogen bonding, is deuterated. Angle-selective ^{2}H-ENDOR spectra are recorded at the g_{xx}, g_{yy} and g_{zz} turning points in the 130-GHz EPR spectrum of the ^{14}N-nitroxide (cf. Figure 14.15, top). Outer extrema, g_{xx} and g_{zz}, in the HF-EPR spectrum yield single-crystal like ENDOR spectra corresponding to B_o oriented along $(\theta,\phi) = (90°,0°)$ and $(0°,0°)$, respectively, in the nitroxide axis system. However, the intermediate turning point in the HF-EPR spectrum corresponds to B_o oriented not only along g_{yy} (i.e., $\theta,\phi = 90°, 90°$) but also along all directions satisfying the relation $g(\theta,\phi) = g_{yy}$. As we illustrate in Figure 14.17, these orientations lie in two planes containing the y-axis that are tilted at angle $\theta = \xi_z$ to the z-axis, where $\sin^2 \xi_z = (g_{yy} - g_{zz})/(g_{xx} - g_{zz})$ from Eq. 2.23 when $g(\xi_z, 0) = g_{yy}$. Values of ξ_z are in the range 45°–52° depending on spin label and environmental polarity (see Tables A2.1–A2.3). The resulting ENDOR spectrum is a two-dimensional powder pattern extending from $(\theta,\phi) = (90°, 90°)$ to $(\xi_z, 0)$, whose detailed shape depends on orientation of the hyperfine/quadrupole axes to those of the nitroxide.

The principal element of the ^{2}H-hyperfine tensor and ^{2}H-quadrupole tensor, determined by fitting the orientation-selective ENDOR spectra, is: $A_{zz}^{dip} = 1.92\,\mathrm{MHz}$ and $2P_{zz} = 0.27\,\mathrm{MHz}$, respectively (see Figure 14.16 for axes). Taking explicit account of the unpaired electron spin densities in the oxygen and nitrogen $2p\pi$-orbitals (see McConnell and Strathdee 1959), the hydrogen-bond length deduced from A_{zz}^{dip} is $r_{\mathrm{H-O}} = 0.174 \pm 0.006\,\mathrm{nm}$, with $\rho_\pi^{\mathrm{O}} = \rho_\pi^{\mathrm{N}} = 0.475$ and $\phi_D = 60°$ (Smirnova et al. 2007). For comparison, the offset point-dipole model of Section 14.14.2 gives $r_{\mathrm{H-O}} = 0.191\,\mathrm{nm}$, with the same parameters. However, both models predict a much smaller non-axiality $A_{xx}^{dip} - A_{yy}^{dip}$ than that found by fitting the experimental spectra (0.88 MHz from experiment, and 0.24 MHz from point-dipole model). The ^{2}H-quadrupole coupling depends on the length $r_{\mathrm{D-O}}$ of the hydrogen bond (Soda and Chiba 1969). DFT simulations for neutral benzosemiquinone, where the C–O radical is closest to our N–O case, yield the calibration: $2P_{zz}(\mathrm{Hz}) = 3.38 \times 10^5 - 474/r_{\mathrm{D-O}}^3\,(\mathrm{nm}^3)$ (Sinnecker et al. 2004). This leads to $r_{\mathrm{D-O}} = 0.191\,\mathrm{nm}$, which is close to the point-dipole estimate.

It is worthwhile noting here that Mims-ENDOR is the pulse-ENDOR variant best suited to determining weak proton hyperfine couplings such as those described for conformational analysis from CW-ENDOR in Sections 14.11–14.13.

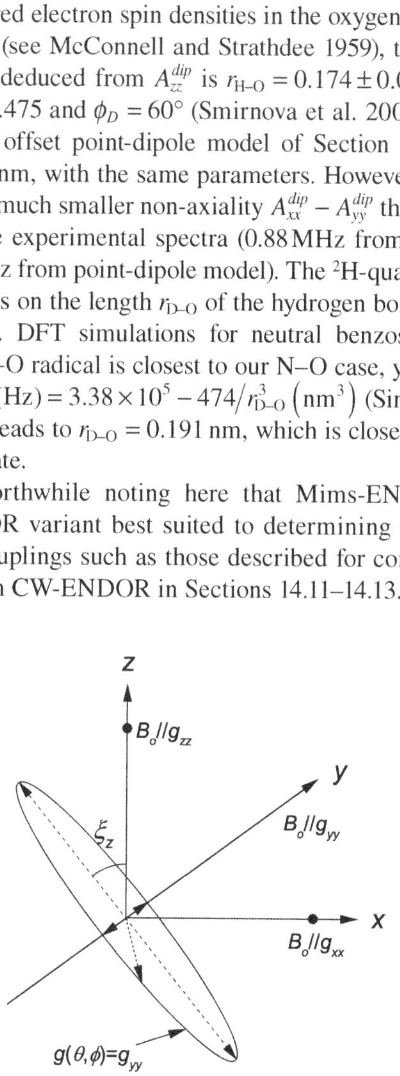

FIGURE 14.17 Orientation selection in high-field EPR/ENDOR. B_x- and B_z-regions of HF-EPR spectrum in Figure 14.15 (B_o along g_{xx}, g_{zz} respectively) produce single-crystal like ENDOR spectra. In B_y-region of HF-EPR spectrum, B_o lies not only along g_{yy} but in two planes containing g_{yy}, tilted to z-axis at angle ξ_z satisfying $g(\xi_z, 0) = g_{yy}$. (Second plane, not shown, is tilted in the positive x-direction.) ENDOR spectrum is two-dimensional powder pattern when $B_o \parallel B_y$.

14.15 ELDOR-DETECTED NMR

In ELDOR-detected NMR (see Figure 14.18), a selective inversion pulse (HTA) at the pumping microwave frequency $\omega_{e,p}$ excites *forbidden* transitions, thus burning holes in the spectral lineshape. The holes, which correspond to *allowed* transitions offset by the nuclear transition frequencies, are detected by a primary-echo sequence with selective pulses at the microwave observer frequency $\omega_{e,o}$ (Schosseler et al. 1994). The observer frequency is stepped through the hole pattern, hence avoiding spectrometer dead times. This lets us study broad hyperfine lines, and in particular to determine first-order contributions from ^{14}N-quadrupole splittings for nitroxides (Florent et al. 2011).

In practice, the observer frequency $\omega_{e,o}$ is kept fixed, matched to the microwave cavity resonance, and the pumping frequency is swept through the detection frequency. The delay between pump pulse and start of the detection sequence is set long enough to ensure decay of the electron coherence developed during the lengthy HTA pulse (or a two-step phase cycle $+x$, $-x$ is used). ELDOR-detected NMR is best performed with high-field spectrometers, because nuclear Larmor frequencies are then higher. This lets us study nuclei with low gyromagnetic ratios or small hyperfine splittings, because the nuclear frequencies are then resolved from the central blindspot where allowed transitions are excited.

We illustrate the method for the $S = \frac{1}{2}, I = \frac{1}{2}$ system treated in Appendices H.2, H.3. From Eqs. H.20 and H.21, the normalized intensities of the allowed (I_a) and forbidden (I_f) transitions are:

$$I_a = \cos^2 \eta = \frac{\left|\omega_I^2 - \frac{1}{4}(\omega_\alpha - \omega_\beta)^2\right|}{\omega_\alpha \omega_\beta} \tag{14.71}$$

$$I_f = \sin^2 \eta = \frac{\left|\omega_I^2 - \frac{1}{4}(\omega_\alpha + \omega_\beta)^2\right|}{\omega_\alpha \omega_\beta} \tag{14.72}$$

where $\cos \eta$ and $\sin \eta$ are proportional to matrix elements between the states connected by allowed and forbidden transitions, respectively (cf. Eqs. H.9 and H.10). We can view these transition amplitudes as admixture coefficients. For a selective pulse, the effective turning angle β_{eff} is reduced relative to the nominal angle β_o for a non-selective transition by just this factor

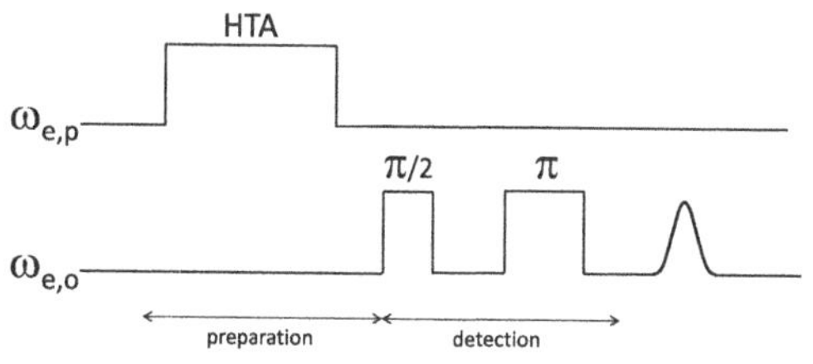

FIGURE 14.18 ELDOR-detected NMR pulse sequence. High turning-angle preparation pulse (HTA) is selective with microwave frequency $\omega_{e,p}$. Primary-echo detection pulses are selective with microwave frequency $\omega_{e,o}$. Echo intensity is recorded as function of pump-frequency offset $\omega_{e,p} - \omega_{e,o}$.

(Schweiger and Jeschke 2001). The effective pulse angles for allowed and forbidden transitions are therefore:

$$\beta_a = \beta_o \cos\eta = \beta_o\sqrt{I_a} \tag{14.73}$$

$$\beta_f = \beta_o \sin\eta = \beta_o\sqrt{I_f} \tag{14.74}$$

where $\beta_o = \omega_1 t_{HTA}$ for microwave field strength $\omega_1 = \gamma_e B_1$ and duration t_{HTA} of the selective pulse (see Section 13.2). Note that, because of the dependence on I_a or I_f, the effective turning angles depend on the orientation θ of the magnetic field according to Eqs. 14.2–14.6.

Figure 14.19 shows the level scheme for the $S=\frac{1}{2}, I=\frac{1}{2}$ system with strong hyperfine coupling: $|A_{hf}| > 2|\omega_I|$ (e.g., a ^{15}N-nitroxide). Applying a selective $\beta_f = \pi$ pumping pulse to the (2,3) transition inverts the polarization of this forbidden transition. As we see in the figure, this abolishes the population difference between states that are connected by allowed transitions, hence nulling the detected primary echo. From Table P.3 in Appendix P, the single-transition operator that corresponds to a pumping pulse of effective angle β_f applied to the forbidden (2,3) transition is: $\beta_f S_x^{(23)} = \beta_f\left(I_x S_x + I_y S_y\right)$. Starting from the equilibrium density matrix, σ_{eq}, we therefore get:

$$\sigma_{eq} = S_z \xrightarrow{\beta_f I_x S_x} \xrightarrow{\beta_f I_y S_y} \sigma_{prep}$$

$$= \tfrac{1}{2}\left(1+\cos\beta_f\right)S_z - \sin\beta_f \cdot S_y^{(23)} + \tfrac{1}{2}\left(1-\cos\beta_f\right)I_z \tag{14.75}$$

at the end of the preparation pulse, where we have used Tables P.1–P.3 and neglect evolution of electron coherence during the pulse. What interests us is the reduction of electron polarization S_z by the pulse. This has amplitude $\sigma_{eq} - \sigma_{prep} = 1 - \frac{1}{2}\left(1+\cos\beta_f\right)$ that is then detected by the primary-echo sequence. For $\beta_f = \pi$ as in the example of Figure 14.19, this becomes $\frac{1}{2}\left(1-\cos\beta_f\right) = 1$, i.e., 100% reduction.

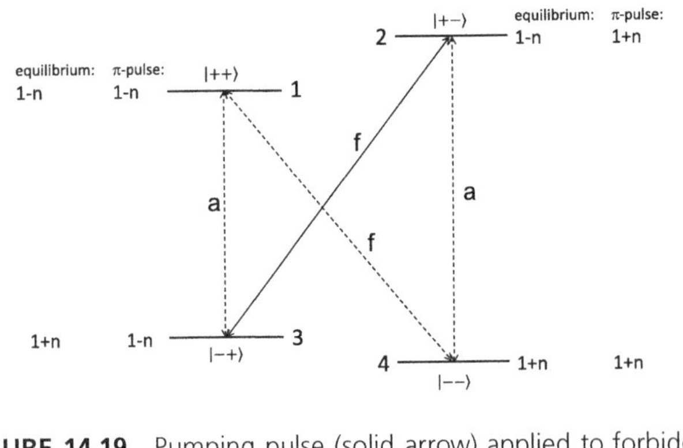

FIGURE 14.19 Pumping pulse (solid arrow) applied to forbidden transition (2,3) for $S=\frac{1}{2}, I=\frac{1}{2}$ with $|A_{hf}| > 2|\omega_I|$. Relative populations $(1 \pm n)$ of four $|M_S, m_I\rangle$ states at equilibrium and after selective π-pulse are indicated; HTA pump-pulse equalizes populations across allowed transitions (*a*). Forbidden transition $|+,-\rangle \rightarrow |-,+\rangle$ (solid line, *f*) is referred to as zero-quantum, and forbidden transition $|+,+\rangle \rightarrow |-,-\rangle$ (dashed line, *f*) as double-quantum. Allowed transitions are (true) single-quantum transitions.

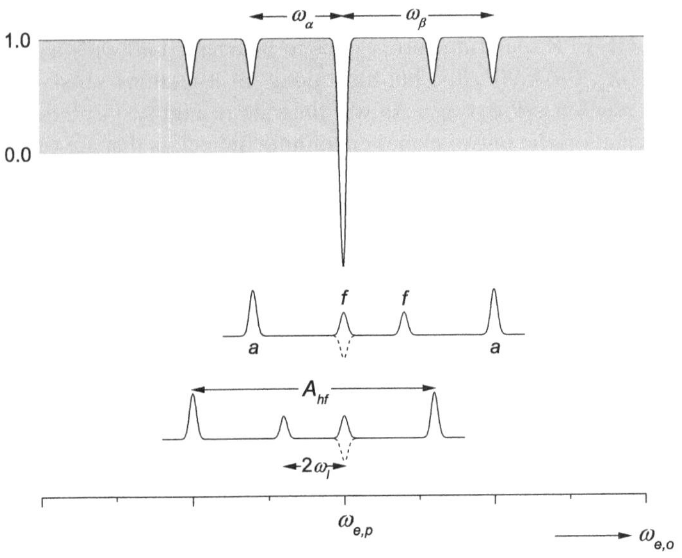

FIGURE 14.20 Hole burning in inhomogeneously broadened EPR line (shaded area with $\Gamma_{inh} \rightarrow \infty$) during ELDOR-detected NMR, for $S=\frac{1}{2}, I=\frac{1}{2}$ with $\frac{1}{2}|A_{hf}| > |\omega_I|$. Two spin packets shifted by $2|\omega_I|$ share a common forbidden-line position (*f*). Pumping microwave π-pulse applied at this position $(\omega_{e,p})$ inverts *z*-magnetization of these two forbidden transitions (dashed lines). This reduces polarization on allowed transitions (*a*) to zero (see populations in Figure 14.19), producing holes in inhomogeneous lineshape at offsets $\pm\omega_\alpha$ and $\mp\omega_\beta$. Central hole includes contributions from pumping on-resonance allowed transitions. Hole pattern is detected by primary echo with selective pulses of frequency $\omega_{e,o}$; but in practice, $\omega_{e,o}$ is fixed and pumping frequency $\omega_{e,p}$ is swept.

Figure 14.20 illustrates the situation when inhomogeneous line broadening is much larger than the hyperfine splitting of individual spin-packets ($\Gamma_{inh} >> A_{hf}, \omega_I$). For the $S=\frac{1}{2}, I=\frac{1}{2}$ system with strong hyperfine coupling, the higher-frequency forbidden hyperfine line of one spin packet superimposes on the lower-frequency forbidden line of a second spin packet that is shifted by $\omega_\alpha + \omega_\beta \approx 2|\omega_I|$. Pumping at this position, with microwave frequency $\omega_{e,p}$ and $\beta_f = \pi$, produces holes of depth I_a in the inhomogeneous lineshape at offset frequencies $\pm\omega_\alpha$ and $\mp\omega_\beta$. Correspondingly, the depth of the central hole is $4I_f$ because both forbidden transitions are inverted. We have ignored, for the moment, pumping of the two spin packets with allowed transitions at frequency $\omega_{e,p}$, because this contributes mostly to the central hole when $I_f << I_a$. The hole pattern is detected with a primary-echo sequence using selective pulses of frequency $\omega_{e,o}$. As a function of $\omega_{e,p} - \omega_{e,o}$, this is a pure nuclear spectrum yielding the ENDOR frequencies ω_α and ω_β directly.

As deduced from the discussion of Eq. 14.75, the depth of each of the side holes produced by pumping the two forbidden transitions is $h\left(\omega_{\alpha,\beta}\right) = \frac{1}{2}\left(1-\cos\beta_f\right)I_a$. The contribution from pumping the two allowed transitions, which so far we have neglected, is correspondingly $h\left(\omega_{\alpha,\beta}\right) = \frac{1}{2}\left(1-\cos\beta_a\right)I_f$. We need to include the latter whenever I_f is appreciable. Using Eqs. 14.73 and 14.74, the depth of the side holes therefore becomes:

$$h(\omega_{\alpha,\beta}) = \tfrac{1}{2}(I_a + I_f) - \tfrac{1}{2}\left(I_a \cos\left(\beta_o\sqrt{I_f}\right) + I_f \cos\left(\beta_o\sqrt{I_a}\right)\right)$$
$$= \tfrac{1}{2}\left(1 - I_a \cos\left(\beta_o\sqrt{I_f}\right) - I_f \cos\left(\beta_o\sqrt{I_a}\right)\right) \quad (14.76)$$

where $I_a + I_f = 1$. Figure 14.21 shows the dependence of the hole depth on intensity of the forbidden transition that we deduce from Eq. 14.76, for different values of the nominal turning angle β_o. Clearly, higher turning angles enhance hole-burning (and the ELDOR-detected NMR signal) for spin packets that have low values of I_f. This is why we refer to the pump pulse as an HTA-pulse.

Note that when $I_f << I_a$, the turning angle for allowed transitions is many times that for forbidden transitions: $\beta_a = \sqrt{I_a/I_f} \times \beta_f$. This means that the magnetization of the allowed transitions nutates around the pumping field during the HTA pulse, and their contribution to the central hole can be anywhere between zero and $4I_a$ depending on the value of β_a. Similar considerations apply to the contribution to each side hole, which lies between zero and I_f.

The turning angle also has an influence on orientation selection in powder EPR lineshapes. This is the same problem as that discussed for ESEEM in Section 14.9. The pseudosecular terms and, therefore, the forbidden-transition intensity, go to zero at the canonical turning points in the EPR powder pattern. In ELDOR-detected NMR, this problem is partly circumvented by improving the efficiency of the HTA pulse, because this emphasizes weak features at expense of the strong (see Cox et al. 2017). The latter reference also includes information on optimizing other features of the ELDOR-detected NMR experiment.

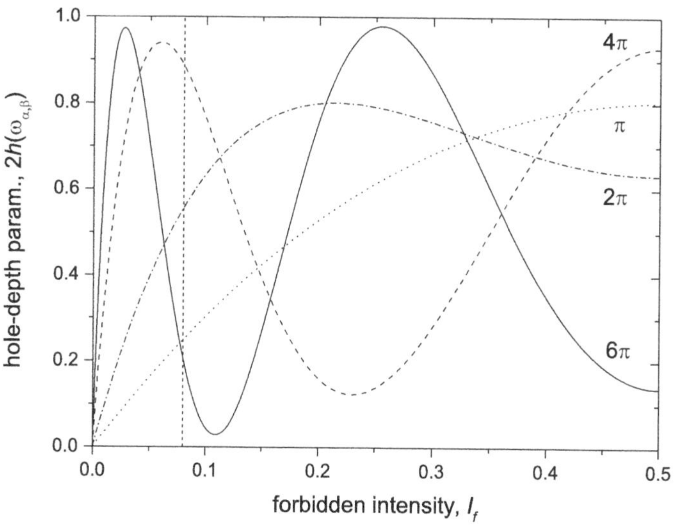

FIGURE 14.21 Hole depth $h(\omega_{\alpha,\beta})$ as function of forbidden-transition intensity I_f, for selective pulse irradiating forbidden transition. From Eq. 14.76 with $I_a + I_f = 1$, for nominal turning angles $\beta_o = \pi$, 2π, 4π and 6π, as indicated. Vertical dashed line indicates upper limit of I_f at 9 GHz, for a point-dipole coupled proton at distance $r = 0.25$ nm.

14.15.1 ELDOR-Detected NMR of ^{15}N-Nitroxides

For ^{15}N-nitroxides, information on the N-hyperfine tensor is available already from the high-field EPR spectrum, although the smaller A_{xx}^{N} and A_{yy}^{N} elements might not be so precisely defined. Also, for $I = \frac{1}{2}$ there is no quadrupole interaction. Nonetheless, ^{15}N-studies reveal several interesting features of the ELDOR-detected NMR experiment (see Cox et al. 2017).

For instance, at the high-field turning point (g_{zz}) of the 94-GHz EPR spectrum we are in the regime of strong hyperfine coupling $\frac{1}{2}\left|A_{zz}^{N}\right| > \left|\omega_I^{N}\right|$, whereas at the low-field turning point (g_{xx}) we are in the weak-coupling limit $\frac{1}{2}\left|A_{xx}^{N}\right| < \left|\omega_I^{N}\right|$ (see description in Appendix H.3). This has consequences for the appearance of the spectrum at different magnetic field settings. We see this from Figure 14.22 that refers to standard experimental conditions with fixed detection frequency $\omega_{e,o}$ and swept pump frequency $\omega_{e,p}$, in contrast to the intuitive situation depicted previously in Figure 14.20.

Upper and lower halves of Figure 14.22 refer to strong and weak coupling, respectively. Dotted lines show EPR spectra in these two situations as described by Figure H.2 of Appendix H.3. Fixed detection is at one of the two allowed EPR transitions, in response to pumping the forbidden EPR transitions. Solid lines show the hole patterns that we observe in an ELDOR-detected

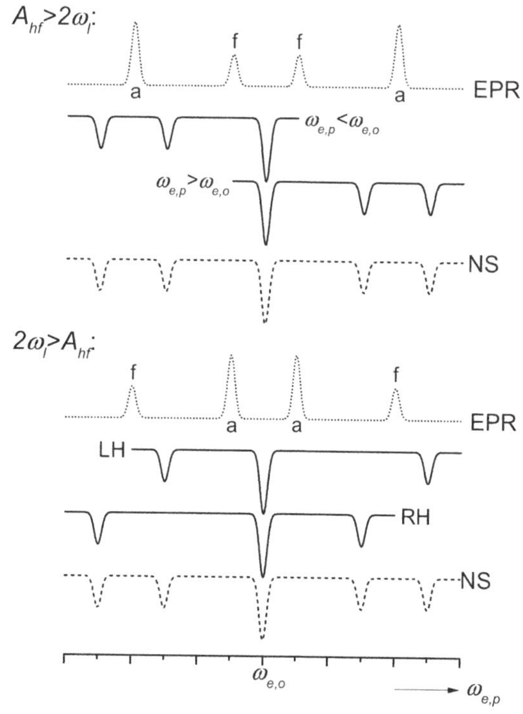

FIGURE 14.22 ELDOR-detected NMR spectrum for $S = \frac{1}{2}, I = \frac{1}{2}$ in strong $\frac{1}{2}|A_{hf}| > |\omega_I|$ (*upper*), and weak $\frac{1}{2}|A_{hf}| < |\omega_I|$ (*lower*), hyperfine-coupling limits. *Solid lines*: fixed observer-frequency $\omega_{e,o}$ detecting one of the two *allowed* transitions (*a*), whilst sweeping pump-frequency $\omega_{e,p}$ that excites *forbidden* transitions (*f*) (cf. Figure 14.20). *Dashed lines*: non-selective detection (NS). *Dotted lines*: EPR spectrum (see Figure H.2).

NMR experiment. For strong hyperfine coupling, the forbidden transitions that are pumped lie entirely to either higher or lower frequency of the fixed detection. The left- and right-hand sides of the ELDOR-detected spectrum then correspond to $\omega_{e,p} < \omega_{e,o}$ and $\omega_{e,p} > \omega_{e,o}$, respectively. For weak hyperfine coupling, however, there is not this clear distinction. Detection at either of the two allowed transitions, left (LH) or right (RH), responds to pumping forbidden transitions that lie both below and above this observation frequency. Therefore, we can distinguish strong and weak coupling depending on whether the ELDOR-detected NMR transitions lie entirely above or below $\omega_{e,o}$, or on both sides of $\omega_{e,o}$ (see Figure 14.22).

Observation at the high-field turning points (g_{zz}) in the 94-GHz EPR spectrum of ^{15}N-PYMeOH (1-oxyl-2,2,5,5-tetramethylpyrrolinyl-3-methanol) yields ELDOR lines lying $\frac{1}{2}A_{zz}^{N} \pm \omega_I^{N}$ above $\omega_{e,o}$, or $-\frac{1}{2}A_{zz}^{N} \mp \omega_I^{N}$ below $\omega_{e,o}$ (for $m_I = \pm\frac{1}{2}$, respectively), which corresponds to strong coupling (Cox et al. 2017). On the other hand, observation at the low-field turning points (g_{xx}) yields lines split about $\omega_{e,o}$ by $\pm\frac{1}{2}A_{xx}^{N} - \omega_I^{N}$ or $\mp\frac{1}{2}A_{xx}^{N} + \omega_I^{N}$ (for $m_I = \pm\frac{1}{2}$), which corresponds to weak coupling (Cox et al. 2017). A quasi-rotation pattern with systematically varied field position B_o, as in Figure 14.15, is complex in the intermediate positions, because of the changeover from strong to weak hyperfine coupling, and also because the turning angle of the HTA pulse varies with field orientation (see Cox et al. 2017).

14.15.2 ELDOR-Detected NMR of ^{14}N-Nitroxides

For ^{14}N-nitroxides (with $I = 1$), the nuclear quadrupole interaction contributes in first order to the ELDOR-detected NMR spectrum. This is not the case for the (high-field) EPR spectrum. We see this already from the nuclear frequencies $\omega_{\alpha,\pm1}$, $\omega_{\beta,\pm1}$ for $I = 1$ that are given by Eqs. 14.28 and 14.29 in Section 14.4. Because the principal axis of ^{14}N-hyperfine and quadrupolar interactions coincides, these equations with $\theta_Q = 0$, i.e., $P = \frac{3}{2}P_{zz}$, give the nuclear frequencies when the static field is parallel to the nitroxide z-axis.

For strong hyperfine coupling, $\frac{1}{2}|A_{zz}^{N}| > |\omega_I^{N}|$, in the high-field (g_{zz}) region of 94-GHz EPR, the nuclear transition frequencies become:

$$\omega_{\alpha,\pm1} = \mp\tfrac{1}{2}A_{zz} - \left(\omega_I \pm \tfrac{3}{2}P_{zz}\right) \tag{14.77}$$

$$\omega_{\beta,\pm1} = \mp\tfrac{1}{2}A_{zz} + \left(\omega_I \mp \tfrac{3}{2}P_{zz}\right) \tag{14.78}$$

where the upper sign is for the $m_I = +1 \leftrightarrow 0$ transition and lower for $m_I = -1 \leftrightarrow 0$. The two ELDOR frequencies for the higher-field ($m_I = -1$) hyperfine line are therefore $\omega_{\alpha,-1}$ and $\omega_{\beta,-1}$, relative to the observer frequency (see Figure 14.23). These are centred about $\frac{1}{2}A_{zz}$ and split by $2\omega_I + 3P_{zz}$. For the lower-field ($m_I = +1$) hyperfine line, the splitting is the same, but the ELDOR frequencies are centred about $-\frac{1}{2}A_{zz}$.

Observation at the outermost high-field turning point $(g_{zz}, m_I = -1)$ in the 94-GHz EPR spectrum of

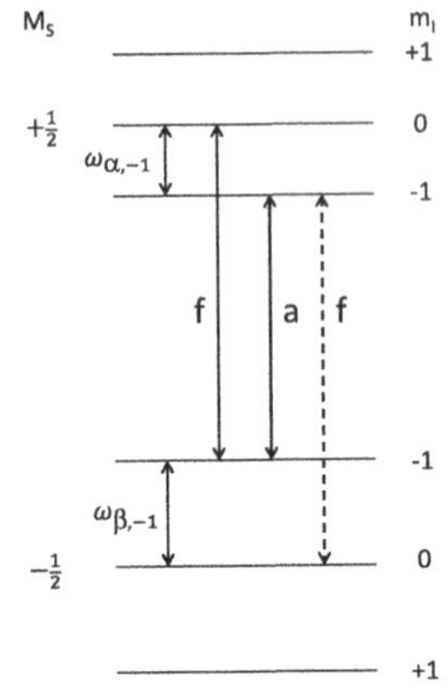

FIGURE 14.23 Energy-level scheme for $S = \frac{1}{2}, I = 1$. EPR (allowed *a*: observe), ELDOR (forbidden *f*: pump) and nuclear $(\omega_{\alpha,-1}, \omega_{\beta,-1})$ transitions for highest-field ($m_I = -1$) hyperfine line. Transitions selected at high-field g_{zz}-position in HF-EPR, with pumping the solid-line *f*-transition.

^{14}N-PYMeOH (1-oxyl-2,2,5,5-tetramethylpyrrolinyl-3-methanol) yields ELDOR lines solely at positive frequencies, lying $\frac{1}{2}A_{zz}^{N} \pm \left(\omega_I^{N} + \frac{3}{2}P_{zz}^{N}\right)$ above $\omega_{e,o}$ (Nalepa et al. 2014). Determining ω_I^{N} from pulse ENDOR at the same field setting then gives us the principal quadrupole element P_{zz}^{N} from the ELDOR-detected NMR splitting (see Table 4.9). As explained in Section 4.5, this provides us with a valuable way to study environmental polarity.

14.16 CONCLUDING SUMMARY

1. *Three-pulse ESEEM:*
   Three-pulse ESEEM records the stimulated-echo intensity as a function of the second time-delay T, for fixed first delay τ. The echo-decay envelope is modulated by the basic nuclear frequencies, ω_α and ω_β. Fourier transforming the modulation gives the nuclear-frequency spectrum. For $I = \frac{1}{2}$:

$$\omega_\alpha = \sqrt{\left(\omega_I + \tfrac{1}{2}A\right)^2 + \tfrac{1}{4}B^2} \tag{14.79}$$

$$\omega_\beta = \sqrt{\left(\omega_I - \tfrac{1}{2}A\right)^2 + \tfrac{1}{4}B^2} \tag{14.80}$$

where for axial hyperfine tensor $(A_\perp, A_\perp, A_\parallel)$:

$$A = A_\parallel \cos^2\theta + A_\perp \sin^2\theta \tag{14.81}$$

$$B = \left(A_\parallel - A_\perp\right)\sin\theta\cos\theta \tag{14.82}$$

For weak hyperfine couplings ($A, B << \omega_I$), the nuclear frequencies are centred about the nuclear Larmor frequency ω_I. For strong hyperfine couplings, they are centred about $\frac{1}{2}A_{eff} = \frac{1}{2}\sqrt{A^2 + B^2}$. For $I = 1$, with weak hyperfine and quadrupole coupling P, the four nuclear frequencies to first-order (i.e., neglecting B) are:

$$\omega_{\alpha,\pm1} = \left|\omega_I + \tfrac{1}{2}A \pm P\right| \quad (14.83)$$

$$\omega_{\beta,\pm1} = \left|\omega_I - \tfrac{1}{2}A \pm P\right| \quad (14.84)$$

where the second subscript defines transitions with $\Delta m_I = \pm 1$.

2. *CW-ENDOR:*
Continuous-wave ENDOR is a double-resonance experiment. We record the nuclear spectrum by sweeping the radiofrequency field and detecting alleviation of EPR saturation by the nuclear transitions that we induce. In proton-ENDOR of nitroxides, the hyperfine splitting about the proton NMR frequency: $\omega^{H}_{\alpha,\beta} = \omega_H \pm \tfrac{1}{2}A^{H}(\theta)$ is given by:

$$A^{H}(\theta) = \left(\frac{\mu_o}{4\pi}\right)\frac{g_e\beta_e\gamma_H}{r^3}\left(3\cos^2\theta - 1\right) + a^{H}_o \quad (14.85)$$

where a^{H}_o is the (small) isotropic hyperfine coupling constant, which becomes negligible for $r > 0.5\,\text{nm}$. Orientation selection at fixed field within the EPR powder pattern from a frozen glass helps identify the $A^{H}_{\perp}$ and $A^{H}_{\parallel}$ hyperfine couplings corresponding to $\theta = 90^\circ$ and $\theta = 0$ orientations. Distances r from the proton to the midpoint of the N–O bond can then be identified for use in conformational analysis.

3. *Pulse-ENDOR:*
Pulse ENDOR employs a selective radiofrequency pulse to invert nuclear polarization. In Davies-ENDOR, we use a selective microwave pulse to burn a hole in the inhomogeneous EPR lineshape. This is refilled by selective nuclear transitions, on sweeping the radiofrequency. We monitor the hole depth with a selective primary-echo sequence. In Mims-ENDOR, we apply the radiofrequency pulse during the second evolution period of a non-selective three-pulse echo sequence, and detect nuclear transitions by reduction in intensity of the stimulated echo. Davies-ENDOR is more suited to larger hyperfine interactions, e.g., directly hydrogen-bonded protons, and Mims-ENDOR (which can have blind spots) to smaller unresolved hyperfine splittings, e.g., more remote protons and hydrogen-bonded deuterium. Pulse-ENDOR benefits considerably from the orientation selection that we get from g-value dispersion in high-field EPR.

4. *ELDOR-detected NMR:*
ELDOR-detected NMR employs two microwave frequencies. One frequency pumps forbidden transitions that burn holes at allowed transitions within the inhomogeneous EPR lineshape. The second frequency detects the allowed transitions with a selective primary-echo sequence. The pump pulse is selective, with a high turning angle that optimizes efficiency for weak forbidden transitions. ELDOR-detected NMR is particularly useful for determining quadrupole couplings of ^{14}N-nitroxides. We have strong coupling, $\tfrac{1}{2}\left|A^{N}_{zz}\right| > \left|\omega^{N}_{I}\right|$ for ^{14}N, and the first-order nuclear transition frequencies, in the high-field (g_{zz}) region of 94-GHz EPR, become:

$$\omega_{\alpha,\pm1} = \mp\tfrac{1}{2}A_{zz} - \left(\omega_I \pm \tfrac{3}{2}P_{zz}\right) \quad (14.86)$$

$$\omega_{\beta,\pm1} = \mp\tfrac{1}{2}A_{zz} + \left(\omega_I \mp \tfrac{3}{2}P_{zz}\right) \quad (14.87)$$

where the upper sign is for the $m_I = +1 \leftrightarrow 0$ transition and lower for $m_I = -1 \leftrightarrow 0$. Knowing ω_I then gives us the quadrupole interaction P_{zz}.

REFERENCES

Astashkin, A.V., Dikanov, S.A., and Tsvetkov, Y.D. 1987. Spectrometer dead time: effect on electron spin echo modulation spectra in disordered systems. *Chem. Phys. Lett.* 136:204–208.

Bartucci, R., Guzzi, R., Marsh, D., and Sportelli, L. 2003. Intramembrane polarity by electron spin echo spectroscopy of labeled lipids. *Biophys. J.* 84:1025–1030.

Bartucci, R., Guzzi, R., Sportelli, L., and Marsh, D. 2009. Intramembrane water associated with TOAC spin-labelled alamethicin: electron spin-echo envelope modulation by D_2O. *Biophys. J.* 96:997–1007.

Bartucci, R., Guzzi, R., Esmann, M., and Marsh, D. 2014. Water penetration profile at the protein-lipid interface in Na,K-ATPase membranes. *Biophys. J.* 107:1375–1382.

Cox, N., Nalepa, A., Lubitz, W., and Savitsky, A. 2017. ELDOR-detected NMR: a general and robust method for electron-nuclear hyperfine spectroscopy? *J. Magn. Reson.* 280:63–78.

Davies, E.R. 1974. New pulse ENDOR technique. *Phys. Lett. A* 47:1–2.

Dikanov, S.A., Shubin, A.A., and Parmon, V.N. 1981. Modulation effects in the electron spin echo resulting from hyperfine interaction with a nucleus of an arbitrary spin. *J. Magn. Reson.* 42:474–487.

Dikanov, S.A. and Tsvetkov, Y.D. 1985. Structural applications of the electron spin echo method. *J. Struct. Chem.* 26:766–804.

Dikanov, S.A. and Tsvetkov, Y.D. 1992. *Electron Spin Echo Envelope Modulation (ESEEM) Spectroscopy*, Boca Raton, FL: CRC Press.

Erilov, D.A., Bartucci, R., Guzzi, R. et al. 2005. Water concentration profiles in membranes measured by ESEEM of spin-labeled lipids. *J. Phys. Chem. B* 109:12003–12013.

Ernst, R.R., Bodenhausen, G., and Wokaun, A. 1987. *Principles of NMR in One and Two Dimensions*, Oxford: Clarendon Press.

Fauth, J.M., Schweiger, A., and Ernst, R.R. 1989. Recovery of broad hyperfine lines in electron spin-echo envelope modulation spectroscopy of disordered systems. *J. Magn. Reson.* 81:262–274.

Florent, M., Kaminker, I., Nagaranjan, V., and Goldfarb, D. 2011. Determination of the ^{14}N quadrupole coupling constant of nitroxide spin probes by W-band ELDOR-detected NMR. *J. Magn. Reson.* 210:192–199.

Gemperle, C. and Schweiger, A. 1991. Pulsed electron-nuclear double resonance methodology. *Chem. Rev.* 91:1481–1505.

Guzzi, R., Bartucci, R., Sportelli, L., Esmann, M., and Marsh, D. 2009. Conformational heterogeneity and spin-labelled -SH groups: Pulsed EPR of Na,K-ATPase. *Biochemistry* 48:8343–8354.

Heine, V. 1960. *Group Theory in Quantum Mechanics: An Introduction to Its Present Usage*, Oxford: Pergamon Press.

Jiang, F., Tsai, S.-W., and Makinen, M.W. 1998. ENDOR determined structure of a complex of α-chymotrypsin with a spin-labeled transition-state inhibitor analogue. *J. Phys. Chem. B* 102:4619–4627.

Linder, M., Höhener, A., and Ernst, R.R. 1980. Orientation of tensorial interactions determined from two-dimensional NMR powder spectra. *J. Chem. Phys.* 73:4959–4970.

Makinen, M.W., Mustafi, D., and Kasa, S. 1998. ENDOR of spin labels for structure determination: from small molecules to enzyme reaction intermediates. In *Biological Magnetic Resonance*, Vol. 14, Spin Labeling. The Next Millenium. ed. Berliner, L.J., 181–249. New York: Plenum Press.

McConnell, H.M. and Strathdee, J. 1959. Theory of anisotropic hyperfine interactions in π-electron radicals. *Mol. Phys.* 2:129–138.

Mims, W.B. 1965. Pulse ENDOR experiments. *Proc. Roy. Soc. London* 283:452–457.

Mims, W.B. 1972. Envelope modulation in spin-echo experiments. *Phys. Rev. B* 5:2409–2419.

Morton, J.R. 1964. Electron spin resonance spectra of oriented radicals. *Chem. Rev.* 64:453–471.

Mustafi, D. and Makinen, M.W. 1994. Catalytic conformation of carboxypeptidase A - structure of a true enzyme reaction intermediate determined by electron-nuclear double resonance. *J. Biol. Chem.* 269:4587–4595.

Mustafi, D., Sachleben, J.R., Wells, G.B., and Makinen, M.W. 1990. Structure and conformation of spin-labeled amino acids in frozen solutions determined by electron nuclear double resonance. 1. Methyl *N*-(2,2,5,5-tetramethyl-1-oxypyrrolinyl-3-carbonyl)-L-alanate, a molecule with a single preferred conformation. *J. Am. Chem. Soc.* 112:2558–2566.

Mustafi, D., Joela, H., and Makinen, M.W. 1991. The effective position of the electronic point dipole of the nitroxyl group of spin labels determined by ENDOR spectroscopy. *J. Magn. Reson.* 91:497–504.

Mustafi, D., Sosa-Peinado, A., and Makinen, M.W. 2001. ENDOR structural characterization of a catalytically competent acylenzyme reaction intermediate of wild-type TEM-1 β-lactamase confirms glutamate-166 as the base catalyst. *Biochemistry* 40:2397–2409.

Mustafi, D., Sosa-Peinado, A., Gupta, V., Gordon, D.J., and Makinen, M.W. 2002. Structure of spin-labeled methylmethanethiolsulfonate in solution and bound to TEM-1 β-lactamase determined by electron nuclear double resonance spectroscopy. *Biochemistry* 41:797–808.

Nalepa, A., Möbius, K., Lubitz, W., and Savitsky, A. 2014. High-field ELDOR-detected NMR study of a nitroxide radical in disordered solids: towards characterization of heterogeneity of microenvironments in spin-labeled systems. *J. Magn. Reson.* 242:203–213.

Narayana, P.A., Bowman, M.K., Kevan, L., Yudanov, V.F., and Tsvetkov, Y.D. 1975. Electron spin echo envelope modulation of trapped radicals in disordered glassy systems - application to molecular structure around excess electrons in γ-irradiated 10 M sodium-hydroxide alkaline ice glass. *J. Chem. Phys.* 63:3365–3371.

Savitsky, A., Dubinskii, A.A., Plato, M. et al. 2008. High-field EPR and ESEEM investigation of the nitrogen quadrupole interaction of nitroxide spin labels in disordered solids: toward differentiation between polarity and proticity matrix effects on protein function. *J. Phys. Chem. B* 112:9079–9090.

Schosseler, P., Wacker, T., and Schweiger, A. 1994. Pulsed ELDOR detected NMR. *Chem. Phys. Lett.* 224:319–324.

Schweiger, A. and Jeschke, G. 2001. *Principles of Pulse Electron Paramagnetic Resonance*, Oxford: Oxford University Press.

Shubin, A.A. and Dikanov, S.A. 1983. The influence of nuclear quadrupole interactions upon electron spin echo modulation induced by weak hyperfine interactions. *J. Magn. Reson.* 52:1–12.

Sinnecker, S., Reijerse, E., Neese, F., and Lubitz, W. 2004. Hydrogen bond geometries from electron paramagnetic resonance and electron-nuclear double resonance parameters: density functional study of quinone radical anion-solvent interactions. *J. Am. Chem. Soc.* 126:3280–3290.

Smirnova, T.I., Smirnov, A.I., Paschenko, S.V., and Poluektov, O.G. 2007. Geometry of hydrogen bonds formed by lipid bilayer nitroxide probes: a high-frequency pulsed ENDOR/EPR study. *J. Am. Chem. Soc.* 129:3476–3477.

Soda, G. and Chiba, T. 1969. Deuteron magnetic resonance study of cupric sulfate pentahydrate. *J. Chem. Phys.* 39:439–455.

Wells, G.B., Mustafi, D., and Makinen, M.W. 1990. Structure and conformation of spin-labeled amino acids in frozen solutions determined by electron-nuclear double resonance. 2. Methyl *N*-(2,2,5,5-tetramethyl-1-oxypyrrolinyl-3-carbonyl)-L-tryptophanate, a molecule with multiple conformations. *J. Am. Chem. Soc.* 112:2566–2574.

Wells, G.B., Mustafi, D., and Makinen, M.W. 1994. Structure at the active site of an acylenzyme of α-chymotrypsin and implications for the catalytic mechanism - an electron-nuclear double resonance study. *J. Biol. Chem.* 269:4577–4586.

Yudanov, V.F., Dikanov, S.A., Grishin, Y.A., and Tsvetkov, Y.D. 1976. Structure of the immediate environment of stable radicals in frozen solutions. *J. Struct. Chem.* 17:387–392.

Distance Measurements

15

15.1 INTRODUCTION

Measurement of distances between nitroxide spin labels is based on the magnetic dipole–dipole interaction between two unpaired electron spins. As we saw in Chapter 9, the strength of this interaction depends inversely on the third power of the distance between the two spins, but it depends also on the angle between the interspin vector and the direction of the static magnetic field.

We can use conventional CW-EPR spectra, if the distance between the spins is short (≤ 1.8 nm). The upper limit arises when the dipolar interaction no longer has a perceptible influence on the powder lineshape of an individual nitroxide. Only for relatively short distances (<1 nm) can we reliably resolve dipolar splittings in the powder lineshape (see Figure 9.4). Otherwise, we need to use convolution/deconvolution to extract the dipolar powder pattern from the overall lineshape. Alternatively, we can evaluate moments of the composite lineshape to measure the extent of dipolar line broadening.

For longer distances, we use pulsed electron–electron double resonance, commonly known as DEER. The pump pulse of the ELDOR sequence reverses the sign of the dipolar field experienced by the spins whose transverse magnetization we detect. Other sources of inhomogeneous broadening are removed by the echo detection sequence. So we can measure much weaker dipolar interactions, i.e., longer distances, than with CW-EPR. The longest interspin distance that we can measure is limited by intrinsic T_2-relaxation and corresponds to ca. 8 nm. Measurement of short distances by pulse methods is limited by the available microwave power, i.e., by the maximum dipolar splitting that we can excite. A typical lower bound is >1.8 nm. In principle, pulse ELDOR measures both the interspin distance for a specific nitroxide pair and the background signal from more remote spin labels. The latter gives us useful information on the macroscopic distribution of the spins: three dimensional, two dimensional, or a linear chain.

15.2 DIPOLAR PAIR SPECTRA

Determination of distances is most straightforward when a pair of spins has strong dipolar coupling that is greater than the nitrogen hyperfine coupling. Isotropic Heisenberg spin exchange then does not enter. The dipolar splitting is resolved directly in the EPR spectrum of the resulting triplet spin state (see Section 9.2). For this, the two nitroxides must be relatively close. From Eq. 9.14 for like spins, the dipolar splitting $\pm\Delta B_{dd}$ of the resonance magnetic field in the EPR spectrum from a pair of identical nitroxides is:

$$\pm\Delta B_{dd}(\theta) = \frac{3}{4}\cdot\frac{\mu_o}{4\pi}\left(\frac{g\beta_e}{r_{12}^3}\right)\left(1-3\cos^2\theta\right) \tag{15.1}$$

where θ is the angle that the static magnetic field makes with interspin vector, $\mathbf{r}_{12}$. The dipolar splitting ranges from $\Delta B_{dd}^{\parallel} = -\frac{3}{2}(\mu_o/4\pi)g\beta_e/r_{12}^3$ for $\theta = 0°$ to $\Delta B_{dd}^{\perp} = +\frac{3}{4}(\mu_o/4\pi)g\beta_e/r_{12}^3$ for $\theta = 90°$. Best resolution therefore comes from angular-dependent measurements on oriented systems. Otherwise, we get the Pake powder pattern that is characteristic of randomly oriented systems (see Section 9.3 and Figures 9.3 and 9.4).

Figure 15.1 gives an example of the dipole–dipole splitting in the EPR spectrum from a pair of spin-labelled steroids (cholestane) in aligned phospholipid bilayer membranes. The spin-label concentration is increased to 8 mol% so that there is an appreciable population of nearest-neighbour spin-label pairs, P. These give the resonance lines, indicated by arrows, in the outer wings of the more intense spectrum that comes from isolated spin labels, I. We see no

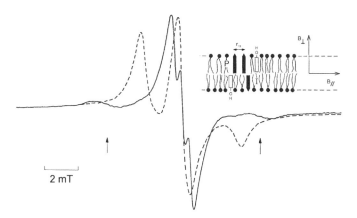

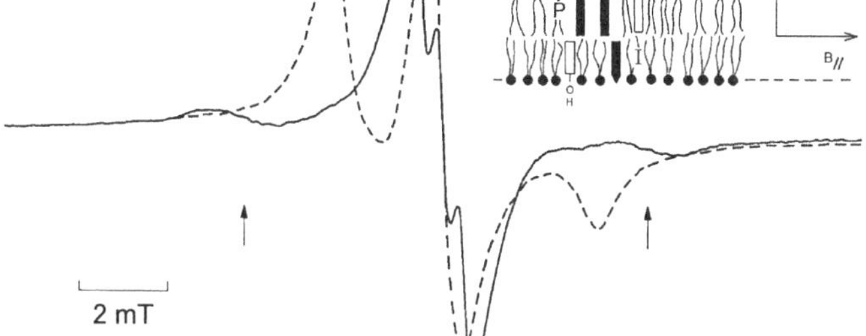

FIGURE 15.1 9-GHz EPR spectra of 8 mol% 3-DOXYL cholestane in aligned membranes of dipalmitoyl phosphatidylcholine + 20 mol% cholesterol. *Solid line*: magnetic field normal to membrane plane ($B_\perp$); *dashed line*: magnetic field parallel to membrane plane ($B_\parallel$). Data from Marsh and Smith (1973). Arrows indicate EPR lines from interacting spin-label pairs with the $B_\perp$-configuration, where all pair-vectors $\mathbf{r}_{12}$ are equivalent. *Inset*: spin-label pair, P, and isolated spin label, I, in aligned membrane.

hyperfine structure in the pair lines, because the nitrogen coupling constant is halved and relative intensities modified, for triplet states (see Section 9.8). The pair spectrum is well resolved only when the magnetic field is oriented perpendicular to the plane of the membrane ($B_\perp$) and, consequently, all interspin vectors $\mathbf{r}_{12}$ have the same (perpendicular) orientation to $\mathbf{B}_o$. The separation of the pair lines $2|\Delta B_{dd}^{\perp}|$ gives a separation between adjacent steroid molecules in the membrane of $r_{12} = 0.69\,\text{nm}$, which is found to be sensitive to lipid chain unsaturation, cholesterol content, temperature and the chain-melting transition (Marsh and Smith 1972; Marsh and Smith 1973; Marsh 1974).

15.3 DIPOLAR CONVOLUTION AND DECONVOLUTION

First, we must remember that convolution of a dipolar powder pattern with an underlying spectrum of the isolated spin label is valid only if the two lineshapes are mutually independent. Clearly, this holds for a random distribution of monoradical spin labels, e.g., in a frozen glass, but for a rigid biradical it decidedly does not. Often we are interested in two spin labels attached at specific sites on a macromolecule (Steinhoff et al. 1991; Steinhoff et al. 1997). For this case, the spin-label attachments must have sufficient configurational flexibility to ensure that the nitroxide x,y,z-axes approximate an isotropic orientational distribution relative to the interspin vector $\mathbf{r}_{12}$. Mostly, there is a distribution of interspin distances in these situations, and the resulting r_{12}-distribution that we derive may partially compensate shortcomings in the assumption of independent orientational distributions. Hustedt et al. (2006) simulated this scenario by assuming a random distribution of nitroxide orientations restricted within a cone at the point of attachment. It turns out that we need cone half-angles over 40° for the spectra to be independent of cone-axis orientation to the interspin axis. Generally, the results of dipolar convolution provide a reasonable estimate for average distances in the tether-in-a-cone model but are less robust for the distance distribution. Tethers are rigid in the model, but segmental flexibility of real tethers means that only the cone angle, and not its lateral extent, is significant. Convolution methods are described in Appendix T and in Section 2.6 of Chapter 2.

For independent lineshapes, dipolar broadening results from convolution of the dipolar absorption envelope, $D(\Delta B_{dd}) = D(B-B')$, such as that given at the top of Figure 9.3 in Chapter 9, with the normal first-derivative EPR spectrum, $f_o(B')$, of the spin label without dipolar interaction (Steinhoff et al. 1991; Steinhoff et al. 1997):

$$F_{dd}(B) = \int f_o(B')D(B-B')\cdot \mathrm{d}B' \tag{15.2}$$

We evaluate the integral by using the convolution theorem for Fourier transforms – see Section 2.6 and Appendix T. Only one of the two lineshapes should be the first-derivative to produce a first-derivative EPR spectrum by convolution (see Eq. 2.40 in Section 2.6). If we include a distribution of interspin distances, the dipolar envelope must be weighted by this distribution. Often we assume a Gaussian distribution (Steinhoff et al. 1997), leading to a net dipolar powder pattern that is given by:

$$D(B-B') = \frac{1}{\sigma_{r12}\sqrt{2\pi}} \int_{r_{min}}^{\infty} D\left(\Delta B_{dd}(r_{12})\right) \exp\left(-\frac{\left(r_{12} - \langle r_{12}\rangle\right)^2}{2\sigma_{r12}^2}\right) \cdot \mathrm{d}r_{12} \tag{15.3}$$

where $D\left(\Delta B_{dd}(r_{12})\right)$ is the dipolar envelope for fixed interspin distance r_{12}, and $\langle r_{12}\rangle$ and $\sigma_{r_{12}}$ are the mean and standard deviation, respectively, of the distribution. We need the lower integration limit, r_{min}, in Eq. 15.3 to prevent approach beyond the contact distance and/or to exclude dipolar powder patterns that do not fit within the sweep width of the experimental spectrum (see Section 15.8 below).

Figure 15.2 shows the effect of convoluting the powder spectrum of a non-interacting spin pair with dipolar powder patterns corresponding to decreasing interspin separations. For a spin separation of $r_{12} = 2.0$ nm, the only effect of the dipolar coupling is a slight broadening of the central hyperfine manifold.

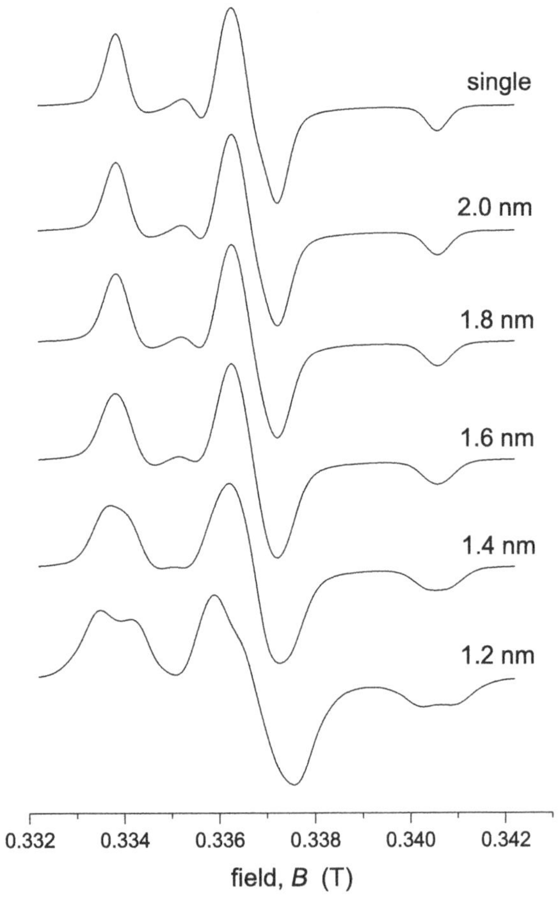

FIGURE 15.2 Convolution of dipolar absorption powder envelope (see top of Figure 9.3) with powder spectrum for a non-interacting nitroxide (*single*). Dipolar coupling increases from top to bottom. Separation, r_{12} (nm), of the pair of interacting *unlike* spins is indicated.

We see this as a slight decrease in line height when the spectra are normalized to the double-integrated intensity, instead of to line height as in the figure. With decreasing distance between spins, dipolar broadening increases progressively up to $r_{12} = 1.6$ nm, which again we notice most in the central hyperfine manifold. Beyond this, the shape of the outer hyperfine peaks changes to give a pronounced dipolar splitting for $r_{12} = 1.2$ nm. This becomes somewhat smeared out when we include a distribution of interspin distances. Note that the maximal extent of the dipolar envelopes used for the convolutions in Figure 15.2 ranges from $\Delta B_{dd}^{\parallel} = \pm 0.24$ mT for $r_{12} = 2.0$ nm to $\Delta B_{dd}^{\parallel} = \pm 1.08$ mT for $r_{12} = 1.2$ nm, which is considerably smaller than the overall spectral anisotropy of the non-interacting spin label. We therefore classify the interacting spins as *unlike*, and the interspin distances quoted correspond to this condition, i.e., $2\Delta B_{dd}^{\perp} = (\mu_o/4\pi) g\beta_e / r_{12}^3$ (see Eq. 9.19 in Chapter 9).

Figure 15.3 shows data obtained from convolution, for α-helical peptides $(Ala_4Lys)_4Ala$ with MTSSL spin labels attached at two differently spaced cysteine residues substituted in the sequence (Banham et al. 2008). Dipolar patterns with maximum extents ranging from ±0.24 to ±1.26 mT were required for fitting. The mean separations determined assuming *unlike* spin labels, $\langle r_{dd} \rangle$ are compared with those deduced from molecular modelling, $\langle r_{12} \rangle$, for each pair of substitutions. We see a reasonable correlation in both absolute values and trends of the mean distances. Also, the standard deviations of the distributions agree reasonably well with those deduced from molecular modelling. Distributions with mean distances greater than 2 nm cannot be determined by this method, because the resultant dipolar broadening is too small (cf. Figure 15.2).

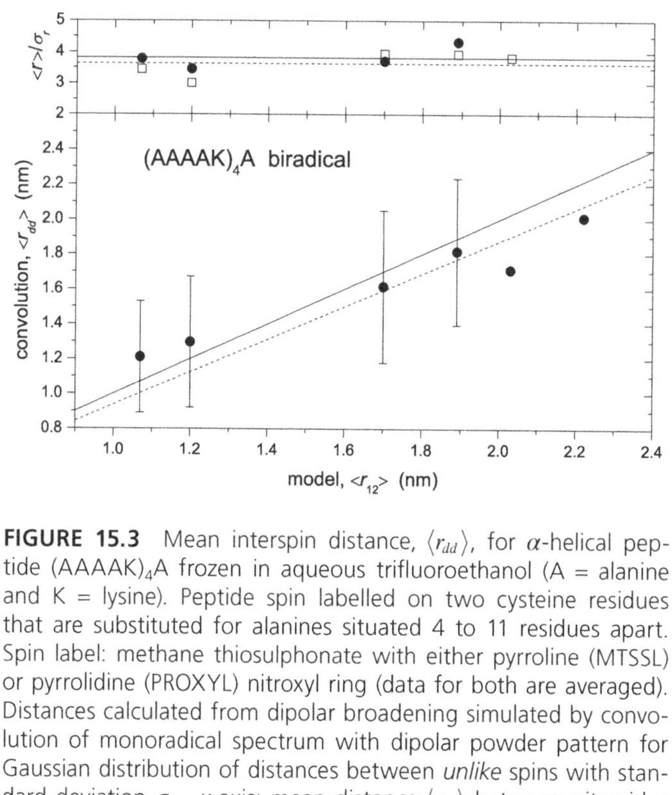

FIGURE 15.3 Mean interspin distance, $\langle r_{dd} \rangle$, for α-helical peptide $(AAAAK)_4A$ frozen in aqueous trifluoroethanol (A = alanine and K = lysine). Peptide spin labelled on two cysteine residues that are substituted for alanines situated 4 to 11 residues apart. Spin label: methane thiosulphonate with either pyrroline (MTSSL) or pyrrolidine (PROXYL) nitroxyl ring (data for both are averaged). Distances calculated from dipolar broadening simulated by convolution of monoradical spectrum with dipolar powder pattern for Gaussian distribution of distances between *unlike* spins with standard deviation σ_{dd}. *x*-axis: mean distance $\langle r_{12} \rangle$ between nitroxides from molecular modelling. *Lower panel*: solid line, $\langle r_{dd} \rangle = \langle r_{12} \rangle$; dotted line: linear regression constrained to pass through origin, $\langle r_{dd} \rangle = (0.94 \pm 0.04) \times \langle r_{12} \rangle$. *Upper panel*: solid line, mean experimental value of $\langle r_{dd} \rangle / \sigma_{r_{dd}} = 3.8$ (solid circles); dashed line, $\langle r_{12} \rangle / \sigma_{r_{12}} = 3.6$ from modelling (open squares). Data from Banham et al. (2008).

15.4 DIPOLAR DECONVOLUTION

In principle, we can find the interspin distance distribution from Fourier deconvolution to give the dipolar broadening envelope, which we then fit with a sum of individual envelopes, $D(\Delta B_{dd}(r_{12}))$, for different values of r_{12}. The convolution theorem (see Eq. 2.41 in Section 2.6) implies that the Fourier transform of the spectrum from a double-labelled system is given by the Fourier transform of the spectrum from the single labels multiplied by that of the dipolar broadening envelope:

$$\mathcal{F}(F_{dd}(B)) = \mathcal{F}(f_o(B)) \times \mathcal{F}(D(B)) \tag{15.4}$$

Thus, we obtain the net dipolar envelope from spectra of the single and doubly labelled systems by using the relation:

$$D(B) = \mathcal{F}^{-1}\left[\frac{\mathcal{F}(F_{dd}(B))}{\mathcal{F}(f_o(B))}\right] \tag{15.5}$$

where $\mathcal{F}^{-1}$ represents the reverse Fourier transform. Deconvolution performed on the usual, first-derivative EPR spectra produces the absorption dipolar powder pattern, as also does deconvolution of the integrated EPR spectra (see Eq. 2.40 in Section 2.6). Unfortunately, deconvolution is an inherently unstable algorithm, because dividing by the Fourier transform (which is an oscillatory function decaying to zero) produces potential infinities, especially when we take noise into account. Ways to handle this are discussed by Rabenstein and Shin (1995), Altenbach et al. (2001) and Banham et al. (2008).

Figure 15.4 illustrates the Fourier deconvolution procedure. Because we are dealing solely with real spectra, only the real (cosine) part of the Fourier transform is needed. The upper panel in the figure shows the ratio of the Fourier transforms of double- and single-labelled systems. We take the top spectrum from Figure 15.2, at a total scan width of 20 mT, for the single label. For the double label, we take its convolution with a dipolar powder pattern corresponding to $r_{12} = 1.2$ nm (bottom spectrum in Figure 15.2). Both these spectra are simulations and are completely free of noise. Nevertheless, the Fourier signal, which is shown by the solid line in the top panel of Figure 15.4, contains spikes that arise from the numerical instabilities in deconvolution. The dashed line in the top panel shows the effect of adding just 0.3% of uncorrelated random noise to each spectrum. The Fourier signal quickly becomes swamped by the instabilities arising from this noise, which we must null by using a low-pass filter. The lower panel in Figure 15.4 shows

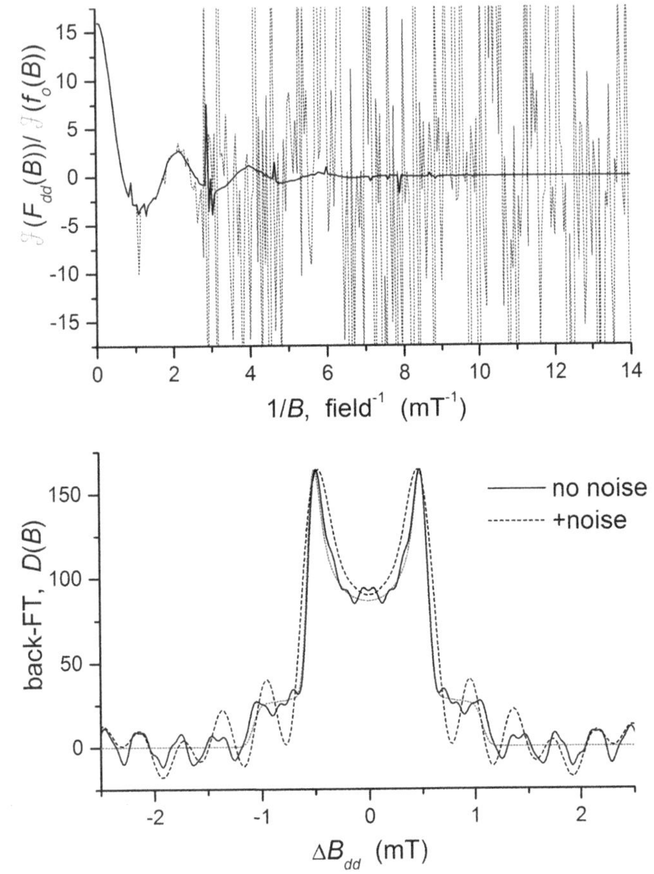

FIGURE 15.4 Deconvolution of first-derivative powder spectrum $F_{dd}(B)$ for dipolar-coupled spin-label pair (bottom spectrum in Figure 15.2). Single spin label spectrum $f_o(B)$ is top in Figure 15.2. *Top panel*: ratio of real Fourier transforms of F_{dd} and f_o (solid line) and of corresponding spectra with 0.3% random noise added (dashed line). Full transform extends out to 50 mT^{-1}. *Bottom panel*: deconvoluted dipolar spectrum from reverse Fourier transform of solid line in top panel (solid line) and of dashed line zeroed beyond 2.6 mT^{-1} with low-pass filter (dashed line). Dotted line: dipolar powder pattern used for original convolution in Figure 15.2.

the dipolar spectrum, $D(\Delta B_{dd})$, that we obtain by back transforming the Fourier signal from the upper panel. The dipolar powder pattern from the noise-free spectra (solid line) contains ripples that arise from the spike-instabilities in Fourier space. We could have edited these out to produce a result much closer to the original Pake pattern, but that still contains some residual ripple from the numerical procedures. With noise included, however, editing numerical spikes is possible only to a limited extent: exclusively on the residual signal recovered after low-pass filtering. In any case, the back-transformed spectrum is dominated by oscillations that arise from truncation by the filter (dashed line in the lower panel). These occur because the Fourier transform of a step-function is a sinc-function (see Appendix T). Clearly, we must reduce noise in the original experimental spectra as much as possible by signal averaging, for deconvolution to be successful.

The dipolar powder pattern that we produce by deconvolution can then be fit from a library of single Pake doublets to give the distribution of interspin distances, r_{12}, from which we then obtain the mean and standard deviation (Altenbach et al. 2001; Banham et al. 2008).

15.5 RANDOM DISTRIBUTION OF SPINS

Consider the distribution of nearest-neighbour distances, r_{nn}, between spins in a random spatial distribution of spin labels, as with a frozen glass. The probability, $w(r_{nn})\cdot dr_{nn}$, of finding a nearest-neighbour distance between r_{nn} and $r_{nn} + dr_{nn}$ is equal to the probability that there is no nearest neighbour within a distance less than r_{nn}, multiplied by the probability that the nearest neighbour lies within $r_{nn} + dr_{nn}$:

$$w(r_{nn}) = \left(1 - \int_0^{r_{nn}} w(r_{nn})\cdot dr_{nn}\right) 4\pi r_{nn}^2 n_{SL} \tag{15.6}$$

where n_{SL} is the average number of spin labels per unit volume. Solving Eq. 15.6 for $w(r_{nn})$, we get the normalized distribution function (Chandrasekhar 1943):

$$w(r_{nn}) = \exp\left(-\tfrac{4}{3}\pi r_{nn}^3 n_{SL}\right) 4\pi r_{nn}^2 n_{SL} \tag{15.7}$$

The mean nearest-neighbour distance $\langle r_{nn}\rangle$ is then given by:

$$\langle r_{nn}\rangle = \int_0^{\infty} r_{nn} w(r_{nn})\cdot dr_{nn} = \frac{\Gamma\left(\frac{4}{3}\right)}{\left(\frac{4}{3}\pi n_{SL}\right)^{1/3}} = 0.55396/n_{SL}^{1/3} \tag{15.8}$$

where the gamma function is defined by the integral: $\Gamma(x) = \int_0^{\infty} e^{-t} t^{x-1}\,dt$. For spin-label concentration c_M in moles/litre, the mean nearest-neighbour distance in nm is: $\langle r_{nn}(\text{nm})\rangle = 0.6560 c_M^{-1/3}$. Figure 15.5 shows the normalized distribution for a dimensionless nearest-neighbour distance, r_{nn}, that is normalized to the mean separation, $\langle r_{nn}\rangle$. As shown by Steinhoff et al. (1997), we can fit the nearest-neighbour distribution with a normalized Gaussian function. The mean nearest-neighbour distance obtained from the Gaussian is practically identical with that for $w(r)$, and the ratio of the mean to the standard deviation of the best fitting Gaussian is: $\langle r_{nn}\rangle/\sigma_{r_{nn}} = 2.61 \pm 0.02$.

By an analogous method, we find that the normalized nearest-neighbour distribution for a random spatial surface distribution of spin labels is given by:

$$w_s(r_{nn}) = \exp\left(-\pi r_{nn}^2 n_{SL}^s\right) 2\pi r_{nn} n_{SL}^s \tag{15.9}$$

where n_{SL}^s is the surface density of spins, per unit area. The mean nearest-neighbour distance in the two-dimensional case is then:

$$\langle r_{nn}\rangle_s = \tfrac{1}{2}\left(n_{SL}^s\right)^{-1/2} \tag{15.10}$$

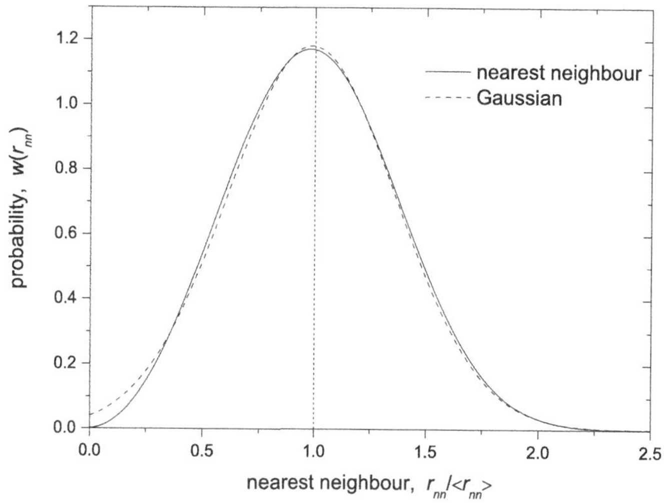

FIGURE 15.5 Distribution function $w(r_{nn})$ for nearest-neighbour distances (solid line) in random spatial distribution of spins (Eq. 15.7). *x*-axis: separation r_{nn} between nearest neighbours normalized to its mean value $\langle r_{nn}\rangle$. *Dashed line*: least-squares fit of normalized Gaussian distribution to $w(r_{nn})$, giving mean value $r_{nn}/\langle r_{nn}\rangle = 0.985 \pm 0.002$ and standard deviation $\sigma_{r_{nn}} = 0.377 \pm 0.001$.

Not surprisingly, this surface distribution (i.e., Eq. 15.9) is not well represented by a Gaussian function.

Figure 15.6 shows the mean distances $\langle r_{dd}\rangle$ between TEMPO spin probes at different concentrations in a frozen glycerol–water glass (Steinhoff et al. 1997). These values come from simulating the EPR lineshapes by convoluting a dipolar powder lineshape for a Gaussian distribution of interspin distances with the

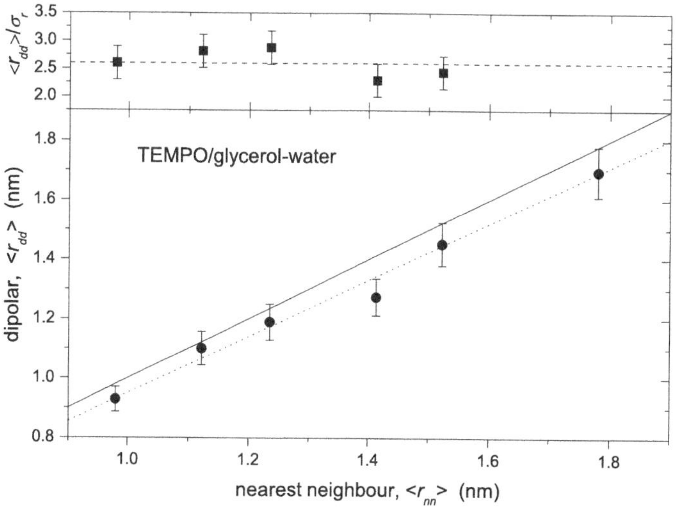

FIGURE 15.6 Mean experimental interspin distance $\langle r_{dd}\rangle$ (circles), calculated from dipolar broadening by convolution using dipolar powder pattern for Gaussian distribution of interspin distances with standard deviation $\sigma_{r_{dd}}$ (squares). *x*-axis: mean nearest-neighbour distance $\langle r_{nn}\rangle$ from Eq. 15.8, for random distribution of TEMPO spin probe in glycerol–water glass. *Lower panel*: solid line, $\langle r_{dd}\rangle = \langle r_{nn}\rangle$; dotted line, linear regression constrained to pass through origin: $\langle r_{dd}\rangle = (0.95 \pm 0.01) \times \langle r_{nn}\rangle$. *Upper panel*, dashed line: $\langle r_{dd}\rangle/\sigma_{r_{dd}} = 2.6$. Data from Steinhoff et al. (1997).

experimental lineshape obtained at low spin-label concentration. They are compared with the average nearest-neighbour distances $\langle r_{nn}\rangle$ that we calculate from the spin-label concentration by using Eq. 15.8. To within the experimental error, the agreement of $\langle r_{dd}\rangle$ with $\langle r_{nn}\rangle$ is good: from linear regression $\langle r_{dd}\rangle$ is only 5% smaller. Also, the ratio of the mean separation to the Gaussian standard deviation is close to the value of $\langle r_{dd}\rangle/\sigma_{r_{dd}} = 2.6$ that fits the random distribution given by Eq. 15.7. We obtain the values of $\langle r_{dd}\rangle$ in Figure 15.6 by assuming like spins; they would be ca. 15% smaller for unlike spins (see Section 9.4 in Chapter 9). On the other hand, the dipolar broadening contains contributions from spin labels more distant than nearest neighbours, neglecting which underestimates the nearest-neighbour distance. From simulating a random particle distribution, Steinhoff et al. (1997) find that the average distance to next-nearest neighbours is 1.5× that to nearest neighbours. To allow for this, we need to increase $\langle r_{dd}\rangle$ by 12%. This suggests that assuming like spins compensates for neglecting neighbours beyond the nearest. We treat dilute random distributions of spins later, by using the statistical theory given in Section 15.9.

15.6 ABSOLUTE-VALUE FIRST MOMENT AND MEAN DIPOLAR SPLITTING

A simplified approach to obtaining distances from deconvolution uses the mean splitting, $\langle 2|\Delta B_{dd}|\rangle$, averaged over the entire powder pattern (Rabenstein and Shin 1995). Figure 15.7 gives the definition of the dipolar splitting, $2|\Delta B_{dd}|$, and powder intensities, $I(\Delta B_{dd})$, in the deconvoluted spectrum. For a single Pake powder pattern, the mean splitting is given by:

$$\langle 2|\Delta B_{dd}|\rangle = \int_{-2\Delta B_{dd}^o}^{\Delta B_{dd}^o} 2|\Delta B_{dd}| \cdot I(\Delta B_{dd})\,\mathrm{d}\Delta B_{dd} = \frac{4}{3\sqrt{3}} 2\Delta B_{dd}^o \quad (15.11)$$

where $I(\Delta B_{dd})$ is the normalized dipolar semi-envelope that is given by Eq. 9.16 of Chapter 9 (see Figure 9.3) and $2\Delta B_{dd}^o = (\mu_o/4\pi) g\beta_e/r_{12}^3$ for unlike spins. With a distribution of distances r_{12}, the contributions of the different Pake patterns to the mean splitting are additive. On using Eq. 15.11, the experimental mean splitting $\langle 2|\Delta B_{dd}|\rangle$ then gives us the weighted average of ΔB_{dd}^o i.e., the mean value $\langle 1/r_{12}^3\rangle$. Only for narrow distributions can the latter give a good approximation to the mean distance $\langle r_{12}\rangle$. In general, the effective value $r_{eff,3}$ that we deduce from $\langle 1/r_{12}^3\rangle$ will differ from the true value of $\langle r_{12}\rangle$, by being biased towards shorter distances (see Section 15.8, below).

When the distribution of interspin distances is relatively broad, we can resolve only the initial positive part of the deconvolution in Fourier space, that without any oscillation, before the signal becomes swamped by noise (cf. Figure 15.4). Under these conditions, a Gaussian fit in Fourier space is sometimes used (Xiao and Shin 2000; Banham et al. 2008). This automatically

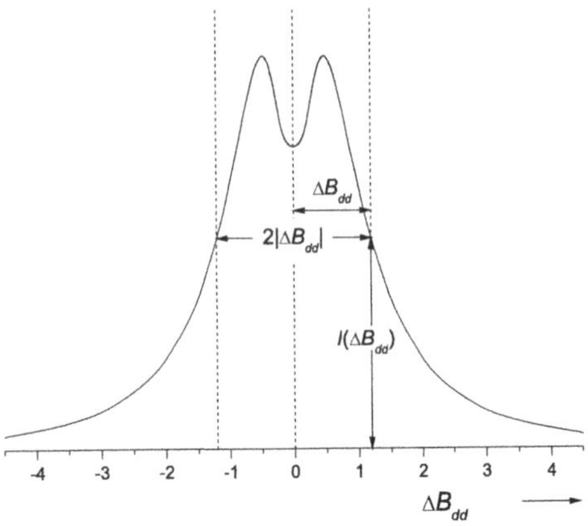

FIGURE 15.7 Spectral splittings ΔB_{dd} and intensities $I(\Delta B_{dd})$ used to define mean dipolar splitting (or absolute first moment) $\langle 2|\Delta B_{dd}|\rangle$ in deconvoluted dipolar powder absorption spectrum (cf. Eq. 15.11).

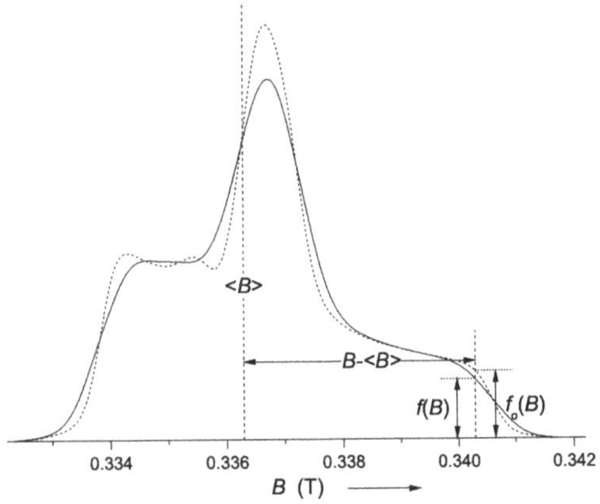

FIGURE 15.8 Dipolar second moment for anisotropic spin-label spectrum. Absorption EPR spectra for nitroxides with (solid line) and without (dotted line) dipolar broadening are normalized to integrated area. $\langle B\rangle \equiv \int B\cdot f(B)\mathrm{d}B$: mean resonance field, or first moment. Dipolar contribution to second moment: $\langle \Delta B_{dd}^2\rangle = \langle \Delta B^2\rangle - \langle \Delta B_o^2\rangle$, for $\langle \Delta B^2\rangle$ and $\langle \Delta B_o^2\rangle$ from Eq. 15.12 for spectra with and without dipolar broadening, respectively.

forces the resulting dipolar powder pattern also to be Gaussian, and a numerical back-transform is unnecessary because this result is analytical (see Table T.1 in Appendix T). The mean splitting, or twice the one-sided first moment, of a Gaussian dipolar spectrum is: $\langle 2|\Delta B_{dd}|\rangle = \sigma_B/\sqrt{2\pi}$, where $\sigma_B(=1/\sigma_t)$ is the standard deviation of the Gaussian spectrum and σ_t is the standard deviation of the Gaussian fit in Fourier space. To within the Gaussian approximation, we can use this result immediately with Eq. 15.11 to get the mean value $\langle 1/r_{12}^3\rangle$.

15.7 SECOND MOMENT OF DIPOLAR LINESHAPE

The second moment of the EPR lineshape from a system of interacting spins has two important advantages: (1) an analytical expression is available for an arbitrary number of spins interacting by the dipole–dipole interaction, and (2) Heisenberg exchange interaction does not contribute to the second moment of *like* spins, because of exchange narrowing, although it does contribute for unlike spins (Van Vleck 1948). We define the second moment $\langle \Delta B^2\rangle$ of a symmetrical EPR lineshape *f*(*B*) by:

$$\langle \Delta B^2\rangle = \frac{\int_{-\infty}^{\infty} (B-\langle B\rangle)^2 f(B)\cdot \mathrm{d}B}{\int_{-\infty}^{\infty} f(B)\cdot \mathrm{d}B} \tag{15.12}$$

where $B=\langle B\rangle$ is the mean resonance field about which the line is centred. Figure 15.8 illustrates the general definition of the second moment for an anisotropic lineshape (see discussion surrounding Eq. U.12 in Appendix U). The denominator in Eq. 15.12 ensures that the lineshape is normalized. For an isolated pair of spins, we can calculate the second moment of the dipolar powder lineshape directly from Eq. 15.12, by using the Pake pattern that is given in Eq. 9.16. This is exactly analogous to calculation of the mean splitting in Section 15.6 (cf. Eq. 15.11):

$$\langle \Delta B_{dd}{}^2\rangle = \int_{-2\Delta B_{dd}^o}^{\Delta B_{dd}^o} (\Delta B_{dd})^2\cdot I(\Delta B_{dd})\mathrm{d}\Delta B_{dd} = \frac{4}{5}\left(\Delta B_{dd}^o\right)^2 \tag{15.13}$$

where $I(\Delta B_{dd})$ is the normalized dipolar semi-envelope, $\Delta B_{dd}^o = \frac{3}{4}(\mu_o/4\pi)g\beta_e/r_{12}^3$ for like spins and $\Delta B_{dd}^o = \frac{1}{2}(\mu_o/4\pi)g\beta_e/r_{12}^3$ for unlike spins. However, the method introduced by Van Vleck (1948) is far more general, because it applies to an unrestricted number of mutually interacting pairs of spins.

For a general system of interacting spins, we show in Appendix U that the second moment is given by (see Eq. U.18):

$$\langle \Delta B^2\rangle = -\frac{\mathrm{Tr}\left([\mathcal{H}_{dd},S_x]^2\right)}{(g\beta_e)^2\,\mathrm{Tr}\left(S_x{}^2\right)} \tag{15.14}$$

where $\mathcal{H}_{dd}$ is the dipolar Hamiltonian and $S_x = \sum_j s_{jx}$ is the *x*-component of the total spin. Here Tr(..) is the trace of the corresponding operator, which is equal to the sum of the diagonal elements in a convenient representation. The *x*-component of the spin operators enters into the calculation because EPR transitions are induced by a microwave magnetic field that is perpendicular to the static field, which specifies the direction of *z*-quantization. For like spins, we get $\mathcal{H}_{dd}$ by generalizing Eq. 9.12 of Chapter 9, and for unlike spins by generalizing Eq. 9.18.

Evaluating the two traces in Eq. 15.14 leads to the second moment contributed by the dipolar interaction between *like* spins (see Appendix U, Eq. U.27):

$$\left\langle \Delta B_{dd}^2 \right\rangle = \frac{3}{4}\left(\frac{\mu_o}{4\pi}\right)^2 s(s+1)g^2\beta_e^2 \sum_j \frac{\left(1-3\cos^2\theta_{jk}\right)^2}{r_{jk}^6} \qquad (15.15)$$

where $s = \frac{1}{2}$ for all k, which is a dummy index. In a powder sample, we need to sum the angular terms over a sphere, finally giving:

$$\left\langle \Delta B_{dd}^2 \right\rangle_{SS} = \frac{3}{5}\left(\frac{\mu_o}{4\pi}\right)^2 s(s+1)g^2\beta_e^2 \sum_j 1/r_{jk}^6 \qquad (15.16)$$

as the second moment of the dipolar broadening for *like* spins. The dipolar Hamiltonian for *unlike* spins is smaller by a factor of 2/3 than that for like spins (compare Eqs. 9.13 and 9.19 in Chapter 9), which becomes a factor of 4/9 in the second moment. The dipolar second moment in a powder sample of *unlike* spins is therefore:

$$\left\langle \Delta B_{dd}^2 \right\rangle_{SS'} = \frac{4}{15}\left(\frac{\mu_o}{4\pi}\right)^2 s'(s'+1)g^2\beta_e^2 \sum_k 1/r_{jk}^6 \qquad (15.17)$$

where $s' = \frac{1}{2}$ for all k unlike spins. Because the spins are distinguishable, a contribution $\left\langle \Delta B^2 \right\rangle_{ex} = \frac{1}{3}s'(s'+1)\sum\left(J_{12,k}/g\beta_e\right)^2$ must be added to Eq. 15.17 to allow for weak Heisenberg exchange coupling, $J_{12,k}$, between unlike spins (see, e.g., Abragam and Bleaney 1970).

For a distribution of interspin distances, we see immediately from Eqs. 15.16 and 15.17 that the second moment does not yield the mean interspin distance directly but instead gives the weighted mean value $\left\langle 1/r_{12}^6 \right\rangle$, which replaces the sum over k in the two equations above. The inverse root-mean sixth power, $r_{eff,6} = \left\langle 1/r_{12}^6 \right\rangle^{-1/6}$ is an effective mean distance that is biased towards distances shorter than the true mean $\langle r_{12} \rangle$, to an extent that depends on the distribution width. The situation is analogous to that for the mean dipolar splitting or absolute first moment, which we discussed already in Section 15.6. For a Gaussian distribution, we give an explicit treatment for both moments in Section 15.8, which directly follows this one.

Figure 15.9 shows dipolar distances deduced from the second moments of the spectra of α-helical $(AAAAK)_4A$ biradicals with labelled residues at various distances apart. We obtain the dipolar contribution to the second moment by subtracting $\left\langle \Delta B^2 \right\rangle$ for the singly-labelled peptide from the corresponding value for the doubly-labelled peptide (see legend to Figure 15.8). The open circles in Figure 15.9 are the original experimental data for the effective mean $r_{eff,6}$ (Banham et al. 2008) but adjusted to the assumption of unlike spins that we used for convolution analysis of the same spin-labelled peptides in Figure 15.3. As expected, these values of $\left\langle 1/r_{12}^6 \right\rangle^{-1/6}$ lie consistently below the mean distances $\langle r_{12} \rangle$ that are deduced for these peptides from molecular modelling. Correcting these values by using Eq. 15.21 for a Gaussian distribution (see the following Section 15.8) results in mean values that lie closer to the predictions of molecular modelling. When compared with aggregate experimental values of $\langle r_{12} \rangle$ that are obtained from convolution and deconvolution (Banham et al. 2008; and see Figure 15.3), the agreement improves further. Nonetheless, the shortest mean distance is still larger than expected (Marsh 2014). The main disadvantage of second-moment calculations is that they overemphasize data in the outer wings of the spectra, where signal-to-noise ratio is lowest. Of course, this preferentially affects the widest splittings, i.e., the shortest distances (cf. Figure 15.10 in the Section immediately following).

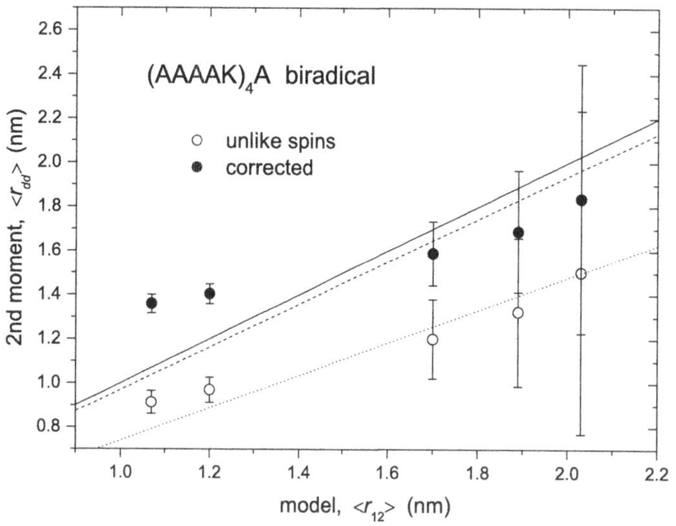

FIGURE 15.9 Mean interspin distance $\langle r_{dd} \rangle$ for α-helical peptide $(AAAAK)_4A$, as in Figure 15.3 but obtained from second moment $\left\langle \Delta B^2 \right\rangle$. Effective distances $\langle r_{dd} \rangle \equiv r_{eff,6}$ from Eq. 15.17 (open circles). Values corrected for Gaussian distribution (Eq. 15.21) with $\sigma_{r_{12}} = 0.38$ nm – cf. Figure 15.3 (solid circles). *x*-axis: mean distance $\langle r_{12} \rangle$ between nitroxides from molecular modelling. *Solid line*: $\langle r_{dd} \rangle = \langle r_{12} \rangle$; *dotted line*: unweighted linear regression constrained to pass through origin for open circles; *dashed line*: regression for corrected values (solid circles). Experimental data from Banham et al. (2008).

15.8 GAUSSIAN DISTANCE DISTRIBUTIONS

Frequently, a Gaussian distribution of interspin distances is assumed not only in dipolar convolution of CW spectra (Steinhoff et al. 1997) but also in some analyses of pulsed dipolar EPR (Fajer et al. 2007; Brandon et al. 2012). As seen already in Section 15.5, the Gaussian distribution is a good approximation for the nearest-neighbour distances in a random dispersion.

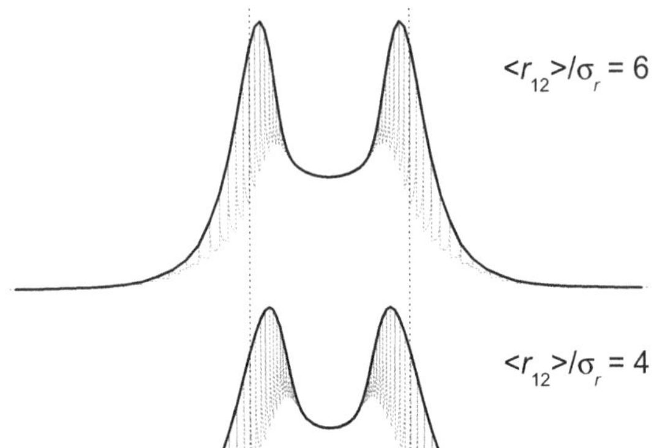

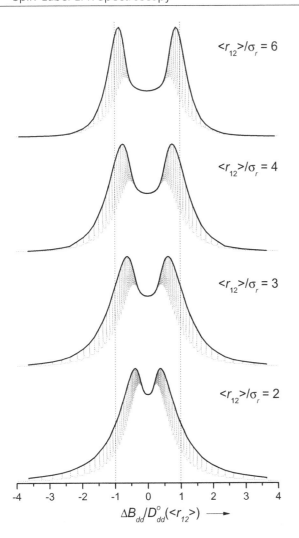

FIGURE 15.10 Dipolar powder patterns for Gaussian distribution of interspin distances, r_{12}. *Dashed lines*: superpositions of dipolar envelopes for each interspin distance. To give true spectrum, these must be convoluted with the intrinsic lineshape. Ratio of mean separation $\langle r_{12}\rangle$ to standard deviation $\sigma_{r_{12}}$ decreases from top to bottom. Perpendicular peaks of envelope corresponding to mean separation occur at $\pm D_{dd}^{o}(\langle r_{12}\rangle)$ (vertical dotted lines).

Empirically, it also describes the distance distributions found in the geometrical tether-in-a-cone model for biradicals (Hustedt et al. 2006) and in molecular modelling of MTSSL-labelled α-helical peptides (Banham et al. 2008).

The probability that the separation between spins has a value between r_{12} and $r_{12} + dr_{12}$ is given by:

$$p(r_{12})\cdot dr_{12} = \frac{1}{\sigma_{r_{12}}\sqrt{2\pi}}\exp\left(-\frac{\left(r_{12}-\langle r_{12}\rangle\right)^2}{2\sigma_{r_{12}}^2}\right)\cdot dr_{12} \qquad (15.18)$$

where $\langle r_{12}\rangle$ and $\sigma_{r_{12}}$ are the mean and standard deviation, respectively, of the Gaussian distribution. The full dipolar powder pattern is then a sum of normalized Pake doublets that are weighted by Eq. 15.18 for each value of the dipolar interaction $D_{dd}^{o}(r_{12})$. Figure 15.10 shows the resulting spectral envelopes for different values of the distribution width. As $\sigma_{r_{12}}$ increases, the apparent doublet peaks move together and the width at half-height of the overall envelope decreases. This happens for two reasons. Firstly, for a given increment in r_{12}, the Pake doublets become more closely spaced as r_{12} increases, because $D_{dd}^{o}(r_{12})$ depends on $1/r_{12}^3$. Secondly, for a fixed integrated intensity, the line height of a Pake doublet increases with decreasing splitting D_{dd}^{o} (cf. Eq. 9.16). Even for a relatively small distribution width, given by $\sigma_{r_{12}} = 0.17\times\langle r_{12}\rangle$ at the top of Figure 15.10, the peak splitting is less than that corresponding to the mean separation $\langle r_{12}\rangle$, and there is little evidence for the parallel shoulders that are characteristic of a single Pake doublet (cf. Figure 9.3). The latter features appear only when the distribution becomes very narrow.

Note that the total extent of the envelopes reaches beyond the range displayed in Figure 15.10: we see that they do not quite return to the baseline. Pake doublets extending out to $\Delta B_{dd} = \pm 16 D_{dd}^{o}(\langle r_{12}\rangle)$ are summed for these envelopes, which includes distances down to $r_{12} \approx 0.4\langle r_{12}\rangle$. Nonetheless, the range displayed covers most of that over which the line height is appreciable. This illustrates the difficulty in determining the full extent of the shorter side of the distribution function from the dipolar powder pattern, even if the wings are relatively free of noise. In practice, the minimum distance is determined by the sweep width, SW, of the experimental spectrum: $r_{min}^3 \approx 3(\mu_o/4\pi)g\beta_e/SW$, which amounts to $r_{min} \approx 0.65$ nm for $SW = 20$ mT. This lower cut-off accounts for the one-sided nature of some distance distributions that are deduced from experimental dipolar powder patterns.

The width of the distance distribution determines the extent to which the effective separation between spins $r_{eff,n}$ that are deduced from the mean width (see Eq. 15.11), or from the second moment (see Eqs. 15.16 and 15.17), of the dipolar powder pattern differ from the true mean $\langle r_{12}\rangle$. The effective value from the mean width depends on the weighted average $\langle 1/r_{12}^3\rangle$ and that from the second moment on the average $\langle 1/r_{12}^6\rangle$ (cf. Sections 15.6 and 15.7). We get these mean values by integrating over the probability distribution, $p(r_{12})$:

$$\left\langle \frac{1}{r_{12}^n}\right\rangle = \int_{r_{min}}^{\infty} \frac{1}{r_{12}^n} p(r_{12})\cdot dr_{12} \qquad (15.19)$$

where $n = 3$ or 6 and r_{min} is some suitable cut-off that we choose to prevent the integral from diverging. Figure 15.11 shows values of $1/r_{12}^3$ that are weighted by Gaussian distance distributions (i.e., Eq. 15.18) with different widths. At values of $r_{12} - \langle r_{12}\rangle$ more negative than those shown in the figure, the ordinate begins to diverge because of the inverse power dependence on r_{12}. For the largest distribution width shown $(\sigma_{r_{12}} = 0.25\langle r_{12}\rangle)$, this is already evident from the incipient minimum at the far left. (This numerical instability is more extreme when it comes to evaluating $\langle 1/r_{12}^6\rangle$.) Restricting the integration in Eq. 15.19 to the range displayed in the figure, the values of $r_{eff,3} = \langle 1/r_{12}^3\rangle^{-1/3}$ are given by the inset to Figure 15.11. Even for a narrow distribution width $\sigma_{r_{12}} = 0.1\langle r_{12}\rangle$, the effective distance $r_{eff,3}$ is still somewhat lower than the true mean because of the bias introduced by the $1/r_{12}^3$

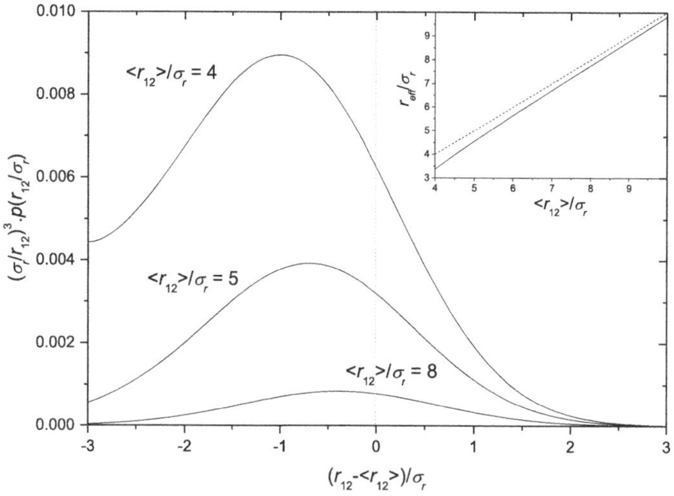

FIGURE 15.11 Dependence of weighted values of $1/r_{12}^3$ on inter-spin distance r_{12}, for Gaussian distribution $p(r_{12})$ of r_{12} with standard deviation σ_{r12} and mean value $\langle r_{12}\rangle$. *Inset*: dependence of effective mean separation $r_{eff} = \langle 1/r_{12}^3\rangle^{-1/3}$ on true mean value (solid line); dashed line: $r_{eff} = \langle r_{12}\rangle$. Mean value $\langle 1/r_{12}^3\rangle$ is obtained by integrating $(1/r_{12}^3)\,p(r_{12})$ over range plotted. All distances normalized to σ_{r12}.

weighting in Eq. 15.19. We show a second-order polynomial fit to the relation between $\langle r_{12}\rangle$ and $r_{eff,3}$ in the insert to Figure 15.11. This gives the following empirical calibration of the true mean value in terms of the effective value that we obtain experimentally from the mean splitting, as described in Section 15.6:

$$\frac{\langle r_{12}\rangle}{\sigma_{r12}} = 1.193 \pm 0.051 + (0.789 \pm 0.016)\frac{r_{eff,3}}{\sigma_{r12}} + (0.0113 \pm 0.0012)\left(\frac{r_{eff,3}}{\sigma_{r12}}\right)^2 \qquad (15.20)$$

where the range of fitting is $\langle r_{12}\rangle/\sigma_{r12} = 4$ to 10. A similar treatment gives the following calibration for the effective distance $r_{eff,6}$ that we get from the second moment, as described in Section 15.7:

$$\frac{\langle r_{12}\rangle}{\sigma_{r12}} = 1.759 \pm 0.052 + (0.723 \pm 0.016)\frac{r_{eff,6}}{\sigma_{r12}} + (0.0138 \pm 0.0011)\left(\frac{r_{eff,6}}{\sigma_{r12}}\right)^2 \qquad (15.21)$$

where the range of fitting is $\langle r_{12}\rangle/\sigma_{r12} = 5$ to 10.

These results emphasize that we expect effective distances deduced both from the second moment of the absorption spectrum, and from the mean splitting in a dipolar deconvolution, to be consistently smaller than the true mean distance. Corrections to the effective values deduced from double-labelled $(AAAAK)_4A$ α-helical peptides that are made by using Eqs. 15.20 and 15.21 (see solid circles in Figure 15.9) fully support this view (Marsh 2014).

15.9 STATISTICAL THEORY OF DIPOLAR BROADENING FOR DILUTE SPINS

We conclude the treatment of CW-lineshapes by describing the dipolar lineshape calculated from the statistical theory for dilute spins (Abragam 1961). This method is important for calculating phase-memory times from spin echoes (see Section 13.14 in Chapter 13). In Section 15.5, we considered nearest-neighbour distances for a random distribution of spins; here, we deal with the entire spin population. Lattice calculations for dilute spin occupancy ($f < 0.01$) yield ratios of fourth to second spectral moments $\langle \Delta\omega^4\rangle/\langle \Delta\omega^2\rangle^2 \gg 3$, which we can fit by assuming a truncated Lorentzian lineshape, but are inconsistent with a Gaussian lineshape (Abragam 1961, and see Appendix U). As we shall see, the statistical theory predicts a (truncated) Lorentzian lineshape. Derivations are given by Abragam (1961) and by Mims (1968). For convenience and in uniformity with the treatment of spin-echo phase-memory times (see Chapter 13), we use the angular resonance frequency ω in this section, instead of the resonance field, B.

The secular dipolar Hamiltonian for two unlike spins contains only $s_{1z}s'_{2z}$ terms, and the shift in Larmor frequency $\Delta\omega_{dd}$ of the central spin by the various surrounding spins is (cf. Eq. 9.19 in Chapter 9):

$$\hbar\Delta\omega_{dd} = \frac{\mu_o}{4\pi} g^2\beta_e^2 \sum_j \frac{\left(1-3\cos^2\theta_j\right)}{r_j^3} s'_{jz} \equiv \hbar\sum_j \Delta\omega_{d,j} \qquad (15.22)$$

where r_j is the distance between the central spin and the jth surrounding spin, θ_j is the angle between the static B_o field and the interspin vector, and s'_{jz} is the z-component of the jth surrounding spin. The sum in the first identity of Eq. 15.22 therefore determines the net local magnetic field from the surrounding spins. In the high-temperature limit, both values of $s'_{jz} = \pm\frac{1}{2}$ are equally likely, which we can allow for by letting r_j range from $-\infty$ to $+\infty$, and replacing the total volume V by $2V$ to restore normalization. For like spins, we need a multiplying factor of $\frac{3}{2}$ in Eq. 15.22 (cf. Eq. 9.13 in Chapter 9).

The probability that the jth spin is situated at position (r_j, θ_j) is simply $\mathrm{d}V_j/V$, where $\mathrm{d}V_j = 2\pi r_j^2 \sin\theta_j \mathrm{d}r_j \mathrm{d}\theta_j$. For N surrounding spins in total, the dipolar-broadened lineshape is therefore given by:

$$I(\Delta\omega) = \int_{V^N} \delta\left(\Delta\omega - \Delta\omega_{d,1} \cdots - \Delta\omega_{d,N}\right) \cdot \frac{\mathrm{d}V_1}{V} \cdots \frac{\mathrm{d}V_N}{V} \qquad (15.23)$$

where the delta-function limits the integral to those configurations of surrounding spins that produce a net dipolar shift in Larmor frequency $\sum_j \Delta\omega_{d,j}$ that is equal to $\Delta\omega$. Using the relation $\delta(x) = (1/2\pi)\int_{-\infty}^{+\infty} \exp(i\rho x)\,\mathrm{d}\rho$ and dropping the subscript j from the N similar definite integrals in Eq. 15.23, the lineshape then becomes:

$$I(\Delta\omega) = \frac{1}{2\pi}\int_{-\infty}^{+\infty} \exp(i\rho\Delta\omega)\cdot \mathrm{d}\rho\left[\frac{1}{2V}\int_{V} \exp\left(-i\rho\Delta\omega_d\right)\mathrm{d}V\right]^N \tag{15.24}$$

with the definition:

$$\int_V \exp\left(-i\rho\Delta\omega_d\right)\mathrm{d}V \equiv 2\pi\int_0^{\pi} \sin\theta\cdot \mathrm{d}\theta \times \int_{-\infty}^{\infty} \exp\left(-i\rho\frac{\mu_o}{4\pi}\frac{g^2\beta_e^2}{2\hbar}\frac{1-3\cos^2\theta}{r^3}\right) r^2\,\mathrm{d}r \tag{15.25}$$

where we include the extension of r to negative values and the renormalization of V. We avoid the divergence as $r \to \infty$ and $N \to \infty$ by rewriting Eq. 15.24 as:

$$I(\Delta\omega) = \frac{1}{2\pi}\int_{-\infty}^{+\infty} \exp(i\rho\Delta\omega)\left[1-\frac{V'(\rho)}{V}\right]^N \mathrm{d}\rho \tag{15.26}$$

with the definition:

$$V'(\rho) \equiv \pi\int_0^{\pi} \sin\theta\,\mathrm{d}\theta\int_{-\infty}^{\infty}\left[1-\exp\left(-i\rho\frac{\mu_o}{4\pi}\frac{g^2\beta_e^2}{2\hbar}\frac{1-3\cos^2\theta}{r^3}\right)\right] r^2\,\mathrm{d}r \tag{15.27}$$

We evaluate the integral over r in Eq. 15.27 by substituting $x = 1/r^3$, which gives: $I_1 = \frac{1}{3}\int_{-\infty}^{+\infty}\left(1-e^{-ibx}\right)/x^2\cdot \mathrm{d}x = \pi|b|/3 \equiv \pi\left(\mu_o/4\pi\right)\left(g^2\beta_e^2/6\hbar\right)|\rho|\left|3\cos^2\theta - 1\right|$. The integral over θ is then given by:

$$V'(\rho) = \pi^2\left(\frac{\mu_o}{4\pi}\right)\frac{g^2\beta_e^2}{6\hbar}|\rho|\int_0^{\pi}\left|3\cos^2\theta-1\right|\sin\theta\cdot \mathrm{d}\theta$$

$$= \frac{4\pi^2}{9\sqrt{3}}\left(\frac{\mu_o}{4\pi}\right)\frac{g^2\beta_e^2}{\hbar}|\rho| \tag{15.28}$$

Taking the thermodynamic limit $N \to \infty$, $V \to \infty$, by making the usual statistical-mechanical replacement (see Eq. 13.89), the lineshape given by Eq. 15.26 becomes:

$$I(\Delta\omega) = \frac{1}{2\pi}\int_{-\infty}^{+\infty} \exp(i\rho\Delta\omega)\exp\left(-n_{SL}V'(\rho)\right)\cdot \mathrm{d}\rho \tag{15.29}$$

where $n_{SL} \equiv N/V$ is the number of spin labels per unit volume. This is the Fourier transform of a reflected exponential function (because of $|\rho|$ in Eq. 15.28). Therefore, the dipolar broadened lineshape for dilute spins is Lorentzian (see Table T.1 in Appendix T):

$$I(\Delta\omega) = \frac{\Delta\omega_{1/2}/\pi}{(\Delta\omega)^2 + \left(\Delta\omega_{1/2}\right)^2} \tag{15.30}$$

From Eq. 15.28, the half-width at half-height of the Lorentzian lineshape is given by:

$$\Delta\omega_{1/2} = \frac{4\pi^2}{9\sqrt{3}}\left(\frac{\mu_o}{4\pi}\right)\frac{g^2\beta_e^2}{\hbar}n_{SL} \tag{15.31}$$

which is directly proportional to the spin concentration, n_{SL}. This contrasts with the situation at higher concentrations ($f > 0.1$), where we expect that the second moment $\langle\Delta\omega^2\rangle$ is directly proportional to the spin concentration (see Eqs. U.28 and U.30 in Appendix U). Equation 15.31 is for unlike spins, which is the case when inhomogeneous broadening is much larger than the dipolar broadening $\Delta\omega_{1/2}$ (see Section 9.4 in Chapter 9, and Marsh 2014). For like spins, the Lorentzian linewidth is 3/2 times greater than that given by Eq. 15.31 (Abragam 1961).

Equation 15.30 predicts a Lorentzian lineshape over the entire frequency range, without cut-off. This implies, amongst other things, that the second moment of the dipolar broadened lineshape diverges (see Appendix U). The divergence arises because we integrated over all r, which allows the spins to come vanishingly near to each other. In practice, the distance of closest approach limits the Lorentzian broadening with a high-frequency cut-off (Abragam 1961; and see also Section 15.13).

15.10 ECHO-DETECTED ELDOR (DEER)

We can extend dipolar spectroscopy to weaker interactions, i.e., longer distances, by using pulsed-ELDOR with selective pump and observe pulses within the common powder pattern of the interacting spins (see Figure 15.12). Detection with a spin-echo sequence removes sources of inhomogeneous broadening other than spin–spin interaction, which is preserved by reversing the dipolar field with the pump pulse. The shortest distance that we can measure by DEER is approximately: $r_{min} = 1.8$ nm, which is limited by the maximum width of the dipolar powder pattern that we can excite. The longest distance accessible to DEER extends to 8 nm, or even 10 nm for perdeuterated macromolecules, and is limited by homogeneous T_2-relaxation.

15.10.1 Three-Pulse DEER

Figure 15.13 shows pulse sequences for the DEER (Double Electron-Electron Resonance) experiment. We concentrate first on the simpler three-pulse variant of DEER. This consists

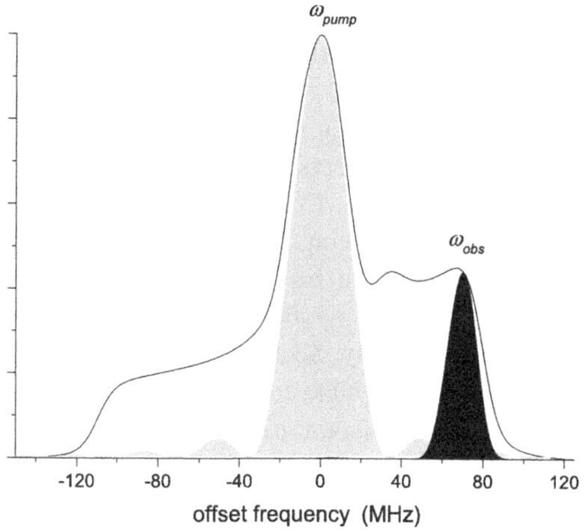

FIGURE 15.12 Excitation profiles for detection (ω_{obs}, black) and pump (ω_{pump}, grey) pulses in powder absorption spectrum from interacting ^{14}N-nitroxides, when using DEER double-resonance sequence. *x*-axis: offset frequency of microwave carrier for fixed magnetic field.

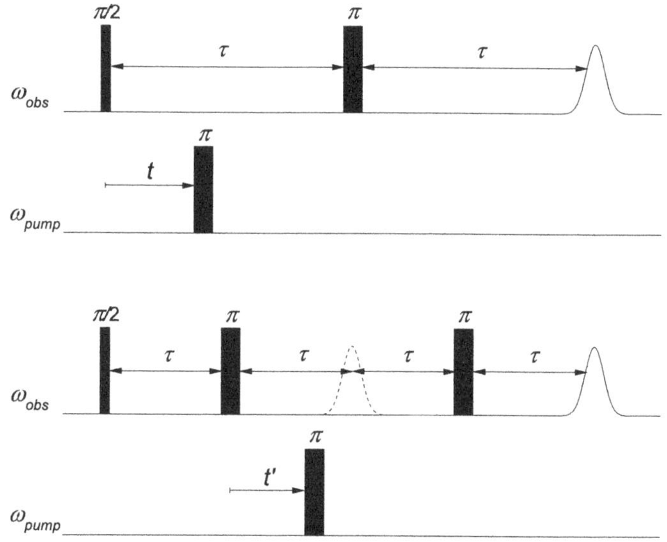

FIGURE 15.13 Pulse sequences for three-pulse DEER with primary-echo detection (upper) and for four-pulse DEER with refocused-echo detection (lower). Time t (or t') for pump pulse (ω_{pump}) is incremented, and interpulse spacing τ for observer echo sequence (ω_{obs}) is fixed. τ-delay for generating refocused echo in four-pulse DEER can differ from that generating original echo.

of a two-pulse primary echo sequence (see Section 13.4) for the observe frequency ω_{obs}, with a single π-pulse at the pump frequency ω_{pump}. The latter comes at variable time t after the observer π/2-pulse, within the evolution period τ between the π/2- and π-pulses of the echo sequence. Spin-echo modulation by dipolar interaction is monitored by incrementing the time t, for fixed τ in the observer channel.

We denote the observed spins as A-spins and the pumped spins as B-spins. Clearly these two sets of spins are distinguishable, because they are selected by pulses with different carrier frequencies. In the frame rotating at angular frequency ω_{obs}, the spin Hamiltonian of the A-spins is:

$$\mathcal{H}_{obs}/\hbar = \Delta\omega_A S_{z,A} + \omega_{AB} S_{z,B} S_{z,A} \tag{15.32}$$

where $\Delta\omega_A = \omega_A - \omega_{obs}$ is the resonance offset frequency of the A-spins. We retain only the secular part of the spin–spin interaction, because $\omega_{pump} - \omega_{obs}$ is much larger than the off-diagonal terms (see Section 9.4). The angular frequency of spin–spin interaction ω_{AB} is given by (cf. Eq. 9.19 for unlike spins):

$$\hbar\omega_{AB} = \frac{\mu_o}{4\pi} g^2 \beta_e^2 \frac{\left(1 - 3\cos^2\theta_{AB}\right)}{r_{AB}^3} + J_{AB} \tag{15.33}$$

where r_{AB} is the length of the vector joining A- and B-spins and θ_{AB} is the angle that this vector makes with the magnetic field. J_{AB} is the Heisenberg spin-exchange interaction, which is likely to be small for the distances that interest us (except for conjugated biradicals – see Weber et al. 2002; Margraf et al. 2009). For a spin-echo sequence alone, dephasing by secular spin–spin interactions would be refocused in the echo. But with DEER, the additional pump pulse reverses dephasing by the $\omega_{AB} S_{z,B}$ term, causing modulation of the echo amplitude.

We can describe DEER echo modulation by using the vector model (Larsen and Singel 1993, and see Section 13.4). After the initial π/2-pulse, dephasing of the A-spin magnetization in the x,y-plane occurs at a rate $\Delta\omega_A + M_{S,B}\omega_{AB}$, where $M_{S,B} = \pm\frac{1}{2}$ is the magnetic quantum number of the B-spins. The pump π-pulse at time t flips the B-spins, which reverses the sign of $M_{S,B}$; and the A-spins now dephase at rate $\Delta\omega_A - M_{S,B}\omega_{AB}$, relative to the situation before the pump pulse. The observe π-pulse at time τ reverses the dephasing, which now takes place at angular frequency $-\Delta\omega_A + M_{S,B}\omega_{AB}$. The echo occurs as usual at time 2τ, when the total accumulated phase is:

$$\begin{aligned}\varphi(2\tau,t) &= \left(\Delta\omega_A + M_{S,B}\omega_{AB}\right)t + \left(\Delta\omega_A - M_{S,B}\omega_{AB}\right)(\tau - t) \\ &\quad + \left(-\Delta\omega_A + M_{S,B}\omega_{AB}\right)\tau = 2M_{S,B}\omega_{AB}t = \pm\omega_{AB}t\end{aligned} \tag{15.34}$$

From Eq. 13.18, we see that modulation of the echo amplitude is determined by the accumulated phase according to:

$$E_{DEER}(t) = \cos(\varphi(2\tau,t)) = \cos(\omega_{AB}t) \tag{15.35}$$

which for $J_{AB} \approx 0$ immediately yields the dipolar coupling frequency.

We also arrive at Eq. 15.35 by using the product-operator method, as described in Sections P.1 and P.2 of Appendix P. The sequence is based on the accumulated phase given by Eq. 15.34.

Up to the second observe pulse, the analogue of Eq. P.11 for the A-spins is:

$$S_z \xrightarrow{\pi/2 S_x} -S_y \xrightarrow{\left(\left(\Delta\omega_A \mp \frac{1}{2}\omega_{AB}\right)\tau \pm \omega_{AB}t\right)S_z}$$
$$-S_y\cos\left(\left(\Delta\omega_A \mp \tfrac{1}{2}\omega_{AB}\right)\tau \pm \omega_{AB}t\right) \quad (15.36)$$
$$+S_x\sin\left(\left(\Delta\omega_A \mp \tfrac{1}{2}\omega_{AB}\right)\tau \pm \omega_{AB}t\right) = \sigma(\tau)$$

The second observe pulse along x reverses the phase of S_y, and subsequent free precession for further time τ produces the echo (cf. Eq. P.12):

$$\sigma(\tau) \xrightarrow{\pi S_x} S_y\cos\left(\left(\Delta\omega_A \mp \tfrac{1}{2}\omega_{AB}\right)\tau \pm \omega_{AB}t\right)$$
$$+S_x\sin\left(\left(\Delta\omega_A \mp \tfrac{1}{2}\omega_{AB}\right)\tau \pm \omega_{AB}t\right) \quad (15.37)$$
$$\xrightarrow{\left(\Delta\omega_A \pm \frac{1}{2}\omega_{AB}\right)\tau S_z} S_y\cos(\pm\omega_{AB}t)$$

This modulation of the refocused S_y-magnetization agrees with Eq. 15.35 from the vector model. We can extend the product-operator method to include the non-vanishing width of the exciting pulses.

15.10.2 Four-Pulse DEER

Three-pulse DEER suffers from a considerable dead time caused by the finite lengths of the observe and pump pulses, which overlap when they approach at $t \approx 0$. This problem is avoided with four-pulse DEER where we detect a refocused echo, instead of the primary echo (see Figure 15.13). A second observer π-pulse, applied at time τ after the primary echo, generates a refocused echo after further time delay τ. The maximum of the time-domain signal then occurs for $t' = \tau$, when the pump pulse is at the position of the primary echo (Pannier et al. 2000). The total accumulated phase at time 4τ of the refocused echo is (cf. Eq. 15.34):

$$\varphi(4\tau,t) = \left(\Delta\omega_A \pm \tfrac{1}{2}\omega_{AB}\right)\tau - \left(\Delta\omega_A \pm \tfrac{1}{2}\omega_{AB}\right)t'$$
$$-\left(\Delta\omega_A \mp \tfrac{1}{2}\omega_{AB}\right)(2\tau - t') + \left(\Delta\omega_A \mp \tfrac{1}{2}\omega_{AB}\right)\tau$$
$$= \pm\omega_{AB}(\tau - t') \quad (15.38)$$

Correspondingly, the modulation of the echo amplitude becomes (cf. Eq. 15.35):

$$E_{4pDEER}(t) = \cos\left(\omega_{AB}(\tau - t')\right) \quad (15.39)$$

for four-pulse DEER. In these ideal cases, where we have complete inversion of the B-spins in both three-pulse and four-pulse DEER, we see from Eqs. 15.35 and 15.39 that the echo amplitude oscillates at frequency ω_{AB} with a modulation depth that changes from +1 to −1.

We now no longer assume ideal π/2- and π-pulses, in order to investigate the effects of their finite widths, $t_{p,1}$ and $t_{p,2}$ respectively. In Section 13.2 of Chapter 13, we saw that the signal amplitude following a nominal π/2-pulse is scaled by a factor $p_{\pi/2} = \sin\beta_1$ (see Eq. 13.4), where $\beta_1 = \omega_1 t_{p,1}$ is the actual pulse angle and $\omega_1 = \gamma_e B_1$ is the strength of the microwave field. Correspondingly, Section 13.3 shows that the effectiveness of a nominal π-pulse in reversing the magnetization direction is given by the factor $p_\pi = \frac{1}{2}(1 - \cos\beta_2) = \sin^2(\beta_2/2)$, where β_2 is the true pulse turning angle. For a pulse of length $t_{p,2}$, we find from Eq. 13.13 that $p_\pi = (\omega_1/\omega_{eff})^2 \sin^2(\omega_{eff}t_{p,2}/2)$, where ω_{eff} is related to the resonance offset $\Delta\omega_o$ by Eq. 13.10: $\omega_{eff} = \sqrt{(\Delta\omega_o)^2 + \omega_1^2}$. Similarly, for a nominal π/2-pulse of length $t_{p,1}$, we get $p_{\pi/2} = (\omega_1/\omega_{eff})\sin(\omega_{eff}t_{p,1})$ (cf. Eqs. 13.14 and 13.11).

For a pump pulse of duration $t_{p,2}$, the fraction of B-spins inverted f_B is given simply by the overlap integral between the excitation profile $p_\pi(\omega_{eff,B})$ and the normalized inhomogeneous EPR lineshape $g(\Delta\omega_B)$ (cf. Eqs. 13.111 and 13.113):

$$f_B = \langle p_B \rangle_g = \int_{-\infty}^{\infty} (\omega_{1,B}/\omega_{eff,B})^2 \sin^2(\omega_{eff,B}t_{p,2}/2)\, g(\Delta\omega_B)\cdot \mathrm{d}\Delta\omega_B \quad (15.40)$$

where $\omega_{eff,B} = \sqrt{(\Delta\omega_B)^2 + \omega_{1,B}^2}$, and $\omega_{1,B}$ allows for the pump pulse having a higher B_1-field than the detection pulses. For the primary-echo observer sequence that excites the A-spins in three-pulse DEER, the excitation function is the product of the profiles for both pulses and the fraction of spins excited becomes (cf. Larsen and Singel 1993):

$$f_A = \int_{-\infty}^{\infty} (\omega_{1,A}/\omega_{eff,A})^3 \sin(\omega_{eff,A}t_{p,1})\sin^2(\omega_{eff,A}t_{p,2}/2)$$
$$\times g(\Delta\omega_A)\cdot \mathrm{d}\Delta\omega_A \quad (15.41)$$

where $\omega_{eff,A} = \sqrt{(\Delta\omega_A)^2 + \omega_{1,A}^2}$. Equation 15.41 comes directly from the fractional efficiency $p_{\pi/2} \times p_\pi = \sin\beta_1 \sin^2(\beta_2/2)$ for ESE (see Section 13.4). For 4-pulse DEER with refocused-echo detection, the corresponding excitation function becomes (Pannier et al. 2000):

$$f_A = \int_{-\infty}^{\infty} (\omega_{1,A}/\omega_{eff,A})^5 \sin(\omega_{eff,A}t_{p,1})\sin^4(\omega_{eff,A}t_{p,2}/2)$$
$$\times g(\Delta\omega_A)\cdot \mathrm{d}\Delta\omega_A \quad (15.42)$$

which derives from the factor $p_{\pi/2} \times p_\pi^2 \sin\beta_1 \sin^4(\beta_2/2)$ for the three-pulse refocused echo. Typical excitation profiles for three-pulse DEER are shown in Figure 15.12. The pump pulse is applied at the absorption maximum in the $m_I = 0$ manifold, to maximize the number of B-spins that are flipped and hence optimize the echo modulation amplitude. The observer pulses are applied close to the high-frequency shoulder of the $m_I = +1$ manifold, well away from the pump pulse. A 12-ns pumping π-pulse is optimum because the first minimum of the excitation profile occurs close to the observer position in a 9-GHz powder pattern (Jeschke and Polyhach 2007). A protocol for performing DEER experiments is given by Schiemann et al. (2007).

15.11 PROCESSING AND ANALYSING 4-PULSE DEER SIGNALS

The product rule is central to all DEER analysis. Derived by Mims (1972) for ESEEM with multiple nuclei, this states that independent contributions to the echo amplitude multiply together.

15.11.1 Background DEER Signal

Invariably, a background signal from non-specific intermolecular spin–spin interactions superimposes on the dipolar modulation from any biradical pair or monoradical cluster. At practicable spin-label concentrations, the background signal is quite strong and must be removed from the composite DEER signal. For a homogeneous distribution of spin labels, this is the contribution of instantaneous diffusion to the echo signal. We treated this already in Section 13.15. From Eqs. 13.110 and 13.93, we know that the intermolecular dipolar signal decays exponentially with a rate constant that depends linearly on the spin concentration, n_{SL}:

$$E_{4p,\mathrm{inter}}(t') = \exp\left(-k_{\mathrm{inter}} n_{SL} f_B \left|\tau - t'\right|\right) \quad (15.43)$$

where k_{inter} is related directly to twice the dipolar half-width $\Delta\omega_{1/2}$ given in Eq. 15.31:

$$k_{\mathrm{inter}} = \frac{8\pi^2}{9\sqrt{3}}\left(\frac{\mu_o}{4\pi}\right)\frac{g^2\beta_e^2}{\hbar} \quad (15.44)$$

and $f_B \equiv \langle p_B \rangle_g$ is the fraction of total electron spins inverted by the pumping pulse, which is given by Eq. 15.40. Correspondingly, the intramolecular signal from the biradical (or monoradical complex) consists of a fraction $1 - f_B$ that is not modulated, and a fraction f_B that is modulated at frequency ω_{AB} (see Eq. 15.39):

$$E_{4p,\mathrm{intra}}(t') = 1 - f_B + f_B \cos\left(\omega_{AB}(\tau - t')\right) \quad (15.45)$$

i.e., f_B functions as a modulation depth. The total DEER signal is the product of the intra- and intermolecular signals:

$$E_{4p,DEER}(t') = E_{4p,\mathrm{intra}}(t') E_{4p,\mathrm{inter}}(t') \quad (15.46)$$

which we get from Eqs. 15.43 and 15.45.

Figure 15.14 shows how to process four-pulse DEER time-domain signals for a localized A–B pair of spins, as in a rigid biradical. We remove the intermolecular signal from the upper trace in the top panel by fitting the exponential background (dashed line), and then dividing by it. (Subtracting the background, instead of dividing, leads to a frequency-domain signal that still is convoluted with the intermolecular broadening.) The lower trace shows the resulting time-domain signal. Fourier transformation (after zero-filling) then yields the dipolar spectrum in the frequency domain, which we show in the bottom panel. If there is no orientation selection of interdipole vectors, the result is a classic Pake powder pattern. We get this when the interspin vector $\mathbf{r}_{AB}$ is uncorrelated with the orientation of the nitroxide axes. A correlation is most likely to occur in rigid molecules. For instance, if $\mathbf{r}_{AB}$ is perpendicular to the nitroxide z-axis, the parallel shoulders will not appear in the dipolar powder pattern (Pannier et al. 2000). Orientational correlations tend to disappear with increasing conformational flexibility. This can be accompanied by an increasing distribution of interspin distances, although less dramatically than for the more closely spaced spins considered earlier in the chapter.

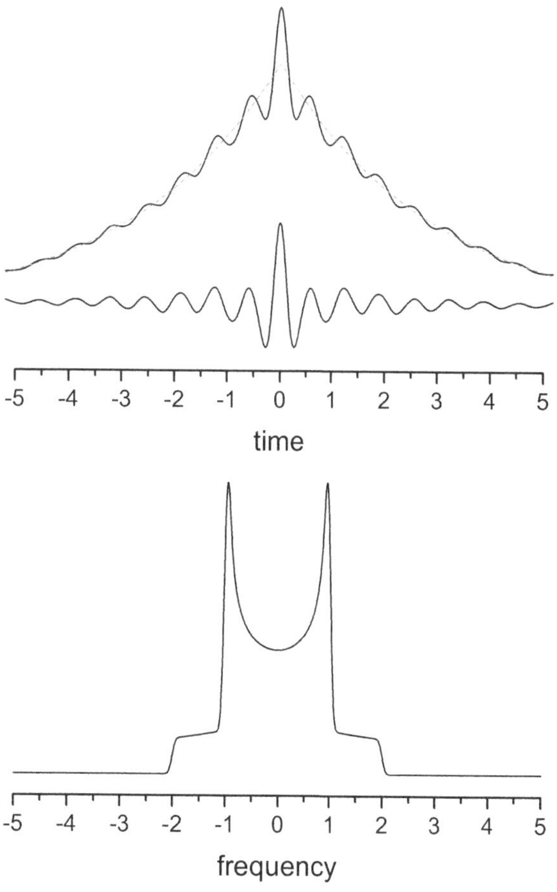

FIGURE 15.14 Processing 4-pulse DEER time-domain signals (upper panel). Fitting and subtracting exponential component (Eq. 15.43; dashed line) that corresponds to intermolecular interactions produces dipolar modulation trace. Fourier transform yields Pake powder spectrum (lower panel), if there is no orientation selection by the pulses.

Of course, Figure 15.14 is an idealized situation. In practice, the DEER signal is attenuated by intrinsic T_2-relaxation of the nitroxide. This limits the upper distance that we can measure; at least one or two dipolar oscillations must occur within the T_2-decay time. Consequently, we perform DEER experiments at low temperatures ~50 K, which is a compromise with the lengthening T_1 that reduces the repetition rate (Jeschke and Polyhach 2007). Also, as we saw in Section 13.14, deuteration of both solvent and the macromolecule under study can increase T_2 considerably; in the latter case, allowing measurement of distances up to 10 nm or more (Jeschke et al. 2004a; Ward et al. 2010; El Mkami et al. 2014). As we saw in Section 13.15 and the top panel of Figure 15.14, instantaneous diffusion also decreases the phase-memory time (or effective T_2). Thus, it is important to decrease the total spin-label concentration as far as possible, whilst still giving sufficiently strong signals (Jeschke 2002).

15.11.2 Spin-Pair Distributions and Spin Counting

Now we turn to distributions (or clusters) of spin pairs. For a distribution of B-spins, relative to a particular A-spin, the normalized DEER intensity as a function of dipolar evolution time $t = \tau - t'$ is the product of individual contributions (Jeschke 2002; Bowman et al. 2004; and see Eq. 15.45):

$$E_{DEER}(t) = \prod_p \left(1 - f_{B,p}\left(1 - \cos\omega_{AB,p}t\right)\right) \tag{15.47}$$

where $f_{B,p}$ is the dipolar modulation depth (see Eq. 15.40) and the product is taken over all B-spins p that are pumped. For $f_{B,p} << 1$, which may apply approximately in nitroxide DEER, we expand the product in Eq. 15.47 retaining only the constant term and those linear in $f_{B,p}$:

$$E_{DEER,lin}(t) = 1 - \sum_p f_{B,p}\left(1 - \cos\omega_{AB,p}t\right) \tag{15.48}$$

This concentrates on those B-spins that are sufficiently close to the central A-spin as to constitute a well-defined distribution of spin pairs (Jeschke 2002). If we neglect any angular correlations between interspin vectors, which holds for flexible molecules, $f_{B,p}$ no longer depends on p. To allow for a general distribution of interspin distances, we must additionally average over all A spins. We indicate this by introducing the constant modulation-depth factor $f_{A,B}$. Up to separations $r = r_{max}$, the contribution from this spin-pair population is then:

$$E_{DEER,pair}(t) = 1 - f_{A,B}\int_{r_{min}}^{r_{max}} P(r)\left(1 - \int_0^{\pi/2}\cos\left(\omega_{AB}(r,\theta)t\right)\sin\theta\,d\theta\right)dr \tag{15.49}$$

where $P(r)$ is the distribution of interspin distances, and $\omega_{AB}(r,\theta)$ is given by Eq. 15.33 with $r_{AB} \equiv r$, $\theta_{AB} \equiv \theta$ (and $J_{AB} = 0$). The distribution $P(r)$ is related to the radial distribution function $G(r)$ by $P(r) = 4\pi r^2 G(r)$. Generalizing to spin clusters, it is normalized such that:

$$\int_{r_{min}}^{r_{max}} P(r)\,dr = \bar{n} - 1 \tag{15.50}$$

where $\bar{n}$ is the average number of spins in the cluster. For a biradical $\bar{n} = 2$, of course. DEER contributions beyond $r = r_{max}$ are factorized out as a normalized background signal $B(t)$. The total normalized DEER signal thus becomes (cf. Eq. 15.46):

$$E_{4pDEER}(t) = E_{DEER,pair}(t)B(t) \tag{15.51}$$

When the dipolar oscillations of $E_{DEER,pair}(t)$ have decayed to zero, we see from Eqs. 15.49 and 15.50 that the total signal becomes $E_{4pDEER}(t) = \left(1 - f_{A,B} \times (\bar{n} - 1)\right)B(t)$. This lets us determine both the total modulation depth $\Delta = f_{A,B} \times (\bar{n} - 1)$ and the functional form of $B(t)$ by fitting the tail of the decay, as illustrated in Figure 15.15. Determination of Δ allows spin counting to obtain $\bar{n}$ for oligomers or aggregates (Bode et al. 2007; Jeschke et al. 2009). To do this, we must calibrate the value of $f_{A,B}$ for the particular experimental setup by using a standard biradical (i.e., with $\bar{n} = 2$). A more general expression for the modulation depth comes from the binomial product in Eq. 15.47, without using the linear approximation:

$$1 - \Delta = \left(1 - f_{A,B}\right)^{\bar{n}-1} \tag{15.52}$$

which is valid for all $f_{A,B}$ and reduces to $\Delta = f_{A,B} \times (\bar{n} - 1)$ when $f_{A,B} << 1$. Note that spin counting requires a two-step phase cycle

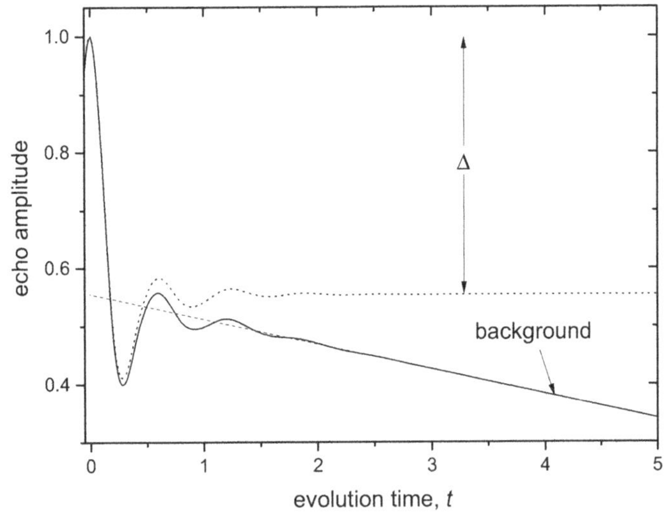

FIGURE 15.15 Separation of dipolar modulation from background (dashed line) in normalized 4-pulse DEER signal (solid line). Background signal $(1-\Delta)B(t)$ fitted – beyond decay of modulation – by (stretched) exponential. Dotted line: base-line corrected dipolar signal.

of the first observe pulse (+x, −x with subtraction on detection) to eliminate receiver offsets.

The background signal may be an exponential decay, as for the intermolecular contribution given by Eq. 15.43. More generally, we expect a stretched exponential (Milov and Tsvetkov 1997):

$$E_{4p,\text{inter}}(t) = \exp\left(-K_{\text{inter}} t^{D/3}\right) \tag{15.53}$$

where D is the fractal dimension of the space in which the remote B-spins are contained, e.g., $D = 2$ for a membrane or $D = 1$ for an extended linear chain, and K_{inter} is a second fitting parameter. Stretched-exponential fits with effective values $D > 3$ can arise in 3-dimensional systems, if excluded-volume effects are important (see Section 15.13). Alternatively, we can use an experimental background signal from a singly labelled system, but this still must be fitted with an analytical function to avoid instabilities from noise, such as we described in Section 15.4 on deconvolution.

15.11.3 Tikhonov Regularization and Gaussian Fitting

Once we have the background signal, we get $E_{DEER,pair}(t) = E_{4pDEER}(t)/B(t)$ from Eq. 15.51, as shown by the dotted line in Figure 15.15. Finally, the modulated part of the echo becomes:

$$f(t) = \frac{E_{4pDEER}(t) - B(t)}{\Delta \times B(t)} + 1$$

$$= \int_{r_{\min}}^{r_{\max}} P_{norm}(r)\left(\int_0^{\pi/2} \cos\left(\omega_{AB}(r,\theta)t\right)\sin\theta\,\mathrm{d}\theta\right)\mathrm{d}r \tag{15.54}$$

which is normalized to $f(0) = 1$ and oscillates about the zero baseline to which it decays. Here, the probability distribution $P_{norm}(r) \equiv P(r)/(\bar{n} - 1)$ is normalized to unity by virtue of Eq. 15.50. Determining the distribution function from this integral equation is a moderately ill-posed problem; because small contributions from noise and imperfections in separating the background factor can produce large errors in $P(r)$. Most frequently, the problem is treated by Tikhonov regularization (Jeschke et al. 2004b; Jeschke et al. 2006; Chiang et al. 2005). We write Eq. 15.54 as:

$$f(t) = \int_{r_{\min}}^{r_{\max}} K(r,t)P_{norm}(r)\,\mathrm{d}r \tag{15.55}$$

where the kernel is defined from Eq. 15.33 by:

$$K(r,t) = \int_0^1 \cos\left(\left(1 - 3z^2\right)\frac{\mu_o}{4\pi}\frac{g^2\beta_e^2}{\hbar r^3}t\right)\mathrm{d}z \tag{15.56}$$

with $z = \cos\theta$. In discrete form needed for numerical solution, this becomes a system of linear simultaneous equations:

$$f(t_i) = \frac{r_{\max} - r_{\min}}{N}\sum_{j=1}^{N} K(r_j,t_i)P_{norm}(r_j) \tag{15.57}$$

which we then write as matrices: $\mathbf{f} = \mathbf{K}\cdot\mathbf{P}$. Straightforward least-squares minimization of $\rho = \|\mathbf{K}\cdot\mathbf{P} - \mathbf{f}\|^2$ typically yields an unphysically large number of sharp peaks that reflect the inherent instability of the solution. Tikhonov regularization smooths the distribution by using the square norm of the second derivative $\eta = \|\mathrm{d}^2\mathbf{P}/\mathrm{d}r^2\|^2$ as criterion. The target function for minimization is:

$$G_\alpha[\mathbf{P}] = \|\mathbf{K}\cdot\mathbf{P} - \mathbf{f}\|^2 + \alpha\|\mathrm{d}^2\mathbf{P}/\mathrm{d}r^2\|^2 \tag{15.58}$$

where the Lagrange multiplier α is the regularization parameter. This yields the regularized solution (Chiang et al. 2005):

$$\mathbf{P} = \left|\mathbf{K}^{\mathrm{T}}\mathbf{K} + \alpha\mathbf{D}^{\mathrm{T}}\mathbf{D}\right|^{-1}\mathbf{K}^{\mathrm{T}}\mathbf{f} \tag{15.59}$$

where $\mathbf{D}$ is the second-derivative operator written as a matrix and superscript-T indicates the transpose. The optimum value of α comes from the elbow point between the two limbs in the L-shaped parametric plot of $\ln\eta(\alpha)$ against $\ln\rho(\alpha)$. This is a compromise between noise instabilities and over-smoothing.

Tikhonov regularization cannot perform well for a mixture of sharp and broad distributions. Either the broad component is too bumpy, or the sharp component is too broad. Then fitting with a specific model, using standard least-squares minimization, could be preferable. Whilst susceptible to trapping in local minima, this approach does not need prior base-line correction, because it is included in the model used for fitting. Brandon et al. (2012) use a sum of Gaussians for the specific distribution function and demonstrate that this performs well for relatively sharp distributions (see also Stein et al. 2015). They also discuss excluded-volume effects for the background function. Note that Gaussian distributions of the positions of the two spins in three dimensions produce a Rice distribution of interspin distances (Kohler et al. 2011):

$$P(r) = \sqrt{\frac{2}{\pi}}\frac{r}{\sigma_{rice}R}\exp\left(-\frac{R^2 + r^2}{2\sigma_{rice}^2}\right)\sinh\left(\frac{Rr}{\sigma_{rice}^2}\right) \tag{15.60}$$

where R is the separation of the centres of the Gaussian distributions with standard deviations σ_1, σ_2 and $\sigma_{rice}^2 = \sigma_1^2 + \sigma_2^2$. Fitting such a distribution with a simple Gaussian function yields the most probable distance but an incorrect value for R, because the maximum in Eq. 15.60 is shifted upward from $r = R$ (see Figure 15.16).

Semi-continuous distance distributions, such as we get in systems doped with a spin probe (e.g., micelles), give echo decays almost without oscillations that are not easy to distinguish from the background signal. Here, specific models are most useful (Bode et al. 2011; Kattnig and Hinderberger 2013). These distributions, together with effects of excluded volume, are the subject of the next two sections.

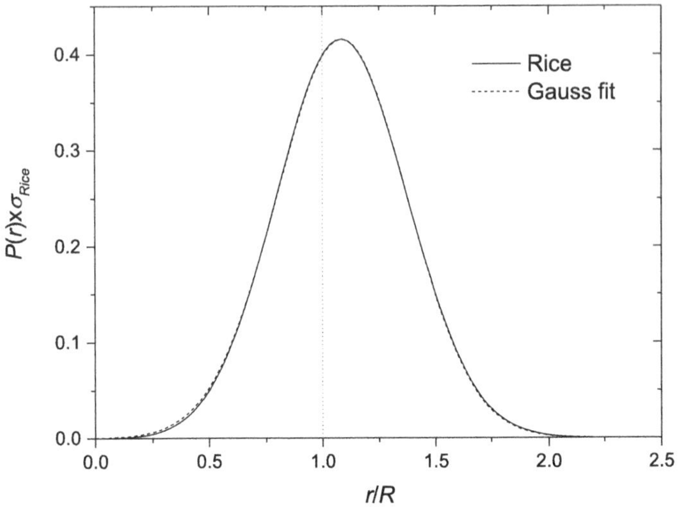

FIGURE 15.16 Rice distribution (Eq. 15.60; solid line) with $\sigma_{Rice} = 0.3R$, where R is distance between centres of two Gaussian spatial distributions. Dashed line: fit with Gaussian distribution in r, giving most probable distance $r = 1.086R$.

15.12 EXPLICIT DISTANCE DISTRIBUTIONS IN SHELLS OF UNIFORM DENSITY

We treat probability functions $P(r)$ for points separated by distance r, when the points distribute uniformly: (i) on a spherical surface, (ii) throughout a sphere, or (iii) in a spherical shell. These can be used, together with trial background decays, for conventional least-squares fitting of the echo time-domain signal.

15.12.1 Spherical Surface and Volume Distance-Distributions

On the surface of a sphere of radius R, the probability that two points are distance r apart, or less, is the surface area of the cap defined by the chord r from the pole, divided by the total surface area $4\pi R^2$ (see Figure 15.17). The area of the cap is πr^2, and the cumulative probability up to separation r is thus: $\int P(r)\cdot dr = r^2/4R^2$. Differentiating with respect to r, the distribution function for uniform density on the surface of a sphere therefore becomes:

$$P_A(r) = \frac{r}{2R^2} \tag{15.61}$$

where $r \leq 2R$ and the subscript A denotes surface area.

For a volume distribution in a sphere of radius R, we have:

$$P_V(r) = \frac{1}{\frac{4}{3}\pi R^3}\int_0^R 4\pi\rho^2 p(r;\rho)\cdot d\rho = \frac{3}{R^3}\int_0^R p(r;\rho)\rho^2\cdot d\rho \tag{15.62}$$

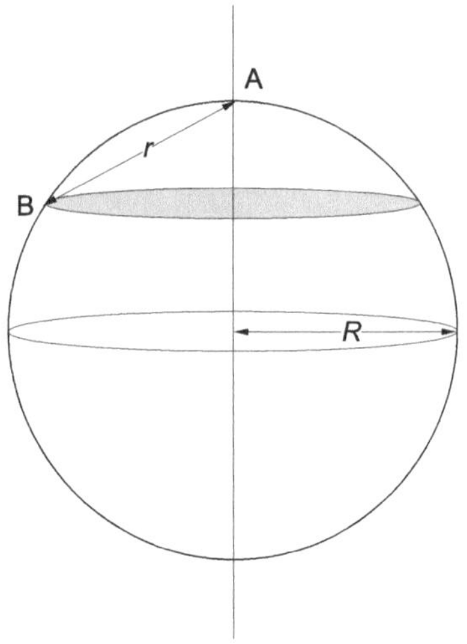

FIGURE 15.17 Uniform distribution of points on surface of sphere of radius R. Area of spherical cap is πr^2, where r is length of chord from pole to edge of cap.

where $p(r;\rho)$ is the conditional probability that points A and B are separated by r, when point A is at distance ρ from the centre of the sphere (see Figure 15.18). When $r \leq R$, we must distinguish between two ranges for ρ. For $\rho \leq R - r$, point B can take up all positions on the surface of a sphere of radius r that is centred on point A. The conditional probability is then simply:

$$p(r;\rho) = \frac{4\pi r^2}{\frac{4}{3}\pi R^3} = \frac{3r^2}{R^3} \tag{15.63}$$

For $R - r \leq \rho \leq R$, the shell of radius r intersects the sphere of radius R and the cap defined by the chord d is inaccessible to point B (see Figure 15.18 left). As in the previous example, the surface area of the inaccessible cap is $\pi d^2 = 2\pi r^2(1+\cos\theta)$,

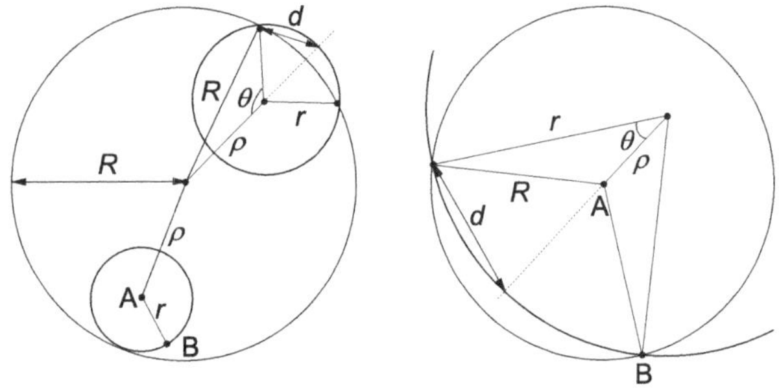

FIGURE 15.18 Uniform distribution of points throughout the volume of sphere of radius R. Point A is at radial position ρ, where $0 \leq \rho \leq R$. Point B is at distance r from point A. *Left panel* ($r \leq R$): $\rho \leq R - r$, point B lies on spherical surface of radius r totally enclosed in the sphere; $\rho \geq R - r$, spherical surface containing point B intersects sphere at angle θ, and spherical cap of area πd^2 is inaccessible to point B, where $d^2 = 2r^2(1+\cos\theta)$. *Right panel* ($R \leq r \leq 2R$): $\rho \leq r - R$, spherical surface of radius r always lies outside the sphere; $\rho \geq r - R$, spherical surface containing point B intersects sphere at angle θ, and only spherical cap of area πd^2 is accessible to point B, where $d^2 = 2r^2(1-\cos\theta)$.

where θ is the excluded angle. We must subtract this from the total area of the shell, $4\pi r^2$. The conditional probability then becomes (cf. Parry and Fischbach 2000):

$$p(r;\rho) = \frac{2\pi r^2(1-\cos\theta)}{\frac{4}{3}\pi R^3} = \frac{3r^2}{2R^3}\left(1 - \frac{\rho^2 + r^2 - R^2}{2\rho r}\right) \quad (15.64)$$

where we get $\cos\theta$ from the triangle that includes the angle θ. Performing the integration in Eq. 15.62 over the two ranges separately, we get the distance distribution:

$$P_V(r) = \frac{3r^2}{R^3} - \frac{9r^3}{4R^4} + \frac{3r^5}{16R^6} \quad (15.65)$$

for $r \le R$. The remaining range of values permitted for r is $R \le r \le 2R$. Here, point B would lie outside the sphere if $\rho \le R - r$; and ρ is allowed only the range $R - r \le \rho \le R$ (see Figure 15.18 right). The shell of radius r then always intersects the sphere at some point, and the conditional probability becomes the same as that given by Eq. 15.64 (Parry and Fischbach 2000). Integrating this over the new allowed range of ρ then produces a result identical to Eq. 15.65, which therefore holds for all separations: $0 \le r \le 2R$. Parry and Fischbach (2000) also treat distance distributions in uniform ellipsoids.

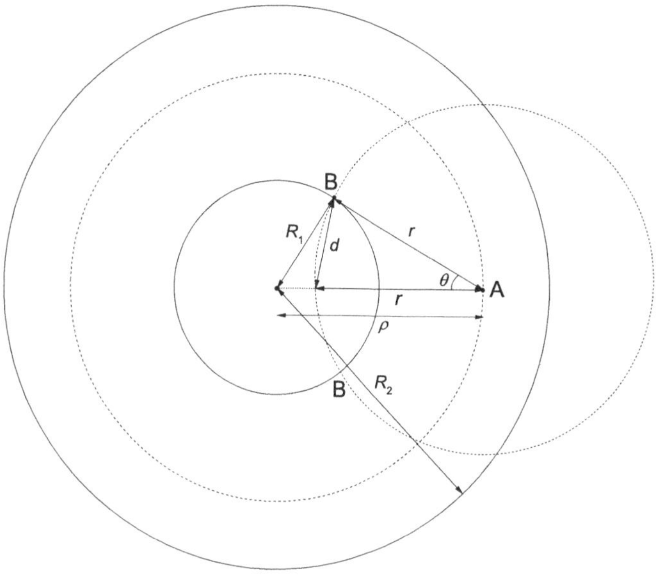

FIGURE 15.19 Distribution of distances r between external point A and points B at uniform density within sphere of radius R_1. ρ is distance of point A from centre of sphere. Probability $P(r)$ is proportional to area πd^2 of spherical cap at radius r, where $d^2 = 2r^2(1-\cos\theta)$. By combinatory arguments, we can extend the model to give distance distribution of points at uniform density in spherical shell of inner radius R_1 and outer radius R_2 (see text).

15.12.2 Sphere-within-Shell Distances

Kattnig and Hinderberger (2013) use a combinatory approach to calculate the distance distribution of points at uniform density in a spherical shell. We begin with a simpler situation shown in Figure 15.19. Points A are confined to a spherical shell of outer radius R_2 that surround a concentric inner sphere of radius R_1, to which points B are confined. The distribution of distances r between two different types of spins A and B (e.g., nitroxides and paramagnetic ions) could correspond to this situation. Consider a point A at distance ρ from the centre of the two spheres. We seek first the distribution of separations r from points B distributed homogeneously in the sphere. The probability is the area $\pi d^2 = 2\pi r^2(1-\cos\theta)$ of the spherical cap of radius r that lies within the sphere:

$$P_\rho(r) = \frac{\pi d^2}{\frac{4}{3}\pi R_1^3} = \frac{3r}{4R_1^3\rho}\left(R_1^2 - (\rho - r)^2\right) \quad (15.66)$$

where we get $\cos\theta$ from the larger triangle that includes angle θ (see Figure 15.19). Here, r is restricted to the range $\rho - R_1 \le r \le \rho + R_1$. Correspondingly, ρ is restricted to $r - R_1 \le \rho \le r + R_1$. Eq. 15.66 with $\rho = R_1$, immediately gives us the distance distribution when points A are restricted uniformly to the surface of the inner sphere.

For points A distributed uniformly on the spherical surface of radius ρ, we must multiply $P_\rho(r)$ by the surface area $4\pi\rho^2$. Then we get the distribution of distances between points A at uniform density in the shell and points B at the same density in the sphere, by integrating ρ over the shell thickness:

$$P_{sh,V}(r) = \frac{1}{\frac{4}{3}\pi\left(R_2^3 - R_1^3\right)} \int_{R_1}^{R_2} 4\pi\rho^2 P_\rho(r)\cdot \mathrm{d}\rho$$
$$= \frac{9r}{4\left(R_2^3 - R_1^3\right)R_1^3} \int_{\max(R_1, r-R_1)}^{\min(R_2, r+R_1)} \rho\left(R_1^2 - (\rho - r)^2\right)\cdot \mathrm{d}\rho \quad (15.67)$$

which is normalized by the volume of the shell. In substituting $P_\rho(r)$ to give the second identity, we take account of restrictions on the range of ρ in Eq. 15.66. This yields four separate integration ranges, depending on the value of r. Using the scaled distance $x = r/R_2$ and ratio of radii $f = R_1/R_2$, distributions for the different ranges become (Kattnig and Hinderberger 2013):

$$P_{sh,V}(x) = \frac{3}{16f^3\left(1-f^3\right)R_2} \times \left(12f^2x^3 - x^5\right)$$
$$0 \le x \le \min(2f, 1-f) \quad (15.68)$$

$$P_{sh,V}(x) = \frac{3}{16f^3\left(1-f^3\right)R_2}$$
$$\times\left(8\left(1-f^3\right)x^2 - 3\left(1-f^2\right)^2 x - 6\left(1-f^2\right)x^3\right)$$
$$\left(f \ge \tfrac{1}{3}\right) \quad 1 - f \le x \le 2f \quad (15.69)$$

$$P_{sh,V}(x) = \frac{3}{16f^3(1-f^3)R_2} \times 16f^3x^2$$

$$\left(f \le \tfrac{1}{3}\right) \quad 2f \le x \le 1-f \qquad (15.70)$$

$$P_{sh,V}(x) = \frac{3}{16f^3(1-f^3)R_2} \times \left(x^5 - 6(1+f^2)x^3 + 8(1+f^3)x^2 - 3(1-f^2)^2 x\right)$$

$$\max(1-f, 2f) \le x < 1+f \qquad (15.71)$$

For instance, Eq. 15.68 comes from integrating over the range $R_1 \le \rho \le r + R_1$ and Eq. 15.69 from the range $R_1 \le \rho \le R_2$. At distances greater than $x = 1 + f$, then $P_{sh,V}(x) = 0$.

15.12.3 Distance Distribution within Spherical Shell

Now we are in a position to construct the distance distribution for a shell by using Eq. 15.65, together with Eqs. 15.68–15.71, as building blocks. We divide a homogeneously filled sphere of radius R_2 into a sphere of radius $R_1 = fR_2$ and a shell of thickness $R_2 - R_1 = (1-f)R_2$. The probability that a given particle is located in the smaller sphere is the ratio of volumes f^3, and hence the probability to be located in the outer shell is $1 - f^3$. The probability that both particle A and particle B are located in the inner sphere is f^6. These particles have the distance distribution P_{V,R_1}, which is given by Eq. 15.65 with $R = R_1$. The probability that B is located in the sphere and A is in the shell is $f^3(1-f^3)$. These particles have the distance distribution $P_{sh,V}$, which is given by Eqs. 15.68–15.71. Building up in this way, we express the distance distribution within the whole of the outer sphere as:

$$P_{V,R_2} = f^6 P_{V,R_1} + 2f^3(1-f^3)P_{sh,V} + (1-f^3)^2 P_{sh} \qquad (15.72)$$

where $P_{sh}(r)$ is the distance distribution we require, i.e., for particles that are confined entirely to the shell. Note that we use the obvious symmetry: $P_{sh,V} \equiv P_{V,sh}$. Rearranging Eq. 15.72 we get the shell distribution:

$$P_{sh}(r) = \frac{P_{V,R_2} - f^6 P_{V,R_1}}{(1-f^3)^2} - \frac{2f^3 P_{sh,V}}{(1-f^3)} \qquad r < \max(2R_1, R_2 - R_1) \qquad (15.73)$$

in terms of distributions that we know already. Eq. 15.72, and hence Eq. 15.73, hold for distances $r < \max(2R_1, R_2 - R_1)$, i.e., $x < \max(2f, 1-f)$. Beyond this, the term P_{V,R_1} no longer contributes to Eq. 15.72, and Eq. 15.73 is replaced by:

$$P_{sh}(r) = \frac{P_{V,R_2}}{(1-f^3)^2} - \frac{2f^3 P_{sh,V}}{(1-f^3)} \qquad r > \max(2R_1, R_2 - R_1) \qquad (15.74)$$

Figure 15.20 shows interparticle distance distributions within spherical shells specified by the ratio $f = R_1/R_2$. For $f = 0$, this corresponds to a filled sphere of radius R_2 (Eq. 15.65) and for $f = 1$ to a spherical surface of radius R_2 (Eq. 15.61). For a shell of width equal to the inner radius, i.e., $f = 0.5$, the explicit expressions are relatively simple:

$$P_{sh}(x) = \frac{12}{49R_2}\left(14x^2 - 15x^3 + 2x^5\right) \qquad 0 \le x \le \tfrac{1}{2} \qquad (15.75)$$

$$P_{sh}(x) = \frac{81}{98R_2}x \qquad \tfrac{1}{2} \le x \le 1 \qquad (15.76)$$

$$P_{sh}(x) = \frac{3}{98R_2}\left(27x - 16x^2 + 24x^3 - 8x^5\right) \qquad 1 \le x \le \tfrac{3}{2} \qquad (15.77)$$

$$P_{sh}(x) = \frac{12}{49R_2}\left(16x^2 - 12x^3 + x^5\right) \qquad \tfrac{3}{2} \le x \le 2 \qquad (15.78)$$

These agree with the results of Tu and Fischbach (2002) for a similar two-shell model but where the two regions have different explicit uniform densities. Expressions that correspond to Eqs. 15.75–15.78 for general f are given in Appendix A15.1 at the end of the chapter.

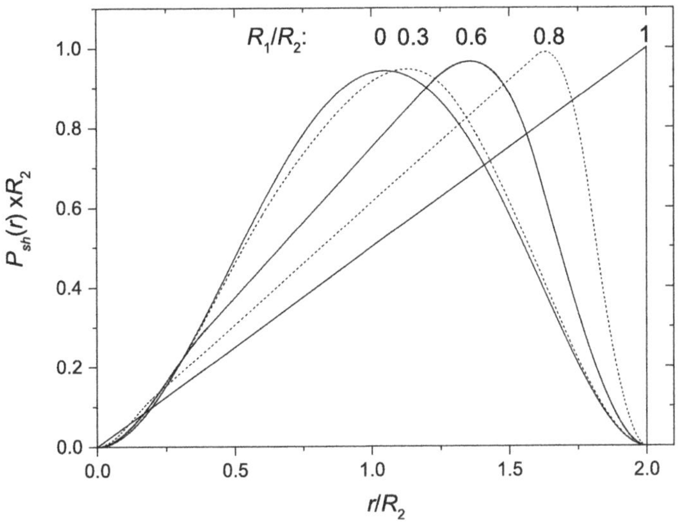

FIGURE 15.20 Distance distributions for particles at uniform density within spherical shells of outer and inner radii R_2 and R_1, respectively. From Eq. 15.73 $\left[\text{for } r/R_2 < \max(2R_1/R_2, 1 - R_1/R_2)\right]$ or Eq. 15.74 $\left[\text{for } r/R_2 > \max(2R_1/R_2, 1 - R_1/R_2)\right]$, and Eqs. 15.68–15.71, for different ratios R_1/R_2, as indicated. Surface distribution $(R_1/R_2 = 1)$ is given by Eq. 15.61; distribution within entire volume of sphere $(R_1/R_2 = 0)$ by Eq. 15.65.

15.13 EXCLUDED VOLUME IN BACKGROUND DEER SIGNALS

Excluded volume limits the closest distance of approach $R_{\min}$ (see Figure 15.21) and the maximum spread in dipolar frequencies. This affects mostly the dipolar evolution at short times. We can generalize the background decay rate from instantaneous diffusion given by $k_{\text{inter}}t$ in Eq. 15.43 with the time-dependent function $k_{\text{inter}}(t)$ (Kattnig et al. 2013 and cf. Eq. 15.55):

$$B(t) = \exp\left(-n_{SL} f_B k_{\text{inter}}(t)\right) \tag{15.79}$$

From Figure 15.21, we see that the decay function is the radial integral:

$$k_{\text{inter}}(t) = 4\pi \int_{R_{\min}}^{\infty} K(r,t) r^2 \cdot \mathrm{d}r$$

$$= \frac{4\pi}{3}\left(\frac{\mu_o}{4\pi}\right)\frac{g^2\beta_e^2}{\hbar} t \int_0^{u_{\max}} \frac{K(u)}{u^2} \cdot \mathrm{d}u \tag{15.80}$$

where the kernel $K(r,t)$ is given by Eq. 15.56. In the second equality, we make the substitution:

$$u(r,t) = \left(\frac{\mu_o}{4\pi}\right)\frac{g^2\beta_e^2}{\hbar}\frac{t}{r^3} \tag{15.81}$$

and $u = u_{\max}$ is the value for $r = R_{\min}$. Correspondingly, the kernel becomes:

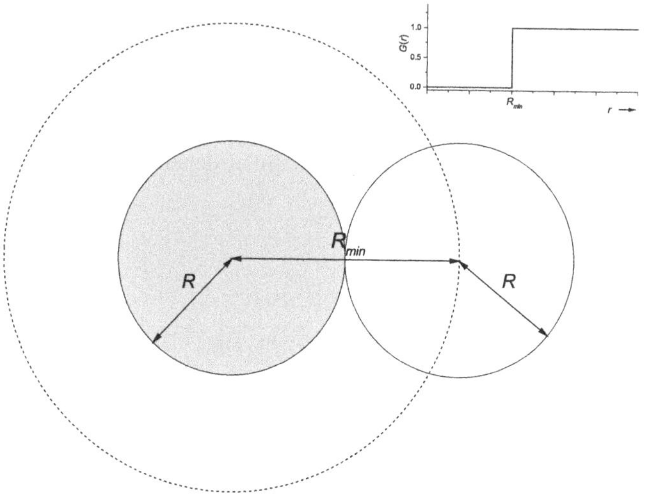

FIGURE 15.21 Excluded volume for nitroxides located at centres of spheres of diameter R. Distance of closest approach is $r = R_{\min}$, and excluded volume is $V_{excl} = \frac{4}{3}\pi R_{\min}^3$. *Inset*: radial distribution function $G(r)$ for spin separation r.

$$K(u) = \int_0^{\pi/2} \left[1 - \cos\left(u(1 - 3\cos^2\theta)\right)\right] \sin\theta \mathrm{d}\theta$$

$$= 2\int_0^1 \sin^2\left(\tfrac{1}{2}u(1-3x^2)\right)\mathrm{d}x \tag{15.82}$$

where $x = \cos\theta$, and we introduce the trignometric identity for half angles. After substituting in Eq. 15.80 and integrating first over u, the decay rate becomes:

$$k_{\text{inter}}(t) = \frac{8\pi}{3}\left(\frac{\mu_o}{4\pi}\right)\frac{g^2\beta_e^2}{\hbar} t \int_0^1 \int_0^{u_{\max}} \frac{\sin^2\left(\frac{1}{2}u\left(1-3x^2\right)\right)}{u^2} \cdot \mathrm{d}u \cdot \mathrm{d}x \tag{15.83}$$

The standard form for the u-integral is:

$$\int_0^q \frac{\sin^2(az)}{z^2} \cdot \mathrm{d}z = a\mathrm{Si}(2aq) - \frac{\sin^2 aq}{q} \tag{15.84}$$

where the sine integral is defined by $\mathrm{Si}(t) = \int_0^t (\sin x)/x \cdot \mathrm{d}x$. Thus from Eq. 15.83, the decay function becomes:

$$k_{\text{inter}}(t) = \frac{4\pi}{3}\left(\frac{\mu_o}{4\pi}\right)\frac{g^2\beta_e^2}{\hbar} t \times \left[\int_0^1 \left(1-3x^2\right)\mathrm{Si}\left(\tfrac{1}{2}d\left(1-3x^2\right)\right) \cdot \mathrm{d}x - \frac{K(u_{\max})}{u_{\max}}\right] \tag{15.85}$$

where we substitute Eq. 15.82 for the second integral. We evaluate the kernel by using Fresnel integrals (Milov et al. 1998; Kattnig et al. 2013):

$$K(u) = 1 - \sqrt{\frac{\pi}{6u}}\left(\cos u \times C\left(\sqrt{\frac{6u}{\pi}}\right) + \sin u \times S\left(\sqrt{\frac{6u}{\pi}}\right)\right) \tag{15.86}$$

where $C(t) = \int_0^t \cos\left(\tfrac{1}{2}\pi x^2\right) \cdot \mathrm{d}x$ and $S(t) = \int_0^t \sin\left(\tfrac{1}{2}\pi x^2\right) \cdot \mathrm{d}x$.

Numerical values of the combined integral in Eq. 15.85 are listed in Appendix A15.2 at the end of the chapter.

Figure 15.22 shows the dependence of the normalized decay rate α on the excluded volume parameter $u_{\max}$. The normalized rate is defined by:

$$k_{\text{inter}}(t) = \frac{8\pi^2}{9\sqrt{3}}\left(\frac{\mu_o}{4\pi}\right)\frac{g^2\beta_e^2}{\hbar} t \times \alpha(u_{\max}) \tag{15.87}$$

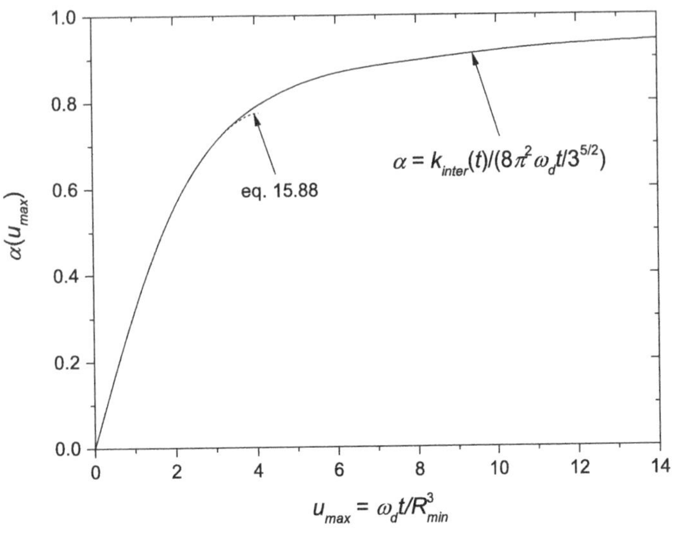

FIGURE 15.22 Factor $\alpha(u_{max})$ by which background decay rate constant $k_{inter} = (8\pi^2/9\sqrt{3})\omega_d$ is reduced, as function of excluded volume parameter $u_{max} = \omega_d t/R^3_{min}$, where $\hbar\omega_d \equiv (\mu_o/4\pi)g^2\beta_e^2$. *Solid line*: $\alpha(u_{max})$ from Eqs. 15.87 and 15.85; *dotted line*: approximate Eq. 15.88. Numerical data from Kattnig et al. (2013).

This yields Eq. 15.45 (where excluded volume is neglected) when $\alpha = 1$. As we see from Figure 15.22, $\alpha(u_{max})$ tends asymptotically to unity as u_{max} increases, i.e., as R_{min} decreases.

Kattnig et al. (2013) give the limiting value of the decay function at short times:

$$k_{inter}(t) = \frac{4\pi}{3}\left(\frac{\mu_o}{4\pi}\right)\frac{g^2\beta_e^2}{\hbar}t\left(\frac{2}{5}u_{max} - \frac{2}{105}u^3_{max} + \frac{212}{225,225}u^5_{max} - \frac{38}{1,072,071}u^7_{max} + \frac{92}{93,532,725}u^9_{max} - \cdots\right) \quad (15.88)$$

where all even powers of u_{max}, i.e., odd powers of t, vanish. Eq. 15.88 shows that excluded volume has its greatest effect at short times, because $dk_{inter}/dt = 0$ at $t = 0$. Furthermore, $k_{inter}(t) \propto t^2$ in this region. This low-t approximation is indicated by the dotted line in Figure 15.22, which holds up to $u_{max} \approx 3.8$. At the other extreme, the limiting value for very long times is (Kattnig et al. 2013):

$$k_{inter}(t) = \frac{4\pi}{3}\left(\frac{\mu_o}{4\pi}\right)\frac{g^2\beta_e^2}{\hbar}t - \frac{4\pi}{3}R^3_{min} \quad (15.89)$$

Thus, for $t \to \infty$ the background decay rate is $k_{inter}(t) \propto t$, as in Eq. 15.43 from the classical approach, but with the absolute rate shifted by the covolume $\frac{4}{3}\pi R^3_{min}$. A stretched exponential therefore does not represent excluded volume effects particularly well. The effective dimensionality varies from $D = 6$ at short times to $D = 3$ at long times, which at least explains background fits that give effective values of $D > 3$.

Kattnig et al. (2013) have explored the effects of crowding and offset of the nitroxide position from the centres of the impinging spheres. Both of these attenuate the effects of excluded volume by decreasing the departures from a simple exponential decay at short times. This possibly accounts for some of the apparent success found in fitting the background with a single exponential.

15.14 ORIENTATION SELECTION IN DEER SIGNALS

In choosing a given position in the nitroxide powder lineshape for the DEER excitation pulses, we select particular orientations of the two interacting nitroxide spins. So far, we assumed that the orientation of the nitroxide axes is uncorrelated with that of the interspin vector. This is likely for flexible biradicals, particularly if the nitroxide is attached flexibly to the molecular backbone. If the biradical is rigid, however, the nitroxide axes have a fixed orientation to the interspin vector. The dipolar interaction recorded in DEER then depends on the position in the nitroxide powder pattern that we choose for the observer frequency. For instance, assume that the nitroxide z-axis is perpendicular to the interspin vector. Observing in the outer edges of the spectrum, where the nitroxide z-axis is parallel to the magnetic field, would then give the dipolar spectrum for $\theta_{AB} = 90°$ with the interspin axis perpendicular to the field. Correspondingly, if the nitroxide z-axis lies along the interspin vector, we get the dipolar spectrum for $\theta_{AB} = 0°$. Clearly, we must take account of any orientation selection when determining interspin distances, otherwise we would get an incorrect distribution. Characteristically, sharp features in a distance distribution can arise artefactually from orientation selection. An important control is to check that the apparent distance distribution does not change with position of the observer frequency in the powder spectrum.

Conversely, for rigid biradicals, analysis of the positional dependence of the observer frequency can give useful information on molecular orientation and configuration. In Eq. 15.48 of Section 15.11, the individual modulation amplitudes $f_{B,p}$ contributing to the DEER signal now depend both on the orientation θ of the interspin vector to the magnetic field $\mathbf{B_o}$, and on the fixed orientations (x,y,z) of the nitroxide axes to the interspin vector $\mathbf{r}_{AB}$ (see Figure 15.23). Including this angular dependence, Eq. 15.49 then becomes:

$$E_{DEER,pair}(t) = 1 - \int_{r_{min}}^{r_{max}} P(r)\left(\int_0^{\pi/2} f_B(\theta)\left[1 - \cos(\omega_{AB}(r,\theta)t)\right]\sin\theta\, d\theta\right)dr \quad (15.90)$$

where the fraction of spins inverted $f_B(\theta)$ is given by Eq. 15.40, and we have not yet averaged over the A spins. Orientation selection affects $f_B(\theta)$ because the Rabi frequency $\omega_{eff,B}$ depends on the resonance offset $\Delta\omega_B$. We therefore see that both modulation period and modulation depth of the echo amplitude depend on the orientation of the nitroxide axes, because these determine

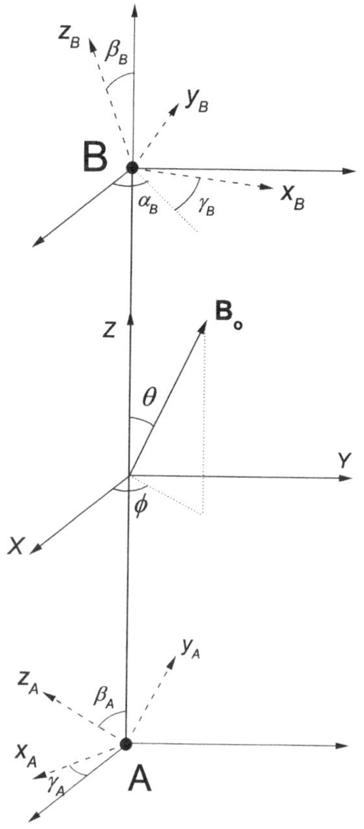

FIGURE 15.23 Orientation of nitroxide coordinate systems, (x_A, y_A, z_A) and (x_B, y_B, z_B), and dipolar axis system (X,Y,Z); for an A–B spin pair. $(\alpha_A, \beta_A, \gamma_A)$ and $(\alpha_B, \beta_B, \gamma_B)$ are Euler angles of nitroxide systems, relative to dipolar axes. (θ,ϕ) are polar angles of magnetic field $\mathbf{B}_o$ in dipolar axis system.

the resonance frequency at different positions of excitation throughout the powder lineshape.

Equation 15.90 for the dipolar pairs refers to a single A spin. Now we must average over all A spins, which we do by using Eq. 15.42 for the fractional magnetization $f_A(\theta)$ excited by the detection sequence. Because the labels A and B are arbitrary, in that both have the same powder pattern, the total refocused echo magnetization becomes (Polyhach et al. 2007; Marko et al. 2009):

$$\lambda(\theta) = \tfrac{1}{2}\left(f_A(\theta,\omega_{obs}) f_B(\theta,\omega_{pump}) + f_A(\theta,\omega_{pump}) f_B(\theta,\omega_{obs})\right) \tag{15.91}$$

which is the orientational intensity function, represented here by $\lambda(\theta)$. The second term on the right is the fraction of A spins that are excited by the pump pulse, multiplied by the fraction of B spins that are inverted by the observer sequence. We now rewrite Eq. 15.54 for the modulated part of the echo by changing the order of integration:

$$f(t) = \int_0^{\pi/2} \lambda(\theta)\left(\int_{r_{min}}^{r_{max}} \cos(\omega_{AB}(r,\theta)t) P_{norm}(r)\,dr\right)\sin\theta\,d\theta \tag{15.92}$$

because here we treat a rigid biradical, and therefore the angular distribution is more significant than the radial distribution. This assumes that we can average independently over r, and over θ (see Marko et al. 2009). Instead of the integral Eq. 15.55, we now get:

$$f(t) = \int_0^1 K(z,t)\lambda(z)\,dz \tag{15.93}$$

where $z = \cos\theta$, and the new kernel is:

$$K(z,t) = \int_{r_{min}}^{r_{max}} \cos\left(\left(1-3z^2\right)\frac{\mu_o}{4\pi}\frac{g^2\beta_e^2}{\hbar r^3}t\right) P_{norm}(r)\cdot dr \tag{15.94}$$

The kernel requires either a model for the radial distribution function $P_{norm}(r)$ or evaluating the raw distance distribution by Tikhonov regularization as described in Section 15.11. In the latter case, we minimize orientational bias in the experimental data by averaging over DEER time traces determined at several different observer positions (Marko et al. 2009; Marko et al. 2010). We then get the orientational intensity function $\lambda(\cos\theta)$ by using this kernel in Eq. 15.93 and again employing Tikhonov regularization as in Section 15.11.

To determine the orientation function by using Eqs. 15.42 and 15.40 for f_A and f_B, we need the angular dependence of the resonance frequencies ω_A and ω_B for electron spins A and B of the dipolar pair. The angular resonance frequency of the m_I-hyperfine manifold from each spin is given by:

$$\hbar\omega_o(\theta,\phi;\alpha\beta\gamma) = g_{eff}(\theta,\phi;\alpha\beta\gamma)\beta_e B_o + A_{eff}(\theta,\phi;\alpha\beta\gamma)m_I \tag{15.95}$$

where (θ,ϕ) are the polar angles of the static field B_o, relative to the interspin vector, and $(\alpha\beta\gamma)$ are the Euler angles of the A or B nitroxide axes, relative to the same dipolar axis system (see Figure 15.23). The angular-dependent effective hyperfine coupling A_{eff} in Eq. 15.95 is related to the principal hyperfine tensor components (A_{xx}, A_{yy}, A_{zz}) by (Marko et al. 2009):

$$\begin{aligned} A_{eff}^2(\theta,\phi;\alpha\beta\gamma) = A_{xx}^2 &+ \left(A_{yy}^2 - A_{xx}^2\right)(\sin\theta\sin\gamma\cos(\alpha-\phi) \\ &+ \sin\theta\cos\gamma\sin(\alpha-\phi) + \cos\theta\cos\gamma\sin\beta)^2 \\ &+ \left(A_{zz}^2 - A_{xx}^2\right)(\sin\theta\sin\beta\sin(\alpha-\phi) + \cos\theta\cos\beta)^2 \end{aligned} \tag{15.96}$$

and similarly for the effective g-value g_{eff}. Near-axiality of the hyperfine tensor and smallness of the Zeeman anisotropy at 9 GHz mean that we can put $\gamma = 0$ in Eq. 15.96. In fact, only the three Euler angles β_A, α_B and β_B (see Figure 15.23) cause significant changes in the DEER signals at 9 GHz (Marko and Prisner 2013).

Clearly, we should get more detailed orientation information from EPR at operating frequencies higher than 9 GHz. This is illustrated in Figure 15.24. At an EPR frequency of 9 GHz,

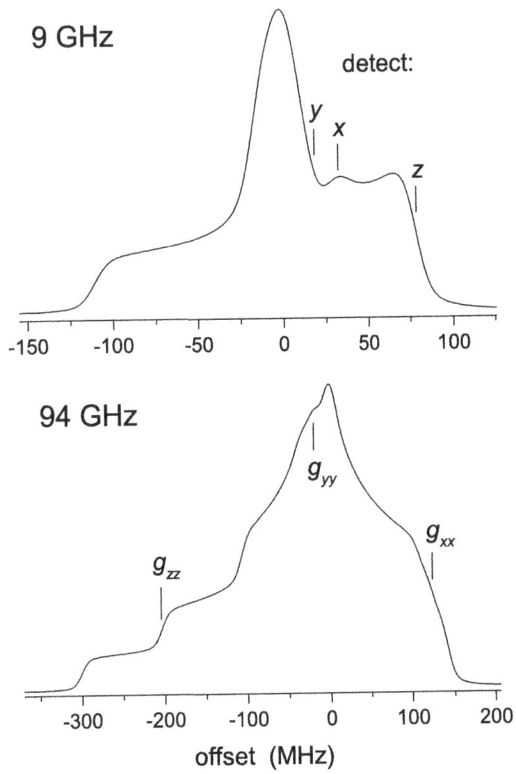

FIGURE 15.24 *Top*: observer positions in $m_I = +1$ manifold that correspond to magnetic field directed along nitroxide *x*-, *y*-, *z*-axes in a 9-GHz powder pattern. Pump pulse is centred on absorption maximum of $m_I = 0$ manifold (see Figure 15.12). *Bottom*: 94-GHz echo-detected absorption powder pattern. g_{xx}, g_{yy} and g_{zz} positions correspond to $m_I = 0$ manifold with magnetic field directed along nitroxide *x*-, *y*- and *z*-axes, respectively. Various combinations of peaks or shoulders from different manifolds allow pairing of pump and detection pulses for same or different nitroxide directions (see Reginsson et al., 2012).

observer frequencies that correspond to the magnetic field oriented along the *x*- and *y*-nitroxide axes of the $m_I = +1$ hyperfine manifold already overlap with the central $m_I = 0$ manifold, where we apply the inverting pump pulse. At ten times higher EPR frequency of 94 GHz, the much wider spectral spread from *g*-value anisotropy allows a more varied choice of pump and observer positions. With appropriate combination of magnetic field and offset frequencies, spin pairs with various differently oriented nitroxide axes can be selected (Reginsson et al. 2012; Polyhach et al. 2007).

Figure 15.25 gives a highly schematic illustration of orientation-selective DEER experiments. Idealized time traces are shown for observer frequencies chosen such that, in a rigid biradical, the interspin vector is inclined at angle θ to the magnetic field direction. When we select $\theta = 0$, the modulation period is longer than for $\theta = 90°$; and the modulation disappears completely at the magic angle $\cos^2\theta = \frac{1}{3}$. Also, the amplitude of the modulation depends on orientational selection.

In general, we get information on molecular orientation by detailed fitting and comparison with libraries of non-degenerate

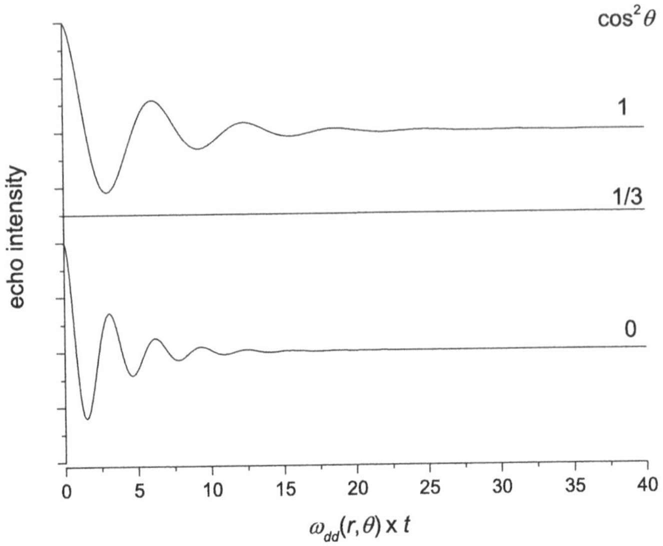

FIGURE 15.25 Schematic orientation-selective DEER. Observer frequency chosen such that interspin vector is oriented at angle $\theta = 0$, 54.7° and 90° (top to bottom), to static magnetic field. For rigid biradical with nitroxide *z*-axis parallel to interspin vector, $\cos^2\theta = 1$ corresponds to outer edge of 9-GHz nitroxide powder pattern.

DEER time traces, which are obtained by simulation according to Eqs. 15.93–15.96 (Abdullin et al. 2015; Tkach et al. 2013). However, Marko et al. (2010) describe an analytical solution for the case where the two nitroxide *z*-axes are parallel, i.e., $\langle\beta_A\rangle = \langle\beta_B\rangle \equiv \beta$, in 9-GHz DEER. For illustration, we concentrate on the $\theta = 0$ orientation of the magnetic field to the interspin vector, because then the azimuthal orientation ϕ does not enter. Assuming that Gaussian inhomogeneous broadening is larger than the excitation width of the detection sequence, Marko et al. (2010) find that the modulation depth Δ for $\theta = 0$ has a Gaussian dependence on observer frequency ω_{obs} at fixed magnetic field B_o:

$$\Delta(\omega_{obs}) = \lambda(\theta = 0, \beta) \times g(\omega_A) \propto \exp\left(-\frac{(\omega_{obs} - \omega_A)^2}{2\sigma^2}\right) \tag{15.97}$$

where $g(\omega_A)$ is the nitroxide powder lineshape, $\omega_A(B_o)$ is the resonance frequency of the A-spins whose transverse magnetization is determined and σ is the standard deviation of the Gaussian. We therefore get the maximum amplitude at the observer frequency where: $\omega_{obs} = \omega_A(B_o)$. From Eqs. 15.95 and 15.96, this is given by:

$$\begin{aligned}\hbar\omega_{obs}^{max} &= g_{eff}(\theta = 0; \beta)\beta_e B_o + A_{eff}(\theta = 0; \beta) \\ &= \left(g_{xx} + (g_{zz} - g_{xx})\cos^2\beta\right)\beta_e B_o + \sqrt{A_{xx}^2 + \left(A_{zz}^2 - A_{xx}^2\right)\cos^2\beta}\end{aligned} \tag{15.98}$$

where we use the intermediate-field approximation for the *g*-value term (see Section 2.2 and Eq. 2.11).

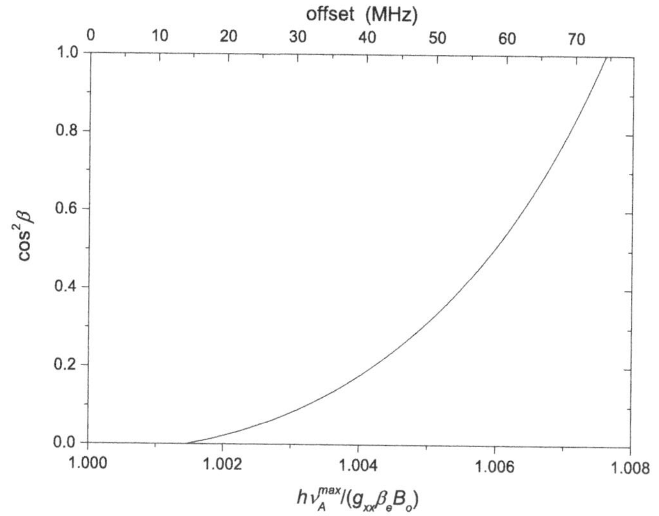

FIGURE 15.26 Dependence of orientation β of nitroxide z-axis to interspin vector, on observer frequency ν_{obs}^{max} that gives maximum modulation amplitude Δ when $\theta = 0$ (magnetic field $\mathbf{B_o}$ parallel to interspin vector). See Eqs. 15.97 and 15.98, for 9-GHz DEER. Offset frequency is referred to x-turning point of $m_I = +1$ manifold: $\nu_{xx} = g_{xx}\beta_e B_o/h$.

We get the orientational intensity functions $\lambda(\cos\theta)$ from Eqs. 15.93 and 15.94 as described above, and from these select the values at $\theta = 0$. Together with the echo powder lineshape $g(\omega_A)$, we then get the modulation depths $\Delta(\omega_{obs})$ according to Eq. 15.97. In this way, (Marko et al. 2010) find that 9-GHz DEER with pairs of rigidly labelled cytidine analogues in DNA produces Gaussian dependences on ω_{obs}. We then get the nitroxide orientation β by solving Eq. 15.98, which is quadratic in $\cos^2\beta$. Figure 15.26 shows the dependence of $\cos^2\beta$ on the normalized observer frequency $\omega_{obs}^{max} = 2\pi\nu_{obs}^{max}$. The offset shown in Figure 15.26 is given relative to frequency $h\nu_A = g_{xx}\beta_e B_o$. This in turn is offset from the pump frequency $h\nu_B$ at the absorption maximum of the powder pattern by approximately A_{xx} (ca. 14 MHz).

15.15 CONCLUDING SUMMARY

1. *Dipolar splittings for like spins:*
   For strong dipolar coupling, the resonance field of two *like* electron spins is split by:

$$2\Delta B_{dd}^{\perp} = \frac{3}{2}\cdot\frac{\mu_o}{4\pi}\left(\frac{g\beta_e}{r_{12}^3}\right) \tag{15.99}$$

   when $\mathbf{r}_{12}$ is perpendicular to the static magnetic field, and $(\mu_o/4\pi)g\beta_e = 1.860$ mT nm³. Here, the dipolar splitting must be greater than the sum of hyperfine and Zeeman anisotropies and is measured from the extrema, or turning points, in the EPR powder spectrum. This condition applies for short distances: $r_{12} \le 0.66$ nm.

2. *Dipolar splittings for unlike spins:*
   For weak dipolar coupling, the dipolar splitting of two *unlike* electron spins is:

$$2\Delta B_{dd}^{\perp} = \frac{\mu_o}{4\pi}\left(\frac{g\beta_e}{r_{12}^3}\right) \tag{15.100}$$

   when $\mathbf{r}_{12}$ is perpendicular to the static magnetic field. In a Pake powder pattern, $2\Delta B_{dd}^{\perp}$ is the separation between the perpendicular discontinuities. Dipolar splittings are extracted either by convolution/deconvolution of the nitroxide powder pattern or from evaluating moments of the powder lineshape. Assuming *unlike* spins is the default situation, whenever dipolar splittings are appreciably narrower than the single-nitroxide powder pattern (i.e., for $r_{12} >> 0.66$ nm).

3. *Mean splittings and moments for distributions:*
   The mean splitting, or absolute-value first moment, of the dipolar powder lineshape is given by:

$$\langle 2|\Delta B_{dd}|\rangle = \frac{4}{3\sqrt{3}}\left(\frac{\mu_o}{4\pi}\right)g\beta_e\left\langle\frac{1}{r_{12}^3}\right\rangle \tag{15.101}$$

   for *unlike* spins. The second moment of the dipolar powder lineshape from *unlike* spins is:

$$\langle \Delta B_{dd}^2\rangle = \frac{1}{5}\left(\frac{\mu_o}{4\pi}\right)^2 g^2\beta_e^2\left\langle\frac{1}{r_{12}^6}\right\rangle \tag{15.102}$$

   which depends on an average over the distance distribution different from that for the absolute-value first moment. Both are biased to shorter distances than the true mean value $\langle r_{12}\rangle$, the shift being larger for the second moment.

4. *Pulse ELDOR for long distances:*
   Long distances (1.8–8 nm) are determined by pulse ELDOR (or DEER). We eliminate inhomogeneous broadening, other than that from electron dipolar interaction, by using a refocused-echo detection sequence. After removing background interaction with distant spins, the modulated part of the time-domain echo signal is:

$$f(t) = \int_{r_{min}}^{r_{max}} K(r,t)P_{norm}(r)\,dr \tag{15.103}$$

   with kernel:

$$K(r,t) = \int_0^1 \cos\left(\left(1-3z^2\right)\frac{\mu_o}{4\pi}\frac{g^2\beta_e^2}{\hbar r^3}t\right)dz \tag{15.104}$$

   where $z = \cos\theta$. We determine the normalized interspin distance distribution $P_{norm}(r_{12})$ either by Tikhonov regularization, or by fitting with a sum of Gaussians.

Regularization is a compromise between broadening narrow components and introducing artefactual spikes in broad components. Fitting has the problem of uniqueness, and of finding a global minimum, but includes the background signal in the fitting procedure. For rigid systems, we check whether orientation selection distorts the distance distribution by using different positions within the powder lineshape for the observer pulse sequence, and if necessary, we average over these.

APPENDIX A15.1: GENERAL EXPRESSIONS FOR DISTANCE DISTRIBUTION IN SPHERICAL SHELLS

The probability distribution for the normalized distance $x = r/R_2$ between two points at uniform density in a spherical shell of inner and outer radii R_1 and R_2 comes from Eqs. 15.65, 15.68–15.71, 15.73 and 15.74. With general values of the ratio $f = R_1/R_2$, the following are the equivalents of Eqs. 15.75–15.78 (which are specifically for $f = \frac{1}{2}$):

$$P_{sh}(x) = \frac{3}{8\left(1-f^3\right)^2 R_2}\left(8\left(1-f^3\right)x^2 - 6\left(1+f^2\right)x^3 + x^5\right)$$

$$0 \le x \le \min(2f, 1-f) \qquad \text{(A15.1)}$$

$$P_{sh}(x) = \frac{9\left(1-f^2\right)^2}{8\left(1-f^3\right)^2 R_2}x \quad (f \ge 1/3) \quad 1-f \le x \le 2f \qquad \text{(A15.2)}$$

$$P_{sh}(x) = \frac{3}{16\left(1-f^3\right)^2 R_2}\left(16\left(1-2f^3\right)x^2 - 12x^3 + x^5\right)$$

$$(f \le 1/3) \quad 2f \le x \le 1-f \qquad \text{(A15.3)}$$

$$P_{sh}(x) = \frac{3}{16\left(1-f^3\right)^2 R_2}\left(6\left(1-f^2\right)^2 x - 16f^3x^2 + 12f^2x^3 - x^5\right)$$

$$\max(2f, 1-f) \le x \le 1+f \qquad \text{(A15.4)}$$

$$P_{sh}(x) = \frac{3}{16\left(1-f^3\right)^2 R_2}\left(16x^2 - 12x^3 + x^5\right)$$

$$1+f \le x \le 2 \qquad \text{(A15.5)}$$

where Eq. 15.73 is used for Eqs. A15.1 and A15.2, and Eq. 15.74 is used for Eqs. A15.3–A.15.5.

APPENDIX A15.2: DECAY FUNCTION α FOR EXCLUDED VOLUME (Eq. 15.87)

u_{max}	α
0	0.00000000000000000
0.15625	0.05162706629556498
0.3125	0.10289575864793842
0.625	0.20297572751410472
0.9375	0.29768857369338311
1.25	0.38497223190865181
1.5625	0.46340692632428295
1.875	0.53227764081851469
2.1875	0.59154366480800064
2.5	0.64172880600106878
3.125	0.71877399707291740
3.75	0.77241540554356620
4.375	0.81056539532051426
5	0.83814761550748048
6.25	0.87152574963314597
6.875	0.88136170843399123
7.5	0.88929899054375268
8.125	0.89662337341543504
8.75	0.90375014193053403
10	0.91679121736453198
11.25	0.92715281382076650
12.5	0.93451470664197596
13.75	0.93970817029148796
15	0.94439511732196284
17.5	0.95297886136061277
18.75	0.95613625951642133
20	0.95858908097957250
21.25	0.96088237885312859
22.5	0.96318267190245481
25	0.96703857507138751
27.5	0.96982178852619641
30	0.97249562098761656
32.5	0.97453759156555443
35	0.97634932000970439
40	0.97928361985960965
45	0.98161658730029958
50	0.98348189110936179

See Kattnig et al. (2013). The function $\alpha(u_{max})$ is defined by Eqs. 15.85 and 15.87, and $u_{max} = \left(\mu_o/4\pi\right)\left(g^2\beta_e^2/\hbar\right)t/R_{min}^3$.

REFERENCES

Abdullin, D., Hagelueken, G., Hunter, R.I., Smith, G.M., and Schiemann, O. 2015. Geometric model-based algorithm for orientation-selective PELDOR data. *Mol. Phys.* 113:544–560.

Abragam, A. 1961. *The Principles of Nuclear Magnetism*, Oxford: Oxford University Press.

Abragam, A. and Bleaney, B. 1970. *Electron Paramagnetic Resonance of Transition Ions*, Oxford: Oxford University Press.

Altenbach, C., Oh, K.J., Trabanino, R.J., Hideg, K., and Hubbell, W.L. 2001. Estimation of inter-residue distances in spin labeled proteins at physiological temperatures: experimental strategies and practical limitations. *Biochemistry* 40:15471–15482.

Banham, J.E., Baker, C.M., Ceola, S. et al. 2008. Distance measurements in the borderline region of applicability of CW EPR and DEER: a model study on a homologous series of spin-labelled peptides. *J. Magn. Reson.* 191:202–218.

Bode, B.E., Margraf, D., Plackmeyer, J. et al. 2007. Counting the monomers in nanometer-sized oligomers by pulsed electron-electron double resonance. *J. Am. Chem. Soc.* 129:6736–6745.

Bode, B.E., Dastvan, R., and Prisner, T.F. 2011. Pulsed electron-electron double resonance (PELDOR) distance measurements in detergent micelles. *J. Magn. Reson.* 211:11–17.

Bowman, M.K., Maryasov, A.G., Kim, N., and DeRose, V.J. 2004. Visualization of distance distribution from pulsed double electron-electron resonance data. *Appl. Magn. Reson.* 26:23–39.

Brandon, S., Beth, A.H., and Hustedt, E.J. 2012. The global analysis of DEER data. *J. Magn. Reson.* 218:93–104.

Chandrasekhar, S. 1943. Stochastic problems in physics and astronomy. *Rev. Mod. Physics* 15:1–89.

Chiang, Y.W., Borbat, P.P., and Freed, J.H. 2005. The determination of pair distance distributions by pulsed ESR using Tikhonov regularization. *J. Magn. Reson.* 172:279–295.

El Mkami, H., Ward, R., Bowman, A., Owen-Hughes, T., and Norman, D.G. 2014. The spatial effect of protein deuteration on nitroxide spin-label relaxation: implications for EPR distance measurements. *J. Magn. Reson.* 248:36–41.

Fajer, P.G., Brown, L., and Song, L. 2007. Practical pulsed dipolar ESR (DEER). In *ESR Spectroscopy in Membrane Biophysics*. Biological Magnetic Resonance, Vol. 27, eds. Hemminga, M.A. and Berliner, L.J., 95–128. New York: Springer.

Hustedt, E.J., Stein, R.A., Sethaphong, L. et al. 2006. Dipolar coupling between nitroxide spin labels: the development and application of a tether-in-a-cone model. *Biophys. J.* 90:340–356.

Jeschke, G. 2002. Determination of the nanostructure of polymer materials by electron paramagnetic resonance spectroscopy. *Macromol. Rapid Commun.* 23:227–246.

Jeschke, G. and Polyhach, Y. 2007. Distance measurements on spin-labelled biomacromolecules by pulsed electron paramagnetic resonance. *Phys. Chem. Chem. Phys.* 9:1895–1910.

Jeschke, G., Bender, A., Paulsen, H., Zimmermann, H., and Godt, A. 2004a. Sensitivity enhancement in pulse EPR distance measurements. *J. Magn. Reson.* 169:1–12.

Jeschke, G., Panek, G., Godt, A., Bender, A., and Paulsen, H. 2004b. Data analysis procedures for pulse ELDOR measurements of broad distance distributions. *Appl. Magn. Reson.* 26:223–244.

Jeschke, G., Chechik, V., Ioita, P. et al. 2006. DeerAnalysis2006- a comprehensive software package for analyzing pulsed ELDOR data. *Appl. Magn. Reson.* 30:473–498.

Jeschke, G., Sajid, M., Schulte, M., and Godt, A. 2009. Three-spin correlations in double electron-electron resonance. *Phys. Chem. Chem. Phys.* 11:6580–6591.

Kattnig, D.R. and Hinderberger, D. 2013. Analytical distance distributions in systems of spherical symmetry with applications to double electron-electron resonance. *J. Magn. Reson.* 230:50–63.

Kattnig, D.R., Reichenwallner, J., and Hinderberger, D. 2013. Modeling excluded volume effects for the faithful description of the background signal in double electron-electron resonance. *J. Phys. Chem. B* 117:16542–16557.

Kohler, S.D., Spitzbarth, M., Diederichs, K., Exner, T.E., and Drescher, M. 2011. A short note on the analysis of distance measurements by electron paramagnetic resonance. *J. Magn. Reson.* 208:167–170.

Larsen, R.G. and Singel, D.J. 1993. Double electron-electron resonance spin-echo modulation - spectroscopic measurement of electron spin pair separations in orientationally disordered solids. *J. Chem. Phys.* 98:5134–5146.

Margraf, D., Cekan, P., Prisner, T.F., Sigurdsson, S.T., and Schiemann, O. 2009. Ferro- and antiferromagnetic exchange coupling constants in PELDOR spectra. *Phys. Chem. Chem. Phys.* 11:6708–6714.

Marko, A., Margraf, D., Yu, H. et al. 2009. Molecular orientation studies by pulsed electron-electron double resonance experiments. *J. Chem. Phys.* 130:064102-1–064102-9.

Marko, A., Margraf, D., Cekan, P., Sigurdsson, S.T., and Schiemann, O. 2010. Analytical method to determine the orientation of rigid spin labels in DNA. *Phys. Rev. E* 81:021911-1–021911-9.

Marko, A. and Prisner, T. 2013. An algorithm to analyze PELDOR data of rigid spin label pairs. *Phys. Chem. Chem. Phys.* 15:619–627.

Marsh, D. 1974. An interacting spin label study of lateral expansion in dipalmitoyl lecithin-cholesterol bilayers. *Biochim. Biophys. Acta* 363:373–386.

Marsh, D. 2014. Intermediate dipolar distances from spin labels. *J. Magn. Reson.* 238:77–81.

Marsh, D. and Smith, I.C.P. 1972. Interacting spin labels as probes of molecular separation within phospholipid bilayers. *Biochem. Biophys. Res. Commun.* 49:916–922.

Marsh, D. and Smith, I.C.P. 1973. An interacting spin label study of the fluidizing and condensing effects of cholesterol on lecithin bilayers. *Biochim. Biophys. Acta* 298:133–144.

Milov, A.D., Maryasov, A.G., and Tsvetkov, Y.D. 1998. Pulsed electron double resonance (PELDOR) and its applications in free radicals research. *Appl. Magn. Reson.* 15:107–143.

Milov, A.D. and Tsvetkov, Y.D. 1997. Double electron-electron resonance in electron spin echo: conformations of spin-labeled poly-4-vinylpyridine in glassy solutions. *Appl. Magn. Reson.* 12:495–504.

Mims, W.B. 1968. Phase memory in electron spin echoes, lattice relaxation effects in $CaWO_4$:Er, Ce, Mn. *Phys. Rev.* 168:370–389.

Mims, W.B. 1972. Envelope modulation in spin-echo experiments. *Phys. Rev. B* 5:2409–2419.

Pannier, M., Veit, S., Godt, A., Jeschke, G., and Spiess, H.W. 2000. Dead-time free measurement of dipole-dipole interactions between electron spins. *J. Magn. Reson.* 142:331–340.

Parry, M. and Fischbach, E. 2000. Probability distribution of distance in a uniform ellipsoid: theory and applications to physics. *J. Math. Phys.* 41:2417–2433.

Polyhach, Y., Godt, A., Bauer, C., and Jeschke, G. 2007. Spin pair geometry revealed by high-field DEER in the presence of conformational distributions. *J. Magn. Reson.* 185:118–129.

Rabenstein, M.D. and Shin, Y.K. 1995. Determination of the distance between 2 spin labels attached to a macromolecule. *Proc. Natl. Acad. Sci USA* 92:8239–8243.

Reginsson, G.W., Hunter, R.I., Cruickshank, P.A.S. et al. 2012. W-band PELDOR with 1 kW microwave power: molecular geometry, flexibility and exchange coupling. *J. Magn. Reson.* 216:175–182.

Schiemann, O., Piton, N., Plackmeyer, J. et al. 2007. Spin labeling of oligonucleotides with the nitroxide TPA and use of PELDOR, a pulse EPR method, to measure intramolecular distances. *Nature Protocols* 2:904–923.

Stein, R.A., Beth, A.H., and Hustedt, E.J. 2015. A straightforward approach to the analysis of double electron-electron resonance data. *Methods Enzymol.* 563:531–567.

Steinhoff, H.J., Dombrowsky, O., Karim, C., and Schneiderhahn, C. 1991. Two dimensional diffusion of small molecules on protein surfaces: an EPR study of the restricted translational diffusion of protein-bound spin labels. *Eur. Biophys. J.* 20:293–303.

Steinhoff, H.J., Radzwill, N., Thevis, W. et al. 1997. Determination of interspin distances between spin labels attached to insulin: comparison of electron paramagnetic resonance data with the x-ray structure. *Biophys. J.* 73:3287–3298.

Tkach, I., Pornsuwan, S., Höbartner, C., et al. 2013. Orientation selection in distance measurements between nitroxide spin labels at 94 GHz EPR with variable dual frequency irradiation. *Phys. Chem. Chem. Phys.* 15:3433–3437.

Tu, S.J. and Fischbach, E. 2002. Random distance distribution for spherical objects: general theory and applications to physics. *J. Phys. A.* 35:6557–6570.

Van Vleck, J.H. 1948. The dipolar broadening of magnetic resonance lines in crystals. *Phys. Rev.* 74:1168–1183.

Ward, R., Bowman, A., Sozudogru, E. et al. 2010. EPR distance measurements in deuterated proteins. *J. Magn. Reson.* 207:164–167.

Weber, A., Schiemann, O., Bode, B., and Prisner, T.F. 2002. PELDOR at S- and X-band frequencies and the separation of exchange coupling from dipolar coupling. *J. Magn. Reson.* 157:277–285.

Xiao, W. and Shin, Y.-K. 2000. EPR spectroscopic ruler, the deconvolution method and its applications. In *Distance Measurements in Biological Systems by EPR*. Biological Magnetic Resonance, Vol. 19, eds. Berliner, L.J., Eaton, G.R., and Eaton, S.S., 249–276. New York: Kluwer Academic Publishers.

Site-Directed Spin Labelling (SDSL)

16

16.1 INTRODUCTION

Site-directed spin-labelling (SDSL) refers to covalent modification of proteins on cysteine residues that are introduced at specific positions in the amino-acid sequence by site-directed mutagenesis. This is a combination of molecular biology and protein chemistry. For specific labelling, we first must replace cysteine residues in the native protein sequence by mutagenesis to residues that do not react with sulphydryl reagents; a conservative replacement is by serine. The most commonly used nitroxide is the methyl thiosulphonate, MTSSL, which reacts with –SH groups by disulphide exchange giving the 1-oxy-2,2,5,5-tetramethylpyrrolin-3-ylmethyl-S–S adduct (see Figure 16.1). This spin label produces a nitroxyl side chain (commonly referred to as R1), which is the least perturbing and is that best characterized in a range of secondary and tertiary protein environments. An alternative form of SDSL is to introduce a nitroxyl amino-acid residue directly into the peptide sequence during chemical synthesis. TOAC (i.e., 1-oxyl-2,2,6,6-tetramethyl-4-aminopiperidin-4-yl carboxylic acid) is a popular choice, particularly in helices because of the helicogenic propensity of this C^{α}-disubstituted amino acid.

SDSL exploits the ability of EPR to distinguish regularities along the protein sequence, e.g., for α-helices or β-strands and for tertiary or quaternary interactions, and also to measure distances between nitroxides at different positions in the structure. Periodicities with sequence position identify secondary structural elements in the protein. Both short and long inter-residue distances then define the tertiary fold, and finally we use long dipolar distances to investigate quaternary assembly and other supramolecular structures.

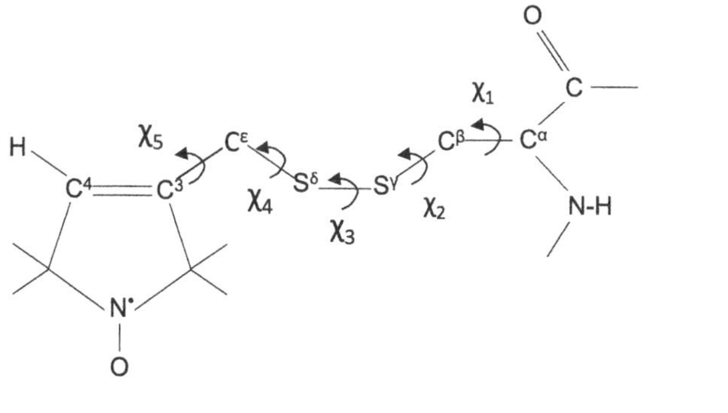

FIGURE 16.1 1-oxy-2,2,5,5-tetramethyl-3-pyrrolin-3-ylmethylsulpho- (MTSSL) S–S adduct of cysteine. R1 side chain with NH, CO and C^{α} of peptide backbone. Dihedral angles χ_1 to χ_5 are labelled. Side-chain atoms have Greek superscripts; nitroxide ring numbering begins at N (≡1).

16.2 SIDE-CHAIN ACCESSIBILITIES AND MOBILITIES

We use mainly two classes of measurement to investigate positional dependence: accessibility to paramagnetic relaxants and spectral lineshapes and linewidths. A third possibility is offered by spin–spin interactions between single label sites in multimeric structures. For accessibilities, we use polar relaxants (such as Ni-complexes, NiEDDA and $Ni(acac)_2$ at neutral pH) and hydrophobic relaxants (such as molecular oxygen) to distinguish water-exposed and lipid-exposed residues in membranes. Charged relaxants, such as chromium trioxalate, additionally probe surface charges and electrostatic potentials. We determine accessibilities from the increase $\Delta P_{1/2} = P_{1/2} - P_{1/2}^{o}$ in half-saturation power $P_{1/2}$ of the spin label by the relaxant, which is directly proportional to relaxant concentration (see Sections 10.6, 11.2 and 11.3). An accessibility parameter Π is defined by (Farahbakhsh et al. 1992):

$$\Pi = \frac{\Delta P_{1/2}/\Delta B_{pp}(0)}{P_{1/2}(\text{DPPH})/\Delta B_{pp}(\text{DPPH})} \quad (16.1)$$

where $\Delta B_{pp}(0)$ is the peak-to-peak central linewidth of the spin label, and $P_{1/2}$(DPPH) and ΔB_{pp}(DPPH) are the half-saturation power and peak-to-peak linewidth, respectively, for a solid sample of 2,2-diphenyl-1-picrylhydrazyl (DPPH). Normalization to the linewidth ΔB_{pp} corrects approximately for the dependence of $P_{1/2}$ on T_2-relaxation rate. More explicit allowance for spin-label rotational dynamics, which is important at moderately high mobilities (Livshits et al. 2003), is described in Sections 11.5 and 11.6. Normalization to the saturation behaviour of a DPPH standard avoids measuring the microwave B_1-field directly but does not allow for differences in resonator Q-factors between samples, which is important for microwave cavities but less so for loop-gap resonators.

Figure 16.2 (left panel) gives a topological map for accessibility to polar and apolar relaxants, specifically the nickel acetylacetonate (20 mM $Ni(acac)_2$) complex and molecular

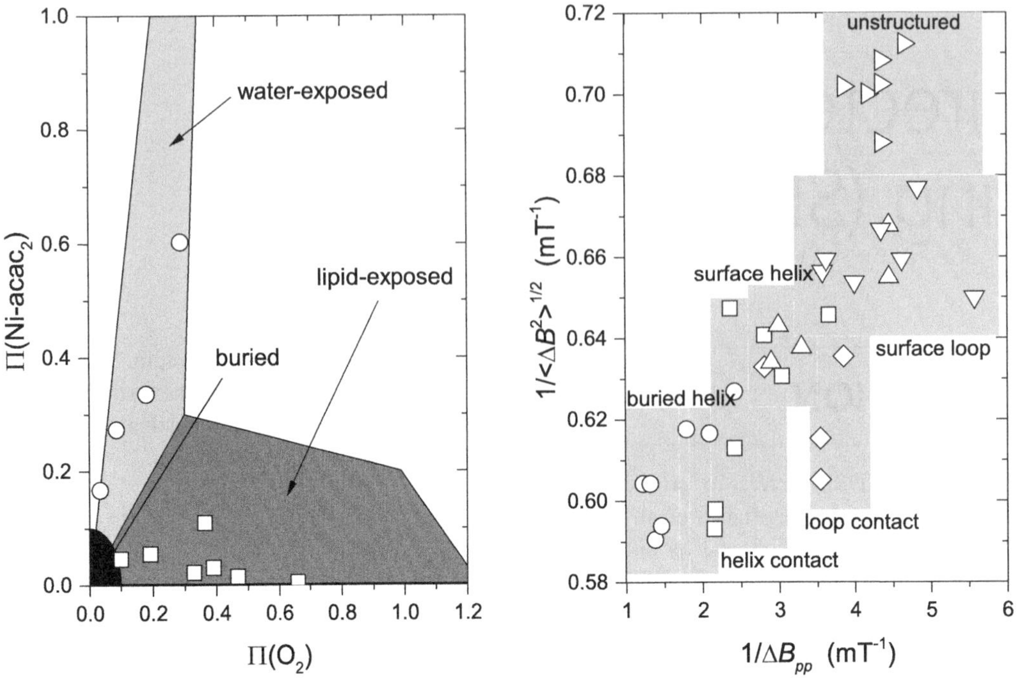

FIGURE 16.2 *Left*: Topographic map for spin-labelled residues in membranous bacteriorhodopsin. Accessibility (Π) to 20 mM nickel acetylacetonate plotted against accessibility to oxygen in equilibrium with air, for surface sites exposed to water (circles) and membrane-embedded sites exposed to lipid (squares). Data from Altenbach et al. (1994); Hubbell and Altenbach (1994). *Right*: Mobility map for spin-labelled residues in aqueous annexin 12. Reciprocal of square root of second moment $\langle \Delta B^2 \rangle$ plotted against reciprocal of central peak-to-peak linewidth $\Delta B_{pp}(m_I = 0)$, for buried residues in helices (circles), helix residues at contact sites (squares), surface residues in helices (triangles) and in loops (inverted triangles), and contact residues in loops (diamonds). Data from Isas et al. (2002). Right-pointing triangles: unstructured regions in flexible linker between C2 domains of synaptotagmin (Huang and Cafiso 2008). Cysteine-substituted residues are labelled with MTSSL, in all cases.

oxygen (O_2 in air). For sites buried in the interior of a helix bundle, accessibilities to both oxygen and $Ni(acac)_2$ are very low. For membrane-embedded sites that are exposed to the lipids, the accessibility to oxygen concentrated in the lipid phase is high, but the $Ni(acac)_2$ accessibility is low. For sites exposed to water at the membrane surface, the $Ni(acac)_2$ accessibility is high and the oxygen accessibility increases with increasing accessibility to $Ni(acac)_2$. Within the membrane, accessibility to oxygen increases with increasing depth, whereas accessibility to $Ni(acac)_2$ decreases. The ratio of the two accessibilities gives us a way to estimate immersion depth in the membrane and to identify loop or contact regions (see next section).

For mobility mapping, two different metrics are often used: the first-derivative linewidth of the central peak and the second moment of the entire absorption lineshape. These, and the alternative absolute-value first moment, are spectral parameters that are valid over the entire motional range (see Chapter 7) and which we can use readily to survey rotational mobility throughout the protein sequence. Figure 16.2 (right panel) shows the correlation between two mobility parameters, the reciprocal central linewidth $1/\Delta B_{pp}(0)$ and reciprocal square-root second moment $1/\langle \Delta B^2 \rangle^{1/2}$, for different cysteine positions labelled with MTSSL in annexin 12. Buried residues in helices display the lowest mobility, and surface residues in interconnecting loops display the highest mobility. Residues at the surface of helices have next to the highest mobility, and those at tertiary contacts have next lowest mobility. Contact residues are grouped into those in loops and those in helices, by differences in central linewidth. Reduced correlation between central linewidth and second moment appears in these and other cases because $\Delta B_{pp}(0)$ emphasizes the more mobile component in two-component spectra, producing the vertically aligned columns in the right panel of Figure 16.2.

16.3 NITROXIDE SCANNING: SECONDARY STRUCTURE

Systematically comparing spin labels introduced sequentially throughout the primary sequence of a protein lets us distinguish residues that are surface-exposed or buried, from both accessibility and mobility measurements. For segments with well-defined secondary structure, the accessibility or mobility parameter, m_i, displays a characteristic periodicity with position, i, in the sequence:

$$m_i = \langle m \rangle + \Delta m \cos\left(\frac{2\pi}{n_P} i + \delta\right) \tag{16.2}$$

where $n_P \cong 3.6$ residues for an α-helix and $n_P \cong 2$ residues for a β-strand, etc. (see Table 16.1). δ is the phase angle that specifies

TABLE 16.1 Protein secondary structures

structure	*residues/ turn, n_P*	*rise/residue, h (nm)*	*helix radius, R_{C^α} (nm)*
antiparallel β-sheet	2	0.34	0.09
α-helix	3.6	0.15	0.23
3_{10}-helix	3.0	0.20	0.19
π-helix	4.3	0.11	0.28

Helix radius is the distance of the C^α-atom from the helix axis.

the azimuthal orientation corresponding to the maxima in m_i, where $\langle m \rangle$ is the mean value and $\pm\Delta m$ the modulation amplitude. The value of $\langle m \rangle$ may also exhibit a slow modulation, which could correspond to the polarity profile for a transmembrane protein, or to overall differences between, for example, neighbouring helices.

Fourier transformation of Eq. 16.2 with respect to position i yields a power spectrum of the angle $\omega = 2\pi/n_P$ between adjacent side chains in the secondary structure. This method of processing lets us filter out low-frequency baseline variations. Also, we can use a sliding window to identify the boundaries of different secondary structural elements better (Mchaourab and Perozo 2000).

16.3.1 Soluble Proteins

Figure 16.3 (top panel) shows the accessibility profile of the polar nickel ethylenediamine-diacetic acid complex, NiEDDA, to site-directed MTSSL-labelled cysteine residues in the water-soluble protein annexin 12. The section scanned includes α-helices D and E and adjacent loops. We see a clear sinusoidal variation in accessibility for the two helical regions, which has a periodicity $n_P = 3.8$ residues close to that expected for α-helices and appears also to extend across the short interconnecting loop. We also see that the mean accessibility is higher for helix D than for helix E, which imposes a slow modulation with a period of $n_P = 23$ residues equal to the total length of the helix-loop-helix motif. Accessibilities of residues in the longer loop region that follows helix E stay consistently high.

Accessibilities of oxygen in samples equilibrated with air follow closely those for NiEDDA (middle panel of Figure 16.3). In particular, the periodic modulation is in phase for these two collisional relaxation reagents. Helix side chains facing the aqueous phase are more accessible than those directed towards the protein interior. The buried internal residues are practically inaccessible to the larger, more polar NiEDDA, whereas a limited accessibility remains for the smaller, apolar oxygen. This is seen most clearly for helix E where Π(NiEDDA) reaches almost zero when $\Pi(O_2)$ still has a finite value. In contrast, oscillations in the interconnecting loop region remain well above zero for both relaxants.

The inverse central linewidth (bottom panel of Figure 16.3) oscillates in-phase with the accessibilities, showing that solvent-exposed helix side chains have higher mobility than internal residues or those at contact sites. The one exception is residue

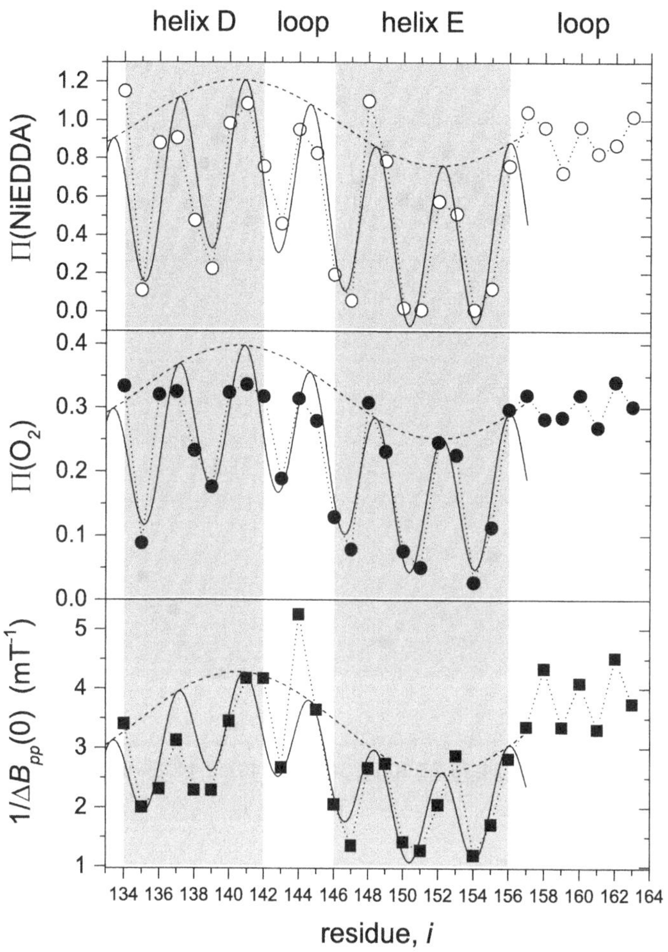

FIGURE 16.3 Profiles of accessibility Π to 3 mM nickel ethylenediaminediacetate (top panel, open circles) and to air-equilibrated oxygen (middle panel, solid circles), and of segmental mobility $1/\Delta B_{pp}(m_I = 0)$ (bottom panel, solid squares), for spin-labelled residues in annexin 12. Cysteine-substituted residues are labelled with MTSSL. Data from Isas et al. (2002). Amino-acid residues, i, are numbered from N- to C-terminal. Shaded areas: α-helices; unshaded areas: connecting loop regions. Solid lines: sinusoidal oscillations with period $n_P = 3.8$ residue (see Eq. 16.2), about mean level varying with period $n_P = 23$ residue (dashed line).

136 that is solvent exposed but hindered by specific interactions. Loop residues mostly are more mobile than helix surface residues. It is notable that the slow 23-residue modulation of the helix-loop-helix motif is more pronounced for the mobility than for the accessibilities. This suggests enhanced backbone dynamics for the shorter, more exposed helix D.

16.3.2 Membrane Proteins

Figure 16.4 shows accessibility scans for the outer, lipid-contacting, transmembrane helix of the KcsA potassium ion channel. Understandably, the profiles for oxygen (top panel) and NiEDDA (middle panel) are now very different, unlike the situation with water-soluble annexin 12. Oxygen concentrates in the hydrophobic interior of the membrane. $\Pi(O_2)$ shows a pronounced

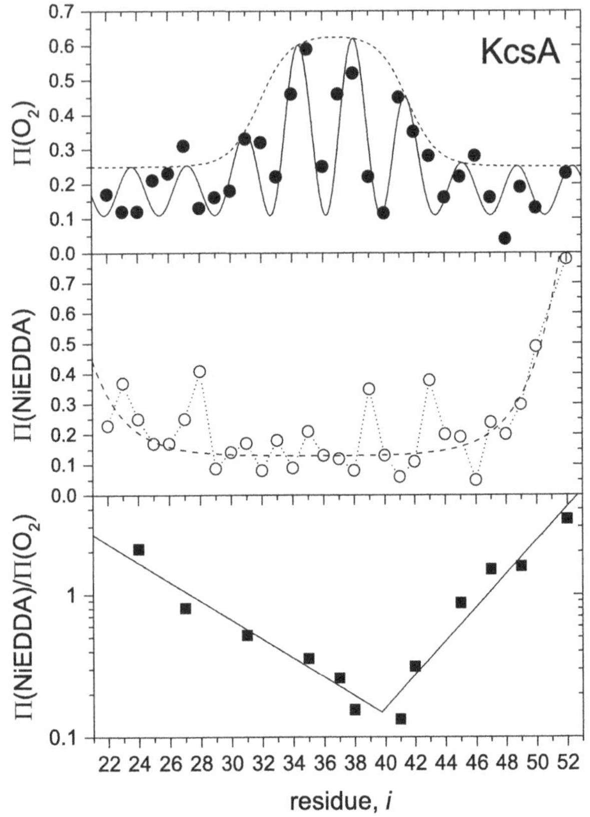

FIGURE 16.4 Profiles of accessibility Π to air-equilibrated oxygen (top panel, solid circles) and to 200 mM nickel ethylenediaminediacetate (middle panel, open circles), for spin-labelled residues in outer transmembrane helix of KcsA K^+-channel. Cysteine-substituted residues are labelled with MTSSL. Data from Perozo et al. (1998). For $\Pi(O_2)$, solid line is sinusoidal oscillation with period of n_P = 3.6 residue (see Eq. 16.2), with maximum envelope given by the transmembrane polarity profile (dashed line – see Eq. 4.44). For Π(NiEDDA), dashed line represents exponential decays from each membrane surface. Bottom panel: accessibility ratio Π(NiEDDA)/$\Pi(O_2)$ on logarithmic scale; solid lines: linear regressions for two halves of the membrane.

3.6-residue α-helical periodicity in this region (solid line), with an envelope determined by the membrane polarity profile described in Section 4.6 of Chapter 4 (dashed line). The lipid-facing surface of the outer helix has high oxygen accessibilities, whereas sites facing the inner channel helix are much less accessible to oxygen. Although electroneutral, the polar NiEDDA complex penetrates only to a very limited extent into the hydrophobic interior of the membrane. Going out from the centre of the membrane, Π(NiEDDA) increases towards the lipid–water interface, and the helix leaves the membrane at residue 50. However, we do not see an unambiguous exit point on the N-terminal side. A few residues show local maxima for Π(NiEDDA), and these are out of phase with those in $\Pi(O_2)$, indicating that these are oriented more towards the channel or the extracellular vestibule contributed by the P-loop. We illustrate this by mapping the accessibility vectors onto the helical wheel plot in Figure 16.5. The resultant Π(NiEDDA) moment is oriented 120° away from that of $\Pi(O_2)$. On the other hand, the mobility given by the inverse

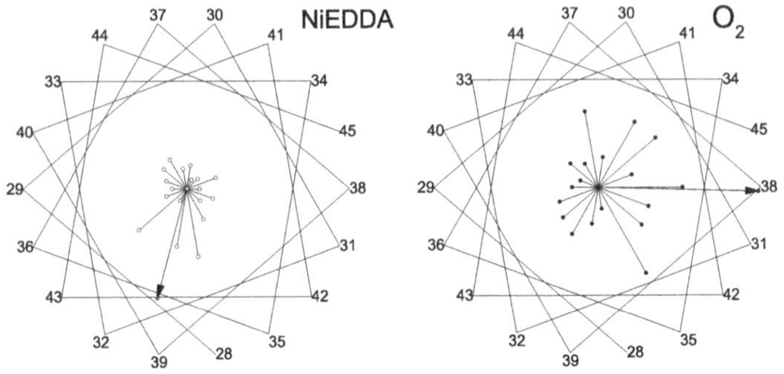

FIGURE 16.5 Vector representation of accessibility Π to oxygen (right) and to NiEDDA (left) from Figure 16.4, projected onto helical wheel for outer transmembrane helix of KcsA. Solid arrow: resultant accessibility moment. View along helical axis from N- to C-terminal.

central linewidth oscillates in phase with $\Pi(O_2)$, and the resultant moment of $1/\Delta B_{pp}(0)$ aligns along that of the oxygen accessibility facing the lipid exterior. Antiphase segregation of local maxima in Π(NiEDDA) and $\Pi(O_2)$ to opposite faces is also a characteristic feature of membrane-associated amphipathic helices, e.g., in annexin 12 (Langen et al. 1998a).

The bottom panel of Figure 16.4 gives the transmembrane profile for the ratio of NiEDDA accessibility to that of oxygen, on a logarithmic scale. We choose only residues that are exposed to the lipids, and for these the log-ratio decreases approximately linearly with distance into the membrane. This implies a linear dependence on penetration depth for the difference in free energy of transfer, $\mu^o_{w/m}(O_2) - \mu^o_{w/m}(\text{NiEDDA})$, of oxygen and NiEDDA from water into the membrane. Altenbach et al. (1994) use this relation to calibrate vertical location d within a membrane, which they determine with lipid-facing spin-labelled residues on helix D of bacteriorhodopsin:

$$d(\text{nm}) = a_1 \log_{10}\left(\frac{\Pi(\text{Niacac}_2)}{\Pi(O_2)}\right) + a_2 \tag{16.3}$$

where the accessibility parameters are for 20 mM nickel acetylacetonate and oxygen in equilibrium with air, and the origin for d is the phosphate position in the lipid polar head group. We list the calibration constants a_1 and a_2 for these and other paramagnetic relaxants in Table 16.2. Note that these estimates become less reliable towards the middle of the membrane, because the concentration of the polar complex there is rather low, and the oxygen profile also becomes rather flat (see Figure 16.4). Therefore, I prefer the depth parameter shown in Figure 16.4, instead of the commonly used reciprocal ratio, $\Pi(O_2)$/Π(NiEDDA), which potentially can diverge in the middle of the membrane.

Essentially, the depth-parameter method combines the approximately exponential decrease in polar relaxant penetration in the outer membrane regions with the effectively exponential part of the sigmoidal oxygen profile within the membrane to cover most of the total membrane thickness. In so far as the paramagnetic metal-induced relaxation is collisional (see Chapter 9), taking the ratio also cancels out local variations in diffusion rates and steric factors which affect

TABLE 16.2 Calibration constants for immersion depths, $d(\text{nm}) = a_1 \log_{10}\left[\Pi(\text{PM})/\Pi(\text{O}_2)\right] + a_2$ (cf. Eq. 16.3), by the collision-gradient method with paramagnetic relaxants, PM, and O_2[a]

PM[b]	O_2	a_1 (nm)	a_2 (nm)[c]	*Ref.*[d]
200 mM NiEDDA	air	−1.11	0.49	1
		−1.11	1.16[e]	2
150 mM NiEDDA	air	−1.27	0.81[f]	3
100 mM NiEDDA	air	−1.11	0.85[g]	4
20 mM $Ni(acac)_2$	air	−0.90	0.60	5
50 mM $Cr(ox)_3$	100%	−0.94	−0.61	6

[a] Calibrations from phosphatidylcholine spin labelled in the *sn*-2 chain, where $d = 0.81, 1.05, 1.4, 1.6$ and 1.86 nm relative to the head-group P-atom for 5-, 7-, 10-, 12- and 16-DOXYL PC, respectively (Dalton et al. 1987). $d \cong 0$ for TEMPO-stearamide and $d \cong -0.5$ nm for *N*-TEMPO PC. Data from ref. 6 are re-fitted to these distances for *n*-DOXYL PC.

[b] NiEDDA, nickel ethylenediaminediacetate; $Ni(acac)_2$, nickel acetylacetonate; $Cr(ox)_3$, chromium trioxalate.

[c] *Note*: for different relaxant concentrations, a_2 is increased by $\Delta a_2 = a_1 \log_{10}\left[r(\text{O}_2)/r(\text{PM})\right]$ where r is the relaxant concentration relative to that used in the calibration.

[d] References: 1. Oh et al. (1996); 2. Oh et al. (1999); 3. Salwinski and Hubbell (1999); 4. Langen et al. (1998a); 5. Altenbach et al. (1994); 6. Farahbakhsh et al. (1992).

[e] In the presence of bound T-domain of diphtheria toxin.

[f] In the presence of bound colicin E1.

[g] In the presence of bound annexin 12.

the two relaxants equally. Note further that scans for helices are relatively coarse-grained, compared with those for lipid chains (see Chapter 4), being determined by the 3.6-residue periodicity of lipid-facing residues. Also, helix contact sites must be excluded because these hinder accessibility of the polar paramagnetic-ion complex, relative to the smaller oxygen molecule.

The form of Eq. 16.3 is particularly useful because the constant a_1 multiplying the log-ratio depends solely on the shape of the two profiles. It is independent of the concentration and absolute partition coefficients of the two relaxants; all such effects contribute only to the additive constant a_2. For example, $a_1 = -0.94$ nm and $a_2 = -0.61$ nm for 50 mM chromium trioxalate and 100% oxygen, whereas $a_1 = -0.90$ nm and $a_2 = +0.60$ nm for 20 mM nickel acetylacetonate and 20% oxygen (see Table 16.2). Using calibrations for a 200 mM NiEDDA/air couple from the first entry in Table 16.2, the linear-regression gradients in the bottom panel of Figure 16.4 give a depth increment of $h \equiv \partial d/\partial i = 0.13$ nm/residue on the C-terminal side and 0.07 nm/residue on the N-terminal side. This suggests that the outer transmembrane helix of KcsA tilts less (if at all) on the C-terminal side than on the N-terminal side (cf. Table 16.1).

We also can use the log accessibility-ratio for polar and apolar relaxants (sometimes called the contrast function) to identify loop regions and contact regions between helices. Such regions are characterized by a constant contrast function because relative accessibility for polar and apolar relaxants is approximately the same at all sites. We illustrate this with the profile of the contrast function for the helix-loop-helix-loop sequence of annexin 12 shown in Figure 16.6. The contrast function varies periodically in the helical regions but remains rather constant in the loop regions. The periodic modulation in the less exposed helix E is particularly deep.

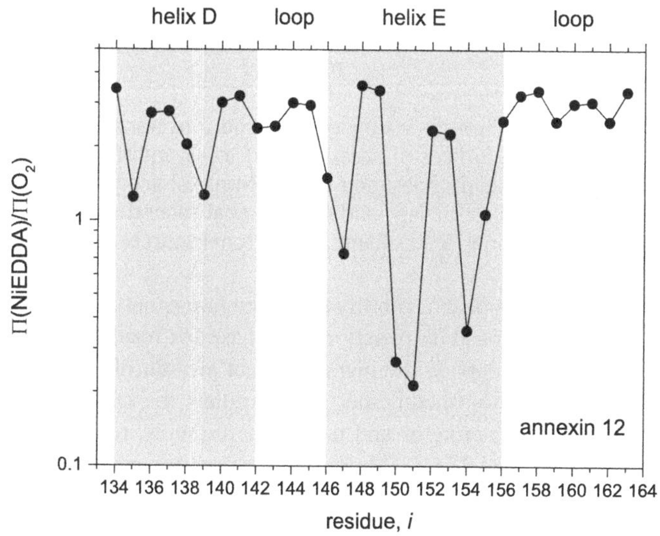

FIGURE 16.6 Profile of accessibility ratio $\Pi(\text{NiEDDA})/\Pi(\text{O}_2)$, on logarithmic scale, from nitroxide scans of annexin 12 (see Figure 16.3).

16.4 β-SHEETS AND β-BARREL PROTEINS

We can identify β-sheet regions in nitroxide scans from an alternation in accessibilities and/or mobilities between odd and even residues, because these are located on opposite faces of the sheet ($n_P = 2$ in Eq. 16.2). Figure 16.7 shows scans for the second strand in the transmembrane β-barrel of the vitamin B_{12} transporter BtuB.

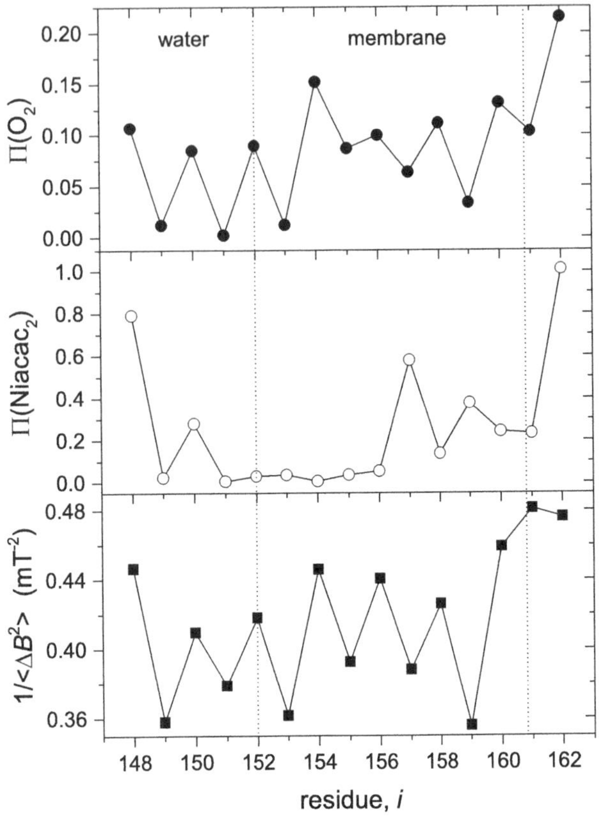

FIGURE 16.7 Nitroxide scanning for strand 2 in transmembrane β-barrel of BtuB. Profiles of accessibility Π to air-equilibrated oxygen (top panel, solid circles) and to 20 mM nickel acetylacetonate (middle panel, open circles), and of reciprocal second moment $1/\langle \Delta B^2 \rangle$ (bottom panel, solid squares). Data from Fanucci et al. (2002).

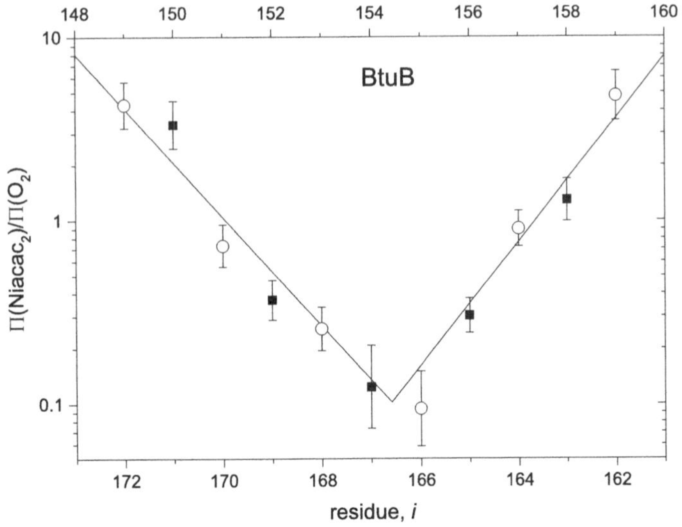

FIGURE 16.8 Depth profiles of accessibility ratio $\Pi(\text{Niacac}_2)/\Pi(O_2)$, on logarithmic scale, for outward-facing residues in second strand (148–160; top axis, solid squares) and third strand (176–162, bottom axis, open circles) of BtuB β-barrel (cf. Figure 16.7). Both plots run from extracellular to periplasmic surface of the protein. Profiles are overlaid to coincide best. Data from Fanucci et al. (2002).

Alternations in accessibility to oxygen (top panel) and nickel acetylacetonate (middle panel) are in phase for residues in the aqueous region outside the membrane but are out of phase for residues within the membrane. This parallels the situation for α-helical soluble proteins and membrane proteins, respectively (cf. Figures 16.3 and 16.4). Alternations in mobility, registered by the reciprocal second moment (bottom panel), are in phase with the oxygen accessibility, because residues exposed to solvent (water or lipid) have higher mobility.

Figure 16.8 shows the transmembrane profile of the nickel/oxygen accessibility ratio for lipid-facing residues of the second (solid squares) and third (open circles) β-strands in the BtuB barrel. Scans for the two strands are aligned to bring sites at similar depths in register. We see a logarithmic depth dependence similar to that noted previously for KcsA in the bottom panel of Figure 16.4. Taking the mean slope of the log accessibility-ratio and the depth calibration for the nickel acetylacetonate/air couple from Table 16.2, we find a transmembrane increment of $h \equiv \partial d/\partial i = 0.28$ nm/residue. This corresponds to a strand tilt of $\beta \approx 36°$, because the increment along a β-strand is $h_\beta = 0.345$ nm/residue (see Table 16.1). Slightly larger average tilts of 38°–39° are found from the crystal structures of the analogous FepA and FhuA transporters, because the tilt is larger for strands on the opposite side of the barrel from those spin labelled (Páli and Marsh 2001). From the absolute values of the accessibility ratio (cf. Table 16.2), we deduce that the polar/apolar interfaces of the membrane are located at residue positions 149/172 and 159/162, respectively, in the two strands.

β-barrels are geometrically constrained structures. Knowing the strand tilt β and number of strands ($N_\beta = 22$, deduced from the sequence), we can calculate the radius of an equivalent circular barrel (Marsh 2000):

$$R = \frac{d_\beta}{2\sin(\pi/N_\beta)\cos\beta} \tag{16.4}$$

where $d_\beta = 0.473$ nm is the strand separation. For BtuB, we obtain a diameter of $2R = 4.0$ nm, compared with a slightly elliptical cross-section of 4.0×4.5 nm found from the crystal structure (Chimento et al. 2003). In principle, we also can estimate the mean twist, θ, of the β-sheet (perpendicular to the strand axis) and coiling, ε, of the β-strands (Marsh 2000; Anbazhagan et al. 2008):

$$\theta = \frac{\theta_o + \dfrac{2\pi}{N_\beta}\left[\left(\dfrac{h_\beta}{d_\beta}\right)^2 \dfrac{\cos\beta}{\tan\beta} + \sin\beta\right]}{\left(\dfrac{h_\beta}{d_\beta}\right)^2 \dfrac{1}{\tan^2\beta} + \dfrac{1}{\cos^2\beta}} \tag{16.5}$$

$$\varepsilon = \left(\frac{h_\beta}{d_\beta}\right)\left(\frac{2\pi}{N_\beta} - \frac{\theta}{\sin\beta}\right)\cos\beta \tag{16.6}$$

where $h_\beta/d_\beta = 0.719 \pm 0.022$ and $\theta_o = -3.5°$ (Páli and Marsh 2001). This yields $\theta = 6.2°$ and $\varepsilon = 3.4°$, for BtuB, which is close to the values predicted for an idealized 22-stranded barrel with a shear number of 24. Average twists and coilings determined from the crystal structures of the analogous FepA and FhuA transporters are $\theta = 6.8°, 6.9°$ and $\varepsilon = 3.5°, 3.8°$, respectively (Páli and Marsh 2001).

16.5 MOBILITY MAPPING

As seen already in Chapter 8, rotational motion is specified both by the angular amplitude (or order parameter) and by the rate of rotation (or correlation time). In principle, the central linewidth and moments of the spectral lineshape are sensitive to both these dynamic parameters. Although easiest to measure, interpretation of the central linewidth is less straightforward. This is because of the limited spectral range at 9 GHz and complications from the non-axial g-tensor in the case of anisotropic rotational diffusion and ordering. As mentioned already, the central linewidth overemphasizes the fastest component in multi-component spectra. On the other hand, moments of the different spectral components in the lineshape are strictly additive.

In this section, we concentrate primarily on the relation of the spectral moments to orientational order parameters and rotational correlation times. We deal specifically with the familiar second moment, but also with the absolute-value first moment. The latter is less susceptible to baseline drift, choice of scan range, and noise in the outer wings of the spectrum than is the second moment (Marsh 2014a). Background information on the method of moments, and some of the salient results, are given in Appendix U.

16.5.1 Angular Amplitudes

For the relation of spectral moments to angular amplitudes of motion, we consider first axial hyperfine anisotropy, in the absence of g-value anisotropy. We illustrate this for a ^{14}N-nitroxide by the absorption powder pattern shown in Figure 16.9. For this symmetric situation (which we also treat in Appendix U), the mean resonance field is given by the centre of the spectrum. From Eq. U.32 in Appendix U, the second moment for axial hyperfine structure is:

$$\left\langle \Delta B^2_{aniso} \right\rangle = \frac{2}{9}\left(\left(\frac{A_{\parallel}}{g\beta_e} \right)^2 + 2\left(\frac{A_{\perp}}{g\beta_e} \right)^2 \right) \tag{16.7}$$

for ^{14}N-nitroxides ($I = 1$). The corresponding expression for ^{15}N-nitroxides $\left(I = \frac{1}{2}\right)$ is:

$$\left\langle \Delta B^2_{aniso} \right\rangle = \frac{1}{12}\left(\left(\frac{A_{\parallel}}{g\beta_e} \right)^2 + 2\left(\frac{A_{\perp}}{g\beta_e} \right)^2 \right) \tag{16.8}$$

where $A_{\parallel}/g\beta_e$ and $A_{\perp}/g\beta_e$ are the principal hyperfine couplings in field units (e.g., mT).

In the motional-narrowing region, the order parameter S_{zz} is given by (see Chapter 8):

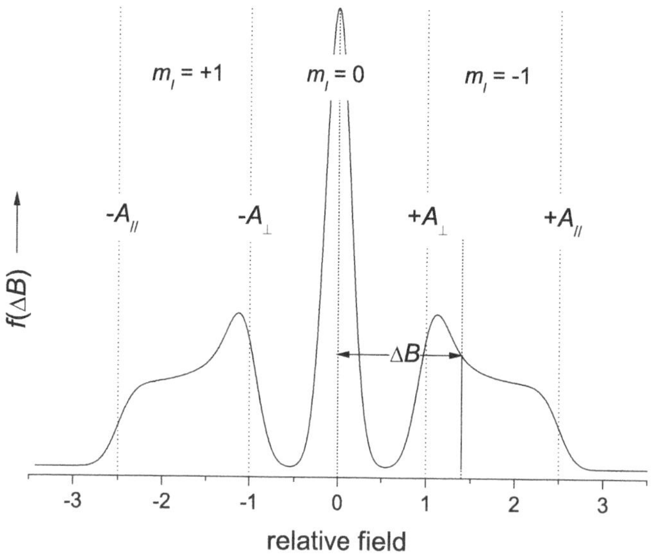

FIGURE 16.9 Absorption powder pattern with axially symmetric hyperfine anisotropy for ^{14}N-nitroxide (I = 1). g-value anisotropy is neglected. ΔB: field separation from centre of spectrum, used for calculating spectral moments. Hyperfine manifolds: $m_I = 0, \pm 1$.

$$S_{zz} \equiv \frac{1}{2}\left(3\left\langle \cos^2\theta \right\rangle - 1\right) = \frac{A_{\parallel} - A_{\perp}}{\Delta A} \tag{16.9}$$

where $\Delta A \equiv A_{zz} - \frac{1}{2}\left(A_{xx} + A_{yy}\right)$, and θ is the angle that the nitroxide z-axis makes with the uniaxial director axis. Angular brackets indicate a time (or ensemble) average. The rotationally invariant isotropic hyperfine coupling is: $a_o \equiv \frac{1}{3}\left(A_{xx} + A_{yy} + A_{zz}\right) = \frac{1}{3}\left(A_{\parallel} + 2A_{\perp}\right)$. Thus, from Eq. 16.7, the second moment is related to the order parameter for axially symmetric rotational diffusion by:

$$\left\langle \Delta B^2_{aniso} \right\rangle = \frac{2}{3}\left(\frac{a_o}{g\beta_e} \right)^2 \left(1 + \frac{2}{9}\left(\frac{\Delta A}{a_o} \right)^2 S_{zz}{}^2 \right) \tag{16.10}$$

for ^{14}N-nitroxides. From Eq. 16.8, the corresponding expression for ^{15}N-nitroxides is given by Eq. 16.10 with the factor $\frac{2}{3}$ replaced by $\frac{1}{4}$.

From Eq. U.10 in Appendix U, the absolute-value first moment for axial hyperfine structure is given by:

$$\left\langle \left| \Delta B_{aniso} \right| \right\rangle = \frac{1}{3}\left(\frac{A_{\parallel}}{g\beta_e} + \frac{A_{\perp}/g\beta_e}{\sqrt{\left(A_{\parallel}/A_{\perp}\right)^2 - 1}} \ln\left(A_{\parallel}/A_{\perp} + \sqrt{\left(A_{\parallel}/A_{\perp}\right)^2 - 1} \right) \right) \tag{16.11}$$

for ^{14}N-nitroxides ($I = 1$). The corresponding expression for ^{15}N-nitroxides $\left(I = \frac{1}{2}\right)$ is:

$$\left\langle \left| \Delta B_{aniso} \right| \right\rangle = \frac{1}{4}\left(\frac{A_{\parallel}}{g\beta_e} + \frac{A_{\perp}/g\beta_e}{\sqrt{\left(A_{\parallel}/A_{\perp}\right)^2 - 1}} \ln\left(A_{\parallel}/A_{\perp} + \sqrt{\left(A_{\parallel}/A_{\perp}\right)^2 - 1} \right) \right) \tag{16.12}$$

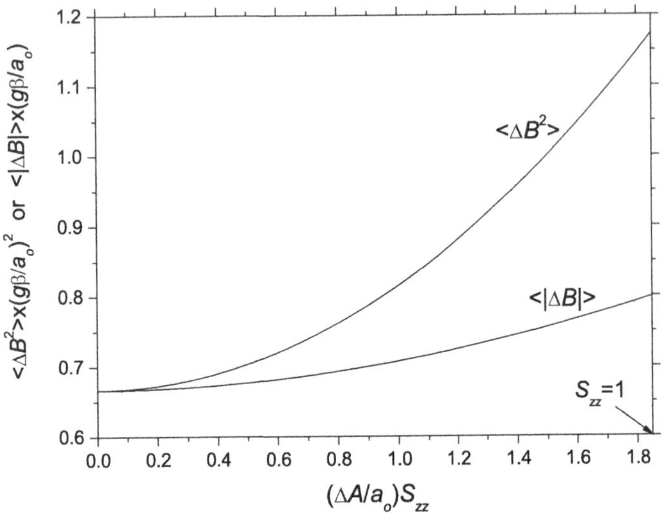

FIGURE 16.10 Dependence of second moment $\langle \Delta B^2_{aniso} \rangle$ and absolute-value first moment $\langle |\Delta B_{aniso}| \rangle$, of ^{14}N-nitroxide absorption powder spectrum on orientational order parameter S_{zz} for axial rotational diffusion (Eqs. 16.10 and 16.13). g-value anisotropy is neglected (cf. Figure 16.9). For each moment, ΔB is scaled to isotropic hyperfine coupling ($a_o \approx 1.50$ mT) in field units. Order parameter is scaled by ratio of hyperfine anisotropy to isotropic coupling: $\Delta A/a_o \approx 1.85$.

– see Appendix U. From Eq. 16.11, together with Eq. 16.9, the absolute first moment is related to the order parameter for axially symmetric rotational diffusion by:

$$\langle |\Delta B_{aniso}| \rangle = \frac{a_o}{3g\beta_e}\left(1+\tfrac{2}{3}\hat{S}_{zz}+\frac{\left(1-\tfrac{1}{3}\hat{S}_{zz}\right)^2}{\sqrt{2\hat{S}_{zz}+\tfrac{1}{3}\hat{S}^2_{zz}}}\times\ln\left(\frac{1+\tfrac{2}{3}\hat{S}_{zz}+\sqrt{2\hat{S}_{zz}+\tfrac{1}{3}\hat{S}^2_{zz}}}{1-\tfrac{1}{3}\hat{S}_{zz}}\right)\right) \tag{16.13}$$

for ^{14}N-nitroxides, where $\hat{S}_{zz} \equiv (\Delta A/a_o)S_{zz}$. The corresponding expression for ^{15}N-nitroxides is given by Eq. 16.13 with the initial factor of $\frac{1}{3}$ replaced by $\frac{1}{4}$.

Figure 16.10 shows the dependence of the two spectral moments, $\langle |\Delta B_{aniso}| \rangle$ and $\langle \Delta B^2_{aniso} \rangle$, on the orientational order parameter S_{zz}. These values are for a ^{14}N-nitroxide. For both moments, ΔB is normalized to the isotropic coupling a_o, and the order parameter S_{zz} is scaled by the ratio of the anisotropy ΔA to a_o. Each moment increases, at a progressively faster rate, as the order parameter increases from the isotropic situation at $S_{zz} = 0$ to the completely ordered state at $S_{zz} = 1$. This corresponds to the change from completely unrestricted rotational amplitude, θ, to no rotation with $\theta = 0$. The increase in second moment with orientational order is greater than that of the absolute first moment, because the second moment preferentially emphasizes larger hyperfine splittings, i.e., greater spectral anisotropy.

16.5.2 Rotational Rates

So far, we considered only the contribution of spectral anisotropy, i.e., angular amplitude of motion, to the spectral moments, without including line broadening. We now treat the other extreme of isotropic rotational diffusion with unrestricted amplitude. Contributions to the spectral moments are principally those of line broadening as the motion slows down. We simulate spectra of this type using the stochastic Liouville equation in Chapter 6 and show spectra going from the fast- to slow-motional regime already in Figure 1.5 of Chapter 1. Figure 16.11 shows the slow-motion EPR absorption spectrum for an ^{14}N-nitroxide that corresponds to an isotropic rotational correlation time $\tau_R = 20$ ns. This illustrates how moments are calculated for spectra without axial symmetry. The centre of gravity $\langle B \rangle$, to which we refer the moments, is determined by the isotropic g-value (see Section U.2 in Appendix U).

Figure 16.12 shows the dependence of the absolute-value first moment and second moment, and of the central peak-to-peak linewidth, on isotropic rotational correlation time $\tau_R \equiv 1/(6D_R)$ for spectra such as those given for an EPR frequency of 9.4 GHz in Chapters 1 and 6. In the fast isotropic motional regime, the moments level off to values that are determined by the isotropic hyperfine splitting, a_o, together with the residual linewidth. Note that only the width of the central $m_I = 0$ manifold contributes to the absolute-value first moment, whereas the linewidths of all three manifolds contribute to the second moment. In the extreme slow-motional regime, the moments also level off, but to values determined by the full spectral anisotropy and residual linewidths, as the spectrum approaches a rigid-limit powder pattern.

Empirically, the sigmoidal dependence of the moments on logarithm of the rotational correlation time in Figure 16.12 is well fit by the logistic equation (see solid lines). The resulting

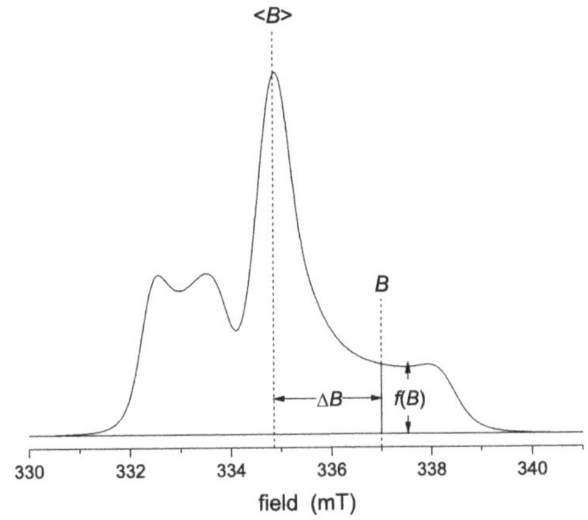

FIGURE 16.11 Definition of moments for non-axial spin-label absorption lineshape. ΔB: field separation from centre of gravity $\langle B \rangle$ of absorption spectrum, determined by isotropic g-value, g_o.

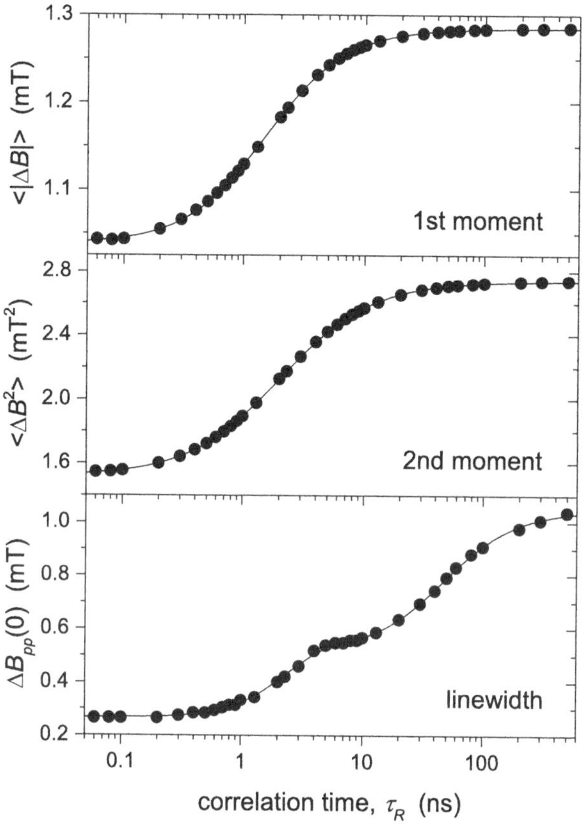

FIGURE 16.12 Dependence of absolute-value first moment $\langle|\Delta B|\rangle$ (top panel) and second moment $\langle\Delta B^2\rangle$ (middle panel) of ^{14}N-nitroxide absorption spectrum, and of central peak-to-peak linewidth $\Delta B_{pp}(m_I = 0)$ in first-derivative spectrum (bottom panel), on correlation time τ_R for isotropic rotational diffusion. Spectra simulated with stochastic-Liouville equation (Figure 1.5 in Chapter 1), and moments referred to field corresponding to isotropic g-value. Solid lines: fits of logistic equation (Eq. 16.14); τ_R-axis is logarithmic.

correlation-time calibrations for the two moments are given by the power law:

$$\tau_R = \tau_{R,o}\left(\frac{M - M_{min}}{M_{max} - M}\right)^{1/p} \qquad (16.14)$$

where M is the measured moment, M_{max} and M_{min} are the limiting values of M at long and short τ_R, respectively, and $\tau_{R,o}$ and the index p are the logistic fitting parameters. For the absolute first moment: $\tau_{R,o} = 1.51$ ns, $M_{max} = 1.285$ mT, $M_{min} = 1.038$ mT and $p = 1.30$; and for the second moment: $\tau_{R,o} = 2.02$ ns, $M_{max} = 2.739$ mT^2, $M_{min} = 1.519$ mT^2 and $p = 1.15$. The smaller value of p indicates sensitivity to a somewhat wider range of correlation times for the second moment, as we see in Figure 16.12. In particular, $\langle\Delta B^2\rangle$ is sensitive to longer correlation times because it emphasizes the linewidths of the outer wings (cf. Section 7.14 on slow motion in Chapter 7).

It is instructive to compare the values of M_{min} that we obtain from Figure 16.10 for fast isotropic rotation with the predictions of Eqs. 16.10 and 16.13 for $S_{zz} = 0$ in the idealized model with zero linewidth. From these equations, we have: $\langle|\Delta B|\rangle = \frac{4}{3}a_o = 1.00$ mT and $\langle\Delta B^2\rangle = \frac{2}{3}a_o^2 = 1.50$ mT^2. The corresponding values of M_{min} include a Gaussian linewidth of $\Delta B_o^G = 0.30$ mT (FWHM), which contributes $\langle|\Delta B|\rangle = \Delta B_o^G/\left(2\sqrt{\pi \ln 2}\right) = 0.10$ mT to the absolute-value first moment, and $\langle\Delta B^2\rangle = \left(\Delta B_o^G\right)^2/(8\ln 2) = 0.016$ mT^2 to the second moment, for a single line. We therefore expect M_{min} to exceed the predictions of Eqs. 16.10 and 16.13 somewhat. The corrected values are thus close to the predictions.

From the bottom panel of Figure 16.12, we see that the central first-derivative linewidths, $\Delta B_{pp}(m_I = 0) \equiv \delta_{pp}$, have a double sigmoidal dependence on the logarithm of the correlation time, joined by a "dead spot" at $\tau_R \approx 7$ ns. In the first region, which corresponds to motional narrowing, the calibration parameters for Eq. 16.14 are: $\tau_{R,o} = 2.3$ ns, $\delta_{max} = 0.589$ mT, $\delta_{min} = 0.266$ mT and $p = 1.8$. In the second region, which corresponds to slow motion for the g-value anisotropy, the calibration parameters are: $\tau_{R,o} = 48$ ns, $\delta_{max} = 1.045$ mT, $\delta_{min} = 0.518$ mT and $p = 1.45$. The latter, extends to longer correlation times than for the two spectral moments, because the Zeeman anisotropy at 9.4 GHz is considerably smaller than the hyperfine anisotropy that dominates the moments.

16.5.3 Order Parameters

We are now in a position to combine the effects of both limited angular amplitudes and dynamical line broadening. We presented such spectra, including slow-motional contributions and covering the full range of axial order parameters from $S_{zz} = 0$ to $S_{zz} = 1$, in Figure 1.6 of Chapter 1. They come from stochastic-Liouville simulations with dynamic parameters and residual linewidths that well reproduce experimental spectra for spin-labelled lipid chains in fluid bilayer membranes (Schorn and Marsh 1996b; Schorn and Marsh 1996a). They also approximate, at least qualitatively, partially ordered environments for spin-labelled side chains in proteins (see, e.g., Columbus et al. 2001; Lopez et al. 2012).

Figure 16.13 shows the dependence of the absolute-value first moment and the second moment on orientational order parameter S_{zz}, for spin-label spectra such as those given in Chapter 1. The rotational diffusion coefficient $D_{R\perp}$ and the residual linewidth are adapted to changes in order parameter (Schorn and Marsh 1997). Qualitatively, the dependence of the moments on order parameter is similar to that predicted by the analytical model (cf. Figure 16.10). However, the absolute values of each moment are larger and the difference between the two moments is less. This results primarily from the slow rotational dynamics, which have their greater effect at larger motional amplitudes (lower order) and are more pronounced for the absolute-value first moment than for the second moment. In contrast, the second moment is preferentially biased towards the degree of order. In this respect, the absolute-value first moment more resembles the central linewidth and scaled mobility parameter (cf. Columbus and Hubbell 2004).

As we see from the solid lines in Figure 16.13, the dependences of the two moments on S_{zz} are well described by second-order polynomials, where the coefficient of the quadratic term is

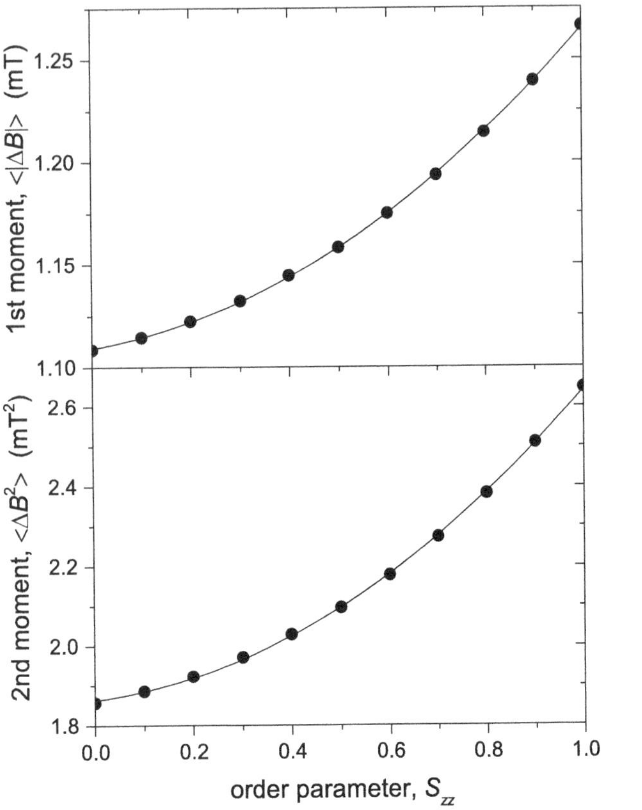

FIGURE 16.13 Dependence of absolute-value first moment $\langle|\Delta B|\rangle$ (top), and second moment $\langle\Delta B^2\rangle$ (bottom), of ^{14}N-nitroxide absorption powder spectrum on orientational order parameter S_{zz} for axial rotational diffusion. 9.4-GHz slow-motion spectra simulated with stochastic-Liouville equation (Figure 1.6 in Chapter 1). Isotropic hyperfine coupling: $a_o \approx 1.47$ mT; hyperfine anisotropy: $\Delta A \approx 2.73$ mT. Simulation parameters from Schorn and Marsh (1997). Residual Gaussian line broadening: 0.08–0.5 mT (FWHM), for $S_{zz} = 0-1.0$. Solid lines: least-squares fits of second-order polynomial.

greater than that of the linear term (cf. Eq. 16.10). Calibration of the order parameter in terms of the measured moment M, therefore comes from solution of a quadratic equation:

$$S_{zz} = \sqrt{aM - c} - b \tag{16.15}$$

where calibration constants a, b and c are related to the polynomial coefficients. For the absolute-value first moment: $a = 8.655\ \text{mT}^{-1}$, $b = 0.1747$ and $c = 9.569$; and for the second moment: $a = 1.604\ \text{mT}^{-2}$, $b = 0.1243$ and $c = 2.971$.

16.6 MTSSL (R1) SIDE-CHAIN ROTAMERS

In principle, the R1-side chain that we produce by modifying cysteine residues with the methyl thiosulphonate spin-label MTSSL contains several bonds about which segmental motion of the nitroxide can occur by rotational isomerism (see Figure 16.1). Only the S–S link, which has partial double-bond character, has a rotational barrier that is high enough to prevent rotameric transitions from taking place within the nitroxide EPR timescale (Jiao et al. 1992). However, interaction of the R1-side chain with the peptide backbone restricts the number of stable conformers and simplifies analysis of EPR lineshapes for spin-labelled α-helices considerably. Evidence for this comes from correlation of x-ray crystal structures from the spin-labelled protein with temperature-dependent solution lineshapes.

Figure 16.14 classifies the allowed C–C rotamers for the R1-side chain, using the trans (**t**), plus (**p**) and minus (**m**) nomenclature suggested for protein structures (Lovell et al. 2000). Stable rotamers correspond to staggered conformations for the threefold potential of carbon sp^3 bonds. They are defined by the dihedral angles: $\chi = 180°, +60°$ and $-60°$ for the **t**, **p** and **m** conformers (i.e., *trans*, *gauche*$^+$ and *gauche*$^-$), respectively, for a right-handed rotation. The dihedral angle illustrated in Figure 16.14 is χ_1; Table 16.3 defines the complete set of dihedrals for the R1-side chain, χ_1 to χ_5. Table 16.4 gives the nomenclature for amino-acid substitution in a protein sequence, using the single-letter code, including R1 for a cysteine residue reacted with MTSSL. Note that the R1 side chain is listed as a ligand (MTN) in protein database (PDB) files.

Crystal structures are available for the R1 spin-label side chain at different sites in α-helices of bacteriophage T4 lysozyme and at a loop site. These locations include surface-exposed sites, tertiary contact sites and a buried site. The S–S bond is clearly resolved for all sites, demonstrating

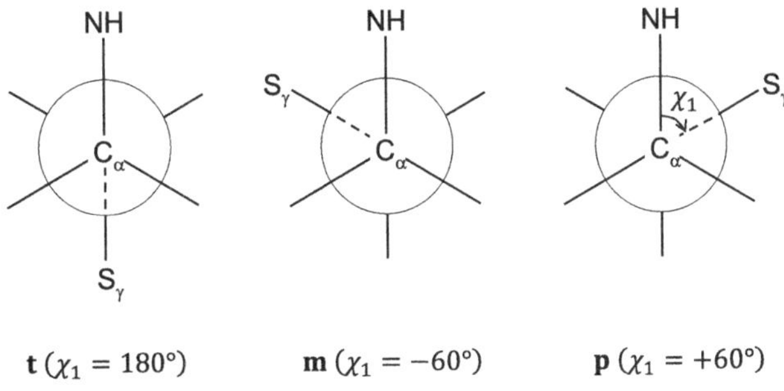

FIGURE 16.14 Newman projections along C$^{\alpha}$ to C$^{\beta}$ direction for cysteine- or R1-side chain (cf. Figure 16.1). **t**, **m** and **p** rotamers defined by dihedral angle χ_1 (=NH–C$^{\alpha}$–C$^{\beta}$–S$^{\gamma}$) about C$^{\alpha}$–C$^{\beta}$ bond.

TABLE 16.3 Dihedral angles for S–S linked methyl thiosulphonate spin-label (MTSSL) side chains

dihedral angle	*bonds*
χ_1	NH–C$^{\alpha}$–C$^{\beta}$–S$^{\gamma}$
χ_2	C$^{\alpha}$–C$^{\beta}$–S$^{\gamma}$–S$^{\delta}$
χ_3	C$^{\beta}$–S$^{\gamma}$–S$^{\delta}$–C$^{\varepsilon}$
χ_4	S$^{\gamma}$–S$^{\delta}$–C$^{\varepsilon}$–C^{3}
χ_5	S$^{\delta}$–C$^{\varepsilon}$–C^{3}–C^{4}

Greek superscripts represent side-chain atoms, and numerical superscripts the nitroxide ring (C$^{\varepsilon}$ ≡ C^{6}) – see Figure 16.1.

TABLE 16.4 Amino-acid single letter code for specifying residue substitutions in site-directed spin labelling

amino acid	*abbreviation*	*single-letter*
alanine	Ala	A
arginine	Arg	R
asparagine	Asn	N
aspartic acid	Asp	D
cysteine	Cys	C
glutamic acid	Glu	E
glutamine	Gln	Q
glycine	Gly	G
histidine	His	H
isoleucine	Ile	I
leucine	Leu	L
lysine	Lys	K
methionine	Met	M
phenylalanine	Phe	F
proline	Pro	P
serine	Ser	S
threonine	Thr	T
tryptophan	Trp	W
tyrosine	Tyr	Y
valine	Val	V
MTSSL-modified Cys	-	R1

Substitutions are specified as: original amino-acid/position-in-sequence/substituting amino-acid, e.g., A41C. R1 (radical 1) is used by default to designate cysteine reacted with methyl thiosulphonate spin-label MTSSL; e.g., A41R1.

the existence of distinct side-chain rotamers that are determined by specific values of the χ_1 and χ_2 dihedral angles (see Table 16.5). As we see from Figure 16.15 (solid circles), the **m,m** χ_1,χ_2-rotamer is strongly preferred; indeed just three rotamers account for nearly all cases. This results from direct interaction of the disulphide with the peptide backbone, particularly from C^{α}-H···S^{δ} and N-H···S^{γ} interactions (Warshaviak et al. 2011). The two exceptions, V75R1 and L118R1 (see Table 16.5), are located at sterically strained sites. The sulphur–backbone association restricts segmental motion of the spin-label side chain to rotations solely about the terminal S^{δ}–C^{ε} and C^{ε}–C^3 bonds. These are affected by transitions between the χ_4- and χ_5-stable rotamers. Such limited internal flexibility implies that the EPR spectra of R1 should reflect motion also of the peptide backbone.

Where resolved, the χ_3 dihedral in Table 16.5 takes the approximate values ±90°, corresponding to one of the two stable conformations about the S–S bond (see also Table A16.1 for MTSSL alone). In the case of RX, complementary S–S conformers allow this bifunctional reagent to bridge between helix residues 115 and 119 of T4 lysozyme (see Table 16.5). For surface-exposed non-interacting sites, dihedrals beyond χ_1 and χ_2 are not resolved in T4 lysozyme, because of the inherent disorder that is associated with χ_4/χ_5 rotational freedom. This includes the loop site A82R1. All dihedral angles are resolved for the buried residue L118R1 and for residue V75R1 which is buried in the interface at a crystal contact site. As noted already, the values of χ_1 and χ_2 for these two residues do not correspond to the usual rotamers; χ_1 for the L118R1 core residue does not even correspond to a stable rotamer. Some water-exposed residues of T4 lysozyme also have a unique complement of χ-angles; in these cases, specific interactions stabilize a single conformation but with χ_1,χ_2 corresponding to one of the usual three low-energy combinations. Some EPR spectra of T4-lysozyme R1-mutants display two components, and certain of the corresponding crystal structures indicate two populations of R1-rotamers. Besides the basal **m,m** rotamer, residue 119R1 also populates the **t,p** rotamer, which is stabilized by interaction with a neighbouring side chain. Residue T115R1 populates both the **m,m** and **t,m** rotamers at 298 K but solely the **m,m** rotamer on cooling to 100 K. This shows the need to study the temperature dependence of the EPR lineshapes before correlating components with the low-temperature x-ray structure.

Loop sites in the effector chaperone Spa15 and the cytolysin repressor CylR2 also have the favoured **m,m** combination (see Table 16.5 and Figure 16.15) and S^{δ}–C^{α} distances of 0.35 nm or less. This again indicates stabilization by the unconventional H-bond with S^{δ}. The two loop sites in Spa15 are buried. Exposed α-helical sites in the B1 immunoglobulin-binding domain of protein G (K28R1 of the GB1-dimer), and a buried loop site in cytochrome P450cam (S190R1) also have the preferred **m,m** rotamers for χ_1,χ_2. Two of the rotamer structures of the exposed β-strand site (N8R1) in GB1 are also **m,m**, and a third is **m,t**. This latter infrequent combination has a S^{δ}–C^{α} distance of 0.46 nm, which precludes forming a H^{α}–S^{δ} H-bond; it is found also for V75R1 in T4 lysozyme, although at a crystal contact (see above). The χ_2 dihedral found for N8R1 at a crystal contact site in GB1 does not correspond to a canonical rotamer, as for buried L118R1 in T4 lysozyme. Also, S48R1 in a loop region of cytochrome P450cam has non-rotameric values for χ_2 (see Table 16.5 and Figure 16.15).

Table 16.5 includes data for lipid-exposed residues at α-helical sites in a membrane protein, the leucine transporter LeuT (open circles in Figure 16.15). Both R1-sites correspond to the **m,m** χ_1,χ_2-rotamer, but all dihedrals are resolved. This is the result of extensive interactions of the nitroxide ring with neighbouring hydrophobic side chains in the low-dielectric environment of the membrane.

We also include data for lipid-exposed residues in a β-barrel membrane protein, the vitamin B_{12} transporter BtuB, in Table 16.5 (solid triangles in Figure 16.15). These differ considerably from the situation with α-helical proteins. The more populated χ_1,χ_2-rotamer is **p,p**, which is not represented in the α-helical proteins, neither water-soluble nor membrane-bound. This rotamer occurs because the C^{α}-H···S^{δ} intra-residue stabilizing interaction is replaced by N_{i+1}-H···S^{δ} and C=O···S^{δ} interactions (Freed et al. 2011). A similar situation arises for the T156R1 site at the barrel surface, although this residue has the **t,m** rotamer that is least frequent of the three found with α-helical proteins. The full complement of dihedral angles χ_1 – χ_5 is resolved at all BtuB sites given in Table 16.5 (as is also true for the α-helical membrane protein LeuT). However, the T156R1 and W371R1 side chains are accommodated differently in hydrophobic pockets, and V10R1 of the Ton box is sterically constrained in a tertiary contact.

TABLE 16.5 Dihedral angles for S–S linked methyl thiosulphonate spin-label (MTSSL) side chains R1, in bacteriophage-T4 lysozyme (T4L), cytolysin repressor (CylR2), effector chaperone (Spa15), cytochrome P450 (cyt P450cam), immunoglobulin-binding domain (GB1), leucine transporter (LeuT), potassium channel (KcsA) and vitamin-B12 transporter (BtuB); and of the bridging side chain RX in T4L

mutant/SL[a]	PDB *id*	*T* (K)	rotamer[b]	χ_1 (°)	χ_2 (°)	χ_3 (°)	χ_4 (°)	χ_5 (°)	*Ref.*[c]
T4L:									
A41R1	2Q9D	100	**t,p**	–175	57	86			1
S44R1 (chain A)	2Q9E	100	**m,m**	–83	–58	–95	76	–86	1
S44R1 (chain B)			**m,m**	–85	–55	–96	71	–78	1
S44R1 (chain C)			**t,m**	173	–96				1
K65R1	-	298	**t,p**	153	89	53			2
V75R1	-	100	**m,t**	–73	173	91	95		2
R80R1	-	298	**m,m**	–74	–66				2
R119R1 (chain A)	-	100	**m,m**	–50	–50				2
R119R1 (chain B)	-	100	**t,p**	175	54				2
A82R1	1ZYT	100	**m,m**	–68	–56	101			3
V131R1	2CUU	100	**m,m**	–69	–60				3
V131R1			**t,p**	175	80				3
V131R1	3G3V	291	**m,m**	–75	–57				3
V131R1			**t,p**	175	83				3
T151R1	3G3X	100	**m,m**	–83	–72				3
T151R1	3G3W	291	**m,m**	–82	–72				3
T115R1	2IGC	100	**m,m**	–81	–57	–92	76	78	4
T115R1/R119A	2OU9	100	**m,m**	–77	–33				4
T115R1	2OU8	298	**m,m**	–94	–28				4
T115R1			**t,m**	163	–63				4
T115R1(4-Br)	2NTG	100	**t,m**	–179	–85	–85	–112	–73	4
L118R1[d]	2NTH[d]	100	**m,p**	–104	32	88	54	107	4
L118R1	5JDT	100	**m,t**	–98	167	–98	–81	89	14
L118R1	5G27	293	**m,t**	–83	152	93	–74	86	14
CylR2:									
T55R1 (chain A)	2XIU	100	**m,m**	–63	–61	–87	–77	161	9
T55R1 (chain B)		100	**m,m**	–74	–53	–97	–72	156	9
Spa15:									
C19R1 (chain A)	2XGA	120	**m,m**	–77	–61	–112	–51	97	10
C19R1 (chain B)		120	**m,m**	–84	–63	–115	–50	97	10
cyt P450cam:									
S48R1 (chain A)	4EK1	100	**p,–107°**	72	–107	93	72	–89	11
S48R1 (chain B)		100	**p,–119°**	75	–119	102	56	–79	11
S190R1		100	**m,m**	–71	–80	–92	77	45	11
GB1 (N8R1/K28R1):									
N8R1 (chain A)	3V3X	100	**m,–120°**	–57	–120	–88	–109	61	12
N8R1 (chain C)		100	**m,m**	–56	–74	–76	–93	77	12
N8R1 (chain D, α)[e]		100	**m,t**	–59	162	72	61	–94	12
N8R1 (chain D, β)[e]		100	**m,m**	–63	–58	–83	–86	74	12
K28R1 (chain B)		100	**m,m**	–61	–63	112	148		12
K28R1 (chain C)		100	**m,m**	–60	–61	96	140	90	12
K28R1 (chain D)		100	**m,m**	–58	–66	97	81	–73	12
LeuT:									
F177R1	3MPN	100	**m,m**	–69	–57	107	103	–24	5
I204R1	3MPQ	100	**m,m**	–69	–59	–87	71	–95	5

(*Continued*)

TABLE 16.5 (*Continued*) Dihedral angles for S–S linked methyl thiosulphonate spin-label (MTSSL) side chains R1, in bacteriophage-T4 lysozyme (T4L), cytolysin repressor (CylR2), effector chaperone (Spa15), cytochrome P450 (cyt P450cam), immunoglobulin-binding domain (GB1), leucine transporter (LeuT), potassium channel (KcsA) and vitamin-B12 transporter (BtuB); and of the bridging side chain RX in T4L

mutant/SL[a]	PDB *id*	*T* (K)	rotamer[b]	χ_1 (°)	χ_2 (°)	χ_3 (°)	χ_4 (°)	χ_5 (°)	*Ref.*[c]
KcsA:									
Y82R1	3STZ	-	**p,m**	59	−89	91	89	−41	13
BtuB:									
V10R1 apo	3M8B	90	**p,p**	56	69	83	67	67	6
V10R1 +Ca^{2+}, B_{12}	3M8D	90	**p,p**	49	60	70	93	46	6
T156R1	3RGM	90	**t,m**	176	−83	−94	−82	−30	7
W371R1	3RGN	90	**p,p**	47	78	64	113	130	7
bifunctional/T4L:									
T115RX-	3L2X	100	**m,m**	−94	−61	−81	−160	105	8
R119RX-			**m,m**	−70	−56	106	122	−90	
TEM-1 β-lactamase:									
E240R1	1BTL[f]	20	**t,t**	−135±30	−155±15	−80±15	−55±10	96±5	15

[a] R1, 1-oxyl-2,2,5,5-tetramethylpyrrolin-3-ylmethylsulpho- (MTSSL) S–S adduct of cysteine; RX, 1-oxyl-2,2,5,5-tetramethylpyrrolinyl-3,4-bis(methylsulpho-) double S–S adduct of cysteine.
[b] Rotamer defined by χ_1, χ_2.
[c] References: 1. Guo et al. (2008); 2. Langen et al. (2000); 3. Fleissner et al. (2009); 4. Guo et al. (2007); 5. Kroncke et al. (2010); 6. Freed et al. (2010); 7. Freed et al. (2011); 8. Fleissner et al. (2011); 9. Gruene et al. (2011); 10. Lillington et al. (2011); 11. Stoll et al. (2012); 12. Cunningham et al. (2012); 13. Raghuraman et al. (2012); 14. Consentius et al. (2016); 15. Mustafi et al. (2002).
[d] Superseded by 5JDT.
[e] α, β are two different conformers.
[f] ENDOR structure; PDB id is for native protein.
Note: further data are given in Table A16.3 of the appendix at the end of the chapter.

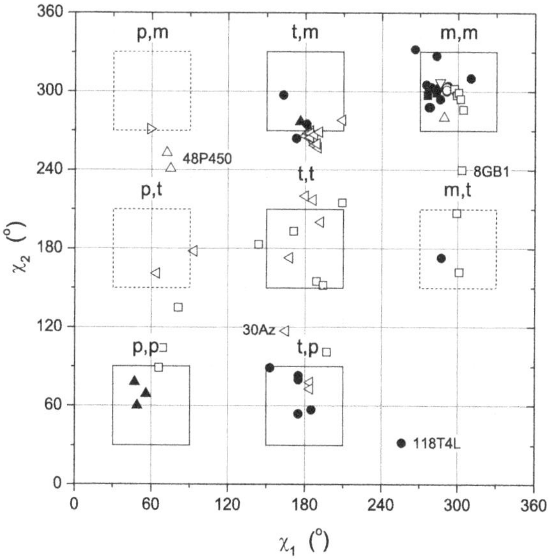

FIGURE 16.15 Rotamer preferences for S–S linked methyl thiosulphonate spin-label (MTSSL) side chains R1; at α-helical and loop sites in T4 lysozyme (solid circles), cytolysin repressor CylR2 (down triangle), effector chaperone Spa15 (solid squares), cytochrome P450cam (open triangles), leucine transporter LeuT (open circles) and KcsA K^+-channel (right triangle); at α-helical and β-strand sites in immunoglobulin-binding domain GB1 (open squares); at β-barrel surface of vitamin-B_{12} transporter BtuB (solid triangles); and at water-exposed β-sheet in azurin (left-pointing triangles). Dihedral angles χ_1 and χ_2 about C^α–C^β and C^β–S^γ bonds plotted from Tables 16.5 and A16.3. Note: $\chi = 300° \equiv -60°$, i.e., **m**.

Table A16.3, in the appendix to this chapter, lists dihedral angles of R1 for two water-exposed β-sheet sites in azurin. These are closer packed than at exposed sites in α-helices. We include this data as left-pointing triangles in Figure 16.15. A range of conformations arises from two crystal forms and inequivalent sites within the unit cell. χ_1, χ_2-Rotamers close to **t,m** are predominant, with none in the **m,m** region, which is that most populated at α-helical sites. Also, some conformations lie in the **p,t** and **t,t** regions that otherwise are populated only by β-strand edge sites of the immunoglobulin-binding domain GB1 (open squares).

16.7 χ_4/χ_5 LINESHAPE SIMULATIONS FOR R1-SIDE CHAINS

As we saw in the previous section, the internal motion of an MTSSL R1-side chain is restricted essentially to rotational isomerism about the last two bonds to which the nitroxide ring is attached (cf. also Fajer et al. 2007). This corresponds to transitions of the χ_4 and χ_5 dihedral angles alone. In the absence of any additional protein motions, the EPR spectrum then retains considerable angular anisotropy. Indeed, lineshapes of R1 at non-interacting, water-exposed sites can display characteristic axial anisotropy with effective $A_{\parallel}$- and $A_{\perp}$-hyperfine splittings

(Columbus et al. 2001; Columbus and Hubbell 2004). These correspond to the canonical χ_4/χ_5-model, upon which backbone motions and/or restrictions from interaction with neighbouring side chains are then superimposed.

Instead of considering dihedral angles explicitly, we treat the χ_4/χ_5-motion with the stochastic-Liouville model for axial ordering in an orienting potential that was introduced in Chapter 8 (Sections 8.11, 8.12). This is an approximation, but then we can incorporate additional motion or additional orientational restriction easily in an *ad hoc* manner. Justification for this approach comes from correlating the orientation of the diffusion axis with that expected for χ_4/χ_5-rotations (see Columbus et al. 2001).

We show the relation of the nitroxide axes (x_M, y_M, z_M) to the rotational diffusion axis Z_R and to the director-axis for ordering Z_D in Figure 16.16. The diffusion axes (X_R, Y_R, Z_R) are related to the nitroxide axes by the Euler angles $(\alpha_D, \beta_D, \gamma_D)$, as defined in Appendix R.2. Most important for water-exposed non-interacting helix sites is the diffusion-tilt angle β_D, whereas α_D and γ_D are relatively insignificant. The protein-fixed director axes are axially symmetric; the Z_R-diffusion axis makes an instantaneous angle θ with the director Z_D. The order parameter for the Z_R-axis is the time (or ensemble) average: $S_{Z_RZ_R} = \frac{1}{2}\left(3\langle\cos^2\theta\rangle - 1\right)$. This leaves open the question of how the director is oriented relative to the protein, which we only can answer by using aligned samples, e.g., single crystals.

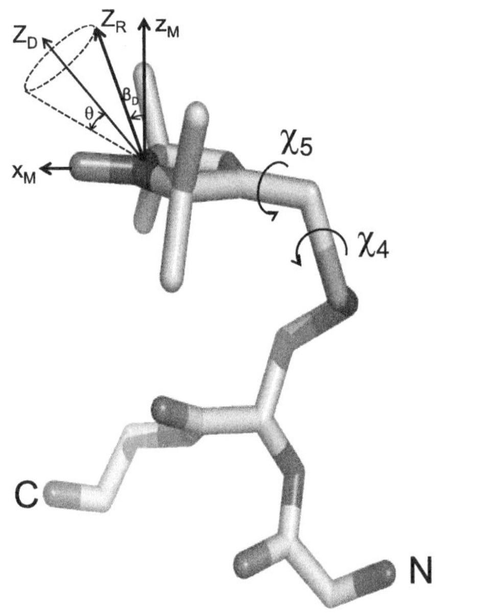

FIGURE 16.16 Axis systems describing rotational motion of R1 side chain in stochastic-Liouville simulations. Principal axes (X_R, Y_R, Z_R) of rotational diffusion tensor are related to nitroxide (x_M, y_M, z_M) axes by Euler angles $(\alpha_D, \beta_D, \gamma_D)$ (see Figure R.2 in Appendix R). Z_R-axis of diffusion tensor performs limited axially-symmetric angular excursions, θ, about director axis Z_D, which is fixed in the protein. For non-aligned sample, spectra must be summed over all orientations ψ of director to static magnetic field, with weighting $\sin\psi$ (i.e., a powder distribution).

16.7.1 Water-Exposed Non-interacting Sites on α-Helices

Residue D72R1 in T4 lysozyme is representative of a non-interacting, water-exposed site on an α-helix with inflexible backbone (Columbus et al. 2001). We show a typical simulated spectrum for this reference state at the top of Figure 16.17. It corresponds to an axially symmetric, partially-averaged powder pattern. The dynamic simulation parameters for this spectrum appear in the first line of Table 16.6. They are characterized by a diffusion tilt $\beta_D = 36°$, an order parameter $S_{Z_RZ_R} = 0.47$, and an effective correlation time: $\bar{\tau}_R = \sqrt[3]{\tau_{R,X}\tau_{R,Y}\tau_{R,Z}} \approx 2$ ns. Note that we take the geometric mean of the correlation times (or elements of the rotational diffusion tensor) because significant changes in the spectral parameters are produced by logarithmic, not linear, changes in τ_R (see Figure 16.12). We use the same principle in spectral fitting, where $\log \tau_R$ is the parameter that is optimized.

Residue V131R1 in T4 lysozyme is a non-interacting, water-exposed site on a short helix that is likely to exhibit backbone flexibility. The second line in Figure 16.17 shows a typical simulated spectrum. The lineshape is similar to that for D72R1 but with reduced anisotropy. Dynamic parameters needed for simulation

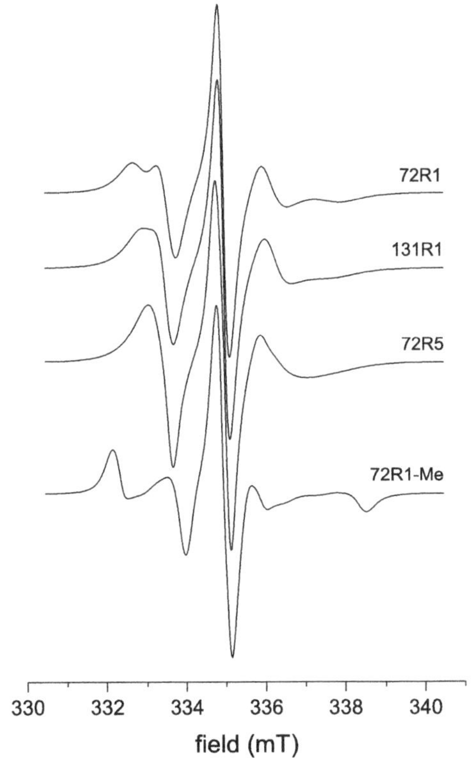

FIGURE 16.17 Simulated spectra for spin labels S–S linked to cysteine mutants of T4 lysozyme at non-interacting, water-exposed α-helix sites (see Table 16.6 for parameters; inhomogeneous broadening, 0.1 mT). 72R1: pyrroline nitroxide (MTSSL) on rigid backbone, $S_{ZZ} = 0.47$, $\bar{\tau}_R = 2.2$ ns, $\beta_D = 36°$. 131R1: pyrroline nitroxide on flexible backbone, $S_{ZZ} = 0.35$, $\bar{\tau}_R = 2.2$ ns, $\beta_D = 36°$. 72R5: pyrrolidine nitroxide on rigid backbone, $S_{ZZ} = 0.47$, $\bar{\tau}_R = 2.2$ ns, $\beta_D = 99°$. 72R1-Me: 4-methyl pyrroline nitroxide on rigid backbone, $S_{ZZ} = 0.81$, $\bar{\tau}_R = 3.7$ ns, $\beta_D = 36°$.

are characterized by a lower order parameter $S_{Z_RZ_R} = 0.35$, but similar correlation time and the same diffusion tilt (see Table 16.6). Backbone flexibility therefore manifests itself by increases in motional amplitude of the Z_R-axis.

We relate the diffusion-tilt angle β_D to the χ_4/χ_5-model by using Figure 16.16, which corresponds to the side-chain configuration in the crystal, with energy minimization for the preferred orientation of the nitroxide ring. Rotation about χ_5 is hindered by interaction of the α-methyl groups and 4-H of the nitroxide ring with S^δ of the disulphide. On the other hand, rotation about χ_4 is relatively free. As shown in Figure 16.16, the S^δ–C^ε axis of χ_4 is oriented at 115° to the C^ε–C^3 axis of χ_5, which lies in the plane of the nitroxide ring (see Table A16.1 at the end of the chapter). The χ_4-axis therefore makes an angle $\approx 115° - 90° = 25°$ with the nitroxide z_M-axis. This is the minimum value; it increases upon rotation about the χ_5-axis. In view of the approximate nature of the model, this orientation correlates reasonably well with the limited diffusion-tilt angle β_D and explains the characteristic axially anisotropic spectra in Figure 16.17. The principal rotation axis ($Z_R \equiv D_{R\parallel}$) is therefore directed approximately along the χ_4-axis, whereas the χ_5-axis corresponds approximately to $D_{R\perp}$.

Further support for the χ_4/χ_5-model comes from simulation of spectra from R5, the saturated pyrrolidine analogue of R1 (see Figure 16.17). Absence of the double bond in R5 means that the C^ε–C^3 χ_5-bond no longer lies in the plane of the nitroxide ring and brings the χ_4-axis approximately perpendicular to the nitroxide z_M-axis (Columbus et al. 2001). Correspondingly, the diffusion-tilt angle β_D obtained from spectral simulation becomes much larger (see Table 16.6).

Increasing the steric bulk of the nitroxide ring by substituting 4-H in R1 with a methyl group restricts the motional amplitude and reduces the rate of motion (see Figure 16.17). The order parameter increases to $S_{Z_RZ_R} = 0.81$, and the effective correlation time at 21 °C increases to $\bar{\tau}_R \approx 4$ ns. This effect of steric hindrance is also consistent with the χ_4/χ_5-model. Temperature-dependent changes arise solely from changes in rotational rate: the order parameter remains constant. The activation energy for rotation increases from 27 ± 1 kJ mol^{-1} for R1 to 42 ± 1 kJ mol^{-1} for 4Me-R1 (Columbus et al. 2001). For comparison, the activation energy for viscous flow of 30% sucrose (used to slow down rotation of the whole protein) is ≈ 25 kJ mol^{-1}. Thus, although the R1 side chain may experience only part of the viscous drag from the aqueous environment, the barrier for bond rotation must be relatively low, in agreement with the χ_4/χ_5 rotameric model.

Systematic nitroxide scans of the basic leucine-zipper motif from transcription factor GCN4 reveal the progressive effect of increasing backbone flexibility in the DNA-binding region on the R1-side chain at water-exposed sites of the helix dimer (Columbus and Hubbell 2004). Very narrow spectra ($\tau_R \approx 0.4$ ns) are found in the highly flexible N-terminal region, whereas spectra at water-exposed sites in the leucine-zipper region of the coiled-coil evidence restricted mobility ($\tau_R \approx 2$ ns). This demonstrates that EPR spectra of the R1-side chain and its congeners can register helix backbone motions directly.

16.7.2 Lipid-Exposed Non-interacting Sites in Transmembrane Helices

Motion of the R1-side chain at lipid-exposed sites on a transmembrane α-helix of the leucine transporter LeuT differs from that of water-exposed helical sites on soluble proteins (see Table 16.6). For the F177R1 residue, rotational diffusion is strongly anisotropic, with the fastest rate about the Y_R-axis, and the degree of order is less than for T4 lysozyme residues. The diffusion-tilt

TABLE 16.6 Dynamic parameters at 21°C from χ_4/χ_5-simulations for S–S linked pyrroline (R1) and pyrrolidine (R5) nitroxides at solvent-exposed α-helical sites of bacteriophage-T4 lysozyme in buffer (T4L; Columbus et al. 2001), and leucine transporter in detergent micelles (LeuT; Kroncke et al. 2010)[a]

mutant/SL[b]	$D_{R,X}$ (s^{-1})	$\tau_{R,X}$ (ns)	$D_{R,Y}$ (s^{-1})	$\tau_{R,Y}$ (ns)	$D_{R,Z}$ (s^{-1})	$\tau_{R,Z}$ (ns)	$S_{Z_RZ_R}$	β_D (°)	α_D (°)[c]
T4L:									
D72R1	9.77×10^7	1.7	6.31×10^7	2.6	7.41×10^7	2.2	0.47	36	4
D72R5	9.77×10^7	1.7	6.31×10^7	2.6	7.41×10^7	2.2	0.47	99	15
V131R1	9.77×10^7	1.7	6.31×10^7	2.6	7.41×10^7	2.2	0.35	36	4
V131R5	9.77×10^7	1.7	6.31×10^7	2.6	7.41×10^7	2.2	0.35	99	15
LeuT:									
F177R1	3.16×10^7	5.3	9.55×10^7	1.8	$<10^7$[d]	>16	0.24	0	−57
F177R5(+)	2.82×10^7	5.9	1.17×10^8	1.4	1.12×10^7	15	0.24	0	−58
I204R1(α)	3.80×10^7	4.4	1.38×10^8	1.2	$<10^7$[d]	>16	0.15	0	−43
I204R1(β)	1.58×10^7	11	3.80×10^7	4.4	6.31×10^6	26	0.38	0	−90
I204R5(+)	2.82×10^7	5.9	1.23×10^8	1.4	1.70×10^7	9.8	0.18	6	−45

[a] Spin-Hamiltonian parameters used for the simulations are: $g_{xx} = 2.0076$, $g_{yy} = 2.0050$, $g_{zz} = 2.0023$, $A_{xx} = 0.62$ mT, $A_{yy} = 0.59$ mT, $A_{zz} = 3.7$ mT (T4L in buffer), and $g_{xx} = 2.0084$, $g_{yy} = 2.0058$, $g_{zz} = 2.0020$, $A_{xx} = 0.51$ mT, $A_{yy} = 0.53$ mT, $A_{zz} = 3.43$ mT (LeuT in detergent).
[b] R1: 1-oxyl-2,2,5,5-tetramethylpyrrolin-3-ylmethylsulpho- (MTSSL) S–S adduct of cysteine; R5(±): (±)-(1-oxyl-2,2,5,5-tetramethylpyrrolidin-3-yl)methylsulpho-S–S adduct of cysteine; α,β: two spectral components for I204R1.
[c] Simulation insensitive to sign of α_D. Taken to be negative (LeuT) for interpretation with anisotropic-rate χ_4/χ_5 model.
[d] Spectra insensitive to $D_{R,Z} < 10^7$ s^{-1}.

angle $\alpha_D = -57°$ is large, such that the Y_R-diffusion axis almost coincides with the orientation of the terminal C^{ε}–C^3 χ_5-bond that is found in the crystal (Kroncke et al. 2010). Thus, fastest rotation is about χ_5, possibly with a slower rotation about χ_4 that is reflected by the value of $D_{R,X}$. Two spectral components are found for residue I204R1 (see Table 16.6). One (α) is similar to that for F177R1, but the other (β), for which $\alpha_D = -90°$, correlates better with the orientation of the side chain in the crystal. The change in geometry from the R1 to R5 side chain leaves the spectral lineshapes relatively unchanged, as we see from the simulation parameters in Table 16.6. Rotation about χ_5 is more restricted for R5, but the different geometry now puts the Y_R-diffusion axis almost collinear with the χ_4-axis, thus explaining the similarity with the spectra of R1. Interestingly, the activation energy associated with the mean diffusion rates $\bar{\tau}_R^{-1}$ is extremely low ~5 kJ mol^{-1} for the lipid-exposed R1-side chains of LeuT, corresponding to a low activation energy for viscous flow in hydrocarbon environments.

16.7.3 Lipid-Exposed Sites on a Transmembrane β-Barrel

Spectra from the R1 spin-label side chain on the lipid-facing surface of the BtuB β-barrel divide into roughly two types (Freed et al. 2011; Xu et al. 2008; Ellena et al. 2011). Residues buried within the lipid bilayer, such as W371R1, have highly immobilized spectra close to the rigid limit. From Table 16.5, we see that W371R1 is the χ_1, χ_2 = **p,p** rotamer, for which the side chain lies in a pocket at the protein surface. On the other hand, residues close to the hydrocarbon–water interface, such as T156R1, display spectra consisting of two components, neither of which is rigidly immobilized.

Dynamic simulation parameters for T156R1 are given in Table 16.7; the proportion of the two components depends on the membrane thickness. For the shortest lipid, DLPC, only the more mobile component is present and the spectrum is characterized by moderately rapid anisotropic rotation about the Z_R-diffusion axis, with no preferential ordering but a large diffusion tilt to the nitroxide z-axis, viz., $S_{Z_RZ_R} = 0$, $\tau_R = 1.7$ ns and $\beta_D = 65°$. Simulated spectra of this type are shown in Figure 1.6 of Chapter 1 for large effective tilt angles, β_{eff} (but with a considerably shorter correlation time, $\tau_{R\parallel} \approx 0.3$ ns). This is the sole spectral component in the crystal, where the R1 side chain is in the **t,m** rotameric state (see Table 16.5) and extends out from the protein surface. The diffusion Z_R-axis is directed roughly along the average of the fourth and fifth bonds such that rotational isomerism of both χ_4 and χ_5 produces the observed spectral averaging (Freed et al. 2011). The slower component for T156R1 is simulated by rotation essentially about the nitroxide x-axis alone: $\beta_D = 0, \gamma_D = 11°$ (see Table 16.7). Spectra simulated for this type of slow anisotropic rotation are shown in the leftmost column of Figure 7.15 in Chapter 7. This rotational mode probably corresponds to a different χ_1, χ_2-rotamer, for which **t,t** is a likely candidate (Freed et al. 2011). The two-component spectra of T156R1 are characterized by a single spin–lattice relaxation time (T_{1e} measured by saturation recovery), which indicates exchange between two rotamers on the microsecond timescale, because this is too rapid for a conformational change of the protein. See also Section 16.9 given below.

The picture that emerges from the BtuB studies is one in which motion of the R1 spin label depends strongly on the solvation environment. In particular, the hydrophobic interior of the membrane favours interactions of the spin label with pockets on the protein surface, whereas more polar regions allow interactions with side chains directed to the solvent interface.

16.7.4 Water-Exposed Residues in β-Sheets

The propensity for interactions with adjacent side chains is greater in β-sheets, because nearest neighbours are less widely separated than in an α-helix.

Because of the pattern of H-bonding between strands, the unit cell in a planar antiparallel β-sheet consists of 4 residues, one pair of which is hydrogen bonded (HB) whereas the other pair is not (NHB) (see Figure 16.18). The side chains of nearest-neighbour residues in adjacent strands are on the same side of the sheet, whereas side chains of nearest neighbours within a strand are on

TABLE 16.7 Dynamic parameters in χ_4/χ_5-simulations for S–S linked pyrroline R1 nitroxides at lipid-facing sites in β-barrel of BtuB vitamin-B12 transporter (Freed et al. 2011)[a]

mutant/lipid[b]	$D_{R,X}$ (s^{-1})	$\tau_{R,X}$ (ns)	$D_{R,Y}$ (s^{-1})	$\tau_{R,Y}$ (ns)	$D_{R,Z}$ (s^{-1})	$\tau_{R,Z}$ (ns)	α_D (°)	β_D (°)	γ_D (°)
T156R1/DLPC	<3 × 10^6	-	1.5 × 10^7	11	9.8 × 10^7	1.7	−23	65	−26
T156R1/DMPC[c]	3.9 × 10^7	4.3	<3 × 10^6	-	<3 × 10^6	-	0	0	11
	<3 × 10^6	-	1.4 × 10^7	12	7.9 × 10^7	2.1	−23	63	−26
T156R1/POPC[c]	1.2 × 10^7	14	<3 × 10^6	-	<3 × 10^6	-	0	0	17
	<3 × 10^6	-	8.3 × 10^6	20	9.8 × 10^7	1.7	−24	58	−18
T156R1/DiErPC[c]	2.2 × 10^7	7.6	<3 × 10^6	-	<3 × 10^6	-	0	0	31
	9.8 × 10^6	17	<3 × 10^6	-	6.9 × 10^7	2.4	−36	50	−13

[a] Spin-Hamiltonian parameters used for the simulations are: $g_{xx} = 2.0085$, $g_{yy} = 2.0059$, $g_{zz} = 2.0021$, $A_{xx} = 0.65$ mT, $A_{yy} = 0.56$ mT, $A_{zz} = 3.50$ mT.
[b] R1, 1-oxyl-2,2,5,5-tetramethyl-3-pyrrolin-3-ylmethylsulpho- (MTSSL) S–S adduct of cysteine; DLPC, dilauroyl phosphatidylcholine; DMPC, dimyristoyl phosphatidylcholine; POPC, palmitoyl-oleoyl phosphatidylcholine; DiErPC, dierucoyl phosphatidylcholine.
[c] Two rows indicate two spectral components.

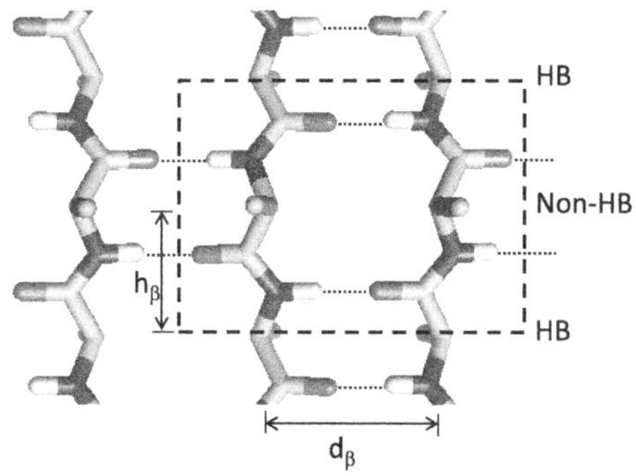

FIGURE 16.18 Adjacent chains in antiparallel β-pleated sheet of poly-L-alanine. Repeating unit (boxed) contains four residues, where only one pair is mutually hydrogen bonded (HB). h_β: rise per residue along strand axis; d_β: strand separation in plane of sheet.

opposite sides of the sheet. Interactions between side chains in the same strand are therefore less likely: the C^α–C^α distance between the i and i+2 residues is $d^{i,i+2}_{C^\alpha-C^\alpha} = 2h_\beta = 0.69$ nm, where h_β is the rise per residue along the strand. The C^α–C^α separation between residues on adjacent strands is smaller for the NHB pair than for the HB pair. This depends both on the separation between strands, d_β, and on the offset Δz between antiparallel strands that is determined by the pattern of hydrogen bonding. From the coordinates of β-polyalanine that we give in Table A16.4 of the Appendix at the end of the chapter, the separations are: $d^{NHB}_{C^\alpha-C^\alpha} = 0.45$ nm and $d^{HB}_{C^\alpha-C^\alpha} = 0.51$ nm, respectively.

Site-directed mutagenesis with the cytosolic retinal binding protein (c-RBP), a soluble flattened β-barrel, indicates little interaction of the R1 spin-label side chain with side chains of residues within the same strand (Lietzow and Hubbell 2004). Strong interactions are found, however, with side chains of residues on adjacent strands. The native HB neighbour of 52R1 is E41, and the EPR spectrum contains an immobile component ($\tau_R \approx 17$ ns), in addition to a more mobile ordered component ($\bar{\tau}_R = 1.3$ ns, $S_{Z_R Z_R} = 0.7$). Substitution of the native glutamic acid with either glutamine or lysine removes the immobile component, which is thought to arise from direct interaction of the nitroxide with the acidic side chain. On the other hand, substitution by aspartate or tryptophan leads to an increased population of the immobile component. Substitution of the native NHB neighbour, isoleucine, with the unbranched leucine removes the immobile component of 52R1, by alleviating interaction with the HB side chain. Other β-branched residues, threonine or valine, at the NHB position increase the population of immobilized component.

For residue N8R1 in a domain-swapped β-strand of the GB1 dimer from the B1 immunoglobulin-binding domain, the CW-EPR spectra at 20°C in 30% Ficoll are rather mobile (Cunningham et al. 2012). They also exhibit low order. Stochastic-Liouville simulations with fixed diffusion tilt $\beta_D = 32°$ and $S_{Z_R Z_R} = 0$ yield a mean correlation time $\bar{\tau}_R \equiv \sqrt[3]{\tau_{R\parallel}\tau^2_{R\perp}} = 2.1$ ns with a diffusion asymmetry $\tau_{R\perp}/\tau_{R\parallel} = 25$. This favours $\chi_1/\chi_2 = \mathbf{m},\mathbf{t}$, without C^α-H$\cdots$$S^\delta$ interaction, as the dominant conformer (cf. Table 16.5). In addition, interstrand interactions between side chains at this water-exposed site must be very limited.

16.8 LINESHAPE SIMULATIONS FROM MOLECULAR-DYNAMICS TRAJECTORIES

Molecular dynamics (MD) simulations can provide additional information on the rotational dynamics of the MTSSL side chain and also of the protein backbone to which it is attached. In principle, we can calculate the entire EPR lineshape from MD trajectories for the orientation of the nitroxide axes. We begin, however, with the more limited task of extracting order parameters S_{zz} for the nitroxide z-axis, and correspondingly of orientation potentials $U(\Omega)$, from MD simulations. After this, we go on to consider Brownian dynamics, and converting angular trajectories to those for tranverse spin magnetization; then leading to requirements imposed on the trajectories by the Fourier transform that yields the EPR spectrum.

16.8.1 Order Parameters from MD

To determine order parameters, we first must find the director for uniaxial ordering. For lipid membranes or liquid crystals this is obvious, but not so for SDSL with a protein. We use MD trajectories for polar angles θ, ϕ that the nitroxide z-axis makes with a fixed direction in the sample (or protein). The director for ordering is then the mode of the resulting angular distribution of nitroxide z-axes (see Figure 8.5). The order parameter is the average over this distribution: $S_{zz} = \frac{1}{2}\left(3\left\langle\cos^2\theta_z\right\rangle - 1\right)$, where as usual θ_z is the inclination of the nitroxide z-axis to the director (see Eq. 8.12). The MD trajectory thus provides information additional to that from the EPR spectrum of an unaligned sample, namely the orientation of the director relative to sample-fixed axes (LaConte et al. 2002).

16.8.2 Orientation Potentials and SLE Simulations

Once determined, we can use the orientation potential either with simpler stochastic dynamics to obtain extended trajectories (Steinhoff and Hubbell 1996; Beier and Steinhoff 2006) or with the stochastic Liouville equation (SLE) to simulate EPR lineshapes as in Section 16.7 (Budil et al. 2006). Assuming thermodynamic equilibrium, we obtain $U(\Omega)$ by fitting the normalized population of orientations, $p(\Omega)$, from MD Ω-trajectories with a Boltzmann distribution (cf. Eq. 8.22 in Section 8.4):

$$U(\Omega) = -k_B T \ln p(\Omega) + const \tag{16.16}$$

where the constant comes from normalizing the distributions, but is unimportant because we always deal with torques: $T_R = -\partial U/\partial\Omega$ that arise from the restoring potential (see Sections 8.11 and 8.12). To avoid generating artifactual torques, the potential surface must be free from digital noise spikes. We achieve this by fitting with analytical functions or by filtering (Beier and Steinhoff 2006). Steinhoff and Hubbell (1996) fit $p(\Omega)$ with products of Gaussians in the three Euler angles $\Omega = (\alpha, \beta, \gamma)$ that relate the nitroxide axis system to the protein-fixed axes.

Budil et al. (2006), on the other hand, use the axis systems shown in Figure 16.16 for the Liouville lineshape simulations of Section 16.7. These relate the diffusion axes by Euler angles $(\alpha_D, \beta_D, \gamma_D)$ to the nitroxide axes, and the principal diffusion axis Z_R (i.e., the molecular long axis that is subject to ordering) by polar angles (θ, ϕ) to the protein-fixed director. Then we need not only the two polar angles of the director relative to the protein axes, as in the order-parameter calculation above, but also the three diffusion tilt angles. Budil et al. (2006) fit both sets to the orientational distribution $p(\theta, \phi)$ from the MD trajectory, simultaneously with the potential $U(\theta, \phi)$. For this purpose, they expand the potential in a series of spherical harmonics $Y_{L,M}(\theta, \phi)$ (cf. Eq. 8.27). Stochastic-Liouville simulations of 250-GHz EPR lineshapes then yield the principal elements of the diffusion tensor. A notable feature of the MD fitting for S44R1 and Q69R1 in T4 lysozyme is that the principal diffusion axis is predicted approximately parallel to the nitroxide N–O bond (x_M), whereas the fastest diffusion rate deduced from the lineshape simulations is approximately about the nitroxide y-axis (y_M) (Budil et al. 2006).

Note that MD is not the only way to get the equilibrium population distribution $p(\Omega)$. We could use some other thermodynamic sampling technique, e.g., Monte-Carlo methods. In early determinations of $U(\Omega)$, MD simulations were performed at high temperature (600 K) *in vacuo* to achieve adequate sampling with relatively short trajectories (Steinhoff and Hubbell 1996; Beier and Steinhoff 2006). Instead, Budil et al. (2006) used Monte-Carlo minimization to locate the potential wells where short MD runs (1 ns) were then used to follow librational fluctuations at ambient temperatures.

16.8.3 Brownian Dynamics Trajectories

To create a stochastic Brownian dynamics trajectory, we introduce the orientation potential into the Langevin equation for the angular velocity $\omega_{R\perp}$ of the nitroxide z-axis, by adding the corresponding torque, $T_{R\perp} = -\partial U/\partial\theta_z$ (cf. Figure 8.3), to Eq. 7.65 of Section 7.12:

$$I^{inert}\left(\partial\omega_{R\perp}/\partial t\right) = -f_{R\perp}\omega_{R\perp}(t) + T_{R\perp}(\Omega(t)) + R_{R\perp}(t) \qquad (16.17)$$

where $f_{R\perp} = k_B T/D_{R\perp}$ is the frictional coefficient. In the limit of high friction, we can neglect inertial motion in Eq. 16.17 and the instantaneous angular velocity becomes:

$$\omega_{R\perp}(t) \equiv \partial\theta_z/\partial t = \left(D_{R\perp}/k_B T\right)\left[-\partial U(\Omega(t))/\partial\theta_z + R_{R\perp}(t)\right] \qquad (16.18)$$

Using Eq. 16.16 to express the torque, the recursive solution for the angular trajectory at the nth time point t_n with step length Δt becomes (Beier and Steinhoff 2006):

$$\theta_z(t_n + \Delta t) = \theta_z(t_n) + D_{R\perp}\Delta t\left[\partial \ln p(\Omega(t))/\partial\theta_z\right] + \Theta_{z,n} \qquad (16.19)$$

where $\Theta_{z,n}$ is a normally distributed random number with mean-square value: $\left\langle\Theta_{z,n}^2\right\rangle = 2D_{R\perp}\Delta t$ (cf. Eq. 7.67). If the orientational potential is fitted by a series of spherical harmonics, or Wigner rotation matrix elements $\mathcal{D}_{0,M}^L$ (cf. Eq. 8.27):

$$U(\Omega) = -\sum_{L,M}\lambda_{L,M}\mathcal{D}_{0,M}^L(\Omega) \qquad (16.20)$$

we can express the derivatives $\partial U/\partial\Omega$ analytically. From Eq. E.3 in Appendix E.1, the dimensionless angular momentum operator is $\mathbf{L} = -i\nabla_\Omega$, where $\nabla_\Omega = \left(\partial/\partial\phi_x, \partial/\partial\phi_y, \partial/\partial\phi_z\right)$. In Cartesian components, we therefore get:

$$L_z = -i\,\partial/\partial\phi_z \qquad (16.21)$$

$$L_\pm = L_x \pm iL_y = \pm\partial/\partial\phi_y - i\,\partial/\partial\phi_x \qquad (16.22)$$

where ϕ_z is the azimuthal rotation about the z-axis of the diffusion tensor and similarly for the x- and y-axes. (Do not confuse this subscript convention with that used for order parameters in Figure 8.3, where θ_z for ordering of the nitroxide z-axis corresponds to the azimuthal rotation: $\phi_{R\perp} \equiv \phi_x$ or ϕ_y.) Because $\mathcal{D}_{0,M}^L$ are eigenfunctions of L_z (see Eqs. R.20 and R.21), the derivatives that we need become (cf. Sezer et al. 2008a):

$$\frac{\partial U}{\partial\phi_x} = -\frac{1}{2}i\sum_{L,M}\lambda_{L,M}\left(C_{L,M}^+\mathcal{D}_{0,M+1}^L + C_{L,M}^-\mathcal{D}_{0,M-1}^L\right)$$
$$\Rightarrow -\sqrt{\frac{3}{2}}i\lambda_{2,0}\left(\mathcal{D}_{0,+1}^2 + \mathcal{D}_{0,-1}^2\right) \qquad (16.23)$$

$$\frac{\partial U}{\partial\phi_y} = -\frac{1}{2}\sum_{L,M}\lambda_{L,M}\left(C_{L,M}^+\mathcal{D}_{0,M+1}^L - C_{L,M}^-\mathcal{D}_{0,M-1}^L\right)$$
$$\Rightarrow -\sqrt{\frac{3}{2}}\lambda_{2,0}\left(\mathcal{D}_{0,+1}^2 - \mathcal{D}_{0,-1}^2\right) \qquad (16.24)$$

$$\frac{\partial U}{\partial\phi_z} = -i\sum_{L,M}\lambda_{L,M}M\mathcal{D}_{0,M}^L \Rightarrow 0 \qquad (16.25)$$

where $C_{L,M}^\pm = \sqrt{L(L+1) - M(M\pm 1)}$, and the expressions on the extreme right are for a Maier–Saupe potential: $U(\Omega) = -\lambda_{2,0}\mathcal{D}_{0,0}^2(\Omega)$ (see Eq. 8.24, where $\lambda_{2,0} \equiv k_B T\,\lambda_2$).

16.8.4 Spin Hamiltonian and Trajectory of Transverse Magnetization

To proceed further, we use the spin Hamiltonian with the Bloch equations (or stochastic Liouville equation) to transform the angular trajectory to that of the resonance frequency (or spin density-matrix), and hence to the time evolution of transverse spin magnetization (DeSensi et al. 2008). The latter is simply the free induction decay (FID) that we know from pulse EPR, where Fourier transformation to the frequency domain gives us the CW-EPR lineshape.

First, we transform the hyperfine and g-tensors from the nitroxide axis system, in which they are diagonal, to the laboratory axis system specified by the magnetic-field direction. We do this by using unitary transformations, as in Eq. B.25 of Appendix B.4: $\mathbf{g}'(t) = \mathbf{R}(t)\cdot\mathbf{g}\cdot\mathbf{R}(t)^{-1}$ and similarly for $\mathbf{A}'(t) = \mathbf{R}(t)\cdot\mathbf{A}\cdot\mathbf{R}(t)^{-1}$. But now the rotation matrices become time dependent:

$$\mathbf{R}(\Omega(t)) = \mathbf{R}_{global}(t)\mathbf{R}_{prot}\mathbf{R}_{\mathrm{NO}}(\Omega_{MD}(t)) \tag{16.26}$$

where $\Omega_{MD}(t)$ is the Euler-angle trajectory of the the nitroxide z-axis in the protein-fixed axis system, $\mathbf{R}_{prot}$ is the fixed rotation from protein-fixed axes to the rotation-diffusion axes of the protein and $\mathbf{R}_{global}(t)$ represents rotational diffusion of the whole protein in the laboratory axis system. If global protein diffusion is isotropic, we do not need $\mathbf{R}_{prot}$. If global diffusion is absent (as in membranes), we must replace $\mathbf{R}_{global}(t)$ by a powder-pattern summation from trajectories with fixed angles of the director to the magnetic field. We get the trajectory for global rotation from Brownian dynamics, by using Eq. 16.18 with $\partial U/\partial\theta_z$ and equivalents. This is conveniently done by using quaternions, instead of the three Euler angles (DeSensi et al. 2008; Sezer et al. 2008a). Because global diffusion is much slower than the motions normally covered by MD simulations, we must extrapolate the MD trajectories to match the longer timescale by using means discussed later. Note that we must eliminate any global rotation and translation from the MD trajectory by periodically checking and resetting the protein coordinates.

We eliminate high-frequency oscillations at the Larmor frequency, by working in a frame that rotates about the static magnetic field B_o at angular frequency $\omega_o = g_o\beta_e B_o$, where g_o is the isotropic g-value. This differs from the rotating frame introduced in Section 6.2, which rotates at frequency ω of the microwave radiation (absent here), and where ω_o stands there for the full resonance frequency. To include slow motion, we need to retain pseudosecular terms that come from off-diagonal elements of the rotated $\mathbf{A}'$-tensor (cf. Section 6.9). The time-dependent Hamiltonian thus becomes:

$$\widehat{\mathcal{H}}(t) = \left(g'_{zz}(t) - g_o\right)\beta_e B_o S_z + A'_{zz}(t)I_z S_z + A'_{xz}(t)I_x S_z + A'_{yz}(t)I_y S_z \tag{16.27}$$

where the crescent hat indicates the rotating frame. As usual, we omit nonsecular terms because they contribute only in second order. From the energy eigenvalues of Eq. 16.27, the electron resonance frequencies become (cf. Eq. H.8 in Appendix H.2):

$$\hbar\widehat{\omega}_{m_I}(t) = \left(g'_{zz}(t) - g_o\right)\beta_e B_o + m_I\sqrt{A'^2_{zz}(t) + A'^2_{xz}(t) + A'^2_{yz}(t)} \tag{16.28}$$

where m_I is the nuclear magnetic quantum number. For the simple case of isotropic rotation: $g'_{zz} = g_{zz}\cos^2\theta + \left(g_{xx}\cos^2\phi + g_{yy}\sin^2\phi\right)\sin^2\theta$ and $\left(A'_{zz}, A'_{xz}, A'_{yz}\right) = \left(A_{zz}\cos\theta, -A_{xx}\sin\theta\cos\phi, A_{yy}\sin\theta\sin\phi\right)$, where θ, ϕ are the polar coordinates of the nitroxide z-axis in the laboratory axis system (cf. Eqs. 2.11 and 2.12).

The Bloch equation for the transverse magnetization $\widehat{M}_+ = \widehat{M}_x + i\widehat{M}_y$ in the rotating frame is simply (see Eq. 6.7 with $B_1 = 0 = 1/T_2$):

$$\frac{\mathrm{d}\widehat{M}_+}{\mathrm{d}t} = -i\widehat{\omega}_{m_I}\widehat{M}_+ \tag{16.29}$$

Recursive solution with Eq. 16.28 and a time-step Δt then gives:

$$\widehat{M}_+(t+\Delta t) = \widehat{M}_+(t)\exp\left(-i\widehat{\omega}_{m_I}(t)\Delta t\right) \tag{16.30}$$

for each m_I-hyperfine line. This yields the required trajectory, when the time dependence of the $\mathbf{g}'$- and $\mathbf{A}'$-tensors is taken from angular trajectories obtained by MD simulation. We take an ensemble average of trajectories with different starting orientations Ω_o:

$$\left\langle\widehat{M}_+(t)\right\rangle = \int P_o(\Omega_o)\widehat{M}_+(\Omega_o, t)\mathrm{d}\Omega_o \tag{16.31}$$

where $P_o(\Omega_o)$ is the equilibrium distribution of starting orientations. The latter includes, if appropriate, the powder distribution ($\propto \sin\theta$).

16.8.5 Fourier Transformation

The trajectory ensemble $\left\langle\widehat{M}_+(t)\right\rangle$ is formally equivalent to the complex FID following a narrow 90°-pulse at frequency ω_o. We get the CW-EPR spectrum $\widehat{M}_+(\omega)$ in the frequency domain from the Fourier transform (see Appendix T.1). The absorption spectrum, centred about the isotropic g-value, then comes from the real part (or cosine transform):

$$v(\omega - \omega_o) = \mathrm{Re}\left[\widehat{M}_+(\omega)\right] = \mathrm{Re}\int_0^T \left\langle\widehat{M}_+(t)\right\rangle\exp(-i\omega t)\cdot\mathrm{d}t \tag{16.32}$$

where T is the total length of the trajectory. To allow for Lorentzian line broadening other than from rotational dynamics, we multiply $\widehat{M}_+(t)$ under the integral in Eq. 16.32 by the T_2-relaxation factor: $\exp\left(-t/T_2^o\right)$. We may also introduce inhomogeneous line broadening by multiplying further with a Gaussian convolution: $\exp\left(-\sigma_G^2 t^2/2\right)$, where σ_G^2 is the Gaussian second moment (see Section 2.6).

16.8.6 Trajectories

The reciprocal relation between time and frequency in a Fourier transform (see Appendix T.1) imposes certain requirements on the MD and magnetization trajectories. To get a frequency resolution equivalent to $\delta B = 0.01$ mT needs a trajectory with total length $T \geq 2\pi/\gamma_e\delta B = 3.6\ \mu\mathrm{s}$, which is not realized by most MD simulations. When the FID is sufficiently relaxed before this time, however, we can fill the remaining $\widehat{M}_+(t)$ trajectory with zeroes, using an apodization window to minimize any step artefact (DeSensi et al. 2008). This is standard practice in Fourier-transform spectroscopy.

Correspondingly, Fourier transformation requires a time increment Δt in the trajectory given by Eq. 16.30 that is short enough to cover the entire spectral width W. This is easily met for spectral widths of 8 mT at 9 GHz and 15 mT at 94 GHz, which correspond to $\Delta t \le 2\pi/(2\gamma_e W) = 2.2$ ns and 1.2 ns, respectively. The factor of two comes from the Nyquist sampling theorem. A further requirement is $\Delta t << \tau_c$, where τ_c is the shortest correlation time in the trajectory. However, trajectories from atomistic MD simulations are sampled at much shorter time steps, typically: $\delta t = 1$ ps. An efficient way to use the MD trajectory is to average the magnetic tensors over the time step $\Delta t >> \delta t$ used for the magnetization trajectory (Sezer et al. 2008a).

If the MD trajectories are too short to yield the desired spectral resolution, we need to extrapolate, e.g., by stochastic dynamics simulations, which are far less time-consuming than MD (Hakansson et al. 2001). Alternatively, we can append the original trajectory, time-reversed to ensure continuity, as many times back-to-back as necessary (Oganesyan et al. 2017). In principle, however, we need MD simulations only until the orientational correlation function is fully relaxed: at time $t \approx 10\tau_c$, where τ_c is the effective correlation time (see bottom panel of Figure 16.19). This offers a different approach to extrapolation, when the system has reached a stationary state with stable residual mean resonance frequency $\langle \hat{\omega}_{m_I} \rangle$ (Oganesyan 2007).

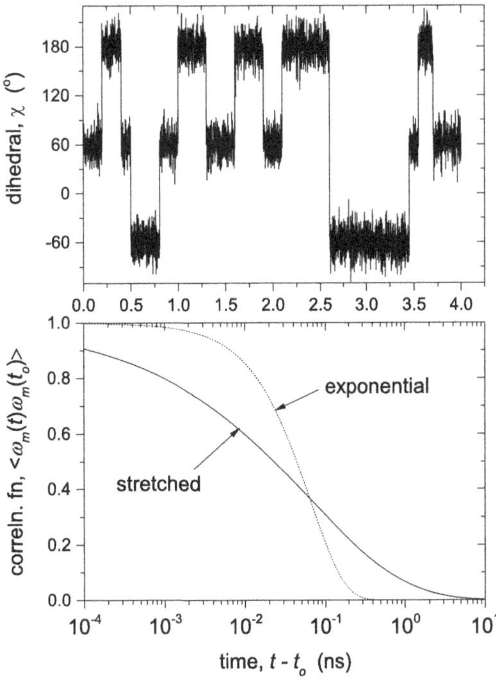

FIGURE 16.19 *Top*: schematic section of MD trajectory for dihedral angle χ that adopts **t** (180°), **p** (+60°) and **m** (−60°) rotamers. *Bottom*: schematic correlation functions $\langle \hat{\omega}_{m_I}(t)\hat{\omega}_{m_I}(t_o) \rangle$ for resonance frequency $\hat{\omega}_{m_I}$; deduced from trajectories for orientation θ_z of nitroxide z-axis to magnetic field. Solid line: stretched exponential (i.e., distribution of correlation times); dashed line: single exponential with $\tau_c = 0.063$ ns.

From the central limit theorem, the fluctuation in resonance frequency $\Delta\hat{\omega}_{m_I} = \hat{\omega}_{m_I} - \langle \hat{\omega}_{m_I} \rangle$ at these long times becomes a Gaussian random function. We thus get the ensemble/time average (see Eq. 13.31 in Section 13.6):

$$\langle \exp(-i\Delta\hat{\omega}_{m_I}) \rangle = \exp\left(-\frac{1}{2}\langle \Delta\hat{\omega}_{m_I}^2 \rangle\right) \tag{16.33}$$

where the second moment comes from integrating the autocorrelation function over time: $\langle \Delta\omega_{m_I}^2 \rangle = \int \langle \Delta\omega_{m_I}(t')\Delta\omega_{m_I}(0) \rangle dt' \equiv 2j_{m_I}(0)$, and the zero-frequency spectral density $j_{m_I}(0)$ is non-vanishing. For times $t > T_{trunc}$ longer than the MD simulation, the trajectory of the transverse magnetization therefore becomes (cf. Eq. 16.30):

$$\langle \hat{M}_+(t) \rangle = \langle \hat{M}_+(T_{trunc}) \rangle \exp\left(-\left(i\langle \Delta\hat{\omega}_{m_I} \rangle + j_{m_I}(0)\right)(t - T_{trunc})\right) \tag{16.34}$$

Both the mean resonance frequency and the correlation function are obtained from the MD trajectory over times $t \le T_{trunc}$ (see Kuprusevicius et al. 2011).

16.8.7 Discrete Markov States

The top panel in Figure 16.19 shows schematically a section from the trajectory for one of the spin-label dihedral angles. Rapid fluctuations take place on a timescale of 1–100 ps within the potential well of a given rotamer, before an abrupt transition takes place to a different stable rotamer. For MTSSL at a surface-exposed site of T4 lysozyme, dihedral angles χ_1 and χ_2 closest to the protein backbone undergo transitions only every 5–10 ns, whereas χ_4 and χ_5 next to the nitroxide ring undergo many more rotamer exchanges within a 50-ns trajectory (Sezer et al. 2009). This is consistent with the χ_4/χ_5 model of Section 16.7, except that we might expect χ_1 and χ_2 transitions still to make some slow-motional contribution to the EPR lineshape. The S–S dihedral χ_3, on the other hand, has a large activation barrier to transition between ±90° rotamers and remains essentially fixed, requiring two separate simulations starting with the different S–S conformers. We then weight these according to their relative populations predicted from energy minimizations.

Trajectories such as the top panel of Figure 16.19 suggest that we can construct a model of MD simulations that is based on jumps between discrete conformers that are specific combinations of $\chi_1 - \chi_5$ rotamers (Sezer et al. 2008b). Essentially, this is the two-site exchange model of Section 6.3, as extended for lipid-chain segments in Section 8.14. The rate equation for evolution of the probability $p_i(t)$ of a particular state i is given by the rate matrix $[W_{ij}]$:

$$\frac{d\mathbf{p}(t)}{dt} = \mathbf{W} \cdot \mathbf{p}(t) \tag{16.35}$$

However, short-term dynamics in MD simulations are likely to be deterministic, not stochastic. Therefore, we must select a lag time τ such that memory of previous orientations is lost. Then we can construct a discrete Markov chain to represent the rotamer jumps. The conditional probability that state i at time t has undergone transition to state j at time $t+\tau$ is given by the transition probability matrix $\left[P_{ij}(\tau)\right]$:

$$\mathbf{p}(t+\tau)=\mathbf{P}(\tau)\cdot\mathbf{p}(t) \tag{16.36}$$

We obtain the elements of $\mathbf{P}(\tau)$ by counting the number of times $N_{ij}(\tau)$ throughout the MD trajectory that an $i \rightarrow j$ transition occurs after lag time τ: $P_{ij}(\tau)=N_{ij}(\tau)\Big/\sum_{j\neq i}N_{ij}(\tau)$. Using the solution of Eq. 16.35, we find from Eq. 16.36 that $\mathbf{P}(\tau)=\exp(\tau\mathbf{W})$. If we write the eigenvalues of $\left[W_{ij}\right]$ as $-1/\tau_i$, the corresponding relaxation timescales become: $\tau_i=-\tau/\ln\lambda_i^P(\tau)$, where λ_i^P are the eigenvalues of $\mathbf{P}(\tau)$. We choose the length of the lag time (~100 ps) by requiring that the τ_i which we obtain are independent of the value of τ (Sezer et al. 2009). Trajectories of arbitrary length can then be generated from the relaxation times in the Markov chain.

The total number of rotamers of the MTSSL side chain is $3\times3\times2\times3\times2=108$, from all possible combinations of $\chi_1-\chi_5$. Some of these are not kinetically significant, and therefore we group the microstates into a smaller number of slowly exchanging macrostates within which constituent microstates exchange rapidly. Table 16.8 lists the five most populated macrostates for MTSSL at position 72 in a surface-exposed helix of T4 lysozyme (i.e., D72R1), which account for 68% of the total population. These come from extensive MD simulations modelled with a total of 37 states that provide a temporal resolution $\Delta t=0.16$ ns, which is more than adequate to cover the full spectral range at high field and narrow enough to allow pre-averaging of the hyperfine and g-tensors (Sezer et al. 2009). The five states in Table 16.8 are characterized by $\chi_1,\chi_2=$ **t,p** or **m,m** rotamers, which are those found most frequently in crystal structures of MTSSL-labelled T4 lysozyme (see Figure 16.15). The first four states are +90° S–S rotamers, whereas the fifth state has $\chi_3=-90°$ and is unlikely to exchange with the other four. Quite remarkably, the (hidden) Markov jump model is able to simulate multifrequency spectra (9, 95 and 170 GHz) of T4 lysozyme labelled at two different surface-exposed helical sites, D72R1 and V131R1, essentially without adjustable parameters (Sezer et al. 2009). The only fitting done was to expand the MD timescale by a factor two to correct for the 2–3× too low viscosity of the non-polarisable TIP3 waters used for simulation, and to take a correlation time of 9.3 ns for global protein rotation that is derived from spectral lineshape fitting.

16.9 SR-EPR: SPIN–LATTICE RELAXATION AND CONFORMATIONAL EXCHANGE

From Chapter 5, we know that the electron spin–lattice relaxation time T_{1e} also depends on the rotational correlation time τ_R (cf. Eq. 5.41) and thus will vary depending on local dynamics at the environment of the nitroxide. From Section 12.14 in Chapter 12, we also see that, with the usual protocols for saturation-recovery measurements of spin–lattice and cross relaxation, ordering is less significant than is rotational diffusion rate. Table 16.9 lists the effective T_{1e}-relaxation times for MTSSL (i.e., R1) at different sites in T4 lysozyme that are determined by saturation-recovery EPR. Values range from 1.9 to 7.5 μs. As we saw in Sections 12.4 and 12.6, these values contain contributions from cross relaxation.

Figure 16.20 shows the dependence of the spin–lattice relaxation rate on the mean correlation time $\bar{\tau}_R$ for local rotational motion, determined from lineshape simulations as in Chapter 6. Short correlation times correspond to faster T_{1e}-relaxation that varies more rapidly with $\bar{\tau}_R$, whereas the relaxation rate is lower at longer correlation times and varies more slowly with $\bar{\tau}_R$. The dependence of ST-EPR recovery rates on τ_R, and their independence of m_I, were discussed in Sections 12.4–12.8 of Chapter 12. Empirically, we can characterize the relaxation rates

TABLE 16.8 Five most-populated discrete Markov states modelling MD trajectories for D72R1 mutant of T4 lysozyme (Sezer et al. 2009)

state	*rotamer*[a]	*population (%)*		χ_1 (°)	χ_2 (°)	χ_3 (°)	χ_4 (°)	χ_5 (°)
1	**t,p**	22.3	20.8[b]	−176	49	93	−82	100
	t,p		1.3[b]	−170	54	102	−65	97
2	**t,p**	15.7	10.6	−168	48	85	−159	−70
	t,p		2.9	−169	54	85	−169	−81
3	**t,p**	13.7	4.0	−169	48	78	−144	20
	t,p		3.8	−162	47	79	−170	88
4	**m,m**	8.7	4.0	−63	−66	94	172	57
	m,m		1.8	−65	−78	84	145	87
5	**m,m**	7.4	1.2	−55	−44	−91	−178	−93
	m,m		1.1	−57	−37	−87	−171	−7

MTSSL dihedral angles $\chi_1-\chi_5$ are given for the two most probable microstates in each state.
[a] Rotamer defined by χ_1,χ_2.
[b] First two microstates in state 1, et seq.

TABLE 16.9 Effective electron spin–lattice relaxation times T_{1e}^{eff} for MTSSL (R1) S–S linked to cysteine residues introduced at different sites in bacteriophage-T4 lysozyme

residue	$\bar{\tau}_R$ (s)	T_{1e}^{eff} (μs)[a]	*Ref.*[b]
K48R1	1.8×10^{-9}	2.2	1
D72R1	2.5×10^{-9}	3.2	1
N81R1	2.3×10^{-9}	2.7	1
E128R1		2.1	2
A129R1		5.5	2
A130R1		2.4	2
V131R1	2.0×10^{-9}	2.3	1
N132R1		1.9	2
L133R1	1.63×10^{-8}	7.5	1
A134R1		3.9	2
L135R1		2.1	2
I150R1	2.8×10^{-9}	2.9	1
T151R1	2.4×10^{-9}	2.7	1

Correlation time of local rotational motion is $\bar{\tau}_R$.

[a] $1/T_{1e}^{eff} \equiv 1/T_{SR,e} = 2\bar{W}_e + 2\bar{W}_x$, which includes cross relaxation, and rates $\bar{W}$ are averaged over all N-hyperfine manifolds m_I.

[b] References: 1. Bridges et al. (2010); 2. Pyka et al. (2005).

$\bar{W} = 1/\left(2T_{1e}^{eff}\right)$ in the two regimes of Figure 16.20 by approximately linear dependences:

$$\bar{W}_f(\text{MHz}) = 0.345 - 0.067 \times \bar{\tau}_R(\text{ns}) \tag{16.37}$$

$$\bar{W}_b(\text{MHz}) = 0.146 - 0.0050 \times \bar{\tau}_R(\text{ns}) \tag{16.38}$$

where $\bar{\tau}_R < 3.2$ ns for Eq. 16.37 (subscript f) and $\bar{\tau}_R > 3.2$ ns for Eq. 16.38 (subscript b).

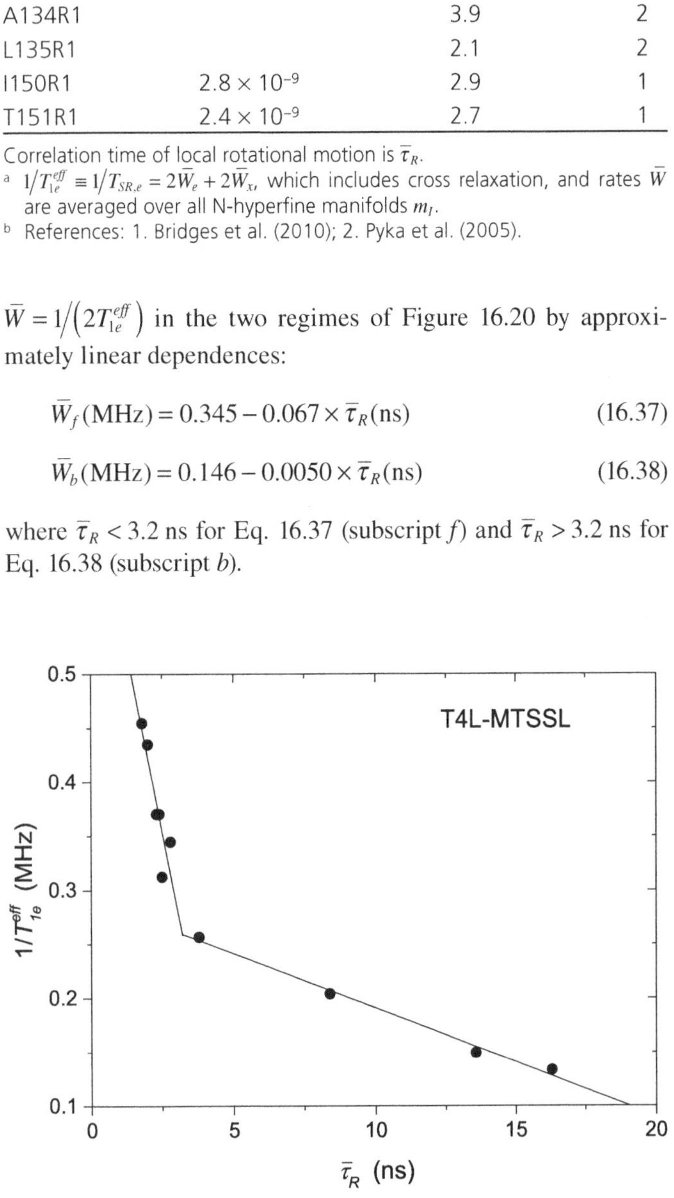

FIGURE 16.20 Dependence of composite spin-lattice relaxation rate $1/T_{1e}^{eff} = 2\bar{W}_e + 2\bar{W}_x$ on rotational correlation time $\bar{\tau}_R$ for S–S linked MTSSL (R1) at different sites in T4 lysozyme at 25°C. Also included are values for 131R1-Me, 131R1-Ph and 131RX135. Data from Bridges et al. (2010).

The distinction between correlation-time regimes lets us determine slow exchange rates in two-component spectra from a single site of labelling by using SR-EPR (Bridges et al. 2010). Kinetic equations for SR-EPR with a two-site model that applies to this type of T_{1e}-exchange spectroscopy are given in Section 12.10 of Chapter 12. The rate constants for biexponential recovery are:

$$\lambda_{1,2} = \bar{W}_b + \bar{W}_f + \tfrac{1}{2}\left(\tau_b^{-1} + \tau_f^{-1}\right) \mp \sqrt{\left(\bar{W}_b - \bar{W}_f\right)^2 + \left(\bar{W}_b - \bar{W}_f\right)\left(\tau_b^{-1} - \tau_f^{-1}\right) + \tfrac{1}{4}\left(\tau_b^{-1} + \tau_f^{-1}\right)^2} \tag{16.39}$$

where τ_b^{-1} and τ_f^{-1} are the rates of transfer from site b to site f, and vice versa, respectively. When exchange is slow, i.e., $k_{ex} \equiv \tau_b^{-1} + \tau_f^{-1} << \left|\bar{W}_b - \bar{W}_f\right|$, the rate constant with appreciable amplitude is $\lambda_2 \approx 2\bar{W}_b + (1 - f_b)k_{ex}$, where f_b is the fractional population of state b. When exchange is fast, i.e., $k_{ex} >> \left|\bar{W}_b - \bar{W}_f\right|$, both components have significant amplitudes and the rate constants are: $\lambda_2 \approx 2(1 - f_b)\bar{W}_b + 2f_b\bar{W}_f + k_{ex}$ and $\lambda_1 \approx 2f_b\bar{W}_b + 2(1 - f_b)\bar{W}_f$. For further details, see Section 12.10.

From the measured exchange rate we can distinguish exchange that occurs only between different MTSSL rotamers from large-scale conformational changes in the protein. Rotamer exchange generally is characterized by rates faster than 2 MHz, whereas functionally significant protein conformational changes are much slower (Bridges et al. 2010).

16.10 TOAC SPIN-LABEL RESIDUE IN PEPTIDES: HELIX ORIENTATIONS

Unlike MTSSL, which is a spin-label side chain, TOAC (1-oxy-2,2,6,6-tetramethyl-4-aminopiperdin-4-yl carboxylic acid) is a nitroxyl amino acid that is incorporated directly in the peptide backbone (see Figure 16.21). The piperidine-N-oxy ring is a special case of a branched side chain. Because of the double substitution at C^α, it is a conservative replacement for α-aminoisobutyric acid (β-alanine), which strongly promotes helix formation.

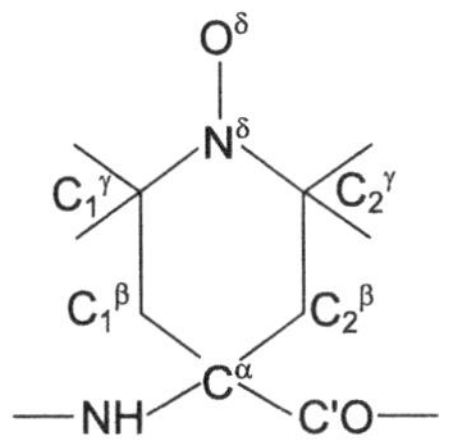

FIGURE 16.21 TOAC amino-acid spin label; peptide chain continues at –NH and C′O–. Nitroxyl ring is a branched side chain. In conventional numbering of piperidine ring: N^δ is position 1; C_2^γ, position 2; C_2^β, position 3; C^α, position 4; C_1^β, position 5; C_1^γ, position 6.

The rigid attachment to the peptide backbone therefore makes TOAC an excellent reporter of helix orientation and dynamics.

For uniaxial motional averaging, the EPR order parameter of TOAC in a regular secondary structure is given by the addition theorem for spherical harmonics (see Eq. 8.31 in Section 8.5):

$$S_{zz} = \langle P_2(\cos\theta)\rangle \cdot P_2(\cos\beta_z) \qquad (16.40)$$

where θ is the angle that the helix axis makes with the director (e.g., membrane normal) and β_z is the inclination of the nitroxide z-axis to the helix axis (see Figure 16.22). $P_2(x) = \frac{1}{2}(3x^2 - 1)$ is a second-order Legendre polynomial, and angular brackets indicate either a time average over the molecular motion or an ensemble average. To determine $\langle P_2(\cos\theta)\rangle$, the orientational order parameter of the helix axis (or, in general, of the molecular diffusion axis), we need to know the diffusion-tilt angle β_z between the spin-label z-axis and the helix axis.

Crystal structures of TOAC in peptides show that the preferred conformation of the spin-label ring is the twist-boat form, of which there are two main conformers: that with ${}^6T_2$ ring puckering and the mirror image ${}^2T_6$ (Flippen-Anderson et al. 1996; Crisma et al. 2005). For ${}^6T_2$, ring position 6 (i.e., C_1^γ) lies above the average plane, and position 2 (i.e., C_2^γ) lies below the plane, as viewed in the projection of Figure 16.21. When the TOAC structures are built into the coordinates of α-poly-L-alanine (see Table A16.5), we find that the ${}^6T_2$ conformer is oriented with the spin-label z-axis close to the helix axis, whereas the z-axis is tilted away from the helix in the ${}^2T_6$ conformer. Table 16.10 lists the orientations β_x, β_y and β_z of the nitroxide x, y, z-axes to the α-helix axis for various TOAC structures. Note that the three β_i are interrelated by the orthonormality condition for direction cosines: $\cos^2\beta_x + \cos^2\beta_y + \cos^2\beta_z = 1$. For the larger ${}^6T_2$ group, the mean values are: $\beta_x = 83.0 \pm 4.2°$ and $\beta_z = 13.3 \pm 2.2°$ ($\beta_y = 79°$); and for the mirror-image conformer are: $\beta_x = 88.5 \pm 6.7°$ and $\beta_z = 64.9 \pm 3.2°$ ($\beta_y = 25°$). We need values of β_x whenever the averaging is not uniaxial, in which case measurements on aligned samples and at higher EPR frequencies can be helpful.

For axially symmetric ordering, the orientation of the nitroxide z-axis in an α-helix is characterized by $P_2(\cos\beta_z) = 0.92 \pm 0.03$ and $P_2(\cos\beta_z) = -0.20 \pm 0.06$, for the two different TOAC ring conformers. The EPR experiment is unable to

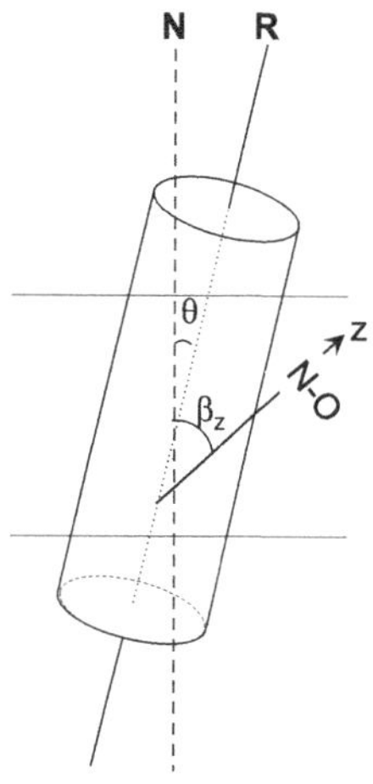

FIGURE 16.22 Orientation of TOAC-labelled helix in a membrane. Principal molecular diffusion axis **R**, inclined at instantaneous angle θ to membrane normal (or director) **N**. Nitroxide z-axis oriented at fixed angle β_z to **R**. Experimental order parameter S_{zz} of nitroxide z-axis is given by Eq. 16.40, where $\langle P_2(\cos\theta)\rangle$ is order parameter of helix diffusion axis.

TABLE 16.10 Angles β_x, β_y, β_z that TOAC nitroxide x, y, z-axes make with the α-helix axis (α-poly-L-alanine), for different TOAC-containing peptide crystal structures (Marsh 2006)

peptide/residue[a]	β_x (°)	β_y (°)		β_z (°)		*pucker*
		$C_1^\gamma N^\delta O^{\delta,b}$	$C_2^\gamma N^\delta O^{\delta,c}$	$C_1^\gamma N^\delta O^{\delta,b}$	$C_2^\gamma N^\delta O^{\delta,c}$	
I/TOAC[1]A	80.0	78.5	81.7	15.3	13.0	${}^6T_2$
/TOAC[1]B	87.4	77.3	77.3	13.0	13.0	
II/TOAC[1]	75.9	82.0	80.6	16.3	17.0	${}^6T_2$
/TOAC[4]	87.8	80.4	76.2	9.8	13.9	
III/TOAC[1]	81.4	78.7	77.3	14.2	15.4	${}^6T_2$
IV/TOAC[4]A	81.2	80.0	84.0	13.4	10.7	${}^6T_2$
/TOAC[4]B	86.5	81.1	79.2	9.6	11.4	-
/TOAC[8]A	83.8	79.5	77.1	12.2	14.4	${}^6T_2$
/TOAC[8]B	65.7	78.7	74.8	27.1	29.2	-
V/TOAC[1]	80.8	26.7	22.5	65.1	69.7	${}^3T_1$
VI/TOAC[1]	92.1	25.4	24.2	64.7	65.9	${}^2T_6$
/TOAC[2]	92.6	25.5	30.5	64.6	59.6	

[a] Peptides and coordinates: I, Z-TOAC-(L-Ala)$_2$-NH*t*Bu, CCDC 123753 (Flippen-Anderson et al. 1996); II, *p*BrBz-TOAC-(L-Ala)$_2$-TOAC-L-Ala-NH*t*Bu, CCDC 123754 (Flippen-Anderson et al. 1996); III, Boc-TOAC-[L-(αMe)Val]$_4$-NH*t*Bu, CCDC 257672 (Crisma et al. 2005); IV, trichogin GA IV *n*Oct-[TOAC^{4,8},Leu-OMe^{11}], CCDC 120048 (Monaco et al. 1999a); V, Ac-TOAC-(Aib)$_3$-L-Trp-Aib-O*t*Bu, CCDC 257673 (Crisma et al. 2005); VI, Fmoc-(TOAC)$_2$-(Aib)$_4$-O*t*Bu, CCDC 257674 (Crisma et al. 2005). A and B indicate two inequivalent molecules in the asymmetric unit.

[b] Nitroxide z-axis defined as normal to (pro-L) $C_1^\gamma N^\delta O^\delta$ plane.

[c] Nitroxide z-axis defined as normal to (pro-D) $C_2^\gamma N^\delta O^\delta$ plane.

determine the sign of the order parameter. From Eq. 16.40, we see that experimental order parameters for uniaxial ordering $|S_{zz}| > 0.26$ are consistent only with the ${}^6T_2$ conformer, for which $P_2(\cos\beta_z) = 0.92$. This seems to be the case for membrane-bound α-helical TOAC peptides. Experimental values of S_{zz} are in the region of 0.56–0.94, corresponding to helix order parameters: $\langle P_2(\cos\theta)\rangle = 0.61–1$ that are consistent with a transmembrane helix (Monaco et al. 1999b; Karim et al. 2004). For TOAC in alamethicin, the tilt of the nitroxide z-axis varies little with temperature and does not depend on chain length of the host lipid membrane, although values of β_z are larger than deduced from crystal structures (Marsh et al. 2007b; Marsh et al. 2007a). This suggests that the principal diffusion axis tilts away from the helix axis of alamethicin.

Incorporation of TOAC into extended β-sheets is unlikely because of strong steric clashes with the side chain from one of the adjacent strands, unless this residue is glycine or the sheet is strongly twisted. Otherwise, possible TOAC label positions will be limited to β-sheet edges, two-stranded anti-parallel β-ribbons or to a single strand/extended chain. For these putative β-strands, the difference in orientation of the TOAC nitroxide z-axis between the two mirror-image conformers is less than for α-helices. For the ${}^6T_2$ conformers, the mean values are $\beta_x = 91.0 \pm 4.8°$ and $\beta_z = 24.9 \pm 3.4°$ ($\beta_y = 65°$), whereas for the mirror-image TOAC conformers: $\beta_x = 84.7 \pm 3.7°$ and $\beta_z = 31.6 \pm 3.3°$ ($\beta_y = 59°$) (Marsh 2006). Note that when the molecular diffusion axis does not coincide with the β-strand direction, we also need the orientation of the β-strand axis to the molecular diffusion axis (cf., Marsh 1997).

16.11 DISTANCES BETWEEN SITE-DIRECTED SPIN LABELS: α-HELICES, β-STRANDS

Measurement of magnetic dipole–dipole interactions between electron spins is covered extensively in Chapter 9. In SDSL, we are concerned mostly with situations where the dipolar splitting is less than the anisotropic spread of the spectrum from a single label (see Figure 9.5). We then must consider the two spins as being unlike (i.e., weak coupling), which is important for calculating their separation (see Section 9.4). Distances between spin labels up to 1.6 nm can be obtained from CW-EPR and in the range 1.8–6 nm from pulsed ELDOR (see Chapter 15). The principal complication that arises in SDSL comes from the flexibility of the spin-label side chain, which gives rise to a distribution of interspin distances. In many cases, dipolar data are processed to yield the distance distribution directly. In other cases, a Gaussian distribution is a reasonable representation (see Chapter 15). On the other hand, inflexible labels (such as TOAC) report the distance between backbone sites directly. A useful alternative is RX (see Table 16.5), the bifunctional version of MTSSL, which reacts with two cysteine residues at positions i and i+4 in an α-helix to provide a rigid anchor.

16.11.1 α-helices

We can estimate the effective mean position of the nitroxide spin, relative to the protein backbone, from systematic SDSL studies on a regular α-helix (Rabenstein and Shin 1995). Such CW-EPR experiments are described in Chapter 15. Assume that the nitroxides are located at a fixed average radial distance R_{NO} from the helix axis, a spin label at residue position n_i has the cylindrical coordinates $(R,\phi,z) \equiv (R_{NO}, 2\pi n_i/n_P, h_\alpha n_i)$. The distance apart of two spin labels at positions n_1 and n_2 in the helix is then:

$$r_{12} = \sqrt{2\left(1-\cos\frac{2\pi\Delta n}{n_P}\right)R_{NO}^2 + (h_\alpha \Delta n)^2} \quad (16.41)$$

where $\Delta n \equiv n_2 - n_1$, $n_P = 3.6$ is the number of residues per turn and $h_\alpha = 0.15$ nm is the vertical rise per residue (see Table 16.1). Solid circles in Figure 16.23 give the mean interspin distances for MTSSL spin labels on two cysteine residues that are positioned increasingly far apart in an aqueous α-helical peptide with sequence $(A_4K)_4A$. We see that the dependence on Δn is described reasonably well by Eq. 16.41, with an effective radial distance of $R_{NO} = 0.74$ nm for each nitroxide. A somewhat larger value results

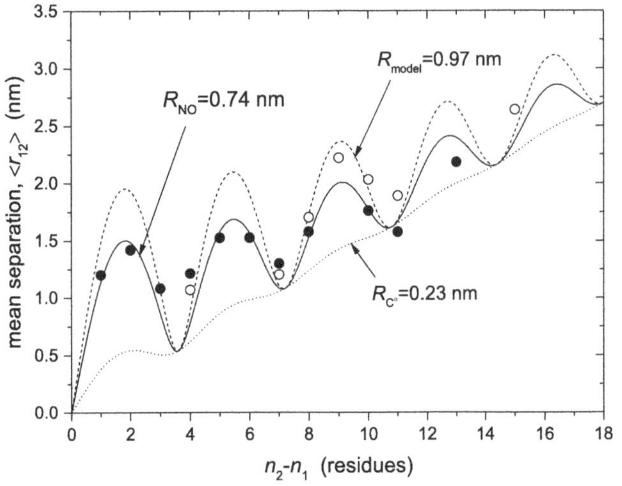

FIGURE 16.23 Dependence of mean separation $\langle r_{12}\rangle$ between nitroxides on number of residues apart $n_2 - n_1$ in an α-helix. *Solid line*: Eq. 16.41 for mean radial distance $R_{NO} = 0.74$ nm of nitroxide from helix-axis; *dashed line*: $R_{NO} = 0.97$ nm; *dotted line*: C^α positions at $R_{C^\alpha} = 0.23$ nm. *Solid circles*: mean separations from dipolar splittings measured by various methods (Rabenstein and Shin 1995; Banham et al. 2008; Marsh 2014b); *open circles*: molecular modelling (Banham et al. 2008); for α-helical peptides $(Ala_4Lys)_4Ala$ spin-labelled with MTSSL on cysteine residues at positions n_1 and n_2 in peptide sequence. Samples frozen in trifluoroethanol/water solutions.

from molecular modelling (open circles). The experimental result corresponds to a fixed distance $\Delta R_{NO} = R_{NO} - R_{C^\alpha} = 0.51$ nm of the nitroxide from the cylinder surface defined by the C^α atoms of the helix backbone (cf. Table 16.1). For reference, the coordinates of an α-polyalanine helix are given in Table A16.5 of the appendix at the end of the chapter.

The reasonable fit with a fixed value of R_{NO} in Figure 16.23 implies that the orientations of the labels at different positions in $(A_4K)_4A$ are similar, as we might expect for similar environments. This must not always be the case. Three separate distances between labelling sites in the first α-helix of bacteriophage-T4 lysozyme (T4L) are longer by 0.5 nm than those between corresponding C^α atoms in the crystal structure (Borbat et al. 2002; and see Table A16.7). This is in spite of the fact that separations differ by four residues and, therefore, correspond to positions almost in vertical register along the helix. The label orientations at all these positions must differ, in this particular case. All dipolar distances measured between pairs of residues in T4L are collected together in Table A16.7, which we give in the appendix at the end of the chapter. There, we compare them with C^α–C^α and C^β–C^β separations of the corresponding residues that come from the crystal structure.

16.11.2 β-sheets

The situation is simpler for β-strands in an antiparallel β-pleated sheet. Side chains of adjacent residues project from opposite sides of the sheet (i.e., $n_P = 2$), with corresponding C^α-atoms situated $h_\beta = 0.34$ nm apart along the strand direction (see Table 16.1 and Figure 16.18).

Dipolar distances between residues in strands S_2, S_4 and S_6 of the α-crystallin domain in eye lens αA-crystallin, measured by CW-EPR, are given in Table A16.8 at the end of the chapter. For residue pairs on the exposed surface of strand S_2 we get separations: $r_{12}(\Delta n = 2) = 0.66$ nm and $r_{12}(\Delta n = 4) = 1.20$ nm, when the residues are two and four apart, respectively. These values are quite close to the predicted vertical increments $h_\beta \times \Delta n$, and to C^α–C^α and C^β–C^β separations deduced from the crystal structure, which are included in Table A16.8. Comparable values for $r_{12}(\Delta n = 2) = 0.63$ and 0.70 nm are found in strands S_4 and S_6, respectively. Therefore, the length of the spin-label side chain is not so significant for distances between residues that are on the same side of the sheet. For spin labels three residues apart, on opposite sides of the sheet in strand S_6 we find that $r_{12}(\Delta n = 3) = 1.70$ nm, however. This exceeds the distance between nitroxides four residues apart, because it directly reflects the length of the side chain perpendicular to the sheet. Using the equivalent of Eq. 16.41 for β-strands: $r_{12}^2 = 2(1 - \cos \pi \Delta n) R_{NO}^2 + (h_\beta \Delta n)^2$, the distance of the nitroxide from the strand axis becomes $R_{NO} = 0.68$ nm, which is an average over the exposed and buried sides of the sheet. This corresponds to an extension $\Delta R_{NO} = R_{NO} - R_{C^\alpha} = 0.59$ nm of the nitroxide from the C^α atoms of the strand backbone (cf. Table 16.1).

16.11.3 Comparison with Crystal Structures

For comparison with the EPR measurements, Table 16.11 gives dimensional information obtained from crystal structures of proteins spin-labelled with MTSSL at solvent-exposed sites. We list separations $r_{C^\alpha\text{-N}}$ and $r_{C^\alpha\text{-O}}$ of the nitroxide N- and O-atoms from the backbone C^α-atoms, together with their differences in cylindrical coordinates (ΔR, $\Delta\phi$ and Δz) referred to the α-helix or β-strand axis. Relative to Tables 16.5 and A16.3, we are restricted here to those x-ray structures where the nitroxide ring is resolved. For these cases, the inherent flexibility of the side chain is inhibited by interactions additional to those found in a simple $(A_4K)_4A$ helix. This occurs more frequently in β-sheets, which therefore are over-represented in Table 16.11. For water-exposed α-helical sites on the surface of T4 lysozyme, values of $\Delta R_{NO} = \frac{1}{2}\left(\Delta R_{C^\alpha\text{-N}} + \Delta R_{C^\alpha\text{-O}}\right)$ are in the range 0.42–0.63 nm, with a value of 0.87 nm for the more highly restricted (4-Br)R1 side chain. The value of ΔR_{NO} deduced for aqueous $(A_4K)_4A$ is within the range for water-exposed sites on T4L; it is higher than the minimum for T4L because side-chain flexibility is not suppressed. For lipid-exposed α-helical sites in the membrane environment of LeuT, we find a wider range of extensions of the side chain: $\Delta R_{NO} = 0.34 - 0.70$ nm. This corresponds to more direct effects of solvation by side chains at the lipid–protein interface. Lipid-exposed residues on the surface of the β-barrel of BtuB are most extended with values of $\Delta R_{NO} = 0.75$ nm, where the side chains are accommodated in hydrophobic pockets between adjacent side chains.

In the case of well-defined side chains in crystals, differences in radial extension from the protein backbone arise mostly from different orientations of the side chain relative to the helix axis (cf. Table 16.5). For water-exposed residues in T4L, the side chain orients along the helix, and $\Delta R_{C^\alpha\text{-N}} \cong \Delta R_{C^\alpha\text{-O}}$ because the N–O bond is parallel to the helix axis. On the other hand, for lipid-exposed residues in BtuB, the side chain orients closer to the normal to the β-sheet and $\Delta R_{C^\alpha\text{-O}} - \Delta R_{C^\alpha\text{-N}} \cong 0.14$ nm, because the N–O bond is perpendicular to the sheet (cf. Table A16.1). However, the side-chain axis is still not completely radial, as we see from the non-zero values of Δz and $\Delta\phi$. The full C^α–NO distance is related to these cylindrical coordinates by:

$$r_{C^\alpha\text{-NO}} = \sqrt{R_{NO}^2 + R_{C^\alpha}^2 + \Delta z^2 - 2R_{NO}R_{C^\alpha}\cos\Delta\phi} \qquad (16.42)$$

where $R_{NO} = \Delta R_{NO} + R_{C^\alpha}$, and we have used the cosine law.

The TOAC spin label in the peptide backbone of an α-helix is a particularly simple situation. Here, the N–O bond is directed radially outward from the C^α atom with $\Delta R_{C^\alpha\text{-N}} \cong r_{C^\alpha\text{-N}}$, the full N–$C^\alpha$ distance, and similarly for the oxygen coordinates (see Table A16.5 at the end of the chapter). In this case, $\Delta R_{NO} \cong 0.36$ nm. The inflexible RX cross linker (bis-MTSSL) also offers a favourable situation with α-helices. It is oriented with the N–O bond directed approximately radially outwards, and its coordinates are simply related to those of the first residue i to which it is linked, although vertically it lies closer to the i+4 residue (see Table 16.11).

TABLE 16.11 C^{α}–nitroxide distances ($r_{C^{\alpha}-N}$ and $r_{C^{\alpha}-O}$) and differences in cylindrical coordinates (ΔR, $\Delta\phi$, Δz) relative to helix- or strand-axis, for S–S linked methyl thiosulphonate spin-label (MTSSL) side chains R1, in bacteriophage-T4 lysozyme (T4L), leucine transporter (LeuT), vitamin-B12 transporter (BtuB), immunoglobulin-binding domain (GB1) and tumor necrosis factor (TNFα); and for TOAC in an α-helix[a]

mutant/SL[b]	*N-atom*				O-atom			
	$r_{C^{\alpha}-N}$ (nm)	$\Delta R_{C^{\alpha}-N}$ (nm)	$\Delta\phi_{C^{\alpha}-N}$ (°)	$\Delta z_{C^{\alpha}-N}$ (nm)	$r_{C^{\alpha}-O}$ (nm)	$\Delta R_{C^{\alpha}-O}$ (nm)	$\Delta\phi_{C^{\alpha}-O}$ (°)	$\Delta z_{C^{\alpha}-O}$ (nm)
				α-helix[c]				
T4L:								
S44R1 (chain A)	0.56	0.43	24	0.32	0.61	0.41	30	0.41
S44R1 (chain B)	0.56	0.43	27	0.30	0.63	0.43	33	0.40
T115R1	0.56	0.42	27	0.34	0.64	0.42	32	0.45
T115R1(4-Br)	0.87	0.76	55	0.00	0.97	0.87	53	−0.01
L118R1 (100K)	0.77	0.58	−10	−0.51	0.89	0.68	−10	−0.58
	0.74	0.58	−7	−0.47	0.87	0.68	−8	−0.55
L118R1 (RT)[d]	0.69	0.51	57	−0.09	0.64	0.52	68	−0.07
LeuT:								
F177R1	0.75	0.64	−50	−0.11	0.87	0.75	−56	−0.10
I204R1	0.52	0.36	17	0.36	0.61	0.33	21	0.48
GB1:								
K28R1 (chain C)	0.72	0.62	−42	−0.12	0.84	0.73	−43	−0.18
K28R1 (chain D)	0.73	0.54	−42	−0.36	0.85	0.65	−44	−0.41
bifunctional/T4L:								
T115RX-	0.79	0.68	5	0.41	0.91	0.80	4	0.44
R119RX-	0.75	0.67	−37	−0.17	0.87	0.79	−39	−0.14
TOAC/α-helix:								
	0.29	0.29	1	−0.03	0.42	0.42	1	−0.03
				β-sheet[e]				
BtuB:								
T156R1	0.83	0.76	35	0.31	0.95	0.89	35	0.28
W371R1	0.80	0.69	61	0.33	0.91	0.83	62	0.30
GB1:								
E15R1 (chain A)[f]	0.88	0.78	−2	0.40	0.96	0.87	−7	0.41
E15R1 (chain B)[f]	0.78	0.74	35	0.19	0.90	0.86	34	0.20
T44R1 (crystal I)	0.82	0.81	2	0.07	0.92	0.92	5	0.10
T44R1 (crystal II)	0.65	0.49	37	0.40	0.75	0.53	37	0.52
N8R1 (chain A)	0.88	0.87	11	−0.13	1.00	0.99	11	−0.13
N8R1 (chain B)	0.83	0.81	−31	−0.04	0.94	0.92	−33	0.00
N8R1 (chain C)	0.79	0.78	−13	−0.10	0.91	0.90	−15	−0.14
N8R1 (chain D)	0.70	0.69	−15	−0.09	0.81	0.79	−18	−0.14
TNFα:								
T77R1	0.62	0.60	27	−0.13	0.67	0.66	22	−0.05

[a] For PBD codes, conformations and references, see Tables 16.5 and A16.3. For TOAC, see Marsh (2006).
[b] R1, 1-oxyl-2,2,5,5-tetramethylpyrrolin-3-ylmethylsulpho- (MTSSL) S–S adduct of cysteine; RX, 1-oxyl-2,2,5,5-tetramethylpyrrolinyl-3,4-bis(methylsulpho-) double S–S adduct of cysteine; TOAC, 1-oxyl-2,2,6,6-tetramethyl-4-aminopiperidin-4-yl carboxylic acid.
[c] Helix axes are located by transforming the PDB-coordinates using three Euler angles and three *x*, *y*, *z* translations, and fitting these transformations to the periodicity $n_{\alpha} = 3.6$ of the entire helix.
[d] Second (major) conformer, not present at 100 K.
[e] Because of the twist, the β-strand axis is found from the local PDB-coordinates of the three *i*, *i* ± 1 residues.
[f] Representative of three other chains.

16.12 ROTAMER LIBRARIES

We see from the discussion of MTSSL rotamers in Section 16.6 that dipolar distances measured by SDSL will depend on the configuration of the R1-side chain, and particularly on distributions in configuration. Distributions in side-chain configuration contribute to dipolar distance distributions that we obtain from DEER by Tikhonov regularization or explicit modelling with Gaussians (see Section 15.11). Distance distributions found experimentally are attributable at least partly to rotamers because the distribution width is narrower for Br-MTSSL, and the bifunctional bis-MTSSL (RX), which are restricted to single rotamers (see Table 16.5).

To allow for effects of spin-label attachment, it helps to construct rotamer libraries that we can use in modelling (Polyhach et al. 2011). Figure 16.24 shows distributions for dihedral angles $\chi_1 - \chi_5$ that are calculated from a 100-ns molecular dynamics trajectory of the R1-side chain alone in implicit solvent, with the attached peptide unit immobilized. (Note that force fields explicitly adapted for the R1 side chain are parameterized in Sezer et al. 2008c). The principal lobes of the distributions in Figure 16.24 correspond to the rotamers identified by x-ray diffraction (cf. Tables 16.5, A16.2, A16.3 and Figure 16.15). For χ_1,χ_2 (and χ_4), these are the **t**, **p**, **m** rotamers that we illustrated for a sp^3-centre in Figure 16.14. The S–S bond, on the other hand, is characterized by $\chi_3 = \pm 90°$, which reflects the partial double-bond character. Likewise, the attachment to the planar pyrroline ring is characterized mainly by values of $\chi_5 \approx \pm 90°$.

For computational economy, the dihedral distributions from molecular dynamics are reduced to a limited number of discrete rotamers (ca. 200) by grouping snapshots in the trajectory. Their frequencies are indicated by the lengths of the solid grey lines shown in Figure 16.24. Completeness of the library is demonstrated by near coincidence with the underlying molecular-dynamics trajectory of distance distributions and DEER time traces predicted for pairs of unrestrained spin labels (Polyhach et al. 2011).

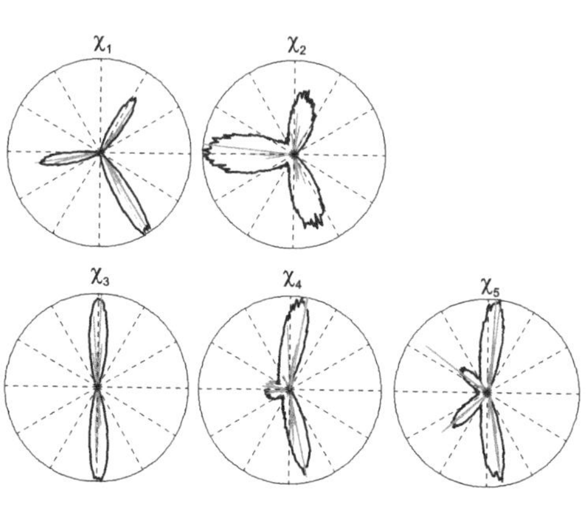

FIGURE 16.24 Rotamer populations in MTSSL side chain. Polar histogram plots for distribution of dihedral angles, $\chi_1 - \chi_5$, from 100-ns molecular dynamics trajectory at 175 K. $\chi = 0$ lies in the positive *x*-direction, and $\chi = 180°$ defines the **t**-rotamer for χ_1,χ_2,χ_4 (cf. Figure 16.14). Solid lines show dihedral-angle frequencies in the rotamer library (Polyhach et al. 2011).

In practice, the relative populations of these different rotamers become modified at the site of labelling. We allow for this by calculating interatomic interactions with the local protein environment, using potentials that are softened empirically to represent residual conformational dynamics (Polyhach et al. 2011). Fewer rotamers are populated appreciably in tighter binding sites than are in more open or exposed ones. For example, 37 rotamers account for 99.5% of the population at the buried site 118R1 in T4 lysozyme, whereas 163 out of 210 rotamers are populated at the exposed loop site 82R1 (Polyhach et al. 2011). The principal application of rotamer libraries is in EPR-assisted molecular modelling. Here, rotamers are accommodated in trial models and predicted distance distributions are then compared with data obtained from DEER measurements, e.g., for docking at the dimer interface of the sodium-proton exchanger NhaA (Polyhach et al. 2011).

A rotamer library is also available for the cysteine adduct of PROXYL-iodoacetamide (i.e., 1-oxyl-2,2,5,5-tetramethylpyrrolidin-3-yl thioacetamide), where the side chain has an additional dihedral angle: $\chi_1 - \chi_6$.

16.13 HELIX ASSEMBLY AND INTERHELICAL DISTANCES: COILED COILS AND α-BUNDLES

An isolated α-helix may be only marginally stable in water but can be stabilized greatly by helix–helix interactions. A motif that occurs in fibrous proteins, contractile proteins, cytoskeletal proteins and DNA-binding leucine zippers is the coiled coil, where two or more α-helices assemble as superhelices. Another frequent motif is the helix bundle, where side chains from adjacent helices more often pack as ridges-in-grooves (Chothia et al. 1981), instead of the knobs-in-holes packing of coiled coils (Crick 1953b). Distance measurements by CW-EPR, as described in Sections 15.3, 15.4 and 15.6–15.8, are well adapted to investigating the close proximities that are involved in helix packing.

16.13.1 Coiled Coils

Soluble coiled-coil proteins are characterized by a heptad amino-acid sequence repeat $(abcdefg)_n$, which contains hydrophobic residues at the first (a) and fourth (d) positions that form the internal contacts between the amphipathic helices (see left panel of Figure 16.25). If the helix period were exactly 3.5 residues per turn, every seventh residue would be in vertical register and the helices could pack as parallel rods. With the natural α-helical period, $2\pi/\omega_\alpha = 3.6$ residues per turn, the apolar stripe of a and d residues in Figure 16.25 (left) coils slowly to the left around the axis of each helix. Vertical register along the axis of the helix bundle is maintained when the axis of each α-helix coils around the central bundle axis. We illustrate this and the resulting knobs-in-holes packing with the two intermeshed helical

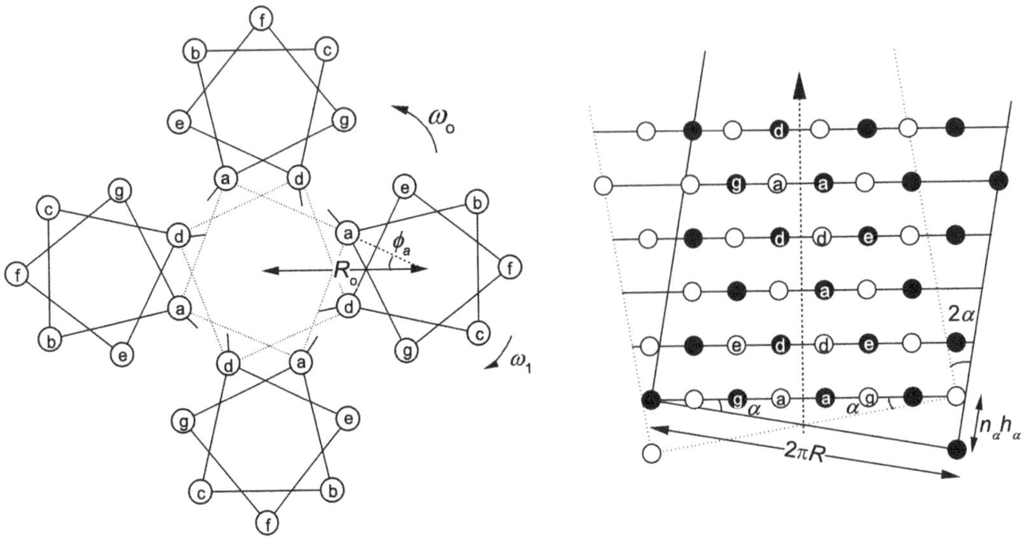

FIGURE 16.25 *Left*: Residue packing in coiled coils. Helical wheels for C^{α}-atoms of backbone show heptad repeats *a-g* in four α-helices forming four-stranded parallel superhelix with internal contact residues *a* and *d*, which are hydrophobic. Additional stabilization can come from salt bridges between complementary ionic residues at positions e and g. Heavy ticks show projected orientations of C^{α}–C^{β} bonds for *a*- and *d*-residues. Helix axes point into the page; rotation ω_1 is right-handed from *a* to *g*. Superhelical rotation ω_o is left-handed. *Right*: Meshing of α-helical nets for two-stranded coiled coil, with $R = 0.5\,\text{nm}$, $h_\alpha = 0.15\,\text{nm}$ and $n_\alpha = 3.6$ residues per turn. Helix with residues given by solid circles faces out of the page and that with open circles faces into the page. Each are tilted by angle α to the superhelix axis (dashed arrow), to produce knobs-in-holes packing of side chains along the line of contact. The two bottom rows show alternating layers of *a-a* and *d-d* contacts between helices. Each *a*- or *d*-residue contacts four residues from the other helix (see upper rows).

nets of side chains shown on the right in Figure 16.25. Here, side chains are placed on the surface of a cylinder at radius R from the axis of the α-helix. Solid circles are on the lower helix with side chains pointing up out of the page, and open circles are from the upper helix with side chains pointing down into the page. To achieve good packing at the helix contact sites, the two nets are tilted at angle α to the superhelix axis, such that a,a and d,d contacts are in vertical register and alternate in layers along the axis. The 3.6-residue periodicity of the α-helix thus translates to a 3.5-residue per turn periodicity about the axis of the superhelix. We see from Figure 16.25 (right) that the tilt α is given by:

$$\tan\alpha = n_\alpha h_\alpha/(2\pi R) \qquad (16.43)$$

where the helix pitch is $n_\alpha h_\alpha = 0.54\,\text{nm}$ (see Table 16.1). For $R \cong 0.5\,\text{nm}$ we get $\alpha \cong 10°$, which clearly depends directly upon our choice of R. But the tilt angle definitely must be much less than 20°, which we would get by taking the cylinder surface at the C^{α}-atoms of the helix backbone (i.e., $R = 0.23\,\text{nm}$). The crossing angle of the two α-helices is $\Omega = 2\alpha$, which we expect to be in the region of 20°. Note that the side chains are not oriented perpendicular to the helix axis; instead they point backwards. Side chains mesh directly as on the right in Figure 16.25, if N- to C-terminus vectors of both helices are pointing in the same direction, i.e., parallel helices. Knobs-in-holes packing of oppositely oriented helices is also possible if the antiparallel helix shifts down along the contact line by ca. 0.1 nm, giving a,d pairing along the contact line. Thus, both parallel and antiparallel helices can assemble as coiled coils.

The helical nets on the right of Figure 16.25 show knobs-in-holes packing for a two-stranded coiled coil. This corresponds to two helical wheels facing one another across a horizontal or vertical diagonal on the left of the figure. From the bold ticks that indicate the directions in which the side chains initially extend from the helix backbone, we see that side-chain packing differs between *a-a* layers and *d-d* layers. This explains the preference for different residues at the two sites. Also, we see that packing patterns and residue pairing in the layers will differ for higher oligomers, such as the four-stranded coiled coil. The C^{α}–C^{β} bond of a knob orients either parallel or perpendicular to the C^{α}–C^{α} vector at the bottom of the hole into which it fits. For four strands, packing is parallel at the *d*-level and perpendicular at the *a*-level, whereas this is reversed for two strands (Harbury et al. 1993).

Because of the tilt, two helices would splay apart away from the contact region, if they did not twist around each other in coiled coils. The locus of each superhelix has x,y,z-coordinates: $H(t) = \left(R_o\cos(\omega_o t),\, R_o\sin(\omega_o t),\, P(\omega_o t/2\pi)\right)$, where R_o and P are the superhelical radius and pitch, respectively. The parametric variable t represents rotation around the superhelix axis at angular frequency ω_o. Discretized in terms of amino-acid residues, the pitch is related to the tilt α of the two α-helices by $h_\alpha \cos\alpha = P(\omega_o/2\pi)$, and the superhelical radius by $h_\alpha \sin\alpha = R_o\omega_o$ (Harbury et al. 1995). Each α-helix has a superhelix for its axis. We use a local x',y',z'-axis system, where z' is tangential to the superhelix and x' lies in a plane perpendicular to the superhelix z-axis. The coordinates of the α-helix backbone describe a circle in the local system: $C'(t) = \left(R_\alpha\cos(\omega_1 t + \phi_a),\, R_\alpha\sin(\omega_1 t + \phi_a),\, 0\right)$, where ϕ_a

is the orientation angle of residue a (see Figure 16.25, left) and $R_\alpha = 0.23$ nm corresponds to the radial position of the C^α-atoms. We use the heptad frequency $\omega_1 = 4\pi/7$ radians per residue (i.e., 102.9° instead of 100°), because we express vertical progression by the parametric variable t that increases along the superhelical axis according to the z-component of $H(t)$. Superhelical and local axis systems are related by the Euler rotation matrix $\hat{R} = (\omega_o t, \alpha, 0)$ (see Figure R.2 in Appendix R.2). The path of the coiled coil then becomes $CC(t) = H(t) + \hat{R}(t) \cdot C'(t)$, in the unprimed superhelix axis system. Explicitly, the coordinates are (Crick 1953a):

$$x(t) = R_o \cos(\omega_o t) + R_\alpha \big(\cos(\omega_o t)\cos(\omega_1 t + \phi_a) - \cos\alpha \sin(\omega_o t)\sin(\omega_1 t + \phi_a)\big) \quad (16.44)$$

$$y(t) = R_o \sin(\omega_o t) + R_\alpha \big(\sin(\omega_o t)\cos(\omega_1 t + \phi_a) + \cos\alpha \cos(\omega_o t)\sin(\omega_1 t + \phi_a)\big) \quad (16.45)$$

$$z(t) = h_\alpha (\cos\alpha) t - R_\alpha \sin\alpha \sin(\omega_1 t + \phi_a) \quad (16.46)$$

which contain only three independent parameters (R_o, ω_o and ϕ_a), because R_α, ω_1 and h_α are known, and the tilt angle is determined by the relation: $h_\alpha \sin\alpha = R_o \omega_o$.

The 33-residue leucine-zipper region of the yeast transcription activator GCN4 forms a parallel, two-stranded coiled coil (O'Shea et al. 1991). Dimerization of GCN4 is needed for DNA binding. The a-position is populated by three Val residues and one Asn; whereas all four d-positions are occupied by Leu residues. Table 16.12 lists the superhelical parameters, where in principle, the crossing angle Ω and pitch P are derived values. Populating all a and d positions with Ile residues alters the packing at the interface between constituent amphipathic helices, which then assemble into a three-stranded coiled coil (Harbury et al. 1993). Formation of the trimer is reflected by the supercoil parameters that we give in Table 16.12. The zipper mutant with all-Leu residues at position a and all-Ile residues at position d reverses the usual distribution of branched and unbranched hydrophobic amino acids found in two-stranded coiled coils. This switches the supercoil assembly to a tetramer (see Figure 16.25, left), with parameters that we include in Table 16.12. Fitting Eq. 16.43 to the tilt angle α of the dimer in Table 16.12 places the cylinder surface for the side chains at effective distance $R = 0.41$ nm from the axis of the α-helix. These results come from x-ray diffraction. Limited EPR results on the leucine zipper region of GCN4 demonstrate strong spin–spin broadening and strong immobilization for the contact d-position at the beginning of the zipper, and moderately high mobility at positions c and f outside the contact zone of the zipper (Columbus and Hubbell 2004). We turn now to a more detailed application of SDSL–EPR distance measurements, namely with coiled coils.

16.13.2 Example of 4-Stranded Coiled Coil: SNARE Complex

A classic example of coiled-coil formation is the SNARE complex that transiently links synaptic vesicles containing the membrane protein VAMP to the presynaptic plasma membrane proteins SNAP-25 and syntaxin-1A (Poirier et al. 1998; Sutton et al. 1998). This primes the vesicle for release of neurotransmitter into the synaptic cleft by exocytosis. All SNARE proteins contain characteristic heptad repeat sequences, but in SNAP-25 these occur in two separate domains at the N- and C-terminal sections, designated respectively SNAP-25(N) and SNAP-25(C). Accordingly, the SNARE complex has the potential to form a four-stranded coiled coil, as depicted on the left in Figure 16.25. Table 16.13 lists the dipolar distances determined by SDSL with pairwise cysteine mutants of SNAP-25 and syntaxin 1A. Further distances were also determined, between the native Cys95 residue in VAMP1 and sites on SNAP-25 and syntaxin (Poirier et al. 1998). Distances were measured by the dipolar deconvolution technique described in Sections 15.3 and 15.4, at low temperatures where all motional effects are suppressed. Because this is CW-EPR, distances greater than 2.5 nm could not be measured. The close proximities (≈0.7 nm) of residues at the N-terminal ends of SNAP-25(N) and SNAP-25(C), and similar proximities at the C-terminal ends, demonstrate the parallel alignment of helices from the two domains in SNAP-25. The intervening sequence (residues 93–135) between the two domains is sufficiently long to allow this type of helix packing. The only arrangement consistent with the observed distances is one where N- to C-terminal vectors of SNAP-25(N), SNAP-25(C), syntaxin and VAMP all point in the same direction. This shows that the SNARE complex is a parallel four-stranded helical bundle. Modelling as a coiled coil, where close contacts were used to assign neighbouring positions within the bundle, was used as a starting point for energy refinement (without EPR distance constraints). This produced a final model of the parallel four-stranded coiled coil that, to within the uncertainty of the conformation of the spin-labelled side chains, was consistent with the distance measurements (Poirier et al. 1998). The model turns out to be remarkably close to the crystal structure that was determined almost simultaneously by x-ray diffraction (Sutton et al. 1998; and see Table 16.13). The structure of the synaptic fusion complex is not so regular as the homotetrameric coiled coil of GCN4 (see Table 16.12), because each of the

TABLE 16.12 Superhelical parameters for GCN4 leucine-zipper n-mer coiled coils

GCN4 *peptide, pad*	R_o (nm)	$2\pi/\omega_o$ (residue/turn)	ϕ_a (°)	$\alpha = \frac{1}{2}\Omega$ (°)	P (nm)	*Ref.*[a]
dimer, pVL[b]	0.49	100	21.6	11.7	14.76	1,2
trimer, pII	0.67	119	20	13.5	17.52	2,3
tetramer, pLI	0.76	139	19.8	13.0	20.54	2

[a] References: 1. O'Shea et al. (1991); 2. Harbury et al. (1993); 3. Harbury et al. (1995).
[b] "pVL" contains 3 Val and one Asn at position *a*, but four Leu at position *d*; pII trimer has four Ile at both *a* and *d*; pLI tetramer has four Leu at *a* and four Ile at *d* (cf. Figure 16.25).

TABLE 16.13 Intra- and inter-helix dipolar distances in nm (bold) for SNAP25(N)/SNAP25(C)/Syntaxin/VAMP four-helix bundle of a SNARE synaptic fusion complex (Poirier et al. 1998)

	SNAP-25(N)							SNAP-25(C)			syntaxin	
	L21	S39	A42	L50	Q53	M71	A74	L160	A185	A199	I195	Y257
L21								(>2.5)		(>2.5)	**1.4**	
S39				**1.0**				**<0.7**				
A42					**1.2**			**1.6**				
L50		*1.57*										
Q53			*1.56*					**2.2**	(>2.5)			
M71									**0.7**	**0.72**	(>2.5)	**1.7**
A74									**1.1**	**0.78**	(>2.5)	**1.5**
L160	*2.77*	*0.60*	*0.97*		*2.08*							
A185					*1.74*	*1.25*	*1.68*					(>2.5)
A199	*7.97*					*1.18*	*1.48*				(>2.5)	**1.9**
I195	*0.51*					*7.37*	*7.79*			*8.18*		
Y257						*2.09*	*1.60*		*3.21*	*1.48*		

Italics: C^{β}–C^{β} distances (nm) from crystal structure (PDB: 1SFC; chains B–D). To compare with EPR distances (bold): reflect about the diagonal.

four different components has its own superhelical character. For instance, the angle α at which each α-helix crosses the overall bundle axis ranges from 5° in syntaxin and SNAP-25(N) up to 25° in VAMP and SNAP-25(C) (Sutton et al. 1998).

The T-SNARES syntaxin-1A and SNAP-25 alone also can form a four-stranded parallel coiled coil, where the V-SNARE VAMP is replaced by a second copy of syntaxin-1A (Zhang et al. 2002). Figure 16.26 shows the dependence on residue position of the dipolar distance between singly labelled syntaxins, for the central heptad of the four-stranded coiled coil. Following Zhang et al. (2002), we use the simplified model given in the top panel of Figure 16.26. The separation between axes of the two syntaxin helices is $2R$; and R is also assumed equal to the radial distance R_{NO} of each nitroxide from its helix axis. The azimuthal offsets of an a-residue, ϕ_1 and ϕ_2, differ between the two syntaxin helices (1 and 2). As defined in Figure 16.26, the origin for ϕ lies along the line joining the two helix axes. When we proceed around the helical wheel, the azimuthal angle of a particular residue increases by $(2\pi/n_\alpha)\Delta n$, where Δn is the number of residues from the chosen reference a-residue and n_α is the number of residues per turn of the helix. The x,y-coordinates of a residue in helix 1 are then $(R\cos(\phi_1+2\pi\Delta n/n_\alpha), R\sin(\phi_1+2\pi\Delta n/n_\alpha))$ with the helix centre as origin. Similarly for helix 2, the coordinates are $(2R+R\cos(\phi_2+2\pi\Delta n/n_\alpha), R\sin(\phi_2+2\pi\Delta n/n_\alpha))$, where we add $2R$ to x because the origin is still the axis of helix 1. Fitting this model to the EPR distances (solid circles), we get the solid line shown in Figure 16.26. The fitted value of $R = 0.81 \pm 0.02$ nm is slightly larger than the radius of this central section in the crystal structure of the SNARE complex: 0.70–0.73 nm (Sutton et al. 1998). For label positions outside the a and d contact sites, this suggests that $R_{NO} \approx 0.8$ nm, which is significantly greater than the radius of the coiled coil. Correspondingly, we get $\Delta R_{NO} = R_{NO} - R_{C^\alpha} = 0.58$ nm compared with 0.51 nm for the isolated $(A_4K)_4A$ helix of Figure 16.23. The fitted values of ϕ_1 and ϕ_2 are such that the a-residues point towards the interior core of the four-stranded coiled coil, which is completed by adding two circles representing SNAP-25(N),(C) immediately below those shown in the top panel of Figure 16.26 (cf. Figure 16.25 left, where the helix axes point into the page).

Before turning to a different mode of helix packing, it is worthwhile to note that monomer SNAREs alone give sharp three-line MTSSL EPR spectra because they are natively unfolded. Only after assembling into coiled coils do they show broad anisotropic spectra, with particularly strong immobilization found at core a and d sites (Poirier et al. 1998; Margittai et al. 2001).

16.13.3 α-Helical Bundles

We mentioned already at the start of this section that a coiled coil is not the only packing motif for α-helices, nor even for a helix bundle. Packing of ridges into grooves offers an alternative to knobs into holes. Whereas the latter (see Figure 16.25) concentrates on

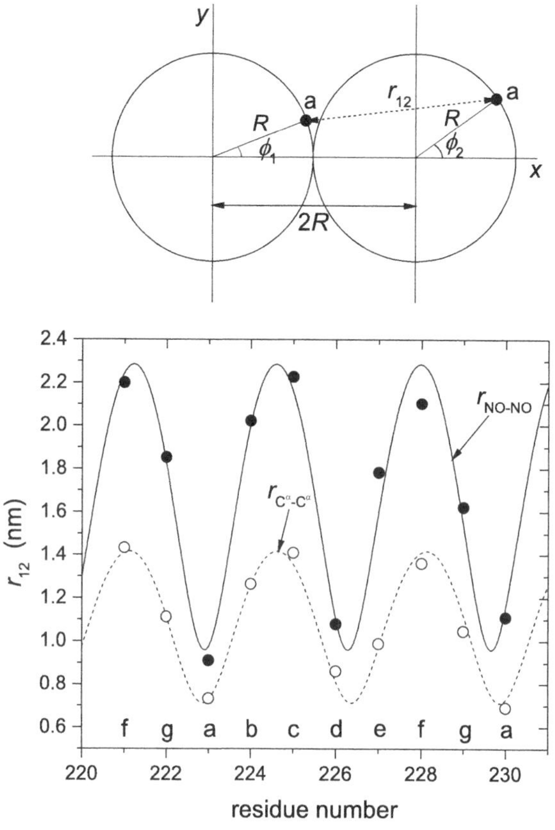

FIGURE 16.26 Dipolar distances r_{12} between residues in adjacent syntaxin-1A chains in four-stranded, parallel coiled-coil assembly with SNAP25 (solid circles; Zhang et al. 2002). Syntaxin-1A is singly labelled with MTSSL. For comparison, open circles are C^α–C^α separations of the backbone in crystal structure of corresponding ternary SNARE complex (PDB: 1SFC) where VAMP2 is surrogate for the second syntaxin-1A. *Solid line*: fits of helical-wheel model in top panel, with $R = 0.81 \pm 0.02$ nm, $\phi_1 = -56 \pm 8°$, $\phi_2 = -104 \pm 8°$ and $n_\alpha = 3.4$ residue/turn, for the dipolar distance. Helix *z*-axis points out of the page: ϕ is right-hand rotation. *Dashed line* for C^α–C^α separation is to guide the eye.

adjacent residues in the sequence, the ridges on the helix surface that are important for packing are those formed by side chains of residues that are $m = 3$ or 4 apart in the sequence. We can visualize the ridges and grooves, and their packing, by using helical nets such as those shown already in Figure 16.25. From reasoning similar to that leading to Eq. 16.43, the ridge (or groove) is inclined at an angle α_m to the helix axis, where (cf. Chothia et al. 1981):

$$\tan\alpha_m = \left(\frac{n_\alpha h_\alpha}{2\pi R}\right)\frac{m}{m - n_\alpha} \tag{16.47}$$

The helix-crossing angle that combines ridges and grooves from residues m apart on one helix with grooves and ridges from residues q apart on the other helix is then simply: $\Omega_{mq} = \alpha_m + \alpha_q$. Intercalation of ±4 ridges/grooves with ±3 grooves/ridges is a predominant packing mode for α-helical bundles. Taking $R = 0.47$ nm, $h_\alpha = 0.151$ nm and $n_\alpha = 3.64$ as mean values from a wide population of proteins analysed, Chothia et al. (1981) predict a crossing angle $\Omega_{34} = \alpha_3 + \alpha_4 = +23°$. The most frequent crossing mode, however, is found when ±4 ridges/grooves on both helices intercalate. This results in a crossing angle $\Omega_{44} = 2\alpha_4 = -52°$ at the point where the helices contact, which is important for other tertiary folds, but is incompatible with a helix bundle. In ridges-in-groves packing, each residue at the interface contacts two residues from the other helix. This differs from knobs-in-holes packing, where a given residue has four close neighbours from the adjacent helix (see Figure 16.25, right). Note that using the present parameters in Eq. 16.43 for knobs-in-holes predicts a crossing angle $\Omega = 2\alpha = 21°$, as opposed to 23° for the 3–4 ridges/grooves. However, the two packing modes differ principally in the relative translations of the two helices. With the same parameters, we predict a crossing angle $\Omega_{14} = \alpha_1 + \alpha_4 = +75°$ (or −105°), for the less frequent 1–4 ridges/grooves packing mode.

16.14 BETA-SHEET TOPOLOGY: INTERSTRAND SEPARATIONS

We saw already in Section 16.7.4 that the distance between residues in register on adjacent strands of β-sheets is rather short. We can expect separations between C^α-atoms in the region of 0.45–0.51 nm (see also Table A16.4). This means a very large dipolar broadening, with intensities well outside the maximum 6–7 mT range of the powder pattern for an isolated nitroxide. Without actually measuring distances, we can use this large qualitative effect on the spectrum to determine the β-sheet topology: i.e., which strands are adjacent, and whether they are arranged parallel or antiparallel.

16.14.1 α-Crystallin Chaperone Fold

AlphaA crystallin is a protein of the eye lens which contains the conserved α-crystallin domain that is characteristic of the chaperone family of small heat-shock proteins. The principal motif of this domain is a seven-strand β-sandwich, as illustrated in Figure 16.27 (Koteiche and Mchaourab 1999; Laganowsky et al. 2010). Strands S_3–S_7 were identified from a twofold periodicity in accessibility to NiEDDA, and in mobility, by using nitroxide-scanning experiments (Berengian et al. 1997; Koteiche et al. 1998; Koteiche and Mchaourab 1999). Note that a periodic exposure to aqueous relaxant already implies that one side of the corresponding sheet is buried in the tertiary structure. Strands S_1 and S_2 were identified by a periodicity of two in spin–spin broadening (Koteiche and Mchaourab 1999). As discussed already in Section 16.11.2, dipolar distances between residues in segment S_2 are characteristic of a β-strand. Spin–spin broadening between residue pairs shows that strand S_3 is adjacent to S_4, which is adjacent to S_5. Also, the strands are antiparallel because dipolar interaction at an N-terminal residue in one strand is strongest with a C-terminal residue in the adjacent strand and vice-versa (Koteiche et al. 1998). Lack of spin–spin broadening with any of the other strands, even those adjacent in the sequence, shows

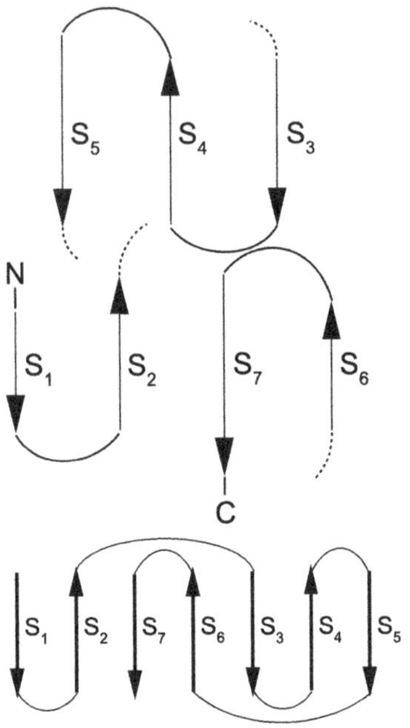

FIGURE 16.27 Topological links between adjacent β-strands in seven-strand β-sandwich domain (S_1–S_7) of αA-crystallin. β-strands indicated by arrows from N- to C-terminus (i.e., with increasing residue number). *Top*: three-stranded sheet; *middle*: four-stranded sheet; *bottom*: double Greek-key topology.

that strands S_3 to S_5 constitute an independent sheet, as depicted in the top panel of Figure 16.27 (Koteiche and Mchaourab 1999). Similarly, spin–spin interactions show that strands S_1 and S_2 form a four-stranded antiparallel sheet with non-contiguous strands S_7 and S_6, in the arrangement shown by the middle panel of Figure 16.27 (Koteiche and Mchaourab 1999). Connections between strands therefore correspond to the double Greek-key topology diagram that we show at the bottom of Figure 16.27.

The two independent sheets pack in a sandwich that buries hydrophobic residues at the interface. The relative arrangement of the two sheets is such that these side chains pack against each other in an aligned fashion. Because the strands twist, those in one sheet must tilt, relative to those in the other, to maintain aligned side-chain packing (Chothia and Janin 1981). The side chains from facing antiparallel strands twist in opposite senses around the strand axis by an amount τ, per residue. This is related to the twist θ per residue of the strand axis itself that we use in Eq. 16.5 of Section 16.4 by: $d_\beta \sin(\tau/2) = h_\beta \sin(\theta/2)$, where d_β is the strand separation within the sheet (see Marsh 2000). The opposing side chains therefore splay apart with a spatial separation given by the bases of the two isosceles triangles: $2D_\beta \sin(\tau/2)$ (per residue), where $D_\beta \cong 1.0$ nm is the separation of the two sheets. This is compensated by tilting the strand axes of the two sheets in the opposite sense by dihedral angle Ω, giving a corresponding separation $2h_\beta \tan(\Omega/2)$ (per residue), where: $h_\beta \tan(\Omega/2) = -D_\beta \sin(\tau/2)$ to cancel the splay. For a typical tilt of $\Omega = -30°$ (Chothia and Janin 1981), we find that $\tau = +11°$, which corresponds to a twist along the strand of $\theta = 14.5°$ per residue.

Aligned packing of the α-crystallin sandwich with modest strand tilt is distinguished from the possibility of orthogonal packing with crossing angle $\Omega \cong 90°$ (Chothia and Janin 1982) by spin labelling pairs of residues where one partner is in the long loop linking S_5 and S_6, or in the shorter loop connecting S_3 and S_4 (Koteiche and Mchaourab 1999). This establishes that the two sheets pack in the relative orientation indicated by the top and middle panels of Figure 16.27, with strand S_6 in one sheet facing strand S_3 in the other sheet, in an antiparallel orientation. Dipolar distances between residues, some of them long-range from DEER, are listed in Table A16.8 of the appendix at the end of the chapter. These are compared with distances between C^α atoms, and between C^β atoms, of the corresponding residues in the crystal structure of the α-crystallin domain of bovine αA-crystallin that was determined only later (Laganowsky et al. 2010). The EPR work was done on the rat sequence, but over the total 96-residue stretch only four residues are different, and two of these changes are conservative. Many nitroxide–nitroxide distances in the α-crystallin domain are close to the C^α–C^α and C^β–C^β distances, in spite of the length of the spin-label side chain. This is most likely because the residues involved are on the same side of the sheet, as mentioned already in Section 16.11.2. Deviations of the dipolar distances from C^α–C^α and C^β–C^β distances are included in the histograms that are shown in Figure 16.28. The β-sheet

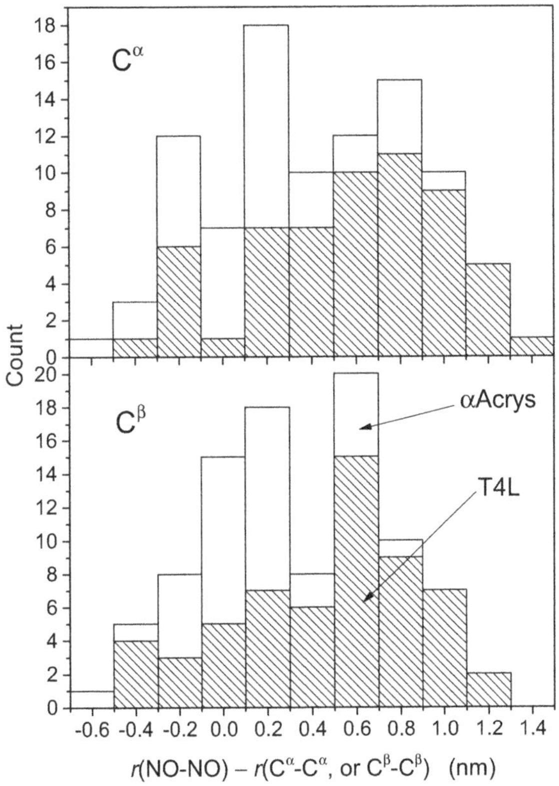

FIGURE 16.28 Distribution of dipolar distances r_{ij} between MTSSL spin labels in double-labelled bacteriophage-T4 lysozyme (T4L) and conserved αA-crystallin domain (αAcrys), relative to separation of C^α-atoms (top panel) or C^β-atoms (bottom panel) of corresponding residues in crystal structure of native T4L (PDB: 2LZM) and αAcrys (PDB: 3L1F). Hatched part of each bar is contribution from T4L and open part that from αAcrys. Data from Tables A16.7 and A16.8.

domain of αA-crystallin has larger populations close to zero than we find with the α-helical protein T4 lysozyme, where the length of the side chain plays a more important role.

16.14.2 Amyloid Fibril Cross-β Structure

A rather special situation arises for the homooligomeric assembly in parallel β-sheets of amyloidogenic protein monomers, when they build fibrils. Here, the core regions of the monomers stack underneath one another, in a parallel in-register orientation, down the length of the fibril. Examples include the microtubulin-associated tau protein (Margittai and Langen 2004), islet amyloid polypeptide IAPP (Jayasinghe and Langen 2004), presynaptic α-synuclein (Chen et al. 2007) and the human prion protein (Cobb et al. 2007), all of which have amyloid cores of >20 residues. Because side chains of adjacent monomers are in register, their separation is only 0.48 nm. Coupling between the like spins of singly labelled proteins in a given sheet then leads to exchange narrowing of the CW-EPR spectrum that collapses the nitrogen hyperfine structure. This is a solid-like effect, because the narrowing persists to low temperatures. At least four closely juxtaposed spin labels are needed to generate this type of exchange narrowing (Ladner et al. 2010). For protein monomers labelled singly and off-centre in a strand, this strong spin exchange is a hallmark of in-register assembly into parallel β-sheets. Dilution with non-paramagnetic spin label abolishes the extensive exchange coupling when less than half of the spin labels are paramagnetic. Conventional spectra of the systems with 40% active spins show strong dipolar broadening, and deconvolution using a reference with only 10% active spins (see Sections 15.3, 15.4) reveals distinct pair populations with separations that are multiples of 0.48 nm, up to fourth-nearest neighbours (Chen et al. 2007). Note that coordinates for a residue in a standard parallel β-pleated sheet are listed in Table A16.6 of the appendix at the end of this chapter.

The parallel, in-register β-sheet cores form a cross-β structure in the amyloid fibrils, where the β-strand axes lie perpendicular to the fibre axis. Core lengths of >20 residues would not fit as a single ≥6.5-nm long strand in a protofilament of 5-nm diameter. Therefore, the cores must pack as two or more strands within each protofilament, with turns or loops folding the strands together. Asymmetry of this fold forces parallel packing between monomers in the corresponding sheets.

16.15 DEER AND SDSL: TERTIARY FOLD, SUBUNITS AND DOCKING

Long-range distance measurements by pulsed ELDOR (i.e., DEER) make an ideal combination with site-directed labelling to investigate macromolecular structure. We can use distances between established secondary structural elements, α-helices and β-sheets, to find the tertiary fold in proteins. On the quaternary level of organization, we can determine the structural arrangement of subunits in homo- and hetero-oligomers. Also, docking of proteins that interact to form supramolecular assemblies can be steered by DEER-determined distances between SDSL-labelled components. For homo-oligomers, spin counting by DEER (see Section 15.11.2) can also yield the size of the oligomer.

16.15.1 Homo-oligomers

We can relate inter-monomer distances directly to the structure of rotationally symmetric homo-oligomers. For a regular N-sided polygon, the distance between the first (1) and subsequent (i) vertices is:

$$r_{1i} = 2R\sin(\pi(i-1)/N) \qquad (16.48)$$

where R is the radial distance from the centre to a vertex (see Figure 16.29). The closest distance is $r_{12} = 2R\sin(\pi/N)$, i.e., that between adjacent monomers; and the longest distance for even N is $2R$, i.e., that between opposite monomers across the polygon. For odd N, all interspin distances are twofold degenerate; whereas for even N, the longest distance is unique.

Complications arise in interpreting DEER distance distributions for oligomers solely in terms of spin pairs, r_{1i}. This is because expansion of the general expression (Eq. 15.47) for the DEER signal introduces multi-spin contributions with dipolar sum and difference frequencies, when $N > 2$ and the individual dipolar depths do not fulfill the condition $f_{B,p} \ll 1$ (Jeschke et al. 2009). Giannoulis et al. (2013) have analysed the situation with homo-oligomers by calculating DEER decays explicitly for regular polygonal arrangements. Corresponding distance distributions were deduced by using Tikhonov regularization (see Eqs. 15.54–15.59 in Section 15.11). With increasing polygon order N, multi-spin effects suppress longer distances and also introduce ghost artefacts. Then we obtain only the nearest-neighbour distance r_{12} with certainty, and the width of this distribution is broadened considerably by multi-spin artefacts.

We get the oligomer size, N, separately by spin counting, using the modulation depth Δ together with Eq. 15.52 (see Section 15.11.2). Only for small oligomers can we deduce

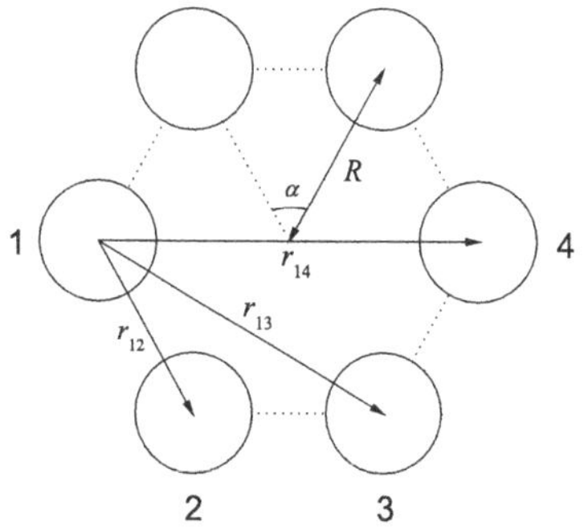

FIGURE 16.29 Geometric relations in a homo-oligomeric N-mer, modelled as regular polygon of outer radius R and internal angle $\alpha = 2\pi/N$. Intermonomer distances are r_{1i}.

N from the distance distribution. For example, long and short interspin distances of $r_{13} = 3.1 \pm 0.1$ nm and $r_{12} = 2.1 \pm 0.1$ nm are found between single R64R1 spin-labelled mutants of the homotetrameric K$^+$-channel KcsA (Endeward et al. 2009). These values are close to the ratio $r_{13}/r_{12} = \sqrt{2}$ expected for a square arrangement. On the other hand, only nearest-neighbour distances r_{12} are deduced reliably from Tikhonov regularization for R1 spin-labelled mutants of the homo-heptameric mechanosensitive channel of small conductance MscS (Pliotas et al. 2012). Fitting the dipolar decay with a regular heptagon model, however, produces all three distances r_{12}, r_{13} and r_{14} for the S147R1 mutant; and these agree with modelling based on the crystal structure (Giannoulis et al. 2013).

By expanding the product in Eq. 15.47 for $p = 1$ to N, we see that the contribution to dipolar modulation from spin pairs alone is proportional to $f_{AB}(1 - f_{AB})^{N-2}$ (von Hagens et al. 2013). This reaches a maximum when the fraction of B–spins inverted is $f_{AB} = 1/(N-1)$. Thus, we can reduce multi-spin contributions by decreasing f_{AB}, both by using lower levels of spin labelling and by dropping the angle of the pump pulse. This helps to determine distances beyond the basal nearest-neighbour value (Ackermann et al. 2017).

A simpler method for determining homooligomer size is just to vary the degree of labelling, F, systematically using EPR intensities. We assume a single labelling site per monomer. The probability of n labels per N-mer then follows binomial statistics:

$$p_n = \frac{N!}{n!(N-n)!} F^n (1-F)^{N-n} \tag{16.49}$$

where F is the total fraction of monomers that are spin labelled and $\sum_0^N p_n = 1$. Instead of distance measurements, we can rely upon CW-EPR intensities at partially saturating radiation power B_1^2, because these are sensitive to weak spin–spin interactions that are comparable to intrinsic spin–lattice relaxation rates (see Sections 11.2 and 11.10). We use a simple model where oligomers containing more than one spin-labelled monomer contribute intensity I'_{ST} that is reduced from the value I^o_{ST} for an isolated spin-labelled monomer. The net normalized intensity of the sample then becomes (Páli et al. 1999):

$$I_{ST} = \left(I^o_{ST} - I'_{ST}\right) \frac{p_1}{\left(1 - p_o\right)} + I'_{ST} \tag{16.50}$$

where the term in the denominator arises because intensities are normalized to the total number of spin labels. Here we write Eq. 16.50 in terms of normalized intensities I_{ST} of the saturation-transfer EPR spectrum (see Eq. 11.69), but they could equally well be conventional CW-EPR intensities at a fixed saturating power. This method was used with ST-EPR to demonstrate the hexameric assembly of a 16-kDa proteolipid (Páli et al. 1999). A hexamer is probably the upper limit for a protein of this size, when the label is at the outer periphery of the oligomer. Labels on the internal surface will give sensitivity to higher values of N. With judicious choice of label position, Langen et al. (1998b) determined membrane assembly of annexin XII trimers simply from quenching of the normal CW-EPR amplitudes by strong spin–spin broadening.

16.15.2 Triangulation and Distance Geometry

To within reflection symmetry, we can triangulate the position in space of a spin label from the distances to three fixed points, whose mutual separations, r_{01}, r_{02} and r_{12}, are known. We can visualize triangulation easily in cylindrical coordinates (R,ϕ,z) as shown in Figure 16.30. The base triangle has vertices located at (0,0,0), $(R_1,0,0)$, $(R_2,\phi_2,0)$, where $R_1 \equiv r_{01}$, $R_2 \equiv r_{02}$ and we get the azimuthal angle ϕ_2 from the cosine law:

$$r_{12}^2 = R_1^2 + R_2^2 - 2R_1R_2\cos\phi_2 \tag{16.51}$$

A general point i then has coordinates (R_i,ϕ_i,z_i) that we obtain from the distances r_{0i}, r_{1i} and r_{2i}, to vertices 0, 1 and 2 of the base triangle:

$$r_{0i}^2 = R_i^2 + z_i^2 \tag{16.52}$$

$$r_{1i}^2 = R_1^2 + r_{0i}^2 - 2R_1R_i\cos\phi_i \tag{16.53}$$

$$r_{2i}^2 = R_2^2 + r_{0i}^2 - 2R_2R_i\cos(\phi_2 - \phi_i) \tag{16.54}$$

From solution of Eqs. 16.52–16.54 we get R_i, ϕ_i and $|z_i|$. To decide on which side of the base triangle the point i lies, we need a reference point outside the basal plane, whose coordinates (R_3,ϕ_3,z_3) with $z_3 \neq 0$ are given by expressions analogous to Eqs. 16.52–16.54. We must remember that triangulated positions based on EPR dipolar distances are those for the N–O group of the nitroxide and depend on the conformation of the MTSSL side chain (cf. Sections 16.11 and 16.12).

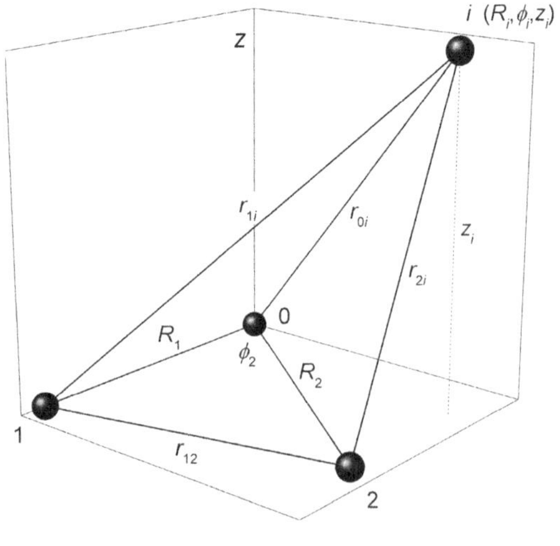

FIGURE 16.30 Triangulation of nitroxide i, with cylindrical coordinates (R_i,ϕ_i,z_i), by using distances r_{0i}, r_{1i} and r_{2i} from nitroxides 0, 1 and 2 of the basal triangle. A fourth reference position, outside the plane of the basal triangle, is needed to specify whether i lies above or below the triangle.

Repeated application of Eqs. 16.52–16.54 leads to accumulation of errors and instability. In general, we can transform sets of interatomic distances r_{ij} into Cartesian atomic coordinates x_i, y_i, z_i stably by using distance geometry (Crippen and Havel 1978; Borbat and Freed 2007). The metric matrix **G** has elements g_{ij} that are the scalar products of the vectors, $\mathbf{v}_i \equiv (x_i, y_i, z_i)$ and $\mathbf{v}_j \equiv (x_j, y_j, z_j)$, from the atom at the origin to atoms i and j:

$$g_{ij} = \mathbf{v}_i \cdot \mathbf{v}_j = \tfrac{1}{2}\left(r_{0i}^2 + r_{0j}^2 - r_{ij}^2\right) \tag{16.55}$$

where the second identity is just the cosine law (e.g., Eq. 16.53). For N atoms in total, the **G**-matrix has dimensions $(N-1)\times(N-1)$. The coordinate vectors $\mathbf{v}_i$ form the rows of a $(N-1)\times 3$ matrix **V**. We proceed by diagonalizing the metric matrix **G** (see Eq. H.14 in Appendix H.2):

$$\Lambda = \mathbf{W}^{\mathrm{T}} \cdot \mathbf{G} \cdot \mathbf{W} \tag{16.56}$$

where $\Lambda = (\lambda_i)$ is a diagonal matrix containing the eigenvalues of **G**, $\mathbf{W} = (\mathbf{w}_i)$ is the column matrix of corresponding eigenvectors $\mathbf{w}_i$, and superscript-T stands for the transpose. If the distances between all pairs of atoms are exact, then only three eigenvalues are non-zero (Braun 1987). Thus, from the inverse of Eq. 16.56, we can write the elements of **G** as:

$$g_{ij} = \mathbf{w}_i \cdot \boldsymbol{\lambda} \cdot \mathbf{w}_j^{\mathrm{T}} \tag{16.57}$$

where $\boldsymbol{\lambda}$ is a diagonal 3×3 matrix of eigenvalues $\lambda_x, \lambda_y, \lambda_z$. We then get the coordinate vectors from combining Eqs. 16.55 and 16.57:

$$\mathbf{v}_i = \boldsymbol{\lambda}^{1/2} \cdot \mathbf{w}_i \tag{16.58}$$

Diagonalization is performed numerically. In practice, measurement inaccuracies produce departures from zero in many of the eigenvalues. Therefore, we take the three largest for $\lambda_x, \lambda_y, \lambda_z$ in Eq. 16.56, together with their corresponding eigenvectors (Borbat and Freed 2007). This produces the optimum fit (Havel et al. 1983). Increased stability is achieved by using the centre of gravity (O), instead of one of the points, as origin. Then $r_{0i}{}^2$ in Eq. 16.55 is replaced by (Crippen and Havel 1978):

$$r_{0i}^2 = \frac{1}{N}\sum_{j=1}^{N} r_{ij}^2 - \frac{1}{N^2}\sum_{j=2}^{N}\sum_{k=1}^{j-1} r_{jk}^2 \tag{16.59}$$

and similarly for r_{0j}^2.

16.15.3 Distance Restraints in Modelling

Less stringently than in distance geometry, we can use DEER distances (not necessarily interconnected) as constraints for molecular modelling (Alexander et al. 2008; Hirst et al. 2011) or molecular dynamics simulations (Islam et al. 2013). Most useful are distances between residues that are far apart in the sequence but close spatially. Distance constraints can accommodate ranges of uncertainty arising from unknown conformations of the spin-label side chain. A cone model (Alexander et al. 2008; Hirst et al. 2011) or a soft square-potential (Bhatnagar et al. 2007) are amongst those used. As we might expect, proper allowance for the conformation of the spin-label side chain is particularly important for interpreting short dipolar distances, such as those from CW-EPR measurements (Sale et al. 2005).

Table A16.7 of the Appendix at the end of this chapter lists the dipolar distances $r_{\mathrm{NO-NO}}$, determined by pulse- and CW-EPR, between 58 pairs of MTSSL spin labels attached to double cysteine-replacement mutants of bacteriophage-T4 lysozyme. These are compared with the distances, $r_{C^\alpha\text{–}C^\alpha}$ and $r_{C^\beta\text{–}C^\beta}$, between the C$^\alpha$- or C$^\beta$-atoms of the corresponding residues in the crystal structure of the native protein. Most dipolar distances are larger than $r_{C^\alpha\text{–}C^\alpha}$ and $r_{C^\beta\text{–}C^\beta}$, but some are smaller although to a lesser extent than the positive excursions. The mean positive difference with respect to C$^\beta$-separations is $\Delta r = r_{\mathrm{NO-NO}} - r_{C^\beta\text{–}C^\beta} = +0.57 \pm 0.31$ nm, and the mean negative difference is $\Delta r = r_{\mathrm{NO-NO}} - r_{C^\beta\text{–}C^\beta} = -0.30 \pm 0.06$ nm, where the ranges quoted are the standard deviations. Figure 16.28 shows distributions of the differences between dipolar distances and separations of C$^\alpha$- or C$^\beta$-atoms in the crystal. The total range is −0.5 to +1.5 nm. With respect to C$^\alpha$, most differences lie in the broad range +0.5 to +1.2 nm; but with respect to C$^\beta$, they peak strongly at +0.6 nm. Distributions such as these form the basis for restraints employed in molecular modelling or molecular dynamics simulations. Note that the simple geometric cone model (Hirst et al. 2011; Alexander et al. 2013) puts the nitroxide at an effective distance of 0.60 nm from the C$^\beta$-atom, along an axis that makes an angle of 120° with the C$^\alpha$–C$^\beta$ bond.

16.15.4 Docking Subunits

If three-dimensional structures of individual interacting proteins (or subunits) are known, we can use intermolecular distances between labelled residues to construct the protein complex. We need six parameters to describe the relative arrangement of two rigid bodies fully. Three Euler angles specify the orientation of one set of molecular axes relative to another (see Appendix R.2); and three components of a translation vector determine the distance apart. Thus, in general, we need at least six independent intermolecular distance measurements. Any symmetry in the docked structure can reduce the total number. For example, the homodimer of the Na^+/H^+ antiporter NhaA is generated from the monomer by using the polar angles (θ, ϕ) of the twofold symmetry axis, relative to the molecular z-axis, and the translation vector (x,y) (Hilger et al. 2007). Again, distance restraints must allow for different conformations of the spin-labelled side chains (cf. Section 16.12). Twice the minimum number of measured distances is advisable to correct for these uncertainties. After docking, we can use energy refinement to alleviate any clashes between side groups at the interface.

An illustrative example is docking coupling protein CheW with histidine kinase CheA of the bacterial chemotaxis receptor complex (Borbat and Freed 2007). Table 16.14 gives the distances (in bold) between three spin-labelled residues in CheW and four in a truncated version of CheA; these are measured with pulse dipolar-EPR spectroscopy. Spin-labelling sites were chosen from an NMR structure of CheW and a crystal structure of CheAΔ289 (see distances in italics in Table 16.14) to triangulate the two proteins in the dimeric complex. The 6×6 metric matrix for the seven nitroxides, with S15R1 as origin, has eigenvalues (93.261, 9.062, 5.099, 2.085, 1.259, −0.465) nm^2, all of which should be ≥0 (Borbat and Freed 2007). The three largest eigenvalues and corresponding eigenvectors are used in Eq. 16.56 to produce coordinates of the docked complex. These were then optimized by using rigid-body refinement, in which pulsed-EPR distances were treated as nuclear-Overhauser (NOE) restraints, as in solution-structure determination by NMR (Brunger et al. 1988). The C^β–C^β distances were constrained to be within −0.5 nm and +0.1 nm of the measured EPR distances (Bhatnagar et al. 2007). These limits are based on comparison of EPR dipolar distances with known crystal structures of proteins without spin labels (see Figure 16.28). For example, the differences for CheAΔ289 in Table 16.14 range from −0.7 nm to −0.1 nm.

16.15.5 Conformational Changes

Functionally significant conformational changes in proteins may be accompanied by large-scale movements, e.g., in channel opening or substrate accessibility. Suitably chosen pairs of spin-labelling sites therefore can be used to monitor different conformational populations from the distribution of inter-spin distances. Ideally, distinct conformations will give rise to resolved components in the distance profiles. Then we can use integrated populations to determine fractional occupancies of conformational states and their equilibrium constants (Liu et al. 2016; Casey and Fanucci 2015). For such quantitative work, it is most important to avoid (or correct for) any orientational selection in the DEER profiles (see Section 15.14).

16.16 CONCLUDING SUMMARY

1. *Spin-labelling sites by mutagenesis:*
   Site-directed spin labelling uses site-directed mutagenesis to place cysteine residues at specific positions in a protein sequence. The cysteine SH group is then spin-labelled covalently with a nitroxide reagent, most frequently by forming a S–S bridge with the methane thiosulphonate derivative MTSSL. Alternatively, the –SH group is alkylated with PROXYL-iodoacetamide, producing a side chain with one further bond. TOAC on the other hand is a helicogenic TEMPO-amino acid that is incorporated stereospecifically at specific positions in peptide sequences by using chemical synthesis.
2. *Cysteine scanning and accessibility to relaxants:*
   Systematic scanning of the spin-label position reveals periodicity in segmental mobility of MTSSL, and in accessibility to paramagnetic relaxants, that is characteristic of α-helix (3.6 residues/turn) and β-sheet (2.0 residues/turn) secondary structures. Differential accessibility to hydrophobic (molecular oxygen) and polar (transition-metal complexes, e.g., NiEDDA) relaxants is especially valuable for determining immersion depths in membranes and also for detecting tertiary folds. Relaxation enhancements are most sensitively detected with saturation measurements, which require lower relaxant concentrations than for line-broadening experiments.
3. *Side-chain rotamers (MTSSL):*
   The configuration of the MTSSL (or R1) side chain is specified by five dihedral angles $\chi_1 - \chi_5$, extending outward from the C^α-atom of the protein backbone to the pyrroline-nitroxide ring. The first two dihedrals χ_1, χ_2 have fixed preferential rotamers, which exchange only slowly. The S–S link is fixed in one of two rotamers with $\chi_3 = \pm 90°$. For the two terminal segments, χ_4 populates three rotamers which exchange rapidly and χ_5 populates two rotamers that exchange somewhat less rapidly.

TABLE 16.14 Intra- and inter-domain dipolar distances (nm) for triangulating the CheW/CheAΔ289 chemoreceptor coupling system (Park et al. 2006)

		CheW			CheAΔ289			
	mutant:	S15R1	S72R1	S80R1	N553R1	S568R1	D579R1	E646R1
CheW	S15R1		**2.7**	**1.82**	**3.7**	**5.45**	**6.1**	**4.37**
	S72R1	*1.49/1.53*		**2.45**	**2.7**	**4.9**	**4.6**	**3.25**
	S80R1	*1.14/1.15*	*1.66/1.67*		**2.6**	**4.7**	**5.45**	**3.95**
CheAΔ289	N553R1					**2.35**	**3.45**	**3.2**
	S568R1				*1.91/1.79*		**3.25**	**3.55**
	D579R1				*2.78/2.68*	*2.53/2.30*		**2.8**
	E646R1				*2.57/2.46*	*3.43/3.22*	*2.44/2.52*	

Italics: C^β–C^β/C^α–C^α distances from PDB 1KOS (CheW) and 1B3Q (CheAΔ289) coordinates. To compare with EPR distances (bold): reflect about the diagonal.

4. *Spectral simulation reflecting side-chain diffusion:* Rotamer analysis of MTSSL at solvent-exposed α-helical sites leads to the χ_4/χ_5-model for spectral simulations with the stochastic Liouville equation. Here, the diffusion tilt $\beta_D \approx 36°$ arises from rapid rotation around χ_4 as principal axis, with order parameter S_{zz} for fluctuations of this axis around χ_5. Spectral simulations from molecular-dynamics trajectories additionally admit slow transitions in the dihedrals χ_1, χ_2. Experiments at high field (EPR frequency ≥ 94 GHz) suppress contributions from these slow transitions.
5. *Peptide-backbone ordering from TOAC:* Angular anisotropy of the spectra from TOAC-containing peptides lets us determine order parameters for the peptide backbone, S_{helix}. This is related to the experimental order parameter by $S_{zz} = S_{helix} \times \frac{1}{2}(3l_z^2 - 1)$, where $\frac{1}{2}(3l_z^2 - 1) =$ 0.92 or −0.2 for the $^6T_2$ or $^2T_6$ conformer of the TOAC ring, respectively. Values of $|S_{zz}| > 0.2$ are compatible only with the first alternative ($^6T_2$).
6. *Linkage/end effects in dipolar distances*: For MTSSL-labelled isolated α-helices $(A_4K)_4A$, the average radial distance of the nitroxide from the helix axis is $R_{NO} = 0.74$ nm (Figure 16.23) and is $\Delta R_{NO} = R_{NO} - R_{C\alpha} = 0.51$ nm from the cylinder surface defined by the C^α backbone atoms. For water-exposed surface α-helices in T4 lysozyme ΔR_{NO} = 0.42–0.63 nm. For MTSSL-labelled β-sheets, the extension of the nitroxide from the strand axis is R_{NO} = 0.68 nm, averaged over exposed and buried sides of the sheet, and $\Delta R_{NO} = R_{NO} - R_{C\alpha} = 0.59$ nm from the C^α atoms of the strand backbone. Tables 16.11, A16.7 and A16.8 give more detailed comparisons of EPR distances with protein crystal structures.

APPENDIX A16: ADDITIONAL CONFORMATIONAL AND DISTANCE DATA

TABLE A16.1 Crystal Structure of *S*-(2,2,5,5-tetramethyl-1-oxyl-Δ³-pyrrolin-3-ylmethyl) methanethiosulphonate, MTSSL (Zielke et al. 2008)[a]

interatomic distance	molecule 1 (nm)	molecule 2 (nm)[b]	mean (nm)[c]
O—N	0.1277 (1)	0.1274 (1)	0.1275 (1)
N—C^2	0.1482 (2)	0.1485 (2)	0.1483 (2)
N—C^5	0.1483 (2)	0.1482 (2)	0.1483 (2)
C^4—C^5	0.1503 (2)	0.1501 (2)	0.1502 (2)
C^3—C^4	0.1330 (2)	0.1336 (2)	0.1333 (3)
C^3—C^ε	0.1493 (2)	0.1494 (2)	0.1493 (2)
C^2—C^3	0.1517 (2)	0.1515 (2)	0.1516 (2)
C^ε—S^δ	0.1835 (1)	0.1837 (1)	0.1836 (1)
S^δ—S^γ	0.20661 (5)	0.20684 (5)	0.2067 (1)
S^γ—O^3	0.1435 (1)	0.1437 (1)	0.1436 (1)
S^γ—O^2	0.1437 (1)	0.1441 (1)	0.1439 (2)
S^γ—C^β	0.1761 (2)	0.1762 (2)	0.1761 (2)
bond angle	molecule 1 (°)	molecule 2 (°)[b]	mean (°)[c]
O—N—C^2	121.9 (1)	122.1 (1)	122.0 (1)
O—N—C^5	123.1 (1)	122.7 (1)	122.9 (2)
C^2—N—C^5	115.0 (1)	115.0 (1)	115.0 (1)
N—C^5—C^4	99.5 (1)	99.7 (1)	99.6 (1)
C^3—C^4—C^5	113.6 (1)	113.5 (1)	113.5 (1)
C^4—C^3—C^ε	125.5 (1)	126.0 (1)	125.8 (3)
C^4—C^3—C^2	112.0 (1)	112.1 (1)	112.0 (1)
C^ε—C^3—C^2	122.5 (1)	121.9 (1)	122.2 (3)
N—C^2—C^3	99.7 (1)	99.8 (1)	99.7 (1)
C^3—C^ε—S^δ	114.4 (1)	115.8 (1)	115.1 (7)
C^ε—S^δ—S^γ	101.6 (1)	100.6 (1)	101.1 (5)
O^3—S^γ—O^2	119.3 (1)	118.9 (1)	119.1 (2)
O^3—S^γ—C^β	108.8 (1)	108.8 (1)	108.8 (1)

(Continued)

TABLE A16.1 (*Continued*) Crystal Structure of *S*-(2,2,5,5-tetramethyl-1-oxyl-Δ^3-pyrrolin-3-ylmethyl) methanethiosulphonate, MTSSL (Zielke et al. 2008)[a]

bond angle	molecule 1 (°)	molecule 2 (°)[b]	mean (°)[c]
O^2—S^γ—C^β	108.0 (1)	107.7 (1)	107.9 (1)
O^3—S^γ—S^δ	109.5 (1)	109.9 (1)	109.7 (2)
O^2—S^γ—S^δ	103.9 (1)	104.2 (1)	104.1 (1)
C^β—S^γ—S^δ	106.6 (1)	106.6 (1)	106.6 (1)
dihedral angle	molecule 1 (°)	molecule 2 (°)[b]	mean (°)[c,d]
O—N—C^5—C^4	178.2 (1)	−177.4 (1)	±177.8 (4)
C^2—N—C^5—C^4	−4.0 (1)	−2.3 (1)	−3.1 (9)
N—C^5—C^4—C^3	1.3 (1)	1.8 (2)	1.6 (2)
C^5—C^4—C^3—C^ε	−179.8 (1)	−178.2 (1)	−179.0 (8)
C^5—C^4—C^3—C^2	1.6 (2)	−0.8 (2)	0.4 (12)
O—N—C^2—C^3	−177.3 (1)	177.1 (1)	∓177.2 (1)
C^5—N—C^2—C^3	4.8 (1)	1.9 (1)	3.4 (15)
C^4—C^3—C^2—N	−3.8 (1)	−0.6 (1)	−2.2 (16)
C^ε—C^3—C^2—N	177.6 (1)	176.9 (1)	177.2 (3)
C^4—C^3—C^ε—S^δ (χ_5)	99.3 (1)	−99.7 (1)	±99.5 (2)
C^2—C^3—C^ε—S^δ	−82.3 (1)	83.2 (1)	∓82.7 (5)
C^3—C^ε—S^δ—S^γ (χ_4)	−83.3 (1)	82.5 (1)	∓82.9 (4)
C^ε—S^δ—S^γ—O^3	−30.0 (1)	29.4 (1)	∓29.7 (3)
C^ε—S^δ—S^γ—O^2	−158.5 (1)	157.8 (1)	∓158.2 (3)
C^ε—S^δ—S^γ—C^β (χ_3)	87.6 (1)	−88.4 (11)	±88.0 (4)

Bond lengths, bond angles and dihedral angles listed for pyrroline ring and S-containing side chain of methane thiosulphonate spin label MTSSL. Atom labelling (see Figure 16.1) includes $S^\gamma O_2$ and $C^\beta H_3$, which differ from the R1 amino-acid side chain.

[a] Atom numbering as in Figure 16.1, but note that S^γ, C^β, O^2 and O^3 belong to the methyl sulphonate, which is the leaving group on forming the S–S bridge with the cysteine –SH in Figure 16.1.

[b] Majority species (92%).

[c] Mean for molecule 1 and majority species of molecule 2 (92%), which are stereoisomers.

[d] Upper signs for molecule 1; lower signs for majority species of molecule 2.

TABLE A16.2 ENDOR-constrained dihedral angles for methyl thiosulphonate spin-label (MTSSL) S–S linked to cysteine (Mustafi et al. 2002)[a]

SL[b]	*rotamer*[c]	χ_1 (°)	χ_2 (°)	χ_3 (°)	χ_4 (°)	χ_5 (°)
CysR1	**t,t/m,m,p**	−135±30	−155±15	−80±15	−55±10	96±5
	t,p/t,m,p[d]	180±30	60±15	160±15	−65±10	95±5

[a] Determined from conformational search with distance constraints from proton ENDOR (see Section 14.13).

[b] R1, 1-oxyl-2,2,5,5-tetramethylpyrrolin-3-ylmethylsulpho- (MTSSL) S–S adduct of cysteine.

[c] Rotamers defined by dihedrals,: $\chi_1, \chi_2/\chi_3, \chi_4, \chi_5$. For χ_1, χ_2, χ_4: $\mathbf{t} = \pm 180 \pm 30°$, $\mathbf{p} = 60 \pm 30°$, $\mathbf{m} = -60 \pm 30°$; but for χ_3, χ_5: $\mathbf{p} = 90 \pm 30°$, $\mathbf{m} = -90 \pm 30°$.

[d] Unlikely solution, because $\chi_3 \neq \pm 90°$ dihedral allowed about S–S bond.

TABLE A16.3 Dihedral angles for S–S linked methyl thiosulphonate spin-label (MTSSL) side chains R1 at β-strand sites in immunoglobulin-binding domain (GB1), tumor necrosis factor (TNFα) and azurin[a]

mutant/SL[b]	*crystal form*	*chain*[c]	*rotamer*[d]	χ_1 (°)	χ_2 (°)	χ_3 (°)	χ_4 (°)	χ_5 (°)
GB1:								
E15R1	I	A	**t,t/(p)**	−171	155	131	120	179
		B	**p,(p)/(p),m**	69	104	133	−79	26
		C		135				
		D	**p,p/(p),m,(p)**	66	89	132	−67	47
		E	**t,t/m**	171	−167	−68	−109	−51
		F	**p,(t)/p,t,m**	81	135	93	175	−63
		G	**t,(t)/(p),-,(p)**	−151	−145	135	129	127
		H	**(t),t/m,t,p**	144	−177	−90	−157	70
T44R1	I	A	**t,(p)/p**	−163	101	72	106	28
	II	A,α[e]	**t,t/p,p,m**	−166	152	69	45	−104
		A,β[e]	**m,t/m,p,(m)**	−61	−153	−75	96	−100
TNFα:								
T77R1			**p,(m)/t,(p),p**	38	−112	177	104	90
azurin:								
T30R1	I	A	**t,(m)/m,m,p**	180	−94	−89	−59	86
		B	**t,m/m,m**	−176	−90	−85	−78	−17
		C		(143)	(−74)	(−86)	(−67)	(−155)
		D	**t,m/m**	−151	−82	−80	(−80)	(−43)
	II	A	**t,(m)/m,t**	−178	−92	−86	171	−6
		B	**t,(m)/m,t**	−173	−99	−87	167	2
		C	**t,(m)/m,t**	−173	−97	−89	169	5
		D	**t**	175	(−136)	(−82)	(79)	(50)
		E	**t,(m)/m,t**	−176	−92	−85	159	5
		F	**t,(m)/m,(t),m**	−173	−101	−85	−148	−81
		G	**t,t**	168	173	(73)	(−164)	(−5)
		H	**p,t/p,p,(p)**	64	161	83	46	49
		I	**(p),t/p,m**	93	178	114	−87	−32
		J	**t,t**	−168	−160	(−99)	(152)	(−1)
		K	**t,(m)/m,t**	−176	−95	−82	162	2
		L	**t,(m)/m,(t),m**	−170	−103	−83	−148	−87
		M	**t,(m)**	−173	−93	(−161)	(146)	(16)
		N	**t**	165	117	(15)	(60)	(18)
		O	**t,(m)/m,(m),p**	−169	−91	−81	−119	51
		P	**t,(m)/m,t**	−171	−100	−84	170	−3

[a] GB1 mutants E15R1 (PDB: 5BMG), and T44R1 crystal forms I and II (PDB: 5BMI, 5BMH) (Cunningham et al. 2016). TNFα mutant T77R1 (PDB: 5UUI) (Carrington et al. 2017). Azurin mutants T21R1 (PDB: 4BWW; Florin et al. 2014), and T30R1 crystal forms I and II (PDB: 5I26, 5I28; Abdullin et al. 2016).

[b] R1, 1-oxyl-2,2,5,5-tetramethylpyrrolin-3-ylmethylsulpho- (MTSSL) S–S adduct of cysteine.

[c] Different "chains" are inequivalent monomers in the unit cell.

[d] Rotamers defined by dihedrals: $\chi_1, \chi_2/\chi_3, \chi_4, \chi_5$. For χ_1, χ_2, χ_4: $\mathbf{t} = \pm 180 \pm 30°$, $\mathbf{p} = 60 \pm 30°$, $\mathbf{m} = -60 \pm 30°$; but for χ_3, χ_5: $\mathbf{p} = 90 \pm 30°$, $\mathbf{m} = -90 \pm 30°$. Brackets denote angles outside the ±30° range.

[e] α, β are two different conformers.

TABLE A16.4 Atomic coordinates for amino-acid residues in antiparallel β-pleated sheet of β-poly-L-alanine (Arnott et al. 1967)

	up chain			*down chain*		
atom	X (nm)	Y (nm)	Z (nm)	X (nm)	Y (nm)	Z (nm)
H-bonded pair:						
N	0.033	0.013	−0.0837	0.440	0.013	0.0837
H (on N)	0.130	−0.010	−0.0857	0.343	−0.010	0.0857
C^{α}	−0.014	0.079	0.0383	0.487	0.079	−0.0383
H (on C^{α})	−0.121	0.070	0.0453	0.594	0.070	−0.0453
C^{β}	0.024	0.228	0.0353	0.449	0.228	−0.0353
C′	0.049	0.013	0.1608	0.424	0.013	−0.1608
O	0.170	−0.013	0.1638	0.303	−0.013	−0.1638
non H-bonded pair:						
N	−0.033	−0.013	0.2608	0.506	−0.013	0.4282
H (on N)	−0.130	0.010	0.2588	0.603	0.010	0.4302
C^{α}	0.014	−0.079	0.3828	0.459	−0.079	0.3602
H (on C^{α})	0.121	−0.070	0.3898	0.352	−0.070	0.2992
C^{β}	−0.024	−0.228	0.3798	0.497	−0.228	0.3092
C′	−0.049	−0.013	0.5053	0.522	−0.013	0.1837
O	−0.170	0.013	0.5083	0.643	0.013	0.1807

Coordinates of H-bonded pair: (X, Y, Z), $(-X+d_{\beta}, Y, -Z)$.
Coordinates of non H-bonded pair: $(-X, -Y, Z+h_{\beta})$, $(X+d_{\beta}, -Y, -Z+h_{\beta})$.
$h_{\beta} = 0.3445$ nm; $d_{\beta} = 0.473$ nm.

TABLE A16.5 Atomic coordinates for amino-acid residue in an α-helix

atom	X (nm)	Y (nm)	Z (nm)	R (nm)	ϕ (°)
α-poly-L-alanine:[a]					
N	0.1375	−0.0711	−0.0906	0.1548	−27.35
H (on N)	0.1459	−0.0490	−0.1878	0.1539	−18.57
C^{α}	0.2288	0	0	0.2288	0
H (on C^{α})	0.2929	−0.0705	0.0485	0.3013	−13.54
C^{β}	0.3139	0.0998	−0.0808	0.3294	17.63
C′	0.1481	0.0760	0.1054	0.1664	27.16
O	0.1777	0.0691	0.2256	0.1906	21.24
standard right-handed α-helix:[b]					
N	0.1360	−0.0744	−0.0873	0.1550	−28.69
H (on N)	0.1485	−0.0559	−0.1847	0.1586	−20.63
C^{α}	0.2280	0	0	0.2280	0
H on C^{α}	0.2989	−0.0699	0.0467	0.3069	−13.17
C^{β}	0.3049	0.1029	−0.0830	0.3218	18.65
C′	0.1480	0.0718	0.1089	0.1645	25.88
O	0.1765	0.0579	0.2288	0.1858	18.15

Cylindrical polar coordinates for adjacent residues: $Z \pm h_{\alpha}$, R, $\phi \pm \omega_{\alpha}$, where $h_{\alpha} = 0.1495$ nm, $\omega_{\alpha} =$ 99.57° for polyalanine; and $h_{\alpha} = 0.15$ nm, $\omega_{\alpha} = 100°$ for a standard α-helix.
[a] Arnott and Dover (1967).
[b] Parry and Suzuki (1969).

TABLE A16.6 Atomic coordinates for amino-acid residue in a parallel β-sheet

atom	X (nm)	Y (nm)	Z (nm)	R (nm)	ϕ (°)
N	0.0245	0.0336	−0.1225	0.0415	53.92
H (on N)	0.0152	0.1318	−0.1388	0.1327	83.43
C^{α}	0.0985	0	0	0.0985	0
H on C^{α}	0.1053	−0.1093	0.0099	0.1518	−46.08
C^{β}	0.2393	0.0593	−0.0077	0.2466	13.93
C′	0.0253	0.0578	0.1213	0.0631	66.35
O	0.0160	0.1803	0.1378	0.1811	84.93

Cylindrical polar coordinates for adjacent residues: $Z \pm h_{\beta}$, R, $\phi \pm \omega_{\beta}$, where $h_{\beta} = 0.325$ nm, $\omega_{\beta} = 180°$.

TABLE A16.7 Dipolar distances $r_{NO\text{-}NO}$ measured between MTSSL labels in bacteriophage T4 lysozyme and C^{α}–C^{α}, C^{β}–C^{β} distances between residues in native crystal structure (PDB: 2LZM)

residues		$r_{C^{\alpha}\text{-}C^{\alpha}}$ (nm)	$r_{C^{\beta}\text{-}C^{\beta}}$ (nm)	$r_{NO\text{-}NO}$[a] (nm)	Δr[b] (nm)	*Ref.*[c]
T59	D159	3.12	3.36	4.19 ± 0.27	0.83	1
K60	S90	3.53	3.65	3.78 ± 0.45	0.13	1
K60	V94	2.71	2.75	2.55 ± 0.31	−0.20	1
K60	T109	2.94	3.05	3.52 ± 0.26	0.47	1
K60	R154	3.35	3.40	3.41 ± 0.20	0.01	1
D61	R80	2.87	2.89	3.40 ± 0.22	0.51	3
D61	P86	3.45	3.71	3.75 ± 0.20	0.04	3
D61	E128	3.79	3.90	4.62 ± 0.24	0.72	2
D61	K135	3.77	4.01	4.72 ± 0.22	0.71	3
E62	T109	2.69	2.71	2.95 ± 0.27	0.24	2
E62	Q123	4.01	4.21	4.23 ± 0.33	0.02	2
E62	A134	3.45	3.59	4.11 ± 0.15	0.52	1
E62	T155	3.44	3.51	4.12 ± 0.15	0.61	2
E64	Q122	3.22	3.33	3.41 ± 0.25	0.08	1
K65	R76	1.68	1.67	2.14 ± 0.28	0.47	3
K65	R80	2.26	2.23	2.65 ± 0.38	0.42	3
K65	P86	2.89	3.13	3.74 ± 0.27	0.61	3
K65	K135	3.43	3.64	4.63 ± 0.22	0.99	3
R80	K135	2.67	2.71	3.68 ± 0.10	0.97	3
A82	V94	2.12	2.4	3.07 ± 0.33	0.67	1
A82	N132	2.13	2.07	2.63 ± 0.35	0.56	1
A82	A134	2.37	2.56	3.39 ± 0.32	0.83	2
A82	T155	2.56	2.80	3.58 ± 0.25	0.78	1
K83	Q123	1.43	1.51	2.05 ± 0.34	0.54	1
K83	T155	2.29	2.42	3.28 ± 0.30	0.86	2
P86	A112	1.14	1.11	1.30 ± 0.51	0.19	1
P86	R119	1.00	1.03	1.50 ± 0.30	0.47	4
Y88	I100	0.89	0.85	<0.60 ± 0.30	<−0.25	4
D89	A93	0.98	1.20	1.60 ± 0.30	0.40	4
D89	R96	0.84	0.93	<0.60 ± 0.30	<−0.33	4
A93	E108	1.91	2.07	2.33 ± 0.41	0.26	2
A93	A112	1.96	2.14	2.61 ± 0.15	0.47	1
A93	Q123	1.62	1.89	2.48 ± 0.23	0.59	1
A93	A134	2.2	2.36	2.91 ± 0.24	0.55	2
A93	R154	1.34	1.59	2.51 ± 0.24	0.92	1
V94	Q123	1.64	1.78	2.40 ± 0.26	0.62	2

(Continued)

TABLE A16.7 (*Continued*) Dipolar distances $r_{NO\text{–}NO}$ measured between MTSSL labels in bacteriophage T4 lysozyme and C^{α}–C^{α}, C^{β}–C^{β} distances between residues in native crystal structure (PDB: 2LZM)

residues		$r_{C^\alpha\text{–}C^\alpha}$ (nm)	$r_{C^\beta\text{–}C^\beta}$ (nm)	$r_{NO\text{–}NO}{}^{a}$ (nm)	Δr^{b} (nm)	*Ref.*[c]
V94	N132	1.96	1.94	3.17 ± 0.13	1.23	1
E108	Q123	2.05	2.18	2.76 ± 0.24	0.58	2
E108	A134	1.85	2.14	3.24 ± 0.12	1.10	2
E108	T155	2.33	2.53	3.52 ± 0.23	0.99	1
T109	A134	1.81	2.06	3.06 ± 0.28	1.00	1
T115	T155	2.06	2.31	2.82 ± 0.24	0.51	1
N116	V131	1.11	1.01	1.90 ± 1.00	0.89	4
N116	A134	1.19	1.24	2.02 ± 0.15	0.78	1
R119	E128	1.04	0.97	1.99 ± 0.23	1.02	1
R119	V131	1.32	1.33	2.23 ± 0.27	0.90	1
M120	V131	1.05	0.9	1.40 ± 0.30	0.50	4
Q123	V131	1.46	1.38	2.23 ± 0.27	0.85	1
D127	T151	0.95	1.15	1.40 ± 0.24	0.25	4
D127	R154	0.59	0.64	0.70 ± 0.30	0.06	4
D127	T155	0.88	1.02	1.21 ± 0.34	0.19	1
V131	I150	0.87	0.89	0.57 ± 0.04	−0.32	4
V131	T151	1.04	1.24	0.90 ± 0.80	−0.34	4
V131	R154	0.95	0.92	0.65 ± 0.40	−0.27	4
A134	T151	1.07	1.08	0.70 ± 0.08	−0.38	4
N140	K147	1.01	1.12	1.30 ± 0.70	0.18	4
N140	T151	1.55	1.69	2.22 ± 0.33	0.53	2

[a] "±" indicates standard deviation of distance distribution. MTSSL labelling produces the 1-oxyl-2,2,5,5-tetramethylpyrrolin-3-ylmethylsulpho- S–S adduct of cysteine, which replaces the native residue indicated.

[b] $\Delta r \equiv r_{NO\text{–}NO} - r_{C^\beta\text{–}C^\beta}$.

[c] References to EPR data: 1. Kazmier et al. (2011); 2. Alexander et al. (2013); 3. Borbat et al. (2002); 4. Alexander et al. (2008).

TABLE A16.8 Dipolar distances $r_{NO\text{–}NO}$ measured between MTSSL labels in α-crystallin domain of rat C131A αA-crystallin, and C^{α}–C^{α}, C^{β}–C^{β} distances between residues in native crystal structure of truncated bovine αA-crystallin (PDB: 3L1F)[a]

residues		$r_{C^\alpha\text{–}C^\alpha}$ (nm)	$r_{C^\beta\text{–}C^\beta}$ (nm)	$r_{NO\text{–}NO}{}^{b}$ (nm)	Δr^{c} (nm)	*Ref.*[d]
R65	V72	0.52	0.47	0.60 ± 0.30	0.13	1,2
R68	S122	1.25	1.42	1.95 ± 0.50	0.53	2
K70	V72	0.65	0.59	0.66 ± 0.15	0.07	1,2
K70	F74	1.29	1.24	1.20 ± 0.40	−0.04	1,2
K70	S142	0.70	0.57	0.60 ± 0.20	0.03	1,2
V72	S142	0.43	0.44	0.55 ± 0.20	0.11	1,2
I73	Y118	0.87	0.60	0.60 ± 0.30	0.00	2
F74	T140	0.47	0.58	0.60 ± 0.20	0.02	1,2
D84	K99	0.54	0.58	0.76 ± 0.06	0.18	3,2
D84	N101	0.90	0.87	0.75 ± 0.80	−0.12	3,2
K88	E95	0.51	0.45	0.60 ± 0.07	0.15	3,2
K88	H97	0.88	0.81	0.70 ± 0.08	−0.11	3,2
L90[e]	F93	0.54	0.60	0.69 ± 0.11	0.09	3,2
L90[e]	K99	1.99	1.88	1.85 ± 0.40	−0.03	3,2
L90[e]	Q126	1.01	1.02	1.20 ± 0.40	0.18	1,2
D92	N123	1.04	1.11	1.20 ± 0.60	0.09	1,2
E95	R117	0.47	0.56	0.76 ± 0.10	0.20	3,2

(*Continued*)

TABLE A16.8 (*Continued*) Dipolar distances r_{NO-NO} measured between MTSSL labels in α-crystallin domain of rat C131A αA-crystallin, and C^{α}–C^{α}, C^{β}–C^{β} distances between residues in native crystal structure of truncated bovine αA-crystallin (PDB: 3L1F)[a]

residues		$r_{C^{\alpha}-C^{\alpha}}$ (nm)	$r_{C^{\beta}-C^{\beta}}$ (nm)	r_{NO-NO}[b] (nm)	Δr[c] (nm)	*Ref.*[d]
N101	R103	0.69	0.78	0.63 ± 0.08	−0.15	1,2
N101	S111	0.44	0.49	0.76 ± 0.10	0.27	3,2
R103	D105	0.70	0.85	0.95 ± 0.80	0.10	1,2
D105	S111	1.42	1.61	1.80 ± 0.90	0.19	1,2
R119	N123	1.17	1.24	1.30 ± 0.60	0.06	1,2
R119	D125	1.36	1.57	2.30 ± 0.30	0.73	1,2
R119	S127	1.63	1.84	2.40 ± 0.60	0.56	1,2
R119	S130	1.58	1.82	2.30 ± 0.50	0.48	1,2
R119	S132	1.88	2.10	2.30 ± 0.40	0.20	1,2
D125	P144	0.59	0.51	0.70 ± 0.40	0.19	1,2
S127	S142	1.13	1.33	2.00 ± 0.40	0.67	2
A128	P144	0.69	0.53	0.40 ± 0.40	−0.13	1,2
A128	I146	1.01	0.89	1.30 ± 0.40	0.41	1,2
S130	P144	0.95	0.93	0.60 ± 0.10	−0.33	1,2
A131	L133	0.71	0.79	0.70 ± 0.10	−0.09	1,2
A131	S134	1.01	1.13	1.70 ± 0.40	0.57	1,2
S132	A135	0.99	1.05	1.60 ± 0.50	0.55	1,2
S132	S142	0.91	0.90	0.60 ± 0.10	−0.30	1,2
S132	P144	1.51	1.51	0.90 ± 0.50	−0.61	1,2

[a] Strands S_1–S_7 in the β-sandwich are residues: 63–67, 69–77, 84–90, 92–104, 108–120, 130–134 and 137–146.

[b] "±" indicates standard deviation of distance distribution. MTSSL labelling produces the 1-oxyl-2,2,5,5-tetramethylpyrrolin-3-ylmethylsulpho- S–S adduct of cysteine, which replaces the native residue indicated.

[c] $\Delta r \equiv r_{NO-NO} - r_{C^{\beta}-C^{\beta}}$.

[d] References to EPR data: 1. Koteiche and Mchaourab (1999); 2. Alexander et al. (2008); 3. Koteiche et al. (1998).

[e] Q90 in bovine αA-crystallin: 96% identity with rat sequence over residues 60–155.

REFERENCES

Abdullin, D., Hagelueken, G., and Schiemann, O. 2016. Determination of nitroxide spin label conformations *via* PELDOR and X-ray crystallography. *Phys. Chem. Chem. Phys.* 18:10428–10437.

Ackermann, K., Pliotas, C., Valera, S., Naismith, J.H., and Bode, B.E. 2017. Sparse labeling PELDOR spectroscopy on multimeric mechanosensitive membrane channels. *Biophys. J.* 113:1968–1978.

Alexander, N., Bortolus, M., Al-Mestarihi, A., Mchaourab, H., and Meiler, J. 2008. De novo high-resolution protein structure determination from sparse spin-labeling EPR data. *Structure* 16:181–195.

Alexander, N.S., Stein, R.A., Koteiche, H.A. et al. 2013. RosettaEPR: rotamer library for spin label structure and dynamics. *PLOS One* e72851:1–14.

Altenbach, C., Greenhalgh, D.A., Khorana, H.G., and Hubbell, W.L. 1994. A collision gradient method to determine the immersion depth of nitroxides in lipid bilayers: Application to spin-labeled mutants of bacteriorhodopsin. *Proc. Natl. Acad. Sci. USA* 91:1667–1671.

Anbazhagan, V., Qu, J., Kleinschmidt, J.H., and Marsh, D. 2008. Incorporation of outer membrane protein OmpG in lipid membranes. Protein-lipid interactions and β-barrel orientation. *Biochemistry* 47:6189–6198.

Arnott, S. and Dover, S.D. 1967. Refinement of bond angles of an α-helix. *J. Mol. Biol.* 30:209–212.

Arnott, S., Dover, S.D., and Elliot, A. 1967. Structure of β-poly-L-alanine: Refined atomic co-ordinates for an anti-parallel beta-pleated sheet. *J. Mol. Biol.* 30:201–208.

Banham, J.E., Baker, C.M., Ceola, S. et al. 2008. Distance measurements in the borderline region of applicability of CW EPR and DEER: A model study on a homologous series of spin-labelled peptides. *J. Magn. Reson.* 191:202–218.

Beier, C. and Steinhoff, H.J. 2006. A structure-based simulation approach for electron paramagnetic resonance spectra using molecular and stochastic dynamics simulations. *Biophys. J.* 91:2647–2664.

Berengian, A.R., Bova, M.P., and Mchaourab, H.S. 1997. Structure and function of the conserved domain in αA-crystallin. Site-directed spin labeling identifies a β-strand located near a subunit interface. *Biochemistry* 36:9951–9957.

Bhatnagar, J., Freed, J.H., and Crane, B.R. 2007. Rigid body refinement of protein complexes with long-range distance restraints from pulsed dipolar ESR. *Meth. Enzymol.* 423:117–133.

Borbat, P.P. and Freed, J.H. 2007. Measuring distances by pulsed dipolar ESR spectroscopy: spin-labeled histidine kinases. *Meth. Enzymol.* 423:52–116.

Borbat, P.P., Mchaourab, H.S., and Freed, J.H. 2002. Protein structure determination using long-distance constraints from double-quantum coherence ESR: study of T4 lysozyme. *J. Am. Chem. Soc.* 124:5304–5314.

Braun, W. 1987. Distance geometry and related methods for protein structure determination from NMR data. *Q. Rev. Biophys.* 19:115–157.

Bridges, M.D., Hideg, K., and Hubbell, W.L. 2010. Resolving conformational and rotameric exchange in spin-labeled proteins using saturation recovery EPR. *Appl. Magn. Reson.* 37:363–390.

Brunger, A.T., Adams, P.D., Clore, G.M. et al. 1988. Crystallography & NMR system: A new software suite for macromolecular structure determination. *Acta Crystallogr. D* 54:905–921.

Budil, D.E., Sale, K.L., Khairy, K.A., and Fajer, P.G. 2006. Calculating slow-motional electron paramagnetic resonance spectra from molecular dynamics using a diffusion operator approach. *J. Phys. Chem. A* 110:3703–3713.

Carrington, B., Myers, W.K., Horanyi, P., Calmiano, M., and Lawson, A.D.G. 2017. Natural conformational sampling of human TNFα visualized by double electron-electron resonance. *Biophys. J.* 113:371–380.

Casey, T.M. and Fanucci, G.E. 2015. Spin labelling and double electron-electron resonance (DEER) to deconstruct conformational ensembles of HIV protease. *Meth. Enzymol.* 564:153–187.

Chen, M., Margittai, M., Chen, J., and Langen, R. 2007. Investigation of α-synuclein fibril structure by site-directed spin labelling. *J. Biol. Chem.* 282:24970–24979.

Chimento, D.P., Mohanty, A.K., Kadner, R.J., and Wiener, M.C. 2003. Substrate-induced transmembrane signaling in the cobalamin transporter BtuB. *Nat. Struct. Biol.* 10:394–401.

Chothia, C. and Janin, J. 1981. Relative orientation of close-packed β-pleated sheets in proteins. *Proc. Natl. Acad. Sci. USA* 78:4146–4150.

Chothia, C. and Janin, J. 1982. Orthogonal packing of β-pleated sheets in proteins. *Biochemistry* 21:3955–3965.

Chothia, C., Levitt, M., and Richardson, D. 1981. Helix to helix packing in proteins. *J. Mol. Biol.* 145:215–250.

Cobb, N.J., Sönnichsen, F.D., Mchaourab, H., and Surewicz, W.K. 2007. Molecular architecture of human prion protein amyloid: A parallel, in-register β-structure. *Proc. Natl. Acad. Sci. USA* 104:18946–18951.

Columbus, L. and Hubbell, W.L. 2004. Mapping backbone dynamics in solution with site-directed spin labeling: GCN4–58 bZip free and bound to DNA. *Biochemistry* 43:7273–7287.

Columbus, L., Kálai, T., Jekö, J., Hideg, K., and Hubbell, W.L. 2001. Molecular motion of spin labeled side chains in α-helices: Analysis by variation of side chain structure. *Biochemistry* 40:3828–3846.

Consentius, P., Gohlke, U., Loll, B. et al. 2016. Tracking transient conformational states of T4 lysozyme at room temperature combining X-ray crystallography and site-directed spin labeling. *J. Am. Chem. Soc.* 138:12868–12875.

Crick, F.H.C. 1953a. The Fourier transform of a coiled coil. *Acta Cryst.* 6:685–689.

Crick, F.H.C. 1953b. The packing of α-helices: simple coiled-coils. *Acta Cryst.* 6:689–697.

Crippen, G.M. and Havel, T.F. 1978. Stable calculation of coordinates from distance information. *Acta Cryst. A* 34:282–284.

Crisma, M., Deschamps, J.R., George, C. et al. 2005. A topographically and conformationally constrained, spin-labeled, α-amino acid: crystallographic characterization in peptides. *J. Peptide Res.* 65:564–579.

Cunningham, T.F., McGoff, M.S., Sengupta, I. et al. 2012. High-resolution structure of a protein spin label in a solvent-exposed β-sheet and comparison with DEER spectroscopy. *Biochemistry* 51:6350–6359.

Cunningham, T.F., Pornsuwan, S., Horne, W.S., and Saxena, S. 2016. Rotameric preferences of a protein spin label at edge-strand β-sheet sites. *Prot. Sci.* 25:1049–1060.

Dalton, L.A., McIntyre, J.O., and Fleischer, S. 1987. Distance estimate of the active center of D-β-hydroxybutyrate dehydrogenase from the membrane surface. *Biochemistry* 26:2117–2130.

DeSensi, S.C., Rangel, D.P., Beth, A.H., Lybrand, T.P., and Hustedt, E.J. 2008. Simulation of nitroxide electron paramagnetic resonance spectra from Brownian trajectories and molecular dynamics simulations. *Biophys. J.* 94:3798–3809.

Ellena, J.F., Lackowicz, P., Mongomery, H., and Cafiso, D.S. 2011. Membrane thickness varies around the circumference of the transmembrane protein BtuB. *Biophys. J.* 100:1280–1287.

Endeward, B., Butterwick, J.A., MacKinnon, R., and Prisner, T.F. 2009. Pulse electron-electron double-resonance determination of spin-label distances and orientations on the tetrameric potassium ion channel KcsA. *J. Am. Chem. Soc.* 131:15246–15250.

Fajer, M.I., Li, H., Yang, W., and Fajer, P.G. 2007. Mapping electron paramagnetic resonance spin label conformations by the simulated scaling method. *J. Am. Chem. Soc.* 129:13840–13846.

Fanucci, G.E., Cadieux, N., Piedmont, C.A., Kadner, R.J., and Cafiso, D.S. 2002. Structure and dynamics of the β-barrel of the membrane transporter BtuB by site-directed spin labeling. *Biochemistry* 41:11543–11551.

Farahbakhsh, Z.T., Altenbach, C., and Hubbell, W.L. 1992. Spin labeled cysteines as sensors for protein-lipid interaction and conformation in rhodopsin. *Photochem. Photobiol.* 56:1019–1033.

Fleissner, M.R., Cascio, D., and Hubbell, W.L. 2009. Structural origin of weakly ordered nitroxide motion in spin-labeled proteins. *Protein Sci.* 18:893–908.

Fleissner, M.R., Bridges, M.D., Brooks, E.K. et al. 2011. Structure and dynamics of a conformationally constrained nitroxide side chain and applications in EPR spectroscopy. *Proc. Natl. Acad. Sci. USA* 108:16241–16246.

Flippen-Anderson, J.L., George, C., Valle, G. et al. 1996. Crystallographic characterization of geometry and conformation of TOAC, a nitroxide spin-labelled $C^{\alpha,\alpha}$-disubstituted glycine, in simple derivatives and model peptides. *Int. J. Pept. Protein Res.* 47:231–238.

Florin, N., Schiemann, O., and Hagelueken, G. 2014. High-resolution crystal structure of spin labelled (T21R1) azurin from *Pseudomonas aeruginosa*: a challenging structural benchmark for *in silico* spin labelling algorithms. *BMC Struct. Biol.* 14:16.

Freed, D.M., Horanyi, P.S., Wiener, M.C., and Cafiso, D.S. 2010. Conformational exchange in a membrane transport protein is altered in protein crystals. *Biophys. J.* 99:1604–1610.

Freed, D.M., Khan, A.K., Horanyi, P.S., and Cafiso, D.S. 2011. Molecular origin of electron paramagnetic resonance line shapes on β-barrel membrane proteins: The local solvation environment modulates spin-label configuration. *Biochemistry* 50:8792–8803.

Giannoulis, A., Ward, R., Branigan, E., Naismith, J.H., and Bode, B.E. 2013. PELDOR in rotationally symmetric homo-oligomers. *Mol. Phys.* 111:2845–2854.

Gruene, T., Cho, M.K., Karyagina, I. et al. 2011. Integrated analysis of the conformation of a protein-linked spin label by crystallography, EPR and NMR spectroscopy. *J. Biomol. NMR* 49:111–119.

Guo, Z., Cascio, D., Hideg, K., Kálai, T., and Hubbell, W.L. 2007. Structural determinants of nitroxide motion in spin-labeled proteins: Tertiary contact and solvent-inaccessible sites in helix G of T4 lysozyme. *Prot. Science* 16:1069–1086.

Guo, Z., Cascio, D., Hideg, K., and Hubbell, W.L. 2008. Structural determinants of nitroxide motion in spin-labeled proteins: Solvent-exposed sites in helix B of T4 lysozyme. *Protein Sci.* 17:228–239.

Hakansson, P., Westlund, P.O., Lindahl, E., and Edholm, O. 2001. A direct simulation of EPR slow-motion spectra of spin labelled phospholipids in liquid crystalline bilayers based on a molecular dynamics simulation of the lipid dynamics. *Phys. Chem. Chem. Phys.* 3:5311–5319.

Harbury, P.B., Zhang, R., Kim, P.S., and Alber, T. 1993. A switch between two-, three- and four-stranded coiled coils in GCN4 leucine zipper mutants. *Science* 262:1401–1407.

Harbury, P.B., Tidor, B., and Kim, P.S. 1995. Repacking protein cores with backbone freedom: Structure prediction for coiled coils. *Proc. Natl. Acad. Sci. USA* 92:8408–8412.

Havel, T.F., Kuntz, I.D., and Crippen, G.M. 1983. The theory and practice of distance geometry. *Bull. Math. Biol.* 45:665–720.

Hilger, D., Polyhach, Y., Padan, E., Jung, H., and Jeschke, G. 2007. High-resolution structure of a Na^+/H^+ antiporter dimer obtained by pulsed electron paramagnetic resonance distance measurements. *Biophys. J.* 93:3675–3683.

Hirst, S.J., Alexander, N., Mchaourab, H.S., and Meiler, J. 2011. RosettaEPR: An integrated tool for protein structure determination from sparse EPR data. *J. Struct. Biol.* 173:506–514.

Huang, H. and Cafiso, D.S. 2008. Conformation and membrane position of the region linking the two C2 domains in synaptotagmin 1 by site-directed spin labeling. *Biochemistry* 47:12380–12388.

Hubbell, W.L. and Altenbach, C. 1994. Investigation of structure and dynamics in membrane proteins using site-directed spin labeling. *Curr. Opin. Struct. Biol.* 4:566–573.

Isas, J.M., Langen, R., Haigler, H.T., and Hubbell, W.L. 2002. Structure and dynamics of a helical hairpin and loop region in annexin 12: A site-directed spin labeling study. *Biochemistry* 41:1464–1473.

Islam, S.M., Stein, R.A., Mchaourab, H.S., and Roux, B. 2013. Structural refinement from restrained-ensemble simulations based on EPR/DEER data: Application to T4 lysozyme. *J. Phys. Chem. B* 117:4740–4754.

Jayasinghe, S. and Langen, R. 2004. Identifying structural features of fibrillar islet amyloid polypeptide using site-directed spin labeling. *J. Biol. Chem.* 279:48420–48425.

Jeschke, G., Sajid, M., Schulte, M., and Godt, A. 2009. Three-spin correlations in double electron-electron resonance. *Phys. Chem. Chem. Phys.* 11:6580–6591.

Jiao, D., Barfield, M., Comariza, J.E., and Hruby, V.J. 1992. Ab initio molecular orbital studies of the rotational barriers and the ^{33}S and ^{13}C chemical shieldings for dimethyl disulfide. *J. Am. Chem. Soc.* 114:3639–3643.

Karim, C.B., Kirby, T.L., Zhang, Z., Nesmelov, Y., and Thomas, D.D. 2004. Phospholamban structural dynamics in lipid bilayers probed by a spin label rigidly coupled to the peptide backbone. *Proc. Natl. Acad. Sci. USA* 101:14437–14442.

Kazmier, K., Alexander, N.S., Meiler, J., and Mchaourab, H.S. 2011. Algorithm for selection of optimized EPR distance restraints for *de novo* protein structure determination. *J. Struct. Biol.* 173:549–557.

Koteiche, H.A., Berengian, A.R., and Mchaourab, H.S. 1998. Identification of protein folding patterns using site-directed spin labeling. Structural characterization of a β-sheet and putative substrate binding regions in the conserved domain of αA-crystallin. *Biochemistry* 37:12681–12688.

Koteiche, H.A. and Mchaourab, H.S. 1999. Folding pattern of the α-crystallin domain in αA-crystallin determined by site-directed spin labeling. *J. Mol. Biol.* 294:561–577.

Kroncke, B.M., Horanyi, P.S., and Columbus, L. 2010. Structural origins of nitroxide side chain dynamics on membrane protein α-helical sites. *Biochemistry* 49:10045–10060.

Kuprusevicius, E., White, G., and Oganesyan, V.S. 2011. Prediction of nitroxide spin label EPR spectra from MD trajectories: application to myoglobin. *Faraday Discuss.* 148:283–298.

LaConte, L.E.W., Voelz, V., Nelson, W., Enz, M., and Thomas, D.D. 2002. Molecular dynamics simulation of site-directed spin labeling: Experimental validation in muscle fibers. *Biophys. J.* 83:1854–1866.

Ladner, C.L., Chen, M., Smith, D.P. et al. 2010. Stacked sets of parallel, in-register β-strands of $β_2$-microglobulin in amyloid fibrils revealed by site-directed spin labeling and chemical labeling. *J. Biol. Chem.* 285:17137–17147.

Laganowsky, A., Benesch, J.L.P., Landau, M. et al. 2010. Crystal structures of truncated alphaA and alphaB crystallins reveal structural mechanisms of polydispersity important for eye lens function. *Prot. Sci.* 19:1031–1043.

Langen, R., Isas, J.M., Hubbell, W.L., and Haigler, H.T. 1998a. A transmembrane form of annexin XII detected by site-directed spin labeling. *Proc. Natl. Acad. Sci. USA* 95:14060–14065.

Langen, R., Isas, J.M., Luecke, H., Haigler, H.T., and Hubbell, W.L. 1998b. Membrane-mediated assembly of annexins studied by site-directed spin labeling. *J. Biol. Chem.* 273:22453–22457.

Langen, R., Oh, K.J., Cascio, D., and Hubbell, W.L. 2000. Crystal structures of spin labeled T4 lysozyme mutants: Implications for the interpretation of EPR spectra in terms of structure. *Biochemistry* 39:8396–8405.

Lietzow, M.A. and Hubbell, W.L. 2004. Motion of spin label side chains in cellular retinol-binding protein: Correlation with structure and nearest-neighbor interactions in an antiparallel β-sheet. *Biochemistry* 43:3137–3151.

Lillington, J.E.D., Lovett, J.E., Johnson, S. et al. 2011. *Shigella flexneri* Spa15 crystal structure verified in solution by double electron-electron resonance. *J. Mol. Biol.* 405:427–435.

Liu, Z.L., Casey, T.M., Blackburn, M.E. et al. 2016. Pulsed EPR characterization of HIV-1 protease conformational sampling and inhibitor-induced population shifts. *Phys. Chem. Chem. Phys.* 18:5819–5831.

Livshits, V.A., Dzikovski, B.G., and Marsh, D. 2003. Anisotropic motion effects in CW non-linear EPR spectra: Relaxation enhancement of lipid spin labels. *J. Magn. Reson.* 162:429–442.

Lopez, C.J., Oga, S., and Hubbell, W.L. 2012. Mapping molecular flexibility of proteins with site-directed spin labeling: A case study of myoglobin. *Biochemistry* 51:6568–6583.

Lovell, S.C., Word, J.M., Richardson, J.S., and Richardson, D.C. 2000. The penultimate rotamer library. *Proteins Struct. Funct. Genet.* 40:389–408.

Margittai, M., Fasshauer, D., Pabst, S., Jahn, R., and Langen, R. 2001. Homo- and heterooligomeric SNARE complexes studied by site-directed spin labeling. *J. Biol. Chem.* 276:13169–13177.

Margittai, M. and Langen, R. 2004. Template-assisted filament growth by parallel stacking of tau. *Proc. Natl. Acad. Sci. USA* 101:10278–10283.

Marsh, D. 1997. Dichroic ratios in polarized Fourier transform infrared for nonaxial symmetry of β-sheet structures. *Biophys. J.* 72:2710–2718.

Marsh, D. 2000. Infrared dichroism of twisted β-sheet barrels. The structure of *E. coli* outer membrane proteins. *J. Mol. Biol.* 297:803–808.

Marsh, D. 2006. Orientation of TOAC amino-acid spin labels in α-helices and β-strands. *J. Magn. Reson.* 180:305–310.

Marsh, D. 2014a. EPR moments for site-directed spin-labelling. *J. Magn. Reson.* 248:66–70.

Marsh, D. 2014b. Intermediate dipolar distances from spin labels. *J. Magn. Reson.* 238:77–81.

Marsh, D., Jost, M., Peggion, C., and Toniolo, C. 2007a. Lipid chainlength dependence for incorporation of alamethicin in membranes: EPR studies on TOAC-spin labelled analogues. *Biophys. J.* 92:4002–4011.

Marsh, D., Jost, M., Peggion, C., and Toniolo, C. 2007b. TOAC spin labels in the backbone of alamethicin: EPR studies in lipid membranes. *Biophys. J.* 92:473–481.

Mchaourab, H. and Perozo, E. 2000. Determination of protein folds and conformational dynamics using spin-labeling EPR spectroscopy. In *Biological Magnetic Resonance*, 19, eds. Berliner, L.J., Eaton, S.S., and Eaton, G.R., 185–247. New York: Kluwer.

Monaco, V., Formaggio, F., Crisma, M. et al. 1999a. Determining the occurrence of a 3_{10}-helix and an α-helix in two different segments of a lipopeptaibol antibiotic using TOAC, a nitroxide spin-labeled C^{α}-tetrasubstituted α-amino acid. *Bioorg. Med. Chem.* 7:119–131.

Monaco, V., Formaggio, F., Crisma, M. et al. 1999b. Orientation and immersion depth of a helical lipopeptaibol in membranes using TOAC as an ESR probe. *Biopolymers* 50:239–253.

Mustafi, D., Sosa-Peinado, A., Gupta, V., Gordon, D.J., and Makinen, M.W. 2002. Structure of spin-labeled methylmethanethiolsulfonate in solution and bound to TEM-1 β-lactamase determined by electron nuclear double resonance spectroscopy. *Biochemistry* 41:797–808.

O'Shea, E.K., Klemm, J.D., Kim, P.S., and Alber, T. 1991. X-ray structure of the GCN4 leucine zipper, a two-stranded, parallel coiled coil. *Science* 254:539–544.

Oganesyan, V.S. 2007. A novel approach to the simulation of nitroxide spin label EPR spectra from a single truncated dynamical trajectory. *J. Magn. Reson.* 188:196–205.

Oganesyan, V.S., Chami, F., White, G.F., and Thomson, A.J. 2017. A combined EPR and MD simulation study of a nitroxyl spin label with restricted internal mobility sensitive to protein dynamics. *J. Magn. Reson.* 274:24–35.

Oh, K.J., Zhan, H., Cui, C. et al. 1996. Organization of diphtheria toxin T domain in bilayers: A site-directed spin labeling study. *Science* 273:810–812.

Oh, K.J., Zhan, H.J., Cui, C. et al. 1999. Conformation of the diphtheria toxin T domain in membranes: A site-directed spin-labeling study of the TH8 helix and TL5 loop. *Biochemistry* 32:10336–10343.

Park, S.Y., Borbat, P.P., Gonzalez-Bonet, G. et al. 2006. Reconstruction of the chemotaxis receptor-kinase assembly. *Nature Struct. Mol. Biol.* 13:400–407.

Parry, D.A.D. and Suzuki, E. 1969. Intrachain potential energy of the α-helix and of a coiled coil strand. *Biopolymers* 7:189–197.

Páli, T., Finbow, M.E., and Marsh, D. 1999. Membrane assembly of the 16-kDa proteolipid channel from *Nephrops norvegicus* studied by relaxation enhancements in spin-label ESR. *Biochemistry* 38:14311–14319.

Páli, T. and Marsh, D. 2001. Tilt, twist and coiling in β-barrel membrane proteins: relation to infrared dichroism. *Biophys. J.* 80:2789–2797.

Perozo, E., Marien Cortes, D., and Cuello, L.G. 1998. Three-dimensional architecture and gating mechanism of a K^+ channel studied by EPR spectroscopy. *Nat. Struct. Biol.* 5:459–469.

Pliotas, C., Ward, R., Branigan, E. et al. 2012. Conformational state of the MscS mechanosensitive channel in solution revealed by pulsed electron-electron double resonance (PELDOR) spectroscopy. *Proc. Nat. Acad. Sci. USA* 109:15983–15984/E2675–E2682.

Poirier, M.A., Xiao, W.Z., Macosko, J.C. et al. 1998. The synaptic SNARE complex is a parallel four-stranded helical bundle. *Nature Struct. Biol.* 5:765–769.

Polyhach, Y., Bordignon, E., and Jeschke, G. 2011. Rotamer libraries of spin labelled cysteines for protein studies. *Phys. Chem. Chem. Phys.* 13:2356–2366.

Pyka, J., Ilnicki, J., Altenbach, C., Hubbell, W.L., and Fronciscz, W. 2005. Accessibility and dynamics of nitroxide side chains in T4 lysozyme measured by saturation recovery EPR. *Biophys. J.* 89:2059–2068.

Rabenstein, M.D. and Shin, Y.-K. 1995. Determination of the distance between two spin labels attached to a macromolecule. *Proc. Natl. Acad. Sci. USA* 92:8239–8243.

Raghuraman, H., Cordero-Morales, J.F., Jogini, V. et al. 2012. Mechanism of Cd^{2+} coordination during slow inactivation in potassium channels. *Structure* 20:1332–1342.

Sale, K., Song, L.K., Liu, Y.S., Perozo, E., and Fajer, P. 2005. Explicit treatment of spin labels in modeling of distance constraints from dipolar EPR and DEER. *J. Am. Chem. Soc.* 127:9334–9335.

Salwinski, L. and Hubbell, W.L. 1999. Structure in the channel forming domain of colicin E1 bound to membranes: The 402–424 sequence. *Prot. Sci.* 8:562–572.

Schorn, K. and Marsh, D. 1996a. Lipid chain dynamics and molecular location of diacylglycerol in hydrated binary mixtures with phosphatidylcholine: Spin label ESR studies. *Biochemistry* 35:3831–3836.

Schorn, K. and Marsh, D. 1996b. Lipid chain dynamics in diacylglycerol-phosphatidylcholine mixtures studied by slow-motional simulations of spin label ESR spectra. *Chem. Phys. Lipids* 82:7–14.

Schorn, K. and Marsh, D. 1997. Extracting order parameters from powder EPR lineshapes for spin-labelled lipids in membranes. *Spectrochim. Acta A* 53:2235–2240.

Sezer, D., Freed, J.H., and Roux, B. 2008a. Simulating electron spin resonance spectra of nitroxide spin labels from molecular dynamics and stochastic trajectories. *J. Chem. Phys.* 128:165106.

Sezer, D., Freed, J.H., and Roux, B. 2008b. Using Markov models to simulate electron spin resonance spectra from molecular dynamics trajectories. *J. Phys. Chem. B* 112:11014–11027.

Sezer, D., Freed, J.H., and Roux, B. 2008c. Parametrization, molecular dynamics simulation, and calculation of electron spin resonance spectra of a nitroxide spin label on a polyalanine α-helix. *J. Phys. Chem. B* 112:5755–5767.

Sezer, D., Freed, J.H., and Roux, B. 2009. Multifrequency electron spin resonance spectra of a spin-labeled protein calculated from molecular dynamics simulations. *J. Am. Chem. Soc.* 131:2597–2605.

Steinhoff, H.J. and Hubbell, W.L. 1996. Calculation of electron paramagnetic resonance spectra from Brownian dynamics trajectories: Application to nitroxide side chains in proteins. *Biophys. J.* 71:2201–2212.

Stoll, S., Lee, Y.T., Zhang, M. et al. 2012. Double electron-electron resonance shows cytochrome P450cam undergoes a conformational change in solution upon binding substrate. *Proc. Natl. Acad. Sci. USA* 109:12888–12893.

Sutton, R.B., Fasshauer, D., Jahn, R., and Brunger, A.T. 1998. Crystal structure of a SNARE complex involved in synaptic exocytosis at 2.4 angstrom resolution. *Nature* 395:347–353.

von Hagens, T., Polyhach, Y., Sajid, M., Godt, A., and Jeschke, G. 2013. Suppression of ghost distances in multiple-spin double electron-electron resonance. *Phys. Chem. Chem. Phys.* 15:5854–5866.

Warshaviak, D.T., Serbulea, L., Houk, K.N., and Hubbell, W.L. 2011. Conformational analysis of a nitroxide side chain in an α-helix with density functional theory. *J. Phys. Chem. B* 115:397–405.

Xu, Q., Kim, M., Ho, K.W.D. et al. 2008. Membrane hydrocarbon thickness modulates the dynamics of a membrane transport protein. *Biophys. J.* 95:2849–2858.

Zhang, F., Chen, Y., Kweon, D.-H., Kim, C.S., and Shin, Y.K. 2002. The four-helix bundle of the neuronal target membrane SNARE complex is neither disordered in the middle nor uncoiled at the C-terminal region. *J. Biol. Chem.* 277:24294–24298.

Zielke, V., Eickmeier, H., Hideg, K., Reuter, H., and Steinhoff, H.J. 2008. A commonly used spin label: *S*-(2,2,5,5-tetramethyl-1-oxyl-Δ^3-pyrrolin-3-ylmethyl) methanethiosulfonate. *Acta Cryst. C* 64:o586–o589.

Fundamental Physical Constants

QUANTITY	VALUE	RELATIVE ERROR
Bohr magneton	$\beta_e = \frac{\|e\|\hbar}{2m_e} = 9.27400968(20)\times10^{-24}\ \text{J T}^{-1}$	$(\pm2.2\times10^{-8})$
$\frac{h}{\beta_e}$	$\frac{h}{\beta_e} = \frac{4\pi m_e}{\|e\|} = 0.714477319(63)\times10^{-10}\ \text{s T}$	$(\pm8.8\times10^{-8})$
	$= 0.0714477319(63)\ \text{T/GHz}$	
free electron g-factor	$g_e = -2(1+a_e) = -2.00231930436153(53)$ a_e: electron magnetic moment anomaly	$(\pm2.6\times10^{-13})$
electron magnetic moment	$\mu_e = -9.28476430(21)\times10^{-24}\ \text{J T}^{-1}$ (fundamental)	$(\pm2.2\times10^{-8})$
electron gyromagnetic ratio	$\gamma_e = \frac{2\|\mu_e\|}{\hbar} = 1.760859708(39)\times10^{11}\ \text{s}^{-1}\ \text{T}^{-1}$	$(\pm2.2\times10^{-8})$
nuclear magneton	$\beta_N = \frac{m_e}{m_p}\beta_e = 5.05078353(11)\times10^{-27}\ \text{J T}^{-1}$	$(\pm2.2\times10^{-8})$
proton magnetic moment	$\mu_p = 1.410606743(33)\times10^{-26}\ \text{J T}^{-1}$	$(\pm2.4\times10^{-8})$
proton gyromagnetic ratio	$\gamma_p = \frac{2\mu_p}{\hbar} = 2.675222005(63)\times10^{8}\ \text{s}^{-1}\ \text{T}^{-1}$	$(\pm2.4\times10^{-8})$
Planck's constant	$h = 6.62606957(29)\times10^{-34}\ \text{J s}$	$(\pm4.4\times10^{-8})$
	$\hbar \equiv h/2\pi = 1.054571726(47)\times10^{-34}\ \text{J s}$	
Avogadro's number	$N_A = 6.02214129(27)\times10^{23}\ \text{mol}^{-1}$	$(\pm4.4\times10^{-8})$
Boltzmann's constant	$k_B = 1.3806488(13)\times10^{-23}\ \text{J K}^{-1}$	$(\pm9.1\times10^{-7})$
ideal gas constant	$R = N_A k_B = 8.3144621(75)\ \text{J mol}^{-1}\ \text{K}^{-1}$	$(\pm9.1\times10^{-7})$
elementary charge	$\|e\| = 1.602176565(35)\times10^{-19}\ \text{C}$	$(\pm2.2\times10^{-8})$
Faraday constant	$F = N_A\|e\| = 9.64853365(21)\times10^{4}\ \text{C mol}^{-1}$	$(\pm2.2\times10^{-8})$
electron rest mass	$m_e = 9.10938291(40)\times10^{-31}\ \text{kg}$	$(\pm4.4\times10^{-8})$
proton rest mass	$m_p = 1.672621777(74)\times10^{-27}\ \text{kg}$	$(\pm4.4\times10^{-8})$
	$m_p/m_e = 1836.15267245(75)$	$(\pm4.1\times10^{-10})$
atomic mass unit (^{12}C/12)	$1\ \text{Da} = 1.660538921(73)\times10^{-27}\ \text{kg}$	$(\pm4.4\times10^{-8})$
velocity of light (vacuum)	$c = 2.99792458\times10^{8}\ \text{m s}^{-1}$	(defined)
permeability (vacuum)	$\mu_o = 4\pi\times10^{-7}\ \text{H m}^{-1}\ (\text{or N A}^{-2})$	(defined)
permittivity (vacuum)	$\varepsilon_o \equiv 1/(\mu_o c^2) = 8.854187817\ldots\times10^{-12}\ \text{F m}^{-1}$	(defined)

Number in brackets is standard uncertainty in last two digits.
From CODATA 2010, National Institute of Standards, USA.

SYMBOLS

a, b, unique and non-unique semi-axes of rotational ellipsoid.
a, b, slow-rotation correlation-time calibration constants for $\langle A_{zz} \rangle$ (Eq. 7.80; Table 7.2).
a_{m_I}, b_{m_I}, slow-rotation correlation-time calibration constants for ΔB_{m_I} $(m_I = \pm 1)$ (Eq. 7.81; Table 7.2).
a_m, b_m slow-motion correlation-time calibration constants for phase-memory time (Eq. 13.64, Table 13.1) (m = low, central, high-field).
$a_m, b_m, \omega'_m, \Gamma_m$, parameters in Heisenberg-exchange coupled lineshapes (Eq. 9.112).
a, secular hyperfine coupling (either a_o or A_{zz}): aI_zS_z.
a_o, isotropic hyperfine coupling constant: $a_o = \frac{1}{3}(A_{xx} + A_{yy} + A_{zz})$.
$a_o^{\varepsilon=1}$, isotropic hyperfine constant extrapolated to $\varepsilon_r = 1$ (Eq. 4.31).
a_o^N, a_o^O, a_o^H, isotropic nitrogen, oxygen (^{17}O) and proton hyperfine coupling constant.
$a_o^{C\alpha}, a_o^{C\beta}, a_o^{C\gamma}$, isotropic ^{13}C hyperfine coupling for α, β and γ C-atoms.
$a_{o,o}^N, a_{o,h}^N$, isotropic hyperfine constant of free and H-bonded nitroxide (Eq. 4.39).
a_N, isotropic nitrogen hyperfine coupling constant.
a_{HE}, apparent (reduced) a_N with Heisenberg exchange.
a_{HE}^{eff}, effective a_N observed with Heisenberg exchange (Figure 9.15).
$a_{HE,dd}^{eff}$, as for a_{HE}^{eff} but including magnetic dipole-dipole interaction (Eq. 9.183).
a_e, effective hyperfine coupling for Gaussian inhomogeneous lineshape (Eq. 2.32).
$\mathbf{A}$, magnetic vector potential (Eq. F.1).
$\mathbf{A}$, Bloch-equations matrix (Eq. 11.5).
$\mathbf{A}_n$, Bloch matrix for nth harmonic of modulation frequency (Eq. 11.27).
A, B, C, coefficients of m_I-dependent, Lorentzian absorption linewidth (Eq. 5.58).
$A(\gamma), B(\gamma), C(\gamma)$, angle-dependent linewidth coefficients for ordered systems (Eqs. 8.130–8.132).
$A_0, A_2, A_4; B_0, B_2, B_4; C_0, C_2, C_4$; constants in field-orientation dependence of linewidth coefficients for ordered systems (Eq. 8.94).
A_{pp}, B_{pp}, C_{pp}, coefficients of m_I-dependent peak-to-peak, first-derivative, Lorentzian linewidth (Eq. 7.32).
A, B, C, D, E, dipolar alphabet: angle- and spin-dependent terms in magnetic dipole–dipole Hamiltonian (Eqs. 9.5–9.9).
A, B, secular (AI_zS_z) and pseudosecular (BI_xS_z) hyperfine interaction (see Eqs. 14.1 and H.6).
A_{eff}, effective hyperfine coupling for $S = \frac{1}{2}$, $I = \frac{1}{2}$: $A_{eff} = \sqrt{A^2 + B^2}$ (see Eqs. 14.2, 14.3).
$A(\theta,\phi)$, angular-dependent effective hyperfine coupling (Eq. 2.12).
$A_{\parallel}^{dip}, A_{\perp}^{dip}$, principal values of point-dipole hyperfine tensor.
$A_{xx}^{dip}, A_{yy}^{dip}, A_{zz}^{dip}$, principal values of dipolar hyperfine tensor.
$A^H(\theta)$, proton hyperfine constant (dipolar plus isotropic – see Eq. 14.51).
$A^{H,dip}$, dipolar contribution to proton hyperfine constant (see Eq. 14.52).
$A_{\parallel}^H, A_{\perp}^H$, principal values of axial proton hyperfine tensor.
A_{xx}, A_{yy}, A_{zz}, principal elements of hyperfine tensor.
$\langle \mathbf{A}' \rangle$, partially rotationally-averaged hyperfine tensor $\mathbf{A}$ (see Eqs. 8.1, 8.5).
$\langle A_{\parallel} \rangle, \langle A_{\perp} \rangle$, principal elements of axially partially-averaged hyperfine tensor (Eqs. 8.6, 8.7).
$\langle A_{zz} \rangle$, half the outer splitting $(2\langle A_{zz} \rangle)$ of slow-motion spectrum (see Eq. 7.80).
A_{zz}^R, half the outer splitting $(2A_{zz}^R)$ of rigid-limit spectrum (see Eq. 7.80).
A_{zz}^N, principal nitrogen hyperfine tensor element.
A_{max}, A_{min}, half the outer/inner hyperfine splitting in partially motionally averaged powder pattern.
A_1, A_2, A_3, saturation-recovery amplitudes (e.g., Eqs. 12.16, 12.35).
A_{HE}, A_{dd}, constants in temperature-dependent line broadening by Heisenberg exchange/dipole–dipole interaction (Eqs. 9.164–9.166).
$A_G(B), A_L(B), A_V(B)$, Gaussian, Lorentzian and Voigt lineshapes (Eqs. 1.25, 1.22, 2.36).
$A_p^{(q)}$, spin operators in perturbation Hamiltonian (Eqs. O.11–O.15; O.35–O.39) – modulated by angular-dependent factors $F^{(q)}(t)$.
b_1, b_2, c_1, c_2, calibration constants for fast anisotropic rotation (Table 7.1; Eqs. 7.55, 7.56).
b, ratio of nuclear to electron spin–lattice relaxation rates: $b \equiv W_n/(2W_e)\,(= T_{1n}/T_{1e})$.
b'', normalized Heisenberg exchange rate: $b'' \equiv W_{HE}/(6W_e)$ $(= \frac{1}{3}T_1^o\tau_{HE}^{-1})$.
b''', normalized rate of paramagnetic spin–lattice relaxation enhancement: $b''' = W_{RL}/W_e$.
$\mathbf{B}$, magnetic field (strictly, magnetic induction: $\mathbf{B} = \mu_o\mu_r\mathbf{H}$).
B_o, static magnetic field.
B_x, B_y, B_z, Cartesian components of static magnetic field.
B_1, microwave (or radiofrequency) magnetic field.
$\hat{B}_{eff}$, effective field in frame rotating about $\mathbf{B}_o$ at $\boldsymbol{\omega}$: $\hat{\mathbf{B}}_{eff} = \mathbf{B}_o - \boldsymbol{\omega}/\gamma_e + \mathbf{B}_1$ (Figure 6.2).
B_{mod}, modulation amplitude of static field.
$\mathbf{B}_{dip}$, magnetic field from an electron-spin dipole (Eq. 3.16).
B_{loc}, local magnetic field.
$B_{m_I}(\theta,\phi)$, angular-dependent resonance field of m_I-hyperfine line (Eq. 2.13).
$B_{\parallel}, B_{\perp}$, resonance field for static field parallel (||), perpendicular (⊥) to principal axis.
$(B_1^2)_{1/2}$, microwave power at half-saturation (Eq. 11.15).
$B(t)$, DEER background signal (Eq. 15.51).
c, normalized cross-relaxation rate: $c \equiv W_x/(2W_{e,0})$ for I = 1; $c \equiv W_x/\bar{W}_e$ for $I = \frac{1}{2}$.
c_M, molar concentration (of ions).
c_R, concentration of paramagnetic relaxant R.

c_N, c_O, admixture coefficients of nitrogen, oxygen atomic orbitals in nitroxide $2p\pi$ molecular orbital (Eqs. 3.1, 3.66).

c_s, s-electron admixture in sp^2-hybrid N–C valence orbitals (Eqs. 3.59, 3.60).

c_{p_x}, p_x-electron admixture in sp^2-hybrid N–C valence orbitals (Eq. K.24).

$c_k^{(m)}$, admixture of atomic orbital on atom k in molecular orbital m.

$c_{k,s}^{(m)}, c_{k,x}^{(m)}, c_{k,y}^{(m)}$, admixture of s-, p_x-, p_y- orbitals on atom k in molecular orbital m (e.g., Eq. 3.69).

$c_{O,s}^{(n)}, c_{O,x}^{(n)}, c_{O,y}^{(n)}$, admixtures of s-, p_x-, p_y- orbitals in oxygen lone-pair sp^2-hybrid $\psi_n^{(O)}$ (Eq. 3.68).

$\left(C_{O,y}^{(n)}\right)^2$, sum of $\left(c_{O,y}^{(n)}\right)^2$ over all $2p_y$ hybrid lone-pair orbitals: $\left(C_{O,y}^{(n)}\right)^2 = \sum_n \left(c_{O,y}^{(n)}\right)^2$ (cf. Eq. 4.37).

$c_m(t)$, admixture coefficient of eigenstate m in time-dependent perturbation theory (Eq. 5.2).

$c_{L,M}(t)$, expansion coefficient in spherical harmonics (Eq. 7.18).

$c_{K,M}^L(t)$, expansion coefficient in Wigner rotation matrices $\mathcal{D}_{K,M}^L$ (Eq. 8.115).

$C(\tau/\tau_c)$, function giving dependence of primary-echo decay on dynamics (Eqs. 13.36, 13.37).

$C(\tau/\tau_c, T/\tau_c)$, function giving dependence of stimulated-echo decay on dynamics (Eqs. 13.124, 13.125).

C, spin-rotation coupling tensor (Eq. 5.33).

C_1, C_2, C_3, correlation-time calibration constants for asymptotic phase-memory time (Eqs. 13.61–13.63).

C_E, constant relating shift in isotropic hyperfine coupling to electric field (Eq. 4.30).

C_{m_I}, coefficients in saturation recovery.

$C_{Ram}, C_{loc}, \Delta_{loc}$, constants characterizing temperature dependence of Raman and local-mode spin–lattice relaxation mechanisms (see Eq. 12.46).

$(C_m(\omega))_{-+}$, $\langle -\frac{1}{2}|\cdots|+\frac{1}{2}\rangle$ matrix element of mth coefficient in expansion of $Z(\Omega,\omega)$ in spherical harmonics $G_m(\Omega)$ (see Eqs. 6.63, 6.64) – for CW-EPR.

$(C_m(t))_{-+}$, $\langle -\frac{1}{2}|\cdots|+\frac{1}{2}\rangle$ matrix element of mth coefficient in expansion of $\chi(\Omega,t)$ in spherical harmonics $G_m(\Omega)$ (Eq. 13.49) – for pulse EPR.

$\left(C_{L,M}\right)^{m_I m_I'} \equiv \left(C_{L,M}\right)_{-+}^{m_I m_I'}$, matrix element $\langle -\frac{1}{2}, m_I|C_{L,M}|+\frac{1}{2}, m_I'\rangle$ of expansion coefficients in $Y_{L,M}(\theta,\phi)$, for axial tensors (see Eq. 6.91).

$\left(C_{K,M}^L\right)^{m_I m_I'} \equiv \left(C_{K,M}^L\right)_{-+}^{m_I m_I'}$, matrix element $\langle -\frac{1}{2}, m_I|C_{K,M}^L|+\frac{1}{2}, m_I'\rangle$ of expansion coefficients in $\mathcal{D}_{K,M}^L(\psi,\theta,\phi)$, for non-axial tensors.

$\mathbf{C}(\omega)$, column matrix of $\left(C_{K,M}^L\right)_{-+}^{m_I m_I'}$ (see Eqs. 6.110, 6.114).

$\mathbf{C}(t)$, column matrix of $\left(C_m\right)_{-+}$ (see Eqs. 13.51, 13.52).

d, normalized difference in spin–lattice relaxation rates: $d = \delta/\overline{W}_e$ for $I = \frac{1}{2}$; $d = \delta/W_{e,0}$ for $I = 1$.

d_1, normalized difference in spin–lattice relaxation rates for I = 1: $d_1 = \delta_1/W_{e,0}$.

d_S, distance of closest approach between spin labels.

d_{RL}, distance of closest approach between paramagnetic relaxant R and spin label L.

d_β, separation of β-strands in a β-sheet.

$d_{K,M}^L(\theta)$, matrix elements of reduced Wigner rotation matrices.

D, electric displacement ($\mathbf{D} = \varepsilon_o\varepsilon_r\mathbf{E}$).

D, second-derivative probability operator $d^2\mathbf{P}/dr^2$ in Tikhonov regularization (see Eq. 15.59).

D, E, zero-field splitting constants in dipolar triplet state: $S = s_1 + s_2 = 1$ (Eq. 9.17).

D, fractal dimension of space containing remote B-spins giving background DEER (Eq. 15.53).

$D(\Delta B_{dd})$, dipolar absorption envelope $D(B - B')$: Pake pattern (see top of Figure 9.3).

D_{dd}, strength of angle-dependent dipolar interaction for two electron spins: $D_{dd} \equiv (\mu_o/4\pi)(g^2\beta_e^2/r_{12}^3)(1 - 3\cos^2\theta)$.

D_{dd}^o, dipolar interaction strength: $D_{dd}^o \equiv (\mu_o/4\pi)(g^2\beta_e^2/r_{12}^3)$.

D_{O_2}, translational diffusion coefficient of dissolved oxygen.

D_R, isotropic rotational diffusion coefficient.

$D_{R\parallel}, D_{R\perp}$, principal elements of diffusion tensor for axial rotation.

$D_{R,x}, D_{R,y}, D_{R,z}$, principal elements of diffusion tensor for anisotropic rotation.

D_T, translational diffusion coefficient.

D_β, separation of β-sheets.

$\mathcal{D}_{K,M}^L(\Omega)$, matrix elements of Wigner rotation matrices.

e, elementary (electron negative) charge.

E, electric field ($\mathbf{E} = -\nabla V$).

$\mathbf{E}_R$, reaction electric field (Eq. 4.1).

$E_{loc,x}$, local electric field component along N–O bond/x-axis (Eq. 4.36).

E_a, Arrhenius activation energy.

E_m, electronic energy level (e.g., $m \equiv \pi^*$).

${}^1E, {}^3E$, singlet ($S = 0$)/triplet ($S = 1$) energy levels: $J_{12} = {}^3E - {}^1E$.

$E_{2p}(2\tau)$, amplitude of two-pulse electron spin echo for pulse separation τ.

$E_{2p,I}(2\tau)$, amplitude of two-pulse electron spin echo for nuclear spin I.

$E_{3p}(2\tau + T)$ or $E_{3p}(\tau, T)$, amplitude of three-pulse electron spin echo with short delay τ and long delay T.

$E_{3p}^\alpha(2\tau + T), E_{3p}^\beta(2\tau + T)$, contributions from $M_S = \pm\frac{1}{2}$ states to amplitude of three-pulse electron spin echo.

$E_{norm}(\tau, T)$, normalized three-pulse electron spin-echo amplitude (see Eq. 14.38).

$E_{DEER}(t), E_{4pDEER}(t)$, echo amplitude in 3-pulse/4-pulse electron–electron double resonance (DEER).

$E_{4p,inter}$, intermolecular dipolar background signal in 4-pulse DEER (Eq. 15.43).

$E_{4p,intra}$, intramolecular 4-pulse DEER signal (Eq. 15.45).

$E_{DEER,lin}$, linearized DEER signal for dilute distribution of B-spins (Eq. 15.48).

$E_{DEER,pair}$, DEER signal for distribution of A-B spin pairs (Eq. 15.49).

f, vector representation of normalized DEER signal $f(t_i)$ (see Eq. 15.57).

$f(t)$, normalized modulated part of DEER signal (Eqs. 15.54, 15.92).

f_A, f_B, fraction of A-, B-spins inverted by a pulse (Eqs. 15.40–15.42).

$f_{A,B}$, (constant) modulation-depth parameter for distribution of A-B spin pairs in DEER (see Eq. 15.49).

$f_{B,p}$, DEER modulation-depth for distribution of B-spins pumped (Eq. 15.47).

f_R, rotational friction coefficient (Eq. 7.6).

f_R^o, rotational friction coefficient for sphere of equal volume (Eq. 7.10).

$f_{R\parallel}, f_{R\perp}$, principal elements of frictional tensor for axially anisotropic rotational diffusion.

$f_{R\parallel,m}^o$, rotational friction coefficient for right circular cylinder (Eq. 11.84).

f_T, translational friction coefficient (Eq. 9.138).

f_T^o, translational friction coefficient for a sphere (Eq. 9.142).

f_{1W}, fraction of nitroxides singly hydrogen-bonded to water.

f_{2W}, fraction of nitroxides doubly hydrogen-bonded to water.

f_k, fractional population of site/state k.

f_m, fractional population of hyperfine state $m_I = m$.

$f_h([\mathrm{OH}])$, fraction of hydrogen-bonded nitroxides.

$f_O(\varepsilon_{r,B}), f_{BW}(\varepsilon_{r,B})$, functional dependence of reaction field on bulk permittivity (Eq. 4.5) according to Onsager/Block–Walker models (Eqs. 4.23, 4.24).

$F(T_\perp^d t)$, integral defining powder lineshape for three-pulse ESEEM (Eq. 14.47).

$F_{R\parallel}, F_{R\perp}$, shape factors for axially anisotropic rotation (Eqs. 7.11 and 7.12).

F_T, shape factors for translational diffusion (Eqs. 9.143–9.145).

F_i, frictional drag on particle i.

$F_{dd}(B)$, EPR lineshape including dipolar broadening (see Eq. 15.2).

$F_q(\theta)$ $(q = 0, 1, 2)$, angular-dependent functions proportional to spherical harmonics, specifically $\mathrm{P}_l^m(\theta)$ (see Eqs. 10.29–10.32).

$F^{(q)}(t)$ $(q = 0, 1, 2)$, time-dependent angular functions related to spherical harmonics, specifically $\mathrm{P}_l^m(\theta,\phi)$. (see Eqs. 5.15–5.18; 8.48–8.51; 9.24–9.26).

$\mathcal{F}\{g(\omega)\}(t)$, Fourier transform of $g(\omega)$ with respect to t (Eq. 2.42).

$g(\omega_o - \omega)$, inhomogeneous lineshape.

$g(\theta,\phi)$, angular-dependent effective g-value (Eq. 2.11).

g_o, isotropic g-value: $g_o = \frac{1}{3}(g_{xx} + g_{yy} + g_{zz})$.

$g_o^{\varepsilon=1}$, isotropic g-value extrapolated to $\varepsilon_r = 1$ (Eq. 4.38).

$g_{o,o}, g_{o,h}$, isotropic g-value of free and H-bonded nitroxide (Eq. 4.41).

g_{xx}, g_{yy}, g_{zz}, principal elements of g-tensor.

$\langle \mathbf{g} \rangle$, partially rotationally-averaged g-tensor $\mathbf{g}$.

$\langle g_\parallel \rangle, \langle g_\perp \rangle$, principal elements of axially partially-averaged g-tensor (Eqs. 8.8 and 8.9)

g_J, Landé g-factor.

g_R, g-factor of paramagnetic relaxant R.

$G(t)$, autocorrelation function (Eq. 5.10).

$G(t',t)$, autocorrelation function $\langle \Delta\omega_o(t')\Delta\omega_o(t) \rangle$ for fluctuations in Larmor frequency (see Eq. 13.35).

$G^{(M)}(t)$, correlation function for dipole-dipole relaxation: $M =$ 0, 1, 2 (Eq. 9.36 rotational; Eq. 9.43 translational).

$G_{LM}(t)$, autocorrelation function for spherical harmonics $\mathrm{Y}_{L,M}$ in isotropic diffusion (Eq. 7.24).

$g_{KM}(t)$, autocorrelation function for second-rank Wigner rotation matrix elements $\mathcal{D}_{K,M}^2$ (Eq. 8.85).

h, Planck's constant.

$\hbar$, Planck's constant divided by 2π.

$h(t)$, switching function that changes sign (± 1) every time a B-spin flips (Eq. 13.85).

$h(\omega_o - \omega')$, inhomogeneous envelope function.

$h(\omega)$, hole-burning depth parameter at position ω.

$h(m_I)$, height of first-derivative m_I-hyperfine line.

h_{pp}, peak-to-peak height of first-derivative EPR line (Figure 9.15).

h_{disp}, height of dispersion component in first-derivative EPR line (Figure 9.15).

h_{NO}, molecular-orbital parameter (see Eq. 3.4).

h_α, h_β, rise per residue in an α-helix, β-strand.

$\mathcal{H}$, Hamiltonian operator.

$\mathcal{H}_{mn}$, matrix element of Hamiltonian operator.

$\mathcal{H}_o$, zero-order (or static) Hamiltonian operator.

$\mathcal{H}_1$, perturbation Hamiltonian operator.

$\mathcal{H}_1^{(q)}$, spin part of perturbation Hamiltonian (see Eq. 5.15).

$\mathcal{H}_1^\ddagger(t)$, time-dependent perturbation Hamiltonian in interaction representation (Eq. O.3).

i_σ, fractional ionic character of a covalent bond.

I, ionic strength.

I, nuclear-spin quantum number.

I_x, I_y, I_z, nuclear-spin operator: x, y, z components.

$I_\pm (\equiv I_x \pm iI_y)$, shift operators for nuclear spin.

I^α, I^β, nuclear polarization operators that select the $m_I = \pm\frac{1}{2}$ hyperfine states $(I^\alpha + I^\beta = \hat{1})$.

$I_i^{(rs)}$ $(i = x,y,z)$, single-transition operators for nuclear-spin transition $|r\rangle$ to $|s\rangle$.

I_{ii}^{inert} $(i = x,y,z)$, principal elements of moment of inertia tensor.

I_a, I_f, intensities of allowed and forbidden transitions for $S = \frac{1}{2}$, $I = \frac{1}{2}$, where $I_a + I_f = 1$ (see Eqs. H.20 and H.21).

I_{ST}, normalized integrated intensity of V_2' saturation-transfer EPR spectrum (Eq. 11.69).

I_{broad}, ESEEM amplitude from D_2O hydrogen-bonded to nitroxide.

I_{narrow}, ESEEM amplitude from D_2O not hydrogen-bonded to nitroxide.

I_{tot}, total D_2O ESEEM amplitude $I_{tot} = I_{broad} + I_{narrow}$.

$j(\omega)$, reduced spectral density: $j(\omega) = \tau/(1 + \omega^2\tau^2)$.

$j_q(\omega)$ $(q = 0, 1, 2, \ldots)$, spectral density for partial motional averaging (Eq. 8.56).

$j_{KM}(0)$, zero-frequency spectral density for second-rank reduced Wigner rotation matrices $d_{K,M}^2$ (Eq. 8.88, for strong jump), or for Wigner rotation matrices $\mathcal{D}_{K,M}^L$ (Eq. 8.119 for Brownian diffusion), with field along director $\gamma = 0$.

$j_{NM}(0,\gamma)$, angle-dependent zero-frequency spectral density for $d_{K,M}^2$ (Eq. 8.92, for strong jump), or for $\mathcal{D}_{K,M}^L$ (Eq. 8.126, for Brownian diffusion).

j_{NM}^L, additive contributions $(L = 0, 2, 4)$ to $j_{NM}(0,\gamma)$ (Eqs. 8.95 and 8.96).

$j^{(q)}(\omega)$, spectral density from autocorrelation function of $F^{(q)}(t)$.
$j^{AA}(\omega)$, compound spectral density for hyperfine interaction (END) (Eq. 5.23).
$j_m^{AA}(0)$, zero-frequency compound spectral density for hyperfine interaction (END) (Eq. A8.10).
$j^{gg}(\omega)$, compound spectral density for Zeeman anisotropy (g-value) (Eqs. 5.27 and A8.8).
$j^{gA}(\omega)$, compound spectral density for hyperfine-Zeeman interaction cross term (Eqs. 5.31 and A8.9).
$j^{SJ}(\omega)$, compound spectral density for spin-rotation interaction (Eqs. 5.36 and 5.39).
$j_{mnm'n'}(\omega_{mn})$, spectral density in Redfield theory (Eq. O.81), where $\hbar\omega_{mn} \equiv E_m - E_n$.
J, exchange integral between electrons in orbitals φ_a, φ_b (Eq. 9.68).
J, electron angular momentum operator (orbit plus spin): $\mathbf{J} = \mathbf{L} + \mathbf{S}$.
J, angular momentum operator of rotating molecule: $\mathbf{J}\hbar = \mathbf{I}^{inert} \cdot \boldsymbol{\omega}_R$.
$J(\Omega)$, flux of spin labels through orientation Ω (Eq. 7.1).
$J(x)$, flux of spin labels in direction x (Eq. 9.133).
$J^{(q)}(\omega)$ ($q = 0, 1, 2, \ldots$), spectral density.
J_{12}, Heisenberg spin-exchange constant, for electron 1 and 2: $J_{12}\mathbf{s}_1 \cdot \mathbf{s}_2$.
J_{RL}, Heisenberg exchange constant, for spin exchange between paramagnetic relaxant R and spin label L.
k, dimensionless echo modulation-depth parameter in ESEEM (see Eqs. 14.20, 14.21).
k, R_o, b, constants in correlation-time calibrations for V_2' saturation-transfer EPR (Eq. 11.83, Table 11.4).
$k_{2,diff}$, diffusion-controlled bimolecular collision rate constant.
k_B, Boltzmann's constant.
k_{coll}, bimolecular collision rate constant.
k_{HE}, bimolecular rate constant for Heisenberg exchange (cf. Eq. 10.50): $\tau_{HE}^{-1} = k_{HE}[SL]$.
k_{RL}, bimolecular rate constant for Heisenberg spin-exchange of paramagnetic relaxant R with spin label L.
k_{inter}, (double) strength of dipolar half-width (Eq. 15.44).
$k_{inter}(t)$, generalized decay rate in DEER background signal (Eqs. 15.79 and 15.80).
K, matrix of kernel $K(r_j, t_i)$ in Tikhonov regularization of DEER signal (see Eq. 15.57).
K, direct Coulomb integral between electrons in orbitals φ_a, φ_b (see Eq. 9.67).
K, intrinsic binding constant for hydrogen bonding to nitroxide.
$K_{A,h}$, effective association constant for H-bonding: $f_h([\mathrm{OH}]) = K_{A,h}[\mathrm{OH}]$.
$K(k)$, elliptic integral (Eq. 2.29).
$K(2\tau/\tau_S), G(2\tau/\tau_S)$, functions determining primary and stimulated echo decays by B-spin flips (Eqs. 13.100, 13.101, 13.143 and 13.144).
$K(r,t)$, kernel in Tikhonov regularization of DEER signal (Eqs. 15.55–15.57; Eqs. 15.93 and 15.94).
K_{inter}, constant in stretched-exponential representation of DEER background signal (see Eq. 15.53).
K_S, surface ion-binding constant.
K_v, constant relating isotropic hyperfine coupling to bulk dielectric constant (Eq. 4.31).
$K_{v,g}$, constant relating isotropic g-value to bulk dielectric constant (Eq. 4.38).
l_x, l_y, l_z, direction cosines to x, y, z axes.
l_x, l_y, l_z, x-, y-, z-components of angular momentum operator.
l, orbital angular momentum of a single electron.
$\mathbf{l}^{(k)}$, orbital angular momentum of electron on atom k.
L, total orbital angular momentum operator.
L_x, L_y, L_z, components of angular momentum operator
L_{max}, K_{max}, M_{max}, cut-off values for non-zero $C_{K,M}^L$ expansion coefficients.
$L''/L, C'/C, H''/H$, diagnostic line-height ratios in low-, central- and high-field regions of V_2' saturation-transfer EPR spectrum (Figure 11.16).
$\mathbf{m}_{el}$, total electric dipole moment (Eq. 4.2).
m_I, nuclear spin projection quantum number.
M_S, electron spin projection quantum number.
M_o, equilibrium z-magnetization, aligned along static field B_o
$M_x(t), M_y(t), M_z(t)$, electron-spin magnetization x, y, z components.
$\tilde{M}_x(t), \tilde{M}_y(t)$, electron-spin magnetization x, y components in frame rotating about static field (z-axis) – see Eq. 6.4.
$\tilde{M}_+(t)$, complex transverse magnetization in rotating frame: $\tilde{M}_x(t) + i\tilde{M}_y(t)$.
$\tilde{M}_k$, complex rotating-frame transverse magnetization of kth angular zone (Eq. 6.36).
$\tilde{\mathbf{M}}$, magnetization vector in rotating frame: $\tilde{\mathbf{M}} \equiv (u, v, M_z)$.
$\tilde{\mathbf{M}}_n$, nth Fourier amplitude of rotating-frame magnetization with field modulation: $\tilde{\mathbf{M}}_n \equiv (u_n, v_n, M_{z,n})$.
$\tilde{\mathbf{M}}_{n,n}$, nth-harmonic component of rotating-frame magnetization depending on nth power of field-modulation amplitude: $\tilde{\mathbf{M}}_{n,n} \equiv (u_{n,n}, v_{n,n}, M_{z,n,n})$.
M_r, molecular mass.
$\bar{n}$, mean number of spins in a cluster (see Eq. 15.50).
n_D, refractive index of neat nitroxide (see Eq. 4.4).
n_i, population difference of $M_S = \pm\frac{1}{2}$ electron spin state i $(n_i = N_i^+ - N_i^-)$.
n_i^o, population difference n_i at (Boltzmann) thermal equilibrium.
n_P, periodicity (in peptide residues) of protein helices (Eq. 16.2).
n_α, n_β, residues per turn in an α-helix, β-strand.
n_R, paramagnetic relaxant concentration in ions per unit volume.
$n_{R,S}$, paramagnetic relaxant surface concentration in ions per unit area.
n_{SL}, spin-label concentration in molecules per unit volume.
n_{SL}^s, spin-label surface density in molecules per unit area.
n_x, n_y, n_z, electron occupation numbers in p_x, p_y, p_z-orbitals.
n_{klm}, population of chains in configuration m with orientation θ_k, ϕ_l.
n_{kl}, n_m, chain population with orientation θ_k, ϕ_l; population in configuration m.

N_A, Avogadro's number.
N_Z, total number of states, $(2S+1)(2I+1)$ (see Eq. 6.47).
$N_i^{\pm}$, population of $M_S = \pm\frac{1}{2}$ electron spin state i.
N_β, number of strands in a β-sheet/barrel.
p, linear momentum vector.
p_x, p_y, p_z, x-, y-, z-components of linear momentum operator.
p, exponent in denominator of saturation curve (see Eq. 11.14).
p_{bond}, electric dipole moment of substituent in nitroxide ring.
p_{el}, permanent electric dipole moment.
p_{ex}, probability of Heisenberg spin exchange on collision (Eqs. 9.92, 10.20 and 10.22).
p_β, fraction of spins flipped by pulse of angle β (cf. Eqs. 13.13 and 13.111).
$p(t;\tau)$, frequency Fourier transform of conditional probability $P(t;\Delta\omega)$ in spectral diffusion (Eq. A13.1).
$p(r,\rho)$, conditional probability that A and B are r apart, when A is distance ρ from centre of sphere (see Eq. 15.62).
$P(\Omega,t)$, direct probability of orientation Ω at time t.
$P(\Omega_o;\Omega,t)$, conditional probability: given orientation Ω_o at $t=0$, we reach Ω at time t.
$P_o(\Omega)$, equilibrium distribution in orientation $\Omega \equiv \theta,\phi$.
$P(\mathbf{r}_o;\mathbf{r},t)$, conditional probability: given position $\mathbf{r}_o$ at $t=0$, we reach $\mathbf{r}$ at time t.
$P_o(\mathbf{r}_o)$, probability of location at position $\mathbf{r}_o$ at time $t=0$.
$P(t;\Delta\omega)$, conditional probability in spectral diffusion: $\equiv P(\omega_o;\omega,t)$ (Section A13.1).
$P_A(r), P_V(r)$, probability of two spins r apart given uniform surface/volume densities (see, e.g., Eq. 15.62).
$P_{sh}(r)$, probability of two spins r apart when confined to a shell at uniform density (Eqs. 15.75–15.78 with $x \equiv r/R_2$).
$P_{norm}(r)$, normalized distribution of interspin distances $P(r)$ in DEER (see Eq. 15.50).
P_{klm}, probability of chain configuration m, with orientation θ_k, ϕ_l to diffusion axes (Eqs. 8.159–8.162).
P, constant in power-dependence of saturation parameter: $P \equiv \gamma_e^2 T_1 T_2^o$ (Eq. 11.8).
P, diagonal part of nuclear quadrupole interaction PI_z^2 for $I \geq 1$.
P_{2p}, value of quadrupole P for ^{14}N $2p$-orbital.
$P_\parallel$, principal element of axial nuclear quadrupole tensor.
P_{xx}, P_{yy}, P_{zz}, principal elements of nuclear quadrupole tensor.
$P_L(\cos\theta)$, Lth order Legendre polynomial with argument $\cos\theta$.
q, fixed electric charge (see Figure 4.6, Section 4.3.1).
q, electric field gradient in units of e (Eq. 3.40).
q_{np}, electric field gradient along symmetry axis of an np-electron, in units of e.
q_{xx}, q_{yy}, q_{zz}, components of electric field gradient in units of e.
Q, electric quadrupole moment in units of e.
Q_2^0, quadrupolar tensor operator (Eq. 3.44).
$Q_{\mathrm{N}}^{\mathrm{N}}, Q_{\mathrm{NO}}^{\mathrm{N}}, Q_{\mathrm{NC}\alpha}^{\mathrm{N}}, Q_{\mathrm{O}}^{\mathrm{O}}, Q_{\mathrm{NO}}^{\mathrm{O}}$, σ–π spin-polarization constants relating isotropic hyperfine coupling to unpaired π-electron spin density (see Eqs. 3.14, 3.30 and 3.35).
r_e, effective vector position of unpaired electron in a nitroxide (see Eq. 14.53).
r_{eff}, effective molecular radius: $v_{mol} = \frac{4}{3}\pi r_{eff}^3$, for reaction field (Eq. 4.1).
r_{NO}, N–O bond length.
r_L, diffusion radius of spin label L.
r_R, diffusion radius of paramagnetic relaxant R.
$r_{2p,\mathrm{N}}^{eff}, r_{2p,\mathrm{O}}^{eff}$, effective offsets of nitrogen, oxygen $2p\pi$-orbitals from nitroxide x,y-plane.
$\langle r_{2p,\mathrm{N}}^{-3}\rangle, 1/r^3$ averaged over nitrogen $2p$-orbital (see Eq. 3.22).
r_{12}, distance between spin 1 and spin 2.
$r_{eff,3}$, effective mean interspin distance deduced from dipolar splitting: $r_{eff,3} = \langle 1/r_{12}^3\rangle^{-1/3}$.
$r_{eff,6}$, effective mean interspin distance deduced from dipolar second moment: $r_{eff,6} = \langle 1/r_{12}^6\rangle^{-1/6}$.
$\langle r_{nn}\rangle$, mean nearest-neighbour distance in a random distribution.
$R(t)$, random frictional torque in Langevin dynamics (Eqs. 7.65–7.67).
$\mathbf{R}(\phi,z)$, rotation matrix for rotation by angle ϕ about the z-axis (Eq. I.20 and cf. Eq. B.25).
$\mathbf{R}(\psi,\theta,\phi)$, rotation matrix for a general rotation by Euler angles ψ, θ, ϕ (Eq. I.21).
$R^{p,o}$, CW-ELDOR reduction factor: ratio of observer (o) absorption in presence and absence of pumping (p) power (Eq. 10.8).
$R_\infty^{p,o}$, CW-ELDOR reduction factor $R^{p,o}$, for pumping transition p and observing transition o, extrapolated to infinite pumping power ($W^{-1} = 0$ in Eq. 10.11).
$(R_{KM})_{LL'}$, matrix elements in expansion of $c_{K,M}^{L'}$ (Eqs. 8.117 and 8.118).
R_{m_I}, coefficients in echo decay for sudden jumps $\pm\Delta\omega_{m_I}$ at rate $1/\tau_c$: $R_{m_I}^2 = 1/\tau_c^2 - \Delta\omega_{m_I}^2$ (Eqs. 13.65, 13.66, 13.133 and 13.134).
$R_{2e}, R_{2x_1}, R_{2x_2}$, rate constants in time evolution of $\langle S_+\rangle$ (Eq. O.60).
$R_{mm',nn'}$, elements of relaxation matrix in Redfield theory (Eqs. O.82–O.84).
$s(t)$, switching function that brings about phase reversal by π-pulse (cf. Eq. 13.30).
$\mathbf{s}_1, \mathbf{s}_2$, operators for individual (interacting) electron spins: $s_1 = s_2 = \frac{1}{2}$ (Sections 9.1, 9.2 and 9.4).
S, elliptical integral (Eqs. 7.15 and 7.16).
S, integrated EPR absorption intensity: $S = \int v(\omega)\mathrm{d}\omega$ (see Eq. 11.11).
$S(\chi_0)$, correction to line-height ratio for inhomogeneous broadening (Eqs. 7.37–7.41).
S, electron-spin quantum number.
S_x, S_y, S_z, electron-spin operator: x, y, z components.
$S_\pm(\equiv S_x \pm iS_y)$, shift operators for electron spin.
S^α, S^β, electron polarization operators that select the $M_S = \pm\frac{1}{2}$ states $\left(S^\alpha + S^\beta = \hat{1}\right)$.
$S_i^{(rs)}$ $(i = x,y,z)$, single-transition operators for electron-spin transition $|r\rangle$ to $|s\rangle$.
S_R, spin of paramagnetic relaxant R.
S_{xx}, S_{yy}, S_{zz}, principal elements of order tensor, i.e., order parameters of x-, y-, z-axes (Eq. 8.12).
$S_{XX}, S_{YY}, S_{ZZ}, S_{XY}, S_{XZ}, S_{YZ}$, elements of segmental order tensor (Eqs. 8.152–8.157).
S_{seg}, S_o, segmental/backbone order parameter (see Eq. 8.32).

$S_{a,b}$, overlap integral between electron orbitals φ_a, φ_b (Eq. 9.66; Eq. K.7) – when little chance of ambiguity, abbreviated to S (Eqs. K.14 and K.15).

$SR(t)$, saturation recovery signal.

t_p, pulse length.

t_P, duration of collision between paramagnetic relaxant and spin label.

T_1, spin–lattice (or longitudinal) relaxation time.

T_1^{eff}, effective spin–lattice relaxation time, defined by the steady-state population difference n_p (Eq. 10.1) of an irradiated transition.

$T_{1,b}^{eff}$, effective spin–lattice relaxation time at site b (cf. Eq. 10.1).

$T_{1,b}^{o}$, intrinsic spin–lattice relaxation time at site b.

T_{1e}, electron spin–lattice relaxation time.

$T_{1,L}$, electron spin–lattice relaxation time of spin label L.

T_{1n}, nuclear spin–lattice relaxation time (see W_n).

T_{1R}, spin–lattice relaxation time of paramagnetic relaxant R.

$T_{1,dd}^{-1}$, magnetic dipole–dipole contribution to spin–lattice relaxation rate.

T_{1HE}^{-1}, rate of spin–lattice relaxation induced by Heisenberg exchange.

T_2, transverse (or spin–spin) relaxation time.

T_2^{sec}, secular contribution to transverse relaxation time (cf. Eq. 5.49).

T_{2e}, electron transverse relaxation time.

T_{2M}, phase-memory time for spin-echo decay, also called T_M (Eq. 13.40).

T_{2M}^{∞}, asymptotic phase-memory time from long-time tails in echo decay (Eq. 13.60).

$T_{2,L}$, transverse relaxation time of spin label L.

$T_{2,R}$, transverse relaxation time of paramagnetic relaxant R.

$T_{2,dd}^{-1}$, rate of transverse relaxation by magnetic dipole–dipole interaction.

$T_{2dd}^{\prime -1}$, rate of transfer of transverse magnetization by non-secular dipole–dipole interaction (see Eq. 9.169).

T_{2HE}^{-1}, rate of transverse relaxation by Heisenberg exchange.

$T_{\parallel}^d$, $T_{\perp}^d$, principal elements of dipolar hyperfine interaction tensor (Eqs. 3.20 and 3.19).

$T_{\parallel,o}^d$, value of $T_{\parallel}^d$ for $\rho_\pi^N = 1$ (see Eq. 3.22).

$(T_1T_2)^{eff}$, effective T_1T_2 relaxation-time product obtained from saturation parameter $P \equiv \gamma_e^2 T_1 T_2^o$.

T_{SE}, stimulated-echo decay time.

T_{SR}, saturation recovery time.

$T_{SR,e}$, long-pulse saturation recovery time.

$T_{SR,n}$, SR-ELDOR recovery time.

T_R, frictional torque (see Eq. 7.6).

$\mathbf{T}_R = (T_{R,x}, T_{R,y}, T_{R,z})$, torque generated by orienting potential: $\mathbf{T}_R = i\mathbf{L}U(\Omega)$ (Eqs. 8.141 and 8.142).

$T_{R\perp}$, torque on nitroxide z-axis: $T_{R\perp} = -\partial U/\partial\theta_z$ (see Eq. 16.17).

$T_{L,M}$, components of irreducible spherical tensor $\mathbf{T}_L$ of rank L: $M = -L, -L+1, \ldots, L-1, L$ (Eq. S.1).

$\langle a_1 L_1 \| T_L \| a_2 L_2 \rangle$, reduced matrix element of irreducible spherical tensor operator $T_{L,M}$, in Wigner-Eckhart theorem (Eq. Q.19).

u, v, rotating-frame transverse magnetization $\widehat{M}_x(t)$, $\widehat{M}_y(t)$ (for B_1-field along x-axis): dispersion/absorption EPR signal (see Eqs. 6.9 and 6.10).

u_k, v_k, dispersion/absorption EPR signal from kth angular zone (Eqs. 6.42 and 6.43).

$u_n, v_n, M_{z,n}$, complex Fourier amplitudes at nth harmonic of the field modulation (Eq. 11.24).

$u_{n,\nu}, v_{n,\nu}, M_{z,n,\nu}$, coefficients in dependence of nth Fourier harmonic on νth power of B_{mod} (Eq. 11.29).

$u_{nk}, v_{nk}, M_{z,nk}$, Fourier amplitudes of nth harmonic for kth angular zone (see Eqs. 11.91–11.93).

u_{max}, excluded volume parameter in DEER background signal: $u_{max}(t) \equiv (\mu_o/4\pi) g^2 \beta_e^2 t / R_{min}^3$ (Eq. 15.87).

$U(\theta_i)$, potential of mean torque/orienting pseudopotential (see Eq. 8.22).

U_1, first harmonic absorption signal, in-phase with the field modulation.

U_1', first harmonic absorption signal, 90° out-of-phase with the field modulation.

$\bar{v}$, partial specific volume.

v_{12}, steric volume occupied by one spin-label in a colliding pair: $\frac{4}{3}\pi(r_1 + r_2)^3$.

V, electric potential.

$V(\mathbf{r})$, electrostatic potential of electron cloud.

V_{loc}, local electrostatic potential.

V_N, V_O, local electric potential at N-, O-atom.

V_{mol}, molecular volume.

$V'(\rho)$, integral in statistical theory of dilute dipolar spins – Section 15.9. (Eqs. 15.26, 15.27 and Eqs. 13.89, 13.90, 13.104, 13.109).

V_o, V_1, V_2, zeroth, first and second harmonic absorption signal, in-phase with the field modulation.

V_1', V_2', first and second harmonic absorption signal, 90° out-of-phase with the field modulation.

$w(r_{nn})$, probability of nearest-neighbour distance r_{nn} in a random distribution (Eq. 15.7).

$w_s(r_{nn})$, probability of nearest-neighbour distance r_{nn} in a random *surface* distribution (Eq. 15.9).

W, rate of transition induced by microwave (or radiofrequency) irradiation.

$W_\uparrow, (W_\downarrow)$, rate of upward (downward) transition in a two-level system.

W_0, rate of $m_I = \pm 1$ to $m_I = \mp 1$ transition in a $I = 1$ system.

W_e, rate of electron spin–lattice transition $(T_{1e}^{-1} = 2W_e)$.

W_{e,m_I}, rate of electron spin–lattice transition for m_I-hyperfine state.

$W_{e,SJ}$, rate of electron spin–lattice relaxation by spin-rotation interaction.

W_{el}, electrostatic energy.

W_{HE}, rate of single Heisenberg spin-exchange transition $(T_{1HE}^{-1} = 2W_{HE})$.

W_n, rate of nuclear spin–lattice transition ($T_{1n}^{-1} = 2W_n$ for $I = \frac{1}{2}$, $T_{1n}^{-1} = W_n$ for $I = 1$).

W_{RL}, rate of paramagnetic spin–lattice relaxation enhancement.

W_x, cross-relaxation rate: $W_x = \frac{1}{2}(W_{x1} + W_{x2})$.

W_{x1}, W_{x2}, individual rates of cross relaxation (double quantum, zero quantum).

$W_{m'm}$ or $W_{m\to m'}$, probability per unit time of transition from state m to state m' (Eqs. 5.8 and 5.52).

$W_{k,k+1}$, probability of transition from kth to $(k+1)$th angular zone (see Eq. 6.37).

$W_{kk'llmm}$, $W_{kkll'mm}$, rate of angular transition $\theta_k \to \theta_{k'}$, $\phi_l \to \phi_{l'}$ for chain in configuration m (Eq. 8.160).

$W_{kkllmm'}$, rate of conformational transition $m \to m'$ for chain at orientation θ_k, ϕ_l (Eq. 8.161).

$W_m(t)$, probability of m B-spin flips in time t (Eq. 13.94).

$W_H(\tau_1,\tau_2)$, $W_L(\tau_1,\tau_2)$, relaxation rates in high-/low-field regions of primary-echo-detected lineshapes (Eq. 13.82).

X_N, X_O, X_C, electronegativity of N-, O- and C-atoms.

X_S, mole fraction of spin label.

$Y_{L,M}(\theta,\phi)$, spherical harmonic function.

z, number of new neighbours encountered in diffusive jump (Eq. 9.150).

z_T, total number of nearest neighbours.

z_L, electric charge of spin label L (in units of e).

z_R, electric charge of paramagnetic relaxant R (in units of e).

$Z(\Omega,\omega)$, reduced spin density-matrix in rotating frame: $Z \equiv Z' + iZ''$ (Eq. 6.59).

$\bar{Z}_{-+}(\omega)$, sum of matrix elements $\langle -\frac{1}{2}|Z(\Omega,\omega)|+\frac{1}{2}\rangle$ over distribution of angles Ω (Eq. 6.61).

$Z^{(n)}(\Omega,\omega)$, nth-harmonic component of reduced spin density-matrix in rotating frame: $Z^{(n)} \equiv Z'^{(n)} + iZ''^{(n)}$ (cf. Eq. 11.95).

Z_p, fractional population (or degeneracy) of site being irradiated/saturated.

α, factor relating Gaussian linewidth to hyperfine multiplet width (Eq. 2.31).

α, Lagrange multiplier in Tikhonov regularization (Eq. 15.58).

α, β, wave functions of $M_S = \pm\frac{1}{2}$ electron spin states.

α,β,γ, Euler angles (Section R.2, Figure R.2).

α_D, electric polarizability of neat nitroxide (see Eq. 4.2).

α_N, α_O, coulomb integrals for nitrogen, oxygen $2p_z$ atomic orbitals (Eqs. 3.4 and 3.3).

$\bar{\alpha}, \Delta\alpha$, mean energy, half difference in energies in molecular orbital theory: $\bar{\alpha} \equiv \frac{1}{2}(\mathcal{H}_{AA} + \mathcal{H}_{BB})$, $\Delta\alpha \equiv \frac{1}{2}(\mathcal{H}_{AA} - \mathcal{H}_{BB})$ (Eqs. K.14 and K.15).

α_{AB}, β_{AB}, γ_{AB}, calibrations in allowed-value equations for fast anisotropic rotation (Eqs. 7.61–7.64).

α_{CB}, β_{CB}, γ_{CB}, calibrations in allowed-value equations for fast anisotropic rotation (Eqs. 7.57–7.60).

$\alpha^{(M)}(t)$, scaling factors for dipolar relaxation: $M = 0, 1, 2$ (Eq. 9.36 rotational; Eq. 9.43 translational).

$\alpha(u_{max})$, normalized decay rate in background DEER signal $u_{max} \equiv (\mu_o/4\pi) g^2\beta_e^2 t/R_{min}^3$ (Eq. 15.87, Figure 15.22).

$\langle\alpha^2\rangle\tau_R$, mean-square amplitude, correlation-time product in librational model for primary echo-detected lineshapes (Section 13.13).

β, tilt angle of a β-strand (Eq. 16.4 for a β-barrel).

β, resonance integral in molecular orbital theory: $\beta \equiv \mathcal{H}_{AB}$ (Eqs. K.14 and K.15).

β_{NO}, resonance integral for nitrogen, oxygen $2p_z$ atomic orbitals (Eq. 3.5).

β_D, diffusion tilt angle, where Euler angles $(\alpha_D, \beta_D, \gamma_D)$ relate diffusion axes to nitroxide axes.

β_x, β_y, β_z, angle of nitroxide x,y,z-axes to helix axis for a TOAC residue.

β_o, nominal turning angle for (equivalent) non-selective pulse $(\beta_o = \omega_1 t_p)$.

β_p, pulse turning/flip angle: $\beta_p = \gamma_e B_1 t_p$.

β_1, β_2, β_3, turning/flip angles of first, second and third pulses in a (echo) sequence.

β_a, β_f, effective turning angles of selective pulses applied to allowed, forbidden transition.

β_e, Bohr magneton: $e\hbar/(2m_e)$ (electron).

β_N, nuclear magneton: $e\hbar/(2M_p)$ (proton).

$\boldsymbol{\gamma}$, electron gyromagnetic-ratio matrix (Eq. 11.28).

γ_e, electron gyromagnetic ratio.

γ_n, γ_I, nuclear gyromagnetic ratio.

γ_p, γ_H, proton gyromagnetic ratio.

γ_2, ε, strengths of nonaxial ordering potential in units of k_BT (Eq. 8.27).

Γ_{inh}, inhomogeneous linewidth.

Γ_R, Markov operator for relaxation processes other than rotational diffusion (see Eq. 11.47).

Γ_Ω, Markov operator for rotational diffusion with angle Ω (see Eqs. 6.51 and 6.52).

$\tilde{\Gamma}_\Omega$, symmetrized Markov operator (Eq. 8.135), for ordering with axial (Eq. 8.136) and nonaxial (Eqs. 8.140–8.142) potentials.

δ, difference in electron spin–lattice relaxation rates: $\delta = \frac{1}{2}(W_{e,+} - W_{e,-})$ for $I = \frac{1}{2}$; $\delta = \frac{1}{2}(W_{e,+1} - W_{e,-1})$ for $I = 1$.

δ_1, difference in electron spin–lattice relaxation rates for $I = 1$: $\delta_1 = \frac{1}{2}(W_{e,+1} + W_{e,-1}) - W_{e,0}$.

δ_{pm}, relative paramagnetic chemical shift – NMR (Eq. 3.29).

δA, off-axial hyperfine anisotropy, e.g., $\delta A = \frac{1}{2}(A_{xx} - A_{yy})$.

δg, off-axial g-value anisotropy, e.g., $\delta g = \frac{1}{2}(g_{xx} - g_{yy})$.

δa_{HE}, additional shift in a_{HE} from cage re-encounters (Eqs. 9.130 and 9.131).

δq^+, excess positive charge on nitroxide N-atom.

Δ, total modulation depth in DEER (see Eq. 15.52).

ΔA, axial hyperfine anisotropy, e.g., $\Delta A = A_{zz} - \frac{1}{2}(A_{xx} + A_{yy})$.

Δg, axial g-value anisotropy, e.g., $\Delta g = g_{zz} - \frac{1}{2}(g_{xx} + g_{yy})$.

$\Delta B(\theta,\phi)$, resonance-field offset (Eqs. 2.23, 14.57).

ΔB_x, ΔB_y, ΔB_z, resonance-field offsets for static field along x,y,z principal axes, relative to field along z (Section 2.4: $\Delta B_z = 0$), or relative to isotropic resonance position (Section 14.14.1).

$\Delta B_{1/2}$, half-width at half-height of absorption EPR spectrum.

$\Delta B_{1/2}^L$, $\Delta B_{1/2}^G$, half-width at half-height of Lorentzian/Gaussian absorption spectrum.

ΔB_{m_I}, $\Delta B_{1/2}^L$ for outer extrema $(m_I = \pm 1)$ of slow-motion spectrum (see Eq. 7.81).

$\Delta B_{m_I}^R$, $\Delta B_{1/2}^L$ for outer extrema $(m_I = \pm 1)$ of rigid-limit spectrum (see Eq. 7.81).

$\Delta\Delta B_{1/2}$, increase in half-width at half-height of absorption EPR spectrum.

ΔB_{pp}, peak-to-peak linewidth of first-derivative EPR spectrum.

ΔB_{pp}^{tot}, total peak-to-peak linewidth of first-derivative spectrum (see Eq. 2.30).

ΔB_{pp}^{L}, ΔB_{pp}^{G}, peak-to-peak linewidth of Lorentzian/Gaussian first-derivative spectrum.
$\Delta\Delta B_{pp}$, increase in peak-to-peak linewidth of first-derivative EPR spectrum.
$\Delta B_{dd}(\theta)$, angle-dependent dipole–dipole splitting for a pair of spin labels (Eq. 9.14).
$\Delta B_{dd}^{\parallel}, \Delta B_{dd}^{\perp}$, dipolar splitting for static field parallel/perpendicular to interspin vector.
$\langle 2|\Delta B_{dd}|\rangle$, mean dipolar splitting (or absolute-value first moment) of a dipolar powder pattern (cf. Eq. 15.11).
$\langle \Delta B_{dd}^{2}\rangle$, second moment of a dipolar powder pattern (cf. Eq. 15.13).
$\langle \Delta B^{n}\rangle$, nth moment of absorption lineshape about mean field position (Eq. U.1).
$\Delta\Delta\omega_{pp}$, exchange-induced increase in peak-to-peak linewidth: $\Delta\Delta\omega_{pp} = \left(2/\sqrt{3}\right)T_{2,HE}^{-1}$.
$\Delta E_{\pi-\pi*}$, energy of π–π* electronic transition.
Δn_i, departure of population difference n_i from Boltzmann equilibrium: $\Delta n_i = n_i - n_i^{o}$.
Δp, change in coherence order (Section 13.17).
$\Delta\omega_o$, offset in electron Larmor frequency.
$\Delta\omega_{d,j}$, shift in Larmor frequency of A-spin by dipolar interaction of jth B-spin: $\Delta\omega_{d,j}^{o}s_{jz}'$ (Eq. 13.84).
$\Delta\omega_{1/2}$, half-width at half-height for Lorentzian dipolar broadening (Eqs. 15.30 and 15.31).
$\langle \Delta\omega_{m_I}^{2}\rangle$, second moment of hyperfine and g-value spectral anisotropy.
$\langle \Delta\omega^{2}\rangle_{HE}$, second moment of Heisenberg exchange frequency ω_{HE}.
ε, coiling of β-strands (Eq. 16.6).
ε, factor relating translational and rotational correlation times in spin diffusion: $\tau_D = \varepsilon\tau_R$.
ε, shielding factor of nitrogen excess charge: $\varepsilon\delta q^{+}$.
$\varepsilon(t)$, perturbation Hamiltonian for B_1 radiation field (Eq. 6.57).
$\varepsilon(r)$, distance-dependent relative permittivity.
ε_o, permittivity of free space.
ε_r, relative (dielectric) permittivity (numerically equal to dielectric constant).
$\varepsilon_{r,B}$, bulk relative permittivity.
ζ_k, spin-orbit coupling constant for electron on atom k.
η, dynamic viscosity.
η, asymmetry parameter in electric field gradient (Eq. 3.46).
η, product-operator transformation for diagonalization of $S = \frac{1}{2}$, $I = \frac{1}{2}$ system: $2\eta I_y S_z$ (see Eqs. 14.8, 14.9).
η_{M_S}; η_α, η_β, angle rotating ($S = \frac{1}{2}$, $I = \frac{1}{2}$) basis vectors to diagonalize hyperfine interaction for α, β, i.e., $M_S = \pm\frac{1}{2}$ -states (see Sections H.2, H.3).
$\eta, \eta(\alpha)$, square norm of second-derivative probability $\left\|d^2\mathbf{P}/dr^2\right\|$ in Tikhonov regularization.
θ, twist of β-strands in β-sheets (Eq. 16.5).
θ, ϕ, angular polar coordinates.
$\theta_x, \theta_y, \theta_z$, angles of static field (Figure 2.2) – or of ordering director (Figure 8.3) – to nitroxide x, y, z-axes.
θ_C, cone half-angle in restricted random walk model (Figure 8.5).
θ_{eff}, angle of effective field $\hat{\mathbf{B}}_{eff}$ in rotating frame to static field $\mathbf{B}_o$ (see Figure 6.2).
θ_{CNC}, nitroxide C–N–C bond angle.
κ, inverse Debye screening length.
κ, exponential decay constant of relative permittivity (Eq. 4.18).
κ, slip coefficient: $f_R(\text{slip}) = \kappa f_R(\text{stick})$, $0 < \kappa < 1$.
λ, spin-orbit coupling constant: $\mathcal{H}_{SOC} = \lambda\mathbf{L}\cdot\mathbf{S}$.
λ, decay length in transmembrane polarity profile (Eq. 4.44).
λ, step-length for diffusive jumps (see Eqs. 9.152, 9.159).
$\lambda(\theta)$, orientational intensity function in DEER (Eqs. 15.91, 15.92).
$\lambda_1, \lambda_2, \lambda_3$, saturation-recovery rate constants (e.g., Eqs. 12.16, 12.35).
λ_2, strength of Maier–Saupe ordering potential in units of k_BT (see Eq. 8.24).
μ_o, permeability of free space.
μ_e, μ_n, magnetic dipole moment operator (electron or nuclear).
μ_L, magnetic moment of spin label L.
μ_R, magnetic moment of paramagnetic relaxant R.
ν, linear frequency.
$\nu_a^{\pm}, \nu_f^{\pm}$, allowed/forbidden transition frequencies of exchange-coupled spin pairs (Eqs. 9.81, 9.82).
ξ, rotation angle – see $\chi^{(I)}(2\xi)$.
ξ_α, ξ_β, rotation angles for $M_S = \pm\frac{1}{2}$ states.
π_c, electron density transferred in covalent bonding.
Π, accessibility parameter to paramagnetic relaxant from CW-saturation (Eq. 16.1).
ρ, inverse axial ratio of rotational ellipsoid: b/a.
$\rho, \rho(\alpha)$, square norm $\left\|\mathbf{K}\cdot\mathbf{P}-\mathbf{f}\right\|^2$ in Tikhonov regularization (Eq. 15.58).
$\rho(T)$, solvent density.
$\rho(\mathbf{r})$, electric charge-density distribution.
ρ_{el}, electric charge density.
ρ_x, ρ_y, rotational diffusion anisotropies, $D_{R,x}/D_{R,z}$, $D_{R,y}/D_{R,z}$.
ρ_1', out-of-phase/in-phase ratio for first harmonic double-integrated intensity (Eqs. 11.65 and 11.68).
ρ_π^{N}, nitroxide unpaired spin density on nitrogen.
ρ_π^{O}, nitroxide unpaired spin density on oxygen.
σ, saturation parameter (Eq. 11.8).
$\boldsymbol{\sigma}$, spin density-matrix operator (cf. Eq. 6.44).
$\sigma(0)$, (equilibrium) spin density-matrix operator, at start of pulse sequence.
$\sigma(t)$, spin density-matrix operator at various times t in a pulse sequence.
$\sigma^{\ddagger}(t)$, spin density-matrix in interaction representation (Eq. N.21).
σ_{mn}, m-,n-element of spin density-matrix (Eq. N.3).
σ_{-+}, off-diagonal spin density-matrix element that produces the transverse magnetization detected in pulse EPR (Eq. 13.26): $\sigma_{-+} \equiv \langle -\frac{1}{2}|\sigma|+\frac{1}{2}\rangle$, et seq.
σ_{echo}, spin density-matrix (product-operator formalism) at end of spin-echo pulse sequence.
σ_{eq}, equilibrium spin density-matrix operator ($\sigma_{eq} = S_z$) before start of pulse sequence.

$\sigma_{prep}, \sigma_{mix}$, spin density-matrix (product-operator formalism) after preparation, mixing periods in pulse sequence.

σ_o, equilibrium spin density-matrix operator (Eq. 6.47).

$\sigma_o(\omega), \sigma_1(\omega), \sigma_o'(\omega), \sigma_1'(\omega)$, summations (Eqs. 9.104, 9.105, 9.107, 9.108) in Heisenberg-exchange lineshapes (Eqs. 9.103, 9.106). See Eqs. 9.176–9.178 for corresponding magnetic-dipolar case.

σ_{RL}, steric accessibility factor for Heisenberg exchange between paramagnetic relaxant *R* and spin label *L*.

σ_{r12}, standard deviation in Gaussian distribution of interspin distances r_{12} (see Eq. 15.18).

σ_{rice}, width parameter in Rice distribution of interspin distances (see Eq. 15.60).

σ_G^2, Gaussian second moment (or variance).

τ, twist about β-strand axis.

τ_0, τ_2, correlation times in strong-jump model with ordering (Eq. 8.88).

$\tau_{2,0}, \tau_{2,2}$, rotational correlation times for second-order spherical tensor (Eq. 7.28: $L = 2$; $K = 0,2$).

$\tau_{L,M}$, anisotropic rotational correlation times for Wigner rotation matrices (Eq. 7.28; Eq.8.137 with ordering).

τ_{KML}, relaxation times in $j_{KM}(0)$ spectral densities for Brownian diffusion with ordering (Eqs. 8.119 and 8.122).

τ_c, correlation time for exponential correlation function (Eq. 5.12).

τ_D, time constant characterizing mutual translational diffusion of spin label and relaxant: $\tau_D \equiv d_{RL}^2/(2D_T)$.

τ_J, correlation time for angular velocity/inertial angular momentum: $\tau_J \equiv I^{inert}/f_R$ (see Eq. 7.71).

$\tau_{J,i}$ $(i = x,y,z)$, correlation time for spin-rotation coupling (Eq. 5.35).

τ_l, lifetime between jumps in sudden-jump model (see Eq. 6.29).

τ_L, (effective) correlation time for different models of rotational diffusion (Eqs. 7.22, 7.74 and 7.75).

τ_P, time constant for lifetime of collisions between spin labels (Eq. 9.90) or with paramagnetic relaxant (Section 10.7).

τ_{RE}, time between cage re-encounters.

τ_R, correlation time for isotropic rotational diffusion.

$\bar{\tau}_R$, geometric-mean correlation time for axial rotational diffusion: $\bar{\tau}_R = \sqrt{\tau_{R\parallel}\tau_{R\perp}}$. For non-axial rotation: $\bar{\tau}_R = \sqrt[3]{\tau_{R,X}\tau_{R,Y}\tau_{R,Z}}$.

$\tau_{R\parallel}$, correlation time for rotation about the unique (||) axis in axial diffusion.

$\tau_{R\perp}$, correlation time for rotation about the non-unique (⊥) axis in axial diffusion.

$\tau_R^{eff}(m_I)$, effective saturation-transfer EPR correlation times from *L″/L* and *H″/H* ratios for axially anisotropic diffusion (Eq. 11.90).

τ_A^{-1}, τ_B^{-1}, rate of transfer from site **A** to **B**, and vice-versa, in two-site exchange (see Eq. 6.14).

τ_{coll}^{-1}, frequency of collisions between spin labels or with paramagnetic relaxant.

τ_{HE}^{-1}, rate of Heisenberg spin-exchange transitions with all electron states: $\tau_{HE}^{-1} = (2I+1)W_{HE}$.

τ_j^{-1}, hopping frequency between adjacent sites (Eqs. 9.152 and 9.159).

τ_k^{-1}, rate constant for transfer/exchange from site *k* (see Eq. 10.49).

τ_{obs}^{-1}, rate of transfer of saturation from observer position (SR-ELDOR).

τ_p^{-1}, rate of transfer of saturation from pump position (SR-ELDOR).

τ_S^{-1}, rate of B-spin flips (see Eq. 13.94).

$\tau_{sd}(\omega)^{-1}$, spectral diffusion rate (Eq. 11.78).

$\varphi(2\tau,t), \varphi(4\tau,t)$, accumulated phase in a 3-pulse/4-pulse DEER experiment (Eqs. 15.34, 15.38).

φ_N, φ_O, $2p_z$ atomic orbitals on nitrogen, oxygen.

$\varphi_{K,M}^L(\Omega)$, normalized eigenfunctions of Schroedinger equation for a symmetric top: $\varphi_{K,M}^L(\Omega) = \sqrt{(2L+1)/8\pi^2}\,\mathcal{D}_{K,M}^L(\Omega)$.

$\bar{\phi}, \phi_o$, mean azimuthal angle $\bar{\phi}$ and amplitude $\pm\phi_o$ for lateral (azimuthal) ordering (Section 8.6).

Φ, total flux across a sphere (Eqs. 9.147, 9.148).

$\chi(\Omega,t)$, reduced spin density-matrix: $\chi(\Omega,t) = \sigma(\Omega,t) - \sigma_o(\Omega)$.

χ_{-+}, matrix element $\langle -\tfrac{1}{2}|\chi(\Omega,t)|+\tfrac{1}{2}\rangle$ of reduced spin density-matrix (cf. Eq. 6.49).

χ_0, ratio of Gaussian to Lorentzian broadening of $m_I = 0$ ^{14}N-hyperfine line, $\Delta B_{pp}^G/\Delta B_{pp}^L(0)$.

$\chi_1, \chi_2, \ldots$, amino acid side-chain dihedral angles.

$\chi^{(l)}(2\xi)$, character for rotation 2ξ of the full rotation group.

${}^1\chi_0; {}^3\chi_{\pm 1}, {}^3\chi_0$, two-electron singlet ($S$ = 0)/triplet (S = 1) spin functions (Eqs. 9.70–9.73).

$\chi(t)$, wave function of exchange-coupled spins (Eqs. 9.85–9.88).

ψ, θ, ϕ, Euler angles (Section R.2, Figure R.2).

$|\psi(0)|^2$, unpaired electron density at the nucleus (see Eq. 3.13).

ψ_L, electrostatic surface potential of spin-label system.

$\psi_{N\text{-}O}, \psi_{N\text{-}C_1}, \psi_{N\text{-}C_2}$, sp^2-hybrid nitrogen valence orbitals (Eqs. 3.58–3.60).

$\psi_m^{(k)}$, (hybrid) atomic orbital on atom *k* in molecular orbital *m*.

$\psi_n^{(O)}$, oxygen lone-pair sp^2-hybrid orbital (Eq. 3.68).

${}^1\psi, {}^3\psi$, two-electron singlet (S = 0)/triplet (S = 1) wave functions (Eqs. 9.64, 9.65).

Ψ, diagnostic line-height ratio for Gaussian/Lorentzian lineshape contributions (Eq. 2.33).

Ψ_m, molecular orbital *m* (Eqs. 3.1, 3.66, 3.69).

$\Psi^{(b)}, \Psi^*$, nitroxide bonding/anti-bonding π-orbital (Eqs. 3.10 and 3.9).

$\Psi_{MO}^{(b)}, \Psi_{MO}^{(a)}$, bonding, antibonding molecular orbitals (Eq. K.17).

ω, dihedral angle Cα–C–N–Cα of peptide backbone.

ω, angular frequency ($2\pi\nu$).

ω_{AB}, angular frequency of spin–spin interaction between A-spin and B-spin (Eq. 15.33).

ω_1, radiation field strength/Rabi frequency: $\omega_1 = \gamma_e B_1$.

ω_{eff}, precession frequency about effective field in rotating frame: $\omega_{eff} = \gamma_e B_{eff}$ (Eq. 13.10).

$\omega_{m'm}$, transition energy between states *m* and *m′* in angular frequency units: $\omega_{m'm} = \omega_{m'} - \omega_m$.

ω_{mw}, microwave angular frequency.

ω_{rf}, angular radiofrequency.

ω_{mod}, static-field modulation angular frequency.

ω_e, electron (angular) Larmor frequency: $\omega_e = \gamma_e B_o = (g_e \beta_e / \hbar) B_o$.

$\omega_{e,p}$, $\omega_{e,o}$, pump and observer microwave (angular) frequencies in ELDOR.

ω_{pump}, ω_{obs}, pump and observer microwave (angular) frequencies in DEER.

$\omega_{obs}^{\max}$, observer frequency giving maximum amplitude in orientation-selective DEER (Eq. 15.98).

ω_{HE}, Heisenberg exchange frequency: $\omega_{HE} \equiv \tau_{HE}^{-1}$.

ω_I, nuclear Larmor (angular) frequency.

ω_I^{N}, nitrogen nuclear Larmor (angular) frequency.

ω_L, (angular) Larmor frequency of spin label L.

$\omega_{m_I}(\theta,\phi)$, angular resonance frequency of m_I-hyperfine line.

$\omega_{e,m}$, complex resonance frequency of $m_I = m$ hyperfine line under Heisenberg exchange: $\omega_{e,m} = \omega'_m + i\Gamma_m$ (eigenvalues of Eq. 9.109).

$\omega_{\max,m}$, position of resonance absorption maximum of $m_I = m$ hyperfine line under Heisenberg exchange and/or dipole–dipole interaction (see Eqs. 9.119, 9.124, 9.182).

ω_p, (angular) irradiation frequency pumping a two-state transition.

ω_Q, strength of nuclear quadrupole interaction: $\omega_Q = 3e^2Q/8I(2I-1) \equiv \frac{3}{4}P_{zz}$.

ω_R, (angular) Larmor frequency of paramagnetic relaxant R.

ω_R, angular velocity of rotation (Eqs. 7.6, 7.65).

ω_α, ω_β, angular frequency of nuclear transition within $M_S = \pm\frac{1}{2}$ states of $S = \frac{1}{2}$, $I = \frac{1}{2}$ spin system (see Eqs. 14.5 and 14.6).

$\omega_{\alpha,\beta}^{\text{H}}$, proton ENDOR frequencies (cf. ω_α, ω_α and see Eq. 14.50).

$\omega_{\alpha,\Delta m_I}$, $\omega_{\beta,\Delta m_I}$, angular frequency of nuclear transition Δm_I within $M_S = \pm\frac{1}{2}$ states of $S = \frac{1}{2}$ system with general I.

$\omega_\pm$, diagonalized nuclear frequencies for $S = \frac{1}{2}$, $I = \frac{1}{2}$: $\frac{1}{2}\omega_+ I'_z + \omega_- I'_z S_z$ (see Eqs. 14.4–14.6).

$\omega^{M_S m_I}$, allowed $(\Delta m_I = 0)$ and forbidden $(\Delta m_I = \pm 1)$ EPR transition frequencies for $M_S = -\frac{1}{2}$ to $M_S = +\frac{1}{2}$ (see Figs. 6.9 and 6.11).

$\omega_p^{(q)}$, angular frequency in spectral density $j^{(q)}\left(\omega_p^{(q)}\right)$ associated with spin operator $A_p^{(q)}$.

Ω, angular orientation: θ,ϕ in polar coordinates; ψ,θ,ϕ in Euler angles.

Ω, crossing angle between packed α-helices or dihedral angle between β-strands in adjacent β-sheets.

Appendices A–M
Fundamentals

APPENDIX A: UNITS AND CONVERSIONS

A.1 SI Units and cgs Units

This book is written in Système International (SI) units – except where I forgot to make the change. Much of the original literature, however, is expressed in centimetre-gram-second (cgs) units. A mixed cgs system is used, the so-called Gaussian system, where magnetic quantities are in emu units and electrostatic quantities are in esu units. The result is that the velocity of light in vacuum c appears in equations that involve both types of electromagnetic quantities, because this relates the emu and esu systems (see Table A.3, given later). Notably, the Bohr magneton ($\beta_e = e\hbar/2m$ in SI units) is $\beta_e = e\hbar/2mc$ in cgs Gaussian units.

One advantage of SI units is the unification of electric and magnetic units. The fundamental SI base units are the metre, kilogram, second and ampere (MKSA). Additionally, the common practical units (volt, coulomb, ohm, farad, etc.) are retained. A further advantage is that SI units are a so-called rationalized system. This means that magnetic permeability and dielectric permittivity are defined such that factors of 4π appear in electromagnetic equations involving radial fields (e.g., magnetic field from a point dipole, or electrostatic potential from a point charge), but not otherwise (e.g., in the isotropic hyperfine coupling, which depends only on the magnetic dipole density of the electron at the nucleus). The result of rationalization is that Maxwell's electromagnetic equations become considerably simpler in SI units.

Table A.1 gives the conversion between relevant SI units and cgs units. Only the emu system is used for electrical units; the relation between emu and esu systems is given later in Table A.3. Mostly, the emu units are related simply by factors of ten to the SI units. Together with use of the Bohr magneton, this means that many of the equations for magnetic resonance remain the same in the two systems. The notable exception is that expressions involving the field from a magnetic dipole contain factors of $\mu_o/4\pi \equiv 10^{-7}$ H m^{-1} in SI units. Table A.2 lists some of the basic equations that differ between the two systems. Many are those directly involving electrostatics. The important differences in magnetism are equations relating to dipolar interactions, where the SI expressions unfortunately are more cumbersome.

A fundamental difference between the two systems is that magnetic permeability and electric permittivity have units in the SI system, but not in the cgs systems. In the SI system, the permeability, μ_o, and permittivity, ε_o, of free space are defined such that the relative permeability, μ_r, and relative permittivity, ε_r, in the SI system are equal to μ and ε in the cgs system, where they are dimensionless. For example, ε_r is numerically equal to the dielectric constant. Note that μ_o, ε_o and the velocity of light in vacuum, c, are all defined quantities in SI units, being related by: $c = 1/\sqrt{\mu_o\varepsilon_o}$.

The mole – fundamentally a cgs unit – is retained as the base unit for amount of substance in the SI system. This means that 1 mole must be multiplied by 10^{-3} to give the molar mass in kg mol^{-1}. Molar concentration becomes 1 M $\equiv 10^3$ mole m^{-3}.

TABLE A.1 Conversion between SI units and cgs units

quantity	SI	cgs (emu)
magnetic induction, *B*	1 tesla (T or Wb m^{-2})	10^4 gauss (G)
magnetic field, *H*	1 A m^{-1}	$4\pi\times10^{-3}$ oersted (Oe)
magnetic flux, Φ	1 weber (Wb)	10^8 maxwell (Mx)
charge, *Q*	1 coulomb (C)	10^{-1} emu
potential, *V*	1 volt (V)	10^8 emu
force, *F*	1 newton (N)	10^5 dynes (dyn)
energy, *E*	1 joule (J or N m)	10^7 ergs
power	1 watt (W or J s^{-1})	10^7 erg s^{-1}
pressure, *P*	1 pascal (Pa or N m^{-2})	10 dyn cm^{-2}
dynamic viscosity, η	1 Pa s	10 poise (P)

Electrical units are given only in emu.

TABLE A.2 Electromagnetic equations in SI system and in cgs Gaussian system

equation	SI *units*	cgs *units*
magnetic induction	$\mathbf{B}=\mu_r\mu_o\mathbf{H}=\mu_o(\mathbf{H}+\mathbf{M})$	$\mathbf{B}=\mu\mathbf{H}=\mathbf{H}+4\pi\mathbf{M}$
magnetic permeability	$\mu_r\mu_o=\mu_r\times\left(4\pi\times10^{-7}\ \mathrm{H\,m^{-1}}\right)$	$\mu=\mu_r$
electrodynamic momentum	$\mathbf{p}_{tot}=\mathbf{p}+e\mathbf{A}$	$\mathbf{p}_{tot}=\mathbf{p}+(e/c)\mathbf{A}$ (esu)
Lorentz force	$\mathbf{F}=e(\mathbf{E}+\mathbf{v}\times\mathbf{B})$	$\mathbf{F}=e\left(\mathbf{E}+\frac{1}{c}\mathbf{v}\times\mathbf{B}\right)$
Bohr magneton	$\beta_e=e\hbar/2m_e$	$\beta_e=e\hbar/2m_ec$ (esu)
nuclear magneton	$\beta_N=e\hbar/2m_p$	$\beta_N=e\hbar/2m_pc$ (esu)
isotropic hyperfine coupling	$a_o=\frac{2\mu_o}{3}g\beta_eg_N\beta_N\lvert\psi(0)\rvert^2$	$a_o=\frac{8\pi}{3}g\beta_eg_N\beta_N\lvert\psi(0)\rvert^2$
dipole–dipole interaction	$E_{dd}=\frac{\mu_o}{4\pi}\left(\frac{\boldsymbol{\mu}_1\cdot\boldsymbol{\mu}_2}{r_{12}^3}-3\frac{(\boldsymbol{\mu}_1\cdot\mathbf{r}_{12})\cdot(\boldsymbol{\mu}_2\cdot\mathbf{r}_{12})}{r_{12}^5}\right)$	$E_{dd}=\frac{\boldsymbol{\mu}_1\cdot\boldsymbol{\mu}_2}{r_{12}^3}-3\frac{(\boldsymbol{\mu}_1\cdot\mathbf{r}_{12})\cdot(\boldsymbol{\mu}_2\cdot\mathbf{r}_{12})}{r_{12}^5}$
dipolar hyperfine coupling	$T_\parallel^d=\frac{4}{5}(\mu_o/4\pi)g_e\beta_eg_N\beta_N\langle r_{2p}^{-3}\rangle\rho_\pi^N$	$T_\parallel^d=\frac{4}{5}g_e\beta_eg_N\beta_N\langle r_{2p}^{-3}\rangle\rho_\pi^N$
dipolar splitting	$\pm\Delta B_{dd}(\theta)=\frac{3}{4}\times\frac{\mu_o}{4\pi}\left(\frac{g\beta_e}{r_{12}^3}\right)(1-3\cos^2\theta)$	$\pm\Delta B_{dd}(\theta)=\frac{3}{4}\left(\frac{g\beta_e}{r_{12}^3}\right)(1-3\cos^2\theta)$
dipolar relaxation	$\frac{1}{T_{1dd}}=\frac{3}{2}\left(\frac{\mu_o}{4\pi}\right)^2\gamma_e^2\hbar^2s(s+1)\left[J^{(1)}(\omega)+J^{(2)}(2\omega)\right]$	$\frac{1}{T_{1dd}}=\frac{3}{2}\gamma_e^4\hbar^2s(s+1)\left[J^{(1)}(\omega)+J^{(2)}(2\omega)\right]$
	$\frac{1}{T_{2dd}}=\frac{3}{8}\left(\frac{\mu_o}{4\pi}\right)^2\gamma_e^4\hbar^2s(s+1)\left[J^{(0)}(0)+10J^{(1)}(\omega)+J^{(2)}(2\omega)\right]$	$\frac{1}{T_{2dd}}=\frac{3}{8}\gamma_e^4\hbar^2s(s+1)\left[J^{(0)}(0)+10J^{(1)}(\omega)+J^{(2)}(2\omega)\right]$
dipolar second moment	$\langle\Delta B_{dd}^2\rangle=\frac{3}{5}\left(\frac{\mu_o}{4\pi}\right)^2s(s+1)g^2\beta_e^2\sum_j 1/r_{jk}^6$	$\langle\Delta B_{dd}^2\rangle=\frac{3}{5}s(s+1)g^2\beta_e^2\sum_j 1/r_{jk}^6$
electric displacement	$\mathbf{D}=\varepsilon_r\varepsilon_o\mathbf{E}=\varepsilon_o\mathbf{E}+\mathbf{P}$	$\mathbf{D}=\varepsilon\mathbf{E}=\mathbf{E}+4\pi\mathbf{P}$
electric permittivity	$\varepsilon_r\varepsilon_o\equiv\varepsilon_r/(\mu_oc^2)$	$\varepsilon=\varepsilon_r$
Gauss theorem	$\nabla\cdot\mathbf{D}=\rho$	$\nabla\cdot\mathbf{D}=4\pi\rho$
Coulomb energy	$W_{el}=\frac{Ze^2}{4\pi\varepsilon_or}$	$W_{el}=\frac{Ze^2}{r}$
reaction field	$\mathbf{E}_\mathbf{R}=f(\varepsilon_{r,B})\frac{-\mathbf{m}_{el}}{4\pi\varepsilon_or_{eff}^3}$	$\mathbf{E}_\mathbf{R}=f(\varepsilon_{r,B})\frac{-\mathbf{m}_{el}}{r_{eff}^3}$
	$E_R(\text{Onsager})=\frac{p_{el}}{4\pi\varepsilon_or_{eff}^3}\frac{2(\varepsilon_{r,B}-1)}{2\varepsilon_{r,B}+1}$	$E_R(\text{Onsager})=\frac{p_{el}}{r_{eff}^3}\frac{2(\varepsilon_{r,B}-1)}{2\varepsilon_{r,B}+1}$
Lorenz–Lorentz	$\frac{n_D^2-1}{n_D^2+2}=\frac{\alpha_D}{4\pi\varepsilon_or_{eff}^3}$	$\frac{n_D^2-1}{n_D^2+2}=\frac{\alpha_D}{r_{eff}^3}$
Poisson equation	$\nabla^2V=-\rho/\varepsilon_o\varepsilon_r$	$\nabla^2V=-4\pi\rho/\varepsilon$
electric potential	$V_1(r,\theta)=\frac{q}{4\pi\varepsilon_o\varepsilon_{r,1}}\sum_{l=0}^{\infty}\left(\frac{d^l}{r^{l+1}}+A_lr^l\right)\mathrm{P}_l(\cos\theta)$	$V_1(r,\theta)=\frac{q}{\varepsilon_1}\sum_{l=0}^{\infty}\left(\frac{d^l}{r^{l+1}}+A_lr^l\right)\mathrm{P}_l(\cos\theta)$
electric field gradient	$eq_{zz}\equiv\frac{\partial^2V}{\partial z^2}=\frac{-e}{4\pi\varepsilon_or^3}\left(\frac{3z^2}{r^2}-1\right)$	$eq_{zz}\equiv\frac{\partial^2V}{\partial z^2}=\frac{-e}{r^3}\left(\frac{3z^2}{r^2}-1\right)$

TABLE A.3 Conversion between emu and esu systems

quantity	SI	emu	esu[a]
current, *I*	1 ampere (A)	10^{-1}	$c/10$
charge, *Q*	1 coulomb (C)	10^{-1}	$c/10$
potential, *V*	1 volt (V)	10^{8}	$10^{8}/c$
resistance, *R*	1 ohm (Ω)	10^{9}	$10^{9}/c^2$
inductance, *L*	1 henry (H)	10^{9}	$10^{9}/c^2$
capacitance, *C*	1 farad (F)	10^{-9}	$10^{-9}\times c^2$

[a] $c = 2.99792458\times10^{8}$ m s^{-1} is the velocity of electromagnetic radiation in vacuum (defined).

A.2 Electromagnetic Units and Conversion

The emu units of the cgs system are related to SI units by:

$$1\text{ coulomb of charge} = 10^{-1}\text{emu}$$

$$1\text{ volt of potential} = 10^{8}\text{emu}$$

The esu units are related to the emu units by:

$$\frac{\text{current (or charge) in esu}}{\text{current (or charge) in emu}} = c$$

$$\frac{\text{potential in esu}}{\text{potential in emu}} = 1/c$$

$$\frac{\text{resistance (inductance, or 1/capacitance) in esu}}{\text{resistance (inductance, or 1/capacitance) in emu}} = 1/c^2$$

Table A.3 lists these relationships.

APPENDIX B: VECTORS, MATRICES AND TENSORS

B.1 Vector Algebra

A vector, such as the electron spin **S**, has both a length and a direction. Expressed in Cartesian components (S_x, S_y, S_z), the vector is given by:

$$\mathbf{S} = S_x\mathbf{i} + S_y\mathbf{j} + S_z\mathbf{k} \tag{B.1}$$

where **i**, **j** and **k** are unit vectors directed along the *x*-, *y*- and *z*-axes, respectively. Alternatively, we can express the components in terms of the direction cosines, *l*, *m* and *n*, of the vector **S**:

$$\mathbf{S} = S(l\mathbf{i} + m\mathbf{j} + n\mathbf{k}) \tag{B.2}$$

where $l \equiv \cos\theta_x$, $m \equiv \cos\theta_y$ and $n \equiv \cos\theta_z$ are the cosines of the angles θ_x, θ_y and θ_z between **S** and the *x*-, *y*- and *z*-axes, respectively; *S* is the absolute magnitude of **S**. Note that only two of the direction cosines are independent because: $l^2 + m^2 + n^2 \equiv \cos^2\theta_x + \cos^2\theta_y + \cos^2\theta_z = 1$. We use direction cosines in Chapter 2 as an alternative either to rotating axes explicitly or to diagonalizing the spin Hamiltonian directly.

The scalar product of two vectors, such as the electron spin **S** and the magnetic field **B**, is given by the length of **B** times the projection of **S** on **B**, or vice-versa:

$$\mathbf{S}\cdot\mathbf{B} = \mathbf{B}\cdot\mathbf{S} = BS\cos\theta \tag{B.3}$$

where θ is the angle between **B** and **S** (see Figure B.1). In terms of Cartesian components, the scalar product is given by:

$$\mathbf{B}\cdot\mathbf{S} = B_xS_x + B_yS_y + B_zS_z \tag{B.4}$$

because we have: $\mathbf{i}\cdot\mathbf{i} = \mathbf{j}\cdot\mathbf{j} = \mathbf{k}\cdot\mathbf{k} = 1$ and $\mathbf{i}\cdot\mathbf{j} = \mathbf{j}\cdot\mathbf{k} = \mathbf{k}\cdot\mathbf{i} = 0$, for the unit vectors directed along the perpendicular axes. If we write the components of the **S** vector in matrix notation as a column, then the scalar product is given by:

$$\mathbf{B}\cdot\mathbf{S} = \begin{pmatrix} B_x & B_y & B_z \end{pmatrix}\begin{pmatrix} S_x \\ S_y \\ S_z \end{pmatrix} \tag{B.5}$$

Identity of Eqs. B.4 and B.5 defines the rule for matrix multiplication: we multiply the elements of a row (**B**) by the corresponding elements of the succeeding column (**S**) and add the resulting products. We often use the scalar product, together with tensors, for depicting spin Hamiltonians in EPR. Note that a scalar, such as the energy *E* of a spin system, has only a value and does not depend on direction.

The vector product ($\mathbf{l} = \mathbf{r}\times\mathbf{p}$) of two vectors, say the position **r** and linear momentum **p** of an orbiting electron, is a vector of length **p** times the projection of **r** on a direction perpendicular to **p**, which is directed perpendicular to the plane containing **r** and **p** with a sense that corresponds to a right-hand screw (i.e., that takes **r** into **p**) – see Figure B.2. If θ is the angle between **r** and **p**, the magnitude of the vector product is $pr\sin\theta$, which is equal to

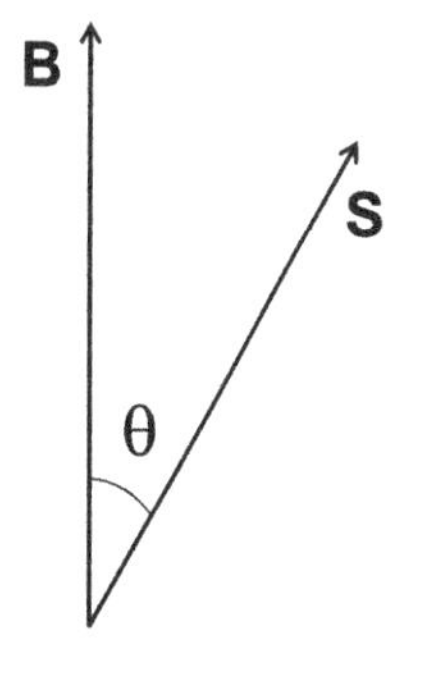

FIGURE B.1 Scalar product of magnetic-field vector, **B**, and spin vector, **S**: $\mathbf{B}\cdot\mathbf{S} = BS\cos\theta$, where *B* and *S* are magnetic-field strength and value of the spin, respectively.

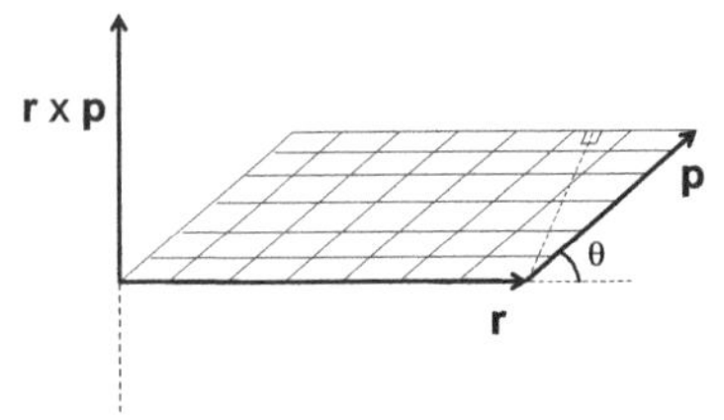

FIGURE B.2 Vector product of position vector, **r**, and linear momentum vector, **p**, is directed perpendicular to the plane defined by vectors **r** and **p**, and has magnitude $rp\sin\theta$. Note that $p\sin\theta$ is the component of **p** along the perpendicular to **r**. The direction of vector product $\mathbf{r}\times\mathbf{p}$ is that of a right-hand screw, which takes **r** into **p**.

the area of the parallelogram with sides **r** and **p**. When the sense of the vector product is changed, the sign reverses:

$$\mathbf{p}\times\mathbf{r} = -\mathbf{r}\times\mathbf{p} \tag{B.6}$$

and the product vector then points down instead of up. The Cartesian components of the vector product are given by:

$$l_x = yp_z - zp_y \tag{B.7}$$

$$l_y = zp_x - xp_z \tag{B.8}$$

$$l_z = xp_y - yp_x \tag{B.9}$$

where x, y and z are the components of **r**. Equations B.7–B.9 follow from the vector products of the unit vectors: $\mathbf{i}\times\mathbf{i} = \mathbf{j}\times\mathbf{j} = \mathbf{k}\times\mathbf{k} = 0$ and $\mathbf{i}\times\mathbf{j} = -\mathbf{j}\times\mathbf{i} = \mathbf{k}$, $\mathbf{k}\times\mathbf{i} = -\mathbf{i}\times\mathbf{k} = \mathbf{j}$, $\mathbf{j}\times\mathbf{k} = -\mathbf{k}\times\mathbf{j} = \mathbf{i}$. We can express the components of the vector product by expanding a determinant with unit vectors in the first row:

$$\mathbf{r}\times\mathbf{p} = \begin{vmatrix} \mathbf{i} & \mathbf{j} & \mathbf{k} \\ x & y & z \\ p_x & p_y & p_z \end{vmatrix}$$

$$= \mathbf{i}\begin{vmatrix} y & z \\ p_y & p_z \end{vmatrix} + \mathbf{j}\begin{vmatrix} z & x \\ p_z & p_x \end{vmatrix} + \mathbf{k}\begin{vmatrix} x & y \\ p_x & p_y \end{vmatrix} \tag{B.10}$$

Note that Eqs. B.7–B.10 for the vector $\mathbf{l} = l_x\mathbf{i} + l_y\mathbf{j} + l_z\mathbf{k}$ define the rules for multiplying out determinants. We should mention that **l**, as defined here, is the orbital angular momentum vector for an electron. The magnitude of the angular momentum about the origin of the vector **r** is r times the component of the linear momentum **p** in a direction that is perpendicular to both **r** and the rotation axis. The direction of the angular momentum vector is parallel to the rotation axis (see Figure B.2).

The triple scalar product $\mathbf{B}\cdot(\mathbf{r}\times\mathbf{p})$, such as that between a magnetic field **B** and the orbital angular momentum $\mathbf{r}\times\mathbf{p}$ of an electron, is the scalar product of **B** with the vector product of **r** and **p**. Reversing the order of the scalar product leaves the value unchanged, whereas reversing the order within the vector product changes the sign:

$$\mathbf{B}\cdot(\mathbf{r}\times\mathbf{p}) = (\mathbf{r}\times\mathbf{p})\cdot\mathbf{B} = -\mathbf{B}\cdot(\mathbf{p}\times\mathbf{r}) = -(\mathbf{p}\times\mathbf{r})\cdot\mathbf{B} \tag{B.11}$$

The triple scalar product is equal to the volume of the parallelopiped formed by the three vectors. Thus, cyclic permutations between the vector and scalar product are also possible:

$$\mathbf{B}\cdot(\mathbf{r}\times\mathbf{p}) = \mathbf{r}\cdot(\mathbf{p}\times\mathbf{B}) = \mathbf{p}\cdot(\mathbf{B}\times\mathbf{r}) \tag{B.12}$$

which we express in the compressed notation: $[\mathbf{B},\mathbf{r},\mathbf{p}] = [\mathbf{r},\mathbf{p},\mathbf{B}] = [\mathbf{p},\mathbf{B},\mathbf{r}]$. In terms of Cartesian components, the triple scalar product then becomes:

$$[\mathbf{r},\mathbf{p},\mathbf{B}] = \begin{vmatrix} x & y & z \\ p_x & p_y & p_z \\ B_x & B_y & B_z \end{vmatrix} \tag{B.13}$$

which we evaluate by multiplying out the determinant. We use the triple scalar product in Appendix F.2 to express the interaction of the orbital magnetic moment with a magnetic field and in Appendix R.1 for determining the relation between the rotation operator and angular momentum.

The triple vector product $\mathbf{a}\times(\mathbf{b}\times\mathbf{c})$ can be expanded in terms of two scalar products:

$$\mathbf{a}\times(\mathbf{b}\times\mathbf{c}) = (\mathbf{a}\cdot\mathbf{c})\mathbf{b} - (\mathbf{a}\cdot\mathbf{b})\mathbf{c} \tag{B.14}$$

We use this relation in the following section, for the curl of a vector product.

B.2 Vector Calculus

An important class of vector is that generated by the action of the gradient differential operator ∇ on a scalar, such as a wave function ψ, electrostatic potential V, or concentration in diffusion. An example used much in magnetic resonance is the quantum mechanical operator for the linear momentum (see Appendix C.3):

$$\hat{\mathbf{p}} \equiv -i\hbar\nabla\psi = (-i\hbar)\left(\mathbf{i}\frac{\partial}{\partial x} + \mathbf{j}\frac{\partial}{\partial y} + \mathbf{k}\frac{\partial}{\partial z}\right)\psi \tag{B.15}$$

where $\hbar$ is Planck's constant divided by 2π and $i = \sqrt{-1}$. (Note that $i^2 = -1$ is not to be confused with the square of the unit vector: $\mathbf{i}\cdot\mathbf{i} = |\mathbf{i}|^2 = +1$.) An indispensable differential operator is the scalar product of the gradient operator with itself:

$$\nabla^2 \equiv \nabla\cdot\nabla = \frac{\partial^2}{\partial x^2} + \frac{\partial^2}{\partial y^2} + \frac{\partial^2}{\partial z^2} \tag{B.16}$$

We call this operator the Laplacian. It appears in the quantum mechanical Schrödinger equation, in the classical diffusion equation, and in the Laplace and Poisson equations of electrostatics. In polar coordinates (r,θ,ϕ), the angular part of the ∇^2 operator appears equivalently in the rotational diffusion equation and in the quantum mechanical expression for angular momentum.

Note that strictly speaking ∇^2 is the divergence of the gradient – see Eq. B.17 immediately following.

The scalar differential (or overall rate of change) of a vector, such as the electric displacement **D**, is the divergence operator that is the scalar product with ∇:

$$\mathrm{div}\mathbf{D} \equiv \nabla \cdot \mathbf{D} = \frac{\partial D_x}{\partial x} + \frac{\partial D_y}{\partial y} + \frac{\partial D_z}{\partial z} \tag{B.17}$$

We use the divergence to express the equation of continuity for a general flux density, e.g., $\nabla \cdot \mathbf{D} = 0$. This forms the basis for Gauss' theorem in electrostatic calculations (see Section 4.2.2 in Chapter 4).

The vector differential of a vector, such as the magnetic vector potential **A**, is the curl (or rotation) operator which is the vector product with ∇:

$$\mathrm{curl}\mathbf{A} \equiv \nabla \times \mathbf{A} = \begin{vmatrix} \mathbf{i} & \mathbf{j} & \mathbf{k} \\ \frac{\partial}{\partial x} & \frac{\partial}{\partial y} & \frac{\partial}{\partial z} \\ A_x & A_y & A_z \end{vmatrix}$$

$$= \mathbf{i}\left(\frac{\partial A_z}{\partial y} - \frac{\partial A_y}{\partial z}\right) + \mathbf{j}\left(\frac{\partial A_x}{\partial z} - \frac{\partial A_z}{\partial x}\right) + \mathbf{k}\left(\frac{\partial A_y}{\partial x} - \frac{\partial A_x}{\partial y}\right) \tag{B.18}$$

We use the curl to introduce the magnetic field into quantum mechanics, via the identity: $\mathbf{B} = \mathrm{curl}\mathbf{A}$, in Appendix F.1. The curl of a vector product $\mathbf{a} \times \mathbf{b}$ is given by the product law with separate derivatives of **a** and **b**. These are then expanded by the expression for the triple vector product (Eq. B.14):

$$\nabla \times (\mathbf{a} \times \mathbf{b}) = \nabla_{\mathbf{a}} \times (\mathbf{a} \times \mathbf{b}) + \nabla_{\mathbf{b}} \times (\mathbf{a} \times \mathbf{b})$$

$$= (\mathbf{b} \cdot \nabla)\mathbf{a} - (\nabla \cdot \mathbf{a})\mathbf{b} + (\nabla \cdot \mathbf{b})\mathbf{a} - (\mathbf{a} \cdot \nabla)\mathbf{b} \tag{B.19}$$

where $\mathbf{b} \cdot \nabla \equiv \nabla_{\mathbf{a}} \cdot \mathbf{b}$ and $\mathbf{a} \cdot \nabla \equiv \nabla_{\mathbf{b}} \cdot \mathbf{a}$. Note that: $\mathbf{a} \cdot \nabla \equiv a_x \partial/\partial x + a_y \partial/\partial y + a_z \partial/\partial z$, etc. We use Eq. B.19 for deriving the dipolar field in Appendix F.4.

B.3 Matrices

We introduced row and column matrices already in Eq. B.5 for the scalar product of two vectors. A square matrix combines both rows and columns. We can use it to depict a series of simultaneous linear equations that contain terms similar to those appearing on the right of Eq. B.4. For instance, the transformation from one set of coordinate axes given by the vector **x** to a new set **x′** uses the rotation matrix **R**: $\mathbf{x}' = \mathbf{R}\mathbf{x}$. Explicitly, rotation of the x, y, z system by an angle ϕ about the z-axis is given by:

$$\begin{pmatrix} x' \\ y' \\ z' \end{pmatrix} = \begin{pmatrix} \cos\phi & \sin\phi & 0 \\ -\sin\phi & \cos\phi & 0 \\ 0 & 0 & 1 \end{pmatrix} \begin{pmatrix} x \\ y \\ z \end{pmatrix} \tag{B.20}$$

– see Eqs. I.4 and I.20 in Appendix I.

An important application of matrices is to the eigenvalue equations of quantum mechanics, particularly the Schrödinger equation: $\mathcal{H}\psi = E\psi$ (see Appendices C and D). If we express the wave function, ψ, in terms of a complete orthogonal set of basis functions, φ_n, the Schrödinger equation in matrix form becomes:

$$\begin{pmatrix} \mathcal{H}_{11} - E_1 & \mathcal{H}_{12} & \mathcal{H}_{13} \\ \mathcal{H}_{21} & \mathcal{H}_{22} - E_2 & \mathcal{H}_{23} \\ \mathcal{H}_{31} & \mathcal{H}_{32} & \mathcal{H}_{33} - E_3 \end{pmatrix} \begin{pmatrix} \varphi_1 \\ \varphi_2 \\ \varphi_3 \end{pmatrix} = 0 \tag{B.21}$$

where $\mathcal{H}_{ij}$ are the matrix elements of the Hamiltonian $\mathcal{H}$ referred to the basis vectors φ_n (see Appendix C.2). For simplicity, we assume here that the three basis functions, φ_1, φ_2 and φ_3, form a complete set. Because the Hamiltonian represents a real observable, viz., the energy of the system, it must be Hermitian and the off-diagonal matrix elements are related by: $\mathcal{H}_{21} = \mathcal{H}_{12}^*$, et seq., where the asterisk indicates a complex conjugate. (If $z = a + ib$, where a and b are real and $i = \sqrt{-1}$, the complex conjugate is $z^* = a - ib$.) We determine the eigenvalues of the Schrödinger equation, (i.e., the energy levels, E_i) by bringing the Hamiltonian matrix of Eq. B.21 into a diagonal form. To do this, we refer the Hamiltonian to suitable combinations of the φ_n, which then become the eigenfunctions. We explain how later, in Appendix G.

B.4 Tensors

If a vector property of a spin system, say the magnetic moment **μ**, depends not only on the magnitude of the spin components (S_x, S_y, S_z) but also on their vectorial direction **S**, then the constant of proportionality, the g-value $(\mu = \beta_e \mathbf{g} \cdot \mathbf{S})$, does not simply have a single value. Instead, the g-factor is a tensor that we depict by a square matrix:

$$\begin{pmatrix} \mu_x \\ \mu_y \\ \mu_z \end{pmatrix} = \beta_e \begin{pmatrix} g_{xx} & 0 & 0 \\ 0 & g_{yy} & 0 \\ 0 & 0 & g_{zz} \end{pmatrix} \begin{pmatrix} S_x \\ S_y \\ S_z \end{pmatrix} \tag{B.22}$$

where the g-tensor is diagonal for a suitable choice of the molecular axes, but with any other than this principal-axis system, some off-diagonal elements are non-zero. The energy of the magnetic moment in a magnetic field **B** is given by the scalar product (cf. Eq. B.3): $\mathcal{H}_{mag} = -\boldsymbol{\mu} \cdot \mathbf{B} = \beta_e \mathbf{B} \cdot \mathbf{g} \cdot \mathbf{S}$, which written in matrix form according to Eq. B.5 becomes:

$$\mathcal{H}_{mag} = \beta_e \begin{pmatrix} B_x & B_y & B_z \end{pmatrix} \begin{pmatrix} g_{xx} & 0 & 0 \\ 0 & g_{yy} & 0 \\ 0 & 0 & g_{zz} \end{pmatrix} \begin{pmatrix} S_x \\ S_y \\ S_z \end{pmatrix} \tag{B.23}$$

where we refer all vectors to the principal axis system of the g-tensor, viz. the nitroxide x,y,z-axes. We can refer tensors to other axis systems by using rotation matrices, **R**, of the type introduced in Eq. B.20. Referred to the axis system $\mathbf{x}'$, the spin vector is simply $\mathbf{S}' = \mathbf{R}\cdot\mathbf{S}$.

We write the magnetic field vector as a row in Eq. B.23 and, therefore, it is the transpose, $\mathbf{B}^T$, of the column vector $\mathbf{B}$. We obtain the transpose of a matrix by exchanging rows for columns, in order. The transpose of a square matrix corresponds to reflecting the elements about the leading diagonal. In the new coordinate system, the transpose of $\mathbf{B}$ is: $\mathbf{B}'^T = (\mathbf{R}\cdot\mathbf{B})^T = \mathbf{B}^T\cdot\mathbf{R}^T = \mathbf{B}^T\cdot\mathbf{R}^{-1}$ because the transpose of a product is equal to the product of the individual transposes but in the reverse order (we can demonstrate this by direct multiplication). For rotation matrices (or any other unitary matrix) the inverse is equal to the transpose: $\mathbf{R}^T = \mathbf{R}^{-1}$. We see this from the form of Eq. B.20, where the transpose reverses the sense of the rotation by exchanging ϕ with $-\phi$. Thus we can develop the Hamiltonian Eq. B.23 in operator form:

$$\mathcal{H}_{mag} = \beta_e\mathbf{B}^T\cdot\mathbf{g}\cdot\mathbf{S} \equiv \beta_e\mathbf{B}^T\cdot\mathbf{R}^{-1}\mathbf{R}\cdot\mathbf{g}\cdot\mathbf{R}^{-1}\mathbf{R}\mathbf{S}$$

$$= \beta_e\mathbf{B}'^T\cdot\left(\mathbf{R}\cdot\mathbf{g}\cdot\mathbf{R}^{-1}\right)\cdot\mathbf{S}' \tag{B.24}$$

The g-tensor is therefore given in the new (primed) coordinate system by:

$$\mathbf{g}' = \mathbf{R}\cdot\mathbf{g}\cdot\mathbf{R}^{-1} \tag{B.25}$$

We use this type of unitary transformation extensively in magnetic resonance, as an alternative way to diagonalize the spin Hamiltonian. Unitary transformations are particularly well adapted to handling time-dependent problems, e.g., Eq. D.7 for the Heisenberg representation in Appendix D, or the interaction representation for the spin density-matrix operator in Eq. N.21 of Appendix N.

Coordinate transformations like Eq. B.25 are used in perturbation theory, where we must rotate the tensors of the perturbation Hamiltonian to the static field direction. Important applications are in relaxation calculations (Chapter 6) and in hyperfine spectroscopy, ESEEM and ENDOR (Chapter 14). We illustrate here with the axial hyperfine interaction:

$$\mathcal{H}_{hfs} = \mathbf{I}'^T\cdot\mathbf{A}\cdot\mathbf{S}' = A_{\parallel}I_{z'}S_{z'} + A_{\perp}I_{x'}S_{x'} \tag{B.26}$$

where the principal z'-axis of the hyperfine interaction is inclined at an angle θ to the static magnetic field, which lies along the (unprimed) z-direction. The rotation matrix is then (see Eqs. I.2–I.4 in Appendix I):

$$\mathbf{R}(\theta, y) = \begin{pmatrix} \cos\theta & \sin\theta \\ -\sin\theta & \cos\theta \end{pmatrix} \tag{B.27}$$

In the magnetic-field axis system (unprimed), the hyperfine Hamiltonian is given by:

$$\mathcal{H}_{hfs} = \mathbf{I}^T\cdot\left(\mathbf{R}\cdot\mathbf{A}\cdot\mathbf{R}^{-1}\right)\cdot\mathbf{S} \tag{B.28}$$

For perturbation theory, we retain only the S_z-terms and the resulting pseudosecular Hamiltonian becomes:

$$\mathcal{H}_{hfs} = \left(A_{\parallel}\cos^2\theta + A_{\perp}\sin^2\theta\right)I_zS_z + \left(A_{\parallel} - A_{\perp}\right)\sin\theta\cos\theta\cdot I_xS_z \tag{B.29}$$

We give a more general version of this transformation that includes both g-value anisotropy and non-secular terms in Chapter 5 (viz. Eq. 5.15).

In ESEEM or ENDOR, we need to include also the nuclear electric quadrupole interaction for nuclei with spins $I \geq 1$. Expressed in its principal axis system (double-primed), the spin Hamiltonian for an axial nuclear quadrupole interaction is given by Eq. 3.45 of Chapter 3:

$$\mathcal{H}_Q = \frac{3e^2qQ}{4I(2I-1)}I_{z''}^2 \tag{B.30}$$

where the quadrupole principal axis z'' is inclined at angle θ_Q to the magnetic field direction. We now must relate $I_{z''}$ to the nuclear spin components in the magnetic field axis system (see Eq. B.27):

$$I_{z''} = I_z\cos\theta_Q + I_x\sin\theta_Q \tag{B.31}$$

Substituting in Eq. B.30 and retaining only terms in I_z^2, we get:

$$\mathcal{H}_Q = \frac{3e^2qQ}{8I(2I-1)}\left(3\cos^2\theta_Q - 1\right)I_z^2 \tag{B.32}$$

where we use the replacement: $I_x^2 \rightarrow \frac{1}{2}\left(I^2 - I_z^2\right)$ that is appropriate to axial symmetry (i.e., $I_y^2 = I_x^2$), to get the full contribution from the I_z^2 spin operator.

APPENDIX C: QUANTUM MECHANICAL BASICS

We use quantum mechanics to describe the quantization of experimental observables. The most important of these are the energy of the system, and also – from our point of view – magnetic moments and spin. Our formulation must take into account the wave-particle duality of electrons and other elementary particles, on the one hand, and of electromagnetic radiation, i.e., photons, on the other. We do this by using Schrödinger's wave mechanics. In the limit of macroscopic systems, we must obtain the results of classical physics. We call this the correspondence principle, which we get by notionally letting Planck's constant, h, tend to zero. The correspondence principle is important for choosing the form of the quantum mechanical operators.

C.1 Basic Postulates

In this section, we introduce the Schrödinger wave-mechanical formalism (although the Heisenberg operator form of quantum mechanics is useful for appreciating certain aspects of the dynamics of spins in magnetic resonance). The wave-mechanical method depends on a series of postulates that we summarize below.

1. Each quantum-mechanical state, e.g., of an electron, is represented by a state function, or wave function, ψ. The square of the absolute value of the wave function, $|\psi|^2 = \psi^* \psi$, at a particular point is the probability density (per unit volume) of finding the particle at that point. (Here the asterisk indicates the complex conjugate: for complex variable $z = a + ib$, where a and b are real and $i = \sqrt{-1}$, the complex conjugate is $z^* = a - ib$.) Usually, we normalize the wave function to unit probability by integrating over all space: i.e., $\int \psi^* \psi \cdot \mathrm{d}v = 1$, where $\mathrm{d}v = \mathrm{d}x\mathrm{d}y\mathrm{d}z$ is the volume element.
2. Each observable of a quantum system, e.g., the energy or angular momentum, is represented by an operator, $\hat{A}$ (e.g., the differentiation operation: $\mathrm{d}/\mathrm{d}x$).
3. The only allowed values of an observable $\hat{A}$, e.g., the energy or spin projection, are the eigenvalues a_n of the operator equation:

$$\hat{A}\varphi_n = a_n \varphi_n \tag{C.1}$$

where φ_n are the eigenfunctions of the operator $\hat{A}$. For the eigenvalues a_n to be real, the operator $\hat{A}$ must be Hermitian, i.e., equal to its Hermitian conjugate $\hat{A}^\dagger$. This condition is defined by: $\int \varphi_m^* \hat{A} \varphi_n \cdot \mathrm{d}v = \int \varphi_n^* \hat{A}^* \varphi_m \cdot \mathrm{d}v$, where the integral is over all space. The eigenfunctions of a Hermitian operator are orthogonal, i.e., $\int \varphi_m^* \varphi_n \cdot \mathrm{d}v = 0\,(m \neq n)$.

4. The expectation value of an observable $\hat{A}$ for a quantum state ψ is given by:

$$\langle \hat{A} \rangle = \frac{\int \psi^* \hat{A} \psi \cdot \mathrm{d}v}{\int \psi^* \psi \cdot \mathrm{d}v} \tag{C.2}$$

If the state function ψ is not an eigenfunction φ_n of $\hat{A}$, we can expand it as a linear combination of these eigenfunctions:

$$\psi = \sum_n c_n \varphi_n \tag{C.3}$$

The expectation value of $\hat{A}$ then becomes:

$$\langle \hat{A} \rangle = \frac{\sum_n |c_n|^2 a_n}{\sum_n |c_n|^2} \tag{C.4}$$

i.e., $|c_n|^2$ is the weight of the eigenvalue a_n in state ψ. Normalization of the wave function ψ means that $\sum_n |c_n|^2 = 1$.

C.2 Representation of Quantum-Mechanical Operators

At this point, it is useful to introduce the Dirac notation for the eigenfunctions and expectation values. The eigenfunction labelled by a set of quantum numbers, n, is given by:

$$\varphi_n \equiv |n\rangle \tag{C.5}$$

and the complex conjugate is given by:

$$\varphi_n^* \equiv \langle n| \tag{C.6}$$

The expectation value of $\hat{A}$ is then given by:

$$\langle \hat{A} \rangle \equiv \langle \psi | \hat{A} | \psi \rangle = \sum_n c_n^* c_n \langle n | \hat{A} | n \rangle = \sum_n |c_n|^2 a_n^2 \tag{C.7}$$

where we imply integration over all space automatically in $\langle \psi | \hat{A} | \psi \rangle$. We assume the eigenfunctions in Eq. C.7 to be normalized:

$$\langle n | n \rangle = 1 \tag{C.8}$$

and orthogonality of the eigenfunctions is given by:

$$\langle n | m \rangle = 0 \quad n \neq m \tag{C.9}$$

Thus we can summarize orthogonality and normalization in shorthand notation by using the Kronecker δ: $\langle n|m\rangle = \delta_{n,m}$ (where $\delta_{n,m} = 1$ for $n = m$ and is zero otherwise).

In the treatment above, we assume that the eigenfunctions φ_n of the operator $\hat{A}$ are known. If not, we can always expand the wave function ψ in terms of the eigenfunctions, say $u_n \equiv |n\rangle$, of some other operator provided that these constitute a complete orthonormal set for the problem in hand:

$$|\psi\rangle = \sum_n c_n |n\rangle = \sum_n \langle n|\psi\rangle |n\rangle \tag{C.10}$$

where $\langle n|\psi\rangle \equiv \int u_n^* \psi \cdot \mathrm{d}v$ is the overlap integral between ψ and u_n. In EPR, the eigenfunctions that we use for the expansion are those of the electron spin angular momentum (see Appendix E.3). The values $\langle n|\hat{A}|n\rangle \equiv \int u_n^* \hat{A} u_n \cdot \mathrm{d}v$ are the diagonal elements, A_{nn}, in a matrix representation of the operator $\hat{A}$ (see Section B.3 of Appendix B):

$$\hat{\mathbf{A}} = \begin{pmatrix} A_{11} & A_{12} & \cdots & A_{1n} \\ A_{21} & A_{22} & & \\ \vdots & & \ddots & \\ A_{n1} & & & A_{nn} \end{pmatrix} \tag{C.11}$$

where the off-diagonal elements are defined by:

$$A_{mn} \equiv \langle m|\hat{\mathbf{A}}|n\rangle \equiv \int u_m^* \hat{\mathbf{A}} u_n \cdot \mathrm{d}v \tag{C.12}$$

for normalized eigenfunctions. Matrix $\hat{\mathbf{A}}$ is Hermitean, i.e., corresponds to a real observable, if $A_{mn} = A_{nm}^*$. Solution of the eigenvalue equation Eq. C.1 involves reducing the matrix $\hat{\mathbf{A}}$ to a form in which only the diagonal elements, A_{nn}, are non-zero. Therefore, we call this process diagonalization – in our case, diagonalization of the spin Hamiltonian. We give methods for matrix diagonalization later in Appendix G on perturbation theory and illustrate them in Appendix H for various spin Hamiltonians.

C.3 Correspondence and Complementarity Principles

The above completes definition of the operator method. It remains to derive the form of the operators for the different physical variables. The correspondence principle requires that the relations between physical variables in classical physics are satisfied by the corresponding quantum mechanical operators. For example, the z-component of the orbital angular momentum is given by (see Eq. B.9):

$$l_z = xp_y - yp_x \tag{C.13}$$

which applies to classical mechanics and also holds true for the corresponding operators in quantum mechanics.

The complementarity principle is the expression of classical variables, such as the position, x, and the linear momentum, p_x, in terms of quantum mechanical operators. In the Schrödinger representation, the position operator remains simply as the variable x, but the linear momentum is replaced by the differential operator:

$$p_x \rightarrow -i\hbar \frac{\partial}{\partial x} \tag{C.14}$$

where $\hbar$ is Planck's constant divided by 2π (i.e., $\hbar \equiv h/2\pi$). The form of Eq. C.14 ensures that the expectation value for linear momentum of an elementary particle corresponds to the de Broglie wave ($\psi \approx e^{ix/\lambda\!\!\!^{-}}$, where $\lambda\!\!\!^{-}$ is the de Broglie wavelength divided by 2π).

An immediate consequence of the Schrödinger representation is that the operators for position and momentum do not commute:

$$[x, p_x] \equiv xp_x - p_x x = i\hbar \tag{C.15}$$

Observation of the position, x, disturbs subsequent observation of the linear momentum, p_x, and vice-versa. For Gaussian wave packets, this commutation relation leads directly to the uncertainty principle:

$$\Delta p_x \Delta x \geq \hbar \tag{C.16}$$

where the widths Δp_x and Δx of the Gaussian packets are the uncertainties in measurement of momentum and position, respectively.

APPENDIX D: SCHRÖDINGER EQUATION AND HEISENBERG EQUATION OF MOTION

D.1 Schrödinger Equation

The time-independent Schrödinger equation is the eigenvalue equation for the total energy. The Hamiltonian operator for the energy of a particle is:

$$\mathcal{H} = T + V = \frac{\mathbf{p}^2}{2m} + V \tag{D.1}$$

where T is the kinetic energy, V is the potential energy, $\mathbf{p}$ is the linear momentum and m is the mass of the particle. Using the three-dimensional generalization of Eq. C.14 for the operator $\mathbf{p}$, the time-independent Schrödinger equation becomes:

$$\mathcal{H}\psi = \frac{-\hbar^2}{2m}\nabla^2\psi + V\psi = E\psi \tag{D.2}$$

where E is the eigenvalue for the total energy of the particle and ψ is the corresponding eigenfunction. The operator ∇^2 stands for $\partial^2/\partial x^2 + \partial^2/\partial y^2 + \partial^2/\partial z^2$ in Cartesian coordinates (see Eq. B.16 in Appendix B). For $V = 0$, Eq. D.2 is simply the differential equation for a standing wave. Hence this is referred to as the wave equation and ψ is the wave function.

We obtain the time-dependent Schrödinger equation from Eq. D.2 by making the operator substitution:

$$E \rightarrow i\hbar \frac{\partial}{\partial t} \tag{D.3}$$

This operator replacement ensures that a plane wave ($\psi \approx e^{-i\omega t}$, where ω is the angular frequency) satisfies Planck's equation for quantization of the energy of electromagnetic radiation: $E = \hbar\omega$. The time-dependent Schrödinger equation then becomes:

$$i\hbar\frac{\partial\psi}{\partial t} = \mathcal{H}\psi = \frac{-\hbar^2}{2m}\nabla^2\psi + V\psi \tag{D.4}$$

For $V = 0$, this is not identical with the classical differential equation for wave propagation, which is second order in t. The quantum-mechanical matter wave is therefore dispersive, i.e., changes wavelength as it propagates.

D.2 Heisenberg Equation of Motion

A formal solution of the time-dependent Schrödinger equation (Eq. D.4) is:

$$\psi(t) = e^{-i\hat{\mathcal{H}}t/\hbar}|\psi\rangle \tag{D.5}$$

where $|\psi\rangle$ is the normalized eigenfunction of the time-independent Schrödinger equation, i.e., Eq. D.2. Note that an exponential operator is defined by the series expansion: $e^x = 1 + x + x^2/2! + x^3/3! + ...$, where we define the factorials by, e.g.,: $3! = 1 \times 2 \times 3$. The expectation value for an operator $\hat{A}$ is then given from Eq. D.5 by:

$$\langle \hat{A}(t)\rangle = \langle \psi(t)|\hat{A}|\psi(t)\rangle = \langle \psi|e^{i\hat{\mathcal{H}}t/\hbar}\hat{A}e^{-i\hat{\mathcal{H}}t/\hbar}|\psi\rangle \tag{D.6}$$

This allows us to define the time-dependent operator:

$$\hat{A}(t) = e^{i\hat{\mathcal{H}}t/\hbar}\hat{A}e^{-i\hat{\mathcal{H}}t/\hbar} \tag{D.7}$$

Here, we have transferred the time dependence of the wave function in the Schrödinger representation to a time-dependent operator for the observable $\hat{A}$. We call this the Heisenberg representation. In the context of the density matrix and time-dependent perturbations, it becomes the interaction representation (see Appendix N.4). Differentiating Eq. D.7 with respect to time leads to the Heisenberg equation of motion:

$$i\hbar\frac{\partial \hat{A}(t)}{\partial t} = \left[\hat{A}(t), \mathcal{H}\right] \tag{D.8}$$

where the square brackets represent the commutator as introduced in Eq. C.15: $\left[\hat{A}, \mathcal{H}\right] \equiv \hat{A}\mathcal{H} - \mathcal{H}\hat{A}$. We call operators that commute with the Hamiltonian constants of motion, because then: $\partial \hat{A}/\partial t = 0$. Examples are the square S^2, and the z-projection S_z, of the electron spin (see Appendix E.3, later).

The Heisenberg equation of motion is important, because it provides a quantum mechanical justification for the classical equations of motion of the magnetic moment associated with electron or nuclear spins. We use this classical approach frequently to describe magnetic resonance by the Bloch equations. For a static magnetic field, B_z, directed along the z-direction, the spin Hamiltonian is:

$$\mathcal{H}_S = -\hbar\gamma_e B_z S_z \tag{D.9}$$

where γ_e is the gyromagnetic ratio of the electron – see Chapter 1. Note that $\mathcal{H}_S$ commutes with the spin-independent part of the Hamiltonian $\mathcal{H}_o$. The equation of motion for the expectation value of the x-component of the spin is then, from Eq. D.8:

$$\frac{dS_x}{dt} = -\frac{i}{\hbar}\left[S_x, \mathcal{H}_S\right] = i\gamma_e B_z\left[S_x, S_z\right] = \gamma_e B_z S_y \tag{D.10}$$

(see Eq. E.5 in Appendix E for spin commutation rules), and similarly:

$$\frac{dS_y}{dt} = -\gamma_e B_z S_x \tag{D.11}$$

$$\frac{dS_z}{dt} = 0 \tag{D.12}$$

Now Eqs. D.10–D.12 are just the components of the classical equation of motion for the precession of the electron spin magnetization magnetization ($\mathbf{M} \approx \gamma_e\hbar\mathbf{S}$, see Eq. 1.1) about the static field $\mathbf{B}$:

$$\frac{d\mathbf{M}}{dt} = \gamma_e\mathbf{M} \times \mathbf{B} \tag{D.13}$$

as discussed in Section 6.2 of Chapter 6. Because of this correspondence, the phenomenological Bloch equations prove very useful for interpreting spin-label EPR experiments. Now, of course, the macroscopic magnetization $\mathbf{M}$ represents an average over the ensemble of electron spins $\mathbf{S}$.

APPENDIX E: ANGULAR-MOMENTUM/ SPIN OPERATORS AND THEIR MATRIX ELEMENTS

E.1 Orbital Angular Momentum

Classically, the angular momentum, $\mathbf{l}$, of an orbiting electron is given by the vector product:

$$\mathbf{l} = \mathbf{r} \times \mathbf{p} \tag{E.1}$$

where $\mathbf{r}$ is the position and $\mathbf{p}$ the linear momentum of the electron (see Figure B.2). The x,y,z components of the angular momentum are therefore:

$$l_x = yp_z - zp_y \tag{E.2}$$

and cyclic permutations of x,y,z for l_y and l_z (see Eqs. B.7–B.9). We get quantum mechanical expressions for the orbital angular momentum from Eqs. E.2 by using the usual operator replacements for the linear momentum: $p_y \rightarrow -i\hbar(\partial/\partial y)$, et seq. (see Eq. C.14 in Appendix C). In polar coordinates, the angular momentum about the z-axis then takes a particularly simple form:

$$l_z = -i\hbar(x\partial/\partial y - y\partial/\partial x) = -i\hbar\partial/\partial\phi \tag{E.3}$$

where ϕ is the azimuthal angle about the z-axis and similar expressions apply for azimuthal angles about the x- and y-axes.

E.2 Commutation Relations

Because the quantum mechanical angular momentum depends on differential operators such as $\partial/\partial x$, et seq., the order in which we apply operators corresponding to l_x and l_y is important.

They do not commute. We can derive commutation relations between the components of the angular momentum from the well-known relations for the position and linear momentum, such as Eq. C.15 in Appendix C. The result is:

$$\left[l_x, l_y\right] \equiv l_x l_y - l_y l_x = i\hbar l_z \tag{E.4}$$

and cyclic permutations, $x \to y \to z \to x$. The commutation relations given in Eq. E.4 represent a fundamental property of angular momentum in quantum mechanics. In fact, we can take them as a more general definition of angular momentum, including both the electron and nuclear spins.

We therefore define spin angular momentum by commutation relations for the spin components, S_x, S_y and S_z, that are of the form:

$$\left[S_x, S_y\right] = iS_z \tag{E.5}$$

and cyclic permutations. We omit the factor of $\hbar$ on the right of Eq. E.4 in this expression, because spin is usually expressed in units of $\hbar$. This is taken into account, for instance, in the definition of the Bohr magneton $(\beta_e = e\hbar/2m_e)$, which relates the magnetic moment to the spin (see Chapter 1 and Appendix F). Whereas we express all that follows in terms of the electron spin, $\mathbf{S}$ – as is appropriate to EPR – we should remember that all relations apply equally to angular momentum in general, including also nuclear spins.

E.3 Eigenvalues, Raising/Lowering Operators and Matrix Elements

Both the square of the total spin, S^2, and its z-component, S_z, are well defined in quantum mechanics, but the S_x and S_y components are not. That is to say, both S^2 and S_z have well-defined eigenvalues. Because S^2 and S_z commute they also share common eigenfunctions, which we label with the quantum numbers S and M_S that refer to the total spin and its z-projection, respectively. Both operators are diagonal in this representation and their matrix elements (in this case the eigenvalues) are given by:

$$\langle S, M_S | S^2 | S, M_S \rangle = S(S+1) \tag{E.6}$$

$$\langle S, M_S | S_z | S, M_S \rangle = M_S \tag{E.7}$$

where the magnetic quantum number is $M_S = -S, -S+1, \ldots, S-1, +S$ for a given S, and the spin quantum number is a positive integer or half-integer. In particular, for a single electron: $S = \frac{1}{2}$ and $M_S = -\frac{1}{2}, +\frac{1}{2}$.

Eqs. E.6 and E.7 follow from the definition of spin angular momentum that we give in Eq. E.5. The definition of orbital angular momentum, Eq. E.1, is more restrictive. This results in eigenvalues of l^2 and l_z analogous to those for S^2 and S_z, except that the quantum number l for the total orbital angular momentum can take only whole-integer values. Half-integer values of angular momentum are allowed only for the spin. The eigenfunctions for whole-integer quantum numbers, i.e., for the orbital angular momentum, are the well-known spherical harmonics, $Y_{l,m_l}(\theta,\phi)$ (see Table K.1 in Appendix K).

The S_x- and S_y-components of the spin have only off-diagonal matrix elements. They connect spin states that differ in M_S by ± 1. We define the shift operators, S_+ and S_-, because these only raise or lower, respectively, the value of M_S of the spin state on which they operate:

$$S_\pm = S_x \pm iS_y \tag{E.8}$$

From Eq. E.5, the corresponding commutation relations for the raising and lowering operators are:

$$\left[S_z, S_\pm\right] = \pm S_\pm \tag{E.9}$$

$$\left[S_\pm, S_\mp\right] = \pm 2S_z \tag{E.10}$$

The only non-zero matrix elements of the raising and lowering operators are:

$$\langle S, M_S+1 | S_+ | S, M_S \rangle = \sqrt{S(S+1) - M_S(M_S+1)} \tag{E.11}$$

$$\langle S, M_S-1 | S_- | S, M_S \rangle = \sqrt{S(S+1) - (M_S-1)M_S} \tag{E.12}$$

where for $S = \frac{1}{2}$ both values are +1. We then easily obtain the matrix elements of S_x and S_y from their relation to the shift operators:

$$S_x = \tfrac{1}{2}(S_+ + S_-) \tag{E.13}$$

$$S_y = \frac{-i}{2}(S_+ - S_-) \tag{E.14}$$

It is worthwhile mentioning that we can obtain the central results given in Eqs. E.6 and E.7, and in Eqs. E.11 and E.12, most readily by applying the shift operators, S_+ and S_-, together with the commutation relations in Eqs. E.9 and E.10.

We list matrices for some angular momentum operators that are useful in spin-label EPR in Table E.1.

E.4 Rigid Rotator: A Classical Example

The Hamiltonian for the rotational kinetic energy of a rigid body is:

$$\mathcal{H} = \frac{\hbar^2}{2}\left(\frac{L_x^2}{I_{xx}^{inert}} + \frac{L_y^2}{I_{yy}^{inert}} + \frac{L_z^2}{I_{zz}^{inert}}\right) \tag{E.15}$$

where $\mathbf{L} = (L_x, L_y, L_z)$ is the angular momentum (see Eq. E.1) and $(I_{xx}^{inert}, I_{yy}^{inert}, I_{zz}^{inert})$ are the moments of inertia, about the x,y,z-axes of the rigid body. For a symmetrical top, x- and y-axes are equivalent: $I_{xx}^{inert} = I_{yy}^{inert} \equiv I_\perp^{inert}$ and $I_{zz}^{inert} \equiv I_\parallel^{inert}$. Now, because the square of the total angular momentum is $L^2 = L_x^2 + L_y^2 + L_z^2$, we have: $L_x^2 + L_y^2 = L^2 - L_z^2$. Therefore, the Hamiltonian for the symmetric top becomes:

TABLE E.1 Angular-Momentum Matrices for $S=\frac{1}{2}$, I = 1, and *p*-orbitals

$S=\frac{1}{2}$	S_x		S_y		S_z	
M_S	$\lvert+\frac{1}{2}\rangle$	$\lvert-\frac{1}{2}\rangle$	$\lvert+\frac{1}{2}\rangle$	$\lvert-\frac{1}{2}\rangle$	$\lvert+\frac{1}{2}\rangle$	$\lvert-\frac{1}{2}\rangle$
$\langle+\frac{1}{2}\rvert$	0	+1/2	0	$-i/2$	+1/2	0
$\langle-\frac{1}{2}\rvert$	+1/2	0	$+i/2$	0	0	−1/2

$I=1$	I_x			I_y			I_z		
	$\lvert+1\rangle$	$\lvert 0\rangle$	$\lvert-1\rangle$	$\lvert+1\rangle$	$\lvert 0\rangle$	$\lvert-1\rangle$	$\lvert+1\rangle$	$\lvert 0\rangle$	$\lvert-1\rangle$
$\langle+1\rvert$	0	$1/\sqrt{2}$	0	0	$-i/\sqrt{2}$	0	+1	0	0
$\langle 0\rvert$	$1/\sqrt{2}$	0	$1/\sqrt{2}$	$i/\sqrt{2}$	0	$-i/\sqrt{2}$	0	0	0
$\langle-1\rvert$	0	$1/\sqrt{2}$	0	0	$i/\sqrt{2}$	0	0	0	−1

$l=1$	l_x			l_y			l_z		
p_i	$\lvert p_x\rangle$	$\lvert p_y\rangle$	$\lvert p_z\rangle$	$\lvert p_x\rangle$	$\lvert p_y\rangle$	$\lvert p_z\rangle$	$\lvert p_x\rangle$	$\lvert p_y\rangle$	$\lvert p_z\rangle$
$\langle p_x\rvert$	0	0	0	0	0	i	0	$-i$	0
$\langle p_y\rvert$	0	0	$-i$	0	0	0	i	0	0
$\langle p_z\rvert$	0	i	0	$-i$	0	0	0	0	0

Note that the p-orbital wave functions are: $\lvert p_x\rangle = \frac{1}{\sqrt{2}}(\lvert-1\rangle - \lvert+1\rangle)$, $\lvert p_y\rangle = \frac{i}{\sqrt{2}}(\lvert-1\rangle + \lvert+1\rangle)$, and $\lvert p_z\rangle = \lvert 0\rangle$, when given in terms of the $\lvert m_l\rangle$ wave functions that are specified by the magnetic quantum number, $m_l = 0, \pm 1$ (see Table K.1 in Appendix K).

$$\mathcal{H} = \frac{\hbar^2}{2}\left(\frac{L_x^2+L_y^2}{I_\perp^{inert}} + \frac{L_z^2}{I_\parallel^{inert}}\right) = \frac{\hbar^2}{2}\left(\frac{L^2}{I_\perp^{inert}} + \left(\frac{1}{I_\parallel^{inert}} - \frac{1}{I_\perp^{inert}}\right)L_z^2\right) \tag{E.16}$$

From the equivalents of Eqs. E.6 and E.7, the eigenvalues of the spherical top are then:

$$E_{L,M} = \frac{\hbar^2}{2I_\perp^{inert}}L(L+1) + \frac{\hbar^2}{2}\left(\frac{1}{I_\parallel^{inert}} - \frac{1}{I_\perp^{inert}}\right)M^2 \tag{E.17}$$

where L, M are the quantum numbers for the total angular momentum and z-projection, respectively. The eigenfunctions are the Wigner rotation matrices with elements $\mathcal{D}_{K,M}^{L}(\phi\theta\psi)$ (see Appendix R.3), where K is the quantum number for rotation about the space-fixed z-axis and M that for rotation about the body-fixed z-axis (Brink and Satchler 1968).

APPENDIX F: MAGNETIC FIELD IN QUANTUM MECHANICS

F.1 Magnetic Vector Potential

We can introduce the magnetic field (strictly, magnetic induction), **B**, into the quantum mechanical Hamiltonian in an *ad hoc* way, via the Bohr magneton, as we do in Chapter 1. A more fundamental approach is to use the magnetic vector potential, **A**, which is defined by:

$$\mathbf{B} = \text{curl}\mathbf{A} = \nabla \times \mathbf{A} \tag{F.1}$$

where curl is the rotation vector operator (∇ is the differential vector operator with Cartesian components: $\partial/\partial x$, $\partial/\partial y$ and $\partial/\partial z$ – see Appendix B.2 for a summary of vector calculus). Equation F.1 defines **A** to within a scalar potential term that is fixed by the additional condition: $\text{div}\mathbf{A} = \nabla\cdot\mathbf{A} = 0$. From classical electrodynamics, the total momentum of an electron with charge $-e$ in a uniform magnetic field becomes:

$$\mathbf{p}_{tot} = \mathbf{p} + e\mathbf{A} \tag{F.2}$$

where **p** is the usual kinetic part of the momentum. We need the form given in Eq. F.2 to satisfy the Lorentz force equation. The Hamiltonian of the electron in a magnetic field is then:

$$\mathcal{H} = \frac{(\mathbf{p}+e\mathbf{A})^2}{2m} = \frac{p^2}{2m} + \frac{e}{2m}(\mathbf{p}\cdot\mathbf{A} + \mathbf{A}\cdot\mathbf{p}) + \frac{e^2A^2}{2m} \tag{F.3}$$

where the first term on the right is the usual kinetic energy and the last is a small diamagnetic term that is proportional to the square of the field. The middle term on the right that is linear in **A** is the paramagnetic energy.

F.2 Orbital Paramagnetism: Bohr Magneton

For a constant field, $\mathbf{B}_o$, we can express the vector potential, which must satisfy Eq. F.1, as:

$$\mathbf{A} = \frac{1}{2}\mathbf{B}_o \times \mathbf{r} \tag{F.4}$$

where $\mathbf{r}$ is the vector position of the electron, relative to the nucleus. For this solenoidal potential: $\nabla \cdot \mathbf{A} = 0$, and therefore $\mathbf{p}$ and $\mathbf{A}$ commute because: $\mathbf{p}\cdot\mathbf{A} - \mathbf{A}\cdot\mathbf{p} = -i\hbar\nabla\cdot\mathbf{A} = 0$, after making the usual operator substitution: $\mathbf{p} \to -i\hbar\nabla$ (see Eq. C.14). The perturbation Hamiltonian for the interaction of the electron with the magnetic field in Eq. F.3 then becomes:

$$\mathcal{H}'_{mag} = \frac{e}{m}\mathbf{A}\cdot\mathbf{p} = \frac{e}{2m}\mathbf{B}_o\times\mathbf{r}\cdot\mathbf{p} = \frac{e}{2m}\mathbf{B}_o\cdot\mathbf{r}\times\mathbf{p} = \frac{e\hbar}{2m}\mathbf{B}_o\cdot\mathbf{l} \tag{F.5}$$

where $\mathbf{l}$ is the quantum mechanical operator for the orbital angular momentum in units of $\hbar$ (see Eq. E.1, et seq. in Appendix E). Note that we have used the properties of the triple scalar product given by Eq. B.12 in Appendix B. Equation F.5 corresponds to the Hamiltonian for a magnetic dipole, μ_l, in a uniform magnetic field $\mathbf{B}_o$:

$$\mathcal{H}'_{mag} = -\mu_l \cdot \mathbf{B}_o \tag{F.6}$$

Hence the magnetic moment associated with the orbital angular momentum of the electron is:

$$\mu_l = -g_l\beta_e\mathbf{l} \tag{F.7}$$

where $\beta_e = e\hbar/2m$ is the Bohr magneton and the g-value associated with the orbital moment is: $g_l = 1$.

F.3 Spin Paramagnetism

This still does not account for the magnetic moment associated with the electron spin. The latter emerges naturally from Dirac's relativistic treatment of quantum mechanics. The magnetic moment of the electron is given there by:

$$\mu_s = -g_e\beta_e\mathbf{s} \tag{F.8}$$

where $g_e = 2$ and the spin quantum number is $s = \frac{1}{2}$, with spin projections $m_s = \pm\frac{1}{2}$ (see also Appendix E for half-integral quantum numbers).

F.4 Magnetic Dipole Interactions

The magnetic field associated with a magnetic dipole moment $\boldsymbol{\mu}$ is the origin of magnetic dipole–dipole interactions, including those in hyperfine anisotropy. The magnetic vector potential at position $\mathbf{r}$ relative to the dipole is:

$$\mathbf{A} = \frac{\mu_o}{4\pi}\frac{\boldsymbol{\mu}\times\mathbf{r}}{r^3} = \frac{\mu_o}{4\pi}\nabla\frac{1}{r}\times\boldsymbol{\mu} \tag{F.9}$$

which derives from the current in a small circular loop. From Eqs. F.1 and F.9, the resulting field is:

$$\mathbf{B}_{dip} = \frac{\mu_o}{4\pi}\nabla\times\left(\nabla\frac{1}{r}\times\boldsymbol{\mu}\right) = \frac{\mu_o}{4\pi}\left((\boldsymbol{\mu}\cdot\nabla)\nabla - \boldsymbol{\mu}\nabla^2\right)\frac{1}{r} \tag{F.10}$$

where we have used Eq. B.19 for the curl of a vector product and two terms drop out because $\boldsymbol{\mu}$ is a constant vector. The final term in Eq. F.10 vanishes because $\nabla^2(1/r) = 0$, as in Laplace's equation. Performing the first derivative, the remaining term becomes:

$$\mathbf{B}_{dip} = -\frac{\mu_o}{4\pi}(\boldsymbol{\mu}\cdot\nabla)\frac{\mathbf{r}}{r^3} = -\frac{\mu_o}{4\pi}\left[\frac{1}{r^3}(\boldsymbol{\mu}\cdot\nabla)\mathbf{r} + \mathbf{r}\left(\boldsymbol{\mu}\cdot\nabla\frac{1}{r^3}\right)\right] \tag{F.11}$$

Further differentiation then gives the dipolar field:

$$\mathbf{B}_{dip} = \frac{\mu_o}{4\pi}\left(\frac{3(\boldsymbol{\mu}\cdot\mathbf{r})\mathbf{r}}{r^5} - \frac{\boldsymbol{\mu}}{r^3}\right) \tag{F.12}$$

This classical expression is entirely analogous to the electric field produced by an electric dipole, derived for two opposite charges. The interaction energy with a second dipole $\boldsymbol{\mu}'$ is then given by the Hamiltonian: $\mathcal{H}_{dd} = -\boldsymbol{\mu}'\cdot\mathbf{B}_{dip}$.

APPENDIX G: QUANTUM MECHANICAL PERTURBATION THEORY

G.1 Unperturbed and Perturbation Hamiltonians

Perturbation theory gives solutions of the Schrödinger equation in the presence of a small energetic perturbation, $\mathcal{H}'$, of the Hamiltonian, when we know the solution for the unperturbed Hamiltonian, $\mathcal{H}_o$. The Schrödinger equation for the unperturbed system:

$$\mathcal{H}_o|n\rangle = E_n|n\rangle \tag{G.1}$$

provides the zero-order wave functions $|n\rangle$, which are a complete, normalized, orthogonal set, i.e., $\langle m|n\rangle = 1$ for $m = n$ and otherwise = 0. The Schrödinger equation for the perturbed system is:

$$\mathcal{H}\psi = (\mathcal{H}_o + \mathcal{H}')\psi = E\psi \tag{G.2}$$

We seek a solution by expressing the eigenfunctions, ψ, as a linear combination of the zero-order functions:

$$\psi = c_1|1\rangle + c_2|2\rangle + ... \tag{G.3}$$

Then we obtain the coefficients, c_i, in this expansion by substituting Eq. G.3 into Eq. G.2 and using the orthogonality properties of the zero-order functions.

G.2 Degenerate-State Perturbation Energies

For simplicity, we assume that the zero-order Hamiltonian has just two eigenfunctions (as in the spin Hamiltonian for $S = \frac{1}{2}$). Multiplying Eq. G.2 on the left by $\langle 1|\times$, or by $\langle 2|\times$, (with the implied integration) yields:

$$c_1\langle 1|\mathcal{H}|1\rangle + c_2\langle 1|\mathcal{H}|2\rangle = c_1 E \tag{G.4}$$

$$c_1\langle 2|\mathcal{H}|1\rangle + c_2\langle 2|\mathcal{H}|2\rangle = c_2 E \tag{G.5}$$

These are two linear simultaneous equations to determine the coefficients c_1 and c_2. We can write these equations in matrix form (see Section B.3 of Appendix B):

$$\begin{pmatrix} \mathcal{H}_{11} & \mathcal{H}_{12} \\ \mathcal{H}_{21} & \mathcal{H}_{22} \end{pmatrix}\begin{pmatrix} c_1 \\ c_2 \end{pmatrix} = E\begin{pmatrix} c_1 \\ c_2 \end{pmatrix} \tag{G.6}$$

where the matrix elements are: $\mathcal{H}_{mn} \equiv \langle m|\mathcal{H}|n\rangle$. Introducing the components of $\mathcal{H}$ from Eq, G.2 and rearranging yields:

$$\begin{pmatrix} (E_1 + \mathcal{H}'_{11} - E) & \mathcal{H}'_{12} \\ \mathcal{H}'_{21} & (E_2 + \mathcal{H}'_{22} - E) \end{pmatrix}\begin{pmatrix} c_1 \\ c_2 \end{pmatrix} = 0 \tag{G.7}$$

where we have also used Eq. G.1. The condition for a non-trivial solution of Eq. G.7 is well known; it is given by the secular equation:

$$\begin{vmatrix} (E_1 + \mathcal{H}'_{11} - E) & \mathcal{H}'_{12} \\ \mathcal{H}'_{21} & (E_2 + \mathcal{H}'_{22} - E) \end{vmatrix} = 0 \tag{G.8}$$

On multiplying out the determinant, solution of Eq. G.8 immediately yields the energy levels, E:

$$E_{\pm} = \frac{E'_1 + E'_2 \pm \sqrt{(E'_1 - E'_2)^2 + 4|\mathcal{H}'_{12}|^2}}{2} \tag{G.9}$$

where:

$$E'_1 = E_1 + \langle 1|\mathcal{H}'|1\rangle \tag{G.10}$$

$$E'_2 = E_2 + \langle 2|\mathcal{H}'|2\rangle \tag{G.11}$$

These are known as the characteristic roots of Eq. G.8. The solution given in Eq. G.9 is an exact diagonalization of the Hamiltonian matrix and, in principle, is independent of the size of the perturbation. If the unperturbed states are degenerate (i.e., $E_1 = E_2$), this diagonalization procedure is unavoidable.

G.3 Non-degenerate Perturbation Energies

For nondegenerate states, we can use the approximation that $|\mathcal{H}_{12}| << |E_1 - E_2|$. To first order, this results in the energies E'_1 and E'_2 defined by Eqs. G.10 and G.11, respectively. The second-order correction to the energy comes from expanding the square root in Eq. G.9, giving the energy correct to second order:

$$E_+ = E_1 + \langle 1|\mathcal{H}'|1\rangle + \frac{|\langle 1|\mathcal{H}'|2\rangle|^2}{E_1 - E_2} \tag{G.12}$$

$$E_- = E_2 + \langle 2|\mathcal{H}'|2\rangle - \frac{|\langle 1|\mathcal{H}'|2\rangle|^2}{E_1 - E_2} \tag{G.13}$$

Clearly, we cannot use this expansion when the zero-order states are degenerate, i.e., $E_1 - E_2 = 0$. Equation G.12 or G.13 has a very straightforward interpretation. The first-order correction to the energy, E_n, is given by the diagonal element, $\langle n|\mathcal{H}'|n\rangle$, of the perturbation matrix. The second-order correction is given by the square of the off-diagonal matrix element, $\langle m|\mathcal{H}'|n\rangle$, divided by the difference in energy, $E_n - E_m$, from that of the state with which it connects.

G.4 Perturbed Wave Functions

Knowing the energies $E_{\pm}$, we can calculate the admixture coefficients c_1 and c_2 for the wave functions. From Eq. G.7, we get the relations:

$$c_1 = \frac{\mathcal{H}'_{12}}{E_- - E'_1} c_2 \tag{G.14}$$

$$c_2 = \frac{\mathcal{H}'_{21}}{E_+ - E'_2} c_1 \tag{G.15}$$

The wave functions are therefore given by:

$$\psi_1 = \frac{1}{\sqrt{N_1}}\left(|1\rangle + \frac{\mathcal{H}'_{21}}{E_+ - E'_2}|2\rangle\right) \tag{G.16}$$

$$\psi_2 = \frac{1}{\sqrt{N_2}}\left(|2\rangle + \frac{\mathcal{H}'_{12}}{E_- - E'_1}|1\rangle\right) \tag{G.17}$$

where N_1 and N_2 are normalization constants. A convenient way to express the normalization is to use the trignometrical identity: $\cos^2\vartheta + \sin^2\vartheta = 1$. We can then write the orthogonal wave functions as:

$$\psi_1 = \cos\vartheta \cdot |1\rangle + \sin\vartheta \cdot |2\rangle \quad \text{(G.18)}$$

$$\psi_2 = \cos\vartheta \cdot |2\rangle - \sin\vartheta \cdot |1\rangle \quad \text{(G.19)}$$

where

$$\tan\vartheta = \frac{\mathcal{H}'_{21}}{E_+ - E'_2} = -\frac{\mathcal{H}'_{12}}{E_- - E'_1} \quad \text{(G.20)}$$

We get an alternative form of Eq. G.20 from the trigonometric relationship for the double angle:

$$\tan 2\vartheta = \frac{2|\mathcal{H}'_{21}|}{E'_1 - E'_2} \quad \text{(G.21)}$$

For non-degenerate states, we can approximate the wave functions by:

$$\psi_1 = |1\rangle + \frac{\mathcal{H}'_{21}}{E_1 - E_2}|2\rangle \quad \text{(G.22)}$$

$$\psi_2 = |2\rangle - \frac{\mathcal{H}'_{12}}{E_1 - E_2}|1\rangle \quad \text{(G.23)}$$

The wave functions in Eqs. G.22 and G.23 are not normalized, but this is unimportant for a small perturbation at this level of approximation.

G.5 Perturbation Theory Summary

For degenerate states, we must diagonalize the perturbation Hamiltonian directly. For the 2 × 2 degenerate case ($E_1 = E_2$), the energies and wave functions are given from Eqs. G.9, G.16 and G.17 by:

$$E_\pm = \tfrac{1}{2}(E'_1 + E'_2) \pm |\mathcal{H}'_{12}| = E_1 + \tfrac{1}{2}(\mathcal{H}'_{11} + \mathcal{H}'_{22}) \pm |\mathcal{H}'_{12}| \quad \text{(G.24)}$$

$$\psi_{1,2} = \frac{1}{\sqrt{2}}(|1\rangle \pm |2\rangle) \quad \text{(G.25)}$$

For non-degenerate states, we can generalize the approximations given in Eqs. G.12 and G.22 by:

$$E_n = E_n^{(0)} + \langle n|\mathcal{H}'|n\rangle + \sum_m \frac{\langle n|\mathcal{H}'|m\rangle\langle m|\mathcal{H}'|n\rangle}{E_n - E_m} \quad \text{(G.26)}$$

$$\psi_n = |n\rangle + \sum_m \frac{\langle m|\mathcal{H}'|n\rangle}{E_n - E_m}|m\rangle \quad \text{(G.27)}$$

where $m \neq n$.

APPENDIX H: SPIN-HAMILTONIAN DIAGONALIZATION

H.1 Anisotropic g-Values

As a simple example of diagonalization of the spin Hamiltonian, we treat the angular variation of the g-values for an axial spin Hamiltonian:

$$\mathcal{H}_{Zeeman} = \beta_e B_o \left(g_\parallel \cos\theta . S_z + g_\perp \sin\theta . S_x\right) \quad \text{(H.1)}$$

where the magnetic field, $B_o(\cos\theta, \sin\theta)$, is inclined at angle θ to the $g_\parallel$ principal axis of the g-tensor. Remembering that $S_x = \frac{1}{2}(S_+ + S_-)$ and taking the matrix elements from Table E.1 in Appendix E, we get the matrix form of the spin Hamiltonian:

M_S	$\lvert +\frac{1}{2}\rangle$	$\lvert -\frac{1}{2}\rangle$
$\langle +\frac{1}{2}\rvert$	$\frac{1}{2}g_\parallel\beta_e B_o\cos\theta$	$\frac{1}{2}g_\perp\beta_e B_o\sin\theta$
$\langle -\frac{1}{2}\rvert$	$\frac{1}{2}g_\perp\beta_e B_o\sin\theta$	$-\frac{1}{2}g_\parallel\beta_e B_o\cos\theta$

(H.2)

Because the diagonal elements can be comparable (when $\theta \approx 90°$), we must diagonalize this spin Hamiltonian exactly. The secular equation (Eq. G.8 of Appendix G) yields the following two solutions for the energy levels:

$$E_\pm = \pm\tfrac{1}{2}\beta_e B_o\sqrt{g_\parallel^2\cos^2\theta + g_\perp^2\sin^2\theta} \quad \text{(H.3)}$$

Therefore, the angular-dependent g-value becomes:

$$g(\theta) = \sqrt{g_\parallel^2\cos^2\theta + g_\perp^2\sin^2\theta} \quad \text{(H.4)}$$

which we define from $E_\pm \equiv \pm\frac{1}{2}g(\theta)\beta_e B_o$. This differs from the axial version of Eq. 2.11 for the angular-dependent g-value in Chapter 2. The difference is that we assume the g-value anisotropy is small for Eq. 2.11 (valid for spin labels), whereas Eq. H.4 is exact. In fact, Eq. H.4 has the same form as the axial version of Eq. 2.12 for the hyperfine coupling. Making the approximation: $(g_\parallel - g_\perp) << g_\perp$, in Eq. H.4 we get:

$$g(\theta) = g_\parallel\cos^2\theta + g_\perp\sin^2\theta \quad \text{(H.5)}$$

which agrees with Eq. 2.11.

We get the eigenfunctions of the spin Hamiltonian in Eq. H.1 by using the eigenvalues of Eq. H.3, as described in Appendix G (viz., Eqs. G.16 and G.17). When expressed in terms of the basis functions $|\pm\frac{1}{2}\rangle$ that refer to the principal axes of the g-tensor, the eigenfunctions depend on the magnetic field direction, θ. We can avoid this by changing the axes for the spin Hamiltonian to the so-called strong-field precession axes. Essentially, this is the approach that we take in Section 2.2 of Chapter 2, as we show explicitly later in Appendix I.1.

H.2 Anisotropic Hyperfine Interaction

As an example of anisotropic hyperfine coupling, we take a spin Hamiltonian that includes just the secular and pseudosecular terms:

$$\mathcal{H}'_{hfs} = \hbar\omega_o S_z + \hbar\omega_I I_z + AI_zS_z + BI_xS_z \tag{H.6}$$

where $\hbar\omega_o S_z$ and $\hbar\omega_I I_z$ are the electron and nuclear Zeeman interactions, AI_zS_z is the secular part of the hyperfine interaction and BI_xS_z is the pseudosecular (i.e., off-diagonal) part. Note that ω_o, A and B depend on the orientation of the static magnetic field. The Hamiltonian in Eq. H.6 is diagonal in the electron spin but is not diagonal in the nuclear spin. For a nuclear spin $I=\frac{1}{2}$, the matrix form of Eq. H.6 in the $|M_S,m_I\rangle$ basis is (see Table E.1 in Appendix E):

M_S, m_I	$\lvert M_S,+\frac{1}{2}\rangle$	$\lvert M_S,-\frac{1}{2}\rangle$
$\langle M_S,+\frac{1}{2}\rvert$	$\hbar\omega_o M_S + \frac{1}{2}(\hbar\omega_I + AM_S)$	$\frac{1}{2}BM_S$
$\langle M_S,-\frac{1}{2}\rvert$	$\frac{1}{2}BM_S$	$\hbar\omega_o M_S - \frac{1}{2}(\hbar\omega_I + AM_S)$

(H.7)

Diagonalizing Eq. H.7, according to Eqs. G.8 and G.9, gives the following energy levels:

$$E_\pm(M_S) = \hbar\omega_o M_S \pm \tfrac{1}{2}\sqrt{(\hbar\omega_I + AM_S)^2 + B^2M_S^2} \tag{H.8}$$

From Eqs. G.18 and G.19, the new eigenfunctions are:

$$|M_S\rangle_+ = \cos\eta_{M_S}|M_S,+\tfrac{1}{2}\rangle + \sin\eta_{M_S}|M_S,-\tfrac{1}{2}\rangle \tag{H.9}$$

$$|M_S\rangle_- = \cos\eta_{M_S}|M_S,-\tfrac{1}{2}\rangle - \sin\eta_{M_S}|M_S,+\tfrac{1}{2}\rangle \tag{H.10}$$

where from Eqs. G.20 or G.21:

$$\tan\eta_{M_S} = \frac{-BM_S}{\hbar\omega_I + AM_S + \sqrt{(\hbar\omega_I + AM_S)^2 + B^2M_S^2}} \tag{H.11}$$

$$\tan 2\eta_{M_S} = \frac{-BM_S}{\hbar\omega_I + AM_S} \tag{H.12}$$

whichever is the more convenient.

Comparing with Eqs. I.2 and I.3 of Appendix I, we see that Eqs. H.9 and H.10 correspond to rotation of the basis system by the angle η_{M_S}. We can diagonalize the spin Hamiltonian by rotating state vectors that are specified in the $m_I=\pm\frac{1}{2}$ nuclear basis states. The rotation matrix is given by (cf. Eq. I.4):

$$U(\eta_{M_S}) = \begin{pmatrix} \cos\eta_{M_S} & \sin\eta_{M_S} \\ -\sin\eta_{M_S} & \cos\eta_{M_S} \end{pmatrix} \tag{H.13}$$

i.e., U is the matrix of eigenvectors of the Hamiltonian. Diagonalization of $\mathcal{H}'_{hfs}$ is accomplished by a unitary transformation, as for rotation:

$$\mathcal{H}'^{d}_{hfs} = U\mathcal{H}'_{hfs}U^{-1} \tag{H.14}$$

where $\mathcal{H}'^{d}_{hfs}$ is then diagonal (see Section B.4, e.g., Eq. B.25).

H.3 Allowed and Forbidden Transitions

Figure H.1 shows the energy levels for a $S=\frac{1}{2}$, $I=\frac{1}{2}$ spin system according to Eq. H.8. We know from Appendix M.3 that the selection rules for an EPR transition are $\Delta M_S=\pm1$ and $\Delta m_I=0$. For nuclear transitions, they are exactly the reverse. From Eq. H.8 for $I=\frac{1}{2}$, the nuclear transition frequencies ω_α, ω_β are therefore given by:

$$\hbar\omega_{\alpha,\beta} = E_+(\pm\tfrac{1}{2}) - E_-(\pm\tfrac{1}{2}) = \sqrt{(\hbar\omega_I \pm \tfrac{1}{2}A)^2 + \tfrac{1}{4}B^2} \tag{H.15}$$

where α stands for the $M_S=+\frac{1}{2}$ electron state and β for $M_S=-\frac{1}{2}$. These are the angular nuclear frequencies that we observe in ENDOR or ESEEM (see Chapter 14).

The angular frequencies of the two *allowed* EPR transitions $\omega_a^{(\pm)}$ then become, from Eq. H.8:

$$\hbar\omega_a^{(\pm)} = E_\pm(+\tfrac{1}{2}) - E_\pm(-\tfrac{1}{2}) = \hbar\left(\omega_o \pm \tfrac{1}{2}(\omega_\alpha - \omega_\beta)\right) \tag{H.16}$$

Correspondingly, the frequencies of the two *forbidden* EPR transitions $\omega_f^{(\pm)}$ are given by:

$$\hbar\omega_f^{(\pm)} = E_\pm(+\tfrac{1}{2}) - E_\mp(-\tfrac{1}{2}) = \hbar\left(\omega_o \pm \tfrac{1}{2}(\omega_\alpha + \omega_\beta)\right) \tag{H.17}$$

– see Figure H.1. The *forbidden* transitions are strictly forbidden only when $B=0$ in Eq. H.6. Otherwise, the pseudosecular terms admix some of the $m_I=-\frac{1}{2}$ state into the $m_I=+\frac{1}{2}$ state and vice-versa (see Eqs. H.9 and H.10). The forbidden transitions then have non-vanishing intensity that depends on the degree of admixture.

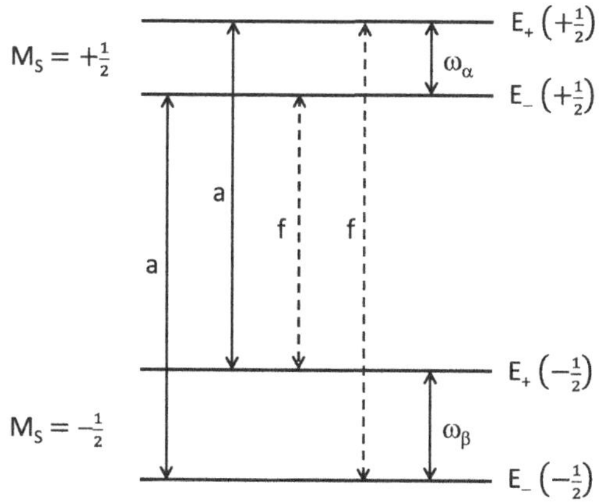

FIGURE H.1 Energy levels for $S=\frac{1}{2}$, $I=\frac{1}{2}$ spin system, including pseudosecular hyperfine interaction according to Eq. H.8. Allowed EPR transitions "a" are indicated by solid lines and forbidden transitions "f" by dashed lines. Nuclear transitions are ω_α and ω_β.

We get the relative intensities from the respective transition probabilities (see Appendix M.2). According to Eqs. M.11 and M.13, the intensity of the allowed transitions is proportional to the square of the perturbation matrix element connecting the two states:

$$I_a^{(\pm)} \propto \left| {}_{\pm}\left\langle -\tfrac{1}{2} \right| S_x \left| +\tfrac{1}{2} \right\rangle_{\pm} \right|^2 = \tfrac{1}{4}\cos^2(\eta_\alpha - \eta_\beta) \tag{H.18}$$

and that of the forbidden transitions is:

$$I_f^{(\pm)} \propto \left| {}_{\mp}\left\langle -\tfrac{1}{2} \right| S_x \left| +\tfrac{1}{2} \right\rangle_{\pm} \right|^2 = \tfrac{1}{4}\sin^2(\eta_\alpha - \eta_\beta) \tag{H.19}$$

where we use Eqs. H.9 and H.10 with η_α for $M_S = +\frac{1}{2}$ and η_β for $M_S = -\frac{1}{2}$; and substitute trigonometric identities for the difference of two angles. (Note that Schweiger and Jescke (2001) define angles η_α and η_β as twice those used here.) For $\eta_\alpha = 0 = \eta_\beta$ (i.e., no pseudosecular terms), we get $I_f^{(\pm)} = 0$ and the forbidden transitions are truly forbidden. From Eqs. H.11 and H.12, the normalized intensities of allowed and forbidden transitions become (Iwasaki and Toriyama 1985):

$$I_a = \cos^2\eta = \frac{\left|\omega_I^2 - \frac{1}{4}(\omega_\alpha - \omega_\beta)^2\right|}{\omega_\alpha\omega_\beta} \tag{H.20}$$

$$I_f = \sin^2\eta = \frac{\left|\omega_I^2 - \frac{1}{4}(\omega_\alpha + \omega_\beta)^2\right|}{\omega_\alpha\omega_\beta} \tag{H.21}$$

where $\eta \equiv \eta_\alpha - \eta_\beta$ and $I_a + I_f = 1$.

We distinguish between strong, $\frac{1}{2}|A_{hf}| > |\omega_I|$, and weak, $\frac{1}{2}|A_{hf}| < |\omega_I|$, hyperfine-coupling limits because these have different relative spectral intensities (see Figure H.2). For strong coupling (upper spectrum), the outer lines are allowed (*a*) and split by the hyperfine coupling $|A_{hf}|$, whereas the inner lines are forbidden (*f*) and split by twice the nuclear Larmor frequency $2|\omega_I|$. For weak coupling (lower spectrum), the situation is exactly the reverse.

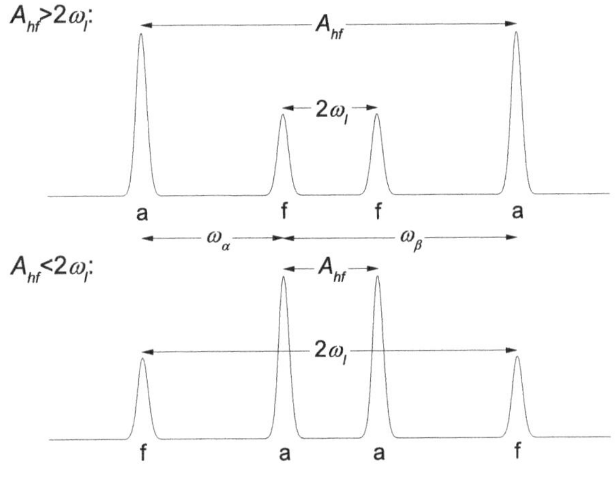

FIGURE H.2 EPR absorption spectra for $S=\frac{1}{2}$, $I=\frac{1}{2}$ spin system, according to Eqs. H.15–H.17, H.20 and H.21. *Upper*: strong hyperfine coupling $\frac{1}{2}|A_{hf}|>|\omega_I|$; *lower*: weak hyperfine coupling $\frac{1}{2}|A_{hf}|<|\omega_I|$.

APPENDIX I: ROTATION OF AXES

I.1 Strong-Field Precession Axes.

An alternative to diagonalizing the spin-Hamiltonian matrix by using the secular equation (as in Appendix H), is for us to rotate the axes to which the spin Hamiltonian is referred, so as to express it immediately in diagonal form. Consider an axial Zeeman Hamiltonian $(g_{xx} = g_{yy})$, as we do in Appendix H.1 but write it here in matrix form (see Section B.4 in Appendix B):

$$\mathcal{H}_{Zeeman} = \beta_e \mathbf{B}^{\mathrm{T}} \cdot \mathbf{g} \cdot \mathbf{S} = \beta_e \begin{pmatrix} B_x & B_z \end{pmatrix} \begin{pmatrix} g_{xx} & 0 \\ 0 & g_{zz} \end{pmatrix} \begin{pmatrix} S_x \\ S_z \end{pmatrix} \tag{I.1}$$

The magnetic field, B, lies in the x-z plane and is inclined at an angle θ to the g_{zz} principal axis of the g-tensor, i.e., $B_x = B_o \sin\theta$ and $B_z = B_o \cos\theta$ (see Figure I.1).

We rotate the g-tensor axis system (x,z), i.e., the nitroxide axis system, by an angle $-\psi$ about its y-axis, as indicated in Figure I.1. In terms of the new axis system (x',z'), coordinates in the original g-tensor system are given by:

$$x = x'\cos\psi + z'\sin\psi \tag{I.2}$$

$$z = -x'\sin\psi + z'\cos\psi \tag{I.3}$$

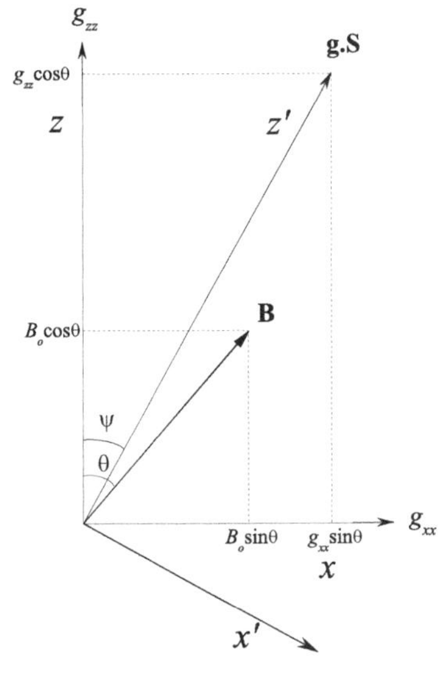

FIGURE I.1 Rotation of axes by angle ψ, from g-tensor axis system (x,z) to strong-field precession axis system (x',z'). (A right-handed rotation about y takes x into z; those indicated are negative rotations.) Coordinates in the two systems are related by Eqs. I.1–I.3. Spectrometer magnetic field, **B**, lies in the x,z-plane and is inclined at angle θ to the nitroxide z-axis. The z-component of the electron spin is directed along z' (i.e., strong-field precession axis, **g·S**) that makes angle ψ with the nitroxide z-axis, where $\tan\psi = (g_{xx}/g_{zz})\tan\theta$.

– see Figure I.1. Rotation of the spin vector, **S**, in the g-tensor axis system is given by the matrix operation that corresponds to the coordinate transformation in Eqs. I.2 and I.3:

$$\mathbf{S} \equiv \begin{pmatrix} S_x \\ S_z \end{pmatrix} = \begin{pmatrix} \cos\psi & \sin\psi \\ -\sin\psi & \cos\psi \end{pmatrix} \begin{pmatrix} S'_x \\ S'_z \end{pmatrix} \equiv \mathbf{R}(\psi)\cdot\mathbf{S}' \tag{I.4}$$

where $\mathbf{S}'$ is the spin vector in the rotated axis system and $\mathbf{R}(\psi)$ is the rotation matrix. ($\mathbf{R}(-\psi)$ is the matrix rotating the unprimed x,z-axes –see Figure I.1.)

We now express the spin Hamiltonian in terms of the rotated spin vector, $\mathbf{S}'$, and require that $\mathcal{H}_{Zeeman}$ is then diagonal. By using Eq. I.4 and the condition that the Hamiltonian (being an energy) must be independent of the axis system used, Eq. I.1 becomes:

$$\mathcal{H}_{Zeeman} = \beta_e \mathbf{B}^{\mathrm{T}}\cdot\mathbf{g}\cdot\mathbf{S} = \beta_e \mathbf{B}^{\mathrm{T}}\cdot\mathbf{g}\cdot\mathbf{R}\cdot\mathbf{S}' = g\beta_e B_o S'_z \tag{I.5}$$

where the last equality specifies that the Hamiltonian is diagonal in the rotated axis system. (On the extreme right, g is no longer a tensor but is simply the angular-dependent g-value.) This means that only the z-element of the row vector $\mathbf{B}^{\mathrm{T}}\cdot\mathbf{g}\cdot\mathbf{R}$ (which multiplies S'_z) is non-vanishing. By multiplying the component matrices that are defined in Eqs. I.1 and I.4, we find that the x-element of $\mathbf{B}^{\mathrm{T}}\cdot\mathbf{g}\cdot\mathbf{R}$ is:

$$B_o\left(g_{xx}\sin\theta\cos\psi - g_{zz}\cos\theta\sin\psi\right) = 0 \tag{I.6}$$

which (because it multiplies S'_x) must be zero to diagonalize the Zeeman Hamiltonian. Thus, the spin Hamiltonian becomes diagonal when the rotation angle, ψ, is given by:

$$\tan\psi = \frac{g_{xx}}{g_{zz}}\tan\theta \tag{I.7}$$

Because of the g-value anisotropy (i.e., $g_{xx} \neq g_{zz}$), the electron spin is not quantized along the magnetic field direction, θ, but along the direction ψ which we call the strong-field precession axis (see Fig. I.1).

Finally, we find from matrix multiplication that the non-zero z-element of $\mathbf{B}^{\mathrm{T}}\cdot\mathbf{g}\cdot\mathbf{R}$ is:

$$B_o\left(g_{xx}\sin\theta\sin\psi + g_{zz}\cos\theta\cos\psi\right) = B_o g(\theta) \tag{I.8}$$

where $g(\theta)$ is the angular-dependent g-value that is defined by the final right-hand equality in Eq. I.5. Substituting for ψ from Eq. I.7 in Eq. I.8 gives the following expression for the g-value:

$$g(\theta) = \sqrt{g_{xx}^2\sin^2\theta + g_{zz}^2\cos^2\theta} \tag{I.9}$$

As it must, this agrees with the result of direct diagonalization that is given by Eq. H.4 in Appendix H, where $g_{\parallel} \equiv g_{zz}$ and $g_{\perp} \equiv g_{xx}$. Note that simply assuming that $\psi = \theta$ in Eq. I.8, immediately gives:

$$g(\theta) = g_{xx}\sin^2\theta + g_{zz}\cos^2\theta \tag{I.10}$$

This is the result (viz., Eq. 2.11) that we obtained in Section 2.2 of Chapter 2 by making the same approximation: that the z-component of the electron spin is directed along the magnetic field direction. To be consistent with Eqs. I.6 and I.7, this approximation requires that the g-value anisotropy is small: $g_{xx} \approx g_{zz}$, which is justified for nitroxide spin labels (see Section 2.2).

I.2 Hyperfine Interaction

We can apply a similar procedure to the hyperfine interaction. We assume an axial hyperfine tensor $\left(A_{xx} = A_{yy}\right)$ and write the spin Hamiltonian for the hyperfine interaction in matrix form:

$$\mathcal{H}'_{hfs} = \mathbf{I}^{\mathrm{T}}\cdot\mathbf{A}\cdot\mathbf{S} = (I_x \quad I_z)\begin{pmatrix} A_{xx} & 0 \\ 0 & A_{zz} \end{pmatrix}\begin{pmatrix} S_x \\ S_z \end{pmatrix} \tag{I.11}$$

As we saw in Section 2.2 of Chapter 2, the strong hyperfine field at the nucleus causes the nuclear spin **I** to be quantized along a direction different from that of the electron spin **S**. Therefore, we must apply a rotation $\mathbf{I} = \mathbf{R}(\chi)\cdot\mathbf{I}''$ to the nuclear spin, different from that which we applied to the electron spin, viz., $\mathbf{S} = \mathbf{R}(\psi)\cdot\mathbf{S}'$. The spin Hamiltonian for the hyperfine interaction then becomes:

$$\begin{aligned}\mathcal{H}_{hfs} &= \mathbf{I}^{\mathrm{T}}\cdot\mathbf{A}\cdot\mathbf{S} = \left(\mathbf{R}(\chi)\mathbf{I}''\right)^{\mathrm{T}}\cdot\mathbf{A}\cdot\mathbf{R}(\psi)\mathbf{S}' \\ &= \mathbf{I}''^{\mathrm{T}}\cdot\mathbf{R}^{-1}(\chi)\mathbf{A}\mathbf{R}(\psi)\cdot\mathbf{S}' = AI''_z S'_z\end{aligned} \tag{I.12}$$

where the last equality specifies that the hyperfine spin Hamiltonian becomes diagonal. On the extreme right, A is no longer a tensor but is simply the angular-dependent scalar hyperfine coupling. The superscript "T" in Eq. I.12 indicates that the leftmost vector **I** is a transpose, i.e., a row vector instead of a column vector – see Eq. I.11. Therefore, the transpose of the product $\mathbf{RI}''$ (which is equal to the product of the individual transposes, in the reverse order) is $\left(\mathbf{RI}''\right)^{\mathrm{T}} = \mathbf{I}''^{\mathrm{T}}\mathbf{R}^{\mathrm{T}} = \mathbf{I}''^{\mathrm{T}}\mathbf{R}^{-1}$, because the transpose of a rotation matrix is equal to its inverse: $\mathbf{R}^{\mathrm{T}}(\chi) \equiv \mathbf{R}(-\chi) = \mathbf{R}^{-1}(\chi)$ (see Eq. I.4).

To diagonalize the hyperfine Hamiltonian, the xz-element of the $\mathbf{R}^{-1}(\chi)\mathbf{A}\mathbf{R}(\psi)$ matrix must be zero, because it multiplies the pseudosecular term $I''_x S'_z$. We ignore the zx-element, which multiplies the nonsecular term $I''_z S'_x$, because it contributes only in second order (cf. Appendix J, which follows). From matrix multiplication, the xz-element is:

$$A_{xx}\sin\psi\cos\chi - A_{zz}\cos\psi\sin\chi = 0 \tag{I.13}$$

Therefore, the angle χ by which we must rotate **I** is given by:

$$\tan\chi = \frac{A_{xx}}{A_{zz}}\tan\psi \tag{I.14}$$

where the angle ψ, which diagonalizes the Zeeman Hamiltonian, is given by Eq. I.7. The zz-element of the $\mathbf{R}^{-1}(\chi)\mathbf{A}\mathbf{R}(\psi)$ matrix, which multiplies $I_z''S_z'$ and gives the diagonal hyperfine interaction, becomes:

$$A_{xx}\sin\psi\sin\chi + A_{zz}\cos\psi\cos\chi = A(\theta) \quad (I.15)$$

Substituting for χ from Eq. I.14 gives the following expression for the angular-dependent hyperfine coupling:

$$A(\theta) = \sqrt{A_{xx}^2\sin^2\psi + A_{zz}^2\cos^2\psi} \quad (I.16)$$

If we make the approximation that **S** lies along the magnetic field direction, as we do in Section 2.2 of Chapter 2, then $\psi = \theta$ and the angular-dependent hyperfine coupling becomes:

$$A(\theta) = \sqrt{A_{xx}^2\sin^2\theta + A_{zz}^2\cos^2\theta} \quad (I.17)$$

which is identical with the axial version of Eq. 2.12 in Section 2.2.

As discussed in Chapter 2, the approximation given by Eq. I.17 is acceptable for nitroxide spin labels. When the g-value anisotropies are larger than for a nitroxide, we must substitute ψ from Eq. I.7 into Eq. I.16. This gives the following expression for the hyperfine coupling:

$$A(\theta) = \frac{1}{g(\theta)}\sqrt{g_{xx}^2A_{xx}^2\sin^2\theta + g_{zz}^2A_{zz}^2\cos^2\theta} \quad (I.18)$$

where $g(\theta)$ is given by Eq. I.9. This equation (Eq. I.18) is the general result for an axial spin Hamiltonian.

I.3 General Rotation Matrices

Already for axial hyperfine structure, rotating the axes is more straightforward than direct diagonalization of the spin Hamiltonian as in Appendix H. It is especially useful when the symmetry is not axial and additional angular relations are needed. This is because sequential rotations about different axes are given simply by the product of the rotation matrices. For example, a right-handed rotation by ψ about the y-axis, followed by a right-handed rotation ϕ about the resulting z'-axis, is given by $\mathbf{R}(\phi,z)\mathbf{R}(\psi,y)$ where:

$$\mathbf{R}(\psi,y) = \begin{pmatrix} \cos\psi & 0 & -\sin\psi \\ 0 & 1 & 0 \\ \sin\psi & 0 & \cos\psi \end{pmatrix} \quad (I.19)$$

$$\mathbf{R}(\phi,z) = \begin{pmatrix} \cos\phi & \sin\phi & 0 \\ -\sin\phi & \cos\phi & 0 \\ 0 & 0 & 1 \end{pmatrix} \quad (I.20)$$

$\mathbf{R}(\psi,y)$ in Eq. I.19 is the 3 × 3 equivalent of Eq. I.4 (but for a positive rotation) and operates on the unprimed column vector (x,y,z).

In general, we need three independent rotations to bring a given vector to an arbitrary new direction. These are usually specified by the Euler angles ϕ, θ and ψ, as described in Figure R.2 of Appendix R. In terms of these Euler angles, a complete rotation of the coordinate axes is given by:

$$\mathbf{R}(\phi,\theta,\psi) \equiv \mathbf{R}(\psi,z'')\mathbf{R}(\theta,y')\mathbf{R}(\phi,z)$$
$$= \begin{pmatrix} \cos\psi\cos\theta\cos\phi - \sin\psi\sin\phi & \cos\psi\cos\theta\sin\phi + \sin\psi\cos\phi & -\cos\psi\sin\theta \\ -\sin\psi\cos\theta\cos\phi - \cos\psi\sin\phi & -\sin\psi\cos\theta\sin\phi + \cos\psi\cos\phi & \sin\psi\sin\theta \\ \sin\theta\cos\phi & \sin\theta\sin\phi & \cos\theta \end{pmatrix} \quad (I.21)$$

Rotation matrices based on the Euler angles are particularly important in relaxation theory, where frequently we use Wigner rotation matrices that possess more symmetrical transformation properties than their Cartesian counterparts (see Appendix R.3). This convention for the rotation matrices and notation for Euler angles agrees with that adopted by Van et al. (1974).

APPENDIX J: SECOND-ORDER HYPERFINE SHIFTS

As an example of second-order nondegenerate perturbation theory, we take the isotropic hyperfine interaction, a_o, for a ^{14}N-nitroxide with isotropic g-value, g_o. In a magnetic field, B, the full spin Hamiltonian (excluding the nuclear Zeeman effect and quadrupole interaction) is given by:

$$\mathcal{H}_S = g_o\beta_e BS_z + a_o\left(I_xS_x + I_yS_y + I_zS_z\right)$$
$$= g_o\beta_e BS_z + a_o\left[I_zS_z + \tfrac{1}{2}(I_+S_- + I_-S_+)\right] \quad \text{(J.1)}$$

where the electron spin, $S = \frac{1}{2}$ (but not the nitrogen nuclear spin, $I = 1$), is quantized along the magnetic field direction, z (cf. Section 2.2). Taking matrix elements from Table E.1 in Appendix E, the spin Hamiltonian in matrix form is:

M_S, m_I	$\lvert+\frac{1}{2},+1\rangle$	$\lvert+\frac{1}{2},0\rangle$	$\lvert-\frac{1}{2},+1\rangle$	$\lvert+\frac{1}{2},-1\rangle$	$\lvert-\frac{1}{2},0\rangle$	$\lvert-\frac{1}{2},-1\rangle$
$\langle+\frac{1}{2},+1\rvert$	$\frac{1}{2}g_o\beta_eB+\frac{1}{2}a_o$	0	0	0	0	0
$\langle+\frac{1}{2},0\rvert$	0	$\frac{1}{2}g_o\beta_eB$	$a_o/\sqrt{2}$	0	0	0
$\langle-\frac{1}{2},+1\rvert$	0	$a_o/\sqrt{2}$	$-\frac{1}{2}g_o\beta_eB-\frac{1}{2}a_o$	0	0	0
$\langle+\frac{1}{2},-1\rvert$	0	0	0	$\frac{1}{2}g_o\beta_eB-\frac{1}{2}a_o$	$a_o/\sqrt{2}$	0
$\langle-\frac{1}{2},0\rvert$	0	0	0	$a_o/\sqrt{2}$	$-\frac{1}{2}g_o\beta_eB$	0
$\langle-\frac{1}{2},-1\rvert$	0	0	0	0	0	$-\frac{1}{2}g_o\beta_eB+\frac{1}{2}a_o$

We choose the ordering of states to bring the Hamiltonian matrix as close as possible to a diagonal form. As in Appendix G.2, we could easily diagonalize the two resulting 2 × 2 submatrices. However, we need not do this here, because the states that are coupled by the off-diagonal hyperfine matrix elements differ in energy by the full Zeeman interaction, $g_o\beta_eB$, which is very much bigger than the hyperfine coupling, a_o.

From Appendix G.3, we know that the second-order perturbation energy is given simply by the off-diagonal matrix element squared, divided by the difference in on-diagonal energies of the states connected by the off-diagonal element (see Eqs. G.12 and G.13). Thus, the energy levels $E(M_S, m_I)$ of the spin Hamiltonian in Eq. J.1 are given by:

$$E\left(\pm\tfrac{1}{2},\pm1\right) = \pm\tfrac{1}{2}g_o\beta_eB + \tfrac{1}{2}a_o \quad \text{(J.2)}$$

$$E\left(\mp\tfrac{1}{2},\pm1\right) = \mp\tfrac{1}{2}g_o\beta_eB - \tfrac{1}{2}a_o \mp \frac{a_o^2}{2g_o\beta_eB} \quad \text{(J.3)}$$

$$E\left(\pm\tfrac{1}{2},0\right) = \pm\tfrac{1}{2}g_o\beta_eB \pm \frac{a_o^2}{2g_o\beta_eB} \quad \text{(J.4)}$$

With the selection rules $\Delta M_S = \pm1$, $\Delta m_I = 0$, the EPR transitions occur at fields B_{m_I} that we get from:

$$h\nu = g_o\beta_eB_{\pm1} \pm a_o + \frac{a_o^2}{2h\nu} \quad \text{(J.5)}$$

$$h\nu = g_o\beta_eB_0 + \frac{a_o^2}{h\nu} \quad \text{(J.6)}$$

where ν is the spectrometer frequency and h is Planck's constant. The second-order downfield shift of the central $(m_I = 0)$ hyperfine line is therefore twice that of each outer $(m_I = \pm1)$ line. This leads to a difference between the two ^{14}N-hyperfine splittings of $a_o^2/h\nu$, where the high-field splitting, $B_{-1} - B_0$, is the larger.

At the common microwave operating frequency of $\nu = 9$ GHz, the second-order correction to the isotropic hyperfine splitting of a ^{14}N-nitroxide is 0.25%: 105 MHz or 4 μT. Second-order hyperfine shifts also affect the determination of isotropic g-values. Measurements on the central ($m_I = 0$) hyperfine line require the correction $\delta g_o = +5\times10^{-5}$ for a ^{14}N-nitroxide at 9 GHz. For the mean of the two outer hyperfine lines, the correction to g_o is half this value. It is clear that second-order hyperfine shifts are significant only for high-resolution measurements in non-viscous solvents, where the lines are narrow. They are even less important for high-field instruments: for a 94-GHz spectrometer, the second-order corrections are one tenth of those quoted for 9 GHz.

APPENDIX K: ATOMIC STRUCTURE AND MOLECULAR BONDING

K.1 Atomic Structure

Here we give a very abbreviated treatment of the electronic structure of atoms, such as we need to describe the molecular orbital structure of nitroxide radicals.

Consider the Hamiltonian, $\mathcal{H}_o$, for an electron in a central-field potential, $U(r)$, that depends only on the radial coordinate, r, of the electron, relative to the nucleus:

$$\mathcal{H}_o = \frac{-\hbar^2}{2m}\nabla^2 + U(r) \quad \text{(K.1)}$$

The central field includes not only the electrostatic attraction of the electron to the positively charged nucleus but also the central part of the Coulombic repulsion from the other electrons in the atom, which partially screens the nuclear charge. Then we treat the non-central part of the Coulombic repulsion later as a perturbation. We choose the central potential by iterative solutions of the radial part of the Schrödinger equation that result finally in self-consistent values of $U(r)$ and the corresponding eigenfunctions.

The Laplacian operator ∇^2 in polar coordinates (r,θ,ϕ) is given by:

$$\nabla^2 = \frac{1}{r^2}\left\{\frac{\partial}{\partial r}\left(r^2\frac{\partial}{\partial r}\right)+\hat{l}^2(\theta,\phi)\right\} \tag{K.2}$$

where $\hat{l}^2$ is the operator for the square of the orbital angular momentum of the electron. Using a central potential, $U(r)$, lets us separate the radial and angular variables. The eigenfunctions of the Schrödinger equation are then:

$$\psi_{n,l,m_l} = R_{nl}(r)\times Y_{l,m_l}(\theta,\phi) \tag{K.3}$$

where n is the principal quantum number that specifies the electron shells: $n=1$, $n=2$, etc. The higher electron shells have higher energy, and their region of maximum density is shifted further out from the nucleus. The number of nodes, i.e., points at which the wave function crosses zero, is $n-1$. The energy levels, E_{nl}, that we determine by the self-consistent field treatment depend not only on n, but also on the angular momentum quantum number, l. For a given electron shell, n, the l quantum number can take integer values from 0 to $n-1$. Only for the hydrogen atom are the energy levels with different l degenerate: this is a special feature of the $1/r$ potential, which is unique to hydrogen.

The angular wave functions are the spherical harmonics, Y_{l,m_l}, that are eigenfunctions of the $\hat{l}^2$ angular momentum operator, which has eigenvalues $l(l+1)$ (see Eq. E.6 in Appendix E). We list the spherical harmonics that are of interest in Table K.1: $l=0$ corresponds to s-electrons and $l=1$ to p-electrons. The spherical harmonics are normalized: $|Y_{l,m_l}|^2 = 1$ and mutually orthogonal $\langle l',m_l'|l,m\rangle = \delta_{ll'}\delta_{m_l m_l'}$. Written out fully, the orthogonality relation for normalized spherical harmonics is:

TABLE K.1 Spherical harmonics, Y_{l,m_l}, and p-orbitals

	polar	*Cartesian*
$Y_{0,0}$	$\frac{1}{\sqrt{4\pi}}$	
$Y_{1,0}$	$\sqrt{\frac{3}{4\pi}}\cos\theta$	$\sqrt{\frac{3}{4\pi}}\left(\frac{z}{r}\right)$
$Y_{1,\pm1}$	$\mp\sqrt{\frac{3}{8\pi}}\sin\theta\exp(\pm i\phi)$	$\mp\sqrt{\frac{3}{8\pi}}\left(\frac{x\pm iy}{r}\right)$
p_z	$Y_{1,0}$	$\sqrt{\frac{3}{4\pi}}\left(\frac{z}{r}\right)$
p_x	$\frac{-1}{\sqrt{2}}(Y_{1,+1}+Y_{1,-1})$	$\sqrt{\frac{3}{4\pi}}\left(\frac{x}{r}\right)$
p_y	$\frac{-i}{\sqrt{2}}(Y_{1,+1}-Y_{1,-1})$	$\sqrt{\frac{3}{4\pi}}\left(\frac{y}{r}\right)$

$$\int_o^{2\pi}\int_o^{\pi} Y^*_{l',m_l'}Y_{l,m_l}\sin\theta\cdot d\theta\cdot d\phi = \delta_{ll'}\delta_{mm'} \tag{K.4}$$

For $m_l \neq 0$, the spherical harmonics contain imaginary parts. The p-orbitals in real space, p_x, p_y and p_z, which we need to discuss chemical bonding, are constructed from normalized linear combinations of the $l=1$ spherical harmonics that are mutually orthogonal, as we see in Table K.1.

Note that any linear combination of eigenfunctions of a given operator must itself also be an eigenfunction of the same operator. For a free atom, it is unimportant which linear combinations of Y_{l,m_l} we take for a given l, provided that they are a complete set. This is because, in the absence of a magnetic field, these eigenstates all have the same energy. In chemical bonding, however, particular p-orbitals are preferred, removing the degeneracy present in the free atom. Fixing the orbitals in the chemical structure causes so-called orbital quenching. For instance, the expectation value of l_z, the z-projection of the orbital angular momentum, is zero for each of the p_x, p_y and p_z orbitals.

To complete this description of atomic structure, we need to introduce the quantum number for the electron spin projection: $m_s = \pm\frac{1}{2}$ for spin-up and spin-down, respectively. The electron wave function is then specified fully by four quantum numbers: $|n,l,m_l,m_s\rangle$. The electronic structure of the atom is built up by filling the orbitals of lower energy according to Pauli's exclusion principle, which requires that no two electrons have all four quantum numbers the same. The exclusion principle results from the more general Pauli Principle for indistinguishable particles of odd-integer spin. This states that the wave function must be anti-symmetric (i.e., change sign) on interchange of any two particles (see Section 9.6 in Chapter 9). The first shell ($n=1$, $l=0$) therefore may contain up to two electrons in the s-orbital, resulting in the $1s^2$ configuration for the filled shell (i.e., the He atom). The second shell ($n=2$; $l=0$, 1) may contain up to eight electrons, two in the s-orbital and six in the p-orbitals, which corresponds to the configuration $1s^22s^22p^6$ for a filled second shell (i.e., the Ne atom).

The electron configurations expressed in the above way assume that all p-electrons, for instance, are degenerate. This is not exactly the case, if we consider the perturbation by the non-central part of the electron–electron repulsion. For electrons with given l, the exchange interaction contributed by the non-central perturbation results in parallel spin orientations being preferred energetically over antiparallel orientations (see Sections 9.6, 9.7). This corresponds to the first of Hund's rules, which requires that as many electrons as possible have parallel spins in the lowest energy term. Nitrogen, for example, has three $2p$ electrons and the ground-state configuration is specified by: $1s^22s^22p_xp_yp_z$. Oxygen has one further $2p$ electron and the ground-state configuration is $1s^22s^22p_xp_y^2p_z$. Of course, for the free O atom, it is arbitrary in which of the $2p$ orbitals the two paired electrons reside. Here, we choose the $2p_y$ orbital because that corresponds more closely to the oxygen lone pair in nitroxides, where x is directed along the N–O bond and the $2p_z$ orbital participates in the π-bond that contains the unpaired electron spin (see Figure 2.1 in Chapter 2).

K.2 Molecular Orbitals and Bonding

In a diatomic molecule A–B, molecular bonding involves sharing of unpaired valence electrons in unfilled orbitals of atom A with those of atom B. The two paired electrons then reside in a bonding molecular orbital Ψ_{MO} that we approximate by a linear combination of the corresponding atomic orbitals ψ_A and ψ_B:

$$\Psi_{MO} = a\psi_A + b\psi_B \tag{K.5}$$

To ensure that the molecular orbital remains normalized, the admixture coefficients are related by:

$$a^2 + b^2 + 2abS_{A,B} = 1 \tag{K.6}$$

where the overlap integral is defined by:

$$S_{A,B} = \langle \psi_A | \psi_B \rangle \equiv \int \psi_A^* \psi_B dv \tag{K.7}$$

with integration, as usual, over all space. The overlap integral is non-zero because orbitals ψ_A and ψ_B have different centres and, unlike atomic orbitals about a single centre, are therefore not orthogonal. We can normalize the molecular orbital in Eq. K.5 by dividing with $N = \sqrt{a^2 + b^2 + 2abS_{A,B}}$. In general, strong overlap accumulates negative electron charge density between the positively charged nuclei and produces strong bonding.

If the atomic orbitals are either *s*-orbitals, or *p*-orbitals that lie along the A–B direction, we get maximum end-on overlap between orbitals which produces a strong σ-bond. If the *p*-orbitals are perpendicular to the bond direction, we get sideways overlap between orbitals and a weaker π-bond (see Figure K.1). An example of a π-bond is that between the N-atom and O-atom, which contains the unpaired electron in a nitroxide radical. The $2p_z$-orbital of the nitrogen that participates in the π-orbital is perpendicular to the N–O bond, which lies in the *x*-direction (see Fig. 2.1 in Chapter 2).

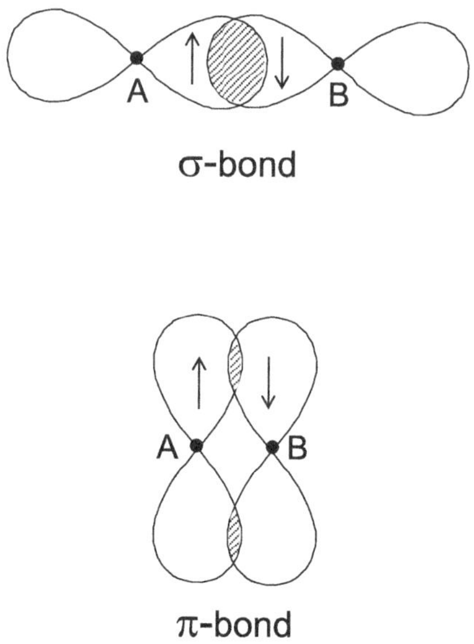

FIGURE K.1 σ-bonding (upper panel) and π-bonding (lower panel) between two *p*-orbitals. The σ-molecular orbital is cylindrically symmetric about the bond axis, whereas the π-molecular orbital is not and retains the mutual nodal plane of the two *p*-orbitals.

We determine the molecular orbitals and their energy levels by using the variation principle, which states that the expectation value of the energy $\langle \Psi_{MO} | \mathcal{H} | \Psi_{MO} \rangle / N^2$ is always greater than, or equal to, the true energy. Therefore, we seek to minimize this energy integral $\bar{E}$ that is given by:

$$\bar{E}\left(a^2 + b^2 + 2abS_{A,B}\right) = \langle \Psi_{MO} | \mathcal{H} | \Psi_{MO} \rangle$$
$$= a^2 \mathcal{H}_{AA} + b^2 \mathcal{H}_{BB} + 2ab\mathcal{H}_{AB} \tag{K.8}$$

where the matrix elements of the one-electron Hamiltonian are: $\mathcal{H}_{AB} = \langle \psi_A | \mathcal{H} | \psi_B \rangle$, etc., and the multiplying factor of N^2 on the left comes from the normalization. Differentiating Eq. K.8 with respect to admixture coefficient a we get:

$$\frac{\partial \bar{E}}{\partial a}\left(a^2 + b^2 + 2abS_{A,B}\right) + \bar{E}\left(2a + 2bS_{A,B}\right)$$
$$= 2a\mathcal{H}_{AA} + 2b\mathcal{H}_{AB} \tag{K.9}$$

From the minimization condition $\partial \bar{E}/\partial a = 0$, we then get:

$$\left(\mathcal{H}_{AA} - \bar{E}\right)a + \left(\mathcal{H}_{AB} - S_{A,B}\bar{E}\right)b = 0 \tag{K.10}$$

Similarly by minimizing with respect to admixture coefficient b we also get:

$$\left(\mathcal{H}_{AB} - S_{A,B}\bar{E}\right)a + \left(\mathcal{H}_{BB} - \bar{E}\right)b = 0 \tag{K.11}$$

We solve these linear simultaneous equations by using the secular equation:

$$\begin{vmatrix} \mathcal{H}_{AA} - \bar{E} & \mathcal{H}_{AB} - S_{A,B}\bar{E} \\ \mathcal{H}_{AB} - S_{A,B}\bar{E} & \mathcal{H}_{BB} - \bar{E} \end{vmatrix} = 0 \tag{K.12}$$

as for degenerate-state perturbation theory described in Appendix G.2. Quadratic solutions for the two energy levels are:

$$\bar{E}^{\pm} = E_o \pm \Delta E \tag{K.13}$$

with

$$E_o = \frac{1}{2\left(1 - S_{A,B}^2\right)}\left(\mathcal{H}_{AA} + \mathcal{H}_{BB} - 2\mathcal{H}_{AB}S_{A,B}\right) = \frac{\bar{\alpha} - \beta S}{1 - S^2} \tag{K.14}$$

$$\Delta E = \frac{1}{1 - S_{A,B}^2}\sqrt{\tfrac{1}{4}\left(\mathcal{H}_{AA} - \mathcal{H}_{BB}\right)^2 + \mathcal{H}_{AB}^2 - \left(\mathcal{H}_{AA} + \mathcal{H}_{BB}\right)\mathcal{H}_{AB}S_{A,B} + \mathcal{H}_{AA}\mathcal{H}_{BB}S_{A,B}^2}$$

$$= \frac{1}{1 - S^2}\sqrt{\left(\beta - \bar{\alpha}S\right)^2 + (\Delta\alpha)^2\left(1 - S^2\right)} \qquad \text{(K.15)}$$

where $\bar{\alpha} \equiv \frac{1}{2}\left(\mathcal{H}_{AA} + \mathcal{H}_{BB}\right)$ and $\Delta\alpha \equiv \frac{1}{2}\left(\mathcal{H}_{AA} - \mathcal{H}_{BB}\right)$ are the mean energy and half the difference in energies, respectively, of the two atoms, $\beta \equiv \mathcal{H}_{AB}$ is the resonance integral, and we have dropped the subscript to the overlap integral S. The resonance integral is responsible for stable bonding, and β is therefore negative (as is $\bar{\alpha}$). The minus sign in Eq. K.13 corresponds to the bonding orbital (b) and the plus sign to the antibonding orbital (a) – see Figure K.2.

We derive the molecular orbital coefficients a and b as in Appendix G.4 for degenerate-state perturbation theory. From Eq. K.10, their ratio is given by:

$$\frac{b}{a} = \frac{\bar{E}^{\pm} - \bar{\alpha} - \Delta\alpha}{\beta - \bar{E}^{\pm}S} \qquad \text{(K.16)}$$

in terms of which the normalized version of Eq. K.5 becomes:

$$\Psi_{MO} = \frac{1}{\sqrt{1 + (b/a)^2 + 2(b/a)S_{A,B}}}\left(\psi_A + (b/a)\psi_B\right) \qquad \text{(K.17)}$$

We get the bonding orbital $\Psi_{MO}^{(b)}$ from Eq. K.17 by taking $\bar{E}^-$ in Eq. K.16 and the antibonding orbital $\Psi_{MO}^{(a)}$ by taking $\bar{E}^+$.

Simplifications ensue if we neglect overlap, i.e., assume that $S = 0$. This is the approach normally taken in Hückel molecular orbital theory, where the parameters are largely empirical. Note that redefining β as: $\beta_{eff} \equiv \mathcal{H}_{AB} - S\bar{\alpha}$ leaves a much smaller S-dependent term, $S\left(\bar{\alpha} - \bar{E}\right)$, in Eq. K.12. This reparameterization therefore partially justifies neglecting overlap (Streitwieser 1961).

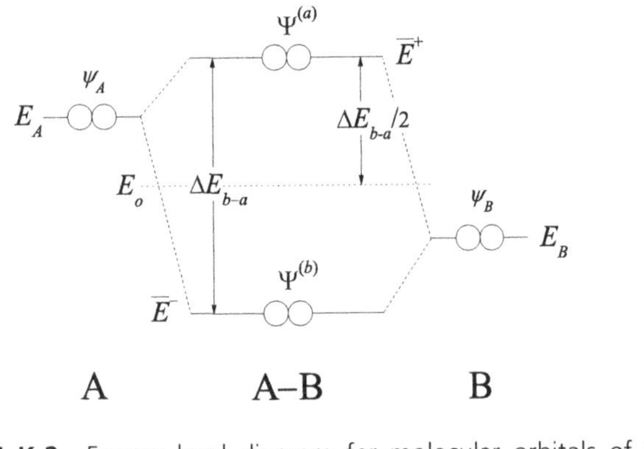

FIGURE K.2 Energy level diagram for molecular orbitals of an A–B bond (see Eqs. K.13–K.15). $\Psi^{(b)}$ and $\Psi^{(a)}$ are bonding and antibonding molecular orbitals, respectively, where constituent atomic orbitals are ψ_A and ψ_B (see Eq. K.5). Energy levels of the isolated atoms are $E_A = \mathcal{H}_{AA} = \langle\psi_A|\mathcal{H}|\psi_A\rangle$ and $E_B = \mathcal{H}_{BB}$, and E_o is given by Eq. K.14. Each energy level can accommodate two electrons with opposite spins.

A different approximation is possible when the electronegativities of A and B are only moderately different, i.e., $|\Delta\alpha| << |\beta|$ as for nitrogen and oxygen. The energy levels and molecular orbital coefficients then become:

$$\bar{E}^{\pm} \approx \frac{\bar{\alpha} \pm \beta}{1 \pm S} \pm \frac{(\Delta\alpha)^2}{2\left(\beta - \bar{\alpha}S\right)} \qquad \text{(K.18)}$$

$$a = \frac{1}{\sqrt{2(1 \pm S)}}\left(\pm 1 + \tfrac{1}{2}(1 \pm S)\frac{\Delta\alpha}{\beta - \bar{\alpha}S}\right) \qquad \text{(K.19)}$$

$$b = \frac{1}{\sqrt{2(1 \pm S)}}\left(1 \mp \tfrac{1}{2}(1 \pm S)\frac{\Delta\alpha}{\beta - \bar{\alpha}S}\right) \qquad \text{(K.20)}$$

– see Mulliken (1935).

Figure K.3 shows the overlap integrals calculated from Slater orbitals for nitrogen and oxygen, as a function of the N–O distance. For bond lengths found in nitroxides, the overlap integrals for π-bonding and for σ-bonding are: $S_{N,O} = 0.146$ ($2p\pi$) and $S_{N,O} = 0.294$ ($2sp^2$ σ-hybrid), respectively. We expect that the resonance integral is proportional to the overlap integral, for separations above and down to the bond distance. A form sometimes assumed is (Wolfsberg and Helmholz 1952; Hoffmann 1963):

$$\beta = K\bar{\alpha}S \qquad \text{(K.21)}$$

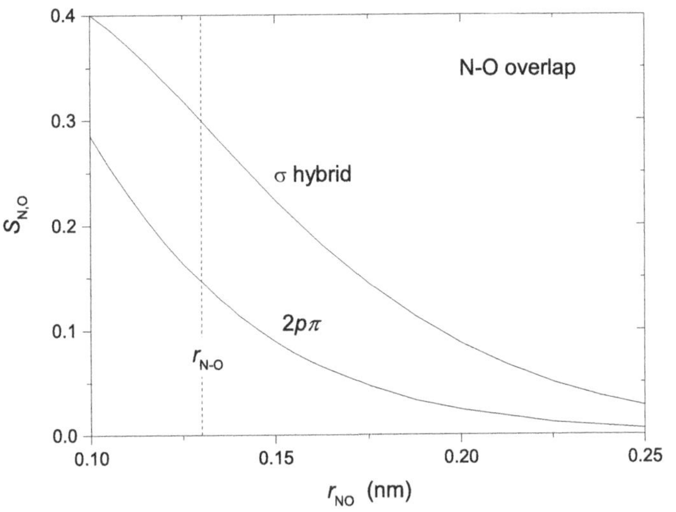

FIGURE K.3 Dependence of N–O overlap integrals for $2p_z\pi$-orbitals and σ-bonding hybrid $2sp^2$-orbitals $\Psi_{\sigma_b} = 0.451|s\rangle_N + 0.545|p_x\rangle_N + 0.451|s\rangle_O + 0.545|p_x\rangle_O$ (see Eq. 3.69), on separation r_{NO} of nitrogen and oxygen atoms. Vertical dashed line indicates the N–O bond distance r_{N-O} in nitroxides. From tabulations for Slater atomic orbitals by Mulliken et al. (1949).

with $K \approx 1.75-2.0$. Values for the Coulomb integrals come from the negative of the corresponding nitrogen and oxygen valence-state ionization potentials, but can be estimated directly from the neutral-atom negativities of Klopman (1965): $\bar{\alpha} = -8.8$ eV and $\Delta\alpha = 0.7$ eV (N minus O). The resonance integral for $2p\pi$-bonding estimated from Eq. K.21 then becomes: $\beta = -2.3$ eV (cf. –2.5 eV estimated from optical absorption in Section 3.2). Correspondingly, the parameter used in Sections 3.2 and 4.2.3 becomes $h_{NO} \equiv \Delta\alpha/\beta = -0.3$ eV, instead of –0.5 eV estimated from Pauling electronegativities. The redefined value of β is, however, $\beta_{eff} \equiv \beta - S\bar{\alpha} = -1.0$ eV.

K.3 Hybridization and Bond Orientation

We achieve directionality in chemical bonding by hybridizing one or more p-orbitals with the corresponding spherically-symmetric s-orbital. This results in orbitals with a single major lobe that points out in a unique direction from the central atom. For example, the two sp_x-hybrid atomic orbitals are given by:

$$\psi_{sp} = \frac{1}{\sqrt{2}}\left(|s\rangle \pm |p_x\rangle\right) \tag{K.22}$$

Each has predominantly a single lobe that points in the direction of one of the two lobes of the original p_x-orbital. This serves to concentrate the electron density in the direction of the atom with which the sp_x-hybrid forms a molecular orbital. Note that Fig. K.1 is a highly idealized depiction of σ-bonding for p-orbitals in most molecules. Unlike with π-bonding, the σ-bond would normally not involve pure p-orbitals alone, but some form of s-p hybrid.

The σ-bonds C_1–NO–C_2 of a nitroxide are close to coplanar, (especially for the five-membered rings – see Table 3.4 in Chapter 3), and lie in the x,y-plane. However, the two C–N bonds lie neither along the direction of the nitrogen $2p_x$-orbital nor along that of the $2p_y$-orbital. To form σ-bonding orbitals, we must take linear combinations of the nitrogen $2s$, $2p_x$ and $2p_y$ orbitals such that they lie along the bond directions. We call this trigonal hybridization. All three hybrid orbitals contain s-electron character; that lying along the N–O bond ψ_{N-O} contains only p_x-electron character, whereas those lying along the C–N bonds ψ_{N-C_1} and ψ_{N-C_2} contain both p_x- and p_y-electron character (see Figure K.4). Let c_s^2 be the fraction of nitrogen s-electron density in each of the two equivalent C–N bonds, then that in the N–O bond must be $1-2c_s^2$ so that the total adds up to unity. The sp^2 hybrid nitrogen orbital that is directed along the N–O bond is therefore:

$$\psi_{N-O} = \sqrt{1-2c_s^2}\,|s\rangle + \sqrt{2c_s^2}\,|p_x\rangle \tag{K.23}$$

where the p_x-electron admixture comes immediately from the normalization condition: $\langle\psi_{N-O}|\psi_{N-O}\rangle = 1$. The two sp^2 nitrogen hybrids that participate in N–C bonding must be of the form:

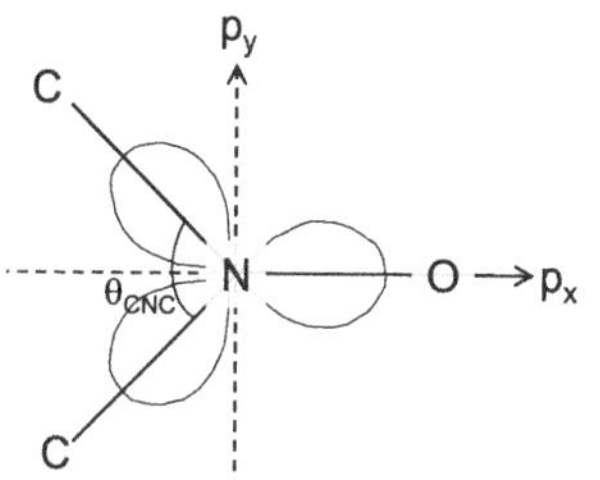

FIGURE K.4 Principal lobes (grey lines) of the three sp^2-hybrid orbitals of the nitrogen atom in a nitroxide radical. The hybrid N-orbitals form σ-bonds in the x-y plane with the O-atom and two C-atoms. The N–O directed sp^2-hybrid contains only the p_x-orbital, whereas the contributions of the p_x-orbital and p_y-orbitals to the N–C directed hybrids are $\cos\frac{1}{2}\theta_{CNC}$ and $\sin\frac{1}{2}\theta_{CNC}$, respectively.

$$\psi_{N-C_{1,2}} = c_s|s\rangle + c_{p_x}|p_x\rangle \pm \sqrt{1-c_s^2-c_{p_x}^2}\,|p_y\rangle \tag{K.24}$$

because, being equivalent, they may differ only in the sign of the admixture coefficients. We get the p_x-admixture coefficient: $c_{p_x} = -\sqrt{\frac{1}{2}-c_s^2}$ from the condition that these orbitals must be orthogonal to ψ_{N-O}: $\langle\psi_{N-C}|\psi_{N-O}\rangle = c_s\sqrt{1-2c_s^2} + c_{p_x}\sqrt{2c_s^2} = 0$. With this value of c_{p_x}, we see also that the two $\psi_{N-C_{1,2}}$ orbitals in Eq. K.24 are mutually orthogonal, as they must be. The required hybrids are therefore:

$$\psi_{N-C_1} = c_s|s\rangle - \sqrt{\frac{1-2c_s^2}{2}}\,|p_x\rangle - \frac{1}{\sqrt{2}}|p_y\rangle \tag{K.25}$$

$$\psi_{N-C_2} = c_s|s\rangle - \sqrt{\frac{1-2c_s^2}{2}}\,|p_x\rangle + \frac{1}{\sqrt{2}}|p_y\rangle \tag{K.26}$$

Because the orbitals behave like vectors (see, e.g., Streitwieser 1961), we can relate the admixture coefficients of p_x and p_y in Eqs. K.25 and K.26 to the C_1–N–C_2 bond angle θ_{CNC}. From Fig. K.4, we see that the admixture coefficients are the projections on the y- and x-axes, and their ratio is given by:

$$\tan\tfrac{1}{2}\theta_{CNC} = \frac{1}{\sqrt{1-2c_s^2}} \tag{K.27}$$

from which we deduce that the s-electron admixture is: $c_s^2 = \cos\theta_{CNC}/(\cos\theta_{CNC}-1)$. Values of the bond angle θ_{CNC} are given for various nitroxides in Table 3.4 of Chapter 3.

Analogues of the above apply to sp^2 hybrid orbitals centred on the oxygen atom. That for the N–O bond is analogous to Eq. K.23 and those for the oxygen lone pairs to Eqs. K.25 and K.26. Note that trigonal sp^2 hybrids that are fully equivalent point towards vertices 1,2,3 of an equilateral triangle, with angles $\theta_{12} = \theta_{23} = \theta_{31} = 120°$ between the orbital axes. The hybrid orbitals corresponding to Eqs. K.23, K.25 and K.26 then become:

$$\psi_1 = \frac{1}{\sqrt{3}}|s\rangle + \sqrt{\frac{2}{3}}|p_x\rangle \tag{K.28}$$

$$\psi_2 = \frac{1}{\sqrt{3}}|s\rangle - \sqrt{\frac{1}{6}}|p_x\rangle - \frac{1}{\sqrt{2}}|p_y\rangle \tag{K.29}$$

$$\psi_3 = \frac{1}{\sqrt{3}}|s\rangle - \sqrt{\frac{1}{6}}|p_x\rangle + \frac{1}{\sqrt{2}}|p_y\rangle \tag{K.30}$$

respectively, i.e., $c_s = 1/\sqrt{3}$ when the wavefunctions are normalized. Taking the rotation matrix from Eq. I.20 in Appendix I, we can generate ψ_2 by rotating ψ_1 by 120° about the z-axis, and ψ_3 with a further 120°-rotation, and so on. Thus, the hybrid orbitals behave as vectors in the way that led to Eq. K.27 above.

APPENDIX L: *g*-VALUES FOR A *p*-ELECTRON

As a particularly instructive example for the application of second-order perturbation theory, we now calculate the g-values for the hypothetical case of a p-electron where the orbital degeneracy is removed by a strong interaction, such as that involved in chemical bonding.

Assume that the unpaired electron is in a p_z-orbital and that the fully occupied orbitals p_x and p_y lie lower in energy by an amount ΔE, which is much larger than the spin-orbit coupling energy (see Figure L.1). This simple hypothetical model is similar to the situation in nitroxide free radicals. Spin-orbit coupling (see Section 3.12) introduces small admixtures of the p_x and p_y orbitals into the p_z-orbital that contains the unpaired electron. Therefore, in addition to the interaction of the orbital and spin magnetic moments with the magnetic field **B**, we must also include spin-orbit coupling in the perturbation Hamiltonian:

$$\mathcal{H}' = \zeta\mathbf{l}\cdot\mathbf{s} + \beta_e\mathbf{B}\cdot(\mathbf{l} + g_e\mathbf{s}) \tag{L.1}$$

where **l** is the orbital angular momentum and **s** is the true spin of the unpaired electron and ζ is the spin-orbit coupling constant.

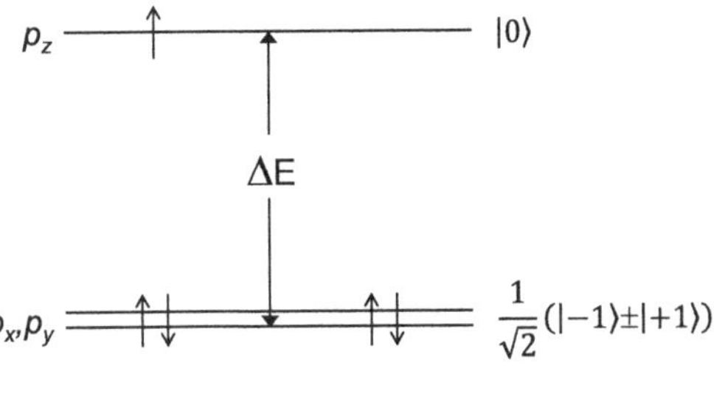

FIGURE L.1 Simple model for calculating g-values of a p-electron. Unpaired electron resides in the p_z-orbital which is higher in energy by $\Delta E >> \varsigma l \cdot s$ than the fully occupied p_x- and p_y-orbitals.

The first-order perturbation energy is (cf. Eqs. G.10 and G.11 of Appendix G):

$$E^{(1)} = \langle p_z|\mathcal{H}'|p_z\rangle = g_e\beta_e\mathbf{B}\cdot\mathbf{s} \tag{L.2}$$

because $\langle p_z|l_j|p_z\rangle = 0$ for $j = x, y, z$ (see Table E.1 of Appendix E). The energy splitting ΔE causes orbital quenching of the unpaired p_z electron (e.g., by chemical bonding), and we obtain only free-spin paramagnetism in first order. The second-order perturbation energy is given by (cf. Eq. G.26 of Appendix G):

$$E^{(2)} = \frac{\sum_{j=x,y}\langle p_z|\mathcal{H}'|p_j\rangle\langle p_j|\mathcal{H}'|p_z\rangle}{\Delta E} \approx \frac{2\beta_e\zeta}{\Delta E}\sum_{j=x,y}\langle p_z|\mathbf{l}\cdot\mathbf{s}|p_j\rangle\langle p_j|\mathbf{B}\cdot\mathbf{l}|p_z\rangle \tag{L.3}$$

where we retain only the term that is linear in the magnetic field **B** in the right-hand equality. To get the g-values, we identify Eqs. L.2 and L.3 for the perturbation energy with the Zeeman part of the spin Hamiltonian:

$$\mathcal{H}_S = \beta_e\left(g_{xx}B_xS_x + g_{yy}B_yS_y + g_{zz}B_zS_z\right) \tag{L.4}$$

Equating with the corresponding B_xs_x, B_ys_y and B_zs_z terms from $E^{(1)} + E^{(2)}$, gives the following expressions for the g-tensor elements:

$$g_{zz} = g_e \tag{L.5}$$

$$g_{xx} = g_e + \frac{2\zeta}{\Delta E}\sum_{j=x,y}\langle p_z|l_x|p_j\rangle\langle p_j|l_x|p_z\rangle \tag{L.6}$$

$$g_{yy} = g_e + \frac{2\zeta}{\Delta E}\sum_{j=x,y}\langle p_z|l_y|p_j\rangle\langle p_j|l_y|p_z\rangle \tag{L.7}$$

The matrix elements of l_x and l_y for p-orbitals are given in Table E.1 of Appendix E. Remember that the p-orbitals are linear combinations of the eigenfunctions $|m_l\rangle$ of l_z (see Table K.1 of Appendix K). Equation L.6 for the g_{xx} tensor element then becomes:

$$g_{xx} = g_e + \frac{2\zeta}{\Delta E} \tag{L.8}$$

Evaluating Eq. L.7 similarly, we find that $g_{yy} = g_{xx}$, as required from the axial symmetry of the model.

APPENDIX M: TIME-DEPENDENT PERTURBATION THEORY AND SELECTION RULES

M.1 Time-Dependent Perturbations

We assume that the time-dependent spin Hamiltonian consists of a static part, $\mathcal{H}_o$, and a time-dependent perturbation, $\mathcal{H}_1 F(t)$:

$$\mathcal{H}(t) = \mathcal{H}_o + \mathcal{H}_1 F(t) \tag{M.1}$$

The eigenstates, u_m, of the stationary system are given by the solutions of the time-independent Schrödinger equation:

$$\mathcal{H}_o u_m = \hbar\omega_m u_m \tag{M.2}$$

where $\hbar\omega_m$ is the energy of the mth spin state. The time-dependent Schrödinger equation (see Appendix D) reads:

$$i\hbar\frac{\partial\psi}{\partial t} = \left(\mathcal{H}_o + \mathcal{H}_1 F(t)\right)\psi(t) \tag{M.3}$$

where the wave function $\psi(t)$ describes the time evolution of the spin system. We solve Eq. M.3 by making a series expansion in the eigenfunctions $u_m \exp(-i\omega_m t)$ of the unperturbed time-dependent Schrödinger equation:

$$\psi(t) = \sum_m c_m(t) u_m \exp(-i\omega_m t) \tag{M.4}$$

To obtain the expansion coefficients, $c_m(t)$, we substitute Eq. M.4 into Eq. M.3 giving:

$$i\hbar\sum_m \frac{\partial c_m}{\partial t} u_m \exp(-i\omega_m t) = \sum_m \mathcal{H}_1 u_m F(t)\exp(-i\omega_m t) c_m(t) \tag{M.5}$$

We then multiply on the left by $u^*_{m'}$ and integrate over all space to exploit the orthogonality properties of the u_m solutions of the static Schrödinger equation: $\int u^*_{m'} u_m \cdot \mathrm{d}^3 r = \delta_{m',m}$ (see Appendix C.1). This gives the rate equation for the expansion coefficients in Eq. M.4:

$$\frac{\partial c_{m'}(t)}{\partial t} = \frac{-i}{\hbar}\sum_m \langle m'|\mathcal{H}_1|m\rangle F(t)\exp(i\omega_{m'm}t)c_m(t) \tag{M.6}$$

where $\omega_{m'm} \equiv \omega_{m'} - \omega_m$, and we give the matrix elements of the perturbation in Dirac notation: $\langle m'|\mathcal{H}_1|m\rangle \equiv \int u^*_{m'}\mathcal{H}_1 u_m \cdot \mathrm{d}^3 r$, with the integration over all space.

For transitions from the system originally in a single state m, we approximate the coefficients on the right of Eq. M.6 in first order by $c_m(t) \approx c_m(0) = 1$ and replace the summation by a single value. The amplitude of spin state m' at time t then becomes:

$$c_{m'}(t) = \frac{-i}{\hbar}\langle m'|\mathcal{H}_1|m\rangle \int_0^t F(t')\exp(i\omega_{m'm}t')\cdot \mathrm{d}t' \tag{M.7}$$

in first-order perturbation theory.

M.2 Spectroscopic Transition Probabilities

An important application of Eq. M.7 is perturbation of the electron Zeeman levels by a microwave magnetic field $\mathbf{B}_1$. The time dependence is then harmonic:

$$F(t) = 2\cos\omega t = \exp(-i\omega t) + \exp(+i\omega t) \tag{M.8}$$

where the second equality gives the two constituent circular polarizations. Integrating Eq. M.7 with this time dependence gives the time evolution for the amplitude of the m' spin state:

$$c_{m'}(t) = \frac{\langle m'|\mathcal{H}_1|m\rangle}{\hbar} \times\left(\frac{\exp\left(i(\omega_{m'm}-\omega)t\right)-1}{\omega_{m'm}-\omega} + \frac{\exp\left(i(\omega_{m'm}+\omega)t\right)-1}{\omega_{m'm}+\omega}\right) \tag{M.9}$$

where $\omega_{m'm}$ is the angular resonance (i.e., transition) frequency. Only the first of the two terms on the right contributes appreciably to the resonance at $\omega = \omega_{m'm}$. The transition probability per unit time, $W_{m'm} = |c_{m'}|^2/t$, is then given by:

$$W_{m'm} = \frac{|\langle m'|\mathcal{H}_1|m\rangle|^2}{\hbar^2 t}\left(\frac{2\left(1-\cos(\omega_{m'm}-\omega)t\right)}{(\omega_{m'm}-\omega)^2}\right) \tag{M.10}$$

The bracketed term on the right peaks sharply at the transition frequency $\omega = \omega_{m'm}$, behaving at long times like a δ-function (see Fig. M.1). If we change the frequency variable from $(\omega_{m'm}-\omega)$ to $x = (\omega_{m'm}-\omega)t/2$, the integral of this function over the entire lineshape becomes $2t\int_{-\infty}^{\infty} x^{-2}\sin^2 x\,\mathrm{d}x = 2\pi\cdot t$. Therefore, we can express the rate of transition in terms of a δ-function:

$$W_{m'm} = \frac{2\pi}{\hbar^2}|\langle m'|\mathcal{H}_1|m\rangle|^2\,\delta(\omega_{m'm}-\omega) \tag{M.11}$$

with the normalizing integral: $\int_{-\infty}^{\infty}\delta(\omega_{m'm}-\omega)d(\omega_{m'm}-\omega) = 1$

Note that, like any other lineshape function, the units of the δ-function are those of its inverse argument,

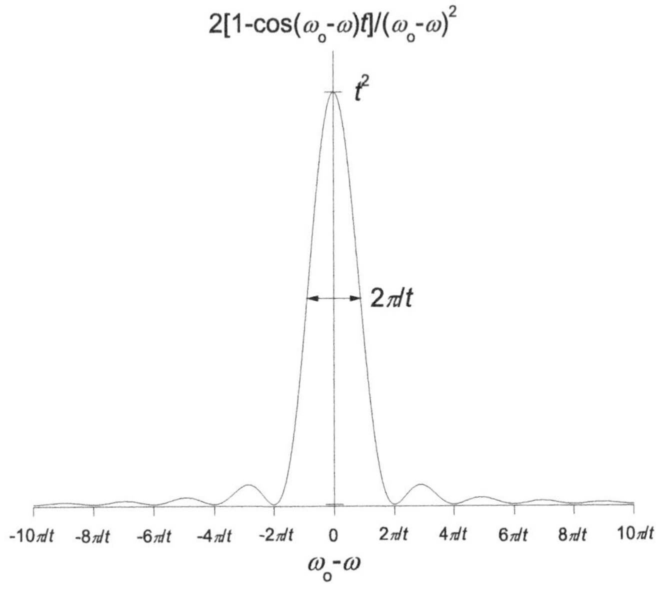

FIGURE M.1 Frequency dependence of the lineshape function: $2\left[1-\cos(\omega_{m'm}-\omega)t\right]/(\omega_{m'm}-\omega)^2$ in Eq. M.10. This asymptotically approaches a δ-function, as $t \to \infty$.

here the inverse angular frequency. In terms of a magnetic field scan, the δ-function in Eq. M.11 is given by: $\delta(\omega_{m'm}-\omega)=\delta(B_{m'm}-B)\times \mathrm{d}B/\mathrm{d}\omega=(1/\gamma_e)\delta(B_{m'm}-B)$, where $\gamma_e \equiv g\beta_e/\hbar$ is the gyromagnetic ratio of the electron.

Equation M.11 is a fundamental result for spectroscopic transitions, which embodies Fermi's well-known golden rule. Note that the time dependence in Eq. M.10 disappears only after integrating over $(\omega_{m'm}-\omega)$. For EPR, this corresponds to integrating over the spectral lineshape. For optical spectroscopy, it could correspond instead to integrating over the frequency of a white-light radiation source. The generalization of Eq. M.11 for inhomogeneously broadened spin-label EPR spectra is thus:

$$W_{m'm}(\omega)=\frac{2\pi}{\hbar^2}\left|\langle m'|\mathcal{H}_1|m\rangle\right|^2 g(\omega) \qquad \text{(M.12)}$$

where $g(\omega)$ is the spectral lineshape function, which we normalize according to $\int g(\omega)\,\mathrm{d}\omega = 1$.

M.3 Selection Rules

In addition to frequency matching, which is given by the δ-function in Eq. M.11, selection rules for a particular type of spectroscopy are determined by the matrix element $\langle m'|\mathcal{H}_1|m\rangle$ of the perturbation. For magnetic dipole transitions in EPR, the perturbation of the electron spin **S** by the microwave $\mathbf{B}_1$-field is given by:

$$\mathcal{H}_1 = g\beta_e \mathbf{B}_1\cdot\mathbf{S} \qquad \text{(M.13)}$$

where we ignore the small g-value anisotropy of the nitroxide. The S_z component of **S** has no matrix elements connecting the $M_S=+\frac{1}{2}$ and $M_S=-\frac{1}{2}$ spin states. To induce transitions in the electron spin orientation, the microwave magnetic field therefore must be perpendicular to the static $\mathbf{B}_o$ field. Conventionally, we put $\mathbf{B}_1$ along the x-axis and then $\langle +\frac{1}{2}|\mathcal{H}_1|-\frac{1}{2}\rangle = g\beta_e B_{1,x}\langle +\frac{1}{2}|S_x|-\frac{1}{2}\rangle=\frac{1}{2}g\beta_e B_{1,x}$. Quite generally, the selection rule for magnetic dipole transitions in EPR is $\Delta M_S=\pm 1$, because S_x only connects M_S-states that differ by ± 1 (see Appendix E.3). This includes the $S=1$ triplet state of a nitroxide biradical. Note that the selection rule for the nitrogen nuclear spin is $\Delta m_I = 0$ in spin-label EPR, because the microwave frequency is far too high to fulfill the matching condition of Eq. M.11 for nuclear transitions.

REFERENCES

Brink, D.M. and Satchler, G.R. 1968. *Angular Momentum*, Oxford: Oxford University Press.

Hoffmann, R. 1963. An extended Hückel theory. I. Hydrocarbons. *J. Chem. Phys.* 39:1397–1412.

Iwasaki, M. and Toriyama, K. 1985. Exact analysis of polycrystalline electron spin echo envelope modulation including mutual nuclear arrangements and quadrupole interactions and its application to methyl radicals in irradiated crystals of lithium acetate dihydrate. *J. Chem. Phys.* 82:5415–5423.

Klopman, G. 1965. Electronegativity. *J. Chem. Phys.* 43:S124–S129.

Mulliken, R.S. 1935. Electronic structures of molecules. XI. Electroaffinity, molecular orbitals and dipole moments. *J. Chem. Phys.* 3:573–585.

Mulliken, R.S., Rieke, C.A., Orloff, D. and Orloff, H. 1949. Formulas and numerical tables for overlap integrals. *J. Chem. Phys.* 17:1248–1267.

Schweiger, A. and Jeschke, G. 2001. *Principles of Pulse Electron Paramagnetic Resonance*, Oxford: Oxford University Press.

Streitwieser, A. 1961. *Molecular Orbital Theory for Organic Chemists*, New York: John Wiley.

Van, S.P., Birrell, G.B., and Griffith, O.H. 1974. Rapid anisotropic motion of spin labels. Models for motion averaging of the ESR parameters. *J. Magn. Reson.* 15:444–459.

Wolfsberg, M. and Helmholz, L. 1952. The spectra and electronic structure of the tetrahedral ions MnO_4^-, CrO_4^{--}, and ClO_4^-. *J. Chem. Phys.* 20:837–843.

Appendices N–V

Specialist Topics

APPENDIX N: SPIN DENSITY-MATRIX

Density-matrix methods extend quantum mechanical treatments of single-spin systems to ensembles of spins. This is analogous with the classical approach to macroscopic magnetization in statistical mechanics. We treat the expansion coefficients, c_n, of the wave function as an operator, which has matrix elements given by the products, $c_n c_m^*$. Clearly the density matrix is Hermitean $\left(c_n c_m^* = c_m c_n^*\right)$, and the on-diagonal elements $|c_n|^2$ are probabilities (or occupations) of the different eigenstates φ_n with energies E_n (see Appendix C.1). We need off-diagonal elements of the density matrix to calculate expectation values of the transverse spin components S_x and S_y, for example.

N.1 Spin-Operator Expectation Values

As usual, we expand the wave function ψ of the spin system in a complete orthonormal set of spin basis functions, u_n:

$$\psi = \sum_n c_n u_n \tag{N.1}$$

which are not necessarily the eigenstates of the system (see Appendix C). The expectation value of a spin operator, say S_y, is then given by:

$$\langle S_y \rangle \equiv \langle \psi | S_y | \psi \rangle = \sum_{n,m} c_m^* c_n \langle m | S_y | n \rangle \tag{N.2}$$

The time evolution is determined solely by the products, $c_n c_m^*$, which form the elements, σ_{nm}, of the spin density-matrix:

$$\sigma_{nm} \equiv \langle n | \sigma | m \rangle = c_n c_m^* \tag{N.3}$$

From the rules for matrix multiplication, the expectation value given in Eq. N.2 then becomes:

$$\langle S_y \rangle = \sum_{n,m} \langle n|\sigma|m \rangle \langle m|S_y|n \rangle$$

$$= \sum_n \langle n|\sigma S_y|n \rangle$$

$$\text{i.e., } \langle S_y \rangle = \mathrm{Tr}\left(\sigma S_y\right) = \mathrm{Tr}\left(S_y \sigma\right) \tag{N.4}$$

where the trace, $\mathrm{Tr}\left(\sigma S_y\right)$, is the sum of the on-diagonal elements for the product of the operators σ and S_y (see Appendix U). This result is particularly useful, because the trace does not depend on which set of basis functions we choose to represent the spin density matrix. The eigenfunctions $\left|M_S = \pm\tfrac{1}{2}\right\rangle$ of the spin projection S_z are usually the most convenient.

The density matrix forms the bridge between quantum mechanics and statistical mechanics. For an ensemble of spin systems, the density matrix elements σ_{nm} are the ensemble averages, $\overline{c_n c_m^*}$, of the products of the expansion coefficients, and the expectation value in Eq. N.4 also becomes the ensemble average $\overline{\langle S_y \rangle}$. For example, the macroscopic magnetization $\left(M_x, M_y, M_z\right)$ is related directly to the ensemble averages of the expectation values of the spin components $\left(\overline{\langle S_x \rangle}, \overline{\langle S_y \rangle}, \overline{\langle S_z \rangle}\right)$:

$$\overline{\langle S_x \rangle} = \mathrm{Tr}(\sigma S_x) = \left\langle +\tfrac{1}{2}\middle|\sigma\middle|-\tfrac{1}{2}\right\rangle\left\langle -\tfrac{1}{2}\middle|S_x\middle|+\tfrac{1}{2}\right\rangle + \left\langle -\tfrac{1}{2}\middle|\sigma\middle|+\tfrac{1}{2}\right\rangle\left\langle +\tfrac{1}{2}\middle|S_x\middle|-\tfrac{1}{2}\right\rangle = \tfrac{1}{2}(\sigma_{+-} + \sigma_{-+}) \tag{N.5}$$

$$\overline{\langle S_y \rangle} = \mathrm{Tr}(\sigma S_y) = \left\langle +\tfrac{1}{2}\middle|\sigma\middle|-\tfrac{1}{2}\right\rangle\left\langle -\tfrac{1}{2}\middle|S_y\middle|+\tfrac{1}{2}\right\rangle + \left\langle -\tfrac{1}{2}\middle|\sigma\middle|+\tfrac{1}{2}\right\rangle\left\langle +\tfrac{1}{2}\middle|S_y\middle|-\tfrac{1}{2}\right\rangle = -\tfrac{1}{2}i(\sigma_{+-} - \sigma_{-+}) \tag{N.6}$$

$$\overline{\langle S_z \rangle} = \mathrm{Tr}(\sigma S_z) = \left\langle +\tfrac{1}{2}\middle|\sigma\middle|+\tfrac{1}{2}\right\rangle\left\langle +\tfrac{1}{2}\middle|S_z\middle|+\tfrac{1}{2}\right\rangle + \left\langle -\tfrac{1}{2}\middle|\sigma\middle|-\tfrac{1}{2}\right\rangle\left\langle -\tfrac{1}{2}\middle|S_z\middle|-\tfrac{1}{2}\right\rangle = \tfrac{1}{2}(\sigma_{++} - \sigma_{--}) \tag{N.7}$$

where the matrix elements of σ are $\sigma_{+-} \equiv \left\langle +\tfrac{1}{2}\middle|\sigma\middle|-\tfrac{1}{2}\right\rangle$, et seq., and we take the non-zero matrix elements of the spin components for $S = \tfrac{1}{2}$ from Table E.1 in Appendix E.

A particularly simple (but extremely important) example is the expectation value of the raising operator $S_+ \equiv S_x + iS_y$. For $S = \tfrac{1}{2}$, this operator has only one non-zero matrix element: $\left\langle +\tfrac{1}{2}\middle|S_+\middle|-\tfrac{1}{2}\right\rangle = 1$. The expectation value is therefore determined by a single off-diagonal element of the spin density matrix:

$$\overline{\langle S_+ \rangle} = \mathrm{Tr}(\sigma S_+) = \left\langle -\tfrac{1}{2}\middle|\sigma\middle|+\tfrac{1}{2}\right\rangle\left\langle +\tfrac{1}{2}\middle|S_+\middle|-\tfrac{1}{2}\right\rangle = \sigma_{-+} \tag{N.8}$$

The $\overline{\langle S_+ \rangle}$ expectation value is crucial because it determines the EPR signal that is induced by right circularly polarized microwave radiation (see Section M.3 in Appendix M). It therefore specifies both the T_2-relaxation time (see Chapter 5) and the EPR lineshape (see Chapter 6).

N.2 Equilibrium Density Matrix

If we take the eigenstates φ_n of the time-independent spin Hamiltonian $\mathcal{H}_o$ as basis states, then $\langle m|\mathcal{H}_o|n\rangle = \delta_{mn}E_n$ where E_n are the energy eigenstates of the spin system. At equilibrium, the diagonal elements of the spin density matrix must correspond to the Boltzmann distribution:

$$\sigma_{nn} = \frac{\exp\left(-E_n/k_BT\right)}{\sum_m \exp\left(-E_m/k_BT\right)} \equiv \frac{\exp\left(-(\mathcal{H}_o)_{nn}/k_BT\right)}{\mathrm{Tr}\left[\exp\left(-\mathcal{H}_o/k_BT\right)\right]} \tag{N.9}$$

where $(\mathcal{H}_o)_{nn} \equiv \langle n|\mathcal{H}_o|n\rangle = E_n$ is the diagonal element of $\mathcal{H}_o$ in its eigenfunction representation and k_B is Boltzmann's constant. The off-diagonal elements of σ are zero in this representation and so are those of the operator $\exp\left(-\mathcal{H}_o/k_BT\right)$ because from series expansion of the exponential we get the identity: $\exp\left(-\mathcal{H}_o/k_BT\right)\varphi_m = \exp\left(-E_m/k_BT\right)\varphi_m$. Similarly, the denominator on the far right of Eq. N.9 is the sum of the diagonal elements of the exponential operator. Therefore, we can write the equilibrium spin density-matrix, σ_o, quite generally in operator form as:

$$\sigma_o = \frac{\exp\left(-\mathcal{H}_o/k_BT\right)}{\mathrm{Tr}\left[\exp\left(-\mathcal{H}_o/k_BT\right)\right]} \tag{N.10}$$

where the statistical mechanical partition function $\mathrm{Tr}\left[\exp\left(-\mathcal{H}_o/k_BT\right)\right]$ does not depend on the representation, because the trace is invariant.

The spin Hamiltonian $\mathcal{H}_o$ corresponds to energies that are of the order of, at most, a few degrees Kelvin. Thus the high-temperature approximation is valid for spin-label EPR applications. The equilibrium spin density-matrix is then given to first order by:

$$\sigma_o \approx \frac{\mathbf{1}-\left(\mathcal{H}_o/k_BT\right)}{\mathrm{Tr}(\mathbf{1})} \tag{N.11}$$

because $\mathrm{Tr}\left(\mathcal{H}_o\right)=0$ for both Zeeman and dipolar components. The partition coefficient then becomes equal to the number of spin states: $\mathrm{Tr}(\mathbf{1}) = 2S+1$, in the high-temperature approximation. The spin Hamiltonian is dominated by the Zeeman term: $\mathcal{H}_o \approx \hbar\omega_o S_z$, where ω_o is the angular resonance frequency. When analyzing EPR experiments, we usually drop the identity operator $\mathbf{1}$ in the numerator, because it remains invariant. Also, we can ignore the constant factors unless the absolute number of spins is important. A convenient short-hand expression for the equilibrium spin density matrix is therefore:

$$\sigma_o = S_z \tag{N.12}$$

where we also absorb the sign, because normally only relative signs are important. We favour this sign convention because it corresponds with the macroscopic magnetization M_z that we often use for describing pulse experiments (see Chapter 13). Equation N.12 therefore forms a convenient starting point for analyzing pulse-EPR experiments by using the product-operator formalism (see Appendix P).

N.3 Time Dependence

The equation of motion for the elements of the spin density matrix is:

$$\frac{\mathrm{d}\sigma_{km}}{\mathrm{d}t} = \frac{\mathrm{d}c_k c_m^*}{\mathrm{d}t} = c_k\frac{\mathrm{d}c_m^*}{\mathrm{d}t} + \frac{\mathrm{d}c_k}{\mathrm{d}t}c_m^* \tag{N.13}$$

We get the time-derivatives of the c_k-coefficients from the time-dependent Schrödinger Equation for spin Hamiltonian $\mathcal{H}$ (Eq. D.4 in Appendix D):

$$i\hbar\frac{\partial\psi}{\partial t} = \mathcal{H}\psi \tag{N.14}$$

which expanded in terms of Eq. N.1 becomes:

$$i\hbar\sum_n \frac{\partial c_n}{\partial t}u_n = \sum_n c_n\mathcal{H}u_n \tag{N.15}$$

Using the orthogonality properties of the u_n basis functions (as in time-dependent perturbation theory, see Appendix M.1), the time evolution of the c_k-coefficient is therefore:

$$i\hbar\frac{\partial c_k}{\partial t} = \sum_n c_n\langle k|\mathcal{H}|n\rangle \tag{N.16}$$

Substituting this result in Eq. N.13, the time dependence of the elements of the density matrix becomes:

$$\begin{aligned}\frac{d\sigma_{km}}{dt} &= \frac{i}{\hbar}\sum_n\left[c_kc_n^*\langle n|\mathcal{H}|m\rangle - \langle k|\mathcal{H}|n\rangle c_nc_m^*\right]\\ &= \frac{i}{\hbar}\sum_n\left(\sigma_{kn}\mathcal{H}_{nm} - \mathcal{H}_{kn}\sigma_{nm}\right)\\ &= \frac{i}{\hbar}\sum_n\langle k|\sigma\mathcal{H} - \mathcal{H}\sigma|m\rangle\end{aligned} \tag{N.17}$$

where we use the fact that, like the Hamiltonian, the density matrix is Hermitean (i.e., $\sigma_{mn} = \sigma_{nm}^*$, where the asterisk means the complex conjugate).

In operator form, the evolution in time of the spin density matrix is therefore:

$$\frac{\mathrm{d}\sigma}{\mathrm{d}t} = \frac{i}{\hbar}\left[\sigma(t),\mathcal{H}(t)\right] \tag{N.18}$$

where square brackets indicate the commutator $[\sigma,\mathcal{H}] \equiv \sigma\mathcal{H} - \mathcal{H}\sigma$. We call this relation the quantum-mechanical Liouville equation (or Liouville–von Neumann equation). Apart from the change of sign, Eq. N.18 for the density matrix is very similar to the Heisenberg equation of motion for a quantum-mechanical observable (see Eq. D.8

in Appendix D). When $\mathcal{H} \equiv \mathcal{H}_o$ is independent of time, the complete formal solution of Eq. N.18 is:

$$\sigma(t) = e^{-i\mathcal{H}_o t/\hbar}\sigma(0)e^{i\mathcal{H}_o t/\hbar} \tag{N.19}$$

where $\sigma(0)$ is the initial constant value of the spin density matrix. Using the eigenfunction representation φ_n of $\mathcal{H}_o$, the matrix elements of $\sigma(t)$ then become:

$$\begin{aligned}\sigma_{mn}(t) &\equiv \int \varphi_m^* e^{-i\mathcal{H}_o t/\hbar}\sigma(0)e^{i\mathcal{H}_o t/\hbar}\varphi_n \mathrm{d}^3\mathbf{r} \\ &= \int \left(e^{i\mathcal{H}_o t/\hbar}\varphi_m\right)^* \sigma(0)e^{i\mathcal{H}_o t/\hbar}\varphi_n \mathrm{d}^3\mathbf{r} \\ &= e^{-i\omega_{mn}t}\sigma_{mn}(0)\end{aligned} \tag{N.20}$$

where $\hbar\omega_{mn} \equiv (E_m - E_n)$. Again, we use the identity: $\exp(-\mathcal{H}_o/k_BT)\varphi_m = \exp(-E_m/k_BT)\varphi_m$ from series expansion of the exponential operator.

N.4 Interaction Representation

We now assume, as in time-dependent perturbation theory, that the spin Hamiltonian is composed of the time-independent part, $\mathcal{H}_o$, and a much smaller time-dependent term, $\mathcal{H}_1(t)$. Introducing the interaction representation, in analogy with Eq. N.19, we now define $\sigma^\ddagger(t)$ by:

$$\sigma(t) = e^{-i\mathcal{H}_o t/\hbar}\sigma^\ddagger(t)e^{i\mathcal{H}_o t/\hbar} \tag{N.21}$$

where $\sigma^\ddagger(t)$ represents the time dependence of the density matrix that arises from the perturbation $\mathcal{H}_1(t)$. We thus remove all the time dependence associated with the rapid precession induced by $\mathcal{H}_o$. Note that the interaction representation for time-dependent perturbations (which is analogous to the Heisenberg representation of Appendix D.2) is often represented by an asterisk. Here, we use the double dagger to avoid confusion with the complex conjugate, which we represent throughout by an asterisk. We can redefine the perturbation Hamiltonian similarly in terms of $\mathcal{H}_1^\ddagger(t)$, where:

$$\mathcal{H}_1^\ddagger(t) = e^{i\mathcal{H}_o t/\hbar}\mathcal{H}_1(t)e^{-i\mathcal{H}_o t/\hbar} \tag{N.22}$$

From Eq. N.18, the resulting equation of motion for $\sigma^\ddagger(t)$ then becomes:

$$\frac{\mathrm{d}\sigma^\ddagger}{\mathrm{d}t} = \frac{i}{\hbar}\left[\sigma^\ddagger, \mathcal{H}_1^\ddagger(t)\right] \tag{N.23}$$

when using the interaction representation. In analogy with Eq. N.20, the matrix elements become:

$$\sigma_{mn}^\ddagger = e^{i\omega_{mn}t}\sigma_{mn} \tag{N.24}$$

$$\mathcal{H}_{mn}^\ddagger = e^{i\omega_{mn}t}\mathcal{H}_{mn} \tag{N.25}$$

in the interaction representation. Alternatively, we can expand the exponential operators in Eq. N.22 to give the Baker–Hausdorff formula:

$$\begin{aligned}\mathcal{H}_1^\ddagger(t) &= \left(1 + \frac{i}{\hbar}\mathcal{H}_o t + \frac{i^2}{2!\hbar^2}\mathcal{H}_o^2 t^2 + \ldots\right) \\ &\quad\times \mathcal{H}_1(t)\left(1 - \frac{i}{\hbar}\mathcal{H}_o t + \frac{i^2}{2!\hbar^2}\mathcal{H}_o^2 t^2 + \ldots\right) \\ &= \mathcal{H}_1 + \frac{i}{\hbar}\left[\mathcal{H}_o, \mathcal{H}_1\right]t \\ &\quad + \frac{1}{\hbar^2}\left(\mathcal{H}_o\mathcal{H}_1\mathcal{H}_o - \tfrac{1}{2}\mathcal{H}_o^2\mathcal{H}_1 - \tfrac{1}{2}\mathcal{H}_1\mathcal{H}_o^2\right)t^2 + \ldots \\ &= \mathcal{H}_1(t) + \frac{i}{\hbar}\left[\mathcal{H}_o, \mathcal{H}_1(t)\right]t - \frac{1}{2\hbar^2}\left[\mathcal{H}_o, \left[\mathcal{H}_o, \mathcal{H}_1(t)\right]\right]t^2 + \ldots\end{aligned} \tag{N.26}$$

where the commutators enter to preserve the order of the operators. For a time-independent Hamiltonian: $\mathcal{H}_o = \hbar\omega_o S_z$, we obtain the following forms for spin operators that can enter into $\mathcal{H}_1$:

$$(S_z)^\ddagger \equiv e^{i\omega_o S_z t}S_z e^{-i\omega_o S_z t} = S_z \tag{N.27}$$

$$(S_\pm)^\ddagger \equiv e^{i\omega_o S_z t}S_\pm e^{-i\omega_o S_z t} = e^{\pm i\omega_o t}S_\pm \tag{N.28}$$

where the double dagger again indicates the interaction representation. When (as here) $\mathcal{H}_o$ is a purely Zeeman Hamiltonian, the interaction representation corresponds to the rotating frame with Larmor frequency ω_o.

To obtain a solution for the time evolution of $\sigma^\ddagger(t)$ we first integrate Eq. N.23 from $t = 0$ to t:

$$\sigma^\ddagger(t) = \sigma^\ddagger(0) + \frac{i}{\hbar}\int_0^t \left[\sigma^\ddagger(t'), \mathcal{H}_1^\ddagger(t')\right]\mathrm{d}t' \tag{N.29}$$

We then get a first-order approximation by assuming that $\sigma^\ddagger(t') = \sigma^\ddagger(0)$ in the integrand:

$$\sigma^\ddagger(t) = \sigma^\ddagger(0) + \frac{i}{\hbar}\int_0^t \left[\sigma^\ddagger(0), \mathcal{H}_1^\ddagger(t')\right]\mathrm{d}t' \tag{N.30}$$

A better approximation comes from substituting Eq. N.30 in the right-hand side of Eq. N.29, which gives the solution correct to second order:

$$\begin{aligned}\sigma^\ddagger(t) &= \sigma^\ddagger(0) + \frac{i}{\hbar}\int_0^t \left[\sigma^\ddagger(0), \mathcal{H}_1^\ddagger(t')\right]\mathrm{d}t' \\ &\quad + \left(\frac{i}{\hbar}\right)^2 \int_0^t \mathrm{d}t' \int_0^{t'} \left[\left[\sigma^\ddagger(0), \mathcal{H}_1^\ddagger(t'')\right], \mathcal{H}_1^\ddagger(t')\right]\mathrm{d}t''\end{aligned} \tag{N.31}$$

We shall find the time derivative of Eq. N.31 more useful, giving:

$$\frac{d\sigma^{\ddagger}(t)}{dt}=\frac{i}{\hbar}\left[\sigma^{\ddagger}(0),\mathcal{H}_1^{\ddagger}(t)\right]+\left(\frac{i}{\hbar}\right)^2\int_0^t\left[\left[\sigma^{\ddagger}(0),\mathcal{H}_1^{\ddagger}(t')\right],\mathcal{H}_1^{\ddagger}(t)\right]dt' \tag{N.32}$$

which is equivalent to normal time-dependent perturbation theory (cf. Appendix M), taken to second order. This is the starting point used to treat spin relaxation or the response of a spin system to electromagnetic radiation.

APPENDIX O: RELAXATION THEORY WITH DENSITY MATRICES

O.1 Spin Relaxation

We begin with Eq. N.32 for evolution of the density matrix under a time-dependent perturbation. The perturbation $\mathcal{H}_1(t)$ that causes relaxation is random with mean value zero. Therefore, the first term in Eq. N.32 vanishes on taking an ensemble average. Correlations between fluctuations in $\mathcal{H}_1(t)$ are stationary, i.e., are independent of t, depending only on the difference $\tau=t-t'$ (see Section 5.2 in Chapter 5). Also, they die out at times much longer than the correlation time $\tau=\tau_c$. Consequently, we can extend the limit of the integral in Eq. N.32 to infinity, if we restrict ourselves to time intervals appreciably longer than τ_c. Further, because the perturbation is small, $\sigma^{\ddagger}$ varies only slowly with time and we can replace $\sigma^{\ddagger}(0)$ with $\sigma^{\ddagger}(t)$ in the integral. Then Eq. N.32 becomes (Abragam 1961):

$$\frac{d\overline{\sigma^{\ddagger}}}{dt}=-\frac{1}{\hbar^2}\int_0^\infty\overline{\left[\mathcal{H}_1^{\ddagger}(t),\left[\mathcal{H}_1^{\ddagger}(t-\tau),\sigma^{\ddagger}(t)\right]\right]}d\tau \tag{O.1}$$

for relaxation by random fluctuations, where the bar indicates an ensemble average over the population of spins.

As in Chapter 5, we write the Hamiltonian for the random perturbation as the product of a time-dependent term $F(t)$ and a time-independent term $\mathcal{H}_1$ that contains the spin operators (see Eq. 5.1):

$$\mathcal{H}_1(t)=\mathcal{H}_1F(t) \tag{O.2}$$

In the interaction representation this becomes:

$$\mathcal{H}_1^{\ddagger}(t)=e^{i\mathcal{H}_o t/\hbar}\mathcal{H}_1(t)e^{-i\mathcal{H}_o t/\hbar}=F(t)\sum_p\mathcal{H}_{1,p}e^{i\omega_p t} \tag{O.3}$$

in analogy with Eq. N.25 for the matrix elements, and see also Eqs. N.26–28. The ω_p are combinations of eigenfrequencies of $\mathcal{H}_o$, and $\mathcal{H}_{1,p}$ are spin operators from the commutator $[\mathcal{H}_o,\mathcal{H}_1]$. Substituting Eq. O.3 in Eq. O.1, for $t>>\tau_c$, we get:

$$\frac{d\overline{\sigma^{\ddagger}}}{dt}=-\frac{1}{\hbar^2}\sum_{p,p'}e^{i(\omega_p-\omega_{p'})t}\left[\mathcal{H}_{1,p'}^{\dagger},\left[\mathcal{H}_{1,p},\sigma^{\ddagger}(t)\right]\right]\times\int_0^\infty\overline{F^*(t)F(t-\tau)}e^{-i\omega_p\tau}d\tau \tag{O.4}$$

where the asterisk stands as usual for the complex conjugate and the single dagger for the Hermitean conjugate. Choosing the Hermitian conjugate for $\mathcal{H}_{1,p'}^{\dagger}(t)$ is somewhat arbitrary (as is p'), but simplifies things later. The rapidly varying exponentials $e^{i(\omega_p-\omega_{p'})t}$ average to zero at long times. This means that only the terms where $\omega_p-\omega_{p'}=0$ contribute appreciably in Eq. O.4 (i.e., terms with $p'=p$). We then get:

$$\frac{d\overline{\sigma^{\ddagger}}}{dt}=-\frac{1}{\hbar^2}\sum_p\left[\mathcal{H}_{1,p}^{\dagger},\left[\mathcal{H}_{1,p},\sigma^{\ddagger}(t)\right]\right]j(\omega_p) \tag{O.5}$$

where the spectral density is:

$$j(\omega)=\int_0^\infty\overline{F^*(t)F(t-\tau)}e^{-i\omega\tau}d\tau \tag{O.6}$$

as defined by Eqs. 5.7 and 5.9 in Chapter 5.

To treat spin–lattice relaxation, we need the equation of motion for the ensemble average of the expectation value $\overline{\langle S_z\rangle}^{\ddagger}=\mathrm{Tr}(S_z\sigma^{\ddagger})$, and for transverse relaxation that of $\overline{\langle S_+\rangle}^{\ddagger}=\mathrm{Tr}(S_+\sigma^{\ddagger})$. For convenience, we take each in the interaction representation, which removes rapidly varying terms. Multiplying both sides of Eq. O.5 by the spin component S_i, and then taking the trace, we get:

$$\frac{d\overline{\langle S_i\rangle}^{\ddagger}}{dt}=-\frac{1}{\hbar^2}\sum_p j(\omega_p)\mathrm{Tr}\left(S_i\left[\mathcal{H}_{1,p}^{\dagger},\left[\mathcal{H}_{1,p},\sigma^{\ddagger}\right]\right]\right) \tag{O.7}$$

where $S_i\equiv S_z$ or S_+. By twice applying the relation $\mathrm{Tr}(A[B,C])=\mathrm{Tr}([A,B]C)$ this becomes (Abragam 1961):

$$\begin{aligned}\frac{d\overline{\langle S_i\rangle}^{\ddagger}}{dt}&=-\frac{1}{\hbar^2}\sum_p j(\omega_p)\mathrm{Tr}\left(\left[\left[S_i,\mathcal{H}_{1,p}^{\dagger}\right],\mathcal{H}_{1,p}\right]\sigma^{\ddagger}\right)\\&=-\frac{1}{\hbar^2}\sum_p j(\omega_p)\overline{\left\langle\left[\left[S_i,\mathcal{H}_{1,p}^{\dagger}\right],\mathcal{H}_{1,p}\right]\right\rangle}^{\ddagger}\end{aligned} \tag{O.8}$$

Equations O.7 and O.8 hold when we can average over $\mathcal{H}_1(t)$ and $\sigma^{\ddagger}(t)$ separately, i.e., for $t>>\tau_c$. We get exponential relaxation from terms on the right-hand side of Eq. O.8 that are linear in $\overline{\langle S_i\rangle}^{\ddagger}$, when all other terms are zero (at least after taking the trace).

In practice, we are dealing with more than one term in the spin Hamiltonian. Then we write Eq. O.2 as:

$$\mathcal{H}_1(t)=\sum_q\mathcal{H}_1^{(q)}F^{(q)}(t) \tag{O.9}$$

TABLE O.1 Commutators and useful relations

operation/operator	*expression*
commutator	$[AB,C]=A[B,C]+[A,C]B$
	$[A,BC]=[A,B]C+B[A,C]$
	$\mathrm{Tr}(A[B,C])=\mathrm{Tr}([A,B]C)$
spin-operator commutators	$[S_z,S_\pm]=\pm S_\pm$
	$[S_\pm,S_\mp]=\pm 2S_z$
	$[S_k,I_l]=0$, where $k,l\equiv\pm,z$
spin-operator relations	$S_\pm S_\mp=S^2-S_z^2\pm S_z$
	$S_zS_\pm=\pm S_\pm+S_\pm S_z$
	$S_+S_-+S_-S_+=2\left(S^2-S_z^2\right)$
trace relations	$\mathrm{Tr}\left(S_\pm S_\mp\right)=\frac{2}{3}S(S+1)$
	$\mathrm{Tr}\left(S_zS_\pm\right)=0$
$S=\frac{1}{2}$ *relations*	$S_zS_\pm=-S_\pm S_z=\pm\frac{1}{2}S_\pm$
	$S_\pm S_\mp=\frac{1}{2}\pm S_z$
	$S_+S_-+S_-S_+=1$

where we arrange the terms such that the $F^{(q)}$ are mutually orthogonal spherical harmonics, $Y_{2,q}$ (see Appendix V.2). We designate the corresponding spectral densities by $j^{(q)}(\omega_p)$. Abragam (1961) shows how the method applies to magnetic dipole–dipole interactions between two spins. This results in Eqs. 9.27, 9.28 and Eqs. 9.34, 9.35 of Chapter 9 for the dipolar T_1-, T_2-relaxation rates of like and unlike spins, respectively.

Table O.1 lists relations useful for evaluating commutators of spin operators and their traces. Further expressions for the traces of spin operators appear in Table U.1 of Appendix U.

O.2 Dipolar Transverse Relaxation

As a first illustration of the spin-operator method, we treat T_2-relaxation induced by dipole–dipole interaction between unlike electron spins S_1 and S_2. Only the final results of such a calculation are given in Chapter 9. The time-independent Hamiltonian for unlike spins is (in angular frequency units):

$$\mathcal{H}_o=\omega_1S_{1z}+\omega_2S_{2z} \tag{O.10}$$

where ω_1 and ω_2 are the angular resonance frequencies of the two spins. The spin operators, $\mathcal{H}_1^{(q)}=\sum_p A_p^{(q)}$, defined in Eq. O.9 then consist of the terms (see Eqs. 9.5–9.9):

$$A_1^{(0)}=\left(\mu_o/4\pi\right)\gamma_e^2\hbar S_{1z}S_{2z} \tag{O.11}$$

$$A_2^{(\pm0)}=-\tfrac{1}{4}\left(\mu_o/4\pi\right)\gamma_e^2\hbar S_{1\pm}S_{2\mp} \tag{O.12}$$

$$A_1^{(\pm1)}=-\tfrac{3}{2}\left(\mu_o/4\pi\right)\gamma_e^2\hbar S_{1\pm}S_{2z} \tag{O.13}$$

$$A_2^{(\pm1)}=-\tfrac{3}{2}\left(\mu_o/4\pi\right)\gamma_e^2\hbar S_{1z}S_{2\pm} \tag{O.14}$$

$$A^{(\pm2)}=-\tfrac{3}{4}\left(\mu_o/4\pi\right)\gamma_e^2\hbar S_{1\pm}S_{2\pm} \tag{O.15}$$

in angular frequency units. The corresponding time-dependent functions are:

$$F^{(0)}(t)=\left(3\cos^2\theta(t)-1\right)/r_{12}^3(t) \tag{O.16}$$

$$F^{(\pm1)}(t)=\sin\theta(t)\cos\theta(t)e^{\mp i\phi(t)}/r_{12}^3(t) \tag{O.17}$$

$$F^{(\pm2)}(t)=\sin^2\theta(t)e^{\mp 2i\phi(t)}/r_{12}^3(t) \tag{O.18}$$

For a constant interspin separation r_{12}, the angular (or ensemble) averages are $\left|F^{(0)}\right|^2=\frac{4}{5}r_{12}^{-6}$, $\left|F^{(\pm1)}\right|^2=\frac{2}{15}r_{12}^{-6}$ and $\left|F^{(\pm2)}\right|^2=\frac{8}{15}r_{12}^{-6}$ (see Eqs. 5.16–5.18). Otherwise we require solution of, e.g., the translational diffusion equation (see Section 9.5.2 of Chapter 9).

From Eqs. O.5, O.6 and O.8, we see that we need the expansions of $A_p^{(q)}$ in the interaction representation (i.e., Eq. O.5) to find the frequencies $\omega_p^{(q)}$ that are associated with the corresponding spectral densities $j^{(q)}\left(\omega_p^{(q)}\right)$. To do this we use Eq. N.26 for expansion of the interaction representation, with $\mathcal{H}_o$ in angular frequency units:

$$\begin{aligned}A_p^{(q)\ddagger}&=A_p^{(q)}+it\left[\mathcal{H}_o,A_p^{(q)}\right]+\frac{(it)^2}{2!}\left[\mathcal{H}_o,\left[\mathcal{H}_o,A_p^{(q)}\right]\right]\\&\quad+\frac{(it)^3}{3!}\left[\mathcal{H}_o,\left[\mathcal{H}_o,\left[\mathcal{H}_o,A_p^{(q)}\right]\right]\right]+\ldots\\&=A_p^{(q)}\left(1+i\omega_p^{(q)}t+\frac{\left(i\omega_p^{(q)}t\right)^2}{2!}+\frac{\left(i\omega_p^{(q)}t\right)^3}{3!}+\ldots\right)\\&=A_p^{(q)}\exp\left(i\omega_p^{(q)}t\right)\end{aligned} \tag{O.19}$$

where the second identity is established by explicit calculation, which then yields the value of $\omega_p^{(q)}$. Given that the series develops in this way, we need only calculate the first commutator, i.e., $\left[\mathcal{H}_o,A_p^{(q)}\right]$. Thus, we find that (cf. Abragam 1961):

$$\left[\mathcal{H}_o,A_1^{(0)}\right]=0 \tag{O.20}$$

$$\left[\mathcal{H}_o,A_2^{(\pm0)}\right]=\pm(\omega_1-\omega_2)A_2^{(\pm0)} \tag{O.21}$$

$$\left[\mathcal{H}_o,A_1^{(\pm1)}\right]=\pm\omega_1A_1^{(\pm1)} \tag{O.22}$$

$$\left[\mathcal{H}_o,A_2^{(\pm1)}\right]=\pm\omega_2A_2^{(\pm1)} \tag{O.23}$$

$$\left[\mathcal{H}_o,A^{(\pm2)}\right]=(\omega_1+\omega_2)A^{(\pm2)} \tag{O.24}$$

which give the various values that we need for $\omega_p^{(q)}$, e.g., $\omega_2^{(\pm0)}=\pm(\omega_1-\omega_2)$ from Eq. O.21.

We now evaluate the commutators in Eq. O.8 that we need for the time evolution of transverse magnetization $\langle S_{1+}\rangle$ of spin 1. The first, involving $A_1^{(0)}$ for the diagonal elements of the perturbation, is straightforward:

$$\left[\left[S_{1+},A_1^{(0)\dagger}\right],A_1^{(0)}\right]=\left(\mu_o/4\pi\right)^2\gamma_e^4\hbar^2S_{2z}^2S_{1+} \tag{O.25}$$

because the unlike spins are only weakly coupled and therefore commute. This gives the secular contribution. Taking the trace and performing ensemble averaging, we get:

$$\mathrm{Tr}\left(\left[\left[S_{1+},A_1^{(0)\dagger}\right],A_1^{(0)}\right]\sigma^{\ddagger}\right)=\tfrac{1}{3}\left(\mu_o/4\pi\right)^2\gamma_e^4\hbar^2S_2(S_2+1)\overline{\langle S_{1+}\rangle}^{\ddagger} \tag{O.26}$$

where we take $\mathrm{Tr}\left(S_z^2\right)=\frac{1}{3}S(S+1)$ from Table U.1 and $\overline{\langle S_{1+}\rangle}^{\ddagger}=\mathrm{Tr}\left(S_{1+}\sigma^{\ddagger}\right)$ is the transverse magnetization in the interaction representation. For independent spin systems, the trace of the product is the product of the traces for the S_1- and S_2-terms taken separately. Substituting Eq. O.26 into Eq. O.8, we then get the secular contribution to the transverse relaxation:

$$\frac{\mathrm{d}\overline{\langle S_{1+}\rangle}^{\ddagger}}{\mathrm{d}t}=-\frac{1}{T_2^{\mathrm{sec}}(S_1)}\overline{\langle S_{1+}\rangle}^{\ddagger} \tag{O.27}$$

where the secular contribution to the T_2-relaxation rate of spin 1 is:

$$\frac{1}{T_2^{\mathrm{sec}}(S_1)}=\frac{1}{3}\left(\frac{\mu_o}{4\pi}\right)^2\gamma_e^4\hbar^2S_2(S_2+1)j^{(0)}(0) \tag{O.28}$$

From Eq. O.20, the spectral density is that at zero frequency, as it must be for a secular contribution. Equation O.28 agrees with the secular contribution to the T_2-relaxation rate of dipolar-coupled unlike spins that we give in Eq. 9.35. Note that we use the spectral densities $J^{(q)}(\omega)=2j^{(q)}(\omega)$ in Chapter 9.

Of the remaining $q=0$ terms, the commutator for $A_2^{(-0)}$ is zero. That for $A_2^{(+0)}$ is rather more involved. Our strategy is to isolate terms containing only S_{1+} for the spin-1 system and to reduce the spin-2 terms to those containing only the eigenoperators S_2^2, S_{2z}^2 and S_{2z}. Using the results of Table O.1, we then get:

$$\left[\left[S_{1+},A_2^{(+0)\dagger}\right],A_2^{(+0)}\right]=\tfrac{1}{8}\left(\mu_o/4\pi\right)^2\gamma_e^4\hbar^2\left[S_{1z}S_{2+},S_{1+}S_{2z}\right]$$

$$=\frac{1}{8}\left(\frac{\mu_o}{4\pi}\right)^2\gamma_e^4\hbar^2\left((S_{2+}S_{2-})S_{1z}S_{1+}-(S_{2-}S_{2+})S_{1+}S_{1z}\right)$$

$$=\frac{1}{8}\left(\frac{\mu_o}{4\pi}\right)^2\gamma_e^4\hbar^2\left(S_2^2-S_{2z}^2\right)S_{1+} \tag{O.29}$$

The final equality in Eq. O.29 uses the relations: $S_\pm S_\mp=S^2-S_z^2\pm S_z$ and $S_zS_+=-S_+S_z=\frac{1}{2}S_+$, where the first (used for spin 2) is quite general, but the second (used for spin 1) applies only when $S=\frac{1}{2}$. Of the $q=1$ terms, the commutator for $A_1^{(-1)}$ is zero, and that for $A_1^{(+1)}$ is straightforward:

$$\left[\left[S_{1+},A_1^{(+1)\dagger}\right],A_1^{(+1)}\right]=\tfrac{9}{2}\left(\mu_o/4\pi\right)^2\gamma_e^4\hbar^2S_{2z}^2S_{1+} \tag{O.30}$$

The remaining $q=1$ commutator, that for the pseudosecular contribution $A_2^{(\pm1)}$, becomes:

$$\left[\left[S_{1+},A_2^{(\pm1)\dagger}\right],A_2^{(\pm1)}\right]=-\frac{9}{4}\left(\frac{\mu_o}{4\pi}\right)^2\gamma_e^4\hbar^2\left[S_{1+}S_{2\mp},S_{1z}S_{2\pm}\right]$$

$$=\frac{9}{4}\left(\frac{\mu_o}{4\pi}\right)^2\gamma_e^4\hbar^2\left(S_2^2-S_{2z}^2\right)S_{1+} \tag{O.31}$$

by using the same relations as for Eq. O.29. Of the $q=2$ terms, the commutator for $A^{(-2)}$ is zero, and that for $A^{(+2)}$ becomes:

$$\left[\left[S_{1+},A^{(+2)\dagger}\right],A^{(+2)}\right]=\frac{9}{8}\left(\frac{\mu_o}{4\pi}\right)^2\gamma_e^4\hbar^2\left[S_{1z}S_{2-},S_{1+}S_{2+}\right]$$

$$=\frac{9}{8}\left(\frac{\mu_o}{4\pi}\right)^2\gamma_e^4\hbar^2\left(S_2^2-S_{2z}^2\right)S_{1+} \tag{O.32}$$

again as in Eq. O.29.

Gathering up the contributions from Eqs. O.29–O.32 and treating them in the same way as for the secular contribution in Eqs. O.26–O.28, we finally get the T_2-relaxation rate of spin 1:

$$\frac{1}{T_2(S_1)}=\frac{1}{3}\left(\frac{\mu_o}{4\pi}\right)^2\gamma_e^4\hbar^2S_2(S_2+1)\times\left(j^{(0)}(0)+\tfrac{1}{4}j^{(0)}(\omega_1-\omega_2)\right.$$
$$\left.+\tfrac{9}{2}j^{(1)}(\omega_1)+9j^{(1)}(\omega_2)+\tfrac{9}{4}j^{(2)}(\omega_1+\omega_2)\right) \tag{O.33}$$

where frequencies for the spectral densities again come from Eqs. O.20–O.24. This result is in agreement with Eq. 9.35 in Chapter 9 for unlike dipolar-coupled spins. With $S_2=\frac{1}{2}$, Eq. O.33 gives the dipolar broadening for spin labels in different hyperfine states. With general S_2, it gives the broadening of a spin label by a different paramagnetic species.

O.3 Nitroxide Transverse Relaxation/Linewidths

The most common spin-label application of motional-narrowing theory is to differential broadening of the nitrogen hyperfine lines by rotational modulation of the A- and g-anisotropies (see Chapter 5). Treatment here is analogous to that of the magnetic dipole–dipole interaction given in the previous section, but now we must include cross terms between the hyperfine and Zeeman anisotropies. The spins involved are those of the electron and nitrogen nucleus, S and I respectively. The two spin systems are independent, i.e., S and I commute $\left([I_j,S_k]=0\right.$, where $\left.j,k\equiv\pm,z\right)$, because their resonance frequencies ω_e and ω_a are very far apart.

The time-independent part of the Hamiltonian for an isolated nitroxide is (see Eq. 5.14 in Chapter 5):

$$\mathcal{H}_o = g_o\beta_e B_o S_z + a_o \mathbf{I}\cdot\mathbf{S} \tag{O.34}$$

where g_o and a_o are the isotropic g-value and hyperfine coupling constant, respectively, and B_o is the static magnetic field oriented along the z-axis. We assume for simplicity that the perturbation Hamiltonian is axial. The spin operators that we need to apply Eq. O.8 are then (see Eq. 5.15):

$$A_1^{(0)} = \tfrac{1}{3}\left(\Delta g\beta_e B_o S_z + \Delta A I_z S_z\right) \tag{O.35}$$

$$A_2^{(\pm 0)} = -\tfrac{1}{12}\Delta A I_\pm S_\mp \tag{O.36}$$

$$A_1^{(\pm 1)} = \tfrac{1}{2}\left(\Delta g\beta_e B_o S_\pm + \Delta A I_z S_\pm\right) \tag{O.37}$$

$$A_2^{(\pm 1)} = \tfrac{1}{2}\Delta A I_\pm S_z \tag{O.38}$$

$$A^{(\pm 2)} = \tfrac{1}{4}\Delta A I_\pm S_\pm \tag{O.39}$$

with $\Delta g = g_{zz} - \tfrac{1}{2}\left(g_{xx}+g_{yy}\right)$ and $\Delta A = A_{zz} - \tfrac{1}{2}\left(A_{xx}+A_{yy}\right)$. The corresponding orthogonal time-dependent functions $F^{(q)}(t)$ are the same as the angular-dependent terms for dipole–dipole interactions, i.e., Eqs. O.16–O.18 with $r_{12} = 1$.

We evaluate the commutators $\left[\mathcal{H}_o, A_p^{(q)}\right]$ needed to specify the frequencies in the spectral densities $j^{(q)}\left(\omega_p^{(q)}\right)$ by keeping just the diagonal terms in Eq. O.34, i.e., $\mathcal{H}_o = g_o\beta_e B_o S_z + a_o I_z S_z$. Then we get (cf. Eqs. O.20–O.24):

$$\left[\mathcal{H}_o, A_1^{(0)}\right] = 0 \tag{O.40}$$

$$\left[\mathcal{H}_o, A_2^{(\pm 0)}\right] = \mp(\omega_e \mp \omega_a) A_2^{(\pm 0)} \tag{O.41}$$

$$\left[\mathcal{H}_o, A_1^{(\pm 1)}\right] = \pm\omega_e A_1^{(\pm 1)} \tag{O.42}$$

$$\left[\mathcal{H}_o, A_2^{(\pm 1)}\right] = \pm\omega_a A_2^{(\pm 1)} \tag{O.43}$$

$$\left[\mathcal{H}_o, A^{(\pm 2)}\right] = \pm(\omega_e \mp \omega_a) A^{(+2)} \tag{O.44}$$

where $\omega_e = g_o\beta_e B_o/\hbar + a_o m_I$ and $\omega_a = a_o S_z = \tfrac{1}{2}a_o$. From Eq. O.41, we then get $\omega_2^{(\pm 0)} = \mp(\omega_e \mp \omega_a)$ and so on.

We evaluate the commutators needed for the time evolution of the transverse electron spin magnetization $\langle S_+\rangle$ as in the previous section (cf. Eqs. O.25, O.29–O.32). For Eqs. O.35 and O.37, which contain both Zeeman and hyperfine terms, we must take care to include the cross terms:

$$\left[\left[S_+, A_1^{(0)\dagger}\right], A_1^{(0)}\right] = \tfrac{1}{9}\left(\left(\Delta g\beta_e\right)^2 B_o^2 + (\Delta A)^2 I_z^2 + 2\Delta g\beta_e B_o \Delta A I_z\right) S_+ \tag{O.45}$$

$$\left[\left[S_+, A_1^{(+1)\dagger}\right], A_1^{(+1)}\right] = \tfrac{1}{2}\left(\left(\Delta g\beta_e\right)^2 B_o^2 + (\Delta A)^2 I_z^2 + 2\Delta g\beta_e B_o \Delta A I_z\right) S_+ \tag{O.46}$$

where Eq. O.45 is the secular contribution (see Eq. O.35). Commutators involving $A_2^{(+0)}$, $A_1^{(-1)}$ and $A^{(-2)}$ are zero and also do not contribute cross-terms. Additionally, cross terms between the Hermitean conjugates $A_2^{(\pm 1)}$ also vanish. The remaining non-zero commutators become as follows:

$$\left[\left[S_+, A_2^{(-0)\dagger}\right], A_2^{(-0)}\right] = \tfrac{1}{72}(\Delta A)^2\left[I_+ S_z, I_- S_+\right] = \tfrac{1}{72}(\Delta A)^2\left(I^2 - I_z^2\right) S_+ \tag{O.47}$$

$$\left[\left[S_+, A_2^{(\pm 1)\dagger}\right], A_2^{(\pm 1)}\right] = -\tfrac{1}{4}(\Delta A)^2\left[I_\mp S_+, I_\pm S_z\right] = \tfrac{1}{4}(\Delta A)^2\left(I^2 - I_z^2\right) S_+ \tag{O.48}$$

$$\left[\left[S_+, A_2^{(+2)\dagger}\right], A_2^{(+2)}\right] = \tfrac{1}{8}(\Delta A)^2\left[I_- S_z, I_+ S_+\right] = \tfrac{1}{8}(\Delta A)^2\left(I^2 - I_z^2\right) S_+ \tag{O.49}$$

where again evaluations proceed as for Eq. O.29. Note that Eq. O.48 is the pseudosecular contribution (see Eq. O.38).

Now we gather together the individual terms from Eqs. O.45–O.49. To identify the contribution for an individual hyperfine, we note that the nuclear operators act solely on the spin eigenfunctions $|I, m_I\rangle$. Then we can replace I^2, I_z^2 and I_z by their eigenvalues $I(I+1)$, m_I^2 and m_I, respectively. Substituting in Eq. O.8 then leads to time evolution of $\overline{\langle S_+\rangle}^{\ddagger} = \mathrm{Tr}\left(S_+\sigma^{\ddagger}\right)$ that is analogous to Eq. O.27 and is characterized by the transverse relaxation rate (cf. Eq. 5.57 in Chapter 5):

$$\frac{1}{T_2(m_I)} = A + Bm_I + Cm_I^2 \tag{O.50}$$

with

$$A = \left(\Delta g\beta_e/\hbar\right)^2\left(\tfrac{1}{9}j^{(0)}(0) + \tfrac{1}{2}j^{(1)}(\omega_e)\right)B_o^2 + \tfrac{1}{4}I(I+1)(\Delta A)^2 \times\left(\tfrac{1}{18}j^{(0)}(\omega_e+\omega_a) + j^{(1)}(\omega_a) + \tfrac{1}{2}j^{(2)}(\omega_e-\omega_a)\right) \tag{O.51}$$

$$B = \left(\Delta g\beta_e/\hbar\right)\Delta A\left(\tfrac{2}{9}j^{(0)}(0) + j^{(1)}(\omega_e)\right)B_o \tag{O.52}$$

$$C = \tfrac{1}{2}(\Delta A)^2\left(\tfrac{2}{9}j^{(0)}(0) - \tfrac{1}{36}j^{(0)}(\omega_e+\omega_a) + j^{(1)}(\omega_e) - \tfrac{1}{2}j^{(1)}(\omega_a) - \tfrac{1}{4}j^{(2)}(\omega_e-\omega_a)\right) \tag{O.53}$$

Note that only one of the two alternatives in Eq. O.48 contributes for a given hyperfine state (at least for $I \le 1$). A rigorous justification comes from applying specific single-transition operators as described later, in Section O.4 (and see also Marsh 2017). The spectral densities for isotropic rotational diffusion with correlation time τ_R are given by (see Section 7.5 in Chapter 7):

$$j^{(q)}(\omega) = \left|F^{(q)}(0)\right|^2 \frac{\tau_R}{1+\omega^2\tau_R^2} \tag{O.54}$$

where the angular averages are: $\left|F^{(0)}\right|^2 = \frac{4}{5}$, $\left|F^{(\pm1)}\right|^2 = \frac{2}{15}$ and $\left|F^{(\pm2)}\right|^2 = \frac{8}{15}$ (see Eqs. 5.16–5.18). The linewidth coefficients of Eqs. O.51–O.53 then become:

$$A = \frac{1}{15}\left(\left(\frac{\Delta g\beta_e B_o}{\hbar}\right)^2\left(\frac{4}{3}+\frac{1}{1+\omega_e^2\tau_R^2}\right)\right.$$
$$\left. + \tfrac{1}{2}I(I+1)(\Delta A)^2\left(\frac{1}{1+\omega_a^2\tau_R^2}+\frac{7}{3}\frac{1}{1+\omega_e^2\tau_R^2}\right)\right)\tau_R \tag{O.55}$$

$$B = \frac{2}{15}\left(\frac{\Delta g\beta_e B_o}{\hbar}\right)\Delta A\left(\frac{4}{3}+\frac{1}{1+\omega_e^2\tau_R^2}\right)\tau_R \tag{O.56}$$

$$C = \frac{1}{30}(\Delta A)^2\left(\frac{8}{3}-\frac{1}{1+\omega_a^2\tau_R^2}-\frac{1}{3}\frac{1}{1+\omega_e^2\tau_R^2}\right)\tau_R \tag{O.57}$$

where we use the approximation $\omega_e \pm \omega_a \cong \omega_e$. Equations O.55–O.57 agree with the axial version of the linewidth coefficients given in Chapter 5 (viz., Eqs. 5.59, 5.61 and 5.62 with $\delta g = \delta A = 0$).

Note that the modified spectral densities of Chapter 5 are related to the conventional spectral densities used here by: $j^{AA}(\omega) = \frac{1}{4}\left((\Delta A)^2 + 3(\delta g)^2\right) j^{(1)}(\omega)$, $j^{gg}(\omega) = \frac{1}{4}\left((\Delta g)^2 + 3(\delta g)^2\right)\left(\beta_e/\hbar\right)^2 j^{(1)}(\omega)$ and $j^{gA}(\omega) = \frac{1}{4}\left(\Delta g\Delta A + 3\delta g\delta A\right)\left(\beta_e/\hbar\right) j^{(1)}(\omega)$.

O.4 Transverse Relaxation and Coherence Transfer for ^{14}N-nitroxides

We now concentrate specifically on ^{14}N-nitroxides. This lets us treat not only transverse relaxation but also transfer of transverse magnetization (i.e., of coherence) between hyperfine states (Marsh 2017). To allow for coherence transfer, we need to use single-transition spin operators $S_+^{(m_I)}$ for particular hyperfine states m_I. For nitrogen nuclear spin $I = 1$ these are (see, e.g., Ernst et al. 1987):

$$S_+^{(0)} = \left(1 - I_z^2\right)S_+ \tag{O.58}$$

$$S_+^{(\pm1)} = \tfrac{1}{2}\left(I_z^2 \pm I_z\right)S_+ \tag{O.59}$$

which correspond to the transverse electron spin operator S_+ for the $m_I = 0, \pm1$ ^{14}N-hyperfine transitions, respectively. To find the time evolution of expectation values for the single-transition operators, we therefore must treat coupled relaxation of the operators $\langle S_+\rangle$, $\langle I_z S_+\rangle$ and $\langle I_z^2 S_+\rangle$.

The double commutators that we need for the time evolution of S_+ are given by Eqs. O.45–O.49, in Section O.3. Substituting in Eq. O.8, we then get:

$$\frac{d\overline{\langle S_+\rangle}^{\ddagger}}{dt} = -R_{2e}\overline{\langle S_+\rangle}^{\ddagger} - R_{2x1}\overline{\langle I_z S_+\rangle}^{\ddagger} - R_{2x2}\overline{\langle I_z^2 S_+\rangle}^{\ddagger} \tag{O.60}$$

where:

$$R_{2e} = \left(\frac{\Delta g\beta_e B_o}{\hbar}\right)^2\left(\tfrac{1}{9}j^{(0)}(0) + \tfrac{1}{2}j^{(1)}(\omega_e)\right)$$
$$+ (\Delta A)^2\left(\tfrac{1}{36}j^{(0)}(\omega_e+\omega_a) + j^{(1)}(\omega_a) + \tfrac{1}{4}j^{(2)}(\omega_e-\omega_a)\right) \tag{O.61}$$

$$R_{2x1} = \left(\Delta g\beta_e B_o/\hbar\right)\Delta A\left(\tfrac{2}{9}j^{(0)}(0) + j^{(1)}(\omega_e)\right) \tag{O.62}$$

$$R_{2x2} = \tfrac{1}{2}(\Delta A)^2\left(\tfrac{2}{9}j^{(0)}(0) + j^{(1)}(\omega_e) - \tfrac{1}{36}j^{(0)}(\omega_e+\omega_a)\right.$$
$$\left. - j^{(1)}(\omega_a) - \tfrac{1}{4}j^{(2)}(\omega_e-\omega_a)\right) \tag{O.63}$$

Note that Eqs. O.61 and O.63 now contain contributions from both pseudosecular terms $A_2^{(\pm1)}$ (see Eq. O.48), because we specify spin operators for the different hyperfine states explicitly here by using Eqs. O.58 and O.59.

Direct evaluation shows that we get the double commutators needed for the evolution of $I_z S_+$ by multiplying Eqs. O.45–O.47 and O.49 on the right with I_z. Only for the pseudosecular term, i.e., Eq. O.48, does this replacement not hold. There we get instead:

$$\left[\left[I_z S_+, A_2^{(\pm1)\dagger}\right], A_2^{(\pm1)}\right] = \tfrac{1}{4}(\Delta A)^2\left(I^2 - \tfrac{1}{2} - I_z^2\right)I_z S_+ \tag{O.64}$$

Following Abragam (1961), we relate terms in $I_z^3 S_+$ to the expectation value $\langle I_z S_+\rangle$, by using the special relation $I_z^3 = I_z$ that holds only for $I = 1$. From Eq. O.8, we then get the time evolution:

$$\frac{d\overline{\langle I_z S_+\rangle}^{\ddagger}}{dt} = -R'_{2e}\overline{\langle I_z S_+\rangle}^{\ddagger} - R'_{2x1}\overline{\langle I_z^2 S_+\rangle}^{\ddagger} \tag{O.65}$$

where

$$R'_{2e} = R_{2e} + R_{2x2} - \tfrac{1}{4}(\Delta A)^2 j^{(1)}(\omega_a) \tag{O.66}$$

$$R'_{2x1} = R_{2x1} \tag{O.67}$$

given in terms of the rate constants R_{2e}, R_{2x1} and R_{2x2} of Eqs. O.61–O.63. Note that $\langle S_+\rangle$ does not couple to $\langle I_z S_+\rangle$.

We turn now to $\langle I_z^2 S_+\rangle$. By evaluating double commutators directly, we find that we must multiply Eqs. O.45–O.47 and O.49 on the right by I_z^2 to get the evolution of $I_z^2 S_+$. Again, this simple replacement does not hold for the pseudosecular term, i.e., Eq. O.48. In that case we get:

$$\left[\left[I_z^2 S_+, A_2^{(\pm 1)\dagger}\right], A_2^{(\pm 1)}\right] = \tfrac{1}{4}(\Delta A)^2\left(I^2 - \tfrac{3}{2} - I_z^2\right) I_z^2 S_+ + \tfrac{1}{8}(\Delta A)^2 I^2 S_+ \tag{O.68}$$

where we must retain contributions from both pseudosecular terms $A_2^{(\pm 1)}$, as with Eq. O.64. In addition to terms in $I_z^3 S_+$, we now also have terms in $I_z^4 S_+$. We relate these to the expectation value $\langle I_z^2 S_+ \rangle$ by using the special relation $I_z^4 = I_z^2$ that holds only for $I = 1$. From Eq. O.8, we then get the time evolution:

$$\frac{d\overline{\langle I_z^2 S_+ \rangle}^{\ddagger}}{dt} = -R''_{2e}\overline{\langle I_z^2 S_+ \rangle}^{\ddagger} - R''_{2x_1}\overline{\langle I_z S_+ \rangle}^{\ddagger} - R''_{2x_2}\overline{\langle S_+ \rangle}^{\ddagger} \tag{O.69}$$

where

$$R''_{2e} = R_{2e} + R_{2x_2} - \tfrac{3}{4}(\Delta A)^2 j^{(1)}(\omega_a) \tag{O.70}$$

$$R''_{2x_1} = R_{2x_1} \tag{O.71}$$

$$R''_{2x_2} = \tfrac{1}{2}(\Delta A)^2 j^{(1)}(\omega_a) \tag{O.72}$$

when expressed using the rate constants given by Eqs. O.61–O.63.

We finally obtain the time evolution of the transverse electron magnetization $\overline{\langle S_+^{(0)} \rangle}^{\ddagger}$, $\overline{\langle S_+^{(\pm 1)} \rangle}^{\ddagger}$ for the individual $m_I = 0, \pm 1$ hyperfine transitions by combining Eqs. O.58–O.60, O.65 and O.69:

$$\frac{d\overline{\langle S_+^{(0)} \rangle}^{\ddagger}}{dt} = -\frac{1}{T_{2e}(0)}\overline{\langle S_+^{(0)} \rangle}^{\ddagger} - \frac{1}{T'_{2e}(0)}\left(\overline{\langle S_+^{(+1)} \rangle}^{\ddagger} + \overline{\langle S_+^{(-1)} \rangle}^{\ddagger}\right) \tag{O.73}$$

$$\frac{d\overline{\langle S_+^{(\pm 1)} \rangle}^{\ddagger}}{dt} = -\frac{1}{T_{2e}(\pm 1)}\overline{\langle S_+^{(\pm 1)} \rangle}^{\ddagger} - \frac{1}{T'_{2e}(\pm 1)}\overline{\langle S_+^{(0)} \rangle}^{\ddagger} \tag{O.74}$$

Here, the direct transverse relaxation rate constants are given by:

$$\frac{1}{T_{2e}(0)} = R_{2e} - R''_{2x_2} \tag{O.75}$$

$$\frac{1}{T_{2e}(\pm 1)} = \tfrac{1}{2}\left(R'_{2e} + R''_{2e} + R''_{2x_2}\right) \pm R_{2x_1} \tag{O.76}$$

where we use Eq. O.71 to replace R''_{2x_1}. (The primed rate constants are those for coherence transfer.)

From Eqs. O.61, O.62, O.66, O.70 and O.72, we find that the transverse electron relaxation times $T_{2e}(0)$ and $T_{2e}(\pm 1)$ given by Eqs. O.75 and O.76 agree with those predicted for $I = 1$ by Eqs. O.50–O.53. This specific example justifies the assumption made for the general case in Section O.3. There only one of the the two alternatives for the pseudosecular contribution is included from Eq. O.48, for a general hyperfine state m_I. As already noted, we retain both pseudosecular terms in the present treatment because the single-transition operators handle the different values of m_I explicitly. We reach a similar conclusion by using single-transition operators specifically for ^{15}N-nitroxides, i.e., $I = \frac{1}{2}$ (Marsh 2017).

The rates of coherence transfer that appear on the right of Eqs. O.73 and O.74 are characterized by:

$$\frac{1}{T'_{2e}(0)} = R_{2e} + R_{2x_2} - R''_{2e} - R''_{2x_2} \tag{O.77}$$

$$\frac{1}{T'_{2e}(\pm 1)} = \tfrac{1}{2} R''_{2x_2} \tag{O.78}$$

Note that coherence transfer to the $m_I = 0$ manifold is identical for both $m_I = \pm 1$ manifolds (Eq. O.78). Also, there is no coherence transfer between $m_I = \pm 1$ manifolds (Eq. O.74). These coherence-transfer terms (Eqs. O.77, O.78) are mostly ignored in conventional treatments of nitroxide transverse relaxation by Redfield theory. Normally, near-degenerate terms differing only by the hyperfine frequency ω_a are omitted from Eq. O.5 by using the condition $p' = p$ (see Section O.1), instead of the less stringent requirement: $\omega_p - \omega_{p'} \approx 0$ (see Galeev and Salikhov 1996; Marsh 2017). This automatically excludes the possibility of transferring transverse electron magnetization between different hyperfine states. Nonetheless, there is experimental evidence for electron coherence transfer between hyperfine states of isolated nitroxides, from non-vanishing dispersion-like distortions of the absorption EPR spectrum in the absence of electron spin–spin interaction (Peric et al. 2012; Bales et al. 2014). See also Section 9.17 in Chapter 9.

Using Eqs. O.61, O.63, O.70 and O.72, the coherence transfer rate constants for each of the three $m_I = 0, \ \pm 1$ ^{14}N-hyperfine lines (i.e., Eqs. O.77 and O.78) become:

$$\frac{1}{T'_{2e}(0)} = \frac{1}{T'_{2e}(\pm 1)} = \tfrac{1}{4}(\Delta A)^2 j^{(1)}(\omega_a) = \frac{1}{30}(\Delta A)^2 \frac{\tau_R}{1 + \omega_a^2 \tau_R^2} \tag{O.79}$$

where the final equality comes from Eq. O.54. Electron coherence transfer is contributed entirely by the pseudosecular terms, which are associated with the spectral density at the frequency $\omega_a \left(= \tfrac{1}{2} a_o\right)$ of the hyperfine coupling. This is anticipated, because pseudosecular terms change only the nitrogen nuclear hyperfine state. Comparing Eq. O.79 with Eqs. O.50, O.51 and O.53 for $I = 1$, we see that the rate of electron coherence transfer is half the pseudosecular contribution to the transverse electron relaxation rate $1/T_{2e}(0)$. Therefore, depending on circumstances, it can be appreciable (see Fig. 5.6).

A corresponding treatment for ^{15}N-nitroxides where $I = \frac{1}{2}$ shows that the expression for the coherence transfer rate constant $1/T'_{2e}\left(\pm\tfrac{1}{2}\right)$ is just half that given for ^{14}N-nitroxides by Eq. O.79 (Marsh 2017).

O.5 Relaxation Matrix

Alternatively, instead of using the commutator method of Sections O.2–O.4, we may express the spin operators in terms of their matrix elements referred to the eigenstates $|m\rangle$ of the unperturbed Hamiltonian $\mathcal{H}_o$. This corresponds to Redfield's approach and leads to the relaxation matrix (Redfield 1957; 1965). Expanding the commutators in Eq. O.1, we get:

$$\frac{d\sigma^{\ddagger}}{dt}=\frac{1}{\hbar^2}\int_0^{\infty}\{\mathcal{H}_1^{\ddagger}(t-\tau)\sigma^{\ddagger}(t)\mathcal{H}_1^{\ddagger}(t)+\mathcal{H}_1^{\ddagger}(t)\sigma^{\ddagger}(t)\mathcal{H}_1^{\ddagger}(t-\tau)$$

$$-\sigma^{\ddagger}(t)\mathcal{H}_1^{\ddagger}(t-\tau)\mathcal{H}_1^{\ddagger}(t)-\mathcal{H}_1^{\ddagger}(t)\mathcal{H}_1^{\ddagger}(t-\tau)\sigma^{\ddagger}(t)\}d\tau \qquad (O.80)$$

whilst postponing ensemble averaging until later. From the rules for matrix multiplication, the contribution of the first operator product $\mathcal{H}_1^{\ddagger}(t-\tau)\sigma^{\ddagger}(t)\mathcal{H}_1^{\ddagger}(t)$ to the $d\sigma^{\ddagger}_{mm'}/dt$ matrix element is given by the sum $\sum_{n,n'}\mathcal{H}^{\ddagger}_{1,mn}(t-\tau)\sigma^{\ddagger}_{nn'}(t)\mathcal{H}^{\ddagger}_{1,n'm'}(t)$. From Eq. N.25, this becomes $\sum_{n,n'}\mathcal{H}_{1,mn}(t-\tau)\mathcal{H}_{1,n'm'}(t)e^{-i\omega_{mn}\tau}e^{i(\omega_{mn}+\omega_{n'm'})t}\sigma^{\ddagger}_{nn'}(t)$, in the original representation for $\mathcal{H}_1(t)$. Now we perform the ensemble averaging and integrate over τ for this term. This results in a contribution $\sum_{n,n'}j_{mnm'n'}(\omega_{mn})e^{i(\omega_{mm'}-\omega_{nn'})t}\sigma^{\ddagger}_{nn'}(t)$, where the spectral density is defined in analogy with Eqs. 5.7, 5.9 by:

$$j_{mnm'n'}(\omega_{mn})=\int_0^{\infty}\overline{\mathcal{H}_{1,mn}(t-\tau)\mathcal{H}^{\dagger}_{1,m'n'}(t)}e^{-i\omega_{mn}\tau}\,d\tau \qquad (O.81)$$

and we have used the identity: $\omega_{mn}+\omega_{n'm'}=\omega_{mm'}-\omega_{nn'}$ for the differences between energy levels (recall the definition: $\hbar\omega_{mn}\equiv E_m-E_n$). Clearly these modified spectral densities are linearly related to the conventional spectral densities $j^{(q)}(\omega)$ that we define in Eq. O.6. Because the Hamiltonian is Hermitean, we write the matrix element of the spin operator as $\mathcal{H}_{1,n'm'}=\mathcal{H}^{\dagger}_{1,m'n'}$, where the single dagger is the Hermitean conjugate. We can treat the second term in the integrand of Eq. O.80 similarly, which then yields the corresponding contribution $\sum_{n,n'}j_{mnm'n'}(\omega_{n'm'})e^{i(\omega_{mm'}-\omega_{nn'})t}\sigma^{\ddagger}_{nn'}(t)$. However, the third term is given by the sum $\sum_{n,n'}\sigma^{\ddagger}_{mn}(t)\mathcal{H}^{\ddagger}_{1,nn'}(t-\tau)\mathcal{H}^{\ddagger}_{1,n'm'}(t)$, which contains the $\sigma^{\ddagger}_{mn}(t)$ element of the density matrix, i.e., no longer $\sigma^{\ddagger}_{nn'}(t)$ over which we make the summation. To produce the latter element we relabel some of the indices and make certain restrictions (Atherton 1973). We replace n by n' which gives $\sigma^{\ddagger}_{mn}\to\sigma^{\ddagger}_{mn'}$. Then we restrict the summation to terms where $n=m$ by making the replacement $\sigma^{\ddagger}_{mn'}\to\delta_{nm}\sigma^{\ddagger}_{nn'}$ ($\delta_{nm}=1$ if $n=m$ and is zero otherwise). Having done this, we must relabel the old n' indices as say p. The third term thus becomes $\delta_{nm}\sum_{n',p}\mathcal{H}_{1,n'p}(t-\tau)\mathcal{H}_{1,pm'}(t)e^{-i\omega_{n'p}\tau}e^{i(\omega_{n'p}+\omega_{pm'})t}\sigma^{\ddagger}_{nn'}(t)$, which contributes $\delta_{nm}\sum_{n',p}j_{n'pm'p}(\omega_{n'p})e^{i(\omega_{mm'}-\omega_{nn'})t}\sigma^{\ddagger}_{nn'}(t)$ to the right-hand side of Eq. O.80. Here $\omega_{n'p}+\omega_{pm'}=\omega_{n'm'}$ and we expand $\omega_{n'm'}$ to $\omega_{mm'}-\omega_{nn'}$ by using the fact that $n=m$. By a similar procedure, the fourth term contributes $\delta_{n'm'}\sum_{n',p}j_{mpnp}(\omega_{pn})e^{i(\omega_{mm'}-\omega_{nn'})t}\sigma^{\ddagger}_{nn'}(t)$.

Collecting together all contributions, the time evolution of the elements of the spin density matrix in the interaction representation (i.e., Eq. O.80) finally becomes:

$$\frac{d\sigma^{\ddagger}_{mm'}}{dt}=\sum_{n,n'}R_{mm',nn'}e^{i(\omega_{mm'}-\omega_{nn'})t}\sigma^{\ddagger}_{nn'}(t) \qquad (O.82)$$

where the relaxation matrix is defined by (Atherton 1973):

$$R_{mm',nn'}=\frac{1}{\hbar^2}\left(j_{mnm'n'}(\omega_{mn})+j_{mnm'n'}(\omega_{n'm'})-\delta_{nm}\sum_p j_{n'pm'p}(\omega_{n'p})-\delta_{n'm'}\sum_p j_{mpnp}(\omega_{pn})\right) \qquad (O.83)$$

Because the random fluctuations are stationary, the spectral densities given by Eq. O.81 are independent of time, and so therefore are the elements of the relaxation matrix. The time-dependent exponential in Eq. O.82 produces rapid oscillations that average to zero at long times, except when $\omega_{nn'}=\omega_{mm'}$ and the exponential becomes unity. Terms satisfying the latter condition therefore dominate, and we often omit those where $\omega_{nn'}\neq\omega_{mm'}$ from the summation. Under these circumstances, $\omega_{mn}=\omega_{m'n'}=-\omega_{n'm'}$ and the real parts of the first two spectral densities are then equal because they are even functions. If we omit imaginary parts of the relaxation matrix, which give rise to small frequency shifts, Eq. O.83 reduces to (cf. Freed and Fraenkel 1963):

$$R_{mm',nn'}=\frac{1}{\hbar^2}\left(2j_{mnm'n'}(\omega_{mn})-\delta_{nm}\sum_p j_{n'pm'p}(\omega_{n'p})-\delta_{n'm'}\sum_p j_{mpnp}(\omega_{pn})\right) \qquad (O.84)$$

Using Eq. N.24, we can rewrite Eq. O.82 in the matrix elements of the original representation:

$$\frac{d\sigma_{mm'}}{dt}=-i\omega_{mm'}\sigma_{mm'}(t)+\sum_{n,n'}R_{mm',nn'}\sigma_{nn'}(t)$$

$$=\frac{i}{\hbar}[\sigma(t),\mathcal{H}_o]+\sum_{n,n'}R_{mm',nn'}\sigma_{nn'}(t) \qquad (O.85)$$

where again we may exclude states for which $\omega_{nn'}\neq\omega_{mm'}$. The first term on the right is the motion of the spins in the static field (cf. Appendix D.2), and the second term represents the relaxation.

Being a perturbation treatment, the restrictions on Redfield theory are those appropriate to motional-narrowing theory, i.e., that $1/R_{mm'nn'} >> \tau_c$ where τ_c is the characteristic motional correlation time. The semi-diagonal elements have a straightforward interpretation. Element $R_{mm,nn}$ is simply the transition probability $W_{nm} = 2j_{mnmn}(\omega_{mn})/\hbar^2$ from state n to state m that corresponds to spin–lattice relaxation processes. The last two terms in Eq. O.83 are then zero. The $R_{mm',mm'}$ element is analogous to $1/T_2$, the transverse relaxation rate. The first two terms in Eq. O.83 then correspond to the secular contribution to the linewidth $1/T_2^{sec} = 2j_{mmm'm'}(0)/\hbar^2$ that comes from random modulation of the energy separation $\hbar\omega_{mm'}$. The last two terms are the lifetime broadening by transitions from all other states p to the states m and m' under observation, which make the contribution: $\frac{1}{2}\sum_p (W_{pm'} + W_{pm})$ to the T_2 relaxation rate. The net transverse relaxation rate for the $m \to m'$ transition is therefore:

$$\frac{1}{T_{2,mm'}} = -R_{mm',mm'} = \frac{1}{T_{2,mm'}^{sec}} + \frac{1}{2}\sum_{p\neq m,m'}(W_{pm'} + W_{pm}) \tag{O.86}$$

Concretely for spin labels: m and m' correspond to the electron spin states $M_S = \pm\frac{1}{2}$ with a fixed nuclear state m_I. Then p corresponds to both electron spin states $M_S = +\frac{1}{2}$ and $M_S = -\frac{1}{2}$, but with different hyperfine states, i.e., different values of m_I (cf. Figure 5.4 in Chapter 5).

Note that Eqs. O.82 and O.85 are incomplete because they do not include a lattice bath. The rates of transitions in both directions are equal: $R_{mm,nn} = R_{nn,mm}$. We can correct for this in an *ad hoc* manner by replacing the spin density-matrix σ in these equations by $\sigma - \sigma_o$, where σ_o is the density matrix at thermal equilibrium (Abragam 1961; Redfield 1965).

O.6 Redfield Theory for a Nitroxide

To illustrate use of the relaxation matrix, we apply Redfield theory to the classic case of a nitroxide undergoing rotational diffusion. We treated this problem already in Section O.3 by using spin-operator methods. However, the traditional approach to motional narrowing in the spin-label literature is to use the relaxation matrices of Redfield theory. The transverse relaxation rate for the $m \to m'$ transition is given by the $R_{mm',mm'}$ element of the relaxation matrix:

$$1/T_{2,mm'} = -R_{mm',mm'} \tag{O.87}$$

where $m \equiv \left|-\frac{1}{2}, m_I\right\rangle$ and $m' \equiv \left|+\frac{1}{2}, m_I\right\rangle$, for a nitroxide. From Eq. O.84, this is determined by the modified spectral densities with $n = m$ and $n' = m'$:

$$R_{mm',mm'} = \frac{1}{\hbar^2}\left(2j_{mmm'm'}(0) - \sum_p j_{m'pm'p}(\omega_{m'p}) - \sum_p j_{mpmp}(\omega_{pm})\right) \tag{O.88}$$

where $\omega_{mm} = 0$ for the first term.

We assume axial symmetry of the perturbation Hamiltonian, which therefore is given by Eqs. O.35–O.39 in Section O.3. For the secular part (i.e., Eq. O.35), we need only the diagonal matrix elements $\left\langle\pm\frac{1}{2}, m_I\right|S_z\left|\pm\frac{1}{2}, m_I\right\rangle = \pm\frac{1}{2}$ and $\left\langle\pm\frac{1}{2}, m_I\right|S_z I_z\left|\pm\frac{1}{2}, m_I\right\rangle = \pm\frac{1}{2}m_I$. The secular contribution to the first modified spectral density in Eq. O.88 then becomes:

$$\begin{aligned} j_{mmm'm'}(0) &= \left\langle-\tfrac{1}{2}, m_I\right|A_1^0\left|-\tfrac{1}{2}, m_I\right\rangle\left\langle+\tfrac{1}{2}, m_I\right|A_1^0\left|+\tfrac{1}{2}, m_I\right\rangle j^{(0)}(0) \\ &= \tfrac{1}{9}\left(\Delta g\beta_e B_o + \Delta A m_I\right)^2\left\langle-\tfrac{1}{2}, m_I\right|S_z\left|-\tfrac{1}{2}, m_I\right\rangle \\ &\quad\times\left\langle+\tfrac{1}{2}, m_I\right|S_z\left|+\tfrac{1}{2}, m_I\right\rangle j^{(0)}(0) \\ &= -\tfrac{1}{36}\left(\Delta g\beta_e B_o + \Delta A m_I\right)^2 j^{(0)}(0) \end{aligned} \tag{O.89}$$

where we use Eq. O.81 to express this in terms of the conventional spectral density $j^{(q)}(\omega)$ defined by Eq. O.54 in Section O.3. The other two modified spectral densities contributed by the secular Hamiltonian are given by $p = m'$ and $p = m$, respectively, in Eq. O.88:

$$\begin{aligned} j_{m'm'm'm'}(0) &= \left\langle-\tfrac{1}{2}, m_I\right|A_1^0\left|-\tfrac{1}{2}, m_I\right\rangle\left\langle-\tfrac{1}{2}, m_I\right|A_1^0\left|-\tfrac{1}{2}, m_I\right\rangle j^{(0)}(0) \\ &= \tfrac{1}{36}\left(\Delta g\beta_e B_o + \Delta A m_I\right)^2 j^{(0)}(0) \end{aligned} \tag{O.90}$$

$$\begin{aligned} j_{mmmm}(0) &= \left\langle+\tfrac{1}{2}, m_I\right|A_1^0\left|+\tfrac{1}{2}, m_I\right\rangle\left\langle+\tfrac{1}{2}, m_I\right|A_1^0\left|+\tfrac{1}{2}, m_I\right\rangle j^{(0)}(0) \\ &= \tfrac{1}{36}\left(\Delta g\beta_e B_o + \Delta A m_I\right)^2 j^{(0)}(0) \end{aligned} \tag{O.91}$$

The total secular contribution to the T_2-relaxation rate therefore becomes:

$$\frac{1}{T_2^{sec}(m_I)} = \frac{1}{9\hbar^2}\left(\Delta g\beta_e B_o + \Delta A m_I\right)^2 j^{(0)}(0) \tag{O.92}$$

in agreement with the secular part of Eqs. O.50–O.53 of Section O.3, and Eqs. 5.59, 5.61, 5.62 of Chapter 5 for the axial approximation. Note that here ΔA is expressed in energy units.

Pseudosecular and nonsecular terms make no contribution to $j_{mmm'm'}(0)$ because they involve a change in at least one of the quantum numbers M_S or m_I. Therefore, we are left with the last two terms in Eq. O.88. We turn first to contributions from $A_1^{(\pm1)}$ given by Eq. O.37, because these are analogous to the secular contributions, except that the electron spin operator is nonsecular, $S_\pm$ instead of S_z. The non-vanishing extended spectral densities in Eq. O.88 are given by $p = m$ and $p = m'$, respectively:

$$\begin{aligned} j_{m'mm'm}(\omega_{m'm}) &= \left\langle+\tfrac{1}{2}, m_I\right|A_1^{(+1)}\left|-\tfrac{1}{2}, m_I\right\rangle \\ &\quad\times\left\langle+\tfrac{1}{2}, m_I\right|A_1^{(+1)}\left|-\tfrac{1}{2}, m_I\right\rangle j^{(1)}(\omega_{m'm}) \\ &= \tfrac{1}{4}\left(\Delta g\beta_e B_o + \Delta A m_I\right)^2\left\langle+\tfrac{1}{2}, m_I\right|S_+\left|-\tfrac{1}{2}, m_I\right\rangle \\ &\quad\times\left\langle+\tfrac{1}{2}, m_I\right|S_+\left|-\tfrac{1}{2}, m_I\right\rangle j^{(1)}(\omega_{m'm}) \\ &= \tfrac{1}{4}\left(\Delta g\beta_e B_o + \Delta A m_I\right)^2 j^{(1)}(\omega_{m'm}) \end{aligned} \tag{O.93}$$

$$j_{mm'mm'}(\omega_{mm'}) = \langle -\tfrac{1}{2}, m_I | A_1^{(-1)} | +\tfrac{1}{2}, m_I \rangle \times \langle -\tfrac{1}{2}, m_I | A_1^{(-1)} | +\tfrac{1}{2}, m_I \rangle j^{(1)}(\omega_{mm'}) = \tfrac{1}{4}(\Delta g \beta_e B_o + \Delta A m_I)^2 \langle -\tfrac{1}{2}, m_I | S_- | +\tfrac{1}{2}, m_I \rangle \times \langle -\tfrac{1}{2}, m_I | S_- | +\tfrac{1}{2}, m_I \rangle j^{(1)}(\omega_{mm'}) = \tfrac{1}{4}(\Delta g \beta_e B_o + \Delta A m_I)^2 j^{(1)}(\omega_{mm'}) \quad (O.94)$$

where $\omega_{m'm} = -\omega_{m'm} = (g_o \beta_e B_o + a_o m_I)/\hbar$. By contrast, the pseudosecular contribution, $A_2^{(\pm 1)}$ given by Eq. O.38, is diagonal in the electron spin, but has only off-diagonal elements of the nuclear spin:

$$j_{m'pm'p}(\omega_{m'p}) = \langle -\tfrac{1}{2}, m_I \pm 1 | A_2^{(\pm 1)} | -\tfrac{1}{2}, m_I \rangle \times \langle -\tfrac{1}{2}, m_I \pm 1 | A_2^{(\pm 1)} | -\tfrac{1}{2}, m_I \rangle j^{(1)}(\omega_{m'p}) = \tfrac{1}{4}(\Delta A)^2 \langle -\tfrac{1}{2}, m_I \pm 1 | S_z I_\pm | -\tfrac{1}{2}, m_I \rangle \times \langle -\tfrac{1}{2}, m_I \pm 1 | S_z I_\pm | -\tfrac{1}{2}, m_I \rangle j^{(1)}(\omega_{m'p}) = \tfrac{1}{16}(\Delta A)^2 (I(I+1) - m_I(m_I \pm 1)) j^{(1)}(\omega_{m'p}) \quad (O.95)$$

$$j_{mpmp}(\omega_{mp}) = \langle +\tfrac{1}{2}, m_I \pm 1 | A_2^{(\pm 1)} | +\tfrac{1}{2}, m_I \rangle \times \langle +\tfrac{1}{2}, m_I \pm 1 | A_2^{(\pm 1)} | +\tfrac{1}{2}, m_I \rangle j^{(1)}(\omega_{mp}) = \tfrac{1}{16}(\Delta A)^2 (I(I+1) - m_I(m_I \pm 1)) j^{(1)}(\omega_{mp}) \quad (O.96)$$

where $\omega_{m'p} = -\omega_{mp} = \mp \tfrac{1}{2} a_o/\hbar$. The remaining terms, from $A_2^{(\pm 0)}$ and $A^{(2\pm)}$ (see Eqs. O.35 and O.39), contain only off-diagonal elements from both electron and nuclear spins, of which the non-vanishing contributions are:

$$j_{m'pm'p}(\omega_{m'p}) = \langle +\tfrac{1}{2}, m_I - 1 | A_2^{(-0)} | -\tfrac{1}{2}, m_I \rangle \times \langle +\tfrac{1}{2}, m_I - 1 | A_2^{(-0)} | -\tfrac{1}{2}, m_I \rangle j^{(0)}(\omega_{m'p}) = \tfrac{1}{144}(\Delta A)^2 \langle +\tfrac{1}{2}, m_I - 1 | S_+ I_- | -\tfrac{1}{2}, m_I \rangle \times \langle +\tfrac{1}{2}, m_I - 1 | S_+ I_- | -\tfrac{1}{2}, m_I \rangle j^{(0)}(\omega_{m'p}) = \tfrac{1}{144}(\Delta A)^2 (I(I+1) - m_I(m_I - 1)) j^{(0)}(\omega_{m'p}) \quad (O.97)$$

$$j_{mpmp}(\omega_{mp}) = \langle -\tfrac{1}{2}, m_I + 1 | A_2^{(+0)} | +\tfrac{1}{2}, m_I \rangle \times \langle -\tfrac{1}{2}, m_I + 1 | A_2^{(+0)} | +\tfrac{1}{2}, m_I \rangle j^{(0)}(\omega_{mp}) = \tfrac{1}{144}(\Delta A)^2 \langle -\tfrac{1}{2}, m_I + 1 | S_- I_+ | +\tfrac{1}{2}, m_I \rangle \times \langle -\tfrac{1}{2}, m_I + 1 | S_- I_+ | +\tfrac{1}{2}, m_I \rangle j^{(0)}(\omega_{mp}) = \tfrac{1}{144}(\Delta A)^2 (I(I+1) - m_I(m_I + 1)) j^{(0)}(\omega_{mp}) \quad (O.98)$$

where $\omega_{m'p} = -\omega_{mp} - a_o/\hbar = (g_o \beta_e B_o + a_o m_I - \tfrac{1}{2} a_o)/\hbar$; and

$$j_{m'pm'p}(\omega_{m'p}) = \langle +\tfrac{1}{2}, m_I + 1 | A^{(2+)} | -\tfrac{1}{2}, m_I \rangle \times \langle +\tfrac{1}{2}, m_I + 1 | A^{(2+)} | -\tfrac{1}{2}, m_I \rangle j^{(2)}(\omega_{m'p}) = \tfrac{1}{16}(\Delta A)^2 \langle +\tfrac{1}{2}, m_I + 1 | S_+ I_+ | -\tfrac{1}{2}, m_I \rangle \times \langle +\tfrac{1}{2}, m_I + 1 | S_+ I_+ | -\tfrac{1}{2}, m_I \rangle j^{(2)}(\omega_{m'p}) = \tfrac{1}{16}(\Delta A)^2 (I(I+1) - m_I(m_I + 1)) j^{(2)}(\omega_{m'p}) \quad (O.99)$$

$$j_{mpmp}(\omega_{mp}) = \langle -\tfrac{1}{2}, m_I - 1 | A^{(-2)} | +\tfrac{1}{2}, m_I \rangle \times \langle -\tfrac{1}{2}, m_I - 1 | A^{(-2)} | +\tfrac{1}{2}, m_I \rangle j^{(2)}(\omega_{mp}) = \tfrac{1}{16}(\Delta A)^2 \langle -\tfrac{1}{2}, m_I - 1 | S_- I_- | +\tfrac{1}{2}, m_I \rangle \times \langle -\tfrac{1}{2}, m_I - 1 | S_- I_- | +\tfrac{1}{2}, m_I \rangle j^{(2)}(\omega_{mp}) = \tfrac{1}{16}(\Delta A)^2 (I(I+1) - m_I(m_I - 1)) j^{(2)}(\omega_{mp}) \quad (O.100)$$

where $\omega_{m'p} = -\omega_{mp} + a_o/\hbar = (g_o \beta_e B_o + a_o m_I + \tfrac{1}{2} a_o)/\hbar$.

Finally, collecting up all terms from Eqs. O.92–O.100 and expressing the total transverse relaxation rate as in Eq. O.50, the linewidth coefficients become:

$$A = (\Delta g \beta_e/\hbar)^2 \left(\tfrac{1}{9} j^{(0)}(0) + \tfrac{1}{2} j^{(1)}(\omega_e)\right) B_o^2 + \tfrac{1}{4} I(I+1)(\Delta A)^2 \left(\tfrac{1}{18} j^{(0)}(\omega_e) + j^{(1)}(\omega_a) + \tfrac{1}{2} j^{(2)}(\omega_e)\right) \quad (O.101)$$

$$B = (\Delta g \beta_e/\hbar) \Delta A \left(\tfrac{2}{9} j^{(0)}(0) + j^{(1)}(\omega_e)\right) B_o \quad (O.102)$$

$$C = \tfrac{1}{2}(\Delta A)^2 \left(\tfrac{2}{9} j^{(0)}(0) - \tfrac{1}{36} j^{(0)}(\omega_e) + j^{(1)}(\omega_e) - \tfrac{1}{2} j^{(1)}(\omega_a) - \tfrac{1}{4} j^{(2)}(\omega_e)\right) \quad (O.103)$$

where $\omega_e \cong (g_o \beta_e B_o + a_o m_I)/\hbar$ and $\omega_a = \tfrac{1}{2} a_o/\hbar$. These coefficients agree with what we give in Chapter 5, in the case of axial symmetry (viz., Eqs. 5.59, 5.61 and 5.62 with $\delta g = \delta A = 0$). They agree with Eqs. O.50–O.53 in Section O.3, which use the same axial approximation.

APPENDIX P: PRODUCT-OPERATOR FORMALISM FOR PULSE EPR

P.1 Single Pulse: Free Precession

Description of pulse EPR by the spin density-matrix operator $\sigma(t)$ begins with Eq. N.19 of Appendix N for the time evolution from its equilibrium value. The latter is given by the spin operator $\sigma(0) = S_z$, in the high-temperature approximation (see Eq. N.12). We view all operations in the frame that rotates about the static B_o-field (z-axis) at the angular frequency ω of the microwave B_1-field, which lies along the x-direction. Assuming that the microwave pulse is sufficiently short and intense, the Hamiltonian $\mathcal{H}_1 = \gamma_e B_1 S_x$ dominates in the rotating frame and is essentially constant for the duration t_p of the pulse. From Eq. N.19, the action of the pulse applied along the x-axis then is given by:

$$\sigma(t_p) = \exp(-i\beta_p S_x)\sigma(0)\exp(i\beta_p S_x)$$
$$= \exp(-i\beta_p S_x) S_z \exp(i\beta_p S_x) \quad \text{(P.1)}$$

where $\beta_p \equiv \gamma_e B_1 t_p$ is the angle by which the magnetization is rotated about the x-axis (see Section 13.2 in Chapter 13). We readily identify the exponentials in Eq. P.1 as rotation operators (cf. Eqs. R.5, R.6 in Appendix R). The action of the pulse therefore corresponds simply to rotations in a spin-operator space. We use a shorthand for Eq. P.1:

$$\sigma(0) \xrightarrow{\beta_p S_x} \sigma(t) \quad \text{(P.2)}$$

where the arrow implies bracketting $\sigma(0)$ by the rotation operators: $R(\beta,x)\sigma(0)R^{-1}(\beta,x)$, i.e., a transformation by the unitary operator $R(\beta,x) \equiv \exp(-i\beta S_x)$. Now we can express the action of a pulse along the x-axis solely by using spin operators:

$$S_z \xrightarrow{\beta_p S_x} S_z \cos\beta_p - S_y \sin\beta_p \quad \text{(P.3)}$$

$$S_y \xrightarrow{\beta_p S_x} S_y \cos\beta_p - S_z \sin\beta_p \quad \text{(P.4)}$$

Of course, an x-axis pulse does not rotate S_x, which remains unchanged by the $\beta_p S_x$ transformation. We get corresponding rotations from a pulse along the y-axis by permuting the x,y,z indices in Eqs. P.3 and P.4 cyclically. The rotational transformations for each transverse spin operator are listed in the first two rows of Table P.1.

After the pulse, we get free precession of the magnetization under the action of the static Zeeman Hamiltonian, $\mathcal{H}_o = \hbar\Delta\omega_o S_z$, by applying Eq. N.19 further to Eq. P.1:

$$\sigma(t) = \exp(-i\Delta\omega_o t S_z)\sigma(t_p)\exp(i\Delta\omega_o t S_z)$$
$$= \exp(-i\Delta\omega_o t S_z)\exp(-i\beta_p S_x) S_z \exp(i\beta_p S_x)\exp(i\Delta\omega_o t S_z) \quad \text{(P.5)}$$

The first equality corresponds to rotation by an angle $\Delta\omega_o t$ about the z-axis. The corresponding spin-operator transformations for free precession are:

$$S_y \xrightarrow{\Delta\omega_o t S_z} S_y \cos\Delta\omega_o t - S_x \sin\Delta\omega_o t \quad \text{(P.6)}$$

$$S_x \xrightarrow{\Delta\omega_o t S_z} S_x \cos\Delta\omega_o t + S_y \sin\Delta\omega_o t \quad \text{(P.7)}$$

Of course, the S_z operator remains unchanged during rotation about the z-axis. We give these transformations for free precession in the third row of Table P.1. The advantage of the spin-operator method for constructing pulse sequences becomes apparent from the second line in Eq. P.5. We can replace the nested unitary transformations by sequential transformations. Equation P.3 gives the action of the pulse, which is followed by free evolution that is given by Eq. P.6:

$$S_z \xrightarrow{\beta_p S_x} S_z \cos\beta_p - S_y \sin\beta_p \xrightarrow{\Delta\omega_o t S_z}$$
$$S_z \cos\beta_p - \sin\beta_p\left(S_y \cos\Delta\omega_o t - S_x \sin\Delta\omega_o t\right) \quad \text{(P.8)}$$

The final expression in Eq. P.8 is the spin density-matrix operator $\sigma(t)$ defined in Eq. P.5. By applying Eq. N.4, the average values of the transverse magnetization components become:

$$\overline{\langle S_x(t)\rangle} = \mathrm{Tr}(S_x\sigma(t)) = \tfrac{1}{2}\sin\beta_p \sin\Delta\omega_o t \quad \text{(P.9)}$$

$$\overline{\langle S_y(t)\rangle} = \mathrm{Tr}(S_y\sigma(t)) = \tfrac{1}{2}\sin\beta_p \cos\Delta\omega_o t \quad \text{(P.10)}$$

TABLE P.1 Effect of rotation transformations $R(\phi,j)S_iR^{-1}(\phi,j)$ and of unitary transformations US_iU^{-1} on spin operators S_i, where $R \equiv \exp(-i\phi S_j)$ and $U \equiv \exp(-i\phi I_k S_j)$, for subscripts $i,j,k \equiv x, y$ or z

R,U	S_x	S_y	S_z
$\exp(-i\phi S_x)$	S_x	$S_y\cos\phi + S_z\sin\phi$	$S_z\cos\phi - S_y\sin\phi$
$\exp(-i\phi S_y)$	$S_x\cos\phi - S_z\sin\phi$	S_y	$S_z\cos\phi + S_x\sin\phi$
$\exp(-i\phi S_z)$	$S_x\cos\phi + S_y\sin\phi$	$S_y\cos\phi - S_x\sin\phi$	S_z
$\exp(-i\phi I_k S_x)$	S_x	$S_y\cos\frac{\phi}{2} + 2I_kS_z\sin\frac{\phi}{2}$	$S_z\cos\frac{\phi}{2} - 2I_kS_y\sin\frac{\phi}{2}$
$\exp(-i\phi I_k S_y)$	$S_x\cos\frac{\phi}{2} - 2I_kS_z\sin\frac{\phi}{2}$	S_y	$S_z\cos\frac{\phi}{2} + 2I_kS_x\sin\frac{\phi}{2}$
$\exp(-i\phi I_k S_z)$	$S_x\cos\frac{\phi}{2} + 2I_kS_y\sin\frac{\phi}{2}$	$S_y\cos\frac{\phi}{2} - 2I_kS_x\sin\frac{\phi}{2}$	S_z

Note: the three bottom rows apply for $I = \frac{1}{2}$ only, where $k \equiv x, y$ or z and $I_k^2 = \frac{1}{4}$. See Eq. P.20 for $I \geq 1$.

The resulting free induction decay (FID) signal therefore arises from transverse magnetization oscillating at the offset frequency $\Delta\omega_o$, with an amplitude that is proportional to $\sin\beta_p$. Note that, quite generally, the transverse x-magnetization $\langle S_x(t)\rangle$ is given by the coefficient of S_x in the final expression for $\sigma(t)$ and likewise for the transverse y-magnetization.

P.2 Pulse Sequences

We now use the primary spin echo ($\pi/2-\tau-\pi-\tau-echo$) as a simple example of a pulse sequence. The first pulse, and subsequent free precession up to the second pulse, is a special case of the FID example that we gave above:

$$S_z \xrightarrow{\pi/2S_x} -S_y \xrightarrow{\Delta\omega_o\tau S_z} -S_y\cos\Delta\omega_o\tau + S_x\sin\Delta\omega_o\tau \tag{P.11}$$

where the $\pi/2$-pulse along the x-direction tips the equilibrium z-magnetization to the transverse y-direction. The second pulse, also along the x-direction, reverses the phase of the transverse magnetization, and subsequent free precession produces the echo:

$$-S_y\cos\Delta\omega_o\tau + S_x\sin\Delta\omega_o\tau \xrightarrow{\pi S_x}$$

$$S_y\cos\Delta\omega_o\tau + S_x\sin\Delta\omega_o\tau \xrightarrow{\Delta\omega_o\tau S_z} S_y \tag{P.12}$$

where the π-pulse inverts the sign of S_y and leaves S_x unchanged. Free precession then refocusses the transverse magnetization to produce an echo along the positive y-axis at time τ after the second pulse. In this ideal case, the whole of the magnetization is directed along the y-axis after time $t=\tau$. Generally this is not the case, e.g., for different pulse angles, and then we identify the echo as that part of the y-magnetization which does not depend on the dephasing $\Delta\omega_o t$ at time $t=\tau$ after the second pulse (see Section 13.5 in Chapter 13).

P.3 Evolution under Product Operators

So far, we treated free precession solely under the influence of the Zeeman Hamiltonian. This contains only a single spin operator S_z (or I_z), and its action is described fully by the first three rows in Table P.1. We also need a method to handle evolution under product operators, such as the secular part of the hyperfine interaction aI_zS_z, where a stands equally for the isotropic hyperfine coupling a_o or the anisotropic tensor element A_{zz}. After applying a single 90°-pulse, the spin operator for the transverse magnetization is $-S_y$. Subsequent evolution under the aI_zS_z part of the Hamiltonian is given by (see Eq. P.6):

$$-S_y \xrightarrow{(aI_zt)S_z} -S_y\cos(aI_zt) + S_x\sin(aI_zt) \tag{P.13}$$

where we bundle I_z into the phase shift, because it does not act on the electron spin. However, because I_z is an operator, we must expand the trignometrical functions as power series:

$$\cos(aI_zt) = 1 - \frac{(at)^2}{2!}I_z^2 + \frac{(at)^4}{4!}I_z^4\ldots \tag{P.14}$$

$$\sin(aI_zt) = I_z\left[at - \frac{(at)^3}{3!}I_z^2 + \frac{(at)^5}{5!}I_z^4\ldots\right] \tag{P.15}$$

For a nuclear spin $I=\frac{1}{2}$, we have $I_z^2=\left(\frac{1}{2}\right)^2$, $I_z^4=\left(\frac{1}{2}\right)^4$, ... (see Appendix E). Substituting these values into Eqs. P.14, P.15, we recover the trignometrical series without operators:

$$\cos(aI_zt) = 1 - \frac{\left(\frac{1}{2}at\right)^2}{2!} + \frac{\left(\frac{1}{2}at\right)^4}{4!}\ldots \equiv \cos\left(\tfrac{1}{2}at\right) \tag{P.16}$$

$$\sin(aI_zt) = 2I_z\left[\tfrac{1}{2}at - \frac{\left(\frac{1}{2}at\right)^3}{3!} + \frac{\left(\frac{1}{2}at\right)^5}{5!}\ldots\right] \equiv 2I_z\sin\left(\tfrac{1}{2}at\right) \tag{P.17}$$

Hence, for $I=\frac{1}{2}$, the transformation with the bilinear spin operator in Eq. P.13 becomes:

$$-S_y \xrightarrow{(at)I_zS_z} -S_y\cos\left(\tfrac{1}{2}at\right) + 2I_zS_x\sin\left(\tfrac{1}{2}at\right) \tag{P.18}$$

which applies equally when I_z is substituted by I_x or I_y. This follows from the eigenvalue relation $I^2=I_x^2+I_y^2+I_z^2=I(I+1)$, and thus we get:

$$I_x^2 = I_y^2 = I_z^2 = \tfrac{1}{3}I(I+1) \tag{P.19}$$

because the x-, y- and z-axes are equivalent in the absence of a static field. In the three bottom rows of Table P.1, we list transformations equivalent to Eq. P.18, for each of the three groups of I_kS_j product operators where $I=\frac{1}{2}$. By using Eq. P.19, we also see how we can extend Eq. P.18 to nuclear spins with $I>\frac{1}{2}$:

$$-S_y \xrightarrow{(at)I_zS_z} -S_y\cos\left(\sqrt{\tfrac{1}{3}I(I+1)}at\right)$$

$$+\frac{1}{\sqrt{\frac{1}{3}I(I+1)}}I_zS_x\sin\left(\sqrt{\tfrac{1}{3}I(I+1)}at\right) \tag{P.20}$$

which reverts to Eq. P.18 for $I=\frac{1}{2}$.

We need to know not only how to transform a single-spin operator by a product operator, but also how to transform a bilinear spin operator I_kS_j by a product operator. This proves to be straightforward. Recalling the definition of the unitary transformation, we have:

$$UI_kS_jU^{-1} = UI_kU^{-1}US_jU^{-1} \tag{P.21}$$

which means that we can perform the transformations separately. We transform S_j with the S-part of U and I_k with the I-part of U. Then we multiply the two separate results, for example:

$$I_yS_x \xrightarrow{\phi I_zS_z} \left(I_y\cos\tfrac{1}{2}\phi - 2I_xS_z\sin\tfrac{1}{2}\phi\right)\left(S_x\cos\tfrac{1}{2}\phi + 2I_zS_y\sin\tfrac{1}{2}\phi\right) \tag{P.22}$$

where we use the bottom rows of Table P.1 for the individual transformations. Clearly, such transformations generate higher-order spin products, including product operators for the same spin.

It turns out that we can substitute bilinear operators for the same spin, e.g., S_xS_y, by single-spin operators. By explicitly multiplying the S_x- and S_y-matrices for a spin $S=\frac{1}{2}$ from Table E.1 in Appendix E, we find that:

$$S_xS_y = -S_yS_x = \tfrac{1}{2}iS_z \tag{P.23}$$

Corresponding relations come from cyclic permutation of the axes:

$$S_yS_z = -S_zS_y = \tfrac{1}{2}iS_x \tag{P.24}$$

$$S_zS_x = -S_xS_z = \tfrac{1}{2}iS_y \tag{P.25}$$

These replacements hold only for $I,\ S=\frac{1}{2}$; we cannot use them for nuclear spins $I>\frac{1}{2}$. Applying these results to the example given in Eq. P.22, we get:

$$I_yS_x \xrightarrow{\phi I_zS_z} I_yS_x\cos^2\tfrac{1}{2}\phi + \left(I_yI_zS_y - I_xS_zS_x\right)\sin\phi$$
$$-4I_xI_zS_zS_y\sin^2\tfrac{1}{2}\phi = I_yS_x\cos^2\tfrac{1}{2}\phi + I_yS_x\sin^2\tfrac{1}{2}\phi = I_yS_x \tag{P.26}$$

As we might anticipate, the transverse product I_yS_x is unchanged by rotation of the two spin components by the same angle about the z-axis. This also holds good for the action of ϕI_zS_z on the products I_xS_x and I_yS_y, and of ϕI_yS_z on I_xS_x and I_zS_x. However, we get a different result when the transformation operator and the spin product have a spin component in common. For example:

$$I_yS_x \xrightarrow{\phi I_yS_z} I_y\left(S_x\cos\tfrac{1}{2}\phi + 2I_yS_y\sin\tfrac{1}{2}\phi\right)$$
$$= I_yS_x\cos\tfrac{1}{2}\phi + \tfrac{1}{2}S_y\sin\tfrac{1}{2}\phi \tag{P.27}$$

where we have used $I_y^2=\frac{1}{4}$ for $I=\frac{1}{2}$ from Eq. P.19. This type of transformation is important because it generates transverse magnetization, viz., a term containing solely S_y, which corresponds to the EPR signal. We summarize transformations for different combinations of bilinear spin operators of both types in Table P.2.

P.4 Evolution under Non-diagonal Hamiltonians

Applying the spin-operator method to evolution/free-precession periods is straightforward, if the Hamiltonian $\mathcal{H}_o$ is diagonal in the spin-operator basis. This restricts $\mathcal{H}_o$ to the secular terms, i.e., those which contain only the spin operators S_z and I_z. If the Hamiltonian contains off-diagonal terms, we must first diagonalize it, before we can apply the new S_z and I_z operators to describe the evolution under $\mathcal{H}_o$. Afterwards, we must apply the reverse transformation to return to the original Cartesian basis. The latter is the laboratory axis system, where the pulses are applied and the transverse magnetization is detected.

We incorporate diagonalization into the spin-operator method by using the unitary transformation, $U\mathcal{H}_oU^{-1}$, that rotates $\mathcal{H}_o$ to axes where it is diagonal (see Eq. H.14 in Appendix H). An important example is the pseudosecular contribution to the hyperfine interaction, BI_xS_z, which produces the ESEEM effect in spin-echo EPR and also is crucial for ELDOR-detected NMR (see Chapter 14). Including the hyperfine interaction, the spin Hamiltonian is:

$$\mathcal{H}_o = \hbar\Delta\omega_oS_z + \hbar\omega_II_z + AI_zS_z + BI_xS_z \tag{P.28}$$

where $\hbar\omega_II_z$ is the nuclear Zeeman interaction and AI_zS_z is the secular (i.e., diagonal) part of the hyperfine interaction. We describe diagonalization of this pseudosecular Hamiltonian for $I=\frac{1}{2}$ in Section H.2 of Appendix H.

The unitary transformation matrix that diagonalizes Eq. P.28 for the M_S electron state is given by Eq. H.13. Because it takes I_x into I_z, this corresponds to rotation about the y-axis, for which we use the exponential operator (see Eq. R.6 in Appendix R):

$$U(\eta_{M_S}) = \exp(-2i\eta_{M_S}I_y) \tag{P.29}$$

where the angle $2\eta_{M_S}$ is given by Eq. H.12. The factor of two appears in the exponential to satisfy Eqs. H.9–H.10, as we explain in Appendix R.1. We must apply Eq. P.29 twice, for the two $M_S=\pm\frac{1}{2}$ spin orientations. To do this, we use sequential product operators:

TABLE P.2 Effect of unitary transformations $UI_jS_iU^{-1}$ on bilinear spin operators I_jS_i, where $U\equiv\exp(-i\phi I_kS_j)$, for subscripts $i,j,k\equiv x,y,z$ plus cyclic permutations

U	I_jS_i	I_iS_i	I_kS_i	I_iS_k
$\exp(-i\phi I_kS_k)$	I_jS_i	I_iS_i	$I_kS_i\cos\frac{\phi}{2}+\frac{1}{2}S_j\sin\frac{\phi}{2}$	$I_iS_k\cos\frac{\phi}{2}+\frac{1}{2}I_j\sin\frac{\phi}{2}$
$\exp(-i\phi I_jS_k)$	$I_jS_i\cos\frac{\phi}{2}+\frac{1}{2}S_j\sin\frac{\phi}{2}$	I_iS_i	I_kS_i	$I_iS_k\cos\frac{\phi}{2}-\frac{1}{2}I_k\sin\frac{\phi}{2}$
$\exp(-i\phi I_kS_j)$	I_jS_i	I_iS_i	$I_kS_i\cos\frac{\phi}{2}-\frac{1}{2}S_k\sin\frac{\phi}{2}$	I_iS_k
$\exp(-i\phi I_jS_j)$	$I_jS_i\cos\frac{\phi}{2}-\frac{1}{2}S_k\sin\frac{\phi}{2}$	I_iS_i	I_kS_i	I_iS_k

Note: $k\neq i,j$ with $i\neq j$; this table applies only for $I=\frac{1}{2}$.

$$U(\eta_\alpha,\eta_\beta)=\exp\left[-2i\left(\eta_\alpha\left(\tfrac{1}{2}+S_z\right)I_y+\eta_\beta\left(\tfrac{1}{2}-S_z\right)I_y\right)\right] \quad \text{(P.30)}$$

where η_α, η_β correspond to the two values α,β of M_S. The polarization operator $\left(\tfrac{1}{2}+S_z\right)\equiv S^\alpha$ in Eq. P.30 selects $M_S=+\tfrac{1}{2}$ and $\left(\tfrac{1}{2}-S_z\right)\equiv S^\beta$ selects $M_S=-\tfrac{1}{2}$ (see later in Section P.5). A more compact form of Eq. P.30 is:

$$U(\xi,\eta)=\exp\left[-i\left(\xi I_y+2\eta S_z I_y\right)\right] \quad \text{(P.31)}$$

where

$$\xi\equiv\eta_\alpha+\eta_\beta \quad \text{(P.32)}$$

$$\eta\equiv\eta_\alpha-\eta_\beta \quad \text{(P.33)}$$

Remember that from Eq. H.12:

$$\tan 2\eta_{\alpha,\beta}=\frac{\mp B}{2\hbar\omega_I\pm A} \quad \text{(P.34)}$$

where the upper signs apply to η_α and the lower to η_β. Note also: Schweiger and Jeschke (2001) define η_α and η_β as twice those that we use here (see Sections H.2, H.3 of Appendix H).

By using a redefined nuclear spin vector, we now can rewrite the pseudosecular Hamiltonian of Eq. P.28 in its diagonal form:

$$\mathcal{H}_o^d=\hbar\left(\Delta\omega_o S_z+\tfrac{1}{2}\omega_+ I'_z+\omega_- I'_z S_z\right) \quad \text{(P.35)}$$

where the prime indicates the transformed nuclear axis system. To satisfy the eigenvalues given by Eq. H.8 in Appendix H, the sum and difference nuclear frequencies in Eq. P.35 become:

$$\tfrac{1}{2}\hbar(\omega_++\omega_-)=\sqrt{\left(\hbar\omega_I+\tfrac{1}{2}A\right)^2+\tfrac{1}{4}B^2} \quad \text{(P.36)}$$

$$\tfrac{1}{2}\hbar(\omega_+-\omega_-)=\sqrt{\left(\hbar\omega_I-\tfrac{1}{2}A\right)^2+\tfrac{1}{4}B^2} \quad \text{(P.37)}$$

We see this most readily by writing out expressions for the $M_S=\pm\tfrac{1}{2}$ energy levels separately. Using the spin-operator formalism, we therefore express the diagonalization procedure as:

$$\hbar\omega_I+AI_zS_z+BI_xS_z\xrightarrow{\xi I_y+2\eta I_yS_z}\tfrac{1}{2}\hbar\omega_+I_z+\hbar\omega_-I_zS_z \quad \text{(P.38)}$$

which we may verify directly by using Tables P.1, P.2 and Eqs. P.32–P.34, P.36, P.37.

Now we can turn to the pulse-EPR experiment. As illustration, we consider the FID generated by a single 90°-pulse. For simplicity, we assume that the pulse is on resonance, i.e., $\Delta\omega_o=0$ in Eq. P.35. The effect of the pulse is the same as in Eq. P.11, but then we must transform the result to the diagonal basis of $\mathcal{H}_o$:

$$S_z\xrightarrow{\pi/2S_x}-S_y\xrightarrow{\xi I_y+2\eta I_yS_z}-S_y\cos\eta+2I_yS_x\sin\eta \quad \text{(P.39)}$$

where ξI_y does not act on the electron spin S_y, and we have used the bottom row of Table P.1. Free precession now takes place under the influence of the diagonal Hamiltonian, which we give in Eq. P.35. Evolution under the nuclear Zeeman term is given by:

$$-S_y\cos\eta+2I_yS_x\sin\eta\xrightarrow{\frac{1}{2}\omega_+tI_z}$$
$$-S_y\cos\eta+2\left(I_y\cos\tfrac{1}{2}\omega_+t-I_x\sin\tfrac{1}{2}\omega_+t\right)S_x\sin\eta \quad \text{(P.40)}$$

where the electron spin components remain unchanged. Subsequent application of the hyperfine term then gives:

$$\xrightarrow{\omega_-tI_zS_z}-\left(S_y\cos\tfrac{1}{2}\omega_-t-2I_zS_x\sin\tfrac{1}{2}\omega_-t\right)\cos\eta$$
$$+2\left(I_yS_x\cos\tfrac{1}{2}\omega_+t-I_xS_x\sin\tfrac{1}{2}\omega_+t\right)\sin\eta \quad \text{(P.41)}$$

where we use Eq. P.27 and equivalent to transform the $\sin\eta$ term (see Table P.2). For detection, we must transform back to the Cartesian (i.e., laboratory) frame, by using the inverse operator: $-\left(\xi I_y+2\eta I_yS_z\right)$. Because we detect the electron spin magnetization, we need consider only the action of the $-2\eta I_yS_z$ term:

$$\xrightarrow{-2\eta I_yS_z}-\left[\left(S_y\cos\eta+2I_yS_x\sin\eta\right)\cos\tfrac{1}{2}\omega_-t\right.$$
$$\left.-2I_zS_x\sin\tfrac{1}{2}\omega_-t\right]\cos\eta+2\left[\left(I_yS_x\cos\eta-\tfrac{1}{2}S_y\sin\eta\right)\cos\tfrac{1}{2}\omega_+t\right.$$
$$\left.-I_xS_x\sin\tfrac{1}{2}\omega_+t\right]\sin\eta \quad \text{(P.42)}$$

where we again use Eq. P.27 and Table P.2. The transverse magnetization comes from the terms that contain S_x or S_y alone and, therefore, is given by:

$$\overline{\langle S_y\rangle}=-\left(\cos^2\eta\cos\tfrac{1}{2}\omega_-t+\sin^2\eta\cos\tfrac{1}{2}\omega_+t\right) \quad \text{(P.43)}$$

with $\overline{\langle S_x\rangle}=0$ because we have assumed that $\Delta\omega_o=0$. Equation P.43 shows that the free induction decay is modulated sinusoidally by the composite nuclear frequencies ω_- and ω_+, with relative amplitudes given by $\tan^2\eta$.

P.5 Single-Transition Operators for Selective Pulses

Soft (i.e., long) pulses are selective because they cover only a narrow range of frequencies (cf. Figure 13.3 in Section 13.3). An ideally selective pulse excites only a single transition. One useful criterion is that the radiation field should be much smaller than the line splittings, e.g., $\omega_1(=\gamma_eB_1)<<a$ where a is the hyperfine coupling. Selective pulses help particularly in extracting unresolved hyperfine structure from inhomogeneously broadened lines. To handle selective pulses by the product-operator method we need single-transition operators, which treat a specific transition $|r\rangle$ to $|s\rangle$ as an isolated two-level subsystem. Single-transition operators $S_i^{(rs)}$ $(i=x,y,z)$ therefore correspond to a (fictitious) spin of $\tfrac{1}{2}$.

We already encountered single-transition operators for specific hyperfine transitions when treating transverse relaxation in Appendix O.4. Selection was achieved there by using the polarization operators for $I=1$ (see Eqs. O.58, O.59). We also used polarization operators for $S=\frac{1}{2}$ to select the $M_S=\pm\frac{1}{2}$ electron spin states in Eq. P.30 of the previous section.

For allowed transitions, we can always define single-transition operators in terms of polarization operators (Gemperle and Schweiger 1991). Here we concentrate on all six transitions for a $S=\frac{1}{2}$, $I=\frac{1}{2}$ spin system, as shown in Figure P.1. Nuclear polarization operators that select the $m_I=+\frac{1}{2}$ and $m_I=-\frac{1}{2}$ hyperfine states are, respectively:

$$I^{\alpha}=\tfrac{1}{2}+I_z \tag{P.44}$$

$$I^{\beta}=\tfrac{1}{2}-I_z \tag{P.45}$$

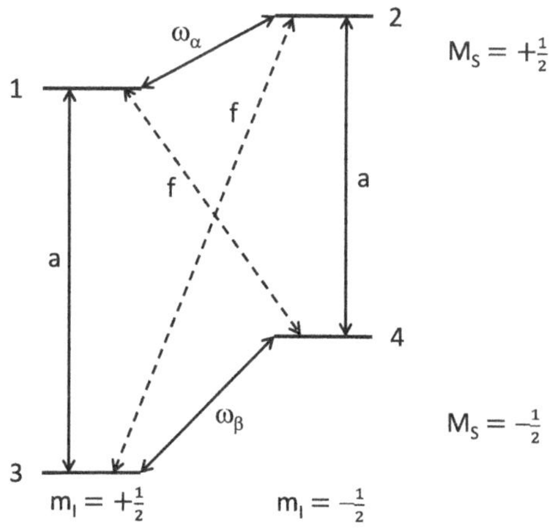

FIGURE P.1 Energy levels and transitions for a $S=\frac{1}{2}$, $I=\frac{1}{2}$ spin system. Allowed and forbidden EPR transitions are "a" and "f", respectively; and ω_α, ω_β are NMR (or ENDOR) transitions. Numbering of levels 1–4 corresponds to that of Schweiger and Jeschke (2001) (and of Ernst et al. 1987).

because $I^{\alpha}=1$ for $m_I=+\frac{1}{2}$, $I^{\alpha}=0$ for $m_I=-\frac{1}{2}$ and vice-versa for I^{β}. Note that where an operator is not specified explicitly, we assume that the constant (e.g., $\frac{1}{2}$) multiplies the unit operator $\hat{1}$. Useful relations between polarization operators are:

$$1=I^{\alpha}+I^{\beta} \tag{P.46}$$

$$I_z=\tfrac{1}{2}\left(I^{\alpha}-I^{\beta}\right) \tag{P.47}$$

We can use Eq. P.46 to split up the Hamiltonian for an x-pulse to operate separately on the two hyperfine transitions:

$$\mathcal{H}_{rf}=\omega_1 S_x=\omega_1 I^{\alpha}S_x+\omega_1 I^{\beta}S_x \tag{P.48}$$

Similarly, we use Eq. P.47 to substitute for I_z in the hyperfine term of the static Hamiltonian $\mathcal{H}_o$.

Electron polarization operators S^{α} and S^{β} are defined similarly to Eqs. P.44, P.45. These lead immediately to single-transition operators for the two nuclear transitions $(\Delta m_I=\pm1,\ \Delta M_S=0)$ that we give in the first two lines of Table P.3. Correspondingly, I^{α} and I^{β} produce single-transition operators for the two allowed EPR transitions $(\Delta M_S=\pm1,\ \Delta m_I=0)$ that we give in lines three and four of Table P.3. The forbidden EPR transitions $(\Delta M_S=\pm1,\ \Delta m_I=\pm1)$ and $(\Delta M_S=\pm1,\ \Delta m_I=\mp1)$ involve simultaneous flips of electron and nuclear spins. We give the corresponding single-transition operators, composed of raising and lowering operators for both spins, in the bottom two rows of Table P.3.

Selective pulses transform operators that correspond to the same single transition (rs), as in the first three lines of Table P.1. For a selective pulse along x, for example, we get:

$$S_z^{(rs)}\xrightarrow{\ \phi\cdot S_x^{(rs)}\ }S_z^{(rs)}\cos\phi-S_y^{(rs)}\sin\phi \tag{P.49}$$

We can verify this directly by applying the relations in Tables P.1 and P.2 to the definitions given in Table P.3. As illustration of the single-transition method, we apply a

TABLE P.3 Single-transition operators $S_i^{(rs)}$ (or $I_i^{(rs)}$)[a] for $S=\frac{1}{2}$, $I=\frac{1}{2}$ with subscripts $i\equiv x,y,z$

transition[b]		r,s	$S_x^{(rs)}$	$S_y^{(rs)}$	$S_z^{(rs)}$
NMR[a]	$++\leftrightarrow+-$	1,2	$I_xS^{\alpha}=I_x\left(\frac{1}{2}+S_z\right)$	$I_yS^{\alpha}=I_y\left(\frac{1}{2}+S_z\right)$	$I_zS^{\alpha}=I_z\left(\frac{1}{2}+S_z\right)$
NMR[a]	$-+\leftrightarrow--$	3,4	$I_xS^{\beta}=I_x\left(\frac{1}{2}-S_z\right)$	$I_yS^{\beta}=I_y\left(\frac{1}{2}-S_z\right)$	$I_zS^{\beta}=I_z\left(\frac{1}{2}-S_z\right)$
allowed EPR	$++\leftrightarrow-+$	1,3	$I^{\alpha}S_x=\left(\frac{1}{2}+I_z\right)S_x$	$I^{\alpha}S_y=\left(\frac{1}{2}+I_z\right)S_y$	$I^{\alpha}S_z=\left(\frac{1}{2}+I_z\right)S_z$
allowed EPR	$+-\leftrightarrow--$	2,4	$I^{\beta}S_x=\left(\frac{1}{2}-I_z\right)S_x$	$I^{\beta}S_y=\left(\frac{1}{2}-I_z\right)S_y$	$I^{\beta}S_z=\left(\frac{1}{2}-I_z\right)S_z$
forbidden EPR	$++\leftrightarrow--$	1,4	$I_xS_x-I_yS_y$	$I_xS_y+I_yS_x$	$\frac{1}{2}\left(I_z+S_z\right)$
			$=\frac{1}{2}\left(I^+S^++I^-S^-\right)$	$=\frac{1}{2}i\left(I^-S^--I^+S^+\right)$	
forbidden EPR	$+-\leftrightarrow-+$	2,3	$I_xS_x+I_yS_y$	$I_xS_y-I_yS_x$	$\frac{1}{2}\left(S_z-I_z\right)$
			$=\frac{1}{2}\left(I^-S^++I^+S^-\right)$	$=\frac{1}{2}i\left(I^+S^--I^-S^+\right)$	

[a] Nuclear transition operators are better designated as $I_i^{(rs)}$ (in the column headings).

[b] Figure P.1 identifies the r,s states and transitions: $M_S,m_I\leftrightarrow M_S',m_I'$; namely, $|1\rangle\equiv|+\frac{1}{2},+\frac{1}{2}\rangle$, $|2\rangle\equiv|+\frac{1}{2},-\frac{1}{2}\rangle$, $|3\rangle\equiv|-\frac{1}{2},+\frac{1}{2}\rangle$, $|4\rangle\equiv|-\frac{1}{2},-\frac{1}{2}\rangle$.

selective pulse of effective angle β_a along x to the allowed EPR transition (1,3) in Figure P.1. We start from thermal equilibrium: $\sigma_{eq} = S_z = \left(I^\alpha + I^\beta\right)S_z = S_z^{(13)} + S_z^{(24)}$, where we use Eq. P.46 to show that the total S_z-polarization is the sum from the two allowed EPR transitions. Applying the $\beta_a S_x^{(13)}$ pulse, which has no effect on $S_z^{(24)}$, we get:

$$S_z = S_z^{(13)} + S_z^{(24)} \xrightarrow{\beta_a S_x^{(13)}} S_z^{(13)} \cos\beta_a - S_y^{(13)} \sin\beta_a + S_z^{(24)}$$

$$= \tfrac{1}{2}\left(1+\cos\beta_a\right)S_z - \sin\beta_a \cdot S_y^{(13)} - \left(1-\cos\beta_a\right)I_z S_z \tag{P.50}$$

by using Eq. P.49 followed with definitions from Table P.3. With an effective pulse angle $\beta_a = \pi/2$, we maximize electron coherence on the (1,3) transition. For $\beta_a = \pi$, on the other hand, the selective pulse creates only two-spin order $2I_zS_z$, which we can view as superposition of nuclear polarization with opposite signs in the $M_S = \pm\frac{1}{2}$ manifolds.

If the pulse is applied to a different transition (rt) that is connected by a common state $|r\rangle$, coherence is transferred according to:

$$S_x^{(rt)} \xrightarrow{\phi \cdot S_x^{(rs)}} S_x^{(rt)} \cos\tfrac{1}{2}\phi - S_y^{(ts)} \sin\tfrac{1}{2}\phi \tag{P.51}$$

The angle of rotation is halved whenever the transformed operator and rotation operator have only one index in common. Coherence is transferred completely from transition (rt) to transition (ts) when the effective pulse angle is $\phi = \pi$.

APPENDIX Q: ADDITION OF ANGULAR MOMENTA AND WIGNER 3j-SYMBOLS

Q.1 Clebsch–Gordon Coefficients

If two angular momentum operators $\mathbf{L}_1$ and $\mathbf{L}_2$ commute, e.g., because they correspond to different electrons or to the orbital and spin momentum for the same electron, then they have simultaneous eigenfunctions (see Appendix E). Here $\mathbf{L}$ stands quite generally for electron orbital (L) or spin (S), or nuclear spin (I) angular momentum, etc. These individual momenta couple to a total angular momentum that is given by $\mathbf{L} = \mathbf{L}_1 + \mathbf{L}_2$. As we know from Appendix E.2, the wavefunctions for the individual momenta, $|L_1, M_1\rangle$ and $|L_2, M_2\rangle$, are eigenfunctions of the operators $\mathbf{L}_1^2$ and L_{1z}, and of $\mathbf{L}_2^2$ and L_{2z}, respectively, where $M_1 = L_1, L_1 - 1, \ldots, -L_1 + 1, -L_1$ and similarly for M_2. Correspondingly, the wave function for the total angular momentum, $|L, M\rangle$, is an eigenfunction of the operators $\mathbf{L}^2$ and L_z. The simple product wavefunctions: $|L_1M_1L_2M_2\rangle \equiv |L_1M_1\rangle|L_2M_2\rangle$ are eigenfunctions of $L_z = L_{1z} + L_{2z}$, with eigenvalues: $M = M_1 + M_2$. However, they are not always eigenfunctions of the $\mathbf{L}^2$ operator. For the latter, we must take linear combinations of the simple product wave functions:

$$|LM\rangle = \sum_{M_1, M_2} C\left(L_1M_1L_2M_2; LM\right)|L_1M_1L_2M_2\rangle \tag{Q.1}$$

where the addition coefficients, $C\left(L_1M_1L_2M_2; LM\right)$, are signed real numbers called Clebsch–Gordan coefficients (see, e.g., Brink and Satchler 1968; Messiah 1965).

We get the largest value of M when M_1 and M_2 have their maximum values, then: $M = L_1 + L_2$. This combination corresponds to a total angular momentum $L = L_1 + L_2$ and is achieved by only one product wave function, hence the vector addition coefficient is: $C\left(L_1L_1L_2L_2; (L_1+L_2)(L_1+L_2)\right) = 1$. There are two possible combinations that give $M = L_1 + L_2 - 1$; these are $M_1 - 1$ and M_2, and M_1 and $M_2 - 1$. Correspondingly, there are two independent wave functions with the form given by Eq. Q.1. One of these belongs to $L = L_1 + L_2$ and the other to $L = L_1 + L_2 - 1$. Similarly, there are three combinations that correspond to $M = L_1 + L_2 - 2$ and these belong to $L = L_1 + L_2$, $L = L_1 + L_2 - 1$ and $L = L_1 + L_2 - 2$. Assuming that $L_1 > L_2$, we can continue this process until we reach the minimum value of M_2, which is $M_2 = -L_2$. The projection quantum number is then: $M = L_1 - L_2$ that belongs to the total angular momentum $L = L_1 - L_2$, for which there is only one combination and the vector addition coefficient is unity. Thus the total angular momentum takes the range of values: $L = L_1 + L_2, L_1 + L_2 - 1, \ldots, |L_1 - L_2| + 1, |L_1 - L_2|$, which is called the triangular rule. For each value of total angular momentum L, the projection is: $M = L, L-1, \ldots, -L+1, -L$. Our problem is to find the Clebsch–Gordan coefficients for the intermediate allowed values of the total angular momentum L.

We start with the simplest example of vector addition. Consider two electron spins $\mathbf{S}_1$ and $\mathbf{S}_2$ with $S_1 = S_2 = \frac{1}{2}$. These couple to a total spin $\mathbf{S}$ that corresponds either to a triplet $(S = S_1 + S_2 = 1)$ or to a singlet $(S = S_1 - S_2 = 0)$ spin state. We discussed this case in Section 9.7 of Chapter 9, for the exchange interaction between two electron spins. We designate the individual spin states with $M_1, M_2 = +\frac{1}{2}$ and $-\frac{1}{2}$ by α_1, α_2 and β_1, β_2, respectively, of which there are four distinct product functions. The symmetric product state $|\alpha_1\alpha_2\rangle \equiv |\alpha_1\rangle|\alpha_2\rangle$ is an eigenfunction of both total spin operators, $\mathbf{S}^2$ and S_z, and has a total spin projection $M_S = M_1 + M_2 = +1$. This eigenfunction therefore belongs to the $S = 1$ triplet state and has the maximum value of M_S. Operating on this state with the lowering operator, $S_-(= S_{1-} + S_{2-})$, gives:

$$S_-|S = 1, M_S = +1\rangle = \sqrt{2}|S = 1, M_S = 0\rangle \tag{Q.2}$$

and operating similarly on the product state yields:

$$\left(S_{1-} + S_{2-}\right)|\alpha_1\alpha_2\rangle = |\beta_1\alpha_2\rangle + |\alpha_1\beta_2\rangle \tag{Q.3}$$

– see Eq. E.12 in Appendix E. Because Eqs. Q.2 and Q.3 are equivalents, the $M_S = 0$ triplet-state wave function is given by the symmetric linear combination of product functions:

$$|S=1, M_S=0\rangle = \frac{1}{\sqrt{2}}\left(|\alpha_1\beta_2\rangle + |\beta_1\alpha_2\rangle\right) \quad (Q.4)$$

The vector addition coefficients that correspond to this case are therefore: $C\left(\frac{1}{2},\pm\frac{1}{2},\frac{1}{2},\mp\frac{1}{2};1,0\right) = 1/\sqrt{2}$. Operating again with S_- on this wave function then yields the third triplet-state wave function: $|S=1, M_S=-1\rangle = |\beta_1\beta_2\rangle$, as it must by symmetry with the $M_S=+1$ case. The corresponding vector addition coefficient for these two cases is $C\left(\frac{1}{2},\pm\frac{1}{2},\frac{1}{2},\pm\frac{1}{2};1,\pm 1\right) = 1$. The fourth wave function of the coupled spin pair must be a further linear combination of $|\alpha_1\beta_2\rangle$ and $|\beta_1\alpha_2\rangle$. The only combination that is orthogonal to the right-hand side of Eq. Q.4 is the antisymmetric wave function:

$$|S=0, M_S=0\rangle = \frac{1}{\sqrt{2}}\left(|\alpha_1\beta_2\rangle - |\beta_1\alpha_2\rangle\right) \quad (Q.5)$$

This corresponds to the singlet spin state with $S = S_1 - S_2 = 0$, which has only one spin projection: $M_S = 0$. In this case, the vector addition coefficients are: $C\left(\frac{1}{2},\pm\frac{1}{2},\frac{1}{2},\mp\frac{1}{2};0,0\right) = \pm 1/\sqrt{2}$.

Q.2 Properties of Clebsch–Gordon Coupling Coefficients

We can apply the method illustrated by the simple example above to derive Clebsch–Gordon coefficients for more complicated systems with $L > \frac{1}{2}$. Analytical expressions are given by Condon and Shortley (1935) and numerical expressions by Heine (1960). Some useful tables are given in Appendix A8.1 to Chapter 8 and Appendix A6.1 to Chapter 6. The algebraic expressions can be rather complex, but non-vanishing $C(L_1M_1L_2M_2;LM)$ are limited by the triangular condition: $L = |L_1 - L_2|, |L_1 - L_2|+1, \ldots, L_1 + L_2 - 1, L_1 + L_2$ and the selection rule: $M = M_1 + M_2$. For the latter reason, we often contract the symbol to $C(L_1L_2L;M_1M_2) \equiv \langle L_1L_2L;M_1M_2\rangle$ where the condition for M is understood, and any permutation of the triangular condition is allowed. As usual, the allowed range of M_1 is $-L_1$ to $+L_1$ in integer steps and similarly for M_2.

Orthonormality of the basis functions $|L_1, M_1\rangle$ and $|L_2, M_2\rangle$ leads to the following orthogonality relations for the $\langle L_1L_2L;M_1M_2\rangle$:

$$\sum_{M_2}\langle L_1L_2L;M-M_2,M_2\rangle\langle L_1L_2L';M-M_2,M_2\rangle = \delta_{LL'} \quad (Q.6)$$

$$\sum_{L}\langle L_1L_2L;M-M_2,M_2\rangle\langle L_1L_2L;M'-M_2',M_2'\rangle = \delta_{M_2M_2'}\delta_{MM'} \quad (Q.7)$$

Clebsch–Gordon coefficients display several useful symmetries that we can write in the form:

$$\begin{aligned}
\langle L_1L_2L;M_1M_2\rangle &= (-1)^{L_1+L_2-L}\langle L_1L_2L;-M_1,-M_2\rangle \\
&= (-1)^{L_1+L_2-L}\langle L_2L_1L;M_2M_1\rangle \\
&= (-1)^{L_1-M_1}\sqrt{\frac{2L+1}{2L_2+1}}\langle L_1LL_2;M_1,-(M_1+M_2)\rangle \\
&= (-1)^{L_1-M_1}\sqrt{\frac{2L+1}{2L_2+1}}\langle LL_1L_2;M_1+M_2,-M_1\rangle \\
&= (-1)^{L_2+M_2}\sqrt{\frac{2L+1}{2L_1+1}}\langle LL_2L_1;-(M_1+M_2),M_2\rangle \\
&= (-1)^{L_2+M_2}\sqrt{\frac{2L+1}{2L_1+1}}\langle L_2LL_1;-M_2,M_1+M_2\rangle
\end{aligned} \quad (Q.8)$$

A special value worth noting is the parity coefficient:

$$\langle L_1L_2L;00\rangle = 0, \quad \text{for } L_1+L_2+L = \text{odd} \quad (Q.9)$$

when L_1, L_2 and L are all integer. If $L_1+L_2+L=$ even, $\langle L_1L_2L;00\rangle$ is given by Eqs. Q.12, Q.16 below. Additionally, for $L_2 = 0$ (and therefore $M_2 = 0$):

$$\langle L_10L;M_10\rangle = \delta_{L_1L}\delta_{M_1M} \quad (Q.10)$$

and for $L = 0$ (and therefore $M = 0$):

$$\langle L_1L_20;M_1M_2\rangle = \frac{(-1)^{L_1-M_1}}{\sqrt{2L_1+1}}\delta_{L_1L_2}\delta_{M_1,-M_2} \quad (Q.11)$$

Q.3 Wigner 3*j*-Symbols

The symmetries given in Eqs. Q.8 are best represented by the Wigner 3*j*-symbols, which are related to the Clebsch–Gordan coefficients by:

$$\begin{aligned}
C(L_1M_1L_2M_2;LM) &\equiv \langle L_1L_2L;M_1M_2\rangle \\
&= (-1)^{L_1-L_2+M}\sqrt{2L+1}\begin{pmatrix} L_1 & L_2 & L \\ M_1 & M_2 & -M \end{pmatrix}
\end{aligned} \quad (Q.12)$$

where $M_1 + M_2 - M = 0$ and the triangular relation for addition of angular momenta is $|L_1 - L_2| \le L \le L_1 + L_2$. Written in terms of 3*j*-symbols, the expansion in product wave functions that is given by Eq. Q.1 becomes:

$$|LM\rangle = (-1)^{L_1-L_2+M}\sqrt{2L+1}\sum_{M_1}\begin{pmatrix} L_1 & L_2 & L \\ M_1 & (M_1+M) & -M \end{pmatrix} \times |L_1M_1L_2(M_1+M)\rangle \quad (Q.13)$$

where we must sum only over M_1. The 3j-symbol is multiplied by $(-1)^{L_1+L_2+L}$ on interchange of adjacent columns:

$$\begin{pmatrix} L_1 & L_2 & L \\ M_1 & M_2 & M \end{pmatrix} = (-1)^{L_1+L_2+L}\begin{pmatrix} L_2 & L_1 & L \\ M_2 & M_1 & M \end{pmatrix} = \begin{pmatrix} L_2 & L & L_1 \\ M_2 & M & M_1 \end{pmatrix} \tag{Q.14}$$

which also means that it is invariant under cyclic permutations of the columns. For the rows we have:

$$\begin{pmatrix} L_1 & L_2 & L \\ M_1 & M_2 & M \end{pmatrix} = (-1)^{L_1+L_2+L}\begin{pmatrix} L_1 & L_2 & L \\ -M_1 & -M_2 & -M \end{pmatrix} \tag{Q.15}$$

Algebraic expressions for 3j-symbols are complicated in the general case, but one special case that is relevant to stochastic-Liouville lineshape simulations is:

$$\begin{pmatrix} L_1 & L_2 & L \\ 0 & 0 & 0 \end{pmatrix} = (-1)^{(L_1+L_2+L)/2}\sqrt{\frac{(L_1+L_2-L)!(L_2+L-L_1)!(L+L_1-L_2)!}{(L_1+L_2+L+1)!}} \times \frac{((L_1+L_2+L)/2)!}{((L_1+L_2-L)/2)!((L_2+L-L_1)/2)!((L+L_1-L_2)/2)!} \tag{Q.16}$$

when L_1+L_2+L is even; otherwise this 3j-symbol is zero. The factorials appearing in Eq. Q.16 are defined as: $L! = 1\times 2\times 3\times\cdots\times L$ with $0! = 1$. Further useful special cases are (Brink and Satchler 1968):

$$\begin{pmatrix} L_1 & L_2 & L_1+L_2 \\ M_1 & M_2 & -(M_1+M_2) \end{pmatrix} = (-1)^{L_1-L_2+M_1+M_2} \times\sqrt{\frac{(2L_1)!(2L_2)!(L_1+L_2+M_1+M_2)!(L_1+L_2-M_1-M_2)!}{(2L_1+2L_2+1)!(L_1+M_1)!(L_1-M_1)!(L_2+M_2)!(L_2-M_2)!}} \tag{Q.17}$$

$$\begin{pmatrix} L_1 & L_2 & L_1+L_2-2 \\ M_1 & M_2 & -(M_1+M_2) \end{pmatrix} = (-1)^{L_1-L_2+M_1+M_2} \times\sqrt{\frac{(2L_1-2)!(2L_2-2)!(L_1+L_2+M_1+M_2-2)!(L_1+L_2-M_1-M_2-2)!}{2(2L_1+2L_2-1)!(L_1+M_1)!(L_1-M_1)!(L_2+M_2)!(L_2-M_2)!}}$$
$$\times\left[(L_1+M_1)(L_1+M_1-1)(L_2-M_2)(L_2-M_2-1)-2(L_1-M_1)(L_1+M_1)(L_2-M_2)(L_2+M_2) +(L_1-M_1)(L_1-M_1-1)(L_2+M_2)(L_2+M_2-1)\right] \tag{Q.18}$$

We give other explicit expressions for cases of relevance to slow-motion EPR simulations in Table Q.1. Tabulations of numerical values are given by Rotenberg et al. (1959).

Q.4 Wigner–Eckhart Theorem

Clebsch–Gordan coefficients, or 3j-symbols, are important for determining the matrix elements of angular momentum operators and generally of spherical tensor operators (which include spherical harmonics). Irreducible spherical tensor operators T_{LM} are those which obey commutation rules given by Eqs. S.2 and S.3 that we present later in Appendix S. An example is the tensor operator for the nuclear quadrupole interaction (see Section 3.10 in Chapter 3). The Wigner–Eckhart theorem states that the matrix elements of such operators are proportional to a Clebsch–Gordan coefficient, where the constant of proportionality is independent of M and the magnetic quantum numbers of the angular momentum states connected. Consider the matrix element between the states $|a_1L_1M_1\rangle$ and $|a_2L_2M_2\rangle$, where a_1 and a_2 are quantum numbers other than those needed to specify the angular momentum, e.g., for the radial wave function. The Wigner–Eckhart theorem then becomes (Brink and Satchler 1968; Messiah 1965):

$$\langle a_1L_1M_1|T_{L,M}|a_2L_2M_2\rangle = (-1)^{2L}\frac{\langle a_1L_1\|T_L\|a_2L_2\rangle}{\sqrt{2L_1+1}}\langle L_2LL_1;M_2M\rangle \equiv (-1)^{L_1-M_1}\langle a_1L_1\|T_L\|a_2L_2\rangle \times\begin{pmatrix} L_1 & L & L_2 \\ -M_1 & M & M_2 \end{pmatrix} \tag{Q.19}$$

where L_1 and L_2 are either I or S and $\langle a_1L_1\|T_L\|a_2L_2\rangle$ is a so-called reduced matrix element that does not depend on any of the M-indices. The phase and scaling factors in Eq. Q.19 are to some extent arbitrary, but conventional. The reduced matrix element is a scalar quantity that we evaluate most easily by determining the matrix element with $M_1 = L_1$ and $M_2 = L_2$. We do this for the nuclear quadrupole interaction in Section 3.10 of Chapter 3. Integration over the product of three Wigner rotation matrices (or spherical harmonics) is also an example of the Wigner–Eckhart theorem; this is given by Eq. R.18 in Appendix R that follows.

TABLE Q.1 3*j*-symbols used in stochastic-Liouville (slow-motion) calculations

3j-symbol	*value*
$\begin{pmatrix} L & 2 & L \\ M & 0 & -M \end{pmatrix}$	$(-1)^{L-M}\dfrac{3M^2-L(L+1)}{\sqrt{L(L+1)(2L+3)(2L+1)(2L-1)}}$
$\begin{pmatrix} L & 2 & L \\ M & 1 & -(M+1) \end{pmatrix}$	$(-1)^{L-M}(2M+1)\sqrt{\dfrac{3(L+M+1)(L-M)}{L(2L+3)(2L+2)(2L+1)(2L-1)}}$
$\begin{pmatrix} L & 2 & L \\ M & 2 & -(M+2) \end{pmatrix}$	$(-1)^{L-M}\sqrt{\dfrac{3(L+M+1)(L+M+2)(L-M-1)(L-M)}{L(2L+3)(2L+2)(2L+1)(2L-1)}}$
$\begin{pmatrix} L & 2 & L+2 \\ M & 0 & -M \end{pmatrix}$	$(-1)^{L-M}\sqrt{\dfrac{3(L+M+1)(L+M+2)(L-M+1)(L-M+2)}{(L+1)(2L+5)(2L+4)(2L+3)(2L+1)}}$
$\begin{pmatrix} L & 2 & L+2 \\ M & 1 & -(M+1) \end{pmatrix}$	$(-1)^{L-M-1}\sqrt{\dfrac{(L+M+1)(L+M+2)(L+M+3)(L-M+1)}{(L+1)(L+2)(2L+5)(2L+3)(2L+1)}}$
$\begin{pmatrix} L & 2 & L+2 \\ M & 2 & -(M+2) \end{pmatrix}$	$(-1)^{L-M}\sqrt{\dfrac{(L+M+1)(L+M+2)(L+M+3)(L+M+4)}{(2L+5)(2L+4)(2L+3)(2L+2)(2L+1)}}$
$\begin{pmatrix} L & 2 & L-2 \\ M & 0 & -M \end{pmatrix}$	$(-1)^{L-M}\sqrt{\dfrac{3(L+M-1)(L+M)(L-M-1)(L-M)}{2L(L-1)(2L-1)(2L-3)(2L+1)}}$
$\begin{pmatrix} L & 2 & L-2 \\ -M & 1 & (M-1) \end{pmatrix}$	$(-1)^{L-M}\sqrt{\dfrac{(L+M-2)(L+M-1)(L+M)(L-M)}{L(L-1)(2L-1)(2L-3)(2L+1)}}$
$\begin{pmatrix} L & 2 & L-2 \\ -M & 2 & (M-2) \end{pmatrix}$	$(-1)^{L-M}\sqrt{\dfrac{(L+M-3)(L+M-2)(L+M-1)(L+M)}{4L(L-1)(2L-1)(2L-3)(2L+1)}}$

Note: $\begin{pmatrix} L & 2 & L' \\ -M & -m & -M' \end{pmatrix} = (-1)^{L+L'}\begin{pmatrix} L & 2 & L' \\ M & m & M' \end{pmatrix}$.

APPENDIX R: ROTATION OPERATORS, EULER ANGLES AND WIGNER ROTATION MATRICES

R.1 Rotation Operators

The vector for an infinitesimal rotation, $\mathbf{d\varphi}$, is directed along the rotation axis and has a length $d\varphi$ that gives the rotation angle. The change $\mathbf{dr}$ in a linear vector $\mathbf{r}$ that is subjected to this infinitesimal rotation is given by the vector product:

$$\mathbf{dr} = \mathbf{d\varphi} \times \mathbf{r} \tag{R.1}$$

– see Figure R.1. Correspondingly, the rate of change in the vector $\mathbf{r}$ is given by:

$$\frac{\mathbf{dr}}{dt} = \omega_R \times \mathbf{r} \tag{R.2}$$

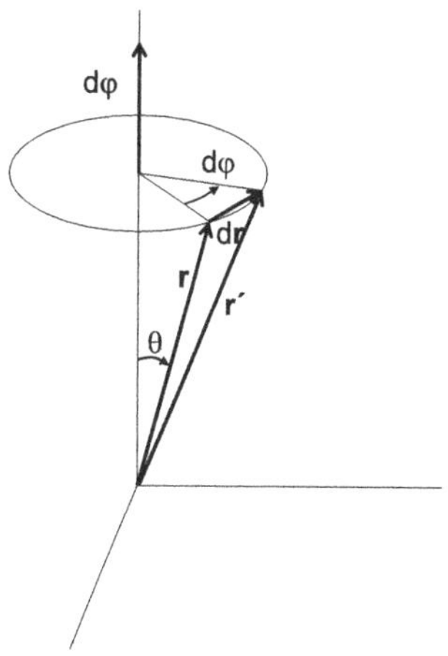

FIGURE R.1 Infinitesimal rotation vector $\mathbf{d\varphi}$ is directed along the rotation axis, in the sense of a right-hand screw. Change $\mathbf{dr}$ in position-vector $\mathbf{r}$ that results from the infinitesimal rotation is perpendicular to both $\mathbf{r}$ and the rotation axis, and is of magnitude: $dr = r\sin\theta \cdot d\phi$, i.e., $\mathbf{dr} = \mathbf{d\varphi} \times \mathbf{r}$ (see Appendix B.1).

where $\omega_R \equiv \mathrm{d}\boldsymbol{\varphi}/\mathrm{d}t$ is the angular velocity vector. Equation R.2, which applies generally to any vector – not only to **r**, is the basis for the transformation to the rotating frame of reference that we use for describing magnetic resonance experiments, e.g., by the Bloch equations (see Section 6.2 in Chapter 6).

Transformation of a wave function $\psi(\mathbf{r})$ by rotation of the position vector **r** is equivalent to the reverse rotation of **r**, whilst leaving the wave function unchanged: $\hat{R}\psi(\mathbf{r}) = \psi\left(\mathbf{R}^{-1}\cdot\mathbf{r}\right)$. For an infinitesimal rotation, the transformed wave function is therefore:

$$\psi(\mathbf{r}-\mathrm{d}\mathbf{r}) = \left(1-(\mathrm{d}\boldsymbol{\varphi}\times\mathbf{r})\cdot\nabla_{\mathbf{r}}\right)\psi(\mathbf{r}) \tag{R.3}$$

where we use the gradient operator $\nabla_{\mathbf{r}}\psi(\mathbf{r})$ $(\equiv \partial\psi/\partial\mathbf{r})$ to give the first-order contribution in a Taylor expansion. The term on the right of Eq. R.3 defines the operator, $\hat{R}(\mathrm{d}\varphi)$, for an infinitesimal rotation:

$$\hat{R}(\mathrm{d}\boldsymbol{\varphi}) = 1-\mathrm{d}\boldsymbol{\varphi}\cdot\left(\mathbf{r}\times\nabla_{\mathbf{r}}\right) = 1-i\mathrm{d}\boldsymbol{\varphi}\cdot\mathbf{L} \tag{R.4}$$

where the triple scalar product is given by: $\left(\mathrm{d}\boldsymbol{\varphi}\times\mathbf{r}\right)\cdot\nabla_{\mathbf{r}} = \mathrm{d}\boldsymbol{\varphi}\cdot\left(\mathbf{r}\times\nabla_{\mathbf{r}}\right)$ – see Eqs. B.11, B.12 in Appendix B – and we use the quantum mechanical operator replacement $(-i\mathbf{r}\times\nabla)$ for the angular momentum $\mathbf{L}$ $(\equiv \mathbf{r}\times\mathbf{p})$ from Appendix E. Note that we can use Eq. R.4 to derive the fundamental commutation relation: $\mathbf{L}\times\mathbf{L} = i\mathbf{L}$ for angular momentum that we introduced by Eq. E.4, and equivalents, in Appendix E. We get the operator for a finite rotation $\boldsymbol{\varphi}$ by n successive applications of the infinitesimal rotation that is given by Eq. R.4:

$$\hat{R}(\boldsymbol{\varphi}) = \lim_{n\to\infty}\left(1-i\frac{\boldsymbol{\varphi}}{n}\cdot\mathbf{L}\right)^n = 1-i\boldsymbol{\varphi}\cdot\mathbf{L}+\frac{(i\boldsymbol{\varphi}\cdot\mathbf{L})^2}{2!}-\frac{(i\boldsymbol{\varphi}\cdot\mathbf{L})^3}{3!}+\cdots \equiv \exp(-i\boldsymbol{\varphi}\cdot\mathbf{L}) \tag{R.5}$$

where letting $n\to\infty$ gives the definition of the exponential operator on the right. We see from Eq. R.4 or R.5 that the rotation operator $\hat{R}$ has the same eigenfunctions, $|L,M\rangle$, as does the angular momentum operator **L**, viz., spherical harmonics.

Note also that Eq. R.5 applies to any angular momentum operator that has the same commutation rules as **L**. With $S=\frac{1}{2}$, for example, a rotation by angle β about the x-axis in real space is given by:

$$R(\beta,x) = \exp\left(-i2\beta S_x\right) \tag{R.6}$$

where the factor two appears in the exponential because S_x operates on the spin functions $|M_S\rangle$, with non-zero matrix elements: $\left\langle\frac{1}{2}\middle|S_x\middle|-\frac{1}{2}\right\rangle = \frac{1}{2}$ and $\left\langle-\frac{1}{2}\middle|S_x\middle|\frac{1}{2}\right\rangle = \frac{1}{2}$. This complication comes about because we are relating a rotation in real space to a rotation in spin space. Generalization of Eq. R.6 to other values of S (or L) is given by the Wigner rotation matrices that we introduce later in this Appendix (see Table R.1).

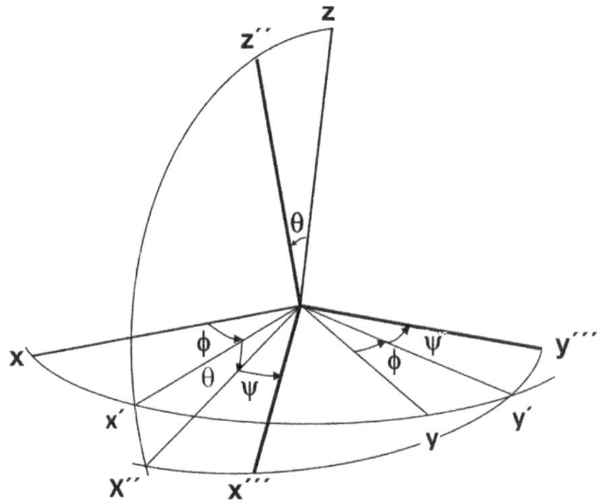

FIGURE R.2 Definition of Euler angles. First rotation by ϕ about the z-axis produces the new axis system x',y',z. Second rotation by θ about the y'-axis produces the new axis system x'',y',z''. Third rotation by ψ about the z''-axis produces the new axis system x''',y''',z''. The same final result is produced by rotating first by ψ about the z-axis, then by θ about the y-axis and finally by ϕ about the z-axis (i.e., all original axes).

R.2 Euler Angles

We need three angles to specify a general rotation; these are the Euler angles ϕ, θ and ψ. First we make a rotation by ϕ about the z-axis, next a rotation by θ about the new y-axis, and then a rotation by ψ about the final z-axis (see Figure R.2). In terms of the rotation matrices given by Eqs. I.19 and I.20 in Appendix I, the general rotation matrix therefore corresponds to the matrix product: $\mathbf{R}(\phi,\theta,\psi) = \mathbf{R}(\psi,z)\mathbf{R}(\theta,y)\mathbf{R}(\phi,z)$, where the rotation matrix to the right automatically generates the new coordinate system for application of the following rotation matrix. To apply the rotation operators defined by Eq. R.5, however, it is more convenient to define all rotations in terms of the original axes. It turns out that to do this, we must merely reverse the order in which we apply the Euler angles (Brink and Satchler 1968; Rose 1957). Consider the second rotation $\hat{R}\left(\theta,y'\right)$ in Figure R.2: we perform this rotation in the axis system that has already undergone a rotation $\hat{R}$. The corresponding rotation in the original axis system is given by the unitary transformation that we use for the g-tensor in Eq. B.25 of Appendix B. If we identify $\hat{R}(\theta,y)$ with the non-rotated matrix **g**, and $\hat{R}\left(\theta,y'\right)$ with **g**′, then Eq. B.25 becomes $\hat{R}(\theta,y) = \hat{R}^{-1}\cdot\hat{R}\left(\theta,y'\right)\cdot\hat{R}$. Operating on the left with $\hat{R}$ shows immediately that the order of rotation is reversed between the two systems, viz., $\hat{R}\cdot\hat{R}(\theta,y) \equiv \hat{R}\left(\theta,y'\right)\cdot\hat{R}$.

R.3 Wigner Rotation Matrices

According to Eq. R.5, we specify a general rotation $\hat{R}(\phi,\theta,\psi)$ by successively applying rotation operators with the three Euler angles in the following order:

$$\hat{R}(\phi,\theta,\psi) = \exp\left(-i\phi L_z\right)\exp\left(-i\theta L_y\right)\exp\left(-i\psi L_z\right) \tag{R.7}$$

TABLE R.1 Reduced Wigner rotation matrices, $d^L_{K,M}(\theta)$

L	K	M	$d^L_{K,M}(\theta)$
0	0	0	1
$\frac{1}{2}$	$\pm\frac{1}{2}$	$\pm\frac{1}{2}$	$\cos(\theta/2)$
	$\mp\frac{1}{2}$	$\pm\frac{1}{2}$	$\pm\sin(\theta/2)$
1	0	0	$\cos\theta$
	0	± 1	$\pm\sqrt{\frac{1}{2}}\sin\theta$
	± 1	0	$\mp\sqrt{\frac{1}{2}}\sin\theta$
	± 1	± 1	$\cos^2(\theta/2)$
	± 1	∓ 1	$\sin^2(\theta/2)$
2	0	0	$\frac{1}{2}\left(3\cos^2\theta-1\right)$
	0	± 1	$\pm\sqrt{\frac{3}{2}}\sin\theta\cos\theta$
	± 1	0	$\mp\sqrt{\frac{3}{2}}\sin\theta\cos\theta$
	± 1	± 1	$\frac{1}{2}(2\cos\theta-1)(1+\cos\theta)$
	± 1	∓ 1	$\frac{1}{2}(2\cos\theta+1)(1-\cos\theta)$
	0	± 2	$\sqrt{\frac{3}{8}}\sin^2\theta$
	± 2	0	$\sqrt{\frac{3}{8}}\sin^2\theta$
	± 1	± 2	$\pm\frac{1}{2}\sin\theta(1+\cos\theta)$
	± 2	± 1	$\mp\frac{1}{2}\sin\theta(1+\cos\theta)$
	± 1	∓ 2	$\mp\frac{1}{2}\sin\theta(1-\cos\theta)$
	± 2	∓ 1	$\mp\frac{1}{2}\sin\theta(1-\cos\theta)$
	± 2	± 2	$\cos^4(\theta/2)$
	± 2	∓ 2	$\sin^4(\theta/2)$

In particular, for integer L:

$$d^L_{K,L}(\theta)=\frac{1}{2^L}\sqrt{\frac{(2L)!}{(L+M)!(L-M)!}}\sin^{L-M}\theta(1+\cos\theta)^M$$

where the angular momentum components, L_z and L_y, are referred to the *original* z- and y-axes. Applying this general rotation to an eigenfunction $|L,M\rangle$ of angular momentum **L** gives the transformation (Rose 1957):

$$\hat{R}|L,M\rangle=\sum_{K=-L}^{+L}\mathcal{D}^L_{K,M}|L,K\rangle \tag{R.8}$$

Exploiting orthogonality of the eigenfunctions by multiplying on the left with $\langle L,K|$, we see that the expansion coefficients are the matrix elements: $\langle L,K|\hat{R}|L,M\rangle=\mathcal{D}^L_{K,M}$. Substituting from Eq. R.7, we then get:

$$\mathcal{D}^L_{K,M}(\phi,\theta,\psi)\equiv\langle L,K|\exp(-i\phi L_z)\exp(-i\theta L_y)\exp(-i\psi L_z)|L,M\rangle$$
$$=\exp(-iK\phi)\times d^L_{K,M}(\theta)\times\exp(-iM\psi) \tag{R.9}$$

These expressions are elements $\mathcal{D}^L_{K,M}$ of the $(2L+1)\times(2L+1)$ Wigner rotation matrices $\mathcal{D}^{(L)}$, which are irreducible representations of the full rotation group. The elements of the reduced rotation matrix, $d^L_{K,M}(\theta)\equiv\langle L,K|\exp(-i\theta L_y)|L,M\rangle$, are real functions that obey the symmetry relations:

$$d^L_{K,M}=d^L_{-M,-K}=(-1)^{K-M}d^L_{M,K}=(-1)^{K-M}d^L_{-K,-M} \tag{R.10}$$

We list the functions $d^L_{K,M}(\theta)$ in Table R.1. The definition of Wigner rotation matrices here is that of Rose (1957), Messiah (1965), Brink and Satchler (1968). That of Edmonds (1957) and of Wigner (1959) corresponds to the rotation $\hat{R}(-\phi,-\theta,-\psi)$ (see Brink and Satchler 1968). This alternative convention is also used sometimes in EPR, e.g., Polnaszek et al. (1973); Glarum and Marshall (1967).

The Wigner rotation matrices have various useful properties (Brink and Satchler 1968). The complex conjugate is given by (see Eqs. R.9, R.10):

$$\mathcal{D}^{L*}_{K,M}(\phi,\theta,\psi)=(-1)^{K-M}\mathcal{D}^L_{-K,-M}(\phi,\theta,\psi) \tag{R.11}$$

Note also that: $\mathcal{D}^{L*}_{K,M}(\phi,\theta,\psi)=\mathcal{D}^L_{M,K}(-\psi,-\theta,-\phi)$. Orthogonality of the rotation matrices is expressed by:

$$\int\mathcal{D}^{L*}_{K,M}(\Omega)\mathcal{D}^{L'}_{K',M'}(\Omega)\cdot d\Omega=\frac{8\pi^2}{2L+1}\delta_{L',L}\delta_{K',K}\delta_{M',M} \tag{R.12}$$

where $d\Omega\equiv\sin\theta\cdot d\theta\cdot d\phi\cdot d\psi$. When K or M is zero, the rotation matrix elements reduce to spherical harmonics:

$$\mathcal{D}^L_{0,M}(0,\theta,\psi)=(-1)^M\sqrt{\frac{4\pi}{2L+1}}Y^*_{L,M}(\theta,\psi) \tag{R.13}$$

$$\mathcal{D}^L_{K,0}(\phi,\theta,0)=\sqrt{\frac{4\pi}{2L+1}}Y^*_{L,K}(\theta,\phi) \tag{R.14}$$

When both K and M are zero, the rotation matrix elements are identical with the Legendre polynomials in $\cos\theta$:

$$\mathcal{D}^L_{0,0}(0,\theta,0)\equiv d^L_{0,0}(\theta)=P_L(\cos\theta) \tag{R.15}$$

We can view the addition of angular momenta in Eq. Q.1 of Appendix Q as rotations of the individual basis functions $|L_1M_1\rangle$ and $|L_2M_2\rangle$ by Wigner matrices. By using Eq. R.8, this leads to the expansion in rotation matrix elements:

$$\mathcal{D}^L_{K,M}=\sum_{K_2,M_2}\langle L_1L_2L;K-K_2,K_2\rangle\langle L_1L_2L;M-M_2,M_2\rangle$$
$$\times\mathcal{D}^{L_1}_{K-K_2,M-M_2}\mathcal{D}^{L_2}_{K_2,M_2} \tag{R.16}$$

where $\langle L_1L_2L;M_1M_2\rangle$ is a Clebsch–Gordan coefficient (see Appendix Q.2). Correspondingly, the inverse relation (contraction) gives the reduction:

$$\mathcal{D}^{L_1}_{K_1,M_1}\mathcal{D}^{L_2}_{K_2,M_2}=\sum_{L=|L_1-L_2|}^{L_1+L_2}\langle L_1L_2L;K_1K_2\rangle\langle L_1L_2L;M_1M_2\rangle$$
$$\times\mathcal{D}^L_{K_1+K_2,M_1+M_2} \tag{R.17}$$

which is known as the Clebsch–Gordan series. We use Eq. R.17 for transforming tensors via intermediate axis systems (see Sections 8.10 and 8.11 in Chapter 8). A relation that is especially useful for evaluating integrals needed in slow-motion EPR simulations (see Section 6.9 in Chapter 6) derives directly from Eq. R.17, together with Eq. R.12:

$$\int \mathcal{D}^{L*}_{K_1+K_2,M_1+M_2}\mathcal{D}^{L_1}_{K_1,M_1}\mathcal{D}^{L_2}_{K_2,M_2}\cdot \mathrm{d}\Omega = \frac{8\pi^2}{2L+1}\langle L_1L_2L;K_1K_2\rangle\langle L_1L_2L;M_1M_2\rangle \quad (R.18)$$

A much more symmetrical form is:

$$\int \mathcal{D}^{L_1}_{K_1,M_1}(\Omega)\mathcal{D}^{L_2}_{K_2,M_2}(\Omega)\mathcal{D}^{L_3}_{K_3,M_3}(\Omega)\cdot \mathrm{d}\Omega = 8\pi^2\begin{pmatrix} L_1 & L_2 & L_3 \\ K_1 & K_2 & K_3 \end{pmatrix}\begin{pmatrix} L_1 & L_2 & L_3 \\ M_1 & M_2 & M_3 \end{pmatrix} \quad (R.19)$$

where $\begin{pmatrix} L_1 & L_2 & L_3 \\ M_1 & M_2 & M_3 \end{pmatrix}$ is a Wigner $3j$-symbol, for which the integer indices are restricted by the conditions $L_3 = |L_1 - L_2|,\ldots,L_1 + L_2$ and $M_1 + M_2 + M_3 = 0$. Equation R.19 is a particular case of the Wigner–Eckhart theorem (i.e., Eq. Q.19 in Appendix Q).

A further important property of Wigner rotation matrices is that they are eigenfunctions of the rigid-rotator Hamiltonian for a symmetric top (see Appendix E.4). Correspondingly, they are also eigenfunctions of the rotational diffusion equation for an axially symmetric rotator (see Section 7.6 in Chapter 7). We therefore have the following relations for the angular momentum operator:

$$L_z\mathcal{D}^L_{K,M} = M\mathcal{D}^L_{K,M} \quad (R.20)$$

$$L_\pm\mathcal{D}^L_{K,M} \equiv \left(L_x \pm iL_y\right)\mathcal{D}^L_{K,M} = \sqrt{L(L+1) - M(M\pm 1)}\,\mathcal{D}^L_{K,M\pm 1} \quad (R.21)$$

$$L^2\mathcal{D}^L_{K,M} \equiv \left(L_x^2 + L_y^2 + L_z^2\right)\mathcal{D}^L_{K,M} = L(L+1)\mathcal{D}^L_{K,M} \quad (R.22)$$

which follow from Eqs. E.7, E.11 and E.12, and E.6, respectively, in Appendix E.

The resultant $(\phi\theta\psi)$ of two consecutive rotations $(\phi_1\theta_1\psi_1)$ followed by $(\phi_2\theta_2\psi_2)$ is given by the closure relation for matrix multiplication:

$$\mathcal{D}^L_{K,M}(\phi\theta\psi) = \sum_{N=-L}^{+L}\mathcal{D}^L_{K,N}(\phi_2\theta_2\psi_2)\mathcal{D}^L_{N,M}(\phi_1\theta_1\psi_1) \quad (R.23)$$

For the reduced Wigner rotation matrix elements, this becomes:

$$d^L_{K,M}(\theta_1+\theta_2) = \sum_{N=-L}^{+L} d^L_{K,N}(\theta_2)d^L_{N,M}(\theta_1) \quad (R.24)$$

These relations are important for transforming spin Hamiltonians (e.g., Eq. S.11) via intermediate axis systems (see Sections 8.10, 8.11).

APPENDIX S: IRREDUCIBLE SPHERICAL TENSORS

S.1 Irreducible Spherical Tensors and Rotational Transformations

Cartesian representations (3×3 matrices) of second-rank tensors, such as the g-tensor, were introduced in Appendix B.4. (First-rank tensors are simply vectors.) These tensors transform under rotations according to Eq. B.25, where **R** is a rotation matrix that is composed of direction cosines. The latter do not display the inherent rotational symmetries that we introduced in the preceding Appendices Q and R. For rotation problems it is better to use spherical tensors, which transform according to the Wigner rotation matrices with elements, $\mathcal{D}^L_{K,M}$ (see Appendix R.3). An irreducible spherical tensor, $\mathbf{T}_L$, of rank L has $2L+1$ components, $T_{L,M}$, where $M = -L, -L+1,\ldots, L-1, L$. These components are transformed by rotation R through Euler angles ϕ,θ,ψ according to (Rose 1957; Brink and Satchler 1968):

$$T'_{L,M} = \sum_{K=-L}^{+L} T_{L,K}\mathcal{D}^L_{K,M}(\phi\theta\psi) \quad (S.1)$$

where $T'_{L,M} = RT_{L,M}R^{-1}$ are the components referred to the new axes (see Eq. B.25). Equation S.1 is the definition of an irreducible spherical tensor, which is irreducible in the sense that we need all $2L+1$ components of $T_{L,K}$ on the right-hand side of Eq. S.1 to express a general rotation. The irreducibility follows directly from the properties of the Wigner rotation matrices in Appendix R.3. It is clear that the $T_{L,M}$ transform in the same way under rotations as do the spherical harmonics $Y_{L,M}$ (see Appendix V.2 and Eqs. R.13, R.14). In fact, viewed as operators, spherical harmonics are irreducible spherical tensors.

Equation S.1 applies equally well to quantum mechanical tensor operators, which transform under rotation according to $T'_{L,M} \equiv \hat{R}T_{L,M}\hat{R}^{-1}$. When $\hat{R}$ represents the infinitesimal rotation given by Eq. R.4 of Appendix R.1, this leads to the commutation relations with general angular momentum operators L_z and $L_\pm$ (Brink and Satchler 1968):

$$\left[L_z, T_{L,M}\right] = M\cdot T_{L,M} \quad (S.2)$$

$$\left[L_\pm, T_{L,M}\right] = \sqrt{(L\pm M+1)(L\mp M)}\cdot T_{L,M\pm 1} \quad (S.3)$$

where $L_\pm = L_x \pm iL_y$.

We can use Eqs. S.2 and S.3 to find the spherical equivalents of Cartesian tensors. For instance, if we take the electron spin (which is a first-rank tensor) as the angular momentum operator (i.e., $\mathbf{L} \equiv \mathbf{S}$), comparing Eq. S.2 with the trivial identity $\left[S_z, S_z\right] = 0$ leads to $S_{1,0} = S_z$. Applying Eq. S.3 then gives:

$$S_{1,\pm 1} = \frac{1}{\sqrt{2}}\left[S_\pm, S_{1,0}\right] = \frac{1}{\sqrt{2}}\left[S_\pm, S_z\right] = \mp\frac{1}{\sqrt{2}}S_\pm \quad (S.4)$$

where we use Eq. E.9 from Appendix E for the commutator between S_z and $S_\pm (\equiv S_x \pm iS_y)$. Because the Cartesian components of any vector transform in the same way under rotation, the expressions for the spherical tensor components $S_{1,M}$ obtained here hold quite generally. The general case for vector **V** is listed in the first two rows of Table S.1.

We can construct a spherical tensor of rank 2, with components $T_{2,M}$, from the product of two vectors **I** and **S**, which are spherical tensors of rank 1 with components $I_{1,K}$ and $S_{1,N}$. The second-rank tensor component with maximum value of M (=2), we can construct in only one way, i.e., with K = 1 and N = 1. Therefore, for this component we have: $T_{2,2} = I_{1,1}S_{1,1}$, which satisfies the commutation relation given by Eq. S.2. We can then construct other components of $T_{2,M}$ from $T_{2,2}$ by applying Eq. S.3:

$$\begin{aligned} T_{2,1} &= \tfrac{1}{2}\left[L_-, I_{1,1}S_{1,1}\right] = \tfrac{1}{2}\left(\left[L_-, I_{1,1}\right]S_{1,1} + I_{1,1}\left[L_-, S_{1,1}\right]\right) \\ &= \tfrac{1}{\sqrt{2}}\left(I_{1,0}S_{1,1} + I_{1,1}S_{1,0}\right) \\ &\equiv -\tfrac{1}{2}\left(I_zS_+ + I_+S_z\right) = -\tfrac{1}{2}\left(I_zS_x + I_xS_z + i\left(I_zS_y + I_yS_z\right)\right) \end{aligned} \tag{S.5}$$

where the expansion of the commutator for the product follows from the differential nature of the angular momentum operator. In the third line of Eq. S.5, we convert the first-rank tensor components to Cartesian vector components by using the results of the preceding paragraph. Expressions of this type, in terms of the components of the spin operators **I** and **S**, are particularly useful for constructing spin Hamiltonians from spherical tensor operators. We recognize immediately that the equivalent of Eq. S.5 expressed in Cartesian tensor components is:

$$T_{2,1} = -\tfrac{1}{2}\left(T_{zx} + T_{xz} + i\left(T_{zy} + T_{yz}\right)\right) \tag{S.6}$$

and correspondingly: $T_{2,2} = I_{1,1}S_{1,1} \equiv \tfrac{1}{2}I_+S_+$ becomes $T_{2,2} = \tfrac{1}{2}\left(T_{xx} - T_{yy} + i\left(T_{xy} + T_{yx}\right)\right)$. These substitutions follow directly from the definition of spherical tensors in Eq. S.1. Applying Eq. S.3 to $T_{2,1}$ yields the next tensor component:

TABLE S.1 Spherical tensors expressed in terms of Cartesian vectors (V_i) and tensors (T_{ij})

Cartesian	*spherical*		
$i,j = x,y,z$	L	M	$T_{L,M}$
V_i	1	0	$V_{1,0} = V_z$
		±1	$V_{1,\pm1} = \mp\tfrac{1}{\sqrt{2}}\left(V_x \pm iV_y\right)$
T_{ij}	0	0	$T_{0,0} = -\tfrac{1}{\sqrt{3}}\left(T_{xx} + T_{yy} + T_{zz}\right)$
	1	0	$T_{1,0} = -\tfrac{1}{\sqrt{2}}i\left(T_{xy} - T_{yx}\right)$
		±1	$T_{1,\pm1} = -\tfrac{1}{2}\left(T_{zx} - T_{xz} \pm i\left(T_{zy} - T_{yz}\right)\right)$
	2	0	$T_{2,0} = \tfrac{1}{\sqrt{6}}\left(3T_{zz} - \left(T_{xx} + T_{yy} + T_{zz}\right)\right)$
		±1	$T_{2,\pm1} = \mp\tfrac{1}{2}\left(T_{xz} + T_{zx} \pm i\left(T_{yz} + T_{zy}\right)\right)$
		±2	$T_{2,\pm2} = \tfrac{1}{2}\left(T_{xx} - T_{yy} \pm i(T_{xy} + T_{yz})\right)$

$$T_{2,0} = \sqrt{\tfrac{2}{3}}\left(I_zS_z - \tfrac{1}{4}\left(I_+S_- + I_-S_+\right)\right) \equiv \sqrt{\tfrac{2}{3}}\left(T_{zz} - \tfrac{1}{2}\left(T_{xx} + T_{yy}\right)\right) \tag{S.7}$$

Repeatedly applying Eq. S.3 then yields the $T_{2,M}$ components with negative values of M.

Table S.1 lists the spherical tensors and their components that derive from a second-rank Cartesian tensor, T_{ij}. There are nine spherical tensor components in all, corresponding to the nine elements of the 3×3 T_{ij} matrix. The Cartesian tensor reduces to three irreducible spherical tensors: a scalar $T_{0,0}$, a vector $T_{1,M}$ and a second-rank tensor $T_{2,M}$. When the Cartesian tensor is expressed in its diagonal, principal axis system, we need only the second-rank irreducible tensor. This is the usual situation for the g- and A-tensors of a spin label, when expressed in the nitroxide molecular coordinate system. In addition, the spherical tensor component $T_{2,\pm1}$ vanishes, but this is not the case for the corresponding product of spin operators **I** and **S**.

Second-order irreducible spherical tensors constructed from the product of two first-order spherical tensors are an example of the general rule for the product of irreducible spherical tensors $T_{L_1,M_1}(A_1)$ and $T_{L_2,M_2}(A_2)$ of ranks L_1 and L_2:

$$T_{L,M}(A_1,A_2) = \sum_{M_2=-L_2}^{+L_2} \langle L_1L_2L; M - M_2, M_2\rangle \times T_{L_1,M-M_2}(A_1)T_{L_2,M_2}(A_2) \tag{S.8}$$

where $\langle L_1L_2L; M_1, M_2\rangle$ is a Clebsch–Gordan coefficient (see Appendix Q.2). The product tensor $T_{L,M}(A_1,A_2)$ satisfies the definition given by Eq. S.1 for an irreducible spherical tensor provided that its rank satisfies the triangular condition: $L = |L_1 - L_2|, |L_1 - L_2| + 1, \ldots, L_1 + L_2 - 1, L_1 + L_2$. We can show this by using Wigner rotation matrices, together with the Clebsch–Gordan series (i.e., Eq. R.17) (Rose 1957).

S.2 Spin Hamiltonian in Irreducible Spherical Tensors

The usefulness of spherical tensors comes from their transformation properties under rotation, which are given by Eq. S.1. We see this best from the anisotropic (i.e., angular-dependent) part of the general spin Hamiltonian for a nitroxide:

$$\mathcal{H}_S = \beta_e \mathbf{B}\cdot\mathbf{g}\cdot\mathbf{S} + \mathbf{I}\cdot\mathbf{A}\cdot\mathbf{S} \tag{S.9}$$

which we give here in an implied Cartesian tensor notation. As already mentioned, the hyperfine and g-tensor are expressed in the nitroxide axis system (x',y',z'), in which they are diagonal (i.e., $g_{ij} = 0$ unless $i = j$ and similarly for A_{ij}) and the z'-axis is the nitrogen $2p\pi$-orbital (see Figure 2.1). On the other hand, the electron spin is quantized along the static magnetic field $\mathbf{B}_o$ and, therefore, is expressed in the laboratory axis system (x,y,z) with $\mathbf{B}_o$ as the z-axis. It is clear from Eq. S.9 that we can express the hyperfine and Zeeman interactions as scalar products, $\mathbf{A}(A_{ij})\cdot\mathbf{T}(IS)$ and $\mathbf{A}(g_{ij})\cdot\mathbf{T}(BS)$ respectively, of two spherical tensors. One second-order tensor $\mathbf{T}_2$

depends only on the spin (and magnetic field) vectors, and the other $\mathbf{A}_2$ depends only on the nitroxide molecular tensors (Freed and Fraenkel 1963; Mehring 1983). Note that our $\mathbf{A}_2$ tensor is given the symbol F throughout the EPR literature. However, definitions of F vary, depending in part on whether the factor $(-1)^M$ is absorbed into F (cf. Eq. S.10 below) and on which convention is used for Wigner rotation matrices (see Appendix R.3). The treatment given here follows that of Mehring (1983) for NMR and avoids these potential ambiguities. In the main text, we give the full spin Hamiltonian (see Eq. S.13 below) by writing out $\mathbf{A}_2$ and $\mathbf{T}_2$ explicitly so as to eliminate possible confusion.

We consider only the anisotropic interactions, which restricts us to second-rank irreducible spherical tensors (cf. Table S.1). The anisotropic part of the hyperfine interaction, for example, is given by the scalar product that we get from rotationally invariant contraction of the spherical tensors defined in Table S.2 (Rose 1957; Tinkham 1964):

$$\mathcal{H}_{hfs} = \sum_{M=-2}^{+2} (-1)^M A'_{2,M} T'_{2,-M} \quad \left(\equiv \sum_{M=-2}^{+2} (-1)^M A'_{2,-M} T'_{2,M} \right) \tag{S.10}$$

where primes indicate that we have chosen the nitroxide coordinate system. This is the natural system for $A'_{2,M}$, but for $T'_{2,-M}$ we must use Eq. S.1 to express the spin vectors in the (unprimed) laboratory coordinate system. Equation S.10 then becomes:

$$\mathcal{H}_{hfs} = \sum_{K,M=-2}^{+2} (-1)^M A'_{2,M} \mathcal{D}^2_{K,-M}(\Omega) T_{2,K} \tag{S.11}$$

TABLE S.2 Irreducible spherical tensor representation of nitroxide spin-Hamiltonian tensors and spin operators

M	$A'_{2,M}$	$T_{2,M}$
hyperfine interaction: $A_{2,M}$, $T_{2,M}(IS)$		
0	$A_{2,0} = \sqrt{\frac{2}{3}}\Delta A$	$T_{2,0} = \sqrt{\frac{2}{3}}\left(I_z S_z - \frac{1}{4}(I_+ S_- + I_- S_+)\right)$
±1	$A_{2,\pm1} = 0$	$T_{2,\pm1} = \mp\frac{1}{2}(I_z S_\pm + I_\pm S_z)$
±2	$A_{2,\pm2} = \delta A$	$T_{2,\pm2} = \frac{1}{2} I_\pm S_\pm$
Zeeman interaction: $\beta_e g_{2,M}$, $T_{2,M}(BS)$		
0	$g_{2,0} = \sqrt{\frac{2}{3}}\Delta g$	$T_{2,0} = \sqrt{\frac{2}{3}} B_o S_z$
±1	$g_{2,\pm1} = 0$	$T_{2,\pm1} = \mp\frac{1}{2} B_o S_\pm$
±2	$g_{2,\pm2} = \delta g$	$T_{2,\pm2} = 0$

Spin-Hamiltonian tensors are referred to the nitroxide axis system (x',y',z'), where the anisotropies are defined by:

$\Delta A = A_{zz} - \frac{1}{2}(A_{xx} + A_{yy})$

$\delta A = \frac{1}{2}(A_{xx} - A_{yy})$

$\Delta g = g_{zz} - \frac{1}{2}(g_{xx} + g_{yy})$

$\delta g = \frac{1}{2}(g_{xx} - g_{yy})$

and strictly speaking the subscripts should be primed.

Spin operators are referred to the laboratory axis system (x,y,z), where the magnetic field defines the z-axis, i.e., $\mathbf{B} \equiv (0,0,B_o)$.

For spin operators, see also Eq. S.8 with $\langle 112;0,M_2 \rangle$ from Table A8.3 in Chapter 8.

where $\Omega = (\phi,\theta,\psi)$ represents the Euler angles that rotate the laboratory axis system, in which we now express the unprimed $T_{2,K}$ components, into the nitroxide axis system. The azimuthal angle of the nitroxide z-axis in the laboratory system is ϕ, that of the magnetic field in the nitroxide system is ψ, and the nitroxide z-axis is tilted at θ to the magnetic field.

Note that Freed and Fraenkel (1963) use a transformation different from that in Eq. S.1. The summation there is over the second subindex of the Wigner rotation matrix in our Eq. S.1, which amounts to substituting $\mathcal{D}^L_{M,K}$ for $\mathcal{D}^L_{K,M}$. This convention is adhered to in the slow-motion simulations reviewed by Freed (1976) and up until the paper by Moro and Freed (1981). It was first changed to our convention in the appendix to Meirovitch et al. (1982). A clear exposition on this point is given by Schneider and Freed (1989). However, they simultaneously change the notation for Wigner rotation matrix elements to $\mathcal{D}^L_{M,K}$ instead of $\mathcal{D}^L_{K,M}$ used previously by them, presumably to maintain consistency with the original convention. This means that their M is now our K and vice-versa. Luckhurst and colleagues, on the other hand, use a consistent convention which agrees with that described here (see Section 8.10 and Appendix A8.2 in Chapter 8), as does that of Nordio and colleagues (see Section 8.11).

The upper part of Table S.2 lists the components of the spherical tensors that we need for the spin Hamiltonian given in Eq. S.11. We neglect non-secular hyperfine terms ($I_\pm S_\mp$ and $I_z S_\pm$), as is usual for relaxation calculations, because they contribute only in second order (see Appendix J). This means that we no longer need consider $K = \pm 2$ in Eq. S.11. Taking into account also that $A_{2,\pm1} = 0$ and $A_{2,-2} = A_{2,2}$ (see Table S.2), we can then write the hyperfine spin Hamiltonian of Eq. S.11 explicitly in spherical tensor components:

$$\begin{aligned} \mathcal{H}_{hfs} = {} & A_{2,0}\mathcal{D}^2_{0,0}T_{2,0} + A_{2,2}\left(\mathcal{D}^2_{0,2} + \mathcal{D}^2_{0,-2}\right)T_{2,0} \\ & + A_{2,0}\left(\mathcal{D}^2_{1,0}T_{2,1} + \mathcal{D}^2_{-1,0}T_{2,-1}\right) + A_{2,2}\left(\mathcal{D}^2_{1,-2}T_{2,1} + \mathcal{D}^2_{-1,-2}T_{2,-1}\right) \\ & + A_{2,2}\left(\mathcal{D}^2_{1,2}T_{2,1} + \mathcal{D}^2_{-1,2}T_{2,-1}\right) \end{aligned} \tag{S.12}$$

where $A_{2,M}$ now stands specifically for the hyperfine tensor components, and the $T_{2,M}$ components are constructed from products of the components of the $\mathbf{I}$ and $\mathbf{S}$ spin vectors.

Clearly from Eq. S.9, the anisotropic Zeeman interaction takes a form identical to Eq. S.11, where $\beta_e \mathbf{g}$ replaces the hyperfine tensor and the magnetic field $\mathbf{B}_o$ replaces the nuclear spin vector. The corresponding spherical tensors for the Zeeman interaction are listed in the lower part of Table S.2. Because the magnetic field defines the axis of quantization, we need only the secular terms. These correspond to the first two terms, those containing $T_{2,0}$, in Eq. S.12. Combining both hyperfine and Zeeman interactions, and substituting the Cartesian equivalents of the spherical tensors that are given in Table S.2, the anisotropy of the full spin Hamiltonian becomes:

$$\begin{aligned} \mathcal{H}_1(\Omega) = {} & \frac{2}{3}\left(\Delta g \beta_e B_o + \Delta A I_z\right)\mathcal{D}^2_{0,0} S_z \\ & + \sqrt{\frac{2}{3}}\left(\delta g \beta_e B_o + \delta A I_z\right)\left(\mathcal{D}^2_{0,2} + \mathcal{D}^2_{0,-2}\right) S_z \end{aligned}$$

$$-\frac{\Delta A}{\sqrt{6}}\left(\mathcal{D}_{1,0}^{2}I_{+}-\mathcal{D}_{-1,0}^{2}I_{-}\right)S_{z}-\frac{\delta A}{2}\left(\mathcal{D}_{1,2}^{2}+\mathcal{D}_{1,-2}^{2}\right)I_{+}S_{z}$$

$$+\frac{\delta A}{2}\left(\mathcal{D}_{-1,2}^{2}+\mathcal{D}_{-1,-2}^{2}\right)I_{-}S_{z} \tag{S.13}$$

where $\Delta g = g_{zz} - \frac{1}{2}(g_{xx} + g_{yy})$ and $\delta g = \frac{1}{2}(g_{xx} - g_{yy})$, and similarly for ΔA and δA. Note that S_z is the only electron spin component present in this Hamiltonian, because all non-secular terms $(S_{\pm})$ are excluded. Equation S.13 is the basis for spin-label relaxation calculations using perturbation theory and for slow-motion simulations using the stochastic Liouville equation (see Chapters 5, 6 and 8).

APPENDIX T: FOURIER TRANSFORMS, CONVOLUTIONS AND CORRELATION FUNCTIONS

T.1 Fourier Transforms

The Fourier transform of a function $f(t)$ in the time domain is a function $F(\omega)$ in the frequency domain that is given by:

$$F(\omega)=\mathcal{F}\{f(t)\}\equiv\int_{-\infty}^{\infty}f(t)\exp(-i\omega t)\cdot \mathrm{d}t \tag{T.1}$$

Correspondingly, the inverse Fourier transform is given by:

$$f(t)=\mathcal{F}^{-1}\{F(\omega)\}\equiv\frac{1}{2\pi}\int_{-\infty}^{\infty}F(\omega)\exp(i\omega t)\cdot \mathrm{d}\omega \tag{T.2}$$

Association of the factor $1/2\pi$ with either the forward or the inverse transform is somewhat arbitrary. The value of the transform at the origin (i.e., zero-frequency) is equal to the integrated area under the function:

$$F(0)=\int_{-\infty}^{\infty}f(t)\cdot \mathrm{d}t \tag{T.3}$$

The integrated area of the power spectrum is given by the Parseval equality:

$$\int_{-\infty}^{\infty}|f(t)|^{2}\,\mathrm{d}t=\frac{1}{2\pi}\int_{-\infty}^{\infty}|F(\omega)|^{2}\,\mathrm{d}\omega \tag{T.4}$$

Scaling of the variables in the transform is given by:

$$\mathcal{F}\{f(at)\}=\frac{1}{|a|}F(\omega/a) \tag{T.5}$$

Shifting the origin of t results in periodic modulation of the transform:

$$\mathcal{F}\{f(t-t_o)\}=\exp(-i\omega t_o)F(\omega) \tag{T.6}$$

and correspondingly periodic modulation of the function $f(t)$ is given by:

$$\mathcal{F}\{f(t)\exp(i\omega_o t)\}=F(\omega-\omega_o) \tag{T.7}$$

The Fourier transform of the derivative function is given by differentiating under the integral in Eq. T.1 and subsequent integration by parts:

$$\mathcal{F}\{\mathrm{d}f/\mathrm{d}t\}=i\omega\mathcal{F}\{f(t)\}=i\omega F(\omega) \tag{T.8}$$

The Fourier transforms of the moments, $t^n f(t)$, of the function $f(t)$ are related to the differentials of its Fourier transform:

$$\mathcal{F}\{t^n f(t)\}=i^n\frac{\mathrm{d}^n}{\mathrm{d}\omega^n}F(\omega) \tag{T.9}$$

which we see from differentiating under the integral sign of $F(\omega)$.

Values of Fourier transform pairs for various functions $f(t)$ are listed in Table T.1. We can generate extensions by using the above rules.

T.2 Convolution

The convolution $f(\omega)*g(\omega)$ of two lineshapes or distributions, $f(\omega)$ and $g(\omega)$, is defined by the integral:

$$f(\omega)*g(\omega)=\int_{-\infty}^{\infty}f(\omega')g(\omega-\omega')\cdot \mathrm{d}\omega' \tag{T.10}$$

TABLE T.1 Fourier transform pairs

	$f(t)$	$F(\omega)=\int_{-\infty}^{\infty}f(t)\exp(-i\omega t)\cdot \mathrm{d}t$
Dirac δ-function	$\delta(t)$	1
	1	$2\pi\delta(\omega)$
reflected exponential	$\exp(-a\lvert t\rvert)$	$\frac{2a}{\omega^2+a^2}$
Lorentzian	$\frac{1}{t^2+a^2}$	$\frac{\pi}{a}\exp(-a\lvert\omega\rvert)$
truncated exponential	$\exp(-at)$ for $t>0$, 0 for $t<0$	$\frac{1}{a+i\omega}=\frac{a}{\omega^2+a^2}-\frac{i\omega}{\omega^2+a^2}$
Gaussian	$\exp(-a^2t^2)$	$\frac{\sqrt{\pi}}{a}\exp\left(-\frac{\omega^2}{4a^2}\right)$
rectangular pulse	1 for $\lvert t\rvert<\tau$, 0 for $\lvert t\rvert>\tau$	$\frac{2\sin(\omega\tau)}{\omega}$
triangular pulse	$1-\frac{\lvert t\rvert}{2\tau}$ for $\lvert t\rvert<2\tau$, 0 for $\lvert t\rvert>2\tau$	$2\tau\left(\frac{\sin(\omega\tau)}{\omega\tau}\right)^2$

We illustrate this in the upper panel of Figure T.1. The convolution operation is commutative, even if the functions are not symmetrical; we can reverse the direction of ω in $f(\omega)$, instead of that in $g(\omega)$:

$$f(\omega)*g(\omega)=g(\omega)*f(\omega)=\int_{-\infty}^{\infty}g(\omega')f(\omega-\omega')\cdot\mathrm{d}\omega' \quad \text{(T.11)}$$

We see this by comparing the upper and lower panels in Figure T.1. The asterisk behaves like a multiplication sign for convolution. The area under a convolution is equal to the product of the areas under the individual functions:

$$\int_{-\infty}^{\infty}(f*g)\cdot\mathrm{d}\omega=\int_{-\infty}^{\infty}f(\omega)\cdot\mathrm{d}\omega\times\int_{-\infty}^{\infty}g(\omega)\cdot\mathrm{d}\omega \quad \text{(T.12)}$$

A most important property is that the Fourier transform of a convolution integral is equal to the product of the Fourier transforms of the component functions:

$$\mathcal{F}\{f(\omega)*g(\omega)\}=\mathcal{F}\{f(\omega)\}\times\mathcal{F}\{g(\omega)\} \quad \text{(T.13)}$$

The derivative of a convolution is equal to the convolution of one function with the derivative of the other:

$$\frac{\mathrm{d}}{\mathrm{d}\omega}(f*g)=f*\left(\frac{\mathrm{d}g}{\mathrm{d}\omega}\right)=\left(\frac{\mathrm{d}f}{\mathrm{d}\omega}\right)*g \quad \text{(T.14)}$$

which we readily see from Eq. T.13 and the expression for the Fourier transform of a derivative (i.e., Eq. T.8).

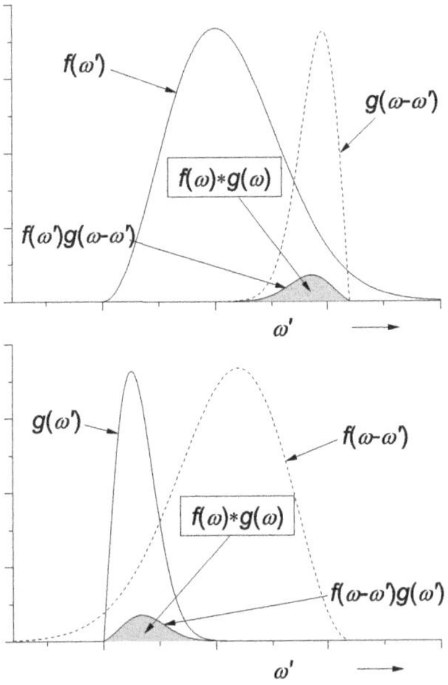

FIGURE T.1 Convolution of envelope function $f(\omega)$ with the lineshape $g(\omega)$. The convolution integral $f(\omega)*g(\omega)$ is the area under the curve $f(\omega')g(\omega-\omega')$ (upper panel), or equivalently under the curve $f(\omega-\omega')g(\omega')$ (lower panel). Note that convolution is commutative: $f(\omega)*g(\omega)\equiv g(\omega)*f(\omega)$. For generality, the lineshapes are illustrated as asymmetric functions.

The mean frequency, or first moment, $\langle\omega\rangle$ of a convolution is equal to sum of the mean frequencies of the component distributions:

$$\langle\omega\rangle_{f*g}=\langle\omega\rangle_f+\langle\omega\rangle_g \quad \text{(T.15)}$$

Correspondingly, the second moment $\langle\omega^2\rangle$ of the convolution is given by:

$$\langle\omega^2\rangle_{f*g}=\langle\omega^2\rangle_f+\langle\omega^2\rangle_g+2\langle\omega\rangle_f\langle\omega\rangle_g \quad \text{(T.16)}$$

The second moments are therefore also additive, if the means are zero. The first and second moments are related directly to the first and second derivatives, $F'(0)$ and $F''(0)$ respectively, of the Fourier transform at the origin:

$$\langle\omega\rangle_f=\frac{\int_{-\infty}^{\infty}\omega f(\omega)\cdot\mathrm{d}\omega}{\int_{-\infty}^{\infty}f(\omega)\cdot\mathrm{d}\omega}=\frac{iF'(0)}{F(0)} \quad \text{(T.17)}$$

$$\langle\omega^2\rangle_f=\frac{\int_{-\infty}^{\infty}\omega^2 f(\omega)\cdot\mathrm{d}\omega}{\int_{-\infty}^{\infty}f(\omega)\cdot\mathrm{d}\omega}=-\frac{F''(0)}{F(0)} \quad \text{(T.18)}$$

which we see from differentiating under the integral sign of the Fourier transform (i.e., Eq. T.9).

T.3 Correlation Functions

The autocorrelation function of a function $f(t)$ in the time domain is defined by the integral:

$$G(\tau)=\langle f(t)f(t-\tau)\rangle=\int_{-\infty}^{\infty}f(t)f(t-\tau)\cdot\mathrm{d}t \quad \text{(T.19)}$$

This is the self-convolution of $f(t)$, but without reversal of the time direction (viz., $f(t)*f(-t)$). Therefore, we obtain the same result if the displacement τ is in either the positive or negative direction:

$$G(\tau)=G(-\tau)=\langle f(t)f(t+\tau)\rangle \quad \text{(T.20)}$$

If $f(t)$ is a complex function, we introduce a complex conjugate into the definition:

$$G(\tau)=\int_{-\infty}^{\infty}f^*(t)f(t-\tau)\cdot\mathrm{d}t=\int_{-\infty}^{\infty}f(t)f^*(t+\tau)\cdot\mathrm{d}t \quad \text{(T.21)}$$

where the order is a matter of convention. The autocorrelation function of $f(t)$ is equal to the square of the Fourier transform of $f(t)$:

$$G(\tau) = \langle f(t) f^*(t+\tau) \rangle = |\mathcal{F}\{f(t)\}|^2 = |F(\omega)|^2 \tag{T.22}$$

i.e., the autocorrelation function is the Fourier transform of the power spectrum.

APPENDIX U: MOMENTS OF EPR LINESHAPES, AND TRACES OF SPIN OPERATORS

U.1 Spectral Moments

The *n*th moment of a normalized absorption lineshape *f(B)* about the mean field position is defined by:

$$\langle \Delta B^n \rangle \equiv \langle (B - \langle B \rangle)^n \rangle = \int_{-\infty}^{\infty} (B - \langle B \rangle)^n f(B) \cdot \mathrm{d}B \tag{U.1}$$

where the mean resonance field position $\langle B \rangle$ is:

$$\langle B \rangle = \int_{-\infty}^{\infty} B \times f(B) \cdot \mathrm{d}B \tag{U.2}$$

Normalization of the lineshape requires that: $\int f(B) \cdot \mathrm{d}B = 1$. For a symmetrical lineshape: $\langle B \rangle = 0$, and also all odd moments are zero.

The *n*th moment is related to the mean *n*th power of the resonance field $\langle B^n \rangle$ plus the lower moments, by binomial expansion of $(B - \langle B \rangle)^n$. We give an example for the second moment later. Similar considerations apply to shifts in the point about which the moment is calculated. If the moment position is shifted up by B_o, the first moment is increased by this amount and the second moment by B_o^2. Beyond the second moment, the lower moments also contribute to the shift.

U.2 First Moment

The first moment of the absorption EPR spectrum is given from Eq. U.1 by:

$$\langle \Delta B \rangle = \int_{-\infty}^{\infty} (B - \langle B \rangle) f(B) \cdot \mathrm{d}B = 0 \tag{U.3}$$

By definition, the first moment about the mean resonance field $\langle B \rangle$ is zero (cf. Eq. U.2). This result is trivial for a symmetrical lineshape, where $\langle B \rangle$ is the centre field. For an anisotropic powder pattern, however, $\langle B \rangle$ is the resonance field that corresponds to the isotropic g-value: $g_o = \langle g \rangle = \frac{1}{3}(g_{xx} + g_{yy} + g_{zz})$. Sometimes referred to as the first-moment "theorem", this is an approximate general result for $S = \frac{1}{2}$ systems with first-order hyperfine structure and is good to better than $\delta g_o = 10^{-4}$ for nitroxides (Hyde and Pilbrow 1980).

Consider, for simplicity, an axial g-tensor $(g_\perp, g_\perp, g_\parallel)$ where θ is the angle between the magnetic field B and the principal ||-axis of the g-tensor. The effective g-value $g(\theta)$ is given by Eq. 2.16 in Chapter 2, where the g-value anisotropy, $\Delta g(\theta) = \langle g \rangle - g(\theta)$, is small compared with the mean g-value, $\langle g \rangle$. Therefore, we can linearize the resonance anisotropy at high EPR fields/frequencies by using the relation: $\delta B/B \approx -\delta g/g$, to give:

$$\Delta B(\theta) = B(\theta) - \langle B \rangle \approx (\Delta B_\parallel - \Delta B_\perp)\cos^2\theta + \Delta B_\perp \tag{U.4}$$

where $\Delta B_\parallel = (\langle g \rangle - g_\parallel)\langle B \rangle / \langle g \rangle$, $\Delta B_\perp = (\langle g \rangle - g_\perp)\langle B \rangle / \langle g \rangle$ and $\langle B \rangle = h\nu / \langle g \rangle \beta_e$ (cf. Eq. 2.23 in Chapter 2). Note that the maximum value of $|\langle g \rangle - g| / \langle g \rangle$ is 1.8×10^{-3} for nitroxides. Applying Eq. 2.15 for axial powder patterns from Section 2.4, the normalized absorption envelope becomes:

$$f(\Delta B) = \frac{1}{2} \frac{1}{\sqrt{\Delta B_\parallel - \Delta B_\perp}} \frac{1}{\sqrt{\Delta B - \Delta B_\perp}} \tag{U.5}$$

From Eqs. U.3 and U.5, the first moment is then given by:

$$\langle \Delta B \rangle = \int_{\Delta B_\perp}^{\Delta B_\parallel} \Delta B \cdot f(\Delta B) \cdot \mathrm{d}\Delta B = \tfrac{1}{3}(\Delta B_\parallel + 2\Delta B_\perp)$$

$$= \left(\langle g \rangle - \tfrac{1}{3}(g_\parallel + 2g_\perp)\right)\langle B \rangle / \langle g \rangle \tag{U.6}$$

Because the first moment is identically zero, Eq. U.6 requires that the mean g-value is given by $\langle g \rangle = \frac{1}{3}(g_\parallel + 2g_\perp)$, i.e., the mean resonance field $\langle B \rangle$ corresponds to the isotropic g-value, g_o.

Ignoring second-order hyperfine shifts (see Appendix J), the $m_I = \pm 1$ hyperfine lines are situated symmetrically about the $m_I = 0$ line, and their contributions to the mean resonance field (and first moment) therefore cancel. The same is true for contributions from a symmetrical intrinsic lineshape that is convoluted with the powder envelope of Eq. U.5. To within the linearization in Eq. U.4, the first moment "theorem" therefore holds good for spin labels with first-order ^{15}N- or ^{14}N-hyperfine structure and inhomogeneous broadening from proton or deuterium hyperfine structure. We can always measure the g_{zz}-element of the g-tensor from a powder spectrum. Thus, determination of g_o from $\langle B \rangle$ immediately gives the sum: $g_{xx} + g_{yy}$, and if we know one of these elements, we can calculate the third. Note that g_{yy} is less dependent on polarity than is g_{xx} (see Chapter 4) and, therefore, will vary less from one system to another.

U.2.1 Absolute-Value First Moment

The first moment $\langle \Delta B \rangle$ about $\langle B \rangle$ vanishes. However, the absolute-value first moment $\langle |\Delta B| \rangle$ (or twice the one-sided first moment) is a finite quantity that is directly related to the spectral anisotropy in a powder pattern, or the broadening of a symmetrical spectrum. Consider the straightforward example of a powder pattern with axially symmetric N-hyperfine anisotropy, i.e., $\mathbf{A} = (A_\perp, A_\perp, A_\parallel)$, at a microwave frequency low enough for us to neglect g-value anisotropy.

We illustrate this for $I = 1$ in Figure U.1. The high-field hyperfine manifold extends from $\Delta B_\perp = A_\perp |m_I|/g\beta_e$ to $\Delta B_\parallel = A_\parallel |m_I|/g\beta_e$, relative to the centre field $\langle B\rangle = h\nu/g\beta_e$, and the low-field manifold is the mirror image. For this symmetrical powder pattern, the absolute first moment for a single hyperfine state, m_I, is given by:

$$\langle|\Delta B_m|\rangle = \int_{-\infty}^{\infty} |B_m - \langle B\rangle| f(B_m)\cdot \mathrm{d}B_m = \int_{\Delta B_\perp}^{\Delta B_\parallel} \Delta B_m f(\Delta B_m)\cdot \mathrm{d}\Delta B_m$$

$$= \int_{-\Delta B_\parallel}^{-\Delta B_\perp} \Delta B_m f(\Delta B_m)\cdot \mathrm{d}\Delta B_m \qquad (U.7)$$

where $\Delta B_m = |B_m - \langle B\rangle|$ and $f(\Delta B_m)$ is normalized over the integration limits indicated. The angular-dependent hyperfine coupling is given by Eq. 2.17 in Chapter 2. The corresponding normalized absorption envelope for the high-field or low-field manifold of the powder pattern is given by (see Eq. 2.18 in Section 2.4):

$$f(\Delta B_m) = \frac{1}{\sqrt{\Delta B_\parallel^2 - \Delta B_\perp^2}}\cdot\frac{\Delta B_m}{\sqrt{\Delta B_m^2 - \Delta B_\perp^2}} \qquad (U.8)$$

From Eqs. U.7 and U.8, the absolute first moment then becomes:

$$\langle|\Delta B_{m_I}|\rangle$$

$$= \frac{1}{2}\left(\frac{A_\parallel}{g\beta_e} + \frac{A_\perp/g\beta_e}{\sqrt{(A_\parallel/A_\perp)^2 - 1}}\ln\left(A_\parallel/A_\perp + \sqrt{(A_\parallel/A_\perp)^2 - 1}\right)\right)|m_I| \qquad (U.9)$$

for one hyperfine manifold of an axially symmetric powder pattern. The absolute first moment of the entire powder pattern is given by taking the average over all $m_I = -I, \ -I+1, \ldots, I-1, I$,

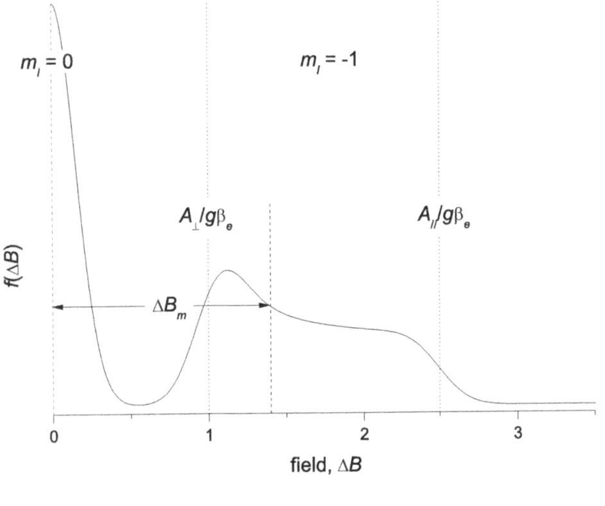

FIGURE U.1 High-field manifold $(m_I = -1)$ in the absorption powder pattern for axially symmetric hyperfine anisotropy of a ^{14}N-nitroxide ($I = 1$). g-value anisotropy is neglected. ΔB_m is the field separation from the centre of the spectrum that is used in calculating the spectral moments according to Eqs. U.7 and U.31. For a ^{15}N-nitroxide $(I = \frac{1}{2})$, the high-field manifold is $m_I = -\frac{1}{2}$ and the central $(m_I = 0)$ manifold is absent.

where I is the nuclear spin. This involves the normalized summation: $\sum_{-I}^{I} |m_I|/(2I+1) = I(I+1)/(2I+1)$ for integer I, (e.g., ^{14}N) and $\sum_{-I}^{I} |m_I|/(2I+1) = (2I+1)/4$ for half-integer I (e.g., ^{15}N). For integer I, the absolute first moment of the whole spectrum then becomes:

$$\langle|\Delta B_{aniso}|\rangle = \frac{1}{2}\left(\frac{A_\parallel}{g\beta_e} + \frac{A_\perp/g\beta_e}{\sqrt{(A_\parallel/A_\perp)^2 - 1}}\right.$$

$$\left.\times \ln\left(A_\parallel/A_\perp + \sqrt{(A_\parallel/A_\perp)^2 - 1}\right)\right)\frac{I(I+1)}{2I+1} \qquad (U.10)$$

and a corresponding substitution of $(2I+1)/4$ for $|m_I|$ in Eq. U.9 gives the result, if I is half-integer.

A particularly useful application of the absolute first moment is calculation of the average splitting $\langle 2|\Delta B_{dd}|\rangle$ in an absorption powder pattern for the magnetic dipole–dipole interaction between two spin labels (Rabenstein and Shin 1995). We treat this in Section 15.6 of Chapter 15, for distance measurements between spin labels. In general, the absolute first moment has advantages over the higher even moments, all of which have positive non-zero values, in that the higher moments specifically emphasize larger splittings or line broadenings and therefore require good signal/noise and baseline in the outer wings of the spectrum.

U.3 Second Moment

From Eq. U.1, the second moment $\langle \Delta B^2\rangle$ of a normalized lineshape $f(B)$ about the mean resonance field $\langle B\rangle$ is given by:

$$\langle \Delta B^2\rangle = \int_{-\infty}^{\infty} (B - \langle B\rangle)^2 f(B)\cdot \mathrm{d}B \qquad (U.11)$$

Expanding $(B - \langle B\rangle)^2 = B^2 - 2B\langle B\rangle + \langle B\rangle^2$, the second moment is related to the mean-square resonance field $\langle B^2\rangle$ by:

$$\langle \Delta B^2\rangle = \langle B^2\rangle - \langle B\rangle^2 \qquad (U.12)$$

It is sometimes more convenient to calculate $\langle B^2\rangle$ than to obtain $\langle \Delta B^2\rangle$ directly. For independent contributions to the line broadening: $\Delta B = \Delta B_1 + \Delta B_2$, the contributions to the second moment are additive:

$$\langle \Delta B^2\rangle = \langle \Delta B_1^2\rangle + \langle \Delta B_2^2\rangle \qquad (U.13)$$

if at least one of the mean values is zero: $\langle \Delta B_1\rangle = 0$ or $\langle \Delta B_2\rangle = 0$. This is the case for symmetrical spectra, including dipolar interactions or first-order hyperfine structure. It also applies specifically to the Gaussian lineshape, where the second moment is equal to the variance σ_G^2, which we recognize from the standard deviation of the normal error distribution.

The power of this method is that we can express moments of EPR lines solely in terms of the traces (i.e., diagonal sums) of spin operators. Then we need not diagonalize the (many-body) Hamiltonian, because the trace of an operator is independent of the basis set that we choose (see Van Vleck 1948). As we shall see below, we evaluate the trace within whichever basis set is most convenient.

The overall shape of an inhomogeneously broadened line arises from a superposition of different transitions that take place between pairs of states $|m\rangle$ and $|m'\rangle$ at resonance field $B_{mm'}$. These transitions are induced by a microwave magnetic field B_{1x} that is oriented perpendicular to the static magnetic field B_{oz}, which defines the direction of z-quantization of the electron spin. The transition probability, and therefore relative intensity, is then proportional to the squared off-diagonal matrix element $|\langle m|S_x|m'\rangle|^2$ of the x-component of the electron spin (see Chapter 1 and Appendix M). We get the mean-square resonance field by summing over all resonant transitions:

$$\langle B^2\rangle = \frac{\sum_{m,m'} B^2_{mm'}|\langle m|S_x|m'\rangle|^2}{\sum_{m,m'}|\langle m|S_x|m'\rangle|^2} \tag{U.14}$$

which then lets us calculate the second moment. Remembering that $|\langle m|S_x|m'\rangle|^2 = \langle m|S_x|m'\rangle\langle m'|S_x|m\rangle$, we can write Eq. U.14 in terms of the sums of diagonal matrix elements, i.e., of traces, of the operators (Van Vleck 1948):

$$\langle B^2\rangle = -\frac{\mathrm{Tr}\left([\mathcal{H}_S,S_x]^2\right)}{(g\beta_e)^2\,\mathrm{Tr}\left(S_x^2\right)} \tag{U.15}$$

where $\mathcal{H}_S$ is the Hamiltonian of the total spin system so that $g\beta_e B_{mm'} = \langle m|\mathcal{H}_S|m\rangle - \langle m'|\mathcal{H}_S|m'\rangle$, and the commutator is defined by $[\mathcal{H}_S,S_x] = (\mathcal{H}_S S_x - S_x\mathcal{H}_S)$. The numerator in Eq. U.15 therefore becomes: $\mathrm{Tr}(\mathcal{H}_S S_x\mathcal{H}_S S_x) + \mathrm{Tr}(S_x\mathcal{H}_S S_x\mathcal{H}_S) - 2\mathrm{Tr}(\mathcal{H}_S S_x^2\mathcal{H}_S)$.

We can write the Hamiltonian as a sum of the Zeeman Hamiltonian $\mathcal{H}_o$ that determines the resonance field $\langle B\rangle$ and a spin-interaction term $\mathcal{H}_1$ that gives rise to the inhomogeneous broadening:

$$\mathcal{H}_S = \mathcal{H}_o + \mathcal{H}_1 = g\beta_e\langle B\rangle S_z + \mathcal{H}_1 \tag{U.16}$$

Then expanding $\mathcal{H}_S$ in the numerator in Eq. U.15, we get:

$$\mathrm{Tr}\left([\mathcal{H}_o,S_x]^2\right) + 2\mathrm{Tr}\left([\mathcal{H}_o,S_x][\mathcal{H}_1,S_x]\right) + \mathrm{Tr}\left([\mathcal{H}_1,S_x]^2\right)$$
$$= \mathrm{Tr}\left([\mathcal{H}_o,S_x]^2\right) + \mathrm{Tr}\left([\mathcal{H}_1,S_x]^2\right)$$

where the cross-term vanishes. This is because the interaction $\mathcal{H}_1$ is composed of spin-projection products of the type S_zI_z or $S_{jz}S_{kz}$, and thus every term involves factors of the type $\mathrm{Tr}(I_z) = 0$ or $\mathrm{Tr}(S_{kz}) = 0$. Then we can write the mean-square resonance field as:

$$\langle B^2\rangle = -\frac{\mathrm{Tr}\left([\mathcal{H}_o,S_x]^2\right)}{(g\beta_e)^2\,\mathrm{Tr}\left(S_x^2\right)} - \frac{\mathrm{Tr}\left([\mathcal{H}_1,S_x]^2\right)}{(g\beta_e)^2\,\mathrm{Tr}\left(S_x^2\right)} = \langle B\rangle^2 - \frac{\mathrm{Tr}\left([\mathcal{H}_1,S_x]^2\right)}{(g\beta_e)^2\,\mathrm{Tr}\left(S_x^2\right)} \tag{U.17}$$

where the first term involves only the Zeeman Hamiltonian and therefore corresponds to a delta-function at $B = \langle B\rangle$. Using Eq. U.12 in Eq. U.17, the second moment becomes:

$$\langle \Delta B^2\rangle = -\frac{\mathrm{Tr}\left([\mathcal{H}_1,S_x]^2\right)}{(g\beta_e)^2\,\mathrm{Tr}\left(S_x^2\right)} \tag{U.18}$$

which is determined solely by the interaction Hamiltonian, $\mathcal{H}_1$. Rules for determining the traces of spin operators, together with values for products of spin-operator components, are given in Table U.1.

TABLE U.1 Rules for traces of spin operators and examples

rule/example	*expression*
addition:	$\mathrm{Tr}(A+B) = \mathrm{Tr}(A) + \mathrm{Tr}(B)$
commutation:	$\mathrm{Tr}(AB) = \mathrm{Tr}(BA)$, even if A and B do not commute
	$\mathrm{Tr}(AB) = \mathrm{Tr}(A)\times\mathrm{Tr}(B)$, if A and B commute, and $A \neq B$
	$\mathrm{Tr}(ABC) = \mathrm{Tr}(CBA) = \mathrm{Tr}(BCA)$, cyclic permutations only
	$\mathrm{Tr}(ABC) \neq \mathrm{Tr}(BAC)$, non-cyclic permutation
commutators:	$\mathrm{Tr}([A,B]) = 0$
	$\mathrm{Tr}(A[B,C]) = \mathrm{Tr}([A,B]C)$
	$\mathrm{Tr}([A,[B,C]]D) = \mathrm{Tr}([B,[A,D]]C) = \mathrm{Tr}([D,A][B,C])$
invariance under rotation:	$\mathrm{Tr}(f(S_x)) = \mathrm{Tr}(f(S_y)) = \mathrm{Tr}(f(S_z))$
examples:	$\mathrm{Tr}(1) = 2S+1$
	$\mathrm{Tr}(S_x) = \mathrm{Tr}(S_y) = \mathrm{Tr}(S_z) = 0$
	$\mathrm{Tr}(S_xS_y) = \mathrm{Tr}(S_xS_z) = \mathrm{Tr}(S_zS_y) = 0$, et seq.
	$\mathrm{Tr}(S_+^2) = \mathrm{Tr}(S_-^2) = 0$
	$\mathrm{Tr}(S_x^2) = \mathrm{Tr}(S_y^2) = \mathrm{Tr}(S_z^2) = \frac{1}{3}\mathrm{Tr}(S^2) = \frac{1}{3}S(S+1)(2S+1)$
	$\mathrm{Tr}(S_+S_-) = \mathrm{Tr}(S_-S_+) = \frac{2}{3}S(S+1)(2S+1)$
	$\mathrm{Tr}(s_{jx}s_{kx}) = \mathrm{Tr}(s_{jy}s_{ky}) = \mathrm{Tr}(s_{jz}s_{kz}) = 0$, for $j \neq k$
	$\mathrm{Tr}(s_{j+}s_{k+}) = \mathrm{Tr}(s_{j-}s_{k-}) = \mathrm{Tr}(s_{j+}s_{k-}) = \mathrm{Tr}(s_{j-}s_{k+}) = 0$, for $j \neq k$
	$\mathrm{Tr}(S_x^2S_y) = 0$, and so on
	$\mathrm{Tr}(S_x^3) = \mathrm{Tr}(S_y^3) = \mathrm{Tr}(S_z^3) = 0$
	$\mathrm{Tr}(S_xS_yS_z) = \frac{1}{6}iS(S+1)(2S+1)$
	$\mathrm{Tr}(S_xS_yS_xS_y) = \frac{1}{15}S(S+1)(S^2+S+2)$
	$\mathrm{Tr}(S_z^4) = \frac{1}{5}\left(S^2(S+1)^2 - \frac{1}{3}S(S+1)\right)(2S+1)$
	$\mathrm{Tr}(S_x^2S_y^2) = \frac{1}{15}\left(S^2(S+1)^2 + \frac{1}{2}S(S+1)\right)(2S+1)$
	$\mathrm{Tr}(s_{jz}^2s_{kz}^2) = \frac{1}{9}s^2(s+1)^2(2s+1)^2$, for $j \neq k$
	$\mathrm{Tr}(s_{jz}s_{kz}s_{jx}s_{kx}) = \mathrm{Tr}(s_{jz}s_{kz}s_{jy}s_{ky}) = 0$, for $j \neq k$

See also Van Vleck (1948) and Van Vleck (1937).

U.3.1 Second Moment for Isotropic Hyperfine Structure

First we treat inhomogeneous broadening from unresolved hyperfine structure contributed by many nuclei, most usually protons. For N equivalent nuclear spins I with hyperfine coupling constant a, the interaction Hamiltonian is:

$$\mathcal{H}_1 = aS_z \sum_{j=1}^{N} I_{jz} \tag{U.19}$$

Choosing a simple product wave function for the electron spin and nuclear spins, we can evaluate the traces in Eq. U.18 within the electron manifold and the nuclear manifolds separately:

$$\langle \Delta B_{hfs}^2 \rangle = -\left(\frac{a}{g\beta_e}\right)^2 \frac{\mathrm{Tr}\left([S_z,S_x]^2\left(\sum I_{jz}\right)^2\right)}{\mathrm{Tr}(S_x^2)}$$

$$= \left(\frac{a}{g\beta_e}\right)^2 \frac{\mathrm{Tr}(S_y^2)}{\mathrm{Tr}(S_x^2)} \times \frac{\mathrm{Tr}\left(\left(\sum I_{jz}\right)^2\right)}{\mathrm{Tr}(1)} \tag{U.20}$$

where the electron spin commutator is $[S_z, S_x] = iS_y$ (see Eq. E.5 in Appendix E) and $\mathrm{Tr}(1) = 2I+1$ in the denominator of the nuclear term allows for the number of spin states in each nuclear manifold $(m_I = -I,\ -I+1, \ldots, I-1, I)$. The invariance of the trace to rotation means that $\mathrm{Tr}(S_y^2) = \mathrm{Tr}(S_x^2)$, (or alternatively one could choose to evaluate each trace in the basis set in which it is diagonal). Expanding the square of the sum in the nuclear term gives:

$$\mathrm{Tr}\left(\left(\sum_j I_{jz}\right)^2\right) = \mathrm{Tr}\left(\sum_{j=1}^{N} I_{jz}^2\right) + \mathrm{Tr}\left(\sum_{j\neq k} I_{jz}I_{kz}\right) = N \times \mathrm{Tr}(I_z^2) \tag{U.21}$$

where the cross terms between different nuclei drop out because $\mathrm{Tr}(I_{jz}I_{kz}) = \mathrm{Tr}(I_{jz})\mathrm{Tr}(I_{kz}) = 0$ for $j \neq k$. For integer nuclear spin, the trace of I_z^2 is given by:

$$\mathrm{Tr}(I_z^2) = \sum_{m_I=-I}^{I} m_I^2 = 2\sum_{m_I=1}^{I} m_I^2 = \tfrac{1}{3} I(I+1)(2I+1) \tag{U.22}$$

from the standard result for the sum of squares of the integers. Explicit calculation shows that the right-hand side of Eq. U.22 also gives the correct value for the corresponding summation with half-integer I. Clearly, contributions to the second moment from different sets, i, of equivalent nuclei are additive. The final result for the contribution to the second moment, $\langle \Delta B^2 \rangle$, from isotropic hyperfine structure is:

$$\langle \Delta B_{hfs}^2 \rangle = \frac{1}{3}\sum_i N_i I_i (I_i+1)\left(\frac{a_i}{g\beta_e}\right)^2 \tag{U.23}$$

where $a_i/g\beta_e$ is the hyperfine coupling in field units (e.g., mT). Note that the number of hyperfine states: $z_I = 2I+1$ does not appear in the final result because it enters into both numerator and denominator in Eq. U.18. This is a general result in every calculation of the second moment, including that for multiple electron spins which now follows.

U.3.2 Second Moment for Dipolar Spin–Spin Interactions

Now we treat a many-body system of interacting electron spins. The second moment has two important advantages when applied to interacting spins: Heisenberg exchange interaction does not contribute to $\langle \Delta B^2 \rangle$ for like spins, and an analytical expression is available for an arbitrary number of spins interacting by the dipole–dipole interaction (Van Vleck 1948).

From generalization of Eqs. 9.5, 9.11 in Chapter 9, the Hamiltonian for *like* spins, $\mathbf{s}_j$, that undergo mutual dipole–dipole interactions is given by:

$$\mathcal{H}_{dd} = \tfrac{1}{2}\frac{\mu_o}{4\pi} g^2\beta_e^2 \sum_{j>k} \frac{1-3\cos^2\theta_{jk}}{r_{jk}^3}\left(3s_{jz}s_{kz} - \mathbf{s}_j \cdot \mathbf{s}_k\right) \tag{U.24}$$

where we can drop the scalar product $\mathbf{s}_j \cdot \mathbf{s}_k$ because it commutes with the x-component of the total spin $S_x = \sum_j s_{jx}$ (cf. Eq. U.18). For the same reason, Heisenberg spin exchange does not contribute to the second moment. We use lower case for the individual spins, to distinguish from the total spin $\mathbf{S} = \sum \mathbf{s}_j$. The total number of spin states is $z_s^N = (2s+1)^N$, where z_s is the number of states for an individual spin $\mathbf{s}$ and N is the total number of spins. (The actual value of z_s^N is not important because it appears in both denominator and numerator of the second moment and cancels out in the end.)

It now remains to find the traces. From the invariance of the trace to rotation, we have: $\mathrm{Tr}(s_{jx}^2) = \mathrm{Tr}(s_{jy}^2) = \mathrm{Tr}(s_{jz}^2)$ for a single spin, and therefore $\mathrm{Tr}(s_{jx}^2) = \frac{1}{3}\mathrm{Tr}(\mathbf{s}_j^2)$. In the diagonal representation, each of the z_s^N matrix elements of the $\mathbf{s}_j^2$ spin operator has the value $s(s+1)$, and hence (Van Vleck 1948, and see Slichter 1990):

$$\mathrm{Tr}(s_{jx}^2) = \tfrac{1}{3}s(s+1)z_s^N \tag{U.25}$$

where $s = \frac{1}{2}$ is the quantum number of an individual spin. For the x-component of the total spin: $\mathrm{Tr}(S_x^2) = \mathrm{Tr}\left(\sum_j s_{jx}^2\right) + \mathrm{Tr}\left(\sum_{j\neq k} s_{jx}s_{kx}\right) = N\mathrm{Tr}(s_{jx}^2)$, where the cross-term falls out because $\mathrm{Tr}(s_{jx}s_{kx}) = \mathrm{Tr}(s_{jx})\mathrm{Tr}(s_{kx}) = 0$ for $j \neq k$. Hence from Eq. U.25 we obtain:

$$\mathrm{Tr}(S_x^2) = \tfrac{1}{3}Ns(s+1)z_s^N \tag{U.26}$$

for the denominator in Eq. U.18.

We turn now to the numerator in Eq. U.18. The spin-containing part of $[\mathcal{H}_{dd},S_x]$ is given by: $\left[\sum_{j,k} s_{jz}s_{kz},\sum_j s_{jx}\right]=\sum_{j,k}\left([s_{jz},s_{jx}]s_{kz}+s_{jz}[s_{kz},s_{kx}]\right)=i\sum_{j,k}(s_{jy}s_{kz}+s_{jz}s_{ky})$, because spin components with $j\neq k$ commute. Taking the square gives: $-\sum_{j,k}\left(s_{jy}^2s_{kz}^2+s_{jz}^2s_{ky}^2\right)$ plus cross-terms of the type: $s_{jy}s_{kz}s_{jz}s_{ky}$, which vanish on taking the trace because they consist of products of terms such as $\mathrm{Tr}(s_{jy})=0$ (cf. Table U.1). The surviving term is then: $\sum_{j,k}\left(\mathrm{Tr}(s_{jy}^2)\mathrm{Tr}(s_{kz}^2)+\mathrm{Tr}(s_{jz}^2)\mathrm{Tr}(s_{ky}^2)\right)=\sum_{j,k}\frac{2}{9}s^2(s+1)^2z_s^N=N\sum_k\frac{2}{9}s^2(s+1)^2z_s^N$, where we assume that all spins are similarly situated when replacing the sum over j by the multiplying factor $N\times$. The summation here serves only to remind us that the angular and radially dependent terms in Eq. U.24 are still to be included.

The second moment contributed by the dipolar interaction of like spins then finally becomes (Van Vleck 1948):

$$\langle\Delta B_{dd}^2\rangle=\frac{3}{4}\left(\frac{\mu_o}{4\pi}\right)^2 s(s+1)g^2\beta_e^2\sum_k\frac{\left(1-3\cos^2\theta_{jk}\right)^2}{r_{jk}^6}\qquad\text{(U.27)}$$

where j is a dummy index because we assume that the environment of every spin is equivalent. In a powder sample we must sum the angular terms over a sphere (see Eq. 5.16), giving the angular average:

$$\langle\Delta B_{dd}^2\rangle_{SS}=\frac{3}{5}\left(\frac{\mu_o}{4\pi}\right)^2 s(s+1)g^2\beta_e^2\sum_j 1/r_{jk}^6\qquad\text{(U.28)}$$

for the second moment of the dipolar broadening with like spins. The dipolar Hamiltonian for *unlike* spins is (see Eq. 9.18):

$$(\mathcal{H}_{dd})_{SS'}=\frac{\mu_o}{4\pi}g^2\beta_e^2\sum_{k\neq j}\frac{1-3\cos^2\theta_{jk}}{r_{jk}^3}s_{jz}s'_{kz}\qquad\text{(U.29)}$$

which differs in the coefficient of the $s_{jz}s_{kz}$ term from that for like spins by a factor 2/3 (cf. Eq. U.24), which then becomes 4/9 in the second moment. The second moment in a powder sample of unlike spins is therefore given by:

$$\langle\Delta B_{dd}^2\rangle_{SS'}=\frac{4}{15}\left(\frac{\mu_o}{4\pi}\right)^2 s'(s'+1)g^2\beta_e^2\sum_k 1/r_{jk}^6\qquad\text{(U.30)}$$

where $s'=\frac{1}{2}$ for all k unlike spins. Note that the summation over $1/r_{jk}^6$ (for both cases) implies that the second moment is directly proportional to the spin concentration, in either a random or a lattice model (Abragam 1961).

U.3.3 Second Moment for Spectral Anisotropy

The spectral anisotropy in powder patterns contributes to the second moment, just as it does to the absolute-value first moment. Consider again the case of axial N-hyperfine anisotropy, when we can neglect g-value anisotropy (see Figure U.1). The normalized powder envelope for one hyperfine manifold is given by Eq. U.8, and the second moment becomes:

$$\langle\Delta B_m^2\rangle=\int_{\Delta B_\perp}^{\Delta B_{//}}\Delta B_m^2\cdot f(\Delta B_m)\cdot\mathrm{d}\Delta B_m=\frac{1}{3}\left(\left(\frac{A_\parallel}{g\beta_e}\right)^2+2\left(\frac{A_\perp}{g\beta_e}\right)^2\right)m_I^2\qquad\text{(U.31)}$$

where $A_\parallel/g\beta_e$ and $A_\perp/g\beta_e$ are the principal hyperfine couplings in field units (e.g., mT). This result applies to a single hyperfine state in an axially symmetric powder pattern. The second moment of the entire powder pattern is given by taking the average over all $m_I=-I,\ -I+1,\ldots,I-1,\ I$, where I is the nuclear spin. The normalized summation that we need is: $\sum_{-I}^{I}m_I^2/(2I+1)=I(I+1)/3$ (cf. Eq. U.22), which holds for both integer and half-integer spins. The second moment of the whole spectrum then becomes:

$$\langle\Delta B_{aniso}^2\rangle=\frac{1}{9}\left(\left(\frac{A_\parallel}{g\beta_e}\right)^2+2\left(\frac{A_\perp}{g\beta_e}\right)^2\right)I(I+1)\qquad\text{(U.32)}$$

where $I=1$ for ^{14}N-spin labels and $I=\frac{1}{2}$ for ^{15}N-spin labels. Unlike for the absolute first moment, the intrinsic linewidth (homogeneous and inhomogeneous) contributes additionally to the second moment.

U.4 Fourth Moment

The fourth moment of a spectral lineshape is given by (Abragam 1961; Van Vleck 1948):

$$\langle\Delta B^4\rangle=\frac{\mathrm{Tr}\left([\mathcal{H}_1,[\mathcal{H}_1,S_x]]^2\right)}{(g\beta_e)^4\,\mathrm{Tr}\left(S_x^2\right)}\qquad\text{(U.33)}$$

We can see this most readily from the proportionality between the moments and derivatives at the origin of the Fourier transform (Abragam 1961):

$$\langle\Delta B^4\rangle=\frac{1}{F(0)}\left(\frac{d^4F(t)}{dt^4}\right)_{t=0}\qquad\text{(U.34)}$$

where $F(0)$ is the Fourier transform of the lineshape, at the origin (cf. Eqs. T.17 and T.18 in Appendix T).

If the interaction Hamiltonian consists of both dipole–dipole and exchange interactions: $\mathcal{H}_1=\mathcal{H}_{ex}+\mathcal{H}_{dd}$, we see that Heisenberg exchange contributes to the fourth moment, although it does not contribute to the second moment. This is because, although $\mathcal{H}_{ex}$ commutes with S_x, it does not commute with all terms in $\mathcal{H}_{dd}$. This is one of the features of exchange narrowing. To produce a broader spectrum in the wings, as indicated by the fourth moment, whilst keeping the second moment and the integrated intensity unchanged, requires narrowing in the centre of the spectrum.

Van Vleck (1948) calculated the fourth moment for dipolar broadening on a simple cubic lattice. With the magnetic field directed along the principal crystal axis, the ratio of the fourth moment to the square of the second moment is $\langle \Delta B_{dd}^4 \rangle / \langle \Delta B_{dd}^2 \rangle^2 = 2.44$. For a Gaussian lineshape (see Eq. 1.25) this ratio is $\langle \Delta B^4 \rangle / \langle \Delta B^2 \rangle^2 = 3$. For a Lorentzian lineshape, the second moment is already infinite, unless integration is truncated at some large but finite value, $\pm \Delta B_{max}$. In the latter case, the second moment is (see, e.g., Abragam 1961): $\langle \Delta B^2 \rangle = 2\Delta B_{max} / (\pi \gamma_e T_2)$ and the ratio of moments is $\langle \Delta B^4 \rangle / \langle \Delta B^2 \rangle^2 = \pi \gamma_e T_2 \Delta B_{max} / 6$. Clearly, the truncated Lorentzian is not a good approximation for dipolar broadening, except for dilute spins, because ΔB_{max} must be much greater than the linewidth $1/(\gamma_e T_2)$ for the model to be realistic.

For a single spin pair with separation r_{12}, the dipolar fourth moment is given by (cf. Eq. 15.13 of Section 15.7 in Chapter 15):

$$\langle \Delta B_{dd}^4 \rangle = \int_{-2\Delta B_{dd}^o}^{\Delta B_{dd}^o} (\Delta B_{dd})^4 \cdot I(\Delta B_{dd}) \cdot d\Delta B_{dd} = \frac{48}{35} (\Delta B_{dd}^o)^4 \quad (U.35)$$

where $\Delta B_{dd}^o = \frac{3}{4} (\mu_o / 4\pi) g \beta_e / r_{12}^3$ for like spins and $\Delta B_{dd}^o = \frac{1}{2} (\mu_o / 4\pi) g \beta_e / r_{12}^3$ for unlike spins. The corresponding expression for the second moment is given by Eq. 15.13 of Chapter 15:

$$\langle \Delta B_{dd}^2 \rangle = \int_{-2\Delta B_{dd}^o}^{\Delta B_{dd}^o} (\Delta B_{dd})^2 \cdot I(\Delta B_{dd}) \cdot d\Delta B_{dd} = \frac{4}{5} (\Delta B_{dd}^o)^2 \quad (U.36)$$

The ratio of moments is then: $\langle \Delta B_{dd}^4 \rangle / \langle \Delta B_{dd}^2 \rangle^2 = (15/7) \times \langle 1/r_{12}^{12} \rangle / \langle 1/r_{12}^6 \rangle^2$, which becomes $15/7 = 2.14$ when there is no distribution in r_{12}.

APPENDIX V: SPHERICAL HARMONICS AND LEGENDRE POLYNOMIALS

V.1 Legendre Polynomials

Legendre polynomials, $P_L(x)$, are solutions to the differential equation:

$$\left((1 - x^2) \frac{d^2}{dx^2} - 2x \frac{d}{dx} \right) \cdot P_L = -L(L+1) \cdot P_L \quad (V.1)$$

where L is a positive integer. The polynomials are given by the series:

$$P_L(x) = \frac{(2L)!}{2^L (L!)^2} \left(x^L - \frac{L(L-1)}{2(2L-1)} x^{L-2} + \frac{L(L-1)(L-2)(L-3)}{2 \times 4(2L-1)(2L-3)} x^{L-4} - \dots \right) \quad (V.2)$$

which is generated by the differential function:

$$P_L(x) = \frac{1}{2^L L!} \frac{d^L (x^2 - 1)^L}{dx^L} \quad (V.3)$$

where factorials are defined by: $L! = 1 \times 2 \times 3 \times \ldots \times (L-2) \times (L-1) \times L$. Explicit functional forms for the Legendre polynomials are listed in Table V.1.

Orthogonality of Legendre polynomials is given by:

$$\int_{-1}^{1} P_L(x) \cdot P_M(x) \cdot dx = \frac{2}{2L+1} \delta_{L,M} \quad (V.4)$$

Recursion relation for Legendre polynomials is:

$$(L+1) P_{L+1} = (2L+1) x \cdot P_L - L \cdot P_{L-1} \quad (V.5)$$

V.2 Spherical Harmonics

Spherical harmonics $Y_{L,M}(\theta, \phi)$ are solutions to the angular parts of Laplace's differential equation:

$$\nabla_\Omega^2 Y_{L,M} \equiv \left(\frac{1}{\sin\theta} \frac{\partial}{\partial\theta} \left(\sin\theta \frac{\partial}{\partial\theta} \right) + \frac{1}{\sin^2\theta} \frac{\partial^2}{\partial\phi^2} \right) Y_{L,M} = -L(L+1) Y_{L,M} \quad (V.6)$$

where ∇_Ω^2 is the Laplacian operator on the surface of a unit sphere. The operator on the left is the square of the orbital angular momentum, $\mathbf{L}^2$, in polar coordinates and simultaneously is the angular part of the Schrödinger equation for a central potential (i.e., a potential without angular dependence). We can

TABLE V.1 Legendre polynomials, $P_L(x)$

L	$P_L(x)$
0	$P_0 = 1$
1	$P_1 = x$
2	$P_2 = \frac{1}{2}(3x^2 - 1)$
3	$P_3 = \frac{1}{2}(5x^3 - 3x)$
4	$P_4 = \frac{1}{8}(35x^4 - 30x^2 + 3)$
5	$P_5 = \frac{1}{8}(63x^5 - 70x^3 + 15x)$
6	$P_6 = \frac{1}{16}(231x^6 - 315x^4 + 105x^2 - 5)$

separate the θ and ϕ parts of Eq. V.6, and $Y_{L,M}(\theta,\phi)$ is therefore the product of a function of θ alone and a function of ϕ alone. The ϕ-dependent differential equation is:

$$\frac{\partial^2}{\partial\phi^2}Y_{L,M} = -M^2 Y_{L,M} \quad \text{(V.7)}$$

where the operator on the left is the square of the z-component, L_z, of the orbital angular momentum. The corresponding θ-dependent differential equation is:

$$\left(\frac{1}{\sin\theta}\frac{\partial}{\partial\theta}\left(\sin\theta\frac{\partial}{\partial\theta}\right)+\frac{M^2}{\sin^2\theta}\right)Y_{L,M} = -L(L+1)Y_{L,M} \quad \text{(V.8)}$$

the solutions of which are the associated Legendre functions, $P_{L,M}(\cos\theta)$, where L is a positive integer and $M=-L, -L+1,\ldots, L-1, L$. The associated Legendre functions are generated by the differential function:

$$P_{L,M}(\cos\theta) = \frac{(-1)^M}{2^L L!}\left(1-\cos^2\theta\right)^{M/2}\frac{d^{L+M}}{d(\cos\theta)^{L+M}}\left(\cos^2\theta-1\right)^L \quad \text{(V.9)}$$

The normalized spherical harmonics are then given by:

$$Y_{L,M}(\theta,\phi) = (-1)^M\sqrt{\frac{2L+1}{4\pi}\frac{(L-M)!}{(L+M)!}}.P_{L,M}(\cos\theta)\cdot e^{iM\phi} \quad \text{(V.10)}$$

This means that $Y_{L,M}$ is the product of $e^{iM\phi}\cdot\sin^{|M|}\theta$ with a polynomial of degree $L-|M|$ in $\cos\theta$ of parity $(-1)^{L-M}$. The choice of phase is that of Condon and Shortley (1935), which gives the complex conjugate:

$$Y_{L,M}(\theta,\phi)^* = (-1)^M Y_{L,-M}(\theta,\phi) \quad \text{(V.11)}$$

For $M=0$, the spherical harmonics are simply Legendre polynomials:

$$Y_{L,0} = \sqrt{\frac{2L+1}{4\pi}}\cdot P_L(\cos\theta) \quad \text{(V.12)}$$

For $M=L$:

$$Y_{L,L} = (-1)^L\sqrt{\frac{2L+1}{4\pi}\frac{(2L)!}{2^{2L}(L!)^2}}\sin^L\theta\cdot e^{iL\phi} \quad \text{(V.13)}$$

Specific functional forms for the spherical harmonics are listed in Table V.2.

Orthogonality of spherical harmonics is given by:

$$\int_0^{2\pi}\int_0^{\pi} Y^*_{L,M}(\theta,\phi)Y_{L',M'}(\theta,\phi)\cdot\sin\theta\cdot d\theta\cdot d\phi = \delta_{M,M'}\delta_{L,L'} \quad \text{(V.14)}$$

Closure relation for spherical harmonics is:

$$\sum_{L=0}^{\infty}\sum_{M=-L}^{+L} Y^*_{L,M}(\theta,\phi)Y_{L,M}(\theta',\phi') = \frac{\delta(\theta-\theta')\delta(\phi-\phi')}{\sin\theta} \equiv \delta(\Omega-\Omega') \quad \text{(V.15)}$$

TABLE V.2 Spherical harmonics, $Y_{L,M}(\theta,\phi)$

L	M	$Y_{L,M}(\theta,\phi)$
0	0	$Y_{0,0} = \frac{1}{\sqrt{4\pi}}$
1	0	$Y_{1,0} = \sqrt{\frac{3}{4\pi}}\cos\theta$
	±1	$Y_{1,\pm1} = -\sqrt{\frac{3}{8\pi}}\sin\theta\cdot e^{\pm i\phi}$
2	0	$Y_{2,0} = \sqrt{\frac{5}{16\pi}}\left(3\cos^2\theta-1\right)$
	±1	$Y_{2,\pm1} = -\sqrt{\frac{15}{8\pi}}\sin\theta\cos\theta\cdot e^{\pm i\phi}$
	±2	$Y_{2,\pm2} = \sqrt{\frac{15}{32\pi}}\sin^2\theta\cdot e^{\pm i2\phi}$
3	0	$Y_{3,0} = \sqrt{\frac{7}{16\pi}}\left(5\cos^3\theta-3\cos\theta\right)$
	±1	$Y_{3,\pm1} = -\sqrt{\frac{21}{64\pi}}\sin\theta\left(5\cos^2\theta-1\right)\cdot e^{\pm i\phi}$
	±2	$Y_{3,\pm2} = \sqrt{\frac{105}{32\pi}}\sin^2\theta\cos\theta\cdot e^{\pm i2\phi}$
	±3	$Y_{3,\pm3} = -\sqrt{\frac{35}{64\pi}}\sin^3\theta\cdot e^{\pm i3\phi}$

where $\Omega \equiv (\theta,\phi)$.

Recursion relation for spherical harmonics is:

$$\cos\theta\cdot Y_{L,M} = \sqrt{\frac{(L+1+M)(L+1-M)}{(2L+1)(2L+3)}}\cdot Y_{L+1,M} + \sqrt{\frac{(L+M)(L-M)}{(2L+1)(2L-1)}}\cdot Y_{L-1,M} \quad \text{(V.16)}$$

Addition theorem for spherical harmonics is:

$$P_L(\cos\Theta) = \frac{4\pi}{2L+1}\sum_{M=-L}^{+L} Y^*_{L,M}(\theta_1,\phi_1)Y_{L,M}(\theta_2,\phi_2) \quad \text{(V.17)}$$

where Θ is the angle between the directions (θ_1,ϕ_1) and (θ_2,ϕ_2). In terms of associated Legendre polynomials $P_{L,M}$, the addition theorem becomes:

$$P_L(\cos\Theta) = P_L(\cos\theta_1)P_L(\cos\theta_2) + 2\sum_{M=1}^{L}\frac{(L-M)!}{(L+M)!}P_{L,M}(\cos\theta_1)P_{L,M}(\cos\theta_2) \times\cos M(\phi_1-\phi_2) \quad \text{(V.18)}$$

which reduces to a simple product of Legendre polynomials P_L, in the case of axial symmetry.

REFERENCES

Abragam, A. 1961. *The Principles of Nuclear Magnetism*, Oxford: Oxford University Press.

Atherton, N.M. 1973. *Electron Spin Resonance. Theory and Applications*, Chichester: Ellis Horwood.

Bales, B.L., Meyer, M., and Peric, M. 2014. EPR line shifts and line shape changes due to Heisenberg spin exchange and dipole-dipole interactions of nitroxide free radicals in liquids: 9. An alternative method to separate the effects of the two interactions employing ^{15}N and ^{14}N. *J. Phys. Chem. A* 118:6154–6162.

Brink, D.M. and Satchler, G.R. 1968. *Angular Momentum*, Oxford: Oxford University Press.

Condon, E.U. and Shortley, G.H. 1935. *Theory of Atomic Spectra*, Cambridge: Cambridge University Press.

Edmonds, A.R. 1957. *Angular Momentum in Quantum Mechanics*, Princeton: Princeton University Press.

Ernst, R.R., Bodenhausen, G., and Wokaun, A. 1987. *Principles of NMR in One and Two Dimensions*, Oxford: Clarendon Press.

Freed, J.H. 1976. Theory of slowly tumbling ESR spectra for nitroxides. In *Spin Labeling, Theory and Applications*, ed. Berliner, L.J., 53–132. New York: Academic Press Inc.

Freed, J.H. and Fraenkel, G.K. 1963. Theory of linewidths in electron spin resonance spectra. *J. Chem. Phys.* 39:326–348.

Galeev, R.T. and Salikhov, K.M. 1996. Theory of dipole widening of magnetic resonance lines in nonviscous liquids. *Chem. Phys. Reports* 15:359-375.

Gemperle, C. and Schweiger, A. 1991. Pulse electron-nuclear double resonance methodology. *Chem. Rev.* 91:1481–1505.

Glarum, S.H. and Marshall, J.H. 1967. Paramagnetic relaxation in liquid-crystal solvents. *J. Chem. Phys.* 46:55-62.

Heine, V. 1960. *Group Theory in Quantum Mechanics*, Oxford: Pergamon Press.

Hyde, J.S. and Pilbrow, J.R. 1980. A moment method for determining isotropic g-values from powder EPR. *J. Magn. Reson.* 41:447–457.

Marsh, D. 2017. Coherence transfer and electron T_1-, T_2-relaxation in nitroxide spin labels. *J. Magn. Reson.* 277:86–94.

Mehring, M. 1983. *Principles of High Resolution NMR in Solids*, Berlin Heidelberg: Springer-Verlag.

Meirovitch, E., Igner, D., Igner, E., Moro, G., and Freed, J.H. 1982. Electron-spin relaxation and ordering in smectic and supercooled nematic liquid crystals. *J. Chem. Phys.* 77:3915–3938.

Messiah, A. 1965. *Quantum Mechanics, Vol. II*, Amsterdam: North Holland Publishing Company.

Moro, G. and Freed, J.H. 1981. Calculation of ESR spectra and related Fokker-Planck forms by the use of the Lanczos algorithm. *J. Chem. Phys.* 74:3757–3772.

Peric, M., Bales, B.L., and Peric, M. 2012. Electron paramagnetic resonance line shifts and line shape changes due to Heisenberg spin exchange and dipole-dipole interactions of nitroxide free radicals in liquids. 8. Further experimental and theoretical efforts to separate the effects of the two interactions. *J. Phys. Chem. A* 116:2855–2866.

Polnaszek, C.F., Bruno, G.V. and Freed, J.H. 1973. ESR line shapes in the slow-motional region: anisotropic fluids. *J. Chem. Phys.* 58:3185–3199.

Rabenstein, M.D. and Shin, Y.K. 1995. Determination of the distance between 2 spin labels attached to a macromolecule. *Proc. Natl. Acad. Sci USA* 92:8239–8243.

Redfield, A.G. 1957. On the theory of relaxation processes. *IBM J. Res. Develop.* 1:19–31.

Redfield, A.G. 1965. The theory of relaxation processes. *Adv. Mag. Res.* 1:1–32.

Rose, M.E. 1957. *Elementary Theory of Angular Momentum*, New York: John Wiley & Sons Inc.

Rotenberg, M., Bivins, R., Metropolis, N., and Wooten, J.K. 1959. *The 3-j and 6-j Symbols*, Cambridge, MA: The Technology Press, MIT.

Schneider, D.J. and Freed, J.H. 1989. Calculating slow motional magnetic resonance spectra: a user's guide. In *Biological Magnetic Resonance*, Vol. 8, Spin-Labeling. Theory and Applications. eds. Berliner, L.J. and Reuben, J., 1–76. New York: Plenum Publishing Corp.

Schweiger, A. and Jeschke, G. 2001. *Principles of Pulse Electron Paramagnetic Resonance*, Oxford: Oxford University Press.

Slichter, C.P. 1990. *Principles of Magnetic Resonance*, 3rd Ed., New York: Springer-Verlag.

Tinkham, M. 1964. *Group Theory & Quantum Mechanics*, New York: McGraw Hill Book Company.

Van Vleck, J.H. 1937. The influence of dipole-dipole coupling on the specific heat and susceptibility of a paramagnetic salt. *J. Chem. Phys.* 5:320–337.

Van Vleck, J.H. 1948. The dipolar broadening of magnetic resonance lines in crystals. *Phys. Rev.* 74:1168–1183.

Wigner, E. 1959. *Group Theory and its Application to the Quantum Mechanics of Atomic Spectra*, New York: Academic Press.

Index

J

L

M

N

O

P

Q

R

U

V

W

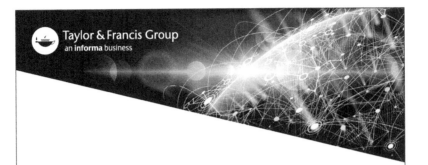

Printed and bound by CPI Group (UK) Ltd, Croydon, CR0 4YY
17/10/2024
01775672-0018